Unit Operations of Chemical Engineering

PARTH PATEL

UNIT OPERATIONS OF CHEMICAL ENGINEERING

SIXTH EDITION

Warren L. McCabe

Late R. J. Reynolds Professor of Chemical Engineering
North Carolina State University

Julian C. Smith

Emeritus Professor of Chemical Engineering
Cornell University

Peter Harriott

Fred H. Rhodes Professor of Chemical Engineering
Cornell University

Boston   Burr Ridge, IL   Dubuque, IA   Madison, WI   New York
San Francisco   St. Louis   Bangkok   Bogotá   Caracas   Kuala
Lumpur   Lisbon   London   Madrid   Mexico City
Milan   Montreal   New Delhi   Santiago   Seoul   Singapore
Sydney   Taipei   Toronto

McGraw-Hill Higher Education

A Division of The **McGraw-Hill** Companies

UNIT OPERATIONS OF CHEMICAL ENGINEERING
International Edition 2001

Exclusive rights by McGraw-Hill Book Co – Singapore, for manufacture and export. This book cannot be re-exported from the country to which it is sold by McGraw-Hill. The International Edition is not available in North America.

Published by McGraw-Hill, an imprint of The McGraw-Hill Companies, Inc., 1221 Avenue of the Americas, New York, NY 10020. Copyright © 2001, 1993, 1985, 1976, 1967, 1956, by The McGraw-Hill Companies, Inc. All rights reserved. No part of this publication may be reproduced or distributed in any form or by any means, or stored in a data base or retrieval system, without the prior written consent of The McGraw-Hill Companies, Inc., including, but not limited to, in any network or other electronic storage or transmission, or broadcast for distance learning.
Some ancillaries, including electronic and print components, may not be available to customers outside the United States.

10   09 08 07 06   05 04 03 02
20   09 08 07 06   05 04 03 02
CTP   SLP

Library of Congress Cataloging-in-Publication Data
McCabe, Warren L. (Warren Lee), 1899-1982
   Unit operations of chemical engineering / Warren L. McCabe, Julian C. Smith, Peter Harriott.—6th ed.
      p. cm
   Includes bibliographical references and index.
   ISBN 0-07-039366-4
   1.   Chemical processes.   I. Smith, Julia C. (Julian Cleveland)
II. Harriott, Peter.   III. Title.
   TP155.7.M3 2001
   660'.2842—dc21   00-055038

www.mhhe.com

When ordering this title, use ISBN 0-07-118173-3

Printed in Singapore

ABOUT THE AUTHORS

JULIAN C. SMITH (B.Chem., Chem.E., Cornell University) is Professor Emeritus of Chemical Engineering at Cornell University, where he joined the faculty in 1946. He was Director of Continuing Engineering Education at Cornell from 1965 to 1971, and Director of the School of Chemical Engineering from 1975 to 1983. He retired from active teaching in 1986. Before joining the faculty at Cornell, he was employed as a chemical engineer by E.I. duPont de Nemours and Co. He has served as a consultant on process development to Du Pont, American Cyanamid, and many other companies, as well as government agencies. He is a member of the American Chemical Society and a Fellow of the American Institute of Chemical Engineers.

PETER HARRIOTT (B. Chem.E., Cornell University, ScD., Massachusetts Institute of Technology) is the Fred H. Rhodes Professor of Chemical Engineering at Cornell University. Before joining the Cornell faculty in 1953, he worked as a chemical engineer for the E.I. duPont de Nemours and Co. and the General Electric Co. In 1966 he was awarded an NSF Senior Postdoctoral Fellowship for study at the Institute for Catalysis in Lyon, France, and in 1988 he received a DOE fellowship for work at the Pittsburgh Energy Technology Center. Professor Harriott is the author of Process Control and a member of the American Chemical Society and the American Institute of Chemical Engineers. He has been a consultant to the U.S. Department of Energy and several industrial firms on problems of mass transfer, reactor design, and air pollution control.

CONTENTS

SECTION III Heat Transfer and Its Applications

SECTION IV **Mass Transfer and Its Applications**

SECTION V   Operations Involving Particulate Solids

PREFACE

This sixth edition of the text on the unit operations of chemical engineering has been extensively revised and updated, with much new material and considerable condensation of some sections. Its basic structure and general level of treatment, however, remain unchanged. It is an introductory text, written for undergraduate students in their junior or senior years who have completed the usual courses in mathematics, physics, chemistry, and an introduction to chemical engineering. An elementary knowledge of material and energy balances and of thermodynamic principles is assumed.

Separate chapters are devoted to each of the principal unit operations, grouped into four sections: fluid mechanics, heat transfer, mass transfer and equilibrium stages, and operations involving particulate solids. One-semester or one-quarter courses may be based on any of these sections or combinations of them. The order of the first 16 chapters has not been changed; later ones, dealing with mass transfer and operations involving solids, have been rearranged in a more logical order.

Nearly all equations have been written for SI units, and the Newton's law conversion factor g_c has been eliminated except in the few instances where it must be included. Symbols for dimensionless groups have been changed to Re for N_{Re}, for example, Pr for N_{Pr}, and so forth. Many new examples and problems have been added, some reflecting the importance of biochemical engineering processes. Material on handling, mixing, and grinding particulate solids has been greatly condensed and dealt with in a single chapter. The number of appendixes is reduced from 22 to 19.

Derivations of the differential equations for continuity and momentum balances, leading to the Navier-Stokes equation, have been added, as well as the differential forms of Fourier's law and Fick's law, emphasizing the analogies among momentum, heat, and mass transfer. The chapter on adsorption has been expanded to include new material on chromatography and ion exchange, and renamed "Fixed-bed Separations." Other new material has been added on viscoelastic fluids, laminar flow in annuli, drag coefficients, affinity laws for pumps, high-efficiency agitators and motionless mixers, plate-type heat exchangers, boiling by submerged tube bundles, cooling towers, aqueous phase extraction, cross-flow filtration, and many other topics.

Many of the problems at the ends of the chapters are new or revised. Most are expressed in SI units. Nearly all the problems can be solved with the aid of a pocket calculator; for a few, a computer solution is preferable.

McGraw-Hill and the authors thank Dr. N. T. Obot for his many suggestions regarding fluid mechanics and heat transfer, and Professor Charles H. Gooding of Clemson University for his detailed and helpful review of the manuscript.

Julian C. Smith
Peter Harriott

Introduction

CHAPTER 1

Definitions and Principles

Chemical engineering has to do with industrial processes in which raw materials are changed or separated into useful products. The chemical engineer must develop, design, and engineer both the complete process and the equipment used; choose the proper raw materials; operate the plants efficiently, safely, and economically; and see to it that products meet the requirements set by the customers. Chemical engineering is both an art and a science. Whenever science helps the engineer to solve a problem, science should be used. When, as is usually the case, science does not give a complete answer, it is necessary to use experience and judgment. The professional stature of an engineer depends on skill in utilizing all sources of information to reach practical solutions to processing problems.

The variety of processes and industries that call for the services of chemical engineers is enormous. Products of concern to chemical engineers range from commodity chemicals such as sulfuric acid and chlorine to high-technology items such as polymeric lithographic supports for the electronics industry, high-strength composite materials, and genetically modified biochemical agents. The processes described in standard treatises on chemical technology and the process industries give a good idea of the field of chemical engineering, as does the 1988 report on the profession by the National Research Council.[1,8][†]

Because of the variety and complexity of modern processes, it is not practicable to cover the entire subject matter of chemical engineering under a single head. The field is divided into convenient, but arbitrary, sectors. This text covers that portion of chemical engineering known as the unit operations.

[†]Superscript numerals in the text correspond to the numbered references at the end of each chapter.

UNIT OPERATIONS

An economical method of organizing much of the subject matter of chemical engineering is based on two facts: (1) Although the number of individual processes is great, each one can be broken down into a series of steps, called operations, each of which in turn appears in process after process; (2) the individual operations have common techniques and are based on the same scientific principles. For example, in most processes solids and fluids must be moved; heat or other forms of energy must be transferred from one substance to another; and tasks such as drying, size reduction, distillation, and evaporation must be performed. The unit operation concept is this: By studying systematically these operations themselves—operations that clearly cross industry and process lines—the treatment of all processes is unified and simplified.

The strictly chemical aspects of processing are studied in a companion area of chemical engineering called reaction kinetics. The unit operations are largely used to conduct the primarily physical steps of preparing the reactants, separating and purifying the products, recycling unconverted reactants, and controlling the energy transfer into or out of the chemical reactor.

The unit operations are as applicable to many physical processes as to chemical ones. For example, the process used to manufacture common salt consists of the following sequence of unit operations: transportation of solids and liquids, transfer of heat, evaporation, crystallization, drying, and screening. No chemical reaction appears in these steps. On the other hand, the cracking of petroleum, with or without the aid of a catalyst, is a typical chemical reaction conducted on an enormous scale. Here the unit operations—transportation of fluids and solids, distillation, and various mechanical separations—are vital, and the cracking reaction could not be utilized without them. The chemical steps themselves are conducted by controlling the flow of material and energy to and from the reaction zone.

Because the unit operations are a branch of engineering, they are based on both science and experience. Theory and practice must combine to yield designs for equipment that can be fabricated, assembled, operated, and maintained. A balanced discussion of each operation requires that theory and equipment be considered together. An objective of this book is to present such a balanced treatment.

Scientific foundations of unit operations

A number of scientific principles and techniques are basic to the treatment of the unit operations. Some are elementary physical and chemical laws such as the conservation of mass and energy, physical equilibria, kinetics, and certain properties of matter. Their general use is described in the remainder of this chapter. Other special techniques important in chemical engineering are considered at the proper places in the text.

UNIT SYSTEMS

The official international system of units is SI (Système International d'Unités). Strong efforts are underway for its universal adoption as the exclusive system for all engineering and science, but older systems, particularly the centimeter-gram-second (cgs) and foot-pound-second (fps) engineering gravitational systems, are still in use and probably will be around for some time. The chemical engineer finds many physiochemical data given in cgs units; that many calculations are most conveniently done in fps units; and that SI units are increasingly encountered in science and engineering. Thus it becomes necessary to be expert in the use of all three systems.

In the following treatment, SI is discussed first, and then the other systems are derived from it. The procedure reverses the historical order, as the SI units evolved from the cgs system. Because of the growing importance of SI, it should logically be given a preference. If, in time, the other systems are phased out, they can be ignored and SI used exclusively.

Physical Quantities

Any physical quantity consists of two parts: a unit, which tells what the quantity is and gives the standard by which it is measured, and a number, which tells how many units are needed to make up the quantity. For example, the statement that the distance between two points is 3 m means all this: A definite length has been measured; to measure it, a standard length, called the meter, has been chosen as a unit; and three 1-m units, laid end to end, are needed to cover the distance. If an integral number of units are either too few or too many to cover a given distance, submultiples, which are fractions of the unit, are defined by dividing the unit into fractions, so that a measurement can be made to any degree of precision in terms of the fractional units. No physical quantity is defined until both the number and the unit are given.

SI Units

The SI system covers the entire field of science and engineering, including electromagnetics and illumination. For the purposes of this book, a subset of the SI units covering chemistry, gravity, mechanics, and thermodynamics is sufficient. The units are derivable from (1) four proportionalities of chemistry and physics; (2) arbitrary standards for mass, length, time, temperature, and the mole; and (3) arbitrary choices for the numerical values of two proportionality constants.

Basic equations

The basic proportionalities, each written as an equation with its own proportionality factor, are

$$F = k_1 \frac{d}{dt}(mu) \tag{1.1}$$

$$F = k_2 \frac{m_a m_b}{r^2} \tag{1.2}$$

$$Q_c = k_3 W_c \tag{1.3}$$

$$T = k_4 \lim_{p \to 0} \frac{pV}{m} \tag{1.4}$$

where[†] $F =$ force
                 $t =$ time
                 $m =$ mass
                 $u =$ velocity
                 $r =$ distance
              $W_c =$ work
              $Q_c =$ heat
                 $p =$ pressure
                 $V =$ volume
                 $T =$ thermodynamic absolute temperature
$k_1, k_2, k_3, k_4 =$ proportionality factors

Equation (1.1) is Newton's second law of motion, showing the proportionality between the resultant of all the forces acting on a particle of mass m and the time rate of increase in momentum of the particle in the direction of the resultant force.

Equation (1.2) is Newton's law of gravitation, giving the force of attraction between two particles of masses m_a and m_b a distance r apart.

Equation (1.3) is one statement of the first law of thermodynamics. It affirms the proportionality between the work performed by a closed system during a cycle and the heat absorbed by that system during the same cycle.

Equation (1.4) shows the proportionality between the thermodynamic absolute temperature and the zero-pressure limit of the pressure-volume product of a definite mass of any gas.

Each equation states that if means are available for measuring the values of all variables in that equation and if the numerical value of k is calculated, then the value of k is constant and depends only on the units used for measuring the variables in the equation.

Standards

By international agreement, standards are fixed arbitrarily for the quantities of mass, length, time, temperature, and the mole. These are five of the *base units* of SI. Currently, the standards are as follows.

The standard of mass is the kilogram (kg), defined as the mass of the international kilogram, a platinum cylinder preserved at Sèvres, France.

The standard of length is the meter (m), defined[5] (since 1983) as the length of the path traveled by light in vacuum during a time interval of $1/299{,}792{,}458^{*}$ of a second.[‡]

The standard of time is the second (s), defined as $9{,}192{,}631.770^{*}$ frequency cycles of a certain quantum transition in an atom of ^{133}Ce.

The standard of temperature is the kelvin (K), defined by assigning the value 273.16^{*} K to the temperature of pure water at its triple point, the unique temperature at which liquid water, ice, and steam can exist at equilibrium.

[†]A list of symbols is given at the end of each chapter.
[‡]The asterisk at the end of a number signifies that the number is exact, by definition.

The mole (abbreviated mol) is defined[7] as the amount of a substance comprising as many elementary units as there are atoms in 12^* g of ^{12}C. The definition of the mole is equivalent to the statement that the mass of one mole of a pure substance in grams is numerically equal to its molecular weight calculated from the standard table of atomic weights, in which the atomic weight of carbon is given as 12.01115. This number differs from 12^* because it applies to the natural isotopic mixture of carbon rather than to pure ^{12}C.

Evaluation of constants

From the basic standards, values of m, m_a, and m_b in Eqs. (1.1) and (1.2) are measured in kilograms, r in meters, and u in meters per second. Constants k_1 and k_2 are not independent but are related by eliminating F from Eqs. (1.1) and (1.2). This gives

$$\frac{k_1}{k_2} = \frac{d(mu)/dt}{m_a m_b / r^2}$$

Either k_1 or k_2 may be fixed arbitrarily. Then the other constant must be found by experiments in which inertial forces calculated by Eq. (1.1) are compared with gravitational forces calculated by Eq. (1.2). In SI, k_1 is fixed at unity and k_2 found experimentally. Equation (1.1) then becomes

$$F = \frac{d}{dt}(mu) \qquad (1.5)$$

The force defined by Eq. (1.5) and also used in Eq. (1.2) is called the *newton* (N). From Eq. (1.5),

$$1 \text{ N} \equiv 1 \text{ kg} \cdot \text{m/s}^2 \qquad (1.6)$$

Constant k_2 is denoted by G and called the *gravitational constant*. Its recommended value is[4]

$$G = 6.6726 \times 10^{-11} \text{ N} \cdot \text{m}^2/\text{kg}^2 \qquad (1.7)$$

Work, energy, and power

In SI, both work and energy are measured in newton-meters, a unit called the *joule* (J), and so

$$1 \text{ J} \equiv 1 \text{ N} \cdot \text{m} = 1 \text{ kg} \cdot \text{m}^2/\text{s}^2 \qquad (1.8)$$

Power is measured in joules per second, a unit called the *watt* (W).

Heat

The constant k_3 in Eq. (1.3) may be fixed arbitrarily. In SI it, like k_1, is set at unity. Equation (1.3) becomes

$$Q_c = W_c \qquad (1.9)$$

Heat, like work, is measured in joules.

Temperature

The quantity pV/m in Eq. (1.4) may be measured in $(N/m^2)(m^3/kg)$, or J/kg. With an arbitrarily chosen gas, this quantity can be determined by measuring p and V of m kg of gas while it is immersed in a thermostat. In this experiment, only constancy of temperature, not magnitude, is needed. Values of pV/m at various pressures and at constant temperature can then be extrapolated to zero pressure to obtain the limiting value required in Eq. (1.4) at the temperature of the thermostat. For the special situation in which the thermostat contains water at its triple point, the limiting value is designated by $(pV/m)_0$. For this experiment Eq. (1.4) gives

$$273.16 = k_4 \lim_{p \to 0} \left(\frac{pV}{m} \right)_0 \tag{1.10}$$

For an experiment at temperature T K, Eq. (1.4) can be used to eliminate k_4 from Eq. (1.10), giving

$$T \equiv 273.16 \frac{\lim_{p \to 0}(pV/m)_T}{\lim_{p \to 0}(pV/m)_0} \tag{1.11}$$

Equation (1.11) is the definition of the Kelvin temperature scale from the experimental pressure-volume properties of a real gas.

Celsius temperature

In practice, temperatures are expressed on the Celsius scale, in which the zero point is set at the ice point, defined as the equilibrium temperature of ice and air-saturated water at a pressure of one atmosphere. Experimentally, the ice point is found to be 0.01K below the triple point of water, and so it is at 273.15K. The Celsius temperature (°C) is defined by

$$T°C \equiv T K - 273.15 \tag{1.12}$$

On the Celsius scale, the experimentally measured temperature of the steam point, which is the boiling point of water at a pressure of 1 atm, is 100.00°C.

Decimal units

In SI, a single unit is defined for each quantity, but named decimal multiples and submultiples also are recognized. They are listed in Table 1.1. Time may be expressed in the nondecimal units: minutes (min), hours (h), or days (d).

Standard gravity

For certain purposes, the acceleration of free fall in the earth's gravitational field is used. From deductions based on Eq. (1.2), this quantity, denoted by g, is nearly constant. It varies slightly with latitude and height above sea level. For precise calculations, an arbitrary standard g_n has been set, defined by

$$g_n \equiv 9.80665^* \text{ m/s}^2 \tag{1.13}$$

TABLE 1.1
SI and cgs prefixes for multiples and submultiples

Factor	Prefix	Abbreviation	Factor	Prefix	Abbreviation
10^{12}	tera	T	10^{-1}	deci	d
10^9	giga	G	10^{-2}	centi	c
10^6	mega	M	10^{-3}	milli	m
10^3	kilo	k	10^{-6}	micro	μ
10^2	hecto	h	10^{-9}	nano	n
10^1	deka	da	10^{-12}	pico	p
			10^{-15}	femto	f
			10^{-18}	atto	a

Pressure units

The natural unit of pressure in SI is the newton per square meter. This unit, called the *pascal* (Pa), is inconveniently small, and a multiple, called the *bar,* also is used. It is defined by

$$1 \text{ bar} \equiv 1 \times 10^5 \text{ Pa} = 1 \times 10^5 \text{ N/m}^2 \qquad (1.14)$$

A more common empirical unit for pressure, used with all systems of units, is the *standard atmosphere* (atm), defined by

$$1 \text{ atm} \equiv 1.01325^* \times 10^5 \text{ Pa} = 1.01325 \text{ bars} \qquad (1.15)$$

CGS Units

The older cgs system can be derived from SI by making certain arbitrary decisions. The standard for mass is the gram (g), defined by

$$1 \text{ g} \equiv 1 \times 10^{-3} \text{ kg} \qquad (1.16)$$

The standard for length is the centimeter (cm), defined by

$$1 \text{ cm} \equiv 1 \times 10^{-2} \text{ m} \qquad (1.17)$$

Standards for time, temperature, and the mole are unchanged.
As in SI, constant k_1 in Eq. (1.1) is fixed at unity. The unit of force is called the *dyne* (dyn), defined by

$$1 \text{ dyn} \equiv 1 \text{ g} \cdot \text{cm/s}^2 \qquad (1.18)$$

The unit for energy and work is the *erg,* defined by

$$1 \text{ erg} \equiv 1 \text{ dyn} \cdot \text{cm} = 1 \times 10^{-7} \text{ J} \qquad (1.19)$$

Constant k_3 in Eq. (1.3) is not unity. A unit for heat, called the *calorie* (cal), is used to convert the unit for heat to ergs. Constant $1/k_3$ is replaced by J, which denotes the quantity called the mechanical equivalent of heat and is measured in joules per calorie. Equation (1.3) becomes

$$W_c = JQ_c \tag{1.20}$$

Two calories are defined.[7] The *thermochemical calorie* (cal), used in chemistry, chemical engineering thermodynamics, and reaction kinetics, is defined by

$$1 \text{ cal} \equiv 4.1840^* \times 10^7 \text{ ergs} = 4.1840^* \text{ J} \tag{1.21}$$

The *international steam table calorie* (cal_{IT}), used in heat power engineering, is defined by

$$1 \text{ cal}_{IT} \equiv 4.1868^* \times 10^7 \text{ ergs} = 4.1868^* \text{ J} \tag{1.22}$$

The calorie is so defined that the specific heat of water is approximately 1 cal/g $\cdot$ °C. The standard acceleration of free fall in cgs units is

$$g_n \equiv 980.665 \text{ cm/s}^2 \tag{1.23}$$

Gas Constant

If mass is measured in kilograms or grams, constant k_4 in Eq. (1.4) differs from gas to gas. But when the concept of the mole as a mass unit is used, k_4 can be replaced by the universal gas constant R, which, by Avogadro's law, is the same for all gases. The numerical value of R depends only on the units chosen for energy, temperature, and mass. Then Eq. (1.4) is written

$$\lim_{p \to 0} \frac{pV}{nT} = R \tag{1.24}$$

where n is the number of moles. This equation applies also to mixtures of gases if n is the total number of moles of all the molecular species that make up the volume V. The accepted experimental value of R is[6]

$$R = 8.31447 \text{ J/K} \cdot \text{mol} = 8.31447 \times 10^7 \text{ ergs/K} \cdot \text{mol} \tag{1.25}$$

Values of R in other units for energy, temperature, and mass are given in Table 1.2. Although the mole is defined as a mass in grams, the concept of the mole is easily extended to other mass units. Thus, the kilogram mole (kg mol) is the usual molecular or atomic weight in kilograms, and the pound mole (lb mol) is that in avoirdupois pounds. When the mass unit is not specified, the gram mole (g mol) is intended. Molecular weight M is a pure number.

Standard molar volume. From Table 1.2, the volume of 1 kg mol of gas at standard conditions (1 atm, 0°C), is $82.056 \times 10^{-3} \times 273 = 22.4 \text{ m}^3$, or 22.4 (L/g mol). In fps units, the standard volume at 1 atm and 32°F is 359 ft^3/lb mol.

TABLE 1.2
Values of the gas constant R

Temperature	Mass	Energy	R
Kelvins	kg mol	J	8,314.47
		cal_{IT}	1.9859×10^3
		cal	1.9873×10^3
		m^3-atm	82.056×10^{-3}
	g mol	cm^3-atm	82.056
Degrees Rankine	lb mol	Btu	1.9858
		$ft \cdot lb_f$	1,545.3
		$hp \cdot h$	7.8045×10^{-4}
		kWh	5.8198×10^{-4}

FPS Engineering Units

In some countries a nondecimal gravitational unit system has long been used in commerce and engineering. The system can be derived from SI by making the following decisions.

The standard for mass is the avoirdupois pound (lb), defined by

$$1 \text{ lb} = 0.45359237^* \text{ kg} \tag{1.26}$$

The standard for length is the inch (in.), defined as 2.54^* cm. This is equivalent to defining the foot (ft) as

$$1 \text{ ft} \equiv 2.54 \times 12 \times 10^{-2} \text{ m} = 0.3048^* \text{ m} \tag{1.27}$$

The standard for time remains the second (s).

The thermodynamic temperature scale is called the Rankine scale, in which temperatures are denoted by degrees Rankine and defined by

$$1°R \equiv \frac{1}{1.8} \text{ K} \tag{1.28}$$

The ice point on the Rankine scale is $273.15 \times 1.8 = 491.67°R$.

The analog of the Celsius scale is the Fahrenheit scale, in which readings are denoted by degrees Fahrenheit. It is derived from the Rankine scale by setting its zero point exactly 32°F below the ice point on the Rankine scale, so that

$$T°F \equiv T°R - (491.67 - 32) = T°R - 459.67 \tag{1.29}$$

The relation between the Celsius and Fahrenheit scales is given by the exact equation

$$T°F = 32 + 1.8°C \tag{1.30}$$

From this equation, temperature differences are related by

$$\Delta T°C = 1.8 \, \Delta T°F = \Delta T \text{ K} \tag{1.31}$$

The steam point is 212.00°F.

Pound force

The fps system is characterized by a gravitational unit of force, called the *pound force* (lb_f). The unit is so defined that a standard gravitational field exerts a force of one pound on a mass of one avoirdupois pound. The standard acceleration of free fall in fps units is, to five significant figures,

$$g_n = \frac{9.80665 \text{ m/s}^2}{0.3048 \text{ m/ft}} = 32.174 \text{ ft/s}^2 \tag{1.32}$$

The pound force is defined by

$$1 \text{ lb}_f \equiv 32.174 \text{ lb} \cdot \text{ft/s}^2 \tag{1.33}$$

Then Eq. (1.1) gives

$$F \text{ lb}_f \equiv \frac{d(mu)/dt}{32.174} \quad \text{lb} \cdot \text{ft/s}^2 \tag{1.34}$$

Equation (1.1) can also be written with $1/g_c$ in place of k_1:

$$F = \frac{d(mu)/dt}{g_c} \tag{1.35}$$

Comparison of Eqs. (1.34) and (1.35) shows that to preserve both numerical equality and consistency of units in these equations, it is necessary to define g_c, called *Newton's law proportionality factor for the gravitational force unit*, by

$$g_c \equiv 32.174 \text{ lb} \cdot \text{ft/s}^2 \cdot \text{lb}_f \tag{1.36}$$

The unit for work and mechanical energy in the fps system is the *foot-pound force* ($ft \cdot lb_f$). Power is measured by an empirical unit, the *horsepower* (hp), defined by

$$1 \text{ hp} \equiv 550 \text{ ft} \cdot \text{lb}_f/\text{s} \tag{1.37}$$

The unit for heat is the *British thermal unit* (Btu), defined by the implicit relation

$$1 \text{ Btu/lb} \cdot {}^{\circ}\text{F} \equiv 1 \text{ cal}_{IT}/\text{g} \cdot {}^{\circ}\text{C} \tag{1.38}$$

As in the cgs system, constant k_3 in Eq. (1.3) is replaced by $1/J$, where J is the mechanical equivalent of heat, equal to 778.17 ft $\cdot$ lb_f/Btu.

The definition of the Btu requires that the numerical value of specific heat be the same in both systems, and in each case the specific heat of water is approximately 1.0.

Conversion of Units

Since three unit systems are in common use, it is often necessary to convert the magnitudes of quantities from one system to another. This is accomplished by using conversion factors. Only the defined conversion factors for the base units are required since conversion factors for all other units can be calculated from them. Interconversions between SI and the cgs system are simple. Both use the same standards for time, temperature, and the mole, and only the decimal conversions defined by Eqs. (1.16) and (1.17) are needed. Both SI and the fps system also use the second as the standard for time; the three conversion factors defined for mass, length, and temperature by Eqs. (1.26), (1.27), and (1.28), respectively, are sufficient for all conversions of units between these two systems.

Example 1.1 demonstrates how conversion factors are calculated from the exact numbers used to set up the definitions of units in SI and the fps system. In conversions involving g_c in fps units, the use of the exact numerical ratio 9.80665/0.3048 in place of the fps number 32.1740 is recommended to give maximum precision in the final calculation and to take advantage of possible cancellations of numbers during the calculation.

EXAMPLE 1.1. Using only exact definitions and standards, calculate factors for converting (a) newtons to pounds force, (b) British thermal units to IT calories, (c) atmospheres to pounds force per square inch, and (d) horsepower to kilowatts.

Solution

(a) From Eqs. (1.6), (1.26), and (1.27),

$$1\text{ N} = 1\text{ kg} \cdot \text{m/s}^2 = \frac{1\text{ lb} \cdot \text{ft/s}^2}{0.45359237 \times 0.3048}$$

From Eq. (1.32)

$$1\text{ lb} \cdot \text{ft/s}^2 = \frac{0.3048}{9.80665}\text{ lb}_f$$

and so
$$1\text{ N} = \frac{0.3048}{9.80665 \times 0.45359237 \times 0.3048}\text{ lb}_f$$

$$= \frac{1}{9.80665 \times 0.45359237}\text{ lb}_f = 0.224809\text{ lb}_f$$

In Appendix 1 it is shown that to convert newtons to pound force, one should multiply by 0.224809. Clearly, to convert from pounds force to newtons, multiply by 1/0.224809 = 4.448221.

(b) From Eq. (1.38)

$$1\text{ Btu} = 1\text{ cal}_{IT}\frac{1\text{ lb}}{1\text{ g}}\frac{1°\text{F}}{1°\text{C}}$$

$$= 1\text{ cal}_{IT}\frac{1\text{ lb}}{1\text{ kg}}\frac{1\text{ kg}}{1\text{ g}}\frac{1°\text{F}}{1°\text{C}}$$

From Eqs. (1.16), (1.26), and (1.31)

$$1\text{ Btu} = 1\text{ cal}_{IT}\frac{0.45359237 \times 1000}{1.8} = 251.996\text{ cal}_{IT}$$

(c) From Eqs. (1.6), (1.14), and (1.15)

$$1\text{ atm} = 1.01325 \times 10^5\text{ kg} \cdot \text{m/s}^2 \cdot \text{m}^2$$

From Eqs. (1.26), (1.27), and (1.36), since 1 ft = 12 in.,

$$1 \text{ atm} = 1.01325 \times 10^5 \times \frac{1 \text{ lb/s}^2}{0.45359237} \frac{0.3048}{\text{ft}}$$

$$= \frac{1.01325 \times 10^5 \times 0.3048}{32.174 \times 0.45359237 \times 12^2} \text{ lb}_f/\text{in.}^2$$

$$= 14.6959 \text{ lb}_f/\text{in.}^2$$

(d) From Eqs. (1.33) and (1.37)

$$1 \text{ hp} = 550 \text{ ft} \cdot \text{lb}_f/\text{s} = 550 \times 32.174 \text{ ft}^2 \cdot \text{lb/s}^3$$

Using Eqs. (1.26) and (1.27) gives

$$1 \text{ hp} = 550 \times 32.174 \times 0.45359237 \times 0.3048^2$$

$$= 745.70 \text{ J/s}$$

Substituting from Eq. (1.8) and dividing by 1,000,

$$1 \text{ hp} = 0.74570 \text{ kW}$$

Although conversion factors may be calculated as needed, it is more efficient to use tables of the common factors. A table for the factors used in this book is given in Appendix 1.

Units and Equations

Although Eqs. (1.1) to (1.4) are sufficient for the description of unit systems, they are but a small fraction of the equations needed in this book. Many such equations contain terms that represent properties of substances, and these are introduced as needed. All new quantities are measured in combinations of units already defined, and all are expressible as functions of the five base units for mass, length, time, temperature, and the mole.

Precision of calculations

In the above discussion, the values of experimental constants are given with the maximum number of significant digits consistent with present estimates of the precision with which they are known, and all digits in the values of defined constants are retained. In practice, such extreme precision is seldom necessary, and defined and experimental constants can be truncated to the number of digits appropriate to the problem at hand, although the advent of the digital computers make it possible to retain maximum precision at small cost. The engineer should use judgment in setting a suitable level of precision for the particular problem to be solved.

General equations

Except for the appearance of the proportionality factors g_c and J, the equations for all three unit systems are alike. In this text, equations are written for SI units, with a reminder to use g_c and J when working examples in cgs or fps units.

Dimensionless equations and consistent units

Equations derived directly from the basic laws of the physical sciences consist of terms that either have the same units or can be written in the same units by using the definitions of derived quantities to express complex units in terms of the five base ones. Equations meeting this requirement are called *dimensionally homogeneous equations*. When such an equation is divided by any one of its terms, all units in each term cancel and only numerical magnitudes remain. These equations are called *dimensionless equations*.

A dimensionally homogeneous equation can be used as it stands with any set of units provided that the same units for the five base units are used throughout. Units meeting this requirement are called *consistent units*. No conversion factors are needed when consistent units are used.

For example, consider the usual equation for the vertical distance Z traversed by a freely falling body during time t when the initial velocity is u_0:

$$Z = u_0 t + \tfrac{1}{2} g t^2 \tag{1.39}$$

Examination of Eq. (1.39) shows that the units in each term reduce to that for length. Dividing the equation by Z gives

$$1 = \frac{u_0 t}{Z} + \frac{g t^2}{2Z} \tag{1.40}$$

A check of each term in Eq. (1.40) shows that the units in each term cancel and each term is dimensionless. A combination of variables for which all dimensions cancel in this manner is called a *dimensionless group*. The numerical value of a dimensionless group for given values of the quantities contained in it is independent of the units used, provided they are consistent. Both terms on the right-hand side of Eq. (1.40) are dimensionless groups.

Dimensional equations

Equations derived by empirical methods, in which experimental results are correlated by empirical equations without regard to dimensional consistency, usually are not dimensionally homogeneous and contain terms in several different units. Equations of this type are *dimensional equations,* or dimensionally nonhomogeneous equations. In these equations there is no advantage in using consistent units, and two or more length units, for example, inches and feet, or two or more time units, for example, seconds and minutes, may appear in the same equation. For example, a formula for the rate of heat loss from a horizontal pipe to the atmosphere by conduction and convection is

$$\frac{q}{A} = 0.50 \frac{\Delta T^{1.25}}{(D_o')^{0.25}} \tag{1.41}$$

where q = rate of heat loss, Btu/h
A = area of pipe surface, ft^2

ΔT = excess of temperature of pipe wall over that of ambient (surrounding atmosphere), °F

D'_o = outside diameter of pipe, in.

Obviously, the units of q/A are not those of the right-hand side of Eq. (1.41), and the equation is dimensional. Quantities substituted in Eq. (1.41) must be expressed in the units as given, or the equation will give the wrong answer. If other units are to be used, the coefficient must be changed. To express ΔT in degrees Celsius, for example, the numerical coefficient must be changed to $0.50 \times 1.8^{1.25} = 1.042$ since there are 1.8 Fahrenheit degrees in 1 Celsius degree of temperature difference.

In this book all equations are dimensionally homogeneous *unless otherwise noted*.

DIMENSIONAL ANALYSIS

Many important engineering problems cannot be solved completely by theoretical or mathematical methods. Problems of this type are especially common in fluid-flow, heat-flow, and diffusion operations. One method of attacking a problem for which no mathematical equation can be derived is that of empirical experimentation. For example, the pressure loss from friction in a long, round, straight, smooth pipe depends on all these variables: the length and diameter of the pipe, the flow rate of the liquid, and the density and viscosity of the liquid. If any one of these variables is changed, the pressure drop also changes. The empirical method of obtaining an equation relating these factors to pressure drop requires that the effect of each separate variable be determined in turn by systematically varying that variable while keeping all others constant. The procedure is laborious, and it is difficult to organize or correlate the results so obtained into a useful relationship for calculations.

There exists a method intermediate between formal mathematical development and a completely empirical study.[2] It is based on the fact that if a theoretical equation does exist among the variables affecting a physical process, that equation must be dimensionally homogeneous. Because of this requirement, it is possible to group many factors into a smaller number of dimensionless groups of variables. The groups themselves rather than the separate factors appear in the final equation.

This method is called *dimensional analysis,* which is an algebraic treatment of the symbols for units considered independently of magnitude. It drastically simplifies the task of fitting experimental data to design equations; it is also useful in checking the consistency of the units in equations, in converting units, and in the scaleup of data obtained in model test units to predict the performance of full-scale equipment.

In making a dimensional analysis, the variables thought to be important are chosen and their dimensions tabulated. If the physical laws that would be involved in a mathematical solution are known, the choice of variables is relatively easy. The fundamental differential equations of fluid flow, for example, combined with the laws of heat conduction and diffusion, suffice to establish the dimensions and dimensionless groups appropriate to a large number of chemical engineering problems. In other situations the choice of variables may be speculative, and testing of the resulting relationships may be needed to establish whether some variables were left out or whether some of those chosen are not needed.

Assuming that the variables are related by a power series, in which the dimension of each term must be the same as that of the primary quantity, an exponential relationship is written in which the exponents relating to any given quantity (for example, length) must be the same on both sides of the equation. The relationship among the exponents is then found algebraically, as shown in Example 1.2.

EXAMPLE 1.2. A steady stream of liquid in turbulent flow is heated by passing it through a long, straight, heated pipe. The temperature of the pipe is assumed to be greater by a constant amount than the average temperature of the liquid. It is desired to find a relationship that can be used to predict the rate of heat transfer from the wall of the liquid.

Solution. The mechanism of this process is discussed in Chap. 12. From the characteristics of the process it may be expected that the rate of heat transfer q/A depends on the quantities listed with their dimensional formulas in Table 1.3. If a theoretical equation for this problem exists, it can be written in the general form

$$\frac{q}{A} = \Psi(D, \bar{V}, \rho, \mu, c_p, k, \Delta T) \qquad (1.42)$$

If Eq. (1.42) is a valid relationship, all terms in the function Ψ must have the same dimensions as those of the left-hand side of the equation, q/A. Let the phrase *the dimensions of* be shown by the use of square brackets. Then any term in the function must conform to the dimensional formula

$$\left[\frac{q}{A}\right] = [D]^a[\bar{V}]^b[\rho]^c[\mu]^d[c_p]^e[k]^f[\Delta T]^g \qquad (1.43)$$

Let an overbar above a symbol denote that it refers to a dimension. Thus $\bar{L}$ refers to the dimension of length. Substituting the dimensions from Table 1.3 gives

$$\bar{H}\bar{L}^{-2}\bar{t} = \bar{L}^a\bar{L}^b\bar{t}^{-b}\bar{M}^c\bar{L}^{-3c}\bar{M}^d\bar{L}^{-d}\bar{H}^e\bar{M}^{-e}\bar{t}^{-e}\bar{H}^f\bar{L}^{-f}\bar{t}^{-f}\bar{T}^{-f}\bar{T}^g \qquad (1.44)$$

TABLE 1.3

Quantities and dimensional formulas for Example 1.2

Quantity	Symbol	Dimensions
Heat flow per unit area	q/A	$\bar{H}\bar{L}^{-2}\bar{t}^{-1}$
Diameter of pipe (inside)	D	$\bar{L}$
Average velocity of liquid	$\bar{V}$	$\bar{L}\bar{t}^{-1}$
Density of liquid	ρ	$\bar{M}\bar{L}^{-3}$
Viscosity of liquid	μ	$\bar{M}\bar{L}^{-1}\bar{t}^{-1}$
Specific heat, at constant pressure, of liquid	c_p	$\bar{H}\bar{M}^{-1}\bar{T}^{-1}$
Thermal conductivity of liquid	k	$\bar{H}\bar{L}^{-1}\bar{t}^{-1}\bar{T}^{-1}$
Temperature difference between wall and fluid	ΔT	$\bar{T}$

Since Eq. (1.43) is assumed to be dimensionally homogeneous, the exponents of the individual primary units on the left-hand side of the equation must equal those on the right-hand side. This gives the following set of equations:

Exponents of $\bar{H}$: $\qquad\qquad 1 = e + f$ (1.45a)

Exponents of $\bar{L}$: $\qquad\qquad -2 = a + b - 3c - d - f$ (1.45b)

Exponents of $\bar{t}$: $\qquad\qquad -1 = -b - d - f$ (1.45c)

Exponents of $\bar{M}$: $\qquad\qquad 0 = c + d - e$ (1.45d)

Exponents of $\bar{T}$: $\qquad\qquad 0 = -e - f + g$ (1.45e)

Here there are seven variables but only five equations. Five of the unknowns may be found in terms of the remaining two. The two letters to be retained must be chosen arbitrarily. The final result is equally valid for all choices, but for this problem it is customary to retain the exponents of the velocity $\bar{V}$ and the specific heat c_p. The letters b and e will be retained and the remaining five eliminated, as follows. From Eq. (1.45a):

$$f = 1 - e \qquad (1.46a)$$

From Eqs. (1.45e) and (1.46a)

$$g = e + f = e + 1 - e = 1 \qquad (1.46b)$$

From Eqs. (1.45c) and (1.46a)

$$d = 1 - b - f = 1 - b - 1 + e = e - b \qquad (1.46c)$$

From Eqs. (1.45d) and (1.46c)

$$c = e - d = e - e + b = b \qquad (1.46d)$$

From Eqs. (1.45b), (1.46a), (1.46c), and (1.46d)

$$a = -2 - b + 3c + d + f$$
$$= -2 - b + 3b + e - b + 1 - e$$
$$= b - 1 \qquad (1.46e)$$

By substituting values from Eqs. (1.46a) through (1.46e) for the letters a, c, d, and f, Eq. (1.43) becomes

$$\left[\frac{q}{A}\right] = [D]^{b-1}[\bar{V}]^b[\rho]^b[\mu]^{e-b}[c_p]^e[k]^{1-e}[\Delta T] \qquad (1.47)$$

Collecting all factors having integral exponents in one group, all factors having exponents b into another group, and those having exponents e into a third gives

$$\left[\frac{qD}{Ak\,\Delta T}\right] = \left[\frac{D\bar{V}\rho}{\mu}\right]^b \left[\frac{c_p\mu}{k}\right]^e \qquad (1.48)$$

The dimensions of each of the three bracketed groups in Eq. (1.48) are zero, and all groups are dimensionless. Any function whatever of these three groups will be dimensionally homogeneous, and the equation will be a dimensionless one. Let such a function be

$$\frac{qD}{Ak\,\Delta T} = \Phi\left(\frac{D\bar{V}\rho}{\mu}, \frac{c_p\mu}{k}\right) \qquad (1.49)$$

or
$$\frac{q}{A} = \frac{k\Delta T}{D}\Phi\left(\frac{D\bar{V}\rho}{\mu}, \frac{c_p\mu}{k}\right)$$
(1.50)

The relationship given in Eqs. (1.49) and (1.50) is the final result of the dimensional analysis. The form of function Φ must be found experimentally, by determining the effects of the groups in the brackets on the value of the group on the left-hand side of Eq. (1.49). The correlations that have been found for this are given in Chap. 12.

Correlating the experimental values of the three groups of variables of Eq. (1.49) is clearly simpler than attempting to correlate the effects of each of the individual factors of Eq. (1.42).

Formation of other dimensionless groups

If a pair of letters other than b and e is selected for retention, three dimensionless groups are again obtained, but one or more differ from the groups of Eq. (1.49). For example, if b and f are kept, the result is

$$\frac{q}{A\bar{V}\rho c_p\Delta T} = \Phi_1\left(\frac{D\bar{V}\rho}{\mu}, \frac{c_p\mu}{k}\right)$$
(1.51)

Other combinations may be found. However, it is unnecessary to repeat the algebra to obtain such additional groups. The three groups in Eq. (1.49) may be combined in any desired manner, by multiplying and dividing them, or reciprocals or multiples of them, together. It is necessary only that each original group be used at least once in finding new groups and that the final assembly contain exactly three groups. For example, Eq. (1.51) is obtained from Eq. (1.49) by multiplying both sides by $(\mu/D\bar{V}\rho)(k/\mu c_p)$

$$\frac{qD}{Ak\,\Delta T}\frac{\mu}{D\bar{V}\rho}\frac{k}{\mu c_p} = \Phi\left(\frac{D\bar{V}\rho}{\mu}, \frac{c_p\mu}{k}\right)\frac{\mu}{D\bar{V}\rho}\frac{k}{\mu c_p} = \Phi_1\left(\frac{D\bar{V}\rho}{\mu}, \frac{c_p\mu}{k}\right)$$

and Eq. (1.51) follows. Note that function Φ_1 is not equal to function Φ. In this way any dimensionless equation may be changed into any number of new ones. This is often useful when it is desired to isolate a single factor in one group. Thus, in Eq. (1.49) c_p appears in only one group, and in Eq. (1.51) k is found in only one. It is shown in Chap. 12 that Eq. (1.51) is more useful for some purposes than Eq. (1.49).

Speculative dimensional analysis. As pointed out by Churchill, dimensional analysis is best looked upon as a speculative process.[3] For one thing, the method does not guarantee that a simple exponential relationship such as Eq. (1.48) applies, although subsequent experiments may well show that such a relationship is entirely satisfactory. For another, the choice of variables should often be considered tentative in making a dimensional analysis. If unnecessary variables are included, the analysis usually shows that they are not needed; if important ones are left out, the analysis may still give a valid result but only for certain limiting or asymptotic conditions as, for example, at very low or very high flow rates. Used correctly, dimensional analysis can be a most useful tool in designing an experimental program.

Named dimensionless groups

Some dimensionless groups occur with such frequency that they have been given names and special symbols. A list of the most important ones is given in Appendix 2.

BASIC CONCEPTS

Underlying the unit operations of chemical engineering are a small number of basic concepts, including the equations of state of gases, material balances, and energy balances. These topics are treated extensively in introductory courses in chemistry and chemical engineering, and because of this they are discussed only briefly here. Illustrative problems, and answers, are included at the end of this chapter for anyone who wishes practice in applying these concepts.

Equations of State of Gases

A pure gas consisting of n mol and held at a temperature T and pressure p will fill a volume V. If any of the three quantities are fixed, the fourth also is determined and only three of these quantities are independent. This can be expressed by the functional equation

$$f(p, T, V, n) = 0 \tag{1.52}$$

Specific forms of this relation are called *equations of state*. Many such equations have been proposed, and several are in common use. The most satisfactory equations of state can be written in the form

$$\frac{pV}{nRT} = 1 + \frac{B}{V/n} + \frac{C}{(V/n)^2} + \frac{D}{(V/n)^3} + \cdots \tag{1.53}$$

This equation, known as the *virial equation,* is well substantiated by molecular gas theory. Coefficients B, C, and D are called the second, third, and fourth virial coefficients, respectively. Each is a function of temperature and is independent of pressure. Additional coefficients may be added, but numerical values for coefficients beyond D are so little known that more than three are seldom used. The virial equation also applies to mixtures of gases. Then the virial coefficients depend on the temperature and composition of the mixture. Rules are available for estimating the values of B, C, and D for mixtures from those of the individual pure gases.[9]

Compressibility factor and molar density

For engineering purposes, Eq. (1.53) often is written

$$z = \frac{p}{\rho_M RT} = 1 + \rho_M B + \rho_M^2 C + \rho_M^3 D \tag{1.54}$$

where z is the compressibility factor and ρ_M the molar density, defined by

$$\rho_M = \frac{n}{V} \tag{1.55}$$

Ideal gas law

Real gases under high pressure require the use of all three virials for accuracy in calculating z from ρ_M and T. As the density is reduced by lowering the pressure, the numerical values of the virial terms fade out, although the values of the coefficients remain unchanged. As the effect of D becomes negligible, the series is truncated by dropping its term, then the term in C, and so on, until at low pressures (about 1 or 2 atm for ordinary gases), all three virials can be neglected. The result is the simple gas law

$$z = \frac{pV}{nRT} = \frac{p}{\rho_M RT} = 1 \tag{1.56}$$

This equation clearly is consistent with Eq. (1.11), which contains the definition of the absolute temperature. The limiting process indicated in Eq. (1.11) rigorously eliminates the virial coefficients to provide a precise definition; Eq. (1.56) covers a useful range of densities for practical calculations and is called the *ideal gas law*.

Partial pressures

A useful quantity for dealing with the individual components in a gas mixture is partial pressure. The partial pressure of a component in a mixture, for example, component A, is defined by the equation

$$p_A \equiv P y_A \tag{1.57}$$

where p_A = partial pressure of component A in mixture
y_A = mole fraction of component A in mixture
P = total pressure on mixture

If all the partial pressures for a given mixture are added, the result is

$$p_A + p_B + p_C + \cdots = P(y_A + y_B + y_C + \cdots)$$

Since the sum of the mole fractions is unity,

$$p_A + p_B + p_C + \cdots = P \tag{1.58}$$

All partial pressures in a given mixture add to the total pressure. This applies to mixtures of both ideal and nonideal gases.

Material balances

The law of conservation of matter states that matter cannot be created or destroyed. This leads to the concept of mass, and the law may be stated in the form that the mass of the materials taking part in any process is constant. It is known now that the law is too restricted for matter moving at velocities near that of light or for substances undergoing nuclear reactions. Under these circumstances energy and mass are interconvertible, and the sum of the two is constant, rather than only 1. In most engineering, however, this transformation is too small to be detected, and in this book it is assumed that mass and energy are independent.

Conservation of mass requires that the materials entering any process either accumulate or leave the process. There can be neither loss nor gain. Most of the processes considered in this book involve neither accumulation nor depletion, and the law of conservation of matter takes the simple form that input equals output. The law is often applied in the form of material balances. The process is debited with everything that enters it and is credited with everything that leaves it. The sum of the credits must equal the sum of the debits. Material balances must hold over the entire process or equipment and over any part of it. They must apply to all the material that enters and leaves the process and to any single material that passes through the process unchanged.

Energy balances

An energy balance may be made for a process, or part of a process, that is separated from the surroundings by an imaginary boundary. As in a mass balance, input across the boundary must equal output plus any accumulation; if conditions are steady and unvarying with time, input equals output.

All forms of energy must be included in an energy balance. In most flow processes some forms, however, such as magnetic, surface, and mechanical stress energies, do not change and need not be considered. The most important forms are kinetic energy, potential energy, enthalpy, heat, and work; in electrochemical processes, electric energy must be added to the list.

Energy balance for single-stream process. As an example of a steady-flow process in which a single stream of material is treated, consider the process shown in Fig. 1.1. The equipment is any device through which the material is passing. Assume the material is flowing through the system at a constant mass rate. Consider the flow of m kg of material. The entering stream has a velocity of u_a m/s and is Z_a m above the horizontal datum above which heights are measured. Its enthalpy (a quantity discussed later) is H_a J/kg. The corresponding quantities for the leaving stream are u_b, Z_b, and H_b. Heat in the amount of Q J is being transferred through the boundaries of the equipment to the material flowing through it during the time m kg of fluid enters the equipment. If the equipment includes a turbine or engine, it may do work, usually by means of a turning shaft, on the outside. If the unit includes a pump, work from the outside must be done on the material, again through the agency of a turning shaft. Work effects of this kind are called *shaft work*. Assume that shaft work equal to W_s J is being done on the outside by the equipment. For this process, the following equation, which is derived in standard texts on technical thermodynamics, applies:[10]

$$m\left[\frac{u_b^2 - u_a^2}{2} + g(Z_b - Z_a) + H_b - H_a\right] = Q - W_s \qquad (1.59)$$

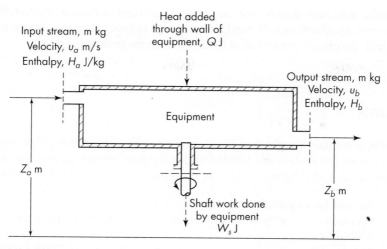

FIGURE 1.1
Diagram for steady-flow process.

or, in fps units,

$$m \left[\frac{u_b^2 - u_a^2}{2g_c J} + \frac{g(Z_b - Z_a)}{g_c J} + H_b - H_a \right] = Q - \frac{W_s}{J} \qquad (1.60)$$

where J, g, and g_c have their usual meanings.

To apply Eq. (1.59) or Eq. (1.60) to a specific situation, a precise choice of the boundaries of the equipment must be made. The inlet and outlet streams must be identified, the inlet and outlet ports located, and rotating shafts noted. All heat-transfer areas between the equipment and its surroundings must be located. The boundaries of the equipment and the cross sections of all shafts and inlet and outlet ports form the *control surface*. This must be a closed envelope, without gaps. Equation (1.59) applies to the equipment and material inside the control surface. For example, the control surface of the process of Fig. 1.1 is bounded by the walls of the equipment and the cross sections of the shaft and inlet and outlet ports, as shown by the dotted lines. The space enclosed by the control surface is called the *control volume*.

The heat effect Q is, by convention, positive when heat flows from outside the control surface into the equipment and negative when heat flows the other way. The shaft work W_s is taken as positive when work is done on the outside of the control surface by the equipment, and negative when work is supplied to the equipment from outside the control surface. Thus work required by a pump located inside the control surface is negative. Both Q and W_s are net effects; if there is more than one heat flow or shaft work, the individual values are added algebraically and the net values of Q and W_s are used in Eqs. (1.59) and (1.60).

Finally, note that there is no term in Eqs. (1.59) and (1.60) for friction. Friction is an internal transformation of mechanical energy to heat and occurs inside the control surface. Its effects are included in the other terms in the equation.

SYMBOLS

In general, quantities are given in SI, cgs, and fps units; quantities given only in either the cgs or fps system are limited to that system; quantities used in only one equation are identified by the number of the equation.

A	Area of heating surface, ft^2 [Eq. (1.41)]
B	Second virial coefficient, equation of state, m^3/kg mol, cm^3/g mol, or ft^3/lb mol
C	Third virial coefficient, equation of state, m^6/(kg mol)^2, cm^6/(g mol)^2, or ft^6(lb mol)^2
c_p	Specific heat, J/g·°C or Btu/lb·°F
D	Diameter, m or ft; also fourth virial coefficient, equation of state, m^9/(kg mol)^3, cm^9/(g mol)^3, or ft^9/(lb mol)^3
D'_o	Outside diameter of pipe, in. [Eq. (1.41)]
F	Force, N, dyn, or lb$_f$
f	Function of
G	Mass velocity, kg/s · m^2 or lb/h · ft^2; also gravitational constant, N · m^2/kg^2, dyn · cm^2/g^2, or lb$_f$ · ft^2/lb^2
g	Acceleration of free fall, m/s^2, cm/s^2, or ft/s^2; g_n, standard value, 9.80665* m/s^2, 980.665 cm/s^2, 32.1740 ft/s^2
g_c	Proportionality factor, $1/k_1$ in Eq. (1.35), 32.1740 ft · lb/lb$_f$ · s^2
$\bar{H}$	Dimension of heat, energy, or work
H	Enthalpy, J/kg
H_a	at entrance
H_b	at exit
h	Heat transfer coefficient, W/m^2 · °C or Btu/h · ft^2 · °F
J	Mechanical equivalent of heat, 4.1868 J/cal$_{IT}$, 778.17 ft · lb$_f$/Btu
k	Thermal conductivity, W/m · °C or Btu/h · ft · °F; also proportionality factor k_1 in Eq. (1.1); k_2 in Eq. (1.2); k_3 in Eq. (1.3); k_4 in Eq. (1.4)
$\bar{L}$	Dimension of length
M	Molecular weight
$\bar{M}$	Dimension of mass
m	Mass, kg, g, or lb; m_a, m_b, masses of particles [Eq. (1.2)]
n	Number of moles
P	Total pressure on mixture
p	Pressure, Pa, dyn/cm^2, or lb$_f$/ft^2; p_A, partial pressure of component A; p_B, partial pressure of component B; p_C, partial pressure of component C
Q	Quantity of heat, J, cal, or Btu; Q_c, heat absorbed by system during cycle
q	Rate of heat transfer, Btu/h [Eq. (1.41)]

R	Gas law constant, 8.31447×10^3 J/K $\cdot$ kg mol, 8.31447×10^7 ergs/K $\cdot$ g mol, or 1.98588 Btu/°R $\cdot$ lb mol
r	Distance between two mass points m_a and m_b [Eq. (1.2)]
T	Temperature, K, °C, °R, or °F; thermodynamic absolute temperature [Eq. (1.11)]
$\bar{T}$	Dimension of temperature
t	Time, s; $\bar{t}$, dimension of time
u	Linear velocity, m/s, cm/s, or ft/s; u_0, initial velocity of falling body
V	Volume, m³, cm³, or ft³
$\bar{V}$	Average fluid velocity, m/s, cm/s, or ft/s
W	Work, J, ergs, or ft $\cdot$ lb$_f$; W_c, work delivered by system during cycle
y	Mole fraction in gas mixture; y_A, mole fraction of component A; y_B, mole fraction of component B; y_C, mole fraction of component C
Z	Height above datum plane, m, cm, or ft
z	Compressibility factor, dimensionless

Greek Letters

ΔT	Temperature difference, °F [Eq. (1.41)]
μ	Absolute viscosity, kg/m $\cdot$ s or lb/ft $\cdot$ s
ρ	Density, kg/m³, g/cm³, or lb/ft³
ρ_M	Molar density, kg mol/m³, g mol/cm³, or lb mol/ft³
Ψ	Function of
Φ, Φ_1	Functions of

PROBLEMS

1.1. Using defined constants and conversion factors for mass, length, time, and temperature, calculate conversion factors for (*a*) foot-pounds force to kilowatthours, (*b*) gallons (1 gal = 231 in.³) to liters (10^3 cm³), (*c*) Btu per pound mole to joules per kilogram mole.
Answers: See App. 1.

1.2. Approximately how many minutes are there in a microcentury?

1.3. The Beattie-Bridgman equation, a famous equation of state for real gases, may be written

$$p = \frac{RT[1 - c/(vT^3)]}{v^2}\left[v + B_0\left(1 - \frac{b}{v}\right)\right] - \frac{A_0}{v^2}\left(1 - \frac{a}{v}\right) \qquad (1.61)$$

where a, A_0, b, B_0, and c are experimental constants and v is the molar volume, 1/g mol. (*a*) Show that this equation can be put into the form of Eq. (1.54), and derive equations for the virial coefficients B, C, and D in terms of the constants in Eq. (1.61). (*b*) For air the constants are $a = 0.01931$, $A_0 = 1.3012$, $b = -0.01101$, $B_0 = 0.04611$, and $c \times 10^{-4} = 66.00$, all in cgs units (atmospheres, liters, gram moles, kelvins, with $R = 0.08206$). Calculate values of the virial coefficients for air in SI units. (c) Calculate z for air at a temperature of 300 K and a molar volume of 0.200 m³/kg mol.

1.4. A mixture of 25 percent ammonia gas and 75 percent air (dry basis) is passed upward through a vertical scrubbing tower, to the top of which water is pumped. Scrubbed gas containing 0.5 percent ammonia leaves the top of the tower, and an aqueous solution containing 10 percent ammonia by weight leaves the bottom. Both entering and leaving gas streams are saturated with water vapor. The gas enters the tower at 37.8°C and leaves at 21.1°C. The pressure of both streams and throughout the tower is 1.02 atm gauge. The air-ammonia mixture enters the tower at a rate of 28.32 m³/min, measured as dry gas at 15.6°C and 1 atm. What percentage of the ammonia entering the tower is not absorbed by the water? How many cubic meters of water per hour are pumped to the top of the tower?
Answers: 1.5%; 2.71 m³/h

1.5. Dry gas containing 75 percent air and 25 percent ammonia vapor enters the bottom of a cylindrical packed absorption tower that is 2 ft in diameter. Nozzles in the top of the tower distribute water over the packing. A solution of ammonia in water is drawn from the bottom of the column, and scrubbed gas leaves the top. The gas enters at 80°F and 760 mm Hg pressure. It leaves at 60°F and 730 mm. The leaving gas contains, on the dry basis, 1.0 percent ammonia. (a) If the entering gas flows through the empty bottom of the column at an average velocity (upward) of 1.5 ft/s, how many cubic feet of entering gas are treated per hour? (b) How many pounds of ammonia are absorbed per hour?
Answers: (a) 16,965 ft³/h; (b) 177 lb

1.6. An evaporator is fed continuously with 25 t (metric tons)/h of a solution consisting of 10 percent NaOH, 10 percent NaCl, and 80 percent H_2O. During evaporation, water is boiled off, and salt precipitates as crystals, which are settled and removed from the remaining liquor. The concentrated liquor leaving the evaporator contains 50 percent NaOH, 2 percent NaCl, and 48 percent H_2O.

Calculate (a) the kilograms of water evaporated per hour, (b) the kilograms of salt precipitated per hour, and (c) the kilograms of concentrated liquor produced per hour.
Answers: (a) 17,600 kg/h; (b) 2,400 kg/h; (c) 5,000 kg/h

1.7. Air is flowing steadily through a horizontal heated tube. The air enters at 40°F and at a velocity of 50 ft/s. It leaves the tube at 140°F and 75 ft/s. The average specific heat of air is 0.24 Btu/lb·°F. How many Btus per pound of air are transferred through the wall of the tube?
Answer: 24.1 Btu/lb

1.8. Check the dimensional consistency of the following empirical equation for heat transfer between a flowing fluid and the surface of a sphere (see Chap. 12):

$$h = 2.0kD_p^{-1} + 0.6D_p^{-0.5}G^{0.5}\mu^{-0.17}c_p^{0.33}k^{0.67}$$

where h = heat transfer coefficient
D_p = diameter of sphere
k = thermal conductivity of fluid
G = mass velocity of fluid

μ = viscosity of fluid
c_p = specific heat of fluid

1.9. In the orifice meters discussed in Chap. 8, a flat disk with a central opening of diameter D_o is set across a pipe of diameter D, and the pressure drop Δp across the opening is measured. It is postulated that Δp is a function of the average fluid velocity in the pipe $\bar{V}$, the density of the fluid ρ, the fluid viscosity μ, and the diameters of the pipe and the opening, D and D_o, respectively. Thus

$$\Delta p = \Phi(\bar{V}, \rho, \mu, D, D_o)$$

Find an acceptable set of dimensionless groups which relate these various factors.
Answers: $\frac{\Delta p}{\rho \bar{V}^2} = \Phi\left(\frac{D_o \bar{V} \rho}{\mu}, \frac{D}{D_o}\right)$

REFERENCES

1. Austin, G. T. *Shreve's Chemical Process Industries.* 5th ed. New York: McGraw-Hill, 1984.
2. Bridgman, P. W. *Dimensional Analysis.* New York: AMS Press, 1978.
3. Churchill, S. W. *Chem. Eng. Education,* **30**(3):158 (1997).
4. *CRC Handbook of Chemistry and Physics.* 69th ed. Boca Raton, FL: CRC Press, 1988, p. F-191.
5. Halladay, D., and R. Resnick. *Fundamentals of Physics.* 3d ed. New York: Wiley, 1988, p. 4.
6. Moldover, M. R., et al. *J. Res. Natl. Bur. Std.* **93**(2):85 (1988).
7. *Natl. Bur. Std. Tech. News Bull.* **55**:3 (March 1971).
8. National Research Council. *Frontiers in Chemical Engineering.* Washington: National Academy Press, 1988.
9. Prausnitz, J. M., R. N. Lichtenthaler, and E. G. de Azevedo. *Molecular Theory of Fluid-Phase Equilibria.* Englewood Cliffs, NJ: Prentice-Hall, 1986.
10. Smith, J. M., H. C. Van Ness, and M. M. Abbott. *Introduction to Chemical Engineering Thermodynamics.* 5th ed. New York: McGraw-Hill, 1996.

SECTION II

Fluid Mechanics

The behavior of fluids is important to process engineering generally and constitutes one of the foundations for the study of unit operations. An understanding of fluids is essential, not only for accurately treating problems in the movement of fluids through pipes, pumps, and all kinds of process equipment but also for the study of heat flow and the many separation operations that depend on diffusion and mass transfer.

The branch of engineering science that has to do with the behavior of fluids—and fluids are understood to include liquids, gases, and vapors—is called *fluid mechanics*. Fluid mechanics in turn is part of a larger discipline called *continuum mechanics*, which also includes the study of stressed solids.

Fluid mechanics has two branches important to the study of unit operations: *fluid statics*, which treats fluids in the equilibrium state of no shear stress, and *fluid dynamics*, which treats fluids when portions of the fluid are in motion relative to other parts.

The chapters of this section deal with those areas of fluid mechanics that are important to unit operations. The choice of subject matter is but a sampling of the huge field of fluid mechanics generally. Chapter 2 treats fluid statics and some of its important applications. Chapter 3 discusses the important phenomena appearing in flowing fluids. Chapter 4 deals with the basic quantitative laws and equations of fluid flow. Chapter 5 treats the flow of incompressible fluids through pipes, Chap. 6 discusses the flow of compressible fluids, and Chap. 7 describes flow past solids immersed in the flowing fluid. Chapter 8 deals with the important engineering tasks of moving fluids through process equipment and of measuring and controlling fluids in flow. Finally, Chap. 9 covers mixing, agitation, and dispersion operations, which in essence are applied fluid mechanics.

CHAPTER 2

Fluid Statics and Its Applications

Nature of fluids

A fluid is a substance that does not permanently resist distortion. An attempt to change the shape of a mass of fluid results in layers of fluid sliding over one another until a new shape is attained. During the change in shape, shear stresses[†] exist, the magnitudes of which depend upon the viscosity of the fluid and the rate of sliding; but when a final shape has been reached, all shear stresses will have disappeared. A fluid in equilibrium is free from shear stresses.

At a given temperature and pressure, a fluid possesses a definite density, which in engineering practice is usually measured in kilograms per cubic meter or pounds per cubic foot. Although the density of all fluids depends on the temperature and pressure, the variation in density with changes in these variables may be small or large. If the density changes only slightly with moderate changes in temperature and pressure, the fluid is said to be *incompressible;* if the changes in density are significant, the fluid is said to be *compressible.* Liquids are generally considered to be incompressible and gases compressible. The terms are relative, however, and the density of a liquid can change appreciably if pressure and temperature are changed over wide limits. Also, gases subjected to small percentage changes in pressure and temperature act as incompressible fluids, and density changes under such conditions may be neglected without serious error.

The pressure in a static fluid is familiar as a surface force exerted by the fluid against a unit area of the walls of its container. Pressure also exists at every point within a volume of fluid. It is a scalar quantity; at any given point its magnitude is the same in all directions.

[†]Shear is the lateral displacement of one layer of material relative to another layer by an external force. *Shear stress* is defined as the ratio of this force to the area of the layer. See Chap. 3.

HYDROSTATIC EQUILIBRIUM

In a stationary mass of a single static fluid, the pressure is constant in any cross section parallel to the earth's surface but varies from height to height. Consider the vertical column of fluid shown in Fig. 2.1. Assume the cross-sectional area of the column is S. At a height Z above the base of the column let the pressure be p and the density be ρ. The resultant of all forces on the small volume of fluid of height dZ and cross-sectional area S must be zero. Three vertical forces are acting on this volume: (1) the force from pressure p acting in an upward direction, which is pS; (2) the force from pressure $p + dp$ acting in a downward direction, which is $(p + dp)S$; (3) the force of gravity acting downward, which is $g\rho S/dZ$. Then

$$+pS - (p + dp)S - g\rho S\,dZ = 0 \tag{2.1}$$

In this equation, forces acting upward are taken as positive and those acting downward as negative. After simplification and division by S, Eq. (2.1) becomes

$$dp + g\rho\,dZ = 0 \tag{2.2}$$

Equation (2.2) cannot be integrated for compressible fluids unless the variation of density with pressure is known throughout the column of fluid. However, it is often satisfactory for engineering calculations to consider ρ to be essentially constant. The density is constant for incompressible fluids and, except for large changes in height, is nearly so for compressible fluids. Integration of Eq. (2.2) on the assumption that ρ is constant gives

$$\frac{p}{\rho} + gZ = \text{const} \tag{2.3}$$

FIGURE 2.1
Hydrostatic equilibrium.

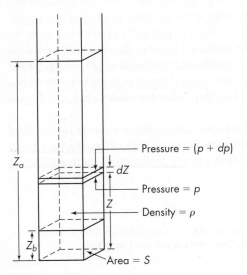

or, between the two definite heights Z_a and Z_b shown in Fig. 2.1,

$$\frac{p_b}{\rho} - \frac{p_a}{\rho} = g(Z_a - Z_b) \tag{2.4}$$

Equation (2.3) expresses mathematically the condition of hydrostatic equilibrium.

Barometric equation

For an ideal gas, the density and pressure are related by the equation

$$\rho = \frac{pM}{RT} \tag{2.5}$$

where M = molecular weight
$\quad\quad T$ = absolute temperature

Substitution from Eq. (2.5) into Eq. (2.2) gives $\quad d\,p + g\,\rho\,d\,z$

$$\frac{dp}{p} + \frac{gM}{RT}dZ = 0 \tag{2.6}$$

Integration of Eq. (2.6) between levels a and b, on the assumption that T is constant, gives

$$\ln\frac{p_b}{p_a} = -\frac{gM}{RT}(Z_b - Z_a)$$

or

$$\frac{p_b}{p_a} = \exp\left[-\frac{gM(Z_b - Z_a)}{RT}\right] \tag{2.7}$$

Equation (2.7) is known as the *barometric equation*.

Methods are available in the literature[3] for estimating the pressure distribution in situations, for example, a deep gas well, in which the gas is not ideal and the temperature is not constant.

Hydrostatic equilibrium in a centrifugal field

In a rotating centrifuge a layer of liquid is thrown outward from the axis of rotation and is held against the wall of the bowl by centrifugal force. The free surface of the liquid takes the shape of a paraboloid of revolution,[2] but in industrial centrifuges the rotational speed is so high and the centrifugal force so much greater than the force of gravity that the liquid surface is virtually cylindrical and coaxial with the axis of rotation. This situation is illustrated in Fig. 2.2, in which r_1 is the radial distance from the axis of rotation to the free liquid surface and r_2 is the radius of the centrifuge bowl. The entire mass of liquid indicated in Fig. 2.2 is rotating as a rigid body, with no sliding of one layer of liquid over another. Under these conditions the pressure distribution in the liquid may be found from the principles of fluid statics.

The pressure drop over any ring of rotating liquid is calculated as follows. Consider the ring of liquid shown in Fig. 2.2 and the volume element of thickness dr

at radius r

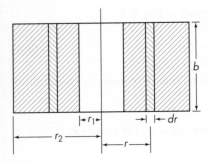

FIGURE 2.2
Single liquid in centrifuge bowl.

dr at a radius r:

$$dF = \omega^2 r \, dm$$

$vol = area \times length$

where dF = centrifugal force
$\quad dm$ = mass of liquid in element
$\quad \omega$ = angular velocity, rad/s

If ρ is the density of the liquid and b the breadth of the ring,

$$dm = 2\pi \rho r b \, dr$$

Eliminating dm gives

$$dF = 2\pi \rho b \omega^2 r^2 \, dr$$

The change in pressure over the element is the force exerted by the element of liquid, divided by the area of the ring:

$$dp = \frac{dF}{2\pi r b} = \omega^2 \rho r \, dr$$

The pressure drop over the entire ring is

$$p_2 - p_1 = \int_{r_1}^{r_2} \omega^2 \rho r \, dr$$

Assuming the density is constant and integrating give

$$p_2 - p_1 = \frac{\omega^2 \rho (r_2^2 - r_1^2)}{2} \tag{2.8}$$

Equation (2.8) strictly applies only when r_1 and r_2 are not greatly different, but for practical systems the error is small.

APPLICATIONS OF FLUID STATICS

Manometers

The manometer is an important device for measuring pressure differences. Figure 2.3 shows the simplest form of manometer. Assume that the shaded portion of the U tube is filled with liquid A having a density ρ_A and that the arms of the U

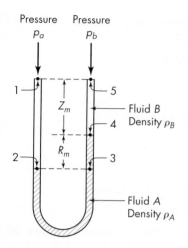

Pressure p_a Pressure p_b

FIGURE 2.3
Simple manometer.

tube above the liquid are filled with fluid B having a density ρ_B. Fluid B is immiscible with liquid A and less dense than A; it is often a gas such as air or nitrogen.

A pressure p_a is exerted in one arm of the U tube and a pressure p_b in the other. As a result of the difference in pressure $p_a - p_b$, the meniscus in one branch of the U tube is higher than in the other, and the vertical distance between the two meniscuses R_m may be used to measure the difference in pressure. To derive a relationship between $p_a - p_b$ and R_m, start at the point 1, where the pressure is p_a; then, as shown by Eq. (2.3), the pressure at point 2 is $p_a + g(Z_m + R_m)\rho_B$. By the principles of hydrostatics, this is also the pressure at point 3. The pressure at point 4 is less than that at point 3 by the amount $gR_m\rho_A$, and the pressure at point 5, which is p_b, is still less by the amount $gZ_m\rho_B$. These statements can be summarized by the equation

$$p_a + g[(Z_m + R_m)\rho_B - R_m\rho_A - Z_m\rho_B] = p_b \tag{2.9}$$

Simplification of this equation gives

$$p_a - p_b = g R_m(\rho_A - \rho_B) \tag{2.10}$$

Note that this relationship is independent of the distance Z_m and of the dimensions of the tube, provided that pressures p_a and p_b are measured in the same horizontal plane. If fluid B is a gas, ρ_B is usually negligible compared to ρ_A and may be omitted from Eq. (2.10).

> EXAMPLE 2.1. A manometer of the type shown in Fig. 2.3 is used to measure the pressure drop across an orifice (see Fig. 8.18). Liquid A is mercury (density 13,590 kg/m³) and fluid B, flowing through the orifice and filling the manometer leads, is brine (density 1,260 kg/m³). When the pressures at the taps are equal, the level of the mercury in the manometer is 0.9 m below the orifice taps. Under operating conditions, the gauge pressure[†] at the upstream tap is 0.14 bar; the pressure at the downstream tap is 250 mm Hg below atmospheric. What is the reading of the manometer in millimeters?

[†]Gauge pressure is pressure measured above the prevailing atmospheric pressure.

Solution. Call atmospheric pressure zero; then the numerical data for substitution in Eq. (2.10) are

$$p_a = 0.14 \times 10^5 = 14{,}000 \text{ Pa}$$

From Eq. (2.4)

$$p_b = Z_b g \rho_A$$
$$= -\frac{250}{1000} \times 9.80665 \times 13{,}590$$
$$= -33{,}318 \text{ Pa}$$

Substituting in Eq. (2.10) gives

$$14{,}000 + 33{,}318 = R_m \times 9.80665 \times (13{,}590 - 1{,}260)$$
$$R_m = 0.391 \text{ m, or } 391 \text{ mm}$$

For measuring small differences in pressure, the *inclined manometer* shown in Fig. 2.4 may be used. In this type, one leg of the manometer is inclined in such a manner that, for a small magnitude of R_m, the meniscus in the inclined tube must move a considerable distance along the tube. This distance is R_m divided by the sine of α, the angle of inclination. By making α small, the magnitude of R_m is multiplied into a long distance R_1, and a large reading becomes equivalent to a small pressure difference; so

$$p_a - p_b = g R_1 (\rho_A - \rho_B) \sin \alpha \qquad (2.11)$$

In this type of pressure gauge, it is necessary to provide an enlargement in the vertical leg so that the movement of the meniscus in the enlargement is negligible within the operating range of the instrument.

Continuous gravity decanter

A gravity decanter of the type shown in Fig. 2.5 is used for the continuous separation of two immiscible liquids of differing densities. The feed mixture enters at one end of the separator; the two liquids flow slowly through the vessel, separate into two layers, and discharge through overflow lines at the other end of the separator.

Provided the overflow lines are so large that frictional resistance to the flow of the liquids is negligible, and provided they discharge at the same pressure as that in

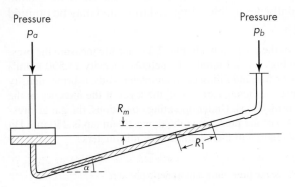

Pressure
p_a

Pressure
p_b

FIGURE 2.4
Inclined manometer.

R_m

R_1

α

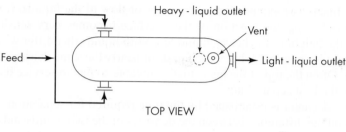

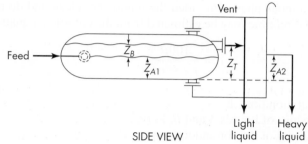

FIGURE 2.5
Continuous gravity decanter for immiscible liquids.

the gas space above the liquid in the vessel, the performance of the decanter can be analyzed by the principles of fluid statics.

For example, in the decanter shown in Fig. 2.5 let the density of the heavy liquid be ρ_A and that of the light liquid be ρ_B. The depth of the layer of heavy liquid is Z_{A1}, and that of the light liquid is Z_B. The total depth of liquid in the vessel Z_T is fixed by the position of the overflow line for the light liquid. Heavy liquid discharges through an overflow leg connected to the bottom of the vessel and rising to a height Z_{A2} above the vessel floor. The overflow lines and the top of the vessel are all vented to the atmosphere.

Since there is negligible frictional resistance to flow in the discharge lines, the column of heavy liquid in the heavy-liquid overflow leg must balance the somewhat greater depth of the two liquids in the vessel. A hydrostatic balance leads to the equation

$$Z_B \rho_B + Z_{A1} \rho_A = Z_{A2} \rho_A \tag{2.12}$$

Solving Eq. (2.12) for Z_{A1} gives

$$Z_{A1} = Z_{A2} - Z_B \frac{\rho_B}{\rho_A} = Z_{A2} - (Z_T - Z_{A1}) \frac{\rho_B}{\rho_A} \tag{2.13}$$

where the total depth of liquid in the vessel is $Z_T = Z_B + Z_{A1}$. From this

$$Z_{A1} = \frac{Z_{A2} - Z_T(\rho_B/\rho_A)}{1 - \rho_B/\rho_A} \tag{2.14}$$

Equation (2.14) shows that the position of the liquid-liquid interface in the separator depends on the ratio of the densities of the two liquids and on the elevations
of the overflow lines.

of the overflow lines. It is independent of the rates of flow of the liquids. It shows that as ρ_A approaches ρ_B, the position of the interface becomes very sensitive to changes in Z_{A2}, the height of the heavy-liquid leg. With liquids that differ widely in density, this height is not critical; but with liquids of nearly the same density, it must be set with care. Often the top of the leg is made movable so that in service it can be adjusted to give the best separation.

The size of a decanter is established by the time required for separation, which in turn depends on the difference between the densities of the two liquids and on the viscosity of the continuous phase. Provided the liquids are clean and do not form emulsions, the separation time may be estimated from the empirical equation[1]

$$t = \frac{100\mu}{\rho_A - \rho_B} \tag{2.15}$$

where t = separation time, h
ρ_A, ρ_B = densities of liquids A and B, kg/m^3
μ = viscosity of the continuous phase, cP

Equation (2.15) is not dimensionless, and the indicated units must be used.

EXAMPLE 2.2. A horizontal cylindrical continuous decanter is to separate 1,500 bbl/d (day) (9.93 m^3/h) of a liquid petroleum fraction from an equal volume of wash acid. The oil is the continuous phase and at the operating temperature has a viscosity of 1.1 cP and a density of 54 lb/ft^3 (865 kg/m^3). The density of the acid is 72 lb/ft^3 (1,153 kg/m^3). Compute (a) the size of the vessel and (b) the height of the acid overflow above the vessel floor.

Solution
(a) The vessel size is found from the separation time. Substitution in Eq. (2.15) gives

$$t = \frac{100 \times 1.1}{1,153 - 865} = 0.38 \text{ h}$$

or 23 min. Since 1 bbl = 42 gal, the rate of flow of each stream is

$$\frac{1,500 \times 42}{24 \times 60} = 43.8 \text{ gal/min}$$

The total liquid holdup is

$$2 \times 43.8 \times 23 = 2,014 \text{ gal}$$

The vessel should be about 95 percent full, so its volume is 2,014/0.95, or 2,120, gal (8.03 m^3).

The length of the tank should be about 5 times its diameter. A tank 4 ft (1.22 m) in diameter and 22 ft (6.10 m) long would be satisfactory; with standard dished heads on the ends, its total volume would be 2,124 gal.

(b) The fraction of the tank volume occupied by the liquid will be 95 percent, and for a horizontal cylinder this means that the liquid depth will be 90 percent of the tank diameter. Thus

$$Z_T = 0.90 \times 4 = 3.6 \text{ ft}$$

If the interface is halfway between the vessel floor and the liquid surface, $Z_{A1} = 1.80$ ft. Solving Eq. (2.14) for Z_{A2}, the height of the heavy-liquid overflow, gives

$$Z_{A2} = 1.80 + (3.60 - 1.80)\tfrac{54}{72} = 3.15 \text{ ft (0.96 m)}$$

Successful operation of a decanter depends on both the sedimentation and the coalescence of the dispersed phase. Equation (2.15) gives poor results if the liquids to be separated are not clean but contain particulates or polymeric films that reduce the rate of coalescence. Such contaminants may also lead to the formation of a dirty layer of uncoalesced droplets—called a "rag"—at the liquid-liquid interface. Coalescing devices such as beds of porous solids, membranes, or high-voltage fields are often necessary for a satisfactory separation. With cleaner liquids the size of a decanter can often be greatly reduced by putting in horizontal or slightly inclined pipes or flat plates, so that the heavy-phase droplets need to fall only a short distance before reaching a layer of the heavy liquid.

Centrifugal decanter

When the difference between the densities of the two liquids is small, the force of gravity may be too weak to separate the liquids in a reasonable time. The separation may then be accomplished in a liquid-liquid centrifuge, shown diagrammatically in Fig. 2.6. It consists of a cylindrical metal bowl, usually mounted vertically, that rotates about its axis at high speed. In Fig. 2.6a the bowl is at rest and contains a quantity of two immiscible liquids of differing densities. The heavy liquid forms a layer on the floor of the bowl beneath a layer of light liquid. If the bowl is now rotated, as in Fig. 2.6b, the heavy liquid forms a layer, denoted as zone A in the figure, next to the inside wall of the bowl. A layer of light liquid, denoted as zone B, forms inside the layer of heavy liquid. A cylindrical interface of radius r_i separates

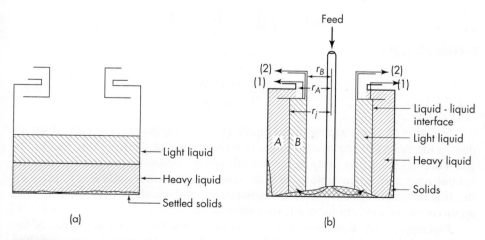

FIGURE 2.6

Centrifugal separation of immiscible liquids: (a) bowl at rest; (b) bowl rotating. Zone A, separation of light liquid from heavy; zone B, separation of heavy liquid from light. (1) Heavy-liquid drawoff. (2) Light-liquid drawoff.

the two layers. Since the force of gravity can be neglected in comparison with the much greater centrifugal force, this interface is vertical. It is called the *neutral zone*.

In operation of the machine, the feed is admitted continuously near the bottom of the bowl. Light liquid discharges at point 2 through ports near the axis of the bowl; heavy liquid passes under a ring, inward toward the axis of rotation, and discharges over a dam at point 1. If there is negligible frictional resistance to the flow of the liquids as they leave the bowl, the position of the liquid-liquid interface is established by a hydrostatic balance and the relative "heights" (radial distances from the axis) of the overflow ports at 1 and 2.

Assume that the heavy liquid, of density ρ_A, overflows the dam at radius r_A, and the light liquid, of density ρ_B, leaves through ports at radius r_B. Then if both liquids rotate with the bowl and friction is negligible, the pressure difference in the light liquid between r_B and r_i must equal that in the heavy liquid between r_A and r_i. The principle is exactly the same as in a continuous gravity decanter.

Thus

$$p_i - p_B = p_i - p_A \tag{2.16}$$

where p_i = pressure at liquid-liquid interface
$\quad\;\; p_B$ = pressure at free surface of light liquid at r_B
$\quad\;\; p_A$ = pressure at free surface of heavy liquid at r_A

From Eq. (2.8)

$$p_i - p_B = \frac{\omega^2 \rho_B (r_i^2 - r_B^2)}{2} \quad \text{and} \quad p_i - p_A = \frac{\omega^2 \rho_A (r_i^2 - r_A^2)}{2}$$

Equating these pressure drops and simplifying lead to

$$\rho_B (r_i^2 - r_B^2) = \rho_A (r_i^2 - r_A^2)$$

Solving for r_i gives

$$r_i = \sqrt{\frac{r_A^2 - (\rho_B/\rho_A) r_B^2}{1 - \rho_B/\rho_A}} \tag{2.17}$$

Equation (2.17) is analogous to Eq. (2.14) for a gravity settling tank. It shows that r_i, the radius of the neutral zone, is sensitive to the density ratio, especially when the ratio is nearly unity. If the densities of the fluids are too nearly alike, the neutral zone may be unstable even if the speed of rotation is sufficient to separate the liquids quickly. The difference between ρ_A and ρ_B should not be less than approximately 3 percent for stable operation.

Equation (2.17) also shows that if r_B is held constant and r_A, the radius of the discharge lip for the heavier liquid, is increased, then the neutral zone is shifted toward the wall of the bowl. If r_A is decreased, the zone is shifted toward the axis; An increase in r_B, at constant r_A, also shifts the neutral zone toward the axis; and a decrease in r_B causes a shift toward the wall. The position of the neutral zone is important practically. In zone A, the lighter liquid is being removed from a mass of heavier liquid; and in zone B, heavy liquid is being stripped from a mass of light liquid. If one of the processes is more difficult than the other, more time should be provided for the more difficult step. For example, if the separation in zone B is more difficult than that in zone A, zone B should be large and zone A small. This is accomplished by moving the neutral zone toward the wall by increasing r_A or decreasing r_B. To obtain a larger time factor in zone A, the opposite adjustments would be made. Many centrifugal separators are so constructed that either r_A or r_B can be varied to control the position of the neutral zone.

Flow through continuous decanters

Equations (2.14) and (2.17) for the interfacial position in continuous decanters are based entirely on hydrostatic balances. As long as there is negligible resistance to flow in the outlet pipes, the position of the interface is the same regardless of the rates of flow of the liquids and of the relative quantities of the two liquids in the feed. The rate of separation is the most important variable, for as mentioned before, it fixes the size of a gravity decanter and determines whether a high centrifugal force is needed. The rates of motion of a dispersed phase through a continuous phase are discussed in Chap. 7.

SYMBOLS

b	Breadth, m or ft
F	Force, N or lb_f
g	Gravitational acceleration, m/s^2 or ft/s^2
g_c	Newton's law proportionality factor, 32.174 ft · lb/lb$_f$ · s^2
M	Molecular weight
m	Mass, kg or lb
p	Pressure, N/m^2 or lb$_f$/ft^2; p_A, at surface of heavy liquid in centrifuge; p_B, at surface of light liquid in centrifuge; p_a, at location a; p_b, at location b; p_i, at liquid-liquid interface; p_1, at free liquid surface; p_2, at wall of centrifuge bowl
R	Gas law constant, 8314.47 J/kg mol · K or 1545 ft · lb$_f$/lb mol · °R
R_m	Reading of manometer, m or ft; R_1, of inclined manometer
r	Radial distance from axis, m or ft; r_A, to heavy-liquid overflow; r_B, to light-liquid overflow; r_i, to liquid-liquid interface; r_1, to free-liquid surface in centrifuge; r_2, to wall of centrifuge bowl
S	Cross-sectional area, m^2 or ft^2
T	Absolute temperature, K or °R
t	Separation time, h
Z	Height, m or ft; Z_{A1}, of layer of heavy liquid in decanter; Z_{A2}, of heavy liquid in overflow leg; Z_B, of layer of light liquid in decanter; Z_T, total depth of liquid; Z_a, Z_b, heights at locations a and b; Z_m, height of pressure connections in manometer above measuring liquid

Greek Letters

α Angle with horizontal
μ Viscosity of continuous phase, cP
ρ Density, kg/m^3 or lb/ft^3; ρ_A, of fluid A; ρ_B, of fluid B
ω Angular velocity, rad/s

PROBLEMS

2.1. A simple U-tube manometer is installed across an orifice meter. The manometer is filled with mercury (specific gravity 13.6), and the liquid above the mercury is carbon tetrachloride (specific gravity 1.6). The manometer reads 200 mm. What is the pressure difference over the manometer in newtons per square meter?

2.2. The temperature of the earth's atmosphere drops about 5°C for every 1000 m of elevation above the earth's surface. If the air temperature at ground level is 15°C and the pressure is 760 mm Hg, at what elevation is the pressure 380 mm Hg? Assume that the air behaves as an ideal gas.

2.3. How much error would be introduced in the answer to Prob. 2.2 if the equation for hydrostatic equilibrium [Eq. (2.4)] were used, with the density evaluated at 0°C and an arithmetic average pressure?

2.4. A continuous gravity decanter is to separate chlorobenzene, with a density of 1,109 kg/m^3, from an aqueous wash liquid having a density of 1,020 kg/m^3. If the total depth in the separator is 1 m and the interface is to be 0.6 m from the vessel floor, (*a*) what should the height of the heavy-liquid overflow leg be; and (*b*) how much would an error of 50 mm in this height affect the position of the interface?

2.5. What should be the volume of the separator in Prob. 2.4 to separate 1,600 kg/h chlorobenzene from 2,100 kg/h of wash liquid? The wash liquid is to be the continuous phase; its viscosity is the same as that of water at the operating temperature of 35°C.

2.6. A centrifuge bowl with 250-mm ID (internal diameter) is turning at 4,000 r/min. It contains a layer of aniline 50 mm thick. If the density of the aniline is 1,002 kg/m^3 and the pressure at the liquid surface is atmospheric, what gauge pressure is exerted on the wall of the centrifuge bowl?

2.7. The liquids described in Prob. 2.4 are to be separated in a tubular centrifuge bowl with an inside diameter of 150 mm, rotating at 8,000 r/min. The free-liquid surface inside the bowl is 40 mm from the axis of rotation. If the centrifuge bowl is to contain equal volumes of the two liquids, what should be the radial distance from the rotational axis to the top of the overflow dam for the heavy liquid?

2.8. A U-tube manometer indicates a pressure drop of 10 in. water across an air filter. The air is at 26°C and a gauge pressure of 60 lb$_f$/in^2. (*a*) What is the pressure drop in lb$_f$/in^2

and in atm? (*b*) What percentage of error is introduced if the density of the air in the manometer leads is neglected?

2.9. A research submarine is designed to operate 3 km below the ocean surface. If the interior pressure is 1 atm, what is the total pressure on a 15-cm-diameter window? The average density of seawater is 1,028 kg/m^3.

2.10. A suspension of very fine sand particles in water is used to separate bituminous coal from denser particles of rock. If the desired density of the suspension is 1,500 kg/m^3, what are the volume fraction and weight fraction of sand that should be used?

REFERENCES

1. Barton, R. L. *Chem. Eng.* **81**(14):111 (1974).
2. Bird, B., W. E. Stewart, and E. N. Lightfoot. *Transport Phenomena.* New York: Wiley, 959, pp. 96–98.
3. Knudsen, J. G., and D. L. Katz. *Fluid Dynamics and Heat Transfer.* New York: McGraw-Hill, 1958, pp. 69–71.
4. Perry, R. H., and D. Green (eds.). *Perry's Chemical Engineers' Handbook.* 7th ed. New York: McGraw-Hill, 1997, p. 10–139.

CHAPTER 3

Fluid Flow Phenomena

The behavior of a flowing fluid depends strongly on whether the fluid is under the influence of solid boundaries. In the region where the influence of the wall is small, the shear stress may be negligible and the fluid behavior may approach that of an ideal fluid, one that is incompressible and has zero viscosity. The flow of such an ideal fluid is called *potential flow* and is completely described by the principles of newtonian mechanics and conservation of mass. The mathematical theory of potential flow is highly developed but is outside the scope of this book. Potential flow has two important characteristics: (1) Neither circulations nor eddies can form within the stream, so that potential flow is also called irrotational flow; and (2) friction cannot develop, so that there is no dissipation of mechanical energy into heat.

Potential flow can exist at distances not far from a solid boundary. A fundamental principle of fluid mechanics, originally stated by Prandtl in 1904,[8] is that, except for fluids moving at low velocities or possessing high viscosities, the effect of the solid boundary on the flow is confined to a layer of the fluid immediately adjacent to the solid wall. This layer is called the *boundary layer,* and shear and shear forces are confined to this part of the fluid. Outside the boundary layer, potential flow survives. Most technical flow processes are best studied by considering the fluid stream as two parts, the boundary layer and the remaining fluid. In some situations such as flow in a converging nozzle, the boundary layer may be neglected; and in others, such as flow through pipes, the boundary layer fills the entire channel, and there is no potential flow.

Within the current of an incompressible fluid under the influence of solid boundaries, four important effects appear: (1) the coupling of velocity-gradient and shear-stress fields, (2) the onset of turbulence, (3) the formation and growth of boundary layers, and (4) the separation of boundary layers from contact with the solid boundary.

In the flow of compressible fluids past solid boundaries, additional effects appear, arising from the significant density changes that are characteristic of compressible fluids. These are considered in Chap. 6, on the flow of compressible fluids.

The velocity field

When a stream of fluid is flowing in bulk past a solid wall, the fluid adheres to the solid at the actual interface between solid and fluid. The adhesion is a result of the force fields at the boundary, which are also responsible for the interfacial tension between solid and fluid. If, therefore, the wall is at rest in the reference frame chosen for the solid-fluid system, *the velocity of the fluid at the interface is zero.* Since at distances away from the solid the velocity is not zero, there must be variations in velocity from point to point in the flowing stream. Therefore, the velocity at any point is a function of the space coordinates of that point, and a velocity field exists in the space occupied by the fluid. The velocity at a given location may also vary with time. When the velocity at each location is constant, the field is invariant with time and the flow is said to be *steady.*

One-dimensional flow. Velocity is a vector, and in general, the velocity at a point has three components, one for each space coordinate. In many simple situations all velocity vectors in the field are parallel or practically so, and only one velocity component, which may be taken as a scalar, is required. This situation, which obviously is much simpler than the general vector field, is called *one-dimensional flow;* an example is steady flow through straight pipe. The following discussion is based on the assumptions of steady one-dimensional flow.

LAMINAR FLOW, SHEAR RATE, AND SHEAR STRESS

Laminar flow

At low velocities fluids tend to flow without lateral mixing, and adjacent layers slide past one another as playing cards do. There are neither cross-currents nor eddies. This regime is called *laminar flow.* At higher velocities turbulence appears and eddies form, which, as discussed later, lead to lateral mixing.

Velocity gradient and rate of shear

Consider the steady one-dimensional laminar flow of an incompressible fluid along a solid plane surface. Figure 3.1a shows the velocity profile for such a stream. The abscissa u is the velocity, and the ordinate y is the distance measured perpendicular from the wall and therefore at right angles to the direction of the velocity. At $y = 0$, $u = 0$, and u increases with distance from the wall but at a decreasing rate. Focus attention on the velocities on two nearby planes, plane A and plane B, a distance Δy apart. Let the velocities along the planes be u_A and u_B, respectively, and assume that $u_B > u_A$. Call $\Delta u = u_B - u_A$. Define the velocity gradient at y_A, du/dy, by

$$\frac{du}{dy} = \lim_{\Delta y \to 0} \frac{\Delta u}{\Delta y} \tag{3.1}$$

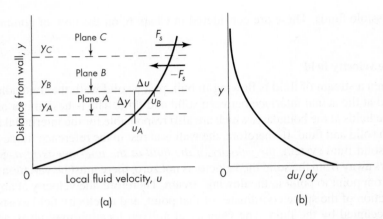

FIGURE 3.1
Profiles of velocity and velocity gradient in layer flow: (*a*) velocity;
(*b*) velocity gradient or rate of shear.

The velocity gradient is clearly the reciprocal of the slope of the velocity profile of Fig. 3.1*a*. The local velocity gradient is also called the *shear rate*, or time rate of shear. The velocity gradient is usually a function of position in the stream and therefore defines a field, as illustrated in Fig. 3.1*b*.

The shear-stress field

Since an actual fluid resists shear, a shear force must exist wherever there is a time rate of shear. In one-dimensional flow the shear force acts parallel to the plane of the shear. For example, at plane C at distance y_C from the wall, the shear force F_s, shown in Fig. 3.1*a*, acts in the direction shown in the figure. This force is exerted by the fluid above plane C on the fluid between plane C and the wall. By Newton's third law, an equal and opposite force $-F_s$ acts on the fluid above plane C from the fluid below plane C. It is convenient to use, not total force F_s, but the force per unit area of the shearing plane, called the *shear stress* and denoted by τ, or

$$\tau = \frac{F_s}{A_s} \tag{3.2}$$

where A_s is the area of the plane. Since τ varies with y, the shear stress also constitutes a field. Shear forces are generated in both laminar and turbulent flow. The shear stress arising from viscous or laminar flow is denoted by τ_v. The effect of turbulence is described later.

RHEOLOGICAL PROPERTIES OF FLUIDS

Newtonian and non-newtonian fluids

The relationships between the shear stress and shear rate in a real fluid are part of the science of rheology. Figure 3.2 shows several examples of the rheological

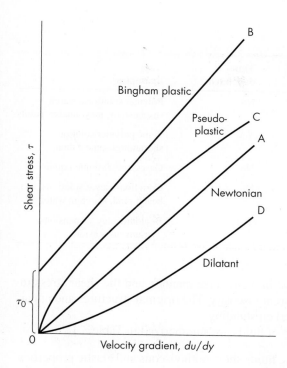

FIGURE 3.2
Shear stress versus velocity gradient for newtonian and non-newtonian fluids.

behavior of fluids. The curves are plots of shear stress versus rate of shear and apply at constant temperature and pressure. The simplest behavior is that shown by curve A, which is a straight line passing through the origin. Fluids following this simple linearity are called newtonian fluids. Gases and most liquids are newtonian. The other curves in Fig. 3.2 represent the rheological behavior of liquids called non-newtonian. Some liquids, for example, sewage sludge, do not flow at all until a threshold shear stress, denoted by τ_0, is attained and then flow linearly, or nearly so, at shear stresses greater than τ_0. Curve B is an example of this relation. Liquids acting this way are called *Bingham plastics*. Line C represents a *pseudoplastic fluid*. The curve passes through the origin, is concave downward at low shear, and becomes nearly linear at high shear. Rubber latex is an example of such a fluid. Curve D represents a *dilatant fluid*. The curve is concave upward at low shear and almost linear at high shear. Quicksand and some sand-filled emulsions show this behavior. Pseudoplastics are said to be *shear rate–thinning* and dilatant fluids *shear rate–thickening.*

Time-dependent flow

None of the curves in Fig. 3.2 depends on the history of the fluid, and a given sample of material shows the same behavior no matter how long the shearing stress has been applied. Such is not the case for some non-newtonian liquids, whose curves of stress versus rate of shear depend on how long the shear has been active. *Thixotropic* liquids break down under continued shear and on mixing give lower shear stress for a given shear rate; that is, their apparent viscosity decreases with

TABLE 3.1
Rheological characteristics of fluids

Designation	Effect of increasing shear rate	Time-dependent?	Examples[2]
Pseudoplastic	Thins	No	Polymer solutions, starch suspensions, mayonnaise, paints
Thixotropic	Thins	Yes	Some polymer solutions, shortening, some paints
Newtonian	None	No	Gases, most simple liquids
Dilatant	Thickens	No	Corn flour–sugar solutions, wet beach sand, starch in water
Rheopectic	Thickens	Yes	Bentonite clay suspensions, gypsum suspensions

time. *Rheopectic* substances behave in the reverse manner, and the shear stress increases with time, as does the apparent viscosity. The original structures and apparent viscosities are usually recovered on standing.

The rheological characteristics of fluids are summarized in Table 3.1.

Viscoelastic fluids. Viscoelastic fluids show both viscous and elastic properties. They exhibit elastic recovery from deformations that occur during flow, but usually only part of the deformation is recovered upon removal of the stress. Examples of viscoelastic fluids are flour dough, napalm, and certain polymer melts.

Viscosity

In a newtonian fluid, the shear stress is proportional to the shear rate, and the proportionality constant is called the viscosity[†]

$$\tau_v = \mu \frac{du}{dy} \tag{3.3}$$

In SI units τ_v is measured in newtons per square meter and μ in kilograms per meter-second or pascal-second. In the cgs system, viscosity is expressed in grams per centimeter-second, and this unit is called the poise (P). Viscosity data are generally reported in millipascal-seconds or in centipoises (cP = 0.01 P = 1 mPa·s), since most fluids have viscosities much less than 1 pascal-second.

In fps units, viscosity is defined using Newton's law conversion factor g_c, and the units of μ are pounds per foot-second or pounds per foot-hour. The defining equation is

$$\tau_v = \frac{\mu}{g_c} \frac{du}{dy} \tag{3.4}$$

Conversion factors among the different systems are given in Table 3.2.

[†]In some texts a minus sign is included in the right-hand side of Eq. (3.3).

TABLE 3.2
Conversion factors for viscosity

Pa · s	P	cP	lb/ft · s	lb/ft · h
1	10	1,000	0.672	2,420
0.1	1	100	0.0672	242
10^{-3}	0.01	1	6.72×10^{-4}	2.42

Viscosity and momentum flux

Although Eq. (3.3) serves to define the viscosity of a fluid, it can also be interpreted in terms of momentum flux. The moving fluid a short distance above the wall possesses some momentum, whereas the fluid immediately adjacent to the wall, where the velocity is zero, has none. The moving fluid must therefore acquire momentum from the faster-moving layer above it, which in turn receives momentum from the next layer up, and so on. Each layer is, in effect, dragged along by the layer above it. In this way x-direction momentum is transferred in the $-y$ direction all the way to the wall, where $u = 0$. Since the wall does not move, the momentum is delivered to the wall as a shear force known as wall shear. The shear stress (shear force per unit area) at the wall is denoted by τ_w.

Momentum is thus transferred from a region of high fluid velocity to one of low velocity, much as heat flows from a region of high temperature to one at a lower temperature. The *rate* of momentum transfer per unit area, or *momentum flux,* is governed by the velocity gradient du/dy. Equation (3.3) therefore states that the momentum flux normal to the direction of fluid flow is proportional to the velocity gradient, with the viscosity as the proportionality factor. The velocity gradient may be thought of as the "driving force" for momentum transfer. The units of momentum flux are $(kg \cdot m/s)/m^2 \cdot s$ or $kg/m \cdot s^2$, the same as the units for τ, since 1 N/m^2 equals 1 $kg/m \cdot s^2$.

Momentum transfer is analogous to conductive heat transfer resulting from a temperature gradient, where the proportionality factor between the heat flux and temperature gradient is called the *thermal conductivity*. This is shown by Fourier's law, on page 291. In laminar flow, momentum is transferred by viscous action as a result of the velocity gradient, and the viscosity may be regarded as the conductivity of momentum transferred by this mechanism. Momentum transfer is also analogous to the transfer of material by molecular diffusion, where the proportionality factor is the diffusivity of mass. This is summarized in Fick's law, page 514.

Viscosities of gases and liquids

The viscosity of a newtonian fluid depends primarily on the temperature and molecular structure and to a minor extent on pressure, except at very high pressures. Gas viscosities at room temperature are generally between 0.005 and 0.02 cP. There is no simple correlation with molecular weight. At 20°C the viscosity is 0.018 cP for air, 0.014 cP for carbon dioxide, 0.007 cP for benzene vapor, and 0.009 cP for hydrogen.

Gas viscosities *increase* with temperature, as predicted by kinetic theory. Values are given for many substances in App. 8 and Ref. 7. For approximate calculations the effect of temperature can be estimated by using the exponential equation

$$\frac{\mu}{\mu_0} = \left(\frac{T}{273}\right)^n \tag{3.5}$$

where μ = viscosity at absolute temperature T, K
μ_0 = viscosity at 0°C (273 K)
n = constant for a particular gas

Exponent $n \cong 0.65$ for air; it is approximately 0.9 for carbon dioxide, 0.8 for butane, and 1.0 for steam.

The viscosity of a gas is almost independent of pressure in the region where the ideal gas laws apply. At very high pressures the viscosity increases with pressure, especially in the neighborhood of the critical point.[9]

The viscosities of liquids are generally much greater than those of gases and cover several orders of magnitude. The viscosity usually increases with molecular weight and decreases significantly when the temperature is raised. For example, the viscosity of water falls from 1.79 cP at 0°C to 0.28 cP at 100°C. The viscosity of a liquid increases with pressure, but the effect is generally insignificant at pressures less than 40 atm. Data for common liquids over a range of temperatures are given in App. 9.

The absolute viscosities of liquids vary over an enormous range of magnitudes, from about 0.1 cP for liquids near their boiling point to as much as 10^6 P for polymer melts. Most extremely viscous materials are non-newtonian and possess no single viscosity independent of shear rate.

Kinematic viscosity. The ratio of the absolute viscosity to the density of a fluid μ/ρ is often useful. This property is called the *kinematic viscosity* and designated by ν. In SI, the unit for ν is square meters per second. In the cgs system, the kinematic viscosity is called the stoke (St), defined as 1 cm^2/s. The fps unit is square feet per second. Conversion factors are

$$1 \ m^2/s = 10^4 \ St = 10.7639 \ ft^2/s$$

For liquids, kinematic viscosities vary with temperature over a somewhat narrower range than absolute viscosities. For gases, the kinematic viscosity increases more rapidly with temperature than does the absolute viscosity.

Rate of shear versus shear stress for non-newtonian fluids

Bingham plastics, like that represented by curve B in Fig. 3.2, follow a rheological equation of the type

$$\tau_v = \tau_0 + K \frac{du}{dy} \tag{3.6}$$

where K is a constant. Over some range of shear rates, dilatant and pseudoplastic fluids often follow a power law, also called the *Ostwald–de Waele* equation,

$$\tau_v = K' \left(\frac{du}{dy}\right)^{n'} \tag{3.7}$$

TABLE 3.3
Flow property indexes for pseudoplastic fluids[2]

Fluid	n'	$K' \times 10^{-3}$
1.5% Carboxymethylcellulose in water	0.554	3.13
3.0% Carboxymethylcellulose in water	0.566	9.31
4.0% Paper pulp in water	0.575	20.02
14.3% Clay in water	0.350	0.173
25% Clay in water	0.185	1.59
Applesauce	0.645	0.500
Banana purée	0.458	6.51
Tomato concentrate	0.59	0.2226

where K' and n' are constants called the *flow consistency index* and the *flow behavior index*, respectively. Such fluids are known as *power law fluids*. For pseudoplastics (curve C) $n' < 1$, and for dilatant fluids (curve D) $n' > 1$. Clearly $n' = 1$ for newtonian fluids. Values of n' and K' for some pseudoplastic fluids are given in Table 3.3.

TURBULENCE

It has long been known that a fluid can flow through a pipe or conduit in two different ways. At low flow rates the pressure drop in the fluid increases directly with the fluid velocity; at high rates it increases much more rapidly, roughly as the square of the velocity. The distinction between the two types of flow was first demonstrated in a classic experiment by Osborne Reynolds, reported in 1883.[10] A horizontal glass tube was immersed in a glass-walled tank filled with water. A controlled flow of water could be drawn through the tube by opening a valve. The entrance to the tube was flared, and provision was made to introduce a fine filament of colored water from the overhead flask into the stream at the tube entrance. Reynolds found that, at low flow rates, the jet of colored water flowed intact along with the mainstream and no cross-mixing occurred. The behavior of the color band showed clearly that the water was flowing in parallel straight lines and that the flow was laminar. When the flow rate was increased, a velocity, called the *critical velocity*, was reached at which the thread of color became wavy and gradually disappeared, as the dye spread uniformly throughout the entire cross section of the stream of water. This behavior of the colored water showed that the water no longer flowed in laminar motion but moved erratically in the form of cross-currents and eddies. This type of motion is *turbulent flow*.

Reynolds number and transition from laminar to turbulent flow

Reynolds studied the conditions under which one type of flow changes to the other and found that the critical velocity, at which laminar flow changes to turbulent flow, depends on four quantities: the diameter of the tube and the viscosity, density, and average linear velocity of the liquid. Furthermore, he found that these four factors can be combined into one group and that the change in the kind of flow occurs at a definite value of the group. The grouping of variables so found was

$$\text{Re} = \frac{D\bar{V}\rho}{\mu} = \frac{D\bar{V}}{\nu} \tag{3.8}$$

where D = diameter of tube
$\bar{V}$ = average velocity of liquid [Eq. (4.9)]
μ = viscosity of liquid
ρ = density of liquid
ν = kinematic viscosity of liquid

The dimensionless group of variables defined by Eq. (3.8) is called the *Reynolds number* Re. It is one of the named dimensionless groups listed in App. 2. Its magnitude is independent of the units used, provided the units are consistent.

Additional observations have shown that the transition from laminar to turbulent flow actually may occur over a wide range of Reynolds numbers. In a pipe, flow is always laminar at Reynolds numbers below 2,100, but laminar flow can persist up to Reynolds numbers well above 24,000 by eliminating all disturbances at the inlet.[1] If the laminar flow at such high Reynolds numbers is disturbed, however, say by a fluctuation in velocity, the flow quickly becomes turbulent. Disturbances under these conditions are amplified, whereas at Reynolds numbers below 2,100 all disturbances are damped and the flow remains laminar. At some flow rates a disturbance may be neither damped nor amplified; the flow is then said to be neutrally stable. Under ordinary conditions, the flow in a pipe or tube is turbulent at Reynolds numbers above about 4,000. Between 2,100 and 4,000 a *transition region* is found where the flow may be either laminar or turbulent, depending upon conditions at the entrance of the tube and on the distance from the entrance.

Reynolds number for non-newtonian fluids

Since non-newtonian fluids do not have a single-valued viscosity independent of shear rate, Eq. (3.8) for the Reynolds number cannot be used. The definition of a Reynolds number for such fluids is somewhat arbitrary; a widely used definition for power law fluids is

$$\text{Re}_n = 2^{3-n'} \left(\frac{n'}{3n'+1} \right)^{n'} \frac{D^{n'}\rho \bar{V}^{2-n'}}{K'} \tag{3.9}$$

The basis of this somewhat complicated definition is discussed on page 113. The onset of turbulence occurs at Reynolds numbers above 2,100 with pseudoplastic fluids, for which $n' < 1$.

Nature of turbulence

Because of its importance in many branches of engineering, turbulent flow has been extensively investigated in recent years, and a large literature has accumulated on this subject.[1,3] Refined methods of measurement have been used to follow in detail the actual velocity fluctuations of the eddies during turbulent flow, and the results of such measurements have shed much qualitative and quantitative light on the nature of turbulence.

Turbulence may be generated in other ways than by flow through a pipe. In general, it can result either from contact of the flowing stream with solid boundaries or from contact between two layers of fluid moving at different velocities. The first kind of turbulence is called *wall turbulence* and the second kind *free turbulence.* Wall turbulence appears when the fluid flows through closed or open channels or past solid shapes immersed in the stream. Free turbulence appears in the flow of a jet into a mass of stagnant fluid or when a boundary layer separates from a solid wall and flows through the bulk of the fluid. Free turbulence is especially important in mixing, which is the subject of Chap. 9.

Turbulent flow consists of a mass of eddies of various sizes coexisting in the flowing stream. Large eddies are continually formed. They break down into smaller eddies, which in turn evolve still smaller ones. Finally, the smallest eddies disappear. At a given time and in a given volume, a wide spectrum of eddy sizes exists. The size of the largest eddy is comparable with the smallest dimension of the turbulent stream; the diameter of the smallest eddies is 10 to 100 μm. Smaller eddies than this are rapidly destroyed by viscous shear. Flow within an eddy is laminar. Since even the smallest eddies contain about 10^{12} molecules, all eddies are of macroscopic size, and turbulent flow is not a molecular phenomenon.

Any given eddy possesses a definite amount of mechanical energy, much like that of a small spinning top. The energy of the largest eddies is supplied by the potential energy of the bulk flow of the fluid. From an energy standpoint, turbulence is a transfer process in which large eddies, formed from the bulk flow, pass their energy of rotation along a continuous series of smaller eddies. This mechanical energy is not appreciably dissipated into heat during the breakup of large eddies into smaller ones, but is passed along almost quantitatively to the smallest eddies. It is finally converted to heat when the smallest eddies are obliterated by viscous action. Energy conversion by viscous action is called *viscous dissipation.*

Deviating velocities in turbulent flow

A typical picture of the variations in the instantaneous velocity at a given point in a turbulent flow field is shown in Fig. 3.3. This velocity is really a single component of the actual velocity vector, all three components of which vary rapidly in magnitude and direction. Also, the instantaneous pressure at the same point fluctuates rapidly and simultaneously with the fluctuations of velocity. Oscillographs showing these fluctuations provide the basic experimental data on which modern theories of turbulence are based.

Although at first sight turbulence seems to be structureless and randomized, studies[14] of oscillographs like that in Fig. 3.3 show that this is not quite so. The randomness and unpredictability of the fluctuations, which are nonetheless constrained between definite limits, exemplify the behavior of certain mathematical "chaotic" nonlinear functions.[6] Quantitative characterization of turbulence, however, is commonly done by statistical analysis of the frequency distributions.

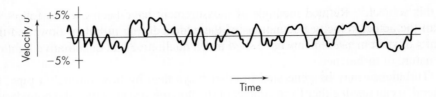

FIGURE 3.3
Velocity fluctuations in turbulent flow. The percentages are based on the constant velocity. [*After F. L. Wattendorf and A. M. Kuethe, Physics,* **5,** *153 (1934).*]

The instantaneous local velocities at a given point can be measured by laser-Doppler anemometers, which are capable of following the rapid oscillations. Local velocities can be analyzed by splitting each component of the total instantaneous velocity into two parts, one a constant part that is the time average, or mean value, of the component in the direction of flow of the stream, and the other, called the *deviating velocity,* the instantaneous fluctuation of the component around the mean. The net velocity is that measured by ordinary flowmeters, such as a pitot tube, which are too sluggish to follow the rapid variations of the fluctuating velocity. The split of a velocity component can be formalized by the following method. Let the three components (in cartesian coordinates) of the instantaneous velocity in directions x, y, and z be u_i, v_i, and w_i, respectively. Assume also that the x is oriented in the direction of flow of the stream and that components v_i and w_i are the y and z components, respectively, both perpendicular to the direction of bulk flow. Then the equations defining the deviating velocities are

$$u_i = u + u' \qquad v_i = v' \qquad w_i = w' \tag{3.10}$$

where u_i, v_i, w_i = instantaneous total velocity components in x, y, and z directions, respectively
$\qquad u$ = constant net velocity of stream in x direction
u', v', w' = deviating velocities in x, y, and z directions, respectively

Terms v and w are omitted in Eqs. (3.10) because there is no net flow in the directions of the y and z axes in one-dimensional flow, and so v and w are zero.

The deviating velocities u', v', and w' all fluctuate about zero as an average. Figure 3.3 is actually a plot of the deviating velocity u'; a plot of the instantaneous velocity u_i, however, would be identical in appearance, since the ordinate would everywhere be increased by the constant quantity u.

For pressure,

$$p_i = p + p' \tag{3.11}$$

where p_i = variable local pressure
$\qquad p$ = constant average pressure as measured by ordinary manometers or pressure gauges
$\qquad p'$ = fluctuating part of pressure due to eddies

Because of the random nature of the fluctuations, the time averages of the fluctuating components of velocity and pressure vanish when averaged over a time period t_0 of the order of a few seconds. Therefore,

$$\frac{1}{t_0} \int_0^{t_0} u' \, dt = 0 \qquad \frac{1}{t_0} \int_0^{t_0} w' \, dt = 0$$

$$\frac{1}{t_0} \int_0^{t_0} v' \, dt = 0 \qquad \frac{1}{t_0} \int_0^{t_0} p' \, dt = 0$$

(3.12)

The reason these averages vanish is that for every positive value of a fluctuation there is an equal negative value, and the algebraic sum is zero.

Although the time averages of the fluctuating components themselves are zero, this is not necessarily true of other functions or combinations of these components. For example, the time average of the mean square of any one of these velocity components is not zero. This quantity for component u' is defined by

$$\frac{1}{t_0} \int_0^{t^0} (u')^2 \, dt = \overline{(u')^2}$$

(3.13)

Thus the mean square is not zero, since u' takes on a rapid series of positive and negative values, which, when squared, always give a positive product. Therefore $\overline{(u')^2}$ is inherently positive and vanishes only when turbulence does not exist.

In laminar flow there are no eddies; the deviating velocities and pressure fluctuations do not exist; the total velocity in the direction of flow u_i is constant and equal to u; and v_i and w_i are both zero.

Statistical nature of turbulence

The distribution of deviating velocities at a single point reveals that the value of the velocity is related to the frequency of occurrence of that value, and that the relationship between frequency and value is gaussian and therefore follows the error curve characteristic of completely random statistical quantities. This result establishes turbulence as a statistical phenomenon, and the most successful treatments of turbulence have been based upon its statistical nature.[3]

By measuring u', v', and w' at different places and over varying time periods, two kinds of data are obtained: (1) The three deviating velocity components at a single point can be measured, each as a function of time, and (2) the values of a single deviating velocity (for example, u') can be measured at different positions over the same time period. Figure 3.4 shows values of u' measured simultaneously at two points separated by vertical distance y. Data taken at different values of y show that the correspondence between the velocities at the two stations varies from a very close relationship at very small values of y to complete independence when y is large. This is to be expected, because when the distance between the measurements is small with respect to the size of an eddy, it is a single eddy that is being measured, and the deviating velocities found at the two stations are strongly correlated. This

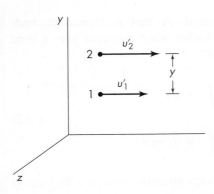

FIGURE 3.4
Fluctuating velocity components in measurement of scale of turbulence.

means that when the velocity at one station changes either in direction or in magnitude, the velocity at the other station acts in practically the same way (or exactly the opposite way). At larger separation distances the measurements are being made on separate eddies, and the correlation disappears.

When the three components of the deviating velocities are measured at the same point, in general any two of them are also found to be correlated and a change in one is accompanied by a change in the other two.

These observations are quantified by defining correlation coefficients.[5] One such coefficient, which corresponds to the situation shown in Fig. 3.4, is defined as follows:

$$R_{u'} = \frac{\overline{u_1' u_2'}}{\sqrt{\overline{(u_1')^2} \, \overline{(u_2')^2}}} \tag{3.14}$$

where u_1' and u_2' are the values of u' at stations 1 and 2, respectively. Another correlation coefficient that applies at a single point is defined by

$$R_{u'v'} = \frac{\overline{u'v'}}{\sqrt{\overline{(u')^2} \, \overline{(v')^2}}} \tag{3.15}$$

where u' and v' are measured at the same point at the same time.

Intensity and scale of turbulence

Turbulent fields are characterized by two average parameters. The first measures the intensity of the field and refers to the speed of rotation of the eddies and the energy contained in an eddy of a specific size. The second measures the size of the eddies. Intensity is measured by the root mean square of a velocity component.

It is usually expressed as a percentage of the mean velocity or as

$$100 \sqrt{\overline{(u')^2}}/u.$$

Very turbulent fields, such as those immediately below turbulence-producing grids, may reach an intensity of 5 to 10 percent. In unobstructed flow, intensities are less and of the order of 0.5 to 2 percent. A different intensity usually is found for each component of velocity.

The scale of turbulence is based on correlation coefficients such as $R_{u'}$, measured as a function of the distance between stations. By determining the values of $R_{u'}$, as a function of y, the scale L_y of the eddy in the y direction is calculated by the integral

$$L_y = \int_0^\infty R_{u'} \, dy \tag{3.16}$$

Each direction usually gives a different value of L_y, depending upon the choice of velocity components used in the definition. For air flowing in pipes at 12 m/s, the scale is about 10 mm, and this is a measure of the average size of the eddies in the pipe.

Isotropic turbulence

Although correlation coefficients generally depend upon the choice of component, in some situations this is not true, and the root-mean-square components are equal for all directions at a given point. In this situation the turbulence is said to be *isotropic,* and

$$\overline{(u')^2} = \overline{(v')^2} = \overline{(w')^2}$$

Nearly isotropic turbulence exists when there is no velocity gradient, as at the centerline of a pipe or beyond the outer edge of a boundary layer. Nearly isotropic turbulence is also found downstream of a grid placed in the flow. Turbulent flow near a boundary is anisotropic, but the anisotropy occurs mainly with the larger eddies. Small eddies, especially those near obliteration from viscous action, are practically isotropic.

Reynolds stresses

It has long been known that shear forces much larger than those occurring in laminar flow exist in turbulent flow wherever there is a velocity gradient across a shear plane. The mechanism of turbulent shear depends upon the deviating velocities in anisotropic turbulence. Turbulent shear stresses are called *Reynolds stresses.* They are measured by the correlation coefficients of the type $R_{u'v'}$ defined in Eq. (3.15).

To relate Reynolds stresses to correlations of deviating velocities, the momentum principle may be used. Consider a fluid in turbulent flow moving in a positive x direction, as shown in Fig. 3.5. Plane S is parallel to the flow. The instantaneous velocity in the plane is u_i, and the mean velocity is u. Assume that u increases with y, the positive direction measured perpendicular to the layer S, so that the velocity gradient du/dy is positive. An eddy moving toward the wall has a negative value of v', and its movement represents a mass flow rate $\rho(-v')$ into the fluid below plane S. The velocity of the eddy in the x direction is u_i, or $u + u'$; if each such eddy crossing plane S is slowed down to the mean velocity u, the rate of momentum transfer per unit area is $\rho(-v')u'$. This momentum flux, after time averaging for all eddies, is

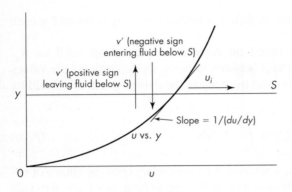

FIGURE 3.5
Reynolds stress.

a turbulent shear stress or Reynolds stress given by the equation

$$\tau_t = \overline{\rho u' v'} \tag{3.17}$$

Eddy viscosity

By analogy with Eq. (3.4), the relationship between shear stress and velocity gradient in a turbulent stream is used to define an eddy viscosity E_v:

$$\tau_t = E_v \frac{du}{dy} \tag{3.18}$$

Quantity E_v is analogous to μ, the absolute viscosity. Also, in analogy with the kinematic viscosity v the quantity ε_M, called the *eddy diffusivity of momentum,* is defined as $\varepsilon_M = E_v/\rho$.

The total shear stress in a turbulent fluid is the sum of the viscous stress and the turbulent stress, or

$$\tau = (\mu + E_v) \frac{du}{dy} \tag{3.19}$$

$$\tau = (v + \varepsilon_M) \frac{d(\rho u)}{dy} \tag{3.20}$$

Although E_v and ε_M are analogous to μ and v, respectively, in that all these quantities are coefficients relating shear stress and the velocity gradient, there is a basic difference between the two kinds of quantities. The viscosities μ and v are true properties of the fluid and are the macroscopic result of averaging motions and momenta of myriad molecules. The eddy viscosity E_v and the eddy diffusivity ε_M are not just properties of the fluid but depend on the fluid velocity and the geometry of the system. They are functions of all factors that influence the detailed patterns of turbulence and the deviating velocities, and they are especially sensitive to location in the turbulent field and the local values of the scale and intensity of the turbulence. Viscosities can be measured on isolated samples of fluid and presented in tables or charts of physical properties, as in Apps. 8 and 9. Eddy viscosities and diffusivities are determined (with difficulty, and only by means of special instruments) by experiments on the flow itself.

BOUNDARY LAYERS

Flow in boundary layers

A boundary layer is defined as that part of a moving fluid in which the fluid motion is influenced by the presence of a solid boundary. As a specific example of boundary layer formation, consider the flow of fluid parallel with a thin plate, as shown in Fig. 3.6. The velocity of the fluid upstream from the leading edge of the plate is uniform across the entire fluid stream. The velocity of the fluid at the interface between the solid and fluid is zero. The velocity increases with distance from the plate, as shown in Fig. 3.6. Each of these curves corresponds to a definite value of x, the distance from the leading edge of the plate. The curves change slope rapidly near the plate; they also show that the local velocity approaches asymptotically the velocity of the bulk of the fluid stream.

In Fig. 3.6 the dashed line OL is so drawn that the velocity changes are confined between this line and the trace of the wall. Because the velocity lines are asymptotic with respect to distance from the plate, it is assumed, in order to locate the dashed line definitely, that the line passes through all points where the velocity is 99 percent of the bulk fluid velocity u_∞. Line OL represents an imaginary surface that separates the fluid stream into two parts: one in which the fluid velocity is constant and the other in which the velocity varies from zero at the wall to a velocity substantially equal to that of the undisturbed fluid. This imaginary surface separates the fluid that is directly affected by the plate from that in which the local velocity is constant and equal to the initial velocity of the approach fluid. The zone, or layer, between the dashed line and the plate constitutes the boundary layer.

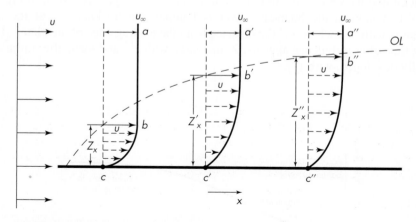

FIGURE 3.6
Prandtl boundary layer: x, distance from leading edge; u_∞, velocity of undisturbed stream; Z_x, thickness of boundary layer at distance x; u, local velocity; abc, $a'b'c'$, $a''b''c''$, curves of velocity versus distance from wall at points c, c', c''; OL, outer limit of boundary layer. (The vertical scale is greatly exaggerated.)

The formation and behavior of the boundary layer are important, not only in the flow of fluids but also in the transfer of heat, discussed in Chap. 12, and mass, discussed in Chap. 17.

Laminar and turbulent flow in boundary layers

The fluid velocity at the solid-fluid interface is zero, and the velocities close to the solid surface are, of necessity, small. Flow in this part of the boundary layer very near the surface therefore is essentially laminar. Actually it is laminar most of the time, but occasionally eddies from the main portion of the flow or the outer region of the boundary layer move very close to the wall, temporarily disrupting the velocity profile. These eddies may have little effect on the average velocity profile near the wall, but they can have a large effect on the profiles of temperature or concentration when heat or mass is being transferred to or from the wall. This effect is most pronounced for mass transfer in liquids.

Farther away from the surface the fluid velocities, though less than the velocity of the undisturbed fluid, may be fairly large, and flow in this part of the boundary layer may become turbulent. Between the zone of fully developed turbulence and the region of laminar flow is a transition, or buffer, layer of intermediate character. Thus a turbulent boundary layer is considered to consist of three zones: the viscous sublayer, the buffer layer, and the turbulent zone. The existence of a completely viscous sublayer is questioned by some, since mass-transfer studies suggest that some eddies penetrate all the way through the boundary layer and reach the wall.

Near the leading edge of a flat plate immersed in a fluid of uniform velocity, the boundary layer is thin, and the flow in the boundary layer is entirely laminar. As the layer thickens, however, at distances farther from the leading edge, a point is reached where turbulence appears. The onset of turbulence is characterized by a sudden rapid increase in the thickness of the boundary layer, as shown in Fig. 3.7.

When flow in the boundary layer is laminar, the thickness Z_x of the layer increases with $x^{0.5}$, where x is the distance from the leading edge of the plate.[12] For a short time after turbulence appears, Z_x increases with $x^{1.5}$ and then, after turbulence is fully developed, with $x^{0.8}$.

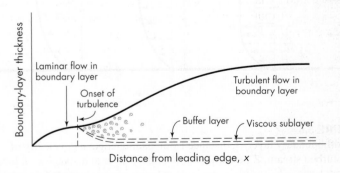

FIGURE 3.7
Development of turbulent boundary layer on a flat plate.
(The vertical scale is greatly exaggerated.)

The initial, fully laminar part of the boundary layer may grow to a moderate thickness of perhaps 2 mm with air or water moving at moderate velocities. Once turbulence begins, however, the thickness of the laminar part of the boundary layer diminishes considerably, typically to about 0.2 mm.

Transition from laminar to turbulent flow; Reynolds number. The factors that determine the point at which turbulence appears in a laminar boundary layer are coordinated by the dimensionless Reynolds number defined by the equation

$$Re_x = \frac{xu_\infty\rho}{\mu}$$ (3.21)

where x = distance from leading edge of plate
u_∞ = bulk fluid velocity
ρ = density of fluid
μ = viscosity of fluid

With parallel flow along a plate, turbulent flow first appears at a critical Reynolds number between about 10^5 and 3×10^6. The transition occurs at the lower Reynolds numbers when the plate is rough and the intensity of turbulence in the approaching stream is high, and at the higher values when the plate is smooth and the intensity of turbulence in the approaching stream is low.

Boundary layer formation in straight tubes

Consider a straight, thin-walled tube with fluid entering it at a uniform velocity. As shown in Fig. 3.8, a boundary layer begins to form at the entrance to the tube, and as the fluid moves through the first part of the channel, the layer thickens. During this stage the boundary layer occupies only part of the cross section of the tube, and the total stream consists of a core of fluid flowing in rodlike manner at constant velocity and an annular boundary layer between the wall and the core. In the boundary layer the velocity increases from zero at the wall to the constant velocity existing in the core. As the stream moves farther down the tube, the boundary layer occupies an increasing portion of the cross section. Finally, at a point well downstream from the entrance, the boundary layer reaches the center of the tube, the rodlike core disappears, and the boundary layer occupies the entire cross section of the stream. At this point the velocity distribution in the tube reaches its final form, as shown by the last curve at the right of Fig. 3.8, and remains unchanged during the remaining length of the tube. Such flow with an unchanging velocity distribution is called *fully developed flow.*

Boundary layer

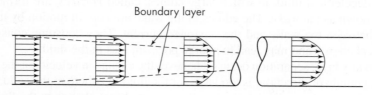

FIGURE 3.8
Development of boundary layer flow in pipe.

Transition length for laminar and turbulent flow. The length of the entrance region of the tube necessary for the boundary layer to reach the center of the tube and for fully developed flow to be established is called the *transition length.* Since the velocity varies not only with length of tube but also with radial distance from the center of the tube, flow in the entrance region is two-dimensional.

The approximate length of straight pipe necessary for completion of the final velocity distribution is, for laminar flow,

$$\frac{x_t}{D} = 0.05 \text{ Re} \tag{3.22}$$

where x_t = transition length
  D = diameter of pipe

Equation (3.22), originally proposed by Nikuradse, was verified experimentally by Rothfus and Prengle.[11] Equation (3.22) shows that for a 50-mm- (2-in.-) ID pipe and a Reynolds number of 1,500, the transition length is 3.75 m (12.3 ft). If the fluid entering the pipe is turbulent and the velocity in the tube is above the critical, the transition length is nearly independent of the Reynolds number and is about 40 to 50 pipe diameters, with little difference between the distribution at 25 diameters and that at greater distances from the entrance. For a 50-mm-ID pipe, 2 to 3 m of straight pipe is sufficient when flow is all turbulent. If the fluid entering the tube is in laminar flow and becomes turbulent on entering the tube, a longer transition length, as large as 100 pipe diameters, is needed.

Boundary layer separation and wake formation

In the preceding paragraphs the growth of boundary layers has been discussed. Now consider what happens at the far side of a submerged object, where the fluid leaves the solid surface.

At the trailing edge of a flat plate that is parallel to the direction of flow, the boundary layers on the two sides of the plate have grown to a maximum thickness. For a time after the fluid leaves the plate, the layers and velocity gradients persist. Soon, however, the gradients fade out, the boundary layers intermingle and disappear, and the fluid once more moves with a uniform velocity. This is shown in Fig. 3.9a.

Suppose, now, the plate is turned at right angles to the direction of flow, as in Fig. 3.9b. A boundary layer forms as before in the fluid flowing over the upstream face. When the fluid reaches the edge of the plate, however, its momentum prevents it from making the sharp turn around the edge, and it separates from the plate and proceeds outward into the bulk of the fluid. Behind the plate is a backwater zone of strongly decelerated fluid, in which large eddies, called *vortices,* are formed. This zone is known as the *wake.* The eddies in the wake are kept in motion by the shear stresses between the wake and the separated current. They consume considerable mechanical energy and may lead to a large pressure loss in the fluid.

Boundary layer separation occurs whenever the change in velocity of the fluid, in either magnitude or direction, is too large for the fluid to adhere to the solid surface. It is most frequently encountered when there is an abrupt change in the flow channel, such as a sudden expansion or contraction, a sharp bend, or an obstruction

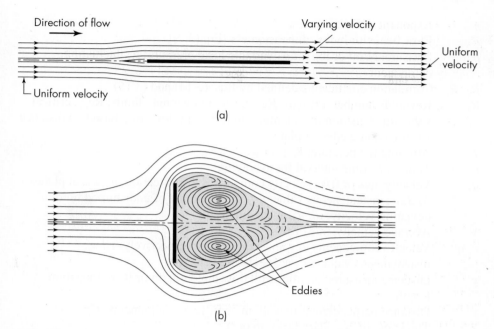

FIGURE 3.9
Flow past flat plate: (*a*) flow parallel with plate; (*b*) flow perpendicular to plate.

around which the fluid must flow. As discussed in Chap. 5, page 122, separation may also occur from the velocity decrease in a smoothly diverging channel. Because of the large energy losses resulting from the formation of a wake, it is often desirable to minimize or prevent boundary layer separation. In some cases this can be done by suction, i.e., by drawing part of the fluid into the solid surface at the area of potential separation. Most often, however, separation is minimized by avoiding sharp changes in the cross-sectional area of the flow channel and by streamlining any objects over which the fluid must flow. For some purposes, such as the promotion of heat transfer or mixing in a fluid, boundary layer separation may be desirable.

SYMBOLS

A	Area m^2 or ft^2; A_s, of plane on which shear force acts
D	Diameter, m or ft
E_v	Eddy viscosity, Pa · s or lb/ft · s, P
F_s	Shear force, N or lb$_f$
g_c	Newton's law proportionality factor, 32.174 ft · lb/lb$_f$ · s^2
K	Constant in Eq. (3.6)
K'	Flow consistency index, g/m · s$^{2-n'}$ or lb/ft · s$^{2-n'}$ [Eq. (3.7)]
L_y	Scale of turbulence, m or ft

n Exponent in Eq. (3.5)

n' Flow behavior index, dimensionless [Eq. (3.7)]

p Pressure, N/m^2 or lb_f/ft^2; p_i, variable local pressure; p', fluctuating component

$R_{u'}, R_{u'v'}$ Correlation coefficients defined by Eqs. (3.14) and (3.15)

Re   Reynolds number, $DV\rho/\mu$; Re_n, for non-newtonian fluids; $Re_{n,c}$, critical value for transition from laminar to turbulent flow; Re_x, based on distance x from leading edge of plate

T Absolute temperature, K

t Time, s; t_0, time interval for averaging

u Velocity, m/s or ft/s; velocity component in x direction; u_A, u_B, at planes A, B; u_i, instantaneous value; u_∞, bulk velocity of undisturbed fluid; u', deviating velocity; u'_1, u'_2, at stations 1, 2

$\bar{V}$ Average velocity, m/s or ft/s

v, w Velocity components in y and z directions, respectively; v_i, w_i, instantaneous values; v', w', deviating velocities

x Distance measured parallel with flow direction, m or ft; x_t, transition length

y Distance perpendicular to wall, m or ft; y_A, y_C, at planes A, C

Z_x Thickness of boundary layer, m or ft

Greek Letters

ε_M Eddy diffusivity of momentum, m^2/s or ft^2/s

μ Viscosity, absolute, Pa · s or lb/ft · s; μ_0, at $T = 273$ K

ν Kinematic viscosity μ/ρ, m^2/s or ft^2/s

ρ Density, kg/m^3 or lb/ft^3

τ Shear stress, N/m^2 or lb_f/ft^2; τ_t, turbulent shear stress; τ_v, laminar shear stress; τ_w, stress at wall; τ_0, threshold stress for Bingham plastic

PROBLEMS

3.1. For the following situations of steady flow, determine whether flow is laminar or turbulent: (a) water at 10°C flowing at an average velocity of 2 m/s in a 100-mm pipe; (b) air at 2-atm pressure and 180°F flowing at 50 ft/s in a 12-in. duct; (c) oil with a specific gravity of 0.78 and a viscosity of 20 cP flowing at 5 ft/s in a 2-in. pipe; (d) polymer melt with a density of 900 kg/m^3 and a viscosity of 1 Pa · s flowing at 0.2 m/s in a 15-mm tube.

3.2. The thickness of the laminar boundary layer on a flat plate Z_x is approximately given by the equation $Z_x = 5.5[\mu x/(u_\infty \rho)]^{1/2}$. Show that at the transition to the turbulent flow the Reynolds number based on this thickness, instead of on x as in Eq. (3.21), is close to the transition Reynolds number for flow in a pipe.

3.3. Use the nomograph in App. 8 to determine the value of n in the equation for gas viscosity, Eq. (3.5), for carbon monoxide and for helium over the ranges 0 to 300°C and 300 to 600°C.

3.4. (a) Estimate the transition length at the entrance to a 15-mm tube through which 100 percent glycerol at 60°C is flowing at a velocity of 0.3 m/s. The density of glycerol is 1,240 kg/m³. (b) Repeat part (a) for 100 percent n-propyl alcohol entering a 3-in. pipe at 30°C and a velocity of 7 ft/s. The density of n-propyl alcohol is 50 lb/ft³.

3.5. (a) Estimate the Reynolds number for flow in an automobile exhaust pipe if the 2-L 4-cycle engine is operating at 3,000 rpm. (b) If the catalytic converter has 4 times the cross section of the exhaust pipe, how small must the channels in the converter be to get laminar flow?

3.6. Plot the viscosity of water, hexane, and 100 percent glycerol on semilog coordinates, and discuss how nearly the viscosity follows an Arrhenius relationship ($\mu = Ae^{-E/RT}$).

3.7. Plot the viscosity at moderate pressures of the halogen gases at 20°C versus their molecular weight. Make a similar plot for some of the normal paraffins, and discuss the trends shown by these plots.

3.8. Air at 30°C and 5 bars, is flowing inside a $\frac{1}{2}$-in. Schedule 40 steel pipe (see App. 3). If the flow is at 4.0 ft³/min at standard temperature and pressure (0°C and 1 atm), is the flow likely to be laminar or turbulent?

3.9. Crude oil is pumped at 1.5 m/s through a pipeline 1 m in diameter. Above what value of the oil viscosity would laminar flow exist? Give your answer in both SI and cgs units.

3.10. The *apparent viscosity* of a non-newtonian liquid at a given shear rate is the value indicated by a viscometer operating on the liquid at that shear rate. It is the viscosity that would be indicated by the viscometer if the liquid were newtonian. (a) Calculate the apparent viscosity of a 4 percent suspension of paper pulp in water at shear rates du/dy of $10\ \text{s}^{-1}$ and $1{,}000\ \text{s}^{-1}$. (b) Repeat for a 25 percent suspension of clay in water.

REFERENCES

1. Garde, R. J. *Turbulent Flow.* New York: Wiley, 1994.
2. Geankoplis, C. J. *Transport Processes and Unit Operations.* 3d ed. Englewood Cliffs, NJ: Prentice-Hall, 1993, pp. 155–157.
3. Hinze, J. O. *Turbulence.* 2d ed. New York: McGraw-Hill, 1975.
4. Kirchhoff, R. H. *Potential Flows: Computer Graphic Solutions.* New York: Marcel Dekker, 1985.
5. Knudsen, J. G., and D. L. Katz. *Fluid Dynamics and Heat Transfer.* New York: McGraw-Hill, 1958, pp. 115–120.
6. Langford, W. F. *Chaos, Fractals and Dynamics.* eds. P. Fischer and W. R. Smith. New York: Marcel Dekker, 1985, pp. 94–95.

7. Perry, R. H., and D. W. Green, eds. *Perry's Chemical Engineers' Handbook.* 7th ed. New York: McGraw-Hill, 1997, p. 2-320.
8. Prandtl, L. *Proc. 3rd Int. Math. Congress,* 1904.
9. Reid, R. C., J. M. Prausnitz, and B. E. Poling. *The Properties of Gases and Liquids.* 4th ed. New York: McGraw-Hill, 1987, p. 417.
10. Reynolds, O. *Phil. Trans. Royal Soc.,* London, Ser. A: 174 (1883).
11. Rothfus, R. R., and R. S. Prengle. *Ind. Eng. Chem.* 44:1683 (1952).
12. Schlichting, H. *Boundary Layer Theory.* 7th ed. New York: McGraw-Hill, 1979, p. 42.
13. Streeter, V. L., and E. B. Wylie. *Fluid Mechanics.* 8th ed. New York: McGraw-Hill, 1985.
14. Wattendorf, F. L., and A. M. Kuethe. *Physics* 5:153 (1934).

CHAPTER 4

Basic Equations of Fluid Flow

In applications of fluid mechanics, the most useful equations are based on the principles of mass balance or continuity; the balances of linear momentum and angular momentum; and the mechanical energy balance. The equations may be written in differential form, showing conditions at a point within a volume element of fluid, or in integrated form applicable to a finite volume or mass of fluid.

Differential equations and shell balances

To be useful in solving engineering problems, the differential equations must be integrated. In some simple cases they can be integrated mathematically, but more often they must be integrated numerically by computer. This would be needed, for example, in as straightforward a situation as the steady flow of air in a large duct containing baffles, or in a highly complex problem such as the transient flow of a high-viscosity non-newtonian polymer melt through a mold.

Useful equations can be derived for some well-defined systems by making a macroscopic shell balance, in which the flows through the boundaries of the entire system are used instead of those through a differential volume element. The system can be as small as a short length of pipe or as large as an entire processing plant.

In this chapter some of the fundamental differential equations are derived, as are many integrated equations based on shell balances. More extensive treatment of the differential equations is found in texts dealing with applied fluid mechanics and transport processes.[1–3,7]

MASS BALANCE IN A FLOWING FLUID; CONTINUITY

In any element of fluid (or in any bounded system), the equation for a mass balance is simply

(Rate of mass flow in) − (Rate of mass flow out) = (Rate of mass accumulation)

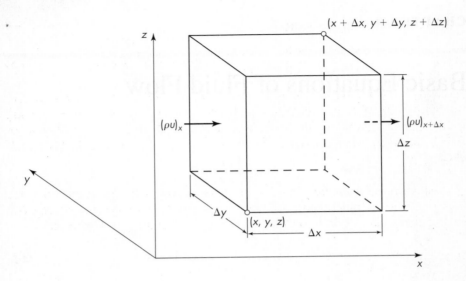

FIGURE 4.1
Region of volume $\Delta x\ \Delta y\ \Delta z$ *fixed in space* through which a fluid is flowing.

For a small volume element $\Delta x\ \Delta y\ \Delta z$, fixed in space as shown in Fig. 4.1, the mass balance is found as follows. For a fluid of density ρ the mass flux in the x direction at the face x is $(\rho u)_x$; at the face $x + \Delta x$ the flux is $(\rho u)_{x+\Delta x}$, where u is the fluid velocity in the x direction. *Flux* is defined as the rate of flow of any quantity per unit area; hence the rate of mass flow entering the element in the x direction is $(\rho u)_x\ \Delta y\ \Delta z$, and that leaving the element is $(\rho u)_{x+\Delta x}\ \Delta y\ \Delta z$. Similar relations may be written for the y and z directions, where v and w are the fluid velocities in the y and z directions, respectively. The rate of accumulation in the volume element is $\Delta x\ \Delta y\ \Delta z\ (\partial \rho / \partial t)$. Thus

$$[(\rho u)_x - (\rho u)_{x+\Delta x}]\ \Delta y\ \Delta z + [(\rho v)_y - (\rho v)_{y+\Delta y}]\ \Delta x\ \Delta z$$
$$+ [(\rho w)_z - (\rho w)_{z+\Delta z}]\ \Delta x\ \Delta y = \Delta x\ \Delta y\ \Delta z\ \frac{\partial \rho}{\partial t} \qquad (4.1)$$

Dividing through by $\Delta x\ \Delta y\ \Delta z$ gives

$$\frac{(\rho u)_x - (\rho u)_{x+\Delta x}}{\Delta x} + \frac{(\rho v)_y - (\rho v)_{y+\Delta y}}{\Delta y} + \frac{(\rho w)_z - (\rho w)_{z+\Delta z}}{\Delta z} = \frac{\partial \rho}{\partial t} \qquad (4.2)$$

Taking the limit as Δx, Δy, and Δz approach zero gives the differential equation of conservation of mass in a fluid

$$\frac{\partial \rho}{\partial t} = -\left[\frac{\partial (\rho u)}{\partial x} + \frac{\partial (\rho v)}{\partial y} + \frac{\partial (\rho w)}{\partial z}\right] = -(\nabla \cdot \rho \mathbf{V}) \qquad (4.3)$$

Equation (4.3) is also known as the *equation of continuity*. The quantity $\nabla \cdot \rho \mathbf{V}$ on the right-hand side denotes the *divergence* of the mass velocity vector $\rho \mathbf{V}$. Carrying out the partial differentiation and rearranging give

$$\frac{\partial \rho}{\partial t} + u\frac{\partial \rho}{\partial x} + v\frac{\partial \rho}{\partial y} + w\frac{\partial \rho}{\partial z} = -\rho\left(\frac{\partial u}{\partial x} + \frac{\partial v}{\partial y} + \frac{\partial w}{\partial z}\right) \qquad (4.4)$$

Equation (4.4) may be written

$$\frac{D\rho}{Dt} = -\rho\left(\frac{\partial u}{\partial x} + \frac{\partial v}{\partial y} + \frac{\partial w}{\partial z}\right) = -\rho(\nabla \cdot \mathbf{V}) \tag{4.5}$$

where $D\rho/Dt$ is the *substantial derivative* or the *derivative following the motion.* This is the rate of density change that would be noted by an observer moving downstream at the velocity of the fluid. Note that $\partial\rho/\partial t$ in Eq. (4.3) is the rate of change observed from a fixed point. At steady state $\partial\rho/\partial t = 0$ in Eqs. (4.3) and (4.4).

Continuity equation for a fluid of constant density. Often in engineering the fluid is almost incompressible, and its density ρ may be considered constant without appreciable error. In such a case $D\rho/Dt = 0$ and Eq. (4.4) becomes

$$\nabla \cdot \mathbf{V} = \left(\frac{\partial u}{\partial x} + \frac{\partial v}{\partial y} + \frac{\partial w}{\partial z}\right) = 0 \tag{4.6}$$

One-dimensional flow. In discussing fluid flow it is helpful to visualize, in the fluid stream, fluid paths called streamlines. A streamline is an imaginary path in a mass of flowing fluid so drawn that at every point the vector of the net velocity along the streamline u is tangent to the streamline. There is no net flow across such a line. In turbulent flow eddies do cross and recross the streamline, but as shown in Chap. 3, the net flow from such eddies in any direction other than that of the flow is zero. Flow along a streamline is therefore one-dimensional, and a single term for velocity is all that is needed.

A stream tube is a tube of large or small cross section and of any convenient cross-sectional shape that is entirely bounded by streamlines. A stream tube can be visualized as an imaginary pipe in the mass of flowing fluid through the walls of which there is no net flow. If the tube has a differential cross-sectional area dS, the velocity through the tube can also be denoted by the single term u.

The mass flow through the differential area is

$$d\dot{m} = \rho u \, dS \tag{4.7}$$

To find the total flow through an impermeable[†] conduit of cross-sectional area S, Eq. (4.7) is integrated across the entire cross section. In general, the local velocity u varies across the cross section. If the fluid is being heated or cooled, the fluid density also varies, but usually the variation is small and can be neglected. The flow rate through the entire cross section is

$$\dot{m} = \rho \int_S u \, dS \tag{4.8}$$

where ρ is constant across the cross section.

[†] If the walls of the conduit are permeable, as in the membrane tubes discussed in Chap. 26, these equations do not apply.

The average velocity $\bar{V}$ of the entire stream flowing through cross-sectional area S is defined by

$$\bar{V} \equiv \frac{\dot{m}}{\rho S} = \frac{1}{S} \int_S u \, dS \tag{4.9}$$

Velocity $\bar{V}$ also equals the total volumetric flow rate of the fluid, divided by the cross-sectional area of the conduit; in fact, it is usually calculated this way. It may be considered to be the *flux of volume,* m³/m² · s or ft³/ft² · s. Thus

$$\bar{V} = \frac{q}{S} \tag{4.10}$$

where q is the volumetric flow rate.

Shell balance for mass flow

Average velocities are useful in making a shell balance or flow through a pipe or a piping system. Consider the flow through a conduit of cross-sectional area S_a at the entrance and area S_b at the exit, in which the local fluid velocity varies within the cross section. The average velocity and density at the entrance are $\bar{V}_a$ and ρ_a; at the exit they are $\bar{V}_b$ and ρ_b. At steady state the mass flow in equals the mass flow out, and the continuity equation becomes

$$\dot{m} = \rho_a \bar{V}_a S_a = \rho_b \bar{V}_b S_b = \rho \bar{V} S \tag{4.11}$$

For the important special case where the flow is through channels of circular cross section

$$\dot{m} = \tfrac{1}{4}\pi D_a^2 \rho_a \bar{V}_a = \tfrac{1}{4}\pi D_b^2 \rho_b \bar{V}_b$$

from which

$$\frac{\rho_a \bar{V}_a}{\rho_b \bar{V}_b} = \left(\frac{D_b}{D_a}\right)^2 \tag{4.12}$$

where D_a and D_b are the diameters of the channel at the upstream and downstream stations, respectively.

Mass velocity

Equation (4.9) can be written

$$\bar{V}\rho = \frac{\dot{m}}{S} \equiv G \tag{4.13}$$

This equation defines the mass velocity G, calculated by dividing the mass flow rate by the cross-sectional area of the channel. In practice, the mass velocity is expressed in kilograms per square meter per second, pounds per square foot per second, or pounds per square foot per hour. The advantage of using G is that it is independent of temperature and pressure when the flow is steady (constant $\dot{m}$) and the cross section is unchanged (constant S). This fact is especially useful when compressible fluids are considered, for both $\bar{V}$ and ρ vary with temperature and pressure. Also certain relationships appear later in this book in which $\bar{V}$ and ρ are associated as their product, so that the mass velocity represents the net effect of both variables. The mass velocity G can also be described as the mass current density or mass flux, where flux is defined generally as any quantity passing through a unit area in unit time. The average velocity $\bar{V}$, as shown by Eq. (4.10), can be described as the volume flux of the fluid.

EXAMPLE 4.1. Crude oil, specific gravity 60°F/60°F = 0.887, flows through the piping shown in Fig. 4.2. Pipe A is 2-in. (50-mm) Schedule 40, pipe B is 3-in. (75-mm) Schedule 40, and each of pipes C is $1\frac{1}{2}$-in. (38-mm) Schedule 40. An equal quantity of liquid flows through each of the pipes C. The flow through pipe A is 30 gal/min (6.65 m³/h). Calculate (a) the mass flow rate in each pipe, (b) the average linear velocity in each pipe, and (c) the mass velocity in each pipe.

Solution. Dimensions and cross-sectional areas of standard pipe are given in App. 3. Cross-sectional areas needed are, for 2-in. pipe, 0.0233 ft²; for 3-in. pipe, 0.0513 ft²; and for $1\frac{1}{2}$-in. pipe, 0.01414 ft².
(a) The density of the fluid is

$$\rho = 0.887 \times 62.37 = 55.3 \text{ lb/ft}^3$$

Since there is 7.48 gal in 1 ft³ (App. 1), the total volumetric flow rate is

$$q = \frac{30 \times 60}{7.48} = 240.7 \text{ ft}^3/\text{h}$$

The mass flow rate is the same for pipes A and B and is the product of the density and the volumetric flow rate, or

$$\dot{m} = 240.7 \times 55.3 = 13{,}300 \text{ lb/h}$$

The mass flow rate through each of pipes C is one-half the total or 13,300/2 = 6,650 lb/h (0.8379 kg/s).
(b) Use Eq. (4.10). The velocity through pipe A is

$$\bar{V}_A = \frac{240.7}{3{,}600 \times 0.0233} = 2.87 \text{ ft/s}$$

through pipe B is

$$\bar{V}_B = \frac{240.7}{3{,}600 \times 0.0513} = 1.30 \text{ ft/s}$$

and through each of pipes C is

$$\bar{V}_C = \frac{240.7}{2 \times 3{,}600 \times 0.01414} = 2.36 \text{ ft/s}$$

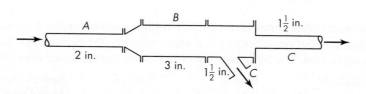

FIGURE 4.2
Piping system for Example 4.1.

(c) Use Eq. (4.13). The mass velocity through pipe A is

$$G_A = \frac{13,300}{0.0233} = 571,000 \text{ lb/ft}^2 \cdot \text{h} \, (744 \text{ kg/m}^2 \cdot \text{s})$$

through pipe B is

$$G_B = \frac{13,300}{0.0513} = 259,000 \text{ lb/ft}^2 \cdot \text{h} \, (351 \text{ kg/m}^2 \cdot \text{s})$$

and through each of pipes C is

$$G_C = \frac{13,300}{2 \times 0.01414} = 470,000 \text{ lb/ft}^2 \cdot \text{h} \, (637 \text{ kg/m}^2 \cdot \text{s})$$

EXAMPLE 4.2. Air at 20°C and 2-atm absolute pressure enters a finned-tube steam heater through a 50-mm tube at an average velocity of 15 m/s. It leaves the heater through a 65-mm tube at 90°C and 1.6 atm absolute. What is the average air velocity at the outlet?

Solution. Let subscript a refer to the heater inlet and b to the outlet. Use Eq. (4.12). The quantities needed are

$$D_a = 0.05 \text{ m} \qquad D_b = 0.065 \text{ m} \qquad p_a = 2 \text{ atm} \qquad p_b = 1.6 \text{ atm}$$
$$T_a = 20 + 273.16 = 293.16 \text{ K} \qquad T_b = 90 + 273.16 = 363.16 \text{ K}$$

Density is found from Eq. (1.56), where $\rho = 1/V$ and V is the volume of 1 kg of air. The number of moles n equals $1/M$, where M is the molecular weight of air. The densities at the inlet and outlet are then $\rho_a = Mp_a/(RT_a)$ and $\rho_b = Mp_b/(RT_b)$. Thus

$$\frac{\rho_a}{\rho_b} = \frac{p_a T_b}{p_b T_a}$$

Substituting in Eq. (4.11) gives

$$\bar{V}_b = \frac{\bar{V}_a \rho_a D_a^2}{\rho_b D_b^2} = \frac{\bar{V}_a p_a T_b D_a^2}{p_b T_a D_b^2}$$

$$= \frac{15 \times 2 \times 0.05^2 \times 363.16}{1.6 \times 0.065^2 \times 293.16}$$

$$= 13.74 \text{ m/s}$$

DIFFERENTIAL MOMENTUM BALANCE; EQUATIONS OF MOTION

A momentum balance may be made on a volume element in much the same way as a mass balance, but since velocity is a vector quantity, the derivation is considerably more complicated. The basic concept of the momentum balance is as follows:

| Rate of momentum accumulation | = | Rate of momentum entering | − | Rate of momentum leaving | + | Sum of forces acting on the system | (4.14) |

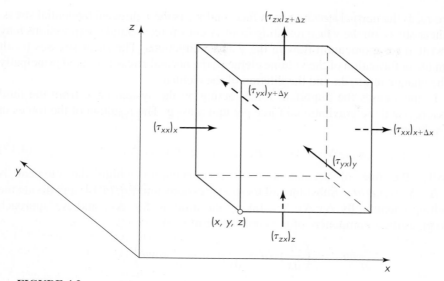

FIGURE 4.3
Volume element $\Delta x\, \Delta y\, \Delta z$ with arrows indicating the direction in which the
x component of momentum is transported through the surfaces.

In Fig. 4.3 the fluid is flowing through all six faces of the volume element in any
arbitrary direction. Since velocity is a vector, Eq. (4.14) has components in each of
the coordinate directions x, y, and z. First we consider only the x component of each
term in Eq. (4.14); the y and z components may be treated analogously.

Consider the flow rates of the x component of momentum into and out of the
volume element in Fig. 4.3. Momentum enters and leaves the volume element partly
by convection from flow of the bulk fluid, and partly by viscous action as a result of
the velocity gradients. The rate at which the x component of momentum enters the
face at x by convection is $(\rho uu)_x\, \Delta y\, \Delta z$, and the rate at which it leaves at $x + \Delta x$
is $(\rho uu)_{x+\Delta x}\, \Delta y\, \Delta z$. That entering at y through the face $\Delta x\, \Delta z$ has a rate
$(\rho vu)_y\, \Delta x\, \Delta z$. Similar expressions may be written for the three other faces. Thus
there is convective flow of x momentum through all six faces of the element. The
net convective flow into the volume element is

$$\Delta y\, \Delta z[(\rho uu)_x - (\rho uu)_{x+\Delta x}] + \Delta x\, \Delta z[(\rho vu)_y - (\rho vu)_{y+\Delta y}]$$
$$+ \Delta x\, \Delta y[(\rho wu)_z - (\rho wu)_{z+\Delta z}] \qquad (4.15)$$

Similarly, the rate at which the x component of momentum enters the face at x
by molecular transport is $(\tau_{xx})_x\, \Delta y\, \Delta z$, and the rate at which it leaves at $x + \Delta x$ is
$(\tau_{xx})_{x+\Delta x}\, \Delta y\, \Delta z$. The rate at which it enters at y is $(\tau_{yx})_y\, \Delta x\, \Delta z$, and similar
expressions may be written for the other three faces. Summing up these six
contributions gives the net flow of x momentum into the volume element by viscous
action:

$$\Delta y\, \Delta z[(\tau_{xx})_x - (\tau_{xx})_{x+\Delta x}] + \Delta x\, \Delta z[(\tau_{yx})_y - (\tau_{yx})_{y+\Delta y}]$$
$$+ \Delta x\, \Delta z[(\tau_{zx})_z - (\tau_{zx})_{z+\Delta z}] \qquad (4.16)$$

Here τ_{xx} is the normal stress on the x face, and τ_{yx} is the x-directed tangential stress, or shear stress, on the y face resulting from viscous forces. Similar expressions may be written for momentum flow in the y and z directions. The shear stresses result from the deformation of the volume element; the normal stress is related principally to the change in $\partial u/\partial x$ and the dilation of the element.[2a,5]

In most cases the important forces acting on the system arise from the fluid pressure p and the gravitational force per unit mass g. The resultant of the forces in the x direction is

$$\Delta y\,\Delta z(p_x - p_{x+\Delta x}) + \rho g_x\,\Delta x\,\Delta y\,\Delta z \tag{4.17}$$

Finally, the rate of accumulation of x momentum within the element is $\Delta x\,\Delta y\,\Delta z\,(\partial\rho u/\partial t)$. Substitute all these expressions into Eq. (4.14) and divide the resulting equation by $\Delta x\,\Delta y\,\Delta z$. Taking the limit as Δx, Δy, and Δz approach zero gives the x component of the equation of motion:

$$\frac{\partial}{\partial t}\rho u = -\left(\frac{\partial}{\partial x}\rho uu + \frac{\partial}{\partial y}\rho vu + \frac{\partial}{\partial z}\rho wu\right)$$

$$-\left(\frac{\partial}{\partial x}\tau_{xx} + \frac{\partial}{\partial y}\tau_{yx} + \frac{\partial}{\partial z}\tau_{zx}\right) - \frac{\partial p}{\partial x} + \rho g_x \tag{4.18}$$

This equation may be rearranged with the help of the equation of continuity [Eq. (4.3)] to give

$$\rho\frac{Du}{Dt} = -\frac{\partial p}{\partial x} - \left(\frac{\partial\tau_{xx}}{\partial x} + \frac{\partial\tau_{yx}}{\partial y} + \frac{\partial\tau_{zx}}{\partial z}\right) + \rho g_x \tag{4.19}$$

Similar equations may be derived for the y and z components. Adding the three components vectorially gives

$$\rho\frac{DV}{Dt} = -\nabla p - [\nabla \cdot \tau] + \rho\mathbf{g} \tag{4.20}$$

The stress at any point depends on the velocity gradients and the rheological properties of the fluid. For newtonian fluids, for example, the x-direction components of the stress tensor are[2a]

$$\tau_{xx} = -2\mu\frac{\partial u}{\partial x} + \left(\frac{2}{3}\mu - \kappa\right)(\nabla \cdot \mathbf{V}) \tag{4.21}$$

$$\tau_{xy} = \tau_{yx} = -\mu\left(\frac{\partial u}{\partial y} + \frac{\partial v}{\partial x}\right) \tag{4.22}$$

$$\nabla \cdot \mathbf{V} = \frac{\partial u}{\partial x} + \frac{\partial v}{\partial y} + \frac{\partial w}{\partial z} \tag{4.23}$$

where κ is the *bulk viscosity*. There is much uncertainty about the value of κ. It is zero for monatomic gases and probably of minor importance for dense gases and liquids.[2a] Equations for stresses in the y and z directions (and all the equations in both cylindrical and spherical coordinate systems) are given by Bird, Stewart, and Lightfoot.[2c]

The general equations of motion for a newtonian fluid with varying density and viscosity are exemplified by the following equation for the x direction, obtained by substitution of Eqs. (4.21) and (4.22) into Eq. (4.19)

$$\rho \frac{Du}{Dt} = -\frac{\partial p}{\partial x} + \frac{\partial}{\partial x}\left[2\mu \frac{\partial u}{\partial x} - \left(\frac{2}{3}\mu - \kappa\right)(\nabla \cdot V)\right]$$
$$+ \frac{\partial}{\partial y}\left[\mu\left(\frac{\partial u}{\partial y} + \frac{\partial v}{\partial x}\right)\right]$$
$$+ \frac{\partial}{\partial z}\left[\mu\left(\frac{\partial w}{\partial x} + \frac{\partial u}{\partial z}\right)\right] + \rho g_x \tag{4.24}$$

$$\rho \frac{Dv}{Dt} = -\frac{\partial p}{\partial y} + \frac{\partial}{\partial x}\left[\mu\left(\frac{\partial v}{\partial x} + \frac{\partial u}{\partial y}\right)\right]$$
$$+ \frac{\partial}{\partial y}\left[2\mu\frac{\partial v}{\partial y} - \left(\frac{2}{3}\mu - \kappa\right)(\nabla \cdot V)\right]$$
$$+ \frac{\partial}{\partial z}\left[\mu\left(\frac{\partial w}{\partial y} + \frac{\partial v}{\partial z}\right)\right] + \rho g_y \tag{4.25}$$

$$\rho \frac{Dw}{Dt} = -\frac{\partial p}{\partial z} + \frac{\partial}{\partial x}\left[\mu\left(\frac{\partial w}{\partial x} + \frac{\partial u}{\partial z}\right)\right] + \frac{\partial}{\partial y}\left[\mu\left(\frac{\partial w}{\partial y} + \frac{\partial v}{\partial z}\right)\right]$$
$$+ \frac{\partial}{\partial z}\left[2\mu\frac{\partial w}{\partial z} - \left(\frac{2}{3}\mu - \kappa\right)(\nabla \cdot V)\right] + \rho g_z \tag{4.26}$$

Navier-Stokes equations. Equations (4.24) through (4.26) are used in their complete form only in setting up highly complicated flow problems. In most situations restricted forms suffice. For a fluid of constant density and viscosity, the equations of motion, known as the Navier-Stokes equations, are

$$\rho\left(\frac{\partial u}{\partial t} + u\frac{\partial u}{\partial x} + v\frac{\partial u}{\partial y} + w\frac{\partial u}{\partial z}\right)$$
$$= \mu\left(\frac{\partial^2 u}{\partial x^2} + \frac{\partial^2 u}{\partial y^2} + \frac{\partial^2 u}{\partial z^2}\right) - \frac{\partial p}{\partial x} + \rho g_x \tag{4.27}$$

$$\rho\left(\frac{\partial v}{\partial t} + u\frac{\partial v}{\partial x} + v\frac{\partial v}{\partial y} + w\frac{\partial v}{\partial z}\right)$$
$$= \mu\left(\frac{\partial^2 v}{\partial x^2} + \frac{\partial^2 v}{\partial y^2} + \frac{\partial^2 v}{\partial z^2}\right) - \frac{\partial p}{\partial y} + \rho g_y \tag{4.28}$$

$$\rho\left(\frac{\partial w}{\partial t} + u\frac{\partial w}{\partial x} + v\frac{\partial w}{\partial y} + w\frac{\partial w}{\partial z}\right)$$
$$= \mu\left(\frac{\partial^2 w}{\partial x^2} + \frac{\partial^2 w}{\partial y^2} + \frac{\partial^2 w}{\partial z^2}\right) - \frac{\partial p}{\partial z} + \rho g_z \tag{4.29}$$

In vector form these equations become

$$\rho \frac{DV}{Dt} = -\nabla p + \mu \nabla^2 V + \rho \boldsymbol{g} \tag{4.30}$$

Euler's equation. For constant density and zero viscosity, as in potential flow, the equation of motion, known as the Euler equation, is

$$\rho \frac{DV}{Dt} = -\nabla p + \rho \boldsymbol{g} \tag{4.31}$$

EXAMPLE 4.3. A newtonian fluid is confined between two broad parallel vertical plates, separated by a distance B, as shown in Fig. 4.4. The plate on the left is stationary; that on the right is moving vertically upward at a constant velocity v_0. Assuming that the flow is laminar, find the equation for the steady-state velocity profile in the fluid.

Solution. Use the Navier-Stokes equation for the y coordinate, Eq. (4.28). At steady state $\partial v/\partial t = 0$ and flow is in the y direction only. Hence the velocities u and z are both zero. From the continuity equation [Eq. (4.6)], $\partial u/\partial y = 0$. Also, $\partial v/\partial z = 0$ and $\rho g_y = -\rho g$. The partial derivatives become derivatives, and Eq. (4.28) becomes

$$\mu \frac{d^2 v}{dx^2} - \frac{dp}{dy} - \rho g = 0 \tag{4.32}$$

The pressure gradient dp/dy is constant, because Eqs. (4.27) and (4.29) show that p is independent of x and z. Integrating Eq. (4.32) gives

$$\frac{dv}{dx} - \frac{x}{\mu} \left(\frac{dp}{dy} + \rho g \right) = C_1 \tag{4.33}$$

Integrating again yields

$$v - \frac{x^2}{2\mu} \left(\frac{dp}{dy} + \rho g \right) = C_1 x + C_2 \tag{4.34}$$

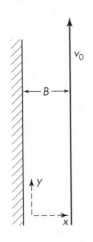

FIGURE 4.4
Flow between vertical plates in Example 4.3.

The boundary conditions are as follows: at $x = 0$, $v = 0$ and at $x = B$, $v = v_0$. Solving for the constants shows that $C_1 = v_0/B - [B/(2\mu)](dp/dy + \rho g)$ and $C_2 = 0$. Substituting in Eq. (4.34) gives

$$v = -\frac{1}{2\mu}\left(\frac{dp}{dy} + \rho g\right)(Bx - x^2) + v_0\frac{x}{B} \tag{4.35}$$

Couette flow

When the plates in a system like that in Example 4.3 are horizontal (or in any situation where the force of gravity may be neglected), the fluid velocity varies linearly with distance from the stationary plate, and the velocity gradient is constant. The viscosity is then related to the shear stress F_s/A by the equation

$$\mu = \frac{F_s B}{A v_0} \tag{4.36}$$

where A is the area of each plate. Flow under these conditions is known as *Couette flow*.

Layer flow with one free surface is discussed in the following section.

MACROSCOPIC MOMENTUM BALANCES

An overall momentum balance can be written for the control volume shown in Fig. 4.5, assuming that the flow is steady and unidirectional in the x direction. In accordance with Eq. (4.14), the sum of the forces acting on the fluid in the x direction equals the increase in the momentum flow rate of the fluid, or

$$\sum F = \dot{M}_b - \dot{M}_a \tag{4.37}$$

Momentum of total stream; momentum correction factor

The momentum flow rate $\dot{M}$ of a fluid stream having a mass flow rate $\dot{m}$ and all moving at a velocity u equals $\dot{m}u$. If u varies from point to point in the cross section of the stream, however, the total momentum flow does not equal the product of the mass flow rate and the average velocity, or $\dot{m}\bar{V}$; in general it is somewhat greater than this.

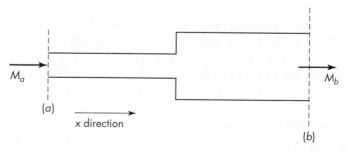

(a)

x direction

(b)

FIGURE 4.5
Momentum balance.

The necessary correction factor is best found from the convective momentum flux, that is, the momentum carried by the moving fluid through a unit cross-sectional area of the channel in unit time. This is the product of the linear velocity normal to the cross section and the mass velocity (or mass flux). For a differential cross-sectional area dS, then, the momentum flux is

$$\frac{d\dot{M}}{dS} = (\rho u)u = \rho u^2 \tag{4.38}$$

The momentum flux of the whole stream, for a constant-density fluid, is

$$\frac{\dot{M}}{S} = \frac{\rho \int_S u^2 \, dS}{S} \tag{4.39}$$

The momentum correction factor β is defined by the relation

$$\beta \equiv \frac{\dot{M}/S}{\rho \bar{V}^2} \tag{4.40}$$

Substituting from Eq. (4.39) gives

$$\beta = \frac{1}{S} \int_S \left(\frac{u}{\bar{V}}\right)^2 dS \tag{4.41}$$

To find β for any given flow situation, the variation of u with position in the cross section must be known.

Thus Eq. (4.37) may be written

$$\sum F = \dot{m}(\beta_b \bar{V}_b - \beta_a \bar{V}_a) \tag{4.42}$$

In using this relation, care must be taken to identify and include in $\sum F$ all force components acting on the fluid in the direction of the velocity component in the equation. Several such forces may appear: (1) pressure change in the direction of flow; (2) shear stress at the boundary between the fluid stream and the conduit or (if the conduit itself is considered to be part of the system) external forces acting on the solid wall; (3) if the stream is inclined, the appropriate component of the force of gravity. Assuming one-dimensional flow in the x direction, a typical situation is represented by the equation

$$\sum F = p_a S_a - p_b S_b + F_w - F_g \tag{4.43}$$

where p_a, p_b = inlet and outlet pressures, respectively
$\quad\quad S_a, S_b$ = inlet and outlet cross sections, respectively
$\quad\quad F_w$ = net force of wall of channel on fluid
$\quad\quad F_g$ = component of force of gravity (written for flow in upward direction)

Layer flow with free surface

In one form of layer flow the liquid layer has a free surface and flows under the force of gravity over an inclined or vertical surface. If such flow is in steady state, with fully developed velocity gradients, the thickness of the layer is constant.

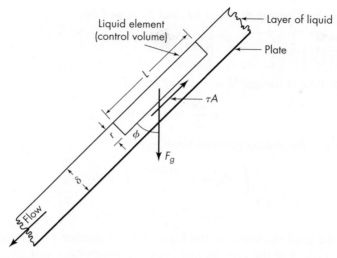

FIGURE 4.6
Forces on liquid element in layer flow.

Conversely there is so little drag at the free liquid surface that the shear stress there can be ignored. If the flow is laminar and the liquid surface is flat and free from ripples, the fluid motion can be analyzed mathematically.

Consider a layer of a newtonian liquid flowing in steady flow at constant rate and thickness over a flat plate, as shown in Fig. 4.6. The plate is inclined at an angle ϕ with the vertical. The breadth of the layer in the direction perpendicular to the plane of the figure is b, and the thickness of the layer in the direction perpendicular to the plate is δ. Isolate a control volume as shown in Fig. 4.6. The upper surface of the control volume is in contact with the atmosphere, the two ends are planes perpendicular to the plate at a distance L apart, and the lower surface is the plane parallel with the wall at a distance r from the upper surface of the layer.

Since the layer is in steady flow with no acceleration, by the momentum principle the sum of all forces on the control volume is zero. The possible forces acting on the control volume in a direction parallel to the flow are the pressure forces on the ends, the shear forces on the upper and lower faces, and the component of the force of gravity in the direction of flow. Since the pressure on the outer surface is atmospheric, the pressures on the control volume at the ends of the volume are equal and oppositely directed. They therefore vanish. Also, by assumption, the shear on the upper surface of the element is neglected. The two forces remaining are therefore the shear force on the lower surface of the control volume and the component of gravity in the direction of flow. Then

$$F_g \cos \phi - \tau A = 0 \tag{4.44}$$

where F_g = gravity force
   τ = shear stress on lower surface of control volume
   A = area of lower surface of control volume

From this equation, noting that $A = bL$ and $F_g = \rho r L b g$,

$$\rho r L b g \cos \phi = \tau L b$$

or
$$\tau = \rho r g \cos \phi \tag{4.45}$$

Since the flow is laminar, $\tau = -\mu \, du/dr$ and

$$-\mu \frac{du}{dr} = g\rho r \cos \phi \tag{4.46}$$

Rearranging and integrating between limits give

$$\int_0^u du = -\frac{g\rho \cos \phi}{\mu} \int_\delta^r r \, dr$$

$$u = \frac{\rho g \cos \phi}{2\mu}(\delta^2 - r^2) \tag{4.47}$$

where δ is the total thickness of the liquid layer. Equation (4.47) shows that in laminar flow on a plate the velocity distribution is parabolic.

Consider now a differential element of cross-sectional area dS, where $dS = b \, dr$. The differential mass flow rate $d\dot{m}$ through this element equals $\rho u b \, dr$. The total mass flow rate of the fluid then is

$$\dot{m} = \int_0^\delta \rho u b \, dr \tag{4.48}$$

Substituting from Eq. (4.47) into Eq. (4.48) and integrating give

$$\frac{\dot{m}}{b} = \frac{\delta^3 \rho^2 g \cos \phi}{3\mu} = \Gamma \tag{4.49}$$

where $\Gamma \equiv \dot{m}/b$ and is called the liquid loading. The units of Γ are kilograms per second per meter of width or pounds per second per foot of width.

Rearrangement of Eq. (4.49) gives, for the thickness of the layer,

$$\delta = \left(\frac{3\mu\Gamma}{\rho^2 g \cos \phi}\right)^{1/3} \tag{4.50}$$

The Reynolds number for flow down a flat plate is defined by the equation.

$$\mathrm{Re} = \frac{4 r_H \bar{V} \rho}{\mu} = 4\delta \frac{\dot{m}}{\rho L_p \delta} \frac{\rho}{\mu} = \frac{4\Gamma}{\mu} \tag{4.51}$$

where r_H = hydraulic radius, defined by Eq. (5.10). For flow of a liquid down either the inside or the outside of a pipe, the layer thickness is usually a very small fraction of the pipe diameter and the Reynolds number is the same as for a flat plate, Eq. (4.51).

Equation (4.50) for the thickness of a falling laminar film was first presented by Nusselt,[4] who used the result to predict heat-transfer coefficients for condensing vapors. Measurements of film thickness on a vertical surface ($\cos \phi = 1$) show that Eq. (4.50) is approximately correct for Re $\approx$ 1,000, but the thickness actually varies with about 0.45 power of the Reynolds number, and the layers are thinner than predicted at low Re and thicker than predicted above Re = 1,000. The deviations may be due to ripples or waves in the films, which are apparent even at quite low Reynolds numbers.

The transition from laminar to turbulent flow is not as easily detected as in pipe flow since the film is very thin and the ripples make it difficult to observe turbulence in the film. A critical Reynolds number of 2,100 has often been used for layer flow, but film thickness measurements[6] indicate a transition at Re ≈ 1,200. Above this point, the thickness increases with about the 0.6 power of the flow rate.

Angular-momentum equation

Analysis of the performance of rotating fluid-handling machinery such as pumps, turbines, and agitators is facilitated by the use of force moments and angular momentum. The moment of a force $\mathbf{F}$ about point O is the vector product of $\mathbf{F}$ and the position vector $\mathbf{r}$ of a point on the line of action of the vector from O. When a force, say F_θ, acts at right angles to the position vector, at a radial distance r from point O, the moment of the force equals the torque T, or

$$F_\theta r = T \tag{4.52}$$

The *angular momentum* (also called the *moment of momentum*) of an object moving about a center of rotation is the vector product of the position vector and the tangential momentum vector of the object (its mass times its tangential component of velocity). Figure 4.7 shows the rotation for a situation involving two-dimensional flow: Fluid at point P is moving about a point O at a velocity V, which has radial and tangential components u_r and u_θ, respectively. The angular momentum of a mass m of fluid at point P is therefore rmu_θ.

Suppose that Fig. 4.7 represents part of the impeller of a centrifugal pump or turbine through which fluid is flowing at a constant mass rate $\dot{m}$. It enters at point Q near the center of rotation at a radial distance r_1 from point O and leaves at radial

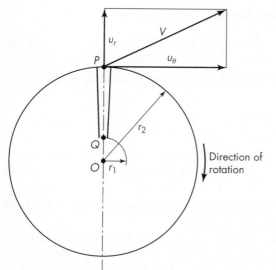

FIGURE 4.7
Angular momentum of flowing liquid.

distance r_2. Its tangential velocities at these points are $u_{\theta 1}$ and $u_{\theta 2}$, respectively. The tangential force F_θ on the fluid at point P is proportional to the rate of change of angular momentum of the fluid; hence the torque is given, from Eq. (4.52), by the relation

$$T = F_\theta r_2 = \dot{m}(r_2 u_{\theta 2} - r_1 u_{\theta 1}) \tag{4.53}$$

Equation (4.53) is the angular momentum equation for steady two-dimensional flow. It is analogous to Eq. (4.42), the momentum equation. It is assumed in deriving Eq. (4.53) that at any given radial distance r all the fluid is moving with the same velocity, so $\beta_1 = \beta_2 = 1$. Applications of Eq. (4.53) are given in Chaps. 8 and 9.

MECHANICAL ENERGY EQUATION

An equation that describes the energy interconversions that occur in a flowing fluid may be derived by forming the scalar product of the local velocity V with the equation of motion. Applying this procedure to Eq. (4.20) yields a general equation which states that the rate of increase in kinetic energy per unit mass equals the net rate of input of kinetic energy by convection less the following: (1) rate of work done by the pressure of the surroundings; (2) rate of reversible conversion to internal energy; (3) rate of work done by viscous forces; (4) irreversible conversion to internal energy; (5) rate of work done by gravity (this term may be positive or negative).[2b] As we will see, mechanical energy may also be added to the fluid by a pump or blower.

Here the derivation is initially restricted to the unidirectional flow of fluids of constant density and zero viscosity, using the Euler equation.

Energy equation for potential flow; Bernoulli equation without friction

The x component of the Euler equation [Eq. (4.31)] is

$$\rho\left(\frac{\partial u}{\partial t} + u\frac{\partial u}{\partial x} + v\frac{\partial u}{\partial y} + w\frac{\partial u}{\partial z}\right) = -\frac{\partial p}{\partial x} + \rho g_x \tag{4.54}$$

For unidirectional flow v and w are zero. Multiplying the remaining terms by the velocity u gives

$$\rho u\left(\frac{\partial u}{\partial t} + u\frac{\partial u}{\partial x}\right) = -u\frac{\partial p}{\partial x} + \rho u g_x$$

or

$$\rho\left[\frac{\partial(u^2/2)}{\partial t} + u\frac{\partial(u^2/2)}{\partial x}\right] = -u\frac{\partial p}{\partial x} + \rho u g_x \tag{4.55}$$

This is the mechanical energy equation for unidirectional potential flow of fluids of constant density when the flow rate varies with time.

Consider now a volume element of a stream tube within a larger stream of fluid, as shown in Fig. 4.8, flowing at a steady rate. Assume that the cross section of the tube increases continuously in the direction of flow, and that the axis of the tube is

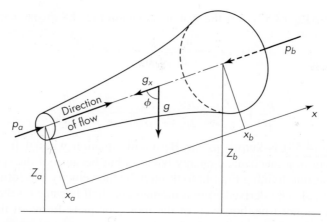

FIGURE 4.8
Potential flow through inclined stream tube.

straight and inclined upward at an angle ϕ from the vertical. Let the pressure, fluid velocity, and elevation at the entrance be p_a, u_a, and Z_a, respectively, and those at the outlet be p_b, u_b, and Z_b. Set the x axis parallel with the axis of the tube.

Since flow is steady, the left-hand term in Eq. (4.55) vanishes. There is no variation in fluid velocity across the cross section, so the flow is unidirectional and velocity u is a function only of x. For substitution in Eq. (4.55), since gravity acts in the negative x direction, $g_x = -g \cos \phi$. If Z is the elevation at any cross section along the tube, then $Z = Z_a + x \cos \phi$, $dZ = \cos \phi \, dx$, and $\cos \phi = dZ/dx$. The partial differentials become total differentials. Hence, from Eq. (4.55),

$$u \frac{d(\rho u^2/2)}{dx} + u\frac{dp}{dx} + \rho u g \cos \phi = 0 \qquad (4.56)$$

Thus for steady flow it is possible to divide through by the velocity u. By doing this and also dividing through by ρ and substituting for $\cos \phi$, Eq. (4.56) becomes

$$\frac{d(u^2/2)}{dx} + \frac{1}{\rho} \frac{dp}{dx} + g \frac{dZ}{dx} = 0 \qquad (4.57)$$

Equation (4.57) is the point form of the Bernoulli equation without friction. Although derived for the special case of an expanding cross section and upward flow, this equation is applicable to constant or contracting cross sections and horizontal or downward flow (the sign of the differential dZ corrects for change in direction).

When the cross section and density are constant, u does not change with position, the term $d(u^2/2)/dx$ is zero, and Eq. (4.57) becomes identical with Eq. (2.2) for a stationary fluid. In unidirectional potential flow at constant velocity, then, the magnitude of the velocity does not affect the pressure drop in the tube; the pressure drop depends only on the rate of change of elevation. In a straight horizontal tube, in consequence, there is *no* pressure drop in steady constant-velocity potential flow.

Integrating Eq. (4.57) over the system shown in Fig. 4.8 gives

$$\frac{p_a}{\rho} + gZ_a + \frac{u_a^2}{2} = \frac{p_b}{\rho} + gZ_b + \frac{u_b^2}{2} \tag{4.58a}$$

In fps units,

$$\frac{p_a}{\rho} + \frac{gZ_a}{g_c} + \frac{u_a^2}{2g_c} = \frac{p_b}{\rho} + \frac{gZ_b}{g_c} + \frac{u_b^2}{2g_c} \tag{4.58b}$$

Equation (4.58) is known as the Bernoulli equation without friction. It is a particular form of a mechanical energy balance, but because of the special conditions which permit dividing Eq. (4.56) through by the velocity u to form Eq. (4.57), Eq. (4.58) can also be derived from a momentum shell balance on the element in Fig. 4.8. This is not true, however, of the more complete forms of the Bernoulli equation which are discussed later in this chapter.

Each term in Eq. (4.58) is a scalar and has the dimensions of energy per unit mass, representing a mechanical energy effect based on a unit mass of the flowing fluid. Terms gZ and $u^2/2$ are the potential and kinetic energy, respectively, of a unit mass of fluid; and p/ρ represents the mechanical work done by forces, external to the stream, on the fluid in pushing it into the tube or the work recovered from the fluid leaving the tube. Equation (4.58) shows that in the absence of friction, when the velocity u is reduced, either the height above datum Z or the pressure p or both must increase. When the velocity increases, it does so only at the expense of Z or p. If the height is changed, compensation must be found in a change of either pressure or velocity.

The Bernoulli equation has a greater range of validity than its derivation implies. Although in the derivation the assumption was made that the stream tube is straight, the principle of conservation of energy permits the extension of the equation to potential flow taking place in curved stream tubes. If the tube is curved, the direction of the velocity changes and in the Bernoulli equation the scalar speed, rather than the vector velocity, is used. In all real situations there are some friction losses in the fluid and some variation of velocity within a cross section of the tube, but in some cases these are small enough to be ignored. In other situations, by the use of correction factors the equation can be modified for use in boundary layer flow, where velocity variations within a cross section occur and friction effects are active. These corrections are discussed in the following sections.

To apply the Bernoulli equation to a specific problem, it is essential to identify the streamline or stream tube and to choose definite upstream and downstream stations. Stations a and b are chosen on the basis of convenience and are usually taken at locations where the most information about pressures, velocities, and heights is available.

EXAMPLE 4.4. Brine, specific gravity 60°F/60°F = 1.15, is draining from the bottom of a large open tank through a 50-mm pipe. The drainpipe ends at a point 5 m below the surface of the brine in the tank. Considering a streamline starting at the surface of the brine in the tank and passing through the center of the drain line to the point of discharge, and assuming that friction along the streamline is negligible, calculate the velocity of flow along the streamline at the point of discharge from the pipe.

Solution. To apply Eq. (4.58), choose station a at the brine surface and station b at the end of the streamline at the point of discharge. Since the pressure at both stations is atmospheric, p_a and p_b are equal, and $p_a/\rho = p_b/\rho$. At the surface of the brine, u_a is negligible, and the term $u_a^2/2$ is dropped. The datum for measurement of heights can be taken through station b, so $Z_b = 0$ and $Z_a = 5$ m. Substitution in Eq. (4.58) gives

$$5g = \tfrac{1}{2}u_b^2$$

and the velocity on the streamline at the discharge is

$$u_b = \sqrt{5 \times 2 \times 9.80665} = 9.90 \text{ m/s}$$

Note that this velocity is independent of density and of pipe size.

Bernoulli equation: correction for effects of solid boundaries

Most fluid flow problems encountered in engineering involve streams that are influenced by solid boundaries and therefore contain boundary layers. This is especially true in the flow of fluids through pipes and other equipment, where the entire stream may be in boundary layer flow.

To extend the Bernoulli equation to cover these practical situations, two modifications are needed. The first, usually of minor importance, is a correction of the kinetic energy term for the variation of local velocity u with position in the boundary layer; the second, of major importance, is the correction of the equation for the existence of fluid friction, which appears whenever a boundary layer forms.

Also, the usefulness of the corrected Bernoulli equation in solving problems of flow of incompressible fluids is enhanced if provision is made in the equation for the work done on the fluid by a pump.

Kinetic energy of stream

The term $u^2/2$ in Eq. (4.58) is the kinetic energy of a unit mass of fluid all of which is flowing at the same velocity u. When the velocity varies across the stream cross section, the kinetic energy is found in the following manner. Consider an element of cross-sectional area dS. The mass flow rate through this is $\rho u \, dS$. Each unit mass of fluid flowing through area dS carries kinetic energy in amount $u^2/2$, and the energy flow rate through area dS is therefore

$$d\dot{E}_k = (\rho u \, dS)\frac{u^2}{2} = \frac{\rho u^3 \, dS}{2}$$

where $\dot{E}_k$ represents the time rate of flow of kinetic energy. The total rate of flow of kinetic energy through the entire cross section S is, assuming constant density within the area S,

$$\dot{E}_k = \frac{\rho}{2} \int_S u^3 \, dS \tag{4.59}$$

The total rate of mass flow is given by Eqs. (4.8) and (4.13), and the kinetic energy per unit mass of flowing fluid, which replaces $u^2/2$ in the Bernoulli equation, is

$$\frac{\dot{E}_k}{\dot{m}} = \frac{\tfrac{1}{2}\int_S u^3 \, dS}{\int_S u \, dS} = \frac{\tfrac{1}{2}\int_S u^3 \, dS}{\bar{V} S} \tag{4.60}$$

Kinetic energy correction factor. It is convenient to eliminate the integral of Eq. (4.60) by a factor operating on $\bar{V}^2/2$ to give the correct value of the kinetic energy as calculated from Eq. (4.60). This factor, called the kinetic energy correction factor, is denoted by α and is defined by

$$\frac{\alpha \bar{V}^2}{2} \equiv \frac{\dot{E}_k}{\dot{m}} = \frac{\int_S u^3 \, dS}{2\bar{V}S}$$

$$\alpha = \frac{\int_S u^3 \, dS}{\bar{V}^3 S} \tag{4.61}$$

If α is known, the average velocity can be used to calculate the kinetic energy from the average velocity by using $\alpha \bar{V}^2/2$ in place of $u^2/2$. To calculate the value of α from Eq. (4.61), the local velocity must be known as a function of location in the cross section, so that the integral in the equation can be evaluated. The same knowledge of velocity distribution is needed to calculate the value of $\bar{V}$ by Eq. (4.9). As shown in Chap. 5, α is 2.0 for laminar flow and is about 1.05 for highly turbulent flow.

Correction of Bernoulli equation for fluid friction

Friction manifests itself by the disappearance of mechanical energy. In frictional flow the quantity

$$\frac{p}{\rho} + \frac{u^2}{2} + gZ$$

is not constant along a streamline, as called for by Eq. (4.58), but always decreases in the direction of flow; and in accordance with the principle of conservation of energy, an amount of heat equivalent to the loss in mechanical energy is generated. Fluid friction can be defined as any conversion of mechanical energy to heat in a flowing stream.

For incompressible fluids, the Bernoulli equation is corrected for friction by adding a term to the right-hand side of Eq. (4.58). Thus, after introducing the kinetic energy correction factors α_a and α_b, Eq. (4.58) becomes

$$\frac{p_a}{\rho} + gZ_a + \frac{\alpha_a \bar{V}_a^2}{2} = \frac{p_b}{\rho} + gZ_b + \frac{\alpha_b \bar{V}_b^2}{2} + h_f \tag{4.62}$$

The units of h_f and those of all other terms in Eq. (4.62) are energy per unit mass. The term h_f represents all the friction generated per unit mass of fluid (and therefore all the conversion of mechanical energy to heat) that occurs in the fluid between stations a and b. It differs from all other terms in Eq. (4.62) in two ways:

1. The mechanical terms represent conditions *at* specific locations, namely, the inlet and outlet stations a and b, whereas h_f represents the loss of mechanical energy at all points *between* stations a and b.
2. Friction is not interconvertible with the mechanical energy quantities.

The sign of h_f, as defined by Eq. (4.62), is always positive. It is zero, of course, in potential flow.

Friction appears in boundary layers because the work done by shear forces in maintaining the velocity gradients in both laminar and turbulent flow is eventually converted to heat by viscous action. Friction generated in unseparated boundary layers is called *skin friction*. When boundary layers separate and form wakes, additional energy dissipation appears within the wake, and friction of this type is called *form friction* since it is a function of the position and shape of the solid.

In a given situation both skin friction and form friction may be active in varying degrees. In the case of Fig. 3.9a, the friction is entirely skin friction; in that of Fig. 3.9b, the friction is largely form friction, because of the large wake, and skin friction is relatively unimportant. The total friction h_f in Eq. (4.62) includes both types of friction loss.

EXAMPLE 4.5. Water with a density of 998 kg/m³ (62.3 lb/ft³) enters a 50 mm (1.969-in.) pipe fitting horizontally, as shown in Fig. 4.9, at a steady velocity of 1.0 m/s (3.28 ft/s) and a gauge pressure of 100 kN/m² (14.48 lb$_f$/in²). It leaves the fitting horizontally, at the same elevation, at an angle of 45° with the entrance direction. The diameter at the outlet is 20 mm (0.787 in.). Assuming the fluid density is constant, the kinetic energy and momentum correction factors at both entrance and exit are unity, and the friction loss in the fitting is negligible, calculate (a) the gauge pressure at the exit of the fitting and (b) the forces in the x and y directions exerted by the fitting on the fluid.

Solution

(a) $\bar{V}_a = 1.0$ m/s. From Eq. (4.12),

$$\bar{V}_b = \bar{V}_a \left(\frac{D_a}{D_b}\right)^2 = 1.0 \left(\frac{50}{20}\right)^2 = 6.25 \text{ m/s}$$

$$p_a = 100 \text{ kN/m}^2$$

The outlet pressure p_b is found from Eq. (4.62). Since $Z_a = Z_b$ and h_f may be neglected, Eq. (4.62) becomes

$$\frac{p_a - p_b}{\rho} = \frac{\bar{V}_b^2 - \bar{V}_a^2}{2}$$

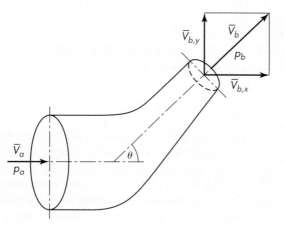

FIGURE 4.9
Flow through reducing fitting, viewed from the top, Example 4.5.

from which

$$p_b = p_a - \frac{\rho(\bar{V}_b^2 - \bar{V}_a^2)}{2} = 100 - \frac{998(6.25^2 - 1.0^2)}{1{,}000 \times 2}$$

$$= 100 - 18.99 = 81.01 \text{ kN/m}^2 \ (11.75 \text{ lb}_f/\text{in.}^2)$$

(b) The forces acting on the fluid are found by combining Eqs. (4.42) and (4.43). For the x direction, since $F_g = 0$ for horizontal flow, this gives

$$\dot{m}(\beta_b \bar{V}_{b,x} - \beta_a \bar{V}_{a,x}) = p_a S_{a,x} - p_b S_{b,x} + F_{w,\,x} \qquad (4.63)$$

where $S_{a,x}$ and $S_{b,x}$ are the projected areas of S_a and S_b on planes normal to the initial flow direction. (Recall that pressure p is a scalar quantity.) Since the flow enters in the x direction, $\bar{V}_{a,x} = \bar{V}_a$ and

$$S_{a,x} = S_a = \frac{\pi}{4}0.050^2 = 0.001964 \text{ m}^2$$

From Fig. 4.9

$$\bar{V}_{b,x} = \bar{V}_b \cos\theta = 6.25 \cos 45° = 4.42 \text{ m/s}$$

Also $S_{b,x} = S_b \sin\theta = \frac{\pi}{4}0.020^2 \sin 45° = 0.000222 \text{ m}^2$

From Eq. (4.11)

$$\dot{m} = \bar{V}_a \rho S_a = 1.0 \times 998 \times 0.001964 = 1.960 \text{ kg/s}$$

Substituting in Eq. (4.63) and solving for $F_{w,x}$, assuming $\beta_a = \beta_b = 1$, give

$$F_{w,x} = 1.96(4.42 - 1.0) - 100{,}000 \times 0.001964 + 81{,}010 \times 0.000222$$

$$= 6.7 - 196.4 + 18.0 = -171.7 \text{ N} \ (-38.6 \text{ lb}_f)$$

Similarly for the y direction, $\bar{V}_{a,y} = 0$ and $S_{a,y} = 0$, and

$$\bar{V}_{b,y} = \bar{V}_b \sin\theta = 4.42 \text{ m/s} \qquad S_{b,y} = S_b \cos\theta = 0.000222 \text{ m}^2$$

Hence $F_{w,y} = \dot{m}(\beta_b \bar{V}_{b,y} - \beta_a \bar{V}_{a,y}) - p_a S_{a,y} + p_b S_{b,y}$

$$= 1.96(4.42 - 0) - 0 + 81.01 \times 0.000222 \times 1{,}000$$

$$= 8.66 + 17.98 = 26.64 \text{ N} \ (5.99 \text{ lb}_f)$$

Pump work in Bernoulli equation

A pump is used in a flow system to increase the mechanical energy of the flowing fluid, the increase being used to maintain flow, provide kinetic energy, offset friction losses, and—sometimes—increase the potential energy. Assume that a pump is installed between stations a and b linked by Eq. (4.62). Let W_p be the work done by the pump per unit mass of fluid. Since the Bernoulli equation is a balance of mechanical energy only, account must be taken of friction occurring within the pump. In an actual pump not only are all the sources of fluid friction active, but mechanical friction occurs as well, in bearings and seals or stuffing boxes. The mechanical energy supplied to the pump as negative shaft work must be discounted by these friction losses to give the net mechanical energy actually available to the flowing fluid. Let h_{fp} be the total friction in the pump per unit mass of fluid. Then the net work to the fluid is $W_p - h_{fp}$. In practice, in place of h_{fp} a pump efficiency denoted by η is used, defined by the equation

$$W_p - h_{fp} \equiv \eta W_p$$

or

$$\eta = \frac{W_p - h_{fp}}{W_p} \tag{4.64}$$

The mechanical energy delivered to the fluid is, then, ηW_p, where $\eta < 1$. Equation (4.62) corrected for pump work is

$$\frac{p_a}{\rho} + g Z_a + \frac{\alpha_a \bar{V}_a^2}{2} + \eta W_p = \frac{p_b}{\rho} + g Z_b + \frac{\alpha_b \bar{V}_b^2}{2} + h_f \tag{4.65}$$

Equation (4.65) is a final working form of the Bernoulli equation for problems on the flow of incompressible fluids.

EXAMPLE 4.6. In the equipment shown in Fig. 4.10, a pump draws a solution of specific gravity 1.84 from a storage tank through a 3-in. (75-mm) Schedule 40 steel pipe. The efficiency of the pump is 60 percent. The velocity in the suction line is 3 ft/s (0.914 m/s). The pump discharges through a 2-in. (50-mm) Schedule 40 pipe to an overhead tank. The end of the discharge pipe is 50 ft (15.2 m) above the level of the solution in the feed tank. Friction losses in the entire piping system are 10 ft·lb$_f$/lb (29.9 J/kg). What pressure must the pump develop? What is the power delivered to the fluid by the pump?

Solution. Use Eq. (4.65). Take station a at the surface of the liquid in the tank and station b at the discharge end of the 2-in. pipe. Take the datum plane for elevations through station a. Since the pressure at both stations is atmospheric, $p_a = p_b$. The velocity at station a is negligible because of the large diameter of the tank in comparison with that of the pipe. For turbulent flows the kinetic energy factor α can be taken

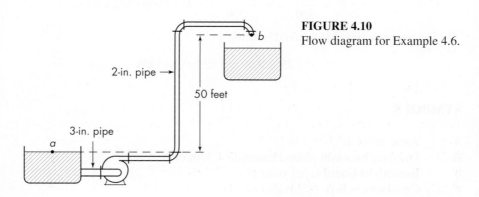

FIGURE 4.10
Flow diagram for Example 4.6.

as 1.0 with negligible error. Equation (4.65) in fps units becomes

$$W_p\eta = \frac{g}{g_c}Z_b + \frac{\bar{V}_b^2}{2g_c} + h_f$$

By App. 3, the cross-sectional areas of the 3- and 2-in. pipes are 0.0513 and 0.0233 ft^2, respectively. The velocity in the 2-in. pipe is

$$\bar{V}_b = \frac{3 \times 0.0513}{0.0233} = 6.61 \text{ ft/s}$$

Then

$$0.60W_p = 50\frac{g}{g_c} + \frac{6.61^2}{64.34} + 10 = 60.68$$

and

$$W_p = \frac{60.68}{0.60} = 101.1 \text{ ft·lb}_f/\text{lb}$$

The pressure developed by the pump can be found by writing Eq. (4.65) over the pump itself. Station a is in the suction connection, and station b is in the pump discharge. The difference in level between suction and discharge can be neglected, so $Z_a = Z_b$, and Eq. (4.65) becomes

$$\frac{p_b - p_a}{\rho} = \frac{\bar{V}_a^2 - \bar{V}_b^2}{2g_c} + W_p\eta$$

The pressure developed by the pump is

$$p_b - p_a = 1.84 \times 62.37\left(\frac{3^2 - 6.61^2}{2 \times 32.17} + 60.68\right)$$

$$= 6,902 \text{ lb}_f/\text{ft}^2 \text{ or } \frac{6,902}{144} = 47.9 \text{ lb}_f/\text{in.}^2 \text{ (330 kN/m}^2\text{)}$$

The power delivered to the pump is the product of W_p and the mass flow rate, divided by the conversion factor 1 hp = 550 ft·lb$_f$/s. The mass flow rate is

$$\dot{m} = 0.0513 \times 3 \times 1.84 \times 62.37 = 17.66 \text{ lb/s}$$

and the power is

$$P = \frac{\dot{m}W_p}{550} = \frac{17.66 \times 101.1}{550} = 3.25 \text{ hp (2.42 kW)}$$

The power delivered to the fluid is then

$$3.25 \times 0.60 = 1.95 \text{ hp (1.45 kW)}$$

SYMBOLS

A	Area, m^2 or ft^2
B	Distance between plates, Example 4.3, m or ft
b	Breadth of liquid layer, m or ft
C_1, C_2	Constants in Eqs. (4.33) and (4.34)

D Diameter of circular channel, m or ft; D_a, at station a; D_b, at station b

E_k Kinetic energy of fluid, J or ft·lb$_f$; $\dot{E}_k$, time rate of flow of kinetic energy, J/s or ft·lb$_f$/s

F Force, N or lb$_f$; F_g, component of gravity force; F_w, net force of channel wall on fluid; $F_{w,x}$, component of F_w in x direction; $F_{w,y}$, component in y direction; F_θ, tangential force or force component

G Mass velocity, kg/m²·s or lb/ft²·s

g Gravitational force vector, N or lb$_f$; g_x, component in x direction

g Acceleration of gravity, m/s² or ft/s²

g_c Newton's law proportionality factor, 32.174 ft·lb/lb$_f$·s²

h Friction loss, J/kg or ft·lb$_f$/lb; h_f, friction loss in conduit between stations a and b; h_{fp}, total friction loss in pump

L Length, m or ft

M Molecular weight; also momentum, kg·m/s or ft·lb/s; $\dot{M}$, time rate of flow of momentum, kg·m/s² or ft·lb/s²; $\dot{M}_a$, at station a; $\dot{M}_b$, at station b

m Mass, kg or lb; $\dot{m}$, mass flow rate, kg/s or lb/s

n Number of moles

P Power, kW or hp

p Pressure, N/m² or lb$_f$/ft²; p_a, at station a; p_b, at station b

q Volumetric flow rate, m³/s or ft³/s

Re   Reynolds number, dimensionless

r Radial distance, m or ft; r_1, at station 1; r_2, at station 2; also, distance from liquid surface, m or ft

r_H Hydraulic radius, m or ft

S Cross-sectional area, m² or ft²; S_a, at station a; S_b, at station b; $S_{a,x}$, $S_{b,x}$, projections of S_a and S_b on planes normal to x axis; $S_{a,y}$, $S_{b,y}$, projections on planes normal to y axis

T Torque, N·m or ft·lb$_f$; also temperature, °C or °F; T_a, temperature at station a; T_b, at station b

t Time, s

u Velocity or velocity component in x direction, m/s or ft/s; u_a, at station a; u_b, at station b; u_r, in radial direction; u_θ, in tangential direction

$\mathbf{V}$ Total velocity vector, m/s or ft/s

V Scalar value of velocity, m/s or ft/s; $\bar{V}$, average velocity; $\bar{V}_a$, at station a; $\bar{V}_b$, at station b; also volume, m³ or ft³

v Velocity or velocity component in y direction, m/s or ft/s; v_0, velocity of moving plate in Example 4.3

W_p Pump work per unit mass of fluid, J/kg or ft·lb$_f$/lb

w Velocity or velocity component in z direction, m/s or ft/s

Z Height above datum plane, m or ft; Z_a, at station a; Z_b, at station b

Greek Letters

α Kinetic energy correction factor defined by Eq. (4.61); α_a, at station a; α_b, at station b

β Momentum correction factor defined by Eq. (4.40); β_a, at station a; β_b, at station b; β_1, at station 1; β_2, at station 2

Γ Liquid loading in layer flow, kg/m·s or lb/ft·s

δ Thickness of liquid layer, m or ft

η Overall efficiency of pump, dimensionless

θ Angle of discharge pipe, Fig. 4.9

κ Bulk viscosity of fluid, mPa·s or lb/ft·s

μ Viscosity, mPa·s or lb/ft·s

ρ Density, kg/m³ or lb/ft³; ρ_a, at station a; ρ_b, at station b

τ Stress, N/m² or lb$_f$/ft²; τ_{xx}, normal stress on the x face of volume element; τ_{yx}, x-directed shear stress on the y face; τ_{zx}, x-directed shear stress on the z face

ϕ Angle with vertical

PROBLEMS

4.1. A liquid is flowing in steady flow through a 75-mm pipe. The local velocity varies with distance from the pipe axis as shown in Table 4.1. Calculate (a) average velocity $\bar{V}$, (b) kinetic energy correction factor α, and (c) momentum correction factor β.

TABLE 4.1
Data for Prob. 4.1

Local velocity u, m/s	Distance from pipe axis, mm	Local velocity u, m/s	Distance from pipe axis, mm
1.042	0	0.919	22.50
1.033	3.75	0.864	26.25
1.019	7.50	0.809	30.00
0.996	11.25	0.699	33.75
0.978	15.00	0.507	35.625
0.955	18.75	0	37.50

4.2. For a given fluid and a given distance between plates in a system like that shown in Fig. 4.4, velocity v_0 must be large enough to counteract the effect of gravity; otherwise, some of the fluid will flow downward. In a particular system, the distance between plates is 1 mm, and the fluid is an oil with a density of 900 kg/m³ and a viscosity of 50 mPa·s. The pressure drop dp/dy is negligible compared with the term ρg. (a) What is the minimum upward velocity of the moving plate so that all the fluid moves upward? (b) If v_0 is set at this minimum value, what is the fluid velocity midway between the plates? (c) What is the shear rate in the fluid at the stationary plate, at the moving plate, and midway between them? Use the equations in Example 4.3.

4.3. (a) A water tank is 30 ft in diameter, and the normal depth is 25 ft. The outlet is a 4-in. horizontal pipe at the bottom. If this pipe is sheared off close to the tank, what is the initial flow rate of water from the tank? (Neglect friction loss in the short stub of pipe.)

(*b*) How long will it take for the tank to be empty? (*c*) Calculate the average flow rate and compare it with the initial flow rate.

4.4. Water at 20°C is pumped at a constant rate of 9 m^3/h from a large reservoir resting on the floor to the open top of an experimental absorption tower. The point of discharge is 5 m above the floor, and friction losses in the 50-mm pipe from the reservoir to the tower amount to 2.5 J/kg. At what height in the reservoir must the water level be kept if the pump can deliver only 0.1 kW?

4.5. Water enters a 100-mm-ID 90° elbow, positioned in a horizontal plane, at a velocity of 6 m/s and a pressure of 70 kN/m^2 gauge. Neglecting friction, what are the magnitude and the direction of the force that must be applied to the elbow to keep it in position without moving?

4.6. A vertical cylindrical reactor 2.5 m in diameter and 4 m high is cooled by spraying water on the top and allowing the water to flow down the outside wall. The water flow rate is 0.15 m^3/min, and the average water temperature is 40°C. Estimate the thickness of the layer of water.

REFERENCES

1. Bennett, C. O., and J. E. Myers. *Momentum, Heat, and Mass Transfer.* 3d ed. New York: McGraw-Hill, 1982.
2. Bird, R. B., W. E. Stewart, and E. N. Lightfoot. *Transport Phenomena.* New York: Wiley, 1960; (*a*) p. 79, (*b*) p. 81, (*c*) pp. 84–91.
3. Geankoplis, C. J. *Transport Processes and Unit Operations.* 3d ed. Englewood Cliffs, NJ: Prentice-Hall, 1993.
4. Nusselt, W. *VDIZ,* **60:**541, 569 (1916).
5. Perry, R. H., and D. W. Green, eds. *Perry's Chemical Engineers' Handbook.* 7th ed. New York: McGraw-Hill, 1997, p. **6**–7.
6. Portalski, S. *Chem. Eng. Sci.* **18:**787 (1963).
7. Streeter, V. L., and E. B. Wylie. *Fluid Mechanics.* 8th ed. New York: McGraw-Hill, 1985.

CHAPTER 5

Incompressible Flow in Pipes and Channels

Chemical engineers in industrial practice are often concerned with the flow of fluids through pipes, tubes, and channels with a noncircular cross section. Usually the pipes are filled with the moving fluid, but some problems involve flow in partially filled pipes, in layers down inclined or vertical surfaces, through beds of solids, or in agitated vessels. This chapter deals with the steady flow of incompressible fluids through closed pipes and channels. Layer flow is discussed in Chap. 4; the flow of compressible fluids is covered in Chap. 6, flow through beds of solids in Chap. 7, and flow in agitated vessels in Chap. 9.

SHEAR STRESS AND SKIN FRICTION IN PIPES

$$Eq \; 4.43 \quad \Sigma F = P_a S_a - P_b S_b + F_w - F_g$$

Shear-stress distribution

Consider the steady flow of fluid of constant density in fully developed flow through a horizontal pipe. Visualize a disk-shaped element of fluid, concentric with the axis of the tube, of radius r and length dL, as shown in Fig. 5.1. Assume the element is isolated as a free body. Let the fluid pressure on the upstream and downstream faces of the disk be p and $p + dp$, respectively. Since the fluid possesses a viscosity, a shear force opposing flow will exist on the rim of the element. Apply the momentum equation (4.43) between the two faces of the disk. Since the flow is fully developed, $\beta_b = \beta_a$, and $\bar{V}_b = \bar{V}_a$, so that $\sum F = 0$. The quantities for substitution in Eq. (4.43) are

$$S_a = S_b = \pi r^2 \qquad p_a = p \qquad p_a S_a = \pi r^2 p \qquad p_b S_b = (\pi r^2)(p + dp)$$

The shear force F_s acting on the rim of the element is the product of the shear stress and the cylindrical area, or $(2\pi r \, dL)\tau$. [Force F_s equals $-F_w$ in Eq. (4.43).] Since the channel is horizontal, F_g is zero. Substituting these quantities into Eq. (4.43) gives

$$\sum F = \pi r^2 p - \pi r^2 (p + dp) - (2\pi r \, dL)\tau = 0$$

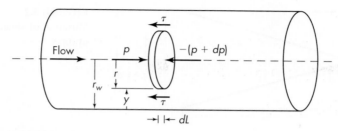

FIGURE 5.1
Fluid element in steady flow through pipe.

Simplifying this equation and dividing by $\pi r^2 \, dL$ give

$$\frac{dp}{dL} + \frac{2\tau}{r} = 0 \qquad (5.1)$$

In steady flow, either laminar or turbulent, the pressure at any given cross section of a stream tube is constant, so that dp/dL is independent of r. Equation (5.1) can be written for the entire cross section of the tube by taking $\tau = \tau_w$ and $r = r_w$, where τ_w is the shear stress at the wall of the conduit and r_w is the radius of the tube. Equation (5.1) then becomes

$$\frac{dp}{dL} + \frac{2\tau_w}{r_w} = 0 \qquad (5.2)$$

Subtracting Eq. (5.1) from Eq. (5.2) gives

$$\frac{\tau_w}{r_w} = \frac{\tau}{r} \qquad (5.3)$$

Also, when $r = 0$, $\tau = 0$. The simple linear relation between τ and r in Eq. (5.3) is shown graphically in Fig. 5.2. Note that this linear relationship applies in both laminar and turbulent flow, and to both newtonian and non-newtonian fluids.

Relation between skin friction and wall shear

Equation (4.62) can be written over a definite length L of the complete stream. In Chap. 4, Δp was defined as $p_b - p_a$, but usually (though not always) $p_a > p_b$ and thus $p_b - p_a$ is usually negative. The term Δp is commonly used for pressure *drop*, that is, $p_a - p_b$, and this terminology is employed in this and subsequent chapters. Here, then, $p_b = p_a - \Delta p$, $Z_b - Z_a = 0$, and the two kinetic energy terms cancel. Also, the only kind of friction is skin friction between the wall and the fluid stream, and the only pressure drop is that resulting from the skin friction. Denote the skin friction by h_{fs} and the pressure drop by Δp_s. Then Eq. (4.62) becomes

$$\frac{p_a}{\rho} = \frac{p_a - \Delta p_s}{\rho} + h_{fs}$$

or

$$\frac{\Delta p_s}{\rho} = h_{fs} \qquad (5.4)$$

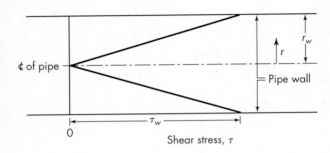

FIGURE 5.2
Variation of shear stress in pipe.

For a definite length L of pipe, dp/dL in Eq. (5.2) becomes $\Delta p_s/L$. Eliminating Δp_s from Eqs. (5.2) and (5.4) gives the following relation between h_{fs} and τ_w:

$$h_{fs} = \frac{2}{\rho}\frac{\tau_w}{r_w}L = \frac{4}{\rho}\frac{\tau_w}{D}L \tag{5.5}$$

where D is the diameter of the pipe.

The friction factor

Another common parameter, especially useful in the study of turbulent flow, is the *Fanning friction factor*,[†] denoted by f and defined as the ratio of the wall shear stress to the product of the density and the velocity head $\bar{V}^2/2$:

$$f \equiv \frac{\tau_w}{\rho\bar{V}^2/2} = \frac{2\tau_w}{\rho\bar{V}^2} \tag{5.6}$$

Relations between skin friction parameters

The four common quantities used to measure skin friction in pipes, h_{fs}, Δp_s, τ_w, and f, are related by the equations

$$h_{fs} = \frac{2}{\rho}\frac{\tau_w}{r_w}L = \frac{\Delta p_s}{\rho} = 4f\frac{L}{D}\frac{\bar{V}^2}{2} \tag{5.7}$$

from which

$$f = \frac{\Delta p_s\, D}{2L\rho\bar{V}^2} \tag{5.8}$$

and

$$\frac{\Delta p_s}{L} = \frac{2f\rho\bar{V}^2}{D} \tag{5.9}$$

Equation (5.9) is the equation usually used to calculate skin friction loss in straight pipe. The subscript s is used in Δp_s and h_{fs} to call attention to the fact that in Eqs. (5.7) through (5.9) these quantities, when they are associated with the

---

[†]Another friction factor common in the fluid mechanics literature and called the *Blasius* or *Darcy friction factor* is $4f$.

Fanning friction factor, relate *only to skin friction*. If other terms in the Bernoulli equation are present or if form friction is also active, $p_a - p_b$ differs from Δp_s. If boundary layer separation occurs, h_f is greater than h_{fs}.

Flow in noncircular channels

In evaluating skin friction in channels of noncircular cross section, the diameter in the Reynolds number and in Eq. (5.8), the definition of the friction factor, is taken as an *equivalent diameter* D_{eq}, defined as 4 times the hydraulic radius. The hydraulic radius is denoted by r_H and in turn is defined as the ratio of the cross-sectional area of the channel to the wetted perimeter of the channel:

$$r_H \equiv \frac{S}{L_p} \qquad (5.10)$$

where S = cross-sectional area of channel
     L_p = perimeter of channel in contact with fluid

Thus, for the special case of a circular tube, the hydraulic radius is

$$r_H = \frac{\pi D^2 / 4}{\pi D} = \frac{D}{4}$$

The equivalent diameter is $4r_H$, or simply D.

An important special case is the annulus between two concentric pipes. Here the hydraulic radius is

$$r_H = \frac{\pi D_o^2 / 4 - \pi D_i^2 / 4}{\pi D_i + \pi D_o} = \frac{D_o - D_i}{4} \qquad (5.11)$$

where D_i and D_o are the inside and outside diameters of the annulus, respectively. The equivalent diameter of an annulus is therefore the difference of the diameters. Also the equivalent diameter of a square duct with a width of side b is $4(b^2/4b) = b$. For flow between parallel plates, when the distance between them b is much smaller than the width of the plates, the equivalent diameter $D_{eq} = 2b$.

The defining equations for friction factor [Eq. (5.7)] and Reynolds number [Eq. (3.8)] can be generalized by substituting $2r_H$ for r or D_{eq} for D. The hydraulic radius is especially useful with turbulent flow. It is much less useful with laminar flow, but in many laminar flow situations the fluid flow relationships can be calculated mathematically, as shown in the following section.

LAMINAR FLOW IN PIPES AND CHANNELS

Equations (5.1) through (5.9) apply to both laminar and turbulent flow provided the fluid is incompressible and the flow is steady and fully developed. Because the shear-stress viscosity relationship for laminar flow is simple, derivations from these equations can be made most readily for laminar flow.

Laminar flow of newtonian fluids

The treatment is especially straightforward for a newtonian fluid, for which quantities such as the velocity distribution, the average velocity, and the momentum and kinetic energy correction factors are readily calculated.

Velocity distribution. The relation between the local velocity and position in the stream is found as follows. In circular channels, because of symmetry about the axis of the tube, the local velocity u depends only on the radius r. Consider a thin ring of radius r and width dr, forming an element of cross-sectional area dS. Then

$$dS = 2\pi r \, dr \tag{5.12}$$

The velocity distribution is found by using the definition of viscosity [Eq. (3.3)], written as

$$\mu = -\frac{\tau}{du/dr} \tag{5.13}$$

The minus sign in the equation accounts for the fact that in a pipe u decreases as r increases. Eliminating τ from Eqs. (5.3) and (5.9) provides the following ordinary differential equation relating u and r:

$$\frac{du}{dr} = -\frac{\tau_w}{r_w \mu} r \tag{5.14}$$

Integration of Eq. (5.14) with the boundary condition $u = 0$, $r = r_w$ gives

$$\int_0^u du = -\frac{\tau_w}{r_w \mu} \int_{r_w}^r r \, dr$$

$$u = \frac{\tau_w}{2r_w \mu} \left(r_w^2 - r^2 \right) \tag{5.15}$$

The maximum value of the local velocity is denoted by $u_{\max}$ and is located at the center of the pipe. The value of $u_{\max}$ is found from Eq. (5.15) by substituting 0 for r, giving

$$u_{\max} = \frac{\tau_w r_w}{2\mu} \tag{5.16}$$

Dividing Eq. (5.15) by Eq. (5.16) gives the following relationship for the ratio of the local velocity to the maximum velocity:

$$\frac{u}{u_{\max}} = 1 - \left(\frac{r}{r_w} \right)^2 \tag{5.17}$$

Equation (5.17) shows that in laminar flow the velocity distribution with respect to the radius is a parabola with the apex at the centerline of the pipe. The distribution is shown as the dashed line in Fig. 5.3.

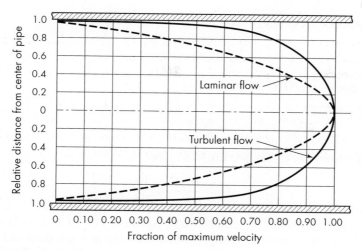

FIGURE 5.3
Velocity distribution in pipe, fully developed flow of newtonian
fluid, for laminar flow and for turbulent flow at Re = 10,000.

$$\bar{V} \equiv \frac{\dot{m}}{\rho S} = \frac{1}{S} \int u \, dS$$

Average velocity. Substitution of dS from Eq. (5.12), u from Eq. (5.15), and πr_w^2 for S into Eq. (4.9) gives

$$\tau_w = \frac{\Delta p_s}{L} \frac{r_w}{2}$$

$$\bar{V} = \frac{\tau_w}{r_w^3 \mu} \int_0^{r_w} (r_w^2 - r^2) r \, dr = \frac{\tau_w r_w}{4\mu} \tag{5.18}$$

Comparison of Eqs. (5.16) and (5.18) shows that

$$\frac{\bar{V}}{u_{max}} = 0.5 \tag{5.19}$$

The average velocity is precisely one-half the maximum velocity.

Kinetic energy correction factor. The kinetic energy factor α is calculated from Eq. (4.61), using Eqs. (5.8) for dS, (5.15) for u, and (5.18) for $\bar{V}$. The final result is $\alpha = 2.0$. The proper term for kinetic energy in the Bernoulli equation [Eq. (4.62) or (4.65)] for laminar flow in a pipe is therefore $\bar{V}^2$.

Momentum correction factor. Again, to obtain the value of β for laminar flow, the defining equation (4.41) is used. The result is $\beta = \frac{4}{3}$.

Hagen-Poiseuille equation

For practical calculations, Eq. (5.18) is transformed by eliminating τ_w in favor of Δp_s by the use of Eq. (5.7) and using the pipe diameter in place of the pipe radius. The result is

$$\bar{V} = \frac{\Delta p_s}{L} \frac{r_w}{2} \frac{r_w}{4\mu} = \frac{\Delta p_s D^2}{32 L \mu}$$

Solving for Δp_s gives

$$\Delta p_s = \frac{32 L \bar{V} \mu}{D^2} \tag{5.20}$$

and since $\Delta p_s = 4\tau_w/(DL)$,

$$\tau_w = \frac{8 \bar{V} \mu}{D} \tag{5.21}$$

Substituting from Eq. (5.21) into Eq. (5.6) gives

$$f = \frac{16 \mu}{D \bar{V} \rho} = \frac{16}{\mathrm{Re}} \tag{5.22}$$

Equation (5.20) is the Hagen-Poiseuille equation. One of its uses is in the experimental measurement of viscosity, by measuring the pressure drop and volumetric flow rate through a tube of known length and diameter. From the flow rate, $\bar{V}$ is calculated by Eq. (4.9) and μ is calculated by Eq. (5.20). In practice, corrections for kinetic energy and entrance effects are necessary.

Laminar flow of non-newtonian liquids

Because of the difference in the relation between shear stress and velocity gradient, the shape of the velocity profile for non-newtonian liquids differs from that of a newtonian liquid. In the more complicated situations of non-newtonian flow, the shape of the profile is determined experimentally. For the simpler cases such as the power law model [Eq. (3.7)] or the Bingham model [Eq. (3.6)], the same methods used for determining the flow parameters of a newtonian fluid can be used for non-newtonian fluids in these categories.

For fluids following the power law model, the velocity variation with radius follows the formula

$$u = \left(\frac{\tau_w}{r_w K'}\right)^{1/n'} \frac{r_w^{1+1/n'} - r^{1+1/n'}}{1 + 1/n'} \tag{5.23}$$

Velocity profiles defined by Eq. (5.23) when $n' = 0.5$ (a pseudoplastic fluid), $n' = 1.0$ (a newtonian fluid), and $n' = 2.0$ (a dilatant fluid) are shown in Fig. 5.4. In all cases K' is assumed to be the same. The curve for the dilatant fluid is narrower and more pointed than a true parabola; that for the pseudoplastic fluid is blunter and flatter.

The pressure difference for the flow of a power law fluid is found by the methods used in deriving Eq. (5.20) for a newtonian fluid. The result is

$$\Delta p_s = 2K' \left(\frac{3n' + 1}{n'}\right)^{n'} \frac{\bar{V}^{n'}}{r_w^{n'+1}} L \tag{5.24}$$

Equation (5.24) corresponds to Eq. (5.20) for a newtonian fluid.

The behavior of fluids following the Bingham plastic flow model is somewhat more complicated. The general shape of the curve of u versus r is shown in

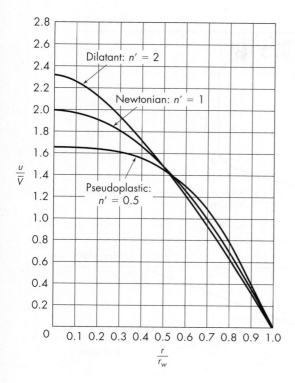

FIGURE 5.4
Velocity profiles in the laminar flow of newtonian and non-newtonian liquids.

Fig. 5.5a. In the central portion of the tube there is no velocity variation with the radius, and the velocity gradient is confined to an annular space between the central portion and tube wall. The center portion is moving in plug flow. In this region the shear stress that would be generated in other types of flow is too small to overcome the threshold shear τ_0. The shear diagram is shown in Fig. 5.5b. For the velocity variation in the annular space between the tube wall and the plug, the following equation applies:

$$u = \frac{1}{K}(r_w - r)\left[\frac{\tau_w}{2}\left(1 + \frac{r}{r_w}\right) - \tau_0\right] \tag{5.25}$$

where K is a constant. The boundary between the plug and the remaining fluid is found by differentiating Eq. (5.25) and setting the velocity gradient equal to zero, or more simply by reading the value from Fig. 5.5b. The result is

$$r_c = \frac{\tau_0}{\tau_w}r_w \tag{5.26}$$

The velocity in the central core u_c, the speed at which the plug is moving, is found by substituting the value of r_c from Eq. (5.26) for r in Eq. (5.25) and rearranging. This gives

$$u_c = \frac{\tau_0}{2Kr_c}(r_w - r_c)^2 \tag{5.27}$$

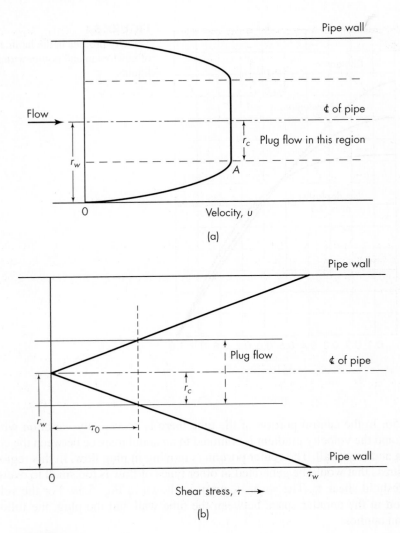

FIGURE 5.5
(*a*) Velocity profile. (*b*) Shear diagram for Bingham plastic flow.

An interesting observation with some non-newtonian mixtures[8] is that at high shear they appear to violate the zero-velocity boundary condition at the wall. For multiphase fluids such as suspensions and fiber-filled polymers, this effect is believed to be the result of a thin layer near the wall that is depleted of particulates or fibers and thus has a lower viscosity than the bulk of the fluid. This gives the appearance of a "slip" at the wall. Empirical formulas are often used in the literature to account for such wall effects.[13]

Laminar flow in an annulus

The relation between local velocity and radial position for the laminar flow of a newtonian fluid through an annular space may be found by the same method as that used for flow in a pipe. The resulting equation, as derived by Bird, Stewart, and Lightfoot,[1] is

$$u = \frac{\Delta p_s r_0^2}{4\mu L} \left[1 - \left(\frac{r}{r_0}\right)^2 + \frac{1 - \kappa^2}{\ln(1/\kappa)} \ln \frac{r}{r_0} \right] \tag{5.28}$$

where r_o = radius of outer wall of annulus
κ = ratio r_i/r_o
r_i = radius of inner wall of annulus

For annular flow the Reynolds number is

$$\text{Re} = \frac{(D_o - D_i)\bar{V}\rho}{\mu} \tag{5.29}$$

where D_i and D_o are the inner and outer diameters of the annulus, respectively. The friction factor may be calculated from Eq. (5.28) as it was in the derivation of Eq. (5.22). For laminar flow in an annulus it is related to the Reynolds number by the equation[5]

$$f = \frac{16}{\text{Re}} \phi_a \tag{5.30}$$

where ϕ_a is the function of D_i/D_o shown in Fig. 5.6. The value of ϕ_a is 1 for circular cross sections and 1.5 for parallel planes. Equations for laminar flow through other cross sections are given in various texts.[5] Flow of viscous non-newtonian liquids through channels of complex shape, as in injection molding

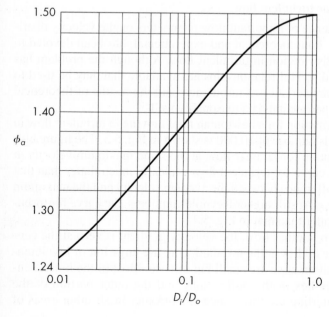

FIGURE 5.6
Value of ϕ_a in Eq. (5.30).

operations, may be analyzed by methods of numerical approximation using Eq. (4.20).

TURBULENT FLOW IN PIPES AND CHANNELS

In turbulent flow through a closed channel, the velocity at the interface between fluid and the solid wall is zero, and (except very infrequently) there are no velocity components normal to the wall. Within a thin volume immediately adjacent to the wall, the velocity gradient is essentially constant and the flow is viscous most of the time. This volume is called the *viscous sublayer*. Formerly it was assumed that this sublayer had a definite thickness and was always free from eddies, but measurements have shown velocity fluctuations in the sublayer caused by occasional eddies from the turbulent fluid moving into this region. Very close to the wall, eddies are infrequent, but there is no region that is completely free of eddies. Within the viscous sublayer only viscous shear is important, and eddy diffusion, if present at all, is minor.

The viscous sublayer occupies only a very small fraction of the total cross section. It has no sharp upper boundary, and its thickness is difficult to define. A transition layer exists immediately adjacent to the viscous sublayer in which both viscous shear and shear due to eddy diffusion exist. The transition layer, which is sometimes called a *buffer layer,* also is relatively thin. The bulk of the cross section of the flowing stream is occupied by entirely turbulent flow called the *turbulent core*. In the turbulent core, viscous shear is negligible in comparison with that from eddy viscosity.

Velocity distribution for turbulent flow

Because of the dependence of important flow parameters on the velocity distribution, considerable study, both theoretical and experimental, has been devoted to determining the velocity distribution in turbulent flow. Although the problem has not been completely solved, useful relationships are available that may be used to calculate the important characteristics of turbulence; and the results of theoretical calculations check with experimental data reasonably well.

A typical velocity distribution for a newtonian fluid moving in turbulent flow in a smooth pipe at a Reynolds number of 10,000 is shown in Fig. 5.3. The figure also shows the velocity distribution for laminar flow at the same maximum velocity at the center of the pipe. The curve for turbulent flow is clearly much flatter than that for laminar flow, and the difference between the average velocity and the maximum velocity is considerably less. At still higher Reynolds numbers, the curve for turbulent flow would be even flatter than that in Fig. 5.3.

In turbulent flow, as in laminar flow, the velocity gradient is zero at the centerline. It is known that the eddies in the turbulent core are large but of low intensity, and those in the transition zone are small but intense. Most of the kinetic energy content of the eddies lies in the buffer zone and the outer portion of the turbulent core. At the centerline the turbulence is isotropic. In all other areas of the turbulent flow regime, turbulence is anisotropic; otherwise, there would be no shear.

It is customary to express the velocity distribution in turbulent flow not as velocity versus distance but in terms of dimensionless parameters defined by the following equations:

$$u^* \equiv \bar{V}\sqrt{\frac{f}{2}} = \sqrt{\frac{\tau_w}{\rho}} \tag{5.31}$$

$$u^+ \equiv \frac{u}{u^*} \tag{5.32}$$

$$y^+ \equiv \frac{yu^*\rho}{\mu} = \frac{y}{\mu}\sqrt{\tau_w\rho} \tag{5.33}$$

where u^* = friction velocity
  u^+ = velocity quotient, dimensionless
  y^+ = distance, dimensionless
  y = distance from wall of tube

Note that y^+ may be considered to be a Reynolds number based on the friction velocity and the distance from the wall. The relationship between y, r, and r_w, the radius of the tube, is

$$r_w = r + y \tag{5.34}$$

Equations relating u^+ to y^+ are called universal velocity distribution laws.

Universal velocity distribution equations

Since the viscous sublayer is very thin, $r \approx r_w$, and Eq. (5.14) can be written, with the substitution of $-dy$ for dr, as

$$\frac{du}{dy} = \frac{\tau_w}{\mu} \tag{5.35}$$

Substituting u^* from Eq. (5.31), u^+ from Eq. (5.32), and y^+ from Eq. (5.33) into Eq. (5.35) gives

$$\frac{du^+}{dy^+} = 1$$

Integrating, with the lower limits $u^+ = y^+ = 0$, gives, for the velocity distribution in the laminar sublayer,

$$u^+ = y^+ \tag{5.36}$$

An empirical equation for the so-called buffer layer is

$$u^+ = 5.00 \ln y^+ - 3.05 \tag{5.37}$$

For the turbulent core a number of correlations have been proposed. An equation proposed by Prandtl,[10] with empirical constants, is

$$u^+ = 2.5 \ln y^+ + 5.5 \tag{5.38}$$

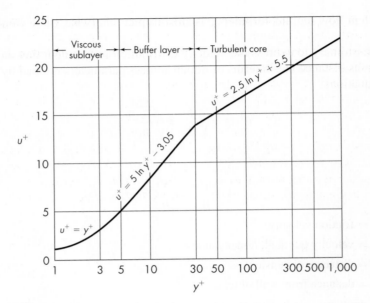

FIGURE 5.7
Universal velocity distribution; turbulent flow of newtonian fluid in smooth pipe.

Figure 5.7 is a semilogarithmic plot of Eqs. (5.36), (5.37), and (5.38). From the two intersections of the three lines representing the equations, the ranges covered by the equations are as follows:

Equation (5.36), for the viscous sublayer: $y^+ < 5$
Equation (5.37), for the buffer zone: $5 < y^+ < 30$
Equation (5.38), for the turbulent core: $30 < y^+$

Limitations of universal velocity distribution laws

The universal velocity equations have a number of limitations. It is certain that the buffer zone has no independent existence and that there is no discontinuity between the buffer zone and the turbulent core. Also, there is doubt as to the reality of the existence of a truly viscous sublayer. The equations do not apply well for Reynolds numbers from the critical to approximately 10,000, and it is known that a simple y^+-u^+ relationship is not adequate for the turbulent core near the buffer zone or in the buffer zone itself. Finally, Eq. (5.38) calls for a finite velocity gradient at the centerline of the pipe, although it is known that the gradient at this point must be zero.

Much research has been devoted to improving the velocity distribution equations and eliminating or reducing some of their deficiencies. This work is reported in advanced texts[3,5,11] but is beyond the scope of this book.

Flow quantities for turbulent flow in smooth round pipes

With a velocity distribution relation at hand, the important flow quantities can be calculated by the usual methods. The quantities of interest are the average

velocity in terms of the maximum velocity at the center of the pipe; relations link-
ing the flow-resistance parameters τ_w and f with the average velocity, the maximum
velocity, and the Reynolds number; the kinetic energy correction factor α; and the
momentum correction factor β.

Calculation of the flow quantities requires an integration with pipe radius from
the centerline of the pipe to the wall. Strictly, the integration should be conducted in
three parts, the first over the range $y^+ = 0$ to $y^+ = 5$, using Eq. (5.36); the second
from $y^+ = 5$ to $y^+ = 30$, using Eq. (5.37); and the third from $y^+ = 30$ to its value at
the center of the pipe, using Eq. (5.38). Because of the thinness of the layers covered by the first two integrations, they may be neglected; and for approximate calculations a single integration based on Eq. (5.38) over the entire range $r = 0$ to
$r = r_w$ is justified, even though this predicts a finite value of velocity at the
pipe wall.

Average velocity. Equation (5.38) may be written for the centerline of the
pipe as

$$u_c^+ = 2.5 \ln y_c^+ + 5.5 \tag{5.39}$$

where u_c^+ and y_c^+ are the values of u^+ and y^+ at the centerline, respectively. Also,
by Eqs. (5.32) and (5.33)

$$u_c^+ = \frac{u_{\max}}{u^*} \tag{5.40}$$

$$y_c^+ = \frac{r_w u^*}{\nu} \tag{5.41}$$

where $u_{\max}$ is the maximum velocity at the centerline and $\nu = \mu/\rho$.

Subtracting Eq. (5.39) from Eq. (5.38) yields

$$u^+ = u_c^+ + 2.5 \ln \frac{y^+}{y_c^+} \tag{5.42}$$

The average velocity $\bar{V}$ is, from Eq. (4.9) after substituting πr_w^2 for S and $2\pi r \, dr$
for dS,

$$\bar{V} = \frac{2\pi}{\pi r_w^2} \int_0^{r_w} ur \, dr \tag{5.43}$$

From Eq. (5.34) $r = r_w - y$ and $dr = -dy$. Also, when $r = 0$, $y = r_w$; and when
$r = r_w$, $y = 0$. Equation (5.43) becomes, after elimination of r,

$$\bar{V} = \frac{2}{r_w^2} \int_0^{r_w} u(r_w - y) \, dy \tag{5.44}$$

Equation (5.44) can be written in dimensionless parameters by substituting u from
Eq. (5.32), y from Eq. (5.33), and u^+ from Eq. (5.42). This gives

$$\bar{V} = \frac{5\nu^2}{r_w^2 u^*} \int_0^{y_c^+} \left(0.4 u_c^+ + \ln \frac{y^+}{y_c^+} \right) (y_c^+ - y^+) \, dy^+ \tag{5.45}$$

Formal integration of Eq. (5.45)[†] gives

$$\frac{\bar{V}}{u^*} = \frac{1}{\sqrt{f/2}} = u_c^+ - 3.75 \tag{5.46}$$

Substituting u_c^+ from Eq. (5.40) and u^* from Eq. (5.31) into Eq. (5.46) gives

$$\frac{\bar{V}}{u_{max}} = \frac{1}{1 + 3.75\sqrt{f/2}} \tag{5.47}$$

Equation (5.47) gives values that are somewhat too high because the low velocities of the fluid in the layers near the wall, some 2 percent of the total volume flow, are not properly taken into account.

The Reynolds number–friction factor law for smooth tubes. The equations at hand can be used to derive an important relation between Re and f for turbulent flow in smooth round pipes. This equation is derived by appropriate substitutions into Eq. (5.39). From the defining equation of y_c^+ [Eq. (5.41)] and from the definition of u^* [Eq. (5.31)]

$$y_c^+ = \frac{r_w \bar{V}}{\nu}\sqrt{\frac{f}{2}} = \frac{D\bar{V}\sqrt{f/2}}{2\nu} = \frac{\text{Re}}{2}\sqrt{\frac{f}{2}} = \text{Re}\sqrt{\frac{f}{8}} \tag{5.48}$$

From Eq. (5.46)

$$u_c^+ = \frac{1}{\sqrt{f/2}} + 3.75 \tag{5.49}$$

Substituting u_c^+ from Eq. (5.49) and y_c^+ from Eq. (5.48) into Eq. (5.39) gives, after rearrangement, the von Kármán equation

$$\frac{1}{\sqrt{f/2}} = 2.5 \ln\left(\text{Re}\sqrt{\frac{f}{8}}\right) + 1.75 \tag{5.50}$$

Equation (5.50) agrees well with experiment. In the range of Reynolds numbers $10^4 < \text{Re} < 10^6$ it predicts friction factors within 2 percent of those read from Fig. 5.10.

The kinetic energy and momentum correction factors. Values of α and β for turbulent flow are closer to unity than those for laminar flow. Equations for these correction factors are readily obtained, however, by integrating Eqs. (4.41) and (4.61), which define them, and using the logarithmic velocity law. The equations so found are

$$\alpha = 1 + 0.78f(15 - 15.9\sqrt{f}) \tag{5.51}$$

$$\beta = 1 + 3.91f \tag{5.52}$$

[†]Substitution of $x = y^+/y_c^+$ and use of standard integral tables suffice for this integration. Also, for the lower limit, $x \ln x = 0$ and $x^2 \ln x = 0$ when $x = 0$. From Eq. (5.31), $\bar{V}/u^* = 1/\sqrt{f/2}$.

For turbulent flow the error is usually very small if α and β are assumed to be unity. For a Reynolds number of 10^4, for example, the friction factor for a smooth tube is 0.0079, α is 1.084, and β is 1.031. For Re $= 10^6$ the values are $f = 0.0029$, $\alpha = 1.032$, and $\beta = 1.011$.

The kinetic energy correction factor may be important in applying Bernoulli's theorem between stations when one is in laminar flow and the other in turbulent flow. Also factors α and β are of some importance in certain types of compact heat-exchange equipment, where there are many changes in size of the fluid channel and where the tubes or heat-transfer surfaces themselves are short.[4] In most practical situations both are taken as unity in turbulent flow.

Relations between maximum velocity and average velocity

The quantities Re and ratio $\bar{V}/u_{\max}$ are useful in relating the average velocity to the maximum velocity in the center of the tube as a function of flow conditions; for example, an important method of measuring fluid flow is the pitot tube (page 232), which can be used to measure $u_{\max}$, and this relationship is then used to determine the average velocity from this single observation.

Experimentally measured values of $\bar{V}/u_{\max}$ as a function of the Reynolds number are shown in Fig. 5.8, which covers the range from laminar flow to turbulent flow. For laminar flow the ratio is exactly 0.5, in accordance with Eq. (5.19). The ratio changes rapidly from 0.5 to about 0.7, when laminar flow changes to turbulent, and then increases gradually to 0.87 when Re $= 10^6$.

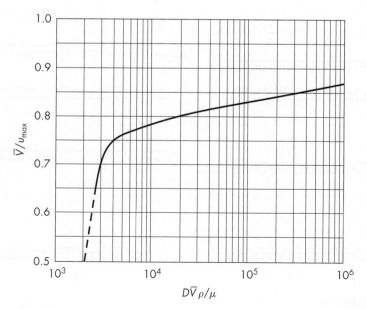

FIGURE 5.8
$\bar{V}/u_{\max}$ versus Re $= D\bar{V}\rho/\mu$.

Effect of roughness

The discussion thus far has been restricted to smooth tubes without defining smoothness. It has long been known that in turbulent flow a rough pipe leads to a larger friction factor for a given Reynolds number than a smooth pipe does. If a rough pipe is smoothed, the friction factor is reduced. When further smoothing brings about no further reduction in the friction factor for a given Reynolds number, the tube is said to be *hydraulically smooth*. Equation (5.50) refers to a hydraulically smooth tube.

Figure 5.9 shows several idealized kinds of roughness. The height of a single unit of roughness is denoted by k and is called the *roughness parameter*. From dimensional analysis, f is a function of both Re and the relative roughness k/D, where D is the diameter of the pipe. For a given kind of roughness, for example, that shown in Fig. 5.9a and b, it can be expected that a different curve of f versus Re would be found for each magnitude of the relative roughness and also that for other types of roughness, such as those shown in Fig. 5.9c and d, a different family of curves of Re versus f would be found for each type of roughness. Experiments on artificially roughened pipe have confirmed these expectations. It has also been found that all clean, new commercial pipes seem to have the same type of roughness and that each material of construction has its own characteristic roughness parameter.

Old, foul, and corroded pipe can be very rough, and the character of the roughness differs from that of clean pipe.

Roughness has no appreciable effect on the friction factor for laminar flow unless k is so large that the measurement of the diameter becomes uncertain.

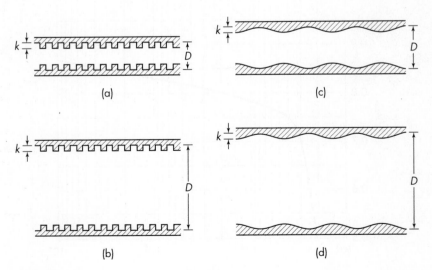

(a)　　(c)

(b)　　(d)

FIGURE 5.9
Types of roughness.

The friction factor chart

For design purposes, the friction characteristics of round pipe, both smooth and rough, are summarized by the friction factor chart (Fig. 5.10), which is a log-log plot of f versus Re. For laminar flow Eq. (5.22) relates the friction factor to the Reynolds number. A log-log plot of Eq. (5.22) is a straight line with a slope of -1. This plot line is shown on Fig. 5.10 for Reynolds numbers less than 2,100.

For turbulent flow the lowest line represents the friction factor for smooth tubes and is consistent with Eq. (5.50). A much more convenient empirical equation for this line is the relation

$$f = 0.046 \, \text{Re}^{-0.2} \tag{5.53}$$

This applies over a range of Reynolds numbers from about 50,000 to 1×10^6. Another equation, applicable over a range of Reynolds numbers from 3,000 to 3×10^6, is

$$f = 0.0014 + \frac{0.125}{\text{Re}^{0.32}} \tag{5.54}$$

The other curved lines in the turbulent flow range represent the friction factors for various types of commercial pipe, each of which is characterized by a different value of k. The parameters for several common metals are given in the figure. Clean wrought-iron or steel pipe, for example, has a k value of 1.5×10^{-4} ft, regardless of the diameter of the pipe. Drawn copper and brass pipe may be considered hydraulically smooth.

Figure 5.10 is useful for calculating h_{fs} from a known pipe size and flow rate, but it cannot be used directly to determine the flow rate for a given pressure drop, since Re is not known until $\bar{V}$ is determined. However, since f changes only slightly with Re for turbulent flow, a trial-and-error solution converges quickly, as shown in Example 5.1.

EXAMPLE 5.1. Water is flowing at 50°F through a long horizontal plastic pipe, 3 in. in inside diameter, at a velocity of 8 ft/s. (a) Calculate the pressure drop in $\text{lb}_f/\text{in.}^2$ per 100 ft of pipe. (b) If the pressure drop must be limited to 2 $\text{lb}_f/\text{in.}^2$ per 100 ft of pipe, what is the maximum allowable velocity of the water?

Solution

(a) Use Eq. (5.9), which in fps units is

$$\frac{\Delta p_s}{L} = \frac{2 f \rho \bar{V}^2}{D g_c} \tag{5.55}$$

The properties of water at 50°F, from App. 6, are $\rho = 62.42$ lb/ft^3 and $\mu = 1.310$ cP. Also $D = \frac{3}{12} = 0.25$ ft, $\bar{V} = 8$ ft/s, and $L = 100$ ft.

The Reynolds number is

$$\text{Re} = \frac{0.25 \times 8 \times 62.42}{1.310 \times 6.72 \times 10^{-4}} = 1.4 \times 10^5$$

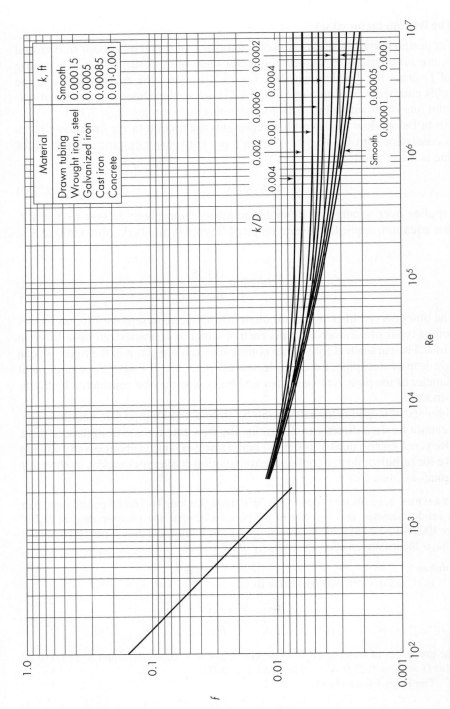

FIGURE 5.10
Friction factor chart.

From Fig. 5.10, $f = 0.0041$. Substituting in Eq. (5.55) gives

$$\Delta p_s = \frac{100 \times 2 \times 0.0041 \times 62.42 \times 8^2}{0.25 \times 32.174}$$

$$= 407.2 \text{ lb}_f/\text{ft}^2 \text{ or } 2.828 \text{ lb}_f/\text{in.}^2$$

(b) Here $\Delta p_s = 2 \times 144 = 288 \text{ lb}_f/\text{ft}^2$. Transposing Eq. (5.55) gives

$$f = \frac{\Delta p_s \, D g_c}{2 L \rho \bar{V}^2} = \frac{288 \times 0.25 \times 32.174}{2 \times 100 \times 62.42 \bar{V}^2} = \frac{0.18556}{\bar{V}^2}$$

The Reynolds number is

$$\mathrm{Re} = \frac{0.25 \bar{V} \times 62.42}{1.310 \times 6.72 \times 10^{-4}} = 1.773 \times 10^4 \bar{V}$$

The maximum allowable velocity $\bar{V}$ must be found by trial, as follows:

$\bar{V}$, ft/s	$f_{\text{calc.}}$	Re $\times 10^5$	f from Fig. 5.10
6	0.0051	1.06	0.0043
7	0.0038	1.24	0.0042
6.6	0.0043	1.17	0.0043

The maximum velocity is therefore 6.6 ft/s.

Reynolds numbers and friction factor for non-newtonian fluids

Equation (5.8), the relation of friction factor to pressure drop, and Eq. (5.24), the equation for the friction pressure loss of a power law fluid, may be used to calculate the friction factor for pseudoplastic fluids. If Δp_s is eliminated from these two equations, the equation for f can be written

$$f = \frac{2^{n'+1} K'}{D^{n'} \rho \bar{V}^{2-n'}} \left(3 + \frac{1}{n'} \right)^{n'} \tag{5.56}$$

From this equation a Reynolds number Re_n for non-newtonian fluids can be defined, on the assumption that for laminar flow

$$f = \frac{16}{\mathrm{Re}_n} \tag{5.57}$$

Combining Eqs. (5.56) and (5.57) yields

$$\mathrm{Re}_n = 2^{3-n'} \left(\frac{n'}{3n' + 1} \right)^{n'} \frac{D^{n'} \rho \bar{V}^{2-n'}}{K'} \tag{5.58}$$

This is the definition of the Reynolds number Re_n given in Eq. (3.9). This Reynolds number reduces to the Reynolds number for a newtonian fluid when $n' = 1$, and it reproduces the linear portion of the logarithmic plot of f versus Re, with a slope of -1, for the laminar flow of newtonian fluids.

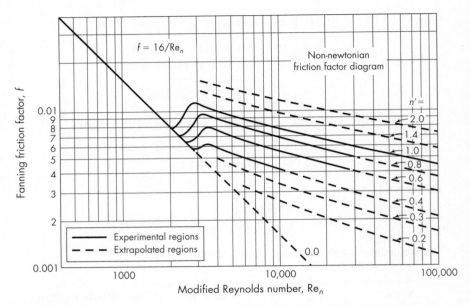

FIGURE 5.11
Friction factor chart, power law fluid. (*After D. W. Dodge and A. B. Metzner.*[2])

Figure 5.11 is a friction factor chart in which f is plotted against Re_n for the flow of power law fluids in smooth pipes. A series of lines depending on the magnitude of n' is needed for turbulent flow. For these lines the following equation, analogous to Eq. (5.50), for newtonian fluids, has been suggested[3]:

$$\frac{1}{\sqrt{f}} = \frac{1.74}{(n')^{0.75}} (\ln Re_n f^{1-0.5n'}) - \frac{0.4}{(n')^{1.2}} \tag{5.59}$$

Figure 5.11 shows that for pseudoplastic fluids ($n' < 1$), laminar flow persists to higher Reynolds number than with newtonian fluids.

Drag reduction in turbulent flow

Dilute solutions of polymers in water or other solvents sometimes give the peculiar effect of a reduction in drag in turbulent flow. The phenomenon was first noted by Toms[12] and has prompted many theoretical studies and some practical applications. As shown in Fig. 5.12, the friction factor can be significantly below the normal value for turbulent flow with only a few parts per million (ppm) of polymer in water, and at 50 to 100 ppm, the drag reduction may be as much as 70 percent. Similar effects have been shown for some polymers in organic solvents.

Drag reduction is found generally with dilute solutions of high-molecular-weight linear polymers and is believed to be related to the extension of these flexible molecules at high turbulent shear stress near the wall. The extended molecules increase the local viscosity, which damps the small eddies and leads to

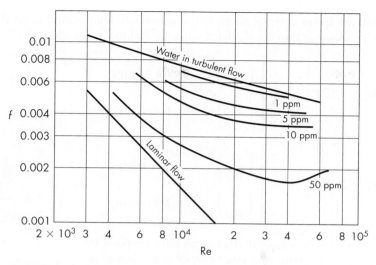

FIGURE 5.12
Friction factors for turbulent flow of dilute solutions of polyethylene oxide (MW $\approx 10^6$). [*After R. W. Patterson and F. H. Abernathy.*[7]]

increased thickness of the viscous sublayer.[6] With a thicker sublayer at the same total flow, the values of $(du/dy)_w$ and τ_w are reduced, giving lower pressure drop. The apparent solution viscosity, as measured in laminar flow, may still be very close to that of the solvent, with only slight departures from newtonian behavior; but careful measurements show non-newtonian behavior, including viscoelastic effects.

The major applications of drag reduction have been to increase the flow of water in a line of fixed size. Only a few parts per million of polyethylene oxide, a cheap nontoxic polymer, can double the capacity of a fire hose or a line carrying cooling water. In a long pipeline or with repeated use, degradation of the polymer under the high shear reduces its effectiveness.

Nonisothermal flow

When liquids flow through a smooth tube which is being cooled or heated, the friction factor is increased or decreased because of the effect of temperature on the viscosity near the wall. When the liquid is being cooled, the higher viscosity near the wall increases the velocity gradient there, leading to increased wall friction. The effect is especially pronounced when the viscosity is a strong function of temperature. For gases, since the viscosity increases with temperature, the opposite is true. For ordinary engineering practice the following simple method is empirically justified for both gases and liquids.

1. The Reynolds number is calculated on the assumption that the fluid temperature equals the *mean bulk temperature,* which is defined as the arithmetic average of the inlet and outlet temperatures.

2. The friction factor corresponding to the mean bulk temperature is divided by a factor ψ, which in turn is calculated from the following equations.[9a]

$$\text{For Re} > 2{,}100: \quad \psi = \begin{cases} \left(\dfrac{\mu}{\mu_w}\right)^{0.17} & \text{for heating} & (5.60a) \\[2ex] \left(\dfrac{\mu}{\mu_w}\right)^{0.11} & \text{for cooling} & (5.60b) \end{cases}$$

$$\text{For Re} < 2{,}100: \quad \psi = \begin{cases} \left(\dfrac{\mu}{\mu_w}\right)^{0.38} & \text{for heating} & (5.61a) \\[2ex] \left(\dfrac{\mu}{\mu_w}\right)^{0.23} & \text{for cooling} & (5.61b) \end{cases}$$

where μ = viscosity of fluid at mean bulk temperature
$\quad\ \mu_w$ = viscosity at temperature of wall of conduit

Equations (5.60) and (5.61) are based on data for values of μ/μ_w between 0.1 and 10 and should not be used for values outside these limits.

In rough commercial pipe, rather than in smooth tubes, flow is commonly in "complete turbulence" where the friction factor is independent of the Reynold number. Under these circumstances f should not be corrected for the effect of wall temperature.[9a]

Viscous dissipation. Frictional or viscous dissipation of mechanical energy can result in heating a fluid, especially highly viscous liquids in small channels. For incompressible flow under adiabatic conditions through channels of constant cross section, the energy from the decrease in pressure Δp_s is converted to thermal energy by the relation

$$\frac{Q}{V} = c_v \rho \, \Delta T = \Delta p_s$$

where Q/V = heat generated per unit volume, J/m^3
$\quad c_v$ = specific heat of liquid, J/kg · °C
$\quad \rho$ = liquid density, kg/m^3
$\quad \Delta T$ = temperature rise, °C

Hence

$$\Delta T = \frac{\Delta p_s}{c_v \rho} \tag{5.62}$$

For the flow of polymers the temperature rise is about 4°C for a pressure drop of 10 MPa; for hydrocarbons it is about 6°C per 10 MPa.[9a]

Turbulent flow in noncircular channels

As discussed earlier, relationships for turbulent flow in pipes may be applied to noncircular channels by substituting the equivalent diameter D_{eq} (or 4 times the hydraulic radius r_H) for the diameter D in the relevant equations.

FRICTION FROM CHANGES IN VELOCITY OR DIRECTION

Whenever the velocity of a fluid is changed, in either direction or magnitude, friction is generated in addition to the skin friction resulting from flow through the straight pipe. Such friction includes form friction resulting from vortices that develop when normal streamlines are disturbed and when boundary layer separation occurs. Often these effects cannot be calculated precisely, and it is necessary to rely on empirical data.

Friction loss from sudden expansion of cross section

If the cross section of the pipe is suddenly enlarged, the fluid stream separates from the wall and issues as a jet into the enlarged section. The jet then expands to fill the entire cross section of the larger conduit. The space between the expanding jet and the conduit wall is filled with fluid in vortex motion characteristic of boundary layer separation, and considerable friction is generated within this space. This effect is shown in Fig. 5.13.

The friction loss h_{fe} from a sudden expansion of cross section is proportional to the velocity head of the fluid in the small conduit and can be written

$$h_{fe} = K_e \frac{\bar{V}_a^2}{2} \tag{5.63}$$

where K_e is a proportionality factor called the *expansion loss coefficient* and $\bar{V}_a$ is the average velocity in the smaller, or upstream, conduit. In this case the calculation of K_e can be made theoretically and a satisfactory result obtained. The calculation utilizes the continuity equation (4.11), the steady-flow momentum-balance equation (4.42), and the Bernoulli equation (4.62). Consider the control volume defined by sections AA and BB and the inner surface of the larger downstream conduit between these sections, as shown in Fig. 5.13. Gravity forces do not appear because the pipe is horizontal, and wall friction is negligible because the wall is relatively short and there is almost no velocity gradient at the wall between the sections. The only

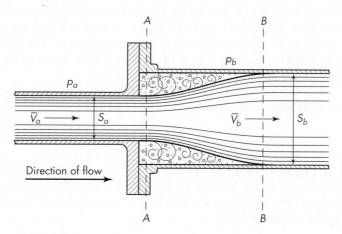

FIGURE 5.13
Flow at sudden enlargement of cross section.

forces, therefore, are pressure forces on sections AA and BB. The momentum equation gives

$$p_a S_a - p_b S_b = \dot{m}(\beta_b \bar{V}_b - \beta_a \bar{V}_a) \tag{5.64}$$

Since $Z_a = Z_b$, Eq. (4.62) may be written for this situation as

$$\frac{p_a - p_b}{\rho} = \frac{\alpha_b \bar{V}_b^2 - \alpha_a \bar{V}_a^2}{2} + h_{fe} \tag{5.65}$$

For usual flow conditions, $\alpha_a = \alpha_b = 1$, and $\beta_a = \beta_b = 1$, and these correction factors are disregarded. Also, elimination of $p_a - p_b$ between Eqs. (5.64) and (5.65) yields, since $\dot{m}/S_b = \rho \bar{V}_b$,

$$h_{fe} = \frac{(\bar{V}_a - \bar{V}_b)^2}{2} \tag{5.66}$$

From Eq. (4.11), $\bar{V}_b = \bar{V}_a(S_a/S_b)$ since ρ is constant, and Eq. (5.66) can be written

$$h_{fe} = \frac{\bar{V}_a^2}{2}\left(1 - \frac{S_a}{S_b}\right)^2 \tag{5.67}$$

Comparison of Eqs. (5.63) and (5.67) shows that

$$K_e = \left(1 - \frac{S_a}{S_b}\right)^2 \tag{5.68}$$

A correction for α and β should be made if the type of flow between the two sections differs. For example, if the flow in the larger pipe is laminar and that in the smaller pipe turbulent, α_b should be taken as 2 and β_b as $\frac{4}{3}$ in Eqs. (5.64) and (5.65).

Friction loss from sudden contraction of cross section.

When the cross section of the conduit is suddenly reduced, the fluid stream cannot follow around the sharp corner and the stream breaks contact with the wall of the conduit. A jet is formed, which flows into the stagnant fluid in the smaller section. The jet first contracts and then expands to fill the smaller cross section, and downstream from the point of contraction the normal velocity distribution eventually is reestablished. The cross section of minimum area at which the jet changes from a contraction to an expansion is called the *vena contracta*. The flow pattern of a sudden contraction is shown in Fig. 5.14. Section CC is drawn at the vena contracta. Vortices appear as shown in the figure.

The friction loss from sudden contraction is proportional to the velocity head in the smaller conduit and can be calculated by the equation

$$h_{fc} = K_c \frac{\bar{V}_b^2}{2} \tag{5.69}$$

where the proportionality factor K_c is called the *contraction loss coefficient* and $\bar{V}_b$ is the average velocity in the smaller, or downstream, section. Experimentally, for laminar flow, $K_c < 0.1$, and the contraction loss h_{fc} is negligible. For turbulent flow, K_c is given by the empirical equation

$$K_c = 0.4\left(1 - \frac{S_b}{S_a}\right) \tag{5.70}$$

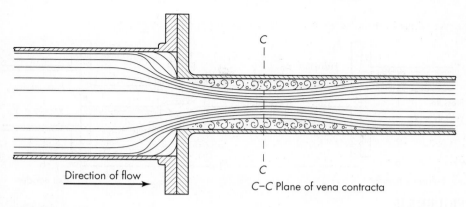

FIGURE 5.14
Flow at sudden contraction of cross section.

where S_a and S_b are the cross-sectional areas of the upstream and downstream conduits, respectively.

Effect of fittings and valves

Fittings and valves disturb the normal flow lines and cause friction. In short lines with many fittings, the friction loss from the fittings may be greater than that from the straight pipe. The friction loss h_{ff} from fittings is found from an equation similar to Eqs. (5.63) and (5.69):

$$h_{ff} = K_f \frac{\bar{V}_a^2}{2} \tag{5.71}$$

where K_f = loss factor for fitting
$\bar{V}_a$ = average velocity in pipe leading to fitting

Factor K_f is found by experiment and differs for each type of connection. A short list of factors is given in Table 5.1.

TABLE 5.1
Loss coefficients for standard threaded pipe fittings[9b]

Fitting	K_f
Elbow, standard	
45°	0.35
90°	0.75
Tee	
Straight through	0.4
Used as elbow	1.0
Return bend, 180°	1.5
Gate valve	
Half open	4.5
Wide open	0.17
Angle valve, wide open	2.0
Globe valve, wide open	6.0

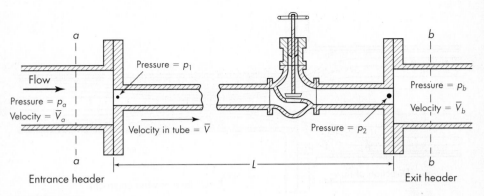

FIGURE 5.15
Flow of incompressible fluid through typical assembly.

Form friction losses in the Bernoulli equation. Form friction losses are incorporated in the h_f term of Eq. (4.65). They are combined with the skin friction losses of the straight pipe to give the total friction loss. Consider, for example, the flow of incompressible fluid through the two enlarged headers, the connecting tube, and the open globe valve shown in Fig. 5.15. Let $\bar{V}$ be the average velocity in the tube, D the diameter of the tube, and L the length of the tube. The skin friction loss in the straight tube is, by Eq. (5.7), $4f(L/D)(\bar{V}^2/2)$; the contraction loss at the entrance to the tube is, by Eq. (5.69), $K_c(\bar{V}^2/2)$; the expansion loss at the exit of the tube is, by Eq. (5.63), $K_e(\bar{V}^2/2)$; and the friction loss in the globe valve is, by Eq. (5.71), $K_f(\bar{V}^2/2)$. When skin friction in the entrance and exit headers is neglected, the total friction is

$$h_f = \left(4f\frac{L}{D} + K_c + K_e + K_f\right)\frac{\bar{V}^2}{2} \tag{5.72}$$

To write the Bernoulli equation for this assembly, take station a in the inlet header and station b in the outlet header. Because there is no pump between stations a and b, $W_p = 0$; also α_a and α_b can be taken as 1.0, the kinetic energy term can be canceled, and Eq. (4.65) becomes

$$\frac{p_a - p_b}{\rho} + g(Z_a - Z_b) = \left(4f\frac{L}{D} + K_c + K_e + K_f\right)\frac{\bar{V}^2}{2} \tag{5.73}$$

EXAMPLE 5.2. Crude oil having a specific gravity of 0.93 and a viscosity of 4 cP is draining by gravity from the bottom of a tank. The depth of liquid above the drawoff connection in the tank is 6 m. The line from the drawoff is 3-in. Schedule 40 pipe. Its length is 45 m, and it contains one ell and two gate valves. The oil discharges into the atmosphere 9 m below the drawoff connection of the tank. What flow rate, in cubic meters per hour, can be expected through the line?

Solution. The quantities needed are

$$\mu = 0.004 \text{ kg/m} \cdot \text{s} \qquad L = 45 \text{ m}$$

$$D = \frac{3.068}{12} = 0.256 \text{ ft (App. 3)} = 0.078 \text{ m}$$

$$\rho = 0.93 \times 998 = 928 \text{ kg/m}^3$$

For fittings, from Table 5.1,

$$\sum K_f = 0.75 + 2 \times 0.17 = 1.09$$

From Eq. (4.62), assuming $\alpha_b = 1$ and since $p_a = p_b$ and $\bar{V}_a = 0$,

$$\frac{\bar{V}_b^2}{2} + h_f = g(Z_a - Z_b) = 9.80665(6 + 9) = 147.1 \text{ m}^2/\text{s}^2 \qquad (5.74)$$

Use Eq. (5.72). There is no final expansion loss, since the stream does not expand upon discharge, and $K_e = 0$. From Eq. (5.70), since S_a is very large, $K_c = 0.4$. Hence

$$h_f = \left(4f\frac{L}{D} + K_c + \sum K_f \right) \frac{\bar{V}_b^2}{2}$$

$$= \left(\frac{4 \times 45 f}{0.078} + 0.4 + 1.09 \right) \frac{\bar{V}_b^2}{2} = (2308 f + 1.49)\frac{\bar{V}_b^2}{2}$$

From Eq. (5.74),

$$\frac{V_b^2}{2} + h_f = \frac{\bar{V}_b^2}{2}(1 + 2{,}308 f + 1.49) = 147.1$$

$$V_b^2 = \frac{147.1 \times 2}{2.49 + 2{,}308 f} = \frac{294.2}{2.49 + 2{,}308 f}$$

Use Fig. 5.10 to find f. For this problem

$$\text{Re} = \frac{0.078 \times 928 \bar{V}_b}{0.004} = 18{,}096 \, \bar{V}_b$$

$$\frac{k}{D} = 0.00015 \times \frac{0.3048}{0.078} = 0.00059$$

Trials give the following:

$\bar{V}_{b,\text{est}}$, m/s	Re $\times 10^{-4}$	f (from Fig. 5.10)	$\bar{V}_{b,\text{cal}}$, m/s
4.00	7.23	0.0056	4.37
4.37	7.91	0.0055	4.40
4.40	7.96	0.0055	4.40

The cross-sectional area of the pipe is 0.0513 ft², or 0.00477 m² (App. 3), and the flow rate is $4.40 \times 3600 \times 0.00477 = 75.6 \text{ m}^3/\text{h}$.

Velocity heads

As shown by Eq. (5.73), the friction loss in a complicated flow system can be expressed as a number of *velocity heads,* defined as $\bar{V}^2/2$ [or $\bar{V}^2/(2g_c)$ in fps units]. It is a measure of the momentum loss resulting from flow through the pipe or fitting. In a tee used as an elbow, for example, all the momentum in one direction is lost as the stream turns 90°, whereas in a standard 90° elbow only three-fourths of it is lost. In an open globe valve, illustrated in Fig. 5.15, the flow makes two 90° turns and passes through a construction, expanding again on the other side. These changes in velocity and direction lead to a friction loss of 6 velocity heads.

A rapid practical method of estimating friction in straight pipe is as follows. By setting $4f(L/D) = 1.0$, it follows from Eq. (5.7) that a length equal to a definite number of pipe diameters generates a friction loss equal to 1 velocity head. Since, in turbulent flow, f varies from about 0.01 to about 0.002, the number of pipe diameters equivalent to a velocity head is from $1/(4 \times 0.01) = 25$ to $1/(4 \times 0.002) = 125$, depending upon the Reynolds number. For ordinary practice, 50 pipe diameters is assumed for this factor. Thus, if the pipe in the system of Fig. 5.15 is standard 2-in. steel (actual ID 2.07 in.) and is 100 ft long, the skin friction is equivalent to $(100 \times 12)/(2 \times 50) = 12$ velocity heads. In this case, the friction from the single fitting and the expansion and contraction are negligible in comparison with that in the pipe. In other cases, where the pipes are short and the fittings and expansion and contraction losses numerous, the friction loss in the pipes only may be negligible.

Separation from velocity decrease

Boundary layer separation can occur even where there is no sudden change in cross section if the cross section is continuously enlarged. For example, consider the flow of a fluid stream through the trumpet-shaped expander shown in Fig. 5.16. Because of the increase of cross section in the direction of flow, the velocity of the fluid decreases, and by the Bernoulli equation, the pressure must increase. Consider two stream filaments, one, *aa*, very near the wall, and the other, *bb*, a short distance from the wall. The pressure increase over a definite length of conduit is the same for both filaments, because the pressure throughout any single cross section is uniform. The loss in velocity head is, then, the same for both filaments. The initial velocity head of filament *aa* is less than that of filament *bb*, however, because filament *aa* is nearer to the wall. A point is reached, at a definite distance along the conduit, where the velocity of filament *aa* becomes zero but where the velocities of filament *bb* and of all other filaments farther from the wall than *aa* are still positive. This point is

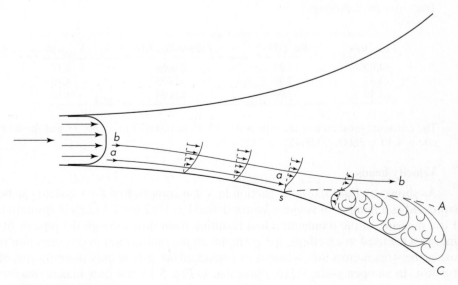

FIGURE 5.16
Separation of boundary layer in diverging channel.

point s in Fig. 5.16. Beyond point s the velocity at the wall changes sign, a backflow of fluid between the wall and filament aa occurs, and the boundary layer separates from the wall. In Fig. 5.16, several curves are drawn of velocity u versus distance from the wall y, and it can be seen how the velocity near the wall becomes zero at point s and then reverses in sign. The point s is called a *separation point*. Line sA is called the *line of zero tangential velocity*.

The vortices formed between the wall and the separated fluid stream, beyond the separation point, cause excessive form friction losses. Separation occurs in both laminar and turbulent flow. In turbulent flow, the separation point is farther along the conduit than in laminar flow. Separation can be prevented if the angle between the wall of the conduit and the axis is made small. The maximum angle that can be tolerated in a conical expander without separation is $7°$.

Minimizing expansion and contraction losses

A contraction loss can be nearly eliminated by reducing the cross section gradually rather than suddenly. For example, if the reduction in cross section shown in Fig. 5.14 is obtained by a conical reducer or by a trumpet-shaped entrance to the smaller pipe, the contraction coefficient K_c can be reduced to approximately 0.05 for all values of S_b/S_a. Separation and vena contracta formation do not occur unless the decrease in cross section is sudden.

An expansion loss can also be minimized by substituting a conical expander for the flanges shown in Fig. 5.13. The angle between the diverging walls of the cone must be less than $7°$, however, or separation may occur. For angles of $35°$ or more, the loss through a conical expander can become greater than that through a sudden expansion for the same area ratio S_a/S_b because of the excessive form friction from the vortices caused by the separation.

SYMBOLS

b	Width of square duct, m or ft; also spacing between parallel plates, m or ft
c_v	Specific heat at constant volume, J/kg·°C or Btu/lb·°F
D	Diameter, m or ft; D_{eq}, equivalent diameter of noncircular channel; D_i, inside diameter of annulus; D_o, outside diameter of annulus
F	Force, N or lb_f; F_g, gravity force; F_s, shear force; F_w, net force of channel wall on fluid
f	Fanning friction factor, dimensionless
g_c	Newton's law proportionality factor, 32.174 $ft·lb/lb_f·s^2$
h_f	Friction loss, J/kg or ft·lb_f/lb; h_{fc}, from sudden contraction; h_{fe}, from sudden expansion; h_{ff}, in flow through fitting or valve; h_{fs}, skin friction
K	Constant in Eqs. (5.25) and (5.27)
K_c	Contraction loss coefficient, dimensionless
K_e	Expansion loss coefficient, dimensionless
K_f	Loss factor for fitting or valve, dimensionless
K'	Flow consistency index, $kg/m·s^{2-n'}$ or $lb/ft·s^{2-n'}$
k	Roughness parameter, m or ft
L	Length, m or ft

L_p Wetted perimeter, m or ft
$\dot{m}$ Mass flow rate, kg/s or lb/s
n' Flow behavior index, dimensionless
p Pressure, N/m^2 or lb$_f$/ft^2; p_a, at station a; p_b, at station b
Q Heat generated by viscous dissipation, J
Re   Reynolds number, dimensionless; Re$_n$, modified Reynolds number for non-newtonian fluids, defined by Eq. (5.58)
r Radial distance from pipe axis, m or ft; r_c, radius of cylinder of fluid in plug flow; r_w, radius of pipe
r_H Hydraulic radius of conduit, m or ft
S Cross-sectional area, m^2 or ft^2; S_a, at station a; S_b, at station b
T Temperature,°C or °F
u Net or time-average local fluid velocity in x direction, m/s or ft/s; u_c, velocity of cylinder of plastic fluid in plug flow; u_{max}, maximum local velocity
u^* Friction velocity, $\bar{V}/\sqrt{f/2}$
u^+ Dimensionless velocity quotient, u/u^*; u_c^+, at pipe axis
V Volume
$\bar{V}$ Average fluid velocity in x direction; $\bar{V}_a$, $\bar{V}_b$, at stations a and b
W_p Pump work, J/kg or ft·lb$_f$/lb
y Radial distance from pipe wall, m or ft
y^+ Dimensionless distance, yu^*/v; y_c^+, at pipe axis
Z Height above datum plane, m or ft; Z_a, at station a; Z_b, at station b

Greek Letters

α Kinetic energy correction factor, dimensionless; α_a, at station a; α_b, at station b
β Momentum correction factor, dimensionless; β_a, at station a; β_b, at station b
Δp Pressure loss, $p_a - p_b$; Δp_s, from skin friction
ΔT Temperature rise from viscous dissipation, °C
κ Ratio r_i/r_o in annular flow [Eq. (5.28)]
μ Absolute viscosity, P or lb/ft·s; μ_w, at temperature of pipe wall
v Kinematic viscosity, μ/ρ, m^2/s or ft^2/s
ρ Density, kg/m^3 or lb/ft^3
τ Shear stress, N/m^2 or lb$_f$/ft^2; τ_w, at pipe wall; τ_0, threshold stress in plastic fluid
ϕ_a Factor in Eq. (5.30) for laminar flow in annulus, dimensionless
ψ Temperature correction factor for skin friction, dimensionless

PROBLEMS

5.1. Prove that the flow of a liquid in laminar flow between infinite parallel flat plates is given by

$$p_a - p_b = \frac{12\mu\bar{V}L}{b^2}$$

where L = length of plate in direction of flow
$\quad\quad b$ = distance between plates

Neglect end effects.

5.2. For flow of water in a smooth 60-mm pipe at Re $= 5 \times 10^4$, what is the thickness of the viscous sublayer? What fraction of the cross-sectional area of the pipe does this represent? About what fraction of the flow is in the viscous sublayer?

5.3. A newtonian fluid is in laminar flow in a rectangular channel with a large aspect ratio. Derive the relationship between local and maximum velocity, and determine the ratio $u_{max}/\bar{V}$.

5.4. Calculate the power required per meter of width of stream to force lubricating oil through the gap between two horizontal flat plates under the following conditions:

> Distance between plates, 6 mm
> Flow rate of oil per meter of width, 100 m^3/h
> Viscosity of oil, 25 cP
> Density of oil, 0.88 g/cm^3
> Length of plates, 3 m

Assume that the plates are very wide in comparison with the distance between them and that end effects can be neglected.

5.5. What would be the temperature rise in the oil from viscous dissipation under the conditions of Prob. 5.4 if the specific heat of the oil is 0.5 cal/g·°C and its density is 820 kg/m^3?

5.6. A liquid with a specific gravity of 2.6 and a viscosity of 2.0 cP flows through a smooth pipe of unknown diameter, resulting in a pressure drop of 0.183 lb$_f$/in.^{2} for 1.73 mi. What is the pipe diameter in inches if the mass rate of flow is 7,000 lb/h?

5.7. Water flows through an 8-in. steel pipe at an average velocity of 6 ft/s. Downstream the pipe divides into an 8-in. main and a 2-in. bypass. The equivalent length of the bypass is 22 ft; the length of the 8-in. pipe in the bypassed section is 16 ft. Neglecting entrance and exit losses, what fraction of the total water flow passes through the bypass?

5.8. A steel pipe 2 ft in diameter carries water at about 15 ft/s. If the pipe has a roughness of 0.0003 ft, could the capacity be increased by inserting a smooth plastic liner that reduces the inside diameter to 1.9 ft? Calculate the change in pressure drop for the same flow and the change in capacity for a fixed pressure drop.

5.9. Water at 60°F is pumped from a reservoir to the top of a mountain through a 6-in. Schedule 120 pipe at an average velocity of 12 ft/s. The pipe discharges into the atmosphere at a level 3,000 ft above the level in the reservoir. The pipeline itself is 4,500 ft long. If the overall efficiency of the pump and the motor driving it is 70 percent and the cost of electric energy to the motor is 4 cents per kilowatthour, what is the hourly energy cost for pumping this water?

5.10. A reverse-osmosis unit for purifying brackish water has about 900,000 hollow fibers that permit the diffusion of water but reject most of the salt. The fibers are 85 μm in outside diameter, 42 μm in inside diameter, and about 3 ft long. The average flow through the tubes is 2,000 gal of water every 24 h when the feed pressure is 400 psig. What is the pressure drop within an individual fiber from the feed end to the discharge end?

5.11. Chlorine gas leaving a reactor at 200°C and 1.5 atm with a flow of 4 lb/min is to be compressed and recycled to the reactor, where the pressure is 1.6 atm. The length of pipe needed is about 80 ft, and the line will have six elbows and two globe valves. Suggest a pipe size and calculate the velocity and pressure drop.

5.12. Exhaust gas from a power plant passes through a 15- by 20-ft rectangular duct at an average velocity of 50 ft/s. The total length of duct is 250 ft, and there are two 90° bends. The gas is at 180°F and about 1 atm, and the properties are similar to those of air. Calculate the pressure drop in the duct and the power required to overcome pressure losses.

5.13. A centrifugal pump takes brine from the bottom of a supply tank and delivers it into the bottom of another tank. The brine level in the discharge tank is 150 ft above that in the supply tank. The line between the tanks is 600 ft of 4-in. Schedule 40 pipe. The flow rate is 400 gal/min. In the line are two gate valves, four standard tees, and four ells. What is the energy cost for running this pump for one 24-h day? The specific gravity of brine is 1.18, the viscosity of brine is 1.2 cP, and the energy cost is $400 per horsepower-year on a basis of 300 d/yr. The overall efficiency of pump and motor is 60 percent.

5.14. Cooling water for a chemical plant must be pumped from a river 2,500 ft from the plant site. Preliminary design calls for a flow of 600 gal/min and 6-in. steel pipe. Calculate the pressure drop and the annual pumping cost if power costs 3 cents per kilowatt hour. Would the use of an 8-in. pipe reduce the power cost enough to offset the increased pipe cost? Use $15/ft of length for the installed cost of 6-in. pipe and $20/ft for 8-in. pipe. Annual charges are 20 percent of the installed cost.

5.15. A fan draws air at rest and sends it through a 200- by 300-mm rectangular duct 45 m long. The air enters at 15°C and 750 mm Hg absolute pressure at a rate of 0.6 m³/s. What is the theoretical power required?

5.16. The laminar flow of a certain aqueous polystyrene sulfonate solution can be represented by the Ostwald–de Waele model with an exponent of 0.500. The solution is contained in a tank that is a vertical cylinder. Solution flows out through a horizontal tube attached to the bottom of the tank. (a) If the time for the height of solution in the tank to fall from 10.00 to 9.90 m is 30.0 min, how much time will it take to half-empty the tank (from a height of 10.0 m down to a height of 5.0 m)? (b) If the tank has a diameter of 2.0 m and the pipe has a diameter of 0.050 m with a length of 200 m, what is the flow consistency index of the solution? The solution has a density of 1,200 kg/m³.

5.17. A portion of a fire protection system can be modeled as a smooth tube with a diameter of 0.12 m, a length of 600 m, and a total drop in elevation of 12 m. End effects can be neglected. How much polyethylene oxide would have to be added to water at 25°C to increase the flow rate by a factor of 1.25? (See Fig. 5.12.)

REFERENCES

1. Bird, R. B., W. E. Stewart, and E. N. Lightfoot. *Transport Phenomena.* New York: Wiley, 1960, pp. 51–53.
2. Dodge, D. W., and A. B. Metzner. *AIChE J.* **5:**189 (1959).
3. Garde, R. J. *Turbulent Flow.* New York: Wiley, 1994.
4. Kays, W. M., and A. L. London. *Compact Heat Exchangers.* 2d ed. New York: McGraw-Hill, 1964.
5. Knudsen, J. G., and D. L. Katz. *Fluid Dynamics and Heat Transfer.* New York: McGraw-Hill, 1958, pp. 97, 101–105, 158–171.
6. Lumley, J. A. *Physics of Fluids* **20**(10):S64 (1977).
7. Patterson, R. W., and F. H. Abernathy. *J. Fluid Mech.* **51:**177 (1972).
8. Pearson, J. R. A. *Mechanics of Polymer Processing.* London: Elsevier, 1985, pp. 191–194.
9. Perry, R. H., and D. W. Green, eds. *Perry's Chemical Engineers' Handbook.* 7th ed. New York: McGraw-Hill, 1997, (*a*) p. **6**-12; (*b*) p. **6**-18.
10. Prandtl, L. *VDIZ.* **77:**105 (1933).
11. Schlichting, H. *Boundary Layer Theory.* 7th ed. New York: McGraw-Hill, 1979.
12. Toms, B. A. *Proc. Intern. Congr. Rheology, Holland.* Amsterdam: North-Holland, 1949, p. **II**-135.
13. Vinogradov, G. V., G. B. Froishteter, K. K. Trilisky, and E. L. Smorodinsky. *Rheologica Acta* **14:**765 (1975).

CHAPTER 6

Flow of Compressible Fluids

Many important applications of fluid dynamics require that density variations be taken into account. The complete field of compressible fluid flow has become very large, and it covers wide ranges of pressure, temperature, and velocity. Chemical engineering practice involves a relatively small area from this field. For incompressible flow the basic parameter is the Reynolds number, a parameter also important in some applications of compressible flow. In compressible flow at ordinary densities and high velocities, a more basic parameter is the Mach number. At very low densities, where the mean free path of the molecules is appreciable in comparison with the size of the equipment or solid bodies in contact with the gas, other factors must be considered. This type of flow is not treated in this text.

DEFINITIONS AND BASIC EQUATIONS

The Mach number, denoted by Ma, is defined as the ratio of u, the speed of the fluid, to a, the speed of sound in the fluid under conditions of flow

$$\mathrm{Ma} \equiv \frac{u}{a} \tag{6.1}$$

By speed of the fluid is meant the magnitude of the relative velocity between the fluid and a solid bounding the fluid or immersed in it, whether the solid is considered to be stationary and the fluid flowing past it or the fluid is assumed to be stationary and the solid moving through it. The former situation is the more common in chemical engineering, and the latter is of great importance in aeronautics, for the motion of missiles, rockets, and other solid bodies through the atmosphere. By definition the Mach number is unity when the speed of fluid equals that of sound in the same fluid at the pressure and temperature of the fluid. Flow is called *subsonic*,

128

sonic, or *supersonic* according to whether the Mach number is less than unity, at or near unity, or greater than unity, respectively. The most interesting problems in compressible flow lie in the high-velocity range, where Mach numbers are comparable with unity or where flow is supersonic.

In this chapter the following simplifying assumptions are made. Although they may appear restrictive, many actual engineering situations may be adequately represented by the mathematical models obtained within the limitations of the assumptions.

1. The flow is steady.
2. The flow is one-dimensional.
3. Velocity gradients within a cross section are neglected, so that $\alpha = \beta = 1$ and $\bar{V} = u$.
4. Friction is restricted to wall shear.
5. Shaft work is zero.
6. Gravitational effects are negligible, and mechanical potential energy is neglected.
7. The fluid is an ideal gas of constant specific heat.

The following basic relations are used:

1. The continuity equation
2. The steady-flow total-energy balance
3. The mechanical energy balance with wall friction
4. The equation for the velocity of sound
5. The equation of state of the ideal gas

Each of these equations must be put into a suitable form.

Continuity equation $eq. 4.7 \Rightarrow d\dot{m} = Sud\delta$

For differentiation, the right-hand side of Eq. (4.7) may be written in logarithmic form:

$$\ln \rho + \ln S + \ln u = \text{const}$$

Differentiating this equation gives

$$\frac{d\rho}{\rho} + \frac{dS}{S} + \frac{du}{u} = 0 \qquad (6.2)$$

Total-energy balance

Consider a fluid in steady flow through a system, entering at station a with velocity u_a and enthalpy H_a and leaving at station b with velocity u_b and enthalpy H_b. For the flow of m kilograms of material, heat in the amount of Q joules must be added through the boundaries of the system to the material flowing through it. Provided there is no significant change in elevation between stations a and b and no work is done by or on the system, the steady-flow total-energy equation (1.59) may be written omitting the terms for potential energy and shaft work. The heat added

$Eq. 1.59 \Rightarrow m \left[\dfrac{u_b^2 - u_a^2}{2} + g(z_b - z_a) + H_b - H_a \right] = Q - W_s$

to the fluid is then given by the equation

$$\frac{Q}{m} = H_b - H_a + \frac{u_b^2}{2} - \frac{u_a^2}{2} \tag{6.3}$$

This equation written differentially is

$$\frac{dQ}{m} = dH + d\left(\frac{u^2}{2}\right) \tag{6.4}$$

Mechanical energy balance

Equation (4.62) may be written over a short length of conduit in the following differential form:

$$\frac{dp}{\rho} + d\left(\frac{\alpha \bar{V}^2}{2}\right) + g\,dZ + dh_f = 0 \tag{6.5}$$

In the light of the assumptions, this equation is simplified by omitting the potential energy terms, noting that $\alpha_a = \alpha_b = 1.0$, $u = \bar{V}$, and restricting the friction to wall shear. Equation (6.5) then becomes

$$\frac{dp}{\rho} + d\left(\frac{u^2}{2}\right) + dh_{fs} = 0 \tag{6.6}$$

From Eq. (5.7), with $D = 4r_H$,

$$5.7\,hf_x = 4f\,\frac{L}{D}\frac{\bar{V}^2}{2}$$

$$dh_{fs} = \frac{u^2}{2} \cdot \frac{f\,dL}{r_H} \tag{6.7}$$

Eliminating dh_{fs} from Eqs. (6.6) and (6.7) gives the form of the mechanical energy equation suitable for treatment of compressible flow:

$$\frac{dp}{\rho} + d\left(\frac{u^2}{2}\right) + \frac{u^2}{2}\frac{f\,dL}{r_H} = 0 \tag{6.8}$$

Velocity of sound

The velocity of sound through a continuous material medium, also called the *acoustical velocity,* is the velocity of a very small compression-rarefaction wave moving adiabatically and frictionlessly through the medium. Thermodynamically, the motion of a sound wave is a constant-entropy, or isentropic, process. The magnitude of the acoustical velocity in any medium is shown in physics texts to be

$$a = \sqrt{\left(\frac{dp}{d\rho}\right)_S} \tag{6.9}$$

where the subscript S calls attention to the isentropic restraint on the process.

Ideal gas equations

Subject to assumptions 1 to 6, Eqs. (6.2) to (6.9) apply to any fluid. In fact, they may be used for incompressible flow simply by assuming that the density ρ is constant. To apply them to compressible flow, it is necessary that the density be related to temperature and pressure. The simplest relation, and one of considerable engineering utility, is the ideal gas law [Eq. (1.56)], which for the present purpose may be written in the form

$$p = \frac{R}{M} \rho T \tag{6.10}$$

where R = molar gas law constant, in units of mechanical energy per mole per degree absolute
   M = molecular weight

The gas may be either pure or a mixture, but if it is not pure, the composition should not change. Equation (6.10) may be written logarithmically and then differentiated to give

$$\frac{dp}{p} = \frac{d\rho}{\rho} + \frac{dT}{T} \tag{6.11}$$

Since the specific heat c_p is assumed to be independent of temperature, the enthalpy of the gas at temperature T is

$$H = H_0 + c_p(T - T_0) \tag{6.12}$$

where H = enthalpy per unit mass at temperature T
   H_0 = enthalpy at arbitrary temperature T_0

The differential form of Eq. (6.12) is

$$dH = c_p\, dT \tag{6.13}$$

Acoustic velocity and Mach number of ideal gas. For an ideal gas, an isentropic path follows the equations

$$p\rho^{-\gamma} = \text{const} \tag{6.14}$$

$$Tp^{-(1-1/\gamma)} = \text{const} \tag{6.15}$$

where γ is the ratio of c_p, the specific heat at constant pressure, to c_v, the specific heat at constant volume. For an ideal gas,

$$\gamma \equiv \frac{c_p}{c_v} = \frac{c_p}{c_p - R/M} \tag{6.16}$$

Since, by assumption, c_p is independent of temperature, so are c_v and γ.

The quantity $(dp/d\rho)_S$ can be calculated by differentiating the logarithmic form of Eq. (6.14), giving

$$\frac{dp}{p} - \gamma \frac{d\rho}{\rho} = 0 \qquad \text{and} \qquad \left(\frac{dp}{d\rho}\right)_S = \gamma \frac{p}{\rho}$$

Substituting into Eq. (6.9) yields

$$a = \sqrt{\frac{\gamma p}{\rho}} = \sqrt{\frac{\gamma T R}{M}} \tag{6.17}$$

Equation (6.10) is used to establish the second equality in Eq. (6.17), which shows that the acoustic velocity of an ideal gas is a function of temperature only. From Eqs. (6.1) and (6.17) the square of the Mach number of an ideal gas is

$$\text{Ma}^2 = \frac{\rho u^2}{\gamma p} = \frac{u^2}{\gamma T R / M} \tag{6.18}$$

The asterisk condition

The state of the fluid moving at its acoustic velocity is important in some processes of compressible-fluid flow. The condition where $u = a$ and $\text{Ma} = 1$ is called the *asterisk condition*, and the pressure, temperature, density, and enthalpy are denoted by p^*, T^*, ρ^*, and H^* at this state.

Stagnation temperature

The stagnation temperature of a high-speed fluid is defined as the temperature the fluid would attain were it brought to rest adiabatically without the development of shaft work. The relation between the actual fluid temperature, the actual fluid velocity, and the stagnation temperature is found by using the total-energy equation (6.3) and the enthalpy equation (6.12). Let station a in Eq. (6.3) refer to the state of the flowing gas at temperature T and enthalpy H, and station b and reference state 0 in Eq. (6.12) to the stagnation condition. Stagnation is denoted by the subscript s. Then $u_a = u$, the velocity of the gas, and $u_b = u_s = 0$. Since the process is adiabatic and $Q = 0$, Eq. (6.3) becomes

$$H_s - H = \frac{u^2}{2} = H_0 - H \tag{6.19}$$

Eliminating $H_0 - H$ from Eq. (6.19) by using Eq. (6.12) gives, for the stagnation temperature T_s,

$$T_s = T + \frac{u^2}{2c_p} \tag{6.20}$$

The *stagnation enthalpy* H_s is defined by the equation

$$H_s = H + \frac{u^2}{2} \tag{6.21}$$

Equation (6.3) can be written

$$\frac{Q}{m} = H_{sb} - H_{sa} = (T_{sb} - T_{sa})c_p \tag{6.22}$$

where H_{sa} and H_{sb} are the stagnation enthalpies at states a and b, respectively. For an adiabatic process, $Q = 0$, $T_{sa} = T_{sb}$, and the stagnation temperature is constant.

PROCESSES OF COMPRESSIBLE FLOW

The flow processes to be considered in this chapter are shown diagrammatically in Fig. 6.1. It is assumed that a very large supply of gas at specified temperature and pressure and at zero velocity and Mach number is available. The origin of the gas is called the *reservoir*, and the temperature and pressure of the gas in the reservoir are called *reservoir conditions*. The reservoir temperature is a stagnation value, which does not necessarily apply at other points in the flow system.

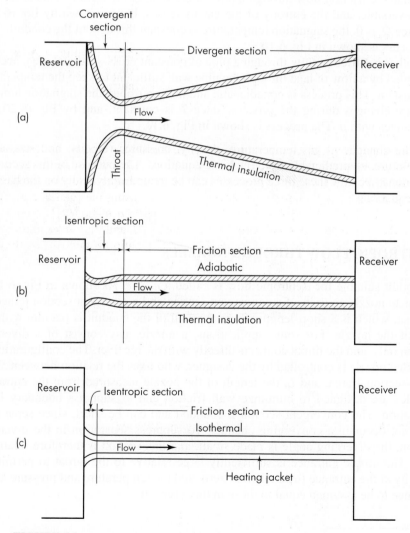

FIGURE 6.1
(*a*) Isentropic flow in a convergent-divergent nozzle. (*b*) Adiabatic friction flow. (*c*) Isothermal friction flow.

From the reservoir the gas is assumed to flow, without friction loss at the entrance, into and through a pipe. The gas leaves the pipe at definite temperature, velocity, and pressure and goes into an exhaust receiver, in which the pressure may be independently controlled at a constant value less than the reservoir pressure.

Within the pipe any one of the following processes may occur:

1. Isentropic expansion. In this process the cross-sectional area of the conduit must change, and the process is described as one of variable area. Because the process is adiabatic, the stagnation temperature does not change in the conduit. Such a process is shown diagrammatically in Fig. 6.1a.
2. Adiabatic friction flow through a pipe of constant cross section. This process is irreversible, and the entropy of the gas increases; but as shown by Eq. (6.22), since $Q = 0$, the stagnation temperature is constant throughout the conduit. This process is shown in Fig. 6.1b.
3. Isothermal friction flow through a pipe of constant cross-sectional area, accompanied by a flow of heat through the pipe wall sufficient to keep the temperature constant. This process is nonadiabatic and nonisentropic; the stagnation temperature changes during the process, since T is constant, and by Eq. (6.20), T_s changes with u. The process is shown in Fig. 6.1c.

The changes in gas temperature, density, pressure, velocity, and stagnation temperature are predictable from the basic equations. The purpose of this section is to demonstrate how these three processes can be treated analytically on the basis of such equations.[†]

ISENTROPIC FLOW THROUGH NOZZLES

A conduit suitable for isentropic flow is called a *nozzle*. As shown in Fig. 6.1a, a complete nozzle consists of a convergent section and a divergent section joined by a throat, which is a short length where the wall of the conduit is parallel with the axis of the nozzle. For some applications, a nozzle may consist of a divergent section only, and the throat connects directly with the receiver. The configuration of an actual nozzle is controlled by the designer, who fixes the relation between S, the cross-sectional area, and L, the length of the nozzle measured from the entrance. Nozzles are designed to minimize wall friction and to suppress boundary layer separation. The convergent section is rounded and can be short, since separation does not occur in a converging channel. To suppress separation in the divergent section, the diverging angle is made small, and this section is therefore relatively long. The nozzle entrance is sufficiently large relative to the throat to permit the velocity at the entrance to be taken as zero, and the temperature and pressure at the entrance to be assumed equal to those in the reservoir.

[†]Generalized treatments, including general heat transfer to and from the gas, injection of gas into the conduit, variations in specific heat and molecular weight, chemical reactions, the drag of internal bodies, and change of phase, are given in Refs. 1 and 3.

The purpose of the convergent section is to increase the velocity and decrease the pressure of the gas. At low Mach numbers the process conforms essentially to the usual Bernoulli relation for incompressible flow [Eq. (4.58)]. In the convergent section flow is always subsonic, but it may become sonic at the throat. Mach numbers greater than unity cannot be generated in a convergent nozzle. In the divergent section, the flow may be subsonic or supersonic. The purpose of the divergent section differs sharply in the two situations. In subsonic flow the purpose of the section is to reduce the velocity and regain pressure, in accordance with the Bernoulli equation. An important application of these nozzles is the measurement of fluid flow, which is discussed in Chap. 8. In supersonic flow, the usual purpose of the divergent section is to obtain Mach numbers greater than unity for use in experimental devices such as wind tunnels.

Flow through a given nozzle is controlled by fixing the reservoir and receiver pressures. For a given flow through a specific nozzle, a unique pressure exists at each point along the axis of the nozzle. The relation is conveniently shown as a plot of p/p_0 versus L, where p_0 is the reservoir pressure and p the pressure at point L. Figure 6.2 shows how the pressure ratio varies with distance and how changes in receiver pressure at constant reservoir pressure affect the pressure distribution. The pressures at the throat and in the receiver are denoted by p_t and p_r, respectively.

If p_r and p_0 are equal, no flow occurs and the pressure distribution is represented by line aa'. If the receiver pressure is slightly below the reservoir pressure, flow occurs and a pressure distribution such as that shown by line abc is established. Pressure recovery in the convergent section is shown by line bc. The maximum velocity occurs at the throat. If the receiver pressure is further reduced, the flow rate and the velocity throughout the nozzle increase. A limit is attained when the velocity at the throat becomes sonic. This case is shown by line ade, where $p_t = p^*$, $u_t = a$, and Ma = 1. The ratio p^*/p_0 is called the *critical pressure ratio* and is denoted by r_c. Flow is subsonic at all other points on line ade.

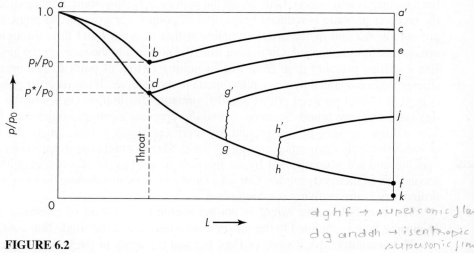

FIGURE 6.2
Variation of pressure ratio with distance from nozzle inlet.

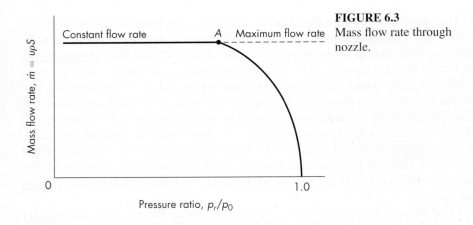

FIGURE 6.3
Mass flow rate through nozzle.

As the receiver pressure is decreased from that of point a' to that of point e, the mass flow rate through the nozzle increases. The flow rate is not affected by reduction of pressure below that corresponding to critical flow. Figure 6.3 shows how mass flow rate varies with the pressure ratio p_r/p_0. The flow rate attains its maximum at point A, which is reached when the pressure ratio in the throat is reduced to its critical value. Further reduction in pressure p_r does not change the flow rate.

The reason for this is as follows: When the velocity in the throat is sonic and the cross-sectional area of the conduit is constant, sound waves cannot move upstream into the throat and the gas in the throat has no way of receiving a message from downstream. Further reduction of the receiver pressure cannot be transmitted to the throat.

If the receiver pressure is reduced to the level shown by point f in Fig. 6.2, the pressure distribution is represented by the continuous line $adghf$. This line is unique for a given gas and nozzle. Only along the path $dghf$ is supersonic flow possible. If the receiver pressure is reduced below that of point f, for example, to point k, the pressure at the end of the nozzle remains at that of point f and flow through the nozzle remains unchanged. On issuing from the nozzle into the receiver the gas suffers a sudden pressure drop from that of point f to that of point k. The pressure change is accompanied by wave phenomena in the receiver. If the receiver pressure is held at a level between points e and f, pressure distribution curves of the type $dgg'i$ and $dhh'j$ are found. Sections dg and dh represent isentropic supersonic flow. The sudden pressure jumps gg' and hh' represent shock waves, where the flow changes suddenly from supersonic to subsonic. Shocks are thermodynamically irreversible and are accompanied by an increase in entropy in accordance with the second law of thermodynamics. Curves $g'i$ and $h'j$ represent subsonic flows in which ordinary pressure recovery is taking place.

The area below line $adghf$ is not accessible to any kind of adiabatic flow. Isentropic flow is confined to the subsonic area $adea'a$ and the single line $dghf$.

The qualitative discussions of Figs. 6.2 and 6.3 apply to the flow of any compressible fluid. Quantitative relations are most easily found for ideal gas flow.

Equations for isentropic flow

The phenomena occurring in the flow of ideal gas through nozzles are described by equations derivable from the basic equations given earlier in this chapter.

Change in gas properties during flow. The density and temperature paths of the gas through any isentropic flow are given by Eqs. (6.14) and (6.15). The constants are evaluated from the reservoir condition. This gives

$$\frac{p}{\rho^{\gamma}} = \frac{p_0}{\rho_0^{\gamma}} \tag{6.23}$$

$$\frac{T}{p^{1-1/\gamma}} = \frac{T_0}{p_0^{1-1/\gamma}} \tag{6.24}$$

These equations apply to both frictionless subsonic and supersonic flow, but they must not be used across a shock front.

Velocity in nozzle. In the absence of friction, the mechanical energy balance [Eq. (6.6)] becomes simply

$$\frac{dp}{\rho} = -d\left(\frac{u^2}{2}\right) \tag{6.25}$$

Eliminating ρ from Eq. (6.25) by substitution from Eq. (6.23) and integrating from a lower limit based on the reservoir, where $p = p_0$, $\rho = \rho_0$, and $u = 0$, gives

$$\int_0^u d\left(\frac{u^2}{2}\right) = -\frac{p_0^{1/\gamma}}{\rho_0} \int_{p_0}^p \frac{dp}{p^{1/\gamma}}$$

Integrating and substituting the limits yield, after rearrangement,

$$u^2 = \frac{2\gamma p_0}{(\gamma - 1)\rho_0}\left[1 - \left(\frac{p}{p_0}\right)^{1-1/\gamma}\right] \tag{6.26}$$

A Mach number form of Eq. (6.26) is convenient. It is derived by substituting u^2 from Eq. (6.2) into the first equality of Eq. (6.18) and eliminating ρ/ρ_0 by substitution from Eq. (6.23). This gives

$$\text{Ma}^2 = \frac{2}{\gamma - 1}\frac{p_0}{p}\frac{\rho}{\rho_0}\left[1 - \left(\frac{p}{p_0}\right)^{1-1/\gamma}\right] = \frac{2}{\gamma - 1}\left[\left(\frac{p_0}{p}\right)^{1-1/\gamma} - 1\right] \tag{6.27}$$

Solved explicitly for the pressure ratio, Eq. (6.27) becomes

$$\frac{p}{p_0} = \frac{1}{\{1 + [(\gamma - 1)/2]\text{Ma}^2\}^{1/(1-1/\gamma)}} \tag{6.28}$$

The critical pressure ratio, denoted by r_c, is found from Eq. (6.28) by substituting p^* for p and 1.0 for Ma:

$$r_c = \frac{p^*}{p_0} = \left(\frac{2}{\gamma + 1}\right)^{1/(1-1/\gamma)} \tag{6.29}$$

The mass velocity is found by calculating the product of u and ρ, using Eqs. (6.23) and (6.26),

$$G = u\rho = \sqrt{\frac{2\gamma\rho_0 p_0}{\gamma - 1}} \left(\frac{p}{p_0}\right)^{1/\gamma} \sqrt{1 - \left(\frac{p}{p_0}\right)^{1-1/\gamma}} \tag{6.30}$$

Effect of cross-sectional area

The relation between the change in cross-sectional area, velocity, and Mach number is useful in correlating the various cases of nozzle flow. Substitution of ρ from Eq. (6.25) into the continuity equation (6.2) gives

$$\frac{du}{u} + \frac{dS}{S} - \left(\frac{d\rho}{dp}\right)_S u\, du = 0 \tag{6.31}$$

Subscript S is used to call attention to the fact that the flow is isentropic. From Eq. (6.9)

$$\left(\frac{dp}{d\rho}\right)_S = a^2 \tag{6.32}$$

Eliminating $(dp/d\rho)_S$ from Eqs. (6.31) and (6.32) gives

$$\frac{du}{u}\left(1 - \frac{u^2}{a^2}\right) + \frac{dS}{S} = 0$$

and, by substituting Ma from Eq. (6.1),

$$\frac{du}{u}(\mathrm{Ma}^2 - 1) = \frac{dS}{S} \tag{6.33}$$

Equation (6.33) shows that for subsonic flow, where Ma < 1, the velocity increases with decreasing cross section (converging conduit) and decreases with increasing cross section (diverging conduit). This corresponds to the usual situation of incompressible flow. Lines *abc*, *ade*, *g'i*, and *h'j* of Fig. 6.2 represent examples. For supersonic flow, where Ma > 1, the velocity increases with increasing cross section, as in the diverging section of the nozzle. This conforms to line *dghf* of Fig. 6.2. The apparent anomaly of supersonic flow is a result of the variation in density and velocity along an isentropic path. Since the mass flow rate is the same at all points in the nozzle, by continuity the cross-sectional area of the nozzle must vary inversely with the mass velocity $u\rho$. The velocity steadily increases with Mach number, and the density decreases. However, at Ma $= 1$, the value of G goes through a maximum. In the subsonic regime the velocity increases faster than the density decreases, the mass velocity increases, and S decreases. In the supersonic regime, the increase in velocity is overcome by a sharper decrease in density, the mass velocity decreases, and S increases to accommodate the total mass flow. This behavior of G is demonstrated by studying the first and second derivatives of Eq. (6.30) in the usual manner for investigating maxima and minima.

EXAMPLE 6.1. Air enters a convergent-divergent nozzle at a temperature of 555.6 K (1000°R) and a pressure of 20 atm. The throat area is one-half that of the discharge of the divergent section. (a) Assuming the Mach number in the throat is 0.8, what are the values of the following quantities at the throat: pressure, temperature, linear velocity, density, and mass velocity? (b) What are the values of p^*, T^*, u^*, and G^* corresponding to reservoir conditions? (c) Assuming the nozzle is to be used supersonically, what is the maximum Mach number at the discharge of the divergent section? For air $\gamma = 1.4$ and $M = 29$.

Solution

(a) The pressure at the throat is calculated from Eq. (6.28):

$$\frac{p_t}{20} = \frac{1}{\{1 + [(1.4 - 1)/2]0.8^2\}^{1/(1-1/1.4)}} = 0.656$$

$$p_t = 13.12 \text{ atm}$$

From Eq. (6.10), since $R = 82.056 \times 10^{-3}$ atm $\cdot$ m³/kg mol $\cdot$ K,

$$\rho_0 = \frac{20 \times 29}{82.056 \times 10^{-3} \times 555.6} = 12.72 \text{ kg/m}^3 \text{ (0.795 lb/ft}^3)$$

Substituting p_0/ρ_0 from Eq. (6.10) into Eq. (6.26) gives, for the velocity in the throat,

$$u_t = \sqrt{\frac{2\gamma R T_0}{M(\gamma - 1)}\left[1 - \left(\frac{p_t}{p_0}\right)^{1-1/\gamma}\right]}$$

$$= \sqrt{\frac{2 \times 1.4 \times 82.056 \times 10^{-3} \times 555.6}{29(1.4 - 1)}\sqrt{1 - 0.656^{1-1/1.4}}}$$

$$= 1.1175 \text{ (m}^3 \cdot \text{atm/kg)}^{1/2}$$

These units are converted to meters per second as follows. From App. 1, 1 atm = 1.01325×10^5 N/m². By definition, 1 N = 1 kg $\cdot$ m/s². Hence

$$u_t = 1.1175\sqrt{1.01325 \times 10^5} = 355.7 \text{ m/s } (1,167 \text{ ft/s})$$

The density at the throat is, from Eq. (6.23),

$$\rho_t = \rho_0\left(\frac{p_t}{p_0}\right)^{1/\gamma} = 12.72 \times 0.656^{1/1.4} = 9.41 \text{ kg/m}^3 \text{ (0.587 lb/ft}^3)$$

The mass velocity at the throat is

$$G_t = u_t \rho_t = 355.7 \times 9.41 = 3348 \text{ kg/m}^2 \cdot \text{s } (686 \text{ lb/ft}^2 \cdot \text{s})$$

[The mass velocity can also be calculated directly by use of Eq. (6.30).] The temperature at the throat is, from Eq. (6.24),

$$T_t = T_0\left(\frac{p_t}{p_0}\right)^{1-1/\gamma} = 555.6 \times 0.656^{1-1/1.4} = 492.5 \text{ K (886.5°R)}$$

(b) From Eq. (6.29)

$$r_c = \frac{p^*}{p_0} = \left(\frac{2}{1.4 + 1}\right)^{1/(1-1/1.4)} = 0.528 \qquad p^* = 20 \times 0.528 = 10.56 \text{ atm}$$

From Eqs. (6.24) and (6.29)

$$T^* = 555.6\left(\frac{2}{1.4+1}\right) = 463 \text{ K } (833.4°\text{R})$$

From Eq. (6.23)

$$\left(\frac{\rho_0}{\rho^*}\right)^\gamma = \frac{p_0}{p^*} \qquad \frac{p^*}{\rho_0} = \left(\frac{p^*}{p_0}\right)^{1/\gamma}$$

$$\rho^* = 12.72 \times 0.528^{1/1.4} = 8.06 \text{ kg/m}^3 \ (0.503 \text{ lb/ft}^3)$$

From Eq. (6.30)

$$G^* = \sqrt{\frac{2 \times 1.4 \times 12.72 \times 20 \times 1.01325 \times 10^5}{1.4-1}} 0.528^{1/1.4}\sqrt{1-0.528^{1-1/1.4}}$$

$$= 3{,}476 \text{ kg/m}^2 \cdot \text{s } (712 \text{ lb/ft}^2 \cdot \text{s})$$

$$u^* = \frac{G^*}{\rho^*} = \frac{3{,}476}{8.06} = 431 \text{ m/s } (1{,}415 \text{ ft/s})$$

(c) Since, by continuity, $G \propto 1/S$, the mass velocity at the discharge is

$$G_r = \frac{3{,}476}{2} = 1{,}738 \text{ kg/m}^2 \cdot \text{s } (356 \text{ lb/ft}^2 \cdot \text{s})$$

From Eq. (6.30)

$$1{,}738 = \sqrt{\frac{2 \times 1.4 \times 12.72 \times 20 \times 1.01325 \times 10^5}{0.4}\left[1-\left(\frac{p_r}{p_0}\right)^{1-1/1.4}\right]\left(\frac{p_r}{p_0}\right)^{1/1.4}}$$

$$\left(\frac{p_r}{p_0}\right)^{1/1.4}\sqrt{1-\left(\frac{p_r}{p_0}\right)^{1-1/1.4}} = 0.1294$$

This equation is solved for p_r/p_0 to give

$$\frac{p_r}{p_0} = 0.0939$$

From Eq. (6.27) the Mach number at the discharge is

$$\text{Ma}_r = \sqrt{\frac{2}{1.4-1}\left(\frac{1}{0.0939^{1-1/1.4}}-1\right)} = 2.20$$

ADIABATIC FRICTION FLOW

Flow through straight pipes of constant cross section is adiabatic when heat transfer through the pipe wall is negligible. The process is shown diagrammatically in Fig. 6.1b. The typical situation is a long pipe into which gas enters at a given pressure and temperature and flows at a rate determined by the length and diameter of the pipe and the pressure maintained at the outlet. In long lines and with a low exit pressure, the speed of the gas may reach the sonic velocity. It is not possible, however, for a gas to pass through the sonic barrier from the direction of either subsonic or supersonic flow; if the gas enters the pipe at a Mach number greater than 1, the Mach number will decrease but will not become less than 1. If an attempt is made, by maintaining a constant discharge pressure and lengthening the pipe, to force the gas to change from subsonic to supersonic flow or from supersonic to subsonic, the mass flow rate will decrease to prevent such a change. This effect is called choking.

The friction parameter

The basic quantity that measures the effect of friction is the friction parameter fL/r_H. This arises from the integration of Eq. (6.8). In adiabatic friction flow, the temperature of the gas changes. The viscosity also varies, and the Reynolds number and friction factor are not actually constant. In gas flow, however, the effect of temperature on viscosity is small, and the effect of Reynolds number on the friction factor f is still less. Also, unless the Mach number is nearly unity, the temperature change is small. It is satisfactory to use an average value for f as a constant in calculations. If necessary, f can be evaluated at the two ends of the conduit and an arithmetic average used as a constant.

Friction factors in supersonic flow are not well established. Apparently, they are approximately one-half those in subsonic flow for the same Reynolds number.[2]

In all the integrated equations in the next section, it is assumed that the entrance to the conduit is rounded to form an isentropic convergent nozzle. If supersonic flow in the conduit is required, the entrance nozzle must include a divergent section to generate a Mach number greater than 1.

Equations for adiabatic friction flow

Equation (6.8) is multiplied by ρ/p, giving

$$\frac{dp}{p} + \frac{\rho}{p}u\,du + \frac{\rho u^2}{2p}\frac{f\,dL}{r_H} = 0 \tag{6.34}$$

It is desired to obtain an integrated form of this equation. The most useful integrated form is one containing the Mach number as the dependent variable and the friction parameter as an independent variable. To accomplish this, the density factor is eliminated from Eq. (6.34) using Eq. (6.18), and relationships between Ma, dp/p, and du/u are found from Eqs. (6.2) and (6.11). Quantity dT/T, when it appears, is eliminated by using Eqs. (6.4), (6.13), and (6.18). The results are

$$\frac{dp}{p} = -\frac{1 + (\gamma - 1)\text{Ma}^2}{1 + [(\gamma - 1)/2]\text{Ma}^2}\frac{d\text{Ma}}{\text{Ma}} \tag{6.35}$$

Also

$$\frac{du}{u} = \frac{dp}{p} + 2\frac{d\text{Ma}}{\text{Ma}} \tag{6.36}$$

Substitution from Eqs. (6.35) and (6.36) into Eq. (6.34) and rearrangement give the final differential equation

$$f\frac{dL}{r_H} = \frac{2(1 - \text{Ma}^2)\,d\text{Ma}}{\gamma\,\text{Ma}^3\{1 + [(\gamma - 1)/2]\text{Ma}^2\}} \tag{6.37}$$

Formal integration of Eq. (6.37) between an entrance station a and exit station b gives

$$\int_{L_a}^{L_b} \frac{f}{r_H}\,dL = \int_{\text{Ma}_a}^{\text{Ma}_b} \frac{2(1 - \text{Ma}^2)\,d\text{Ma}}{\gamma\,\text{Ma}^3\{1 + [(\gamma - 1)/2]\text{Ma}^2\}}$$

$$\frac{\bar{f}}{r_H}(L_b - L_a) = \frac{\bar{f}L}{r_H}$$

$$= \frac{1}{\gamma}\left(\frac{1}{\text{Ma}_a^2} - \frac{1}{\text{Ma}_b^2} - \frac{\gamma + 1}{2}\ln\frac{\text{Ma}_b^2\{1 + [(\gamma - 1)/2]\text{Ma}_a^2\}}{\text{Ma}_a^2\{1 + [(\gamma - 1)/2]\text{Ma}_b^2\}}\right) \tag{6.38}$$

where $\bar{f}$ is the arithmetic average value of the terminal friction factors, $(f_a + f_b)/2$, and $L = L_b - L_a$.

Property equations. For calculating the changes in pressure, temperature, and density, the following equations are useful.

In adiabatic friction flow the ratio of the inlet and outlet pressures is found by direct integration of Eq. (6.35) between the limits p_a, p_b, and Ma_a, Ma_b to give

$$\frac{p_a}{p_b} = \frac{\text{Ma}_b}{\text{Ma}_a}\sqrt{\frac{1 + [(\gamma - 1)/2]\text{Ma}_b^2}{1 + [(\gamma - 1)/2]\text{Ma}_a^2}} \tag{6.39}$$

The temperature ratio is calculated from Eq. (6.20), noting that $T_{0a} = T_{0b}$; so

$$T_a + \frac{u_a^2}{2c_p} = T_b + \frac{u_b^2}{2c_p} \tag{6.40}$$

From Eq. (6.40) and the temperature form of Eq. (6.18) for the Mach number of an ideal gas,

$$T_a + \frac{\gamma R T_a \text{Ma}_a^2}{2Mc_p} = T_b + \frac{\gamma R T_b \text{Ma}_b^2}{2Mc_p} \tag{6.41}$$

From Eq. (6.16)

$$\frac{c_p M}{R} = \frac{\gamma}{\gamma - 1} \tag{6.42}$$

Substituting $c_p M/R$ from Eq. (6.42) into Eq. (6.41) and solving for the temperature ratio give

$$\frac{T_a}{T_b} = \frac{1 + [(\gamma - 1)/2]\text{Ma}_b^2}{1 + [(\gamma - 1)/2]\text{Ma}_a^2} \tag{6.43}$$

The density ratio is calculated from the gas equation of state (6.10) and the pressure and temperature ratios given by Eqs. (6.39) and (6.43), respectively,

$$\frac{\rho_a}{\rho_b} = \frac{p_a T_b}{p_b T_a} = \frac{\text{Ma}_b}{\text{Ma}_a} \sqrt{\frac{1 + [(\gamma - 1)/2]\text{Ma}_a^2}{1 + [(\gamma - 1)/2]\text{Ma}_b^2}} \tag{6.44}$$

Maximum conduit length. To ensure that the conditions of a problem do not call for the impossible phenomenon of a crossing of the sonic barrier, an equation is needed giving the maximum value of $\bar{f}L/r_H$ consistent with a given entrance Mach number. Such an equation is found from Eq. (6.38) by choosing the entrance to the conduit as station a and identifying station b as the asterisk condition, where $\text{Ma} = 1.0$. Then the length $L_b - L_a$ represents the maximum length of conduit that can be used for a fixed value of Ma_a. This length is denoted by L_{max}. Equation (6.38) then gives

$$\frac{\bar{f}L_{\text{max}}}{r_H} = \frac{1}{\gamma}\left(\frac{1}{\text{Ma}_a^2} - 1 - \frac{\gamma + 1}{2} \ln \frac{2\{1 + [(\gamma - 1)/2]\text{Ma}_a^2\}}{\text{Ma}_a^2(\gamma + 1)}\right) \tag{6.45}$$

Corresponding equations for p/p^*, T/T^*, and ρ/ρ^* are found from Eqs. (6.39), (6.43), and (6.44).

Mass velocity

To calculate the Reynolds number for evaluating the friction factor, the mass velocity is needed. From Eq. (6.18) and the definition of G

$$\text{Ma}^2 = \frac{(\rho u)^2}{\rho^2 \gamma T R/M} = \frac{G^2}{\rho^2 \gamma T R/M} = \frac{G^2}{\rho \gamma p}$$

and

$$G = \rho \text{Ma}\sqrt{\frac{\gamma T R}{M}} = \text{Ma}\sqrt{\rho \gamma p} \tag{6.46}$$

Since, for constant-area flow, G is independent of length, the mass velocity can be evaluated at any point where the gas properties are known. Normally the conditions at the entrance to the conduit are used.

EXAMPLE 6.2. Air flows from a reservoir through an isentropic nozzle into a long, straight pipe. The pressure and temperature in the reservoir are 20 atm and 1000°R (555.6 K), respectively, and the Mach number at the entrance of the pipe is 0.05. (*a*) What is the value of $\bar{f}L_{\text{max}}/r_H$? (*b*) What are the pressure, temperature, density, linear velocity, and mass velocity when $L_b = L_{\text{max}}$? (*c*) What is the mass velocity when $\bar{f}L_{\text{max}}/r_H = 400$?

Solution

(*a*) Values for substitution in Eq. (6.45) are $\gamma = 1.4$ and $\text{Ma}_a = 0.05$. Then

$$\frac{\bar{f}L_{\text{max}}}{r_H} = \frac{1}{1.4}\left(\frac{1}{0.05^2} - 1 - \frac{1.4 + 1}{2} \ln \frac{2\{1 + [(1.4 - 1)/2]0.05^2\}}{(1.4 + 1)0.05^2}\right) = 280$$

(b) The pressure at the end of the isentropic nozzle p_a is given by Eq. (6.28):

$$p_a = \frac{20}{\{1 + [(1.4 - 1)/2]0.05^2\}^{1.4/(1.4-1)}} = \frac{20}{1.0016} \approx 20 \text{ atm}$$

The pressure, temperature, and density change in the nozzle are negligible, and except for linear velocity, the reservoir conditions also pertain to the pipe entrance. From Example 6.1, the density of air at 20 atm and 1000°R is 0.795 lb/ft³. The acoustic velocity is, from Eq. (6.17), including g_c in the numerator,

$$a_a = \sqrt{1.4 \times 32.174 \times 1,000 \left(\frac{1,545}{29}\right)} = 1,550 \text{ ft/s (472.4 m/s)}$$

The velocity at the entrance of the pipe is

$$u_a = 0.05 \times 1,550 = 77.5 \text{ ft/s (23.6 m/s)}$$

When $L_b = L_{max}$, the gas leaves the pipe at the asterisk condition, where $Ma_b = 1.0$. From Eq. (6.43)

$$\frac{1,000}{T^*} = \frac{24}{2\{1 + [(1.4 - 1)/2]0.05^2\}} = 1.2$$

$$T^* = 834°R \text{ (463.3 K)}$$

From Eq. (6.44)

$$\frac{0.795}{\rho^*} = \frac{1}{0.05}\sqrt{\frac{2\{1 + [(1.4 - 1)/2]0.05^2\}}{2.4}}$$

$$\rho^* = 0.0435 \text{ lb/ft}^3 \text{ (0.697 kg/m}^3)$$

From Eq. (6.39)

$$\frac{20}{p^*} = \frac{1}{0.05}\sqrt{1.2}$$

$$p^* = 0.913 \text{ atm}$$

The mass velocity through the entire pipe is

$$G = 0.795 \times 77.5 = 0.0435u^* = 61.61 \text{ lb/ft}^2 \cdot \text{s (300.8 kg/m}^2 \cdot \text{s)}$$

$$a = u^* = 1416 \text{ ft/s (431.6 m/s)}$$

Since the exit velocity is sonic, u^* can also be calculated from Eq. (6.17) using $T = T^* = 834°R$ (463.3 K):

$$a = u^* = 1,550\sqrt{\frac{834}{1,000}} = 1,416 \text{ ft/s (431.6 m/s)}$$

(c) Using Eq. (6.45) with $\bar{f} L_{max}/r_H = 400$ gives

$$400 = \frac{1}{1.4}\left(\frac{1}{Ma_a^2} - 1 - \frac{1.4 + 1}{2} \ln \frac{2\{1 + [(1.4 - 1)/2]Ma_a^2\}}{Ma_a^2(1.4 + 1)}\right)$$

This equation must be solved for Ma_a by trial and error. The final result is $Ma_a = 0.04194$. Then

$$u_a = \frac{0.04194}{0.05}(77.5) = 65.0 \text{ ft/s (19.8 m/s)}$$

$$G = 65.0 \times 0.795 = 51.7 \text{ lb/ft}^2 \cdot \text{s} \, (252.4 \text{ kg/m}^2 \cdot \text{s})$$

ISOTHERMAL FRICTION FLOW

The temperature of the fluid in compressible flow through a conduit of constant cross section may be kept constant by a transfer of heat through the conduit wall. Long, small, uninsulated pipes in contact with air transmit sufficient heat to keep the flow nearly isothermal. Also, for small Mach numbers, the pressure pattern for isothermal flow is nearly the same as that for adiabatic flow for the same entrance conditions, and the simpler equations for isothermal flow may be used.

The basic equation for isothermal flow is simple. It is obtained by introducing the mass velocity into the mechanical energy balance [Eq. (6.8)] and integrating directly. Multiplying Eq. (6.8) by ρ^2 gives

$$\rho \, dp + \rho^2 u \, du + \frac{\rho^2 u^2 f \, dL}{2r_H} = 0 \tag{6.47}$$

Since $\rho u = G$, $u \, du = -(G^2 \rho^{-3}) \, d\rho$, and $\rho = Mp/(RT)$, Eq. (6.47) can be written

$$\frac{M}{RT} p \, dp - G^2 \frac{d\rho}{\rho} + \frac{G^2 f \, dL}{2r_H} = 0 \tag{6.48}$$

Rearranging Eq. (6.48) and integrating between stations a and b give

$$p_a^2 - p_b^2 = \frac{G^2 RT}{M} \left[2 \ln \frac{\rho_a}{\rho_b} + \frac{f(L_b - L_a)}{r_H} \right] \tag{6.49}$$

where r_H is the hydraulic radius of the pipe or channel.

The density ratio ρ_a/ρ_b may be used in place of p_a/p_b in Eq. (6.49). Equation (6.49) can also be used when the temperature change over the conduit is small. Then, in place of T, an arithmetic average temperature may be used. For example, adiabatic flow at low Mach numbers (below about 0.3) follows the equation closely.

EXAMPLE 6.3. Air at 1.7 atm gauge and 15°C enters a horizontal 75-mm steel pipe that is 70 m long. The flow rate of the entering air q is 0.265 m³/s. Assuming isothermal flow, what is the pressure at the discharge end of the pipe?

Solution. Use Eq. (6.49). The quantities needed are

$$D = 0.075 \text{ m} \qquad r_H = \frac{0.075}{4} = 0.01875 \text{ m}$$

$$\mu = 0.0174 \text{ cP (App. 8)} = 1.74 \times 10^{-5} \text{ kg/m} \cdot \text{s}$$

$$\rho_a = \frac{29}{22.4} \times \frac{2.7}{1} \times \frac{273}{288} = 3.31 \text{ kg/m}^3$$

$$A = \frac{0.075^2 \pi}{4} = 0.00442 \text{ m}^2$$

Then
$$G = \frac{q\rho_a}{A} = \frac{0.265 \times 3.31}{0.00442} = 198.5 \text{ kg/m}^2 \cdot \text{s}$$

$$\text{Re} = 0.075 \times \frac{198.5}{1.74 \times 10^{-5}} = 8.56 \times 10^5$$

$$\frac{k}{D} = 0.00015 \times \frac{0.3048}{0.075} = 0.00061$$

$$f = 0.0044 \text{ (Fig. 5.10)}$$

Let $(p_a + p_b)/2 = \bar{p}$, so

$$p_a^2 - p_b^2 = (p_a - p_b)(p_a + p_b) = 2\bar{p}(p_a - p_b)$$

From Eq. (6.49)

$$p_a - p_b = \frac{RTG^2}{\bar{p}M}\left(\frac{fL}{2r_H} + \ln\frac{p_a}{p_b}\right)$$

where $p_a = 2.7$ atm
$M = 29$
$L = 70$ m
$R = 82.056 \times 10^{-3}$ m$^3 \cdot$ atm/kg mol $\cdot$ K (Table 1.2)
$T = 15 + 273 = 288$ K

By trial, $\bar{p}$ is found to be 1.982 atm and $p_b = 1.264$ atm.
   Check: Since 1 atm = 101,325 N/m^2 = 101,325 kg/m $\cdot$ s^2 [Eqs. (1.14) and (1.15)],

$$p_b = 2.7 - \frac{82.056 \times 10^{-3} \times 288 \times 198.6^2}{1.982 \times 29 \times 101,325}\left(\frac{0.0044 \times 70}{2 \times 0.01875} + \ln\frac{2.7}{1.264}\right)$$

$$= 1.264 \text{ atm abs or } 0.264 \text{ atm gauge}$$

The average pressure is

$$\bar{p} = \frac{2.7 + 1.264}{2} = 1.982 \text{ atm}$$

Solution of Eq. (6.49) for G and differentiation with respect to pressure p_b yields a maximum mass velocity $G_{\max} = p_b\sqrt{M/(RT)}$, an exit velocity $\bar{V}_{b,\max} = \sqrt{RT/M}$, and an exit Mach number $\text{Ma}_b = 1/\sqrt{\gamma}$. This apparent choking condition, however, is not physically meaningful for isothermal flow because at such high velocities and high rates of expansion, isothermal conditions are not possible.[2]

Heat transfer in isothermal flow

The steady-flow energy equation (6.22) and Eq. (6.20) for the stagnation temperature combine to give, after noting that $T_a = T_b$,

$$\frac{Q}{m} = \frac{u_b^2 - u_a^2}{2} \tag{6.50}$$

Substituting G/ρ for u gives the mass-velocity form of Eq. (6.50),

$$\frac{Q}{m} = \frac{G^2}{2}\left(\frac{1}{\rho_b^2} - \frac{1}{\rho_a^2}\right) \tag{6.51}$$

In these equations Q/m is the heat flow into the gas in joules per kilogram. Equations (6.49) to (6.51) are used for subsonic flow only.

SYMBOLS

A	Area, m^2 or ft^2
a	Acoustic velocity in fluid, m/s or ft/s; a_a, at pipe entrance
c	Specific heat, J/g · °C or Btu/lb · °F; c_p, at constant pressure; c_v, at constant volume
D	Diameter, m or ft
f	Fanning friction factor, dimensionless; f_a, at station a; f_b, at station b; $\bar{f}$, average value
G	Mass velocity, kg/m^2 · s or lb/ft^2 · s; G_r, in receiver; G_t, at throat; G^*, value when Ma = 1.0; G_{max}, maximum value
g	Gravitational acceleration, m/s^2 or ft/s^2
g_c	Newton's law proportionality factor, 32.174 ft · lb/lb_f · s^2
H	Enthalpy, J/g or Btu/lb; H_a, at station a; H_b, at station b; H_s, stagnation value; H_{sa}, H_{sb}, at stations a and b; H_0, at reference temperature; H^*, value when Ma = 1.0
h_f	Friction loss, N · m/g or ft · lb_f/lb; h_{fs}, loss from skin friction
k	Roughness parameter, m or ft
L	Length, m or ft, L_a, from entrance to station a; L_b, to station b; L_{max}, length of pipe when Ma = 1.0 at outlet
Ma	Mach number, u/a; Ma_a, at station a; Ma_b, at station b; Ma_r, at nozzle discharge
M	Molecular weight of fluid
m	Mass, kg or lb
$\dot{m}$	Mass flow rate, kg/s or lb/s
p	Pressure, N/m^2 or lb_f/ft^2; p_a, at station a; p_b, at station b; p_r, in receiver; p_t, at throat of convergent-divergent nozzle; p_0, in reservoir; p^*, value when Ma = 1.0; $\bar{p}$, average value
Q	Quantity of heat, J or Btu
R	Gas law constant, 8.314 J/g mol · K or 1,545 ft · lb_f/lb mol · °R
r_c	Critical pressure ratio, p^*/p_0
r_H	Hydraulic radius of conduit, m or ft
S	Cross-sectional area of conduit, m^2 or ft^2
T	Temperature, K or °R; T_a, at station a; T_b, at station b; T_s, stagnation value; T_t, at nozzle throat; T_0, reference value; T^*, value when Ma = 1.0
u	Fluid velocity, m/s or ft/s; u_a, at station a; u_b, at station b; u_s, stagnation value; u_t, at throat of convergent-divergent nozzle; u^*, value when Ma = 1.0
$\bar{V}$	Average fluid velocity; $\bar{V}_{b,max}$, maximum discharge velocity
Z	Height above datum plane, m or ft

Greek Letters

α	Kinetic energy correction factor; α_a, at station a; α_b, at station b
β	Momentum correction factor
γ	Ratio of specific heats, c_p/c_v

ρ Density of fluid, kg/m³ or lb/ft³; ρ_a, ρ_b, at stations a and b; ρ_t, at throat of convergent-divergent nozzle; ρ_0, in reservoir; ρ^*, value when Ma $= 1.0$

μ Absolute viscosity, kg/m · s or lb/ft · s

Subscripts

a At station a
b At station b
S Isentropic flow
s Stagnation value
0 Reference value, reservoir conditions

PROBLEMS

6.1. For the inlet conditions of Example 6.2, what is the maximum length of pipe that may be used if the pipe is 2 in. in diameter? What is it for the inlet conditions of Example 6.3?

6.2. Natural gas consisting essentially of methane is to be transported through a 20-in.-ID pipeline over flat terrain. Each pumping station increases the pressure to 100 lb$_f$/in.² abs, and the pressure drops to 25 lb$_f$/in.² abs at the inlet to the next pumping station 50 mi away. What is the gas flow rate in cubic feet per hour measured at 60°F and 30 in. Hg pressure?

6.3. A divergent-convergent nozzle has the proportions shown in Table 6.1. Air ($\gamma = 1.40$, $M = 29.0$) enters the nozzle from a reservoir in which the pressure is 20 atm abs and the temperature is 550 K. Plot the pressure, the temperature, and the Mach number versus the length of the nozzle when (*a*) the flow rate is a maximum without shock and with subsonic discharge and (*b*) the flow is isentropic and the discharge is supersonic.

TABLE 6.1
Data in Prob. 6.3

Length		Diameter	Length		Diameter
Reservoir	0	∞	Throat	0.30	0.25
	0.025	0.875		0.40	0.28
	0.050	0.700		0.50	0.35
	0.075	0.575		0.60	0.45
	0.100	0.500		0.70	0.56
	0.150	0.375		0.80	0.68
	0.200	0.300		0.90	0.84
			Receiver	1.00	1.00

6.4. A standard 1-in. Schedule 40 horizontal steel pipe is used to conduct chlorine gas. The gas enters the pipe through a rounded entrance at a pressure of 6 atm abs, a temperature of 120°C, and a velocity of 35 m/s. (*a*) What is the maximum possible length of the pipe? (*b*) What are the pressure and stagnation temperature of the gas at the end of the pipe at maximum length? Assume adiabatic flow. For chlorine, $\gamma = 1.36$, and $M = 70.91$.

6.5. Air at 25°C enters a section of 2-in. Schedule 40 steel pipe at a gauge pressure of 310 kN/m² and a flow rate of 1,200 kg/h. Assuming isothermal flow, what is the pressure drop in 60 m of pipe?

REFERENCES

1. Cambel, A. B. *Handbook of Fluid Dynamics.* Ed. V. L. Streeter. New York: McGraw-Hill, 1961, pp. 8-5 to 8-12.
2. Perry, R. H., and D. W. Green, eds. *Perry's Chemical Engineers' Handbook.* 7th ed. New York: McGraw-Hill, 1997, p. 6-22.
3. Shapiro, A. H., and W. R. Hawthorne. *J. Appl. Mech.* 14: A-317 (1947).

CHAPTER 7

Flow past Immersed Bodies

The discussion in Chaps. 4 and 5 centered on the laws of fluid flow and factors that control changes of pressure and velocity of fluids flowing past solid boundaries and was especially concerned with flow through closed pipes and channels. Emphasis during the discussion was placed primarily on the fluid. In many problems, however, the effect of the fluid on the solid is of interest. The fluid may be at rest and the solid moving through it; the solid may be at rest and the fluid flowing past it; or both may be moving. The situation in which the solid is immersed in, and surrounded by, fluid is the subject of this chapter. It is generally immaterial which phase, solid or fluid, is assumed to be at rest, and it is the relative velocity between the two that is important. An exception to this is seen in some situations when the fluid stream has been previously influenced by solid walls and is in turbulent flow. The scale and intensity of turbulence then may be important parameters in the process.

DRAG AND DRAG COEFFICIENTS

The force in the direction of flow exerted by the fluid on the solid is called *drag*. By Newton's third law of motion, an equal and opposite net force is exerted by the body on the fluid. When the wall of the body is parallel with the direction of flow, as in the case of the thin flat plate shown in Fig. 3.9a, the only drag force is the wall shear τ_w. More generally, however, the wall of an immersed body makes an angle with the direction of flow. Then the component of the wall shear in the direction of flow contributes to drag. Another contribution comes from the fluid pressure which acts in a direction normal to the wall; drag comes from the pressure component in the direction of flow. The total drag on an element of area is the sum of the two components. An extreme example is the drag of a flat plate perpendicular to the flow, as shown in Fig. 3.9b, where the drag is due entirely to the pressure component.

150

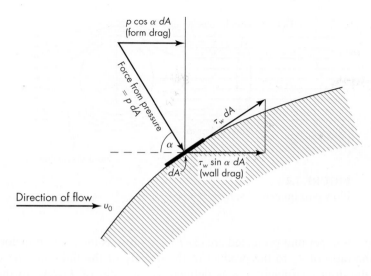

FIGURE 7.1
Wall drag and form drag on immersed body.

Figure 7.1 shows the pressure and shear forces acting on an element of area dA inclined at an angle of $90° - \alpha$ to the direction of flow. The drag from wall shear is $\tau_w \sin \alpha \, dA$, and that from pressure is $p \cos \alpha \, dA$. The total drag on the body is the sum of the integrals of these quantities, each evaluated over the entire surface of the body in contact with the fluid. The total integrated drag from wall shear is called *wall drag,* and the total integrated drag from pressure is called *form drag.*

In potential flow, $\tau_w = 0$, and there is no wall drag. Also, the pressure drag in the direction of flow is balanced by an equal force in the opposite direction, and the integral of the form drag is zero. There is no net drag in potential flow.

The phenomena causing both wall and form drag in actual fluids are complicated, and in general the drag cannot be predicted. For spheres and other regular shapes at low fluid velocities, the flow patterns and drag forces can be estimated by numerical methods using the general momentum balance equations developed in Chap. 4.[15] For irregular shapes and high velocities they are most easily determined by experiment.

Drag coefficients

In treating fluid flow through pipes and channels, a friction factor, defined as the ratio of the shear stress to the product of the velocity head and density, was shown to be useful. An analogous factor, called the *drag coefficient,* is used for immersed solids. Consider a smooth sphere immersed in a flowing fluid and at a distance from the solid boundary of the stream sufficient for the approaching stream to be at a uniform velocity. Define the projected area of the solid body as the area obtained by projecting the body on a plane perpendicular to the direction of flow, as shown in Fig. 7.2. Denote the projected area by A_p. For a sphere, the projected area is that of a great circle, or $(\pi/4)D_p^2$, where D_p is the diameter. If F_D is the total drag,

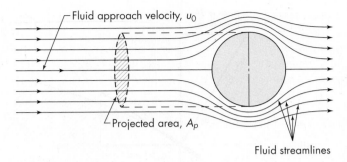

FIGURE 7.2
Flow past immersed sphere.

the average drag per unit projected area is F_D/A_p. Just as the friction factor f is defined as the ratio of τ_w to the product of the density of the fluid and the velocity head, so the drag coefficient C_D is defined as the ratio of F_D/A_p to this same product, or

$$C_D \equiv \frac{F_D/A_p}{\rho u_0^2/2} \tag{7.1}$$

where u_0 is the velocity of the approaching stream (by assumption u_0 is constant over the projected area).

For particles having shapes other than spherical, it is necessary to specify the size and geometric form of the body and its orientation with respect to the direction of flow of the fluid. One major dimension is chosen as the characteristic length, and the other important dimensions are given as ratios to the chosen one. For cylinders the diameter D is taken as the defining dimension, and the length is expressed as the L/D ratio. The orientation between the particle and the stream is also specified. For a cylinder, the angle formed by the axis of the cylinder and the direction of flow is sufficient. Then the projected area is determinate and can be calculated. For a cylinder so oriented that its axis is perpendicular to the flow, A_p is LD_p, where L is the length of the cylinder. For a cylinder with its axis parallel to the direction of flow, A_p is $(\pi/4)D_p^2$, the same as for a sphere of the same diameter.

From dimensional analysis, the drag coefficient of a smooth solid in an incompressible fluid depends upon a Reynolds number and the necessary shape ratios. For a given shape

$$C_D = \phi(\mathrm{Re}_p)$$

The Reynolds number for a particle in a fluid is defined as

$$\mathrm{Re}_p \equiv \frac{G_0 D_p}{\mu} \tag{7.2}$$

where D_p = characteristic length
$\quad G_0 = u_0\rho$

A different C_D-versus-Re_p relation exists for each shape and orientation. The relation must in general be determined experimentally, although a well-substantiated theoretical equation exists for smooth spheres at low Reynolds numbers.

Drag coefficients for compressible fluids increase with an increase in the Mach number when the latter becomes greater than about 0.6. Coefficients in supersonic flow are generally greater than in subsonic flow.

Drag coefficients of typical shapes

In Fig. 7.3, curves of C_D versus Re_p are shown for spheres, long cylinders, and disks. The curves are based on drag measurements for objects held in a fixed position in a flowing stream. The axis of the cylinder and the face of the disk are perpendicular to the direction of flow, and these curves apply only when this orientation is maintained. When a disk or cylinder falls by gravity through a quiescent fluid, it will twist and turn as it falls, giving quite different values for the drag coefficient. Spheres in free fall may rotate and follow a spiral path, giving slightly different drag coefficients than for a fixed sphere. The behavior of drops and bubbles, which can change shape as they move, is discussed in a later section.

From the complex nature of drag, it is not surprising that the variation of C_D with Re_p is more complicated than that of f with Re. The variations in slope of the curves of C_D versus Re_p at different Reynolds numbers are the result of the interplay

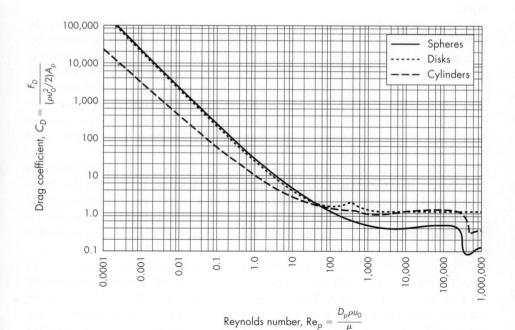

FIGURE 7.3
Drag coefficients for spheres, disks, and cylinders. [*By permission from J. H. Perry and D. W. Green (eds.), Chemical Engineers' Handbook, 7th ed., p. 6-51. Copyright, © 1997, McGraw-Hill Book Company.*]

of the various factors that control form drag and wall drag. Their effects can be followed by discussing the case of the sphere.

For low Reynolds numbers, the drag force for a sphere conforms to a theoretical equation called *Stokes' law,* which may be written

$$F_D = 3\pi\mu u_0 D_p \qquad (7.3)$$

From Eq. (7.3), the drag coefficient predicted by Stokes' law, using Eq. (7.1), is

$$C_D = \frac{24}{\mathrm{Re}_p} \qquad (7.4)$$

In theory, Stokes' law is valid only when Re_p is considerably less than unity. Practically, as shown by the left-hand portion of the graph of Fig. 7.3, Eqs. (7.3) and (7.4) may be used with small error for all Reynolds numbers less than 1. At the low velocities at which the law is valid, the sphere moves through the fluid by deforming it. The wall shear is the result of viscous forces only, and inertial forces are negligible. The motion of the sphere affects the fluid at considerable distances from the body, and if there is a solid wall within 20 or 30 diameters of the sphere, Stokes' law must be corrected for the wall effect. The type of flow treated in this law is called *creeping flow.* The law is especially valuable for calculating the resistance of small particles, such as dust or fogs, moving through gases or liquids of low viscosity or for the motion of larger particles through highly viscous liquids.

As the Reynolds number increases beyond $\mathrm{Re}_p = 1$, the flow pattern behind the sphere becomes different from that in front of the sphere, and the drag coefficient becomes greater than the Stokes' Law limit of $24/\mathrm{Re}_p$. At a Reynolds number of about 20, a zone of recirculating flow develops near the rear stagnation point. The recirculation zone or wake increases in size with increasing Reynolds number, and at $\mathrm{Re}_p = 100$ the wake covers nearly one-half of the sphere. The large eddies or vortices in the wake dissipate considerable mechanical energy and cause the pressure to be much less than the upstream pressure. This makes the form drag quite large relative to the drag caused by wall shear.

At moderate Reynolds numbers of 200 to 300, oscillations develop in the wake and vortices disengage from the wake in a regular fashion, forming in the downstream fluid a series of moving vortices or a "vortex street." A similar vortex street is formed when a fluid flows across wires or long cylinders and is responsible for the hum of telephone wires in the wind. The frequency of vibration can be used to measure flow rate, as described in Chap. 8.

For $\mathrm{Re}_p = 10^3$ to 10^5 the drag coefficient is nearly constant at 0.40 to 0.45, changing only slightly as the point of boundary layer separation slowly shifts toward the nose of the sphere. Figure 7.4a shows the flow pattern for $\mathrm{Re}_p \cong 10^5$, where the boundary layer on the front part of the sphere is still laminar and the angle of separation is 85°. When the front boundary layer becomes turbulent at $\mathrm{Re}_p \cong 300,000$, the separation point moves toward the rear of the sphere and the wake shrinks, as shown in Fig. 7.4b. The remarkable drop in the drag coefficient from 0.45 to 0.10 is the result of this decrease in the size of the wake and the corresponding decrease in form drag.[8]

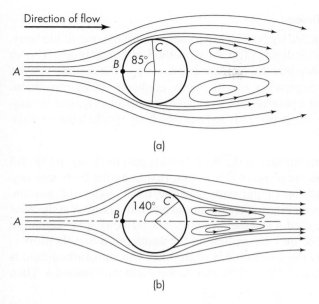

Direction of flow

A

B   85°   C

(a)

A

B   140°   C

(b)

FIGURE 7.4
Flow past single sphere, showing separation and wake formation: (*a*) laminar flow in boundary layer; (*b*) turbulent flow in boundary layer; *B*, stagnation point; *C*, separation point. [*By permission from J. C. Hunsaker and B. G. Rightmire, Engineering Application of Fluid Mechanics, pp. 202–203. Copyright, © 1947, McGraw-Hill Book Company.*]

The Reynolds number at which the attached boundary layer becomes turbulent is called the critical Reynolds number for drag. The curve for spheres shown in Fig. 7.3 applies only when the fluid approaching the sphere is nonturbulent or when the sphere is moving through a stationary static fluid. If the approaching fluid is turbulent, the critical Reynolds number is sensitive to the scale of turbulence and becomes smaller as the scale increases. For example, if the scale of turbulence, defined as $100\sqrt{(u')^2}/u$, is 2 percent, the critical Reynolds number[10] is about 140,000. One method of measuring the scale of turbulence is to determine the critical Reynolds number and use a known correlation between the two quantities.

The curve of C_D versus Re_p for an infinitely long cylinder normal to the flow is much like that for a sphere, but at low Reynolds numbers, C_D does not vary inversely with Re_p because of the two-dimensional character of the flow around the cylinder. For short cylinders, such as catalyst pellets, the drag coefficient falls between the values for spheres and long cylinders and varies inversely with the Reynolds number at very low Reynolds numbers. Disks do not show the drop in drag coefficient at a critical Reynolds number, because once the separation occurs at the edge of the disk, the separated stream does not return to the back of the disk and the wake does not shrink when the boundary layer becomes turbulent. Bodies that show this type of behavior are called *bluff bodies*. For a disk the drag coefficient C_D is approximately unity at Reynolds numbers above 2,000.

The drag coefficients for irregularly shaped particles such as coal, sand, or crushed stone are greater than for spheres of the same nominal size. Data for crushed limestone at $Re_p = 4$ to 400 are shown in Fig. 7.6, page 164. Other tests[5] of irregular solids gave drag coefficients 2 to 3 times those for spheres at Reynolds numbers of 1,000 to 2,000. Results for isometric particles such as cubes and tetrahedra have also been reported.[19]

Form drag and streamlining

Form drag can be minimized by forcing separation toward the rear of the body. This is accomplished by streamlining. The usual method of streamlining is to so proportion the rear of the body that the increase in pressure in the boundary layer, which is the basic cause of separation, is sufficiently gradual to delay separation. Streamlining usually calls for a pointed rear, like that of an airfoil. A typical stream-lined shape is shown in Fig. 7.5. A perfectly streamlined body would have no wake and little or no form drag.

Stagnation point. The streamlines in the fluid flowing past the body in Fig. 7.5 show that the fluid stream in the plane of the section is split by the body into two parts, one passing over the top of the body and the other under the bottom. Streamline AB divides the two parts and terminates at a definite point B at the nose of the body. This point is called a stagnation point. The velocity at a stagnation point is zero. Equation (4.58) may be written on the assumption that the flow is horizontal and friction along the streamline is negligible. The undisturbed fluid is identified with station a in Eq. (4.58) and the stagnation point with station b. Then

$$\frac{p_s - p_0}{\rho} = \frac{u_0^2}{2} \tag{7.5}$$

where p_s = pressure on body at stagnation point
$\quad\ \ p_0$ = pressure in undisturbed fluid
$\quad\ \ u_0$ = velocity of undisturbed fluid
$\quad\ \ \rho$ = density of fluid

The pressure increase $p_s - p_0$ for the streamline passing through a stagnation point is larger than that for any other streamline, because at that point the entire velocity head of the approaching stream is converted to pressure head.

Stagnation pressure

Equation (7.5) may be used for compressible fluids at low Mach numbers but becomes increasingly inaccurate when Ma is larger than about 0.4. The proper pressure to use for p_s then is the isentropic stagnation pressure, defined as the pressure

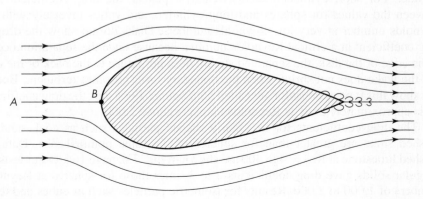

FIGURE 7.5
Streamlined body: AB, streamline to stagnation point B.

of the gas when the stream is brought to rest isentropically, and calculated from the equations of Chap. 6 as follows. From Eq. (6.24)

$$\left(\frac{p_s}{p_0}\right)^{1-1/\gamma} = \frac{T_s}{T_0} \tag{7.6}$$

where p_s = stagnation pressure
$\quad\quad T_s$ = stagnation temperature
$\quad\quad T_0$ = temperature of approaching stream

From Eq. (6.43), using $T_a = T_s$, $T_b = T_0$, $\mathrm{Ma}_a = 0$, and $\mathrm{Ma}_b = \mathrm{Ma}_0$,

$$\frac{T_s}{T_0} = 1 + \frac{\gamma - 1}{2}\mathrm{Ma}_0^2 \tag{7.7}$$

Combining Eqs. (7.6) and (7.7) gives

$$\frac{p_s}{p_0} = \left(1 + \frac{\gamma - 1}{2}\mathrm{Ma}_0^2\right)^{1/(1-1/\gamma)} \tag{7.8}$$

Equation (7.8) can be brought into the same form as Eq. (7.5) by subtracting unity from both sides and expanding the right-hand side of Eq. (7.8) by the binomial theorem. After substitution and simplification, using the relationship for Ma given in Eq. (6.18), and multiplication by p_0/ρ_0, the result is

$$\frac{p_s - p_0}{\rho_0} = \frac{u_0^2}{2}\left(1 + \frac{\mathrm{Ma}_0^2}{4} + \frac{2 - \gamma}{24}\mathrm{Ma}_0^4 + \cdots\right) \tag{7.9}$$

Comparison of Eq. (7.9) with Eq. (7.5) shows that the quantity in the parentheses is the correction factor for converting Eq. (7.5) to a form suitable for compressible fluids in the range $0 \le \mathrm{Ma} < 1.0$.

FLOW THROUGH BEDS OF SOLIDS

In many technical processes, liquids or gases flow through beds of solid particles. Important examples are filtration and the two-phase countercurrent flow of liquid and gas through packed towers. In filtration, the bed of solids consists of small particles that are removed from the liquid by a filter cloth or fine screen. In other equipment, such as ion-exchange or catalytic reactors, a single fluid (liquid or gas) flows through a bed of granular solids. Filtration is discussed in Chap. 29 and packed towers in Chap. 18. The present treatment is restricted to the flow of a single fluid phase through a column of stationary solid particles.

The resistance to the flow of a fluid through the voids in a bed of solids is the resultant of the total drag of all the particles in the bed. Depending on the Reynolds number, $D_p G_0/\mu$, laminar flow, turbulent flow, form drag, separation, and wake formation occur. As in the drag of a single solid particle, there is no sharp transition between laminar and turbulent flow like that occurring in flow through pipes and channels of constant cross section.

The most common methods of calculating the pressure drop through a bed of solids are based on estimates of total drag on the solid boundaries of the tortuous channels through the bed particles. The actual channels are irregular in shape, have a variable cross section and orientation, and are highly interconnected. However, to calculate an equivalent channel diameter, it is assumed that the bed has a set of uniform circular channels whose total surface area and void volume match those of the bed. The total surface area is the surface area per particle times the number of particles, but it is more convenient to base the calculation on the volume fraction particles in the bed and the surface-volume ratio for the particles. This ratio is $6/D_p$ for a sphere, since $s_p = \pi D_p^2$ and $v_p = \frac{1}{6}\pi D_p^3$. For other shapes or irregular particles, the equation for surface-volume ratio includes a sphericity Φ_s, defined as the surface-volume ratio for a sphere of diameter D_p divided by the surface-volume ratio for the particle whose nominal size is D_p. Thus

$$\Phi_s = \frac{6/D_p}{s_p/v_p}$$

or

$$\frac{s_p}{v_p} = \frac{6}{\Phi_s D_p} \tag{7.10}$$

Values of the sphericity for several materials are given in Table 7.1.

The volume fraction particles in the bed is $1 - \varepsilon$, where ε is the porosity or void fraction. If the particles are porous, the pores are generally too small to permit any significant flow through them, so ε is taken to be the external void fraction of the bed and not the total porosity.

To determine the equivalent channel diameter D_{eq}, the surface area for n parallel channels of length L is set equal to the surface-volume ratio times the particle volume $S_0 L(1 - \varepsilon)$, where S_0 is the cross-sectional area of the bed:

$$n\pi D_{eq} L = S_0 L(1 - \varepsilon)\frac{6}{\Phi_s D_p} \tag{7.11}$$

The void volume in the bed is the same as the total volume of the n channels:

TABLE 7.1
Sphericity of miscellaneous materials[†]

Material	Sphericity	Material	Sphericity
Spheres, cubes, short cylinders ($L = D_p$)	1.0	Ottawa sand	0.95
		Rounded sand	0.83
Raschig rings ($L = D_p$)		Coal dust	0.73
$L = D_o, D_i = 0.5D_o$	0.58[‡]	Flint sand	0.65
$L = D_o, D_i = 0.75D_o$	0.33[‡]	Crushed glass	0.65
Berl saddles	0.3	Mica flakes	0.28

[†]By permission, from J. H. Perry (ed.), *Chemical Engineers' Handbook*, 6th ed., p. 5-54, McGraw-Hill Book Company, New York, 1984.
[‡]Calculated value.

$$S_0 L \varepsilon = \tfrac{1}{4} n \pi D_{eq}^2 L \tag{7.12}$$

Combining Eqs. (7.11) and (7.12) gives an equation for D_{eq}:

$$D_{eq} = \frac{2}{3} \Phi_s D_p \frac{\varepsilon}{1 - \varepsilon} \tag{7.13}$$

For the typical void fraction of 0.4, $D_{eq} = 0.44\Phi_s D_p$, or the equivalent diameter is roughly one-half the particle size.

The pressure drop depends on the average velocity in the channels $\bar{V}$, which is proportional to the superficial or empty-tower velocity $\bar{V}_0$ and inversely proportional to the porosity:

$$\bar{V} = \frac{\bar{V}_0}{\varepsilon} \tag{7.14}$$

With the average velocity and channel size now expressed in terms of the measurable parameters $\bar{V}_0$, D_p, and ε, the channel model can be used to predict the form of the correlation for pressure drop. For flow at very low Reynolds numbers, the pressure drop should vary with the first power of the velocity and inversely with the square of the channel size, in accordance with the Hagen–Poiseuille equation for laminar flow in straight tubes, Eq. (5.20). The equations for $\bar{V}$ and D_{eq} are used in Eq. (5.20), and a correction factor λ_1 is added to account for the fact that the channels are actually tortuous and not straight and parallel:

$$\frac{\Delta p}{L} = \frac{32\bar{V}\mu}{D^2} = \frac{32\lambda_1 \bar{V}_0 \mu}{\tfrac{4}{9}\varepsilon \Phi_s^2 D_p^2} \frac{(1 - \varepsilon)^2}{\varepsilon^2} \tag{7.15}$$

or

$$\frac{\Delta p}{L} = \frac{72\lambda_1 \bar{V}_0 \mu}{\Phi_s^2 D_p^2} \frac{(1 - \varepsilon)^2}{\varepsilon^3} \tag{7.16}$$

Several studies have shown that the form of Eq. (7.16) is correct, and experiments give an empirical constant of 150:

$$\frac{\Delta p}{L} = \frac{150\bar{V}_0 \mu}{\Phi_s^2 D_p^2} \frac{(1 - \varepsilon)^2}{\varepsilon^3} \tag{7.17}$$

Equation (7.17) is called the *Kozeny–Carman* equation and is applicable for flow through beds at particle Reynolds numbers up to about 1.0. There is no sharp transition to turbulent flow at this Reynolds number, but the frequent changes in shape and direction of the channels in the bed lead to significant kinetic energy losses at higher Reynolds numbers. The constant 150 corresponds to $\lambda_1 = 2.1$, which is a reasonable value for the tortuosity factor. For a given system, Eq. (7.17) indicates that the flow is proportional to the pressure drop and inversely proportional to the fluid viscosity. This statement is also known as *Darcy's law,* which is often used to describe flow of liquids through porous media.

As the flow rate through a packed bed increases, the slope of the Δp-versus-$\bar{V}_0$ plot gradually increases, and at very high Reynolds numbers, Δp varies with the 1.9

or 2.0 power of the superficial velocity. If the usual equation for turbulent flow in pipes [Eq. (5.7)] is applied to the packed bed, with a correction factor λ_2 to account for the tortuosity, the predicted equation is

$$\frac{\Delta p}{L} = \frac{2 f \rho \bar{V}^2}{D_{eq}} = 2 f \lambda_2 \rho \left(\frac{\bar{V}_0}{\varepsilon}\right)^2 \frac{3(1-\varepsilon)}{2\Phi_s D_p \varepsilon} \qquad (7.18)$$

or

$$\frac{\Delta p}{L} = \frac{3 f \lambda_2 \rho \bar{V}_0^2}{\Phi_s D_p} \frac{1-\varepsilon}{\varepsilon^3} \qquad (7.19)$$

An empirical correlation for pressure drop in packed beds at high Reynolds number ($Re_p > 1,000$) is the Burke–Plummer equation

$$\frac{\Delta p}{L} = \frac{1.75 \rho \bar{V}_0^2}{\Phi_s D_p} \frac{1-\varepsilon}{\varepsilon^3} \qquad (7.20)$$

Although this equation has the same form as Eq. (7.19), the constant 1.75 is much higher than expected based on friction factors for pipe flow. If Re_p is 10^4 and Re based on D_{eq} is therefore about 4,000, the friction factor for smooth pipe is $f \cong 0.01$ (Fig. 5.10). This means that the correction factor λ_2 is $1.75/0.03 = 58$, which is much too large to explain by the tortuosity of the channels or roughness of the particle surfaces. The main contribution to the pressure drop must be the kinetic energy losses caused by changes in channel cross section and flow direction. As the fluid passes between particles, the channel becomes smaller and then larger, and the maximum velocity is much greater than the average velocity. Since the channel area changes rapidly, most of the kinetic energy of the fluid is lost as an expansion loss [see Eq. (5.66)].

To emphasize the magnitude of the kinetic energy losses, the pressure drop from Eq. (7.20) can be divided by $\rho \bar{V}^2/2$ to get the number of velocity heads. Using the average velocity $\bar{V}_0/\varepsilon$ for $\bar{V}$ gives

$$\frac{\Delta p}{(\rho/2)(\bar{V}_0/\varepsilon)^2} = 2 \times 1.75 \left(\frac{1-\varepsilon}{\varepsilon}\right) \frac{L}{\Phi_s D_p} \qquad (7.21)$$

For a typical void fraction of 0.4 and a bed of spheres, the pressure drop corresponds to a loss of 5.25 velocity heads for each layer of particles ($2 \times 1.75 \times 0.6/0.4$). This number is greater than 1.0, because the local velocity at narrow places in the flow channels may be 2 to 3 times the average velocity $\bar{V}$, and much kinetic energy is lost as the fluid passes through each layer of particles.

An equation covering the entire range of flow rates can be obtained by assuming that the viscous losses and the kinetic energy losses are additive. The result is called the *Ergun* equation:

$$\frac{\Delta p}{L} = \frac{150 \bar{V}_0 \mu}{\Phi_s^2 D_p^2} \frac{(1-\varepsilon)^2}{\varepsilon^3} + \frac{1.75 \rho \bar{V}_0^2}{\Phi_s D_p} \frac{1-\varepsilon}{\varepsilon^3} \qquad (7.22)$$

Ergun showed that Eq. (7.22) fitted data for spheres, cylinders, and crushed solids over a wide range of flow rates.[1] He also varied the packing density for some

TABLE 7.2
Void fractions for dumped packings

D_p/D_t	ε for spheres	ε for cylinders
~0	0.34	0.34
0.1	0.38	0.35
0.2	0.42	0.39
0.3	0.46	0.45
0.4	0.50	0.53
0.5	0.55	0.60

materials to verify the $(1 - \varepsilon)^2/\varepsilon^3$ term for the viscous loss part of the equation and the $(1 - \varepsilon)/\varepsilon^3$ term for the kinetic energy part. Note that a small change in ε has a very large effect on Δp, which makes it difficult to predict Δp accurately and to reproduce experimental values after a bed is repacked.

The void fractions for spheres, cylinders, and granular packings generally range from 0.3 to 0.6 depending on the ratio of particle size to tube size and the method of packing. Some data by Leva and Grummer[13] for dumped packings are given in Table 7.2. Vibrating the bed results in void fractions a few percent lower than for dumped packing. Particles with a rough surface had void fractions a few percent larger than for smooth particles. With a distribution of particle sizes, the void fractions are lower than for uniform particles.

For Raschig rings and Berl saddles, which have porosities of 0.60 to 0.75, Eq. (7.22) predicts pressure drops lower than those found experimentally. For these materials and other packings of high surface area and high porosity, the pressure drop should be obtained using the packing factors in Table 18.1 or information from the supplier.

Mixtures of particles

Equation (7.22) can be used for beds consisting of a mixture of different particle sizes by using, in place of D_p, the surface-mean diameter of the mixture $\bar{D}_s$. This mean may be calculated from the number of particles N_i in each size range or from the mass fraction in each size range x_i:

$$\bar{D}_s = \frac{\sum\limits_{i=1}^{n} N_i \bar{D}_{pi}^3}{\sum\limits_{i=1}^{n} N_i \bar{D}_{pi}^2} \qquad (7.23)$$

$$\bar{D}_s = \frac{1}{\sum\limits_{i=1}^{n} (x_i / \bar{D}_{pi})} \qquad (7.24)$$

Compressible fluids. When the density change of the fluid is small—and seldom is the pressure drop large enough to change the density greatly—Eq. (7.22) may be used by calculating the inlet and outlet values of $\bar{V}_0$ and using the arithmetic mean for $\bar{V}_0$ in the equation.

MOTION OF PARTICLES THROUGH FLUIDS

Many processing steps, especially mechanical separations, involve the movement of solid particles or liquid drops through a fluid. The fluid may be gas or liquid, and it may be flowing or at rest. Examples are the elimination of dust and fumes from air or flue gas, the removal of solids from liquid wastes, and the recovery of acid mists from the waste gas of an acid plant.

Mechanics of particle motion

The movement of a particle through a fluid requires an external force acting on the particle. This force may come from a density difference between the particle and the fluid, or it may be the result of electric or magnetic fields. In this section only gravitational or centrifugal forces, which arise from density differences, are considered.

Three forces act on a particle moving through a fluid: (1) the external force, gravitational or centrifugal; (2) the buoyant force, which acts parallel with the external force but in the opposite direction; and (3) the drag force, which appears whenever there is relative motion between the particle and the fluid. The drag force acts to oppose the motion and acts parallel with the direction of movement but in the opposite direction.

In the general case, the direction of movement of the particle relative to the fluid may not be parallel with the direction of the external and buoyant forces, and the drag force then makes an angle with the other two. In this situation, which is called *two-dimensional motion,* the drag must be resolved into components, and this complicates the treatment of particle mechanics. Equations are available for two-dimensional motion,[11] but only the one-dimensional case, where the lines of action of all forces acting on the particle are collinear, is considered in this book.

Equations for one-dimensional motion of particle through fluid

Consider a particle of mass m moving through a fluid under the action of an external force F_e. Let the velocity of the particle relative to the fluid be u. Let the buoyant force on the particle be F_b, and let the drag be F_D. Then the resultant force on the particle is $F_e - F_b - F_D$, the acceleration of the particle is du/dt, and by Eq. (1.35), since m is constant,

$$m \frac{du}{dt} = F_e - F_b - F_D \tag{7.25}$$

The external force can be expressed as a product of the mass and the acceleration a_e of the particle from this force, and

$$F_e = ma_e \tag{7.26}$$

The buoyant force is, by Archimedes's principle, the product of the mass of the fluid displaced by the particle and the acceleration from the external force. The volume of the particle is m/ρ_p, where ρ_p is the density of the particle, and the particle displaces this same volume of fluid. The mass of fluid displaced is $(m/\rho_p)\rho$, where ρ is the density of the fluid. The buoyant force is then

$$F_b = \frac{m\rho a_e}{\rho_p} \qquad (7.27)$$

The drag force, is, from Eq. (7.1),

$$F_D = \frac{C_D u_0^2 \rho A_p}{2} \qquad (7.28)$$

where C_D = dimensionless drag coefficient
   A_p = projected area of particle measured in plane perpendicular to direction of motion of particle
   $u_0 = u$

Substituting the forces from Eqs. (7.26) to (7.28) into Eq. (7.25) gives

$$\frac{du}{dt} = a_e - \frac{\rho a_e}{\rho_p} - \frac{C_D u^2 \rho A_p}{2m} = a_e \frac{\rho_p - \rho}{\rho_p} - \frac{C_D u^2 \rho A_p}{2m} \qquad (7.29)$$

Motion from gravitational force. If the external force is gravity, a_e is g, the acceleration of gravity, and Eq. (7.29) becomes

$$\frac{du}{dt} = g\frac{\rho_p - \rho}{\rho_p} - \frac{C_D u^2 \rho A_p}{2m} \qquad (7.30)$$

Motion in a centrifugal field. A centrifugal force appears whenever the direction of movement of a particle is changed. The acceleration from a centrifugal force from circular motion is

$$a_e = r\omega^2 \qquad (7.31)$$

where r = radius of path of particle
   ω = angular velocity, rad/s

Substituting into Eq. (7.29) gives

$$\frac{du}{dt} = r\omega^2 \frac{\rho_p - \rho}{\rho_p} - \frac{C_D u^2 \rho A_p}{2m} \qquad (7.32)$$

In this equation, u is the velocity of the particle relative to the fluid and is directed outwardly along a radius.

Terminal velocity

In gravitational settling, g is constant. Also, the drag always increases with velocity. Equation (7.30) shows that the acceleration decreases with time and approaches zero. The particle quickly reaches a constant velocity, which is the maximum attainable under the circumstances and which is called the *terminal velocity.* The equation for the terminal velocity u_t is found, for gravitational settling, by taking $du/dt = 0$. Then from Eq. (7.30),

$$u_t = \sqrt{\frac{2g(\rho_p - \rho)m}{A_p \rho_p C_D \rho}} \tag{7.33}$$

In motion from a centrifugal force, the velocity depends on the radius, and the acceleration is not constant if the particle is in motion with respect to the fluid. In many practical uses of centrifugal force, however, du/dt is small in comparison with the other two terms in Eq. (7.32); and if du/dt is neglected, a terminal velocity at any given radius can be defined by the equation

$$u_t = \omega \sqrt{\frac{2r(\rho_p - \rho)m}{A_p \rho_p C_D \rho}} \tag{7.34}$$

Drag coefficient

The quantitative use of Eqs. (7.29) to (7.34) requires that numerical values be available for the drag coefficient C_D. Figure 7.3, which shows the drag coefficient as a function of Reynolds number, indicates such a relationship. Part of the curve of C_D versus Re_p for spheres is reproduced in Fig. 7.6, as well as a curve for crushed limestone particles.[3] These curves, however, apply only under restricted conditions. The particle must be solid; it must be far from other particles and from the vessel walls,

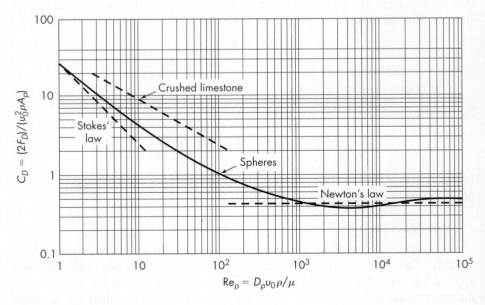

FIGURE 7.6
Drag coefficients for spheres and irregular particles.[3]

so that the flow pattern around the particle is not distorted; and it must be moving at its terminal velocity with respect to the fluid. The drag coefficients for accelerating particles are appreciably greater than those shown in Fig. 7.6, so a particle dropped in a still fluid takes longer to reach terminal velocity than would be predicted[7] using the steady-state values of C_D. Particles injected into a fast-flowing stream also accelerate more slowly than expected, and the drag coefficients in this case are therefore less than the normal values. However, for most processes involving small particles or drops, the time for acceleration to the terminal velocity is still quite small and is often ignored in analysis of the process.[9]

Variations in particle shape can be accounted for by obtaining separate curves of C_D versus Re_p for each shape, as shown in Fig. 7.3 for cylinders and disks. As pointed out earlier, however, the curves for cylinders and disks in Fig. 7.3 apply only to a specified orientation of the particle. In the free motion of nonspherical particles through a fluid, the orientation is constantly changing. This change consumes energy, increasing the effective drag on the particle, and C_D is greater than for the motion of the fluid past a fixed particle. For example, the drag coefficients of crushed lime-stone, as shown in Fig. 7.6, are more than twice as large as those for spheres of the same nominal diameter. As a result, the terminal velocity, especially with disks and other platelike particles, is less than would be predicted from curves for a fixed par-ticle orientation.

In the following treatment the particles are assumed to be spherical, for once the drag coefficients for free-particle motion are known, the same principles apply to any shape.[4,19]

When the particle is at sufficient distance from the boundaries of the container and from other particles, so that its fall is not affected by them, the process is called *free settling*. If the motion of the particle is impeded by other particles, which will happen when the particles are near one another even though they may not actually be colliding, the process is called *hindered settling*. The drag coefficient in hindered settling is greater than that in free settling.

If the particles are very small, brownian movement appears. This is a random motion imparted to the particle by collisions between the particle and the molecules of the surrounding fluid. This effect becomes appreciable at a particle size of about 2 to 3 μm and predominates over the force of gravity with a particle size of 0.1 μm or less. The random movement of the particle tends to suppress the effect of the force of gravity, so settling does not occur. Application of centrifugal force reduces the relative effect of brownian movement.

Motion of spherical particles

If the particles are spheres of diameter D_p,

$$m = \tfrac{1}{6}\pi D_p^3 \rho_p \tag{7.35}$$

and

$$A_p = \tfrac{1}{4}\pi D_p^2 \tag{7.36}$$

Substitution of m and A_p from Eqs. (7.35) and (7.36) into Eq. (7.33) gives the equation for gravity settling of spheres

$$u_t = \sqrt{\frac{4g(\rho_p - \rho)D_p}{3C_D\rho}} \tag{7.37}$$

In the general case, the terminal velocity can be found by trial and error after guessing Re_p to get an initial estimate of C_D. For the limiting cases of very low or very high Reynolds numbers, equations can be used to get u_t directly.

At low Reynolds numbers, the drag coefficient varies inversely with Re_p, and the equations for C_D, F_D, and u_t are

$$C_D = \frac{24}{Re_p} \tag{7.38}$$

$$F_D = 3\pi \mu u_t D_p \tag{7.39}$$

$$u_t = \frac{g D_p^2 (\rho_p - \rho)}{18\mu} \tag{7.40}$$

Equation (7.40), like Eq. (7.3), is a form of Stokes' law, which applies when the particle Reynolds number is less than 1.0. At $Re_p = 1.0$, $C_D = 26.5$ instead of 24, as predicted from Eq. (7.38), and since the terminal velocity depends of the square root of the drag coefficient, Stokes' law is about 5 percent in error at this point. Equation (7.40) can be modified to predict the settling velocity of a small sphere in a centrifugal field by substituting $r\omega^2$ for g.

For $1{,}000 < Re_p < 200{,}000$, the drag coefficient is approximately constant, and the equations are

$$C_D = 0.44 \tag{7.41}$$

$$F_D = 0.055\pi D_p^2 u_t^2 \rho \tag{7.42}$$

$$u_t = 1.75 \sqrt{\frac{g D_p (\rho_p - \rho)}{\rho}} \tag{7.43}$$

Equation (7.43) is Newton's law and applies only for fairly large particles falling in gases or low-viscosity fluids.

As shown by Eqs. (7.40) and (7.43), the terminal velocity u_t varies with D_p^2 in the Stokes' law range, whereas in Newton's law range it varies with $D_p^{0.5}$.

Criterion for settling regime

To identify the range in which the motion of the particle lies, the velocity term is eliminated from the Reynolds number by substituting u_t from Eq. (7.40) to give, for the Stokes' law range,

$$Re_p = \frac{D_p u_t \rho}{\mu} = \frac{D_p^3 g \rho (\rho_p - \rho)}{18\mu^2} \tag{7.44}$$

If Stokes' law is to apply, Re_p must be less than 1.0. To provide a convenient criterion K, let

$$K = D_p \left[\frac{g\rho(\rho_p - \rho)}{\mu^2} \right]^{1/3} \tag{7.45}$$

Then, from Eq. (7.44), $Re_p = \frac{1}{18}K^3$. Setting Re_p equal to 1.0 and solving give $K = 18^{1/3} = 2.6$. If the size of the particle is known, K can be calculated from Eq. (7.45). If K so calculated is less than 2.6, Stokes' law applies.

Substitution for u_t from Eq. (7.43) shows that for Newton's law range $Re_p = 1.75K^{1.5}$. Setting this equal to 1,000 and solving give $K = 68.9$. Thus if K is greater than 68.9 but less than 2,360, Newton's law applies. When K is greater than 2,360, the drag coefficient may change abruptly with small changes in fluid velocity. Under these conditions, as well as in the range between Stokes' law and Newton's law $(2.6 < K < 68.9)$, the terminal velocity is calculated from Eq. (7.37) using a value of C_D found by trial from Fig. 7.6.

EXAMPLE 7.1. (*a*) Estimate the terminal velocity for 80- to 100-mesh particles of limestone ($\rho_p = 2,800$ kg/m^3) falling in water at 30°C. (*b*) How much higher would the velocity be in a centrifugal separator where the acceleration is 50g?

Solution

(*a*) From App. 5,

$$D_p \text{ for 100-mesh} = 0.147 \text{ mm}$$

$$D_p \text{ for 80-mesh} = 0.175 \text{ mm}$$

$$\text{Average diameter } \bar{D}_p = 0.161 \text{ mm}$$

From App. 6, $\mu = 0.801$ cP and $\rho = 62.16$ lb/ft^3 or 995.7 kg/m^3. To find which settling law applies, calculate criterion K [Eq. (7.45)]:

$$K = 0.161 \times 10^{-3} \left[\frac{9.80665 \times 995.7(2,800 - 995.7)}{(0.801 \times 10^{-3})^2} \right]^{1/3}$$

$$= 4.86$$

This is slightly above the Stokes' law range. Assume $Re_p = 5$; then from Fig. 7.6, $C_D \cong 14$, and from Eq. (7.37)

$$u_t = \left[\frac{4 \times 9.80665(2,800 - 995.7)(0.161 \times 10^{-3})}{3 \times 14 \times 995.7} \right]^{1/2}$$

$$= 0.0165 \text{ m/s}$$

Check:

$$Re_p = \frac{0.161 \times 10^{-3} \times 0.0165 \times 995.7}{0.801 \times 10^{-3}} = 3.30$$

Since C_D at $Re_p = 3.30$ is greater than 14, the revised u_t and Re_p will be less than the above values, so guess a lower value of Re_p. Guess

$$Re_p = 2.5 \qquad C_D \cong 20$$

$$u_t = 0.0165 \left(\tfrac{14}{20} \right)^{0.5} = 0.0138 \text{ m/s}$$

$$Re_p = 3.30 \left(\frac{0.0138}{0.0165} \right) = 2.76$$

This is close enough to the value of 2.5, and

$$u_t \cong 0.014 \text{ m/s}$$

(b) Using $a_e = 50g$ in place of g in Eq. (7.45), since only the acceleration changes, $K = 4.86 \times 50^{1/3} = 17.90$. This is still in the intermediate settling range. Estimate $\text{Re}_p = 40$; from Fig. 7.6, $C_D = 4.1$ and

$$u_t = \left[\frac{4 \times 9.80665 \times 50(2,800 - 995.7)(0.161 \times 10^{-3})}{3 \times 4.1 \times 995.7} \right]^{1/2}$$

$$= 0.216 \text{ m/s}$$

Check:

$$\text{Re}_p = \frac{0.161 \times 10^{-3} \times 0.216 \times 995.7}{0.801 \times 10^{-3}} = 43, \text{ close to } 40$$

$$u_t \cong 0.22 \text{ m/s}$$

The calculated terminal velocities are 1.6 to 1.8 times those for a sphere of the same size, but the values are only approximate because the particle shape is unknown.

Hindered settling

In hindered settling, the velocity gradients around each particle are affected by the presence of nearby particles, so the normal drag correlations do not apply. Also, the particles in settling displace liquid, which flows upward and makes the particle velocity relative to the fluid greater than the absolute settling velocity. For a uniform suspension, the settling velocity u_s can be estimated from the terminal velocity for an isolated particle using the empirical equation of Maude and Whitmore[16]

$$u_s = u_t(\varepsilon)^n \tag{7.46}$$

Exponent n changes from about 4.6 in the Stokes' law range to about 2.5 in Newton's law region, as shown in Fig. 7.7. For very small particles, the calculated ratio u_s/u_t is 0.62 for $\varepsilon = 0.9$ and 0.095 for $\varepsilon = 0.6$. With large particles the

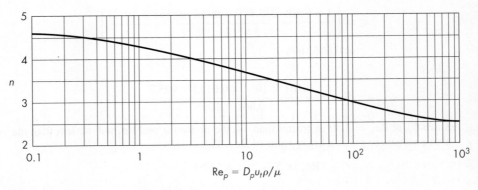

FIGURE 7.7
Plot of exponent n versus Re_p for Eq. (7.46).

corresponding ratios are $u_s/u_t = 0.77$ and 0.28; the hindered settling effect is not as pronounced because the boundary layer thickness is a smaller fraction of the particle size. In any case, Eq. (7.46) should be used with caution, since the settling velocity also depends on the particle shape and size distribution. Experimental data are needed for accurate design of a settling chamber.

If particles of a given size are falling through a suspension of much finer solids, the terminal velocity of the larger particles should be calculated using the density and viscosity of the fine suspension. Equation (7.46) may then be used to estimate the settling velocity with ε taken as the volume fraction of the fine suspension, not the total void fraction. Suspensions of very fine sand in water are used in separating coal from heavy minerals, and the density of the suspension is adjusted to a value slightly greater than that of coal to make the coal particles rise to the surface, while the mineral particles sink to the bottom.

EXAMPLE 7.2. Particles of sphalerite (specific gravity 4.00) are settling under the force of gravity in carbon tetrachloride (CCl_4) at 20°C (specific gravity 1.594). The diameter of the sphalerite particles is 0.004 in. (0.10 mm). The volume fraction of sphalerite in CCl_4 is 0.20. What is the settling velocity of the sphalerite?

Solution. The specific-gravity difference between particles and liquid is $4.00 - 1.594 = 2.406$. The density difference $\rho_p - \rho$ is $62.37 \times 2.406 = 150.06$ lb/ft^3. The density of the CCl_4 is $62.37 \times 1.594 = 99.42$ lb/ft^3. The viscosity of CCl_4 at 20°C, from App. 9, is 1.03 cP. Criterion K, from Eq. (7.45), is

$$K = \frac{0.004}{12}\left[\frac{32.174 \times 99.42 \times 150.06}{(1.03 \times 6.72 \times 10^{-4})^2}\right]^{1/3} = 3.34 \tag{7.47}$$

The settling is almost in the Stokes' law range. The terminal velocity of a free-settling sphalerite particle would be, from Eq. (7.40),

$$u_t = \frac{32.174 \times (0.004/12)^2 \times 150.06}{18 \times 1.03 \times 6.72 \times 10^{-4}}$$

$$= 0.043 \text{ ft/s}$$

The terminal velocity in hindered settling is found from Eq. (7.46). The particle Reynolds number is

$$\text{Re}_p = \frac{0.004 \times 0.043 \times 99.42}{12 \times 1.03 \times 6.72 \times 10^{-4}} = 2.06$$

From Fig. 7.7, $n = 4.1$. From Eq. (7.46), $u_s = 0.043 \times 0.8^{4.1} = 0.017$ ft/s (5.2 mm/s).

Settling and rise of bubbles and drops

Unlike solid particles, dispersed drops of liquid or bubbles of gas may change shape as they move through a continuous phase. Form drag tends to flatten the drops, but the surface tension opposes this force. Because of their large surface energy per unit volume, drops or bubbles smaller than about 0.5 mm are nearly spherical and have about the same drag coefficients and terminal velocities as solid spheres. The coefficient is not exactly the same because skin friction tends to set up circulation patterns inside a falling drop, and movement of the gas-liquid interface makes the total drag somewhat less than that for a rigid sphere. However, impurities that concentrate at the interface inhibit motion of the interface, and the lower drag coefficients are usually noticed only in very pure systems.

Drops from one to a few millimeters in diameter, typical of falling raindrops, are somewhat flattened in the direction of flow and fall more slowly than a sphere of the same volume. (The familiar teardrop shape of the cartoonist is entirely imaginary.) With further increase in size, the drops become flattened ellipsoids or may oscillate from oblate to prolate form. The drag coefficient increases with Reynolds number, and the terminal velocity may go through a maximum with increasing drop size. This is shown in Fig. 7.8 for air bubbles moving relative to turbulently flowing water. The relative velocities are said to be slightly higher than those for quiescent liquid. Various published results, however, for single air bubbles in water do not agree well with one another, probably because of differences in water purity, wall effects, and measurement techniques. A stream of bubbles formed in rapid succession at a central nozzle rises more rapidly than a single bubble, since the bubbles cause an upward flow of liquid in the central region. A similar effect is found for bubbles formed at a vertical electrode in an electrolysis cell. Bubbles in a swarm distributed uniformly over the cross section of the apparatus generally rise more slowly than single bubbles because of the hindered settling effect. In some cases higher average velocities have

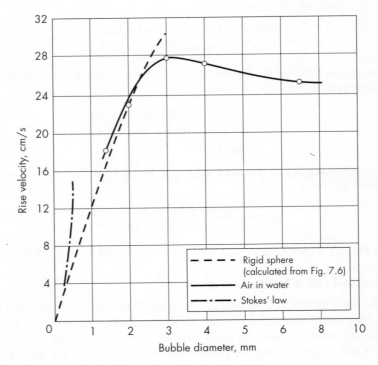

FIGURE 7.8
Rise velocity of air bubbles in water at 70°F. [*By permission, data taken from J. L. L. Baker and B. T. Chao, AIChE J., 11:268 (1965).*]

been found for swarms of bubbles in a small column, but this may have been due to occasional large bubbles or slugs of gas rising up the center.[6] Further work on bubble and drop phenomena is reviewed by Tavlarides et al.[21]

FLUIDIZATION

When a liquid or a gas is passed at very low velocity up through a bed of solid particles, the particles do not move, and the pressure drop is given by the Ergun equation (7.22). If the fluid velocity is steadily increased, the pressure drop and the drag on individual particles increase, and eventually the particles start to move and become suspended in the fluid. The terms *fluidization* and *fluidized bed* are used to describe the condition of fully suspended particles, since the suspension behaves as a dense fluid. If the bed is tilted, the top surface remains horizontal and large objects will either float or sink in the bed depending on their density relative to the suspension. The fluidized solids can be drained from the bed through pipes and valves just as a liquid can, and this fluidity is one of the main advantages of the use of fluidization for handling solids.

Conditions for fluidization

Consider a vertical tube partly filled with a fine granular material such as catalytic cracking catalyst, as shown schematically in Fig. 7.9. The tube is open at the top and has a porous plate at the bottom to support the bed of catalyst and to distribute the flow uniformly over the entire cross section. Air is admitted below the

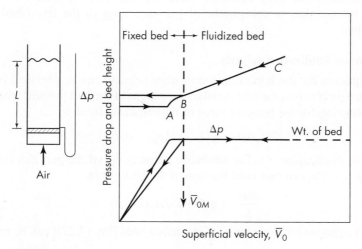

FIGURE 7.9
Pressure drop and bed height versus superficial velocity for a bed of solids.

$$Eq\ 7.17 \Rightarrow \frac{\Delta P}{L} = \frac{150\, \bar{V}_0\, \mu}{\phi_s^2\, D_p^2}\, \frac{(1-\varepsilon)^2}{\varepsilon^3}$$

distributor plate at a low flow rate and passes upward through the bed without causing any particle motion. If the particles are quite small, flow in the channels between the particles will be laminar and the pressure drop across the bed will be proportional to the superficial velocity $\bar{V}_0$ [Eq. (7.17)]. As the velocity is gradually increased, the pressure drop increases, but the particles do not move and the bed height remains the same. At a certain velocity, the pressure drop across the bed counterbalances the force of gravity on the particles or the weight of the bed, and any further increase in velocity causes the particles to move. This is point A on the graph. Sometimes the bed expands slightly with the grains still in contact, since just a slight increase in ε can offset an increase of several percent in $\bar{V}_0$ and keep Δp constant. With a further increase in velocity, the particles become separated enough to move about in the bed, and true fluidization begins (point B).

Once the bed is fluidized, the pressure drop across the bed stays constant, but the bed height continues to increase with increasing flow. The bed can be operated at quite high velocities with very little or no loss of solids, since the superficial velocity needed to support a bed of particles is much less than the terminal velocity for individual particles, as will be shown later.

If the flow rate to the fluidized bed is gradually reduced, the pressure drop remains constant, and the bed height decreases, following the line BC which was observed for increasing velocities. However, the final bed height may be greater than the initial value for the fixed bed, since solids dumped in a tube tend to pack more tightly than solids slowly settling from a fluidized state. The pressure drop at low velocities is then less than that in the original fixed bed. On starting up again, the pressure drop offsets the weight of the bed at point B, and this point, rather than point A, should be considered to give the minimum fluidization velocity $\bar{V}_{OM}$. To measure $\bar{V}_{OM}$, the bed should be fluidized vigorously and allowed to settle with the gas turned off, and the flow rate increased gradually until the bed starts to expand. More reproducible values of $\bar{V}_{OM}$ can sometimes be obtained from the intersection of the graphs of pressure drop in the fixed bed and the fluidized bed.

Minimum fluidization velocity

An equation for the minimum fluidization velocity can be obtained by setting the pressure drop across the bed equal to the weight of the bed per unit area of cross section, allowing for the buoyant force of the displaced fluid:

$$\Delta p = g(1 - \varepsilon)(\rho_p - \rho)L \tag{7.48}$$

At incipient fluidization, ε is the minimum porosity ε_M. (If the particles themselves are porous, ε is the external void fraction of the bed.) Thus

$$\frac{\Delta p}{L} = g(1 - \varepsilon_M)(\rho_p - \rho) \tag{7.49}$$

The Ergun equation for pressure drop in packed beds [Eq. (7.22)] can be rearranged to

$$\frac{\Delta p}{L} = \frac{150\mu \bar{V}_0}{\Phi_s^2 D_p^2} \frac{(1 - \varepsilon)^2}{\varepsilon^3} + \frac{1.75\rho \bar{V}_0^2 (1 - \varepsilon)}{\Phi_s D_p \varepsilon^3} \tag{7.50}$$

Applying Eq. (7.50) to the point of incipient fluidization gives a quadratic equation for the minimum fluidization velocity $\bar{V}_{0M}$:

$$\frac{150\mu\bar{V}_{0M}}{\Phi_s^2 D_p^2}\frac{(1-\varepsilon_M)}{\varepsilon_M^3} + \frac{1.75\rho\bar{V}_{0M}^2}{\Phi_s D_p}\frac{1}{\varepsilon_M^3} = g(\rho_p - \rho) \qquad (7.51)$$

For very small particles, only the laminar flow term of the Ergun equation is significant. With $Re_p < 1$, the equation for minimum fluidization velocity becomes

$$\bar{V}_{0M} \approx \frac{g(\rho_p - \rho)}{150\mu}\frac{\varepsilon_M^3}{1-\varepsilon_M}\Phi_s^2 D_p^2 \qquad (7.52)$$

Many empirical equations state that $\bar{V}_{0M}$ varies with somewhat less than the 2.0 power of the particle size and not quite inversely with the viscosity. Slight deviations from the expected exponents occur because there is some error in neglecting the second term of the Ergun equation and because the void fraction ε_M may change with particle size. For roughly spherical particles, ε_M is generally between 0.40 and 0.45, increasing slightly with decreasing particle diameter. For irregular solids, the uncertainty in ε_M is probably the major error in predicting $\bar{V}_{0M}$ from Eq. (7.51) or (7.52).

Minimum fluidization velocities for particles in air calculated from Eq. (7.51) are shown in Fig. 7.10. Note that the dependence on D_p^2 holds up to particles about 300 μm in size; in many applications of fluidization, the particles are in the range of 30 to 300 μm. However, fluidization is also used for particles larger than 1 mm, as in the fluidized-bed combustion of coal. In the limit of very large sizes, the laminar flow term becomes negligible, and $\bar{V}_{0M}$ varies with the square root of the particle size. The equation for $Re_p > 10^3$ is

$$\bar{V}_{0M} \approx \left[\frac{\Phi_s D_p g(\rho_p - \rho)\varepsilon_M^3}{1.75\rho}\right]^{1/2} \qquad (7.53)$$

The terminal velocity for individual particles falling in still air is also shown in Fig. 7.10. For low Reynolds numbers, u_t and $\bar{V}_{0M}$ both vary with D_p^2, $\rho_p - \rho$, and $1/\mu$, so the ratio $u_t/\bar{V}_{0M}$ depends mainly on the void fraction at minimum fluidization. From Eqs. (7.40) and (7.52),

$$\frac{u_t}{\bar{V}_{0M}} = \frac{gD_p^2(\rho_p - \rho)}{18\mu}\frac{150\mu}{g(\rho_p - \rho)\Phi_s^2 D_p^2}\frac{1-\varepsilon_M}{\varepsilon_M^3}$$

$$= \frac{8.33(1-\varepsilon_M)}{\Phi_s^2\varepsilon_M^3} \qquad (7.54)$$

For spheres, with $\varepsilon_M \approx 0.45$, the terminal velocity is 50 times the minimum fluidization velocity, so a bed that fluidizes at 10 mm/s could probably be operated with velocities up to 400 mm/s with few particles carried out with the exit gas. With a wide distribution of particle sizes there would be more carryover or entrainment of the fines than of average-size particles, but most of the fines can be recovered by filters or cyclone separators and returned to the bed. Some fluidized beds are operated at velocities of 100 times $\bar{V}_{0M}$, with high entrainment but nearly complete recovery of the entrained solids.

$$U_t = 1.75 \sqrt{\frac{gD_p(\rho_p - \rho)}{\rho}}$$

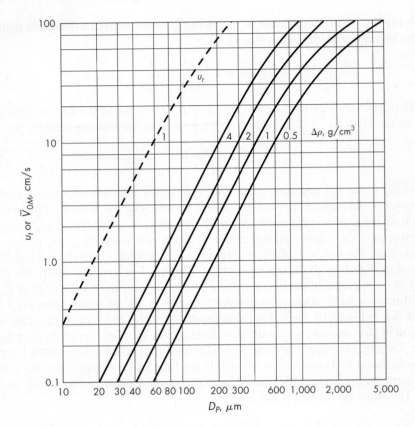

FIGURE 7.10

Minimum fluidization velocity and terminal velocity with air at 20°C and 1 atm ($\varepsilon_M = 0.50$, $\Phi_s = 0.8$, $\Delta\rho = \rho_p - \rho$).

For nonspherical particles Φ_s is less than 1, and Eq. (7.54) might seem to indicate a wider range of fluidization without entrainment. However, the value of ε_M is generally greater for irregular particles than for spheres, and for $\Phi_s = 0.8$ and $\varepsilon_M = 0.5$, the ratio $u_t/\bar{V}_{OM}$ is 52, about the same as that estimated for spheres.

For large particles, the terminal velocity is given by Newton's law [Eq. (7.43)], and this can be compared with $\bar{V}_{OM}$ from Eq. (7.53). For spheres, with Re_p greater than 10^3,

$$\frac{u_t}{\bar{V}_{OM}} = 1.75 \left[\frac{gD_p(\rho_p - \rho)}{\rho}\right]^{1/2} \left[\frac{1.75\rho}{gD_p(\rho_p - \rho)\varepsilon_M^3}\right]^{1/2}$$

$$= \frac{2.32}{\varepsilon_M^{3/2}} \tag{7.55}$$

For $\varepsilon_M = 0.45$, $u_t/\bar{V}_{OM} = 7.7$, which is a much lower ratio than that for fine particles, and entrainment may be more severe when operation is at several times the minimum fluidization velocity. This is a slight disadvantage in the use of coarse particles in a fluidized bed, but the optimum particle size depends on other factors such as chemical reactor efficiency, heat- and mass-transfer rates, grinding cost, or the desired gas velocity.

Types of fluidization

The equations derived for minimum fluidization velocity apply to liquids as well as to gases, but beyond $\bar{V}_{OM}$ the appearance of beds fluidized with liquids or gases is often quite different. When fluidizing sand with water, the particles move farther apart and their motion becomes more vigorous as the velocity is increased, but the average bed density at a given velocity is the same in all sections of the bed. This is called *particulate fluidization* and is characterized by a large but uniform expansion of the bed at high velocities.

Beds of solids fluidized with air usually exhibit what is called aggregative or bubbling fluidization. At superficial velocities somewhat greater than $\bar{V}_{OM}$, most of the gas passes through the bed as bubbles or voids which are almost free of solids, and only a small fraction of the gas flows in the channels between the particles. The particles move erratically and are supported by the fluid, but in the space between bubbles, the void fraction is about the same as that at incipient fluidization. The nonuniform nature of the bed was at first attributed to aggregation of the particles, and the term *aggregative fluidization* was applied; but there is no evidence that the particles stick together, and the term *bubbling fluidization* is a better description of the phenomenon. The bubbles that form behave much as air bubbles in water or bubbles of vapor in a boiling liquid, and the term *boiling bed* is sometimes applied to this type of fluidization. Studies that account for the rates of heat or mass transfer or chemical reaction in a bubbling bed often refer to the *two-phase theory of fluidization,* in which the bubbles are one phase and the dense bed of suspended particles is the second phase.

The behavior of a bubbling fluidized bed depends very strongly on the number and size of the gas bubbles, which are often hard to predict. The average bubble size depends on the nature and size distribution of the particles, the type of distributor plate, the superficial velocity, and the depth of the bed. Bubbles tend to coalesce and grow as they rise through the fluidized bed, and the maximum stable bubble size may be several inches to a few feet in diameter. If a small-diameter column is used with a deep bed of solids, the bubbles may grow until they fill the entire cross section. Successive bubbles then travel up the column separated by slugs of solids. This is called *slugging* and is usually undesirable because of pressure fluctuations in the bed, increased entrainment, and difficulties in scaling up to larger units.

When the superficial gas velocity increases to values much above $\bar{V}_{OM}$, there is a transition from bubbling fluidization to what is called *turbulent fluidization* or fast fluidization. The transition occurs when the bed has expanded so much that there can no longer be a dispersed bubble phase. The gas phase is continuous, and there are small regions of high or low bed density, with rapid density fluctuations at all points in the bed. The velocity for transition to turbulent fluidization is generally in the range of 0.3 to 0.6 m/s (1 to 2 ft/s), but it is difficult to predict the transition velocity because it depends on the particle properties and the average bubble size.

At still higher gas velocities, all particles fed are rapidly entrained with the gas, but can be recovered with a cyclone and returned to the bottom to maintain solids in the unit. This type of system is called a *circulating fluid bed,* even though there is no distinct bed of solids. The volume fraction of solids in the suspension is quite low, often just a few percent. Such a system is quite like an example of pneumatic transport, as discussed later.

The generalization that liquids give particulate fluidization of solids while gases give bubbling fluidization is not completely valid. The density difference is an important parameter, and very heavy solids may exhibit bubbling fluidization with water, while gases at high pressures may give particulate fluidization of fine solids. Also, fine solids of moderate density, such as cracking catalysts, may exhibit particulate fluidization for a limited range of velocities and then bubbling fluidization at high velocities.

Expansion of fluidized beds

With both types of fluidization the bed expands as the superficial velocity increases, and since the total pressure drop remains constant, the pressure drop per unit length decreases as ε increases. Rearranging Eq. (7.48) gives

$$\frac{\Delta p}{L} = g(1 - \varepsilon)(\rho_p - \rho) \tag{7.56}$$

Particulate fluidization. For particulate fluidization the expansion is uniform, and the Ergun equation, which applies to the fixed bed, might be expected to hold approximately for the slightly expanded bed. Assuming the flow between the particles is laminar, using the first term of Eq. (7.50) leads to the following equation for the expanded beds:

$$\frac{\varepsilon^3}{1 - \varepsilon} = \frac{150\bar{V}_0\mu}{g(\rho_p - \rho)\Phi_s^2 D_p^2} \tag{7.57}$$

Note that this equation is similar to Eq. (7.52) for the minimum fluidization velocity, but now $\bar{V}_0$ is the independent variable and ε the dependent variable. Equation (7.57) predicts that $\varepsilon^3/(1 - \varepsilon)$ is proportional to $\bar{V}_0$ for values greater than $\bar{V}_{0M}$. The expanded bed height may be obtained from ε and the values of L and ε at incipient fluidization, using the equation

$$L = L_M \frac{1 - \varepsilon_M}{1 - \varepsilon} \tag{7.58}$$

Data for the fluidization of small glass beads (510 μm) in water[22] are shown in Fig. 7.11. The first data point is for $\varepsilon_M = 0.384$ and $\bar{V}_{0M} = 1.67$ mm/s, and the theoretical line is a straight line from the origin through this point. The actual expansion is slightly less than predicted over much of the range, perhaps because of local variations in void fraction that decrease the hydraulic resistance. Note that the bed height increased nearly linearly with velocity, and the bed height has about doubled at $\bar{V}_0 = 10\bar{V}_{0M}$.

For particulate fluidization of large particles, in water, the expansion of the bed is expected to be greater than that corresponding to Eq. (7.57), since the pressure drop depends partly on the kinetic energy of the fluid, and a greater increase in ε is needed to offset a given percentage increase in $\bar{V}_0$. The expansion

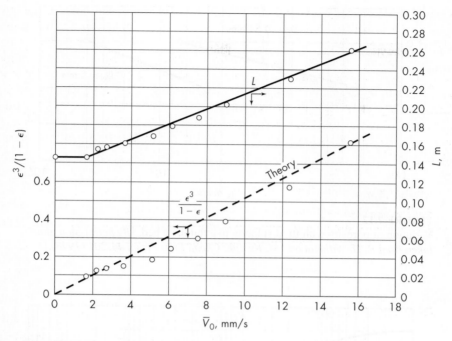

FIGURE 7.11
Bed expansion in particulate fluidization. [*By permission, data taken from R. H. Wilhelm and M. Kwauk, Chem. Eng. Prog.,* **44**:*201 (1948).*]

data can be correlated by the empirical equation proposed by Lewis, Gilliland, and Bauer[14]

$$\bar{V}_0 = \varepsilon^m \tag{7.59}$$

Data for two sizes of glass beads[22] are plotted in Fig. 7.12; and although the data do not fit Eq. (7.59) exactly, a straight line is adequate for engineering estimates of the bed expansion. Data from many investigations show that the slopes of such plots vary from about 0.22 in the laminar region to 0.4 at high Reynolds numbers. These slopes equal $1/m$. A correlation for m given by Leva[12] is shown in Fig. 7.13. To predict the bed expansion, m is estimated using the Reynolds number at the minimum fluidization velocity, and Eq. (7.59) is applied directly or in ratio form. An alternate method is to determine $\bar{V}_{0M}$ and u_t and draw a straight line on a plot such as Fig. 7.12.

EXAMPLE 7.3. A bed of ion-exchange beads 8 ft deep is to be backwashed with water to remove dirt. The particles have a density of 1.24 g/cm³ and an average size of 1.1 mm. What is the minimum fluidization velocity using water at 20°C, and what velocity is required to expand the bed by 25 percent? The beads are assumed to be spherical ($\Phi_s = 1$), and ε_M is taken as 0.40.

Solution. The quantities needed are

$$\mu = 0.01 \text{ P}$$
$$\Delta\rho = 0.24 \text{ g/cm}^3$$

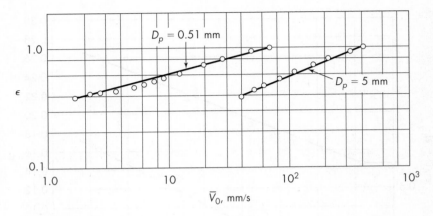

FIGURE 7.12
Variation of porosity with fluid velocity in a fluidized bed. [*By permission, data taken from R. H. Wilhelm and M. Kwauk, Chem. Eng. Prog., **44**:201 (1948).*]

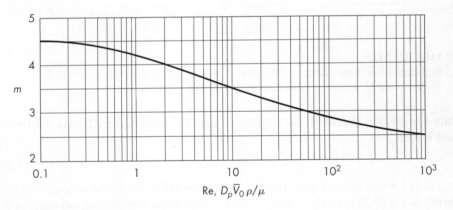

FIGURE 7.13
Exponent m in correlation for bed expansion [Eq. (7.59)]. (*By permission, from M. Leva, Fluidization, p. 89. Copyright, © 1959, McGraw-Hill Book Company.*)

From Eq. (7.51),

$$\frac{150(0.01)\bar{V}_{0M}}{(0.11)^2}\frac{0.6}{0.4^3} + \frac{1.75(1.0)(\bar{V}_{0M})^2}{0.11}\frac{1}{0.4^3} = 980(0.24)$$

$$1,162\bar{V}_{0M} + 248.6\bar{V}_{0M}^2 = 235.2$$

From the quadratic formula, $\bar{V}_{0M} = 0.194$ cm/s or 1.94 mm/s. At $\bar{V}_{0M}$,

$$\text{Re}_p = \frac{0.11(0.194)(1.24)}{0.01} = 2.65$$

From Fig. 7.13, $m \approx 3.9$. From Eq. (7.59)

$$\left(\frac{\varepsilon}{\varepsilon_M}\right)^{3.9} = \frac{\bar{V}_0}{\bar{V}_{0M}}$$

For 25 percent expansion, $L = 1.25 L_M$ or $1 - \varepsilon = (1 - \varepsilon_M)/1.25 = 0.48$. From this, $\varepsilon = 0.52$, and $\bar{V}_0 = 1.94(0.52/0.40)^{3.9} = 5.40$ mm/s.

Bubbling fluidization. For bubbling fluidization, the expansion of the bed comes mainly from the space occupied by gas bubbles, since the dense phase does not expand significantly with increasing total flow. In the following derivation, the gas flow through the dense phase is assumed to be $\bar{V}_{0M}$ times the fraction of the bed occupied by the dense phase, and the rest of the gas flow is to be carried by the bubbles. Thus,

$$\bar{V}_0 = f_b u_b + (1 - f_b)\bar{V}_{0M} \tag{7.60}$$

where f_b = fraction of bed occupied by bubbles
$\quad\quad u_b$ = average bubble velocity

Since all the solid is in the dense phase, the height of the expanded bed times the fraction dense phase must equal the bed height at incipient fluidization

$$L_M = L(1 - f_b) \tag{7.61}$$

Combining Eqs. (7.60) and (7.61) gives

$$\frac{L}{L_M} = \frac{u_b - \bar{V}_{0M}}{u_b - \bar{V}_0} \tag{7.62}$$

When u_b is much greater than $\bar{V}_0$, the bed expands only slightly, even though $\bar{V}_0$ may be several times $\bar{V}_{0M}$.

An empirical equation for bubble velocity in a fluidized bed is[20]

$$u_b \approx 0.7\sqrt{g D_b} \tag{7.63}$$

There is only a small effect of particle size or shape on the coefficient in Eq. (7.63), and although large bubbles are mushroom-shaped rather than spherical, the equation holds quite well with D_b taken as the equivalent spherical diameter. For $D_b = 100$ mm, u_b is 700 mm/s; and if $\bar{V}_{0M} = 10$ mm/s and $\bar{V}_0 = 100$ mm/s, then L/L_M would be 1.15. Doubling the velocity would increase L/L_M to 1.38 if the bubble size were constant; but the bubble size generally increases with gas velocity because of coalescence, and the bed height often increases nearly linearly with velocity. The expansion of the bed is usually in the range of 20 to 50 percent, even at velocities up to 50 times $\bar{V}_{0M}$, in contrast to the large expansions found in particulate fluidization.

Some fine powders fluidized with a gas exhibit particulate fluidization over a limited range of velocities near the minimum fluidization point. With increasing velocity the bed expands uniformly until bubbles start to form, gradually collapses to a minimum height as the velocity is increased past the bubble point, and then expands again as bubble flow becomes predominant. Silica-alumina cracking catalyst

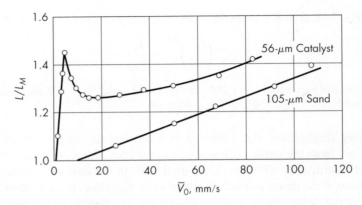

FIGURE 7.14
Expansion of fluidized beds of sand and cracking catalyst.

shows this anomalous behavior, and bed expansion data for a commercial catalyst are contrasted with those for a fine sand in Fig. 7.14. The region of particulate fluidization is found only with quite small or low-density particles. A classification of solids based on these properties is given by Geldart.[2]

Applications of fluidization

Extensive use of fluidization began in the petroleum industry with the development of fluid-bed catalytic cracking. Although the industry now generally uses riser or transport-line reactors for catalytic cracking, rather than fluid beds, the catalyst regeneration is still carried out in fluid-bed reactors, which are as large as 10 m in diameter. Fluidization is used in other catalytic processes, such as the synthesis of acrylonitrile, and for carrying out solid-gas reactions. There is much interest in the fluidized-bed combustion of coal as a means of reducing boiler cost and decreasing the emission of pollutants. Fluidized beds are also used for roasting ores, drying fine solids, and adsorption of gases.

The chief advantages of fluidization are that the solid is vigorously agitated by the fluid passing through the bed, and the mixing of the solids ensures that there are practically no temperature gradients in the bed even with quite exothermic or endothermic reactions. The violent motion of the solids also gives high heat-transfer rates to the wall or to cooling tubes immersed in the bed. Because of the fluidity of the solids, it is easy to pass solids from one vessel to another.

The main disadvantage of gas-solid fluidization is the uneven contacting of gas and solid. Most of the gas passes through the bed as bubbles and directly contacts only a small amount of solid in a thin shell, known as the bubble cloud, around the bubble. A small fraction of the gas passes through the dense phase, which contains nearly all the solid. There is some interchange of gas between the bubbles and the dense phase by diffusion and by turbulent processes such as bubble splitting and coalescence, but the overall conversion of a gaseous reactant is generally much less than with uniform contacting at the same temperature, as in an ideal plug-flow reactor. In turbulent fluidization the average bed density is lower than in bubbling fluidization, but there is better mass transfer between gas and solids, which may improve the performance of the unit as a chemical reactor. Even so, the scaleup of fluidized reactors is often uncertain.

Other disadvantages of fluidized beds, more easily dealt with by proper design, include erosion of vessel internal parts and attrition of the solids, leading to loss of fines. Most fluid-bed reactors have internal or external cyclones to recover fines, but filters and scrubbers are often needed also.

Continuous fluidization; slurry and pneumatic transport

When the velocity of the fluid through the solids becomes large enough, all the particles are entrained in the fluid and carried along with it, to give *continuous fluidization*. As discussed earlier, some gas-solid reactors operate in this fashion, but the principal applications of continuous fluidization are in transporting solids from point to point.

Hydraulic or slurry transport. Particles smaller than about 50 μm in diameter settle very slowly and are readily suspended in a moving liquid. Larger particles are harder to suspend, and when the diameter is 0.25 mm or greater, a fairly large liquid velocity is needed to keep the particles from moving at all, especially in horizontal pipes. The critical velocity $\bar{V}_c$, below which particles will settle out, is typically between 1 and 5 m/s, depending on the density difference between solids and liquid, the particle diameter, the slurry concentration, and the size of the pipe. Critical velocities are larger in big pipe than in small pipe. A semitheoretical general equation for predicting V_c has been proposed by Oroskar and Turian.[17]

The pressure drop in slurries of nonsettling particles may be found from the equations for a homogeneous liquid, with appropriate allowance for the increased density and apparent viscosity. For "settling slurries" there is no single satisfactory correlation; the pressure drop in a horizontal pipe is greater than that in a single-phase fluid of the same density and viscosity as the slurry, especially near the critical velocity, but approaches that in the single-phase liquid as the velocity increases. When the velocity is $3\bar{V}_c$ or greater, the pressure drop in the slurry and that in the equivalent single-phase liquid are equal. The velocity in a long slurry pipeline is typically 1.5 to 2 times $\bar{V}_c$.

Pneumatic conveying. The suspending fluid in a pneumatic conveyor is a gas, usually air, flowing at velocities between 15 and 30 m/s in pipes ranging from 50 to 400 mm in diameter. The principal types of systems are (1) negative-pressure (vacuum) systems, useful in transferring solids from multiple intake points (railroad cars, ships' holds, etc.) to a single delivery point; (2) positive-pressure systems, best for one intake point and one or more points of delivery; (3) vacuum-pressure systems, which combine the advantages of the previous two; and (4) prefluidized systems, which require less air, and consequently less power, than any of the other methods.

Materials handled range in particle size from fine powders to 6.5-mm pellets, and in bulk density from 16 to more than 3,200 kg/m^3. Vacuum systems are typically limited to solids flow rates less than 6,800 kg/h and equivalent conveyor lengths less than 300 m. Pressure systems operate at 1 to 5 atm gauge and are used for free-flowing solids of particles less than 6.5 mm in size, where flow rates greater than 9,000 kg/h are needed. Typically the pressure loss through the system is about 0.5 atm.

Most pneumatic conveyor systems are purchased from the manufacturer; for preliminary design, useful nomographs are given by Perry.[18]

SYMBOLS

A	Area m^2 or ft^2; A_p, projected area of particle
a_e	Acceleration of particle from external force, m/s^2 or ft/s^2
C_D	Drag coefficient, $2F_D/(u_0^2 \rho A_p)$, dimensionless
D	Diameter, m or ft; D_{eq}, equivalent diameter of channels in packed bed; D_p, diameter of spherical particle; also nominal size or characteristic length of a particle; $\bar{D}_{pi}$, average particle diameter in fraction i; $\bar{D}_s$, mean effective diameter for mixture of particles
F	Force, N or lb$_f$; F_D, total drag force; F_b, buoyant force; F_e, external force
f	Fanning friction factor, dimensionless
f_b	Volume fraction of fluidized bed occupied by gas bubbles
G_0	Mass velocity of fluid approaching particle, kg/m$^2 \cdot$s or lb/ft$^2 \cdot$s; also superficial mass velocity in packed bed
g	Gravitational acceleration, m/s^2 or ft/s^2
K	Criterion for settling, defined by Eq. (7.45), dimensionless
L	Length of cylindrical particle, m or ft; also length of channels in packed bed; also total height of packed or fluidized bed; L_M, bed height at incipient fluidization
Ma	Mach number, dimensionless; Ma$_a$, Ma$_b$, at stations a and b; Ma$_0$, of approaching fluid
m	Mass, kg or lb; also exponent in Eq. (7.59)
N_i	Number of particles in each size range
n	Number of channels in packed bed; also exponent in Eq. (7.46)
p	Pressure, N/m^2 or lb$_f$/ft^2; p_s, at stagnation point; p_0, in undisturbed fluid
Re	Reynolds number, dimensionless; Re$_p$, particle Reynolds number, $D_p G_0/\mu$
r	Radius of particle path, m or ft
S	Cross-sectional area, m^2 or ft^2; S_0, of empty tower
s_p	Surface area of single particle, m^2 or ft^2
T	Temperature, K, °C, °F, or °R; T_a, T_b, at stations a and b; T_s, at stagnation point; T_0, of approaching stream
t	Time, s

u Velocity of fluid or particle, m/s or ft/s; u_b, average bubble velocity in fluidized bed; u_s, settling velocity of uniform suspension; u_t, terminal velocity of particle; u_0, velocity of approaching stream; u', fluctuating component

$\bar{V}$ Volumetric average fluid velocity, m/s or ft/s; $\bar{V}_c$, critical velocity in hydraulic transport; $\bar{V}_0$, superficial or empty-tower velocity; $\bar{V}_{OM}$, minimum superficial velocity for fluidization

v_p Volume of single particle, ft^3 or m^3

x_i Volume fraction of particles of size i in bed of mixed particles

Greek Letters

α Angle with perpendicular-to-flow direction

γ Ratio of specific heats, c_p/c_v

Δp Pressure drop in packed or fluidized bed

$\Delta \rho$ Density difference $\rho_p - \rho$

ε Porosity or volume fraction of voids in bed of solids; ε_M, minimum porosity for fluidization

λ_1 Constant in Eqs. (7.15) and (7.16)

λ_2 Constant in Eqs. (7.18) and (7.19)

μ Absolute viscosity, cP or lb/ft·s; μ_s, effective viscosity of suspension

ρ Density, kg/m^3 or lb/ft^3; ρ_p, of particle; ρ_0, of approaching stream

τ_w Shear stress at channel boundary, N/m^2 or lb$_f$/ft^2

Φ_s Sphericity, defined by Eq. (7.10)

ϕ Function

ω Angular velocity, rad/s

PROBLEMS

7.1. A partial oxidation is carried out by passing air with 1.2 mole percent (mol %) hydrocarbon through 40-mm tubes packed with 2 m of 3-mm by 3-mm cylindrical catalyst pellets. The air enters at 350°C and 2.0 atm with a superficial velocity of 1 m/s. What is the pressure drop through the packed tubes? How much would the pressure drop be reduced by using 4-mm pellets? Assume $\varepsilon = 0.40$.

7.2. A catalyst tower 40 ft high and 18 ft in diameter is packed with 1-in.-diameter spheres. Gas enters the top of the bed at a temperature of 450°F and leaves at the same temperature. The pressure at the bottom of the catalyst bed is 30 lb$_f$/in.^{2} abs. The bed porosity is 0.40. If the gas has average properties similar to propane and the time of contact (based on flow in the void space) between the gas and the catalyst is 8 s, what is the inlet pressure?

7.3. The pressure drop for airflow through a column filled with 1-in. ceramic Raschig rings is 0.01 in. water per foot when $G_0 = 80$ lb/ft^2·h and 0.9 in./ft water when

$G_0 = 800$ lb/ft$^2 \cdot$ h, all for a mass velocity of the liquid flowing countercurrently of 645 lb/ft$^2 \cdot$ h. Since the change in pressure drop with liquid rate is slight in the range of liquid mass velocities between 645 and 1,980 lb/ft$^2 \cdot$ h, ignore the liquid holdup and estimate the void fraction if the rings have a wall thickness of $\frac{1}{8}$ in. Use this void fraction and the Ergun equation to predict the pressure drop, and discuss the difference between predicted and experimental values.

7.4. The following data are reported for the flow of air through beds of granular activated carbon. Compare the pressure drops with values predicted using the Ergun equation, and predict the pressure drops for both sizes at air velocities of 100 and 200 ft/min.

$\bar{V}_0$, ft/min	4 × 6 Mesh, Δp, in. H$_2$O/ft	4 × 8 Mesh, Δp, in. H$_2$O/ft
10	0.25	0.32
40	1.40	1.80

7.5. The pressure drop through a particle bed can be used to determine the external surface area and the average particle size. Data for a bed of crushed ore particles show $\Delta p/L = 84$ (lb$_f$/in.^{2})/ft for airflow at a superficial velocity of 0.015 ft/s. The measured void fraction is 0.47, and the estimated sphericity Φ_s is 0.7. Calculate the average particle size and the surface area per unit mass if the solid has a density of 4.1 g/cm^3. How sensitive is the answer to an error of 0.01 in ε?

7.6. A 150-mm-thick demister pad for removing fine droplets of H$_2$SO$_4$ from a gas stream is made of 50-μm fibers randomly oriented in the plane perpendicular to the gas flow. The average porosity is 0.90. Calculate the pressure drop for gas velocities of 0.3 and 0.9 m/s at 90°C and 1 atm using the drag coefficients for individual fibers.

7.7. How long will it take for the spherical particles in Table 7.3 to settle, at their terminal velocities under free-settling conditions, through 2 m of water at 20°C?

7.8. A cyclone separator is used to remove sand grains from an airstream at 150°C. If the cyclone body is 0.6 m in diameter and the average tangential velocity is 16 m/s, what is the radial velocity near the wall of particles 20 and 40 μm in size? How much greater are these values than the terminal velocity in gravity settling?

TABLE 7.3
Data for Prob. 7.7

Substance	Specific gravity	Diameter, mm
Galena	7.5	0.25
		0.025
Quartz	2.65	0.25
		0.025
Coal	1.3	6
Steel	7.7	25

7.9. Urea pellets are made by spraying drops of molten urea into cold gas at the top of a tall tower and allowing the material to solidify as it falls. Pellets 6 mm in diameter are to be made in a tower 25 m high containing air at 20°C. The density of urea is 1,330 kg/m^3. (*a*) What would be the terminal velocity of the pellets, assuming free-settling conditions? (*b*) Would the pellets attain 99 percent of this velocity before they reached the bottom of the tower?

7.10. Spherical particles 1 mm in diameter are to be fluidized with water at twice the minimum velocity. The particles have an internal porosity of 40 percent, an average pore diameter of 10 μm, and a particle density of 1.5 g/cm^3. Prove that the flow through the internal pores is very small compared to the flow between the particles, and that the internal porosity can be neglected in predicting the fluidization behavior.

7.11. Catalyst pellets 5 mm in diameter are to be fluidized with 45,000 kg/h of air at 1 atm and 80°C in a vertical cylindrical vessel. The density of the catalyst particles is 960 kg/m^3; their sphericity is 0.86. If the given quantity of air is just sufficient to fluidize the solids, what is the vessel diameter?

7.12. According to a brochure from Dow Chemical Company, the pressure drop for water flowing through a bed of 20-mesh to 50-mesh Dowex 50-X8 resin is said to be proportional to the flow rate and has a value of 0.80 (lb$_f$/in.^{2})/ft at a flow rate of 10 gal/min · ft^2. (*a*) Predict the pressure drop, using an arithmetic average particle size and a void fraction of 0.35. (*b*) What average particle size or alternate value of void fraction would be needed for agreement with the published pressure drop?

7.13. In backwashing a bed of 20-mesh to 50-mesh Dowex 50-X8 resin, the bed starts to expand when the flow rate reaches 0.4 gal/min · ft^2 and has expanded 45 percent at 6 gal/min · ft^2. Are these values consistent with fluidization theories?

REFERENCES

1. Ergun, S. *Chem. Eng. Prog.* **48:**89 (1952).
2. Geldart, D. *Powder Technology* **7:**285 (1973).
3. Hartman, M., O. Trnka, and K. Svoboda. *Ind. Eng. Chem. Res.* **33:**1979 (1994).
4. Heiss, J. F., and J. Coull. *Chem. Eng. Prog.* **48:**133 (1952).
5. Hottovy, J. D., and N. D. Sylvester. *Ind. Eng. Chem. Proc. Des. Dev.* **18:**433 (1979).
6. Houghton, G., A. M. McLean, and P. D. Ritchie. *Chem. Eng. Sci.* **7:**26 (1957).
7. Hughes, R. R., and E. R. Gilliland. *Chem. Eng. Prog.* **48:**497 (1952).
8. Hunsaker, J. C., and B. G. Rightmire. *Engineering Applications of Fluid Mechanics.* New York: McGraw-Hill, 1947, pp. 202–203.
9. Ingebo, R. D. NACA Tech. Note 3762 (1956).
10. Knudsen, J. G., and D. L. Katz. *Fluid Mechanics and Heat Transfer.* New York: McGraw-Hill, 1958, p. 317.
11. Lapple, C. E., and C. B. Shepherd. *Ind. Eng. Chem.* **32:**605 (1940).
12. Leva, M. *Fluidization.* New York: McGraw-Hill, 1959.
13. Leva, M., and M. Grummer. *Chem. Eng. Prog.* **43:**713 (1947).
14. Lewis, W. K., E. R. Gilliland, and W. C. Bauer. *Ind. Eng. Chem.* **41:**1104 (1949).
15. Masliyah, J. H., and N. Epstein. *J. Fluid Mech.* **44:**493 (1970).

16. Maude, A. D., and R. L. Whitmore. *Br. J. Appl. Phys.* **9:**477 (1958).
17. Oroskar, A. R., and R. M. Turian. *AIChE J.* **26:**550 (1980).
18. Perry, J. H., and D. W. Green (eds.). *Chemical Engineers' Handbook.* 7th ed. New York: McGraw-Hill, 1997, pp. **21**-19 to **21**-27.
19. Pettyjohn, E. S., and E. B. Christiansen. *Chem. Eng. Prog.* **44:**157 (1948).
20. Rowe, P. N. *Fluidization.* Eds. J. F. Davidson and D. Harrison. New York: Academic, 1971, p. 145.
21. Tavlarides, L. L., C. A. Coulaloglou, M. A. Zeitlin, G. E. Klinzing, and B. Gal-Or. *Ind. Eng. Chem.* **62**(11):6 (1970).
22. Wilhelm, R. H., and M. Kwauk. *Chem. Eng. Prog.* **44:**201 (1948).

CHAPTER 8

Transportation and Metering of Fluids

Preceding chapters have dealt with theoretical aspects of fluid motion. The engineer is concerned, also, with practical problems in transporting fluids from one place to another and in measuring their rates of flow. Such problems are the subject of this chapter.

The first part of the chapter deals with the transportation of fluids, both liquids and gases. Solids are sometimes handled by similar methods by suspending them in a liquid to form a pumpable slurry or by conveying them in a high-velocity gas stream. It is cheaper to move fluids than solids, and materials are transported as fluids whenever possible. In the process industries, fluids are nearly always carried in closed channels sometimes square or rectangular in cross section but much more often circular. The second part of the chapter discusses common methods of measuring the flow rate.

PIPE, FITTINGS, AND VALVES

Pipe and tubing

Fluids are usually transported in pipe or tubing, which is circular in cross section and available in widely varying sizes, wall thicknesses, and materials of construction. There is no clear-cut distinction between the terms *pipe* and *tubing*. Generally speaking, pipe is heavy-walled and relatively large in diameter and comes in moderate lengths of 20 to 40 ft (6 to 12 m); tubing is thin-walled and often comes in coils several hundred feet long. Metallic pipe can be threaded; tubing usually cannot. Pipe walls are usually slightly rough; tubing has very smooth walls. Lengths of pipe are joined by screwed, flanged, or welded fittings; pieces of tubing are connected by compression fittings, flare fittings, or soldered fittings. Finally, tubing is usually extruded or cold-drawn, while metallic pipe is made by welding, casting, or piercing a billet in a piercing mill.

Pipe and tubing are made from many materials, including metals and alloys, wood, ceramics, glass, and various plastics. Polyvinyl chloride, or PVC, pipe is extensively used for water lines. In process plants the most common material is low-carbon steel, fabricated into what is sometimes called black-iron pipe. Wrought-iron and cast-iron pipes are also used for a number of special purposes.

Sizes. Pipe and tubing are specified in terms of their diameter and their wall thickness. With steel pipe the standard nominal diameters, in U.S. practice, range from $\frac{1}{8}$ to 30 in. For large pipe, more than 12 in. in diameter, the nominal diameters are the actual outside diameters; for small pipe the nominal diameter does not correspond to any actual dimension. The nominal value is close to the actual inside diameter for 3- to 12-in. pipe, but for very small pipe this is not true. Regardless of wall thickness, the outside diameter of all pipe of a given nominal size is the same, to ensure interchangeability of fittings. Standard dimensions of steel pipe are given in App. 3. Pipe of other materials is also made with the same outside diameters as steel pipe to permit interchanging parts of a piping system. These standard sizes for steel pipe, therefore, are known as IPS (iron pipe size) or NPS (normal pipe size). Thus the designation "2-in. nickel IPS pipe" means nickel pipe having the same outside diameter as standard 2-in. steel pipe.

The wall thickness of pipe is indicated by the *schedule number*, which increases with the thickness. Ten schedule numbers—10, 20, 30, 40, 60, 80, 100, 120, 140, and 160—are in use, but with pipe less than 8 in. in diameter only numbers 40, 80, 120, and 160 are common. For steel pipe the actual wall thicknesses corresponding to the various schedule numbers are given in App. 3; with other alloys the wall thickness may be greater or less than that of steel pipe, depending on the strength of the alloy. With steel at ordinary temperatures the allowable stress is one-fourth the ultimate strength of the metal.

The size of tubing is indicated by the outside diameter. The normal value is the actual outer diameter, to within very close tolerances. Wall thickness is ordinarily given by the BWG (Birmingham wire gauge) number, which ranges from 24 (very light) to 7 (very heavy). Sizes and wall thicknesses of heat-exchanger tubing are given in App. 4.

Selection of pipe sizes. The pipe size selected for a particular installation depends mainly on the cost of the pipe and fittings and the cost of energy needed for pumping the fluid. The cost of the pipe and the annual capital charges increase with about the 1.5 power of the pipe diameter, while the power cost for turbulent flow varies with the -4.8 power of the diameter. Equations have been presented giving the optimum pipe diameter as a function of flow rate and fluid density,[8] but these can be converted to the optimum velocity, which is nearly independent of the flow rate. For turbulent flow of liquids in steel pipes larger than 1 in. (25 mm) in diameter, the optimum velocity is

$$\bar{V}_{\text{opt}} = \frac{12\,\dot{m}^{0.1}}{\rho^{0.36}} \tag{8.1}$$

where $\bar{V}_{opt}$ = optimum velocity, ft/s
  $\dot{m}$ = mass flow rate, lb/s
  ρ = fluid density, lb/ft^3

For water and similar fluids $\bar{V}_{opt}$ is 3 to 6 ft/s (0.9 to 1.8 m/s); for air or steam at low to moderate pressures $\bar{V}_{opt}$ is 20 to 80 ft/s (6 to 24 m/s). For flow in heat-exchanger tubes, the optimum design velocity is often higher than that given by Eq. (8.1) because of improved heat transfer at high fluid velocities.

When flow is by gravity from overhead tanks or when a viscous liquid is being pumped, low velocities are favored, in the range of 0.2 to 0.8 ft/s (0.06 to 0.24 m/s). The relations among pipe size, volumetric flow rate, and fluid velocity are shown in App. 3.

For large complex piping systems the cost of piping may be a substantial fraction of the total investment, and elaborate computer methods of optimizing pipe sizes are justified.

Joints and fittings

The methods used to join pieces of pipe or tubing depend in part on the properties of the material but primarily on the thickness of the wall. Thick-walled tubular products are usually connected by screwed fittings, by flanges, or by welding. Pieces of thin-walled tubing are joined by soldering or by compression or flare fittings. Pipe made of brittle materials such as glass or carbon or cast iron is joined by flanges or bell-and-spigot joints.

When screwed fittings are used, the ends of the pipe are threaded externally with a threading tool. The thread is tapered, and the few threads farthest from the end of the pipe are imperfect, so that a tight joint is formed when the pipe is screwed into a fitting. Tape of polytetrafluoroethylene is wrapped around the threaded end to ensure a good seal. Threading weakens the pipe wall, and the fittings are generally weaker than the pipe itself; when screwed fittings are used, therefore, a higher schedule number is needed than with other types of joints. Screwed fittings are standardized for pipe sizes up to 12 in. (300 mm), but because of the difficulty of threading and handling large pipe, they are rarely used in the field with pipe larger than 3 in. (75 mm).

Lengths of pipe larger than about 2 in. (50 mm) are usually connected by flanges or by welding. Flanges are matching disks or rings of metal bolted together and compressing a gasket between their faces. The flanges themselves are attached to the pipe by screwing them on or by welding or brazing. A flange with no opening, used to close a pipe, is called a *blind flange* or a *blank flange*. For joining pieces of large steel pipe in process piping, especially for high-pressure service, welding has become the standard method. Welding makes stronger joints than screwed fittings do, and since it does not weaken the pipe wall, lighter pipe can be used for a given pressure. Properly made welded joints are leakproof, whereas other types of joints are not. Environmental protection legislation considers flanged and screwed joints to be sources of emission of volatile materials. Almost the only disadvantage of a welded joint is that it cannot be opened without destroying it.

Allowances for expansion

Almost all pipe is subjected to varying temperatures, and in some high-temperature lines the temperature change is very large. Such changes cause the pipe to expand and contract. If the pipe is rigidly fixed to its supports, it may tear loose, bend, or even break. In large lines, therefore, fixed supports are not used; instead the pipe rests loosely on rollers or is hung from above by chains or rods. Provision is also made in all high-temperature lines for taking up expansion, so that the fittings and valves are not put under strain. This is done by bends or loops in the pipe, by packed expansion joints, by bellows or packless joints, and sometimes by flexible metal hose.

Prevention of leakage around moving parts

In many kinds of processing equipment it is necessary to have one part move in relation to another part without excessive leakage of a fluid around the moving member. This is true in packed expansion joints and in valves where the stem must enter the valve body and be free to turn without allowing the fluid in the valve to escape. It is also necessary where the shaft of a pump or compressor enters the casing, where an agitator shaft passes through the wall of a pressure vessel, and in other similar places.

Common devices for minimizing leakage while permitting relative motion are stuffing boxes and mechanical seals. Neither completely stops leakage, but if no leakage whatever of the process fluid can be tolerated, it is possible to modify the device to ensure that only innocuous fluids leak into or escape from the equipment. The motion of the moving part may be reciprocating or rotational or both together; it may be small and occasional, as in a packed expansion joint, or virtually continuous, as in a process pump.

Stuffing boxes. A stuffing box can provide a seal around a rotating shaft and also around a shaft that moves axially. In this it differs from mechanical seals, which are good only with rotating members. The "box" is a chamber cut into the stationary member surrounding the shaft or pipe, as shown in Fig. 8.1a. Often a boss is provided on the casing or vessel wall to give a deeper chamber. The annular space between the shaft and the wall of the chamber is filled with *packing,* consisting of a rope or rings of inert material containing a lubricant such as graphite. The packing, when compressed tightly around the shaft, keeps the fluid from passing out through the stuffing box and yet permits the shaft to turn or move back and forth. The packing is compressed by a follower ring, or gland, pressed into the box by a flanged cap or packing nut. The shaft must have a smooth surface so that it does not wear away the packing; even so, the pressure of the packing considerably increases the force required to move the shaft. A stuffing box, even under ideal conditions, does not completely stop fluid from leaking out; in fact, when the box is operating properly, there should be small leakage. Otherwise the wear on the packing and the power loss in the unlubricated stuffing box are excessive.

When the fluid is toxic or corrosive, means must be provided to prevent it from escaping from the equipment. This can be done by using a *lantern gland* (Fig. 8.1b), which may be looked upon as two stuffing boxes on the same shaft, with two sets of

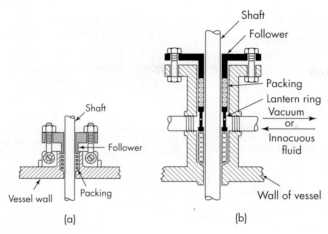

FIGURE 8.1
Stuffing boxes: (*a*) simple form; (*b*) with lantern gland.

packing separated by a lantern ring. The ring is H-shaped in cross section, with holes drilled through the bar of the H in the direction perpendicular to the axis of the shaft. The wall of the chamber of the stuffing box carries a pipe that takes fluid to or away from the lantern ring. By applying vacuum to this pipe, any dangerous fluid that leaks through one set of packing rings is removed to a safe place before it can get to the second set. Or by forcing a harmless fluid such as water under high pressure into the lantern gland, it is possible to ensure that no dangerous fluid leaks out the exposed end of the stuffing box.

Mechanical seals. In a rotary, or mechanical, seal the sliding contact is between a ring of graphite and a polished metal face, usually of carbon steel. A typical seal is shown in Fig. 8.2. Fluid in the high-pressure zone is kept from leaking out around the shaft by the stationary graphite ring held by springs against the face of the rotating metal collar. Stationary U-cup packing of rubber or plastic is set in the space between the body of the seal and the chamber holding it around the shaft; this keeps fluid from leaking past the nonrotating part of the seal and yet leaves the graphite ring free to move axially so that it can be pressed tightly against the collar. Rotary seals require less maintenance than stuffing boxes and have come into wide use in equipment handling highly corrosive fluids.

Valves

A typical processing plant contains thousands of valves of many different sizes and shapes. Despite the variety in their design, however, all valves have a common primary purpose: to slow down or stop the flow of a fluid. Some valves work best in on-or-off service, fully open or fully closed. Others are designed to throttle, to reduce the pressure and flow rate of a fluid. Still others permit flow in one direction only or only under certain conditions of temperature and pressure. A steam trap, which is a special form of valve, allows water and inert gas to pass through while holding back the steam. Finally, by using sensors and automatic control systems to adjust the valve position and thus the flow through the valve, the temperature, pressure, liquid level,

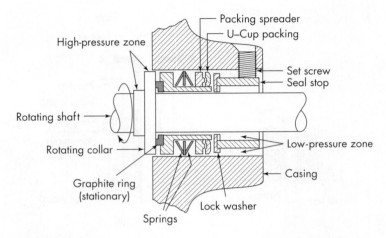

FIGURE 8.2
Mechanical seal.

or other fluid properties can be controlled at points remote from the valve itself.

In all cases, however, the valve initially stops or controls flow. This is done by placing an obstruction in the path of the fluid, an obstruction that can be moved about as desired inside the pipe with little or no leakage of fluid from the pipe to the outside. Where the resistance to flow introduced by an open valve must be small, the obstruction and the opening that can be closed by it are large. For precise control of flow rate, usually obtained at the price of a large pressure drop, the cross-sectional area of the flow channel is greatly reduced, and a small obstruction is set into the small opening.

Gate valves and globe valves. The two most common types of valves, gate valves and globe valves, are illustrated in Fig. 8.3. In a gate valve the diameter of the opening through which the fluid passes is nearly the same as that of the pipe, and the direction of flow does not change. As a result, a wide-open gate valve introduces only a small pressure drop. The disk is tapered and fits into a tapered seat; when the valve is opened, the disk rises into the bonnet, completely out of the path of the fluid. Gate valves are not recommended for controlling flow and are usually left fully open or fully closed.

Globe valves (so called because in the earliest designs the valve body was spherical) are widely used for controlling flow. The opening increases almost linearly with stem position, and wear is evenly distributed around the disk. The fluid passes through a restricted opening and changes direction several times, as can be seen by visualizing the flow through the valve illustrated in Fig. 8.3b. As a result, the pressure drop in this kind of valve is large.

Most automatic control valves are similar to globe valves, but the handwheel is replaced by a spring-diaphragm pneumatic activator or an electric motor, and the valve position depends on a signal from the controller. The simple disk shown in

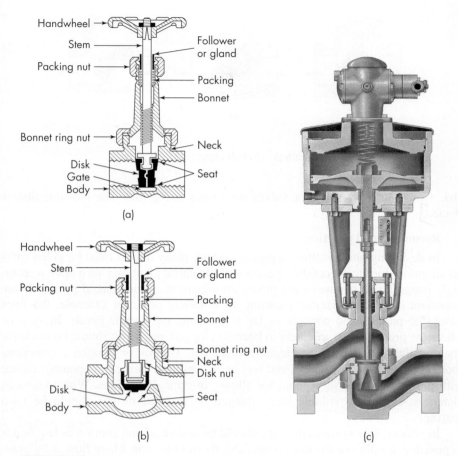

FIGURE 8.3
Common valves: (*a*) gate valve; (*b*) globe valve; (*c*) control valve with pneumatic valve activator.

Fig. 8.3*b* may be replaced by a tapered plug or other shape (Fig. 8.3*c*), designed to give certain flow lift characteristics.

Plug cocks and ball valves. For temperatures below 250° C, metallic plug cocks are useful in chemical process lines. As in a laboratory stopcock, a quarter turn of the stem takes the valve from fully open to fully closed; and when it is fully open, the channel through the plug may be as large as the inside of the pipe itself, and the pressure drop is minimal. In a ball valve the sealing element is spherical, and the problems of alignment and "freezing" of the element are less than with a plug cock. In both plug cocks and ball valves, the area of contact between the moving element and seat is large, and both can therefore be used in throttling service. Ball valves find occasional applications in flow control.

Check valves. A check valve permits flow in one direction only. It is opened by the pressure of the fluid in the desired direction; when the flow stops or tends to reverse, the valve automatically closes by gravity or by a spring pressing against the

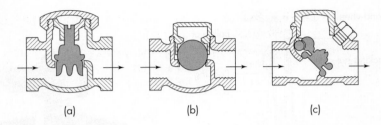

FIGURE 8.4
Check valves: (*a*) lift check; (*b*) ball check; (*c*) swing check.

disk. Common types of check valves are shown in Fig. 8.4. The movable disk is shaded.

Recommended practice

In designing and installing a piping system, many details must be given careful attention, for the successful operation of the entire plant may turn upon a seemingly insignificant feature of the piping arrangement. Some general principles are important enough to warrant mention. In installing pipe, for example, the lines should be parallel and contain, as far as possible, right-angle bends. In systems where the process lines are likely to become clogged, provision should be made for opening the lines to permit cleaning them out. Unions or flanged connections should be generously included, and tees or crosses with their extra openings closed with plugs should be substituted for elbows in critical locations. With hazardous materials, especially volatile ones, flanged or screwed fittings should be used sparingly.

In gravity flow systems the pipe should be oversized and contain as few bends as possible. Fouling of the lines is particularly troublesome where flow is by gravity, since the pressure head on the fluid cannot be increased to keep the flow rate up if the pipe becomes restricted.

Leakage through valves should also be expected. Where complete stoppage of flow is essential, therefore, where leakage past a valve would contaminate a valuable product or endanger the operators of the equipment, a valve or check valve is inadequate. In this situation a blind flange set between two ordinary flanges will stop all flow; or the line can be broken at a union or pair of flanges and the open ends capped or plugged.

Valves should be mounted with their stems up, if possible. They should be accessible and well supported without strain, with suitable allowance for thermal expansion of the adjacent pipe. Room should be allowed for fully opening the valve and for repacking the stuffing box.

PUMPS

This section deals with the transportation of liquids through pipes and channels. Liquid are sometimes moved by gravity from elevated tanks, or from a "blowcase" (a storage vessel pressurized from an external source of compressed gas), but by far the most common devices for the purpose are pumps.

Pumps increase the mechanical energy of the liquid, increasing its velocity, pressure, or elevation—or all three. The two major classes are positive-displacement pumps and centrifugal pumps. Positive-displacement units apply pressure directly to the liquid by a reciprocating piston, or by rotating members which form chambers alternately filled by and emptied of the liquid. Centrifugal pumps generate high rotational velocities, then convert the resulting kinetic energy of the liquid to pressure energy.

In pumps the density of liquid does not change appreciably and may be considered constant.

Developed head

A typical pump application is shown diagrammatically in Fig. 8.5. The pump is installed in a pipeline to provide the energy needed to draw liquid from a reservoir and discharge a constant volumetric flow rate at the exit of the pipeline, Z_b m or ft above the level of the liquid. At the pump itself, the liquid enters the suction connection at station a and leaves the discharge connection at station b. A Bernoulli equation can be written between stations a and b. Equation (4.65) serves for this. Since the only friction is that occurring in the pump itself and is accounted for by the mechanical efficiency η, $h_f = 0$. Then Eq. (4.65) can be written

$$\eta W_p = \left(\frac{p_b}{\rho} + g Z_b + \frac{\alpha_b \bar{V}_b^2}{2} \right) - \left(\frac{p_a}{\rho} + g Z_a + \frac{\alpha_a \bar{V}_a^2}{2} \right) \qquad (8.2a)$$

or in fps units,

$$\eta W_p = \left(\frac{p_b}{\rho} + \frac{g Z_b}{g_c} + \frac{\alpha_b \bar{V}_b^2}{2 g_c} \right) - \left(\frac{p_a}{\rho} + \frac{g Z_a}{g_c} + \frac{\alpha_a \bar{V}_a^2}{2 g_c} \right) \qquad (8.2b)$$

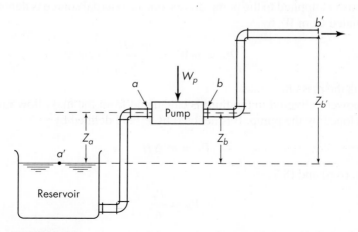

FIGURE 8.5
Pump flow system.

The quantities in the parentheses are called *total heads* and are denoted by H, or

$$H = \frac{p}{\rho} + gZ + \frac{\alpha \bar{V}^2}{2} \tag{8.3a}$$

and

$$H = \frac{p}{\rho} + \frac{gZ}{g_c} + \frac{\alpha \bar{V}^2}{2g_c} \tag{8.3b}$$

In pumps the difference between the heights of the suction and discharge connections is usually negligible, and Z_a and Z_b can be dropped from Eq. (8.2). If H_a is the total suction head, H_b the total discharge head, and $\Delta H = H_b - H_a$, Eq. (8.2) can be written

$$W_p = \frac{H_b - H_a}{\eta} = \frac{\Delta H}{\eta} \tag{8.4}$$

The head H in Eq. (8.3) has the dimensions of work per unit mass (or, equivalently, length squared per time squared). By multiplying through by $1/g$ (or g_c/g in fps units), we obtain

$$\frac{H}{g} = \frac{p}{\rho g} + Z + \frac{\alpha \bar{V}^2}{2g} \tag{8.5a}$$

and

$$\frac{H g_c}{g} = \frac{p g_c}{\rho g} + Z + \frac{\alpha \bar{V}^2}{2g} \tag{8.5b}$$

In Eqs. (8.5a) and (8.5b) each term has the dimension of length. Often the head developed by a pump is expressed in meters or feet of fluid, although here the term *head* refers to $\Delta H/g$ or $\Delta H g_c/g$. Note that in fps units ΔH and $\Delta H g_c/g$ are virtually the same numerically.

Power requirement

The power supplied to the pump drive from an external source is denoted by P_B. It is calculated from W_p by

$$P_B = \dot{m} W_p = \frac{\dot{m} \Delta H}{\eta} \tag{8.6}$$

where $\dot{m}$ is the mass flow rate.

The power delivered to the fluid is calculated from the mass flow rate and the head developed by the pump. It is denoted by P_f and defined by

$$P_f = \dot{m} \Delta H \tag{8.7}$$

From Eqs. (8.6) and (8.7)

$$P_B = \frac{P_f}{\eta} \tag{8.8}$$

Equations (8.2) to (8.8) can also be used for fans by using an average density $\bar{\rho} = (\rho_a + \rho_b)/2$ for ρ.

Suction lift and cavitation

The power calculated by Eq. (8.6) depends on the difference in pressure between discharge and suction and is independent of the pressure level. From energy considerations it is immaterial whether the suction pressure is below atmospheric pressure or well above it, as long as the fluid remains liquid. However, if the suction pressure is only slightly greater than the vapor pressure, some liquid may flash to vapor inside the pump, a process called cavitation, which greatly reduces the pump capacity and causes severe erosion. If the suction pressure is actually less than the vapor pressure, there will be vaporization in the suction line, and no liquid can be drawn into the pump.

To avoid cavitation, the pressure at the pump inlet must exceed the vapor pressure by a certain value, called the *net positive suction head* (NPSH). The required value of NPSH is about 2 to 3 m (5 to 10 ft) for small centrifugal pumps; but it increases with pump capacity, impeller speed, and discharge pressure, and values up to 15 m (50 ft) are recommended for very large pumps. For a pump taking suction from a reservoir, like that shown in Fig. 8.5, the available NPSH is customarily calculated as

$$\text{NPSH} = \frac{1}{g}\left(\frac{p_{a'} - p_v}{\rho} - h_{fs}\right) - Z_a \tag{8.9a}$$

or in fps units,

$$\text{NPSH} = \frac{g_c}{g}\left(\frac{p_{a'} - p_v}{\rho} - h_{fs}\right) - Z_a \tag{8.9b}$$

where $p_{a'}$ = absolute pressure at surface of reservoir
p_v = vapor pressure
h_{fs} = friction in suction line

The velocity head at the pump inlet $\alpha_a \bar{V}_a^2/2$ could be subtracted from the result given by Eq. (8.9) to give a more theoretically correct value of the available NPSH. But this term is usually only about 30 to 60 cm (1 to 2 ft) and is accounted for in the values of the minimum required net positive suction head (NPSHR) specified by the pump manufacturers. Because of the cyclic variations in flow rate, the velocity head (or more properly, in this case, the acceleration head) may be important in specifying positive-displacement pumps.[6]

For the special situation where the liquid is practically nonvolatile ($p_v = 0$), the friction negligible ($h_{fs} = 0$), and the pressure at station a' atmospheric, the maximum possible suction lift can be obtained by subtracting the required NPSH from the barometric head. For cold water, this maximum suction lift is about 10.4 m (34 ft); the maximum practical lift is about 7.6 m (25 ft).

EXAMPLE 8.1. Benzene at $100°\,\text{F}$ ($37.8°\,\text{C}$) is pumped through the system of Fig. 8.5 at the rate of 40 gal/min ($9.09\ \text{m}^3/\text{h}$). The reservoir is at atmospheric pressure. The gauge pressure at the end of the discharge line is 50 $\text{lb}_f/\text{in.}^2$ ($345\ \text{kN/m}^2$). The discharge is 10 ft, and the pump suction is 4 ft above the level in the reservoir. The discharge line is $1\frac{1}{2}$-in. Schedule 40 pipe. The friction in the suction line is known to be 0.5 $\text{lb}_f/\text{in.}^2$ ($3.45\ \text{kN/m}^2$), and that in the discharge line is 5.5 $\text{lb}_f/\text{in.}^2$ ($37.9\ \text{kN/m}^2$). The mechanical efficiency of the pump is 0.60 (60 percent). The density of benzene is 54 lb/ft^3 (865 kg/m^3), and its vapor pressure at 100° F (37.8° C) is 3.8 lb$_f$/in.^{2} (26.2 kN/m^2). Calculate (a) the developed head of the pump and (b) the total power input. (c) If the pump manufacturer specifies a required NPSHR of 10 ft (3.05 m), will the pump be suitable for this service?

Solution

(a) The pump work W_p is found by using Eq. (4.65). The upstream station a' is at the level of the liquid in the reservoir, and the downstream station b' is at the end of the discharge line, as shown in Fig 8.5. When the level in the tank is chosen as the datum of heights and it is noted that $\bar{V}_{a'} = 0$, Eq. (4.65) gives

$$W_p \eta = \frac{p_{b'}}{\rho} + \frac{g Z_{b'}}{g_c} + \frac{\alpha_{b'} \bar{V}_{b'}^2}{2g_c} + h_f - \frac{p_{a'}}{\rho}$$

The exit velocity $\bar{V}_{b'}$ is found by using data from App. 3. For a $1\frac{1}{2}$-in. Schedule 40 pipe, a velocity of 1 ft/s corresponds to a flow rate of 6.34 gal/min, and

$$\bar{V}_{b'} = \frac{40}{6.34} = 6.31 \text{ ft/s}$$

With $\alpha_{b'} = 1.0$, Eq. (4.65) gives

$$W_p \eta = \frac{(14.7 + 50)(144)}{54} + \frac{g}{g_c}(10) + \frac{6.31^2}{2 \times 32.17} + \frac{(5.5 + 0.5)(144)}{54} - \frac{14.7 \times 144}{54}$$
$$= 159.9 \text{ ft} \cdot \text{lb}_f/\text{lb}$$

By Eq. (8.4) $W_p \eta$ is also the developed head, and

$$\Delta H = H_b - H_a = 159.9 \text{ ft} \cdot \text{lb}_f/\text{lb} \ (477.9 \text{ J/kg})$$

(b) The mass flow rate is

$$\dot{m} = \frac{40 \times 54}{7.48 \times 60} = 4.81 \text{ lb/s} \ (2.18 \text{ kg/s})$$

The power input is, from Eq. (8.6),

$$P_B = \frac{4.81 \times 159.9}{550 \times 0.60} = 2.33 \text{ hp} \ (1.74 \text{ kW})$$

(c) Use Eq. (8.9), $p_{a'}/\rho = 14.7 \times 144/54 = 39.2 \text{ ft} \cdot \text{lb}_f/\text{lb}$. The vapor pressure corresponds to a head of

$$\frac{3.8 \times 144}{54} = 10.1 \text{ ft} \cdot \text{lb}_f/\text{lb} \ (30.2 \text{ J/kg})$$

The friction in the suction line is

$$h_f = \frac{0.5 \times 144}{54} = 1.33 \text{ ft} \cdot \text{lb}_f/\text{lb} \ (3.98 \text{ J/kg})$$

The value of the available NPSH from Eq. (8.9), assuming $g/g_c = 1$, is

$$\text{NPSH} = 39.2 - 10.1 - 1.33 - 4 = 23.77 \text{ ft} \ (7.25 \text{ m})$$

The available NPSH is considerably larger than the minimum required value of 10 ft, so the pump should be suitable for the proposed service.

Positive-Displacement Pumps

In the first major class of pumps, a definite volume of liquid is trapped in a chamber, which is alternately filled from the inlet and emptied at a higher pressure through the discharge. There are two subclasses of positive-displacement pumps. In reciprocating pumps, the chamber is a stationary cylinder that contains a piston or plunger; in rotary pumps, the chamber moves from inlet to discharge and back to the inlet.

Reciprocating pumps

Piston pumps, plunger pumps, and diaphragm pumps are examples of reciprocating pumps. In a piston pump, liquid is drawn through an inlet check valve into the cylinder by the withdrawal of a piston and then is forced out through a discharge check valve on the return stroke. Most piston pumps are double-acting with liquid admitted alternately on each side of the piston so that one part of the cylinder is being filled while the other is being emptied. Often two or more cylinders are used in parallel with common suction and discharge headers, and the configuration of the pistons is adjusted to minimize fluctuations in the discharge rate. The piston may be motor-driven through reducing gears, or a steam cylinder may be used to drive the piston rod directly. The maximum discharge pressure for commercial piston pumps is about 50 atm.

For higher pressures, plunger pumps are used. A heavy-walled cylinder of small diameter contains a close-fitting reciprocating plunger, which is merely an extension of the piston rod. At the limit of its stroke the plunger fills nearly all the space in the cylinder. Plunger pumps are single-acting and usually are motor-driven. They can discharge against a pressure of 1,500 atm or more.

In a diaphragm pump, the reciprocating member is a flexible diaphragm of metal, plastic, or rubber. This eliminates the need for packing or seals exposed to the liquid being pumped, a great advantage when handling toxic or corrosive liquids. A typical unit is shown in Fig. 8.6. Diaphragm pumps handle small to moderate amounts of liquid, up to about 100 gal/min, and can develop pressures in excess of 100 atm.

The mechanical efficiency of reciprocating pumps varies from 40 to 50 percent for small pumps to 70 to 90 percent for large ones. It is nearly independent of speed within normal operating limits and decreases slightly with an increase in discharge pressure because of added friction and leakage.

Volumetric efficiency. The ratio of the volume of fluid discharged to the volume swept by the piston or plunger is called the *volumetric efficiency*. In positive-displacement pumps the volumetric efficiency is nearly constant with increasing discharge pressure, although it drops a little because of leakage. Because of the constancy of volume flow, plunger and diaphragm pumps are widely used as "metering pumps," injecting liquid into a process system at controlled but adjustable volumetric rates.

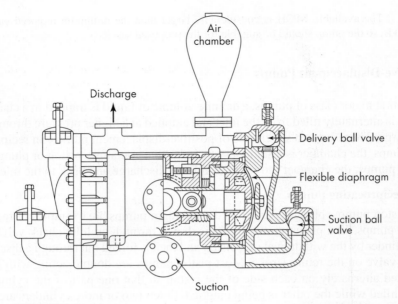

FIGURE 8.6
Diaphragm pump.

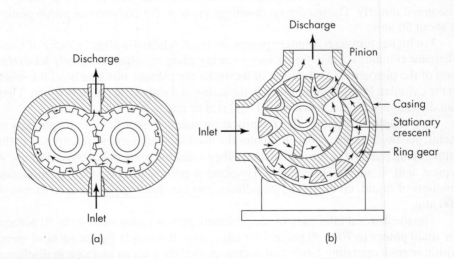

FIGURE 8.7
Gear pumps: (*a*) spur-gear pump; (*b*) internal-gear pump.

Rotary pumps

A wide variety of rotary positive-displacement pumps are available. They bear such names as gear pumps, lobe pumps, screw pumps, cam pumps, and vane pumps. Two examples of gear pumps are shown in Fig. 8.7. Unlike reciprocating pumps, rotary pumps contain no check valves. Close tolerances between the moving and stationary parts minimize leakage from the discharge space back to the suction space; they also limit the operating speed. Rotary pumps operate best on clean, moderately viscous fluids, such as light lubricating oil. Discharge pressures up to 200 atm or more can be attained.

In the spur-gear pump (Fig. 8.7a) intermeshing gears rotate with close clearance inside the casing. Liquid entering the suction line at the bottom of the casing is caught in the spaces between the teeth and the casing and is carried around to the top of the casing and forced out the discharge. Liquid cannot short-circuit back to the suction because of the close meshing of the gears in the center of the pump.

In the internal-gear pump (Fig. 8.7b) a spur gear, or pinion, meshes with a ring gear with internal teeth. Both gears are inside the casing. The ring gear is coaxial with the inside of the casing, but the pinion, which is externally driven, is mounted eccentrically with respect to the center of the casing. A stationary metal crescent fills the space between the two gears. Liquid is carried from inlet to discharge by both gears, in the spaces between the gear teeth and the crescent.

Centrifugal Pumps

In the second major class of pumps, the mechanical energy of the liquid is increased by centrifugal action. A simple, but very common, example of a centrifugal pump is shown in Fig. 8.8. The liquid enters through a suction connection concentric with the axis of a high-speed rotary element called the *impeller*, which carries radial vanes integrally cast in it. Liquid flows outward in the spaces between the vanes and leaves the impeller at a considerably greater velocity with respect to the ground than at the entrance to the impeller. In a properly functioning pump, the space between the vanes is completely filled with liquid flowing without cavitation. The liquid leaving the outer periphery of the impeller is collected in a spiral casing called the *volute* and leaves the pump through a tangential discharge connection. In the

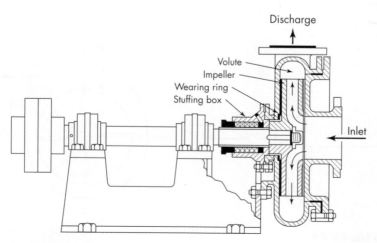

FIGURE 8.8
Single-suction centrifugal pump.

volute the velocity head of the liquid from the impeller is converted to pressure head. The power is applied to the fluid by the impeller and is transmitted to the impeller by the torque of the driveshaft, which usually is driven by a direct-connected motor at constant speed, commonly at 1,750 or 3,450 r/min.

Under ideal conditions of frictionless flow, the mechanical efficiency of a centrifugal pump is, of course, 100 percent, and $\eta = 1$. An ideal pump operating at a given speed delivers a definite discharge rate at each specific developed head. Actual pumps, because of friction and other departures from perfection, fall considerably short of the ideal case.

Centrifugal pumps constitute the most common type of pumping machinery in ordinary plant practice. They come in a number of types other than the simple volute machine shown in Fig. 8.8. A common type uses a double-suction impeller, which accepts liquid from both sides. The impeller itself may be a simple open spider, or it may be enclosed or shrouded. Handbooks, texts on pumps, and especially the catalogs of pump manufacturers show many types, sizes, and designs of centrifugal pumps.

Centrifugal pump theory

The basic equations interrelating the power, developed head, and capacity of a centrifugal pump are derived for the ideal pump from fundamental principles of fluid dynamics. Since the performance of an actual pump differs considerably from that of an ideal one, actual pumps are designed by applying experimentally measured corrections to the ideal situation.

Figure 8.9 shows diagrammatically how the liquid flows through a centrifugal pump. The liquid enters axially at the suction connection, station a. In the rotating eye of the impeller, the liquid spreads out radially and enters the channels between the vanes at station 1. It flows through the impeller, leaves the periphery of the impeller at station 2, is collected in the volute, and leaves the pump discharge at station b.

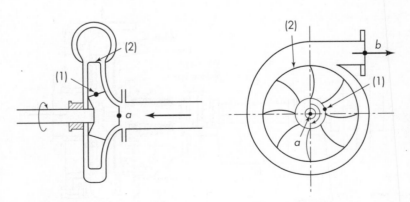

FIGURE 8.9
Centrifugal pump showing Bernoulli stations.

The performance of the pump is analyzed by considering separately the three parts of the total path: first, the flow from station a to station 1; second, the flow through the impeller from station 1 to station 2; and third, the flow through the volute from station 2 to station b. The heart of the pump is the impeller, and the fluid mechanical theory of the second section of the fluid path is considered first.

Figure 8.10 shows a single vane, one of several in an impeller. The vectors represent the various velocities at stations 1 and 2 at the entrance and exit of the vane, respectively. Consider first the vectors at station 2. By virtue of the design of the pump, the tangent to the impeller at its terminus makes an angle β_2 with the tangent to the circle traced out by the impeller tip. Vector v_2 is the velocity of the fluid at point 2 as seen by an observer moving with the impeller, and it is therefore a *relative* velocity. Two idealizations are now accepted. It is assumed, first, that all liquid flowing across the periphery of the impeller is moving at the same speed, so the numerical value (but not the vector direction) is v_2 at all points; second, it is assumed that the angle between the vector v_2 and the tangent is the actual vane angle β_2. This assumption in turn is equivalent to an assumption that there are an infinite number of vanes, of zero thickness, at an infinitesimal distance apart. This ideal state is referred to as *perfect guidance*. Point 2 at the tip of the blades is moving at peripheral velocity u_2 with respect to the axis. Vector V_2 is the resultant velocity of the fluid stream leaving the impeller as observed from the ground. It is called the *absolute velocity of the fluid* and is, by the parallelogram law, the vector sum of relative velocity v_2 and peripheral velocity u_2. The angle between vectors V_2 and u_2 is denoted by α_2.

A comparable set of vectors applies to the entrance to the vanes at station 1, as shown in Fig. 8.10a. In the usual design, α_1 is nearly 90°, and vector V_1 can be considered radial.

Figure 8.10b is the vector diagram for point 2 that shows the relations among the various vectors in a more useful way. It also shows how the absolute velocity vector V_2 can be resolved into components, a radial component denoted by V_{r2} and a peripheral component denoted by V_{u2}.

The power input to the impeller, and therefore the power required by the pump, can be calculated from the angular momentum equation for steady flow, Eq. (4.53). To be consistent with Fig. 8.10, quantity $u_{\theta2}$ in Eq. (4.53) becomes V_{u2}, and $u_{\theta1}$ becomes V_{u1}. Then

$$T = \dot{m}(r_2 V_{u2} - r_1 V_{u1}) \tag{8.10}$$

The momentum correction factors [the β's of Eq. (4.42)] are unity in view of the assumption of perfect guidance. Also, in radial flow, where $\alpha = 90°$, $V_{u1} = 0$. At the entrance, therefore, $r_1 V_{u1} = 0$, the second term in the parentheses of Eq. (8.10) vanishes, and

$$T = \dot{m} r_2 V_{u2} \tag{8.11}$$

Since $P = T\omega$, the power equation for an ideal pump is

$$P_{fr} = \dot{m}\omega r_2 V_{u2} \tag{8.12}$$

where the subscript r denotes a frictionless pump.

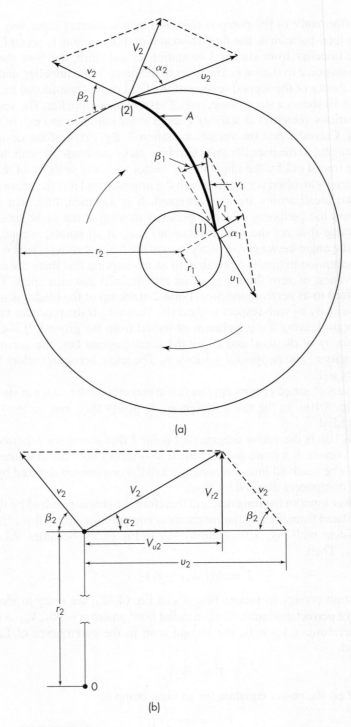

FIGURE 8.10
Velocities at entrance and discharge of vanes in centrifugal pump: (*a*) vectors and vane; (*b*) vector diagram at tip of vane.

Head-flow relations for an ideal pump

From Eq. (8.7), written for an ideal pump, $P = P_{fr} = \dot{m}\Delta H$, and therefore,

$$\Delta H = r_2 V_{u2}\omega \tag{8.13}$$

Since $\omega r_2 = u_2$,

$$\Delta H = u_2 V_{u2} \tag{8.14}$$

From Fig. 8.10b, $V_{u2} = u_2 - V_{r2}/\tan \beta_2$, and

$$\Delta H_r = u_2(u_2 - V_{r2}/\tan \beta_2) \tag{8.15}$$

The volumetric flow rate q_r through the pump is given by

$$q_r = V_{r2} A_p \tag{8.16}$$

where A_p is the total cross-sectional area of the channel around the periphery. Combining Eqs. (8.15) and (8.16) gives

$$\Delta H_r = u_2(u_2 - q_r/A_p \tan \beta_2) \tag{8.17}$$

Since u_2, A_p, and β_2 are constant, Eq. (8.17) shows that the relation between head and volumetric flow is linear. The slope of the head flow rate line depends on the sign of $\tan \beta_2$ and therefore varies with angle β_2. If β_2 is less than $90°$, as is nearly always the case, the line has a negative slope. Flow in a piping system may become unstable if the line is horizontal or has a positive slope.

Head-work relation in an ideal pump

The work done per unit mass of liquid passing through an ideal pump, from Eqs. (8.6) and (8.12), is

$$W_{pr} = \frac{P_{fr}}{\dot{m}} = \omega r_2 V_{u2} \tag{8.18}$$

A Bernoulli equation written between stations 1 and 2, assuming no friction, neglecting $Z_a - Z_b$, and assuming perfect guidance, gives

$$\omega r_2 V_{u2} = \frac{p_2}{\rho} - \frac{p_1}{\rho} + \frac{V_2^2}{2} - \frac{V_1^2}{2} \tag{8.19}$$

Also, Bernoulli equations written between stations a and 1 and b and 2, respectively, are

$$\frac{p_a}{\rho} + \frac{\alpha_a \bar{V}_a^2}{2} = \frac{p_1}{\rho} + \frac{V_1^2}{2} \tag{8.20}$$

$$\frac{p_2}{\rho} + \frac{V_2^2}{2} = \frac{p_b}{\rho} + \frac{\alpha_b \bar{V}_b^2}{2} \tag{8.21}$$

Adding Eqs. (8.19), (8.20), and (8.21) gives

$$\frac{p_b}{\rho} + \frac{\alpha_b \bar{V}_b^2}{2} = \frac{p_a}{\rho} + \frac{\alpha_a \bar{V}_a^2}{2} + \omega r_2 V_{u2} \tag{8.22}$$

Equation (8.22) can be written in the form

$$\Delta H_r = H_b - H_a = \omega r_2 V_{u2} = W_{pr} \tag{8.23}$$

Equation (8.23) is in agreement with Eq. (8.4) for a frictionless or ideal pump, where $\eta = 1.0$, $\Delta H = \Delta H_r$, and $W_p = W_{pr}$.

Actual performance of centrifugal pump

The developed head of an actual pump is considerably less than that calculated from the ideal pump relation of Eq. (8.23). Also, the efficiency is less than unity, and the fluid horsepower is greater than the ideal horsepower. Head losses and power losses are discussed separately.

Loss of head; circulatory flow

A basic assumption in the theory of the ideal pump was that of complete guidance, so the angle between vectors v_2 and u_2 equals the vane angle β_2. The guidance is not perfect in the real pump, and the actual stream of fluid leaves at an angle considerably less than β_2. The physical reason is that the velocity in a given cross section is far from uniform. The effect is the result of an end-to-end circulatory flow of liquid within the impeller channels superimposed on the net flow through the channel. Because of circulation, the resultant velocity V_2' is smaller than the theoretical value. The vector diagram in Fig. 8.11 shows how circulation modifies the theoretical velocities and angles. The full lines and the primed quantities apply to the actual pump, and the dotted lines and the unprimed quantities apply to the theoretical case. The speed of the pump, and hence u_2, and the flow of fluid through the pump, proportional to V_{r2}, are the same in the two cases. It is clear that the angle β_2' and the tangential component V_{u2}' both decrease. By Eq. (8.23), then, the developed head is also decreased.

Performance characteristics

The performance of a given pump is commonly illustrated by plots of actual head, power consumption, and efficiency versus volumetric flow rate. The theoretical head flow rate (usually called *head capacity*) relation is a straight line, in accordance with Eq. (8.17); the actual developed head at given flow rates is considerably less and drops precipitously to zero as the flow rate increases to a certain value. This is known as the *zero-head flow rate;* it is the maximum flow the pump can deliver

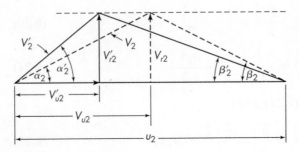

FIGURE 8.11
Vector diagram showing effect of circulation in pump impeller.

under any conditions. The rated or optimum flow rate is, of course, less than this. The maximum head is developed at zero flow, with the discharge line blocked off. (A centrifugal pump should not be operated under such a condition for more than a short time; a positive-displacement pump, never.)

The difference between the theoretical and actual performance results primarily from circulatory flow. Contributing factors to the loss of head are fluid friction in the passages and channels of the pump and shock losses from the sudden change in direction of the liquid leaving the impeller and joining the stream of liquid traveling circumferentially around the casing. Friction is highest at the maximum flow rate; shock losses are a minimum at the rated operating conditions of the pump and become greater as the flow rate is increased or decreased from the rated value.

Power consumption. The fluid power P_f rises with flow rate to a maximum value at or near the rated capacity, then falls slightly. The total or actual power required P_B rises through most of the range of flow rates. The difference between them represents power lost in the pump; it results from fluid friction and shock losses, both of which are conversion of mechanical energy to heat, and by leakage, disk friction, and bearing losses. Leakage is the unavoidable reverse flow from the impeller discharge past the wearing ring to the suction eye; this reduces the volume of the actual discharge from the pump per unit of power expended. Disk friction is the friction between the outer surface of the impeller and the liquid in the space between the impeller and the inside of the casing. Bearing losses constitute the power required to overcome mechanical friction in the bearing and stuffing boxes or seals of the pump.

Efficiency. As shown by Eq. (8.8), the pump efficiency is the ratio of fluid power to the total power consumed. The efficiency rises rapidly with flow rate at low rates, reaches a maximum at or near the rated capacity, then falls as the flow rate approaches the zero-head value.

Characteristic curves. The head-capacity curve is known as the *characteristic curve* of the pump. Such curves for a centrifugal pump with a 5-in. (125-mm) impeller are shown in Fig. 8.12. At the rated speed of 3,450 r/min the rated capacity is 200 gal/min (45.4 m³/h) at a total head of 88 ft (27 m) of fluid. The power required is 5.5 hp (4.1 kW); the efficiency is 80 percent. If the flow rate is reduced to 150 gal/min or increased to 240 gal/min, the efficiency falls to 77 percent. At lower impeller speeds the developed head, power required, and efficiency are all less than those at 3,450 r/min. At 1,750 r/min, for example, the maximum efficiency is only 77 percent at 105 gal/min and a head of 22 ft; the power required at this flow rate is about 0.7 hp. A smaller impeller at the same speed also gives lower flow rates, power requirements, and efficiency.

Affinity laws. When a complete set of performance curves is not available, the characteristics of a particular pump can be predicted from a similar pump and the theoretical equations for an ideal pump. The effects of impeller size and speed on capacity, head, and power are found from Eqs. (8.16), (8.14), and (8.7). The

TABLE 8.1
Affinity laws for pumps

Characteristic	Constant D	Constant n
Capacity	$q \propto n$	$q \propto D$
Head	$\Delta H \propto n^2$	$\Delta H \propto D^2$
Power	$P \propto n^3$	$P \propto D^3$

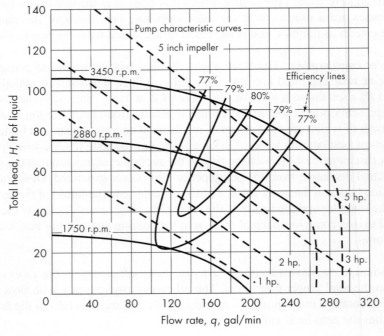

FIGURE 8.12
Characteristic curves of a centrifugal pump operating at various speeds. (*By permission from Perry's Chemical Engineers' Handbook, 7th ed., p. 10-25. Copyright © 1997, McGraw-Hill.*)

relationships are called the *affinity laws;* they are given in Table 8.1, in which D is the diameter of the pump impeller and n is the impeller speed.

Affinity laws are useful when an existing pump must be modified to give a higher or lower head or a different capacity. Changing the impeller size or speed is often less expensive than buying a new pump.

Multistage centrifugal pumps

The maximum head that it is practicable to generate in a single impeller is limited by the peripheral speed reasonably attainable. A so-called high-energy centrifugal pump can develop a head of more than 650 ft (200 m) in a single stage; but generally when a head greater than about 100 ft (30 m) is needed, two or more impellers can be mounted in series on a single shaft and a multistage pump so obtained. The discharge from the first stage provides suction for the second, the discharge from the second provides suction for the third, and so forth. The developed heads of all stages add to give a total head several times that of a single stage.

Leakproof pumps

Because of environmental considerations, leakproof centrifugal pumps are increasingly used for handling dangerous liquids. There are two main types, both of which contain no seals or stuffing boxes. In canned-rotor pumps, a stainless-steel canlike enclosure surrounds the motor rotor to keep the pumped fluid away from the motor. In magnetic-drive pumps, the impeller, which carries magnets, is driven by a magnet-carrying disk on the other side of the casing wall. Both types are less efficient than conventional pumps, but a lower efficiency is often preferable to installing complicated mechanical seals and seal-flushing systems.

Pump priming

Equation (8.23) shows that the theoretical head developed by a centrifugal pump depends on the impeller speed, the radius of the impeller, and the velocity of the fluid leaving the impeller. If these factors are constant, the developed head is the same for fluids of all densities and is the same for liquids and gases. The increase in pressure, however, is the product of the developed head and the fluid density. If a pump develops, say, a head of 100 ft and is full of water, the increase in pressure is $100 \times 62.3/144 = 43\,\text{lb}_f/\text{in.}^2$ (2.9 atm). If the pump is full of air at ordinary density, the pressure increase is about $0.05\,\text{lb}_f/\text{in.}^2$ (0.0035 atm). A centrifugal pump trying to operate on air, then, can neither draw liquid upward from an initially empty suction line nor force liquid along a full discharge line. A pump with air in its casing is *airbound* and can accomplish nothing until the air has been replaced by a liquid. Air can be displaced by priming the pump from an auxiliary priming tank connected to the suction line or by drawing liquid into the suction line by an independent source of vacuum. Also, several types of self-priming pumps are available.

Positive-displacement pumps can compress a gas to a required discharge pressure and are not usually subject to air binding.

FANS, BLOWERS, AND COMPRESSORS

These are machines that move and compress gases. Fans discharge large volumes of gas (usually air) into open spaces or large ducts. They are low-speed machines that generate very low pressures, on the order of 0.04 atm. Blowers are high-speed rotary devices (using either positive displacement or centrifugal force) that develop a maximum pressure of about 2 atm. Compressors, which are also positive-displacement or centrifugal machines, discharge at pressures from 2 atm to several thousand atmospheres. Note that while *pump* generally refers to a device for moving a liquid, the terms *air pump* and *vacuum pump* designate machines for compressing a gas.

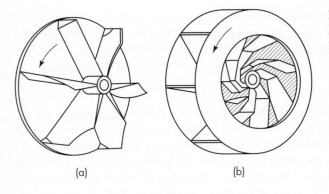

FIGURE 8.13
Impellers for centrifugal
fans.

(a)   (b)

In fans the density of the fluid does not change appreciably and may be assumed constant. In blowers and compressors, however, the density change is too great to justify this assumption, and in discussing these devices compressible flow theory is required.

Fans

Large fans are usually centrifugal, operating on exactly the same principle as centrifugal pumps. Their impeller blades, however, may be curved forward; this would lead to instability in a pump, but not in a fan. Typical fan impellers are shown in Fig. 8.13; they are mounted inside light sheet-metal casings. Clearances are large and discharge heads low, from 5 to 60 in. (130 to 1,500 mm) H_2O. Sometimes, as in ventilating fans, nearly all the added energy is converted to velocity energy and almost none to pressure head. In any case, the gain in velocity absorbs an appreciable fraction of the added energy and must be included in estimating the efficiency and power. The total efficiency, where the power output is credited with both pressure and velocity heads, is about 70 percent.

Since the change in density in a fan is small, the incompressible flow equations used in the discussion of centrifugal pumps are adequate. One difference between pumps and gas equipment recognizes the effect of pressure and temperature on the density of the gas entering the machine. Gas equipment is ordinarily rated in terms of *standard cubic feet*. A volume in standard cubic feet is that measured at a specified temperature and pressure regardless of the actual temperature and pressure of the gas to the machine. Various standards are used in different industries, but a common one is based on a pressure of 30 in. Hg and a temperature of 60° F (520° R). This corresponds to a molal volume of 378.7 ft^3/lb mol.

EXAMPLE 8.2. A centrifugal fan is used to take flue gas at rest and at a pressure of 29.0 in. (737 mm) Hg and a temperature of 200° F (93.3°C) and discharge it at a pressure of 30.1 in. (765 mm) Hg and a velocity of 150 ft/s (45.7 m/s). Calculate the power needed to move 10,000 std ft^3/min (16,990 m^3/h) of gas. The efficiency of the fan is 65 percent, and the molecular weight of the gas is 31.3.

Solution. The actual suction density is

$$\rho_a = \frac{31.3 \times 29.0(460 + 60)}{378.7 \times 30(460 + 200)} = 0.0629 \text{ lb/ft}^3$$

and the discharge density is

$$\rho_b = 0.0629\left(\frac{30.1}{29.0}\right) = 0.0653 \text{ lb/ft}^3$$

The average density of the flowing gas is

$$\bar{\rho} = \tfrac{1}{2}(0.0629 + 0.0653) = 0.0641 \text{ lb/ft}^3$$

The mass flow rate is

$$\dot{m} = \frac{10{,}000 \times 31.3}{378.7 \times 60} = 13.78 \text{ lb/s}$$

The developed pressure is

$$\frac{p_b - p_a}{\bar{\rho}} = \frac{(30.1 - 29)(144)(14.7)}{29.92 \times 0.0641} = 1214 \text{ ft} \cdot \text{lb}_f/\text{lb}$$

The velocity head is

$$\frac{\bar{V}_b^2}{2g_c} = \frac{150^2}{2 \times 32.17} = 349.7 \text{ ft} \cdot \text{lb}_f/\text{lb}$$

From Eq. (8.2b), calling $\alpha_a = \alpha_b = 1.0$, $\bar{V}_a = 0$, and $Z_a = Z_b$,

$$W_p = \frac{1}{\eta}\left(\frac{p_b - p_a}{\bar{\rho}} + \frac{\bar{V}_b^2}{2g_c}\right) = \frac{1{,}214 + 349.7}{0.65} = 2{,}406 \text{ ft} \cdot \text{lb}_f/\text{lb}$$

From Eq. (8.6)

$$P_B = \frac{\dot{m}W_p}{550} = \frac{13.78 \times 2{,}406}{550} = 60.3 \text{ hp } (45.0 \text{ kW})$$

Blowers and Compressors

When the pressure on a compressible fluid is increased adiabatically, the temperature of the fluid also increases. The temperature rise has a number of disadvantages. Because the specific volume of the fluid increases with temperature, the work required to compress a pound of fluid is larger than if the compression were isothermal. Excessive temperatures lead to problems with lubricants, stuffing boxes, and materials of construction. The fluid may be one that cannot tolerate high temperatures without decomposing.

For the isentropic (adiabatic and frictionless) pressure change of an ideal gas, the temperature relation is, using Eq. (6.24),

$$\frac{T_b}{T_a} = \left(\frac{p_b}{p_a}\right)^{1-1/\gamma} \tag{8.24}$$

where T_a, T_b = inlet and outlet absolute temperatures, respectively
p_a, p_b = corresponding inlet and outlet pressures
γ = ratio of specific heats c_p/c_v

For a given gas, the temperature ratio increases with an increase in the compression ratio p_b/p_a. This ratio is a basic parameter in the engineering of blowers and compressors. In blowers with a compression ratio below about 3 or 4, the adiabatic temperature rise is not large, and no special provision is made to reduce it. In compressors, however, where the compression ratio may be as high as 10 or more, the isentropic temperature becomes excessive. Also, since actual compressors are not frictionless, the heat from friction is absorbed by the gas, and temperatures well above the isentropic temperature are attained. Compressors, therefore, are cooled by jackets through which cold water or refrigerant is circulated. In small cooled compressors, the exit gas temperature may approach that at the inlet, and isothermal compression is achieved. In very small ones, air cooling by external fins cast integrally with the cylinder is sufficient. In larger units, where cooling capacity is limited, a path different from isothermal or adiabatic compression, called polytropic compression, is followed.

Positive-displacement blower

A positive-displacement blower is shown in Fig. 8.14a. These machines operate as gear pumps do except that, because of the special design of the "teeth," the clearance is only a few thousandths of an inch. The relative position of the impellers is maintained precisely by heavy external gears. A single-stage blower can discharge gas at 0.4 to 1 atm gauge, a two-stage blower at 2 atm. The blower shown in Fig. 8.14a has two lobes. Three-lobe machines are also common.

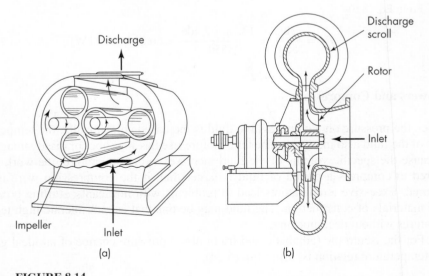

FIGURE 8.14
Typical blowers: (a) positive-displacement two-lobe blower; (b) single-suction centrifugal blower.

Centrifugal blowers

A single-stage centrifugal blower is shown in Fig. 8.14b. In appearance it resembles a centrifugal pump, except that the casing is narrower and the diameters of the casing and discharge scroll are relatively larger than in a centrifugal pump. The operating speed is high—3,600 r/min or more. The ideal power is given by Eq. (8.12). High speeds and large impeller diameters are required because very high heads, in meters or feet of low-density fluid, are needed to generate modest pressure ratios. Thus the velocities appearing in a diagram like Fig. 8.10 are, for a centrifugal blower, approximately 10 times those in a centrifugal pump.

Positive-displacement compressors

Rotary positive-displacement compressors can be used for discharge pressures up to about 6 atm. These devices include sliding-vane, screw-type, and liquid-piston compressors. (See Ref. 7.) For high to very high discharge pressures and modest flow rates, reciprocating compressors are the most common type. An example of a single-stage compressor is shown in Fig. 8.15. These machines operate mechanically in the same way as reciprocating pumps, with the differences that leak prevention is more difficult and the temperature rise is important. The cylinder walls and cylinder heads are cored for cooling jackets using water or refrigerant. Reciprocating compressors are usually motor-driven and are nearly always double-acting.

When the required compression ratio is greater than can be achieved in one cylinder, multistage compressors are used. Between each stage are coolers, which are tubular heat exchangers cooled by water or refrigerant. Intercoolers have sufficient heat-transfer capacity to bring the interstage gas streams to the initial suction temperature. Often an aftercooler is used to cool the high-pressure gas from the final stage.

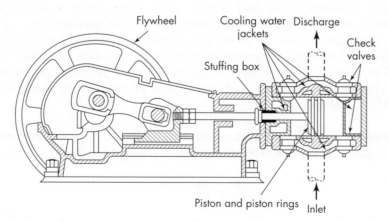

FIGURE 8.15
Reciprocating compressor.

Centrifugal compressors

Centrifugal compressors are multistage units containing a series of impellers on a single shaft rotating at high speeds in a massive casing.[1,4] Internal channels lead from the discharge of one impeller to the inlet of the next. These machines compress enormous volumes of air or process gas—up to 200,000 ft³/min (340,000 m³/h) at the inlet—to an outlet pressure of 20 atm. Smaller-capacity machines discharge at pressures up to several hundred atmospheres. Interstage cooling is needed on the high-pressure units.

Axial-flow machines handle even larger volumes of gas, up to 600,000 ft³/min (1×10^6 m³/h), but at lower discharge pressures of 2 to 10 or 12 atm. In these units the rotor vanes propel the gas axially from one set of vanes directly to the next. Interstage cooling is normally not required.

Equations for blowers and compressors

Because of the change in density during compressible flow, the integral form of the Bernoulli equation is inadequate. Equation (4.65), however, can be written differentially and used to relate the shaft work to the differential change in pressure head. In blowers and compressors the mechanical, kinetic, and potential energies do not change appreciably, and the velocity and static head terms can be dropped. Also, on the assumption that the compressor is frictionless, $\eta = 1.0$ and $h_f = 0$. With these simplifications, Eq. (4.65) becomes

$$dW_{pr} = \frac{dp}{\rho}$$

Integration between the suction pressure p_a and the discharge pressure p_b gives the work of compression of an ideal frictionless gas

$$W_{pr} = \int_{p_a}^{p_b} \frac{dp}{\rho} \tag{8.25}$$

To use Eq. (8.25), the integral must be evaluated, which requires information on the path followed by the fluid in the machine from suction to discharge. The procedure is the same whether the compressor is a reciprocating unit, a rotary positive-displacement unit, or a centrifugal unit, provided only that the flow is frictionless and that in a reciprocating machine the equation is applied over an integral number of cycles, so there is neither accumulation nor depletion of fluid in the cylinders. Otherwise the basic assumption of steady flow, which underlies Eq. (4.65), would not hold.

Adiabatic compression. For uncooled units, the fluid follows an isentropic path. For ideal gases, the relation between p and ρ is given by Eq. (6.14), which may be written

$$\frac{p}{\rho^\gamma} = \frac{p_a}{\rho_a^\gamma}$$

or

$$\rho = \frac{\rho_a}{p_a^{1/\gamma}} p^{1/\gamma} \tag{8.26}$$

Substituting ρ from Eq. (8.26) into Eq. (8.25) and integrating give

$$W_{pr} = \frac{p_a^{1/\gamma}}{\rho_a} \int_{p_a}^{p_b} \frac{dp}{p^{1/\gamma}} = \frac{p_a^{1/\gamma}}{(1 - 1/\gamma)\rho_a}\left(p_b^{1-1/\gamma} - p_a^{1-1/\gamma}\right)$$

By multiplying the coefficient by $p_a^{1-1/\gamma}$ and dividing the terms in the parentheses by the same quantity, this equation becomes

$$W_{pr} = \frac{p_a\gamma}{(\gamma - 1)\rho_a}\left[\left(\frac{p_b}{p_a}\right)^{1-1/\gamma} - 1\right] \tag{8.27}$$

Equation (8.27) shows the importance of the compression ratio p_b/p_a.

Isothermal compression. When cooling during compression is complete, the temperature is constant and the process is isothermal. The relation between p and ρ then is simply

$$\frac{p}{\rho} = \frac{p_a}{\rho_a} \tag{8.28}$$

Eliminating ρ from Eqs. (8.25) and (8.28) and integrating give

$$W_{pr} = \frac{p_a}{\rho_a}\int_{p_a}^{p_b}\frac{dp}{p} = \frac{p_a}{\rho_a}\ln\frac{p_b}{p_a} = \frac{RT_a}{M}\ln\frac{p_b}{p_a} \tag{8.29}$$

For a given compression ratio and suction condition, the work requirement in isothermal compression is less than that for adiabatic compression. This is one reason why cooling is useful in compressors.

A close relation exists between the adiabatic and isothermal cases. By comparing the integrands in the equation above, it is clear that if $\gamma = 1$, the equations for adiabatic and for isothermal compression are identical.

Polytropic compression. In large compressors the path of the fluid is neither isothermal nor adiabatic. The process may still be assumed to be frictionless, however. It is customary to assume that the relation between pressure and density is given by

$$\frac{p}{\rho^n} = \frac{p_a}{\rho_a^n} \tag{8.30}$$

where n is a constant. Use of this equation in place of Eq. (8.26) obviously yields Eq. (8.27) with the replacement of γ by n.

The value of n is found empirically by measuring the density and pressure at two points on the path of the process, for example, at suction and discharge. The value of n is calculated by the equation

$$n = \frac{\ln(p_b/p_a)}{\ln(\rho_b/\rho_a)}$$

This equation is derived by substituting p_b for p and ρ_b for ρ in Eq. (8.30) and taking logarithms.

Compressor efficiency. The ratio of the theoretical work (or fluid power) to the actual work (or total power input) is, as usual, the efficiency and is denoted by η. The maximum efficiency of reciprocating compressors is about 80 to 85 percent; it can be up to 90 percent for centrifugal compressors.

Power equation. The power required by an adiabatic compressor is readily calculated from Eq. (8.27). The *dimensional* formula is

$$P_B = \frac{0.371 T_a \gamma q_0}{(\gamma - 1)\eta} \left[\left(\frac{p_b}{p_a} \right)^{1-1/\gamma} - 1 \right] \tag{8.31a}$$

where P_B = power, kW
    q_0 = volume of gas compressed, std m^3/s, evaluated at 0°C and 760 mm Hg
    T_a = inlet temperature, K

For isothermal compression

$$P_B = \frac{1.97 T_a q_0}{\eta} \ln \frac{p_b}{p_a} \tag{8.32a}$$

The corresponding equation for adiabatic compression in fps units is

$$P_B = \frac{0.0643 T_a \gamma q_0}{520(\gamma - 1)\eta} \left[\left(\frac{p_b}{p_a} \right)^{1-1/\gamma} - 1 \right] \tag{8.31b}$$

where P_B = brake horsepower
    q_0 = volume of gas compressed, std ft^3/min
    T_a = inlet temperature, °R

For isothermal compression, in fps units

$$P_B = \frac{0.341 T_a q_0}{520\eta} \ln \frac{p_b}{p_a} \tag{8.32b}$$

EXAMPLE 8.3. A three-stage reciprocating compressor is to compress 180 std ft^3/min (306 m^3/h) of methane from 14 to 900 lb$_f$/in.^{2} (0.95 to 61.3 atm) abs. The inlet temperature is 80°F (26.7°C). For the expected temperature range the average properties of methane are

$$C_p = 9.3 \text{ Btu/lb mol} \cdot °\text{F (38.9 J/g mol} \cdot °\text{C)} \qquad \gamma = 1.31$$

(a) What is the brake horsepower if the mechanical efficiency is 80 percent? (b) What is the discharge temperature from the first stage? (c) If the temperature of the cooling water is to rise 20°F (11.1°C), how much water is needed in the intercoolers and aftercooler for the compressed gas to leave each cooler at 80°F (26.7°C)? Assume that jacket cooling is sufficient to absorb frictional heat.

Solution
    (a) For a multistage compressor it can be shown that the total power is a minimum if each stage does the same amount of work. By Eq. (8.27) this is equivalent to the use of the same compression ratio in each stage. For a three-stage machine, therefore, the

compression ratio of one stage should be the cube root of the overall compression ratio, 900/14. For one stage

$$\frac{p_b}{p_a} = \left(\frac{900}{14}\right)^{1/3} = 4$$

The power required for each stage is, by Eq. (8.31b),

$$P_B = \frac{(80 + 460)(0.0643) \times 1.31 \times 180}{520(1.31 - 1)(0.80)}(4^{1-1/1.31} - 1) = 24.6 \text{ hp}$$

The total power for all stages is $3 \times 24.6 = 73.8$ hp (55.0 kW).

(b) From Eq. (8.24), the temperature at the exit of each stage is

$$T_b = (80 + 460)4^{1-1/1.31} = 750°\text{R} = 290°\text{F} (143.3°\text{C})$$

(c) Since 1 lb mol = 378.7 std ft^3(see p. 210), the flow rate is

$$\frac{180 \times 60}{378.7} = 28.5 \text{ lb mol/h} (12.9 \text{ kg mol/h})$$

The heat load in each cooler is

$$28.5(290 - 80)(9.3) = 55,660 \text{ Btu/h}$$

The total heat load is $3 \times 55,660 = 166,980$ Btu/h. The cooling water requirement is

$$\frac{166,980}{20} = 8,349 \text{ lb/h} = 16.7 \text{ gal/min} (3.79 \text{ m}^3\text{/h})$$

Vacuum pumps

A compressor that takes suction at a pressure below atmospheric and discharges against atmospheric pressure is called a *vacuum pump*. Any type of blower or compressor—reciprocating, rotary, or centrifugal—can be adapted to vacuum practice by modifying the design to accept very low-density gas at the suction and attain the large compression ratios necessary. As the absolute pressure at the suction decreases, the volumetric efficiency drops and approaches zero at the lowest absolute pressure attainable by the pump. Usually the mechanical efficiency is also lower than that for compressors. The required displacement increases rapidly as the suction pressure falls, so a large machine is needed to move much gas. The compression ratio used in vacuum pumps is higher than that in compressors, ranging up to 100 or more, with a correspondingly high adiabatic discharge temperature. Actually, however, the compression is nearly isothermal because of the low mass flow rate and the effective heat transfer from the relatively large area of exposed metal.

Jet ejectors. An important kind of vacuum pump that does not use moving parts is the jet ejector, shown in Fig. 8.16, in which the fluid to be moved is entrained in a high-velocity stream of a second fluid. The motive fluid and the fluid to be moved may be the same, such as when compressed air is used to move air, but usually they are not. Industrially greatest use is made of steam-jet ejectors, which are valuable

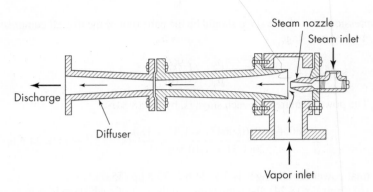

FIGURE 8.16
Steam-jet ejector.

for drawing a fairly high vacuum. As shown in Fig. 8.16, steam at about 7 atm is admitted to a converging-diverging nozzle, from which it issues at supersonic velocity into a diffuser cone. The air or other gas to be moved is mixed with the steam in the first part of the diffuser, lowering the velocity to acoustic velocity or below; in the diverging section of the diffuser the kinetic energy of the mixed gases is converted to pressure energy, so that the mixture can be discharged directly to the atmosphere. Often it is sent to a water-cooled condenser, particularly if more than one stage is used. Otherwise each stage would have to handle all the steam admitted to the preceding stages. As many as five stages are used in industrial processing.

Jet ejectors require very little attention and maintenance and are especially valuable with corrosive gases that would damage mechanical vacuum pumps. For difficult problems the nozzles and diffusers can be made of corrosion-resistant metal, graphite, or other inert material. Ejectors, particularly when multistage, use large quantities of steam and water. They are rarely used to produce absolute pressures below 1 mm Hg. Steam jets are no longer as popular as they once were, because of the great increase in the cost of steam. In many instances where corrosion is not a serious consideration, they have been replaced by mechanical vacuum pumps, which use much less energy for the same service.

Comparison of Devices for Moving Fluids

In all types of fluid-moving equipment, the flow capacity, power requirements, and mechanical efficiency are all highly important. Reliability and ease of maintenance are also highly desirable, often essential. In small units, simplicity and trouble-free operation are usually more important than high mechanical efficiency with its saving of a few kilowatts of power.

Positive-displacement machines. Positive-displacement machines, in general, handle smaller quantities of fluids at higher discharge pressures than do centrifugal machines. Positive-displacement pumps are not subject to air binding and are usually self-priming. In both positive-displacement pumps and blowers, the discharge rate is nearly independent of the discharge pressure, so that these machines are extensively used for controlling and metering flow. Reciprocating devices require considerable maintenance but can produce the highest pressures. They deliver a pulsating stream. Rotary pumps work best on fairly viscous lubricating fluids, discharging a steady stream at moderate to high pressures. They cannot be used with slurries. Rotary blowers usually discharge gas at a maximum pressure of 2 atm from a single stage. The discharge line of a positive-displacement pump cannot be closed without stalling or breaking the pump, so that a bypass line with a pressure relief valve is required.

Centrifugal machines. Centrifugal pumps, blowers, and compressors all deliver fluid at a uniform pressure without shocks or pulsations. They run at higher speeds than positive-displacement machines and are connected to the motor drive directly instead of through a gearbox. The discharge line can be completely closed without damage. Centrifugal pumps can handle a wide variety of corrosive liquids and slurries. Centrifugal blowers and compressors are much smaller, for given capacity, than reciprocating compressors and require less maintenance.

Vacuum devices. For producing vacuum, reciprocating machines are effective for absolute pressures down to about 10 mm Hg. Rotary vacuum pumps can lower the absolute pressure to 0.01 mm Hg and over a wide range of low pressures are cheaper to operate than multistage steam-jet ejectors. For very high vacuums, specialized devices such as diffusion pumps are needed.

MEASUREMENT OF FLOWING FLUIDS

To control industrial processes, it is essential to know the amount of material entering and leaving the process. Because materials are transported in the form of fluids wherever possible, it is important to measure the rate at which a fluid is flowing through a pipe or other channel. Many different types of meters are used industrially. Selection of a meter is based on the applicability of the instrument to the specific problem, its installed cost and costs of operation, the range of flow rates it can accommodate (its *rangeability*), and its inherent accuracy. Sometimes a rough indication of the flow rate is all that is needed; at other times a highly accurate measurement, usually of the mass flow rate, is required for such purposes as controlling reactor feeds or transferring custody of the fluid from one owner to another.

A few types of flowmeters measure the mass flow rate directly, but the majority measure the volumetric flow rate or the average fluid velocity, from which the volumetric flow rate can be calculated. To convert the volumetric rate to the mass flow rate requires that the fluid density under the operating conditions be known. Most meters operate on all the fluid in the pipe or channel and are known as *full-bore meters*. Others, called *insertion meters,* measure the flow rate, or more commonly the fluid velocity, at one point only. The total flow rate, however, can often be inferred with considerable accuracy from this single-point measurement.

Detailed descriptions of commercial flowmeters, listing their advantages and limitations, are available in the literature.[3]

Full-Bore Meters

The most common types of full-bore meters are venturi and orifice meters and variable-area meters such as rotameters. Other full-bore measuring devices include V-element, magnetic, vortex-shedding, turbine, and positive-displacement meters; ultrasonic meters; and mass flow devices such as Coriolis flowmeters.

Venturi meter

A venturi meter is shown in Fig. 8.17. A short conical inlet section leads to a throat section, then to a long discharge cone. Pressure taps at the start of the inlet section and at the throat are connected to a manometer or differential pressure transmitter.

In the upstream cone the fluid velocity is increased and its pressure decreased. The pressure drop in this cone is used to measure the flow rate. In the discharge cone the velocity is decreased and the original pressure largely recovered. The angle of the discharge cone is made small, between 5° and 15°, to prevent boundary layer separation and to minimize friction. Since there is no separation in a contracting cross section, the upstream cone can be made shorter than the downstream cone. Typically 90 percent of the pressure loss in the upstream cone is recovered.

Although venturi meters can be applied to the measurement of gas flow rates, they are most commonly used with liquids, especially large flows of water where, because of the large pressure recovery, a venturi requires less power than other types of meters.

The basic equation for a venturi meter is obtained by writing the Bernoulli equation for incompressible fluids across the upstream cone. If $\bar{V}_a$ and $\bar{V}_b$ are the average upstream and downstream velocities, respectively, and ρ is the density of the fluid, Eq. (4.65) becomes

$$\alpha_b \bar{V}_b^2 - \alpha_a \bar{V}_a^2 = \frac{2(p_a - p_b)}{\rho} \qquad (8.33)$$

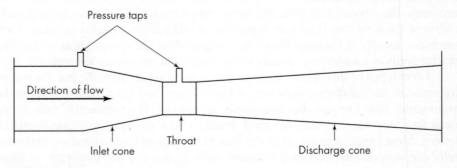

FIGURE 8.17
Venturi meter.

The continuity relation (4.12) can be written, since the density is constant, as

$$\bar{V}_a = \left(\frac{D_b}{D_a}\right)^2 \bar{V}_b = \beta^2 \bar{V}_b \tag{8.34}$$

where D_a = diameter of pipe
D_b = diameter of throat of meter
β = diameter ratio D_b/D_a

If $\bar{V}_a$ is eliminated from Eqs. (8.33) and (8.34), the result is

$$\bar{V}_b = \frac{1}{\sqrt{\alpha_b - \beta^4 \alpha_a}} \sqrt{\frac{2(p_a - p_b)}{\rho}} \tag{8.35}$$

Venturi coefficient. Equation (8.35) applies strictly to the frictionless flow of noncompressible fluids. To account for the small friction loss between locations a and b, Eq. (8.35) is corrected by introducing an empirical factor C_v and writing

$$\bar{V}_b = \frac{C_v}{\sqrt{1 - \beta^4}} \sqrt{\frac{2(p_a - p_b)}{\rho}} \tag{8.36}$$

The small effects of the kinetic energy factors α_a and α_b are also taken into account in the definition of C_v. The coefficient C_v is determined experimentally. It is called the *venturi coefficient, velocity of approach not included*. The effect of the approach velocity $\bar{V}_a$ is accounted for by the term $1/\sqrt{1 - \beta^4}$. When D_b is less than $\frac{1}{4}D_a$, the approach velocity and the term β can be neglected, since the resulting error is less than 0.2 percent.

For a well-designed venturi, the constant C_v is about 0.98 for pipe diameters of 2 to 8 in. and about 0.99 for larger sizes.[5]

Volumetric and mass flow rates. The velocity through the venturi throat $\bar{V}_b$ is not the quantity usually desired. The flow rates of practical interest are the volumetric and mass flow rates through the meter. The volumetric flow rate is calculated by substituting $\bar{V}_b$ from Eq. (8.36) into Eq. (4.10) to get

$$q = \bar{V}_b S_b = \frac{C_v S_b}{\sqrt{1 - \beta^4}} \sqrt{\frac{2(p_a - p_b)}{\rho}} \tag{8.37}$$

where q = volumetric flow rate
S_b = area of throat

The mass flow rate is obtained by multiplying the volumetric flow rate by the density, or

$$\dot{m} = q\rho = \frac{C_v S_b}{\sqrt{1 - \beta^4}} \sqrt{2(p_a - p_b)\rho} \tag{8.38}$$

where $\dot{m}$ is the mass flow rate.

Orifice meter

The venturi meter has certain practical disadvantages for ordinary plant practice. It is expensive, it occupies considerable space, and its ratio of throat diameter to pipe diameter cannot be changed. For a given meter and definite manometer system, the maximum measurable flow rate is fixed, so if the flow range is changed, the throat diameter may be too large to give an accurate reading or too small to accommodate the next maximum flow rate. The orifice meter meets these objections to the venturi but at the price of a larger power consumption.

A standard sharp-edged orifice is shown in Fig. 8.18. It consists of an accurately machined and drilled plate mounted between two flanges with the hole concentric with the pipe in which it is mounted. (Off-center or segmental openings are also used on occasion.) The opening in the plate may be beveled on the downstream side. Pressure taps, one above and one below the orifice plate, are installed and are connected to a manometer or differential pressure transmitter. The positions of the taps are arbitrary, and the coefficient of the meter will depend upon the position of the taps. Three of the recognized methods of placing the taps are shown in Table 8.2. Flange taps are the most common. The taps shown in Fig. 8.18 are vena contracta taps.

The principle of the orifice meter is identical with that of the venturi. The reduction of the cross section of the flowing stream in passing through the orifice increases the velocity head at the expense of the pressure head, and the reduction in pressure between the taps is measured by the manometer. Bernoulli's equation

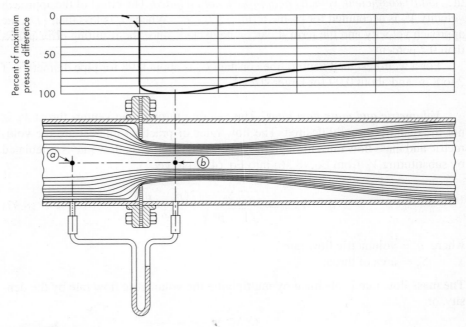

FIGURE 8.18
Orifice meter.

TABLE 8.2
Data on orifice taps

Type of tap	Distance of upstream tap from upstream face of orifice	Distance of downstream tap from downstream face
Flange	1 in. (25 mm)	1 in. (25 mm)
Vena contracta	1 pipe diameter (actual inside)	0.3–0.8 pipe diameter, depending on β
Pipe	$2\frac{1}{2}$ times nominal pipe diameter	8 times nominal pipe diameter

provides a basis for correlating the increase in velocity head with the decrease in pressure head.

One important complication appears in the orifice meter that is not found in the venturi. Because of the sharpness of the orifice, the fluid stream separates from the downstream side of the orifice plate and forms a free-flowing jet in the downstream fluid. A vena contracta forms, as shown in Fig. 8.18. The jet is not under the control of solid walls, as is the case in the venturi, and the area of the jet varies from that of the opening in the orifice to that of the vena contracta. The area at any given point, for example at the downstream tap, is not easily determinable, and the velocity of the jet at the downstream tap is not easily related to the diameter of the orifice. Orifice coefficients are smaller and more variable than those for the venturi, and the quantitative treatment of the orifice meter is modified accordingly.

Extensive and detailed design standards for orifice meters are available in the literature.[2] They must be followed if the performance of a meter is to be predicted accurately without calibration. For approximate or preliminary design, however, it is satisfactory to use an equation similar to Eq. (8.36) as follows:

$$u_o = \frac{C_o}{\sqrt{1 - \beta^4}} \sqrt{\frac{2(p_a - p_b)}{\rho}} \tag{8.39}$$

where u_0 = velocity through orifice
          β = ratio of orifice diameter to pipe diameter
     p_a, p_b = pressures at stations a and b in Fig. 8.18

In Eq. (8.39), C_o is the *orifice coefficient, velocity of approach not included.* It corrects for the contraction of the fluid jet between the orifice and the vena contracta, for friction, and for α_a and α_b. Coefficient C_o is always determined experimentally. It varies considerably with changes in β and with Reynolds number at the orifice Re_o. This Reynolds number is defined by

$$\text{Re}_o = \frac{D_o u_o \rho}{\mu} = \frac{4\dot{m}}{\pi D_o \mu} \tag{8.40}$$

where D_o is the orifice diameter.

Equation (8.39) is useful for design because C_o is almost constant and independent of β provided Re_o is greater than about 30,000. Under these conditions C_o may be taken as 0.61 for both flange taps and vena contracta taps. For process

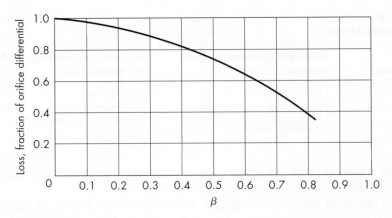

FIGURE 8.19
Overall pressure loss in orifice meters. (*After American Society of Mechanical Engineers.*[2])

applications, β should be between 0.20 and 0.75. If β is less than 0.25, the term $\sqrt{1 - \beta^4}$ differs negligibly from unity. Equations (8.37) and (8.38) for venturi meters may be used for orifice meters by substituting C_o for C_v, u_o for $\bar{V}_b$, and S_o, the cross-sectional area of the orifice, for S_b.

It is especially important that enough straight pipe be provided both upstream and downstream of the orifice to ensure a flow pattern that is normal and undisturbed by fittings, valves, or other equipment. Otherwise the velocity distribution will not be normal, and the orifice coefficient will be affected in an unpredictable manner. Data are available for the minimum length of straight pipe that should be provided upstream and downstream of the orifice to ensure normal velocity distribution.[2] Straightening vanes in the approach line may be used if the required length of pipe is not available upstream of the orifice.

Pressure recovery. Because of the large friction losses from the eddies generated by the reexpanding jet below the vena contracta, the pressure recovery in an orifice meter is poor. The resulting power loss is one disadvantage of the orifice meter. The fraction of the orifice differential that is permanently lost depends on the value of β, and the relationship between the fractional loss and β is shown in Fig. 8.19. For a value of β of 0.5, the lost head is about 73 percent of the orifice differential.

The pressure difference measured by pipe taps, where the downstream tap is eight pipe diameters below the orifice, is really a measurement of permanent loss rather than of the orifice differential.

EXAMPLE 8.4. An orifice meter with flange taps is to be installed in a 100-mm line to measure the flow of water. The maximum flow rate is expected to be 50 m³/h at 15° C. The manometer used to measure the differential pressure is to be filled with mercury, and water is to fill the leads above the surfaces of the mercury. The water temperature will be 15° C throughout. (*a*) If the maximum manometer reading is to be 1.25 m, what diameter, to the nearest millimeter, should be specified for the orifice? (*b*) What will be the power to operate the meter at full load?

Solution

(a) Equation (8.38) is used to calculate the orifice diameter. The quantities to be substituted are

$$q = \frac{50}{3,600} = 0.0139 \text{ m}^3/\text{s}$$

$$\rho = 62.37 \times 16.018 = 999 \text{ kg/m}^3 \qquad (\text{App. 6})$$

$$C_o = 0.61 \qquad g = 9.80665 \text{ m/s}^2$$

From Eq. (2.10),

$$p_a - p_b = 9.80665 \times 1.25 \times (13.6 - 1.0)(999)$$
$$= 154,300 \text{ N/m}^2$$

Substituting these values in Eq. (8.38) gives

$$0.0139 = \frac{0.61 S_o}{\sqrt{1 - \beta^4}} \sqrt{\frac{2 \times 154,300}{999}}$$

from which

$$\frac{S_o}{\sqrt{1 - \beta^4}} = 1.296 \times 10^{-3} = \frac{\pi D_o^2}{4\sqrt{1 - \beta^4}}$$

As a first approximation, call $\sqrt{1 - \beta^4} = 1.0$. Then

$$D_o = 40.6 \text{ mm} \qquad \beta = \frac{40.6}{100} = 0.406$$

and

$$\sqrt{1 - \beta^4} = \sqrt{1 - 0.406^4} = 0.986$$

The effect of this term is negligible in view of the desired precision of the final result. To the nearest millimeter, the throat diameter should be 41 mm.

Check the Reynolds number. The viscosity of water at 15°C, from App. 6, is 1.147 cP or 0.001147 kg/m·s

$$S_o = \frac{\pi D_o^2}{4} = \frac{\pi \times 0.041^2}{4} = 0.00132 \text{ m}^2$$

$$u_o = \frac{q}{S_o} = \frac{0.0139}{0.00132} = 10.53 \text{ m/s}$$

The Reynolds number, from Eq. (8.40), is

$$\text{Re}_o = \frac{0.041 \times 10.53 \times 999}{0.001147} = 376,000$$

The Reynolds number is large enough to justify the value of 0.61 for C_o.

(b) From Fig. 8.19, for $\beta = 0.406$, the permanent loss in pressure is 81 percent of the differential. Since the maximum volumetric flow rate is 0.0139 m³/s, the power required to operate the orifice meter at full flow is

$$P = 0.81 q (p_a - p_b) = \frac{0.81 \times 0.0139 \times 154,300}{1,000} = 1.737 \text{ kW}$$

Flow of compressible fluids through venturis and orifices

The preceding discussion of fluid meters was concerned only with the flow of fluids of constant density. When fluids are compressible, similar equations and discharge coefficients for the various meters may be used. Equation (8.38) for venturi meters is modified to the form

$$\dot{m} = \frac{C_v Y S_b}{\sqrt{1 - \beta^4}} \sqrt{2(p_a - p_b)\rho_a} \tag{8.41}$$

For orifice meters, when β is small, the equation is

$$\dot{m} = 0.61 \, Y S_o \sqrt{2(p_a - p_b)\rho_a} \tag{8.42}$$

In Eqs. (8.41) and (8.42), Y is a dimensionless expansion factor, and ρ_a is the density of the fluid under upstream conditions. For the isentropic flow of an ideal gas through a venturi, Y can be calculated theoretically by integrating Eq. (6.25) between stations a and b and combining the result with the definition of Y implied in Eq. (8.41). The result is

$$Y = \left(\frac{p_b}{p_a}\right)^{1/\gamma} \left\{ \frac{\gamma(1 - \beta^4)[1 - (p_b/p_a)^{1-1/\gamma}]}{(\gamma - 1)(1 - p_b/p_a)[1 - \beta^4(p_b/p_a)^{2/\gamma}]} \right\}^{1/2} \tag{8.43}$$

Equation (8.43) shows that Y is a function of p_b/p_a, β, and γ. This equation cannot be used for orifices because of the vena contracta. An empirical equation in p_b/p_a, β, and γ for standard sharp-edged orifices is[2]

$$Y = 1 - \frac{0.41 + 0.35\beta^4}{\gamma}\left(1 - \frac{p_b}{p_a}\right) \tag{8.44}$$

Equations (8.43) and (8.44) must not be used when p_b/p_a is less than about 0.53, which is the critical pressure ratio at which airflow becomes sonic.

V-element meters

In these meters the flow is restricted by a V-shaped indentation in the side of the pipe or by a metal wedge inserted in the pipe, as shown in Fig. 8.20. They are relatively expensive devices, but their accuracy is high, approximately ±0.5 percent of the measured rate, and they can measure flow rates of hard-to-handle fluids such as liquids containing solid particles or undissolved gases or gases carrying drops of condensate. The flow coefficient is about 0.8; unlike that of orifice meters, it is essentially constant at low flow rates, down to Reynolds numbers as low as 500.

Area meters: Rotameters

In the orifice, nozzle, or venturi, the variation of flow rate through a constant area generates a variable pressure drop, which is related to the flow rate. Another class of meters, called *area meters*, consists of devices in which the pressure drop is constant, or nearly so, and the area through which the fluid flows varies with the flow rate. The area is related, through proper calibration, to the flow rate.

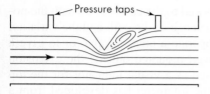

FIGURE 8.20
V-element meter.

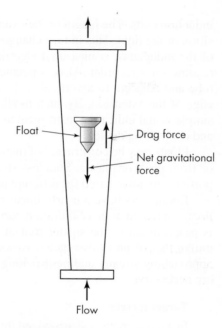

FIGURE 8.21
Principle of a rotameter.

The most important area meter is the *rotameter,* which is shown in Fig. 8.21. It consists essentially of a gradually tapered glass tube mounted vertically in a frame with the large end up. The fluid flows upward through the tapered tube and suspends freely a float (which actually does not float but is completely submerged in the fluid). The float is the indicating element, and the greater the flow rate, the higher the float rides in the tube. The entire fluid stream must flow through the annular space between the float and the tube wall. The tube is marked in divisions, and the reading of the meter is obtained from the scale reading at the reading edge of the float, which is taken at the largest cross section of the float. A calibration curve must be available to convert the observed scale reading to flow rate. Rotameters can be used for either liquid or gas flow measurement.

The bore of a glass rotameter tube is either an accurately formed, plain conical taper or a taper with three beads, or flutes, parallel with the axis of the tube. The tube shown in Fig. 8.21 is a tapered tube. In the first rotameters, angled notches in the top of the float made it rotate, but the float does not rotate in most current designs. For opaque liquids, for high temperatures or pressures, or for other conditions where glass is impracticable, metal tubes are used. Metal tubes are plain tapered. Since in a metal tube the float is invisible, means must be provided for either indicating or transmitting the meter reading. This is accomplished by attaching a rod, called an *extension,* to the top or bottom of the float and using the extension as an armature. The extension is enclosed in a fluid-tight tube mounted on one of the fittings. Since the inside of this tube communicates directly with the interior of the rotameter, no stuffing box for the extension is needed. The tube is surrounded by external induction coils. The length of the extension exposed to the coils varies with the position of the float. This in turn changes the inductance of the coil, and the variation of the inductance is measured electrically to operate a control valve or to give a reading on a recorder. Also, a magnetic follower, mounted outside the extension tube and adjacent to a vertical scale, can be used as a visual indicator for the top edge of the extension. By such modifications the rotameter has developed from a simple visual indicating instrument using only glass tubes into a versatile recording and controlling device.

Floats may be constructed of metals of various densities from lead to aluminum or from glass or plastic. Stainless-steel floats are common. Float shapes and proportions are also varied for different applications.

Rotameters have a nearly linear relationship between flow and position of the float, compared with a calibration curve for an orifice meter, for which the flow rate is proportional to the square root of the reading. The calibration of a rotameter, unlike that of an orifice meter, is not sensitive to the velocity distribution in the approaching stream, and neither long, straight approaches nor straightening vanes are necessary.

Target meters

In a target meter a sharp-edged disk is set at right angles to the direction of flow, as shown in Fig. 8.22, and the drag force exerted on the disk by the fluid is measured. The flow rate is proportional to the square root of this force and to the fluid density. Target meters are rugged and inexpensive and can be used with a variety of fluids, even viscous liquids and slurries. The bar mechanism, however, tends to clog if the solids content of the slurry is high.

Vortex-shedding meters

In a vortex-shedding meter the "target" is a bluff body, often trapezoidal in cross section (Fig. 8.23). This body is designed to create, when flow is turbulent, a

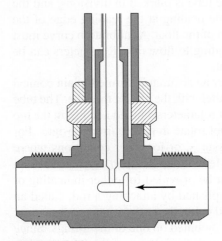

FIGURE 8.22
Target meter.

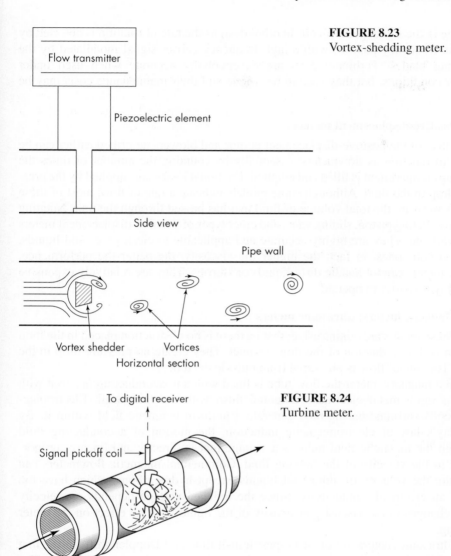

FIGURE 8.23
Vortex-shedding meter.

Flow transmitter

Piezoelectric element

Side view

Pipe wall

Vortex shedder   Vortices
Horizontal section

To digital receiver

Signal pickoff coil →

FIGURE 8.24
Turbine meter.

"vortex street" in its wake. (See Chap. 7, p. 154) Sensors close to the bluff body measure the pressure fluctuations and hence the frequency of the vortex shedding, from which the volumetric flow rate may be inferred. These meters are applicable to many types of fluids, including high-temperature gas and steam. The minimum Reynolds number required for a linear response is fairly high, so the flow rate of highly viscous liquids cannot be measured by this type of instrument.

Turbine meters

In the turbine meter shown in Fig. 8.24, a bladed rotor is suspended axially in the flow stream and spins at a rate proportional to the fluid velocity. In some models the rotor blades are made of a magnetic material that induces an alternating voltage in the signal pickoff coil. In other designs the rate of rotation is detected by a radio-frequency pickoff, with a high-frequency carrier signal modulated by the spinning blades. Turbine meters are exceptionally accurate when used under proper conditions, but they tend to be fragile and their maintenance costs may be high.

Positive-displacement meters

Many of the positive-displacement pumps and blowers described earlier can be made to function as flowmeters, essentially by counting the number of times the moving compartment is filled and emptied. Frictional losses are supplied by the pressure drop in the fluid. Although some models indicate a rate of flow, most of these meters measure the total volume of fluid that has passed through the unit. Nutating disk, oscillating piston, sliding vane, and other types of positive-displacement meters are available. They are highly accurate and applicable to clean gases and liquids, even viscous ones; in fact, the higher the viscosity, the better the performance. These meters cannot handle dirty liquids or slurries. They are relatively expensive and may be costly to operate.

Magnetic meters; ultrasonic meters

These meters are nonintrusive; that is, there is no obstruction placed in the fluid stream and no reduction of the flow channel. They create no pressure drop in the fluid. The rate of flow is measured from outside the tube.

In a magnetic meter the flow tube is lined with a nonconducting material with two or more metal electrodes mounted flush with the liner wall. Electromagnetic coils surrounding the tube generate a uniform magnetic field within it. By Faraday's law of electromagnetic induction, the motion of a conducting fluid through the magnetic field induces a voltage that is directly and linearly proportional to the velocity of the flowing fluid. Commercial magnetic flowmeters can measure the velocity of almost all liquids except hydrocarbons, which have too small an electrical conductivity. Since the induced voltage depends on velocity only, changes in the viscosity or density of the liquid have no effect on the meter reading.

Ultrasonic meters are of two types: transit time and Doppler shift. In the first type a high-frequency pressure wave is beamed at an angle across the pipe. The velocity of the wave is found from its time of transit. When the wave is transmitted in the direction of flow, its velocity is increased, and vice versa. From the change in transit time from that in a quiescent fluid the fluid velocity can be determined. Transit-time meters are applicable to clean fluids only.

Doppler-shift meters, on the other hand, depend on reflections of the pressure wave from suspended particles or bubbles in the stream, which are assumed to be moving at the velocity of the stream. The pressure wave is projected into the fluid at an angle to the direction of flow. The difference between the frequency of the projected wave and that of the reflected wave is proportional to the fluid velocity.

Though not highly accurate, ultrasonic meters are useful in many types of service, including measuring the flow rate of corrosive fluids.

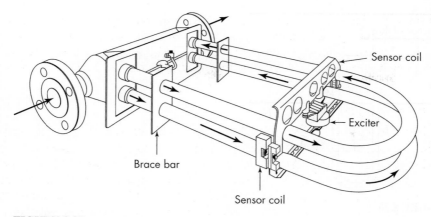

FIGURE 8.25
Typical Coriolis mass flow sensor geometry.

Coriolis meters

An object moving in a rotating system experiences a Coriolis force proportional to its mass and forward velocity and to the angular velocity of the system. This force is perpendicular to the object's direction of travel and to the direction of the angular velocity of the system. In a Coriolis meter (Fig. 8.25) the fluid is passed through two U-shaped curved tubes that are vibrated at their natural frequency. This creates an alternating Coriolis force that produces small elastic deformations in the tubes. From the magnitude of the deformations the mass flow rate may be calculated.

Coriolis meters are highly accurate and directly measure the mass flow rate. Used mostly with small pipes, they are costly to install and operate. Consequently their applications are generally limited to difficult fluids or to situations in which their high accuracy justifies their higher cost.

Insertion Meters

In this type of meter the sensing element, which is small compared to the size of the flow channel, is inserted into the flow stream. A few insertion meters measure the average flow velocity, but the majority measure the local velocity at one point only. The positioning of the sensing element is therefore important if the total flow rate is to be determined. The local measured velocity must bear a constant and known relationship to the average velocity of the fluid.

The point of measurement may be at the centerline of the channel, and the average velocity may be found from the ratio of the average to the maximum velocity. (See Chap. 5.) Alternatively, the sensor may be located at the *critical point* in the channel where the local velocity equals the average velocity. In either case precautions must be taken, usually by providing long calming sections upstream of the meter, to ensure that the velocity profile is fully developed and not distorted.

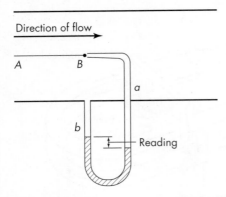

FIGURE 8.26
Principle of pitot tube.

Pitot tube

The pitot tube is a device used to measure the local velocity along a streamline. The principle of the device is shown in Fig. 8.26. The opening of the impact tube a is perpendicular to the flow direction. The opening of the static tube b is parallel to the direction of flow. The two tubes are connected to the legs of a manometer or equivalent device for measuring small pressure differences. The static tube measures the static pressure p_0 since there is no velocity component perpendicular to its opening. The impact opening includes a stagnation point B at which the streamline AB terminates.

The pressure p_s, measured by the impact tube, is the stagnation pressure of the fluid given for ideal gases by Eq. (7.8). Then Eq. (7.9) applies, where p_0 is the static pressure measured by tube b. Solving Eq. (7.9) for u_0 gives

$$u_0 = \left(\frac{2(p_s - p_0)}{p_0\{1 + \mathrm{Ma}^2/4 + [(2 - \gamma)/24]\mathrm{Ma}^4 + \cdots\}} \right)^{1/2} \tag{8.45}$$

Since the manometer of the pitot tube measures the pressure difference $p_s - p_0$, Eq. (8.45) gives the local velocity of the point where the impact tube is located. Normally, only the first Mach number term in the equation is significant.

For incompressible fluids, the Mach number correction factor is unity, and Eq. (8.45) becomes simply

$$u_0 = \sqrt{\frac{2(p_s - p_0)}{\rho}} \tag{8.46}$$

The velocity measured by an ideal pitot tube would conform exactly to Eq. (8.45). Well-designed instruments are in error by not more than 1 percent of theory, but when precise measurements are to be made, the pitot tube should be calibrated and an appropriate correction factor applied. This factor is used as a coefficient before the bracketed terms in Eq. (8.45). It is nearly unity in well-designed pitot tubes.

The disadvantages of the pitot tube are (1) that most designs do not give the average velocity directly and (2) that its readings for gases are extremely small. When it is used for measuring low-pressure gases, some form of multiplying gauge, like that shown in Fig. 2.5, must also be used.

EXAMPLE 8.5. Air at 200°F (93.3°C) is forced through a long, circular flue 36 in. (914 mm) in diameter. A pitot tube reading is taken at the center of the flue at a sufficient distance from flow disturbances to ensure normal velocity distribution. The pitot reading is 0.54 in. (13.7 mm) H_2O, and the static pressure at the point of measurement is 15.25 in. (387 mm) H_2O. The coefficient of the pitot tube is 0.98.

Calculate the flow of air, in cubic feet per minute, measured at 60°F (15.6°C) and a barometric pressure of 29.92 in. (760 mm) Hg.

Solution. Assume the Mach number correction is negligible. The velocity at the center of the flue, which is that measured by the pitot tube, is calculated by Eq. (8.46) using fps units and a coefficient of 0.98. Equation (8.46) becomes

$$u_0 = 0.98 \sqrt{\frac{2g_c(p_s - p_0)}{\rho}} \qquad (8.47)$$

The necessary quantities are as follows. The absolute pressure at the instrument is

$$p = 29.92 + \frac{15.25}{13.6} = 31.04 \text{ in. Hg}$$

Since 1 lb mole occupies 359 ft³ at 32°F and 1 atm, the density of the air is

$$\rho = \frac{29 \times 492 \times 31.04}{359(460 + 200)(29.92)} = 0.0625 \text{ lb/ft}^3$$

From the manometer reading

$$p_s - p_0 = \frac{0.54}{12}(62.37) = 2.81 \text{ lb}_f/\text{ft}^2$$

By Eq. (8.47), the maximum velocity is

$$u_{max} = 0.98 \sqrt{2 \times 32.174 \left(\frac{2.81}{0.0625}\right)} = 52.7 \text{ ft/s}$$

This is sufficiently low for the Mach number correction to be negligible. To obtain the average velocity from the maximum velocity, Fig. 5.8 is used. The Reynolds number, based on the maximum velocity, is calculated as follows. From App. 8, the viscosity of air at 200° F is 0.022 cP, and

$$Re_{max} = \frac{(36/12)(52.7)(0.0625)}{0.022(0.000672)} = 670,000$$

The ratio $\bar{V}/u_{max}$, from Fig. 5.8, is a little greater than 0.86. Using 0.86 as an estimated value gives

$$\bar{V} = 0.86 \times 52.7 = 45.3 \text{ ft/s}$$

The Reynolds number Re is $670{,}000 \times 0.86 = 576{,}000$, and $\bar{V}/u_{max}$ is exactly 0.86 as estimated. The volumetric flow rate is

$$q = 45.3 \left(\frac{36}{12}\right)^2 \left(\frac{\pi}{4}\right)\left(\frac{520}{660}\right)\left(\frac{31.04}{29.92}\right)(60) = 15{,}704 \text{ ft}^3/\text{min} \ (7.41 \text{ m}^3/\text{s})$$

Thermal meters

These meters measure the flow rate directly by measuring the temperature rise in the fluid as it passes over a heating element, or the rate of heat transfer to the stream from a heated surface. Thermal meters are most commonly insertion meters, used to measure gas flow in large ducts.

A typical unit consists of an electrically heated stainless-steel tube adjacent to a similar tube containing a resistance thermometer, and a separate unheated tube also adjacent to a tube with a matched resistance thermometer. Gas passes transversely over the tubes, and the one next to the heating element becomes hotter than the other one. The temperature difference between the tubes is inversely proportional to the mass flow rate of the gas: It is greatest when there is no flow and becomes smaller as the flow rate increases. Flow velocities between 0.08 and 46 m/s (0.25 and 150 ft/s), it is claimed, can be measured with an accuracy of ± 1 percent. While a single measuring point is sufficient when the velocity profile is symmetric, as many as eight points may be provided in a single instrument to give accurate flow rates with an asymmetric velocity profile. Special thermal meters are capable of withstanding temperatures up to 455°C (850°F), high radiation levels, or strong vibration of the duct.

Other insertion meters

Modified forms of magnetic meters, turbine meters, ultrasonic meters, and other types are available as insertion meters. They all have advantages for certain services. Insertion meters are generally cheaper than full-bore meters and are usually the most cost-effective method of measuring flow in large pipes.

SYMBOLS

A	Area, m^2 or ft^2; A_p, cross-sectional area of channels at periphery of pump impeller
C_o	Orifice coefficient, velocity of approach not included
C_p	Molal specific heat at constant pressure, J/g mol $\cdot$ °C or Btu/lb mol $\cdot$ °F
C_v	Venturi coefficient, velocity of approach not included
c_p	Specific heat at constant pressure, J/g $\cdot$ °C or Btu/lb $\cdot$ °F
c_v	Specific heat at constant volume, J/g $\cdot$ °C or Btu/lb $\cdot$ °F
D	Diameter, m or ft; D_a, of pipe; D_b, of venturi throat; D_o, of orifice
g	Gravitational acceleration, m/s^2 or ft/s^2
g_c	Newton's law proportionality factor, 32.174 ft $\cdot$ lb/lb$_f$ $\cdot$ s^2
H	Total head, J/kg or ft $\cdot$ lb$_f$/lb; H_a, at station a; H_b, at station b

h_f	Friction loss, J/kg or ft·lb$_f$/lb; h_{fs}, in pump suction line
IPS	Iron pipe size, standard for steel pipe
M	Molecular weight
Ma	Mach number, dimensionless
m	Mass, kg or lb
$\dot{m}$	Mass flow rate, kg/s or lb/s
NPS	Normal pipe size, standard for steel pipe
NPSH	Net positive suction head; NPSHR, minimum required value
n	Rotational speed, r/s; also exponent in Eq. (8.30)
P	Power, W or ft·lb$_f$/s; P_B, power supplied to pump, kW or hp; P_f, fluid power in pump; P_{fr}, in ideal pump
p	Pressure, atm or lb$_f$/ft^2; p_a, at station a; $p_{a'}$, at station a'; p_b, at station b; $p_{b'}$, at station b'; p_s, impact pressure; p_v, vapor pressure; p_0, static pressure; p_1, p_2, at stations 1 and 2
q	Volumetric flow rate, m^3/s or ft^3/s; q_r, through ideal pump; q_0, compressor capacity, std ft^3/min
R	Gas law constant, 8.314 N·m/g mol·K or 1,545 ft·lb$_f$/lb mol·°R
r	Radius, m or ft; r_1, of impeller at suction; r_2, of impeller at discharge
Re	Reynolds number in pipe, $D\bar{V}\rho/\mu$
Re$_{max}$	Maximum local Reynolds number in pipe, $Du_{max}\rho/\mu$
Re$_o$	Reynolds number at orifice, $D_o u_o \rho/\mu$
S	Cross-sectional area, m^2 or ft^2; S_b, of venturi throat; S_o, of orifice
T	Absolute temperature, K or °R; T_a, at compressor inlet; T_b, at compressor discharge; also torque, J or ft·lb$_f$
u	Local fluid velocity, m/s or ft/s; u_{max}, maximum velocity in pipe; u_o, at orifice; u_0, at impact point of pitot tube; u_1, peripheral velocity at inlet of pump impeller; u_2, at impeller discharge; u_θ, tangential velocity component; $u_{\theta 1}$, $u_{\theta 2}$, at stations 1 and 2
V	Resultant velocity, absolute, in pump impeller, m/s or ft/s; V_{r2}, radial component of velocity V_2; V_u, tangential component; V_{u1}, of velocity V_1; V_{u2}, of velocity V_2; V_1, at suction; V_2, at discharge; V_2', V_{u2}', actual velocities in pump
$\bar{V}$	Average fluid velocity, m/s or ft/s; $\bar{V}_a$, at station a; $\bar{V}_{a'}$, at station a'; $\bar{V}_b$, at station b; $\bar{V}_{b'}$, at station b'; $\bar{V}_{opt}$, optimum velocity in pipe
v	Fluid velocity relative to pump impeller, m/s or ft/s; v_1, at suction; v_2, at discharge
W_p	Pump work, J/kg or ft·lb$_f$/lb; W_{pr}, by ideal pump
Y	Expansion factor, flowmeter
Z	Height above datum plane, m or ft; Z_a, at station a; Z_b, at station b

Greek Letters

α	Kinetic energy correction factor; α_a, at station a; α_b, at station b; $\alpha_{b'}$, at station b'; also angle between absolute and peripheral velocities in pump impeller; α_1, at suction; α_2, at discharge

β Vane angle in pump impeller; β_1, at suction; β_2, at discharge; also β_2', actual angle in pump; also ratio, diameter of orifice or venturi throat to diameter of pipe

γ Ratio of specific heats c_p/c_v

ΔH Head developed by pump; ΔH_r, in frictionless or ideal pump

η Overall mechanical efficiency of pump, fan, or blower

μ Absolute viscosity, cP or lb/ft·s

ρ Density, kg/m^3 or lb/ft^3; ρ_a, at station a; ρ_b, at station b; $\bar{\rho}$, average density $(\rho_a + \rho_b)/2$

ω Angular velocity, rad/s

PROBLEMS

8.1. Make a preliminary estimate of the approximate pipe size required for the following services: (a) a transcontinental pipeline to handle 10,000 std m^3/h of natural gas at an average pressure of 3 atm abs and an average temperature of 20°C; (b) feeding a slurry of p-nitrophenol crystals in water to a continuous centrifugal separator at the rate of 1 t (metric ton)/h of solids. The slurry carries 45 percent solids by weight. For p-nitrophenol $\rho = 1,475$ kg/m^3.

8.2. It is proposed to pump 10,000 kg/h of toluene at 114°C and 1.1 atm abs pressure from the reboiler of a distillation tower to a second distillation unit without cooling the toluene before it enters the pump. If the friction loss in the line between the reboiler and pump is 7 kN/m^2 and the density of toluene is 866 kg/m^3, how far above the pump must the liquid level in the reboiler be maintained to give a net positive suction head of 2.5 m?

8.3. Calculate the power required to drive the pump in Prob. 8.2 if the pump is to elevate the toluene 10 m, the pressure in the second unit is atmospheric, and the friction loss in the discharge line is 35 kN/m^2. The velocity in the pump discharge line is 2 m/s.

8.4. Air entering at 70°F and atmospheric pressure is to be compressed to 4,000 lb$_f$/in.^{2} gauge in a reciprocating compressor at the rate of 125 std ft^3/min. If the compression ratio is the same in each stage, how many stages should be used? What is the theoretical shaft work per standard cubic foot for frictionless adiabatic compression? What is the brake horsepower if the efficiency of each stage is 85 percent? For air, $\gamma = 1.40$.

8.5. What is the discharge temperature of the air from the first stage in Prob. 8.4?

8.6. After the installation of the orifice meter of Example 8.4, the manometer reading at a definite constant flow rate is 45 mm. Calculate the flow through the line in m^3/h measured at 15°C.

8.7. Natural gas having a specific gravity relative to air of 0.60 and a viscosity of 0.011 cP is flowing through a 6-in. Schedule 40 pipe in which is installed a standard sharp-edged orifice equipped with flange taps. The gas is at 100° F and 20 lb$_f$/in.^{2} abs at the upstream tap. The manometer reading is 46.3 in. of water at 60° F. The ratio of specific heats for natural gas is 1.30. The diameter of the orifice is 2.00 in. Calculate the rate of flow of gas through the line in cubic feet per minute based on a pressure of 14.4 $lb_f/in.^2$ and a temperature of 60° F.

8.8. A horizontal venturi meter having a throat diameter of 20 mm is set in a 75-mm-ID pipeline. Water at 15° C is flowing through the line. A manometer containing mercury under water measures the pressure differential over the instrument. When the manometer reading is 500 mm, what is the flow rate in m^3/h? If 12 percent of the differential is permanently lost, what is the power consumption of the meter?

8.9. A V-element meter is used to measure the flow of a 15 percent slurry of ion-exchange beads in water. The slurry is carried in a 3-in. Schedule 40 pipe, and the expected range of flows is 30 to 150 gal/min. The particle density is 1,250 kg/m^3, and the average particle size is 250 μm. (a) What is the expected pressure drop at maximum flow if the V element or wedge extends across two-thirds of the pipe diameter? (b) If the differential pressure transmitter has an accuracy of 0.05 $lb_f/in.^2$, what is the accuracy of the flow measurement at maximum and minimum flows?

8.10. The mass flow rate of flue gas in a 1.2-m by 2-m rectangular duct is to be measured using a thermal meter. The normal gas composition is 76 percent N_2, 3 percent O_2, 14 percent CO_2, and 7 percent H_2O; and the average velocity is 12 m/s at duct conditions of 150°C and 1 atm. (a) If a 5,000-W heating element is centrally mounted in the duct, what is the temperature rise after the heated gas has mixed with the rest of the gas? What is the accuracy of the flow measurement if the upstream and downstream temperatures can be determined to ±0.01°C? (b) If the meter is calibrated for the normal gas composition, what is the effect of a change to 12 percent CO_2?

REFERENCES

1. Dwyer, J. J. *Chem. Eng. Prog.* **70**(10):71 (1974).
2. *Fluid Meters: Their Theory and Applications.* 6th ed. New York: American Society of Mechanical Engineers, 1971, pp. 58–65.
3. Ginesi, D., and G. Grebe. *Chem. Eng.* **94**(9):102 (1987).
4. Haden, R. C. *Chem. Eng. Prog.* **70**(3):69 (1974).
5. Jorissen, A. L. *Trans. ASME* **74**:905 (1952).
6. Neerken, R. F. *Chem. Eng.* **94**(12):76 (1987).
7. Perry, R. H., and D. W. Green (eds.). *Chemical Engineers' Handbook.* 7th ed. New York: McGraw-Hill, 1997, p. **10**-50.
8. Peters, M. S., and K. D. Timmerhaus. *Plant Design and Economics for Chemical Engineers.* 3d ed. New York: McGraw-Hill, 1980.

CHAPTER 9

Agitation and Mixing of Liquids

Many processing operations depend for their success on the effective agitation and mixing of fluids. Though often confused, agitation and mixing are not synonymous. Agitation refers to the induced motion of a material in a specified way, usually in a circulatory pattern inside some sort of container. Mixing is the random distribution, into and through one another, of two or more initially separate phases. A single homogeneous material, such as a tankful of cold water, can be agitated, but it cannot be mixed until some other material (such as a quantity of hot water or some powdered solid) is added to it.

The term *mixing* is applied to a variety of operations, differing widely in the degree of homogeneity of the "mixed" material. Consider, in one case, two gases that are brought together and thoroughly blended and, in a second case, sand, gravel, cement, and water tumbled in a rotating drum for a long time. In both cases the final product is said to be mixed. Yet the products are obviously not equally homogeneous. Samples of the mixed gases—even very small samples—all have the same composition. Small samples of the mixed concrete, on the other hand, differ widely in composition.

This chapter deals with the agitation and mixing of liquids, the dispersion of liquids and gases into other liquids, and the suspension of solids in liquids. Mixing of stiff pastes, elastomers, and dry, solid powders is discussed in Chap. 28.

Purposes of agitation

Liquids are agitated for a number of purposes, depending on the objectives of the processing step. These purposes include

1. Suspending solid particles.
2. Blending miscible liquids, for example, methyl alcohol and water.
3. Dispersing a gas through the liquid in the form of small bubbles.

4. Dispersing a second liquid, immiscible with the first, to form an emulsion or suspension of fine drops.
5. Promoting heat transfer between the liquid and a coil or jacket.

Often one agitator serves several purposes at the same time, as in the catalytic hydrogenation of a liquid. In a hydrogenation vessel, the hydrogen gas is dispersed through the liquid in which solid particles of catalyst are suspended, with the heat of reaction simultaneously removed by a cooling coil or jacket. Heat transfer in agitated vessels is discussed in Chap. 15.

AGITATED VESSELS

Liquids are most often agitated in some kind of tank or vessel, usually cylindrical in form and with a vertical axis. The top of the vessel may be open to the air; more usually it is closed. The proportions of the tank vary widely, depending on the nature of the agitation problem. A standardized design such as that shown in Fig. 9.1, however, is applicable in many situations. The tank bottom is rounded, not flat, to eliminate sharp corners or regions into which fluid currents would not penetrate. The liquid depth is approximately equal to the diameter of the tank. An impeller is mounted on an overhung shaft, that is, a shaft supported from above. The shaft is driven by a motor, sometimes directly connected to the shaft but more often connected to it through a speed-reducing gearbox. Accessories such as inlet and outlet lines, coils,

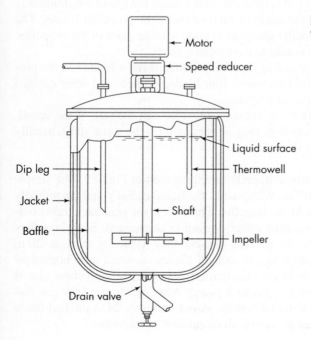

FIGURE 9.1
Typical agitation process vessel.

Motor

Speed reducer

Liquid surface

Dip leg

Thermowell

Jacket

Shaft

Baffle

Impeller

Drain valve

jackets, and wells for thermometers or other temperature-measuring devices are usually included.

The impeller causes the liquid to circulate through the vessel and eventually return to the impeller. Baffles are often included to reduce tangential motion. The flow patterns in agitated vessels are discussed in detail later in this chapter.

Impellers

Impeller agitators are divided into two classes. Those that generate currents parallel with the axis of the impeller shaft are called *axial-flow impellers;* those that generate currents in a radial or tangential direction are called *radial-flow impellers.*

The three main types of impeller for low- to moderate-viscosity liquids are propellers, turbines, and high-efficiency impellers. Each type includes many variations and subtypes which are not considered here. For very viscous liquids, the most widely used impellers are helical impellers and anchor agitators.

Propellers. A propeller is an axial-flow, high-speed impeller for liquids of low viscosity. Small propellers turn at full motor speed, either 1,150 or 1,750 rpm; larger ones turn at 400 to 800 rpm. The direction of rotation is usually chosen to force the liquid downward, and the flow currents leaving the impeller continue until deflected by the floor of the vessel. The highly turbulent swirling column of liquid leaving the impeller entrains stagnant liquid as it moves along, probably considerably more than an equivalent column from a stationary nozzle would. The propeller blades vigorously cut or shear the liquid. Because of the persistence of the flow currents, propeller agitators are effective in very large vessels.

A revolving propeller traces out a helix in the fluid, and if there were no slip between liquid and propeller, one full revolution would move the liquid longitudinally a fixed distance depending on the angle of inclination of the propeller blades. The ratio of this distance to the propeller diameter is known as the *pitch* of the propeller. A propeller with a pitch of 1.0 is said to have *square pitch.*

A typical propeller is illustrated in Fig. 9.2a. Standard three-blade marine propellers with square pitch are most common; four-blade, toothed, and other designs are sometimes employed for special purposes.

Propellers rarely exceed 18 in. in diameter regardless of the size of the vessel. In a deep tank two or more propellers may be mounted on the same shaft, usually directing the liquid in the same direction.

Turbines. Four types of turbine impeller are illustrated in Fig. 9.2. The simple straight-blade turbine shown in Fig. 9.2b pushes the liquid radially and tangentially with almost no vertical motion at the impeller. The currents it generates travel outward to the vessel wall and then flow either upward or downward. Such impellers are sometimes called paddles. In process vessels they typically turn at 20 to 150 rpm. The disk turbine, with multiple straight blades mounted on a horizontal disk (Fig. 9.2c), like the straight-blade impeller, creates zones of high shear rate; it is especially useful for dispersing a gas in a liquid. Also widely used for gas dispersion is the concave-blade CD-6 disk turbine shown in Fig. 9.2d. A pitched-blade turbine (Fig. 9.2e) is used when good overall circulation is important.

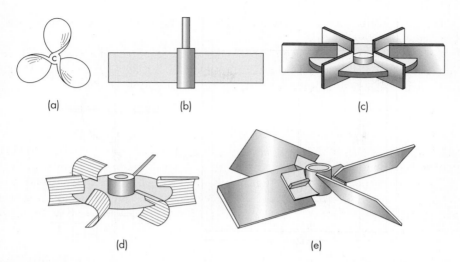

FIGURE 9.2
Impellers for liquids of moderate viscosity: (*a*) three-blade marine propeller;
(*b*) simple straight-blade turbine; (*c*) disk turbine; (*d*) concave-blade CD-6 impeller
(*Chemineer, Inc.*); (*e*) pitched-blade turbine.

"Standard" turbine design. The designer of an agitated vessel has an unusually large number of choices to make as to type and location of the impeller, the proportions of the vessel, the number and proportions of the baffles, and so forth. Each of these decisions affects the circulation rate of the liquid, the velocity patterns, and the power consumed. As a starting point for design in ordinary agitation problems, a turbine agitator of the type shown in Fig. 9.3 is commonly used. Typical proportions are

$$\frac{D_a}{D_t} = \frac{1}{3} \qquad \frac{H}{D_t} = 1 \qquad \frac{J}{D_t} = \frac{1}{12}$$

$$\frac{E}{D_t} = \frac{1}{3} \qquad \frac{W}{D_a} = \frac{1}{5} \qquad \frac{L}{D_a} = \frac{1}{4}$$

The number of baffles is usually 4; the number of impeller blades ranges from 4 to 16 but is generally 6 or 8. Special situations may, of course, dictate different proportions from those listed above; it may be advantageous, for example, to place the agitator higher or lower in the tank, or a much deeper tank may be needed to achieve the desired process result. The listed "standard" proportions, nonetheless, are widely accepted and are the basis of many published correlations of agitator performance.

High-efficiency impellers. Variations of the pitched-blade turbine have been developed to provide more uniform axial flow and better mixing, as well as to reduce the power required for a given flow rate. The high-efficiency impeller HE-3, shown in Fig. 9.4*a*, has three slanted blades that are crimped to decrease the blade angle near the tip. The A310 fluid-foil impeller (Fig. 9.4*b*) uses airfoil-shaped blades

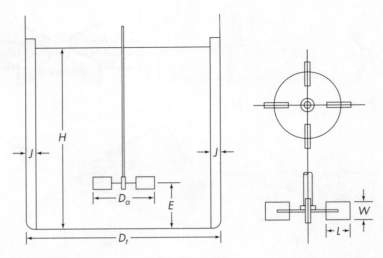

FIGURE 9.3
Measurements of turbine. (*After Rushton et al.*[45])

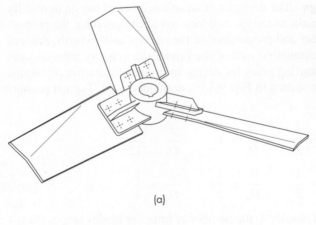

(a)

FIGURE 9.4
High-efficiency impellers:
(*a*) HE-3 impeller (*Chemineer, Inc.*); (*b*) A310 fluid-foil impeller (*Lightnin Division, General Signal*).

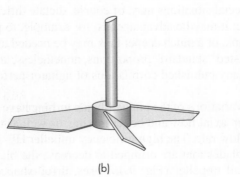

(b)

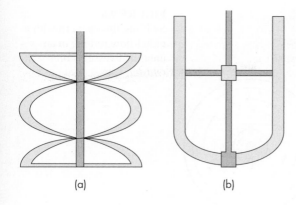

FIGURE 9.5
Impellers for high-viscosity liquids:
(*a*) double-flight helical-ribbon
impeller; (*b*) anchor impeller.

(a)   (b)

which taper so that they are narrower at the tip than at their base. These impellers are widely used to mix low- or moderate-viscosity liquids, but are not recommended for very viscous liquids or for dispersing gases.

Impellers for highly viscous liquids. Well-designed turbine impeller systems can be used with viscosities up to about 50 Pa·s. For viscosities above 20 Pa·s, however, the helical-ribbon impeller shown in Fig. 9.5a is often more effective. The diameter of the helix is very close to the inside diameter of the tank, guaranteeing liquid motion all the way to the tank wall even with very viscous materials. Helical ribbons have been used successfully with viscosities up to 25,000 Pa·s.

To provide good agitation near the floor of the tank, an anchor impeller (Fig. 9.5b) may be used. Because it creates no vertical motion, it is a less effective mixer than a helical ribbon, but it promotes good heat transfer to or from the vessel wall. For this purpose both anchors and helical ribbons may be equipped with scrapers that physically remove liquid from the tank wall.

Flow patterns

The way a liquid moves in an agitated vessel depends on many things: the type of impeller; the characteristics of the liquid, especially its viscosity; and the size and proportions of the tank, baffles, and impeller. The liquid velocity at any point in the tank has three components, and the overall flow pattern in the tank depends on the variations in these three velocity components from point to point. The first velocity component is radial and acts in a direction perpendicular to the shaft of the impeller. The second component is longitudinal and acts in a direction parallel with the shaft. The third component is tangential, or rotational, and acts in a direction tangent to a circular path around the shaft. In the usual case of a vertical shaft, the radial and tangential components are in a horizontal plane, and the longitudinal component is vertical. The radial and longitudinal components are useful and provide the flow necessary for the mixing action. When the shaft is vertical and centrally located in the tank, the tangential component is generally disadvantageous. The tangential flow follows a circular path around the shaft and creates a vortex in the liquid, as shown in Fig. 9.6 for a flat-blade turbine. Exactly the same flow pattern would be observed with a pitched-blade turbine or a propeller. The swirling perpetuates stratification at

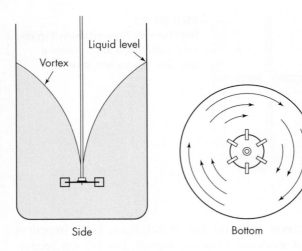

Liquid level

Vortex

Side   Bottom

FIGURE 9.6
Swirling flow pattern with a radial-flow turbine in an unbaffled vessel. (*After Oldshue.*[37])

the various levels without providing longitudinal flow between levels. If solid particles are present, circulatory currents tend to throw the particles to the outside by centrifugal force; from there they move downward and to the center of the tank at the bottom. Instead of mixing, its reverse—concentration—occurs. Since, in circulatory flow, the liquid flows with the direction of motion of the impeller blades, the relative velocity between the blades and the liquid is reduced, and the power that can be absorbed by the liquid is limited. In an unbaffled vessel, circulatory flow is induced by all types of impellers, whether axial flow or radial flow. If the swirling is strong, the flow pattern in the tank is virtually the same regardless of the design of the impeller. At high impeller speeds the vortex may be so deep that it reaches the impeller, and gas from above the liquid is drawn down into the charge. Generally this is undesirable.

Prevention of swirling. Circulatory flow and swirling can be prevented by any of three methods. In small tanks, the impeller can be mounted off center, as shown in Fig. 9.7. The shaft is moved away from the centerline of the tank, then tilted in a plane perpendicular to the direction of the move. In larger tanks, the agitator may be mounted in the side of the tank, with the shaft in a horizontal plane but at an angle with a radius.

In large tanks with vertical agitators, the preferable method of reducing swirling is to install baffles, which impede rotational flow without interfering with radial or longitudinal flow. A simple and effective baffling is attained by installing vertical strips perpendicular to the wall of the tank. Baffles of this type are shown in Fig. 9.1. Except in very large tanks, four baffles are sufficient to prevent swirling and vortex formation. Even one or two baffles, if more cannot be used, have a strong effect on the circulation patterns. For turbines, the width of the baffle need be no more than one-twelfth of the vessel diameter; for propellers, no more than one-eighteenth the tank diameter is needed.[5] For viscous liquids even narrower baffles are generally used, and baffles are not needed at all when $\mu > 10$ Pa·s. Baffles are also not needed with side-entering, inclined, or off-center propellers.

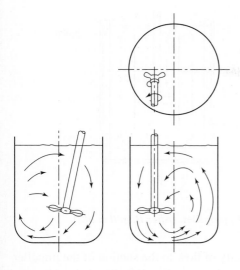

FIGURE 9.7
Flow pattern with off-center propeller. (*After Bissell et al.*[5])

Once the swirling is stopped, the specific flow pattern in the vessel depends on the type of impeller. Propeller agitators usually drive the liquid down to the bottom of the tank, where the stream spreads radially in all directions toward the wall, flows upward along the wall, and returns to the suction of the propeller from the top. Propellers are used when strong vertical currents are desired, for example, when heavy solid particles are to be kept in suspension. They are not ordinarily used when the viscosity of the liquid is greater than about 5 Pa·s. Pitched-blade turbines with 45° down-thrusting blades are also used to provide strong axial flow for suspension of solids. Axial-flow impellers, however, tend to change their discharge flow pattern from axial flow at low liquid viscosities to radial flow when the viscosity is very high.[38]

Flat-blade turbines give good radial flow in the plane of the impeller, with the flow dividing at the wall, to form two separate circulation patterns. One portion flows downward along the wall and back to the center of the impeller from below, and the other flows upward toward the surface and back to the impeller from above. In an unbaffled tank, there are strong tangential flows and vortex formation at moderate stirrer speeds. With baffles present, the vertical flows are increased, and there is more rapid mixing of the liquid.

In a vertical cylindrical tank, the depth of the liquid should be equal to or somewhat greater than the diameter of the tank. If greater depth is desired, two or more impellers are mounted on the same shaft. The lowest impeller is commonly a radial-flow unit such as a straight-blade turbine; the upper ones are usually axial-flow impellers. The lowest impeller is mounted about one impeller diameter above the bottom of the tank.

Draft tubes. The return flow to an impeller of any type approaches the impeller from all directions, because it is not under the control of solid surfaces. The flow to and from a propeller, for example, is essentially similar to the flow of air to and from a fan operating in a room. In most applications of impeller mixers this is not a

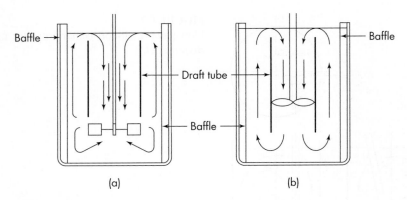

FIGURE 9.8
Draft tubes, baffled tank: (*a*) turbine; (*b*) propeller. (*After Bissell et al.*[5])

limitation, but when the direction and velocity of flow to the suction of the impeller are to be controlled, draft tubes are used, as shown in Fig. 9.8. These devices may be useful when high shear at the impeller itself is desired, as in the manufacture of certain emulsions, or where solid particles that tend to float on the surface of the liquid in the tank are to be dispersed in the liquid. Draft tubes for propellers are mounted around the impeller, and those for turbines are mounted immediately above the impeller. This is shown in Fig. 9.8. Draft tubes add to the fluid friction in the system; and for a given power input, they reduce the rate of flow, so they are not used unless they are required.

Circulation rates

For a processing vessel to be effective, regardless of the nature of the agitation problem, the volume of fluid circulated by the impeller must be great enough to sweep out the entire vessel in a reasonable time. Also, the velocity of the stream leaving the impeller must be sufficient to carry the currents to the remotest parts of the tank. In mixing and dispersion operations, the circulation rate is not the only factor, or even the most important one; turbulence in the moving stream often governs the effectiveness of the operation. Turbulence results from properly directed currents and large velocity gradients in the liquid. Circulation and turbulence generation both consume energy; the relations between power input and the design parameters of agitated vessels are discussed later. Some agitation problems, as will be shown, call for large flows or high average velocities, while others require high local turbulence or power dissipation. Although both flow rate and power dissipation increase with stirrer speed, selection of the type and size of the impeller influences the relative values of flow rate and power dissipation. In general, large impellers moving at medium speed are used to promote flow, and smaller impellers operating at high speed are used where intense turbulence is required.

Flow number. A turbine or propeller agitator is, in essence, a pump impeller operating without a casing and with undirected inlet and output flows. The governing

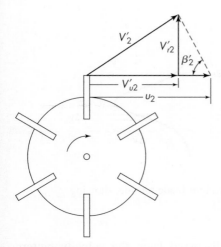

FIGURE 9.9
Velocity vectors at tip of turbine impeller blade.

relations for turbines are similar to those for centrifugal pumps discussed in Chap. 8.[19a] Consider the flat-blade turbine impeller shown in Fig. 9.9. The nomenclature is the same as in Fig. 8.11: u_2 is the velocity of the blade tips; V'_{u2} and V'_{r2} are the actual tangential and radial velocities of the liquid leaving the blade tips, respectively; and V'_2 is the total liquid velocity at the same point. Assume that the tangential liquid velocity is some fraction k of the blade-tip velocity, or

$$V'_{u2} = ku_2 = k\pi D_a n \tag{9.1}$$

since $u_2 = \pi D_a n$. The volumetric flow rate through the impeller, from Eq. (8.16), is

$$q = V'_{r2} A_p \tag{9.2}$$

Here A_p is taken to be the area of the cylinder swept out by the tips of the impeller blades, or

$$A_p = \pi D_a W \tag{9.3}$$

where D_a = impeller diameter
W = width of blades

From the geometry of Fig. 9.9

$$V'_{r2} = (u_2 - V'_{u2}) \tan \beta'_2 \tag{9.4}$$

Substituting for V'_{u2} from Eq. (9.1) gives

$$V'_{r2} = \pi D_a n(1 - k) \tan \beta'_2 \tag{9.5}$$

Figure 9.10 shows the velocity profile of liquid flowing radially from the blade of a standard turbine. The velocities shown are those at the tip of the blade. The radial velocity is a maximum in the plane of the middle of the blade and is much

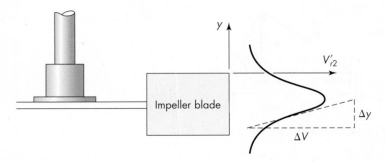

FIGURE 9.10
Typical velocity profile of flow from a straight-blade turbine, showing
definition of shear rate. (*After Oldshue.*[37])

smaller at the upper and lower edges. As discussed later, the velocity pattern
changes with distance from the impeller tip. The volumetric flow rate q is the total
flow leaving the impeller, as measured at the tip of the blades. From Eqs. (9.2) to
(9.4) this volumetric flow rate is therefore

$$q = K\pi^2 D_a^2 n W (1 - k) \tan \beta_2' \tag{9.6}$$

where K is a constant that allows for the fact that the radial velocity is not actually
constant over the width of the blade. For geometrically similar impellers W is proportional to D_a, and K, k, and β_2' are approximately constant. Hence

$$q \propto n D_a^3 \tag{9.7}$$

The ratio of these two quantities is called the *flow number* N_Q, which is defined
by

$$N_Q \equiv \frac{q}{n D_a^3} \tag{9.8}$$

Equations (9.6) to (9.8) imply that N_Q is constant for each type of impeller. For a
standard flat-blade turbine[22] in a baffled vessel, N_Q may be taken as 1.3. This gives
the discharge flow from the tip of the impeller and not the total flow produced. The
high-velocity stream of liquid leaving the tip of the impeller entrains some of the
slowly moving bulk liquid, which slows down the jet but increases the total flow
rate. For flat-blade turbines, the total flow, estimated from the average circulation
time for particles or dissolved tracers,[22] was shown to be

$$q_T = 0.92 n D_a^3 \frac{D_t}{D_a} \tag{9.9}$$

For the typical ratio $D_t/D_a = 3$, q_T is $2.76 n D_a^3$, or 2.1 times the value at the
impeller ($N_Q = 1.3$). Equation (9.9) should be used only for D_t/D_a ratios between
2 and 4.

For axial-flow impellers such as pitched-blade turbines, marine propellers,
or shaped-blade turbines q is the discharge rate in a vertical direction as measured immediately below the impeller. The flow number N_Q may be considered

constant. For the design of baffled agitated vessels the following values are recommended:

For marine propellers[19b] (square pitch)	$N_Q = 0.5$
For a four-blade 45° turbine[19b] $\left(\dfrac{W}{D_a} = \dfrac{1}{6}\right)$	$N_Q = 0.87$
For a disk turbine	$N_Q = 1.3$
For HE-3 high-efficiency impeller	$N_Q = 0.47$

Velocity patterns and velocity gradients

More details of the flow patterns, the local velocities, and the total flow produced by an impeller have been obtained by use of small velocity probes[22] or by photographic measurements of tracer particles.[13] Some of Cutter's results[13] for a 4-in. flat-blade turbine in an 11.5-in. tank are shown in Fig. 9.11. As the fluid leaves the impeller blades, the radial component of the fluid velocity V_r' at the centerline of the impeller is about 0.6 times the tip speed u_2. The radial velocity decreases with vertical distance from the centerline; but as shown in Fig. 9.10, the jet extends beyond the edge of the blades because of entrainment, and integration gives a total flow of $0.75q_B$, where q_B is the flow that would exist if all the fluid were moving at velocity u_2 across the sides of the cylinder swept out by the blades. The entrained flow at this point is therefore 25 percent of the flow coming directly from the blades.

As the jet travels away from the impeller, it slows down because of the increased area for flow and because more liquid is entrained. Along the centerline of the impeller, the velocity drops more or less linearly with radial distance, and the product $V_r'r$ is nearly constant, as was shown by other studies.[22] The total volumetric flow increases with radius to about $1.2q_B$ because of further entrainment and then drops near the vessel wall because the flow has started to divide into the upward and

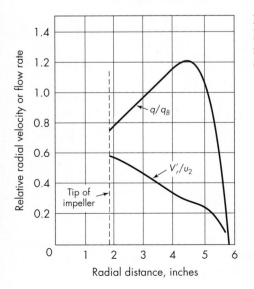

FIGURE 9.11
Radial velocity V_r'/u_2 and volumetric flow rate q/q_B in a turbine-agitated vessel. (*After Cutter.*[13])

downward circulation loops. The maximum flow of $1.2q_B$ compared to the radial discharge velocity of $0.6u_2$ indicates a total flow twice the direct impeller discharge, in agreement with the factor of 2.1 calculated by using Eq. (9.9).

The velocity gradients in an agitated vessel vary widely from one point to another in the fluid. The gradient equals $\Delta V/\Delta y$, as illustrated in Fig. 9.10. It is quite large near the edge of the jet leaving the impeller, since the velocity is high and the jet is relatively narrow. Based on the vertical velocity profiles at the blade tip, the velocity gradient at this point is approximately $0.9u/0.75W$, where $0.9u$ is the resultant of the radial and tangential velocities, and $0.75W$ is one-half the width of the jet leaving the impeller. Since $u = \pi n D_a$ and $W = D_a/5$ for a standard turbine, this corresponds to a velocity gradient of $19n$, which can serve as an estimate of the maximum shear rate in the region near a turbine impeller. As the jet travels away from the impeller, it slows down, and the velocity gradient at the edge of the jet diminishes. Behind the turbine blades there are intense vortices where the local shear rate may be as high as $50n$.[48]

Figure 9.12 shows the fluid currents observed with a six-blade turbine, 6 in. in diameter, turning at 200 rpm in a 12-in. vessel containing cold water.[33] The plane of observation passes through the axis of the impeller shaft and immediately in front of a radial baffle. Fluid leaves the impeller in a radial direction, separates into longitudinal streams flowing upward or downward over the baffle, flows inward toward the impeller shaft, and ultimately returns to the impeller intake. At the bottom of the vessel, immediately under the shaft, the fluid moves in a swirling motion; elsewhere the currents are primarily radial or longitudinal.

The numbers in Fig. 9.12 indicate the scalar magnitude of the fluid velocity at various points as fractions of the velocity of the tip of the impeller blades. Under the

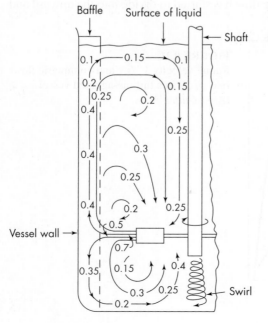

FIGURE 9.12
Velocity patterns in turbine agitator. (*After Morrison et al.*[33])

conditions used, the tip velocity is 4.8 ft/s (1.46 m/s). The velocity in the jet quickly drops from the tip velocity to about 0.4 times the tip velocity near the vessel wall. Velocities at other locations in the vessel are on the order of 0.25 times the tip velocity, although there are two toroidal regions of almost stagnant fluid, one above and one below the impeller, in which the velocity is only 0.10 to 0.15 times the tip velocity.

Increasing the impeller speed increases the tip velocity and the circulation rate. It does not, however, increase the fluid velocity at a given location in the same proportion, for a fast-moving jet entrains much more material from the bulk of the liquid than a slower-moving jet does, and the jet velocity drops very quickly with increasing distance from the impeller.

With shear-thinning or pseudoplastic liquids, a turbine may give a local region of high shear rate around the impeller, but near the wall the shear rate is much lower and the apparent viscosity may be much higher. Hence the velocity near the wall may be very much less than indicated in Fig. 9.12. An axial-flow impeller or helical ribbon may be needed to avoid forming stagnant regions in the vessel.

Power consumption

An important consideration in the design of an agitated vessel is the power required to drive the impeller. When the flow in the tank is turbulent, the power requirement can be estimated from the product of the flow q produced by the impeller and the kinetic energy E_k per unit volume of the fluid. These are

$$q = nD_a^3 N_Q$$

and

$$E_k = \frac{\rho(V_2')^2}{2}$$

Velocity V_2' is slightly smaller than the tip speed u_2. If the ratio V_2'/u_2 is denoted by α, $V_2' = \alpha \pi n D_a$ and the power requirement is

$$P = nD_a^3 N_Q \frac{\rho}{2}(\alpha \pi n D_a)^2$$

$$= \rho n^3 D_a^5 \left(\frac{\alpha^2 \pi^2}{2} N_Q\right) \tag{9.10}$$

In dimensionless form,

$$\frac{P}{n^3 D_a^5 \rho} = \frac{\alpha^2 \pi^2}{2} N_Q \tag{9.11}$$

The left-hand side of Eq. (9.11) is called the power number N_P, defined by

$$N_P \equiv \frac{P}{n^3 D_a^5 \rho} \tag{9.12a}$$

In fps units,

$$N_P \equiv \frac{P g_c}{n^3 D_a^5 \rho} \tag{9.12b}$$

For a standard six-blade turbine, $N_Q = 1.3$; and if α is taken as 0.95, then $N_P = 5.8$. This is in good agreement with observation, as shown later.

Power correlations. To estimate the power required to rotate a given impeller at a given speed, empirical correlations of power (or power number) with the other variables of the system are needed. The form of such correlations can be found by dimensional analysis, given the important measurements of the tank and impeller, the distance of the impeller from the tank floor, the liquid depth, and the dimensions of the baffles if they are used. The number and arrangement of the baffles and the number of blades in the impeller must also be fixed. The variables that enter the analysis are the important measurements of tank and impeller, the viscosity μ and the density ρ of the liquid, and the speed n. Also, unless provision is made to eliminate swirling, a vortex will appear at the surface of the liquid. Some of the liquid must be lifted above the average, or unagitated, level of the liquid surface, and this lift must overcome the force of gravity. Accordingly, the acceleration of gravity g must be considered as a factor in the analysis.

The various linear measurements can all be converted to dimensionless ratios, called *shape factors,* by dividing each by one of their number which is arbitrarily chosen as a basis. The diameter of the impeller D_a and that of the tank D_t are suitable choices for this base measurement, and the shape factors are calculated by dividing each of the remaining measurements by the magnitude of D_a or D_t. Let the shape factors, so defined, be denoted by $S_1, S_2, S_3, \ldots, S_n$. The impeller diameter D_a is then also taken as the measure of the size of the equipment and used as a variable in the analysis, just as the diameter of the pipe was in the dimensional analysis of friction in pipes. Two mixers of the same geometric proportions throughout but of different sizes will have identical shape factors but will differ in the magnitude of D_a. Devices meeting this requirement are said to be geometrically similar or to possess geometric similarity.

When the shape factors are temporarily ignored and the liquid is assumed newtonian, the power P is a function of the remaining variables, or

$$P = \psi(n, D_a, \mu, g, \rho) \qquad (9.13)$$

Application of the method of dimensional analysis gives the result[45]

$$\frac{P}{n^3 D_a^5 \rho} = \psi\left(\frac{n D_a^2 \rho}{\mu}, \frac{n^2 D_a}{g}\right) \qquad (9.14)$$

By taking account of the shape factors, Eq. (9.14) can be written

$$\frac{P}{n^3 D_a^5 \rho} = \psi\left(\frac{n D_a^2 \rho}{\mu}, \frac{n^2 D_a}{g}, S_1, S_2, \ldots, S_n\right) \qquad (9.15)$$

The first dimensionless group in Eq. (9.14), $P/n^3 D_a^5 \rho$, is the power number N_P. The second, $n D_a^2 \rho / \mu$, is a Reynolds number Re; the third, $n^2 D_a / g$, is the Froude number Fr. Equation (9.15) can therefore be written

$$N_P = \psi(\text{Re}, \text{Fr}, S_1, S_2, \ldots, S_n) \qquad (9.16)$$

The three dimensionless groups in Eq. (9.14) may be given simple interpretations.[23] Consider the group $nD_a^2\rho/\mu$. Since the impeller tip speed u_2 equals $\pi D_a n$,

$$\text{Re} = \frac{nD_a^2\rho}{\mu} = \frac{(nD_a)D_a\rho}{\mu} \propto \frac{u_2D_a\rho}{\mu} \tag{9.17}$$

and this group is proportional to a Reynolds number calculated from the diameter and peripheral speed of the impeller. This is the reason for the name of the group. At low Reynolds numbers ($\text{Re} < 10$), viscous flow prevails throughout the vessel, and at $\text{Re} > 10^4$ the flow is turbulent everywhere. A transition region exists at intermediate Reynolds numbers.

The power number N_p is analogous to a friction factor or a drag coefficient. It is proportional to the ratio of the drag force acting on a unit area of the impeller and the inertial stress, that is, the flow of momentum associated with the bulk motion of the fluid.

The Froude number Fr is a measure of the ratio of the inertial stress to the gravitational force per unit area acting on the fluid. It appears in fluid dynamic situations where there is significant wave motion on a liquid surface. It is especially important in ship design. It is not important when baffles are used or when $\text{Re} < 300$. Unbaffled vessels are rarely used at high Reynolds numbers, and hence the Froude number is not included in the following correlations.

Power correlations for specific impellers. Typical plots of N_p versus Re for baffled tanks fitted with centrally located impellers are shown in Fig. 9.13. The top

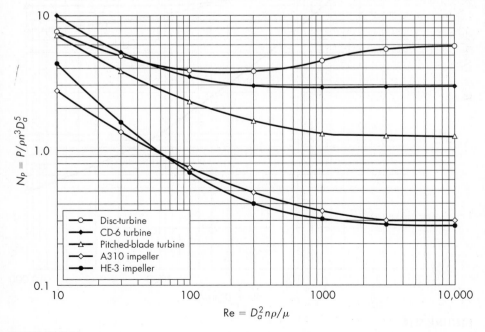

FIGURE 9.13
Power number N_p versus Reynolds number Re for turbines and high-efficiency impellers.

curve is for a six-blade disk turbine with the shape factors of a standard turbine given on page 241. At high Reynolds numbers the curve levels off at a power number of 5.8, in agreement with the value calculated earlier. The curve for the CD-6 concave-blade turbine is similar but levels off at a value of 2.9. The pitched turbine with four blades set at an angle of 45° draws about 70 percent as much power as the standard turbine at low Reynolds numbers, but only about 20 percent as much at high Reynolds numbers. The A310 and HE-3 high-efficiency impellers have much lower power numbers than the turbines, but also have lower flow numbers and are usually operated at higher speeds. The power number for all five impellers is constant when $Re > 10^4$, and it varies inversely with the Reynolds number when $Re < 10$.

Power numbers for a marine propeller and a helical ribbon are shown in Fig. 9.14. For the propeller the power number when $Re = 10^4$ is about 50 percent greater in a baffled tank than in an unbaffled one, but at low Reynolds numbers there

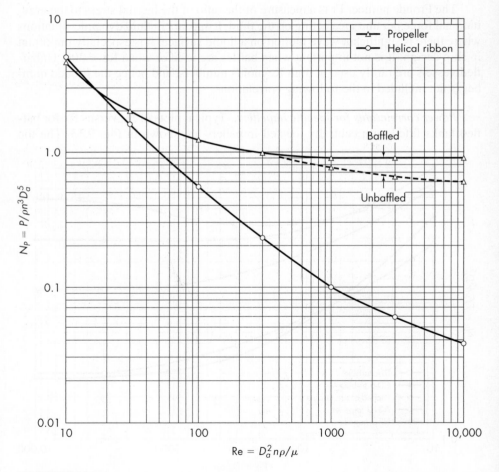

FIGURE 9.14

Power number N_p versus Reynolds number Re for marine propellers (pitch = 1.5:1) and helical ribbons.

is no difference. Baffles are not used with the helical impeller, and N_P decreases rapidly as Re increases. No data are available for a helical impeller for Re $> 10^4$, but these impellers are commonly used only at low Reynolds numbers. Power numbers for an anchor agitator (not shown in the plot) are slightly greater than for a helical impeller over the entire range of Reynolds numbers.

Effect of system geometry. The effects on N_P of the shape factors $S_1, S_2, \ldots, S_n$ in Eq. (9.16) are sometimes small and sometimes very large. Sometimes two or more of the factors are interrelated; that is, the effect of changing S_1, say, may depend on the magnitude of S_2 or S_3. With a flat-blade turbine operating at high Reynolds numbers in a baffled tank, the effects of changing the system geometry may be summarized as follows.[3]

1. Decreasing S_1, the ratio of impeller diameter to tank diameter, increases N_P when the baffles are few and narrow and decreases N_P when the baffles are many and wide. Thus shape factors S_1 and S_5 are interrelated. With four baffles and S_5 equal to $\frac{1}{12}$, as is common in industrial practice, changing S_1 has almost no effect on N_P.
2. The effect of changing S_2, the clearance, depends on the design of the turbine. Increasing S_2 increases N_P for a disk turbine of the type shown in Fig. 9.3. For a pitched-blade turbine increasing S_2 lowers N_P considerably, as shown in Table 9.1; for an open straight-blade turbine it lowers N_P slightly.
3. With a straight-blade open turbine the effect of changing S_4, the ratio of blade width to impeller diameter, depends on the number of blades. For a six-blade turbine N_P increases directly with S_4; for a four-blade turbine N_P increases with $S_4^{1.25}$. With pitched-blade turbines the effect of blade width on power consumption is much smaller than with straight-blade turbines (see Table 9.1).
4. Two straight-blade turbines on the same shaft draw about 1.9 times as much power as one turbine alone, provided the spacing between the two impellers is at least equal to the impeller diameter. Two closely spaced turbines may draw as much as 2.4 times the power of a single turbine.
5. The shape of the tank has relatively little effect on N_P. The power consumed in a horizontal cylindrical vessel, whether baffled or unbaffled, or in a baffled vertical tank with a square cross section is the same as in a vertical cylindrical tank. In an unbaffled square tank the power number is about 0.75 times that in a baffled cylindrical vessel. Circulation patterns, of course, are strongly affected by tank shape, even though power consumption is not.

TABLE 9.1

Effect of blade width and clearance on power consumption of six-blade 45° turbines[10,42]

$W/D_a, S_4$	Clearance, S_2	K_T
0.3	0.33	2.0
0.2	0.33	1.63
0.2	0.25	1.74
0.2	0.17	1.91

Calculation of power consumption. The power delivered to the liquid is computed from Eq. (9.12) after a relationship for N_P is specified. Rearranging Eq. (9.12) gives

$$P = N_P n^3 D_a^5 \rho \tag{9.18}$$

At low Reynolds numbers, the lines of N_P versus Re for both baffled and unbaffled tanks coincide, and the slope of the line on logarithmic coordinates is -1. Therefore

$$N_P = \frac{K_L}{\text{Re}} \tag{9.19}$$

This leads to

$$P = K_L n^2 D_a^3 \mu \tag{9.20}$$

The flow is laminar in this range, and density is no longer a factor. Equations (9.19) and (9.20) can be used when Re is less than 10.

In baffled tanks at Reynolds numbers larger than about 10,000, the power number is independent of the Reynolds number, and viscosity is not a factor. In this range the flow is fully turbulent, and Eq. (9.16) becomes

$$N_P = K_T \tag{9.21}$$

from which

$$P = K_T n^3 D_a^5 \rho \tag{9.22}$$

Magnitudes of constants K_T and K_L for various types of impellers and tanks are shown in Table 9.2.

TABLE 9.2
Values of constants K_L and K_T in Eqs. (9.19) and (9.21) for baffled tanks having four baffles at tank wall, with width equal to 10 percent of tank diameter

Type of impeller	K_L	K_T
Propeller, three blades		
Pitch 1.0[43]	41	0.32
Pitch 1.5[37]	48	0.87
Turbine		
Six-blade disk[37] ($S_3 = 0.25$, $S_4 = 0.2$)	65	5.75
Six pitched blades[42] (45°, $S_4 = 0.2$)	—	1.63
Four pitched blades[37] (45°, $S_4 = 0.2$)	44.5	1.27
Flat paddle, two blades[43] ($S_4 = 0.2$)	36.5	1.70
HE-3 impeller	43	0.28
Helical ribbon	52	—
Anchor[37]	300	0.35

EXAMPLE 9.1. A disk turbine with six flat blades is installed centrally in a vertical baffled tank 2 m in diameter. The turbine is 0.67 m in diameter and is positioned 0.67 m above the bottom of the tank. The turbine blades are 134 mm wide. The tank is filled to a depth of 2 m with an aqueous solution of 50% NaOH at 65°C, which has a viscosity of 12 cP and a density of 1,500 kg/m³. The turbine impeller turns at 90 rpm. What power will be required?

Solution. First calculate the Reynolds number. The quantities needed are

$$D_a = 0.67 \text{ m} \qquad n = \frac{90}{60} = 1.5 \text{ r/s}$$

$$\mu = 0.012 \text{ Pa} \cdot \text{s} \qquad \rho = 1,500 \text{ kg/m}^3$$

Then

$$\text{Re} = \frac{D_a^2 n \rho}{\mu} = \frac{0.67^2 \times 1.5 \times 1,500}{0.012} = 84,169$$

Since Re > 10^4, $N_P = K_T$. From Table 9.2, $K_T = N_P = 5.8$, and from Eq. (9.24)

$$P = 5.8 \times 1.5^3 \times 0.67^5 \times 1,500 = 3,964 \text{ W, or } 3.96 \text{ kW}$$

EXAMPLE 9.2. The agitation system of Example 9.1 is to be used to mix a rubber-latex compound having a viscosity of 120 Pa·s and a density of 1,120 kg/m³. What power will be required?

Solution. The Reynolds number is now

$$\text{Re} = \frac{0.67^2 \times 1.5 \times 1,120}{120} = 6.3$$

This is well within the range of laminar flow. From Table 9.2, $K_L = 65$, and from Eq. (9.20)

$$P = 65 \times 1.5^2 \times 0.67^3 \times 120 = 5,278 \text{ W, or } 5.28 \text{ kW}$$

This power requirement is independent of whether the tank is baffled. There is no reason for baffles in a vessel operated at such a low Reynolds number, as a vortex does not form under these conditions.

Note that a 10,000-fold increase in viscosity increases the power by only about 33 percent over that required in the baffled tank operating on the low-viscosity liquid.

Power consumption in non-newtonian liquids. In correlating power data for non-newtonian liquids, the power number $P/n^3 D_a^5 \rho$ is defined in the same way as for newtonian fluids. The Reynolds number is not easily defined, since the apparent viscosity of the fluid varies with the shear rate (velocity gradient) and the shear rate changes considerably from one point to another in the vessel. Successful correlations have been developed, however, with a Reynolds number defined as in Eq. (9.17), using an average apparent viscosity μ_a calculated from an average shear rate $(du/dy)_{av}$. The Reynolds number is then

$$\text{Re}_n = \frac{n D_a^2 \rho}{\mu_a} \qquad (9.23)$$

For a power law fluid, as shown by Eq. (3.7), the average apparent viscosity can be related to the average shear rate by the equation

$$\mu_a = K'\left(\frac{du}{dy}\right)^{n'-1}_{av} \tag{9.24}$$

Substitution in Eq. (9.23) gives

$$\mathrm{Re}_n = \frac{nD_a^2\rho}{K'(du/dy)^{n'-1}_{av}} \tag{9.25}$$

For a straight-blade turbine in pseudoplastic liquids, it has been shown that the average shear rate in the vessel is directly related to the impeller speed. For a number of pseudoplastic liquids a satisfactory, though approximate, relation is[9,18,27]

$$\left(\frac{du}{dy}\right)_{av} = 11n \tag{9.26}$$

Note that the average shear rate of $11n$ is slightly more than one-half the maximum estimated value of $19n$ (see p. 250). The volumetric average shear rate for the tank is probably much less than $11n$, but the effective value for power consumption depends heavily on shear rates in the region of the stirrer.

Combining Eqs. (9.25) and (9.26) and rearranging give

$$\mathrm{Re}_n = \frac{n^{2-n'}D_a^2\rho}{11^{n'-1}K'} \tag{9.27}$$

Figure 9.15 shows the power number–Reynolds number correlation for a six-blade turbine impeller in pseudoplastic fluids. The dashed curve is taken from Fig. 9.13 and applies to newtonian fluids, for which $\mathrm{Re} = nD_a^2\rho/\mu$. The solid curve is for pseudoplastic liquids, for which Re_n is given by Eqs. (9.23) and (9.27). At Reynolds numbers below 10 and above 100, the results with pseudoplastic

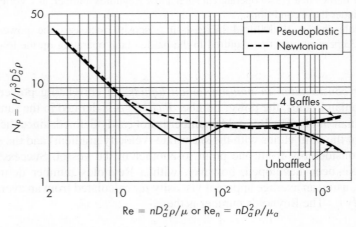

FIGURE 9.15
Power correlation for a six-blade turbine in pseudoplastic liquids.

liquids are the same as with newtonian liquids. In the intermediate range of Reynolds numbers between 10 and 100, pseudoplastic liquids consume less power than newtonian fluids. The transition from laminar to turbulent flow in pseudoplastic liquids is delayed until the Reynolds number reaches about 40, instead of 10 as in newtonian liquids.

The flow patterns in an agitated pseudoplastic liquid differ considerably from those in a newtonian liquid. Near the impeller the velocity gradients are large, and the apparent viscosity of a pseudoplastic liquid is low. As the liquid travels away from the impeller, the velocity gradient decreases, and the apparent viscosity of the liquid rises. The liquid velocity drops rapidly, decreasing the velocity gradients further and increasing the apparent viscosity still more. Even when there is high turbulence near the impeller, therefore, the bulk of the liquid may be moving in slow laminar flow and consuming relatively little power. The toroidal rings of slowly moving liquid indicated in Fig. 9.12 are very strongly marked when the agitated liquid is a pseudoplastic.

BLENDING AND MIXING

Mixing is a much more difficult operation to study and describe than agitation. The patterns of fluid flow and velocity in an agitated vessel are complex but reasonably definite and reproducible. The power consumption is readily measured. The results of mixing studies, on the other hand, are seldom highly reproducible and depend in large measure on how mixing is defined by the particular experimenter. Often the criterion for good mixing is visual, as in the use of interference phenomena to follow the blending of gases in a duct[30] or the color change of an acid-base indicator to determine liquid blending times.[17,35] Other criteria that have been used include the rate of decay of concentration or temperature[25] fluctuations, the variation in the analyses of small samples taken at random from various parts of the mix, the rate of transfer of a solute from one liquid phase to another, and in solid-liquid mixtures, the visually observed uniformity of the suspension.

Blending of miscible liquids

Miscible liquids are blended in relatively small process vessels by propellers, turbines, or high-efficiency impellers, usually centrally mounted, and in large storage and waste treatment tanks by side-entering propellers or jet mixers. In a process vessel, all the liquid is usually well agitated and blending is fairly rapid. In a large storage tank, the agitator may be idle much of the time and be turned on only to blend the stratified layers of liquid that were formed as the tank was being filled. Stratified blending is often very slow.

Blending in process vessels. The impeller in a process vessel produces a high-velocity stream, and the liquid is well mixed in the region close to the impeller because of the intense turbulence. As the stream slows down while entraining other liquid and flowing along the wall, there is some radial mixing, as large eddies break down to smaller ones, but there is probably little mixing in the direction of flow. The fluid completes a circulation loop and returns to the eye of the impeller, where vigorous mixing again occurs. Calculations based on this model show that essentially complete mixing (99 percent) should be achieved if the contents of the tank are circulated about 5 times. The mixing time can then be predicted from the correlations for total flow produced by various impellers. For a standard six-blade turbine, from Eq. (9.9)

$$q_T = 0.92 n D_a^3 \frac{D_t}{D_a} \tag{9.28}$$

$$t_T \approx \frac{5V}{q_T} = 5 \frac{\pi D_t^2 H}{4} \frac{1}{0.92 n D_a^2 D_t} \tag{9.29}$$

or

$$n t_T \left(\frac{D_a}{D_t}\right)^2 \left(\frac{D_t}{H}\right) = \text{const} = 4.3 \tag{9.30}$$

For a given tank and impeller or geometrically similar systems, the mixing time is predicted to vary inversely with the stirrer speed, which is confirmed by experimental studies, provided the Reynolds number is high.[13,32] Figure 9.16 shows the results for several systems plotted as $n t_T$ versus Re. For a turbine in a baffled tank, with $D_a/D_t = \frac{1}{3}$ and $D_t/H = 1$, the value of $n t_T$ for Re > 2,000 is 36, compared to the value predicted from Eq. (9.30) of $9 \times 4.3 = 38.7$.

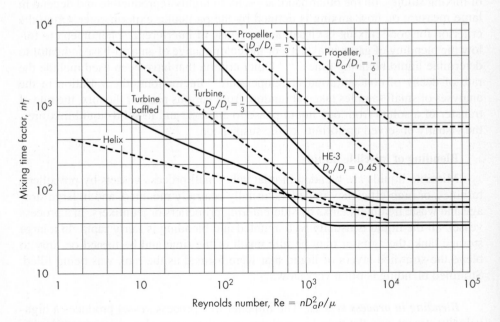

FIGURE 9.16
Mixing times in agitated vessels. Dashed lines are for unbaffled tanks; solids lines are for baffled tanks.

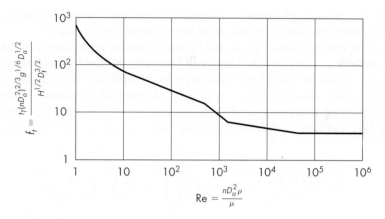

FIGURE 9.17
Correlation of blending times for miscible liquids in a turbine-agitated baffled vessel. (*After Norwood and Metzner.*[35])

The mixing times are appreciably greater when the Reynolds numbers are in the range of 10 to 1000, even though the power consumption is not much different than for the turbulent range. As shown in Fig. 9.16, the mixing time using baffled turbines varies with about the -1.5 power of the stirrer speed in this region and then increases more steeply as the Reynolds number is reduced still further. The data in Fig. 9.16 are for certain ratios of impeller size to tank size. A general correlation given by Norwood and Metzner[35] for turbines is shown in Fig. 9.17. Their mixing time factor can be rearranged to show how it differs from the prediction for the turbulent regime, Eq. (9.30):

$$f_t = \frac{t_T \left(n D_a^2\right)^{2/3} g^{1/6} D_a^{1/2}}{H^{1/2} D_t^{3/2}} = n t_T \left(\frac{D_a}{D_t}\right)^2 \left(\frac{D_t}{H}\right)^{1/2} \left(\frac{g}{n^2 D_a}\right)^{1/6} \tag{9.31}$$

The Froude number in Eq. (9.31) implies some vortex effect, which may be present at low Reynolds numbers, but it is doubtful whether this term should be included for a baffled tank at high Reynolds numbers. When $Re > 10^5$, f_t is almost constant at a value of 5. For $D_a/D_t = \frac{1}{3}$, $D_a/H = 1$, and if we ignore the Froude number, $n t_T$ is about 5, somewhat greater than predicted from Eq. (9.30).

For the HE-3 high-efficiency impeller, the mixing time factors shown in Fig. 9.16 are based on the following empirical correlation[16] for the turbulent regime and a correlation factor for low Reynolds numbers.

$$n t_T = 16.9 \left(\frac{D_t}{D_a}\right)^{1.67} \left(\frac{H}{D_t}\right)^{0.5} \tag{9.32}$$

The mixing times at high Reynolds numbers are somewhat greater than with a standard turbine, but the power required is very much less (see Fig. 9.13). Consequently high-efficiency impellers are usually operated at higher speeds and with higher D_a/D_t ratios than turbines. At the same power input per unit volume, mixing is slightly faster than with turbines in the turbulent mixing regime.

A helical ribbon agitator gives much shorter mixing times for the same power input with very viscous liquids[32] but is slower than the turbine with thin liquids. The mixing times for propellers seem high by comparison with turbines, but of course the power consumption is more than an order of magnitude lower at the same stirrer speed. The propeller data in Fig. 9.16 were taken from a general correlation of Fox and Gex,[17] whose mixing time function differs from both Eqs. (9.30) and (9.31):

$$f'_t = \frac{t_T \left(n D_a^2\right)^{2/3} g^{1/6}}{H^{1/2} D_t} = n t_T \left(\frac{D_a}{D_t}\right)^{3/2} \left(\frac{D_t}{H}\right)^{1/2} \left(\frac{g}{n^2 D_a}\right)^{1/6} \qquad (9.33)$$

Their data were for D_a/D_t of 0.07 to 0.18; the extrapolation to $D_a/D_t = \frac{1}{3}$ for Fig. 9.16 is somewhat uncertain. Because a vortex often forms in an unbaffled vessel the inclusion of the Froude number in Eq. (9.33) is probably justified.

In a pseudoplastic liquid, blending times at Reynolds numbers below about 1,000 are much longer than in newtonian liquids under the same impeller conditions.[18,32] In the regions of low shear, far from the impeller, the apparent viscosity of the pseudoplastic liquid is greater than it is near the impeller. In these remote regions turbulent eddies decay rapidly, and zones of almost stagnant liquid are often formed. Both effects lead to poor mixing and long blending times. An axial-flow impeller with a draft tube can be used to promote good circulation throughout the vessel. At high Reynolds numbers there is little difference in the mixing characteristics of newtonian and pseudoplastic liquids.

When gas bubbles, liquid drops, or solid particles are dispersed in a liquid, the blending time for the continuous phase is increased, even if the comparison is made at the same specific power input.[15] The effect increases with viscosity, and for viscous liquids the blending time can be up to twice the normal value when the gas holdup is only 10 percent.

EXAMPLE 9.3. An agitated vessel 6 ft (1.83 m) in diameter contains a six-blade straight-blade turbine 2 ft (0.61 m) in diameter, set one impeller diameter above the vessel floor, and rotating at 80 rpm. It is proposed to use this vessel for neutralizing a dilute aqueous solution of NaOH at 70° F with a stoichiometrically equivalent quantity of concentrated nitric acid (HNO_3). The final depth of liquid in the vessel is to be 6 ft (1.83 m). Assuming that all the acid is added to the vessel at one time, how long will it take for the neutralization to be complete?

Solution. Figure 9.16 is used. The quantities needed are

$$D_t = 6 \text{ ft} \qquad D_a = 2 \text{ ft} \qquad E = 2 \text{ ft}$$

$$n = \frac{80}{60} = 1.333 \text{ r/s}$$

Density of liquid: $\qquad \rho = 62.3 \text{ lb/ft}^3 \qquad$ (App. 6)

Viscosity of liquid: $\qquad \mu = 6.6 \times 10^{-4} \text{ lb/ft} \cdot \text{s} \qquad$ (App. 6)

The Reynolds number is

$$\mathrm{Re} = \frac{nD_a^2\rho}{\mu} = \frac{1.333 \times 2^2 \times 62.3}{6.60 \times 10^{-4}} = 503{,}000$$

From Fig. 9.16, for Re $= 503{,}000$, $nt_T = 36$. Thus

$$t_T = \frac{36}{1.333} = 27\,\mathrm{s}$$

Stratified blending in storage tanks

For effective blending in a large tank, a side-entering propeller must be oriented precisely with regard to both its angle with the horizontal (for top-to-bottom circulation) and, in the horizontal plane, the angle it makes with the tangent to the tank wall at the point of entry. For optimum results this angle has been found[37] to be between 80° and 83°. The time required for stratified blending depends on the circulation rate but more importantly on the rate of erosion of the interface between the stratified liquid layers. No general correlations are available for stratified blending.

Jet mixers

Circulation in large vessels may also be induced by one or more jets of liquid. Sometimes jets are set in clusters at several locations in the tank. The stream from a single jet maintains its identity for a considerable distance, as seen in Fig. 9.18, which shows the behavior of a circular liquid jet issuing from a nozzle and flowing at high velocity into a stagnant pool of the same liquid. The velocity in the jet issuing from the nozzle is uniform and constant. It remains so in a core, the area of which decreases with distance from the nozzle. The core is surrounded by an expanding turbulent jet, in which the radial velocity decreases with distance from the centerline of the jet. The shrinking core disappears at a distance from the nozzle of $4.3D_j$, where D_j is the diameter of the nozzle. The turbulent jet maintains its

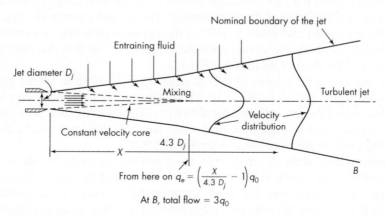

FIGURE 9.18
Flow of a submerged circular jet. (*After Rushton and Oldshue.*[46])

integrity well beyond the point at which the core has disappeared, but its velocity steadily decreases. The radial decrease in velocity in the jet is accompanied by a pressure increase in accordance with the Bernoulli principle. Fluid flows into the jet and is absorbed, accelerated, and blended into the augmented jet. This process is called *entrainment*. An equation applying over distances larger than $4.3D_j$ is

$$q_e = \left(\frac{X}{4.3D_j} - 1 \right) q_0 \tag{9.34}$$

where q_e = volume of liquid entrained per unit time at distance X from nozzle
q_0 = volume of liquid leaving jet nozzle per unit time

In addition to entrainment, strong shear stresses exist at the boundary between the jet and the surrounding liquid. These stresses tear off eddies at the boundary and generate considerable turbulence, which also contributes to the mixing action.

A large flow of liquid alone does not achieve satisfactory mixing. Enough time and space must be provided for the stream to blend thoroughly into the mass of fluid by the mechanism of entrainment.

Correlations are available for the mixing time with a side-entering jet in relatively small process vessels.[17]

Static mixers

Gases or low-viscosity liquids can often be satisfactorily blended by passing them together through a length of open pipe or a pipe containing orifice plates or segmented baffles. Under appropriate conditions the pipe length may be as short as 5 to 10 pipe diameters, but 50 to 100 pipe diameters is recommended.[24]

More difficult mixing tasks are accomplished by static mixers, commercial devices which consist of a series of metal inserts placed in the pipe. One of the main types is the helical-element mixer (Fig. 9.19a) which is mainly used with viscous liquids and pastes. Each element, 1 to 1.5 pipe diameters in length, divides the stream in two, gives it a 180° twist, and delivers it to the next element, which is set at 90° to the trailing edge of the first element. The second element divides the already divided stream and twists it 180° in the opposite direction. Successive elements further subdivide the stream until the striations are so thin that the blending process can be finished by molecular diffusion.

The recommended number of helical elements is 6 for Re = 100 to 1,000, 12 for Re = 10 to 100, and 18 for Re < 10. More elements are needed for very viscous liquids because of the lower molecular diffusivity. The pressure drop per unit length is about 6 times that in the empty pipe when Re < 10, but increases to about 50 to 100 times that in the empty pipe when Re = 2,000.[34]

Another type of static mixer, used for gases and low-viscosity liquids, is the turbulent vortex mixer shown in Fig. 9.19b. Each element of this mixer has four tabs protruding at an angle into the pipe. Counterrotating longitudinal vortices are formed in the wakes of the tabs, providing efficient blending of the liquids or gases after only two to four elements. The pressure drop per unit length is 25 to 40 times the value for the empty pipe at Re = 10^4 to 10^5, but since only a few elements are needed, the overall pressure drop is not large.

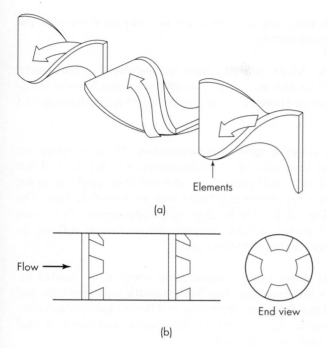

FIGURE 9.19
Static mixers: (*a*) elements of a helical-element mixer; (*b*) turbulent vortex mixer.

Elements

(a)

Flow →

End view

(b)

SUSPENSION OF SOLID PARTICLES

Particles of solids are suspended in liquids for many purposes, perhaps to produce a homogeneous mixture for feeding to a processing unit, to dissolve the solids, to catalyze a chemical reaction, or to promote growth of a crystalline product from a supersaturated solution. Suspension of solids in an agitated vessel is somewhat like fluidization of solids with liquids, as discussed in Chap. 7, in that the particles are separated and kept in motion by the fluid flowing past them. However, the fluid flow pattern created by the agitator has regions of horizontal flow as well as upward and downward flow, and to keep the solids in suspension in a tank generally requires much higher average fluid velocities than would be needed to fluidize the solids in a vertical column.

Degrees of suspension

When solids are suspended in an agitated tank, there are several ways to define the condition of suspension. Different processes require different degrees of suspension, and it is important to use the appropriate definition and correlation in a design or scaleup problem. The degrees of suspension are given below in the order of increasing uniformity of suspension and increasing power input.

Nearly complete suspension with filleting. Most of the solid is suspended in the liquid, with a few percent in stationary fillets of solid at the outside periphery of the bottom or at other places in the tank. Having a small amount of solids not in motion may be permissible in a feed tank to a processing unit, as long as the fillets do not grow and the solids do not cake.[36] For crystallization or a chemical reaction, the presence of fillets would be undesirable.

Complete particle motion. All the particles either are suspended or are moving along the tank bottom. Particles moving on the bottom have a much lower mass-transfer coefficient than suspended particles, which might affect the performance of the unit.[20]

Complete suspension or complete off-bottom suspension. All the particles are suspended off the tank bottom or do not stay on the bottom more than 1 or 2 s. When this condition is just reached, there will generally be concentration gradients in the suspension and there may be a region of clear liquid near the top of the tank. The gradient in solid concentration will have little effect on the performance of the unit as a dissolver or a chemical reactor, and the mass-transfer coefficient will not increase very much with further increases in stirrer speed.

Uniform suspension. At stirrer speeds considerably above those needed for complete suspension, there is no longer any clear liquid near the top of the tank, and the suspension appears uniform. However, there may still be vertical concentration gradients, particularly if the solids have a wide size distribution, and care is needed in getting a representative sample from the tank.

Correlations for suspension

Complete suspension of solids would be satisfactory for most purposes, and the correlations developed to predict conditions for suspension have generally used this criterion. Some of these correlations are discussed here along with guidelines for scaleup. Keep in mind that these correlations give the minimum agitation condition for suspension, and that requirements for dispersion of a gas or good heat transfer to a coil or jacket may indicate higher power inputs for some cases.

The ease with which solids are suspended in a liquid depends on the physical properties of the particles and the liquid and on the circulation patterns in the tank. The terminal velocity of free-settling particles can be calculated by using the drag coefficient curve (Fig. 7.3), as shown previously. It is more difficult to suspend particles that have a high settling velocity, but there is no simple correlation between this velocity and a characteristic velocity in the tank, such as the tip speed of the impeller. The limiting factor is the liquid velocity near the bottom, where the flow is almost horizontal; and to get complete suspension, the velocity in this region is generally many times the settling velocity. Since it is difficult to measure or predict the velocity distribution, investigators have generally measured the critical stirrer speed for complete suspension and developed empirical equations for different agitators.

Zwietering's[51] correlation is based on data for five types of impellers in six tanks from 6 in. to 2 ft in diameter. The critical stirrer speed is given by the dimensionless equation

$$n_c D_a^{0.85} = S v^{0.1} D_p^{0.2} \left(g \frac{\Delta \rho}{\rho} \right)^{0.45} B^{0.13} \tag{9.35}$$

where n_c = critical stirrer speed
D_a = agitator diameter
S = shape factor
v = kinematic viscosity
D_p = average particle size
g = gravitational acceleration
$\Delta\rho$ = density difference
ρ = liquid density
B = 100 × weight of solid/weight of liquid

Typical values of S are given in Table 9.3. Note that for the same impeller size ratio and clearance ratio, the critical speed is about the same for the standard turbine and the flat paddle, and S for the propeller is slightly lower. However, because of the large differences in the power number N_P, the turbine takes about twice as much power as the paddle and 15 to 20 times as much power as the propeller to achieve complete suspension of solids. With the turbine in normal position, slightly less than one-half of the flow produced by the impeller goes down at the wall (see Fig. 9.12), and the average velocity along the bottom of the tank is relatively low. The total flow produced by a propeller of the same size is not quite as great, but it is all directed downward, giving higher velocities and higher shear rates near the bottom. If the only purpose of the stirrer is to suspend solids, a propeller or other axial-flow impeller such as pitched-blade turbine would be chosen rather than a radial-flow impeller. However, a standard turbine may be needed for good gas dispersion or to get higher shear, and then correlations such as Eq. (9.35) could be used in checking conditions needed for solid suspension.

The positive exponent for viscosity in Eq. (9.35) is questionable, since the terminal velocity is lower in more viscous fluids, and viscosity was not varied independently in Zwietering's tests. Other studies[7,50] have shown a slight decrease in the critical speed as the viscosity was increased. A theoretical equation for stirrer speed by Wichterle[50] predicts that the exponents for viscosity, particle size, and density difference all change with particle size and may even change sign. His model is based on comparing the particle terminal velocity and the product of particle size

TABLE 9.3
Shape factor S in Eq. (9.35) for critical stirrer speed

Impeller type	D_t/D_a	D_t/E	S
Six-blade turbine	2	4	4.1
$D_a/W = 5$	3	4	7.5
$N_P = 6.2$	4	4	11.5
Two-blade paddle	2	4	4.8
$D_a/W = 4$	3	4	8
$N_P = 2.5$	4	4	12.5
Three-blade propeller	3	4	6.5
$N_P = 0.5$	4	4	8.5
	4	2.5	9.5

and shear rate near the bottom of the tank. The model does not yet account for the effect of solids concentration, and it requires independent measurement of shear rates, but it should be useful in interpreting existing data and in scaleup.

Power consumption

In discussing scaleup procedures, all workers report that lower stirrer speeds are needed to suspend solids in larger tanks if geometric similarity is maintained. However, small differences in the exponents for D_a in the equation for n_c can lead to large differences in power requirements, since the power per unit volume P/V varies with $n^3 D_a^2$. Some data for suspending sand in water using 45° pitched-blade turbines (downward-pumping) are given in Fig. 9.20. Some of the points shown are direct observations, and others were calculated from the authors' correlations for $D_p = 200\ \mu\text{m}$, $D_a/D_t = \frac{1}{3}$, and $E/D_t = \frac{1}{4}$. The solid lines show the range of tank sizes used, and the dashed lines are extrapolations. The large differences in P/V may be due partly to different criteria for complete suspension or slight differences in geometry.

In the study of Rao et al.,[42] n_c varied with $D_a^{-0.85}$, in agreement with Zwietering, and this indicates P/V varies with $D_a^{-0.55}$ [the exponent equals $3 \times (-0.85) + 2$]. A six-blade turbine was used with $W/D_a = 0.30$, and suspension was judged complete if particles did not stay stationary on the flat bottom for more than 2 s. Chudacek[10] used narrower turbines ($W/D_a = 0.20$) but the same suspension criterion, and his

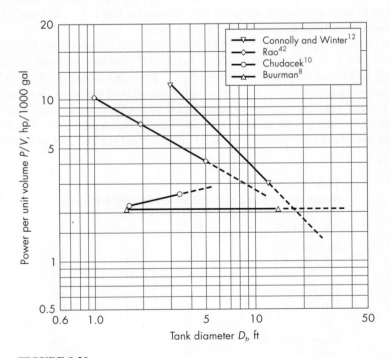

FIGURE 9.20
Power required for complete suspension of solids in agitated tanks using pitched-blade turbines.

correlation indicates that P/V increases somewhat with D_a. However, he also reported conditions for other criteria, and for 98 percent complete suspension, P/V was 30 to 50 percent less than the values in Fig. 9.20, and P/V decreased slightly as D_a increased. The tests of Buurman et al.[8] were made in dished-bottom tanks using four-blade stirrers with $W/D_a = 0.25$. An ultrasound Doppler meter was used to measure velocity near the bottom, and a transition in the velocity marked complete suspension. These tests showed that n_c varied with $D_a^{-2/3}$ and that P/V was independent of D_a. The more negative exponents for D_a in other studies were attributed to incorrect scaling of the stirrer blade thickness.

The strongest dependence of P/V on D_a and the highest power consumptions were reported by Connolly and Winter,[12] who recommended applying constant torque per unit volume, which is equivalent to making nD_a constant and P/V proportional to D_a^{-1}. However, they did not measure the minimum stirrer speed but only showed that their criterion led to nearly uniform suspension in each tank. The power required for uniform suspension is greater than for complete off-bottom suspension, and the ratio is greater for small tank sizes. Buurman found nearly homogeneous suspension at $1.2n_c$ in a 14-ft tank, but $1.9n_c$ was needed for comparable uniformity in the 1.5-ft tank.

If satisfactory suspension is obtained in a small tank, whether judged by visual observations, particle velocities, or mass-transfer rates, the safe scaleup rule is to keep geometric similarity and constant power per unit volume. The ratios $D_a/D_t = \frac{1}{3}$ and $E/D_t = \frac{1}{4}$ are often recommended, though some prefer $D_a/D_t = 0.4$ for solids suspension. The critical speed can be reduced by decreasing the clearance, but it may be hard to start the stirrer if it is in a layer of solids very near the bottom.

EXAMPLE 9.4. An agitated vessel 6 ft (1.8 m) in diameter with a working depth of 8 ft (2.44 m) is used to prepare a slurry of 150-mesh fluorspar in water at 70°F. The solid has a specific gravity of 3.18, and the slurry is 25 percent solids by weight. The impeller is a four-blade pitched-blade turbine 2 ft (0.61 m) in diameter set 1.5 ft above the vessel floor. (*a*) What is the power required for complete suspension? (*b*) What is the critical stirrer speed?

Solution

(*a*) Use the data of Buurman et al. in Fig. 9.20, correcting for differences in physical properties and solids concentration. Assume that the exponents of Eq. (9.35) apply and that n_c varies with $D_p^{0.2} \Delta\rho^{0.45} B^{0.13}$:

	Sand	Fluorspar
$D_p, \mu m$	200	104
$\Delta\rho, g/cm^3$	1.59	2.18
B	11.1	33.3

Change in n_c:

$$\left(\frac{104}{200}\right)^{0.2} \left(\frac{2.18}{1.59}\right)^{0.45} \left(\frac{33.3}{11.1}\right)^{0.13} = 1.157$$

Change in P:

$$(1.157)^3 = 1.55$$

From Fig. 9.20, $P/V = 2.1$ hp/1,000 gal for sand. For fluorspar, $P/V = 2.1 \times 1.55 = 3.3$ hp/1,000 gal:

$$V = \tfrac{1}{4}\pi \times 6^2 \times 8 = 226.2 \text{ ft}^3$$

$$226.2 \times 7.48 = 1,692 \text{ gal}$$

$$P = 3.3 \times \frac{1,692}{1,000} = 5.58 \text{ hp } (4.16 \text{ kW})$$

(b) $D_a = 2$ ft. From Table 9.2, for a four-blade turbine, $K_T = N_P = 1.27$

$$\rho_m = \text{slurry density} = \frac{1}{0.25/3.18 + 0.75} = 1.207 \text{ g/cm}^3 = 75.2 \text{ lb/ft}^3$$

From Eq. (9.12b)

$$n_c^3 = \frac{Pg_c}{N_P \rho D_a^5} = \frac{5.58 \times 550 \times 32.17}{1.27 \times 75.2 \times 2^5} = 32.3$$

$$n_c = 3.18 \text{ r/s}$$

DISPERSION OPERATIONS

When solids are merely suspended in a liquid, the size and surface area of the solid particles exposed to the liquid are fixed, as is the total volume of suspended solids. At high shear rates, however, agglomerates may be broken up, and with fragile or sensitive materials the particles themselves may be degraded, their diameter reduced, and new surface area created. This is especially important in fermentations and similar operations, in which biological cells may be destroyed if the local shear rates in the vessel are too high.[40]

In liquid-liquid and gas-liquid dispersion operations, also the size of the drops or bubbles and the total interfacial area between the dispersed and continuous phases vary with conditions and the degree of agitation. New area must constantly be created against the force of interfacial tension. Drops and bubbles are continually coalescing and being redispersed. In most gas-liquid operations, bubbles rise through the liquid pool and escape from the surface and must be replaced by new ones.

In this dynamic situation, the volume of the dispersed phase held up in the liquid pool depends on the rate of rise of the bubbles and the volumetric feed rate. In liquid-liquid dispersion, the holdup may depend on the rise or fall velocity of the dispersed drops, or it may it be fixed by the ratio of the two phases in the feed. For both gas-liquid and liquid-liquid dispersions, statistical averages are used to characterize the system, since a distribution of drops or bubble sizes is expected, and the holdup and interfacial area may vary with position in the vessel.

Characteristics of dispersed phase; mean diameter

Despite these variations, a basic relationship exists between the holdup Ψ (the volume fraction of dispersed phase in the system), the interfacial area a per unit volume, and the bubble or drop diameter D_p. If the total volume of the dispersion is taken as unity, the volume of dispersed phase, by definition, is Ψ. Let the number of drops or bubbles in this volume be N. Then if all the drops or bubbles were spheres of diameter D_p, their total volume would be given by

$$\frac{\pi N D_p^3}{6} = \Psi \tag{9.36}$$

The total surface area of the drops or bubbles in this volume would be

$$\pi N D_p^2 = a \tag{9.37}$$

Dividing Eq. (9.37) by Eq. (9.36) and rearranging give

$$a = \frac{6\Psi}{D_p} \tag{9.38}$$

The interfacial area per unit volume of dispersion is proportional to the holdup and inversely proportional to the drop size. Usually, a large area is desired to give high rates of mass transfer or reaction. To account for a distribution of drop sizes, an equivalent average diameter $\bar{D}_s$ is used based on the total volume and total area of the dispersed phase. Diameter $\bar{D}_s$ is the volume-surface mean diameter (the Sauter mean diameter), and it can be calculated from Eq. (7.23) if the drop size distribution is known. If the interfacial area and holdup can be measured independently, the average drop size can be determined by using Eq. (9.39).

$$\bar{D}_s = \frac{6\Psi}{a} \tag{9.39}$$

Liquid-liquid dispersions

Various types of equipment can be used to disperse one liquid, say, benzene, in an immiscible liquid such as water. A stirred tank or an in-line mixer can produce drops in the size range of 0.05 to 1.0 mm. Such liquid-liquid dispersions have a high interfacial area, but they are not stable, since the drops will settle (or rise) and coalesce in the absence of agitation. Stable emulsions of very small droplets can be formed in colloid mills or other high-shear devices, but emulsification is not covered in this text.

In a stirred tank, the average drop size depends on a balance between breakup of large drops in regions of high shear and drop coalescence in regions of lower shear. Shear stress at the drop surface tends to deform the drop, and deformation is resisted by the interfacial tension and the viscosity of the dispersed phase. An important dimensionless group is the Weber number, denoted by We, which for a stirred tank is the ratio of the fluid kinetic energy at the impeller tip speed to a surface-tension stress based on D_a

$$\text{We} = \frac{\rho_c(nD_a)^2}{\sigma/D_a} = \frac{\rho_c n^2 D_a^3}{\sigma} \tag{9.40}$$

where ρ_c = density of continuous phase
σ = interfacial tension

Other definitions of the Weber number have been used in other situations.

One of several proposed correlations for the dispersion of liquids with standard six-blade turbines is[31]

$$\frac{\bar{D}_s}{D_a} = 0.058(1 + 5.4\Psi)We^{-0.6} \tag{9.41}$$

The term $1 + 5.4\Psi$ reflects the increase in frequency of coalescence with increasing droplet concentration. Equation (9.41) is based on data for the dispersion of low-viscosity liquids in a small tank. When the dispersed phase is quite viscous, larger drops are produced, because the viscous drop resists deformation. A correction factor $(\mu_d/\mu_c)^{0.1}$ could be added to Eq. (9.41) for a rough estimate of this viscosity effect. There are no data for large tanks, but somewhat greater average drop sizes might be expected, since the increased circulation times would give greater chance for coalescence.

Static mixers can also be used to make liquid-liquid dispersions, and about 10 to 20 elements in series produce an equilibrium dispersion. By considering the surface energy of a drop and the disruptive energy due to turbulent velocity fluctuations, $\bar{D}_s$ is predicted to vary with a fractional power of the Weber number and the friction factor[29]

$$\frac{\bar{D}_s}{D} = CWe^{-0.6}f^{-0.4} \tag{9.42}$$

where $We = \dfrac{\rho_c \bar{V}^2 D}{\sigma}$

D = pipe diameter
$\bar{V}$ = average velocity
f = friction factor $= \dfrac{D\Delta P}{2\rho_c \bar{V}^2 L}$

Data for low-viscosity drops dispersed in water in small Kenics mixers ($D = 0.5$ to 1.0 in.) can be fitted to Eq. (9.42) with $C = 0.35$.[4,29] The friction factor was 0.42 for Re $= 10,000$ to $20,000$. Much larger drops were formed when the drop viscosity was 20 to 200 cP, but the data are scattered and the effects of Re and Ψ are also uncertain. Tests in empty pipes gave $\bar{D}_s$ values several times those for the static mixers, which is consistent with the much lower values of f. The drop-size distributions from the mixers are nearly normal, and the maximum size is about $1.5\bar{D}_s$.

In some cases, nearly uniform drops of moderate size are needed, such as when drops must rise against the flow of water in a counterflow spray extraction column. Drops can be formed one at a time from orifices or formed by the breakup of a liquid jet. For the slow formation of a drop from a submerged circular orifice, the drop diameter can be predicted by equating the net buoyant force on the drop to the opposing drag at the edge of the orifice, which is proportional to the interfacial tension.

$$F_b - F_g = g\frac{\pi D_o^3 \rho}{6}(\rho_c - \rho_d) = \pi D_o \sigma \tag{9.43}$$

where F_b = total buoyant force
$\quad F_g$ = force of gravity
$\quad D_o$ = orifice diameter
$\quad \rho_d$ = density of dispersed phase

Rearranging Eq. (9.43) gives

$$\frac{D_p}{D_o} = \left(\frac{6\sigma}{g D_o^2 (\rho_c - \rho_d)} \right)^{1/3} \tag{9.44}$$

At very low flow rates, the drops formed are slightly smaller than predicted by Eq. (9.44) because a small amount of liquid is left behind when the drop detaches.[11] At high flow rates a jet forms, and the breakup to drops of about the jet diameter occurs by Rayleigh instability. Drop sizes under jetting conditions depend on many factors and can be predicted only approximately.[26,47]

EXAMPLE 9.5. Cyclohexane is dispersed in water at 25°C in a baffled vessel 30 cm in diameter with a normal depth of 35 cm. The agitator is a standard six-blade turbine 10 cm in diameter. (a) If the stirrer speed is 6 r/s and the suspension is 8% cyclohexane by volume, calculate the power consumption and power per unit volume, and estimate the mean droplet size. (b) If the liquid mixture is pumped at 1.2 m/s through a Kenics helical-element mixer that is 2 cm in diameter and has 20 elements, each 3 cm long, estimate the mean droplet size and the power consumption. The friction factor may be taken as 0.42.

Solution
$\quad$ (a) For cyclohexane,

$$\rho = 760 \text{ kg/m}^3$$

$$\sigma = 46 \text{ dyne/cm} = 46 \times 10^{-3} \text{ N/m}$$

For the mixture,

$$\rho = 0.08(760) + 0.92(1,000) = 981 \text{ kg/m}^3$$

$$D_a = 0.1 \text{ m}$$

$$n = 6 \text{ s}^{-1}$$

Assume the viscosity is that of water at 25°C = 0.09 cP = 9×10^4 Pa·s.

$$\text{Re} = \frac{D_a^2 n \rho}{\mu} = \frac{0.1 \times 6 \times 981}{9 \times 10^{-4}} = 6.54 \times 10^5$$

From Table 9.2 $N_P = 5.75$, and from Eq. (9.22),

$$P = 5.75 \times 6^3 \times 0.1^5 \times 981 = 12.2 \text{ W}$$

The tank volume $V = \pi \times 0.3^2 \times 0.35/4 = 2.47 \times 10^{-2}$ m³, and

$$\frac{P}{V} = \frac{12.2 \times 10^{-3}}{2.47 \times 10^{-2}} = 0.494 \text{ kW/m}^3 \text{ or } 2.50 \text{ hp/1,000 gal}$$

This is moderately vigorous agitation. To find the drop size,

$$\text{We} = \frac{\rho_c n^2 D_a^3}{\sigma} = \frac{1,000 \times 6^2 \times 0.1^3}{46 \times 10^{-3}} = 783$$

From Eq. (9.41), for $\Psi = 0.08$,

$$\frac{\bar{D}_s}{D_a} = \frac{0.058[1 + (5.4 \times 0.08)]}{783^{0.6}} = 1.52 \times 10^{-3}$$

Hence $\bar{D}_s = 1.5 \times 10^{-4}$ m, or 0.15 mm

(b) For the Kenics mixer, $\bar{V} = 1.2$ m/s and $f = 0.42$. Thus

$$We = \frac{1,000 \times 1.2^2 \times 0.02}{46 \times 10^{-3}} = 626$$

From Eq. (9.42),

$$\frac{\bar{D}_s}{D} = \frac{0.35}{626^{0.6} \times 0.42^{0.4}} = 1.04 \times 10^{-2}$$

$$\bar{D}_s = 1.04 \times 10^{-2} \times 0.02 = 2.1 \times 10^{-4} \text{ m, or } 0.21 \text{ mm}$$

The actual volume of $\bar{D}_s$ may be higher because the effect of Ψ is not accounted for. For the Kenics mixer, the length $L = 20 \times 0.03 = 0.6$ m. The pressure drop, from Eq. (5.9), is

$$\Delta p = \frac{2 \times 0.42 \times 1.2^2 \times 981 \times 0.6}{0.02} = 3.56 \times 10^4 \text{ Pa or } 5.16 \text{ lb}_f/\text{in.}^2$$

The energy loss (power consumed) equals the volumetric flow rate q times the pressure drop.

$$P = q \, \Delta p = \frac{\pi \times 1.2 \times 0.2^2}{4} \times 3.56 \times 10^4 = 13.4 \text{ W, or } 0.0134 \text{ kW}$$

The volume of the mixer is

$$V = \frac{\pi \times 0.2^2 \times 0.6}{4} = 1.884 \times 10^{-4} \text{ m}^3$$

The power consumed per unit volume is then

$$\frac{P}{V} = \frac{0.0134}{1.884 \times 10^{-4}} = 71.1 \text{ kW/m}^3 \text{ or } 360 \text{ hp/1,000 gal}$$

Although the total power demand is not high, the power consumed per unit volume in the Kenics mixer is much larger than the average value for the stirred tank. The local energy dissipation rate near the impeller in the tank, however, is also much greater than the average for the tank.

Gas-liquid dispersions

Gas can be dispersed in a liquid by passing it through a sparger with multiple orifices, a porous ceramic or metal plate, or an open pipe that discharges gas directly under a turbine impeller. At very low gas rates, bubbles can be formed one at a time at an orifice in nearly quiescent liquid, and the bubble size can be predicted from Eq. (9.44). However, for most applications, moderate to high gas rates are used to increase the gas holdup and interfacial area. The bubbles then interact, and the average size and area depend on the rates of coalescence and redispersion, which are influenced by the level of turbulence as well as the physical properties of the system.

When a gas is dispersed into a pure liquid in a stirred tank, small bubbles are formed in the high-shear region near the tip of the impeller, but they coalesce rapidly in low-shear regions away from the impeller. The average bubble size for air in water is usually in the range of 2 to 5 mm, much greater than that given by Eq. (9.41) for liquid-liquid dispersions. The interfacial area a (which is more important than the average size) increases with a fractional power of the energy dissipation rate and the superficial velocity, and it is greater for systems with low interfacial tension. Many equations of the following type have been proposed, but there is no consensus on the exponents and constants, and Eq. (9.45) is intended to show just the approximate effect of those variables:

$$a = \frac{C(P/V)^{0.4}\bar{V}_s^{0.5}}{\sigma^{0.6}} \tag{9.45}$$

In electrolyte solutions, in multicomponent nonionic solutions, and in solutions with even a small amount of surface-active impurity, coalescence of bubbles may be greatly retarded. The average bubble size in a stirred tank is then smaller than that with pure liquids, and the difference in surface area may be severalfold. For particular systems, such as air with solutions of sodium sulfite, correlations similar to Eq. (9.45) have been presented, but there is no satisfactory general correlation. Experimental data should be obtained for each new system, and equations such as Eq. (9.45) used to guide scaleup. Some generalizations that can be made are that the effects of P/V and $\bar{V}_s$ with coalescing systems are greater than with pure liquids, and that the area at a given power dissipation rate is about the same whether a single gas inlet pipe, a sparger, or a sintered disk is used to introduce the gas.

Effect of gas flow on power consumption. The power consumed by a turbine impeller dispersing a gas is less than that indicated by Fig. 9.13 for agitating liquids only. The ratio of the power required when gas is present to that for the liquid alone depends mainly on the superficial gas velocity and to some extent on the impeller speed, tank size, impeller design and diameter, and the properties of the liquid.[21] Some data for dispersing air in water with standard six-blade turbines $D_a/D_t = \frac{1}{3}$ are shown in Fig. 9.21. The relative power P_g/P_0 drops rapidly to about 0.5 or 0.6 at a velocity of 10 mm/s and then decreases slowly to less than 0.3 at a velocity of 90 mm/s. The work of Pharamond et al.[41] showed that relative power is almost independent of stirrer speed in the low-velocity region where $P_g/P_0 > 0.5$. The relative power is about 10 to 15 percent higher for $D_t = 1.0$ m than for D_t of 0.48 or 0.29 m, but there are no data to show whether this trend continues for still larger tanks.

The study of Dickey[14] covers the higher range of gas velocities, and in this region the relative power generally decreases with increasing stirrer speed as well as with increasing gas velocity. This work and others[6,28,41] show that P_g varies with about the 2.1 to 2.9 power of the stirrer speed compared to the 3.0 power for liquids. The exponent for stirrer speed depends on gas velocity and other variables, and no simple correlation is available. In the region of high gas velocities, P_g/P_0 also depends on the ratio of impeller diameter to tank size. For $D_a/D_t = 0.4$ (data not shown here), the values of P_g/P_0 are lower by about 0.03 to 0.10 than those for

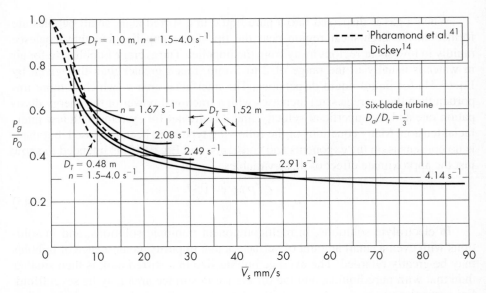

FIGURE 9.21
Power consumption in aerated turbine-agitated vessels.

$D_a/D_t = 0.33$. The main effect of using a larger impeller is that greater volumes of gas can be dispersed at a given stirrer speed.

Results for relative power consumption have often been presented as a function of a dimensionless aeration number $N_{Ae} = q_g/nD_a^3$, where q_g is the total gas flow and nD_a^3 is a measure of the flow rate of liquid from the impeller. The data in Fig. 9.21 are replotted in this form in Fig. 9.22. The curves in Fig. 9.22 are more widely spread because increasing N_{Ae} by increasing the gas flow q_g leads to lower values of P_g/P_0, but increasing N_{Ae} by decreasing the impeller speed n has the opposite effect. This type of plot permits the inclusion of published data for which values of n, D_a, and V_s are not available. The dashed line in Fig. 9.22 is for the CD-6 impeller which has six strongly curved blades with the concave side forward. The test conditions were not specified, but the CD-6 impeller clearly gives a smaller drop in power consumption on gassing than do standard turbines.

The data shown in Figs. 9.21 and 9.22 are for air and water but can be used for other systems if the change in physical properties is not very great. A fourfold increase in viscosity and a 40 percent reduction in surface tension had no significant effect on P_g/P_0, but the relative power was 10 to 20 percent lower with an Na_2SO_4 solution, in which the bubbles were smaller because of reduced coalescence.[21]

The decrease in power with gassing is not just an effect of the lower average density of the gas-liquid dispersion, since the gas holdup is generally 10 percent or less when P_g/P_0 is reduced to 0.5. The decrease in power is associated with the formation of gas pockets behind the turbine blades.[47] Bubbles are captured in the centrifugal field of vortices that form behind the horizontal edges of the blades, and coalescence leads to large cavities that interfere with normal liquid flow.

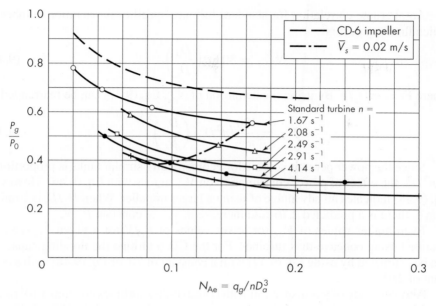

FIGURE 9.22
Relative power consumption in agitated vessels versus aeration number N_{Ae}.

The change in power dissipation with gassing must be allowed for in the design of large units. An agitator drive chosen to handle the torque for a gassed system could be overloaded if the system has to operate occasionally with no gas flow, and a dual-speed drive might be needed. Also, good performance sometimes requires constant power dissipation per unit volume, and scaleup may lead to different values of $\bar{V}_s$ and P_g/P_0.

Gas-handling capacity and flooding of turbine impellers. If the gas throughput to a turbine-agitated vessel is progressively increased, the impeller eventually floods and can no longer disperse the gas effectively. The flooding point is not as distinct a transition as in a packed column, and various criteria for flooding have been proposed. One definition of flooding based on visual inspection is that it occurs when most of the bubbles rise vertically between the turbine blades rather than being dispersed radially from the tips of the blades.[14] The critical gas velocity for this transition $\bar{V}_{s,c}$ was found to be proportional to the power per unit volume dissipated by the stirrer, with a slight effect of tank size. From data for tanks 1.54 and 0.29 m in diameter and velocities up to 75 mm/s, the following dimensional equation was obtained:

$$\bar{V}_{s,c} = 0.114 \left(\frac{P_g}{V} \right) \left(\frac{D_t}{1.5} \right)^{0.17} \tag{9.46}$$

In Eq. (9.46), P_g/V is in kW/m³, D_t in m, and $\bar{V}_{s,c}$ in m/s. If $\bar{V}_{s,c}$ is fixed by design or scaleup requirements, this equation can be used to predict the initial power consumption, since a transition to flooding will occur as the stirrer speed is reduced below a critical value.

An alternate correlation for standard turbines specifies the aeration number at which flooding occurs[2]

$$N_{Ae,f} = 30\,\text{Fr} \left(\frac{D_a}{D_t}\right)^{3.5} \tag{9.47}$$

where $\text{Fr} = n^2 D_a/g$. By using the definition of N_{Ae}, Eq. (9.47) can be rearranged to

$$\bar{V}_{s,c} = \frac{120}{\pi g}(n^3 D_a^2)\left(\frac{D_a}{D_t}\right)^{5.5} \tag{9.48}$$

Since the power per unit volume for geometrically similar tanks is proportional to $n^3 D_a^2$, the effect of power consumption on $\bar{V}_{s,c}$ agrees with Eq. (9.46). However, Eq. (9.48) indicates an advantage of using a larger impeller. When D_a/D_t is $\frac{1}{2}$ rather than $\frac{1}{3}$, there is a predicted 2.76-fold increase in $\bar{V}_{s,c}$ at constant P_g/V.

Even higher values of $\bar{V}_{s,c}$ (or lower values of P_g/V for a given $\bar{V}_{s,c}$) can be obtained using concave-disk turbines. For the CD-6 turbine the flooding capacity[2] can be estimated by using Eq. (9.47) with a constant of 70, or Eq. (9.48) with a constant of 280.

Another study of gas dispersion with concave-disk turbines covered a 10^4 range of tank volumes and showed that the power required to avoid flooding decreases with the $-\frac{1}{12}$ power of the tank volume or the -0.25 power of the tank diameter.[49] This is in reasonable agreement with the -0.17 power indicated by Eq. (9.46) and indicates that Eq. (9.48), which indicates no effect of tank size, may be quite conservative for a large scaleup.

AGITATOR SELECTION AND SCALEUP

Mixer selection

There is not necessarily any direct relation between power consumed and the amount or degree of mixing. When a low-viscosity liquid is swirled about in an unbaffled vessel, the particles may follow circular paths indefinitely and mix little or not at all. Almost none of the energy supplied is used for mixing. If baffles are added, mixing becomes rapid; a larger fraction of the energy is used for mixing and relatively less for circulation.

When the mixing time is critical, the best mixer is the one that mixes in the required time with the smallest amount of power. In many cases a short mixing time is desirable, but not essential, and the mixing time is a compromise arrived at by considering the energy cost for mixing and the capital cost of the mixer. For mixing reagents in a feed tank or blending product from different batches in a storage tank, a relatively small mixer might be used, even if several minutes are required for complete mixing.

Scaleup

A major problem in agitator design is to scale up from a laboratory or pilot-plant agitator to a full-scale unit. The scaleup of vessels for suspending solids has already been discussed. For some other problems generalized correlations such as those shown in Figs. 9.13 to 9.17 are available for scaleup. For many other problems adequate correlations are not available; for these situations various methods of scaleup have been proposed, all based on geometric similarity between the laboratory and plant equipment. It is not always possible, however, to have the large and small vessels geometrically similar. Furthermore, even if geometric similarity is obtainable, dynamic similarity and kinematic similarity are not, so that the results of the scaleup are not always fully predictable. As in most engineering problems, the designer must rely on judgment and experience.

Power consumption in large vessels can be accurately predicted from curves of N_P versus Re, as shown in Figs. 9.13 and 9.14. Such curves may be available in the published literature, or they may be developed from pilot-plant studies using small vessels of the proposed design. With low-viscosity liquids the amount of power consumed by the impeller per unit volume of liquid has been used as a measure of mixing effectiveness, based on the reasoning that increased amounts of power mean higher degree of turbulence and a higher degree of turbulence means better mixing. Studies have shown this to be at least roughly true. In a given mixer, the amount of power consumed can be directly related to the rate of solution of a gas or the rate of certain reactions, such as oxidations, that depend on the intimacy of contact of one phase with another. In a rough qualitative way, it may be said that $\frac{1}{2}$ to 1 hp per 1,000 gal of thin liquid gives "mild" agitation, 2 to 3 hp per 1,000 gal gives "vigorous" agitation, and 4 to 10 hp per 1,000 gal gives "intense" agitation. These figures refer to the power that is actually delivered to the liquid and do not include power used in driving gear reduction units or in turning the agitator shaft in bearings and stuffing boxes. The agitator designed in Example 9.3 would require about $1\frac{1}{2}$ hp per 1,000 gal of liquid and should provide rather mild agitation. (Note that 5 hp per 1,000 gal is equivalent to 1.0 kW/m^3.)

The optimum ratio of impeller diameter to vessel diameter for a given power input is an important factor in scaleup. The nature of the agitation problem strongly influences this ratio: For some purposes the impeller should be small, relative to the size of the vessel; for others it should be large. For dispersing a gas in a liquid, for example, the optimum ratio[44] is about 0.25; for bringing two immiscible liquids into contact, as in liquid-liquid extraction vessels, the optimum ratio[39] is 0.40. For some blending operations the ratio should be 0.6 or even more. In any given operation, since the power input is kept constant, the smaller the impeller, the higher the impeller speed. In general, operations that depend on large velocity gradients rather than on high circulation rates are best accomplished by small, high-speed impellers, as is the case in the dispersion of gases. For operations that depend on high circulation rates rather than steep velocity gradients, a large, slow-moving impeller should be used.

When constant power per unit volume and geometric similarity are maintained in scaling up, the impeller speed changes with $D_a^{-2/3}$, as shown below. The terms inside the square brackets are all constant, so $n^3 D_a^2$ must be constant. The power per unit volume is

$$\frac{P}{V} = \frac{N_P n^3 D_a^5 \rho}{(\pi/4)D_t^2 H} = \left[\frac{4N_P \rho}{\pi} \left(\frac{D_a}{D_t} \right)^2 \left(\frac{D_a}{H} \right) \right] n^3 D_a^2 \tag{9.49}$$

Thus
$$\frac{n_2}{n_1} = \left(\frac{D_{a1}}{D_{a2}}\right)^{2/3}$$
(9.50)

The decrease in impeller speed on scaleup at constant P/V leads to longer mixing times in large tanks. Trying to maintain a constant mixing time is usually impractical, as shown in the following example.

EXAMPLE 9.6. A pilot-plant vessel 1 ft (305 mm) in diameter is agitated by a six-blade turbine impeller 4 in. (102 mm) in diameter. When the impeller Reynolds number is 10^4, the blending time of two miscible liquids is found to be 15 s. The power required is 2 hp per 1,000 gal (0.4 kW/m³) of liquid. (*a*) What power input would be required to give the same blending time in a vessel 6 ft (1,830 mm) in diameter? (*b*) What would be the blending time in the 6-ft (1,830-mm) vessel if the power input per unit volume were the same as in the pilot-plant vessel?

Solution

(*a*) Since the Reynolds number in the pilot-plant vessel is large, the Froude number term in Eq. (9.31) would not be expected to apply, and the correlation in Fig. 9.16 will be used in place of the more complicated relation in Fig. 9.17. From Fig. 9.16, for Reynolds numbers of 10^4 and above, the mixing time factor nt_T is constant; and since time t_T is assumed constant, speed n will be the same in both vessels.

In geometrically similar vessels the power input per unit volume is proportional to P/D_a^3. At high Reynolds numbers, from Eq. (9.22)

$$\frac{P}{D_a^3} = K_T n^3 D_a^2 \rho$$

For a liquid of given density this becomes

$$\frac{P}{D_a^3} = c_2 n^3 D_a^2$$

where c_2 is a constant. From this the ratio of power inputs per unit volume in the two vessels is

$$\frac{P_6/D_{a6}^3}{P_1/D_{a1}^3} = \left(\frac{n_6}{n_1}\right)^3 \left(\frac{D_{a6}}{D_{a1}}\right)^2$$
(9.51)

Since $n_1 = n_6$,

$$\frac{P_6/D_{a6}^3}{P_1/D_{a1}^3} = \left(\frac{D_{a6}}{D_{a1}}\right)^2 = 6^2 = 36$$

The power per unit volume required in the 6-ft vessel is then $2 \times 36 = 72$ hp per 1,000 gal (14.4 kW/m³). This is an impractically large amount of power to deliver to a low-viscosity liquid in an agitated vessel.

(*b*) If the power input per unit volume is to be the same in the two vessels, Eq. (9.50)

$$\frac{n_6}{n_1} = \left(\frac{D_{a1}}{D_{a6}}\right)^{2/3}$$

Since nt_T is constant, $n_6/n_1 = t_{T1}/t_{T6}$, and

$$\frac{t_{T6}}{t_{T1}} = \left(\frac{D_{a6}}{D_{a1}}\right)^{2/3} = 6^{2/3} = 3.30$$

The blending time in the 6-ft vessel would be $3.30 \times 15 = 49.5$ s.

Although it is impractical to achieve the same blending time in the full-scale unit as in the pilot-plant vessel, a moderate increase in blending time in the larger vessel reduces the power requirement to a reasonable level. Such tradeoffs are often necessary in scaling up agitation equipment.

SYMBOLS

A_p Area of cylinder swept out by tips of impeller blades, m^2 or ft^2
a Interfacial area per unit volume, m^{-1} or ft^{-1}
B Solids concentration in suspension [Eq. (9.35)]
C Constant in Eq. (9.42)
D Diameter of pipe, m or ft; D_a, diameter of impeller; D_j, diameter of jet and nozzle; D_o, orifice diameter; D_p, diameter of particles, drops, or bubbles; D_t, tank diameter
$\bar{D}_s$ Volume-surface mean diameter of drops or bubbles, m or ft
E Height of impeller above vessel floor, m or ft
E_k Kinetic energy of fluid, J/m^3 or $ft \cdot lb_f/ft^3$
Fr   Froude number, $n^2 D_a/g$
f Friction factor, dimensionless
f_t Blending time factor, dimensionless, defined by Eq. (9.31); f_t', by Eq. (9.33)
g Gravitational acceleration, m/s^2 or ft/s^2
g_c Newton's law proportionality factor, $32.174\ ft \cdot lb/lb_f \cdot s^2$
H Depth of liquid in vessel, m or ft
J Width of baffles, m or ft
K Constant in Eq. (9.19)
K_L, K_T Constants in Eqs. (9.19) and (9.21), respectively
K' Flow consistency index of non-newtonian fluid
k Ratio of tangential liquid velocity at blade tips to blade-tip velocity
L Length of impeller blades or of static mixer, m or ft
N Number of drops or bubbles per unit volume
N_{Ae} Aeration number, $q_g/n D_a^3$; $N_{Ae,f}$, at flooding
N_P Power number, $P/n^3 D_a^5 \rho$; $N_{P,g}$ at gas redispersion point
N_Q Flow number, $q/n D_a^3$; $N_{Q,g}$, at gas redispersion point
n Rotational speed, r/s; n_c, critical speed for complete solids suspension
n' Flow behavior index of non-newtonian fluid
P Power, kW or $ft \cdot lb_f/s$; P_g, with gas dispersion or at gas redispersion point; P_0, power consumption in ungassed liquid
p Pressure, N/m^2 or lb_f/ft^2
q Volumetric flow rate, m^3/s or ft^3/s; q_B, theoretical flow leaving the impeller blades; q_T, total liquid flow rate; q_e, entrained in jet; q_g, total gas flow rate; q_0, leaving jet nozzle
Re   Agitator Reynolds number, $n D_a^2 \rho/\mu$; Re_n, for pseudoplastic liquid, defined by Eq. (9.25)
r Radial distance from impeller axis, m or ft

S	Shape factor; $S_1 = D_a/D_t$; $S_2 = E/D_t$; $S_3 = L/D_a$; $S_4 = W/D_a$; $S_5 = J/D_t$; $S_6 = H/D_t$; also factor in Eq. (9.35)
t_T	Blending time, s
u	Velocity, m/s or ft/s; u_2, velocity of impeller blade tip
V	Volume, m^3 or ft^3
V'	Resultant velocity, absolute, in impeller, m/s or ft/s; V'_r, radial component; V'_{r2}, radial component of velocity V'_2; V'_{u2}, tangential component of velocity V'_2; V'_2, actual velocity at impeller blade tips
$\bar{V}$	Average velocity of liquid in pipe, m/s or ft/s; $\bar{V}_s$, superficial velocity of gas in agitated vessel; $\bar{V}_{s,c}$, critical velocity at flooding
W	Impeller width, m or ft
We	Weber number, $D\rho\bar{V}^2/\sigma$ or $D_a^3 n^2 \rho_c/\sigma$
X	Distance from jet nozzle, m or ft
y	Coordinate normal to flow direction

Greek Letters

α	Ratio V'_2/u_2
β_2	Angle between impeller blade tips and the tangent to the circle traced out by the impeller tip; β'_2, angle between the actual relative velocity vector of the liquid and the tangent
Δp	Pressure drop, N/m^2 or lb$_f$/ft^2
$\Delta\rho$	Density difference, kg/m^3 or lb/ft^3
μ	Absolute viscosity, P or lb/ft·s; μ_a, apparent viscosity of non-newtonian fluid; μ_c, viscosity of continuous phase in liquid-liquid dispersion; μ_d, of dispersed phase
ν	Kinematic viscosity, m^2/s or ft^2/s
ρ	Density, kg/m^3 or lb/ft^2; ρ_c, of continuous phase in liquid-liquid dispersion; ρ_d, of dispersed phase; ρ_m, of liquid-solid suspension
σ	Interfacial tension, dyn/cm or lb$_f$/ft
Ψ	Volumetric fractional gas or liquid holdup in dispersion, dimensionless
ψ	Function

PROBLEMS

9.1. A tank 1.2 m in diameter and 2 m high is filled to a depth of 1.2 m with a latex having a viscosity of 10 P and a density of 800 kg/m^3. The tank is not baffled. A three-blade 360-mm-diameter propeller is installed in the tank 360 mm from the bottom. The pitch is 1:1 (pitch equals diameter). The motor available develops 8 kW. Is the motor adequate to drive this agitator at a speed of 800 rpm?

9.2. What is the maximum speed at which the agitator of the tank described in Prob. 9.1 may be driven if the liquid is replaced by one having a viscosity of 1 P and the same density?

9.3. What power is required for the mixing operation of Prob. 9.1 if a propeller 360 mm in diameter turning at 15 r/s is used and if four baffles, each 120 mm wide, are installed?

9.4. The propeller in Prob. 9.1 is replaced with a six-blade turbine 400 mm in diameter, and the fluid to be agitated is a pseudoplastic power law liquid having an apparent viscosity of 15 P when the velocity gradient is $10 \ s^{-1}$. At what speed should the turbine rotate to deliver 1 kW/m^3 of liquid? For this fluid $n' = 0.75$ and $\rho = 950 \ \text{kg/m}^3$.

9.5. A mixing time of 29 s was measured for a 4.5-ft baffled tank with a 1.5-ft six-blade turbine and a liquid depth of 4.8 ft. The turbine speed was 75 rpm, and the fluid has a viscosity of 3 cP and a density of 65 lb/ft^3. Estimate the mixing times if an impeller one-quarter or one-half the tank diameter were used with the speeds chosen to give the same power per unit volume.

9.6. A pilot-plant reactor, a scale model of a production unit, is of such size that 1 g charged to the pilot-plant reactor is equivalent to 500 g of the same material charged to the production unit. The production unit is 2 m in diameter and 2 m deep and contains a six-blade turbine agitator 0.6 m in diameter. The optimum agitator speed in the pilot-plant reactor is found by experiment to be 330 rpm. (*a*) What are the significant dimensions of the pilot-plant reactor? (*b*) If the reaction mass has the properties of water at 70°C and the power input per unit volume is to be constant, at what speed should the impeller turn in the large reactor? (*c*) At what speed should it turn if the mixing time is to be kept constant? (*d*) At what speed should it turn if the Reynolds number is held constant? (*e*) Which basis would you recommend for scaleup? Why?

9.7. A stirred tank reactor 3 ft in diameter with a 12-in. flat-blade turbine has been used for a batch reaction in which the blending time of added reagents is considered critical. Satisfactory results were obtained with a stirrer speed of 400 rpm. The same reaction is to be carried out in a tank 7 ft in diameter, for which a 3-ft standard turbine is available. (*a*) What conditions would give the same blending time in the larger tank? (*b*) What would be the percentage change in the power per unit volume? Density $\rho = 60$ lb/ft^3; viscosity $\mu = 5$ cP.

9.8. A six-blade disk turbine ($D_a = 3$ ft) is used to disperse hydrogen gas into a slurry reactor containing methyl linoleate at 90°C and 60 lb$_f$/in.^{2} gauge with 1 percent suspended catalyst particles ($\bar{D}_s = 50 \ \mu$m, $\rho_p = 4$ g/cm^3). The reactor diameter is 9 ft and the depth is 12 ft. The gas flow rate is 1,800 std ft^3/min. The oil viscosity is 1.6 cP, and the density is 0.84 g/cm^3 at 90°C. The reactor is fully baffled. (*a*) What agitator speed is needed to give 5 hp/1,000 gal during the reaction? (*b*) What is the power consumption with gas flow on and with gas flow off?

9.9. For the conditions of Prob. 9.8, estimate the power required for complete suspension of the catalyst.

9.10. A 15 percent slurry of 20-to-28-mesh limestone in water is to be kept in suspension in a 20-ft-diameter tank using a six-blade 45° turbine. (*a*) If $D_a/D_t = \frac{1}{3}$ and $W/D_a = 0.2$, what stirrer speed is required? (*b*) Calculate the stirrer speed and power requirement if $D_a/D_t = 0.4$.

9.11. A reaction in which the product forms a crystalline solid has been studied in a 1-ft-diameter pilot-plant reactor equipped with a 4-in. six-blade turbine with curved blades. At stirrer speeds less than 600 rpm, a solid deposit sometimes forms on the bottom, and this condition must be avoided in the commercial reactor. Density of the liquid is 70 lb/ft^3; viscosity is 3 cP. (*a*) What is the power consumption in the small reactor, and what is recommended for an 8,000-gal reactor if geometric similarity is preserved? (*b*) How much might the required power be lowered by using a different type of agitator or different geometry?

9.12. Gaseous ethylene (C_2H_4) is to be dispersed in water in a turbine-agitated vessel at 110°C and an absolute pressure of 3 atm. The vessel is 3 m in diameter with a maximum liquid depth of 3 m. For a flow rate of 800 m^3/h of ethylene, measured at process conditions, specify (*a*) the diameter and speed of the turbine impeller and (*b*) the power drawn by the agitator. (*c*) What would be the gas flow rate at incipient flooding under these conditions?

9.13. (*a*) If the 6-ft tank described in Example 9.6 is equipped with a large impeller ($D_a/D_t = \frac{2}{3}$ instead of $\frac{1}{3}$), what impeller speed will be needed to give the same power input of 2 hp per 1,000 gal? (*b*) How does the predicted mixing time compare with that for the smaller impeller in the same tank?

9.14. In the scaleup of a stirred-tank process involving a shear-sensitive suspension, it is important to keep the maximum shear rate below a critical value. (*a*) If scaleup requires a 10-fold increase in the tank diameter and if the power per unit volume and geometric ratios are kept constant, by what factor will the shear rate change? (*b*) What could be done to keep the maximum shear rate unchanged?

REFERENCES

1. Bakker, A., and L. E. Gates. *Chem. Eng. Prog.*, **91**(12):25 (1995).
2. Bakker, A., J. M. Smith, and K. J. Myers. *Chem. Eng.*, **101**(12):98 (1994).
3. Bates, R. L., P. L. Fondy, and R. R. Corpstein. *Ind. Eng. Chem. Proc. Des. Dev.*, **2**(4):310 (1963).
4. Berkman, P. D., and R. V. Calabrese. *AIChE J.*, **34**:602 (1988).
5. Bissell, E. S., H. C. Hesse, H. J. Everett, and J. H. Rushton. *Chem. Eng. Prog.*, **43**:649 (1947).
6. Botton, R., D. Cosserat, and J. C. Charpentier. *Chem. Eng. Sci.*, **35**:82 (1980).
7. Bowen, R. L., Jr. *AIChE J.*, **35**:1575 (1989).
8. Buurman, C., G. Resoort, and A. Plaschkes. *Chem. Eng. Sci.*, **41**:2865 (1986).
9. Calderbank, P. H., and M. B. Moo-Young. *Trans. Inst. Chem. Eng. Lond.*, **37**:26 (1959).
10. Chudacek, M. W. *Ind. Eng. Chem. Fund.*, **35**:391 (1986).
11. Clift, R., J. R. Grace, and M. E. Weber. *Bubbles, Drops, and Particles*. New York: Academic Press, 1978.
12. Connolly, J. R., and R. L. Winter. *Chem. Eng. Prog.*, **65**(8):70 (1969).
13. Cutter, L. A. *AIChE J.*, **12**:35 (1966).
14. Dickey, D. S., in M. Moo-Young (ed.). *Advances in Biotechnology*, vol. 1. New York: Pergamon Press, 1981, p. 483.
15. Einsele, A., and R. K. Finn. *Ind. Eng. Chem. Proc. Des. Dev.*, **19**:600 (1980).
16. Fasano, J. B., A. Bakker, and W. R. Penney. *Chem. Eng.*, **101**(8):110 (1994).
17. Fox, E. A., and V. E. Gex. *AIChE J.*, **2**:539 (1956).

18. Godleski, E. S., and J. C. Smith. *AIChE J.*, **8:**617 (1962).
19. Gray, J. B., in V. W. Uhl and J. B. Gray (eds.). *Mixing: Theory and Practice,* vol. 1. New York: Academic, 1969; (*a*) pp. 181–4, (*b*) pp. 207–8.
20. Harriott, P. *AIChE J.*, **8:**93 (1962).
21. Hassan, I. T. M., and C. W. Robinson. *AIChE J.*, **23:**48 (1977).
22. Holmes, D. B., R. M. Voncken, and J. A. Dekker. *Chem. Eng. Sci.,* **19:**201 (1964).
23. Hunsaker, J. C., and B. G. Rightmire. *Engineering Applications of Fluid Mechanics.* New York: McGraw-Hill, 1947, chap. 7.
24. Jacobs, L. J. Paper presented at *Eng. Found. Mixing Res. Conf.,* South Berwick, ME, Aug. 12–17, 1973.
25. Khang, S. J., and O. Levenspiel. *Chem. Eng.,* **83**(21):141 (1976).
26. Meister, B. J., and G. F. Scheele. *AIChE J.*, **15:**689 (1969).
27. Metzner, A. B., R. H. Feehs, H. L. Ramos, R. E. Otto, and J. D. Tuthill. *AIChE J.*, **7:**3 (1961).
28. Michel, B. J., and S. A. Miller. *AIChE J.*, **8:**262 (1962).
29. Middleman, S. *Ind. Eng. Chem. Proc. Des. Dev.,* **13:**78 (1974).
30. Miller, E., S. P. Foster, R. W. Ross, and K. Wohl. *AIChE J.*, **3:**395 (1957).
31. Mlynek, Y., and W. Resnick. *AIChE J.*, **18:**122 (1972).
32. Moo-Young, M., K. Tichar, and F. A. L. Dullien. *AIChE J.*, **18:**178 (1972).
33. Morrison, P. P., H. Olin, and G. Rappe. Chemical Engineering Research Report, Cornell University, June 1962 (unpublished).
34. Myers, K. J., A. Bakker, and D. Ryan. *Chem. Eng. Prog.,* **93**(6):28 (1997).
35. Norwood, K. W., and A. B. Metzner. *AIChE J.*, **6:**432 (1960).
36. Oldshue, J. Y. *Ind. Eng. Chem.,* **61**(9):79 (1969).
37. Oldshue, J. Y. *Fluid Mixing Technology,* Chemical Engineering, New York: McGraw-Hill, 1983, p. 32.
38. Oldshue, J. Y. *Chem. Eng. Prog.,* **85**(5):33 (1989).
39. Overcashier, R. H., H. A. Kingsley, Jr., and R. B. Olney. *AIChE J.*, **2:**529 (1956).
40. Perry, R. H., and D. W. Green (eds.). *Perry's Chemical Engineers' Handbook,* 7th ed., New York: McGraw-Hill, 1997, p. **18**-16.
41. Pharamond, J. C., M. Roustan, and H. Roques. *Chem. Eng. Sci.,* **30:**907 (1975).
42. Rao, K. S. M. S. R., V. B. Rewatkar, and J. B. Joshi. *AIChE J.*, **34:**1332 (1988).
43. Rushton, J. H. *Ind. Eng. Chem.,* **44:**2931 (1952).
44. Rushton, J. H. *Chem. Eng. Prog.,* **50:**587 (1954).
45. Rushton, J. H., E. W. Costich, and H. J. Everett. *Chem. Eng. Prog.,* **46:**395, 467 (1950).
46. Rushton, J. H., and J. Y. Oldshue. *Chem. Eng. Prog.,* **49**(4):165 (1953).
47. Scheele, G. F., and B. J. Meister. *AIChE J.*, **14:**9 (1968).
48. Van't Riet, K., and John M. Smith. *Chem. Eng. Sci.,* **28:**1031 (1973).
49. Von Essen, J. *Chem. Eng.,* **105**(8):80 (1998).
50. Wichterle, K. *Chem. Eng. Sci.,* **43:**467 (1988).
51. Zwietering, Th. N. *Chem. Eng. Sci.,* **8:**244 (1957).

Heat Transfer and Its Applications

Practically all the operations that are carried out by the chemical engineer involve the production or absorption of energy in the form of heat. The laws governing the transfer of heat and the types of apparatus that have for their main object the control of heat flow are therefore of great importance. This section of the book deals with heat transfer and its applications in process engineering.

Nature of heat flow

When two objects at different temperatures are brought into contact, heat flows from the object at the higher temperature to that at the lower temperature. The net flow is always in the direction of the temperature decrease. The mechanisms by which the heat may flow are three: conduction, convection, and radiation.

Conduction. If a temperature gradient exists in a continuous substance, heat can flow unaccompanied by any observable motion of matter. Heat flow of this kind is called conduction, and according to Fourier's law, the heat flux is proportional to the temperature gradient and opposite to it in sign. For one-dimensional heat flow, Fourier's law is

$$\frac{dq}{dA} = -k\frac{dT}{dx} \tag{III.1}$$

where q = rate of heat flow in direction normal to surface
A = surface area
T = temperature
x = distance normal to surface
k = proportionality constant or thermal conductivity

In metals, thermal conduction results from the motion of free electrons, and there is close correspondence between thermal conductivity and electrical

287

conductivity. In solids that are poor conductors of electricity and in most liquids, thermal conduction results from momentum transfer between adjacent vibrating molecules or atoms. In gases, conduction occurs by the random motion of molecules, so that heat is "diffused" from hotter regions to colder ones. The most common example of pure conduction is heat flow in opaque solids such as the brick wall of a furnace or the metal wall of a heat exchanger tube. Conduction of heat in liquids or gases is often influenced by flow of the fluids, and both conductive and convective processes are lumped together under the term convection or convective heat transfer.

Convection. Convection can refer to the flow of heat associated with the movement of a fluid, such as when hot air from a furnace enters a room, or to the transfer of heat from a hot surface to a flowing fluid. The second meaning is more important for unit operations, as it includes heat transfer from metal walls, solid particles, and liquid surfaces. The convective flux is usually proportional to the difference between the surface temperature and the temperature of the fluid, as stated in Newton's law of cooling

$$\frac{q}{A} = h(T_s - T_f) \tag{III.2}$$

where T_s = surface temperature
T_f = bulk temperature of fluid, far from surface
h = heat-transfer coefficient

Note that the linear dependence on the temperature driving force $T_s - T_f$ is the same as that for pure conduction in a solid of constant thermal conductivity, as can be shown by integrating Eq. (III.1). Unlike thermal conductivity, the heat-transfer coefficient is not an intrinsic property of the fluid, but depends on the flow patterns determined by fluid mechanics as well as on the thermal properties of the fluid. If $T_f - T_s > 0$, heat will be transferred from the fluid to the surface.

Natural and forced convection. When currents in a fluid result from buoyancy forces created by density differences, and the density differences are caused by temperature gradients in the fluid, the action is called natural convection. When the currents are due to a mechanical device such as a pump or agitator, the flow is independent of density differences and is called forced convection. Buoyancy forces also exist in forced convection, but usually they have only a small effect.

Radiation. Radiation is a term given to the transfer of energy through space by electromagnetic waves. If radiation is passing through empty space, it is not transformed to heat or any other form of energy, nor is it diverted from its path. If, however, matter appears in its path, the radiation will be transmitted, reflected, or absorbed. It is only the absorbed energy that appears as heat, and this transformation is quantitative. For example, fused quartz transmits practically all the radiation that strikes it; a polished opaque surface or mirror will reflect most of the radiation impinging on it; a black or matte surface will absorb most of the radiation received by it and will transform such absorbed energy quantitatively to heat.

The energy emitted by a black body is proportional to the fourth power of the absolute temperature

$$W_b = \sigma T^4 \tag{III.3}$$

where W_b = rate of radiant energy emission per unit area
σ = Stefan-Boltzmann constant
T = absolute temperature

Monatomic and most diatomic gases are transparent to thermal radiation, and it is quite common to find that heat is flowing through masses of such gases both by radiation and by conduction-convection. Examples are the loss of heat from a radiator or uninsulated steam pipe to the air of a room and heat transfer in furnaces and other high-temperature gas-heating equipment. The two mechanisms are mutually independent and occur in parallel, so that one type of heat flow can be controlled or varied independently of the other. Conduction-convection and radiation can be studied separately and their separate effects added in cases where both are important. In very general terms, radiation becomes important at high temperatures and is independent of the circumstances of the flow of the fluid. Conduction-convection is sensitive to flow conditions and is relatively unaffected by temperature level.

Chapter 10 deals with conduction in solids, Chaps. 11 to 13 with heat transfer to fluids by conduction and convection, and Chap. 14 with heat transfer by radiation. In Chaps. 15 and 16 the principles developed in the preceding chapters are applied to the design of equipment for heating, cooling, condensing, and evaporating.

CHAPTER 10

Heat Transfer by Conduction

Conduction is most easily understood by considering heat flow in homogeneous isotropic solids because in these there is no convection and the effect of radiation is negligible unless the solid is translucent to electromagnetic waves. First, the general law of conduction is discussed; second, situations of steady-state heat conduction, where the temperature distribution within the solid does not change with time, are treated; third, some simple cases of unsteady conduction, where the temperature distribution does change with time, are considered.

BASIC LAW OF CONDUCTION

The basic relation for heat flow by conduction is the proportionality between heat flux and the temperature gradient. It is known as *Fourier's law,*[3] which for steady one-dimensional flow in the x direction has already been given by Eq. (III.1). To repeat,

$$\frac{dq}{dA} = -k\frac{dT}{dx} \tag{10.1}$$

where q = rate of heat flow in direction normal to surface
$\quad A$ = surface area
$\quad T$ = temperature
$\quad x$ = distance measured normal to surface
$\quad k$ = thermal conductivity

The general expressions of Fourier's law for flow in all three directions in an isotropic[†] material are

$$\left(\frac{dq}{dA}\right)_x = -k\,\frac{\partial T}{\partial x} \qquad (10.2)$$

$$\left(\frac{dq}{dA}\right)_y = -k\,\frac{\partial T}{\partial y} \qquad (10.3)$$

$$\left(\frac{dq}{dA}\right)_z = -k\,\frac{\partial T}{\partial z} \qquad (10.4)$$

The partial derivatives in Eqs. (10.1) to (10.4) call attention to the fact that the temperature may vary with both location and time. These equations are the components of the single vector equation

$$\frac{dq}{dA} = -k\nabla T \qquad (10.5)$$

Equation (10.5) states that the flux vector dq/dA is proportional to the temperature gradient ∇T and is oppositely directed. In an isotropic material, therefore, heat flows by conduction in the direction of steepest temperature descent.

Thermal conductivity

The proportionality constant k is a physical property of the substance called the *thermal conductivity*. It, like the newtonian viscosity μ, is one of the so-called transport properties of the material. This terminology is based on the analogy between Eqs. (3.4) and (10.1). In Eq. (3.4) the quantity τ is a rate of momentum flow per unit area, the quantity du/dy is the velocity gradient, and μ is the required proportionality factor. In Eq. (10.1), q/A is the rate of heat flow per unit area, dT/dx is the temperature gradient, and k is the proportionality factor. The minus sign is omitted in Eq. (3.4) because of convention in choosing the direction of the force vector.

In engineering units, q is measured in watts or Btu/h and dT/dx in °C/m or °F/ft. Then the units of k are W/m·°C or Btu/ft^2·h·(°F/ft), which may be written Btu/ft·h·°F.

Fourier's law states that k is independent of the temperature gradient but not necessarily of temperature itself. Experiment does confirm the independence of k for a wide range of temperature gradients, except for porous solids, where radiation between particles, which does not follow a linear temperature law, becomes an important part of the total heat flow. On the other hand, k is a function of temperature, but not a strong one. For small ranges of temperature, k may be considered constant.

[†]In an isotropic material the thermal conductivity k is the same in all directions. This is true for fluids and for most homogeneous solids; the principal nonisotropic materials are noncubic crystals and laminated or fibrous solids such as wood.[1]

For larger temperature ranges, the thermal conductivity can usually be approximated by an equation of the form

$$k = a + bT \tag{10.6}$$

where a and b are empirical constants.

Thermal conductivities of metals cover a wide range of values, from about 17 W/m·°C (10 Btu/ft·h·°F) for stainless steel and 45 W/m·°C (26 Btu/ft·h·°F) for mild steel, to 380 W/m·°C (220 Btu/ft·h·°F) for copper and 415 W/m·°C (240 Btu/ft·h·°F) for silver. The thermal conductivity of metals is generally nearly constant or decreases slightly as the temperature is increased, and the conductivity of alloys is less than that of pure metals. For glass and most nonporous materials, the thermal conductivities are much lower, from about 0.35 to 3.5 W/m·°C (0.2 to 2 Btu/ft·h·°F); for these materials k may either increase or decrease as the temperature rises.

For most liquids k is lower than that for solids, with typical values of about 0.17 W/m·°C (0.1 Btu/ft·h·°F), and k decreases by 3 to 4 percent for a 10°C rise in temperature. Water is an exception, with $k = 0.5$ to 0.7 W/m·°C (0.3 to 0.4 Btu/ft·h·°F), and k goes through a maximum as the temperature is raised.

Gases have the smallest thermal conductivities, with values as low as 0.007 W/m·°C (0.004 Btu/ft·h·°F). For air at 0°C, k is 0.024 W/m·°C (0.014 Btu/ft·h·°F). The thermal conductivity decreases with increasing molecular weight or with decreasing temperature because of changes in the molecular velocity and energy distribution. For monatomic gases k varies with about $T^{0.5}$, but for polyatomic gases k changes more rapidly with temperature. The thermal conductivity of gases is nearly independent of pressure up to about 10 bars; at higher pressures k increases slightly with pressure. Values of k for some solids, liquids, and gases are given in Apps. 6 and 10 through 13. More complete tables are available in the literature.[8]

Solids having low thermal conductivities are used for insulation on pipes, vessels, and buildings. Porous materials such as fiberglass pads or polymer foams act by entrapping air and eliminating convection. Their k values may be nearly as low as that of air itself, and if a high-molecular-weight gas is trapped in a closed-cell foam, k can be less than that for air.

STEADY-STATE CONDUCTION

Simple examples of steady-state conduction are shown in Fig. 10.1. In Fig. 10.1a a flat-walled insulated tank contains a refrigerant at perhaps −10°C, while the air outside the tank is at 28°C. The temperature falls linearly with distance across the layer of insulation as heat flows from the air to the refrigerant. As we will see in a later chapter, there may actually be a temperature drop between the bulk of the air and the outside surface of the insulation, but it is assumed to be negligible in Fig. 10.1a. Figure 10.1b shows a similar tank containing boiling water at 100°C, losing heat to air at 20°C. As before, the temperature profile in the insulation is linear, but heat flows in the opposite direction and x in Eq. (10.1) must be measured

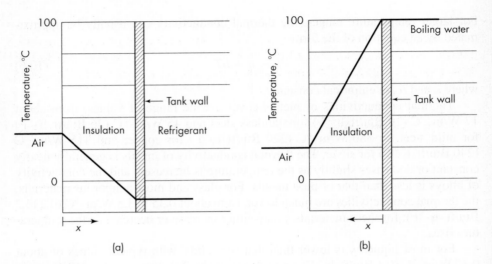

FIGURE 10.1
Temperature gradients outside insulated tanks: (a) heat flow into the tank; (b) heat flow from the tank.

outward from the inside surface. Again, there may be a temperature change in the air near the tank wall; again it is assumed to be negligible.

The rate of heat flow is found as follows, assuming that k is independent of temperature. Since in steady state there can be neither accumulation nor depletion of heat within the slab, q is constant along the path of heat flow. If x is the distance from the hot side, Eq. (10.1) can be written

$$dT = -\frac{q}{kA} \, dx \tag{10.7}$$

Since the only variables in Eq. (10.7) are x and T, direct integration gives

$$\frac{q}{A} = k \frac{T_1 - T_2}{x_2 - x_1} = k \frac{\Delta T}{B} \tag{10.8}$$

where $x_2 - x_1 = B =$ thickness of layer of insulation
$T_1 - T_2 = \Delta T =$ temperature drop across layer

When the thermal conductivity k varies linearly with temperature, in accordance with Eq. (10.6), Eq. (10.8) still can be used rigorously by taking an average value $\bar{k}$ for k, which may be found either by using the arithmetic average of the individual values of k for the two surface temperatures T_1 and T_2, or by calculating the arithmetic average of the temperatures and using the value of k at that temperature.

Equation (10.8) can be written in the form

$$\frac{q}{A} = \frac{\Delta T}{R} \tag{10.9}$$

where R is the thermal resistance of the solid between points 1 and 2. Equation (10.9) is an instance of the general rate principle, which equates a rate to the ratio of a driving force to a resistance. In heat conduction, q is the rate and ΔT is the driving force. The resistance R, as shown by Eq. (10.9) and using $\bar{k}$ for k to account for a linear variation of k with temperature, is $B/\bar{k}$. The reciprocal of a resistance is a heat-transfer coefficient h, as in Newton's law [Eq. (III.2)]. For heat conduction, then, $h = \bar{k}/B$. Both R and h depend on the dimensions of the solid as well as on the thermal conductivity k, which is a property of the material.

EXAMPLE 10.1. A layer of pulverized cork 6 in. (152 mm) thick is used as a layer of thermal insulation in a flat wall. The temperature of the cold side of the cork is 40°F (4.4°C), and that of the warm side is 180°F (82.2°C). The thermal conductivity of the cork at 32°F (0°C) is 0.021 Btu/ft·h·°F (0.036 W/m·°C), and that at 200°F (93.3°C) is 0.032 (0.055). The area of the wall is 25 ft² (2.32 m²). What is the rate of heat flow through the wall in Btu per hour (watts)?

Solution. The arithmetic average temperature of the cork layer is $(40 + 180)/2 = 110°F$. By linear interpolation the thermal conductivity at 110°F is

$$\bar{k} = 0.021 + \frac{(110 - 32)(0.032 - 0.021)}{200 - 32}$$

$$= 0.021 + 0.005 = 0.026 \text{ Btu/ft} \cdot \text{h} \cdot °F$$

Also, $A = 25 \text{ ft}^2$ $\Delta T = 180 - 40 = 140°F$ $B = \dfrac{6}{12} = 0.5 \text{ ft}$

Substituting in Eq. (10.8) gives

$$q = \frac{0.026 \times 25 \times 140}{0.5} = 182 \text{ Btu/h (53.3 W)}$$

Compound resistances in series

Consider a flat wall constructed of a series of layers, as shown in Fig. 10.2. Let the thicknesses of the layers be B_A, B_B, and B_C and the average conductivities of the materials of which the layers are made be $\bar{k}_A$, $\bar{k}_B$, and $\bar{k}_C$, respectively. Also let the area of the compound wall, at right angles to the plane of the illustration, be A. Let ΔT_A, ΔT_B, and ΔT_C be the temperature drops across layers A, B, and C, respectively. Assume, further, that the layers are in excellent thermal contact, so that no temperature difference exists across the interfaces between the layers. Then if ΔT is the total temperature drop across the entire wall,

$$\Delta T = \Delta T_A + \Delta T_B + \Delta T_C \tag{10.10}$$

It is desired, first, to derive an equation for calculating the rate of heat flow through the series of resistances and, second, to show how the rate can be calculated as the ratio of the overall temperature drop ΔT to the overall resistance of the wall. Equation (10.8) can be written for each layer, using $\bar{k}$ in place of k,

$$\Delta T_A = q_A \frac{B_A}{\bar{k}_A A} \qquad \Delta T_B = q_B \frac{B_B}{\bar{k}_B A} \qquad \Delta T_C = q_C \frac{B_C}{\bar{k}_C A} \tag{10.11}$$

Adding Eqs. (10.11) gives

$$\Delta T_A + \Delta T_B + \Delta T_C = \frac{q_A B_A}{A\bar{k}_A} + \frac{q_B B_B}{A\bar{k}_B} + \frac{q_C B_C}{A\bar{k}_C} = \Delta T$$

Temperature drops

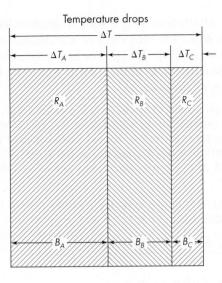

FIGURE 10.2
Thermal resistances in series.

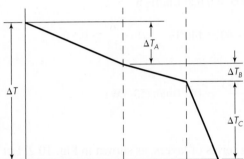

Since, in steady heat flow, all the heat that passes through the first resistance must pass through the second and in turn pass through the third, q_A, q_B, and q_C are equal and all can be denoted by q. Using this fact and solving for q/A give

$$\frac{q}{A} = \frac{\Delta T}{B_A/\bar{k}_A + B_B/\bar{k}_B + B_C/\bar{k}_C} = \frac{\Delta T}{R_A + R_B + R_C} = \frac{\Delta T}{R} \quad (10.12)$$

where R_A, R_B, R_C = resistance of individual layers
R = overall resistance

Equation (10.12) shows that in heat flow through a series of layers the overall thermal resistance equals the sum of the individual resistances.

The rate of flow of heat through several resistances in series clearly is analogous to the current flowing through several electric resistances in series. In an electric circuit the potential drop over any one of several resistances is to the total potential drop in the circuit as the individual resistances are to the total resistance. In the same way the potential drops in a thermal circuit, which are the temperature differences, are to the total temperature drop as the individual thermal resistances are to the total thermal resistance. This can be expressed mathematically as

$$\frac{\Delta T}{R} = \frac{\Delta T_A}{R_A} = \frac{\Delta T_B}{R_B} = \frac{\Delta T_C}{R_C} \tag{10.13}$$

Figure 10.2 also shows the pattern of temperatures and the temperature gradients. Depending on the thickness and thermal conductivity of the layer, the temperature drop in that layer may be a large or small fraction of the total temperature drop; a thin layer of low conductivity may well cause a much larger temperature drop and a steeper thermal gradient than a thick layer of high conductivity.

EXAMPLE 10.2. A flat furnace wall is constructed of a 4.5-in. (114-mm) layer of Sil-o-cel brick, with a thermal conductivity of 0.08 Btu/ft·h·°F (0.138 W/m·°C) backed by a 9-in. (229-mm) layer of common brick, of conductivity 0.8 Btu/ft·h·°F (1.38 W/m·°C). The temperature of the inner face of the wall is 1400°F (760°C), and that of the outer face is 170°F (76.6°C). (a) What is the heat loss through the wall? (b) What is the temperature of the interface between the refractory brick and the common brick? (c) Supposing that the contact between the two brick layers is poor and that a "contact resistance" of 0.50°F·h·ft²/Btu (0.088°C·m²/W) is present, what would be the heat loss?

Solution

(a) Consider 1 ft² of wall ($A = 1$ ft²). The thermal resistance of the Sil-o-cel layer is

$$R_A = \frac{4.5/12}{0.08} = 4.687$$

and that of the common brick is

$$R_B = \frac{9/12}{0.8} = 0.938$$

The total resistance is

$$R = R_A + R_B = 4.687 + 0.938 = 5.625 \, °F·h·ft²/Btu$$

The overall temperature drop is

$$\Delta T = 1400 - 170 = 1230°F$$

Substitution in Eq. (10.12) gives, for the heat loss from 1 ft² of wall,

$$q = \frac{1,230}{5.625} = 219 \, Btu/h \, (64.2 \, W)$$

(b) The temperature drop in one of a series of resistances is to the individual resistance as the overall temperature drop is to the overall resistance, or

$$\frac{\Delta T_A}{4.687} = \frac{1,230}{5.625}$$

from which

$$\Delta T_A = 1025°F$$

The temperature at the interface is $1400 - 1025 = 375°F$ (190.6°C).

(c) The total resistance, which now includes a contact resistance, is

$$R = 5.625 + 0.500 = 6.125$$

The heat loss from 1 ft^2 is

$$q = \frac{1,230}{6.125} = 201 \text{ Btu/h (58.9 W)}$$

Heat flow through a cylinder

Consider the hollow cylinder represented by Fig. 10.3. The inside radius of the cylinder is r_i, the outside radius is r_o, and the length of the cylinder is L. The thermal conductivity of the material of which the cylinder is made is k. The temperature of the outside surface is T_o, and that of the inside surface is T_i. It is desired to calculate the rate of heat flow outward for this case.

Consider a very thin cylinder, concentric with the main cylinder, of radius r, where r is between r_i and r_o. The thickness of the wall of this cylinder is dr; and if dr is small enough with respect to r for the lines of heat flow to be considered parallel, Eq. (10.1) can be applied and written in the form

$$q = -k\frac{dT}{dr}2\pi rL \tag{10.14}$$

since the area perpendicular to the heat flow is equal to $2\pi rL$ and the dx of Eq. (10.1) is equal to dr. Rearranging Eq. (10.14) and integrating between limits gives

$$\int_{r_i}^{r_o} \frac{dr}{r} = \frac{2\pi Lk}{q} \int_{T_o}^{T_i} dT$$

$$\ln r_o - \ln r_i = \frac{2\pi Lk}{q}(T_i - T_o)$$

$$q = \frac{k(2\pi L)(T_i - T_o)}{\ln(r_o/r_i)} \tag{10.15}$$

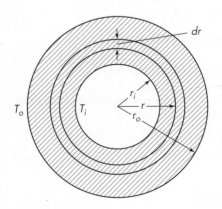

FIGURE 10.3
Flow of heat through thick-walled cylinder.

Equation (10.15) can be used to calculate the flow of heat through a thick-walled cylinder. It can be put in a more convenient form by expressing the rate of flow of heat as

$$q = \frac{k \bar{A}_L (T_i - T_o)}{r_o - r_i} \tag{10.16}$$

This is of the same general form as Eq. (10.8) for heat flow through a flat wall with the exception of $\bar{A}_L$, which must be so chosen that the equation is correct. The term $\bar{A}_L$ can be determined by equating the right-hand sides of Eqs. (10.15) and (10.16) and solving for $\bar{A}_L$:

$$\bar{A}_L = \frac{2\pi L (r_o - r_i)}{\ln(r_o/r_i)} \tag{10.17}$$

Note from Eq. (10.17) that $\bar{A}_L$ is the area of a cylinder of length L and radius $\bar{r}_L$, where

$$\bar{r}_L = \frac{r_o - r_i}{\ln(r_o/r_i)} \tag{10.18}$$

The form of the right-hand side of Eq. (10.18) is important enough to repay memorizing. It is known as the *logarithmic mean,* and in the particular case of Eq. (10.18), $\bar{r}_L$ is called the *logarithmic mean radius.* It is the radius that, when applied to the integrated equation for a flat wall, will give the correct rate of heat flow through a thick-walled cylinder.

The logarithmic mean is less convenient than the arithmetic mean, and the latter can be used without appreciable error for thin-walled tubes, where r_o/r_i is nearly 1. The ratio of the logarithmic mean $\bar{r}_L$ to the arithmetic mean $\bar{r}_a$ is a function of r_o/r_i, as shown in Fig. 10.4. Thus, when $r_o/r_i = 2$, the logarithmic mean is $0.96\bar{r}_a$ and the

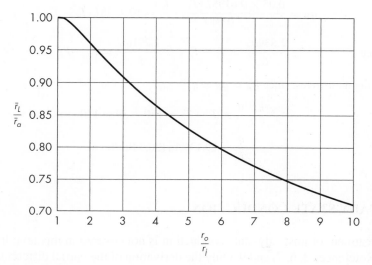

FIGURE 10.4
Relation between logarithmic and arithmetic means.

error in the use of the arithmetic mean is 4 percent. The error is 1 percent where $r_o/r_i = 1.4$.

EXAMPLE 10.3. A tube of 60-mm (2.36-in.) outer diameter (OD) is insulated with a 50-mm (1.97-in.) layer of silica foam, for which the conductivity is 0.055 W/m · °C (0.032 Btu/ft · h · °F), followed with a 40-mm (1.57-in.) layer of cork with a conductivity of 0.05 W/m · °C (0.03 Btu/ft · h · °F). If the temperature of the outer surface of the pipe is 150°C (302°F) and the temperature of the outer surface of the cork is 30°C (86°F), calculate the heat loss in watts per meter of pipe.

Solution. These layers are too thick to use the arithmetic mean radius, and the logarithmic mean radius should be used. For the silica layer

$$\bar{r}_L = \frac{80 - 30}{\ln(80/30)} = 50.97 \text{ mm}$$

and for the cork layer

$$\bar{r}_L = \frac{120 - 80}{\ln(120/80)} = 98.64 \text{ mm}$$

Call silica substance A and cork substance B. From Eq. (10.16)

$$q_A = \frac{k_A \bar{A}_A (T_i - T_x)}{x_A} \qquad q_B = \frac{k_B \bar{A}_B (T_x - T_o)}{x_B}$$

where T_x is the temperature at the interface between the silica and the cork. From Eqs. (10.17) and (10.18),

$$\bar{A}_A = 2\pi (0.05097) L = 0.3203 L \qquad \bar{A}_B = 2\pi (0.09864) L = 0.6198 L$$

Then

$$q_A = \frac{0.055 \times 0.3203 L (T_i - T_x)}{0.050} = 0.3522 L (T_i - T_x)$$

$$q_B = \frac{0.05 \times 0.6198 L (T_x - T_o)}{0.040} = 0.7748 L (T_x - T_o)$$

Hence

$$\frac{2.839 q_A}{L} = T_i - T_x \qquad \frac{1.291 q_B}{L} = T_x - T_o$$

Adding these gives, since $q_A = q_B = q$

$$\frac{4.13 q}{L} = T_i - T_o = 150 - 30 = 120$$

$$\frac{q}{L} = 29.1 \text{ W/m} \ (30.3 \text{ Btu/ft} \cdot \text{h})$$

UNSTEADY-STATE CONDUCTION

A full treatment of unsteady-state conduction is not covered in this text; it is available in References 2, 6, 7, and 9. Only the derivation of the partial differential equations for one-dimensional heat flow, with or without constant surface temperatures, and the results of the integration of the equations for some simple shapes are covered in this section. It is assumed throughout that k is independent of temperature.

One-dimensional heat flow with constant surface temperature

Figure 10.5 represents a section through a large slab of material of thickness $2s$, initially all at a uniform temperature T_a. At the start of heating both surface temperatures are quickly increased to and subsequently held at temperature T_s. The temperature pattern shown in Fig. 10.5 reflects conditions after a relatively short time t_T has elapsed since the start of heating.

Focus attention on the thin layer of thickness dx located at distance x from the left side of the slab. The two sides of the element are isothermal surfaces. The temperature gradient at x is, at a definite instant of time, $\partial T/\partial x$, and the heat input in time interval dt at x is $-kA(\partial T/\partial x)\,dt$, where A is the area of the layer perpendicular to the flow of heat and k is the thermal conductivity of the solid. The gradient at distance $x + dx$ is slightly different from that at x and may be represented as

$$\frac{\partial T}{\partial x} + \frac{\partial}{\partial x}\frac{\partial T}{\partial x}\,dx$$

The heat flow out of the layer at $x + dx$ is, then,

$$-kA\left(\frac{\partial T}{\partial x} + \frac{\partial}{\partial x}\frac{\partial T}{\partial x}\,dx\right)dt$$

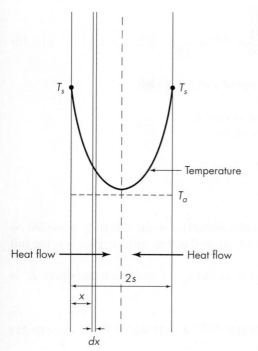

FIGURE 10.5
Unsteady-state conduction in solid slab.

T_s

T_s

Temperature

T_a

Heat flow

Heat flow

$2s$

x

dx

The excess of heat input over heat output, which is the accumulation of heat in layer dx, is

$$-kA\frac{\partial T}{\partial x}dt + kA\left(\frac{\partial T}{\partial x} + \frac{\partial^2 T}{\partial x^2}dx\right)dt = kA\frac{\partial^2 T}{\partial x^2}dx\,dt$$

The accumulation of heat in the layer must increase the temperature of the layer. If c_p and ρ are the specific heat and density, respectively, the accumulation is the product of the mass (volume times density), the specific heat, and the increase in temperature, or $(\rho A\,dx)c_p\,(\partial T/\partial t)\,dt$. Then, by a heat balance,

$$kA\frac{\partial^2 T}{\partial x^2}dx\,dt = \rho c_p A\,dx\,\frac{\partial T}{\partial t}dt$$

or, after division by $\rho c_p A\,dx\,dt$,

$$\frac{\partial T}{\partial t} = \frac{k}{\rho c_p}\frac{\partial^2 T}{\partial x^2} = \alpha\frac{\partial^2 T}{\partial x^2} \tag{10.19}$$

The term α in Eq. (10.19) is called the *thermal diffusivity* of the solid and is a property of the material. It has the dimensions of area divided by time.

General solutions of unsteady-state conduction equations are available for certain simple shapes such as the infinite slab, the infinitely long cylinder, and the sphere. For example, the integration of Eq. (10.19) for the heating or cooling of an infinite slab of known thickness from both sides by a medium at constant surface temperature gives

$$\frac{T_s - \bar{T}_b}{T_s - T_a} = \frac{8}{\pi^2}\left(e^{-a_1 \text{Fo}} + \frac{1}{9}e^{-9a_1 \text{Fo}} + \frac{1}{25}e^{-25a_1 \text{Fo}} + \cdots\right) \tag{10.20}$$

where $T_s =$ constant average temperature of surface of slab
$T_a =$ initial temperature of slab
$\bar{T}_b =$ average temperature of slab at time t_T
$\text{Fo} =$ Fourier number, defined as $\alpha t_T/s^2$
$\alpha =$ thermal diffusivity
$t_T =$ time of heating or cooling
$s =$ one-half slab thickness
$a_1 = (\pi/2)^2$

Equation (10.20) may also be used for a slab heated from one side only, provided no heat is transferred at the other side and $\partial T/\partial x = 0$ at that surface. Here s is the full slab thickness.

For an infinitely long solid cylinder of radius r_m the average temperature $\bar{T}_b$ is given by the equation[6]

$$\frac{T_s - \bar{T}_b}{T_s - T_a} = 0.692e^{-5.78\text{Fo}} + 0.131e^{-30.5\text{Fo}} + 0.0534e^{-74.9\text{Fo}} + \cdots \tag{10.21}$$

where Fo $= \alpha t_T/r_m^2$. For a sphere of radius r_m the corresponding equation is[2]

$$\frac{T_s - \bar{T}_b}{T_s - T_a} = 0.608e^{-9.87\text{Fo}} + 0.152e^{-39.5\text{Fo}} + 0.0676e^{-88.8\text{Fo}} + \cdots \quad (10.22)$$

When Fo is greater than about 0.1, only the first term of the series in Eqs. (10.20) to (10.22) is significant, and the other terms may be ignored. Under these conditions the time required to change the temperature from T_a to $\bar{T}_b$ can be found by rearranging Eq. (10.20), with all except the first term of the series omitted, to give for the slab

$$t_T = \frac{1}{\alpha}\left(\frac{2s}{\pi}\right)^2 \ln \frac{8(T_s - T_a)}{\pi^2(T_s - \bar{T}_b)} \quad (10.23)$$

For the infinite cylinder the corresponding equation, found from Eq. (10.21), is

$$t_T = \frac{r_m^2}{5.78\alpha} \ln \frac{0.692(T_s - T_a)}{T_s - \bar{T}_b} \quad (10.24)$$

For a sphere, from Eq. (10.22),

$$t_T = \frac{r_m^2}{9.87\alpha} \ln \frac{0.608(T_s - T_a)}{T_s - \bar{T}_b} \quad (10.25)$$

Figure 10.6 is a plot of Eqs. (10.21) to (10.23). The ordinate of this figure is known as the *unaccomplished temperature change*, that is, the fraction of the total possible temperature change that remains to be accomplished at any time. Except at very low values of Fo, Eqs. (10.23) to (10.25) apply and all three semilogarithmic plots are straight lines.

Heat flow with variable surface temperature

Equations (10.20) to (10.25) apply only when the surface temperature is constant, which implies negligible thermal resistance between the surface and the fluid. When this resistance is significant, the surface temperature changes with time, as sketched in Fig. 10.7. The temperature gradient in the fluid near the surface is shown as a straight line, which implies heat transfer by conduction through a thin stagnant layer and perfectly mixed fluid outside this layer. With fluid flowing past the surface, the actual gradient is slightly curved; but a linear gradient is shown for convenience, and this does not affect the subsequent derivations. There is no accumulation of energy at the interface, so the heat flux to the surface equals the flux into the solid, and the change in gradient at the interface reflects the difference in conductivities.

In this example, the thermal conductivity of the solid is assumed to be less than that of the fluid, so the temperature gradient is steeper in the solid than in the fluid. With increasing time, the gradient in the solid and the heat flux decrease, and the surface temperature increases, since a lower driving force is needed to transfer heat through the fluid layer.

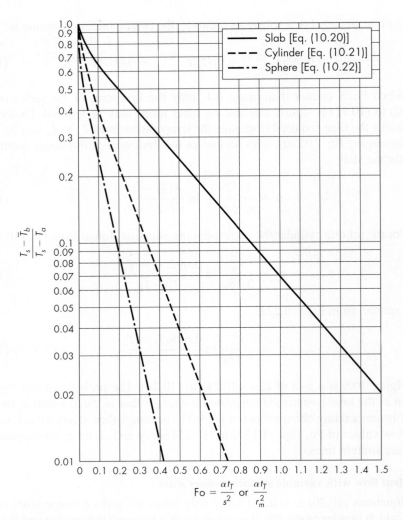

FIGURE 10.6
Average temperatures during unsteady-state heating or cooling of a large slab, an infinitely long cylinder, or a sphere.

When there is an external resistance, Eq. (10.19) for conduction in the solid still applies, but a new boundary condition is needed. The heat flux into the solid is equated to the rate of heat transfer through the fluid layer, as given by Newton's law for convective heat transfer

$$\frac{q}{A} = h(T_f - T_s) = -k\left(\frac{\partial T}{\partial x}\right)_{x=0} \tag{10.26}$$

Equations (10.26) and (10.19) can be solved to get the temperature distribution in the solid and then the average solid temperature and the total heat transferred. The results are often given in charts or tables with the Fourier number and the Biot number as parameters. The Fourier number is a dimensionless time, and the

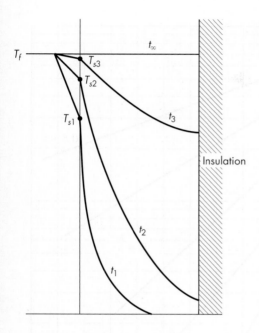

FIGURE 10.7
Temperature gradients in and near a slab being heated from one side.

Biot number, denoted by Bi, is a measure of the relative importance of the external and internal resistances.

For a flat slab

$$\text{Bi} = \frac{hs}{k} \tag{10.27}$$

For a sphere or cylinder

$$\text{Bi} = \frac{hr_m}{k} \tag{10.28}$$

Charts showing local and average temperatures in slabs, cylinders, and spheres were developed by M. P. Heisler[5] and others[4] and are available in heat-transfer texts and handbooks.[6,7,9] Figure 10.8 shows the average temperature change with time for a sphere for intermediate values of Biot number. When the Biot number is very large, the external resistance can be neglected, and the equations for a constant surface temperature apply.

When the Biot number is small, most of the resistance is external to the solid, and the rate of heat transfer can be approximated by using an effective coefficient for the solid phase and an overall heat-transfer coefficient. For a sphere, the effective internal coefficient is about $5k/r_m$, which corresponds to an average conduction distance of $r_m/5$ (a shell of this thickness has one-half the sphere volume). For a long cylinder, the effective coefficient is $3k/r_m$, and for a slab it is $2k/s$.

For a sphere, the unsteady-state heat balance is

$$\rho c_p \left(\frac{4}{3} \pi r_m^3 \right) \frac{d\bar{T}_b}{dt} = U \left(4\pi r_m^2 \right) (T_f - \bar{T}_b) \tag{10.29}$$

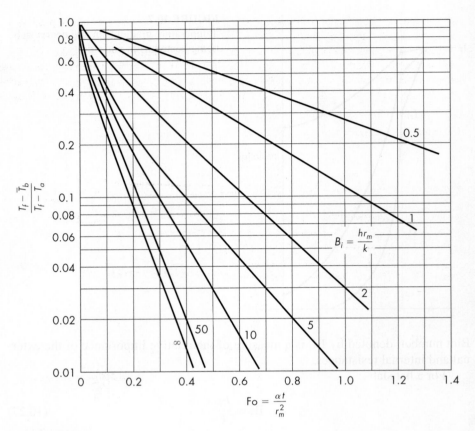

FIGURE 10.8
Change with time of the average temperature of a sphere, with external thermal resistance.

where

$$\frac{1}{U} \cong \frac{1}{h} + \frac{r_m}{5k}$$

(10.30)

Rearranging Eq. (10.29) gives

$$\frac{d\bar{T}_b}{T_f - \bar{T}_b} = \frac{3U\,dt}{\rho c_p r_m}$$

(10.31)

Integration leads to

$$\ln \frac{T_f - \bar{T}_b}{T_f - T_a} = \frac{-3Ut}{\rho c_p r_m}$$

(10.32)

Similar equations can be developed for approximating heat transfer to other shapes at low Biot numbers. For a long cylinder, the 3 in Eq. (10.31) is replaced by 2, and for a slab, $3/r_m$ in Eq. (10.31) becomes $1/s$.

EXAMPLE 10.4. Plastic spheres 5 mm in size and originally at 80°C are cooled in a fluidized bed where the air temperature is 30°C. The density of the solid is 1,100 kg/m³, the thermal conductivity is 0.13 W/m·°C, and the specific heat is 1,700 J/kg·°C. The external heat-transfer coefficient is 50 W/m²·°C. (a) How long will it take for the average solid temperature to reach 35°C? (b) What fraction of the resistance to heat transfer is in the external film? (c) Repeat part (a) for a constant surface temperature of 30°C.

Solution

(a) From Eq. (10.30), the overall resistance is

$$\frac{1}{U} = \frac{1}{50} + \frac{2.5 \times 10^{-3}}{5 \times 0.13} = 0.0238$$

$$U = 41.9 \text{ W/m}^2 \cdot {}^\circ\text{C}$$

From Eq. (10.28),

$$\text{Bi} = \frac{50(2.5 \times 10^{-3})}{0.13} = 0.96$$

$$\frac{T_f - \bar{T}_b}{T_f - T_a} = \frac{30 - 35}{30 - 80} = 0.1$$

By interpolation from Fig. 10.8,

$$\text{Fo} \cong 1.06 = \frac{\alpha t}{r_m^2}$$

$$\alpha = \frac{k}{\rho c_p} = \frac{0.13}{1,100 \times 1,700} = 6.95 \times 10^{-8}$$

$$t = \frac{1.06(2.5 \times 10^{-3})^2}{6.95 \times 10^{-8}} = 95 \text{ s}$$

An approximate solution can be obtained by using Eq. (10.32) as rearranged for cooling.

$$\ln \frac{\bar{T}_b - T_f}{T_a - T_f} = \frac{-3Ut}{\rho c_p r_m}$$

From Eq. (10.30),

$$\frac{1}{U} = 50 + \frac{2.5 \times 10^{-3}}{5(0.13)} = 0.0238$$

$$U = 41.9 \text{ W/m}^2 \cdot {}^\circ\text{C}$$

$$t = -\ln\left(\frac{5}{50}\right) \frac{1,100(1.7 \times 10^3)(2.5 \times 10^{-3})}{3 \times 41.9} = 86 \text{ s}$$

(b) Fraction of external resistance:

$$\frac{U}{h} = \frac{41.9}{50} = 0.84$$

(c) From Fig. 10.6 or Fig. 10.8, for Bi $= \infty$ and Fo $= 0.19$,

$$t = \frac{0.19(2.5 \times 10^{-3})^2}{6.95 \times 10^{-8}} = 17 \text{ s}$$

Total heat transferred

The total heat Q_T transferred to the solid in time t_T through a unit area of surface is often of interest. From the definition of average temperature, the heat required to raise the temperature of a unit mass of solid from T_a to $\bar{T}_b$ is $c_p(\bar{T}_b - T_a)$. For a slab of thickness $2s$ and density ρ, the total surface area (both sides) of a unit mass is $1/s\rho$. The total heat transferred per unit area is therefore given by

$$\frac{Q_T}{A} = s\rho c_p(\bar{T}_b - T_a) \tag{10.33}$$

The corresponding equation for an infinitely long cylinder is

$$\frac{Q_T}{A} = \frac{r_m \rho c_p(\bar{T}_b - T_a)}{2} \tag{10.34}$$

For a sphere the equation is

$$\frac{Q_T}{A} = \frac{r_m \rho c_p(\bar{T}_b - T_a)}{3} \tag{10.35}$$

EXAMPLE 10.5. A flat slab of plastic initially at 70°F (21.1°C) is placed between two platens at 250°F (121.1°C). The slab is 1.0 in. (2.54 cm) thick. (a) How long will it take to heat the slab to an average temperature of 210°F (98.9°C)? (b) How much heat, in Btu, will be transferred to the plastic during this time per square foot of surface? The density of the solid is 56.2 lb/ft³ (900 kg/m³), the thermal conductivity is 0.075 Btu/ft·h·°F (0.13 W/m·°C), and the specific heat is 0.40 Btu/lb·°F (1.67 J/g·°C).

Solution

(a) The quantities for use with Fig. 10.6 are

$$k = 0.075 \text{ Btu/ft} \cdot \text{h} \cdot \text{°F} \qquad \rho = 56.2 \text{ lb/ft}^3 \qquad c_p = 0.40 \text{ Btu/lb} \cdot \text{°F}$$

$$s = \frac{0.5}{12} = 0.0417 \text{ ft} \qquad T_s = 250\text{°F} \qquad T_a = 70\text{°F} \qquad \bar{T}_b = 210\text{°F}$$

Then

$$\frac{T_s - \bar{T}_b}{T_s - T_a} = \frac{250 - 210}{250 - 70} = 0.222 \qquad \alpha = \frac{k}{\rho c_p} = \frac{0.075}{56.2 \times 0.40} = 0.00335$$

From Fig. 10.6, for a temperature-difference ratio of 0.222,

$$\text{Fo} = 0.52 = \frac{0.00335 t_T}{0.0417^2} \qquad t_T = 0.27 \text{ h} = 16 \text{ min}$$

(b) Substitution in Eq. (10.33) gives heat flow per total surface area

$$\frac{Q_T}{A} = 0.0417 \times 56.2 \times 0.40(210 - 70) = 131 \text{ Btu/ft}^2 \text{ (1,487 kJ/m}^2)$$

Semi-infinite solid

Sometimes solids are heated or cooled in such a way that the temperature changes in the solid are confined to the region near one surface. Consider, for example, a very thick flat wall of a chimney, initially all at a uniform temperature T_a. Suppose that the inner surface of the wall is suddenly heated to, and held at, a high temperature T_s, perhaps by suddenly admitting hot flue gas to the chimney. Temperatures inside the chimney wall will change with time, rapidly near the hot surface and more slowly farther away. If the wall is thick enough, there will be no measurable change in the temperature of the outer surface for a considerable time. Under these conditions the heat may be considered to be "penetrating" a solid of essentially infinite thickness. Figure 10.9 shows the temperature patterns in such a wall at various times after exposure to the hot gas, indicating the sharp discontinuity in temperature at the hot surface immediately after exposure and the progressive changes at interior points at later times.

For this situation integration of Eq. (10.19) with the appropriate boundary conditions gives, for temperature T at any point a distance x from the hot surface, the equation

$$\frac{T_s - T}{T_s - T_a} = \frac{2}{\sqrt{\pi}} \int_0^Z e^{-Z^2} \, dZ \tag{10.36}$$

where $Z = x/2\sqrt{\alpha t}$, dimensionless
α = thermal diffusivity
x = distance from surface
t = time after change in surface temperature, h

The function in Eq. (10.36) is known as the *Gauss error integral* or *probability integral*. Equation (10.36) is plotted in Fig. 10.10.

Equation (10.36) indicates that at any time after the surface temperature is changed there will be some change in temperature at all points in the solid, even points far removed from the hot surface. The actual change at such distant points,

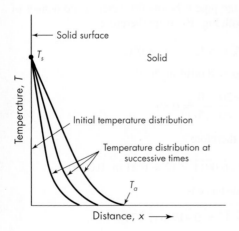

FIGURE 10.9
Temperature distributions in unsteady-state heating of semi-infinite solid.

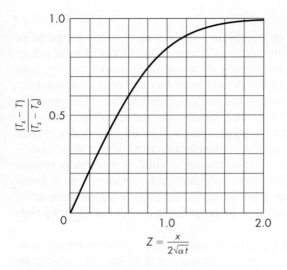

FIGURE 10.10
Unsteady-state heating or cooling of semi-infinite solid.

$$Z = \frac{x}{2\sqrt{\alpha t}}$$

however, is negligibly small. Beyond a certain distance from the hot surface not enough heat has penetrated to affect the temperature significantly. This *penetration distance* x_p is arbitrarily defined as that distance from the surface at which the temperature change is 1 percent of the initial change in surface temperature. That is, $(T - T_a)/(T_s - T_a) = 0.01$ or $(T_s - T)/(T_s - T_a) = 0.99$. Figure 10.10 shows that the probability integral reaches a value of 0.99 when $Z = 1.82$, from which

$$x_p = 3.64\sqrt{\alpha t} \tag{10.37}$$

EXAMPLE 10.6. A sudden cold wave drops the atmospheric temperature to $-20°C$ ($-4°F$) for 12 h. (*a*) If the ground was initially all at 5°C (41°F), how deep would a water pipeline have to be buried to be in no danger of freezing? (*b*) What is the penetration distance under these conditions? The thermal diffusivity of soil is 0.0011 m²/h (0.0118 ft²/h).

Solution

(*a*) Assume that the surface of the ground quickly reaches and remains at $-20°C$. Unless the temperature at the location of the pipe is below 0°C, there is no danger of freezing. The quantities required for use with Fig. 10.10 are therefore

$$T_s = -20°C \qquad T_a = 5°C \qquad T = 0°C$$

$$t = 12\,\text{h} \qquad \alpha = 0.0011\,\text{m}^2/\text{h}$$

$$\frac{T_s - T}{T_s - T_a} = \frac{-20 - 0}{-20 - 5} = 0.80$$

From Fig. 10.10, $Z = 0.91$. The depth x is therefore

$$x = 0.91 \times 2\sqrt{\alpha t} = 0.91 \times 2\sqrt{0.0011 \times 12} = 0.21\,\text{m (0.69 ft)}$$

(*b*) From Eq. (10.37) the penetration distance is

$$x_p = 3.64\sqrt{0.0011 \times 12} = 0.419\,\text{m (1.37 ft)}$$

To find the total heat transferred to a semi-infinite solid in a given time, it is necessary to find the temperature gradient and heat flux at the hot surface as a function of time. The temperature gradient at the surface is found by differentiating Eq. (10.36) to give

$$\left(\frac{\partial T}{\partial x}\right)_{x=0} = -\frac{T_s - T_a}{\sqrt{\pi \alpha t}} \tag{10.38}$$

The heat flow rate at the surface is therefore

$$\left(\frac{q}{A}\right)_{x=0} = -k\left(\frac{\partial T}{\partial x}\right)_{x=0} = \frac{k(T_s - T_a)}{\sqrt{\pi \alpha t}} \tag{10.39}$$

After substitution of dQ/dt for q, Eq. (10.39) can be integrated to give the total quantity of heat transferred per unit area Q_T/A in time t_T, as follows:

$$\frac{Q_T}{A} = \frac{k(T_s - T_a)}{\sqrt{\pi \alpha}} \int_0^{t_T} \frac{dt}{\sqrt{t}} = 2k(T_s - T_a)\sqrt{\frac{t_T}{\pi \alpha}} \tag{10.40}$$

SYMBOLS

A	Area, m^2 or ft^2; $\bar{A}_L$, logarithmic mean
a	Constant in Eq. (10.6)
a_1	$(\pi/2)^2$
B	Thickness of slab, m or ft; B_A, B_B, B_C, of layers A, B, C, respectively
Bi	Biot number, dimensionless; hs/k for slab; hr_m/k for cylinder or sphere
b	Constant in Eq. (10.6)
c_p	Specific heat at constant pressure, J/g · °C or Btu/lb · °F
e	Base of Naperian logarithms, 2.71828 ⋯
Fo	Fourier number, dimensionless; $\alpha t_T/s^2$ for slab; $\alpha t_T/r_m^2$ for cylinder or sphere
h	Individual or surface heat-transfer coefficient, W/m · °C or Btu/ft^2 · h · °F
k	Thermal conductivity, W/m · °C or Btu/ft · h · °F; k_A, k_B, k_C, of layers A, B, C, respectively; $\bar{k}$, average value
L	Length of cylinder, m or ft
Q	Quantity of heat, J or Btu; Q_T, total quantity transferred
q	Heat flow rate, W or Btu/h; q_A, q_B, q_C, in layers A, B, C, respectively
R	Thermal resistance, m^2 · °C/W or ft^2 · °F · h/Btu; R_A, R_B, R_C, of layers A, B, C, respectively
r	Radial distance or radius, m or ft; r_i, inside radius; r_m, radius of solid cylinder or sphere; r_o, outside radius; $\bar{r}_L$, logarithmic mean; $\bar{r}_a$, arithmetic mean
s	Half-thickness of slab, m or ft
T	Temperature, °C or °F; T_a, initial temperature; $\bar{T}_b$, average temperature at end of time t_T; T_f, of bulk of fluid; T_i, of inside surface; T_o, of outside

surface; T_s, of surface; T_x, at solid-solid interface; T_1, T_2, at locations 1, 2, respectively

t ⠀⠀Time, s or h; t_T, time required to heat or cool

U ⠀⠀Overall heat-transfer coefficient, W/m·°C or Btu/ft²·h·°F

x ⠀⠀Distance from surface, m or ft; x_1, x_2, at locations 1, 2, respectively; x_p, penetration distance in semi-infinite solid

y ⠀⠀Distance, m or ft

Z ⠀⠀$x/2\sqrt{\alpha t}$, dimensionless

Greek Letters

α ⠀⠀Thermal diffusivity, $k/\rho c_p$, m²/s or ft²/h

ΔT ⠀⠀Overall temperature drop; ΔT_A, ΔT_B, ΔT_C, in layers A, B, C, respectively

ρ ⠀⠀Density, kg/m³ or lb/ft³

PROBLEMS

10.1. A furnace wall consists of 200 mm of refractory fireclay brick, 100 mm of kaolin brick, and 6 mm of steel plate. The fire side of the refractory is at 1150°C, and the outside of the steel is at 30°C. An accurate heat balance over the furnace shows the heat loss from the wall to be 300 W/m². It is known that there may be thin layers of air between the layers of brick and steel. To how many millimeters of kaolin are these air layers equivalent? See App. 11 for thermal conductivities.

10.2. A standard 1-in. Schedule 40 steel pipe carries saturated steam at 250°F. The pipe is lagged (insulated) with a 2-in. layer of 85 percent magnesia pipe covering, and outside this magnesia there is a $\frac{1}{2}$-in. layer of cork. The inside temperature of the pipe wall is 249°F, and the outside temperature of the cork is 90°F. Thermal conductivities, in Btu/ft·h·°F, are: for steel, 26; for magnesia, 0.034; for cork, 0.03. Calculate (a) the heat loss from 100 ft of pipe in Btu per hour; (b) the temperatures at the boundaries between metal and magnesia and between magnesia and cork.

10.3. Derive the equation for steady-state heat transfer through a spherical shell of inner radius r_1 and outer radius r_2. Arrange the result for easy comparison with the solution for a thick-walled cylinder.

10.4. A very long, wide sheet of plastic 4 mm thick and initially at 20°C is suddenly exposed on both sides to an atmosphere of steam at 102°C. (a) If there is negligible thermal resistance between the steam and the surfaces of the plastic, how long will it take for the temperature at the centerline of the sheet to change significantly? (b) What would be the bulk average temperature of the plastic at this time? For the plastic, $k = 0.138$ W/m·°C and $\alpha = 0.00035$ m²/h.

10.5. A long steel rod 1 in. in diameter is initially at a uniform temperature of 1200°F. It is suddenly immersed in a quenching bath of oil at 150°F. In 4 min its average temperature drops to 250°F. How long would it take to lower the temperature from 1200 to 250°F (a) if the rod were $2\frac{1}{2}$ in. in diameter, (b) if it were 5 in. in diameter? For steel, $k = 26$ Btu/ft·h·°F; $\rho = 486$ lb/ft^3; $c_p = 0.11$ Btu/lb·°F.

10.6. Steel spheres 3 in. in diameter heated to 700°F are to be cooled by immersion in an oil bath at 125°F. If there is negligible thermal resistance between the oil and the steel surfaces, (a) calculate the average temperature of the spheres 10 s and 1 and 6 min after immersion. (b) How long would it take for the unaccomplished temperature change to be reduced to 1 percent of the initial temperature difference? The steel has the same thermal properties as in Prob. 10.5.

10.7. Under the conditions described in Example 10.6, what is the average rate of heat loss per unit area from the ground to the air during the 12-h period? The thermal conductivity of soil is 0.7 W/m·°C.

10.8. For the same initial temperatures T_s and T_a and at the same Fourier number, would the average temperature of a sphere be higher or lower than that of a cylinder or a slab? (See Fig. 10.6.) What are the physical reasons that would lead you to expect this?

10.9. The heat-transfer rate to the jacket of an agitated polymerization kettle is 7.4 kW/m^2 when the polymerization temperature is 50°C and the water in the jacket is at 20°C. The kettle is made of stainless steel with a wall 12 mm thick, and there is a thin layer of polymer ($k = 0.16$ W/m·°C) left on the wall from previous runs. (a) What is the temperature drop across the metal wall? (b) How thick would the polymer deposit have to be to account for the rest of the temperature difference? (c) By what factor could the heat flux be increased by using a stainless-clad reactor with a 3-mm stainless-steel layer bonded to a 9-mm mild-steel shell?

10.10. (a) Compare the thermal conductivities and thermal diffusivities of air and water at 100°F. (b) Calculate the penetration distances in a stagnant mass of air and one of water, at 50°F and 1 atm, each of which is exposed for 10 s to a hot metal surface at 100°F. Comment on the difference.

10.11. An oil storage tank has 20-mm steel walls covered with 50 mm of fiberglass insulation. If the oil temperature is kept at 150°C, (a) what is the rate of heat loss when the outside temperature is 20°C and the external air coefficient is 20 W/m^2·°C? (b) How much would the heat loss be reduced by doubling the thickness of the insulation?

10.12. An incinerator with $\frac{1}{2}$-in. steel walls is lined with 4 in. of firebrick to protect the steel and has 3 in. of magnesia insulation on the outside. When the gas temperature is 1400°F, the inside wall of the firebrick is at 1200°F as measured by an optical pyrometer. (a) What is the temperature of the steel wall? (b) How much more insulation would be needed to raise the inside wall temperature to 1300°F?

10.13. Particles of coal about 50 micrometers (μm) in size are injected into a stream of air at 300°C that carries them to a boiler. How long will it take to heat the particles from

25°C to within 5°C of the air temperature? Assume that the external coefficient h is given by $hD_p/k = 2.0$, where D_p is the particle diameter.

REFERENCES

1. Bird, R. B., W. E. Stewart, and E. N. Lightfoot. *Transport Phenomena.* New York: Wiley, 1960, p. 245.
2. Carslaw, H. S., and J. C. Jaeger. *Conduction of Heat in Solids.* New York: Wiley, 1969.
3. Fourier, J. *The Analytical Theory of Heat,* trans. by A. Freeman. New York: Dover, 1955.
4. Grober H., S. Erk, and U. Grigull. *Fundamentals of Heat Transfer.* New York: McGraw-Hill, 1961.
5. Heisler, M. P. *Trans. ASME,* **68:**493 (1946); **69:**227 (1947).
6. McAdams, W. H. *Heat Transmission,* 3rd ed. New York: McGraw-Hill, 1954.
7. Mills, A. F. *Heat Transfer.* Homewood, IL: Irwin, 1992.
8. Reid, R. C., J. M. Prausnitz, and B. E. Poling. *The Properties of Gases and Liquids,* 4th ed. New York: McGraw-Hill, 1987.
9. Thomas, L. C. *Heat Transfer.* Englewood Cliffs, NJ: Prentice-Hall, 1992.

CHAPTER 11

Principles of Heat Flow in Fluids

Heat transfer from a warmer fluid to a cooler fluid, usually through a solid wall separating the two fluids, is common in chemical engineering practice. The heat transferred may be latent heat accompanying a phase change such as condensation or vaporization, or it may be sensible heat from the rise or fall in the temperature of a fluid without any phase change. Typical examples are reducing the temperature of a fluid by transfer of sensible heat to a cooler fluid, the temperature of which is increased thereby; condensing steam using cooling water; and vaporizing water from a solution at a given pressure by condensing steam at a higher pressure. All such cases require that heat be transferred by conduction and convection.

TYPICAL HEAT-EXCHANGE EQUIPMENT

To establish a basis for the specific discussion of heat transfer to and from flowing fluids, consider the simple tubular condenser of Fig. 11.1. It consists essentially of a bundle of parallel tubes A, the ends of which are expanded into tube sheets B_1 and B_2. The tube bundle is inside a cylindrical shell C and is provided with two channels D_1 and D_2, one at each end, and two channel covers E_1 and E_2. Steam or other pure vapor is introduced through nozzle F into the shell-side space surrounding the tubes, condensate is withdrawn through connection G, and any noncondensable gas that might be present at start-up or that might enter with the inlet vapor is removed through vent K. Connection G leads to a trap, which is a device that allows liquid to flow but holds back vapor. The fluid to be heated is pumped through connection H into channel D_2. It flows through the tubes into the other channel D_1 and is discharged through connection J. The two fluids are physically separated but are in thermal contact with the thin metal tube walls separating them. Heat flows through the tube walls from the condensing vapor to the cooler fluid in the tubes.

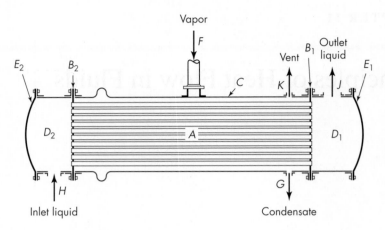

FIGURE 11.1

Single-pass tubular condenser: A, tubes; B_1, B_2, tube sheets; C, shell; D_1, D_2, channels; E_1, E_2, channel covers; F, vapor inlet; G, condensate outlet; H, cold-liquid inlet; J, warm-liquid outlet; K, noncondensed gas vent.

If the vapor entering the condenser is a single component, not a mixture, and is not superheated, and if the condensate is not subcooled below its condensing temperature, then the temperature throughout the shell side of the condenser is constant. The reason for this is that the temperature of the condensing vapor is fixed by the pressure of the shell-side space, and the pressure in that space is constant. The temperature of the fluid in the tubes increases continuously as the fluid flows through the tubes.

The temperatures of the condensing vapor and of the liquid are plotted against the tube length in Fig. 11.2. The horizontal line represents the temperature of the condensing vapor, and the curved line below it represents the rising temperature of the tube-side fluid. In Fig. 11.2, the inlet and outlet fluid temperatures are T_{ca} and T_{cb}, respectively, and the constant temperature of the vapor is T_h. At a length L from the entrance end of the tubes, the fluid temperature is T_c, and the local difference between the temperatures of vapor and fluid is $T_h - T_c$. This temperature difference is called a *point temperature difference* and is denoted by ΔT. The point temperature difference at the inlet of the tubes is $T_h - T_{ca}$, denoted by ΔT_1, and that at the exit end is $T_h - T_{cb}$, denoted by ΔT_2. The terminal point temperature differences ΔT_1 and ΔT_2 are called the *approaches*.

The change in temperature of the fluid $T_{cb} - T_{ca}$ is called the *temperature range* or, simply, the *range*. In a condenser there is but one range, that of the cold fluid being heated.

In this text the symbol ΔT is used exclusively to signify a temperature *difference* between two objects or two fluids. It does *not* denote the temperature change in a given fluid.

A second example of simple heat-transfer equipment is the double-pipe exchanger shown in Fig. 11.3. It is assembled of standard metal pipe and standardized return bends and return heads, the latter equipped with stuffing boxes. One fluid

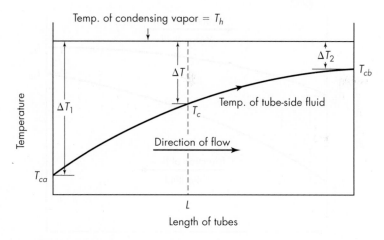

FIGURE 11.2
Temperature–length curves for condenser.

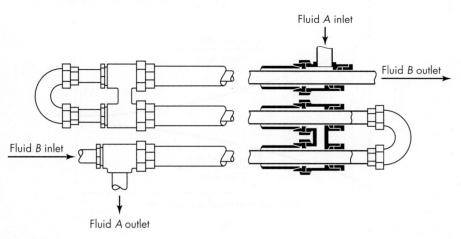

FIGURE 11.3
Double-pipe heat exchanger.

flows through the inside pipe, and the second fluid flows through the annular space between the outside and the inside pipe. The function of a heat exchanger is to increase the temperature of a cooler fluid and decrease that of a hotter fluid. In a typical exchanger, the inner pipe may be $1\frac{1}{4}$ in. and the outer pipe $2\frac{1}{2}$ in., both IPS. Such an exchanger may consist of several passes arranged in a vertical stack. Double-pipe exchangers are useful when not more than 100 to 150 ft^2 of surface is required. For larger capacities, more elaborate shell-and-tube exchangers, containing up to thousands of square feet of area, and described on pages 432 to 436, are used.

Countercurrent and parallel flows

The two fluids enter at different ends of the exchanger, shown in Fig. 11.3, and pass in opposite directions through the unit. This type of flow is that commonly used

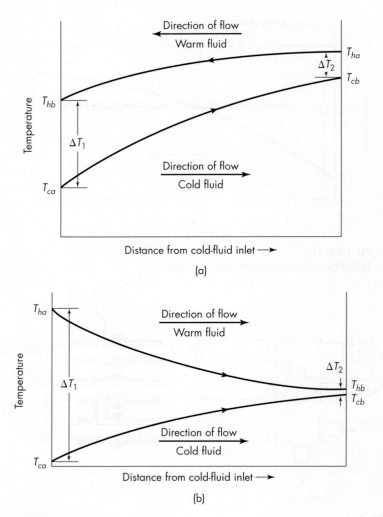

FIGURE 11.4
Temperatures in (*a*) countercurrent flow and (*b*) parallel flow.

and is called *counterflow* or *countercurrert flow*. The temperature–length curves for this case are shown in Fig. 11.4*a*. The four terminal temperatures are denoted as follows:

Temperature of entering hot fluid T_{ha}
Temperature of leaving hot fluid T_{hb}
Temperature of entering cold fluid T_{ca}
Temperature of leaving cold fluid T_{cb}

The approaches are

$$T_{ha} - T_{cb} = \Delta T_2 \qquad \text{and} \qquad T_{hb} - T_{ca} = \Delta T_1 \qquad (11.1)$$

The warm-fluid and cold-fluid ranges are $T_{ha} - T_{hb}$ and $T_{cb} - T_{ca}$, respectively.

If the two fluids enter at the same end of the exchanger and flow in the same direction to the other end, the flow is called *parallel*. The temperature–length curves for parallel flow are shown in Fig. 11.4b. Again, subscript a refers to the entering fluids and subscript b to the leaving fluids. The approaches are $\Delta T_1 = T_{ha} - T_{ca}$ and $\Delta T_2 = T_{hb} - T_{cb}$.

Parallel flow is rarely used in a single-pass exchanger such as that shown in Fig. 11.3 because, as inspection of Fig. 11.4a and b will show, it is not possible with this method of flow to bring the exit temperature of one fluid nearly to the entrance temperature of the other, and the heat that can be transferred is less than that possible in countercurrent flow. In the multipass shell-and-tube exchangers described on pages 435 and 436, counterflow is usually desirable, but parallel flow is used in some passes. Parallel flow is used in special situations in which it is important to change the temperature of one fluid very rapidly, such as when quenching a hot fluid from a chemical reactor to stop further reactions.

In some exchangers one fluid flows across banks of tubes at right angles to the axis of the tubes. This is known as *crossflow*. An automobile radiator and the condenser in a home refrigerator are examples of crossflow heat exchangers.

ENERGY BALANCES

Quantitative attack on heat-transfer problems is based on energy balances and estimations of rates of heat transfer. Rates of transfer are discussed later in this chapter. Many, perhaps most, heat-transfer devices operate under steady-state conditions, and only this type of operation is considered here.

Enthalpy balances in heat exchangers

In heat exchangers there is no shaft work, and mechanical, potential, and kinetic energies are small in comparison with the other terms in the energy-balance equation. Thus, for one stream through the exchanger

$$\dot{m}(H_b - H_a) = q \tag{11.2}$$

where $\dot{m}$ = flow rate of stream
          $q = Q/t$ = rate of heat transfer into stream
          H_a, H_b = enthalpies per unit mass of stream at entrance and exit, respectively

Equation (11.2) can be written for each stream flowing through the exchanger.

A further simplification in the use of the heat-transfer rate q is justified. One of the two fluid streams, that outside the tubes, can gain or lose heat by transfer with the ambient air if the fluid is colder or hotter than the ambient. Heat transfer to or from the ambient is not usually desired in practice, and it is usually reduced to a small magnitude by suitable insulation. It is customary to neglect it in comparison with the heat transfer through the walls of the tubes from the warm fluid to the cold fluid, and q is interpreted accordingly.

Accepting the above assumptions, Eq. (11.2) can be written for the warm fluid as

$$\dot{m}_h(H_{hb} - H_{ha}) = q_h \tag{11.3}$$

and for the cold fluid as

$$\dot{m}_c(H_{cb} - H_{ca}) = q_c \tag{11.4}$$

where $\dot{m}_c, \dot{m}_h$ = mass flow rates of cold fluid and warm fluid, respectively
H_{ca}, H_{ha} = enthalpy per unit mass of entering cold fluid and entering warm fluid, respectively
H_{cb}, H_{hb} = enthalpy per unit mass of leaving cold fluid and leaving hot fluid, respectively
q_c, q_h = rates of heat addition to cold fluid and warm fluid, respectively

The sign of q_c is positive, but that of q_h is negative because the warm fluid loses, rather than gains, heat. The heat lost by the warm fluid is gained by the cold fluid, and

$$q_c = -q_h$$

Therefore, from Eqs. (11.3) and (11.4),

$$\dot{m}_h(H_{ha} - H_{hb}) = \dot{m}_c(H_{cb} - H_{ca}) = q \tag{11.5}$$

Equation (11.5) is called the *overall enthalpy balance*.

If only sensible heat is transferred and constant specific heats are assumed, the overall enthalpy balance for a heat exchanger becomes

$$\dot{m}_h c_{ph}(T_{ha} - T_{hb}) = \dot{m}_c c_{pc}(T_{cb} - T_{ca}) = q \tag{11.6}$$

where c_{pc} = specific heat of cold fluid
c_{ph} = specific heat of warm fluid

Enthalpy balances in total condensers

For a condenser

$$\dot{m}_h \lambda = \dot{m}_c c_{pc}(T_{cb} - T_{ca}) = q \tag{11.7}$$

where $\dot{m}_h$ = rate of condensation of vapor
λ = latent heat of vaporization of vapor

Equation (11.7) is based on the assumption that the vapor enters the condenser as saturated vapor (no superheat) and the condensate leaves at condensing temperature without being further cooled. If either of these sensible-heat effects is important, it must be accounted for by an added term in the left-hand side of Eq. (11.7). For example, if the condensate leaves at a temperature T_{hb} that is less than T_h, the condensing temperature of the vapor, Eq. (11.7) must be written

$$\dot{m}_h[\lambda + c_{ph}(T_h - T_{hb})] = \dot{m}_c c_{pc}(T_{cb} - T_{ca}) \tag{11.8}$$

where c_{ph} is now the specific heat of the condensate.

HEAT FLUX AND HEAT-TRANSFER COEFFICIENTS

Heat-transfer calculations are based on the area of the heating surface and are expressed in watts per square meter or Btu per hour per square foot of surface through which the heat flows. The rate of heat transfer per unit area is called the *heat flux*. In many types of heat-transfer equipment the transfer surfaces are constructed from tubes or pipe. Heat fluxes may then be based on either the inside area or the outside area of the tubes. Although the choice is arbitrary, it must be clearly stated, because the numerical magnitude of the heat fluxes will not be the same for both.

Average temperature of fluid stream

When a fluid is being heated or cooled, the temperature will vary throughout the cross section of the stream. If the fluid is being heated, the temperature of the fluid is a maximum at the wall of the heating surface and decreases toward the center of the stream. If the fluid is being cooled, the temperature is a minimum at the wall and increases toward the center. Because of these temperature gradients throughout the cross section of the stream, it is necessary, for definiteness, to state what is meant by the temperature of the stream. It is agreed that it is the temperature that would be attained if the entire fluid stream flowing across the section in question were withdrawn and mixed adiabatically to a uniform temperature. The temperature so defined is called the *average* or *mixing-cup stream temperature*. The temperatures plotted in Fig. 11.4 are all average stream temperatures.

Overall Heat-Transfer Coefficient

As shown in Chap. 10, Eqs. (10.8) and (10.12), the heat flux through layers of solids in series is proportional to a driving force, the overall temperature difference ΔT. This also applies to heat flow through liquid layers and solids in series. In a heat exchanger the driving force is taken as $T_h - T_c$, where T_h is the average temperature of the hot fluid and T_c is that of the cold fluid. The quantity $T_h - T_c$ is the *overall local temperature difference* ΔT. It is clear from Fig. 11.4 that ΔT can vary considerably from point to point along the tube; and therefore, since the heat flux is proportional to ΔT, the flux also varies with tube length. It is necessary to start with a differential equation by focusing attention on a differential area dA through which a differential heat flow dq occurs under the driving force of a local value of ΔT. The local flux is then dq/dA and is related to the local value of ΔT by the equation

$$\frac{dq}{dA} = U\Delta T = U(T_h - T_c) \tag{11.9}$$

The quantity U, defined by Eq. (11.9) as a proportionality factor between dq/dA and ΔT, is called the *local overall heat-transfer coefficient*.

To complete the definition of U for a tabular exchanger, it is necessary to specify the area. If A is taken as the outside tube area A_o, then U becomes a coefficient based on that area and is written U_o. Likewise, if the inside area A_i is chosen, the coefficient is also based on that area and is denoted by U_i. Since ΔT and dq are independent of the choice of area, it follows that

$$\frac{U_o}{U_i} = \frac{dA_i}{dA_o} = \frac{D_i}{D_o} \tag{11.10}$$

where D_i and D_o are the inside and outside tube diameters, respectively.

In a plate-type heat exchanger the areas on both sides are the same, and there is only one value of U.

Integration over total surface; logarithmic mean temperature difference

To apply Eq. (11.9) to the entire area of a heat exchanger, the equation must be integrated. This can be done formally where certain simplifying assumptions are accepted. The assumptions are that (1) the overall coefficient U is constant, (2) the specific heats of the hot and cold fluids are constant, (3) heat exchange with the ambient is negligible, and (4) the flow is steady and either parallel or counter-current, as shown in Fig. 11.4.

The most questionable of these assumptions is that of a constant overall coefficient. The coefficient does in fact vary with the temperatures of the fluids, but its change with temperature is gradual, so that when the temperature ranges are moderate, the assumption of constant U is not seriously in error.

Assumptions 2 and 4 imply that if T_c and T_h are plotted against q, as shown in Fig. 11.5, straight lines are obtained. Since T_c and T_h vary linearly with q, ΔT does likewise and $d(\Delta T)/dq$, the slope of the graph of ΔT versus q, is constant. Therefore

$$\frac{d(\Delta T)}{dq} = \frac{\Delta T_2 - \Delta T_1}{q_T} \tag{11.11}$$

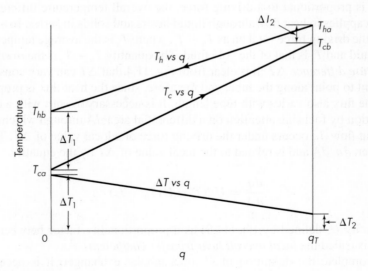

FIGURE 11.5
Temperature versus heat flow rate in countercurrent flow.

where ΔT_1, ΔT_2 = approaches

q_T = rate of heat transfer in entire exchanger

Elimination of dq from Eqs. (11.9) and (11.11) gives

$$\frac{d(\Delta T)}{U \Delta T \, dA} = \frac{\Delta T_2 - \Delta T_1}{q_T} \tag{11.12}$$

The variables ΔT and A can be separated, and if U is constant, the equation can be integrated over the limits A_T and 0 for A and ΔT_2 and ΔT_1, where A_T is the total area of the heat-transfer surface. Thus

$$\int_{\Delta T_1}^{\Delta T_2} \frac{d(\Delta T)}{\Delta T} = \frac{U(\Delta T_2 - \Delta T_1)}{q_T} \int_0^{A_T} dA$$

or

$$\ln \frac{\Delta T_2}{\Delta T_1} = \frac{U(\Delta T_2 - \Delta T_1)}{q_T} A_T \tag{11.13}$$

Equation (11.13) can be written

$$q_T = U A_T \frac{\Delta T_2 - \Delta T_1}{\ln(\Delta T_2/\Delta T_1)} = U A_T \, \overline{\Delta T_L} \tag{11.14}$$

where

$$\overline{\Delta T_L} = \frac{\Delta T_2 - \Delta T_1}{\ln(\Delta T_2/\Delta T_1)} \tag{11.15}$$

Equation (11.15) defines the *logarithmic mean temperature difference* (LMTD). It is of the same form as Eq. (10.18) for the logarithmic mean radius of a thick-walled tube. When ΔT_1 and ΔT_2 are nearly equal, their arithmetic average can be used for $\overline{\Delta T_L}$ within the same limits of accuracy given for Eq. (10.18), as shown in Fig. 10.4.

Equation (11.14) is one of the most important equations for heat transfer to and from fluids. It can be used to predict the performance of a certain heat exchanger, which may require trial-and-error calculations since $\overline{\Delta T_L}$ is not known. It can also be used to calculate the area required for a new exchanger when the flows and temperature approaches are specified. Finally, it is used in the following form to determine the overall coefficient from measured values of q_T and $\overline{\Delta T_L}$:

$$U = \frac{q_T}{A_T \, \overline{\Delta T_L}} \tag{11.16}$$

If one of the fluids is at constant temperature, as in a condenser, no difference exists among countercurrent flow, parallel flow, or multipass flow, and Eq. (11.15) applies to all of them. In countercurrent flow, ΔT_2, the warm-end approach, may be less than ΔT_1, the cold-end approach. In this case, for convenience and to eliminate negative numbers and logarithms, the subscripts in Eq. (11.15) may be interchanged.

Variable overall coefficient

When the overall coefficient varies regularly, the rate of heat transfer may be predicted from Eq. (11.17), which is based on the assumption that U varies linearly with the temperature drop over the entire heating surface[1]:

$$q_T = A_T \frac{U_2 \, \Delta T_1 - U_1 \, \Delta T_2}{\ln(U_2 \, \Delta T_1 / U_1 \, \Delta T_2)} \qquad (11.17)$$

where U_1, U_2 = local overall coefficients at ends of exchanger
   $\Delta T_1, \Delta T_2$ = temperature approaches at corresponding ends of exchanger

Equation (11.17) calls for use of a logarithmic mean value of the $U \, \Delta T$ cross product, where the overall coefficient at one end of the exchanger is multiplied by the temperature approach at the other. The derivation of this equation requires that assumptions 2 to 4 given earlier on page 322 be accepted.

LMTD not always valid

The LMTD is not always the correct mean temperature difference to use. It should *not* be used when U changes appreciably or when ΔT is not a linear function of q. As an example, consider an exchanger used to cool and condense a super-heated vapor, with the temperature diagram shown in Fig. 11.6. The ΔT driving force is a linear function of q while the vapor is being cooled, but ΔT is a different

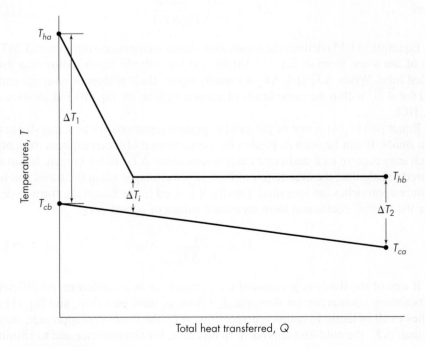

FIGURE 11.6
Temperature profiles in cooling and condensing superheated vapor.

linear function of q in the condensing section of the exchanger. Furthermore, U is not the same in the two parts of the exchanger. The cooling and condensing sections must be sized separately by using the appropriate values of q, U, and LMTD rather than some kind of average U and an overall LMTD.

The LMTD is also incorrect when heat is transferred to or from a reacting fluid in a jacketed reactor. Figure 11.7 shows the temperature profiles for an exothermic reaction in a water-cooled reactor—the lower line shows the temperature of the coolant, the upper line that of the reacting mixture. Because of heat generated by the reaction, the reactant temperature rises rapidly near the reactor inlet, and then, as the reaction slows, the reactant temperature drops. The ΔT's at both the reactor inlet and reactor outlet are relatively small. Clearly the average temperature drop is much greater than the drop at either end of the reactor and cannot be found from the logarithmic mean of the terminal ΔT's.

For the reactor example, or for other cases where one or more of the four assumptions listed on page 322 is not valid, Eq. (11.9) can be integrated by evaluating local values of U, ΔT, and q at several intermediate points in the exchanger. Graphical or numerical evaluation of the area under a plot of $1/U \Delta T$ versus q, between the limits of zero and q_T, will then give the area A_T of the heat-transfer surface required.

Multipass exchangers

In multipass shell-and-tube exchangers the flow pattern is complex, with parallel, countercurrent, and crossflow all present. Under these conditions, even when the overall coefficient U is constant, the LMTD cannot be used. Calculation procedures for multipass exchangers are given in Chap. 15.

Individual Heat-Transfer Coefficients

The overall coefficient depends on many variables including the physical properties of the fluids and of the solid wall, the flow rates, and the exchanger dimensions. The only logical way to predict the overall coefficient is to use correlations for the individual resistances of the solid and the fluid layers and to add these resistances to find the overall resistance, which is the reciprocal of the overall coefficient. Consider the

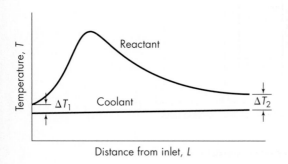

FIGURE 11.7
Temperature patterns in jacketed tubular reactor.

local overall coefficient at a specific point in the double-pipe exchanger shown in Fig. 11.3. For definiteness, assume that the warm fluid is flowing through the inside pipe and that the cold fluid is flowing through the annular space. Assume also that the Reynolds numbers of the two fluids are sufficiently large to ensure turbulent flow and that both surfaces of the inside tube are clear of dirt or scale. If now a plot is prepared, as shown in Fig. 11.8, with temperature as the ordinate and distance perpendicular to the wall as the abscissa, several important facts become evident. In the figure, the metal wall of the tube separates the warm fluid on the right from the cold fluid on the left. The change in temperature with distance is shown by the line $T_a T_b T_{wh} T_{wc} T_e T_g$. The temperature profile is thus divided into three separate parts,

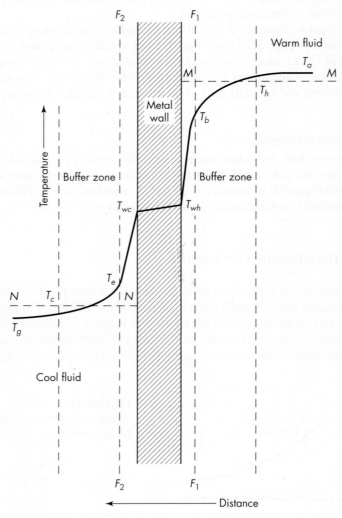

FIGURE 11.8
Temperature gradients in forced convection.

one through each of the two fluids and the other through the metal wall. The overall effect, therefore, should be studied in terms of these individual parts.

It was shown in Chap. 5 that in turbulent flow through conduits three zones exist, even in a single fluid, so that the study of one fluid is itself complicated. In each fluid shown in Fig. 11.8 there is a thin sublayer at the wall, a turbulent core occupying most of the cross section of the stream, and a buffer zone between them. The velocity gradients were described in Chap. 5. The velocity gradient is large near the wall, small in the turbulent core, and in rapid change in the buffer zone. It has been found that the temperature gradient in a fluid being heated or cooled when flowing in turbulent flow follows much the same course. The temperature gradient is large at the wall and through the viscous sublayer, small in the turbulent core, and in rapid change in the buffer zone. Basically, the reason for this is that heat must flow through the viscous sublayer by conduction, which calls for a steep temperature gradient in most fluids because of the low thermal conductivity, whereas the rapidly moving eddies in the core are effective in equalizing the temperature in the turbulent zone. In Fig. 11.8 the dashed lines F_1F_1 and F_2F_2 represent the boundaries of the viscous sublayers and of the buffer zone.

The average temperature of the warm stream is somewhat less than the maximum temperature T_a and is represented by the horizontal line MM, which is drawn at temperature T_h. Likewise, line NN, drawn at temperature T_c, represents the average temperature of the cold fluid.

The overall resistance to the flow of heat from the warm fluid to the cold fluid is a result of three separate resistances operating in series. In general, as shown in Fig. 11.8, the wall resistance is small in comparison with that of the fluids. The fluid resistances are generally computed by using correlations for individual heat-transfer coefficients or film coefficients, which are the reciprocals of the resistances.

The film coefficient for the warm fluid is defined by the equation

$$h = \frac{dq/dA}{T_h - T_w} \tag{11.18}$$

For the cold fluid the terms in the denominator are reversed to make h positive

$$h = \frac{dq/dA}{T_w - T_c} \tag{11.19}$$

where dq/dA = local heat flux, based on area in contact with fluid
T_h = local average temperature of warm fluid
T_c = local average temperature of cold fluid
T_w = temperature of wall in contact with fluid

The reciprocals of these coefficients, $1/h_h$ and $1/h_c$, are the thermal resistances. For conduction through a solid, such as a metal wall of thickness x_w and thermal conductivity k, the thermal resistance equals x_w/k. Appropriately corrected for changes in area, the individual resistances may be added to give the overall resistance $1/U$.

A second expression for h is derived from the assumption that heat transfer very near the wall occurs only by conduction, and the heat flux is given by Eq. (10.1), noting that the normal distance x may be replaced by y, the distance normal to the wall in Fig. 11.8. Thus

$$\frac{dq}{dA} = -k\left(\frac{dT}{dy}\right)_w \tag{11.20}$$

The subscript w calls attention to the fact that the gradient must be evaluated at the wall. Eliminating dq/dA from Eqs. (11.19) and (11.20) gives

$$h = -k\frac{(dT/dy)_w}{T - T_w} \tag{11.21}$$

In Eq. (11.21), T is the average fluid temperature, which is T_h for the warm side and T_c for the cool side. The denominator is changed to $T_w - T$ for the cool side to make h positive.

Equation (11.21) can be put into a dimensionless form by multiplying by the ratio of an arbitrary length to the thermal conductivity. The choice of length depends on the situation. For heat transfer at the inner surface of a tube, the tube diameter D is the usual choice. Multiplying Eq. (11.21) by D/k gives

$$\text{Nu} = \frac{hD}{k} = -D\frac{(dT/dy)_w}{T - T_w} \tag{11.22}$$

On the cold-fluid side of the tube wall $T < T_w$, and the denominator in Eqs. (11.21) and (11.22) becomes $T_w - T$. The dimensionless group hD/k is called a Nusselt number Nu. That shown in Eq. (11.22) is a local Nusselt number based on diameter. The physical meaning of the Nusselt number can be seen by inspection of the right-hand side of Eq. (11.22). The numerator $(dT/dy)_w$ is, of course, the gradient at the wall. The factor $(T - T_w)/D$ can be considered the average temperature gradient across the entire pipe, and the Nusselt number is the ratio of these two gradients.

Another interpretation of the Nusselt number can be obtained by considering the gradient that would exist if all the resistance to heat transfer were in a laminar layer of thickness x in which heat transfer was only by conduction. The heat-transfer rate and coefficient follow from Eqs. (10.1) and (11.18)

$$\frac{dq}{dA} = \frac{k(T - T_w)}{x} \tag{11.23}$$

$$h = \frac{k}{x} \tag{11.24}$$

From the definition of the Nusselt number,

$$\frac{hD}{k} = \text{Nu} = \frac{k}{x}\frac{D}{k} = \frac{D}{x} \tag{11.25}$$

The Nusselt number is the ratio of the tube diameter to the equivalent thickness of the laminar layer. Sometimes x is called the film thickness, and it is generally slightly greater than the thickness of the laminar boundary layer because there is some resistance to heat transfer in the buffer zone.

Equation (11.18), when applied to the two fluids of Fig. 11.8, becomes, for the inside of the tube (the warm side in Fig. 11.8),

$$h_i = \frac{dq/dA_i}{T_h - T_{wh}} \tag{11.26}$$

and for the outside of the tube (the cold side), from Eq. (11.19)

$$h_o = \frac{dq/dA_o}{T_{wc} - T_c} \tag{11.27}$$

where A_i, and A_o are the inside and outside areas of the tube, respectively.

The cold fluid could, of course, be inside the tubes and the warm fluid outside. Coefficients h_i and h_o refer to the *inside* and the *outside* of the tube, respectively, and not to a specific fluid.

Calculation of overall coefficients from individual coefficients

The overall coefficient is constructed from the individual coefficients and the resistance of the tube wall in the following manner. The rate of heat transfer through the tube wall is given by the differential form of Eq. (10.16),

$$\frac{dq}{d\bar{A}_L} = \frac{k_m(T_{wh} - T_{wc})}{x_w} \tag{11.28}$$

where $T_{wh} - T_{wc}$ = temperature difference through tube wall
$\quad k_m$ = thermal conductivity of wall
$\quad x_w$ = tube wall thickness
$\quad dq/d\bar{A}_L$ = local heat flux, based on logarithmic mean of inside and outside areas of tube

If Eqs. (11.26) to (11.28) are solved for the temperature differences and the temperature differences added, the result is

$$(T_h - T_{wh}) + (T_{wh} - T_{wc}) + (T_{wc} - T_c) = T_h - T_c = \Delta T$$

$$= dq\left(\frac{1}{dA_i\, h_i} + \frac{x_w}{d\bar{A}_L\, k_m} + \frac{1}{dA_o\, h_o}\right) \tag{11.29}$$

Assume that the heat-transfer rate is arbitrarily based on the outside area. If Eq. (11.29) is solved for dq, and if both sides of the resulting equation are divided by dA_o, the result is

$$\frac{dq}{dA_o} = \frac{T_h - T_c}{\dfrac{1}{h_i}\left(\dfrac{dA_o}{dA_i}\right) + \dfrac{x_w}{k_m}\left(\dfrac{dA_o}{d\bar{A}_L}\right) + \dfrac{1}{h_o}} \tag{11.30}$$

Now
$$\frac{dA_o}{dA_i} = \frac{D_o}{D_i} \qquad \text{and} \qquad \frac{dA_o}{d\bar{A}_L} = \frac{D_o}{\bar{D}_L}$$

where D_o, D_i, and $\bar{D}_L$ are the outside, inside, and logarithmic mean diameters of the tube, respectively. Therefore

$$\frac{dq}{dA_o} = \frac{T_h - T_c}{\frac{1}{h_i}\left(\frac{D_o}{D_i}\right) + \frac{x_w}{k_m}\left(\frac{D_o}{\bar{D}_L}\right) + \frac{1}{h_o}} \tag{11.31}$$

Comparing Eq. (11.9) with Eq. (11.31) shows that

$$U_o = \frac{1}{\frac{1}{h_i}\left(\frac{D_o}{D_i}\right) + \frac{x_w}{k_m}\left(\frac{D_o}{\bar{D}_L}\right) + \frac{1}{h_o}} \tag{11.32}$$

If the inside area A_i is chosen as the base area, division of Eq. (11.29) by dA_i gives for the overall coefficient

$$U_i = \frac{1}{\frac{1}{h_i} + \frac{x_w}{k_m}\left(\frac{D_i}{\bar{D}_L}\right) + \frac{1}{h_o}\left(\frac{D_i}{D_o}\right)} \tag{11.33}$$

Resistance form of overall coefficient

A comparison of Eqs. (10.12) and (11.32) suggests that the reciprocal of an overall coefficient can be considered to be an overall resistance composed of three resistances in series. The total, or overall, resistance is given by the equation

$$R_o = \frac{1}{U_o} = \frac{D_o}{D_i h_i} + \frac{x_w}{k_m}\frac{D_o}{\bar{D}_L} + \frac{1}{h_o} \tag{11.34}$$

The individual terms on the right-hand side of Eq. (11.34) represent the individual resistances of the two fluids and of the metal wall. The overall temperature drop is proportional to $1/U$, and the temperature drops in the two fluids and the wall are proportional to the individual resistances, or, for the case of Eq. (11.34),

$$\frac{\Delta T}{1/U_o} = \frac{\Delta T_i}{D_o/D_i h_i} = \frac{\Delta T_w}{(x_w/k_m)(D_o/\bar{D}_L)} = \frac{\Delta T_o}{1/h_o} \tag{11.35}$$

where ΔT = overall temperature drop
ΔT_i = temperature drop through inside fluid
ΔT_w = temperature drop through metal wall
ΔT_o = temperature drop through outside fluid

Equation (11.35) can be used to determine the wall temperatures T_{wh} and T_{wc} by calculating ΔT_i or ΔT_o and adding these values to, or subtracting them from, T_i or T_o.

Fouling factors

In actual service, heat-transfer surfaces do not remain clean. Scale, dirt, and other solid deposits form on one or both sides of the tubes, provide additional resistances to heat flow, and reduce the overall coefficient. The effect of such deposits is taken into account by adding a term $1/(dA\,h_d)$ to the term in parentheses in Eq. (11.29) for each scale deposit. Thus, assuming that scale is deposited on both the inside and the outside surface of the tubes, Eq. (11.29) becomes, after correction for the effects of scale,

$$\Delta T = dq\left(\frac{1}{dA_i\,h_{di}} + \frac{1}{dA_i\,h_i} + \frac{x_w}{d\bar{A}_L\,k_m} + \frac{1}{dA_o\,h_o} + \frac{1}{dA_o h_{do}}\right) \quad (11.36)$$

where h_{di} and h_{do} are the *fouling factors* for the scale deposits on the inside and outside tube surfaces, respectively. The following equations for the overall coefficients based on outside and inside areas, respectively, follow from Eq. (11.36):

$$U_o = \frac{1}{D_o/D_i h_{di} + D_o/D_i h_i + (x_w/k_w)(D_o/\bar{D}_L) + 1/h_o + 1/h_{do}} \quad (11.37)$$

and

$$U_i = \frac{1}{1/h_{di} + 1/h_i + (x_w/k_m)(D_i/\bar{D}_L) + D_i/D_o h_o + D_i/D_o h_{do}} \quad (11.38)$$

The actual thicknesses of the deposits are neglected in Eqs. (11.37) and (11.38).

Numerical values of fouling factors are given in Reference 3 corresponding to satisfactory performance in normal operation, with reasonable service time between cleanings. They cover a range of approximately 600 to 11,000 W/m² · °C (100 to 2,000 Btu/ft² · h · °F). Fouling factors for ordinary industrial liquids fall in the range of 1,700 to 6,000 W/m² · °C (300 to 1000 Btu/ft² · h · °F). Fouling factors are usually set at values that also provide a safety factor for design.

EXAMPLE 11.1. Methyl alcohol flowing in the inner pipe of a double-pipe exchanger is cooled with water flowing in the jacket. The inner pipe is made from 1-in. (25-mm) Schedule 40 steel pipe. The thermal conductivity of steel is 26 Btu/ft·h·°F (45 W/m·°C). The individual coefficients and fouling factors are given in Table 11.1. What is the overall coefficient, based on the outside area of the inner pipe?

Solution. The diameters and wall thickness of 1-in. Schedule 40 pipe, from App. 3, are

$$D_i = \frac{1.049}{12} = 0.0874\text{ ft} \qquad D_o = \frac{1.315}{12} = 0.1096\text{ ft} \qquad x_w = \frac{0.133}{12} = 0.0111\text{ ft}$$

TABLE 11.1
Data for Example 11.1

	Coefficient	
	Btu/ft² · h · °F	W/m² · °C
Alcohol coefficient h_i	180	1,020
Water coefficient h_o	600	1,700
Inside fouling factor h_{di}	1,000	5,680
Outside fouling factor h_{do}	500	2,840

The logarithmic mean diameter $\bar{D}_L$ is calculated as in Eq. (10.18), using diameter in place of radius:

$$\bar{D}_L = \frac{D_o - D_i}{\ln(D_o/D_i)} = \frac{0.1096 - 0.0874}{\ln(0.1096/0.0874)} = 0.0983 \text{ ft}$$

The overall coefficient is found from Eq. (11.37):

$$U_o = \cfrac{1}{\cfrac{0.1096}{0.0874 \times 1,000} + \cfrac{0.1096}{0.0874 \times 180} + \cfrac{0.0111 \times 0.1096}{26 \times 0.0983} + \cfrac{1}{600} + \cfrac{1}{500}}$$

$$= 80.9 \text{ Btu/ft}^2 \cdot \text{h} \cdot {}^\circ\text{F} \ (459 \text{ W/m}^2 \cdot {}^\circ\text{C})$$

Special cases of the overall coefficient

Although the choice of area to be used as the basis of an overall coefficient is arbitrary, sometimes one particular area is more convenient than others. Suppose, for example, that one individual coefficient h_i is large numerically in comparison with the other h_o, and that fouling effects are negligible. Also, assuming the term representing the resistance of the metal wall is small in comparison with $1/h_o$, the ratios D_o/D_i and $D_o/\bar{D}_L$ have so little significance that they can be disregarded, and Eq. (11.32) can be replaced by the simpler form

$$U_o = \frac{1}{1/h_o + x_w/k_m + 1/h_i} \tag{11.39}$$

In such a case, it is advantageous to base the overall coefficient on that area that corresponds to the largest resistance, or the lowest value of h.

For large-diameter thin-walled tubes, flat plates or any other case in which a negligible error is caused by using a common area for A_i, $\bar{A}_L$, and A_o, Eq. (11.39) can be used for the overall coefficient, and U_i and U_o are identical.

Sometimes one coefficient, say, h_o, is so very small in comparison with both x_w/k and the other coefficient h_i that the term $1/h_o$ is very large compared with the other terms in the resistance sum. When this is true, the larger resistance is called the *controlling resistance,* and it is sufficiently accurate to equate the overall coefficient to the small individual coefficient, or in this case, $h_o = U_o$.

Classification of individual heat-transfer coefficients

The problem of predicting the rate of heat flow from one fluid to another through a retaining wall reduces to the problem of predicting the numerical values of the individual coefficients of the fluids concerned in the overall process. A wide variety of individual cases are met in practice, and each type of phenomenon must be considered separately. The following classification is followed in this text:

1. Heat flow to or from fluids inside tubes, without phase change
2. Heat flow to or from fluids outside tubes, without phase change
3. Heat flow from condensing fluids
4. Heat flow to boiling liquids

TABLE 11.2
Magnitudes of heat-transfer coefficients

	Range of values of h	
Type of processes	W/m$^2 \cdot$ °C	Btu/ft$^2 \cdot$ h $\cdot$ °F
Steam (dropwise condensation)	30,000–100,000	5,000–20,000
Steam (film-type condensation)	6,000–20,000	1,000–3,000
Boiling water	1,700–50,000	300–9,000
Condensing organic vapors	1,000–2,000	200–400
Water (heating or cooling)	300–20,000	50–3,000
Oils (heating or cooling)	50–1,500	10–300
Steam (superheating)	30–100	5–20
Air (heating or cooling)	1–50	0.2–10

To convert from Btu/ft$^2 \cdot$ h $\cdot$ °F to W/m$^2 \cdot$ °C, multiply by 5.6783.

Source: By permission of author and publisher from W. H. McAdams, *Heat Transmission,* 3rd ed., p. 5. Copyright by author, 1954, McGraw-Hill Book Company.

Magnitude of heat-transfer coefficients

The ranges of values covered by the coefficient h vary greatly, depending upon the character of the process.[2] Some typical ranges are shown in Table 11.2.

SYMBOLS

A	Area m^2 or ft^2; A_T, total area of heat-transfer surface; A_i, of inside of tube; A_o, of outside of tube; $\bar{A}_L$, logarithmic mean
c_p	Specific heat at constant pressure, J/g $\cdot$ °C or Btu/lb $\cdot$ °F; c_{pc}, of cool fluid; C_{ph}, of warm fluid
D	Diameter, m or ft; D_i, inside diameter of tube; D_o, outside diameter of tube; $\bar{D}_L$, logarithmic mean
H	Enthalpy, J/g or Btu/lb; H_a, at entrance; H_b, at exit; H_{ca}, H_{cb}, of cool fluid; H_{ha}, H_{hb}, of warm fluid
h	Individual or surface heat-transfer coefficient, W/m$^2 \cdot$ °C or Btu/ft$^2 \cdot$ h $\cdot$ °F; h_i, for inside of tube; h_o, for outside of tube
h_d	Fouling factor, W/m$^2 \cdot$ °C or Btu/ft$^2 \cdot$ h $\cdot$ °F; h_{di}, inside tube; h_{do}, outside tube
k	Thermal conductivity, W/m $\cdot$ °C or Btu/ft $\cdot$ h $\cdot$ °F; k_m, of tube wall
L	Length, m or ft
LMTD	Logarithmic mean temperature difference
$\dot{m}$	Mass flow rate, kg/h or lb/h; $\dot{m}_c$, of cool fluid; $\dot{m}_h$, of warm fluid
Nu	Nusselt number, hD/k, dimensionless
Q	Quantity of heat, J or Btu
q	Heat flow rate, W or Btu/h; q_T, total in exchanger; q_c, to cool fluid; q_h, to warm fluid
R	Overall thermal resistance, $1/U$, m$^2 \cdot$ °C/W or ft$^2 \cdot$ h $\cdot$ °F/Btu; R_o, based on outside surface area

r_m Radius of spherical particle, m or ft

T Temperature, °F or °C; T_a, at inlet, or initial value; T_b, at outlet; T_c, of
cool fluid; T_{ca}, at cool-fluid inlet; T_{cb}, at cool-fluid outlet; T_h, of warm
fluid; T_{ha}, at warm-fluid inlet; T_{hb}, at warm-fluid outlet; T_s, of surface;
T_w, of tube wall; T_{wc}, on cool-fluid side; T_{wh}, on warm-fluid side; $\bar{T}_b$, bulk
average temperature of solid sphere

t Time, h or s

U Overall heat-transfer coefficient, W/m² · °C or Btu/ft² · h · °F; U_i, based on
inside surface area; U_o, based on outside surface area; U_1, U_2, at ends of
exchanger

x Film thickness, m or ft [Eqs. (11.23) to (11.25)]

x_w Thickness of tube wall, m or ft

y Distance into fluid normal to tube wall, m or ft, measured in direction of
heat flow

Greek Letters

α Thermal diffusivity, $k/\rho c_p$, m²/s or ft²/h

ΔT Overall temperature difference $T_h - T_c$, °C or °F; ΔT_i, between tube wall
and fluid inside tube; ΔT_o, between tube wall and fluid outside tube; ΔT_w,
through the tube wall; ΔT_1, ΔT_2, at ends of exchanger; $\overline{\Delta T_L}$, logarithmic
mean

λ Latent heat of vaporization, J/g or Btu/lb

ρ Density of spherical particle, kg/m³ or lb/ft³

PROBLEMS

11.1. Calculate the overall heat-transfer coefficients based on both inside and outside areas
for the following cases.
Case 1   Water at 10°C flowing in a $\frac{3}{4}$-in. 16 BWG condenser tube and saturated steam
at 105°C condensing on the outside. $h_i = 12$ kW/m² · °C. $h_o = 14$ kW/m² · °C. $k_m = 120$ W/m · °C.
Case 2   Benzene condensing at atmospheric pressure on the outside of a 25-mm steel
pipe and air at 15°C flowing within at 6 m/s. The pipe wall is 3.5 mm thick. $h_i = 20$ W/m² · °C. $h_o = 1200$ W/m² · °C. $k_m = 45$ W/m · °C.
Case 3   Dropwise condensation from steam at a pressure of 50 lb$_f$ /in.² gauge on a
1-in. Schedule 40 steel pipe carrying oil at 100°F. $h_o = 14{,}000$ Btu/ft² · h · °F. $h_i = 130$ Btu/ft² · h · °F. $k_m = 26$ Btu/ft · h · °F.

11.2. Calculate the temperatures of the inside and outside surfaces of the metal pipe or tubing in cases 1 to 3 of Prob. 11.1.

11.3. Aniline is to be cooled from 200 to 150°F in a double-pipe heat exchanger having a
total outside area of 70 ft². For cooling, a stream of toluene amounting to 8,600 lb/h at

a temperature of 100°F is available. The exchanger consists of $1\frac{1}{4}$-in. Schedule 40 pipe in 2-in. Schedule 40 pipe. The aniline flow rate is 10,000 lb/h. (*a*) If flow is counter-current, what are the toluene outlet temperature, the LMTD, and the overall heat-transfer coefficient? (*b*) What are they if flow is parallel?

11.4. In the exchanger described in Prob. 11.3, how much aniline can be cooled if the over-all heat-transfer coefficient is 70 Btu/ft²·h·°F?

11.5. Carbon tetrachloride flowing at 19,000 kg/h is to be cooled from 85 to 40°C using 13,500 kg/h of cooling water at 20°C. The film coefficient for carbon tetrachloride, outside the tubes, is 1,700 W/m²·°C. The wall resistance is negligible, but h_i, on the water side, including fouling factors, is 11,000 W/m²·°C. (*a*) What area is needed for a counterflow exchanger? (*b*) By what factor would the area be increased if parallel flow were used to get more rapid initial cooling of the carbon tetrachloride?

11.6. In a double-pipe heat exchanger used to cool a slurry, the overall coefficient U_i is 1.84 kW/m²·°C. Water flows in the inner pipe, which is 1-in. Schedule 40 steel pipe; the slurry flows in the outer 2-in. pipe. The inside coefficient is estimated to be 4.8 kW/m²·°C based on a published correlation for turbulent flow. (*a*) Calculate the outside coefficient and the percentage of the total resistance in the outside film, the wall, and the inside film. (*b*) If the correlation for h_i has a possible error of ±10 percent and U_i is accurate to only 10 percent, what is the maximum possible error in h_o?

11.7. A tubular condenser with 120 $\frac{3}{4}$-in. BWG 16 copper tubes 8 ft long condenses 22,100 lb/h of saturated steam at 120°C using 870 gal/min of cooling water. The mea-sured inlet and exit water temperatures are 20 and 46°C. (*a*) How good is the heat bal-ance? (*b*) What is the overall heat-transfer coefficient based on the outside area of the tubes?

11.8. In a counterflow heat exchanger, the hot stream is cooled from 120 to 30°C while the cold stream temperature changes from 20 to 60°C. If the same exchanger were oper-ated with parallel flow, what would be the exit temperatures of the two streams?

11.9. An organic vapor is to be condensed at 80°C and 1.2 atm in a water-cooled condenser, and the condensed organic liquid cooled to 35°C in the same condenser. Cooling water is available at 25°C, and the exit temperature of the water should be about 45°C. Sketch the temperature profiles in the condenser cooler, and explain how you would calculate the required area.

REFERENCES

1. Colburn, A. P. *Ind. Eng. Chem.*, **25**:873 (1933).
2. McAdams, W. H. *Heat Transmission*, 3rd ed. New York: McGraw-Hill, 1954, p. 5.
3. Perry, J. H. (ed.). *Chemical Engineers' Handbook*, 6th ed. New York: McGraw-Hill, 1984, p. **10**-43.

CHAPTER 12

Heat Transfer to Fluids Without Phase Change

In a great many applications of heat exchange, heat is transferred between fluid streams without any phase change in the fluids. This is especially important in heat recovery operations, such as when the hot effluent from an exothermic reactor is used to preheat the incoming cooler feed. Other examples include the transfer of heat from a stream of hot gas to cooling water, and the cooling of a hot liquid stream by air. In such situations the two streams are separated by a metal wall, which constitutes the heat-transfer surface. The surface may consist of tubes or other channels of constant cross section, of flat plates, or in such devices as jet engines and advanced power machinery, of special shapes designed to pack a maximum area of transfer surface into a small volume.

Most fluid-to-fluid heat transfer is accomplished in steady-state equipment, but thermal regenerators, in which a bed of solid shapes is alternately heated by a hot fluid and the hot shapes then used to warm a colder fluid, are also used, especially in high-temperature heat transfer. Regenerator performance is discussed in Chap. 15.

BOUNDARY LAYERS

Regimes of heat transfer

A fluid being heated or cooled may be flowing in laminar flow, in turbulent flow, or in the transition range between laminar and turbulent flow. Also, the fluid may be flowing in forced or natural convection. In some instances more than one flow type may occur in the same stream; for instance, in laminar flow at low velocities, natural convection may be superimposed on forced laminar flow.

The direction of flow of the fluid may be parallel to that of the heating surface, so that boundary layer separation does not occur; or the direction of flow may be perpendicular or at an angle to the heating surface, and then boundary layer separation often occurs.

At ordinary velocities the heat generated from fluid friction is usually negligible in comparison with the heat transferred between the fluids. In most cases friction heating may be neglected. It may be important, however, in operations involving very viscous fluids such as the injection molding of polymers. Friction heating of crude oil in the Alaska pipeline helps keep the oil above the ambient temperature. This decreases the viscosity and lowers the pumping cost. In gas flow at high velocities, at Mach numbers above a few tenths, friction heat becomes appreciable and cannot be ignored. At very high velocities, friction heating may become of controlling importance.

Because the conditions of flow at the entrance to a tube differ from those well downstream from the entrance, the velocity field and the associated temperature field may depend on the distance from the tube entrance. Also, in some situations the fluid flows through a preliminary length of unheated or uncooled pipe, so that the fully developed velocity field is established before heat is transferred to the fluid, and the temperature field is created within an existing velocity field.

Finally, the properties of the fluid—viscosity, thermal conductivity, specific heat, and density—are important parameters in heat transfer. Each of these, especially viscosity, is temperature-dependent. Since the temperature varies from point to point in a flowing stream undergoing heat transfer, a problem appears in the choice of temperature at which the properties should be evaluated. For small temperature differences between fluid and wall and for fluids with weak dependence of viscosity on temperature, the problem is not acute. But for highly viscous fluids such as heavy petroleum oils or where the temperature difference between the tube wall and the fluid is large, the variations in fluid properties within the stream become large, and the difficulty of calculating the heat-transfer rate is increased.

Because of the various effects noted above, the entire subject of heat transfer to fluids without phase change is complex and in practice is treated as a series of special cases rather than as a general theory. All cases considered in this chapter do, however, have a phenomenon in common: In all the formation of a thermal boundary layer, analogous to the hydrodynamic Prandtl boundary layer described in Chap. 3, takes place; it profoundly influences the temperature field and so controls the rate of heat flow.

Thermal boundary layer

Consider a flat plate immersed in a stream of fluid that is in steady flow parallel to the plate, as shown in Fig. 12.1a. Assume that the stream approaching the plate does so at velocity u_0 and temperature T_∞ and that the surface of the plate is maintained at a constant temperature T_w. Assume that T_w is greater than T_∞, so that the fluid is heated by the plate. As described in Chap. 3, a boundary layer develops within which the velocity varies from $u = 0$ at the wall to $u = u_0$ at the outer boundary of the layer. This boundary layer, called the *hydrodynamic boundary layer*, is shown by line OA in Fig. 12.1a. The penetration of heat by transfer from the plate to the fluid changes the temperature of the fluid near the surface of the plate, and a temperature gradient is generated. The temperature gradient also is confined to a

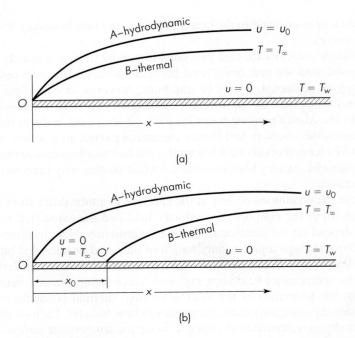

FIGURE 12.1
Thermal and hydrodynamic boundary layers on flat plate:
(a) entire plate heated; (b) unheated length $= x_0$.

layer next to the wall, and within the layer the temperature varies from T_w at the wall to T_∞ at its outside boundary. This layer, called the *thermal boundary layer,* is shown as line OB in Fig. 12.1a. As drawn, lines OA and OB show that the thermal boundary layer is thinner than the hydrodynamic layer at all values of x, where x is the distance from the leading edge of the plate.

Prandtl number

The relationship between the thicknesses of the two boundary layers at a given point along the plate depends on the dimensionless Prandtl number, which is the ratio of the diffusivity of momentum ν or μ/ρ to the thermal diffusivity α or $k/\rho c_p$. Thus

$$\text{Pr} \equiv \frac{\nu}{\alpha} = \frac{c_p \mu}{k} \tag{12.1}$$

When the Prandtl number is greater than unity, which is true for most liquids, the thermal boundary layer is thinner than the hydrodynamic layer because of the relatively low rate of heat conduction. This situation is illustrated in Fig. 12.1a.

The Prandtl number for water at 70°C is about 2.5; for viscous liquids and concentrated solutions it may be as large as 600. Prandtl numbers for liquids change appreciably with temperature because of the change in viscosity. With a high-viscosity fluid the hydrodynamic boundary layer extends farther from the surface of the plate, which can perhaps be understood intuitively. Imagine moving a flat plate through a very viscous liquid such as glycerol: Fluid at a considerable distance from the plate will be set in motion, which means a thick boundary layer.

The Prandtl number of a gas is usually close to 1.0 (0.69 for air, 1.06 for steam), and the two boundary layers have about the same thickness. The Prandtl number for gases is almost independent of temperature because the viscosity and thermal conductivity both increase with temperature at about the same rate. Prandtl numbers for gases and liquids are given in Apps. 16 and 17.

Liquid metals have very low Prandtl numbers, ranging from about 0.01 to 0.04, because of the high thermal conductivity. The temperature gradients extend well beyond the hydrodynamic boundary layer, and special correlations are needed to predict the rate of heat transfer.

In Fig. 12.1a it is assumed that the entire plate is heated and that both boundary layers start at the leading edge of the plate. If the first section of the plate is not heated and if the heat-transfer area begins at a definite distance x_0 from the leading edge, as shown by line $O'B$ in Fig. 12.1b, a hydrodynamic boundary layer already exists at x_0, where the thermal boundary layer begins to form.

The sketches in Fig. 12.1 exaggerate the thickness of the boundary layers for clarity. The actual thicknesses are usually a few percent of the distance from the leading edge of the plate.

In flow through a tube, it has been shown in Chap. 3 that the hydrodynamic boundary layer thickens as the distance from the tube entrance increases, and finally the layer reaches the center of the tube. The velocity profile so developed, called *fully developed flow,* establishes a velocity distribution that is unchanged with additional pipe length. The thermal boundary layer in a heated or cooled tube also reaches the center of the tube at a definite length from the entrance of the heated length of the tube, and the temperature profile is considered fully developed at this point since it covers the entire pipe. Unlike the velocity profile, however, the temperature profile flattens as the length of the tube increases; and in very long pipes the entire fluid stream reaches the temperature of the tube wall, the temperature gradients disappear, and heat transfer ceases.

HEAT TRANSFER BY FORCED CONVECTION IN LAMINAR FLOW

In laminar flow, heat transfer occurs only by conduction, as there are no eddies to carry heat by convection across an isothermal surface. The problem is amenable to mathematical analysis based on the partial differential equations for continuity, momentum, and energy. Such treatments are beyond the scope of this book and are given in standard treatises on heat transfer.[6a] Mathematical solutions depend on the boundary conditions established to define the conditions of fluid flow and heat transfer. When the fluid approaches the heating surface, it may have an already completed hydrodynamic boundary layer or a partially developed one. Or the fluid may approach the heating surface at a uniform velocity, and both boundary layers may be initiated at the same time. A simple flow situation where the velocity is assumed constant in all cross sections and tube lengths is called *plug* or *rodlike flow*.

Independent of the conditions of flow, (1) the heating surface may be isothermal; or (2) the heat flux may be equal at all points on the heating surface, in which case the average temperature of the fluid varies linearly with tube length. Other combinations of boundary conditions are possible.[6a] The basic differential equation for the several special cases is the same, but the final integrated relationships differ.

Most of the simpler mathematical derivations are based on the assumptions that the fluid properties are constant and temperature-independent and that flow is truly laminar with no crosscurrents or eddies. These assumptions are valid when temperature changes and gradients are small, but with large temperature changes the simple model is not in accord with physical reality for two reasons. First, variations in viscosity across the tube distort the usual parabolic velocity distribution profile of laminar flow. Thus, if the fluid is a liquid and is being heated, the layer near the wall has a lower viscosity than the layers near the center and the velocity gradient at the wall increases. A crossflow of liquid toward the wall is generated. If the liquid is being cooled, the reverse effect occurs. Second, since the temperature field generates density gradients, natural convection may set in, which further distorts the flow lines of the fluid. The effect of natural convection may be small or large, depending on a number of factors to be discussed in the section on natural convection.

In this section three types of heat transfer in laminar flow are considered: (1) heat transfer to a fluid flowing along a flat plate, (2) heat transfer in plug flow in tubes, and (3) heat transfer to a fluid stream that is in fully developed flow at the entrance to the tube. In all cases, the temperature of the heated length of the plate or tube is assumed to be constant, and the effect of natural convection is ignored.

Laminar flow heat transfer to flat plate

Consider heat flow to the flat plate shown in Fig. 12.1b. The conditions are assumed to be as follows:

Velocity of fluid approaching plate and at and beyond the edge of the boundary layer OA: u_0.
Temperature of fluid approaching plate and at and beyond the edge of the thermal boundary layer $O'B$: T_∞.
Temperature of plate: from $x = 0$ to $x = x_0$, $T = T_\infty$; for $x > x_0$, $T = T_w$, where $T_w > T_\infty$.
The following properties of the fluid are constant and temperature-independent: density ρ, conductivity k, specific heat c_p, and viscosity μ.

Detailed analysis of the situation yields the equation[2]

$$\left(\frac{dT}{dy}\right)_w = \frac{0.332(T_w - T_\infty)}{\sqrt[3]{1 - (x_0/x)^{3/4}}} \sqrt[3]{\frac{c_p\mu}{k}} \sqrt{\frac{u_0\rho}{\mu x}} \qquad (12.2)$$

where $(dT/dy)_w$ is the temperature gradient at the wall. From Eq. (11.21), the relation between the local heat-transfer coefficient h_x at any distance x from the leading edge and the temperature gradient at the wall is

$$h_x = \frac{k}{T_w - T_\infty}\left(\frac{dT}{dy}\right)_w \qquad (12.3)$$

Eliminating $(dT/dy)_w$ gives

$$h_x = \frac{0.332k}{\sqrt[3]{1-(x_0/x)^{3/4}}} \sqrt[3]{\frac{c_p\mu}{k}} \sqrt{\frac{u_0\rho}{\mu x}}$$

This equation can be put into a dimensionless form by multiplying by x/k, giving

$$\frac{h_x x}{k} = \frac{0.332}{\sqrt[3]{1-(x_0/x)^{3/4}}} \sqrt[3]{\frac{c_p\mu}{k}} \sqrt{\frac{u_0 x\rho}{\mu}} \qquad (12.4)$$

The left-hand side of this equation is, from Eq. (11.22), a Nusselt number corresponding to the distance x, or Nu_x. The second group is the Prandtl number Pr, and the third group is a Reynolds number corresponding to distance x, denoted by Re_x. Equation (12.4) then can be written

$$\mathrm{Nu}_x = \frac{0.332}{\sqrt[3]{1-(x_0/x)^{3/4}}} \sqrt[3]{\mathrm{Pr}} \sqrt{\mathrm{Re}_x} \qquad (12.5)$$

The local Nusselt number can be interpreted as the ratio of the distance x to the thickness of the thermal boundary layer, since conduction through a layer of thickness y would give a coefficient k/y. Thus

$$\mathrm{Nu}_x = \frac{h_x x}{k} = \frac{k}{y}\frac{x}{k} = \frac{x}{y} \qquad (12.6)$$

When the plate is heated over its entire length, as shown in Fig. 12.1a, $x_0 = 0$ and Eq. (12.5) becomes

$$\mathrm{Nu}_x = 0.332 \sqrt[3]{\mathrm{Pr}} \sqrt{\mathrm{Re}_x} \qquad (12.7)$$

Equation (12.7) gives the local value of the Nusselt number at distance x from the leading edge. More important in practice is the average value of Nu over the entire length of the plate x_1, defined as

$$\mathrm{Nu} = \frac{h x_1}{k} \qquad (12.8)$$

where

$$h = \frac{1}{x_1}\int_0^{x_1} h_x\, dx$$

Equation (12.4) can be written for a plate heated over its entire length, since $x_0 = 0$, as

$$h_x = \frac{C}{\sqrt{x}}$$

where C is a constant containing all factors other than h_x and x. Then

$$h = \frac{C}{x_1}\int_0^{x_1}\frac{dx}{\sqrt{x}} = \frac{2C}{x_1}\sqrt{x_1} = \frac{2C}{\sqrt{x_1}} = 2h_{x_1} \qquad (12.9)$$

The average coefficient is clearly twice the local coefficient at the end of the plate, and Eq. (12.7) gives

$$\mathrm{Nu} = 0.664 \sqrt[3]{\mathrm{Pr}} \sqrt{\mathrm{Re}_{x_1}} \qquad (12.10)$$

These equations are valid only for Prandtl numbers of 1.0 or greater, since the derivation assumes a thermal boundary layer no thicker than the hydrodynamic layer. However, they can be used for gases with $\text{Pr} \approx 0.7$ with little error. The equations are also restricted to cases where the Nusselt number is fairly large, say, 10 or higher, since axial conduction, which was neglected in the derivation, has a significant effect for thick boundary layers.

Laminar flow heat transfer in tubes

The simplest situation of laminar flow heat transfer in tubes is defined by the following conditions. The velocity of the fluid throughout the tube and at all points in any cross section of the stream is constant, so that $u = u_0 = \bar{V}$, that is, plug flow; the wall temperature is constant; and the properties of the fluid are independent of temperature. Mathematically this model is identical to that of heat flow by conduction into a solid rod at constant surface temperature, using as heating time the period of passage of a cross section of the fluid stream at velocity $\bar{V}$ through a tube of length L. This time period is $t_T = L/\bar{V}$. Equation (10.21), then, can be used for plug flow of a fluid by substituting $L/\bar{V}$ for t_T in the Fourier number, which becomes

$$\text{Fo} = \frac{\alpha t_T}{r_m^2} = \frac{4k t_T}{c_p \rho D^2} = \frac{4kL}{c_p \rho D^2 \bar{V}} \tag{12.11}$$

Graetz and Peclet numbers

Two other dimensionless groups are commonly used in place of the Fourier number in treating heat transfer to fluids. The *Graetz number* is defined by the equation

$$\text{Gz} \equiv \frac{\dot{m} c_p}{kL} \tag{12.12}$$

where $\dot{m}$ is the mass flow rate. Since $\dot{m} = (\pi/4)\rho \bar{V} D^2$,

$$\text{Gz} = \frac{\pi}{4} \frac{\rho \bar{V} c_p D^2}{kL} \tag{12.13}$$

The Graetz number can also be calculated from the Reynolds and Prandtl numbers and the D/L ratio[†]

$$\text{Gz} = \frac{\pi}{4}\rho \bar{V} D^2 \frac{c_p}{kL}\frac{\mu}{\mu} = \frac{\pi}{4}\,\text{Re}\,\text{Pr}\,\frac{D}{L} \tag{12.14}$$

The *Peclet number* Pe is defined as the product of the Reynolds number and the Prandtl number, or

$$\text{Pe} \equiv \text{Re}\,\text{Pr} = \frac{D\bar{V}\rho}{\mu}\frac{c_p \mu}{k} = \frac{\rho \bar{V} c_p D}{k} = \frac{D\bar{V}}{\alpha} \tag{12.15}$$

[†]In some references the Graetz number is defined as $\text{Gz} = (D/L)\text{Pe}$ without the $\pi/4$ factor, so the definition should be checked when using published correlations.

The choice among these groups is arbitrary. They are related by the equations

$$Gz = \frac{\pi D}{4L}Pe = \frac{\pi}{Fo} \tag{12.16}$$

In the following discussion the Graetz number is used.

Exit temperature in plug flow

Equation (10.21) becomes, for plug flow,

$$\frac{T_w - \bar{T}_b}{T_w - T_a} = 0.692e^{-5.78\pi/Gz} + 0.131e^{-30.5\pi/Gz} + 0.0534e^{-74.9\pi/Gz} + \cdots \tag{12.17}$$

Here T_a and $\bar{T}_b$ are the inlet and average outlet fluid temperatures, respectively.

Plug flow is not a realistic model for newtonian fluids, but it does apply to highly pseudoplastic liquids ($n' \approx 0$) or to plastic liquids having a high value of the yield stress τ_0.

Fully developed flow

With a newtonian fluid in fully developed flow, the actual velocity distribution at the entrance to the heated section and the theoretical distribution throughout the tube are both parabolic. For this situation the appropriate boundary conditions lead to the development of another theoretical equation, of the same form as Eq. (12.17). This is[7d]

$$\frac{T_w - \bar{T}_b}{T_w - T_a} = 0.81904e^{-3.657\pi/Gz} + 0.09760e^{-22.31\pi/Gz} + 0.01896e^{-53\pi/Gz} + \cdots \tag{12.18}$$

Because of distortions in the flow field from the effects of temperature on viscosity and density, Eq. (12.18) does not give accurate results. The heat-transfer rates are usually larger than those predicted by Eq. (12.18), and empirical correlations have been developed for design purposes. These correlations are based on the Graetz number, but they give the film coefficient or the Nusselt number rather than the change in temperature, since this permits the fluid resistance to be combined with other resistances in determining an overall heat-transfer coefficient.

The Nusselt number for heat transfer to a fluid inside a pipe is the film coefficient multiplied by D/k:

$$Nu \equiv \frac{h_i D}{k} \tag{12.19}$$

The film coefficient h_i is the average value over the length of the pipe and is calculated as follows for the case of constant wall temperature:

$$h_i = \frac{\dot{m}c_p(\bar{T}_b - T_a)}{\pi DL \,\overline{\Delta T_L}} \tag{12.20}$$

Since

$$\overline{\Delta T_L} = \frac{(T_w - T_a) - (T_w - \bar{T}_b)}{\ln[(T_w - T_a)/(T_w - \bar{T}_b)]} \tag{12.21}$$

$$h_i = \frac{\dot{m}c_p}{\pi DL}\ln[(T_w - T_a)/(T_w - \bar{T}_b)] \tag{12.22}$$

and

$$\mathrm{Nu} = \frac{\dot{m}c_p}{\pi kL}\ln[(T_w - T_a)/(T_w - \bar{T}_b)] \tag{12.23}$$

or

$$\mathrm{Nu} = \frac{\mathrm{Gz}}{\pi}\ln[(T_w - T_a)/(T_w - \bar{T}_b)] \tag{12.24}$$

Using Eqs. (12.24) and (12.18), theoretical values of the Nusselt number for parabolic flow can be obtained, and these values are shown in Fig. 12.2. At low Graetz numbers, only the first term of Eq. (12.18) is significant, and the Nusselt number approaches a limiting value of 3.66. It is difficult to get an accurate measurement of the heat-transfer coefficient at low Graetz numbers, since the final temperature difference is very small. For example, at $\mathrm{Gz} = 1.0$, the ratio of exit to inlet driving forces is only 8.3×10^{-6}.

For Graetz numbers greater than 20, the theoretical Nusselt number increases with about the one-third power of Gz. Data for air and for moderate-viscosity liquids follow a similar trend, but the coefficients are about 20 percent greater than predicted from theory. An empirical equation for moderate Graetz numbers (greater than 20) is

$$\mathrm{Nu} \approx 2.0\,\mathrm{Gz}^{1/3} \tag{12.25}$$

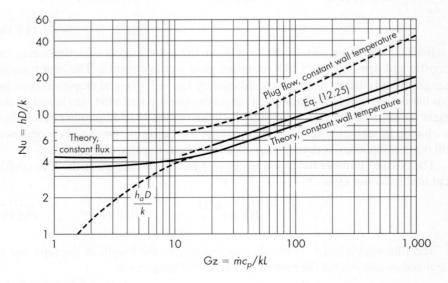

FIGURE 12.2
Heat transfer for laminar flow in tubes with a parabolic velocity profile. (It does not include effects of natural convection or viscosity gradients.)

The increase in film coefficient with increasing Graetz number or decreasing length is a result of the change in shape of the temperature profile. For short lengths, the thermal boundary layer is very thin, and the steep temperature gradient gives a high local coefficient. With increasing distance from the entrance, the boundary layer becomes thicker and eventually reaches the center of the pipe, giving a nearly parabolic temperature profile. The local coefficient is approximately constant from that point on, but the average coefficient continues to decrease with increasing length until the effect of the high initial coefficient is negligible. In practice, the change in local coefficient with length is usually not calculated, and the length-average film coefficient is used in obtaining the overall coefficient.

Correction for heating or cooling

For viscous liquids with large temperature drops, a modification of Eq. (12.25) is required to account for differences between heating and cooling. When a liquid is being heated, the lower viscosity near the wall makes the velocity profile more like that for plug flow, with a very steep gradient near the wall and little gradient near the center. This leads to a higher rate of heat transfer, as can be shown by comparing temperature approaches calculated from Eqs. (12.17) and (12.18). When a viscous liquid is cooled, the velocity gradient at the wall is decreased, giving a lower rate of heat transfer. A dimensionless, but empirical, correction factor ϕ_v accounts for the difference between heating and cooling:

$$\phi_v \equiv \left(\frac{\mu}{\mu_w}\right)^{0.14} \tag{12.26}$$

This factor is added to Eq. (12.25) to give the final equation for laminar flow heat transfer:

$$\text{Nu} = 2\left(\frac{\dot{m}c_p}{kL}\right)^{1/3}\left(\frac{\mu}{\mu_w}\right)^{0.14} = 2\,\text{Gz}^{1/3}\phi_v \tag{12.27}$$

In Eqs. (12.26) and (12.27), μ is the viscosity at the arithmetic mean temperature of the fluid $(T_a + \bar{T}_b)/2$, and, μ_w is the viscosity at the wall temperature T_w. For liquids $\mu_w < \mu$ and $\phi_v > 1.0$ when the liquid is being heated, and $\mu_w > \mu$ and $\phi_v < 1.0$ when the liquid is being cooled.

The coefficients presented in Eqs. (12.19) to (12.27) are based on a logarithmic mean driving force $\overline{\Delta T_L}$. Some workers have presented correlations for a coefficient h_a based on the arithmetic mean driving force $\overline{\Delta T_a}$. When the Graetz number is 10 or larger, the ratio of inlet to exit driving forces is less than 2.0, and there is little difference between $\overline{\Delta T_L}$ and $\overline{\Delta T_a}$ or between h and h_a. However, at low Graetz numbers the temperature approach becomes very small, and the coefficient h_a varies inversely with the Graetz number, as shown in Fig. 12.2. There is no apparent advantage to using h_a, and h is recommended for design calculations.

The equations and experimental results discussed up to this point are for constant wall temperature and would apply for heating a fluid with a condensing vapor. If a counterflow exchanger is used, the wall temperature will change along the length of the exchanger, and this will affect the film coefficient for laminar flow. If the two streams have about the same flow rate and heat capacity, the temperature driving force and the heat flux will be nearly constant. The theoretical equation for constant heat flux and parabolic flow gives a limiting Nusselt number of 4.36, compared to 3.66 for constant T_w. At high Graetz numbers, the predicted coefficient for constant flux is also higher than for constant wall temperature, but there are not enough experimental results to develop a separate equation for the case of constant flux.

Effect of inserts on heat transfer

Inserts such as twisted tapes, similar to those in the static mixer shown in Fig. 9.19a, are often used to improve heat transfer to and from viscous solutions. With laminar flow such devices may increase the heat-transfer coefficient by 4 times or more as the heated liquid near the tube wall is moved toward the center and mixed with cooler fluid. Liquid flowing through tubes in turbulent flow is generally well mixed; hence inserts like twisted tapes are much less effective with turbulent flow than they are with laminar flow.[10]

Heat transfer to non-newtonian liquids in laminar flow

For heat transfer to and from liquids that follow the power law relation [Eq. (3.7)], Eq. (12.27) is modified to[11]

$$\frac{h_i D}{k} = 2\delta^{1/3} \left(\frac{\dot{m} c_p}{kL} \right)^{1/3} \left(\frac{m}{m_w} \right)^{0.14} \tag{12.28}$$

where $\delta = (3n' + 1)/4n'$
$m = K' 8^{n'-1}$, at arithmetic mean temperature
$m_w =$ value of m at T_w
$K' =$ flow consistency index
$n' =$ flow behavior index

For shear-thinning fluids ($n' < 1$), the non-newtonian behavior makes the velocity profile more like that for plug flow and increases the heat-transfer coefficient. When $n' = 0.1$, the coefficient for large Graetz numbers is about 1.5 times that for parabolic flow ($n' = 1.0$). The limit of $n' = 0$ corresponds to true plug flow, and the coefficients can be as large as twice those for parabolic flow. Figure 12.2 shows the Nusselt number for plug flow as a dashed line, the slope of which approaches 0.5 at high Graetz numbers.

HEAT TRANSFER BY FORCED CONVECTION IN TURBULENT FLOW

Perhaps the most important situation in heat transfer is the heat flow in a stream of fluid in turbulent flow in a closed channel, especially in tubes. Turbulence is encountered at Reynolds numbers greater than about 2,100, and since the rate of heat transfer is greater in turbulent flow than in laminar flow, most equipment is operated in the turbulent range.

The earliest approach to this case was based on empirical correlations of test data guided by dimensional analysis. The equations so obtained still are much used in design. Subsequently, theoretical study has been given to the problem. A deeper understanding of the mechanism of turbulent flow heat transfer has been achieved, and improved equations applicable over wider ranges of conditions have been obtained.

Dimensional analysis method

Dimensional analysis of the heat flow to a fluid in turbulent flow in a long, straight pipe yields the dimensionless relationship

$$\frac{hD}{k} = \Phi\left(\frac{D\bar{V}\rho}{\mu}, \frac{c_p\mu}{k}\right) = \Phi\left(\frac{DG}{\mu}, \frac{c_p\mu}{k}\right) \tag{12.29}$$

Here the mass velocity G is used in place of its equal $\bar{V}\rho$. Dividing both sides of Eq. (12.29) by the product $(DG/\mu)(c_p\mu/k)$ gives an alternate relationship

$$\frac{h}{c_pG} = \Phi_1\left(\frac{DG}{\mu}, \frac{c_p\mu}{k}\right) \tag{12.30}$$

The three groups in Eq. (12.29) are the Nusselt, Reynolds, and Prandtl numbers, respectively. The left-hand group in Eq. (12.30) is called the *Stanton number* St. The four groups are related by the equation

$$\text{St Re Pr} = \text{Nu} \tag{12.31}$$

Thus, only three of the four are independent.

Empirical equations

To use Eq. (12.29) or (12.30), the function Φ or Φ_1 must be known. One empirical correlation for long tubes with sharp-edged entrances is the Dittus-Boelter equation

$$\text{Nu} = \frac{h_iD}{k} = 0.023\,\text{Re}^{0.8}\,\text{Pr}^n \tag{12.32}$$

where n is 0.4 when the fluid is being heated and 0.3 when it is being cooled.

Using different values of n in Eq. (12.32) is one way of allowing for the higher coefficients found when liquids are heated than when they are cooled. However, the ratio of the coefficients for heating and cooling, according to Eq. (12.32), equals $\text{Pr}^{0.1}$ and does not depend on conditions at the wall of the pipe. A better relationship for turbulent flow[9] is known as the Sieder-Tate equation; it uses the same correction factor ϕ_v as for laminar flow

$$\text{Nu} = \frac{h_iD}{k} = 0.023\,\text{Re}^{0.8}\,\text{Pr}^{1/3}\,\phi_v^{0.14} \tag{12.33}$$

The viscosity ratio term ϕ_v is not very important for low-viscosity liquids such as water, but for viscous oils, where the wall and bulk viscosities may differ by 10-fold, the coefficient for heating may be twice that for cooling.[†]

[†]Some authors use 0.027 instead of 0.023 in Eq. (12.33) and give a viscosity ratio exponent of 0.25 for heating and 0.11 for cooling.[8]

An alternate form of Eq. (12.33) is obtained by dividing both sides by Re Pr and transposing to give what is called the Colburn equation.

$$\text{St Pr}^{2/3}\, \phi_v^{0.14} = \frac{0.023}{\text{Re}^{0.2}} \tag{12.34}$$

In using these equations the physical properties of the fluid, except for μ_w, are evaluated at the bulk fluid temperature T. For gases, the viscosity increases with temperature, but the change is small and the viscosity term in Eqs. (12.33) and (12.34) is usually ignored. These equations should not be used when the Reynolds number is below 6,000 or for molten metals, which have very low Prandtl numbers.

Effect of tube length. Near the tube entrance, where the temperature gradients are still forming, the local coefficient h_x is greater than h_∞ for fully developed flow. At the entrance itself, where there is no previously established temperature gradient, h_x is infinite. Its value drops rapidly toward h_∞ in a comparatively short length of tube. Dimensionally, the effect of tube length is accounted for by another dimensionless group x/D, where x is the distance from the tube entrance. The local coefficient approaches h_∞ asymptotically with an increase in x, but it is practically equal to h_∞ when x/D is about 50. The average value of h_x over the tube length is denoted by h_i. The value of h_i is found by integrating h_x over the length of the tube. Since $h_x \to h_\infty$ as $x \to \infty$, the relation between h_i and h_∞ is of the form[6b]

$$\frac{h_i}{h_\infty} = 1 + \psi\left(\frac{L}{D}\right) \tag{12.35}$$

An equation for short tubes with sharp-edged entrances, where the velocity at the entrance is uniform over the cross section, is

$$\frac{h_i}{h_\infty} = 1 + \left(\frac{D}{L}\right)^{0.7} \tag{12.36}$$

The effect of tube length on h_i fades out when L/D becomes greater than about 50.

Average value of h_i in turbulent flow

Since the temperature of the fluid changes from one end of the tube to the other and the fluid properties μ, k, and c_p are all functions of temperature, the local value of h_i also varies from point to point along the tube. This variation is independent of the effect of tube length.

The effect of fluid properties can be shown by condensing Eq. (12.33) to read, assuming $\mu/\mu_w = 1$,

$$h_i = 0.023 \frac{G^{0.8} k^{2/3} c_p^{1/3}}{D^{0.2} \mu^{0.47}} \tag{12.37}$$

For gases the effect of temperature on h_i is small. At constant mass velocity in a given tube, h_i varies with $k^{2/3} c_p^{1/3} \mu^{-0.47}$. The increase in thermal conductivity and heat capacity with temperature offsets the rise in viscosity, giving a slight increase in h_i. For example, for air h_i increases about 6 percent when the temperature changes from 50 to 100°C.

For liquids the effect of temperature is much greater than for gases because of the rapid decrease in viscosity with rising temperature. The effects of k, c_p, and μ in Eq. (12.37) all act in the same direction, but the increase in h_i with temperature is due mainly to the effect of temperature on viscosity. For water, for example, h_i increases about 50 percent over a temperature range from 50 to 100°C. For viscous oils the change in h_i may be two- or threefold for a 50°C increase in temperature.

In practice, unless the variation in h_i over the length of the tube is more than about 2:1, an average value of h_i is calculated and used as a constant in calculating the overall coefficient U. This procedure neglects the variation of U over the tube length and allows the use of the LMTD in calculating the area of the heating surface. The average value of h_i is computed by evaluating the fluid properties c_p, k, and μ at the average fluid temperature, defined as the arithmetic mean between the inlet and outlet temperatures. The value of h_i calculated from Eq. (12.33), using these property values, is called the *average coefficient*. For example, assume that the fluid enters at 30°C and leaves at 90°C. The average fluid temperature is $(30 + 90)/2 = 60°C$, and the values of the properties used to calculate the average value of h_i are those at 60°C.

For larger changes in h_i, two procedures can be used: (1) The values of h_i at the inlet and outlet can be calculated, corresponding values of U_1 and U_2 found, and Eq. (11.17) used. Here the effect of L/D on the entrance value of h_i is ignored. (2) For even larger variations in h_i, and therefore in U, the tube can be divided into sections and an average U used for each section. Then the lengths of the individual sections can be added to account for the total length of the tube.

Estimation of wall temperature T_w

To evaluate μ_w, the viscosity of the fluid at the wall, temperature T_w must be found. The estimation of T_w requires an iterative calculation based on the resistance equation (11.35). If the individual resistances can be estimated, the total temperature drop ΔT can be split into the individual temperature drops by the use of this equation and an approximate value for the wall temperature found. To determine T_w in this way, the wall resistance $(x_w/k_m)(D_o/\bar{D}_L)$ can usually be neglected, and Eq. (11.35) used as follows.

From the first two members of Eq. (11.35)

$$\Delta T_i = \frac{D_o/D_i h_i}{1/U_o} \Delta T \tag{12.38}$$

Substituting $1/U_o$ from Eq. (11.32) and neglecting the wall resistance term give

$$\Delta T_i = \frac{1/h_i}{1/h_i + D_i/D_o h_o} \Delta T \tag{12.39}$$

In qualitative terms Eq. (12.39) may be written

$$\Delta T_i = \frac{\text{inside resistance}}{\text{overall resistance}} \Delta T$$

TABLE 12.1
Data for Example 12.1

	Heat-transfer coefficient	
	Btu/ft²·h·°F	W/m²·°C
For cooling water h_i	700	3,970
For toluene h_o	500	2,840

Use of Eq. (12.39) requires preliminary estimates of the coefficients h_i and h_o. To estimate h_i, Eq. (12.33) can be used, neglecting ϕ_v. The calculation of h_o will be described later. The wall temperature T_w is then obtained from the following equations:

For heating:
$$T_w = T + \Delta T_i \qquad (12.40)$$

For cooling:
$$T_w = T - \Delta T_i \qquad (12.41)$$

where T is the average fluid temperature.

If the first approximation is not sufficiently accurate, a second calculation of T_w based on the results of the first can be made. Unless the factor ϕ_v is quite different from unity, however, the second approximation is unnecessary.

EXAMPLE 12.1. Toluene is being condensed at 230°F (110°C) on the outside of $\frac{3}{4}$-in. (19-mm) BWG 16 copper condenser tubes through which cooling water is flowing at an average temperature of 80°F (26.7°C). Individual heat-transfer coefficients are given in Table 12.1. Neglecting the resistance of the tube wall, what is the tube wall temperature?

Solution. From App. 4, $D_i = 0.620$ in.; $D_o = 0.750$ in. Hence, from Eq. (12.39)

$$\Delta T_i = \frac{1/700}{1/700 + 0.620/(0.750 \times 500)}(230 - 80) = 69.5°F$$

Since the water is being heated by the condensing toluene, the wall temperature is found from Eq. (12.40) as follows:

$$T_w = 80 + 69.5 = 149.5°F \ (65.3°C)$$

Cross sections other than circular

To use Eq. (12.33) or (12.34) for cross sections other than circular, it is only necessary to replace the diameter D in both Reynolds and Nusselt numbers by the equivalent diameter D_e, defined as 4 times the hydraulic radius r_H. The method is the same as that used in calculating friction loss.

EXAMPLE 12.2. Benzene is cooled from 141 to 79°F (60.6 to 21.1°C) in the inner pipe of a double-pipe exchanger. Cooling water flows countercurrently to the benzene, entering the jacket at 65°F (18.3°C) and leaving at 75°F (23.9°C). The exchanger consists of an inner pipe of $\frac{7}{8}$-in. (22.2-mm) BWG 16 copper tubing jacketed with $1\frac{1}{2}$-in. (38.1-mm) Schedule 40 steel pipe. The linear velocity of the benzene is 5 ft/s (1.52 m/s). Neglecting the resistance of the wall and scale films and assuming $L/D > 150$ for both

TABLE 12.2
Data for Example 12.2

Property	Value at average fluid temperature	
	Benzene	Water[†]
Density ρ, lb/ft³	53.1	62.3
Viscosity μ, lb/ft · h	1.16[‡]	$2.42 \times 0.982 = 2.34$
Thermal conductivity k, Btu/ft · h · °F	0.089[§]	0.346
Specific heat c_p, Btu lb · °F	0.435[¶]	1.000

[†]Appendix 6.
[‡]Appendix 9.
[§]Appendix 13.
[¶]Appendix 15.

pipes, compute the film coefficients of the benzene and water and the overall coefficient based on the outside area of the inner pipe.

Solution. The average temperature of the benzene is $(141 + 79)/2 = 110°F$; that of the water is $(65 + 75)/2 = 70°F$. The physical properties at these temperatures are given in Table 12.2. The diameters of the inner tube are

$$D_{it} = \frac{0.745}{12} = 0.0621 \text{ ft} \qquad D_{ot} = \frac{0.875}{12} = 0.0729 \text{ ft}$$

The inside diameter of the jacket is, from App. 3

$$D_{ij} = \frac{1.610}{12} = 0.1342 \text{ ft}$$

The equivalent diameter of the annular jacket space is found as follows. The cross-sectional area is $(\pi/4)(0.1342^2 - 0.0729^2)$, or 0.00997 ft^2. The wetted perimeter is $\pi(0.1342 + 0.0729)$. The hydraulic radius is

$$r_H = \frac{(\pi/4)(0.1342^2 - 0.0729^2)}{\pi(0.1342 + 0.0729)} = \frac{1}{4}(0.1342 - 0.0729) = \frac{1}{4} \times 0.0613 \text{ ft}$$

The equivalent diameter $D_e = 4r_H = 0.0613 \text{ ft}$.

First the velocity of the water must be computed from the heat flow and the temperature rise in the water. The heat flow, in turn, is found from the mass flow rate of the benzene $\dot{m}_b$, given by

$$\dot{m}_b = \bar{V}_b \rho_b S$$

where S is the inside sectional area of the copper tube. From App. 4, for a $\frac{7}{8}$-in. BWG 16 tube, $S = 0.00303 \text{ ft}^2$. Thus

$$\dot{m}_b = 5 \times 53.1 \times 0.00303 = 0.804 \text{ lb/s}$$

The rate of heat flow q, found by multiplying the mass flow, specific heat, and temperature change of the benzene, is

$$q = 0.804 \times 0.435(141 - 79)$$
$$= 21.68 \text{ Btu/s}$$

The mass flow rate of the water $\dot{m}_w$ is then

$$\dot{m}_w = \frac{21.68}{1.000(75-65)} = 2.168 \text{ lb/s}$$

The water velocity $\bar{V}_w$ is

$$\bar{V}_w = \frac{2.168}{0.00997 \times 62.3} = 3.49 \text{ ft/s}$$

The Reynolds number and Prandtl number of each stream are computed next:

Benzene: $\text{Re} = \dfrac{D_{it}\bar{V}\rho}{\mu} = \dfrac{0.0621 \times 5 \times 3,600 \times 53.1}{1.16} = 5.12 \times 10^4$

$\text{Pr} = \dfrac{c_p\mu}{k} = \dfrac{0.435 \times 1.16}{0.089} = 5.67$

Water: $\text{Re} = \dfrac{D_e\bar{V}\rho}{\mu} = \dfrac{0.0613 \times 3.49 \times 3,600 \times 62.3}{2.34} = 2.05 \times 10^4$

$\text{Pr} = \dfrac{1.00 \times 2.34}{0.346} = 6.76$

Preliminary estimates of the coefficients are obtained from Eq. (12.34), omitting the correction for viscosity ratio:

Benzene: $h_i = \dfrac{0.023 \times 5 \times 3,600 \times 53.1 \times 0.435}{(5.12 \times 10^4)^{0.2} \times 5.67^{2/3}} = 344 \text{ Btu/ft}^2 \cdot \text{h} \cdot {}^\circ\text{F}$

Water: $h_o = \dfrac{0.023 \times 3.49 \times 3,600 \times 62.3 \times 1.000}{(2.05 \times 10^4)^{0.2} \times 6.76^{2/3}} = 691 \text{ Btu/ft}^2 \cdot \text{h} \cdot {}^\circ\text{F}$

In these calculations use is made of the fact that $G = \bar{V}\rho$.

The temperature drop over the benzene resistance, from Eq. (12.39), is

$$\Delta T_i = \frac{1/344}{1/344 + 0.0621/(0.0729 \times 691)}(110 - 70) = 28.1^\circ\text{F}$$

$$T_w = 110 - 28.1 = 81.9^\circ\text{F}$$

The viscosities of the liquids at T_w are now found.

$$\mu_w = \begin{cases} 1.45 \text{ lb/ft} \cdot \text{h} & \text{for benzene} \\ 0.852 \times 2.42 = 2.06 \text{ lb/ft} \cdot \text{h} & \text{for water} \end{cases}$$

The viscosity correction factors ϕ_v, from Eq. (12.26), are

$$\phi_v = \begin{cases} \left(\dfrac{1.16}{1.45}\right)^{0.14} = 0.969 & \text{for benzene} \\[3mm] \left(\dfrac{2.34}{2.06}\right)^{0.14} = 1.018 & \text{for water} \end{cases}$$

The corrected coefficients are

Benzene: $h_i = 344 \times 0.969 = 333 \text{ Btu/ft}^2 \cdot \text{h} \cdot {}^\circ\text{F} \ (1,891 \text{ W/m}^2 \cdot {}^\circ\text{C})$

Water: $h_o = 691 \times 1.018 = 703 \text{ Btu/ft}^2 \cdot \text{h} \cdot {}^\circ\text{F} \ (3,992 \text{ W/m}^2 \cdot {}^\circ\text{C})$

The temperature drop over the benzene resistance and the wall temperature become

$$\Delta T_i = \frac{1/333}{1/333 + 0.0621/(0.0729 \times 703)}(110 - 70) = 28.5°\text{F}$$

$$T_w = 110 - 28.5 = 81.5°\text{F}$$

This is so close to the wall temperature calculated previously that a second approximation is unnecessary.

The overall coefficient is found from Eq. (11.32), neglecting the resistance of the tube wall:

$$\frac{1}{U_o} = \frac{0.0729}{0.0621 \times 333} + \frac{1}{703} = 0.00495$$

$$U_o = \frac{1}{0.00495} = 202 \text{ Btu/ft}^2 \cdot \text{h} \cdot °\text{F} \ (1,147 \text{ W/m}^2 \cdot °\text{C})$$

Effect of roughness

For equal Reynolds numbers the heat-transfer coefficient in turbulent flow is somewhat greater for a rough tube than for a smooth one. The effect of roughness on heat transfer is much less than on fluid friction, and economically it is usually more important to use a smooth tube for minimum friction loss than to rely on roughness to yield a larger heat-transfer coefficient. The effect of roughness on h_i is neglected in practical calculations.

Heat transfer at high velocities

When a fluid flows through a tube at high velocities, temperature gradients appear even when there is no heat transfer through the wall and $q = 0$. In injection molding, for example, polymer melt flows into the cavity at a very high velocity, and the steep velocity gradients in the viscous liquid cause heat to be generated in the fluid by what is called *viscous dissipation*. In the high-velocity flow of compressible gases in pipes, friction at the wall raises the temperature of the fluid at the wall above the average fluid temperature.[4c] The temperature difference between wall and fluid causes a flow of heat from wall to fluid, and a steady state is reached when the rate of heat generation from friction at the wall equals the rate of heat transfer back into the fluid stream. The constant wall temperature thus attained is called the *adiabatic wall temperature*. Further treatment of this subject is beyond the scope of this text. The effect becomes appreciable for Mach numbers above approximately 0.4, and appropriate equations must be used in this range of velocities instead of Eqs. (12.33) and (12.34).

Transfer by Turbulent Eddies and Analogy Between Transfer of Momentum and Heat

On pages 104 to 110 the distribution of velocity and its accompanying momentum flux in a flowing stream in turbulent flow through a pipe was described. Three rather ill-defined zones in the cross section of the pipe were identified. In the first, immediately next to the wall, eddies are rare and momentum flow occurs almost entirely by viscosity; in the second, a mixed regime of combined viscous and turbulent momentum transfer occurs; in the main part of the stream, which occupies the bulk of the cross section of the stream, only the momentum flow generated by the Reynolds stresses of turbulent flow is important. The three zones are called the *viscous sublayer,* the *buffer zone,* and the *turbulent core,* respectively.

In heat transfer at the wall of the tube to or from the fluid stream, the same hydrodynamic distributions of velocity and of momentum fluxes still persist, and in addition, a temperature gradient is superimposed on the turbulent-laminar velocity field. In the following treatment both gradients are assumed to be completely developed and the effect of tube length negligible.

Throughout the stream of fluid, heat flow by conduction occurs in accordance with the equation

$$\frac{q_c}{A} = -k\frac{dT}{dy} \tag{12.42}$$

where q_c = rate of heat flow by conduction
k = thermal conductivity
A = area of isothermal surface
dT/dy = temperature gradient across isothermal surface

The isothermal surface is a cylinder concentric with the axis of the pipe and located a distance y from the wall, or r from the center of the pipe. Here $r + y = r_w$, where r_w is the radius of the pipe.

In addition to conduction, the eddies of turbulent flow carry heat by convection across each isothermal area. Although both mechanisms of heat flow may occur wherever a temperature gradient exists $(dT/dy \neq 0)$, their relative importance varies greatly with distance from the wall. At the tube wall itself eddies very rarely exist, and the heat flux is entirely due to conduction. Equation (12.42) written for the wall is

$$\left(\frac{q}{A}\right)_w = -k\left(\frac{dT}{dy}\right)_w \tag{12.43}$$

where $(q/A)_w$ = total heat flux at wall
$(dT/dy)_w$ = temperature gradient at wall

These quantities are identical with those in Eqs. (11.20) and (11.21).

Within the viscous sublayer, heat flows mainly by conduction, but eddies are not completely excluded from this zone, and some convection does occur. The relative importance of turbulent heat flux compared with conductive heat flux increases rapidly with distance from the wall. In ordinary fluids, having Prandtl numbers above about 0.6, conduction is entirely negligible in the turbulent core, but it may be significant in the buffer zone when the Prandtl number is in the neighborhood of unity. Conduction is negligible in this zone when the Prandtl number is large.

The situation is analogous to momentum flux, where the relative importance of turbulent shear to viscous shear follows the same general pattern. Under certain ideal conditions, the correspondence between heat flow and momentum flow is exact, and at any specific value of r/r_w the ratio of heat transfer by conduction to that by turbulence equals the ratio of momentum flux by viscous forces to that by Reynolds stresses. In the general case, however, the correspondence is only approximate and may be greatly in error. The study of the relationship between heat and momentum flux for the entire spectrum of fluids leads to the *analogy theory,* and the equations so derived are called *analogy equations.* A detailed treatment of the theory is beyond the scope of this book, but some of the more elementary relationships are considered.

Since eddies continually cross an isothermal surface from both directions, they carry heat between layers on either side of the surface, which are at different average temperatures. At a given point the temperature fluctuates rapidly about the constant mean temperature at that point, depending on whether a "hot" or a "cold" eddy is crossing through the point. The temperature fluctuations form a pattern with respect to both time and place, just like the fluctuations in velocity and pressure described on pages 53 to 55. The instantaneous temperature T_i at the point can be split into two parts, the constant average temperature T at the point and the fluctuating or deviating temperature T', or

$$T_i = T + T' \tag{12.44}$$

The time average of the deviating temperature T', denoted by $\bar{T}'$, is zero, and the time-average value of the total instantaneous temperature, denoted by $\bar{T}_i$, is T. The average temperature T is that measured by an ordinary thermometer. To measure T_i and so find T' requires special sensing devices that can follow rapid temperature changes.

The eddy diffusivity of heat

When there is no temperature gradient across the isothermal surface, all eddies have the same temperature independent of the point of origin $dT/dy = 0$, and no net heat flow occurs. If a temperature gradient exists, an analysis equivalent to that leading to Eq. (3.17) shows that the eddies carry a net heat flux from the higher temperature to the lower, in accordance with the equation

$$\frac{q_t}{A} = -c_p \rho \overline{v'T'} \tag{12.45}$$

where v' is the deviating velocity across the surface and the overbar indicates the time average of the product $v'T'$. Although the time averages $\bar{v}'$ and $\bar{T}'$ individually are zero, the average of their product is not, because a correlation exists between these deviating quantities when $dT/dy \neq 0$, in the same way that the deviating velocities u' and v' are correlated when a velocity gradient du/dy exists.

On page 58 an eddy diffusivity for momentum transfer ε_M was defined. A corresponding eddy diffusivity for heat transfer ε_H can be defined by

$$\frac{q_t}{c_p \rho A} \equiv -\varepsilon_H \frac{dT}{dy} = -\overline{v'T'} \tag{12.46}$$

The subscript t refers to the fact that Eq. (12.46) applies to turbulent convection heat transfer. Since conduction also takes place, the total heat flux at a given point, denoted by q, is, from Eqs. (12.42) and (12.46),

$$\frac{q}{A} = \frac{q_c}{A} + \frac{q_t}{A} = -k\frac{dT}{dy} - c_p\rho\varepsilon_H\frac{dT}{dy}$$

or

$$\frac{q}{A} = -c_p\rho(\alpha + \varepsilon_H)\frac{dT}{dy} \tag{12.47}$$

where α is the thermal diffusivity $k/c_p\rho$. The equation for the total momentum flux corresponding to Eq. (12.47) is Eq. (3.20), written as

$$\frac{\tau}{\rho} = (\nu + \varepsilon_M)\frac{du}{dy} \tag{12.48}$$

where ν is the kinematic viscosity μ/ρ.

Eddy diffusivities

The diffusivity of momentum ν and the thermal diffusivity α are true fluid properties; their values depend on the temperature and pressure. The Prandtl number, which is their ratio, is also a fluid property. The eddy diffusitives for momentum and heat ε_M and ε_H, respectively, are not properties of the fluid but depend on the conditions of flow, especially on all factors that affect turbulence. For simple analogies, it is sometimes assumed that ε_M and ε_H are both constants and equal; but when determined by actual velocity and temperature measurements, both are found to be functions of the Reynolds number, the Prandtl number, and position in the tube cross section. Precise measurement of the eddy diffusivities is difficult, and not all reported measurements agree. Results are given in standard treatises.[6c] The ratio $\varepsilon_H/\varepsilon_M$ also varies but is more nearly constant than the individual quantities. The ratio is denoted by ψ. For ordinary liquids, where Pr > 0.6, ψ is close to 1 at the tube wall and in boundary layers generally and approaches 2 in turbulent wakes. For liquid metals ψ is low near the wall, passes through a maximum of about unity at $y/r_w \approx 0.2$, and decreases toward the center of the pipe.[7c]

The Reynolds analogy

The simplest and oldest analogy equation is that of Reynolds, which is derived for flow at high Reynolds numbers in straight round tubes. It can be derived from Eqs. (12.47) and (12.48) after several assumptions are made, most of which are questionable, but it is best regarded as an empirical finding for gases.

From the empirical relationship of Eq. (5.53),

$$f = 0.046\,\text{Re}^{-0.2} \tag{12.49}$$

From Eq. (12.34), assuming Pr = 1.0 and $\phi_v = 1.0$,

$$\text{St} = 0.023\,\text{Re}^{-0.2} \tag{12.50}$$

From these equations

$$\frac{h}{c_p G} \equiv St = \frac{f}{2} \tag{12.51}$$

This is the usual form of the Reynolds analogy equation. It agrees fairly well with experimental data for most gases, which have Prandtl numbers of about unity, provided the temperature drop $T_w - T$ is not large.

The Colburn analogy: Colburn j factor

Comparison of Eq. (12.49) with Eq. (12.34) shows that

$$St\, Pr^{2/3}\, \phi_v^{0.14} \equiv j_H = \frac{f}{2} \tag{12.52}$$

Equation (12.52) is a statement of the Colburn analogy between heat transfer and fluid friction. The factor j_H, defined as $(h/c_p G)(c_p \mu/k)^{2/3}(\mu_w/\mu)^{0.14}$, is called the *Colburn j factor*. It is used in a number of other semiempirical equations for heat transfer. While the Reynolds analogy [Eq. (12.51)] applies only to fluids for which the Prandtl number is close to unity, the Colburn analogy [Eq. (12.52)] applies over a range of Prandtl numbers from 0.6 to about 100.

Equation (12.34) can be written in j-factor form as follows:

$$j_H = 0.023\, Re^{-0.2} \tag{12.53}$$

More accurate analogy equations

A number of more elaborate analogy equations connecting friction and heat transfer in pipes, along flat plates, and in annular spaces have been published. They cover wider ranges of Reynolds and Prandtl numbers than Eq. (12.52) and are of the general form

$$St = \frac{f/2}{\Phi(Pr)} \tag{12.54}$$

where $\Phi(Pr)$ is a complicated function of the Prandtl number. One example, by Friend and Metzner,[3] applying to fully developed flow in smooth pipe is

$$St = \frac{f/2}{1.20 + 11.8\sqrt{f/2}(Pr-1)(Pr)^{-1/3}} \tag{12.55}$$

The friction factor f used in this equation may be that given by Eq. (12.49) or, for a wider range of Reynolds numbers from 3,000 to 3×10^6, by Eq. (5.54)

$$f = 0.00140 + \frac{0.125}{Re^{0.32}} \tag{12.56}$$

Equation (12.55) is said to apply over a range of Prandtl numbers from 0.5 to 600. For turbulent flow with very high Prandtl numbers, Eq. (12.55) indicates that the heat-transfer coefficient increases with the 0.85 to 0.87 power of the Reynolds numbers, rather than the 0.8 power given by Eq. (12.32) or (12.33).

All analogy equations connecting f and h have an important limitation. They apply only to wall, or skin, friction and must not be used for situations in which form drag appears.

HEAT TRANSFER IN TRANSITION REGION BETWEEN LAMINAR AND TURBULENT FLOW

Equation (12.34) applies only for Reynolds numbers greater than 6,000 and Eq. (12.27) only for Reynolds numbers less than 2,100. The range of Reynolds numbers between 2,100 and 6,000 is called the *transition region,* and no simple equation applies here. A graphical method therefore is used. The method is based on graphs of Eqs. (12.27) and (12.34) on a common plot of the Colburn j factor versus Re, with lines of constant values of L/D. To obtain an equation for the laminar flow range, it is necessary to transform Eq. (12.27) in the following manner. Substituted for the Graetz number, using Eqs. (12.15) and (12.16), is the quantity $(\pi D/4L)\text{RePr}$. The result is

$$\text{Nu} = 2\left(\frac{\pi D}{4L}\text{Re Pr}\right)^{1/3}\left(\frac{\mu}{\mu_w}\right)^{0.14}$$

This relation is multiplied by $(1/\text{Re})(1/\text{Pr})$ to give the j factor. The final equation can be written

$$\frac{h_i}{c_p G}\left(\frac{c_p\mu}{k}\right)^{2/3}\left(\frac{\mu_w}{\mu}\right)^{0.14} = j_H = 1.86\left(\frac{D}{L}\right)^{1/3}\left(\frac{DG}{\mu}\right)^{-2/3} \tag{12.57}$$

Equation (12.57) shows that for each value of the length-diameter ratio L/D, a logarithmic plot of the left-hand side versus Re gives a straight line with a slope of $-\frac{2}{3}$. The straight lines on the left-hand portion of Fig. 12.3 are plots of this equation for a few values of L/D. The lines terminate at a Reynolds number of 2,100.

Equation (12.57) should not be used for $L/D > 100$, since it may give coefficients smaller than the limiting values shown in Fig. 12.2.

Equation (12.34), when plotted for long tubes on the same coordinates, gives a straight line with a slope of -0.20 for Reynolds numbers above 6,000. This line is drawn in the right-hand region of Fig. 12.3.

The curved lines between Reynolds numbers of 2,100 and 6,000 represent the transition region. The effect of L/D is pronounced at the lower Reynolds numbers in this region and fades out as a Reynolds number of 6,000 is approached.

Figure 12.3 is a summary chart that can be used for the entire range of Reynolds numbers from 1,000 to 30,000. Beyond its lower and upper limits, Eqs. (12.27) and (12.34), respectively, can be used.

EXAMPLE 12.3. A light motor oil with the characteristics given below and in Table 12.3 is to be heated from 150 to 250°F (65.5 to 121.1°C) in a $\frac{1}{4}$-in. (6.35-mm) Schedule 40 pipe 15 ft (4.57 m) long. The pipe wall is at 350°F (176.7°C). How much oil can be heated in this pipe, in pounds per hour? What coefficient can be expected?

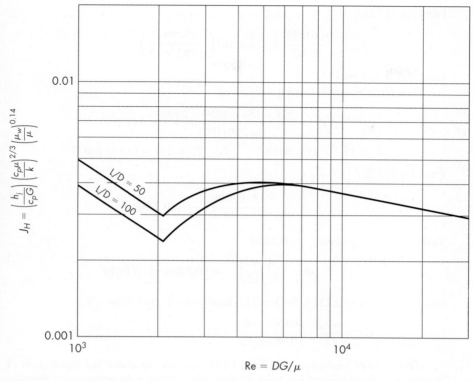

FIGURE 12.3
Heat transfer in transition range (*By permission of author and publisher, from W. H. McAdams, Heat Transmission, 3rd ed. Copyright by author, 1954, McGraw-Hill Book Company.*)

The properties of the oil are as follows: The thermal conductivity is 0.082 Btu/ft·h·°F (0.142 W/m·°C). The specific heat is 0.48 Btu/lb·°F (2.01 J/g·°C).

Solution. Assume the flow is laminar and that the Graetz number is large enough for Eq. (12.27) to apply. Data for substitution into Eq. (12.27) are

$$\mu = \frac{6.0 + 3.3}{2} = 4.65 \, \text{cP} \qquad \mu_w = 1.37 \, \text{cP} \qquad D = \frac{0.364}{12} = 0.0303 \, \text{ft} \qquad \text{(App. 3)}$$

$$\phi_v = \left(\frac{\mu}{\mu_w}\right)^{0.14} = \left(\frac{4.65}{1.37}\right)^{0.14} = 1.187 \qquad k = 0.082 \qquad c_p = 0.48$$

TABLE 12.3
Data for Example 12.3

Temperature		Viscosity, cP
°F	°C	
150	65.5	6.0
250	121.1	3.3
350	176.7	1.37

From Eq. (12.27)

$$\frac{0.0303h}{0.082} = 2 \times 1.187 \left(\frac{0.48\dot{m}}{0.082 \times 15} \right)^{1/3}$$

From this, $h = 4.69\dot{m}^{1/3}$

Data for substitution into Eq. (12.20) are

$$\overline{\Delta T_L} = \frac{350 - 150 - (350 - 250)}{\ln(200/100)} = 144°F$$

$$L = 15 \qquad D = 0.0303 \qquad \bar{T}_b - T_a = 250 - 150 = 100°F$$

From Eq. (12.20)

$$h = \frac{0.48 \times 100\dot{m}}{\pi 0.0303 \times 15 \times 144} = 0.233\dot{m}$$

Then

$$4.69\dot{m}^{1/3} = 0.233\dot{m}$$

$$\dot{m} = \left(\frac{4.69}{0.233} \right)^{3/2} = 90.3 \text{ lb/h (41.0 kg/h)}$$

and

$$h = 0.233 \times 90.3 = 21.0 \text{ Btu/ft}^2 \cdot \text{h} \cdot °F \text{ (119 W/m}^2 \cdot °C)$$

$$Gz = \frac{\dot{m}c_p}{kL} = \frac{90.3 \times 0.48}{0.082 \times 15} = 35.2$$

This is large enough so that Eq. (12.27) applies. To check the assumption of laminar flow, the maximum Reynolds number, which exists at the outlet end of the pipe, is calculated:

$$Re = \frac{DG}{\mu} = \frac{D\dot{m}}{\pi(D^2/4)\mu} = \frac{4 \times 90.3}{\pi \times 0.0303 \times 3.3 \times 2.42}$$

$$= 475$$

This is well within the laminar range.

HEAT TRANSFER TO LIQUID METALS

Liquid metals are used for high-temperature heat transfer, especially in nuclear reactors. Liquid mercury, sodium, and a mixture of sodium and potassium called NaK are commonly used as carriers of sensible heat. Mercury vapor is also used as a carrier of latent heat. Temperatures of 800°C and above are obtainable by using such metals. Molten metals have good specific heats, low viscosities, and high thermal conductivities. Their Prandtl numbers are therefore very low in comparison with those of ordinary fluids.

Equations such as (12.34) and (12.55) do not apply at Prandtl numbers below about 0.5, because the mechanism of heat flow in a turbulent stream differs from that in fluids of ordinary Prandtl numbers. In the usual fluid, heat transfer by conduction is limited to the viscous sublayer when Pr is unity or more and occurs in the buffer zone only when the number is less than unity. In liquid metals, heat transfer by conduction is important throughout the entire turbulent core and may predominate over convection throughout the tube.

Much study has been given to liquid-metal heat transfer in recent years, primarily in connection with its use in nuclear reactors. Design equations, all based on heat-momentum analogies, are available for flow in tubes, in annuli, between plates, and outside bundles of tubes. The equations so obtained are of the form

$$\text{Nu} = \alpha + \beta(\bar{\psi}\ \text{Pe})^{\gamma} \tag{12.58}$$

where α, β, and γ are constants or functions of geometry and of whether the wall temperature or the flux is constant and $\bar{\psi}$ is the average value of $\varepsilon_H/\varepsilon_M$ across the stream. For circular pipes, $\alpha = 7.0$, $\beta = 0.025$, and $\gamma = 0.8$. For other shapes, more elaborate functions are needed. A correlation for $\bar{\psi}$ is given by the equation[1]

$$\bar{\psi} = 1 - \frac{1.82}{\text{Pr}(\varepsilon_M/v)_m^{1.4}} \tag{12.59}$$

The quantity $(\varepsilon_M/v)_m$ is the maximum value of this ratio in the pipe, which is reached at a value of $y/r_w = \frac{5}{9}$. Equation (12.58) becomes, then,

$$\text{Nu} = 7.0 + 0.025\left[\text{Pe} - \frac{1.82\ \text{Re}}{(\varepsilon_M/v)_m^{1.4}}\right]^{0.8} \tag{12.60}$$

A correlation for $(\varepsilon_M/v)_m$ as a function of the Reynolds number is given in Fig. 12.4.

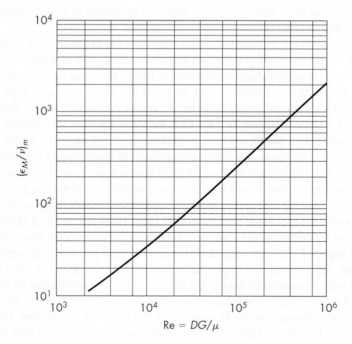

FIGURE 12.4
Values of $(\varepsilon_M/v)_m$ for fully developed turbulent flow of liquid metals in circular tubes.

The critical Peclet number

For a given Prandtl number, the Peclet number is proportional to the Reynolds number, because $Pe = Pr\,Re$. At a definite value of Pe the bracketed term in Eq. (12.60) becomes zero. This situation corresponds to the point where conduction controls and the eddy diffusion no longer affects the heat transfer. Below the critical Peclet number, only the first term in Eq. (12.60) is needed, and $Nu = 7.0$.

For laminar flow at uniform heat flux, by mathematical analysis $Nu = \frac{48}{11} = 4.37$. This has been confirmed by experiment.

HEATING AND COOLING OF FLUIDS IN FORCED CONVECTION OUTSIDE TUBES

The mechanism of heat flow in forced convection outside tubes differs from that of flow inside tubes, because of differences in the fluid flow mechanism. As has been shown on pages 61 and 118, no form drag exists inside tubes except perhaps for a short distance at the entrance end, and all friction is wall friction. Because of the lack of form friction, there is no variation in the local heat transfer at different points in a given circumference, and a close analogy exists between friction and heat transfer. An increase in heat transfer is obtainable at the expense of added friction simply by increasing the fluid velocity. Also, a sharp distinction exists between laminar and turbulent flow, which calls for different treatment of heat-transfer relations for the two flow regimes.

On the other hand, as shown on pages 150 to 155, in the flow of fluids across a cylindrical shape, boundary layer separation occurs, and a wake develops that causes form friction. No sharp distinction is found between laminar and turbulent flow, and a common correlation can be used for both low and high Reynolds numbers. Also, the local value of the heat-transfer coefficient varies from point to point around a circumference. In Fig. 12.5 the local value of the Nusselt number is plotted radially for all points around the circumference of the tube. At low Reynolds numbers, Nu_θ is a maximum at the front and back of the tube and a minimum at the sides. In practice, the variations in the local coefficient h_θ are often of no importance, and average values based on the entire circumference are used.

Radiation may be important in heat transfer to outside tube surfaces. Inside tubes, the surface cannot see surfaces other than the inside wall of the same tube, and heat flow by radiation does not occur. Outside tube surfaces, however, are necessarily in sight of external surfaces, if not nearby, at least at a distance, and the surrounding surfaces may be appreciably hotter or cooler than the tube wall. Heat flow by radiation, especially when the fluid is a gas, is appreciable in comparison with heat flow by conduction and convection. The total heat flow is then a sum of two independent flows, one by radiation and the other by conduction and convection. The relations given in the remainder of this section have to do with conduction and convection only. Radiation, as such and in combination with conduction and convection, is discussed in Chap. 14.

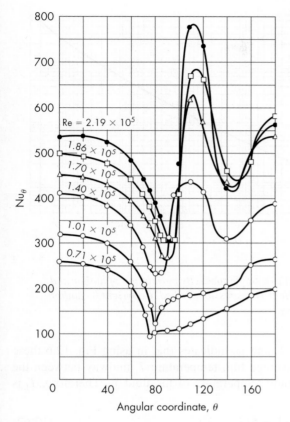

FIGURE 12.5
Local Nusselt number for airflow normal to a circular cylinder. (*Adapted with permission from W. H. Giedt, Trans. ASME, **71:**375, 1949.*)

Fluids flowing normal to a single tube

The variables affecting the coefficient of heat transfer to a fluid in forced convection outside a tube are D_o, the outside diameter of the tube; c_p, μ, and k, the specific heat at constant pressure, the viscosity, and the thermal conductivity, respectively, of the fluid; and G, the mass velocity of the fluid approaching the tube. Dimensional analysis gives, then, an equation of the type of Eq. (12.29):

$$\frac{h_o D_o}{k} = \psi_0\left(\frac{D_o G}{\mu}, \frac{c_p \mu}{k}\right) \tag{12.61}$$

Here, however, ends the similarity between the two types of process—the flow of heat to fluids inside tubes and the flow of heat to fluids outside tubes—and the functional relationships in the two cases differ.

For any one gas for which the Prandtl number is nearly independent of temperature, the Nusselt number is a function only of the Reynolds number. Experimental data for air are plotted in this way in Fig. 12.6. The effect of radiation is not included in this curve and must be calculated separately. The slope of the plot in Fig. 12.6 increases from about 0.4 to 0.7 as Re increases from 10 to 10^5, so no simple exponential equation fits the data.

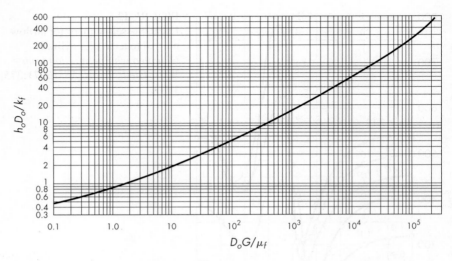

FIGURE 12.6
Heat transfer to air flowing normal to a single tube. (*By permission of author and publisher, from W. H. McAdams, Heat Transmission, 3rd ed. Copyright by author, 1954, McGraw-Hill Book Company.*)

The subscript f on the terms k_f and μ_f indicates that in using Fig. 12.6 these terms must be evaluated at the average film temperature T_f midway between the wall temperature and the mean bulk temperature of the fluid $\bar{T}$. Therefore, T_f is given by the equation

$$T_f = \tfrac{1}{2}(T_w + \bar{T}) \tag{12.62}$$

Figure 12.6 can be used for both heating and cooling.

For heating and cooling liquids flowing normal to single cylinders the following equation is used[4a]:

$$\frac{h_o D_o}{k_f}\left(\frac{c_p \mu_f}{k_f}\right)^{-0.3} = 0.35 + 0.56\left(\frac{D_o G}{\mu_f}\right)^{0.52} \tag{12.63}$$

This equation can also be used for gases from $Re = 1$ to $Re = 10^4$, but it gives lower values of the Nusselt number than Fig. 12.6 at higher Reynolds numbers. Equation (12.63) is plotted in *j*-factor form in Fig. 17.6, in the section of Chap. 17 dealing with the analogies between heat and mass transfer.

Heat-transfer data for flow normal to tubes or rods of noncircular cross section are given in the literature.[4b] Banks of tubes across which the fluid flows are common in industrial exchangers. Problems of heat flow in tube banks are discussed in Chap. 15.

Flow past single spheres

For heat transfer between a flowing fluid and the surface of a single sphere, the following equation is recommended:

$$\frac{h_o D_p}{k_f} = 2.0 + 0.60\left(\frac{D_p G}{\mu_f}\right)^{0.50}\left(\frac{c_p \mu_f}{k_f}\right)^{1/3} \tag{12.64}$$

where D_p is the diameter of the sphere. In a completely stagnant stream the Nusselt number $h_o D_p / k_f$ is equal to 2.0. A plot of Eq. (12.64) is given in Fig. 17.7.

Heat transfer in packed beds

Data for heat transfer between fluids and the particles in packed beds can be obtained from Fig. 17.7 or from Eq. (17.74) by replacing Sc with Pr and Sh with Nu. Heat transfer through a packed bed and to the wall of a packed-tube reactor is discussed in Chap. 15.

NATURAL CONVECTION

As an example of natural convection, consider a hot, vertical plate in contact with the air in a room. The temperature of the air in contact with the plate will be that of the surface of the plate, and a temperature gradient will exist from the plate out into the room. At the bottom of the plate, the temperature gradient is steep, as shown by the full line marked "Z = 10 mm" in Fig. 12.7. At distances above the bottom of the plate, the gradient becomes less steep, as shown by the full curve marked "Z = 240 mm" of Fig. 12.7. At a height of about 600 mm from the bottom of the plate, the temperature–distance curves approach an asymptotic condition and do not change with further increase in height.

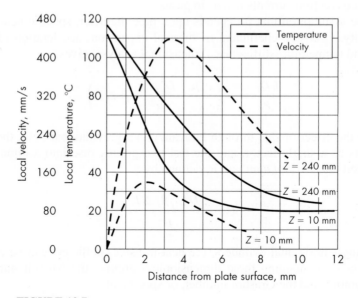

FIGURE 12.7
Velocity and temperature gradients, natural convection from heated vertical plate. (*By permission of author and publisher, from W. H. McAdams, Heat Transmission, 3rd ed. Copyright by author, 1954, McGraw-Hill Book Company.*)

The density of the heated air immediately adjacent to the plate is less than that of the unheated air at a distance from the plate, and the buoyancy of the hot air causes an unbalance between the vertical layers of air of differing density. The unbalanced forces generate a circulation by which hot air near the plate rises and cold air flows toward the plate from the room to replenish the rising airstream. A velocity gradient near the plate is formed. Since the velocities of the air in contact with the plate and out in the room are both zero, the velocity is a maximum at a definite distance from the wall. The velocity reaches its maximum a few millimeters from the surface of the plate. The dashed curves in Fig. 12.7 show the velocity gradients for heights of 10 and 240 mm above the bottom of the plate. For tall plates, an asymptotic condition is approached.

The temperature difference between the surface of the plate and the air in the room at a distance from the plate causes a transfer of heat by conduction into the current of gas next to the wall, and the stream carries the heat away by convection in a direction parallel to the plate.

The natural convection currents surrounding a hot, horizontal pipe are more complicated than those adjacent to a vertical heated plate, but the mechanism of the process is similar. The layers of air immediately next to the bottom and sides of the pipe are heated and tend to rise. The rising layers of hot air, one on each side of the pipe, separate from the pipe at points short of the top center of the pipe and form two independent rising currents with a zone of relatively stagnant and unheated air between them.

Natural convection in liquids follows the same pattern, because liquids are also less dense hot than cold. The buoyancy of heated liquid layers near a hot surface generates convection currents just as in gases.

On the assumption that h depends upon pipe diameter, specific heat, thermal conductivity, viscosity, coefficient of thermal expansion, acceleration of gravity, density, and temperature difference, dimensional analysis gives

$$\frac{hD_o}{k} = \Phi\left(\frac{c_p\mu}{k}, \ \frac{D_o^3\rho^2 g}{\mu^2}, \beta\,\Delta T\right) \tag{12.65}$$

Since the effect of β is through buoyancy in a gravitational field, the product $g\beta\,\Delta T$ acts as a single factor, and the last two groups fuse into a dimensionless group called the *Grashof number* Gr, defined by

$$\mathrm{Gr} = \frac{D_o^3\rho_f^2\beta_g\,\Delta T_o}{\mu_f^2} \tag{12.66}$$

For single horizontal cylinders, the heat-transfer coefficient can be correlated by an equation containing three dimensionless groups, the Nusselt number, the Prandtl number, and the Grashof number, or specifically,

$$\frac{hD_o}{k_f} = \Phi\left(\frac{c_p\mu_f}{k_f}, \ \frac{D_o^3\rho_f^2\beta g\,\Delta T_o}{\mu_f^2}\right) \tag{12.67}$$

where h = average heat-transfer coefficient, based on entire pipe surface
D_o = outside pipe diameter
k_f = thermal conductivity of fluid
c_p = specific heat of fluid at constant pressure
ρ_f = density of fluid
β = coefficient of thermal expansion of fluid
g = acceleration of gravity
ΔT_o = average difference in temperature between outside of pipe and fluid distant from wall
μ_f = viscosity of fluid

Unlike in the equations for laminar or turbulent flow in pipes, where μ, ρ, and k are evaluated at the bulk temperature, the fluid properties μ_f, ρ_f, and k_f are evaluated at the mean film temperature [Eq. (12.62)]. Radiation is not accounted for in this equation.

The coefficient of thermal expansion β is a property of the fluid, defined as the fractional increase in volume at constant pressure of the fluid per degree of temperature change, or mathematically,

$$\beta = \frac{(\partial v/\partial T)_p}{v} \tag{12.68}$$

where v = specific volume of fluid
$(\partial v/\partial T)_p$ = rate of change of specific volume with temperature at constant pressure

For liquids, β can be considered constant over a definite temperature range and Eq. (12.68) written as

$$\beta = \frac{\Delta v/\Delta T}{\bar{v}} \tag{12.69}$$

where $\bar{v}$ is the average specific volume. In terms of density,

$$\beta = \frac{1/\rho_2 - 1/\rho_1}{(T_2 - T_1)(1/\rho_1 + 1/\rho_2)/2} = \frac{\rho_1 - \rho_2}{\bar{\rho}_a(T_2 - T_1)} \tag{12.70}$$

where $\bar{\rho}_a = (\rho_1 + \rho_2)/2$
ρ_1 = density of fluid at temperature T_1
ρ_2 = density of fluid at temperature T_2

For an ideal gas, since $v = RT/p$,

$$\left(\frac{\partial v}{\partial T}\right)_p = \frac{R}{p}$$

and by using Eq. (12.68),

$$\beta = \frac{R/p}{RT/p} = \frac{1}{T} \tag{12.71}$$

The coefficient of thermal expansion of an ideal gas equals the reciprocal of the absolute temperature.

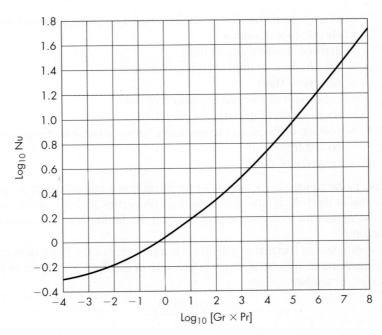

FIGURE 12.8

Heat transfer between single horizontal cylinders and fluids in natural convection.

In Fig. 12.8 is shown a relationship, based on Eq. (12.67), which satisfactorily correlates experimental data for heat transfer from a single horizontal cylinder to liquids or gases. The range of variables covered by the single line of Fig. 12.8 is very great.

For magnitudes of log Gr Pr of 4 or more, the line of Fig. 12.8 follows closely the empirical equation[7b]

$$Nu = 0.53(Gr\,Pr)_f^{0.25} \tag{12.72}$$

Natural convection to air from vertical shapes and horizontal planes

Equations for heat transfer in natural convection between fluids and solids of definite geometric shape are of the form[7a]

$$\frac{hL}{k_f} = b\left[\frac{L^3 \rho_f^2 g \beta_f\,\Delta T}{\mu_f^2}\left(\frac{c_p \mu}{k}\right)_f\right]^n \tag{12.73}$$

where b, n = constants

L = height of vertical surface or length of horizontal square surface

Properties are taken at the mean film temperature. Equation (12.73) can be written

$$Nu_f = b(Gr\,Pr)_f^n \tag{12.74}$$

Values of the constants b and n for various conditions are given in Table 12.4.

TABLE 12.4
Values of constants in Eq. (12.74)

System	Range of Gr Pr	b	n
Vertical plates, vertical cylinders	10^4–10^9	0.59	0.25
	10^9–10^{12}	0.13	0.333
Horizontal plates:			
Heated, facing upward or	10^5–2×10^7	0.54	0.25
cooled, facing down	2×10^7–3×10^{10}	0.14	0.333
Cooled, facing upward or			
heated, facing down	3×10^5–3×10^{10}	0.27	0.25

Source: By permission of author and publishers, from W. H. McAdams, *Heat Transmission,* 3rd ed., pp. 172, 180. Copyright by author, 1954, McGraw-Hill Book Company.

Effects of natural convection in laminar flow heat transfer

In laminar flow at low velocities, in large pipes, and at large temperature drops, natural convection may occur to such an extent that the usual equations for laminar flow heat transfer must be modified. The effect of natural convection in tubes is found almost entirely in laminar flow, as the higher velocities characteristic of flow in the transition and turbulent regimes overcome the relatively gentle currents of natural convection.

The effect of natural convection on the coefficient of heat transfer to fluids in laminar flow through horizontal tubes can be accounted for by multiplying the coefficient h_i, computed from Eq. (12.57) or Fig. 12.3, by the factor[5]

$$\phi_n = \frac{2.25(1 + 0.010\, Gr^{1/3})}{\log Re} \tag{12.75}$$

Natural convection also occurs in vertical tubes, increasing the rate of heat flow, when the fluid flow is upward, to above that found in laminar flow only. The effect is marked at values of Gr between 10 and 10,000 and depends[7e] on the magnitude of the quantity $Gr\, Pr\, D/L$.

EXAMPLE 12.4. Air at 1 atm pressure is passed through a horizontal 2-in. (51-mm) Schedule 40 steam-jacketed steel pipe at a velocity of 1.5 ft/s (0.457 m/s) and an inlet temperature of 68°F (20°C). The pipe wall temperature is 220°F (104.4°C). If the outlet air temperature is to be 188°F (86.7°C), how long must the heated section be?

Solution. To establish the flow regime, the Reynolds number based on the average temperature is calculated. The quantities needed are

$$\bar{T} = \frac{68 + 188}{2} = 128°F \qquad D = \frac{2.067}{12} = 0.1723\,\text{ft} \qquad \text{(App. 3)}$$

$$\mu \text{ (at } 128°F) = 0.019\,\text{cP} \qquad \text{(App. 8)}$$

$$\rho \text{ (at } 68°F) = \frac{29}{359}\left(\frac{492}{68 + 460}\right) = 0.0753\,\text{lb/ft}^3$$

$$\bar{V}\rho = G = 1.5 \times 0.0753 \times 3{,}600 = 406.4\,\text{lb/ft}^2 \cdot \text{h}$$

$$Re = \frac{DG}{\mu} = \frac{0.1723 \times 406.4}{0.019 \times 2.42} = 1{,}522$$

Hence flow is laminar, and Eq. (12.27) applies. The results may later require correction for the effect of natural convection using Eq. (12.75). To use Eq. (12.27), the following quantities are needed:

$$c_p \text{ (at } 128°F) = 0.25 \text{ Btu/lb} \cdot °F \qquad \text{(App. 14)}$$

$$k \text{ (at } 128°F) = 0.0163 \text{ Btu/ft} \cdot h \cdot °F \qquad \text{(App. 12)}$$

$$\text{(By linear interpolation)}$$

$$\mu_w \text{ (at } 220°F) = 0.021 \text{ cP} \qquad \text{(App. 8)}$$

The internal cross-sectional area of pipe is

$$S = 0.02330 \text{ ft}^2 \qquad \text{(App. 3)}$$

The mass flow rate is

$$\dot{m} = GS = 406.4 \times 0.02330 = 9.47 \text{ lb/h}$$

The heat load is

$$q = \dot{m}c_p(\bar{T}_b - T_a) = 9.47 \times 0.25(188 - 68) = 284.1 \text{ Btu/h}$$

The logarithmic mean temperature difference is

$$\Delta T_1 = 220 - 188 = 32°F \qquad \Delta T_2 = 220 - 68 = 152°F$$

$$\overline{\Delta T_L} = \frac{152 - 32}{\ln(152/32)} = 77.0°F$$

The heat-transfer coefficient $h = q/A\,\overline{\Delta T_L}$. From App. 3, for 2-in. Schedule 40 pipe, $A = 0.541L$. Hence

$$h = \frac{284.1}{0.541L \times 77} = \frac{6.820}{L}$$

Also, from Eq. (12.27), the heat-transfer coefficient is

$$h = \frac{2k}{D}\left(\frac{\dot{m}c_p}{kL}\right)^{1/3}\left(\frac{\mu}{\mu_w}\right)^{0.14}$$

$$= \frac{2 \times 0.0163}{0.1723}\left(\frac{9.47 \times 0.25}{0.0163L}\right)^{1/3}\left(\frac{0.019}{0.021}\right)^{0.14} = \frac{0.9813}{L^{1/3}}$$

Equating the two relationships for h gives

$$\frac{0.9813}{L^{1/3}} = \frac{6.820}{L}$$

from which $L = 18.32$ ft (5.58 m).

This result is now corrected for the effect of natural convection by using Eq. (12.75). This requires calculation of the Grashof number, for which the additional quantities needed are

$$\beta \text{ (at } 128°F) = \frac{1}{460 + 128} = 0.0017°R^{-1}$$

$$\Delta T = 220 - 128 = 92°F \qquad \rho \text{ (at } 128°F) = 0.0676 \text{ lb/ft}^3$$

The Grashof number is therefore

$$\text{Gr} = \frac{D^3 \rho^2 g \beta \, \Delta T}{\mu^2}$$

$$= \frac{0.1723^3 \times 0.0676^2 \times 32.174 \times 0.0017 \times 92}{(0.019 \times 6.72 \times 10^{-4})^2} = 0.7192 \times 10^6$$

Hence, from Eq. (12.75),

$$\phi_n = \frac{2.25[1 + 0.01(0.7192 \times 10^6)^{1/3}]}{\log 1{,}522} = 1.34$$

This factor is used to correct the value of L. Hence $L = 18.32/1.34 = 13.7$ ft (4.17 m).

SYMBOLS

A	Area, m^2 or ft^2
b	Constant in Eq. (12.73)
C	Constant
c_p	Specific heat at constant pressure, J/g $\cdot$ °C or Btu/lb $\cdot$ °F
D	Diameter, m or ft; D_e, equivalent diameter, $4r_H$; D_i, inside diameter; D_{ij}, inside diameter of jacket; D_{it}, inside diameter of inner tube; D_o, outside diameter; D_{ot}, outside diameter of inner tube; D_p, of spherical particle; $\bar{D}_L$, logarithmic mean
Fo	Fourier number, $4kL/c_p \rho D^2 \bar{V}$, dimensionless
f	Fanning friction factor, dimensionless
G	Mass velocity, kg/m$^2 \cdot$s, lb/ft$^2 \cdot$s, or lb/ft$^2 \cdot$h
Gr	Grashof number, $D^3 \rho^2 g \beta \, \Delta T/\mu^2$, dimensionless
Gz	Graetz number, $\dot{m}c_p/kL$, dimensionless
g	Gravitational acceleration, m/s^2 or ft/s^2
h	Individual heat-transfer coefficient, W/m$^2 \cdot$°C or Btu/ft$^2 \cdot$h$\cdot$°F; h_a, based on arithmetic mean temperature drop; h_i, average over inside of tube; h_o, for outside of tube or particle; h_x, local value; h_{x1}, at trailing edge of plate; h_∞, for fully developed flow in long pipes; h_θ, local value outside tube
j_H	Colburn j factor, St $(\text{Pr})^{2/3} \phi_v$, dimensionless
K'	Flow consistency index of non-newtonian fluid
k	Thermal conductivity, W/m $\cdot$°C or Btu/ft$\cdot$h$\cdot$°F; k_f, at mean film temperature; k_m, of tube wall
L	Length or height, m or ft
LMTD	Logarithmic mean temperature difference
m	Parameter in Eq. (12.28), $K'8^{n'-1}$; m_w, value at T_w
$\dot{m}$	Mass flow rate, kg/h or lb/h
Nu	Nusselt number, hD/k, dimensionless; Nu_f, at mean film temperature; Nu_x, local value on flat plate; Nu_θ, local value on outside of tube
n	Exponent in Eqs. (12.32) and (12.73)
n'	Flow behavior index of non-newtonian fluid, dimensionless

Pe   Peclet number, $\rho \bar{V} c_p D / k$, dimensionless

Pr   Prandtl number, $c_p \mu / k$, dimensionless

p Pressure, N/m^2 or lb_f/ft^2

q Heat flow rate, W or Btu/h; q_c, by conduction; q_t, by turbulent convection

R Gas law constant

Re   Reynolds number, DG/μ, dimensionless; Re_x, local value on flat plate $u_0 x\rho/\mu$; Re_{x1}, at trailing edge of plate

r Radius, m or ft; r_H, hydraulic radius of channel; r_m, of tube; r_w, radius of pipe

S Cross-sectional area of tube, m^2 or ft^2

St   Stanton number, $h/c_p G$, dimensionless

T Temperature, °C or °F; T_a, at inlet; T_b, at outlet; T_f, mean film temperature; T_i, instantaneous value; T_w, at wall or plate; T_∞, of approaching fluid; $\bar{T}$, average fluid temperature in tube; $\bar{T}_b$, bulk average fluid temperature at outlet; $\bar{T}_i$, time average of instantaneous values; T', fluctuating component; $\bar{T}'$, time average of fluctuating component

t_T Total time of heating or cooling, s or h

U Overall heat-transfer coefficient, $W/m^2 \cdot °C$ or $Btu/ft^2 \cdot h \cdot °F$; U_o, based on outside area; U_1, U_2, at ends of exchanger

u Fluid velocity, m/s or ft/s; u_0, of approaching fluid; u', fluctuating component

$\bar{V}$ Volumetric average fluid velocity, m/s or ft/s

v Specific volume, m^3/kg or ft^3/lb for liquids, m^3/kg mol or ft^3/lb mol for gases; $\bar{v}$, average value

v' Fluctuating component of velocity in y direction; $\bar{v}'$, time-average value

x Distance from leading edge of plate or from tube entrance, m or ft; x_w, wall thickness; x_0, at start of heated section; x_1, length of plate

y Radial distance from wall, m or ft; also, boundary layer thickness

Z Height, m or ft

Greek Letters

α Thermal diffusivity, $k/\rho c_p$, m^2/h or ft^2/h; also constant in Eq. (12.58)

β Coefficient of volumetric expansion, $1/°R$ or $1/K$; also constant in Eq. (12.58); β_f, at mean film temperature

γ Constant in Eq. (12.58)

ΔT Temperature drop, °C or °F; ΔT_i, from inner wall of pipe to fluid; ΔT_o, from outside surface to fluid distant from wall; $\overline{\Delta T_a}$, arithmetic mean temperature drop; $\overline{\Delta T_L}$, logarithmic mean temperature drop

δ Parameter in Eq. (12.28), $(3n' + 1)/4n'$

ε Turbulent diffusivity, m^2/h or ft^2/h; ε_H, of heat; ε_M, of momentum

θ Angular position on outside of tube

μ Absolute viscosity, $kg/m \cdot s$, $lb/ft \cdot s$ or $lb/ft \cdot h$; μ_f, average value of fluid film; μ_w, value at wall temperature

v Kinematic viscosity, m^2/h or ft^2/h

ρ Density, kg/m^3 or lb/ft^3; ρ_f, of fluid film; $\bar{\rho}_a$, arithmetic average value

τ Shear stress, N/m^2 or lb$_f$/ft^2; τ_w, shear stress at pipe wall; τ_0, yield stress of plastic fluid

Φ, Φ_1 Function

ϕ_n Natural convection factor [Eq. (12.75)]

ϕ_v Viscosity correction factor, $(\mu/\mu_w)^{0.14}$

ψ Function in Eq. (12.35); also ratio of turbulent diffusivities, $\varepsilon_H/\varepsilon_M$; $\bar{\psi}$, average value

ψ_0 Function in Eq. (12.61)

PROBLEMS

12.1. Glycerin is flowing at the rate of 700 kg/h through a 30-mm-ID pipe. It enters a heated section 2.5 m long, the walls of which are at a uniform temperature of 115°C. The temperature of the glycerin at the entrance is 15°C. (*a*) If the velocity profile is parabolic, what will be the temperature of the glycerin at the outlet of the heated section? (*b*) What would the outlet temperature be if flow were rodlike? (*c*) How long would the heated section have to be, to heat the glycerin essentially to 115°C?

12.2. Oil at 50°F is heated in a horizontal 2-in. Schedule 40 steel pipe 60 ft long having a surface temperature of 120°F. The oil flow rate is 150 gal/h at inlet temperature. What will be the oil temperature as it leaves the pipe and after mixing? What is the average heat-transfer coefficient? Properties of the oil are given in Table 12.5.

12.3. Oil is flowing through a 75-mm-ID iron pipe at 1 m/s. It is being heated by steam outside the pipe, and the steam-film coefficient may be taken as 11 kW/m$^2 \cdot$°C. At a particular point along the pipe, the oil is at 50°C, its density is 880 kg/m^3, its viscosity is 2.1 cP, its thermal conductivity is 0.135 W/m·°C, and its specific heat is 2.17 J/g·°C. What is the overall heat-transfer coefficient at this point based on the inside area of the pipe? If the steam temperature is 120°C, what is the heat flux at this point based on the outside area of the pipe?

12.4. Kerosene is heated by hot water in a shell-and-tube heater. The kerosene is inside the tubes, and the water is outside. The flow is countercurrent. The average temperature of the kerosene is 110°F, and the average linear velocity is 8 ft/s. The properties of the kerosene at 110°F are: specific gravity, 0.805; viscosity, 1.5 cP; specific heat,

TABLE 12.5
Data for Prob. 12.2

	60°F	120°F
Specific gravity, 60°F/60°F	0.79	0.74
Thermal conductivity, Btu/ft·h·°F	0.072	0.074
Viscosity, cP	18	8
Specific heat, Btu/lb·°F	0.75	0.75

0.583 Btu/lb·°F; and thermal conductivity, 0.0875 Btu/ft·h·°F. The tubes are low-carbon steel $\frac{3}{4}$-in. OD by BWG 16. The heat-transfer coefficient on the shell side is 300 Btu/ft²·h·°F. Calculate the overall coefficient based on the outside area of the tube.

12.5. Assume that the kerosene of Prob. 12.4 is replaced with water at 110°F and flowing at a velocity of 8 ft/s. What percentage increase in overall coefficient may be expected if the tube surfaces remain clean?

12.6. Both surfaces of the tube of Prob. 12.5 become fouled with deposits from the water. The fouling factors are 330 on the inside and 200 on the outside surfaces, both in Btu/ft²·h·°F. What percentage decrease in overall coefficient is caused by the fouling of the tube?

12.7. From the Colburn analogy, how much would the heat-transfer coefficient inside a 1-in. Schedule 40 steel pipe differ from that inside a 1-in. BWG 16 copper tube if the same fluid were flowing in each and the Reynolds number in both cases were 4×10^4?

12.8. Water must be heated from 15 to 50°C in a simple double-pipe heat exchanger at a rate of 3,500 kg/h. The water is flowing inside the inner tube with steam condensing at 110°C on the outside. The tube wall is so thin that the wall resistance may be neglected. Assume that the steam-film coefficient h_0 is 11 kW/m²·°C. What is the length of the shortest heat exchanger that will heat the water to the desired temperature? Average properties of water are as follows:

$$\rho = 993 \text{ kg/m}^3 \qquad k = 0.61 \text{ W/m·°C} \qquad \mu = 0.78 \text{ cP} \qquad c_p = 4.19 \text{ J/g·°C}$$

Hint: Find the optimum diameter for the tube.

12.9. Since the Prandtl number and the heat capacity of air are nearly independent of temperature, Eq. (12.34) seems to indicate that h_i for air increases with $\mu^{0.2}$ (a) Explain this anomaly and determine the approximate dependence of h_i on temperature, using $h_i \propto T^n$. (b) How does h_i for air vary with temperature if the linear velocity, rather than the mass velocity, is kept constant?

12.10. Air is flowing through a steam-heated tubular heater under such conditions that the steam and wall resistances are negligible in comparison with the air-side resistance. Assuming that each of the following factors is changed in turn but that all other original factors remain constant, calculate the percentage variation in $q/\Delta T_L$ that accompanies each change. (a) Double the pressure on the gas, but keep fixed the mass flow rate of the air. (b) Double the mass flow rate of the air. (c) Double the number of tubes in the heater. (d) Halve the diameter of the tubes.

12.11. A sodium-potassium alloy (78 percent K) is to be circulated through $\frac{1}{2}$-in.-ID tubes in a reactor core for cooling. The liquid-metal inlet temperature and velocity are to be 580°F and 32 ft/s. If the tubes are 3 ft long and have an inside surface temperature of 720°F, find the coolant temperature rise and the energy gain per pound of liquid metal. Properties of NaK (78 percent K) are as follows:

$$\rho = 45 \text{ lb/ft}^3 \qquad k = 179 \text{ Btu/ft·h·°F} \qquad \mu = 0.16 \text{ cP} \qquad c_p = 0.21 \text{ Btu/lb·°F}$$

12.12. In a catalytic cracking regenerator, catalyst particles at 600°C are injected into air at 700°C in a fluidized bed. Neglecting the chemical reaction, how long will it take for a 50-μm particle to be heated to within 5°C of the air temperature? Assume the heat-transfer coefficient is the same as for a spherical particle falling at its terminal velocity.

12.13. In a pilot plant, a viscous oil is being cooled from 200 to 110°C in a 1.0-in. jacketed pipe with water flowing in the jacket at an average temperature of 30°C. To get greater cooling of the oil, it has been suggested that the exchanger be replaced with one having a greater inside diameter (1.5 in.) but the same length. (*a*) If the oil is in laminar flow in the 1.0-in. pipe, what change in exit temperature might result from using the larger exchanger? (*b*) Repeat part (*a*), assuming the oil is in turbulent flow.

12.14. In the manufacture of nitric acid, air containing 10 percent ammonia is passed through a pack of fine-mesh wire screens of Pt/Rh alloy. (*a*) Calculate the heat-transfer coefficient for air at 500°C flowing at a superficial velocity of 20 ft/s past wires 0.5 mm in diameter. (*b*) If the surface area of the wire screen is 3.7 cm²/cm² of cross section, what is the temperature change in air, initially at 500°C, flowing through one screen if the surface of the wires is at 900°C?

12.15. Water at 15°C is flowing at right angles across a heated 25-mm-OD cylinder, the surface temperature of which is 120°C. The approach velocity of the water is 1 m/s. (*a*) What is the heat flux, in kilowatts per square meter, from the surface of the cylinder to the water? (*b*) What would be the flux if the cylinder were replaced by a 25-mm-OD sphere, also with a surface temperature of 120°C?

12.16. Water is heated from 15 to 65°C in a steam-heated horizontal 50-mm-ID tube. The steam temperature is 120°C. The average Reynolds number of the water is 450. The individual coefficient of the water is controlling. By what percentage would natural convection increase the total rate of heat transfer over that predicted for purely laminar flow? Compare your answer with the increase indicated in Example 12.4.

12.17. A large tank of water is heated by natural convection from submerged horizontal steam pipes. The pipes are 3-in. Schedule 40 steel. When the steam pressure is atmospheric and the water temperature is 80°F, what is the rate of heat transfer to the water in Btu per hour per foot of pipe length?

12.18. (*a*) Calculate the overall coefficient U for heat transfer through a vertical glass window from a room at 70°F to still air at 0°F. Assume that a single pane of glass is $\frac{1}{8}$ in. thick and 4 ft high. (*b*) Calculate U for a thermopane window with a $\frac{1}{2}$-in. airspace between the two panes. For this glass $k = 0.4$ Btu/h·ft·°F.

12.19. How does U for a thermopane window depend on the spacing between the panes?

REFERENCES

1. Dwyer, O. E. *AIChE J.,* **9:**261 (1963).
2. Eckert, E. R. G., and J. F. Gross. *Introduction to Heat and Mass Transfer.* New York: McGraw-Hill, 1963, pp. 110–4.

3. Friend, W. L., and A. B. Metzner. *AIChE J.,* **4:**393 (1958).
4. Gebhart, B. *Heat Transfer,* 2nd ed. New York: McGraw-Hill, 1971; (*a*) p. 272, (*b*) p. 274, (*c*) p. 283.
5. Kern, D. Q., and D. F. Othmer. *Trans. AIChE,* **39:**517 (1943).
6. Knudsen, J. G., and D. L. Katz. *Fluid Dynamics and Heat Transfer.* New York: McGraw-Hill, 1958; (*a*) pp. 361–90, (*b*) pp. 400–3, (*c*) p. 439.
7. McAdams, W. H. *Heat Transmission,* 3rd ed. New York: McGraw-Hill, 1954; (*a*) pp. 172, 180, (*b*) p. 177, (*c*) p. 215, (*d*) p. 230, (*e*) p. 234.
8. Mills, A. F. *Heat Transfer.* Boston: Irwin Press, 1992.
9. Sieder, E. N., and G. E. Tate. *Ind. Eng. Chem.,* **28:**1429 (1936).
10. Webb, R. L. *Principles of Enhanced Heat Transfer.* New York: Wiley, 1994.
11. Wilkinson, W. L. *Non-Newtonian Fluids.* London: Pergamon, 1960, p. 104.

CHAPTER 13

Heat Transfer to Fluids with Phase Change

Processes of heat transfer accompanied by phase change are more complex than simple heat exchange between fluids. A phase change involves the addition or subtraction of considerable quantities of heat at constant or nearly constant temperature. The rate of phase change may be governed by the rate of heat transfer, but it is often influenced by the rate of nucleation of bubbles, drops, or crystals and by the behavior of the new phase after it is formed. This chapter covers condensation of vapors and boiling of liquids. Crystallization is discussed in Chap. 27.

HEAT TRANSFER FROM CONDENSING VAPORS

The condensation of vapors on the surfaces of tubes cooler than the condensing temperature of the vapor is important when vapors such as those of water, hydrocarbons, and other volatile substances are processed. Some examples will be met later in this text, in discussing the unit operations of evaporation, distillation, and drying.

The condensing vapor may consist of a single substance, a mixture of condensable and noncondensable substances, or a mixture of two or more condensable vapors. Friction losses in a condenser are normally small, so that condensation is essentially a constant-pressure process. The condensing temperature of a single pure substance depends only on the pressure, and therefore the process of condensation of a pure substance is isothermal. Also, the condensate is a pure liquid. Mixed vapors, condensing at constant pressure, condense over a temperature range and yield a condensate of variable composition until the entire vapor stream is condensed, when the composition of the condensate equals that of the original uncondensed vapor.[†]

[†]Exceptions to this statement are found in the condensation of azeotropic mixtures, which are discussed in a later chapter.

A common example of the condensation of one constituent from its mixture with a second noncondensable substance is the condensation of water from a mixture of steam and air.

Condensation of mixed vapors and condensation in the presence of noncondensing gases are discussed briefly later in this chapter. The following discussion is limited to the condensation of a single volatile substance on a cold tube.

Dropwise and film-type condensation

A vapor may condense on a cold surface in one of two ways, which are well described by the terms *dropwise* and *film type*. In film condensation, which is more common than dropwise condensation, the liquid condensate forms a film, or continuous layer, of liquid that flows over the surface of the tube under the action of gravity. It is the layer of liquid interposed between the vapor and the wall of the tube that provides the resistance to heat flow and therefore fixes the magnitude of the heat-transfer coefficient.

In dropwise condensation the condensate begins to form at microscopic nucleation sites. Typical sites are tiny pits, scratches, and dust specks. The drops grow and coalesce with their neighbors to form visible fine drops like those often seen on the outside of a cold-water pitcher in a humid room. The fine drops, in turn, coalesce into rivulets, which flow down the tube under the action of gravity, sweep away condensate, and clear the surface for more droplets. During dropwise condensation, large areas of the tube surface are covered with an extremely thin film of liquid of negligible thermal resistance. Because of this the heat-transfer coefficient at these areas is very high; the average coefficient for dropwise condensation may be 5 to 8 times that for film-type condensation. On long tubes, condensation on some of the surface may be film condensation and the remainder dropwise condensation.

The most important and extensive observations of dropwise condensation have been made on steam, but it has also been observed in ethylene glycol, glycerin, nitrobenzene, isoheptane, and some other organic vapors.[26] Liquid metals usually condense in the dropwise manner. The appearance of dropwise condensation depends upon the wetting or nonwetting of the surface by the liquid, and fundamentally, the phenomenon lies in the field of surface chemistry. Much of the experimental work on the dropwise condensation of steam is summarized in the following paragraphs.[8]

1. Film-type condensation of water occurs on tubes of the common metals if both the steam and the tube are clean, in the presence or absence of air, on rough or on polished surfaces.
2. Dropwise condensation is obtainable only when the cooling surface is not wetted by the liquid. In the condensation of steam it is often induced by contamination of the vapor with droplets of oil. It is more easily maintained on a smooth surface than on a rough surface.
3. The quantity of contaminant or promoter required to cause dropwise condensation is minute, and apparently only a monomolecular film is necessary.
4. Effective drop promotors are strongly adsorbed by the surface, and substances that merely prevent wetting are ineffective. Some promoters are especially

effective on certain metals, for example, mercaptans on copper alloys; other promoters, such as oleic acid, are quite generally effective. Some metals, such as steel and aluminum, are difficult to treat to give dropwise condensation.

5. The average coefficient obtainable in pure dropwise condensation may be as high as 115 kW/m² · °C (20,000 Btu/ft² · °F).

Although attempts are sometimes made to realize practical benefits from these large coefficients by artificially inducing dropwise condensation, this type of condensation is so unstable and the difficulty of maintaining it so great that the method is not common. Also the resistance of the layer of steam condensate even in film-type condensation is ordinarily small in comparison with the resistance inside the condenser tube, and the increase in the overall coefficient is relatively small when dropwise condensation is achieved. For normal design, therefore, film-type condensation is assumed.

Coefficients for film-type condensation

The basic equations for the rate of heat transfer in film-type condensation were first derived by Nusselt.[17,21,23] The Nusselt equations are based on the assumption that the vapor and liquid at the outside boundary of the liquid layer are in thermodynamic equilibrium, so that the only resistance to the flow of heat is that offered by the layer of condensate flowing downward in laminar flow under the action of gravity. It is also assumed that the velocity of the liquid at the wall is zero, that the velocity of the liquid at the outside of the film is not influenced by the velocity of the vapor, and that the temperatures of the wall and the vapor are constant. Superheat in the vapor is neglected, the condensate is assumed to leave the tube at the condensing temperature, and the physical properties of the liquid are taken at the mean film temperature.

Vertical tubes. In film-type condensation, the Nusselt theory shows that the condensate film starts to form at the top of the tube and that the thickness of the film increases rapidly near the top of the tube and then more and more slowly in the remaining length. The heat is assumed to flow through the condensate film solely by conduction, and the local coefficient h_x is therefore given by

$$h_x = \frac{k_f}{\delta} \tag{13.1}$$

where δ is the local film thickness.

The local coefficient therefore changes inversely with the film thickness. The variations of both h_x and δ with distance from the top of the tube are shown for a liquid like methanol in Fig. 13.1.[17]

Film thickness δ is typically two or three orders of magnitude smaller than the tube diameter; it can therefore be found, for flow either inside or outside a tube, from the equation for a flat plate, Eq. (4.50). Since there is a temperature gradient in the film, the properties of the liquid are evaluated at the average film temperature T_f, given later by Eq. (13.11). For condensation on a vertical surface, for which $\cos \beta = 1$, Eq. (4.50) becomes

$$\delta = \left(\frac{3\Gamma \mu_f}{\rho_f^2 g} \right)^{1/3} \tag{13.2}$$

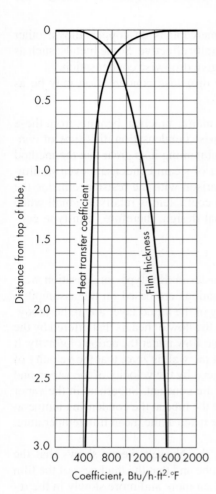

FIGURE 13.1
Film thickness and local coefficients, descending film of condensate. (*By permission, from D. Q. Kern, Process Heat Transfer. Copyright, 1950, McGraw-Hill Book Company.*)

where Γ is the condensate loading, the mass rate per unit length of periphery.

Substitution for δ in Eq. (13.1) gives for the local heat-transfer coefficient, at a distance L from the top of the vertical surface, the equation

$$h_x = k_f \left(\frac{\rho_f^2 g}{3 \Gamma \mu_f} \right)^{1/3}$$

(13.3)

Equation (13.3) applies to condensation either inside or outside a tube. Pure vapors are usually condensed on the outside of tubes; for this situation, for vertical tubes, the local coefficient is given by the relations

$$h_x = \frac{dq}{\Delta T_o \, dA_o} = \frac{\lambda \, d\dot{m}}{\Delta T_o \, \pi D_o \, dL}$$

(13.4)

where λ = heat of vaporization

$\dot{m}$ = local flow rate of condensate

Since $\dot{m}/\pi D_o = \Gamma$, Eq. (13.4) may be written

$$h_x = \frac{\lambda \, d\Gamma}{\Delta T_o \, dL} \tag{13.5}$$

The *average* coefficient h for the entire tube is defined by

$$h \equiv \frac{q_T}{A_o \, \Delta T_o} = \frac{\dot{m}_T \lambda}{\pi D_o L_T \Delta T_o} = \frac{\Gamma_b \lambda}{L_T \, \Delta T_o} \tag{13.6}$$

where q_T = total rate of heat transfer
  $\dot{m}_T$ = total rate of condensation
  L_T = total tube length
  Γ_b = condensate loading at bottom of tube

Eliminating h_x from Eqs. (13.3) and (13.5) and solving for ΔT_o give

$$\Delta T_o = \left(\frac{3\Gamma \mu_f}{\rho_f^2 g}\right)^{1/3} \frac{\lambda \, d\Gamma}{k_f \, dL} \tag{13.7}$$

Substituting ΔT_o from Eq. (13.7) into Eq. (13.6) gives

$$h = \frac{\Gamma_b k_f}{L_T}\left(\frac{\rho_f^2 g}{3\mu_f}\right)^{1/3} \frac{dL}{\Gamma^{1/3} \, d\Gamma} \tag{13.8}$$

Rearranging Eq. (13.8) and integrating between limits lead to

$$h \int_0^{\Gamma_b} \Gamma^{1/3} \, d\Gamma = \frac{\Gamma_b k_f}{L_T}\left(\frac{\rho_f^2 g}{3\mu_f}\right)^{1/3} \int_0^{L_T} dL$$

from which

$$h = \frac{4k_f}{3}\left(\frac{\rho_f^2 g}{3\Gamma_b \mu_f}\right)^{1/3} \tag{13.9}$$

Comparing Eqs. (13.9) and (13.3) shows that the average coefficient for a vertical tube, provided flow is laminar, is $\frac{4}{3}$ times the local coefficient at the bottom of the tube.

Equation (13.9) can be rearranged to include the exit Reynolds number for the condensate film $4\Gamma_b/\mu_f$.

$$h\left(\frac{\mu_f^2}{k_f^3 \rho_f^2 g}\right)^{1/3} = 1.47\left(\frac{4\Gamma_b}{\mu_f}\right)^{-1/3} = 1.47\,\mathrm{Re}^{-1/3} \tag{13.10}$$

On the assumption that the temperature gradient is constant across the film and that $1/\mu$ varies linearly with temperature, the reference temperature for evaluating μ_f, k_f, and ρ_f is given by the equation[21]

$$T_f = T_h - \frac{3(T_h - T_w)}{4} = T_h - \frac{3\,\Delta T_o}{4} \tag{13.11}$$

where T_f = reference temperature

T_h = temperature of condensing vapor
T_w = temperature of outside surface of tube wall

Equation (13.10) is often used in an equivalent form, in which the term Γ_b has been eliminated by combining Eqs. (13.6) and (13.10) to give

$$h = 0.943 \left(\frac{k_f^3 \rho_f^2 g \lambda}{\Delta T_o L \mu_f} \right)^{1/4} \tag{13.12}$$

Equations (13.10) and (13.11) were derived on the assumption of laminar flow of the condensate, and experimental data agree with the theory, but only for low Reynolds numbers. When the Reynolds number $4\Gamma_b/\mu_f$ is greater than about 30, ripples or waves appear on the surface of the condensate film and increase the rate of heat transfer. The average film thickness is still in reasonable agreement with the laminar flow theory, up to Re $\cong 1,200$ (see p. 80), but as Re is increased, the heat-transfer coefficient becomes considerably greater than predicted. When Re is greater than 1,800, the film becomes turbulent and local heat-transfer rates are much higher than predicted from Eq. (13.3).

A correlation for the average coefficient as a function of the Reynolds number at the bottom of the tube is given in Fig. 13.2. At very low Reynolds numbers the graph agrees with Eq. (13.10). For the wavy laminar and turbulent regions the empirical equation is

$$\text{Nu}' = \frac{h}{k_f} \left(\frac{\mu_f^2}{\rho_f^2 g} \right)^{1/3} = (\text{Re}^{-0.44} + 5.82 \times 10^{-6} \text{ Re}^{0.8} \text{ Pr}^{1.3})^{1/2} \tag{13.13}$$

The second term in Eq. (13.13) can be interpreted as a type of Nusselt number with a characteristic length of $(\mu_f^2/\rho_f^2 g)^{1/3}$, but this length is not the same as the film thickness δ.

In the laminar region, where Eq. (13.10) holds, the Prandtl number has no effect on h, since heat is transferred only by conduction through the film, and the heat capacity c_p is not a factor. In the wavy laminar region small ripples cause eddies which improve the heat transfer near the liquid surface, and the Prandtl number has a small effect. In this region Nu' varies with about $\text{Re}^{-0.2}$ compared with $\text{Re}^{-1/3}$ for true laminar flow. In the fully turbulent region the average coefficient increases with about the 0.4 power of the Reynolds number, with a greater dependence on Pr than for turbulent flow in pipes. The 0.65 power of Pr fits the experimental data, but may reflect some influence of the vapor velocity.

If there is a high vapor velocity in the downward direction, the condensate film becomes thinner because of shear stress at the liquid-vapor interface, and the heat-transfer rates are increased. Counterflow of vapor and liquid tends to reduce the heat-transfer rate, and can lead to flooding of the tube. The effects of vapor velocity on the film coefficient have been predicted by Dukler.[9]

Horizontal tubes. Corresponding to Eqs. (13.10) and (13.12) for vertical tubes, the following equations apply to single horizontal tubes:

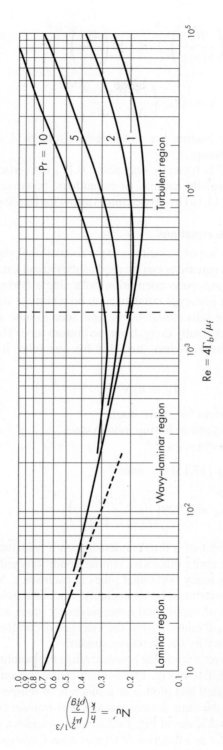

FIGURE 13.2
Film coefficient for condensation on vertical surfaces.

$$h\left(\frac{\mu_f^2}{k_f^3\rho_f^2 g}\right)^{1/3} = 1.51\left(\frac{4\Gamma'}{\mu_f}\right)^{-1/3} \tag{13.14}$$

and
$$h = 0.729\left(\frac{k_f^3\rho_f^2 g\lambda}{\Delta T_o D_o\mu_f}\right)^{1/4} \tag{13.15}$$

where Γ' is the condensate loading per unit *length* of tube $\dot{m}/L$ and all the other symbols have the usual meaning.

For horizontal tubes, Γ' is typically only about $0.1\Gamma_b$ for vertical tubes, and the flow of the condensate is usually laminar. Some authors, however, recommend increasing h found from Eq. (13.14) by 20 percent to allow for rippling when Re > 40.

Practical use of Nusselt equations

In the absence of high vapor velocities, experimental data check Eqs. (13.14) and (13.15) well, and these equations can be used as they stand for calculating heat-transfer coefficients for film-type condensation on a single horizontal tube. Also, Eq. (13.15) can be used for film-type condensation on a vertical stack of horizontal tubes, where the condensate falls cumulatively from tube to tube and the total condensate from the entire stack finally drops from the bottom tube. The average coefficient h_N for the stack of tubes is less than that for one tube; it is given by the equation[24a]

$$h_N = h_1 N^{-1/4} \tag{13.16}$$

where h_N = average coefficient for entire stack
$\quad\quad h_1$ = condensing coefficient for top tube in stack
$\quad\quad N$ = number of tubes in stack

For a stack of tubes Eq. (13.15) becomes[14]

$$h_N = 0.729\left(\frac{k_f^3\rho_f^2 g\lambda}{N\,\Delta T_o D_o\mu_f}\right)^{1/4} \tag{13.17}$$

In general, the coefficient of a film condensing on a horizontal tube is larger than that on a vertical tube under otherwise similar conditions unless the tubes are very short or there are very many horizontal tubes in each stack. Vertical tubes are preferred when the condensate must be appreciably subcooled below its condensation temperature. Mixtures of vapors and noncondensing gases are usually cooled and condensed *inside* vertical tubes, so that the inert gas is continually swept away from the heat-transfer surface by the incoming stream.

For a given substance over a moderate pressure range the quantity $(k_f^3\rho_f^2 g/\mu_f^2)^{1/3}$ is a function of temperature. Use of Fig. 13.2 is facilitated if this quantity, which can be denoted by ψ_f, is calculated and plotted as a function of temperature for a given substance. Quantity ψ_f has the same dimensions as a heat-transfer coefficient, so that both the ordinate and abscissa scales of Fig. 13.2 are dimensionless. Appendix 6 gives the magnitude of ψ_f for water as a function of temperature. Corresponding tables can be prepared for other substances when desired.

EXAMPLE 13.1. A shell-and-tube condenser with vertical $\frac{3}{4}$-in. (19-mm) BWG 16 copper tubes has chlorobenzene condensing at atmospheric pressure in the shell. The latent heat of condensation of chlorobenzene is 324.9 J/g. The tubes are 1.52 m long. Cooling water at an average temperature of 79°C is flowing in the tubes. The water-side coefficient is 4,540 W/m² · °C. (*a*) What is the coefficient of the condensing chlorobenzene? (*b*) What would the coefficient be in a horizontal condenser with the same number of tubes if the average number of tubes in a vertical stack were 6? Neglect fouling factors and tube wall resistance.

Solution

(*a*) A trial-and-error solution may be necessary since the condensate film coefficient depends on the Reynolds number, which in turn depends on the rate of condensation. Also, the wall temperature must be estimated to determine the physical properties of the condensate.

The quantities that may be specified directly are

$$\lambda = 324.9 \text{ J/g} \qquad g = 9.8 \text{ m/s}^2 \qquad L = 1.52 \text{ m} \qquad h_i = 4,540 \text{ W/m}^2 \cdot {}^\circ\text{C}$$

The condensing temperature T_f is 132°C, so the wall temperature T_w must lie between 79 and 132°C. It is probably closer to 79°C because the resistance of the organic condensate is usually greater than that of flowing water. As a first approximation, T_w is taken as 90°C. From Eq. (13.11),

$$T_f = 132 - 3/4(132 - 90) = 100{}^\circ\text{C}$$

$$\rho = 1,106 \text{ kg/m}^3 \text{ at } 20{}^\circ\text{C}$$

$$\rho_f = 1,018 \text{ kg/m}^3 \text{ at } 100{}^\circ\text{C} \qquad \text{(assumes 1\% decrease per 10°C)}$$

$$k_f = 0.144 \text{ W/m} \cdot {}^\circ\text{C at } 10{}^\circ\text{C} \qquad \text{(App. 13)}$$

$$k_f = 0.11 \text{ W/m} \cdot {}^\circ\text{C at } 100{}^\circ\text{C} \qquad \text{(assumes 3\% decrease per 10°C)}$$

$$\mu_f = 0.32 \text{ cP at } 100{}^\circ\text{C} = 3.2 \times 10^{-4} \text{ Pa} \cdot \text{s} \qquad \text{(App. 9)}$$

$$c_p = 0.37 \text{ cal/g} \cdot {}^\circ\text{C at } 100{}^\circ\text{C} = 1,550 \text{ J/kg} \cdot {}^\circ\text{C} \qquad \text{(App. 15)}$$

Then
$$\text{Pr} = \frac{1,550(3.2 \times 10^{-4})}{0.11} = 4.51$$

For a first estimate, assume Re $= 2,000$. From Fig. 13.2,

$$\text{Nu}' = \frac{h}{k_f} \left(\frac{\mu_f^2}{\rho_f^2 g} \right)^{1/3} = 0.23$$

[*Note:* This result may be checked by substituting the values for Re and Pr into Eq. (13.13). This gives Nu' = 0.231.] Then

$$h = 0.23 \times 0.11 \left[\frac{1018^2 \times 9.8}{(3.2 \times 10^{-4})^2} \right]^{1/3} = 1,171 \text{ W/m}^2 \cdot {}^\circ\text{C}$$

$$\frac{1}{U} = \frac{1}{1,171} + \frac{1}{4,540} = 1.07 \times 10^{-3}$$

$$U = 931 \text{ W/m}^2 \cdot {}^\circ\text{C}$$

Check the wall temperature, using Eq. (12.38).

$$\Delta T_i = \frac{1/4,540}{1/931}(53) = 11°C$$

$$T_w = 79 + 11 = 90°C \qquad \text{(as assumed)}$$

From App. 4, the area per tube is

$$A = 0.1963 \text{ ft}^2/\text{ft} \times 0.3048 \times 1.52 = 0.0909 \text{ m}^2$$

Condensate per tube:

$$\frac{931 \times 0.0909(132 - 79)}{324.9} = 13.8 \text{ g/s}$$

$$\Gamma_b = \frac{13.8 \times 10^{-3}}{\pi(0.019)} = 0.231 \text{ kg/s} \cdot \text{m}$$

$$\text{Re} = \frac{4\Gamma_b}{\mu_f} = \frac{4 \times 0.231}{3.2 \times 10^{-4}} = 2,890$$

From Fig. 13.2, Nu′ is slightly more than 0.23. [Solving Eq. (13.13) with Re = 2,890 gives Nu′ = 0.233.] Either way, the first estimate of Re is close enough.[†]

$$h = 1,171 \text{ W/m}^2 \cdot °C \ (206 \text{ Btu/h} \cdot \text{ft}^2 \cdot °F)$$

(b) For a horizontal condenser, use Eq. (13.17) and assume the wall temperature to be about 90°C. For $N = 6$,

$$h = 0.729 \left(\frac{0.144^3 \times 1,018^2 \times 9.8 \times 324,900}{6 \times 53 \times 0.019 \times 3.2 \times 10^{-4}} \right)^{1/4}$$

$$= 1,095 \text{ W/m}^2 \cdot °C \ (193 \text{ Btu/h} \cdot \text{ft}^2 \cdot °F)$$

This is only slightly less than the value for a vertical tube, so no adjustment of the wall temperature is needed.

[†]The distance for conduction implied in Nu′ is

$$\left(\frac{\mu_f^2}{\rho_f^2 g} \right)^{1/3} = \left[\frac{(3.2 \times 10^{-4})^2}{1,018^2 \times 9.8} \right]^{1/3} = 2.16 \times 10^{-5} \text{ m} = 21.6 \ \mu\text{m}$$

The thickness corresponding to the value of h is found from

$$h = \frac{k_f}{x} = 1,171 \text{ W/m}^2 \cdot °C$$

The equivalent thickness is

$$x = \frac{0.11}{1,171} = 9.4 \times 10^{-5} \text{ m} = 94 \ \mu\text{m}$$

The film thickness predicted for laminar flow, from Eq. (4.50), is

$$\delta = \left(\frac{3\mu\Gamma}{\rho^2 g \cos \beta} \right)^{1/3} = \left(\frac{3 \times 3.2 \times 10^{-4} \times 0.233}{1,018^2 \times 9.8 \times 1} \right)^{1/3} = 2.79 \times 10^{-4} \text{ m} = 279 \ \mu\text{m}$$

Thus turbulence makes h more than 3 times greater than that for conduction through a laminar film.

Condensation of superheated vapors

If the vapor entering a condenser is superheated, both the sensible heat of superheat and the latent heat of condensation must be transferred through the cooling surface. For steam, because of the low specific heat of the superheated vapor and the large latent heat of condensation, the heat of superheat is usually small in comparison with the latent heat. For example, 50°C of superheat represents only 100 J/g, as compared with approximately 2,300 J/g for latent heat. In the condensation of organic vapors, such as petroleum fractions, the superheat may be appreciable in comparison with the latent heat. When the heat of superheat is important, either it can be calculated from the degrees of superheat and the specific heat of the vapor and added to the latent heat; or if tables of thermal properties are available, the total heat transferred per pound of vapor can be calculated by subtracting the enthalpy of the condensate from that of the superheated vapor.

The effect of superheat on the rate of heat transfer depends upon whether the temperature of the surface of the tube is higher or lower than the condensation temperature of the vapor. If the temperature of the tube is lower than the temperature of condensation, the tube is wet with condensate, just as in the condensation of saturated vapor, and the temperature of the outside boundary of the condensate layer equals the saturation temperature of the vapor at the pressure existing in the equipment. The situation is complicated by the presence of a thermal resistance between the bulk of the superheated vapor and the outside of the condensate film and by the existence of a temperature drop, which is equal to the degrees of superheat in the vapor, across that resistance. Practically, however, the net effect of these complications is small, and it is satisfactory to assume that the entire heat load, which consists of the heats of superheat and of condensation, is transferred through the condensate film; that the temperature drop is that across the condensate film; and that the coefficient is the average coefficient for condensing vapor as read from Fig. 13.2. The procedure is summarized by the equation

$$q = hA(T_h - T_w) \tag{13.18}$$

where q = total heat transferred, including latent heat and superheat
A = area of heat-transfer surface in contact with vapor
h = coefficient of heat transfer, from Fig. 13.2
T_h = saturation temperature of vapor
T_w = temperature of tube wall

When the vapor is highly superheated and the exit temperature of the cooling fluid is close to that of condensation, the temperature of the tube wall may be greater than the saturation temperature of the vapor, condensation cannot occur, and the tube wall will be dry. The tube wall remains dry until the superheat has been reduced to a point where the tube wall becomes cooler than the condensing temperature of the vapor, and condensation takes place. The equipment can be considered as two sections, one a desuperheater and the other a condenser. In calculations, the two sections must be considered separately. The desuperheater is essentially a gas cooler. The logarithmic mean temperature difference applies, and the heat-transfer coefficient is that for cooling a fixed gas. The condensing section is treated by the methods described in the previous paragraphs.

Because of the low individual coefficient on the gas side, the overall coefficient in the desuperheater section is small, and the area of the heating surface in that section is large in comparison with the amount of heat removed. This situation should be avoided in practice. Superheat can be eliminated more economically by injection of a spray of liquid directly into the superheated vapor, since small drops evaporate very rapidly, cooling the vapor to the saturation temperature. The desuperheating section is thereby eliminated, and condensation occurs with high coefficients.

Condensation of mixed vapors

If the vapor contains two or more volatile components (unless it is an azeotropic mixture), the condensation temperature is no longer constant at a given pressure. Concentration gradients exist in both the vapor and liquid phases as the higher-boiling-point component or components tend to condense, enriching the vapor in lower-boiling-point material. If the coolant temperature is low enough, all the vapor may eventually be condensed; the composition of the condensate will then be the same as that of the original vapor. In other cases some of the low-boiling-point material may not be condensed and must be vented from the condenser.

Because of the concentration gradients in the vapor there is mass transfer[†] of the higher-boiling-point components to the condensate surface through the uncondensed lower-boiling-point components of the vapor. This occurs in parallel with heat transfer to remove sensible heat from the vapor as is cools. In general the rates of heat and mass transfer, and the corresponding heat- and mass-transfer coefficients, vary throughout the condenser and must be estimated at several points. The required area is then found by an iterative procedure. For binary systems a design method was developed by Colburn and Drew.[6] For cases where the controlling resistance for heat and mass transfer in the vapor is the sensible-heat removal from the cooling vapor, a procedure is outlined in Reference 24b.

Effect of noncondensables

When a multicomponent mixture contains a noncondensing gas, the rate of condensation is seriously reduced. As in the condensation of a mixture of condensable vapors, there is mass transfer of one or more components in the vapor phase; but here the condensing molecules must diffuse through a film of noncondensing gas which does not move toward the condensate surface. As condensation proceeds, the relative amount of this inert gas in the vapor phase increases significantly.

The partial pressure of the condensing vapor is less than the total pressure, which lowers the equilibrium condensation temperature. In addition, the partial pressure of the condensing vapor at the condensate surface must be less than it is in the bulk vapor-gas phase, to provide the driving force for mass transfer through the

---

[†]Mass transfer is discussed in Chaps. 17 through 19.

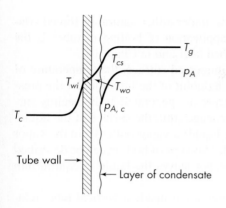

FIGURE 13.3
Profiles of partial pressure and temperature in a condenser when noncondensables are present: p_A, partial pressure of condensable vapor in bulk gas phase; $p_{A,c}$, partial pressure at surface of condensate layer; T_g, temperature of bulk gas phase; T_{cs}, temperature at condensate surface; T_{wi}, temperature inside tube wall; T_{wo}, temperature outside tube wall; T_c, temperature of coolant.

gas film. This further lowers the condensing temperature, and usually the change in temperature due to mass transfer is greater than the change in the equilibrium temperature.

Even a small amount of gas can have a large effect on the rate of condensation. Less than 1 percent air in steam can reduce the condensation rate by more than one-half,[4] and 5 percent of inert gas can decrease the steam condensation rate by a factor[18] of 5. Whenever air or other noncondensable gas is present in the feed, a fraction of the incoming gas must be vented from the condenser. If the vapor fed to a nonvented condenser, for instance, contains 0.1 percent air and 99 percent of the vapor is condensed, then the remaining vapor will contain 10 percent air and the condensation rate will be quite low in the last part of the condenser.

Figure 13.3 shows the profiles of temperature and partial pressure in such a condenser. The condensing temperature drops as the composition of the gas-vapor mixture, and hence its dew point, changes as condensation proceeds. Rigorous methods of solving the general problem are based on equating the heat flow to the condensate surface at any point to the heat flow away from the surface. This involves trial-and-error solutions for the point temperature of the condensate surface and from these an estimation of the point values of the heat flux $U \Delta T$. The values of $1/(U \Delta T)$ for each point are plotted against the heat transferred to that point and the area of the condenser surface found by numerical integration.[7]

A condenser for mixtures of vapors and noncondensable gases is illustrated in Fig. 15.15, p. 458.

HEAT TRANSFER TO BOILING LIQUIDS

Heat transfer to a boiling liquid is a necessary step in evaporation, distillation, and steam generation, and it may also be used to control the temperature of a chemical reactor. The boiling liquid may be contained in a vessel equipped with a heating surface fabricated from horizontal or vertical tubes, in which steam or other vapor is condensed or hot fluid is circulated to supply the heat needed to boil the liquid.

Alternatively, the liquid to be boiled may flow, under either natural or forced convection, inside heated tubes. An important application of boiling in tubes is the evaporation of water from solution, as discussed later and in Chap. 16.

When boiling is accomplished by a hot immersed surface, the temperature of the mass of the liquid is the same as the boiling point of the liquid under the pressure existing in the equipment. Bubbles of vapor are generated at the heating surface, rise through the mass of liquid, and disengage from the surface of the liquid. Vapor accumulates in a vapor space over the liquid; a vapor outlet from the vapor space removes the vapor as fast as it is formed. This type of boiling can be described as *pool boiling of saturated liquid* since the vapor leaves the liquid in equilibrium with the liquid at its boiling temperature.

When a liquid is boiled under natural circulation inside a vertical tube, relatively cool liquid enters the bottom of the tube and is heated as it flows upward at a low velocity. The liquid temperature rises to the boiling point under the pressure prevailing at that particular level in the tube. Vaporization begins, and the upward velocity of the two-phase liquid-vapor mixture increases enormously. The resulting pressure drop causes the boiling point to fall as the mixture proceeds up the tube and vaporization continues. Liquid and vapor emerge from the top of the tubes at very high velocity.

With forced circulation through horizontal or vertical tubes, the liquid may also enter at a fairly low temperature and be heated to its boiling point, changing into vapor near the discharge end of the tube. Sometimes a flow control valve is placed in the discharge line beyond the tube so that the liquid in the tube may be heated to a temperature considerably above the boiling point corresponding to the downstream pressure. Under these conditions there is no boiling in the tube: The liquid is merely heated, as a liquid, to a high temperature, and flashes into vapor as it passes through the valve. Natural- and forced-circulation boilers are called *calandrias;* they are discussed later.

In some types of forced-circulation equipment, the temperature of the mass of the liquid is below that of its boiling point, but the temperature of the heating surface is considerably above the boiling point of the liquid. Bubbles form on the heating surface, but on release from the surface are absorbed by the mass of the liquid. This type of heat transfer is called *subcooled boiling,* even though the fluid leaving the heat exchanger is entirely liquid.

Pool boiling of saturated liquid

Consider a horizontal, electrically heated wire immersed in a vessel containing a boiling liquid. Assume that q/A, the heat flux, and ΔT, the difference between the temperature of the wire surface T_w and that of the boiling liquid T, are measured. Start with a very low temperature drop ΔT. Now raise T_w and increase the temperature drop by steps, measuring q/A and ΔT at each step, until very large values of ΔT are reached. A plot of q/A versus ΔT on logarithmic coordinates will give a curve of the type shown in Fig. 13.4. This curve can be divided into four segments. In the first segment, at low temperature drops, the line AB is straight and has a slope of 1.25. This is consistent with the equation

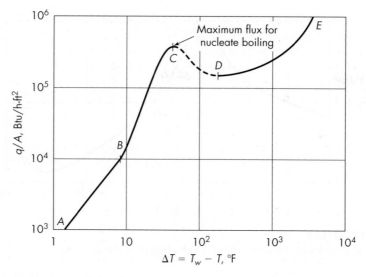

FIGURE 13.4
Heat flux versus temperature drop, boiling water at 212°F on an electrically heated wire: *AB*, natural convection; *BC*, nucleate boiling; *CD*, transition boiling; *DE*, film boiling. (*After McAdams et al.*[22])

$$\frac{q}{A} = a\,\Delta T^{1.25} \tag{13.19}$$

where a is a constant. The second segment, line BC, is also approximately straight, but its slope is greater than that of line AB. The slope of line BC depends upon the specific experiment; it usually lies between 3 and 4. The second segment terminates at a definite point of maximum flux, which is point C in Fig. 13.4. The temperature drop corresponding to point C is called the *critical temperature drop,* and the flux at point C is the *peak flux.* In the third segment, line CD in Fig. 13.4, the flux decreases as the temperature drop rises and reaches a minimum at point D. Point D is called the *Leidenfrost point.* In the last segment, line DE, the flux again increases with ΔT and, at large temperature drops, surpasses the previous maximum reached at point C.

Because, by definition, $h = (q/A)/\Delta T$, the plot of Fig. 13.4 is readily convertible into a plot of h versus ΔT. This curve is shown in Fig. 13.5. A maximum and a minimum coefficient are evident in Fig. 13.5. They do not, however, occur at the same values of the temperature drop as the maximum and minimum fluxes indicated in Fig. 13.4. The coefficient is normally a maximum at a temperature drop slightly lower than that at the peak flux; the minimum coefficient occurs at a much higher temperature drop than that at the Leidenfrost point. The coefficient is proportional to $\Delta T^{0.25}$ in the first segment of the line in Fig. 13.4 and to between ΔT^2 and ΔT^3 in the second segment.

Each of the four segments of the graph in Fig. 13.5 corresponds to a definite mechanism of boiling. In the first section, at low temperature drops, the mechanism

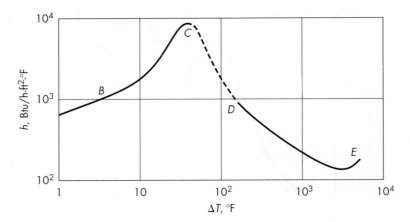

FIGURE 13.5
Heat-transfer coefficients versus ΔT, boiling of water at 1 atm.

is that of heat transfer to a liquid in natural convection, and the variation of h with ΔT agrees with that given by Eq. (12.73). Bubbles form on the surface of the heater, are released from it, rise to the surface of the liquid, and are disengaged into the vapor space; but they are too few to disturb appreciably the normal currents of free convection.

At larger temperature drops, lying between 9 and 45°F (5 and 25°C) in the case shown in Fig. 13.5, the rate of bubble production is large enough for the stream of bubbles moving up through the liquid to increase the velocity of the circulation currents in the mass of liquid, and the coefficient of heat transfer becomes greater than that in undisturbed natural convection. As ΔT is increased, the rate of bubble formation increases and the coefficient increases rapidly.

The action occurring at temperature drops below the critical temperature drop is called *nucleate boiling,* in reference to the formation of tiny bubbles, or vaporization nuclei, on the heating surface. During nucleate boiling, the bubbles occupy but a small portion of the heating surface at a time, and most of the surface is in direct contact with liquid. The bubbles are generated at localized active sites, usually small pits or scratches on the heating surface. As the temperature drop is raised, more sites become active, improving the agitation of the liquid and increasing the heat flux and the heat-transfer coefficient.

Eventually, however, so many bubbles are present that they tend to coalesce and cover portions of the heating surface with a layer of insulating vapor. This layer has a highly unstable surface, from which miniature "explosions" send jets of vapor away from the heating element into the bulk of the liquid. This type of action is called *transition boiling.* In this region, corresponding to segment CD in Fig. 13.4, increasing the temperature drop increases the thickness of the layer of vapor and reduces the number of explosions that occur in a given time. The heat flux and the heat-transfer coefficient both fall as the temperature drop is raised.

Near the Leidenfrost point another distinct change in mechanism occurs. The hot surface becomes covered with a quiescent film of vapor, through which heat is transferred by conduction and (at very high temperature drops) by radiation. The random explosions characteristic of transition boiling disappear and are replaced by the slow and orderly formation of bubbles at the interface between the liquid and the film of hot vapor. These bubbles detach themselves from the interface and rise through the liquid. Virtually all the resistance to heat transfer is offered by the vapor sheath covering the heating element. As the temperature drop increases, the heat flux rises, slowly at first and then more rapidly as radiation heat transfer becomes important. The boiling action in this region is known as *film boiling*.

Film boiling is not usually desired in commercial equipment because the heat-transfer rate is low for such a large temperature drop. Heat-transfer apparatus should be so designed and operated that the temperature drop in the film of boiling liquid is smaller than the critical temperature drop, although with cryogenic liquids this is not always feasible.

The effectiveness of nucleate boiling depends primarily on the ease with which bubbles form and free themselves from the heating surface. The layer of liquid next to the hot surface is superheated by contact with the wall of the heater. The super-heated liquid tends to form vapor spontaneously and so relieve the superheat. It is the tendency of superheated liquid to flash into vapor that provides the impetus for the boiling process. Physically, the flash can occur only by forming vapor-liquid interfaces in the form of small bubbles. It is not easy, however, to form a small bub-ble in a superheated liquid, because at a given temperature the vapor pressure in a very small bubble is less than that in a large bubble or that from a plane liquid surface. A very small bubble can exist in equilibrium with superheated liquid, and the smaller the bubble, the greater the equilibrium superheat and the smaller the ten-dency to flash. By taking elaborate precautions to eliminate all gas and other impu-rities from the liquid and to prevent shock, it is possible to superheat water by several hundred degrees Fahrenheit without formation of bubbles.

A second difficulty appears if the bubble does not readily leave the surface once it is formed. The important factor in controlling the rate of bubble detachment is the interfacial tension between the liquid and the heating surface. If this interfacial ten-sion is large, the bubble tends to spread along the surface and blanket the heat-transfer area, as shown in Fig. 13.6c, rather than leaving the surface to make room for other bubbles. If the interfacial tension between liquid and solid is low, the bubble will pinch off easily, in the manner shown in Fig. 13.6a. An example of intermediate interfacial tension is shown in Fig. 13.6b.

The high rate of heat transfer in nucleate boiling is primarily the result of the turbulence generated in the liquid by the dynamic action of the bubbles.[11]

The coefficient obtained during nucleate boiling is sensitive to a number of variables, including the nature of the liquid, the type and condition of the heating surface, the composition and purity of the liquid, the presence or absence of agita-tion, and the temperature or pressure. Minor changes in some variables cause major changes in the coefficient. The reproducibility of check experiments is poor.

Qualitatively, the effects of some variables can be predicted from a con-sideration of the mechanisms of boiling. A roughened surface provides centers for nucleation that are not present on a polished surface. Thus roughened surfaces usu-ally give larger coefficients than smooth surfaces. This effect, however, is due in part

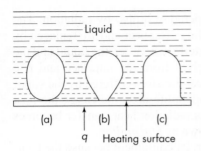

FIGURE 13.6
Effect on bubble formation of interfacial tension between liquid and heating surface. (*After Jakob and Fritz.*[15])

to the fact that the total surface of a rough tube is larger than that of a smooth surface of the same projected area. A very thin layer of scale may increase the coefficient of the boiling liquid, but even a thin scale will reduce the overall coefficient by adding a resistance that reduces the overall coefficient more than the improved boiling liquid coefficient increases it. Gas or air adsorbed on the surface of the heater or contaminants on the surface often facilitate boiling by either the formation or the disengaging of bubbles. A freshly cleaned surface may give a higher or lower coefficient than the same surface after it has been stabilized by a previous period of operation. This effect is associated with a change in the condition of the heating surface. Agitation increases the coefficient by increasing the velocity of the liquid across the surface, which helps to sweep away bubbles.

Curves similar to those in Figs. 13.4 and 13.5 would be obtained when studying pool boiling on a horizontal tube or pipe, although the critical temperature drop and the Leidenfrost point would probably not be the same as for a small-diameter wire.

Maximum flux and critical temperature drop

The maximum flux $(q/A)_{max}$ depends mainly on the nature of the boiling liquid and on the pressure, and somewhat on the type of heating surface. The maximum flux for water is several times that for most organic liquids because of the differences in the heat of vaporization and other physical properties. For both water and organic liquids, the maximum heat flux itself reaches a maximum at an absolute pressure about one-third of the thermodynamic critical pressure p_c and decreases toward zero at very low pressure and at pressures approaching the critical pressure. If the maximum flux is divided by the critical pressure of the boiling substance, a curve is obtained, shown qualitatively in Fig. 13.7, which is much the same for many pure substances and mixtures. The corresponding critical temperature drop (not to be confused with "critical temperature") also varies with pressure, from large values at low pressures to very small ones near the critical pressure. At pressures and temperatures above the critical values, of course, there is no difference between liquid and vapor phases, and "vaporization" has no meaning. At atmospheric pressure the critical temperature drop for water is usually between 30 and 50°C (54 and 90°F). That for organic liquids may be higher or lower, depending largely on how close the operating temperature is to the thermodynamic critical temperature.

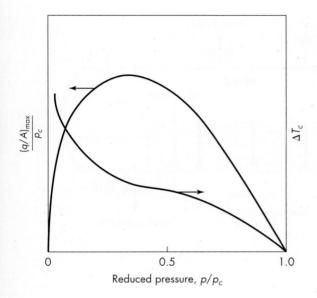

FIGURE 13.7
Qualitative variation of maximum heat flux and critical temperature difference with reduced pressure.

The critical temperature drop can be exceeded in industrial boilers unless precautions are taken. If the source of heat is another fluid, such as condensing steam or hot liquids, the only penalty for exceeding the critical temperature drop is a decrease in flux to a level between that at the peak and that at the Leidenfrost point. If the heat is supplied by an electric heater, exceeding the critical temperature drop may burn out the heater, as the boiling liquid cannot absorb heat fast enough at a large temperature drop, and the heater immediately becomes very hot.

The peak flux at the critical temperature drop is large. For water it is in the range of 350 to 1,250 kW/m² (~100,000 to 400,000 Btu/ft²·h) depending on the purity of water, the pressure, and the type and condition of the heating surface. For organic liquids, the peak flux lies in the range of 125 to 400 kW/m² (40,000 to 130,000 Btu/ft²·h). These limits apply to boiling under atmospheric pressure on a horizontal tube or a flat horizontal surface.

Many correlations have been proposed for estimating the peak flux from the properties of the fluid, based on various models of the physical phenomena. In one approach it is assumed that near the critical temperature drop the streams of bubbles characteristic of nucleate boiling are progressively replaced by jets of vapor leaving the heat-transfer surface. These must, of course, be accompanied by streams of liquid flowing toward the surface. At the peak value of the heat flux, the countercurrent flows of vapor and liquid reach a limiting condition, the process becomes unstable, and the jets of vapor collapse to form a continuous vapor sheath. The phenomenon is analogous to the flooding in a packed tower described in Chap. 18.[29] An equation based on this model was developed by Zuber[28] and modified by Lienhard and Dhir[19] to fit data for a flat plate.

$$(q/A)_{max} = 0.15 \lambda \rho_V^{1/2} [\sigma g (\rho_L - \rho_V)]^{1/4} \qquad (13.20)$$

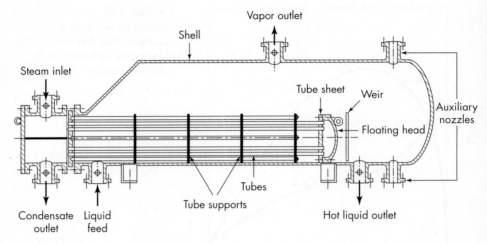

FIGURE 13.8
Kettle-type reboiler.

where σ is the interfacial tension between liquid and vapor, ρ_L and ρ_V are the densities of liquid and vapor, respectively, and the other symbols have their usual meanings.

For horizontal cylinders[25] the constant 0.15 in Eq. (13.20) should be replaced by 0.12.

The peak flux increases with λ and ρ_V because a higher heat of vaporization or a higher density means a smaller vapor volume for the same amount of heat transferred. If the pressure is increased, the effect of greater vapor density is partially offset by the reduction in the heat of vaporization and interfacial tension. These decrease with increasing temperature and eventually lead to lower fluxes, as shown in Fig. 13.7.

Kettle reboilers

A typical kettle-type reboiler, shown in Fig. 13.8, contains a bundle of horizontal tubes submerged in the liquid and arranged to provide some clearance between the lowest tubes and the reboiler shell. Vapor generated at the lower rows of tubes rises and affects the heat-transfer rate from the upper tubes. In the nucleate boiling region, where the ΔT is small, the average heat flux for the bundle is greater than that for a single tube because the increased vapor flow raises the velocity of the liquid-vapor mixture past the tubes. However, vapor blanketing of the tube bundle occurs at lower temperature differences, and the maximum heat flux is much lower than that for a single tube. Boiling curves for a normal paraffin, for a single tube and for a tube bundle, are given in Fig. 13.9, based on data from the Heat Transfer Research Institute.[10] Note that the maximum flux for the tube bundle is only about one-fourth that for the single tube. The critical temperature difference is also smaller for

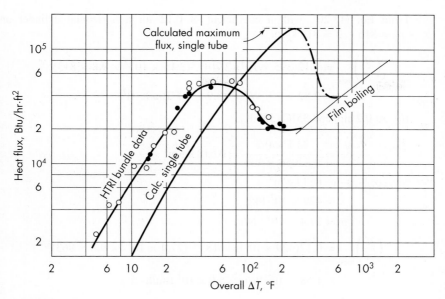

FIGURE 13.9
Boiling heat flux, single tube versus tube bundle. (*After Fair and Klip.*[10])

the tube bundle, although the actual values of ΔT_c are not known, since only the overall temperature difference was measured.

EXAMPLE 13.2. Benzene is boiling at atmospheric pressure in a kettle-type boiler with horizontal tubes. (*a*) Estimate the maximum heat flux for a single tube. The physical properties of benzene are given in Table 13.1. (*b*) Predict the maximum flux for pressures of 0.2 and 5 atm.

Solution
   (*a*) At 80°C (353 K), ρ_V is calculated from the ideal gas law.

$$\rho_V = \frac{Mp}{RT} = \frac{78.1 \times 1.0}{0.082056 \times 353} = 2.70 \text{ kg/m}^3$$

By linear interpolation, $\sigma = 21.5$ dyn/cm $= 21.5 \times 10^{-3}$ N/m.

TABLE 13.1
Properties of benzene

$p\prime$, atm	0.2	1	2	5	10
T, °C	35	80	104	142	179
ρ_L, kg/m³	864	816	788	740	690
λ, kJ/kg	429	395	379	349	314

$\sigma = 31.7$ dyn/cm at 0°C, 29.0 at 20°C, and 18.8 at 100°C

From Eq. (13.20), with a coefficient of 0.12 for a horizontal cylinder and $\rho_L - \rho_V = 816 - 3 = 813 \text{ kg/m}^3$,

$$\left(\frac{q}{A}\right)_{max} = 0.12(395 \times 10^3)(2.70)^{1/2}(2.15 \times 10^{-3} \times 9.8 \times 813)^{1/4}$$

$$= 2.82 \times 10^5 \text{ W/m}^2 \ (8.9 \times 10^4 \text{ Btu/h} \cdot \text{ft}^2)$$

(b) At 0.2 atm and 35°C (308 K),

$$\rho_V = \frac{78.1 \times 0.2}{0.82056 \times 308} = 0.618 \text{ kg/m}^3$$

$$\rho_L - \rho_V = 863 \text{ kg/m}^3$$

$$\lambda = 429 \times 10^3 \text{ J/kg}$$

$$\sigma = 27.4 \times 10^{-3} \text{ N/m}$$

$$\left(\frac{q}{A}\right)_{max} = 0.12(429 \times 10^3)(0.618)^{1/2}(27.4 \times 10^{-3} \times 9.8 \times 863)^{1/4}$$

$$= 1.58 \times 10^5 \text{ W/m}^2 \ (5.0 \times 10^4 \text{ Btu/h} \cdot \text{ft}^2)$$

At 5 atm and 142°C (415 K),

$$\lambda = 349 \times 10^3 \text{ J/kg}$$

$$\rho_V = 11.5 \text{ kg/m}^3$$

$$\rho_L - \rho_V = 740 - 11.5 = 728.5 \text{ kg/m}^3$$

By extrapolation,

$$\sigma \cong 13.7 \times 10^{-3} \text{ N/m}$$

$$\left(\frac{q}{A}\right)_{max} = 0.12(349 \times 10^3)(11.5)^{1/2}(13.7 \times 10^{-3} \times 9.8 \times 728.5)^{1/4}$$

$$= 4.48 \times 10^5 \text{ W/m}^2 \ (1.42 \times 10^5 \text{ Btu/h} \cdot \text{ft}^2)$$

The maximum flux increases 1.78-fold as the pressure rises from 0.2 to 1.0 atm, and 1.59-fold from 1.0 to 5 atm. Further calculations indicate a maximum flux of $4.8 \times 10^5 \text{ W/m}^2$ at a pressure of about 10 atm, which is 0.21 times the critical pressure p_c. The uncertainty in the extrapolated values of σ limits the accuracy of the result.

Minimum heat flux and film boiling

When film boiling is established, undulations of a characteristic wavelength form in the interface between liquid and vapor. These undulations grow into bubbles, which leave the interface at regularly spaced intervals. The diameter of the bubbles is approximately one-half the wavelength of the undulations. Consideration of the dynamics of this process leads to the following equation for the minimum heat flux necessary for stable film boiling on a horizontal plate[12]

$$\left(\frac{q}{A}\right)_{min} = \frac{\pi \lambda \rho_V}{24}\left[\frac{\sigma g(\rho_L - \rho_V)}{(\rho_L + \rho_V)^2}\right]^{1/4} \tag{13.21}$$

where $(q/A)_{min}$ is the minimum heat flux.

Film boiling is a more orderly process than either nucleate boiling or transition boiling and has been subjected to considerable theoretical analysis. Since the heat-transfer rate is governed solely by the vapor film, the nature of the heating surface has no effect in film boiling. For film boiling on a submerged horizontal tube, the following equation applies with considerable accuracy over a wide range of conditions.[2]

$$h_o \left[\frac{\lambda_c \mu_V \, \Delta T}{k_V^3 \rho_V (\rho_L - \rho_V) \lambda' g} \right]^{1/4} = 0.59 + 0.069 \frac{\lambda_c}{D_o} \tag{13.22}$$

where
$\quad h_o$ = heat-transfer coefficient
$\quad \mu_V$ = viscosity of vapor
$\quad \Delta T$ = temperature drop across vapor film
$\quad k_V$ = thermal conductivity of vapor
$\quad \rho_L, \rho_V$ = densities of liquid and vapor, respectively
$\quad D_o$ = outside diameter of heating tube

In Eq. (13.22), λ' is the average difference in enthalpy between the liquid and the superheated vapor. It is given by the following equation, which is Hsu and Westwater's modification[13] of Bromley's approximation to the rigorous solution.[3]

$$\lambda' = \lambda \left(1 + \frac{0.34 c_p \, \Delta T}{\lambda} \right)^2 \tag{13.23}$$

where λ = latent heat of vaporization
$\quad c_p$ = specific heat of vapor at constant pressure

The term λ_c in Eq. (13.22) is the wavelength of the smallest wave that can grow in amplitude on a flat horizontal interface. It is related to the properties of the fluid by the equation

$$\lambda_c = 2\pi \left[\frac{\sigma}{g(\rho_L - \rho_V)} \right]^{1/2} \tag{13.24}$$

where σ is the interfacial tension between liquid and vapor. Equation (13.22) does not include the effect of heat transfer by radiation.

In Eqs. (13.20) through (13.24) the vapor properties are evaluated at the arithmetic average of the wall temperature and the boiling point of the liquid. The liquid properties ρ_L, λ, and σ are evaluated at the boiling point.

Note the similarity between Eq. (13.22) for film boiling and Eq. (13.15) for laminar flow condensation. In both cases, h depends on the $\frac{3}{4}$ power of the thermal conductivity, the $\frac{1}{4}$ power of the latent heat of vaporization, and the $-\frac{1}{4}$ power of the film viscosity and the temperature difference.

Equations have also been developed for film boiling from submerged vertical tubes,[13] but they have less general validity than Eq. (13.22). Vapor disengages from

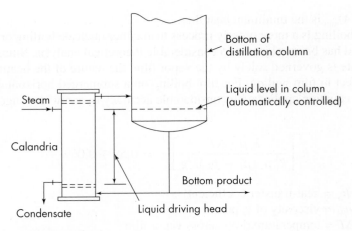

FIGURE 13.10
Calandria reboiler.

a vertical surface in a more complicated fashion than from a horizontal surface, and the theoretical analysis of the process is correspondingly more difficult.

Thermosiphon reboilers

Shell-and-tube units known as calandrias, natural-circulation or thermosiphon reboilers, are often the most economical vaporizers for distillation and evaporation operations. Unlike the boilers discussed so far, they contain no pool of boiling liquid; instead, liquid enters the bottom of the unit and is partially vaporized. The reduction in density causes the vapor-liquid mixture to rise and draw in additional feed liquid. Liquid and vapor leave the top of the unit at high velocity; they are then separated and the liquid is recycled. Typically some 10 to 30 percent of the liquid is vaporized per pass.

Thermosiphon reboilers may contain vertical tubes with the liquid vaporizing inside the tubes, or be horizontal, with the boiling liquid outside the tubes.

A typical vertical unit is shown in Fig. 13.10. The velocity of the liquid entering the tubes is commonly about 1 m/s. The heat-transfer rate depends on the properties of the vaporizing liquid, especially its reduced pressure p/p_c. Typical overall coefficients for steam-heated calandrias are given in Table 13.2. When the absolute pressure is 1 atm or greater, the coefficients are insensitive to the liquid *driving head* indicated in Fig. 13.10. This head is defined as the distance from the bottom tube

TABLE 13.2
Typical overall coefficients in calandria boilers

Service	Overall coefficient U	
	W/m² · °C	Btu/ft² · h · °F
Heavy organic chemicals	570–900	100–160
Light hydrocarbons	900–1,250	160–220
Water, aqueous solutions	1,250–2,000	220–350

sheet to the level of the liquid in the column. Under these pressure conditions the liquid level is typically maintained near the level of the top tube sheet to ensure that the heat-transfer surface is completely wetted and to establish reasonably high circulation rates.

When operation is under vacuum, however, the performance of the reboiler is sensitive to changes in the liquid driving head, especially in the distillation of multicomponent mixtures. The optimum liquid level for vacuum service is midway between the tube sheets, with about 50 percent of the liquid vaporized per pass.[16]

In horizontal thermosiphon reboilers, liquid enters at the center of the bottom of a vessel containing horizontal heated tubes, and it flows horizontally under a transverse baffle toward the ends of the vessel. The partially vaporized mixture rises past the ends of the baffle into the upper section of the vessel, where more vapor is formed. The mixture flows toward the center and leaves through an outlet in the top of the vessel. Heat-transfer rates and design procedures for both vertical and horizontal thermosiphon reboilers are discussed by Fair and Klip.[10]

Forced-circulation reboilers

Sometimes the liquid to be boiled, especially if viscous, is mechanically pumped through the reboiler, usually inside the tubes. In such forced-circulation units, boiling of the liquid is largely suppressed. Superheated liquid leaves the top of the boiler and is later flashed (partially vaporized) by reducing the pressure. Forced-circulation evaporators are discussed in Chap. 16.

Subcooled boiling: Enhancement of peak flux

Subcooled boiling can be demonstrated by pumping a gas-free liquid upward through a vertical annular space consisting of a transparent outer tube and an internal heating element and by observing the effect on the liquid of a gradual increase in heat flux and temperature of the heating element. It is observed that when the temperature of the element exceeds a definite magnitude, which depends on the conditions of the experiment, bubbles form, just as in nucleate boiling, and then condense in the adjacent cooler liquid. Under these conditions a small change in temperature drop causes an enormous increase in heat flux. Fluxes greater than 150×10^6 W/m^2 (50×10^6 Btu/ ft$^2 \cdot$ h) have been reported.[12] Subcooled boiling is important for heat-transfer equipment that must pack great capacity into a small space.

Other methods of obtaining high fluxes, in excess of the normal peak flux in pool boiling, include the use of porous coatings on the heating surface,[1] finned tubes of various designs,[27] and for liquids of low electrical conductivity, the application of high-voltage electric fields.[20]

SYMBOLS

A Area, m^2 or ft^2; A_o, outside of tube

a Constant in Eq. (13.19)

c_p Specific heat at constant pressure, J/g·°C or Btu/lb·°F
D Diameter, m or ft; D_i, inside diameter of tube; D_o, outside diameter
g Gravitational acceleration, m/h^2 or ft/h^2
h Individual heat-transfer coefficient, $W/m^2 \cdot °C$ or $Btu/ft^2 \cdot h \cdot °F$; h_N, average
        for stack of tubes; h_o, outside tube; h_x, local value; h_1, for top tube in stack
k Thermal conductivity, $W/m \cdot °C$ or $Btu/ft \cdot h \cdot °F$; k_V, of vapor; k_f, of
        condensate film
L Length, m or ft; L_T, total length of tube
M Molecular weight of boiling substance
$\dot{m}$ Mass flow rate, kg/h or lb/h; $\dot{m}_T$, total condensate flow rate from stack
        of tubes
N Number of tubes in vertical stack
Nu$'$ Nusselt number, defined by Eq. (13.13), dimensionless
Pr   Prandtl number of condensing liquid, dimensionless
p Pressure, atm or $lb_f/in.^2$; p_c, critical pressure; $p\prime$, vapor pressure
q Rate of heat transfer, J/s or Btu/h; q_T, total rate of heat transfer in
        condenser tube
R Gas law constant, 0.082056 $m^3 \cdot atm/kg \, mol \cdot K$
Re   Reynolds number of condensing film, $4\Gamma/\mu_f$, dimensionless
T Temperature, °C or °F; T_f, average temperature of condensate film; T_h,
        saturation temperature of vapor; T_w, wall or surface temperature
U Overall heat-transfer coefficient, $W/m^2 \cdot °C$ or $Btu/ft^2 \cdot h \cdot °F$
x Equivalent thickness of condensing film, found from heat-transfer
        coefficient, μm

Greek Letters

β Angle with vertical
Γ Condensate loading, kg/m·h or lb/ft·h; Γ_b, at bottom of vertical tube; Γ',
        per unit length of horizontal tube
ΔT Temperature drop, °C or °F; ΔT_o, drop across condensate film, $T_h - T_w$;
        ΔT_c, critical temperature drop
δ Thickness of condensate film, m or ft
λ Heat of vaporization, J/g or Btu/lb; λ', average difference in enthalpy
        between boiling liquid and superheated vapor, defined by Eq. (13.23)
λ_c Wavelength of smallest wave that can grow on flat horizontal surface, m or
        ft [Eq. (13.24)]
μ Absolute viscosity, P or lb/ft·h; μ_V, of vapor; μ_f, of condensate film
ρ Density, kg/m^3 or lb/ft^3; ρ_L, of liquid; ρ_V, of vapor ρ_f, of condensate film
σ Interfacial tension between liquid and vapor, N/m or lb_f/ft
ψ_f Condensation parameter, $(k_f^3 \rho_f^2 g/\mu_f^2)^{1/3}$, $W/m^2 \cdot °C$ or $Btu/ft^2 \cdot h \cdot °F$

PROBLEMS

13.1. A 1-in. BWG 14 copper condenser tube, 3 m long, is to condense ethyl alcohol at atmospheric pressure. Cooling water inside the tube keeps the metal surface at an essentially constant temperature of 25°C. (*a*) How much vapor, in kilograms per hour, will condense if the tube is vertical? (*b*) How much will condense if the tube is horizontal?

13.2. A vertical tubular condenser is to be used to condense 2,100 kg/h of ethyl alcohol, which enters at atmospheric pressure. Cooling water is to flow through the tubes at an average temperature of 30°C. The tubes are 31-mm OD and 27-mm ID. The water-side coefficient is 2,800 W/m²·°C. Fouling factors and the resistance of the tube wall may be neglected. If the available tubes are 3 m long, how many tubes will be needed? Data are as follows:

Boiling point of alcohol: $T_h = 78.4°C$

Heat of vaporization: $\lambda = 856$ J/g

Density of liquid: $\rho_f = 769$ kg/m³

13.3. A horizontal shell-and-tube condenser is to be used to condense saturated ammonia vapor at 145 lb$_f$/in.² abs ($T_h = 82°F$). The condenser has 19 steel tubes (1.5-in. OD, 1.3-in. ID) 14 ft long through which cooling water is flowing. The tubes are arranged hexagonally on 2-in. centers. The latent heat of ammonia at these conditions may be taken as 500 Btu/lb. The cooling water enters at 70°F. Determine the capacity of the condenser for these conditions.

13.4. A study of heat transmission from condensing steam to cooling water in a single-tube condenser gave results for both clean and fouled tubes. For each tube the overall coefficient U was determined at a number of water velocities. The experimental results were represented by the following empirical equations:

$$\frac{1}{U_o} = \begin{cases} 0.00092 + \dfrac{1}{268\bar{V}^{0.8}} & \text{fouled tube} \\[3mm] 0.00040 + \dfrac{1}{268\bar{V}^{0.8}} & \text{clean tube} \end{cases}$$

where U_o = overall heat-transfer coefficient, Btu/ft²·h·°F
    $\bar{V}$ = water velocity, ft/s

The tubes were 0.902-in. ID and 1.000-in. OD, made of admiralty metal, for which $k = 63$ Btu/ft·h·°F. From these data calculate (*a*) the steam-film coefficient (based on steam-side area), (*b*) the water-film coefficient when the water velocity is 1 ft/s (based on water-side area), and (*c*) value of h_{do} for the scale in the fouled tube, assuming the clean tube was free of deposit.

13.5. What heat-transfer coefficient would be expected for natural convection to water at 212°F and 1 atm, outside a 1-in. horizontal pipe with a surface temperature of 213°F? Compare with Fig. 13.5 and comment on the difference.

13.6. A 25-mm-OD copper tube is to be used to boil water at atmospheric pressure. (*a*) Estimate the maximum heat flux obtainable as the temperature of the copper surface is increased. (*b*) If the temperature of the copper surface is 210°C, calculate the boiling film coefficient and the heat flux. The interfacial tension of water at temperatures above 80°C is given by

$$\sigma = 78.38(1 - 0.0025T)$$

where σ = interfacial tension, dyn/cm
$\quad\quad T$ = temperature, °C

13.7. Steam containing 2 percent air is condensed at atmospheric pressure inside 25-mm tubes in a water-cooled condenser. The vertical tubes are 3 m long; the coefficient and temperature of the cooling water are 2500 W/m$^2 \cdot$°C and 30°C, respectively. Calculate the condensation rate expected if no air was present, and use this value to estimate the Reynolds number of the vapor at the tube inlet. If 96 percent of the steam were condensed, what would be the vapor Reynolds number at the tube outlet? Calculate the equilibrium condensation temperature of the exit vapor, and show why the actual temperature at the vapor-liquid interface in the condenser would be lower than this value.

13.8. (*a*) Calculate the heat-transfer coefficient for film boiling of water at 1 atm on submerged horizontal tubes $\frac{1}{2}$, 1, and 2 in. in diameter, with a surface temperature of 180°C. (*b*) How does the effect of tube diameter compare with the effect of tube diameter for condensing steam? (*c*) Compare the equations for film boiling and for film condensation on horizontal tubes, pointing out the similarities and the differences.

13.9. Benzene vapor is condensed at atmospheric pressure in a shell-and-tube condenser with horizontal $\frac{7}{8}$-in. BWG 16 copper tubes. Predict the film coefficient for the top row of tubes and the average coefficient for a stack of 10 tubes.

13.10. At a low overall ΔT the boiling heat flux for a tube bundle is about 4 times that for a single tube, but at a high ΔT the flux is much lower than that for a single tube. Give a logical explanation of the difference.

REFERENCES

1. Bergles, A. E., and M.-C. Chyu. *AIChE Symp. Series,* **77**(208):73 (1981).
2. Breen, B. P., and J. W. Westwater. *Chem. Eng. Prog.,* **58**(7):67 (1962).
3. Bromley, L. A. *Ind. Eng. Chem.,* **44**:2966 (1952).
4. Cengel, Y. A. *Heat Transfer—A Practical Approach.* New York: McGraw-Hill, 1998, p. 478.
5. Chen, S. L., F. A. Gerner, and C. L. Tien. *Exp. Heat Transfer,* **1**:93 (1987).
6. Colburn, A. P., and T. B. Drew. *Trans. AIChE,* **33**:196 (1937).
7. Colburn, A. P., and O. A. Hougen. *Ind. Eng. Chem.,* **26**:1178 (1934).
8. Drew, T. B., W. M. Nagle, and W. Q. Smith. *Trans. AIChE,* **31**:605 (1935).
9. Dukler, A. E. *Chem. Eng. Prog. Symp. Ser.,* **56**(30):1 (1960).
10. Fair, J. R., and A. Klip. *Chem. Eng. Prog.,* **79**(3):86 (1983).
11. Forster, H. K., and N. Zuber. *AIChE J.,* **1**:531 (1955).
12. Gebhart, B. *Heat Transfer,* 2nd ed. New York: McGraw-Hill, 1971, pp. 424–6.
13. Hsu, Y. Y., and J. W. Westwater. *Chem. Eng. Prog. Symp. Ser.,* **56**(30):15 (1960).

14. Incropera, F. P., and D. P. DeWitt. *Fundamentals of Heat Transfer.* New York: Wiley, 1981, p. 492.
15. Jakob, M., and W. Fritz. *VDI-Forschungsh.,* **2:**434 (1931).
16. Johnson, D. L., and Y. Yukawa. *Chem. Eng. Prog.,* **75**(7):47 (1979).
17. Kern, D. Q. *Process Heat Transfer.* New York: McGraw-Hill, 1950, pp. 313ff.
18. Lienhard, J. H. *A Heat-Transfer Textbook.* Englewood Cliffs, NJ: Prentice-Hall, 1981, p. 403.
19. Lienhard, J. H., and V. K. Dhir. NASA Cr-2270, July 1973.
20. Markels, M. Jr., and R. L. Durfee. *AIChE J.,* **10:**106 (1964).
21. McAdams, W. H. *Heat Transmission,* 3rd ed. New York: McGraw-Hill, 1954, pp. 330ff.
22. McAdams, W. H., J. N. Addoms, P. M. Rinaldo, and R. S. Day. *Chem. Eng. Prog.,* **44:**639 (1948).
23. Nusselt, W. *VDIZ.,* **60:**541, 569 (1916).
24. Perry, R. H., and D. W. Green (eds.). *Perry's Chemical Engineers' Handbook,* 7th ed. New York: McGraw-Hill, 1997; (*a*) p. **5**-20, (*b*) p. **11**-12.
25. Sun, K. H., and J. H. Lienhard. *Int. J. Heat Mass Transfer,* **13:**1425 (1970).
26. Welch, J. F., and J. W. Westwater. *Proc. Int. Heat Transfer Conf. Lond., 1961–1962,* p. 302 (1963).
27. Yilmaz, S., and J. W. Westwater. *AIChE Symp. Ser.,* **77**(208):74 (1981).
28. Zuber, N. *Trans. ASME,* **80:**711 (1958).
29. Zuber, N., J. W. Westwater, and M. Tribus. *Proc. Int. Heat Transfer Conf. Lond., 1961–1962,* p. 230 (1963).

CHAPTER 14

Radiation Heat Transfer

Radiation, which may be considered to be energy streaming through space at the speed of light, may originate in various ways. Some types of material will emit radiation when they are treated by external agencies, such as electron bombardment, electric discharge, or radiation of definite wavelengths. Radiation due to these effects will not be discussed here. All substances at temperatures above absolute zero emit radiation that is independent of external agencies. Radiation that is the result of temperature only is called thermal radiation, and this discussion is restricted to radiation of this type.

Fundamental facts concerning radiation

Radiation moves through space in straight lines, or beams, and only substances in sight of a radiating body can intercept radiation from that body. The fraction of the radiation falling on a body that is reflected is called the *reflectivity* ρ. The fraction that is absorbed is called the *absorptivity* α. The fraction that is transmitted is called the *transmissivity* τ. The sum of these fractions must be unity, or

$$\alpha + \rho + \tau = 1 \tag{14.1}$$

Radiation as such is not heat, and when transformed to heat on absorption, it is no longer radiation. In practice, however, reflected or transmitted radiation usually falls on other absorptive bodies and is eventually converted to heat, perhaps after many successive reflections.

The maximum possible absorptivity is unity, attained only if the body absorbs all radiation incident upon it and reflects or transmits none. A body that absorbs all incident radiation is called a *blackbody*.

The complex subject of thermal radiation transfer has received much study in recent years and is covered in a number of texts.[6,8,10] The following introductory treatment discusses the following topics: emission of radiation, absorption by opaque solids, radiation between surfaces, radiation to and from semitransparent materials, and combined heat transfer by conduction-convection and radiation.

EMISSION OF RADIATION

The radiation emitted by any given mass of material is independent of that being emitted by other material in sight of, or in contact with, the mass. The *net* energy gained or lost by a body is the difference between the energy emitted by the body and that absorbed by it from the radiation reaching it from other bodies. Heat flow by conduction and convection may also be taking place independently of the radiation.

When bodies at different temperatures are placed in sight of one another inside an enclosure, the hotter bodies lose energy by emission of radiation faster than they receive energy by absorption of radiation from the cooler bodies, and the temperatures of the hotter bodies decrease. Simultaneously the cooler bodies absorb energy from the hotter ones faster than they emit energy, and the temperatures of the cooler bodies increase. The process reaches equilibrium when all the bodies reach the same temperature, just as in heat flow by conduction and convection. The conversion of radiation to heat on absorption and the attainment of temperature equilibrium through the net transfer of radiation justify the usual practice of calling radiation *heat*.

Wavelength of radiation

Known electromagnetic radiations cover an enormous range of wavelengths, from the short cosmic rays having wavelengths of about 10^{-11} cm to longwave broadcasting waves having lengths of 1,000 m or more.

Radiation of a single wavelength is called *monochromatic*. An actual beam of radiation consists of many monochromatic beams. Although radiation of any wavelength from zero to infinity is, in principle, convertible to heat on absorption by matter, the portion of the electromagnetic spectrum that is of importance in heat flow lies in the wavelength range between 0.5 and 50 μm. Visible light covers a wavelength range of about 0.38 to 0.78 μm, and thermal radiation at ordinary industrial temperatures has wavelengths in the infrared spectrum, which includes waves just longer than the longest visible waves. At temperatures above about 500°C heat radiation in the visible spectrum becomes significant, and the phrases *red heat* and *white heat* refer to this fact. The higher the temperature of the radiating body, the shorter the predominant wavelength of the thermal radiation emitted by it.

For a given temperature, the rate of thermal radiation varies with the state of aggregation of the substance. Monatomic and diatomic gases such as oxygen, argon, and nitrogen radiate weakly, even at high temperatures. Under industrial conditions, these gases neither emit nor absorb appreciable amounts of radiation. Polyatomic gases, including water vapor, carbon dioxide, ammonia, and sulfur dioxide, and hydrocarbons emit and absorb radiation appreciably at furnace temperatures but only in certain bands of wavelength. Solids and liquids emit radiation over the entire spectrum.

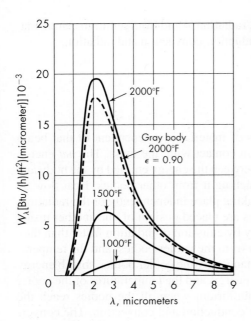

FIGURE 14.1
Energy distribution in spectra of blackbodies and gray bodies.

Emissive power

The monochromatic energy emitted by a radiating surface depends on the temperature of the surface and on the wavelength of the radiation. At constant surface temperature, a curve can be plotted showing the rate of energy emission as a function of the wavelength. Typical curves of this type are shown in Fig. 14.1. Each curve rises steeply to a maximum and decreases asymptotically to zero emission at very large wavelengths. The unit chosen for measuring the monochromatic radiation is based on the fact that, from a small area of a radiating surface, the energy emitted is "broadcast" in all directions through any hemisphere centered on the radiation area. The monochromatic radiation emitted in this manner from unit area in unit time divided by the wavelength is called the *monochromatic radiating power* W_λ. The ordinates in Fig. 14.1 are values of W_λ.

For the entire spectrum of the radiation from a surface, the total radiating power W is the sum of all the monochromatic radiations from the surface, or mathematically,

$$W = \int_0^\infty W_\lambda \, d\lambda \tag{14.2}$$

Graphically, W is the entire area under any of the curves in Fig. 14.1 from wavelengths of zero to infinity. Physically, the total radiating power is the total radiation of all wavelengths emitted by unit area in unit time in all directions through a hemisphere centered on the area.

Blackbody radiation; emissivity

As shown later [Eq. (14.10)], a blackbody has the maximum attainable emissive power at any given temperature and is the standard to which all other radiators are referred. The ratio of the total emissive power W of a body to that of a blackbody W_b is, by definition, the *emissivity* ε of the body. Thus, *at same temperature*

$$\varepsilon \equiv \frac{W}{W_b} \tag{14.3}$$

The *monochromatic emissivity* ε_λ is the ratio of the monochromatic emissive power to that of a blackbody at the same wavelength, or *and temperature*

$$\varepsilon_\lambda \equiv \frac{W_\lambda}{W_{b,\lambda}} \tag{14.4}$$

If the monochromatic emissivity of a body is the same for all wavelengths, the body is called a *gray body*.

Emissivities of solids

Emissivities of solids are tabulated in standard references.[7,9] Emissivity usually increases with temperature. Emissivities of polished metals are low, in the range of 0.03 to 0.08. Those of most oxidized metals range from 0.6 to 0.85; those of non-metals such as refractories, paper, boards, and building materials, from 0.65 to 0.95; and those of paints, other than aluminum paint, from 0.80 to 0.96.

Practical source of blackbody radiation

No actual substance is a blackbody, although some materials, such as certain grades of carbon black, do approach blackness. An experimental equivalent of a blackbody is an isothermal enclosure containing a small peephole. If a sight is taken through the peephole on the interior wall of the enclosure, the effect is the same as viewing a blackbody. The radiation emitted by the interior of the walls or admitted from outside through the peephole is completely absorbed after successive reflections, and the overall absorptivity of the interior surface is unity.

Laws of blackbody radiation

A basic relationship for blackbody radiation is the Stefan–Boltzmann law, which states that the total emissive power of a blackbody is proportional to the fourth power of the absolute temperature, or

$$W_b = \sigma T^4 \tag{14.5}$$

where σ is a universal constant depending only upon the units used to measure T and W_b. The Stefan–Boltzmann law is an exact consequence of the laws of thermodynamics and electromagnetism.

The distribution of energy in the spectrum of a blackbody is known accurately. It is given by Planck's law

$$W_{b,\lambda} = \frac{2\pi \mathbf{h} c^2 \lambda^{-5}}{e^{\mathbf{h}c/\mathbf{k}\lambda T} - 1} \tag{14.6}$$

where $W_{b,\lambda}$ = monochromatic emissive power of blackbody
   h = Planck's constant
   c = speed of light
   λ = wavelength of radiation
   k = Boltzmann's constant
   T = absolute temperature

Equation (14.6) can be written

$$W_{b,\lambda} = \frac{C_1 \lambda^{-5}}{e^{C_2/\lambda T} - 1} \tag{14.7}$$

where C_1 and C_2 are constants. The units of the various quantities and magnitudes of the constants in Eqs. (14.5) and (14.6) are given in the list of symbols, page 427.

Plots of $W_{b,\lambda}$ versus λ from Eq. (14.6) are shown as solid lines in Fig. 14.1 for blackbody radiation at temperatures of 1000, 1500, and 2000°F. The dotted line shows the monochromatic radiating power of a gray body of emissivity 0.9 at 2000°F.

Planck's law can be shown to be consistent with the Stefan–Boltzmann law by substituting $W_{b,\lambda}$ from Eq. (14.6) into Eq. (14.2) and integrating.

At any given temperature, the maximum monochromatic radiating power is attained at a definite wavelength, denoted by λ_{max}. Wien's displacement law states that λ_{max} is inversely proportional to the absolute temperature, or

$$T\lambda_{max} = C \tag{14.8}$$

The constant C is 2,890 when λ_{max} is in micrometers and T is in kelvins, or 5,200 when T is in degrees Rankine.

Wien's law also can be derived from Planck's law [Eq. (14.6)] by differentiating with respect to λ, equating the derivative to zero, and solving for λ_{max}.

ABSORPTION OF RADIATION BY OPAQUE SOLIDS

When radiation falls on a solid body, a definite fraction ρ may be reflected, and the remaining fraction $1 - \rho$ enters the solid to be either transmitted or absorbed. Most solids (other than glasses, certain plastics, quartz, and some minerals) absorb radiation of all wavelengths so readily that, except in thin sheets, the transmissivity τ is zero, and all nonreflected radiation is completely absorbed in a thin surface layer of the solid. The absorption of radiation by an opaque solid is therefore a surface phenomenon, not a volume phenomenon, and the interior of the solid is not of interest in the absorption of radiation. The heat generated by the absorption can flow into or through the mass of an opaque solid only by conduction.

Reflectivity and absorptivity of opaque solids

Since the transmissivity of an opaque solid is zero, the sum of the reflectivity and the absorptivity is unity, and the factors that influence reflectivity affect absorptivity in the opposite sense. In general, the reflectivity of an opaque solid depends on the temperature and character of the surface, the material of which the surface is made, the wavelength of the incident radiation, and the angle of incidence. Two main types of reflection are encountered, specular and diffuse. The first is characteristic of smooth surfaces such as polished metals; the second is found in reflection from rough surfaces or from dull, or matte, surfaces. In specular reflection, the reflected beam makes a definite angle with the surface, and the angle of incidence equals the angle of reflection. The reflectivity from these surfaces approaches unity, and the absorptivity approaches zero. Matte, or dull, surfaces reflect diffusely in all directions, there is no definite angle of reflection, and the absorptivity can approach unity. Rough surfaces, in which the scale of roughness is large in comparison with the wavelength of the incident radiation, will reflect diffusely even if the radiation from the individual units of roughness is specular. Reflectivities of rough surfaces may be either large or small, depending upon the reflective characteristic of the material itself. Most industrial surfaces of interest to the chemical engineer give diffuse reflection, and in treating practical cases, the important simplifying assumption can usually be made that reflectivity and absorptivity are independent of the angle of incidence. This assumption is equivalent to the *cosine law* [see Eq. (14.15)], which states that for a perfectly diffusing surface the intensity (or brightness, in the case of visible light) of the radiation leaving the surface is independent of the angle from which the surface is viewed. This is true whether the radiation is emitted by the surface, giving *diffuse radiation,* or is reflected by it, giving *diffuse reflection.*

The reflectivity may vary with the wavelength of the incident radiation, and the absorptivity of the entire beam is then a weighted average of the monochromatic absorptivities and depends upon the entire spectrum of the incident radiation.

The absorptivity of a gray body, like the emissivity, is the same for all wavelengths. If the surface of the gray body gives diffuse radiation or reflection, its monochromatic absorptivity is also independent of the angle of incidence of the radiant beam. The total absorptivity equals the monochromatic absorptivity and is also independent of the angle of incidence.

Kirchhoff's law

An important generalization concerning the radiating power of a substance is Kirchhoff's law, which states that, at temperature equilibrium, the ratio of the total radiating power of any body to its absorptivity depends only upon the temperature of the body. Thus, consider any two bodies in temperature equilibrium with common surroundings. Kirchhoff's law states that

$$\frac{W_1}{\alpha_1} = \frac{W_2}{\alpha_2} \tag{14.9}$$

where W_1, W_2 = total radiating powers of two bodies
α_1, α_2 = absorptivities of two bodies

This law applies to both monochromatic and total radiation.

If the first body referred to in Eq. (14.9) is a blackbody, $\alpha_1 = 1$, and

$$W_1 = W_b = \frac{W_2}{\alpha_2} \qquad (14.10)$$

where W_b denotes the total radiating power of a blackbody. Thus

$$\alpha_2 = \frac{W_2}{W_b} \qquad (14.11)$$

But, by definition, the emissivity of the second body ε_2 is

$$\varepsilon_2 = \frac{W_2}{W_b} = \alpha_2 \qquad (14.12)$$

Thus, when any body is at temperature equilibrium with its surroundings, its emissivity and absorptivity are equal. This relationship may be taken as another statement of Kirchhoff's law. In general, except for blackbodies or gray bodies, absorptivity and emissivity are not equal if the body is not in thermal equilibrium with its surroundings.

The absorptivity and emissivity, monochromatic or total, of a blackbody are both unity. The cosine law also applies exactly to a blackbody, as the reflectivity is zero for all wavelengths and all angles of incidence.

Kirchhoff's law applies to volumes as well as to surfaces. Since absorption by an opaque solid is effectively confined to a thin layer at the surface, the radiation emitted from the surface of the body originates in this same surface layer. Radiating substances absorb their own radiation, and radiation emitted by the material in the interior of the solid is also absorbed in the interior and does not reach the surface.

Because the energy distribution in the incident radiation depends upon the temperature and character of the originating surface, the absorptivity of the receiving surface may also depend upon these properties of the originating surface. Kirchhoff's law does not, therefore, always apply to nonequilibrium radiation. If, however, the receiving surface is gray, a constant fraction, independent of wavelength, of the incident radiation is absorbed by the receiving surface, and Kirchhoff's law applies whether or not the two surfaces are at the same temperature.

The majority of industrial surfaces, unfortunately, are not gray, and their absorptivities vary strongly with the nature of the incident radiation. Figure 14.2 shows how the absorptivity of various solids varies with the peak wavelength of the incident radiation and thus with the temperature of the source.[5a] A few solids, such as slate, are almost truly gray, and their absorptivities are almost constant. For polished metallic surfaces the absorptivity α_2 rises with the absolute temperature of the source T_1 and also that of surface T_2 according to the equation

$$\alpha_2 = k_1 \sqrt{T_1 T_2} \qquad (14.13)$$

where k_1 is a constant. For most surfaces, however, the absorptivity follows a curve like that indicated for paper, wood, cloth, etc. Such surfaces have high absorptivities for radiation of long wavelengths originating from sources below about 1000°F (540°C); as the source temperature rises above this value, the absorptivity falls,

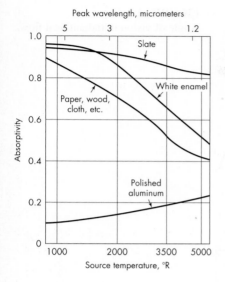

FIGURE 14.2
Absorptivities of various solids versus source temperature and peak wavelength of incident radiation. (*By permission of author and publisher, from H. C. Hottel, p. 62 in W. H. McAdams, Heat Transmission. Copyright by author, 1954, McGraw-Hill Book Company.*)

sometimes very markedly. With a few materials the absorptivity rises once again when the source temperature is very high.

RADIATION BETWEEN SURFACES

The total radiation for a unit area of an opaque body of area A_1, emissivity ε_1, and absolute temperature T_1 is

$$\frac{q}{A_1} = \sigma \varepsilon_1 T_1^4 \tag{14.14}$$

Most surfaces, however, that emit radiation also receive radiation from other surfaces at different temperatures. Some of this incoming radiation is absorbed and must be allowed for in determining the total flux of radiant energy. For example, a steam line in a room is surrounded by the walls, floor, and ceiling of the room, all of which are radiating to the pipe, and although the pipe loses more energy than it absorbs from its surroundings, the net loss by radiation is less than that calculated from Eq. (14.14). Even when a surface is radiating to a clear night sky, the radiated energy is partially absorbed by the water and carbon dioxide in the atmosphere, and part of this absorbed energy is radiated back to the surface.

In furnaces and other high-temperature equipment, where radiation is particularly important, the usual objective is to obtain a controlled rate of net heat exchange between one or more hot surfaces, called *sources,* and one or more cold surfaces, called *sinks.* In many cases the hot surface is a flame, but exchange of energy between surfaces is common, and a flame can be considered to be a special form of translucent surface. The following treatment is limited to the radiant energy transfer between opaque surfaces in the absence of any absorbing medium between them.

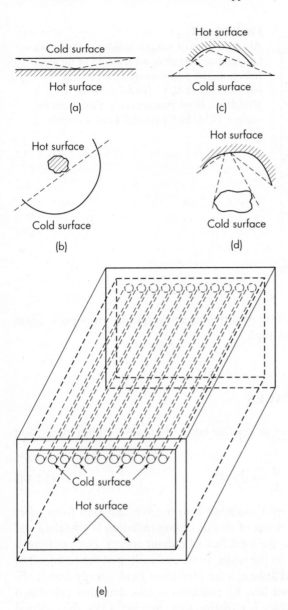

FIGURE 14.3
Angle of vision in radiant heat flow.

The simplest type of radiation between two surfaces occurs where each surface can see only the other, for example, where the surfaces are very large parallel planes, as shown in Fig. 14.3a, and where both surfaces are black. The energy emitted per unit area by the first plane is σT_1^4; that emitted by the second plane is σT_2^4. Assume that $T_1 > T_2$. All the radiation from each of the surfaces falls on the other surface and is completely absorbed. Since the areas of the two surfaces are equal, the net loss of energy per unit area by the first plane and the net gain by the second are $\sigma T_1^4 - \sigma T_2^4$, or $\sigma(T_1^4 - T_2^4)$.

Actual engineering problems differ from this simple situation in the following ways: (1) One or both of the surfaces of interest see other surfaces. In fact, an element of surface in a concave area sees a portion of its own surface. (2) No actual surface is exactly black, and the emissivities of the surfaces must often be considered.

Angle of vision

Qualitatively, the interception of radiation from an area element of a surface by another surface of finite size can be visualized in terms of the angle of vision, which is the solid angle subtended by the finite surface at the radiating element. The solid angle subtended by a hemisphere is 2π steradians (sr). This is the maximum angle of vision that can be subtended at any area element by a plane surface in sight of the element. It will be remembered that the total radiating power of an area element is defined to take this fact into account. If the angle of vision is less than 2π sr, only a fraction of the radiation from the area element will be intercepted by the receiving area and the remainder will pass on to be absorbed by other surfaces in sight of the remaining solid angle. Some of the hemispherical angle of vision of an element of a concave surface is subtended by the originating surface itself.

Figure 14.3 shows several typical radiating surfaces. Figure 14.3a shows how, in two large parallel planes, an area element on either plane is subtended by a solid angle of 2π sr by the other. The radiation from either plane cannot escape being intercepted by the other. A point on the hot body of Fig. 14.3b sees only the cold surface, and the angle of vision is again 2π sr. Elements of the cold surface, however, see, for the most part, other portions of the cold surface, and the angle of vision for the hot body is small. This effect of self-absorption is also shown in Fig. 14.3c, where the angle of vision of an element of the hot surface subtended by the cold surface is relatively small. In Fig. 14.3d, the cold surface subtends a small angle at the hot surface, and the bulk of the radiation from the hot surface passes on to some undetermined background. Figure 14.3e shows a simple muffle furnace, in which the radiation from the hot floor, or source, is intercepted partly by the row of tubes across the top of the furnace, which form the sink, and partly by the refractory walls and the refractory ceiling behind the tubes. The refractory in such assemblies is assumed to absorb and emit energy at the same rate, so the net energy effect at the refractory is zero. The refractory ceiling absorbs the energy that passes between the tubes and reradiates it to the backs of the tubes.

If attention is to be focused on the net energy received by the cold surface, the words *hot* and *cold* in Fig. 14.3 may be interchanged, and the same qualitative conclusions hold.

Square-of-the-distance effect

The energy from a small surface that is intercepted by a large one depends only upon the angle of vision. It is independent of the distance between the surfaces. The energy received per *unit area* of the receiving surface, however, is inversely proportional to the square of the distance between the surfaces, as shown by the following discussion.

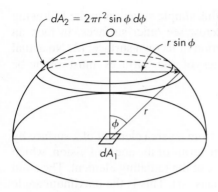

$dA_2 = 2\pi r^2 \sin\phi \, d\phi$

$r\sin\phi$

ϕ

r

dA_1

FIGURE 14.4
Diffuse radiation to a hemispherical surface.

The rate of energy received per unit area of the receiving surface is called the intensity I of the radiation. For diffuse radiation the cosine law can be used to find how the intensity of radiation varies with distance and orientation of the receiving surface from the emitting surface. Consider the element of emitting surface dA_1 shown in Fig. 14.4 at the center of a hemispherical surface A_2 with a radius r. The ring-shaped element of the receiving surface dA_2 has an area $2\pi r^2 \sin\phi \, d\phi$, where ϕ is the angle between the normal to dA_1 and the radius joining dA_1 and dA_2. The intensity at the point directly above dA_1 is denoted by dI_0; that at any other point above A_1 is dI. By the cosine law of diffuse radiation

$$dI = dI_0 \cos\phi \tag{14.15}$$

The relation between emissive power W_1 of the emitting surface and the intensity is found as follows. The rate of energy reception by the element of area dA_2, dq_{dA_2}, is, by Eq. (14.15),

$$dq_{dA_2} = dI \, dA_2 = dI_0 \cos\phi \, dA_2 \tag{14.16}$$

Since $dA_2 = (2\pi r)(r\sin\phi) \, d\phi$,

$$dq_{dA_2} = dI_0 \, 2\pi r^2 \sin\phi \cos\phi \, d\phi \tag{14.17}$$

The rate of emission from area dA_1 must equal the rate at which energy is received by the total area A_2, since all the radiation from dA_1 impinges on some part of A_2. The rate of reception by A_2 is found by integrating dq_{dA_2} over area A_2. Hence,

$$W_1 \, dA_1 = \int_{A_2} dq_{dA_2} = \int_0^{\pi/2} 2\pi \, dI_0 \, r^2 \sin\phi \cos\phi \, d\phi$$

$$= \pi \, dI_0 \, r^2 \tag{14.18}$$

Thus

$$dI_0 = \frac{W_1}{\pi r^2} \, dA_1 \tag{14.19}$$

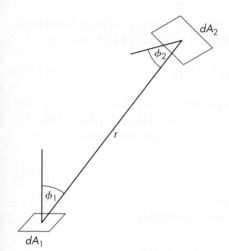

FIGURE 14.5
Differential areas for radiation.

By substitution from Eq. (14.15)

$$dI = \frac{W_1}{\pi r^2} dA_1 \cos \phi \tag{14.20}$$

Quantitative calculation of radiation between black surfaces

The above considerations can be treated quantitatively by setting up a differential equation for the net radiation between two elementary areas and integrating the equation for definite types of arrangement of the surfaces. The two plane elements of area dA_1 and dA_2 in Fig. 14.5 are separated by a distance r and are set at any arbitrary orientation to each other that permits a connecting straight line to be drawn between them. In other words, element dA_1 must see element dA_2; at least some of the radiation from dA_1 must impinge upon dA_2. Angles ϕ_1 and ϕ_2 are the angles between the connecting straight line and the normals to dA_1 and dA_2, respectively.

Since the line connecting the area elements is not normal to dA_2 as it is in Fig. 14.4, the rate of energy reception by element dA_2 of radiation originating at dA_1 is

$$dq_{dA_1 \rightarrow dA_2} = dI_1 \cos \phi_2 \, dA_2 \tag{14.21}$$

where dI_1 is the intensity, at area dA_2, of radiation from area dA_1. From Eqs. (14.20) and (14.21), since element dA_1 is black,

$$dq_{dA_1 \rightarrow dA_2} = \frac{W_1}{\pi r^2} dA_1 \cos \phi_1 \cos \phi_2 \, dA_2$$

$$= \frac{\sigma T_1^4}{\pi r^2} \cos \phi_1 \cos \phi_2 \, dA_1 \, dA_2 \tag{14.22}$$

Similarly, for radiation from dA_2 which impinges on dA_1

$$dq_{dA_2 \rightarrow dA_1} = \frac{\sigma T_2^4}{\pi r^2} \cos \phi_1 \cos \phi_2 \, dA_1 \, dA_2 \tag{14.23}$$

The net rate of transfer dq_{12} between the two area elements is then found from the difference between the rates indicated in Eqs. (14.22) and (14.23), to give

$$dq_{12} = \sigma \frac{\cos \phi_1 \, \cos \phi_2 \, dA_1 \, dA_2}{\pi r^2} (T_1^4 - T_2^4) \qquad (14.24)$$

The integration of Eq. (14.24) for a given combination of finite surfaces is usually a lengthy multiple integration based on the geometry of the two planes and their relation to each other. The resulting equation for any of these situations can be written in the form

$$q_{12} = \sigma A F (T_1^4 - T_2^4) \qquad (14.25)$$

where q_{12} = net radiation between two surfaces
$\quad\quad A$ = area of either of two surfaces, chosen arbitrarily
$\quad\quad F$ = dimensionless geometric factor

The factor F is called the view factor or angle factor; it depends upon the geometry of the two surfaces, their spatial relationship with each other, and the surface chosen for A.

If surface A_1 is chosen for A, Eq. (14.25) can be written

$$q_{12} = \sigma A_1 F_{12} (T_1^4 - T_2^4) \qquad (14.26)$$

If surface A_2 is chosen,

$$q_{12} = \sigma A_2 F_{21} (T_1^4 - T_2^4) \qquad (14.27)$$

Comparing Eqs. (14.26) and (14.27) gives

$$A_1 F_{12} = A_2 F_{21} \qquad (14.28)$$

Factor F_{12} may be regarded as the fraction of the radiation leaving area A_1 that is intercepted by area A_2. If surface A_1 can see only surface A_2, the view factor F_{12} is unity. If surface A_1 sees a number of other surfaces and if its entire hemispherical angle of vision is filled by these surfaces, then

$$F_{11} + F_{12} + F_{13} + \cdots = 1.0 \qquad (14.29)$$

The factor F_{11} covers the portion of the angle of vision subtended by other portions of body A_1. If the surface of A_1 cannot see any portion of itself, F_{11} is zero. The net radiation associated with an F_{11} factor is, of course, zero.

In some situations the view factor may be calculated simply. For example, consider a small blackbody of area A_2 having no concavities and surrounded by a large black surface of area A_1. The factor F_{21} is unity, as area A_2 can see nothing but area A_1. The factor F_{12} is, by Eq. (14.28),

$$F_{12} = \frac{F_{21} A_2}{A_1} = \frac{A_2}{A_1} \qquad (14.30)$$

By Eq. (14.29),

$$F_{11} = 1 - F_{12} = 1 - \frac{A_2}{A_1} \qquad (14.31)$$

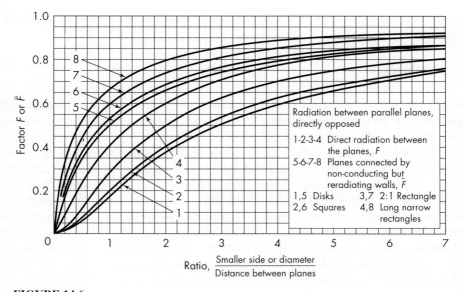

FIGURE 14.6
View factor and interchange factor, radiation between opposed parallel disks, rectangles, and squares.

The factor F has been determined by Hottel[3] for a number of important special cases. Figure 14.6 shows the F factor for equal parallel planes directly opposed. Line 1 is for disks, line 2 for squares, line 3 for rectangles having a ratio of length to width of 2:1, and line 4 for long, narrow rectangles. In all cases, the factor F is a function of the ratio of the side or diameter of the planes to the distance between them.

Allowance for refractory surfaces

When the source and sink are connected by refractory walls in the manner shown in Fig. 14.3e, factor F can be replaced by an analogous factor, called the *interchange factor $\bar{F}$*, and Eqs. (14.26) and (14.27) written as

$$q_{12} = \sigma A_1 \bar{F}_{12}(T_1^4 - T_2^4) = \sigma A_2 \bar{F}_{21}(T_1^4 - T_2^4) \qquad (14.32)$$

The interchange factor $\bar{F}$ has been determined accurately for some simple situations.[4] Lines 5 to 8 of Fig. 14.6 give values of $\bar{F}$ for directly opposed parallel planes connected by refractory walls. Line 5 applies to disks, line 6 to squares, line 7 to 2:1 rectangles, and line 8 to long, narrow rectangles.

An approximate equation for $\bar{F}$ in terms of F is

$$\bar{F}_{12} = \frac{A_2 - A_1 F_{12}^2}{A_1 + A_2 - 2A_1 F_{12}} \qquad (14.33)$$

Equation (14.33) applies where there is but one source and one sink, where neither area A_1 nor A_2 can see any part of itself. It is based on the assumption that the temperature of the refractory surface is constant. This last is a simplifying assumption, as the local temperature of the refractory usually varies between those of the source and the sink.

Nonblack surfaces

The treatment of radiation between nonblack surfaces, in the general case where absorptivity and emissivity are unequal and both depend upon wavelength and angle of incidence, is obviously complicated. Several important special cases can, however, be treated simply.

A simple example is a small body that is not black surrounded by a black surface. Let the areas of the enclosed and surrounding surfaces be A_1 and A_2, respectively, and let their temperatures be T_1 and T_2, respectively. The radiation from surface A_2 falling on surface A_1 is $\sigma A_2 F_{21} T_2^4$. Of this, the fraction α_1, the absorptivity of area A_1 for radiation from surface A_2, is absorbed by surface A_1. The remainder is reflected to the black surroundings and completely reabsorbed by area A_2. Surface A_1 emits radiation in amount $\sigma A_1 \varepsilon_1 T_1^4$, where ε_1 is the emissivity of surface A_1. All this radiation is absorbed by surface A_2, and none is returned by another reflection. The emissivity ε_1 and absorptivity α_1 are not in general equal, because the two surfaces are not at the same temperature. The net energy loss by surface A_1 is

$$q_{12} = \sigma \varepsilon_1 A_1 T_1^4 - \sigma A_2 F_{21} \alpha_1 T_2^4 \tag{14.34}$$

But by Eq. (14.28), $A_2 F_{21} = A_1$, and after elimination of $A_2 F_{21}$, Eq. (14.34) becomes

$$q_{12} = \sigma A_1 \left(\varepsilon_1 T_1^4 - \alpha_1 T_2^4 \right) \tag{14.35}$$

If surface A_1 is gray, $\varepsilon_1 = \alpha_1$ and

$$q_{12} = \sigma A_1 \varepsilon_1 \left(T_1^4 - T_2^4 \right) \tag{14.36}$$

In general, for gray surfaces, Eqs. (14.26) and (14.27) can be written

$$q_{12} = \sigma A_1 \mathscr{F}_{12} \left(T_1^4 - T_2^4 \right) = \sigma A_2 \mathscr{F}_{21} \left(T_1^4 - T_2^4 \right) \tag{14.37}$$

where $\mathscr{F}_{12}$ and $\mathscr{F}_{21}$ are the *overall interchange factors* and are functions of ε_1 and ε_2. The value of σ is 5.672×10^{-8} W/m$^2 \cdot$K^4, or 0.1713×10^{-8} Btu/ft$^2 \cdot$h$\cdot °$R^4.

Two large parallel planes. In simple cases factor $\mathscr{F}$ can be calculated directly. Consider two large gray parallel planes at absolute temperatures T_1 and T_2, as indicated in Fig. 14.7, with emissivities ε_1 and ε_2, respectively. The energy radiated from a unit area of surface 1 equals $\sigma T_1^4 \varepsilon_1$. Part of this energy is absorbed by surface 2, and part is reflected. The amount absorbed, as shown in Fig. 14.7a, equals $\sigma T_1^4 \varepsilon_1 \varepsilon_2$. Part of the reflected beam is reabsorbed by surface 1, and part is re-reflected to surface 2. Of this re-reflected beam an amount $\sigma T_1^4 \varepsilon_1 \varepsilon_2 (1 - \varepsilon_1)(1 - \varepsilon_2)$ is absorbed. Successive reflections and absorptions lead to the following equation for the total amount of radiation originating at surface 1 that is absorbed by surface 2:

$$q_{1 \to 2} = \sigma T_1^4 \varepsilon_1 \varepsilon_2 [1 + (1 - \varepsilon_1)(1 - \varepsilon_2) + (1 - \varepsilon_1)^2 (1 - \varepsilon_2)^2 + \cdots]$$

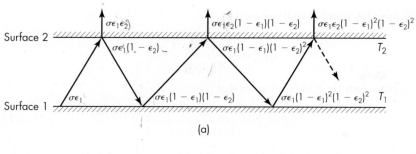

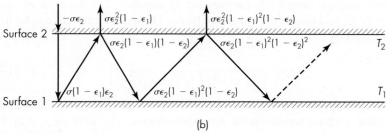

FIGURE 14.7
Evaluation of overall interchange factor for large gray parallel planes: (*a*) energy originating at surface 1, which is absorbed by unit area of surface 2; (*b*) energy originating at surface 2, which is reabsorbed by surface 2.

Some of the energy originating at surface 2, as shown in Fig. 14.7*b*, is reflected by surface 1 and returns to surface 2, where part of it is absorbed. The amount of this energy, per unit area, is

$$q_{2\to2} = -\sigma T_2^4\big[\varepsilon_2 - \varepsilon_2^2(1 - \varepsilon_1) - \varepsilon_2^2(1 - \varepsilon_1)^2(1 - \varepsilon_2) - \cdots\big]$$

The total amount of energy absorbed by a unit area of surface 2 is therefore

$$q_{12} = q_{1\to2} + q_{2\to2}$$

$$= \sigma T_1^4\varepsilon_1\varepsilon_2[1 + (1 - \varepsilon_1)(1 - \varepsilon_2) + (1 - \varepsilon_1)^2(1 - \varepsilon_2)^2 + \cdots]$$

$$- \sigma T_2^4\big\{\varepsilon_2 - \varepsilon_2^2(1 - \varepsilon_1)[1 + (1 + \varepsilon_1)(1 - \varepsilon_2) + \cdots]\big\}$$

Let
$$y = (1 - \varepsilon_1)(1 - \varepsilon_2)$$

Then

$$q_{12} = \sigma T_1^4\varepsilon_1\varepsilon_2(1 + y + y^2 + \cdots) - \sigma T_2^4\big[\varepsilon_2 - \varepsilon_2^2(1 - \varepsilon_1)(1 + y + y^2 + \cdots)\big]$$

But since $y < 1$,

$$1 + y + y^2 + \cdots = \frac{1}{1 - y}$$

Thus
$$q_{12} = \sigma T_1^4\varepsilon_1\varepsilon_2\frac{1}{1 - y} - \sigma T_2^4\left[\varepsilon_2 - \varepsilon_2^2(1 - \varepsilon_1)\frac{1}{1 - y}\right]$$

Substituting for y and simplifying give

$$q_{12} = \frac{\sigma(T_1^4 - T_2^4)}{1/\varepsilon_1 + 1/\varepsilon_2 - 1}$$

Comparison with Eq. (14.37) shows that

$$\mathscr{F}_{12} = \frac{1}{1/\varepsilon_1 + 1/\varepsilon_2 - 1} \tag{14.38}$$

One gray surface completely surrounded by another. Let the area of the enclosed body be A_1 and that of the enclosure be A_2. The overall interchange factor for this case is given by

$$\mathscr{F}_{12} = \frac{1}{1/\varepsilon_1 + (A_1/A_2)[(1/\varepsilon_2) - 1]} \tag{14.39}$$

Equation (14.39) applies strictly to concentric spheres or concentric cylinders, but it can be used without serious error for other shapes. The case of a gray body surrounded by a black one can be treated as a special case of Eq. (14.39) by setting $\varepsilon_2 = 1.0$. Under these conditions $\mathscr{F}_{12} = \varepsilon_1$.

For gray surfaces in general the following approximate equation may be used to calculate the overall interchange factor:

$$\mathscr{F}_{12} = \frac{1}{1/\bar{F}_{12} + [(1/\varepsilon_1) - 1] + (A_1/A_2)[(1/\varepsilon_2) - 1]} \tag{14.40}$$

where ε_1 and ε_2 are the emissivities of source and sink, respectively. If no refractory is present, F is used in place of $\bar{F}$.

Gebhart[1a] describes a direct method for calculating $\mathscr{F}$ in enclosures of gray surfaces where more than two radiating surfaces are present. Problems involving nongray surfaces are discussed in the literature.[5b]

EXAMPLE 14.1. A chamber for heat-curing large aluminum sheets, lacquered black on both sides, operates by passing the sheets vertically between two steel plates 150 mm apart. One of the plates is at 300°C, and the other, exposed to the atmosphere, is at 25°C. (*a*) What is the temperature of the lacquered sheet? (*b*) What is the heat transferred between the walls when equilibrium has been reached? Neglect convection effects. Emissivity of steel is 0.56; emissivity of lacquered sheets is 1.0.

Solution

(*a*) Let subscript 1 refer to hot plate, 2 to lacquered sheets, and 3 to cold plate:

$$\varepsilon_1, \varepsilon_3 = 0.56 \qquad \varepsilon_2 = 1.0$$

$$T_1 = 573 \text{ K} \qquad T_3 = 298 \text{ K}$$

From Eq. (14.37)

$$q_{12} = \sigma A_1 \mathscr{F}_{12}(T_1^4 - T_2^4)$$

$$q_{23} = \sigma A_2 \mathscr{F}_{23}(T_2^4 - T_3^4)$$

At equilibrium $q_{12} = q_{23}$. From Eq. (14.38)

$$\mathscr{F}_{12} = \frac{1}{1/0.56 + 1/1.0 - 1} = 0.56 = \mathscr{F}_{23}$$

Since $A_1 = A_2$,

$$\left(\frac{T_1}{100}\right)^4 - \left(\frac{T_2}{100}\right)^4 = \left(\frac{T_2}{100}\right)^4 - \left(\frac{T_3}{100}\right)^4$$

$$5.73^4 - \left(\frac{T_2}{100}\right)^4 = \left(\frac{T_2}{100}\right)^4 - 2.98^4$$

$$T_2 = 490.4\,\text{K} = 217.4°\text{C}$$

(b) From Eq. (14.37) the heat flux is

$$\frac{q_{12}}{A} = 5.672 \times 0.56(5.73^4 - 4.904^4) = 1,587\,\text{W/m}^2\ (503\,\text{Btu/h} \cdot \text{ft}^2)$$

Check:

$$\frac{q_{23}}{A} = 5.672 \times 0.56(4.904^4 - 2.98^4)$$

$$= 1,587\,\text{W/m}^2$$

Note: If the lacquered sheet is removed, $q_{13} = 3{,}174\,\text{W/m}^2\ (1{,}006\,\text{Btu/h} \cdot \text{ft}^2)$.

RADIATION TO SEMITRANSPARENT MATERIALS

Many substances of industrial importance are to some extent transparent to the passage of radiant energy. Solids such as glass and some plastics, thin layers of liquid, and many gases and vapors are semitransparent materials. Their transmissivity and absorptivity depend on the length of the path of the radiation and also on the wavelength of the beam.

Attentuation: absorption length

As shown below, a material may have very different absorptivities for radiation of different wavelengths. To classify a given material quantitatively as to its ability to absorb or transmit radiation of wavelength λ, it is necessary to define an *absorption length* (or *optical path length*) L_λ in the material. This length is the distance of penetration into the material at which the incident radiation has been attenuated a given amount; that is, the intensity of the radiation has been reduced to a given fraction of the intensity of the incident beam. The fraction usually used is $1/e$, where e is the base of natural logarithms.

The attenuation of an incident radiant beam with a monochromatic intensity $I_{0,\lambda}$ is shown in Fig. 14.8. At distance x from the receiving surface the intensity is

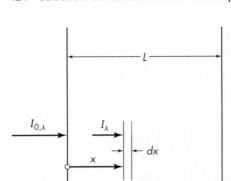

FIGURE 14.8
Attenuation of radiant beam in absorbing material.

reduced to I_λ. By assumption, the attentuation per unit length dI_λ/dx at any given value of x is proportional to the intensity I_λ at that location, or

$$-\frac{dI_\lambda}{dx} = \mu_\lambda I_\lambda \tag{14.41}$$

where μ_λ is the *absorption coefficient* for radiation of wavelength λ. Separating the variables in Eq. (14.41) and integrating between limits, with the boundary condition that $I_\lambda = I_{\lambda,0}$ at $x = 0$, give

$$\frac{I_\lambda}{I_{0,\lambda}} = e^{-\mu_\lambda x} \tag{14.42}$$

The absorption length L_λ is the value of x such that the attenuation is $1/e$. Setting the left-hand side of Eq. (14.42) equal to e^{-1} and setting x equal to L_λ give

$$L_\lambda = \frac{1}{\mu_\lambda} \tag{14.43}$$

The absorption length is therefore the reciprocal of the absorption coefficient.

If the total thickness L of the material is many times larger than L_λ, the material is said to be opaque to radiation of that wavelength. Most solids are opaque to thermal radiation of all wavelengths. If L is less than a few multiples of L_λ, however, the material is said to be transparent or semitransparent. The absorption length L_λ not only varies with wavelength of the incident radiation but also may vary with the temperature and density of the absorbing material. This is especially true of absorbing gases.

Radiation to layers of liquid or solid

In moderately thick layers nearly all solids and liquids are opaque and totally absorb whatever radiation passes into them. In thin layers, however, most liquids and some solids absorb only a fraction of the incident radiation and transmit the rest, depending on the thickness of the layer and the wavelength of the radiation. Thin layers of water (0.01 mm) transmit most of the radiation of wavelengths between 1 and 8 μm, except for absorption peaks at 3 and 6 μm. Layers a few millimeters thick, however, are transparent to visible light (0.38 to 0.78 μm) but absorb virtually all radiant energy with wavelengths greater than 1.5 μm.[2] For heat-transfer purposes, therefore, such layers of water may be considered to have an absorptivity of 1.0.

Layers of solids such as thin films of plastic behave similarly, but the absorption peaks are usually less well marked. Ordinary glass is also transparent to radiation of short wavelengths and opaque to that of longer wavelengths. This is the cause of the *greenhouse effect*, in which the contents of a glass-walled enclosure exposed to sunlight become hotter than the surroundings outside the enclosure. Radiation from the sun's surface, at about 5500 K (10,000°R), is chiefly shortwave and passes readily through the glass; radiation from inside the enclosure, from surfaces at, say, 30°C (86°F), is of longer wavelength and cannot pass through the glass. The interior temperature rises until convective losses from the enclosure equal the input of radiant energy.

Radiation to absorbing gases

Monatomic and diatomic gases, such as hydrogen, oxygen, helium, argon, and nitrogen, are virtually transparent to infrared radiation. More complex polyatomic molecules, including water vapor, carbon dioxide, and organic vapors, absorb radiation fairly strongly, especially radiation of specific wavelengths. The fraction of the incident radiation absorbed by a given amount of a gas or vapor depends on the length of the radiation path and on the number of molecules encountered by the radiation during its passage, that is, on the density of the gas or vapor. Thus the absorptivity of a given gas is a strong function of its partial pressure and a weaker function of its temperature.

If an absorbing gas is heated, it radiates to the cooler surroundings, at the same wavelengths favored for absorption. The emissivity of the gas is also a function of temperature and pressure. Because of the effect of path length, the emissivity and absorptivity of gases are defined arbitrarily in terms of a specific geometry. The emissivities of carbon dioxide and water vapor at 1 atm and charts for correcting them for changes in pressure are given by Hottel.[5c] When both carbon dioxide and water vapor are present, the total radiation is less than that calculated from the two gases separately, since each gas is somewhat opaque to radiation from the other. Correction charts to allow for this interaction are available.[5c]

COMBINED HEAT TRANSFER BY CONDUCTION-CONVECTION AND RADIATION

The total heat loss from a hot body to its surroundings often includes appreciable losses by conduction-convection and radiation. For example, a hot pipeline in a room loses heat nearly equally by each of the two mechanisms. Since the two types of heat transfer occur in parallel, the total loss is, assuming black surroundings,

$$\frac{q_T}{A} = \frac{q_c}{A} + \frac{q_r}{A} = h_c(T_w - T) + \sigma\varepsilon_w(T_w^4 - T^4) \qquad (14.44)$$

where q_T/A = total heat flux
$\qquad q_c/A$ = heat flux by conduction-convection
$\qquad q_r/A$ = heat flux by radiation
$\qquad h_c$ = convective heat-transfer coefficient
$\qquad \varepsilon_w$ = emmissivity of surface
$\qquad T_w$ = temperature of surface
$\qquad T$ = temperature of surroundings

Equation (14.44) is sometimes written

$$\frac{q_T}{A} = (h_c + h_r)(T_w - T) \qquad (14.45)$$

where h_r is a *radiation heat-transfer coefficient*, defined by

$$h_r \equiv \frac{q_r}{A(T_w - T)} \qquad (14.46)$$

This coefficient depends strongly on the absolute magnitude of T_w and to some extent on the temperature difference $T_w - T$. However, when the temperature difference is small, the value of h_r can be approximated from a simple equation using only one temperature. Expansion of the fourth-power term in Eq. (14.44) gives

$$\frac{q_r}{A} = \sigma\varepsilon_w\left(T_w^4 - T^4\right) = \sigma\varepsilon_w\left(T_w^2 + T^2\right)(T_w + T)(T_w - T) \qquad (14.47)$$

If $T_w - T$ is very small, T can be replaced by T_w in all but one term of Eq. (14.47) to give

$$\frac{q_r}{A} \approx \sigma\varepsilon_w\left(2T_w^2\right)(2T_w)(T_w - T) = \sigma\varepsilon_w\left(4T_w^3\right)(T_w - T) \qquad (14.48)$$

From the definition of h_r, Eq. (14.46),

$$h_r \approx 4\sigma\varepsilon_w T_w^3 \qquad (14.49)$$

If the temperature difference $T_w - T$ is more than a few degrees but less than 20 percent of the absolute temperature T_w, the arithmetic average of T_w and T can be used to improve the accuracy of Eq. (14.49).

Equations (14.44) and (14.49) apply to a small area completely surrounded by a surface of much larger area, so that only the emissivity of the small area influences the heat flux. For surfaces of nearly equal areas, the term ε_w should be replaced by $1/[(1/\varepsilon_1) + (1/\varepsilon_2) - 1]$, as indicated by Eq. (14.39).

Radiation in film boiling

In film boiling at a very hot surface a major fraction of the heat transfer occurs by radiation from the surface to the liquid. Equation (14.44) applies to this situation, since the surrounding liquid, as discussed earlier, has an absorptivity of unity. When radiation is active, the film of vapor sheathing the heating element is thicker than it would be if radiation were absent, and the convective heat-transfer coefficient is lower than it would be otherwise. For film boiling at the surface of a submerged horizontal tube, Eq. (13.22) predicts the convective heat-transfer coefficient h_o in the absence of radiation. When radiation is present, the convective coefficient is changed to h_c, which must be found by trial from the equation[1b]

$$h_c = h_o \left(\frac{h_o}{h_c + h_r} \right)^{1/3}$$ (14.50)

where h_o is found from Eq. (13.22) and h_r from Eq. (14.46) or (14.49). Substitution of h_c and h_r into Eq. (14.45) then gives the total rate of heat transfer to the boiling liquid.

SYMBOLS

A	Area, m² or ft²; A_1, of surface 1; A_2, of surface 2
C	Constant in Eq. (14.8), 2,890 μm · K or 5,200 μm · °R; C_1, constant in Eq. (14.7), 3.742×10^{-16} W · m²; C_2, constant in Eq. (14.7), 1.439 cm · K
c	Speed of light, 2.998×10^8 m/s or 9.836×10^8 ft/s
D	Diameter, side of cube, or distance between planes, m or ft
F	View factor or angle factor, dimensionless; F_{11}, F_{12}, F_{13}, for radiation from surface 1 to surfaces 1, 2, 3, respectively; F_{21}, from surface 2 to surface 1
$\bar{F}$	Interchange factor for systems involving refractory surfaces, dimensionless; $\bar{F}_{12}$, from surface 1 to surface 2; $\bar{F}_{21}$, from surface 2 to surface 1
$\mathscr{F}$	Overall interchange factor, dimensionless; $\mathscr{F}_{12}$, from surface 1 to surface 2; $\mathscr{F}_{21}$, from surface 2 to surface 1
$\mathbf{h}$	Planck's constant, 6.626×10^{-34} J · s
h	Individual heat-transfer coefficient, W/m² · °C or Btu/ft² · h · °F; h_c, for convection in the presence of radiation; h_r, for radiation; h_o, for boiling liquid in the absence of radiation
I	Radiation intensity, W/m² or Btu/ft² · h; I_0, at point on normal to radiating surface; I_1, at surface 2 of radiation from surface 1
I_λ	Monochromatic intensity in absorbing material, W/m² or Btu/ft² · h; $I_{0,\lambda}$, at surface of material
$\mathbf{k}$	Boltzmann's constant, 1.380×10^{-23} J/K
k_1	Constant in Eq. (14.13)
L	Thickness of material, radius of hemisphere, m or ft
L_λ	Absorption length, m or ft
q	Heat flow rate, W or Btu/h; q_T, total; q_c, by conduction-convection; q_r, by radiation; q_{12}, net exchange between surfaces 1 and 2; $q_{1\rightarrow2}$, radiation originating at surface 1 that is absorbed by surface 2; $q_{2\rightarrow2}$, radiation originating at surface 2 that returns to surface 2 and is absorbed

r	Radius of hemisphere or length of straight line connecting area elements of radiating surfaces, m or ft
T	Temperature, K or °R; T_w, of wall to surface; T_1, of surface 1; T_2, of surface 2
W	Total radiating power, W/m^2 or $Btu/ft^2 \cdot h$; W_b, of blackbody; W_1, of surface 1; W_2, of surface 2
W_λ	Monochromatic radiating power, $W/m^2 \cdot \mu m$ or $Btu/ft^2 \cdot h \cdot \mu m$; $W_{b,\lambda}$, of blackbody
x	Distance from surface of absorbing material, m or ft
y	Variable, set equal to $(1 - \varepsilon_1)(1 - \varepsilon_2)$

Greek Letters

α	Absorptivity, dimensionless; α_1, of surface 1; α_2, of surface 2; α_λ, for wavelength λ
ε	Emissivity, dimensionless; ε_w, of wall; ε_1, of surface 1; ε_2, of surface 2
ε_λ	Monochromatic emissivity, dimensionless
λ	Wavelength, μm; λ_{max}, wavelength at which $W_{b,\lambda}$ is a maximum
μ_λ	Absorption coefficient, m^{-1} or ft^{-1}
ρ	Reflectivity, dimensionless
σ	Stefan–Boltzmann constant, $5.672 \times 10^{-8} W/m^2 \cdot K^4$ or 0.1713×10^{-8} $Btu/ft^2 \cdot h \cdot °R^4$
τ	Transmissivity, dimensionless
ϕ	Angle with normal to surface; ϕ_1, to surface 1; ϕ_2, to surface 2

PROBLEMS

14.1. Determine the net heat transfer by radiation between two surfaces A and B, expressed as watts per square meter of area B, if the temperatures of A and B are 500 and 200°C, respectively, and the emissivities of A and B are 0.90 and 0.25, respectively. Both surfaces are gray. (*a*) Surfaces A and B are infinite parallel planes 3 m apart. (*b*) Surface A is a spherical shell 3 m in diameter, and surface B is a similar shell concentric with A and 0.3 m in diameter. (*c*) Surfaces A and B are flat parallel squares 2 by 2 m, one exactly above the other, 2 m apart. (*d*) Surfaces A and B are concentric cylindrical tubes with diameters of 300 and 275 mm, respectively. (*e*) Surface A is an infinite plane, and surface B is an infinite row of 100-mm-OD tubes set on 200-mm centers. (*f*) Same as part (*e*) except that 200 mm above the centerlines of the tubes is another infinite plane having an emissivity of 0.90, which does not transmit any of the energy incident upon it. (*g*) Same as part (*f*) except that surface B is a double row of 100-mm-OD tubes set on equilateral 200-mm centers. For part (*e*) $F = 0.66$; for (*f*), $F = 0.88$; for (*g*), $F = 0.98$.

14.2. The black flat roof of a building has an emissivity of 0.9 and an absorptivity of 0.8 for solar radiation. The sun beats down at midday with an intensity of 300 $Btu/ft^2 \cdot h$. (*a*) If the temperature of the air and of the surroundings is 68°F, if the wind velocity is negligible, and if no heat penetrates the roof, what is the equilibrium temperature of the roof? For the rate of heat transfer by conduction-convection use $q/A = 0.38(\Delta T)^{1.25}$, where ΔT is the temperature drop between roof and air in degrees Fahrenheit. (b) What fraction of the heat from the roof is lost by radiation?

14.3. The roof of Prob. 14.2 is painted with an aluminum paint, which has an emissivity of 0.9 and an absorptivity for solar radiation of 0.5. What is the equilibrium temperature of the painted roof?

14.4. A 3-in. Schedule 40 iron pipeline carries steam at 6 atm gauge. The line is uninsulated and is 70 m long. The surrounding air is at 25°C. The emissivity of the pipe wall is 0.70. How many kilograms of steam will condense per hour? What percentage of the heat loss is from conduction-convection?

14.5. A radiant heating system is installed in the plaster ceiling of a room 18 ft long by 18 ft wide by 8 ft high. The temperature of the concrete floor is maintained at 65°F. Assume that no heat flows through the walls, which are coated with a reradiating material. The temperature of the air passing through the room is held at 65°F. If the required heat supply to the floor is 3,500 Btu/h, calculate the necessary temperature at the ceiling surface. How much heat is transferred to the air, in Btu per hour? Emissivity of plaster is 0.93; absorptivity of concrete is 0.63. The convective heat-transfer coefficient between the ceiling and the air is given by the equation $h_c = 0.20(\Delta T)^{1/4}$ Btu/ft$^2 \cdot$h$\cdot$°F.

14.6. On a clear night, when the effective blackbody temperature of space is −70°C, the air is at 15°C and contains water vapor at a partial pressure equal to that of ice or liquid water at 0°C. A very thin film of water, initially at 15°C, is placed in a very shallow well-insulated pan, placed in a spot sheltered from the wind with a full view of the sky. If $h_c = 2.6$ W/m$^2 \cdot$°C, state whether ice will form, supporting the conclusion with suitable calculations.

14.7. Air leaves a heat exchanger at about 300°C and 1.5 atm, and the temperature is measured using a thermocouple inside a $\frac{1}{2}$-in.-diameter thermowell mounted normal to the airflow. If the gas velocity is 25 ft/s and the pipe wall temperature is 270°C, what error in temperature measurement does radiation cause? (Ignore conduction along the axis of the thermowell.)

14.8. In an uninsulated house, there is a 90-mm air gap between the plaster wall and the wood siding. When the inside wall is at 18°C and the outer wall at −9°C, what is the heat loss in watts per square meter by radiation and by natural convection? By what factor would the heat loss be reduced by covering the inside wall with aluminum foil? Would it be better to put the aluminum foil halfway between the two walls? (Correlations for natural convection at vertical surfaces give a film coefficient of 3.9 W/m$^2 \cdot$°C for each wall.)

14.9. In boiling water at 1 atm pressure outside a stainless-steel tube with a surface temperature of 410°F, the heat-transfer coefficient h in the absence of radiation is 32 Btu/h$\cdot$ft$^2 \cdot$°F. If the emissivity of the stainless steel is 0.8, will radiation significantly augment the rate of boiling (i.e., by more than 5 percent)? Assume that the vapor film is transparent to radiation and the boiling liquid is opaque.

REFERENCES

1. Gebhardt, B. *Heat Transfer,* 2nd ed. New York: McGraw-Hill, 1971; (*a*) pp. 150ff., (*b*) p. 421.
2. Grober, H., S. Erk, and U. Grigull. *Fundamentals of Heat Transfer,* 3rd ed. New York: McGraw-Hill, 1961, p. 442.
3. Hottel, H. C. *Mech. Eng.,* **52:**699 (1930).
4. Hottel, H. C. *Notes on Radiant Heat Transmission,* rev. ed. Cambridge, MA: Department of Chemical Engineering, Massachusetts Institute of Technology, 1951.
5. Hottel, H. C., in W. H. McAdams. *Heat Transmission,* 3rd ed. New York: McGraw-Hill, 1954; (*a*) p. 62, (*b*) pp. 77ff, (*c*) p. 86.
6. Hottell, H. C., and A. F. Sarofim. *Radiative Transfer.* New York: McGraw-Hill, 1967.
7. McAdams, W. H. *Heat Transmission,* 3rd ed. New York: McGraw-Hill, 1954, pp. 472ff.
8. Mills, A. F. *Heat Transfer.* Homewood, IL: Irwin, 1992, p. 487.
9. Perry, R. H., and D. W. Green (eds.). *Perry's Chemical Engineers' Handbook,* 7th ed. New York: McGraw-Hill, 1997, p. 5-28.
10. Siegel, R., and J. R. Howell. *Thermal Radiation Heat Transfer,* 3rd ed. New York: McGraw-Hill, 1992.

CHAPTER 15

Heat-Exchange Equipment

In industrial processes heat energy is transferred by a variety of methods, including conduction in electric-resistance heaters; conduction-convection in exchangers, boilers, and condensers; radiation in furnaces and radiant heat dryers; and by special methods such as dielectric heating. Often the equipment operates under steady-state conditions, but in many processes it operates cyclically, as in regenerative furnaces and agitated process vessels.

This chapter deals with equipment types that are of greatest interest to a process engineer: tubular and plate exchangers; extended-surface equipment; mechanically aided heat-transfer devices; condensers and vaporizers; and packed-bed reactors or regenerators. Evaporators are described in Chap. 16. Information on all types of heat-exchange equipment is given in engineering texts and handbooks.[21,22,24]

General design of heat-exchange equipment

The design and testing of practical heat-exchange equipment are based on the general principles given in Chaps. 11 to 14. From material and energy balances, the required heat-transfer rate is calculated. Then, using the overall coefficient and the average ΔT, the required heat-transfer area is determined, and in cyclic equipment, the cycle time. In simple devices these quantities can be evaluated easily and with considerable accuracy, but in complex processing units the evaluation may be difficult and subject to considerable uncertainty. The final design is nearly always a compromise, based on engineering judgment, to give the best overall performance in light of the service requirements.

Sometimes the design is governed by considerations that have little to do with heat transfer, such as the space available for the equipment or the pressure drop that can be tolerated in the fluid streams. Tubular exchangers are, in general, designed in accordance with various standards and codes, such as the Standards of the Tubular Exchanger Manufacturers Association (TEMA)[33] and the ASME-API Unfired Pressure Vessel Code.[1]

In designing an exchanger many decisions—some of them arbitrary—must be made to specify the materials of construction, tube diameter, tube length, baffle spacing, number of passes, and so forth. Compromises must also be made. For example, a high fluid velocity inside small tubes leads to improved heat-transfer coefficients and a small required area, but increases the friction losses and pumping costs. The design of an individual exchanger may be optimized by a formal procedure to balance the heat-transfer area, and hence the price of the equipment and the fixed costs, against the cost of energy to pump the fluids.[25] In processing plants, however, the exchangers are components of a complex network of heat-transfer equipment, and it is the network, not the individual units, that is optimized to give minimum investment and operating costs.[7]

SHELL-AND-TUBE HEAT EXCHANGERS

Tubular heat exchangers are so important and so widely used in the process industries that their design has been highly developed. Standards devised and accepted by TEMA are available covering in detail the materials, methods of construction, technique of design, and dimensions for exchangers.[33] The following sections describe the more important types of exchanger and cover the fundamentals of their engineering, design, and operation.

Single-pass 1-1 exchanger

The simple double-pipe exchanger shown in Fig. 11.3 is inadequate for flow rates that cannot readily be handled in a few tubes. If several double pipes are used in parallel, the weight of metal required for the outer tubes becomes so large that the shell-and-tube construction, such as that shown in Fig. 15.1, where one shell serves for many tubes, is more economical. This exchanger, because it has one shell-side pass and one tube-side pass, is a 1-1 exchanger.

In an exchanger the shell-side and tube-side heat-transfer coefficients are of comparable importance, and both must be large if a satisfactory overall coefficient is to be attained. The velocity and turbulence of the shell-side liquid are as important as those of the tube-side fluid. To promote crossflow and raise the average

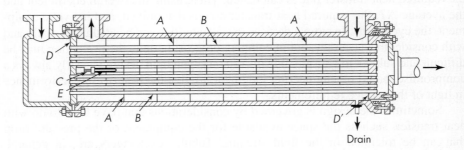

Drain

FIGURE 15.1
Single-pass 1-1 counterflow heat exchanger: A, baffles; B, tubes; C, guide rods; D, D', tube sheets; E, spacer tubes.

velocity of the shell-side fluid, baffles are installed in the shell. In the construction shown in Fig. 15.1, baffles A consist of circular disks of sheet metal with one side cut away. Common practice is to cut away a segment having a height equal to one-fourth the inside diameter of the shell. Such baffles are called *25 percent baffles.* The baffles are perforated to receive the tubes. To minimize leakage, the clearances between baffles and shell and tubes should be small. The baffles are supported by one or more guide rods C, which are fastened between the tube sheets D and D' by setscrews. To fix the baffles in place, short sections of tube E are slipped over rod C between the baffles. In assembling such an exchanger, it is necessary to do the tube sheets, support rods, spacers, and baffles first and then to install the tubes.

The stuffing box shown at the right-hand end of Fig. 15.1 provides for expansion. This construction is practicable only for small shells.

Tubes and tube sheets. As described in Chap. 8, tubes are drawn to definite wall thickness in terms of BWG and true outside diameter (OD), and they are available in all common metals. Tables of dimensions of standard tubes are given in App. 4. Standard lengths of tubes for heat-exchanger construction are 8, 12, 16, and 20 ft. Tubes are arranged in a triangular or square layout, known as *triangular pitch* or *square pitch* (pitch is the distance between centers of adjacent tubes). Triangular pitch is used unless the shell side tends to foul badly, because more heat-transfer area can be packed into a shell of given diameter than with square pitch. If the center-to-center distance between tubes is too small, tubes in triangular pitch cannot be cleaned by running a brush between rows, whereas tubes in square pitch are readily cleaned. Also, square pitch gives a lower shell-side pressure drop than triangular pitch.

TEMA standards specify a minimum pitch of 1.25 times the outside diameter of the tubes for triangular pitch and a minimum cleaning lane of $\frac{1}{4}$ in. for square pitch.

Shell and baffles. Shell diameters are standardized. For shells up to and including 23 in. the diameters are fixed in accordance with American Society for Testing and Materials (ASTM) pipe standards. Standard inside diameters are 8, 10, 12, $13\frac{1}{4}$, $15\frac{1}{4}$, $17\frac{1}{4}$, $19\frac{1}{4}$, $21\frac{1}{4}$, and $23\frac{1}{4}$ in., then 25, 27 in., and so on in 2-in. increments.[14b] These shells are constructed of rolled plate.

The distance between baffles (center to center) is the *baffle pitch,* or baffle spacing. It should not be less than one-fifth the diameter of the shell or more than the inside diameter of the shell.

Tubes are usually attached to the tube sheets by grooving the holes circumferentially and rolling the tube ends into the holes by means of a rotating tapered mandrel, which stresses the metal of the tube beyond the elastic limit, so the metal flows into the grooves. In high-pressure exchangers, the tubes are welded or brazed to the tube sheet after rolling.

Alternate designs

Shell-and-tube exchangers with segmented plate baffles may have vibration problems caused by the fluid flowing at high velocity across the tubes. In the *ROD baffle exchanger* developed by Phillips Petroleum Company, metal rods rather than sheet-metal baffles are used to support the tubes, and flow in the shell is mainly

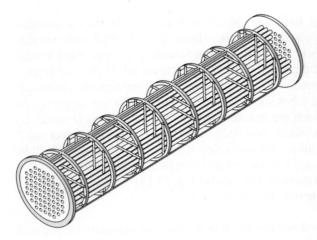

FIGURE 15.2
ROD baffle exchanger,
cutaway view. (*Phillips Petro-leum Co.*)

parallel to the tube axis. The tubes are arranged in square pitch, and rods with a diameter equal to the clearance between tube rows are attached to ring supports and placed between alternate tubes in both horizontal and vertical directions. The normal rod diameter is $\frac{1}{4}$ in., and each tube is supported on all four sides at several points along the exchanger, as shown in Fig. 15.2.

Correlations for the outside film coefficient have been developed using the hydraulic diameter for the Reynolds and Nusselt numbers and allowing for the effects of baffle spacing and leakage around the tube bundle. Flow across the rods leads to vortex formation, and the coefficients for turbulent flow are about 1.5 times those predicted for the same Reynolds number using the Dittus-Boelter equation [Eq. (12.32)]. The coefficients are not as high as those for a segmentally baffled exchanger with close baffle spacing, but the lower pressure drop and reduced vibration failure make the ROD baffle exchanger preferred for many applications.

Another design that requires no baffles uses tubes that are twisted into a helical shape with an oval cross section, so that each tube is supported over its entire length by multiple contact points with adjacent tubes. The end sections are kept circular to permit mounting in standard tube sheets. The twisted tubes give improved heat-transfer coefficients inside and outside because of greater turbulence, and the decrease in required surface area may more than offset the higher cost per square foot. This design eliminates tube vibration and may also reduce the rate of fouling.

With both the twisted tube and ROD baffle exchangers, flow distribution on the shell side is a problem in large-diameter units. With a single inlet pipe, tubes near the inlet would get more than the average flow, and those opposite the inlet would get little flow for an appreciable distance down the exchanger. Flow distribution is improved by enlarging the shell at the ends of the exchanger to make annular zones where fluid enters or leaves radially at lower velocity.

Multipass exchangers

The 1-1 exchanger has limitations, because when the tube-side flow is divided evenly among all the tubes, the velocity may be quite low, giving a low heat-transfer coefficient. If the number of tubes is reduced and the length increased so that the velocity is sufficiently high, the tube length required may be impractical. Using

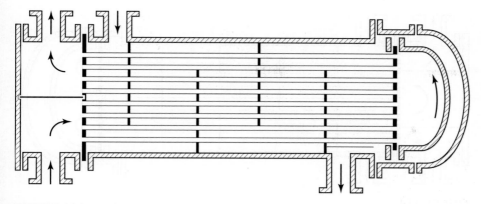

FIGURE 15.3
A 1-2 parallel-counterflow exchanger.

multipass construction with two, four, or more tube passes permits the use of standard tube lengths while ensuring a high velocity and a high tube-side coefficient. The disadvantages are that (1) the construction of the exchanger is slightly more complicated, (2) some sections in the exchanger have parallel flow, which limits the temperature approach, and (3) the friction loss is greatly increased. For example, the average velocity in the tubes of a four-pass exchanger is 4 times that in a single-pass exchanger having the same number and size of tubes and operated at the same liquid flow rate. The tube-side coefficient of the four-pass exchanger is approximately $4^{0.8} = 3.03$ times that for the single-pass exchanger, or even more if the velocity in the single-pass unit is sufficiently low to give laminar flow. The pressure drop per unit length is $4^{1.8}$ times greater, and the length is increased by 4 times; consequently the total friction loss is $4^{2.8} = 48.5$ times that in the single-pass unit, not including the additional expansion and contraction losses. The most economic design calls for such a velocity in the tubes that the increased cost of power for pumping is offset by the decreased cost of the apparatus.

An even number of tube-side passes are used in multipass exchangers. The shell side may be either single-pass or multipass. A common construction is the 1-2 parallel-counterflow exchanger, in which the shell-side liquid flows in one pass and the tube-side liquid in two passes. Such an exchanger is shown in Fig. 15.3. In multipass exchangers, floating heads are frequently used, and the bulge in the shell of the condenser in Fig. 11.1 and the stuffing box shown in Fig. 15.1 are unnecessary. The tube-side liquid enters and leaves through the same head, which is divided by a baffle to separate the entering and leaving tube-side streams.

The 1-2 exchanger is normally arranged so that the cold fluid and the hot fluid enter at the same end of the exchanger, giving parallel flow in the first tube pass and counterflow in the second. This permits a close temperature approach, at least at the exit end of the exchanger.

2-4 Exchanger

The 1-2 exchanger has an important limitation. Because of the parallel-flow pass, the exchanger is unable to bring the exit temperature of one fluid very near to

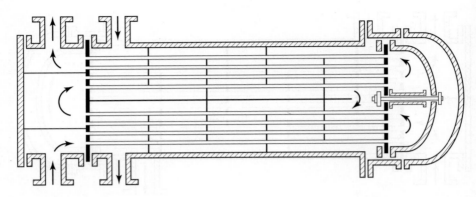

FIGURE 15.4
A 2-4 exchanger.

the entrance temperature of the other. Another way of stating the same limitation is that the heat recovery of a 1-2 exchanger is inherently poor.

A better recovery can be obtained by adding a longitudinal baffle to give two shell passes. A 2-2 exchanger of this kind closely approximates the performance of a double-pipe exchanger, but even with two tube-side passes the total tube length may be insufficient for good heat transfer. More common is the 2-4 exchanger, which has two shell-side and four tube-side passes. This type of exchanger also gives higher velocities and a larger overall heat-transfer coefficient than a 1-2 exchanger having two tube-side passes and operating with the same flow rates. An example of a 2-4 exchanger is shown in Fig. 15.4.

Temperature patterns in multipass exchangers

Temperature–length curves for a 1-2 exchanger are shown in Fig. 15.5a using the following temperature designations:

Inlet temperature of hot fluid T_{ha}
Outlet temperature of hot fluid T_{hb}
Inlet temperature of cold fluid T_{ca}
Outlet temperature of cold fluid T_{cb}
Intermediate temperature of cold fluid T_{ci}

Curve T_{ha}-T_{hb} applies to the shell-side fluid, which is assumed to be the hot fluid.[†] Curve T_{ca}-T_{ci} applies to the first pass of the tube-side liquid, and curve T_{ci}-T_{cb} to the second pass of the tube-side liquid. In Fig. 15.5a curves T_{ha}-T_{hb} and T_{ca}-T_{ci} taken together are those of a parallel-flow exchanger, and curves T_{ha}-T_{hb} and T_{ci}-T_{cb} taken together correspond to a countercurrent exchanger. The curves for a 2-4 exchanger are given in Fig. 15.5b. The dotted lines refer to the shell-side fluid and the solid lines to the tube-side fluid. Again it is assumed that the hotter fluid is in the

---

[†] For reasons of safety and heat economy, the hot fluid is usually put inside the tubes and the cooler fluid put in the shell.

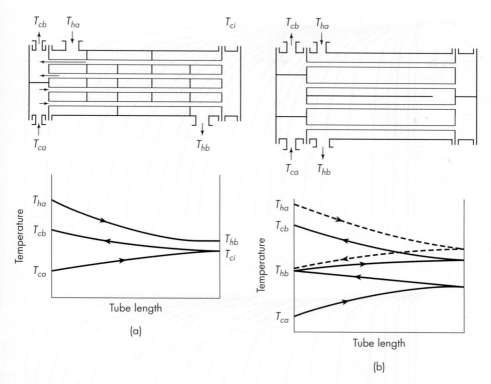

FIGURE 15.5
Temperature–length curves: (*a*) 1-2 exchanger; (*b*) 2-4 exchanger.

shell. The hotter pass of the shell-side fluid is in thermal contact with the two hottest tube-side passes and the cooler shell-side pass with the two coolest tube-side passes. The exchanger as a whole approximates a true countercurrent unit more closely than is possible with a 1-2 exchanger.

Correction of LMTD in multipass exchangers

In multipass exchangers which have more tube passes than shell passes, the flow is countercurrent in some sections and parallel in others. The LMTD, as given by Eq. (11.15), does not apply in this case, and it is customary to define a correction factor F_G, which is so determined that when it is multiplied by the LMTD for countercurrent flow, the product is the true average temperature drop.

Figure 15.6*a* and *b* shows factor F_G for 1-2 and 2-4 exchangers, respectively, derived on the assumptions that the overall heat-transfer coefficient is constant and that all elements of a given fluid stream have the same thermal history in passing through the exchanger.[24a] Each curved line in the figure corresponds to a constant value of the dimensionless ratio Z, defined as

$$Z = \frac{T_{ha} - T_{hb}}{T_{cb} - T_{ca}} \tag{15.1}$$

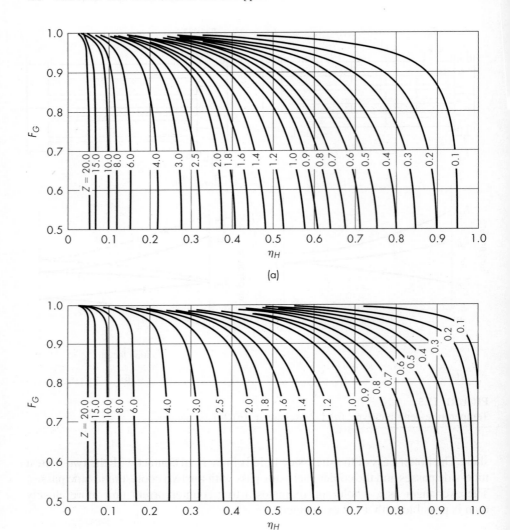

FIGURE 15.6
Correction of LMTD: (*a*) 1-2 exchangers; (*b*) 2-4 exchangers. (*From R. A. Bowman, A. C. Mueller, and W. M. Nagle, Trans. ASME,* **61**:283, 1940. *Courtesy of American Society of Mechanical Engineers.*)

and the abscissas are values of the dimensionless ratio η_H, defined as

$$\eta_H = \frac{T_{cb} - T_{ca}}{T_{ha} - T_{ca}} \tag{15.2}$$

The factor Z is the ratio of the fall in temperature of the hot fluid to the rise in temperature of the cold fluid. The factor η_H is the *heating effectiveness,* or the ratio of the actual temperature rise of the cold fluid to the maximum possible temperature

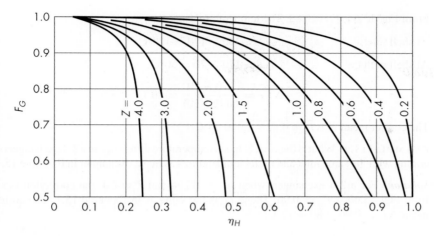

FIGURE 15.7
Correction of LMTD for crossflow. (*From R. A. Bowman, A. C. Mueller, and W. M. Nagle, Trans. ASME,* **61**:283, 1940. *Courtesy of American Society of Mechanical Engineers.*)

rise obtainable if the warm-end approach were zero (based on countercurrent flow). From the numerical values of η_H and Z, factor F_G is read from Fig. 15.6, interpolating between lines of constant Z where necessary, and multiplied by the LMTD for counterflow to give the true mean temperature drop.

Factor F_G is always less than unity. The mean temperature drop, and therefore the capacity of the exchanger, is less than that of a countercurrent exchanger having the same LMTD. When F_G is less than about 0.8, the exchanger should be redesigned with more passes or larger temperature differences; otherwise the heat-transfer surface is inefficiently used, and there is danger that small changes in conditions may cause the exchanger to become inoperable. When F_G is less than 0.75, it falls rapidly as η_H increases, so that operation is sensitive to small changes. In this region, also, any deviations from the basic assumptions on which the charts are based become important, especially that of a uniform thermal history for all elements of fluid. Leakage through and around the baffles may partially invalidate this assumption.

Other combinations of shell-side passes and tube-side passes are used, but the 1-2 and 2-4 are the most common. As the number of shell passes increases, with an even number of tube passes (usually 2 times the number of shell passes), F_G increases and the exchanger performance approaches that for true counterflow.[33]

EXAMPLE 15.1. In the 1-2 exchanger sketched in Fig. 15.5a, the values of the temperatures are $T_{ca} = 70°C$; $T_{cb} = 120°C$; $T_{ha} = 240°C$; $T_{hb} = 120°C$. What is the correct mean temperature drop in this exchanger?

Solution. The correction factor F_G is found from Fig. 15.6a. For this case, from Eqs. (15.1) and (15.2),

$$\eta_H = \frac{120 - 70}{240 - 70} = 0.294 \qquad Z = \frac{240 - 120}{120 - 70} = 2.4$$

From Fig. 15.6a, $F_G = 0.82$. The temperature drops are

At shell inlet: $\qquad\qquad\qquad\qquad \Delta T = 240 - 120 = 120°C$

At shell outlet: $\qquad\qquad\qquad\qquad \Delta T = 120 - 70 = 50°C$

$$\overline{\Delta T}_L = \frac{120 - 50}{\ln(120/50)} = 80°C$$

The correct mean is $\overline{\Delta T} = 0.82 \times 80 = 66°C$.

EXAMPLE 15.2. What is the correct mean temperature difference in a 2-4 exchanger operating with the same inlet and outlet temperatures as in the exchanger in Example 15.1?

Solution. For a 2-4 exchanger, when $\eta_H = 0.294$ and $Z = 2.4$, the correction factor from Fig. 15.6b is $F_G = 0.96$. The $\overline{\Delta T}_L$ is the same as in Example 15.1. The correct mean $\overline{\Delta T} = 0.96 \times 80 = 77°C$.

Heat-transfer coefficients in shell-and-tube exchangers

The heat-transfer coefficient h_i for the tube-side fluid in a shell-and-tube exchanger can be calculated from Eq. (12.33) or (12.34). The coefficient for the shell-side h_o cannot be so calculated because the direction of flow is partly parallel to the tubes and partly across them, and because the cross-sectional area of the stream and the mass velocity of the stream vary as the fluid crosses the tube bundle back and forth across the shell. Also, leakage between baffles and shell and between baffles and tubes short-circuits some of the shell-side liquid and reduces the effectiveness of the exchanger. An approximate but generally useful equation for predicting shell-side coefficients is the *Donohue equation*[6] [Eq. (15.6)], which is based on a weighted average mass velocity G_e of the fluid flowing parallel with the tubes and that flowing across the tubes. The mass velocity G_b parallel with the tubes is the mass flow rate divided by the free area for flow in the baffle window S_b. (The baffle window is the portion of the shell cross section not occupied by the baffle.) This area is the total area of the baffle window less the area occupied by the tubes, or

$$S_b = f_b \frac{\pi D_s^2}{4} - N_b \frac{\pi D_o^2}{4} \qquad\qquad (15.3)$$

where f_b = fraction of cross-sectional area of shell occupied by baffle window (commonly 0.1955)

D_s = inside diameter of shell

N_b = number of tubes in baffle window

D_o = outside diameter of tubes

In crossflow the mass velocity passes through a local maximum each time the fluid passes a row of tubes. For correlating purposes the mass velocity G_c for crossflow is based on the area S_c for transverse flow between the tubes in the row at or closest to the centerline of the exchanger. In a large exchanger S_c can be estimated from the equation

$$S_c = P D_s \left(1 - \frac{D_o}{p}\right) \qquad\qquad (15.4)$$

where p = center-to-center distance between tubes
  P = baffle pitch

The mass velocities are then

$$G_b = \frac{\dot{m}}{S_b} \quad \text{and} \quad G_c = \frac{\dot{m}}{S_c} \tag{15.5}$$

The Donohue equation is

$$\frac{h_o D_o}{k} = 0.2 \left(\frac{D_o G_e}{\mu} \right)^{0.6} \left(\frac{c_p \mu}{k} \right)^{0.33} \left(\frac{\mu}{\mu_w} \right)^{0.14} \tag{15.6}$$

where $G_e = \sqrt{G_b G_c}$. This equation tends to give conservatively low values of h_o, especially at low Reynolds numbers. More elaborate methods of estimating shell-side coefficients are available for the specialist.[24b] In j-factor form Eq. (15.6) becomes

$$\frac{h_o}{c_p G_e} \left(\frac{c_p \mu}{k} \right)^{2/3} \left(\frac{\mu_w}{\mu} \right)^{0.14} = j_H = 0.2 \left(\frac{D_o G_e}{\mu} \right)^{-0.4} \tag{15.7}$$

After the individual coefficients are known, the total area required is found in the usual way from the overall coefficient using an equation similar to Eq. (11.14). As discussed previously, the LMTD must often be corrected for the departure from true counterflow.

EXAMPLE 15.3. A tubular exchanger with 35-in. (889-mm) ID contains eight hundred and twenty-eight $\frac{3}{4}$-in. (19-mm) OD tubes 12 ft (3.66 mm) long on a 1-in. (25-mm) square pitch. Standard 25 percent baffles are spaced 12 in. (305 mm) apart. Liquid benzene at an average bulk temperature of 60°F (15.6°C) is being heated in the shell side of the exchanger at the rate of 100,000 lb/h (45,360 kg/h). If the outside surfaces of the tubes are at 140°F (60°C), estimate the individual heat-transfer coefficient of the benzene.

Solution. The shell-side coefficient is found from the Donohue equation [Eq. (15.6)]. The sectional areas for flow are first calculated from Eqs. (15.3) and (15.4). The quantities needed are

$$D_o = \frac{0.75}{12} = 0.0625 \text{ ft} \qquad D_s = \frac{35}{12} = 2.9167 \text{ ft}$$

$$p = \frac{1}{12} = 0.0833 \text{ ft} \qquad P = 1 \text{ ft}$$

From Eq. (15.4), the area for crossflow is

$$S_c = 2.9167 \times 1 \left(1 - \frac{0.0625}{0.0833} \right) = 0.7292 \text{ ft}^2$$

The number of tubes in the baffle window is approximately equal to the fractional area of the window f times the total number of tubes. For a 25 percent baffle, $f = 0.1955$. Hence

$$N_b = 0.1955 \times 828 = 161.8, \text{ say, } 161 \text{ tubes}$$

The area for flow in the baffle window, from Eq. (15.3), is

$$S_b = 0.1955 \frac{\pi \times 2.9167^2}{4} - 161 \frac{\pi \times 0.0625^2}{4} = 0.8123 \text{ ft}^2$$

The mass velocities are, from Eq. (15.5),

$$G_c = \frac{100,000}{0.7292} = 137,137 \text{ lb/ft}^2 \cdot \text{h} \qquad G_b = \frac{100,000}{0.8123} = 123,107 \text{ lb/ft}^2 \cdot \text{h}$$

$$G_e = \sqrt{G_b G_c} = \sqrt{137,137 \times 123,107} = 129,933 \text{ lb/ft}^2 \cdot \text{h}$$

The additional quantities needed for substitution in Eq. (15.6) are

$$\mu \text{ at } 60°\text{F} = 0.70 \text{ cP} \qquad \mu \text{ at } 140°\text{F} = 0.38 \text{ cP} \qquad \text{(App. 9)}$$

$$c_p = 0.41 \text{ Btu/lb} \cdot °\text{F} \qquad \text{(App. 15)}$$

$$k = 0.092 \text{ Btu/ft} \cdot \text{h} \cdot °\text{F} \qquad \text{(App. 13)}$$

From Eq. (15.6)

$$\frac{h_o D_o}{k} = 0.2 \left(\frac{0.0625 \times 129,933}{0.70 \times 2.42} \right)^{0.6} \left(\frac{0.41 \times 0.70 \times 2.42}{0.092} \right)^{0.33} \left(\frac{0.70}{0.38} \right)^{0.14}$$

$$= 68.59$$

Hence

$$h_o = \frac{68.59 \times 0.092}{0.0625} = 101 \text{ Btu/ft}^2 \cdot \text{h} \cdot °\text{F} \ (573 \text{ W/m}^2 \cdot °\text{C})$$

Crossflow exchangers

In some exchangers, such as air heaters, the shell is rectangular and the number of tubes in each row is the same. Flow is directly across the tubes, and baffles are not needed. Figure 15.7 shows the factor F_G for crossflow exchangers, derived on the assumption that neither stream mixes with itself during flow through the exchanger.[5] The quantities Z and η_H are given by Eqs. (15.1) and (15.2), and as before, F_G is so defined that when it is multiplied by the counterflow LMTD, the product is the correct mean temperature drop.

For the shell-side heat-transfer coefficient in a crossflow exchanger, the following equation is recommended.[3]

$$\frac{h_o D_o}{k} = 0.287 \left(\frac{D_o G}{\mu} \right)^{0.61} \left(\frac{c_p \mu}{k} \right)^{0.33} F_a \qquad (15.8)$$

where G is the mass velocity outside the tubes, based on the minimum area for flow in any tube row, and F_a is an "arrangement factor" that depends on Re and the tube spacing p. The other symbols are the same as in Eq. (15.6). Typical values of F_a are given in Table 15.1.

Heat-transfer units

One method of characterizing heat-exchanger performance is to determine the number of heat-transfer units N_H. For a two-fluid exchanger, N_H is defined as the change in temperature of one stream divided by the average temperature difference,

TABLE 15.1
Arrangement factor F_a for crossflow with square pitch[†]

	F_a			
p/D_o	Re = 2,000	Re = 8,000	Re = 20,000	Re = 40,000
1.25	0.85	0.92	1.03	1.02
1.5	0.94	0.90	1.06	1.04
2.0	0.95	0.85	1.05	1.02

[†]After Ref 3.

or driving force, in the exchanger. Transfer units can be based on either stream, but generally the stream with the lower capacity (flow rate times specific heat) is chosen. The temperature change is always taken as a positive number. When the specific heat and the overall heat-transfer coefficient are constant, the average driving force is the logarithmic mean temperature difference $\overline{\Delta T_L}$. If the cold stream has the lower capacity and thus the greater change in temperature, N_H is defined as

$$N_H = \frac{T_{cb} - T_{ca}}{\overline{\Delta T_L}} \tag{15.9}$$

Since $Q = \dot{m}_c c_{pc}(T_{cb} - T_{ca}) = U A \overline{\Delta T_L}$, an alternate definition is

$$N_H = \frac{U A}{\dot{m}_c c_{pc}} \tag{15.10}$$

In a multipass or crossflow exchanger the effective average driving force is often less than the logarithmic mean temperature difference, and the term F_G must be included in the denominator of the right-hand side of Eq. (15.9).

When N_H is large, the temperature of the stream with the lower capacity can be brought quite close to the inlet temperature of the other stream. The simplest case occurs when the temperature on one side is constant, such as when fluid in the tubes is heated by steam condensing on the outside. To find the temperature change as a function of N_H, the basic equation for heat transfer is integrated as follows:

$$\dot{m}_c c_{pc} \, dT_c = U \, dA \, (T_h - T_c)$$

$$\int_{T_{ca}}^{T_{cb}} \frac{dT_c}{T_h - T_c} = \int_0^A \frac{U \, dA}{\dot{m}_c c_{pc}}$$

$$\ln \frac{T_h - T_{cb}}{T_h - T_{ca}} = -\frac{U A}{\dot{m}_c c_{pc}} = -N_H$$

$$T_h - T_{cb} = (T_h - T_{ca})e^{-N_H} \tag{15.11}$$

Thus if $N_H = 3.0$, the cold stream temperature is brought to within 5 percent of the initial difference between the steam temperature and that of the cold fluid.

Another limiting case occurs when the hot and cold streams have equal capacities, so that $\dot{m}_c c_{pc} = \dot{m}_h c_{ph}$. The driving force is then constant and equal to $T_{ha} - T_{cb}$.

$$N_H = \frac{T_{cb} - T_{ca}}{T_{ha} - T_{cb}} \tag{15.12}$$

Equation (15.12) can be manipulated to find the temperature rise of the cold stream as a function of the inlet temperatures.

$$T_{cb} - T_{ca} = (T_{ha} - T_{ca})\left(\frac{N_H}{1 + N_H}\right) \tag{15.13}$$

The effectiveness of the heat exchanger ε is defined as the actual temperature change divided by the maximum possible change.

$$\varepsilon = \frac{T_{cb} - T_{ca}}{T_{ha} - T_{ca}} = \frac{N_H}{1 + N_H} \tag{15.14}$$

If $N_H = 3$ and the capacities are equal, the cold stream is brought 75 percent of the way to the inlet temperature of the hot stream.

When both streams change temperature but by different amounts, the effectiveness is defined as the temperature change for the stream with the lower capacity divided by the maximum possible change, and the effectiveness depends on the number of transfer units and the capacity ratio R_c:

$$R_c = \frac{\dot{m}c_p \text{ (lower-capacity stream)}}{\dot{m}c_p \text{ (higher-capacity stream)}}$$

A plot of ε versus N_H with R_c as a parameter is shown in Fig. 15.8. The line for $R_c = 0$ corresponds to the case for a steam-heated exchanger, where $\varepsilon = 1 - e^{-N_H}$ The line for $R_c = 1.0$ is based on Eq. (15.14). The plots for other values of R_c permit estimating the exit temperatures for a 1-1 counterflow exchanger without a trial-and-error calculation.

EXAMPLE 15.4. A counterflow exchanger with one tube pass and one shell pass is used to recover heat from an oil stream at 110°C. The exchanger and fluid properties are given below. Estimate the outlet oil temperature.

$$\dot{m}_h = 3{,}000 \text{ kg/h} \qquad T_{ha} = 110°C \qquad c_{ph} = 2{,}300 \text{ J/kg} \cdot °C$$
$$\dot{m}_c = 2{,}400 \text{ kg/h} \qquad T_{ca} = 25°C \qquad c_{pc} = 4{,}180 \text{ J/kg} \cdot °C$$
$$UA = 1.65 \times 10^7 \text{ W/°C}$$

Solution

$$R_c = \frac{3{,}000(2{,}300)}{2{,}400(4{,}180)} = 0.688$$

$$N_H = \frac{UA}{\dot{m}_h c_{ph}} = \frac{1.65 \times 10^7}{6.9 \times 10^6} = 2.39$$

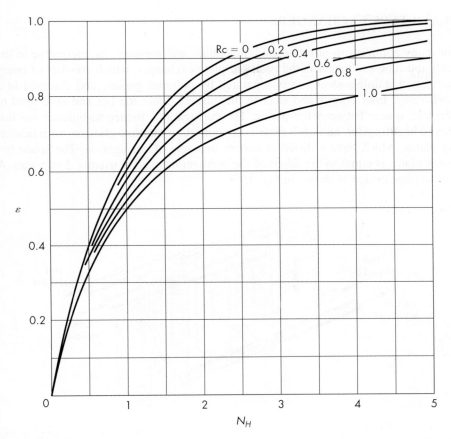

FIGURE 15.8

Effectiveness ε of 1-1 counterflow exchanger versus number of transfer units N_H.

From Fig. 15.8, $\varepsilon \cong 0.78$

$$T_{ha} - T_{hb} = 0.78(110 - 25) = 66.3°C$$

$$T_{hb} = 110 - 66.3 = 43.7°C.$$

Check:

$$Q = 3,000(2,300)(66.3) = 4.57 \times 10^8 \text{ W}$$

$$T_{cb} = 25 + \frac{Q}{2,400 \times 4,180} = 70.6°C$$

$$\Delta T_1 = 43.7 - 25 = 18.7°C$$

$$\Delta T_2 = 110 - 70.6 = 39.4°C$$

$$\overline{\Delta T_L} = 27.8°C$$

$$Q = UA \cdot \overline{\Delta T_L} = 1.65 \times 10^7(27.8) = 4.58 \times 10^8 \text{ W, close enough}$$

PLATE-TYPE EXCHANGERS

For many applications at moderate temperature and pressure, an alternative to the shell-and-tube exchanger is the gasketed plate exchanger, which consists of many corrugated stainless-steel sheets separated by polymer gaskets and clamped in a steel frame. Inlet portals and slots in the gaskets direct the hot and cold fluid to alternate spaces between the plates. The corrugations induce turbulence for improved heat transfer, and each plate is supported by multiple contacts with adjoining plates, which have a different pattern or angle of corrugation. The space between plates is equal to the depth of the corrugations and is usually 2 to 5 mm. A typical plate design is shown in Fig. 15.9.

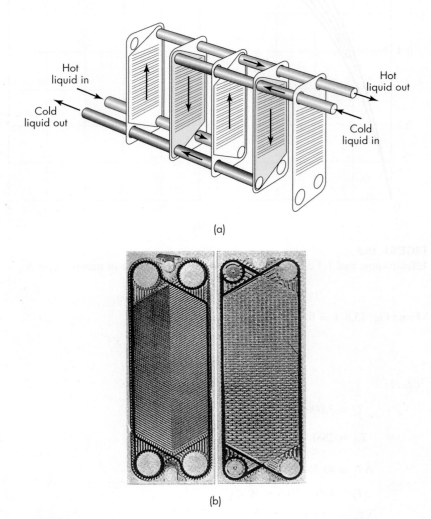

(a)

(b)

FIGURE 15.9
Plate heat exchanger: (a) general layout, (b) detail of plate design.

For a liquid-liquid exchanger, the usual fluid velocity is 0.2 to 1.0 m/s, and because of the small spacing, the Reynolds number is often less than 2,100. However, the corrugations give the flow turbulent characteristics at Reynolds numbers of 100 to 400, depending on the plate design. Evidence for turbulent flow is that the heat-transfer coefficient varies with the 0.6 to 0.8 power of the flow rate, and the pressure drop depends on the 1.7 to 2.0 power of the flow rate. The heat-transfer correlation for a common plate design is[2]

$$\text{Nu} = \frac{h D_e}{k} = 0.37 \, \text{Re}^{0.67} \, \text{Pr}^{0.33} \tag{15.15}$$

The pressure drop is given by the Fanning equation with the following friction factor:

$$f = 2.5 \, \text{Re}^{-0.3} \tag{15.16}$$

In Eq. (15.15), h is based on the nominal area of the corrugated plate. (Other correlations may be based on the corrugated area.) The equivalent diameter is 4 times the hydraulic radius, which for most exchangers is twice the plate spacing. If the hot and cold flows are equal and are introduced at opposite ends of the plates, performance close to true counterflow can be obtained; in the end sections, fluid is heated or cooled from only one side, which slightly reduces the efficiency, and the zigzag flow path also deviates from the ideal pattern. The correction factor decreases[20] as the number of heat-transfer units increases and is about 0.95 when $N_H = 3.0$. When the flows are not balanced, two or more passes can be used for the stream with the lower flow rate; but then the exchanger has some parallel and some counterflow sections, and the LMTD correction factor F_G may be 0.7 to 0.9.

With water or aqueous solutions on both sides, the overall coefficient for a clean plate-type exchanger may be 3,000 to 6,000 W/m$^2 \cdot$K (500 to 1,000 Btu/h$\cdot$ft$^2 \cdot$°F), several times the normal value for a shell-and-tube exchanger. Because of high shear rates, the fouling factors experienced are much lower than those for shell-and-tube exchangers, and the designer may just add 10 percent to the calculated area to allow for fouling.[15] The units can easily be taken apart for thorough cleaning.

Plate exchangers were originally used mainly in the dairy and food processing industries, but they have now found many other applications. New designs and better gaskets permit operation at up to 200°C and 25 atm. Exchangers with plate areas of 2 m^2 and a total area of 1,500 m^2 are available.

EXAMPLE 15.5. Several plate-type exchangers will be used to produce chilled water for air conditioning in the Cornell University Lake-Source-Cooling project. In each unit, up to 4,000 gal/min of water will be cooled from 60 to 44°F using an equal flow of Cayuga Lake water at 41°F. (a) If the plate spacing is 3 mm and the nominal water velocity is 0.5 m/s, what is the overall coefficient for a clean exchanger? (b) How many heat-transfer units will the exchanger have, and what is the estimated height of the heat-transfer section of the plates? (c) Predict the pressure drop for flow between the plates. (d) How many plates are needed if the plates are 1 m wide?

Solution

(a) Use 50°F as the average water temperature for each side. From App. 6,

$$\rho = 62.42 \text{ lb/ft}^3 = 1,000 \text{ kg/m}^3$$

$$\mu = 1.31 \text{ cP} = 1.31 \times 10^{-3} \text{ Pa} \cdot \text{s}$$

$$k = 0.333 \text{ Btu/h} \cdot \text{ft} \cdot {}^\circ\text{F} = 0.576 \text{ W/m} \cdot \text{K}$$

$$c_p = 1.00 \text{ Btu/lb} \cdot {}^\circ\text{F} = 4,184 \text{ J/kg} \cdot \text{K}$$

$$D_e = 2 \times \text{spacing} = 6 \times 10^{-3} \text{ m}$$

$$\text{Re} = \frac{6 \times 10^{-3}(0.5)(1,000)}{1.31 \times 10^{-3}} = 2,290$$

$$\text{Pr} = \frac{4,184(1.31 \times 10^{-3})}{0.576} = 9.52$$

$$\text{Nu} = 0.37(2,290)^{0.67}(9.52)^{0.33} = 139$$

$$h_i \cong h_o = \frac{139(0.576)}{6 \times 10^{-3}} = 1.33 \times 10^4 \text{ W/m}^2 \cdot \text{K} \ (2,350 \text{ Btu/h} \cdot \text{ft}^2 \cdot {}^\circ\text{F})$$

Assume 0.7-mm stainless-steel plates with $k_m = 9.4$ Btu/h·ft·°F or $k_m = 16.3$ W/m·K.

$$h_{\text{wall}} = \frac{16.3}{7 \times 10^{-4}} = 2.33 \times 10^4 \text{ W/m} \cdot \text{K}$$

$$\frac{1}{U} = \frac{2}{1.33 \times 10^4} + \frac{1}{2.33 \times 10^4}$$

$$U = 5.17 \times 10^3 \text{ W/m}^2 \cdot \text{K} \ (911 \text{ Btu/h} \cdot \text{ft}^2 \cdot {}^\circ\text{F})$$

(b) For the warm stream:

$$T_{ha} - T_{hb} = 60 - 44 = 16{}^\circ\text{F}$$

For the cold stream:

$$T_{cb} - T_{ca} = 57 - 41 = 16{}^\circ\text{F}$$

$$N_H = \int \frac{dT}{\Delta T} = \frac{16}{3} = 5.33$$

This is an unusually large number of transfer units, and the LMTD correction factor F_G is about 0.9.[20]

For plates with a width w and length L for the heat-transfer section, the heat-transfer area per channel is

$$A = 2wL$$

The cold flow per channel is

$$\dot{m}_c = (0.5 \text{ m/s})(3 \times 10^{-3} \ w)(1,000) = 1.5w \text{ kg/s}$$

$$Q = 1.5wc_p(T_{cb} - T_{ca}) = U(2wL)(0.9 \ \Delta T)$$

$$L = \frac{1.5(4,184)}{2(0.9)(5.17 \times 10^3)}\left(\frac{16}{3}\right) = 3.6 \text{ m}$$

Since the maximum plate height for standard exchangers is only about 4 m, including inlet and exit ports, either a lower velocity should be used or two exchangers placed in series, keeping the counterflow arrangement. Then the height of each heat-transfer section would be one-half as great.

$$\text{Revised } L = \frac{3.6}{2} = 1.8 \text{ m}$$

Allowing for inlet and exit ports, the total plate height would be about 3 m.

(c)

$$\frac{\Delta p}{L} = 2f\frac{u^2\rho}{D_e}$$

$$f = \frac{2.5}{(2,290)^{0.3}} = 0.245$$

$$\Delta p = \frac{2(0.245)(0.5)^2(1,000)(3.6)}{6 \times 10^{-3}} = 7.35 \times 10^4 \text{ N/m}^2 \text{ (10.6 lb/in.}^2)$$

The overall pressure drop will be greater because of losses in the feed and discharge piping.

(d) Cold flow:

$$\dot{m}_c = \frac{4,000 \text{ gal/min}}{264.7 \times 60} = 0.252 \text{ m}^3/\text{s}$$

For $w = 1$ m, the flow per channel is 1.5×10^{-3} m³/s.

$$\text{Number of channels} = \frac{0.252}{1.5 \times 10^{-3}} = 147$$

$$\text{Total number of channels } n = 2 \times 147 = 294$$

$$\text{Number of plates } n + 1 = 295$$

If two exchanger sections are placed in series to keep the total plate height less than 4 m, the total number of plates for each unit is $2 \times 295 = 590$.

EXTENDED-SURFACE EQUIPMENT

Difficult heat-exchange problems arise when one of two fluid streams has a much lower heat-transfer coefficient than the other. A typical case is heating a fixed gas, such as air, by means of condensing steam. The individual coefficient for the steam is typically 100 to 200 times that for the airstream; consequently, the overall coefficient is essentially equal to the individual coefficient for the air, the capacity of a unit area of heating surface will be low, and many meters or feet of tube will be required to provide reasonable capacity. Other variations of the same problem are found in heating or cooling viscous liquids or in treating a stream of fluid at low flow rate, because of the low rate of heat transfer in laminar flow.

To conserve space and to reduce the cost of the equipment in these cases, certain types of heat-exchange surfaces, called *extended surfaces,* have been developed in which the outside area of the tube is multiplied, or extended, by fins, pegs, disks, and other appendages and the outside area in contact with the fluid thereby

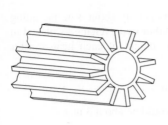

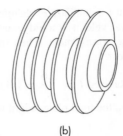

FIGURE 15.10
Types of extended surface:
(a) longitudinal fins; (b) transverse fins.

(a)

(b)

made much larger than the inside area. The fluid stream having the lower coefficient is brought into contact with the extended surface and flows outside the tubes, while the other fluid, having the high coefficient, flows through the tubes. The quantitative effect of extending the outside surface can be seen from the overall coefficient, written in the following form, in which the resistance of the tube wall is neglected:

$$U_i = \frac{1}{1/h_i + A_i/A_o h_o} \tag{15.17}$$

Equation (15.17) shows that if h_o is small and h_i large, the value of U_i will be small; but if the area A_o is made much larger than A_i, the resistance $A_i/A_o h_o$ becomes small and U_i increases just as if h_o were increased, with a corresponding increase in capacity per unit length of tube or unit of inside area.

Types of extended surface

Two common types of extended surfaces are available, examples of which are shown in Fig. 15.10. Longitudinal fins are used when the direction of flow of the fluid is parallel to the axis of the tube; transverse fins are used when the direction of flow of the fluid is across the tubes. Spikes, pins, studs, or spines are also used to extend surfaces, and tubes carrying these can be used for either direction of flow. In all types, it is important that the fins be in tight contact with the tube, both for structural reasons and to ensure good thermal contact between the base of the fin and the wall.

Fin efficiency

The outside area of a finned tube consists of two parts, the area of the fins and the area of the bare tube not covered by the bases of the fins. A unit area of fin surface is not so efficient as a unit area of bare tube surface because of the added resistance to the heat flow by conduction through the fin to the tube. Thus, consider a single longitudinal fin attached to a tube, as shown in Fig. 15.11, and assume that the heat is flowing to the tube from the fluid surrounding the fin. Let the temperature of the fluid be T and that of the bare portion of the tube T_w. The temperature at the base of the fin will also be T_w. The temperature drop available for heat transfer to the bare tube is $T - T_w$, or ΔT_o. Consider the heat transferred to the fin at the tip, the point farthest from the tube wall. To reach the wall of the tube, this heat must flow by conduction through the entire length of the fin, from tip to base. Other increments of heat, entering the fin at points intermediate between tip and base, also must flow through a part of the fin length. A temperature gradient will be necessary, therefore, from the tip of the fin to the base, and the tip will be warmer than the base. If T_F is the temperature of the fin at a distance x from the base, the temperature drop available for heat transfer

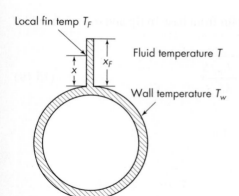

Local fin temp T_F

Fluid temperature T

Wall temperature T_w

FIGURE 15.11
Tube and single longitudinal fin.

from fluid to fin at that point will be $T - T_F$. Since $T_F > T_w$, $T - T_F < T - T_w = \Delta T_o$, and the efficiency of any unit area away from the fin base is less than that of a unit area of bare tube. The difference between $T - T_F$ and ΔT_o is zero at the base of the fin and is a maximum at the tip of the fin. Let the average value of $T - T_F$, based on the entire fin area, be denoted by $\overline{\Delta T_F}$. The efficiency of the fin is defined as the ratio of $\overline{\Delta T_F}$ to ΔT_o and is denoted by η_F. An efficiency of unity (or of 100 percent) means that a unit area of fin is as effective as a unit area of bare tube as far as the temperature drop is concerned. Any actual fin will have an efficiency smaller than 100 percent.

Calculations for extended-surface exchangers

Consider, as a basis, a unit area of tube. Let A_F be the area of the fins and A_b the area of the bare tube. Let h_o be the heat-transfer coefficient of the fluid surrounding the fins and tube. Assume that h_o is the same for both fins and tube. An overall coefficient, based on the inside area A_i, can be written

$$U_i = \frac{1}{A_i/[h_o(\eta_F A_F + A_b)] + x_w D_i/k_m \bar{D}_L + 1/h_i} \tag{15.18}$$

To use Eq. (15.18), it is necessary to know the values of the fin efficiency η_F and of the individual coefficients h_i and h_o. The coefficient h_i is calculated by the usual method. The calculation of the coefficient h_o will be discussed later.

The fin efficiency η_F can be calculated mathematically, on the basis of certain reasonable assumptions, for fins of various types.[9] For example, the efficiency of longitudinal fins is given in Fig. 15.12, in which η_F is plotted as a function of the

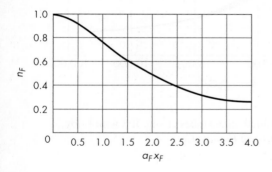

FIGURE 15.12
Fin efficiency, longitudinal fins.

quantity $a_F x_F$, where x_F is the height of the fin from base to tip and a_F is defined by the equation

$$a_F = \sqrt{\frac{h_o L_p / S}{k_m}} \tag{15.19}$$

where h_o = coefficient outside tube
$\quad k_m$ = thermal conductivity of metal in fin
$\quad L_p$ = perimeter of fin
$\quad S$ = cross-sectional area of fin

The product $a_F x_F$ is dimensionless.

Fin efficiencies for other types of extended surface are available.[9] Figure 15.12 shows that the fin efficiency is nearly unity when $a_F x_F < 0.5$. Extended surfaces are neither efficient nor necessary if the coefficient h_o is large. Also, fins increase the pressure drop.

The coefficient h_o cannot be accurately found by the use of the equations normally used for calculating the heat-transfer coefficients for bare tubes. The fins change the flow characteristics of the fluid, and the coefficient for an extended surface differs from that for a smooth tube. Individual coefficients for extended surfaces must be determined experimentally and correlated for each type of surface, and such correlations are supplied by the manufacturer of the tubes. A typical correlation for longitudinal finned tubes is shown in Fig. 15.13. The quantity D_e is the equivalent diameter, defined as usual as 4 times the hydraulic radius, which is, in turn, the cross section of the fin-side space divided by the total perimeter of fins and tube calculated as in Example 15.6.

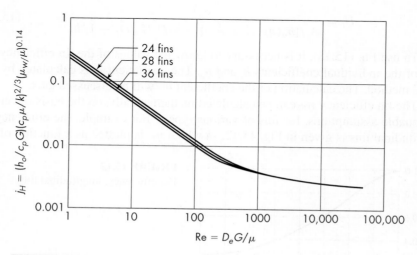

FIGURE 15.13
Heat-transfer coefficients, longitudinal finned tubes; $1\frac{1}{2}$-in. IPS with $\frac{1}{2}$- by 0.035-in. fins in 3-in. IPS shell. (*Brown Fintube Co.*)

EXAMPLE 15.6. Air is heated in the shell of an extended-surface exchanger. The inner pipe is $1\frac{1}{2}$-in. IPS Schedule 40 pipe carrying 28 longitudinal fins $\frac{1}{2}$ in. high and 0.035 in. thick. The shell is 3-in. Schedule 40 steel pipe. The exposed outside area of the inner pipe (not covered by the fins) is 0.416 ft^2 per linear foot; the total surface area of the fins and pipe is 2.830 ft^2/ft. Steam condensing at 250°F inside the inner pipe has a film coefficient of 1,500 Btu/ft^2·h·°F. The thermal conductivity of steel is 26 Btu/ft·h·°F. The wall thickness of the inner pipe is 0.145 in. If the mass velocity of the air is 5,000 lb/h·ft^2 and the average air temperature is 130°F, what is the overall heat-transfer coefficient based on the inside area of the inner pipe? Neglect fouling factors.

Solution. The film coefficient h_o of the air is found from Fig. 15.13. To use this correlation, the Reynolds number of the air must first be calculated as follows. The viscosity of air at 130°F is 0.046 lb/ft·h (App. 8). The equivalent diameter of the shell space is

$$\text{ID of shell (App. 3)} = \frac{3.068}{12} = 0.2557 \text{ ft}$$

$$\text{OD of inner pipe (App. 3)} = \frac{1.900}{12} = 0.1583 \text{ ft}$$

The cross-sectional area of the shell space is

$$\frac{\pi(0.2557^2 - 0.1583^2)}{4} - \frac{28 \times 0.5 \times 0.035}{144} = 0.0282 \text{ ft}^2$$

The perimeter of the airspace is

$$\pi 0.2557 + 2.830 = 3.633 \text{ ft}$$

The hydraulic radius is

$$r_H = \frac{0.0282}{3.633} = 0.00776 \text{ ft}$$

The equivalent diameter is

$$D_e = 4 \times 0.00776 = 0.0310 \text{ ft}$$

The Reynolds number of air is therefore

$$\text{Re} = \frac{0.0310 \times 5,000}{0.046} = 3.37 \times 10^3$$

From Fig. 15.13, the heat-transfer factor is

$$j_H = \frac{h_o}{c_p G}\left(\frac{c_p \mu}{k}\right)^{2/3}\left(\frac{\mu}{\mu_w}\right)^{-0.14} = 0.0031$$

The quantities needed to solve for h_o are

$$c_p = 0.25 \text{ Btu/lb·°F} \qquad \text{(App. 14)}$$
$$k = 0.0162 \text{ Btu/ft·h·°F} \qquad \text{(App. 12)}$$

In computing μ_w, the resistance of the wall and the steam film are considered negligible, so $T_w = 250°F$ and $\mu_w = 0.0528$ lb/ft · h:

$$\left(\frac{\mu}{\mu_w}\right)^{0.14} = \left(\frac{0.046}{0.0528}\right)^{0.14} = 0.981 \qquad Pr = \frac{c_p \mu}{k} = \frac{0.25 \times 0.046}{0.0162} = 0.710$$

$$h_o = \frac{0.0031 \times 0.25 \times 5,000 \times 0.981}{0.710^{2/3}} = 4.78 \text{ Btu/ft}^2 \cdot \text{h} \cdot °F$$

For rectangular fins, disregarding the contribution of the ends of the fins to the perimeter, $L_p = 2L$, and $S = L y_F$, where y_F is the fin thickness and L is the length of the fin. Then, from Eq. (15.19),

$$a_F x_F = x_F \sqrt{\frac{h_o(2L/L y_F)}{k_m}} = x_F \sqrt{\frac{2h_o}{k_m y_F}} = \frac{0.5}{12} \sqrt{\frac{2 \times 4.78}{26(0.035/12)}} = 0.467$$

From Fig. 15.12, $\eta_F = 0.93$.

The overall coefficient is found from Eq. (15.18). The additional quantities needed are

$$D_i = \frac{1.610}{12} = 0.1342 \text{ ft} \qquad \text{(App. 3)}$$

$$\bar{D}_L = \frac{0.1583 - 0.1342}{\ln(0.1583/0.1342)} = 0.1454 \text{ ft}$$

$$A_i = \pi(0.1342) \times 1.0 = 0.422 \text{ ft}^2/\text{lin ft}$$

$$A_F + A_b = 2.830 \text{ ft}^2/\text{lin ft}$$

$$A_F = 2.830 - 0.416 = 2.414 \text{ ft}^2/\text{lin ft}$$

$$x_w = \frac{1.900 - 1.610}{2 \times 12} = 0.0121 \text{ ft}$$

$$U_i = \cfrac{1}{\cfrac{0.422}{4.78(0.93 \times 2.414 + 0.416)} + \cfrac{0.0121 \times 0.1342}{26 \times 0.1454} + \cfrac{1}{1,500}}$$
$$= 29.2 \text{ Btu/ft}^2 \cdot \text{h} \cdot °F \ (166 \text{ W/m}^2 \cdot \text{C})$$

The overall coefficient, when based on the small inside area of the inner pipe, may be much larger than the air-film coefficient based on the area of the extended surface.

Air-cooled exchangers

As cooling water has become scarcer and pollution controls more stringent, the use of air-cooled exchangers has increased. These consist of bundles of horizontal finned tubes, typically 25 mm (1 in.) in diameter and 2.4 to 9 m (8 to 30 ft) long, through which air is circulated by a large fan. Hot process fluids in the tubes, at temperatures from 100 to 400°C (212 to 750°F) or more, can be cooled to about 20°C above the dry-bulb temperature of the air. Heat-transfer areas, based on the outside surface of the tubes, range from 50 to 500 m² (500 to 5,000 ft³); the fins multiply this by a factor of 7 to 20. Air flows between the tubes at velocities of 3 to 6 m/s (10 to 20 ft/s). The pressure drop and power consumption are low, but sometimes to reduce the fan noise to an acceptable level, the fan speed must be lower than that for minimum power consumption. In air-cooled condensers the tubes are usually inclined. Detailed design procedures are given in the literature.[12,18]

SCRAPED-SURFACE EXCHANGERS

Heat transfer to and from viscous liquids, especially food products and other heat-sensitive liquids, is often accomplished in scraped-surface exchangers. Typically these are double-pipe exchangers with a fairly large central tube 100 to 300 mm (4 to 12 in.) in diameter jacketed with steam or cooling liquid. The inside surface of the central tube is wiped by two or more longitudinal blades mounted on a rotating shaft.

The viscous liquid is passed at low velocity through the central tube. Portions of this liquid adjacent to the heat-transfer surface are essentially stagnant, except when disturbed by the passage of the scraper blade. Heat is transferred to the viscous liquid by unsteady-state conduction. If the time between disturbances is short, as it usually is, the heat penetrates only a small distance into the stagnant liquid, and the process is exactly analogous to unsteady-state heat transfer to a semi-infinite solid.

Heat-transfer coefficients in scraped-surface exchangers[13]

Assume that the bulk temperature of the liquid at some location along the exchanger is T, and the temperature of the heat-transfer surface is T_w. Assume for the present that $T_w > T$. Consider a small element of area of the heat-transfer surface over which the blade has just passed. Any liquid that was previously on this surface element has been removed by the blade and replaced by other liquid at temperature T. Heat flows from the surface to the liquid during the time interval t_T, which is the time until the next scraper blade passes the surface element, removes the liquid, and redeposits new liquid on the surface.

From Eq. (10.40), the total amount of heat Q_T transferred during time interval t_T is given by

$$\frac{Q_T}{A} = 2k(T_w - T)\sqrt{\frac{t_T}{\pi\alpha}}$$

where k = thermal conductivity of liquid
   α = thermal diffusivity of liquid
   A = area of heat-transfer surface

The heat-transfer coefficient averaged over each time interval is, by definition,

$$h_i \equiv \frac{Q_T}{t_T A(T_w - T)} \tag{15.20}$$

Substitution from Eq. (10.40) into Eq. (15.20), noting that $\alpha = k/\rho c_p$, gives

$$h_i = 2\sqrt{\frac{k\rho c_p}{\pi t_T}} \tag{15.21}$$

The time interval between the passage of successive blades over a given element of area is

$$t_T = \frac{1}{nB} \tag{15.22}$$

where n = agitator speed, r/h
$\quad$ B = number of blades carried by shaft

Combining Eqs. (15.21) and (15.22) gives, for the heat-transfer coefficient,

$$h_i = 2\sqrt{\frac{k\rho c_p n B}{\pi}} \tag{15.23}$$

Equation (15.23) shows that the heat-transfer coefficient on a scraped surface depends on the thermal properties of the liquid and the agitator speed and implies that it does not depend on the viscosity of the liquid or its velocity through the exchanger. Actually, although Eq. (15.23) gives a good approximation in many cases, it is somewhat of an oversimplification. The liquid at the heat-transfer surface is not all well mixed with the bulk of the fluid, as assumed, especially with viscous liquids, but is partly redeposited behind the scraper blades. Hence the coefficient for a viscous liquid is lower than predicted by Eq. (15.23) and is somewhat affected by changes in liquid viscosity; it also is a function of the liquid velocity in the longitudinal direction and of the diameter and length of the exchanger. An empirical equation for the heat-transfer coefficient incorporating these variables is[30]

$$\frac{h_j D_a}{k} = 4.9 \left(\frac{D_a \bar{V} \rho}{\mu}\right)^{0.57} \left(\frac{c_p \mu}{k}\right)^{0.47} \left(\frac{D_a n}{\bar{V}}\right)^{0.17} \left(\frac{D_a}{L}\right)^{0.37} \tag{15.24}$$

where $\bar{V}$ = bulk average longitudinal velocity
$\quad$ L = length of exchanger
$\quad$ D_a = diameter of scraper (also equal to inside diameter of shell)

Equation (15.24) applies to a small high-speed unit known as a Votator. Data for laminar flow heat transfer in large slow-speed exchangers are given in Reference 28. Scraped-surface devices for evaporating viscous liquids are described in Chap. 16.

CONDENSERS AND VAPORIZERS

As discussed in Chaps. 11 and 13, heat-transfer devices used to liquefy vapors by removing their latent heat are called *condensers*. The latent heat is removed by absorbing it in a cooler liquid, called the *coolant*. Since the temperature of the coolant obviously is increased in a condenser, the unit also acts as a heater; but functionally it is the condensing action that is important, and the name reflects this fact. Condensers fall into two classes. In the first, called shell-and-tube condensers, the condensing vapor and coolant are separated by a tubular heat-transfer surface. In the second, called contact condensers, the coolant and vapor streams, both of which are usually water, are physically mixed and leave the condenser as a single stream.

In industrial practice, liquids are vaporized in kettle-type reboilers or calandrias, as discussed in Chap. 13. The heat of vaporization is most commonly supplied by condensing steam.

Shell-and-tube condensers

The condenser shown in Fig. 11.1 is a single-pass unit, since the entire stream of cooling liquid flows through all the tubes in parallel. In large condensers, this type of flow has a serious limitation. The number of tubes is so large that in single-pass flow the velocity through the tubes is too small to yield an adequate heat-transfer coefficient, and the unit is uneconomically large. Also, because of the low coefficient, long tubes are needed if the cooling fluid is to be heated through a reasonably large temperature range, and such long tubes are not practicable.

To obtain larger velocities, higher heat-transfer coefficients, and shorter tubes, the multipass principle used in heat exchangers may also be used for the coolant in a condenser. An example of a two-pass condenser is shown in Fig. 15.14.

Provision for thermal expansion. Because of the differences in temperature existing in condensers, expansion strains may be set up that are sufficiently severe to buckle the tubes or pull them loose from the tube sheets. The most common method of avoiding damage from expansion is the use of the floating-head construction, in which one of the tube sheets (and therefore one end of the tubes) is structurally independent of the shell. This principle is used in the condenser of Fig. 15.14. The figure shows how the tubes may expand or contract, independent of the shell. A perforated plate is set over the vapor inlet to prevent cutting of the tubes by drops of liquid that may be carried by the vapor.

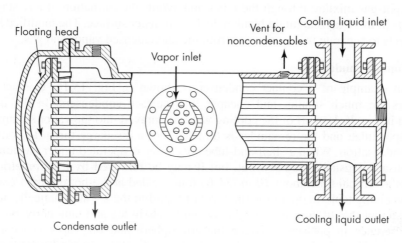

FIGURE 15.14
Two-pass floating-head condenser.

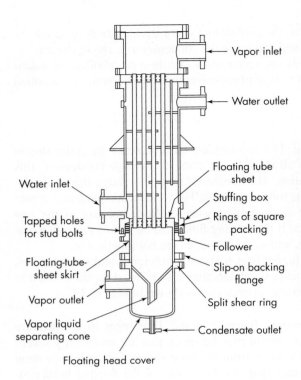

FIGURE 15.15
Outside packed dehumidifying cooler-condenser.

Dehumidifying condensers

A condenser for mixtures of vapors and noncondensable gases is shown in Fig. 15.15. It is set vertically, not horizontally as are most condensers for vapor containing no noncondensable gas; also, vapor is condensed inside the tubes, not outside, and the coolant flows through the shell. This provides a positive sweep of the vapor-gas mixture through the tubes and avoids the formation of any stagnant pockets of inert gas that might blanket the heat-transfer surface. The modified lower head acts to separate the condensate from the uncondensed vapor and gas.

Contact condensers

An example of a contact condenser is shown in Fig. 15.16. Contact condensers are much smaller and cheaper than surface condensers. In the design shown in Fig. 15.16, part of the cooling water is sprayed into the vapor stream near the vapor inlet, and the remainder is directed into a discharge throat to complete the condensation. When a shell-and-tube condenser is operated under vacuum, the condensate is usually pumped out, but it may be removed by a barometric leg. This is a vertical tube about 10 m (34 ft) long, sealed at the bottom by a condensate-receiving tank. In operation the level of liquid in the leg automatically adjusts itself so that the difference in head between levels in leg and tank corresponds to the difference in pressure between the atmosphere and the vapor space in the condenser. Then the liquid flows down the leg as fast as it is condensed, without breaking the vacuum. In a direct-contact condenser the pressure regain in the

Sodexho

PARTHKUMAR D PATEL

WEEK ENDING DATE	WKS	CHECK NUMBER	EMPL ID
01/12/07	1	0066649256	******1047

RETAIN FOR TAX PURPOSES

DIV.	MAIN	SUB	SUB UNITS
07	348	51	

PAYMENTS

OCCUP.	DESCR.	RATE	HRS/DAYS	AMOUNT
900018	RATE 1	6000	102	6120
			TOTAL	6120

YTD. FED. WAGES		TAX DEDUCTIONS		
6120	DESCR.	AMOUNT	YTD	
YTD. FICA WAGES	FIT	102	102	
6120	FICA	468	468	
EXEMPTIONS	ST AL	57	57	
FED | ST				
0 | 0				
EXEMPT STATUS				
FED | ST				
00 | 00	TOTAL	627		

HOURS VESTED	DEDUCTIONS-BACKCHARGES-MEMO ITEMS		
0	DESCR.	DEDN. AMT.	MEMO/BK CHG AMT.
HOURS UNVESTED			
0			
HOURS SICK			
0			
	TOTAL	00	

		TOTAL	

NET AMOUNT
$*****54.93

Sodexho

PARTHKUMAR D PATEL

WEEK ENDING DATE	WKS	CHECK NUMBER	EMPL ID
02/09/07	1	0066945106	*****1047

RETAIN FOR TAX PURPOSES

DIV.	MAIN	SUB	SUB UNITS
07	348	51	

PAYMENTS				
OCCUP.	DESCR.	RATE	HRS/DAYS	AMOUNT
900018	RATE 1	6000	163	9780
			TOTAL	9780

YTD. FED. WAGES
43320
YTD. FICA WAGES
43320

EXEMPTIONS	
FED	ST
0	0

EXEMPT STATUS	
FED	ST
00	00

HOURS VESTED
0
HOURS UNVESTED
0
HOURS SICK
0

TAX DEDUCTIONS		
DESCR.	AMOUNT	YTD
FIT	468	1782
FICA	748	3313
ST AL	160	642
TOTAL	1376	

DEDUCTIONS-BACKCHARGES-MEMO ITEMS		
DESCR.	DEDN. AMT.	MEMO/BK CHG AMT.
	TOTAL	00

	TOTAL	

NET AMOUNT
$*****84.04

AMANDEEPKAUR C MOHEL

WEEK ENDING DATE	WKS	CHECK NUMBER	EMPL ID
02/09/07	1	0066945127	*****0843

	DIV.	MAIN	SUB	SUB UNITS
	07	348	52	

PAYMENTS

OCCUP.	DESCR.	RATE	HRS/DAYS	AMOUNT
900007	RATE 1	6000	194	11640
			TOTAL	11640

YTD. FED. WAGES
53400

YTD. FICA WAGES
53400

EXEMPTIONS	
FED	ST
0	0

EXEMPT STATUS	
FED	ST
00	00

HOURS VESTED
0

HOURS UNVESTED
0

HOURS SICK
0

TAX DEDUCTIONS

DESCR.	AMOUNT	YTD
FIT	654	2826
FICA	891	4085
ST AL	212	965
TOTAL	1757	

DEDUCTIONS-BACKCHARGES-MEMO ITEMS

DESCR.	DEDN. AMT.	MEMO/BK CHG AMT.
TOTAL	00	

		TOTAL	

NET AMOUNT
$*****98.83

PARTHKUMAR D PATEL

WEEK ENDING DATE	WKS	CHECK NUMBER	EMPL ID
03/02/07	1	0067110893	*****1047

RETAIN FOR TAX PURPOSES

DIV.	MAIN	SUB	SUB UNITS
07	348	51	

PAYMENTS

OCCUP.	DESCR.	RATE	HRS/DAYS	AMOUNT
900018	RATE 1	6000	127	7620
			TOTAL	7620

YTD. FED. WAGES
70020

YTD. FICA WAGES
70020

EXEMPTIONS	
FED	ST
0	0

EXEMPT STATUS	
FED	ST
00	00

HOURS VESTED
0

HOURS UNVESTED
0

HOURS SICK
0

TAX DEDUCTIONS

DESCR.	AMOUNT	YTD
FIT	252	2922
FICA	583	5356
ST AL	99	1046
TOTAL	934	

DEDUCTIONS-BACKCHARGES-MEMO ITEMS

DESCR.	DEDN. AMT.	MEMO/BK CHG AMT.
TOTAL	00	

			TOTAL	

NET AMOUNT
$*****66.86

PARTHKUMAR D PATEL

WEEK ENDING DATE	WKS	CHECK NUMBER	EMPL ID
02/02/07	1	0066841835	＊＊＊＊＊1047

RETAIN FOR TAX PURPOSES

DIV.	MAIN	SUB	SUB UNITS
07	348	51	

PAYMENTS

OCCUP.	DESCR.	RATE	HRS/DAYS	AMOUNT
900018	RATE 1	6000	190	11400
			TOTAL	11400

YTD. FED. WAGES			
33540			

TAX DEDUCTIONS

DESCR.	AMOUNT	YTD
FIT	630	1314
FICA	872	2565
ST AL	205	482
TOTAL	1707	

YTD. FICA WAGES
33540

EXEMPTIONS
FED	ST
0	0

EXEMPT STATUS
FED	ST
00	00

HOURS VESTED
0

HOURS UNVESTED
0

HOURS SICK
0

DEDUCTIONS-BACKCHARGES-MEMO ITEMS

DESCR.	DEDN. AMT.	MEMO/BK CHG AMT.
TOTAL	00	

		TOTAL

NET AMOUNT
$＊＊＊＊＊96.93

Vapor inlet

Water inlet (spray) →

Water inlet (jet) →

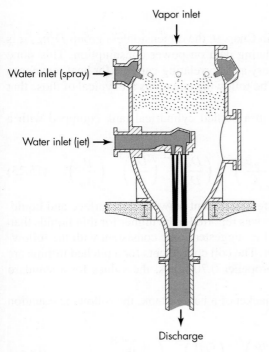

Discharge

FIGURE 15.16
Contact condenser. (*Schutte and Koerting Div., Ketema, Inc.*)

downstream cone of the venturi is often sufficient to eliminate the need for a barometric leg.

Vaporizers

In the kettle-type reboiler shown in Fig. 13.8, page 396, the horizontal shell contains a relatively small tube bundle, two-pass on the tube side, with a floating head and tube sheet. The tube bundle is submerged in a pool of boiling liquid, the depth of which is set by the height of an overflow weir. Feed enters the liquid pool from the bottom. Vapor escapes from the top of the shell; any unvaporized liquid spills over the weir and is withdrawn from the bottom of the shell. The heating fluid, usually steam, enters the tubes as shown; steam condensate is removed through a trap. The auxiliary nozzles shown in Fig. 13.8 are for inspection, draining, or insertion of instrument sensing elements.

The layout and operation of vertical shell-and-tube vaporizers, known as calandrias, natural-circulation or thermosiphon reboilers, are discussed in Chap. 13, page 400, and illustrated in Fig. 13.10.

HEAT TRANSFER IN AGITATED VESSELS

Heat-transfer surfaces, which may be in the form of heating or cooling jackets or coils of pipe immersed in the liquid, are often used in the agitated vessels described in Chap. 9.

Heat-transfer coefficients

In an agitated vessel, as shown in Chap. 9, the dimensionless group $D_a^2 n\rho/\mu$ is a Reynolds number useful in correlating data on power consumption. This same group has been found to be satisfactory as a correlating variable for heat transfer to jackets or coils in an agitated tank. The following equations are typical of those that have been offered for this purpose.

For heating or cooling liquids in a baffled cylindrical tank equipped with a helical coil and a turbine impeller,

$$\frac{h_c D_c}{k} = 0.17\left(\frac{D_a^2 n\rho}{\mu}\right)^{0.67}\left(\frac{c_p\mu}{k}\right)^{0.37}\left(\frac{D_a}{D_t}\right)^{0.1}\left(\frac{D_c}{D_t}\right)^{0.5}\left(\frac{\mu}{\mu_w}\right)^{b} \tag{15.25}$$

where h_c is the individual heat-transfer coefficient between coil surface and liquid. The exponent b on the viscosity ratio was reported to be higher for thin liquids than for viscous oils,[23] but a value of 0.24 is suggested to be consistent with the following equation for the jacket coefficient. The coil coefficients for a pitched turbine are about 0.85 times, and those for a propeller 0.70 times, the values for a standard turbine of the same diameter.[16]

For heat transfer to or from the jacket of a baffled tank, the following equation applies if a standard turbine is used.[32]

$$\frac{h_j D_t}{k} = 0.76\left(\frac{D_a^2 n\rho}{\mu}\right)^{2/3}\left(\frac{c_p\mu}{k}\right)^{1/3}\left(\frac{\mu}{\mu_w}\right)^{0.24} \tag{15.26}$$

where h_j is the coefficient between the liquid and jacketed inner surface of the vessel. The jacket coefficients for a pitched turbine are 0.9 times, and for a propeller 0.6 times, the values for a standard turbine.[16] The standard turbine, however, has a much higher power consumption (see Table 9.2).

When the liquid is very viscous, an anchor agitator is used, which sweeps with close clearance at fairly low speeds over the entire heat-transfer surface. Data for anchor agitators are well correlated by the equation[34b]

$$\frac{h_j D_t}{k} = K\left(\frac{D_a^2 n\rho}{\mu}\right)^{a}\left(\frac{c_p\mu}{k}\right)^{1/3}\left(\frac{\mu}{\mu_w}\right)^{0.18} \tag{15.27}$$

where $K = 1.0$, $a = \frac{1}{2}$ for $10 < \text{Re} < 300$; where $K = 0.36$, $a = \frac{2}{3}$ for $300 < \text{Re} < 40{,}000$.

Equations of this type are generally not applicable to situations differing significantly from those for which the equations were derived. Equations for various types of agitators and arrangements of heat-transfer surface are given in the literature.[34a]

With anchor agitators, scrapers are sometimes attached to the anchor arms to prevent degradation of the liquid in contact with the heated surface. This is especially useful with food products and similar heat-sensitive materials. Scrapers give a modest increase in the heat-transfer coefficient when used with newtonian liquids and may raise the coefficient by as much as 5 times with non-newtonian liquids.[34b] Heat-transfer coefficients may be calculated from Eq. (15.23) for scraped-surface exchangers.

When a liquid is heated in a stirred tank by condensing a vapor in the jacket, the heat-transfer coefficient can be determined from the correlations for film condensation given in Fig. 13.2. When steam is used, the controlling resistance is usually that of the liquid in the tank. When a heating or cooling liquid is passed through the jacket without a phase change, however, the film coefficient is much lower, and the jacket-side resistance may be controlling. With a simple open jacket, the liquid velocity is so low (often less than 0.1 ft/s) that heat transfer is predominantly by natural convection. For cooling water at 100°F, with $\Delta T = 100°F$, and an open jacket 5 ft high, the predicted coefficient for natural convection is only 90 Btu/h·ft²·°F (500 W/m²·K). The mixing caused by natural convection also makes the average temperature in the jacket close to that at the exit, so the temperature difference at the exit should be used for ΔT rather than the logarithmic mean.

Several methods can be used to increase the coefficient for liquid in the jacket. The liquid can be introduced at high velocity through several nozzles, all pointing in the same direction, to induce swirling flow in the jacket (Fig. 15.17a). The swirl velocity is typically 1 to 4 ft/s (0.3 to 1.2 m/s) and can be predicted from the nozzle velocity and the jacket dimensions.[4,17] For 2 ft/s and a jacket width of 3 in., the film coefficient for water at 100°F is 450 Btu/h·ft²·°F (2,500 W/m²·K), as estimated from Eq. (12.32). Another approach is to weld a spiral baffle strip to the outside of the vessel before adding the jacket (Fig. 15.17b). The fluid then flows in a spiral

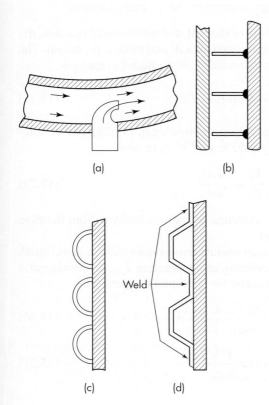

FIGURE 15.17
Jacket designs: (a) agitation nozzles; (b) spiral battle; (c) half-pipe coil; (d) dimple jacket.

(a)

(b)

Weld

(c)

(d)

path with a rectangular cross section, and coefficients several times higher than those with an open jacket can be obtained. However, the gap between the baffle and the outer wall of the jacket leads to some leakage or bypass flow between loops, which decreases the flow along the wall and the rate of heat transfer.

For higher jacket pressures and good control of the flow pattern, the jacket may be formed by welding a half-pipe coil to the vessel outer wall (see Fig. 15.17c). The area for direct heat exchange is reduced, but there is some heat transfer from the area between pipes because of conduction along the vessel wall.[8] The jacket may be divided into two or more zones with separate inlets and exits to permit higher velocity and better heat transfer without increasing the pressure drop.[31] The dimple jacket reactor, Fig. 15.17d, can also be used at high pressures, but the frictional pressure drop is greater than that for other designs, and heat-transfer data are not readily available.

Transient heating or cooling in agitated vessels.

Consider a well-agitated vessel containing m kg or lb of liquid of specific heat c_p. It contains a heat-transfer surface of area A heated by a constant-temperature medium such as condensing steam at temperature T_s. If the initial temperature of the liquid is T_a, its temperature T_b at any time t_T can be found as follows. The basic relation for unsteady-state heat transfer is

$$\text{Rate of accumulation of energy} = \text{energy input} - \text{energy output}$$

For a batch of liquid with no flow in or out of the tank and no chemical reaction, the only energy input is the heat transferred through area A, and there is no output. The accumulation term is the rate of change of enthalpy of the liquid in the tank:

$$mc_p \frac{dT}{dt} = UA(T_s - T) \tag{15.28}$$

If U is constant (usually a reasonable assumption), Eq. (15.28) can be integrated between the limits $t = 0$, $T = T_a$ and $t = t_T$, $T = T_b$ to give

$$\ln \frac{T_s - T_a}{T_s - T_b} = \frac{UA t_T}{mc_p} \tag{15.29}$$

Equation (15.29) is often used to determine overall coefficients from the time required to heat a known mass of liquid.

If the heat-transfer medium is not at a constant temperature but is a liquid (such as cooling water) of specific heat c_{pc} entering at temperature T_{ca} and flowing at a constant rate $\dot{m}_c$, the corresponding equation for the liquid temperature is

$$\ln \frac{T_a - T_{ca}}{T_b - T_{ca}} = \frac{\dot{m}_c c_{pc}}{mc_p} \frac{K_1 - 1}{K_1} t_T \tag{15.30}$$

where

$$K_1 = \exp \frac{UA}{\dot{m}_c c_{pc}} \tag{15.31}$$

Equations for other situations involving transient heat transfer are available in the literature.[14a]

HEAT TRANSFER IN PACKED BEDS

Many catalytic reactions are carried out in multitubular reactors that are similar to shell-and-tube exchangers. The solid catalyst particles are packed in the tubes, and the reactant gases enter and leave through headers at the ends of the reactor. For an exothermic reaction, the heat of reaction is removed by a circulating coolant or a boiling fluid on the shell side. For an endothermic reaction, the energy needed for the reaction is transferred from hot fluid in the shell to the catalyst particles in the tube. The limiting heat-transfer coefficient is usually on the tube side, and the tube size and mass flow rate are often chosen to ensure a nearly constant reaction temperature or to prevent the maximum catalyst temperature from exceeding a safe value. In the following discussion, an exothermic reaction is used as an example, because this is the more common case and because too low an overall coefficient can lead to an uncontrollable rise in reactor temperature or a "runaway" reaction.

Temperature and velocity profiles

The radial temperature profile for an exothermic reaction in a packed tube has the shape shown in Fig. 15.18a. There is a steep gradient near the inside wall and a nearly parabolic temperature profile over the rest of the catalyst bed. The velocity profile (Fig. 15.18b) has a peak near the wall, since the particles are packed more loosely in this region than in the rest of the tube. The temperature and velocity profiles for an empty tube with turbulent flow and a homogeneous reaction would have almost all the gradient near the wall.

Heat-transfer coefficients

For a simple one-dimensional treatment of packed tubes, the heat-transfer coefficient is based on a radial average temperature of the gas, where $\bar{T}$ is the temperature that would result from mixing all the gas flowing through the tube at a given

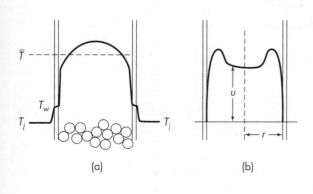

FIGURE 15.18
Temperature and velocity profiles in a packed-tube reactor.

(a)   (b)

distance along the tube. Thus

$$dq = U \, dA \, (\bar{T} - T_j) \tag{15.32}$$

where $dA = \pi D_i \, dL$, and

$$\frac{1}{U} = \frac{1}{h_i} + \frac{1}{h_o D_o / D_i} + \frac{x_w}{k_m \bar{D}_L / D_i}$$

In this simple treatment, the gas and solid temperatures are assumed to be the same, even though, with an exothermic reaction, the catalyst particle must be hotter than the surrounding gas. The difference between gas and solid temperatures can be calculated by using correlations in Chap. 17; this difference is generally only a few degrees compared to a typical driving force $\bar{T} - T_j$ of 20 to 30°C.

The presence of solid particles makes the inside coefficient much greater than that for an empty tube at the same flow rate, since the actual gas velocity between the particles is up to several times the superficial velocity. For air in tubes packed with spheres, the coefficients are 5 to 10 times those for an empty pipe. The coefficients increase with about the 0.6 power of the flow rate and decrease more with increasing tube size than for an empty tube.

The coefficients for a packed tube are highest when the ratio D_p/D_i is about 0.15 to 0.2, as shown by the results in Fig. 15.19. For very small particles, the turbulent mixing in the bed is depressed, and there is a large resistance to heat transfer in the central region, which leads to a temperature profile similar to that for laminar flow. For very large particles, there is rapid mixing and almost no gradient in the center of the tube, but there is a thick region of high void fraction near the wall; most of the resistance to heat transfer is in this region in this case. The dip in the curves at $D_p/D_i \approx 0.3$ was attributed to an increase in the void fraction.[26]

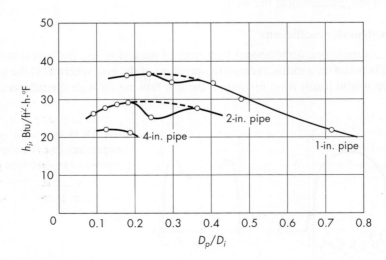

FIGURE 15.19
Heat-transfer coefficients for tubes packed with alumina spheres.
Airflow, 3,000 lb/ft$^2 \cdot$h (4.07 kg/m$^2 \cdot$s).

To predict the rate of heat transfer for different particle and tube sizes, gas flow rates, and gas properties, the coefficient h_i is split into two parts to account for the resistance in the region very near the wall and for the resistance in the rest of the packed bed:

$$\frac{1}{h_i} = \frac{1}{h_{\text{bed}}} + \frac{1}{h_w} \tag{15.33}$$

The bed coefficient is obtained from an effective thermal conductivity k_e. The following equation applies if the temperature profile in the bed is parabolic:

$$h_{\text{bed}} = \frac{4k_e}{r} \tag{15.34}$$

The effective bed conductivity has a static or zero-flow term, which is usually about $5k_g$ when the particles are a porous inorganic material such as alumina, silica gel, or an impregnated catalyst, and k_g is the thermal conductivity of the gas. The turbulent flow contribution to the conductivity is proportional to the mass flow rate and particle diameter, and the factor 0.1 in the following equation agrees with the theory for turbulent diffusion in packed beds.[19]

$$\frac{k_e}{k_g} \approx 5 + 0.1\text{Re}_p\,\text{Pr} \tag{15.35}$$

Note that the particle diameter is used in calculating the Reynolds number for Eq. (15.35), but only the gas properties are used in calculating the Prandtl number. The bed coefficient is obtained by using Eq. (15.35) and the gas conductivity to get k_e, and then Eq. (15.34) gives the value of h_{bed}.

The coefficient h_w can be estimated from the following empirical equation, which was determined by subtracting the calculated bed resistance from the measured overall resistance.[27]

$$\text{Nu}_w \equiv \frac{h_w D_p}{k_g} = 1.94(\text{Re}_p)^{0.5}\,(\text{Pr})^{0.33} \tag{15.36}$$

Equation (15.36) in combination with the equations for h_{bed} explains why the combined coefficient h_i goes through a maximum when D_p/D_i is between 0.1 and 0.2. When D_p/D_i is small, the bed resistance is more important, and increasing D_p increases Re_p and h_{bed}. With large D_p/D_i the wall film controls, and increasing D_p leads to a lower h_w and a lower h_i, since, as shown by Eq. (15.36), h_w varies with $D_p^{-0.5}$. Equation (15.36) was based on results for spheres, but it gives a fairly good fit to the data for cylinders and ring-shaped packings.[27] For packed tubes operating at 200°C or higher, radiation between particles and from the particles to the wall becomes significant, and predicted overall coefficients should be corrected for this effect.[29]

Regenerators

In a regenerator, heat transfer between two streams is carried out by passing the hot and cold fluids alternately through a bed of solids which has appreciable heat storage capacity. The hot fluid gives up heat to the solids, which gradually get warmer; but before equilibrium is reached, the flows are switched, and the cold fluid then removes heat from the bed. In one type of regenerator, two identical beds are used, as in an adsorber-desorber system. (See Fig. 25.1.) A second type uses a rotating bed in the shape of a thick wheel, with cold fluid flowing axially through a sector (usually 180°) of the bed and warm fluid flowing in the opposite direction through the other sector. In rotary regenerators, the bed is often a matrix of rods, screens, or corrugated plates, which gives a large surface area but also a higher void fraction and lower pressure drop than a bed of particles.

Regenerators offer the advantage of high surface area per unit volume and low cost compared to shell-and-tube exchangers. They are also easy to clean, and the packing can be readily replaced. The main problem with rotary units is that a little fluid leaks under the baffles that separate the hot and cold sectors. There is also some mixing of the streams because some of the fluid in the void spaces is carried past the baffles into the other sector. For preheating air with hot combustion gases, the slight leakage of combustion gases into the air, and vice versa, is not a major problem, and rotary regenerators are widely used in electric power plants. They are also used with incinerators, blast furnaces, and gas-turbine engines. Regenerators are generally not suitable for liquids because the thermal capacity of the liquid in the pores would be comparable to that of the solid matrix.

The effectiveness of a regenerator depends on the number of heat-transfer units and the cycle time. For equal flow capacities and negligible resistance in the solid (small Biot numbers Bi), the film coefficients can be combined to get an effective overall coefficient U.

$$\frac{1}{U} = \frac{1}{h_c} + \frac{1}{h_h} \tag{15.37}$$

The number of transfer units is based on the total surface area of the two beds or the rotating wheel.

$$N_H = \frac{UA}{C} \tag{15.38}$$

where $C = \dot{m}_c c_{pc} = \dot{m}_h c_{ph}$. The thermal capacity of the solid (its mass m times its specific heat c_s) is divided by t, the time for heating or cooling (one-half the cycle time) to give a thermal capacity in the same units as the flow capacity of the streams.

$$C_R = \frac{mc_s}{t} \tag{15.39}$$

Then the capacity ratio R_R is used with N_H to characterize the performance.

$$R_R = \frac{C_R}{C} \tag{15.40}$$

The computed effectiveness for an ideal regenerator (no leakage, no mixing, no axial conduction) is shown in Fig. 15.20.[21] For the limiting case of a very short cycle time, $R_R \to \infty$, the temperature of the solid is a function only of axial position and not of time. The effectiveness is then the same as that of an ideal 1-1 counter-flow heat exchanger (see Fig. 15.8). When $R_R = 1$, the solid has just enough ca-

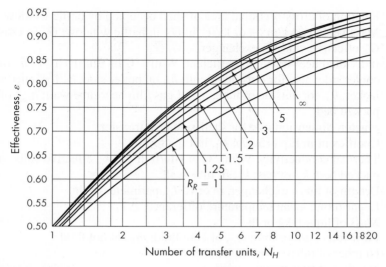

FIGURE 15.20
Effectiveness of a balanced-counterflow regenerator. (*From Mills.*[22])

pacity to absorb all the heat from the hot gas in t s. There would be no point in operating at lower values of R_R, and typical designs have $R_R = 1.5$ to 5.

SYMBOLS

A	Area, m^2 or ft^2; A_F, of fin; A_b, of bare tube; A_i, of inside of tube; A_o, of outside of tube
a	Exponent in Eq. (15.27)
a_F	Fin factor [Eq. (15.19)]
B	Number of scraper blades
Bi	Biot number, dimensionless; $h r_m/k$ for sphere
b	Exponent in Eq. (15.25)
C	Thermal capacity of regenerator gases; C_R, of regenerator solids
c_p	Specific heat at constant pressure, J/g·°C or Btu/lb·°F; c_{pc}, of cool liquid; c_{ph}, of hot liquid
D	Diameter, m or ft; D_a, of impeller or scraper; D_c, outside diameter of coil tubing; D_e, equivalent diameter of noncircular channel; D_i, inside diameter of tube; D_o, of outside of tube; D_p, of particle; D_s, inside diameter of exchanger shell; D_t, of agitated vessel; $\bar{D}_L$, logarithmic mean of inside and outside diameters of tube
F_a	Tube arrangement factor for crossflow [Eq. (15.8)]
F_G	Correction factor for average temperature difference in crossflow or multipass exchangers, dimensionless
f	Fanning friction factor, dimensionless

f_b Fraction of cross-sectional area of shell occupied by baffle window

G Mass velocity, kg/m$^2 \cdot$ s or lb/ft$^2 \cdot$ h; G_b, in baffle window; G_c, in crossflow; G_e, effective value in exchanger, $\sqrt{G_b G_c}$

h Individual heat-transfer coefficient, W/m$^2 \cdot$ °C or Btu/ft$^2 \cdot$ h $\cdot$ °F; h_c, on cooler side; also for outside of coil; h_h, on hotter side; h_i, for inner wall of jacket; h_o, for outside of tube; h_w, of gas film near tube wall; h_{bed}, of packed bed

j_H j factor, dimensionless, for shell-side heat transfer

K Coefficient in Eq. (15.27); K_1, in Eq. (15.30)

k Thermal conductivity, W/m $\cdot$ °C or Btu/ft $\cdot$ h $\cdot$ °F; k_e, effective value for packed bed; k_g, of gas; k_m, of tube wall

L Length of fin or exchanger, m or ft; L_p, perimeter of fin

LMTD   Logarithmic mean temperature difference, $\overline{\Delta T_L}$

m Mass of liquid, kg or lb

$\dot{m}$ Flow rate, kg/h or lb/h; $\dot{m}_c$, of cool fluid; $\dot{m}_h$, of hot fluid

N_b Number of tubes in baffle window

N_H Number of heat-transfer units

Nu   Nusselt number, hD_e/k, dimensionless

n Number of channels in plate exchanger; also impeller or scraper speed, r/s or r/h

P Baffle pitch, m or ft

Pr   Prandtl number, $c_p \mu / k$, dimensionless

p Center-to-center distance between tubes, m or ft; also, pressure, N/m^2 or lb$_f$/ft^2

Q Quantity of heat, J or Btu; Q_T, total amount transferred during time interval t_T

q Rate of heat transfer, W or Btu/h

R_C Thermal capacity ratio; R_R, for regenerator

Re   Reynolds number, DG/μ or $D_a^2 n\rho/\mu$; Re$_p$, for packed bed, $D_p G/\mu$

r_H Hydraulic radius, m or ft

S Cross-sectional area, m^2 or ft^2; S_b, area for flow in baffle window; S_c, area for crossflow in exchanger shell

T Temperature, °C or °F; T_F, at distance x from base of fin; T_a, initial value; T_b, final value; T_{ca}, at cool-fluid inlet; T_{cb}, at cool-fluid outlet; T_{ci}, intermediate cool-fluid temperature; T_{ha}, at warm-fluid inlet; T_{hb}, at warm-fluid outlet; T_j, in jacket; T_r, reduced temperature; T_s, temperature of constant-temperature heating fluid; T_w, of surface or bare portion of finned tube; $\overline{T}$, average value in packed bed

t Time, s or h; t_T, length of time interval

U Overall heat-transfer coefficient, W/m$^2 \cdot$ °C or Btu/ft$^2 \cdot$ h $\cdot$ °F; U_i, based on inside area

u Fluid velocity, m/s or ft/s

$\overline{V}$ Average fluid velocity in longitudinal direction, m/s or ft/s

w Width of plates in plate exchanger

x_F Fin height, m or ft; x_w, thickness of tube wall

y_F Fin thickness, m or ft

Z Ratio of temperature ranges in crossflow or multipass exchanger, dimensionless [Eq. (15.1)]

Greek Letters

α Thermal diffusivity, m^2/h or ft^2/h
Δp Pressure drop, N/m^2 or lb_f/ft^2
ΔT Temperature difference, °C or °F; ΔT_o, between fluid and wall of finned tube; $\overline{\Delta T}$, corrected overall average value; $\overline{\Delta T_F}$, average difference between fluid and fins; $\overline{\Delta T_L}$, logarithmic mean value
ε Effectiveness of heat exchanger, dimensionless
η_F Fin efficiency, $\overline{\Delta T_F}/\Delta T_o$
η_H Heating effectiveness, dimensionless [Eq. (15.2)]
μ Absolute viscosity, cP or lb/ft·h; μ_w, at wall or surface temperature
ρ Density, kg/m^3 or lb/ft^3

PROBLEMS

15.1. Air is blown at a rate of 3 m^3/s (measured at 0°C and 1 atm) at right angles to a tube bank 10 pipes and 10 spaces wide and 10 rows deep. The length of each pipe is 3.5 m. The tubes are on triangular centers, and the center-to-center distance is 75 mm. It is desired to heat the air from 20 to 40°C at atmospheric pressure. What steam pressure must be used? The pipes are 25-mm-OD steel pipe.

15.2. Crude oil at the rate of 150,000 kg/h is to be heated from 20 to 57°C by heat exchange with the bottom product from a distillation unit. The product at 129,000 kg/h is to be cooled from 146 to 107°C. There is available a tubular exchanger with steel tubes with an inside shell diameter of $23\frac{1}{4}$ in., having one pass on the shell side and two passes on the tube side. It has 324 tubes, $\frac{3}{4}$-in. OD, BWG 14, 12 ft long arranged on a 1-in.-square pitch and supported by baffles with a 25 percent cut, spaced at 9-in. intervals. Would this exchanger be suitable; that is, what is the allowable fouling factor? The average properties are shown in Table 15.2.

15.3. A vertical-tube two-pass heater is used for heating gas oil. Saturated steam at 100 $lb_f/in.^2$ gauge is used as a heating medium. The tubes are 1 in. OD by BWG 16

TABLE 15.2
Data for Prob. 15.2

	Product, outside tubes	Crude, inside tubes
c_p, J/g·°C	2.20	1.99
μ, cP	5.2	2.9
ρ, kg/m^3	867	825
k, W/m·°C	0.119	0.137

TABLE 15.3
Data for Prob. 15.4

Temperature, °F	Thermal conductivity, Btu/ft · h · °F	Kinematic viscosity, 10^5 ft²/s	Density, lb/ft³	Specific heat, Btu/lb · °F
100	0.0739	36.6	55.25	0.455
120	0.0737	21.8	54.81	0.466
140	0.0733	14.4	54.37	0.477
160	0.0728	10.2	53.92	0.487
180	0.0724	7.52	53.48	0.498
200	0.0719	5.70	53.03	0.508
220	0.0711	4.52	52.58	0.519
240	0.0706	3.67	52.13	0.530
260	0.0702	3.07		0.540

and are made of mild steel. The oil enters at 60°F and leaves at 150°F. The viscosity–temperature relation is exponential. The viscosity at 60°F is 5.0 cP and at 150°F is 1.8 cP. The oil is 37° API (specific gravity 0.840) at 60°F. The flow of oil is 150 bbl/h (1 bbl = 42 gal). Assume the steam condenses in film condensation. The thermal conductivity of the oil is 0.078 Btu/ft·h·°F, and the specific heat is 0.480 Btu/lb·°F. The velocity of the oil in the tubes should be approximately 4 ft/s. Calculate the length of tubes needed for this heater.

15.4. A petroleum oil having the properties given in Table 15.3 is to be heated in a horizontal multipass heater with steam at 60 lb$_f$/in.² gauge. The tubes are to be steel, $\frac{3}{4}$-in. OD by BWG 16, and their maximum length is 15 ft. The oil enters at 100°F, leaves at 180°F, and enters the tubes at about 4 ft/s. The total flow rate is 200 gal/min. Assuming complete mixing of the oil after each pass, how many passes are required?

15.5. Compare the coefficients predicted by the Donohue equation [Eq. (15.6)] with those found from Eq. (12.63) for flow normal to a single cylinder. Can the difference be reconciled by considering the maximum and average mass velocities in the shell side of the exchanger?

15.6. A shell-and-tube exchanger is used to cool an aqueous stream from 95 to 30°C using cooling water at 20°C. If the flow rate of the cooling water is twice that of the process stream, would a higher overall coefficient be obtained with cooling water in the tubes or in the shell?

15.7. A steam-heated natural-circulation calandria is to be designed to boil 5,000 kg/h of chlorobenzene at atmospheric pressure. (*a*) Approximately how much heat-transfer surface will be required? (*b*) How much area would be required if the average pressure in the calandria were 0.5 atm abs? The normal boiling point of chlorobenzene is 132.0°C; its critical temperature is 359.2°C.

15.8. Liquid styrene at 60°C is being heated in a 2-m-diameter steam-jacketed kettle equipped with a standard six-blade turbine. (*a*) Calculate the film coefficient for the inside wall h_j if the stirrer speed is 140 rpm. (*b*) If a pitched-blade turbine were used with the same power input, how would the coefficient compare with that for the standard turbine?

15.9. A turbine-agitated vessel 2 m in diameter contains 5,500 kg of a dilute aqueous solution. The agitator is $\frac{2}{3}$ m in diameter and turns at 140 rpm. The vessel is jacketed with steam condensing at 110°C; the heat-transfer area is 12 m². The steel walls of the vessel are 10 mm thick. If the solution is at 40°C and the heat-transfer coefficient of the condensing steam is 10 kW/m² · °C, what is the rate of heat transfer between steam and liquid?

15.10. Under the conditions given in Prob. 15.9, how long would it take to heat the vessel contents (*a*) from 20 to 60°C, (*b*) from 60 to 100°C?

15.11. A gas-phase exothermic reaction is carried out in a multitube reactor with the catalyst in 1-in. tubes and boiling water in the jacket. The feed temperature and the jacket temperature are 240°C. The average reactor temperature rises to 250°C a short distance from the inlet and then gradually decreases to 241°C at the reactor exit. The resistance to heat transfer is about equally divided between the bed and the film at the wall. If the tube diameter were increased to 1.5 in. with the same catalyst, what should the jacket temperature be to keep the peak reactor temperature at 250°C? Sketch the temperature profiles for the two cases. What pressure steam would be generated for the two cases?

15.12. Air for a pilot-plant reactor is passed through a 50-mm pipe equipped with electric heaters strapped to the outside. By what factor could the overall heat-transfer coefficient be increased by filling the pipe with 12-mm particles of alumina? The Reynolds number based on the empty pipe is 12,000.

REFERENCES

1. American Society of Mechanical Engineers. *Boiler and Pressure Vessel Code.* New York: ASME, 1995.
2. APV Corporation. *Heat Transfer Handbook.* Goldsboro, NC: 2000.
3. Babcock and Wilcox Co. *Steam—Its Generation and Use,* 40th ed. New York: Babcock and Wilcox, 1992.
4. Bollinger, D. H. *Chem. Eng.,* **89**(19):95 (1982).
5. Bowman, R. A., A. C. Mueller, and W. M. Nagle. *Trans. ASME,* **62**:283 (1940).
6. Donohue, D. A. *Ind. Eng. Chem.,* **41**:2499 (1949).
7. Douglas, J. M. *Conceptual Design of Chemical Processes.* New York: McGraw-Hill, 1988, chap. 8.
8. Fogg, R. M., and V. W. Uhl. *Chem. Eng. Prog.,* **69**(7):76 (1973).
9. Gardner, K. A. *Trans. ASME,* **67**:621 (1945).
10. Gentry, C. C., R. K. Young, and W. M. Small. *AIChE Symp. Ser.,* **80**(236):104 (1984).
11. Gentry, C. C. *Chem. Eng. Prog.,* **86**(7):48 (1990).
12. Gianolio, E., and F. Cuti. *Heat Trans. Eng.,* **3**(1):38 (1981).
13. Harriott, P. *Chem. Eng. Prog. Symp. Ser.,* **29**:137 (1959).
14. Kern, D. Q. *Process Heat Transfer.* New York: McGraw-Hill, 1950; (*a*) pp. 626–37; (*b*) p. 841.
15. Kerner, J. *Chem. Eng.,* **100**(11):177 (1993).
16. Kung, D. M., and P. Harriott. *Ind. Eng. Chem. Res.,* **26**:1654 (1987).
17. Lehrer, I. H. *Heat Trans. Eng.,* **2**(3-4):95 (1981).

18. Lerner, J. E. *Hydrocarbon Proc.,* **51**(2):93 (1972).
19. Li, C. H., and B. A. Finlayson. *Chem. Eng. Sci.,* **32:**1055 (1977).
20. Marriott, J. *Chem. Eng.,* **78**(8):127 (1971).
21. McAdams, W. H. *Heat Transmission,* 3rd ed. New York: McGraw-Hill, 1954.
22. Mills, A. F. *Heat Transfer,* Homewood, IL: Irwin, 1992, p. 751.
23. Oldshue, J. Y., and A. T. Gretton. *Chem. Eng. Prog.,* **50:**615 (1954).
24. Perry, R. H., and D. W. Green (eds.). *Perry's Chemical Engineers' Handbook,* 7th ed. New York: McGraw-Hill, 1997; (*a*) p. **11**-4ff, (*b*) p. **11**-9.
25. Peters, M. S., and K. D. Timmerhaus. *Plant Design and Economics for Chemical Engineers,* 3rd ed. New York: McGraw-Hill, 1980, pp. 678–95.
26. Peters, P. E., M.S. thesis, Cornell University, Ithaca, NY, 1982.
27. Peters, P. E., R. S. Schiffino, and P. Harriott. *Ind. Eng. Chem. Res.,* **27:**226 (1988).
28. Ramdas, V., V. W. Uhl, M. W. Osborne, and J. R. Ortt. *Heat Trans. Eng.,* **1**(4):38 (1980).
29. Schotte, W. *AIChE J.,* **6:**63 (1960).
30. Skelland, A. H. P. *Chem. Eng. Sci.,* **7:**166 (1958).
31. Steve, E. H. *Chem. Eng.,* **105**(1):92 (1998).
32. Strek, F., and S. Masiuk. *Intl. Chem. Eng.,* **7:**693 (1967).
33. Tubular Exchangers Manufacturers Association. *Standards of the TEMA,* 7th ed. New York: TEMA, 1988.
34. Uhl, V. W., and J. B. Gray. *Mixing,* vol. 1. New York: Academic, 1966; (*a*) p. 284; (*b*) pp. 298–303.
35. Walas, F. M. *Chemical Process Equipment,* Stoneham, MA: Butterworths, 1988, chap. 8.

CHAPTER 16

Evaporation

Heat transfer to a boiling liquid has been discussed generally in Chap. 13. A special case occurs so often that it is considered an individual operation. It is called *evaporation* and is the subject of this chapter.

The objective of evaporation is to concentrate a solution consisting of a non-volatile solute and a volatile solvent. In the overwhelming majority of evaporations the solvent is water. Evaporation is conducted by vaporizing a portion of the solvent to produce a concentrated solution of thick liquor. Evaporation differs from drying in that the residue is a liquid—sometimes a highly viscous one—rather than a solid; it differs from distillation in that the vapor usually is a single component, and even when the vapor is a mixture, no attempt is made in the evaporation step to separate the vapor into fractions; it differs from crystallization in that emphasis is placed on concentrating a solution rather than forming and building crystals. In certain situations, for example, in the evaporation of brine to produce common salt, the line between evaporation and crystallization is far from sharp. Evaporation sometimes produces a slurry of crystals in a saturated mother liquor. In this book such processes are considered in Chap. 27, which is devoted to crystallization.

Normally, in evaporation the thick liquor is the valuable product, and the vapor is condensed and discarded. In one specific situation, however, the reverse is true. Mineral-bearing water often is evaporated to give a solid-free product for boiler feed, for special process requirements, or for human consumption. This technique is often called *water distillation,* but technically it is evaporation. Large-scale evaporation processes have been developed and used for recovering potable water from seawater. Here the condensed water is the desired product. Only a fraction of the total water in the feed is recovered, and the remainder is returned to the sea.

Liquid characteristics

The practical solution of an evaporation problem is profoundly affected by the character of the liquor to be concentrated. It is the wide variation in liquor characteristics (which demands judgment and experience in designing and operating evaporators) that broadens this operation from simple heat transfer to a separate art. Some of the most important properties of evaporating liquids are as follows.

Concentration. Although the thin liquor fed to an evaporator may be sufficiently dilute to have many of the physical properties of water, as the concentration increases, the solution becomes more and more individualistic. The density and viscosity increase with solid content until either the solution becomes saturated or the liquor becomes too viscous for adequate heat transfer. Continued boiling of a saturated solution causes crystals to form; these must be removed or the tubes clog. The boiling point of the solution may also rise considerably as the solid content increases, so that the boiling temperature of a concentrated solution may be much higher than that of water at the same pressure.

Foaming. Some materials, especially organic substances, foam during vaporization. A stable foam accompanies the vapor out of the evaporator, causing heavy entrainment.

Temperature sensitivity. Many fine chemicals, pharmaceutical products, and foods are damaged when heated to moderate temperatures for relatively short times. In concentrating such materials special techniques are needed to reduce both the temperature of the liquid and the time of heating.

Scale. Some solutions deposit scale on the heating surface. The overall coefficient then steadily diminishes, until the evaporator must be shut down and the tubes cleaned.

Materials of construction. Whenever possible, evaporators are made of some kind of steel. Many solutions, however, attack ferrous metals or are contaminated by them. Special materials such as copper, nickel, stainless steel, aluminum, impervious graphite, and lead are then used. Since these materials are expensive, high heat-transfer rates become especially desirable to minimize the first cost of the equipment.

Many other liquid characteristics must be considered by the designer of an evaporator. Some of these are the specific heat, heat of concentration, freezing point, gas liberation on boiling, toxicity, explosion hazards, radioactivity, and necessity for sterile operation. Because of the variation in liquor properties, many different evaporator designs have been developed. The choice for any specific problem depends primarily on the characteristics of the liquid.

Single- and multiple-effect operation

Most evaporators are heated by steam condensing on metal tubes. Except in some special horizontal-tube evaporators,[11] the material to be evaporated flows inside the tubes. Usually the steam is at a low pressure, below 3 atm abs; often the boiling liquid is under moderate vacuum, at pressures down to about 0.05 atm abs. Reducing the boiling temperature of the liquid increases the temperature difference between the steam and the boiling liquid and thus increases the heat-transfer rate in the evaporator.

When a single evaporator is used, the vapor from the boiling liquid is condensed and discarded. This method is called *single-effect evaporation,* and although it is simple, it utilizes steam ineffectively. To evaporate 1 kg of water from a solution calls for 1 to 1.3 kg of steam. If the vapor from one evaporator is fed into the steam chest of a second evaporator and the vapor from the second is then sent to a condenser, the operation becomes double-effect. The heat in the original steam is reused in the second effect, and the evaporation achieved by a unit mass of steam fed to the first effect is approximately doubled. Additional effects can be added in the same manner. The general method of increasing the evaporation per kilogram of steam by using a series of evaporators between the steam supply and the condenser is called *multiple-effect evaporation.*

TYPES OF EVAPORATORS

The chief types of steam-heated tubular evaporators are

1. Long-tube vertical evaporators
   a. Upward flow (climbing-film)
   b. Downward flow (falling-film)
   c. Forced circulation

2. Agitated-film evaporators

Once-through and circulation evaporators

Evaporators may be operated either as once-through or as circulation units. In once-through operation, the feed liquor passes through the tubes only once, releases the vapor, and leaves the unit as thick liquor. All the evaporation is accomplished in a single pass. The ratio of evaporation to feed is limited in single-pass units; thus these evaporators are well adapted to multiple-effect operation, where the total amount of concentration can be spread over several effects. Agitated-film evaporators are always operated once-through; falling-film and climbing-film evaporators can also be operated in this way.

Once-through evaporators are especially useful for heat-sensitive materials. By operating under high vacuum, the temperature of the liquid can be kept low. With a single rapid passage through the tubes, the thick liquor is at the evaporation temperature but a short time and can be quickly cooled as soon as it leaves the evaporator.

In circulation evaporators a pool of liquid is held within the equipment. Incoming feed mixes with the liquid from the pool, and the mixture passes through the tubes. Unevaporated liquid discharged from the tubes returns to the pool, so that only part of the total evaporation occurs in one pass. All forced-circulation evaporators are operated in this way; climbing-film evaporators are usually circulation units.

The thick liquor from a circulation evaporator is withdrawn from the pool. All the liquor in the pool must therefore be at the maximum concentration. Since the liquid entering the tubes may contain several parts of thick liquor for each part of feed, its viscosity is high and the heat-transfer coefficient tends to be low.

Circulation evaporators are not well suited to concentrating heat-sensitive liquids. With a reasonably good vacuum the temperature of the bulk of the liquid may be nondestructive, but the liquid is repeatedly exposed to contact with hot tubes. Some of the liquid, therefore, may be heated to an excessively high temperature. Although the average residence time of the liquid in the heating zone may be short, part of the liquid is retained in the evaporator for a considerable time. Prolonged heating of even a small part of a heat-sensitive material such as a food can ruin the entire product.

Circulation evaporators, however, can operate over a wide range of concentration between feed and thick liquor in a single unit, and are well adapted to single-effect evaporation. They may operate either with natural circulation, with the flow through the tubes induced by density differences, or with forced circulation, with flow provided by a pump.

Long-tube evaporators with upward flow

A typical long-tube vertical evaporator with upward flow of the liquid is shown in Fig. 16.1a. The essential parts are (1) a tubular exchanger with steam in the shell

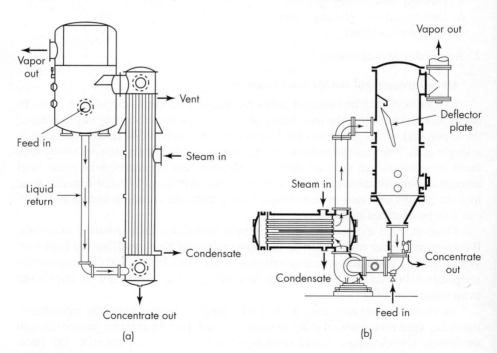

FIGURE 16.1
Evaporators: (*a*) vertical climbing-film, long-tube unit; (*b*) forced-circulation unit with separate two-pass horizontal heating element.

and liquid to be concentrated in the tubes, (2) a separator or vapor space for removing entrained liquid from the vapor, and (3) when operated as a circulation unit, a return leg for the liquid from the separator to the bottom of the exchanger. Inlets are provided for feed liquid and steam, and outlets are provided for vapor, thick liquor, steam condensate, and noncondensable gases from the steam.

The tubular heater operates in the same way as the natural-circulation calandria described in Chap. 13, p. 400. The tubes, however, are larger than those in a calandria: They are typically 25 to 50 mm (1 to 2 in.) in diameter and 3 to 10 m (10 to 32 ft) long. Dilute feed enters the system and mixes with the liquid draining from the separator. Concentrated liquor is withdrawn from the bottom of the heater; the remaining liquor is partially vaporized as it rises through the tubes. The mixture of liquid and vapor from the top of the tubes flows into the separator, where its velocity is greatly reduced. To aid in eliminating liquid droplets, the vapor impinges on and then passes around sets of baffle plates before leaving the separator. The evaporator shown in Fig. 16.1a can be operated only as a circulation unit.

Long-tube vertical evaporators are especially effective in concentrating liquids that tend to foam. Foam is broken when the high-velocity mixture of liquid and vapor impinges against the vapor-head baffle.

Falling-film evaporators[9,12]

Concentration of highly heat-sensitive materials such as orange juice requires a minimum time of exposure to a heated surface. This can be done in once-through falling-film evaporators, in which the liquid enters at the top, flows downstream inside the heated tubes as a film, and leaves from the bottom. The tubes are large— 50 to 250 mm (2 to 10 in.) in diameter. Vapor evolved from the liquid is usually carried downward with the liquid and leaves from the bottom of the unit. In appearance these evaporators resemble long, vertical, tubular exchangers with a liquid–vapor separator at the bottom and a distributor for the liquid at the top.

The chief problem in a falling-film evaporator is that of distributing the liquid uniformly as a film inside the tubes. This is done by a set of perforated metal plates above a carefully leveled tube sheet, by inserts in the tube ends to cause the liquid to flow evenly into each tube, or by "spider" distributors with radial arms from which the feed is sprayed at a steady rate on the inside surface of each tube. Still another way is to use an individual spray nozzle inside each tube.

When recirculation is allowable without damaging the liquid, distribution of liquid to the tubes is facilitated by a moderate recycling of liquid to the tops of the tubes. This provides a larger volume of flow through the tubes than is possible in once-through operation.

For good heat transfer the Reynolds number $4\Gamma/\mu$ of the falling film should be greater than 2,000 at all points in the tube.[5] During evaporation the amount of liquid is continuously reduced as it flows downward, and too great a reduction can lead to dry spots near the bottom of the tube. Thus the amount of concentration that can be done in a single pass is limited.

Falling-film evaporators, with no recirculation and short residence times, handle sensitive products that can be concentrated in no other way. They are also well adapted to concentrating viscous liquids.

Forced-circulation evaporators

In a natural-circulation evaporator[3] the liquid enters the tubes at 0.3 to 1.2 m/s (1 to 4 ft/s). The linear velocity increases greatly as vapor is formed in the tubes, so that in general the rates of heat transfer are satisfactory. With viscous liquids, however, the overall coefficient in a natural-circulation unit may be uneconomically low. Higher coefficients are obtained in forced-circulation evaporators, an example of which is shown in Fig. 16.1b. Here a centrifugal pump forces liquid through the tubes at an entering velocity of 2 to 5.5 m/s (6 to 18 ft/s). The tubes are under sufficient static head to ensure that there is no boiling in the tubes; the liquid becomes superheated as the static head is reduced during flow from the heater to the vapor space, and it flashes into a mixture of vapor and spray in the outlet line from the exchanger just before entering the body of the evaporator. The mixture of liquid and vapor impinges on a deflector plate in the vapor space. Liquid returns to the pump inlet, where it meets incoming feed; vapor leaves the top of the evaporator body to a condenser or to the next effect. Part of the liquid leaving the separator is continuously withdrawn as concentrate.

In the design shown in Fig. 16.1b the exchanger has horizontal tubes and is two-pass on both tube and shell sides. In others, vertical single-pass exchangers are used. In both types the heat-transfer coefficients are high, especially with thin liquids, but the greatest improvement over natural-circulation evaporation is seen with viscous liquids. With thin liquids the improvement with forced circulation does not warrant the added pumping costs over natural circulation; but with viscous material the added costs are justified, especially when expensive metals must be used. An example is caustic soda concentration, which must be done in nickel equipment. In multiple-effect evaporators producing a viscous final concentrate, the first effects may be natural-circulation units and the later ones, handling viscous liquid, forced-circulation units. Because of the high velocities in a forced-circulation evaporator, the residence time of the liquid in the tubes is short—about 1 to 3 s—so that moderately heat-sensitive liquids can be concentrated in them. They are also effective in evaporating salting liquors or those that tend to foam.

Agitated-film evaporator

The principal resistance to overall heat transfer from the steam to the boiling liquid in an evaporator is on the liquid side. One way of reducing this resistance, especially with viscous liquids, is by mechanical agitation of the liquid film, as in the evaporator shown in Fig. 16.2. This is a modified falling-film evaporator with a single jacketed tube containing an internal agitator. Feed enters at the top of the jacketed section and is spread out into a thin, highly turbulent film by the vertical blades of the agitator. Concentrate leaves from the bottom of the jacketed section; vapor rises from the vaporizing zone into an unjacketed separator, which is somewhat larger in diameter than the evaporating tube. In the separator the agitator blades throw entrained liquid outward against stationary vertical plates. The droplets coalesce on these plates and return to the evaporating section. Liquid-free vapor escapes through outlets at the top of the unit.

The chief advantage of an agitated-film evaporator is its ability to give high rates of heat transfer with viscous liquids. The product may have a viscosity as high

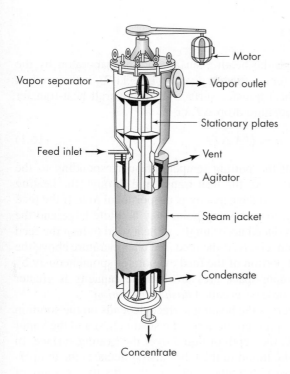

FIGURE 16.2
Agitated-film evaporator.

Motor

Vapor separator → Vapor outlet

Stationary plates

Feed inlet → Vent

Agitator

Steam jacket

Condensate

Concentrate

as 1,000 P at the evaporation temperature. For moderately viscous liquids the heat-transfer coefficient may be estimated from Eq. (15.23). As in other evaporators, the overall coefficient falls as the viscosity rises, but in this design the decrease is slow. With highly viscous materials the coefficient is appreciably greater than in forced-circulation evaporators and much greater than in natural-circulation units. The agitated-film evaporator is particularly effective with such viscous heat-sensitive products as gelatin, rubber latex, antibiotics, and fruit juices. Its disadvantages are the high cost; the internal moving parts, which may need considerable maintenance; and the small capacity of single units, which is far below that of multi-tubular evaporators.

PERFORMANCE OF TUBULAR EVAPORATORS

The principal measures of the performance of a steam-heated tubular evaporator are the capacity and the economy. *Capacity* is defined as the number of kilograms of water vaporized per hour. *Economy* is the number of kilograms vaporized per kilogram of steam fed to the unit. In a single-effect evaporator the economy is nearly always less than 1, but in multiple-effect equipment it may be considerably greater. The steam consumption, in kilograms per hour, is also important. It equals the capacity divided by the economy.

Evaporator Capacity

The rate of heat transfer q through the heating surface of an evaporator, by the definition of the overall heat-transfer coefficient given in Eq. (11.9), is the product of three factors: the area of the heat-transfer surface A, the overall heat-transfer coefficient U, and the overall temperature drop ΔT, or

$$q = U A \, \Delta T \qquad (16.1)$$

If the feed to the evaporator is at the boiling temperature corresponding to the absolute pressure in the vapor space, all the heat transferred through the heating surface is available for evaporation and the capacity is proportional to q. If the feed is cold, the heat required to heat it to its boiling point may be quite large and the capacity for a given value of q is reduced accordingly, as heat used to heat the feed is not available for evaporation. Conversely, if the feed is at a temperature above the boiling point in the vapor space, a portion of the feed evaporates spontaneously by adiabatic equilibration with the vapor-space pressure and the capacity is greater than that corresponding to q. This process is called *flash evaporation.*

The actual temperature drop across the heating surface depends on the solution being evaporated, the difference in pressure between the steam chest and the vapor space above the boiling liquid, and the depth of liquid over the heating surface. In some evaporators the velocity of the liquid in the tubes also influences the temperature drop because the frictional loss in the tubes increases the effective pressure of the liquid. When the solution has the characteristics of pure water, its boiling point can be read from steam tables if the pressure is known, as can the temperature of the condensing steam. In actual evaporators, however, the boiling point of a solution is affected by two factors, boiling point elevation and liquid head.

Boiling-point elevation and Dühring's rule

The vapor pressure of aqueous solutions is less than that of water at the same temperature. Consequently, for a given pressure the boiling point of the solutions is higher than that of pure water. The increase in boiling point over that of water is known as the *boiling-point elevation* (BPE) of the solution. It is small for dilute solutions and for solutions of organic colloids but may be as large as 80°C (144°F) for concentrated solutions of inorganic salts. The BPE must be subtracted from the temperature drop that is predicted from the steam tables.

For strong solutions the BPE is best found from an empirical rule known as *Dühring's rule,* which states that the boiling point of a given solution is a linear function of the boiling point of pure water at the same pressure. Thus if the boiling point of the solution is plotted against that of water at the same pressure, a straight line results. Different lines are obtained for different concentrations. Over wide ranges of pressure the rule is not exact, but over a moderate range the lines are very nearly straight, though not necessarily parallel. Figure 16.3 is a set of Dühring lines for solutions of sodium hydroxide in water.[8] The use of this figure may be illustrated by an example. If the pressure over a 40% solution of sodium hydroxide is such that water boils at 200°F (93.3°C), by reading up from the x axis at 200°F to

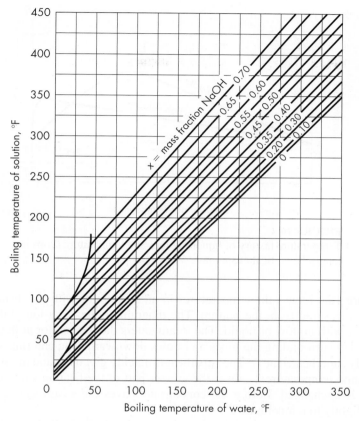

FIGURE 16.3
Dühring lines, system sodium hydroxide–water. (*After McCabe.*[8])

the line for 40% solution and then horizontally to the y axis, it is found that the boiling point of the solution at this pressure is 250°F (121.1°C). The BPE for this solution at this pressure is therefore 50°F (27.8°C).

Effect of liquid head and friction on temperature drop

If the depth of liquid in an evaporator is appreciable, the boiling point corresponding to the pressure in the vapor space is the boiling point of the surface layer of liquid only. A mass of liquid at a distance Z m or ft below the surface is under a pressure of the vapor space plus a head of Z m or ft of liquid and therefore has a higher boiling point. In addition, when the velocity of the liquid is large, frictional loss in the tubes further increases the average pressure of the liquid. In any actual evaporator, therefore, the average boiling point of the liquid in the tubes is higher than the boiling point corresponding to the pressure in the vapor space. This increase in boiling point lowers the average temperature drop between the steam and the liquid and reduces the capacity. The amount of reduction cannot be estimated quantitatively with precision, but the qualitative effect of liquid head, especially with high liquor levels and high liquid velocities, should not be ignored.

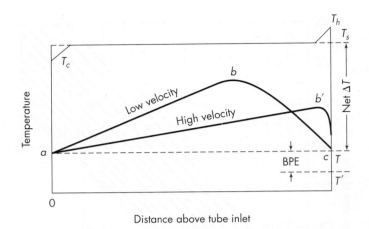

FIGURE 16.4
Temperature history of liquor in tubes and temperature drops in
long-tube vertical evaporator.

Figure 16.4 relates the temperatures in an evaporator with the distance along
the tube, measured from the bottom. The diagram applies to a long-tube vertical
evaporator with upflow of liquid. The steam enters the evaporator at the top of the
steam jacket surrounding the tubes and flows downward. The entering steam may
be slightly superheated at T_h. The superheat is quickly given up, and the steam drops
to saturation temperature T_s. Over the greater part of the heating surface this temperature is unchanged. Before the condensate leaves the steam space, it may be
cooled slightly to temperature T_c.

The temperature history of the liquor in the tubes is shown by lines abc and $ab'c$
in Fig. 16.4. The former applies at low velocities, about 1 m/s, and the latter at
high velocities, above 3 m/s, both velocities based on the flow entering the bottom
of the tube.[2] It is assumed that the feed enters the evaporator at about the boiling
temperature of the liquid at vapor-space pressure, denoted by T. Then the liquid entering the tube is at T, whether the flow is once-through or circulatory. At high
velocities, the fluid in the tube remains liquid practically to the end of the tube and
flashes into a mixture of liquid and vapor in the last few inches of the tube. The
maximum liquid temperature occurs at point b', as shown in Fig. 16.4, almost at the
exit from the tube.

At lower velocities, the liquid flashes at a point nearer the center of the tube and
reaches its maximum temperature, as shown by point b in Fig. 16.4. Point b divides
the tube into two sections, a nonboiling section below point b and a boiling section
above this point.

At both high and low velocities the vapor and concentrated liquid reach equilibrium at the pressure in the vapor space. If the liquid has an appreciable BPE, this
temperature T is greater than T', the boiling point of pure water at the vapor-space
pressure. The difference between T and T' is the BPE.

The temperature drop, corrected for BPE, is $T_s - T$. The true temperature drop,
corrected for both boiling elevation and static head, is represented by the average

distance between T_s and the variable liquid temperature. Although some correlations are available[2] for determining the true temperature drop from the operating conditions, usually this quantity is not available to the designer, and the net temperature drop, corrected for BPE only, is used.

Pressure changes

When the velocity inside an evaporator tube is such that boiling starts inside the tube, the liquid in the nonboiling section moves slowly and the pressure drop from friction is small. In the boiling section, however, the mixture of vapor and liquid has a high velocity, and the friction loss is large. Thus the pressure in the tube falls slowly in the lower part of the tubes (the nonboiling section) and much more rapidly in the upper section, where the velocity is high.

Heat-transfer coefficients

As shown by Eq. (16.1), the heat flux and the evaporator capacity are affected by changes both in the temperature drop and in the overall heat-transfer coefficient. The temperature drop is fixed by the properties of the steam and the boiling liquid and except for the effect of hydrostatic head is not a function of the evaporator construction. The overall coefficient, on the other hand, is strongly influenced by the design and method of operation of the evaporator.

As shown in Chap. 11 [Eq. (11.37)], the overall resistance to heat transfer between the steam and the boiling liquid is the sum of five individual resistances: the steam-film resistance; the two scale resistances, inside and outside the tubes; the tube wall resistance; and the resistance from the boiling liquid. The overall coefficient is the reciprocal of the overall resistance. In most evaporators the fouling factor of the condensing steam and the resistance of the tube wall are very small, and they are usually neglected in evaporator calculations. In an agitated-film evaporator the tube wall is fairly thick, so that its resistance may be a significant part of the total.

Steam-film coefficients. The steam-film coefficient is characteristically high, even when condensation is filmwise. Promoters are sometimes added to the steam to give dropwise condensation and a still higher coefficient. Since the presence of noncondensable gas seriously reduces the steam-film coefficient, provision must be made to vent noncondensables from the steam chest and to prevent leakage of air inward when the steam is at a pressure below atmospheric.

Liquid-side coefficients. The liquid-side coefficient depends to a large extent on the velocity of the liquid over the heated surface. In most evaporators, and especially those handling viscous materials, the resistance of the liquid side controls the overall rate of heat transfer to the boiling liquid. In natural-circulation evaporators the liquid-side coefficient for dilute aqueous solutions is between 1,500 and 3,000 W/m$^2 \cdot$°C (300 and 600 Btu/ft$^2 \cdot$h$\cdot$°F).

Forced circulation gives high liquid-side coefficients even though boiling inside the tubes is suppressed by the high static head. The liquid-side coefficient in a forced-circulation evaporator may be estimated by Eq. (12.33) for heat transfer to a nonboiling liquid if its constant 0.023 is changed[1] to 0.028.

TABLE 16.1
Typical overall coefficients in evaporators

Type	Overall coefficient U	
	W/m² · °C	Btu/ft² · h · °F
Long-tube vertical evaporators		
Natural circulation	1,000–2,500	200–500
Forced circulation	2,000–5,000	400–1,000
Agitated-film evaporator, newtonian liquid, viscosity		
1 cP	2,000	400
1 P	1,500	300
100 P	600	120

The formation of scale on the tubes of an evaporator adds a thermal resistance equivalent to a fouling factor.

Overall coefficients

Because of the difficulty of measuring the high individual film coefficients in an evaporator, experimental results are usually expressed in terms of overall coefficients. These are based on the net temperature drop corrected for boiling-point elevation. The overall coefficient, of course, is influenced by the same factors influencing individual coefficients; but if one resistance (say, that of the liquid film) is controlling, large changes in the other resistances have almost no effect on the overall coefficient.

Typical overall coefficients for various types of evaporators[7,10] are given in Table 16.1. These coefficients apply to conditions under which the various evaporators are ordinarily used. A small accumulation of scale reduces the coefficients to a small fraction of the clean-tube values. An agitated-film evaporator gives a seemingly low coefficient with a liquid having a viscosity of 100 P, but this coefficient is much larger than would be obtained in any other type of evaporator which could handle such a viscous material at all.

Evaporator Economy

The chief factor influencing the economy of an evaporator system is the number of effects. By proper design the enthalpy of vaporization of the steam to the first effect can be used one or more times, depending on the number of effects. The economy also is influenced by the temperature of the feed. If the temperature is below the boiling point in the first effect, the heating load uses a part of the enthalpy of vaporization of the steam and only a fraction is left for evaporation; if the feed is at a temperature above the boiling point, the accompanying flash contributes some evaporation over and above that generated by the enthalpy of vaporization in the steam. Quantitatively, evaporator economy is entirely a matter of enthalpy balances.

Enthalpy balances for single-effect evaporator

In a single-effect evaporator, the latent heat of condensation of the steam is transferred through a heating surface to vaporize water from a boiling solution.

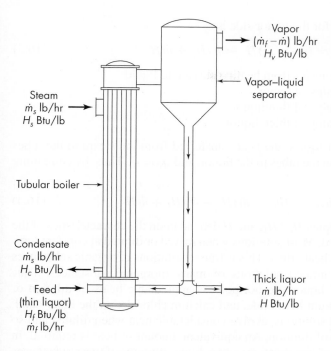

Vapor
$(\dot{m}_f - \dot{m})$ lb/hr
H_v Btu/lb

Vapor–liquid
separator

Steam
$\dot{m}_s$ lb/hr
H_s Btu/lb

Tubular boiler →

Condensate
$\dot{m}_s$ lb/hr
H_c Btu/lb

Feed →
(thin liquor)
H_f Btu/lb
$\dot{m}_f$ lb/hr

Thick liquor
$\dot{m}$ lb/hr
H Btu/lb

FIGURE 16.5
Material and enthalpy
balances in evaporator.

Two enthalpy balances are needed, one for the steam and one for the vapor or liquid side.

Figure 16.5 shows diagrammatically a vertical-tube, single-effect evaporator. The rate of steam flow and of condensate is $\dot{m}_s$, that of the thin liquor, or feed, is $\dot{m}_f$, and that of the thick liquor is $\dot{m}$. The rate of vapor flow to the condenser, assuming that no solids precipitate from the liquor, is $\dot{m}_f - \dot{m}$. Also, let T_s be the condensing temperature of the steam, T the boiling temperature of the liquid in the evaporator, and T_f the temperature of the feed.

It is assumed that there is no leakage or entrainment, that the flow of noncondensables is negligible, and that heat losses from the evaporator need not be considered. The steam entering the steam chest may be superheated, and the condensate usually leaves the steam chest somewhat subcooled below its boiling point. Both the superheat and the subcooling of the condensate are small, however, and it is acceptable to neglect them in making an enthalpy balance. The small errors made in neglecting them are approximately compensated by neglecting heat losses from the steam chest.

Under these assumptions the difference between the enthalpy of the steam and that of the condensate is simply λ_s, the latent heat of condensation of the steam. The enthalpy balance for the steam side is

$$q_s = \dot{m}_s(H_s - H_c) = \dot{m}_s \lambda_s \tag{16.2}$$

where q_s = rate of heat transfer through heating surface from steam
  H_v = specific enthalpy of steam
  H_c = specific enthalpy of condensate
  λ_s = latent heat of condensation of steam
  $\dot{m}_s$ = rate of flow of steam

The enthalpy balance for the liquor side is

$$q = (\dot{m}_f - \dot{m})H_v - \dot{m}_f H_f + \dot{m}H \qquad (16.3)$$

where q = rate of heat transfer from heating surface to liquid
 H_v = specific enthalpy of vapor
 H_f = specific enthalpy of thin liquor
 H = specific enthalpy of thick liquor

In the absence of heat losses, the heat transferred from the steam to the tubes equals that transferred from the tubes to the liquor, and $q_s = q$. Thus, by combining Eqs. (16.2) and (16.3),

$$q = \dot{m}_s \lambda_s = (\dot{m}_f - \dot{m})H_v - \dot{m}_f H_f + \dot{m}H \qquad (16.4)$$

The liquor-side enthalpies H_v, H_f, and H depend upon the characteristics of the solution being concentrated. Most solutions when mixed or diluted at constant temperature do not give much heat effect. This is true of solutions of organic substances and of moderately concentrated solutions of many inorganic substances. Thus sugar, salt, and papermill liquors do not possess appreciable heats of dilution or mixing. Sulfuric acid, sodium hydroxide, and calcium chloride, on the other hand, especially in concentrated solutions, evolve considerable heat when diluted and so possess appreciable heats of dilution. An equivalent amount of heat is required, in addition to the latent heat of vaporization, when dilute solutions of these substances are concentrated to high densities.

Enthalpy balance with negligible heat of dilution

For solutions having negligible heats of dilution, the enthalpy balances over a single-effect evaporator can be calculated from the specific heats and temperatures of the solutions. The heat-transfer rate q on the liquor side includes q_f, the heat transferred to the thin liquor to change its temperature from T_f to the boiling temperature T, and q_v, the heat to accomplish the evaporation. That is,

$$q = q_f + q_v \qquad (16.5)$$

If the specific heat of the thin liquor is assumed constant over the temperature range from T_f to T, then

$$q_f = \dot{m}_f c_{pf}(T - T_f) \qquad (16.6)$$

Also,
$$q_v = (\dot{m}_f - \dot{m})\lambda_v \qquad (16.7)$$

where c_{pf} = specific heat of thin liquor
 λ_v = latent heat of vaporization from thick liquor

If the boiling-point elevation of the thick liquor is negligible, $\lambda_v = \lambda$, the latent heat of vaporization of water at the pressure in the vapor space. When the boiling-point elevation is appreciable, the vapor leaving the solution is superheated by an amount, in degrees, equal to the boiling-point elevation, and λ_v differs slightly from λ. In practice, however, it is nearly always sufficiently accurate to use λ, which may be read directly from steam tables (see App. 7).

Substitution from Eqs. (16.6) and (16.7) into Eq. (16.5) gives the final equation for the enthalpy balance over a single-effect evaporator when the heat of dilution is negligible:

$$q = \dot{m}_f c_{pf}(T - T_f) + (\dot{m}_f - \dot{m})\lambda \qquad (16.8)$$

If the temperature T_f of the thin liquor is greater than T, the term $c_{pf}\dot{m}_f(T - T_f)$ is negative and is the net enthalpy brought into the evaporator by the thin liquor. This item is the *flash evaporation*. If the temperature T_f of the thin liquor fed to the evaporator is less than T, the term $\dot{m}_f c_{pf}(T - T_f)$ is positive and for a given evaporation additional steam will be required to provide this enthalpy. The term $\dot{m}_f c_{pf}(T - T_f)$ is therefore the heating load. In words, Eq. (16.8) states that the heat from the condensing steam is utilized (1) to vaporize water from the solution and (2) to heat the feed to the boiling point; if the feed enters above the boiling point in the evaporator, part of the evaporation is from flash.

Enthalpy balance with appreciable heat of dilution; enthalpy–concentration diagram

If the heat of dilution of the liquor being concentrated is too large to be neglected, an enthalpy–concentration diagram is used for the values of H_f and H in Eq. (16.4). In an enthalpy–concentration diagram the enthalpy, in Btu per pound or joules per gram of solution, is plotted against concentration, in mass fraction or weight percentage of solute.[8] Isotherms drawn on the diagrams show the enthalpy as a function of concentration at constant temperature.

Figure 16.6 is an enthalpy–concentration diagram for solutions of sodium hydroxide and water. Concentrations are in mass fraction of sodium hydroxide, temperatures in degrees Fahrenheit, and enthalpies in Btu per pound of solution. The enthalpy of water is referred to the same datum as in the steam tables, namely, liquid water at 32°F (0°C), so enthalpies from the figure can be used with those from the steam tables when liquid water or steam is involved in the calculations. In finding data for substitution into Eq. (16.4), values of H_f and H are taken from Fig. 16.6, and the enthalpy H_v of the vapor leaving the evaporator is obtained from the steam tables.

The curved boundary lines on which the isotherms of Fig. 16.6 terminate represent conditions of temperature and concentration under which solid phases form. These are various solid hydrates of sodium hydroxide. The enthalpies of all single-phase solutions lie above this boundary line. The enthalpy–concentration diagram can also be extended to include solid phases.

The isotherms on an enthalpy–concentration diagram for a system with no heat of dilution are straight lines. Enthalpy–concentration diagrams can be constructed, of course, for solutions having negligible heats of dilution, but they are unnecessary in view of the simplicity of the specific-heat method described in the last section.

Single-effect calculations

The use of material balances, enthalpy balances, and the capacity equation (16.1) in the design of single-effect evaporators is shown in Example 16.1.

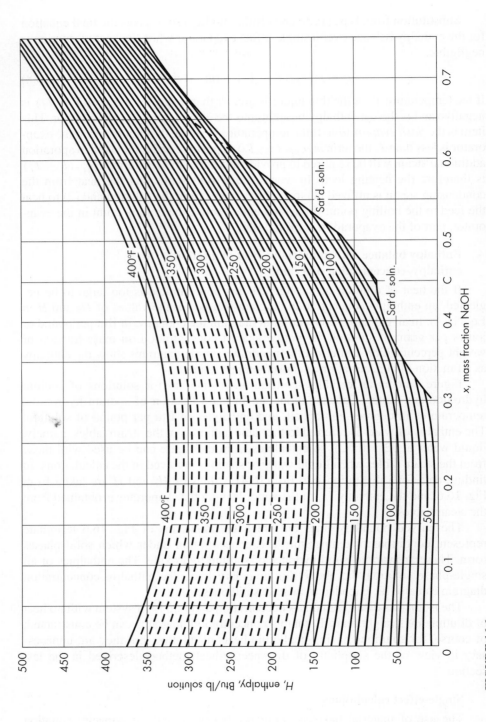

FIGURE 16.6
Enthalpy-concentration diagram, system sodium hydroxide–water. Dashed lines represent extrapolated data. (*After McCabe*[8])

EXAMPLE 16.1. A single-effect evaporator is to concentrate 20,000 lb/h (9,070 kg/h) of a 20% solution of sodium hydroxide to 50% solids. The gauge pressure of the steam is to be 20 $lb_f/in.^2$ (1.37 atm); the absolute pressure in the vapor space is to be 100 mm Hg (1.93 $lb_f/in.^2$). The overall coefficient is estimated to be 250 $Btu/ft^2 \cdot h \cdot °F$ (1,400 $W/m^2 \cdot °C$). The feed temperature is 100°F (37.8°C). Calculate the amount of steam consumed, the economy, and the heating surface required.

Solution. The amount of water evaporated is found from a material balance. The feed contains $\frac{80}{20} = 4$ lb of water per pound of solid; the thick liquor contains $\frac{50}{50} = 1$ lb of water per pound of solid. The quantity evaporated is $4 - 1 = 3$ lb of water per pound of solid, or

$$3 \times 20,000 \times 0.20 = 12,000 \text{ lb/h}$$

The flow rate of thick liquor $\dot{m}$ is

$$20,000 - 12,000 = 8,000 \text{ lb/h } (3,630 \text{ kg/h})$$

Steam consumption Since with strong solutions of sodium hydroxide the heat of dilution is not negligible, the rate of heat transfer is found from Eq. (16.4) and Fig. 16.6. The vaporization temperature of the 50% solution at a pressure of 100 mm Hg is found as follows.

$$\text{Boiling point of water at 100 mm Hg} = 124°F \quad \text{(App. 7)}$$

$$\text{Boiling point of solution} = 197°F \quad \text{(Fig. 16.3)}$$

$$\text{Boiling-point elevation} = 197 - 124 = 73°F$$

The enthalpies of the feed and thick liquor are found from Fig. 16.6:

$$\text{Feed, 20\% solids, 100°F: } H_f = 55 \text{ Btu/lb}$$

$$\text{Thick liquor, 50\% solids, 197°F:} \quad H = 221 \text{ Btu/lb}$$

The enthalpy of the vapor leaving the evaporator is found from steam tables. The enthalpy of superheated water vapor at 197°F and 1.93 $lb_f/in.^2$ is 1,149 Btu/lb; this is the H_v of Eq. (16.4).

The heat of vaporization of steam λ_s at a gauge pressure of 20 $lb_f/in.^2$ is, from App. 7, 939 Btu/lb.

The rate of heat transfer and the steam consumption can now be found from Eq. (16.4):

$$q = (20,000 - 8,000)(1,149) + 8,000 \times 221 - 20,000 \times 55 = 14,456,000 \text{ Btu/h}$$

$$\dot{m}_s = \frac{14,456,000}{939} = 15,400 \text{ lb/h } (6,990 \text{ kg/h})$$

Economy The economy is $12,000/15,400 = 0.78$.

Heating surface The condensation temperature of the steam is 259°F. The heating area required is

$$A = \frac{14,456,000}{250(259 - 197)} = 930 \text{ ft}^2 \text{ } (86.4 \text{ m}^2)$$

If the enthalpy of the vapor H_v were based on saturated vapor at the pressure in the vapor space, instead of on superheated vapor, the rate of heat transfer would be 14,036,000 Btu/h (4,115.7 kW) and the heating area would be 906 ft^2 (84.2 m^2). Thus the approximation would introduce an error of only about 3 percent.

Multiple-effect evaporators

Figure 16.7 shows three long-tube natural-circulation evaporators connected to form a triple-effect system. Connections are made so that the vapor from one effect serves as the heating medium for the next. A condenser and air ejector establish a vacuum in the third effect in the series and withdraw noncondensables from the system. The first effect of a multiple-effect evaporator is the effect to which the raw steam is fed and in which the pressure in the vapor space is the highest. The last effect is that in which the vapor-space pressure is a minimum. In this manner the pressure difference between the steam and the condenser is spread across two or more effects in the multiple-effect system. The pressure in each effect is lower than that in the effect from which it receives steam and higher than that of the effect to which it supplies vapor. Each effect, in itself, acts as a single-effect evaporator, and each has a temperature drop across its heating surface corresponding to the pressure drop in that effect. Every statement that has so far been made about a single-effect evaporator applies to each effect of a multiple-effect system. Arranging a series of evaporator bodies into a multiple-effect system is a matter of interconnecting piping, and not of the structure of the individual units. The numbering of the effects is independent of the order in which liquor is fed to them—they are always numbered in the direction of decreasing pressure. In Fig. 16.7 dilute feed enters the first effect, where it is partly concentrated; it flows to the second effect for additional concentration and then to the third effect for final concentration. Thick liquor is pumped out of the third effect.

In steady operation the flow rates and evaporation rates are such that neither solvent nor solute accumulates or depletes in any of the effects. The temperature, concentration, and flow rate of the feed are fixed, the pressures in steam inlet and condenser established, and all liquor levels in the separate effects maintained. Then all internal concentrations, flow rates, pressures, and temperatures are automatically kept constant by the operation of the process itself. The concentration of the thick liquor can be changed only by changing the rate of flow of the feed. If the thick liquor is too dilute, the feed rate to the first effect is reduced; and if the thick liquor is too concentrated, the feed rate is increased. Eventually the concentration in the last effect and in the thick-liquor discharge will reach a new steady state at the desired level.

The heating surface in the first effect will transmit per hour an amount of heat given by the equation

$$q_1 = A_1 U_1 \Delta T_1 \tag{16.9}$$

If the part of this heat that goes to heat the feed to the boiling point is neglected for the moment, it follows that practically all this heat must appear as latent heat in the vapor that leaves the first effect. The temperature of the condensate leaving the second effect is very near the temperature T_1 of the vapors from the boiling liquid in the first effect. Therefore, in steady operation practically all the heat that was expended in creating vapor in the first effect must be given up when this same vapor condenses in the second effect. The heat transmitted in the second effect, however, is given by the equation

$$q_2 = A_2 U_2 \Delta T_2 \tag{16.10}$$

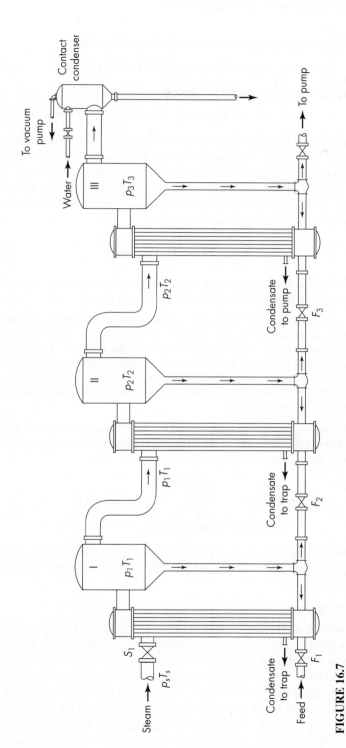

FIGURE 16.7

Triple-effect evaporator: I, II, III, first, second, and third effects; F_1, F_2, F_3, feed or liquor control valves; S_1, steam valve; p_s, p_1, p_2, p_3, pressures; T_s, T_1, T_2, T_3, temperatures.

As has just been shown, q_1 and q_2 are nearly equal, and therefore

$$A_1 U_1 \, \Delta T_1 = A_2 U_2 \, \Delta T_2 \tag{16.11}$$

This same reasoning may be extended to show that, roughly,

$$A_1 U_1 \, \Delta T_1 = A_2 U_2 \, \Delta T_2 = A_3 U_3 \, \Delta T_3 \tag{16.12}$$

It should be understood that Eqs. (16.11) and (16.12) are only approximate equations that must be corrected by the addition of terms which are, however, relatively small compared to the quantities involved in the expressions above.

In ordinary practice the heating areas in all the effects of a multiple-effect evaporator are equal. This is to obtain economy of construction. Therefore, from Eq. (16.12) it follows that since $q_1 = q_2 = q_3 = q$,

$$U_1 \, \Delta T_1 = U_2 \, \Delta T_2 = U_3 \, \Delta T_3 = \frac{q}{A} \tag{16.13}$$

From this it follows that the temperature drops in a multiple-effect evaporator are approximately inversely proportional to the heat-transfer coefficients.

> **EXAMPLE 16.2.** A triple-effect evaporator is concentrating a liquid that has no appreciable elevation in boiling point. The temperature of the steam to the first effect is 108°C, and the boiling point of the solution in the last effect is 52°C. The overall heat-transfer coefficients, in $W/m^2 \cdot °C$, are 2,500 in the first effect, 2,000 in the second effect, and 1,500 in the third effect. (At the solution becomes more concentrated, the viscosity increases and the overall coefficient is reduced.) At what temperatures will the liquid boil in the first and second effects?
>
> **Solution.** The total temperature drop is $108 - 52 = 56°C$. As shown by Eq. (16.13), the temperature drops in the several effects will be approximately inversely proportional to the coefficients. Thus, for example,
>
> $$\Delta T_1 = \frac{\dfrac{1}{2,500}}{\dfrac{1}{2,500} + \dfrac{1}{2,000} + \dfrac{1}{1,500}} \, 56 = 14.3°C$$
>
> In the same manner $\Delta T_2 = 17.9°C$ and $\Delta T_3 = 23.8°C$. Consequently the boiling point in the first effect will be $108 - 14.3 = 93.7°C$, and that in the second effect, 75.8°C.

Methods of feeding

The usual method of feeding a multiple-effect evaporator is to pump the thin liquid into the first effect and send it in turn through the other effects, as shown in Fig. 16.8a. This is called *forward feed*. The concentration of the liquid increases from the first effect to the last. This pattern of liquid flow is the simplest. It requires a pump for feeding dilute solution to the first effect, since this effect is often at about atmospheric pressure, and a pump to remove thick liquor from the last effect. The transfer from effect to effect, however, can be done without pumps, since the flow is in the direction of decreasing pressure, and control valves in the transfer line are all that is required.

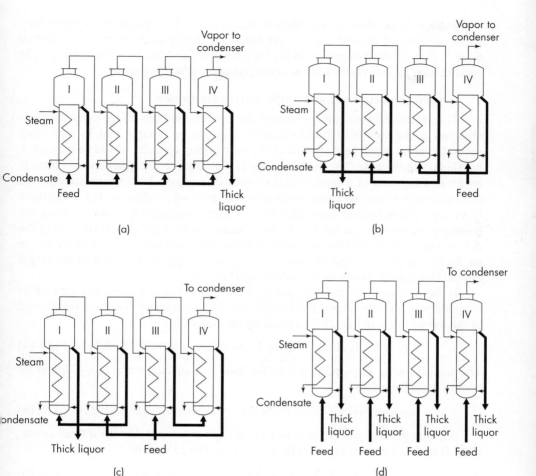

FIGURE 16.8
Patterns of liquor flow in multiple-effect evaporators: (*a*) forward feed; (*b*) backward feed; (*c*) mixed feed; (*d*) parallel feed. (──────) Liquor streams. (────────) Steam and vapor condensate streams.

Another common method is *backward feed,* in which dilute liquid is fed to the last effect and then pumped through the successive effects to the first, as shown in Fig. 16.8*b*. This method requires a pump between each pair of effects in addition to the thick-liquor pump, since the flow is from low pressure to high pressure. Backward feed often gives a higher capacity than forward feed when the thick liquor is viscous, but it may give a lower economy than forward feed when the feed liquor is cold.

Other patterns of feed are sometimes used. In *mixed feed* the dilute liquid enters an intermediate effect, flows in forward feed to the end of the series, and is then pumped back to the first effects for final concentration, as shown in Fig. 16.8*c*. This eliminates some of the pumps needed in backward feed and yet permits the final evaporation to be done at the highest temperature. In crystallizing evaporators, where a slurry of crystals and mother liquor is withdrawn, feed may be admitted directly to each effect to give what is called *parallel feed*, as shown in Fig. 16.8d. In parallel feed there is no transfer of liquid from one effect to another.

Capacity and economy of multiple-effect evaporators

The increase in economy through the use of multiple-effect evaporation is obtained at the cost of reduced capacity. It might be thought that by providing several times as much heating surface the evaporating capacity would be increased, but this is not the case. The total capacity of a multiple-effect evaporator is usually no greater than that of a single-effect evaporator having a heating surface equal to one of the effects and operating under the same terminal conditions, and, when there is an appreciable boiling-point elevation, is often considerably smaller. When the boiling-point elevation is negligible, the effective overall ΔT equals the sum of the ΔT's in each effect, and the amount of water evaporated *per unit area of surface* in an N-effect multiple-effect evaporator is approximately $(1/N)$th that in the single effect. This can be shown by the following analysis.

If the heating load and the heat of dilution are neglected, the capacity of an evaporator is directly proportional to the rate of heat transfer. The heat transferred in the three effects in Fig. 16.7 is given by the equations

$$q_1 = U_1 A_1 \Delta T_1 \qquad q_2 = U_2 A_2 \Delta T_2 \qquad q_3 = U_3 A_3 \Delta T_3 \qquad (16.14)$$

The total capacity is proportional to the total rate of heat transfer q_T, found by adding these equations.

$$q_T = q_1 + q_2 + q_3 = U_1 A_1 \Delta T_1 + U_2 A_2 \Delta T_2 + U_3 A_3 \Delta T_3 \qquad (16.15)$$

Assume that the surface area is A m^2 in each effect and that the overall coefficient U is also the same in each effect. Then Eq. (16.15) can be written

$$q_T = U A (\Delta T_1 + \Delta T_2 + \Delta T_3) = U A \, \Delta T \qquad (16.16)$$

where ΔT is the total temperature drop between the steam in the first effect and the vapor in the last effect.

Suppose now that a single-effect evaporator with a surface area A is operating with the same total temperature drop. If the overall coefficient is the same as in each effect of the triple-effect evaporator, the rate of heat transfer in the single effect is

$$q_T = U A \, \Delta T$$

This is exactly the same equation as that for the multiple-effect evaporator. No matter how many effects are used, provided the overall coefficients are the same, the capacity will be no greater than that of a single effect having an area equal to that of each effect in the multiple unit. The boiling-point elevation tends to make the capacity of a multiple-effect evaporator less than that of the corresponding single effect. Offsetting this are the changes in overall coefficients in a multiple-effect evaporator. In a single-effect unit producing 50% NaOH, for example, the overall coefficient U for this viscous liquid would be small. In a triple-effect unit, the coefficient in the final effect would be the same as that in the single effect; but in the other effects, where the NaOH concentration is much lower than 50 percent, the coefficients would be greater. Thus the average coefficient for the triple-effect evaporator would be greater than that for the single effect. In some cases this overshadows the effect of boiling-point elevation, and the capacity of a multiple-effect unit is actually greater than that of a single effect.

Effect of liquid head and boiling-point elevation. The liquid head and the boiling-point elevation influence the capacity of a multiple-effect evaporator even more than they do that of a single effect. The reduction in capacity caused by the liquid head, as before, cannot be estimated quantitatively. The liquid head reduces the temperature drop available in each effect of a multiple-effect evaporator, as does the boiling-point elevation.

Consider an evaporator that is concentrating a solution with a large boiling-point elevation. The vapor coming from this boiling solution is at the solution temperature and is therefore superheated by the amount of the boiling-point elevation. As discussed in Chap. 13, pages 387 and 388, superheated steam is essentially equivalent to saturated steam at the same pressure when used as a heating medium. The temperature drop in any effect, therefore, is calculated from the temperature of saturated steam at the pressure of the steam chest, and not from the temperature of the boiling liquid in the previous effect. This means that the boiling-point elevation in any effect is lost from the total available temperature drop. This loss occurs in every effect of a multiple-effect evaporator, and the resulting loss of capacity is often important.

The influence of these losses in temperature drop on the capacity of a multiple-effect evaporator is shown in Fig. 16.9. The three diagrams in this figure represent the temperature drops in a single-effect, double-effect, and triple-effect evaporator. The terminal conditions are the same in all three; that is, the steam pressure in the

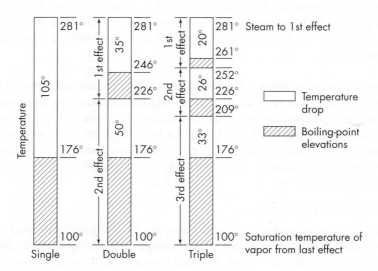

FIGURE 16.9
Effect of boiling-point elevation on capacity of evaporators.

first effect and the saturation temperature of the vapor evolved from the last effect are identical in all three evaporators. Each effect contains a liquid with a boiling-point elevation. The total height of each column represents the total temperature spread from the steam temperature to the saturation temperature of the vapor from the last effect.

Consider the single-effect evaporator. Of the total temperature drop of 181°, the shaded part represents the loss in temperature drop due to boiling-point elevation. The remaining temperature drop, 105°, the actual driving force for heat transfer, is represented by the unshaded part. The diagram for the double-effect evaporator shows two shaded portions because there is a boiling-point elevation in each of the two effects, and the residual unshaded part, totaling 85°, is smaller than in the diagram for the single effect. In the triple-effect evaporator there are three shaded portions since there is a loss of temperature drop in each of three effects, and the total net available temperature drop, 79°, is correspondingly smaller.

In extreme cases of a large number of effects or very high boiling-point elevations, the sum of the boiling-point elevations in a proposed evaporator could be greater than the total temperature drop available. Operation under such conditions is impossible. The design or the operating conditions of the evaporator would have to be revised to reduce the number of effects or increase the total temperature drop.

The economy of a multiple-effect evaporator is not influenced by boiling-point elevations if minor factors, such as the temperature of the feed and changes in the heats of vaporization, are neglected. A kilogram of steam condensing in the first effect generates about a kilogram of vapor, which condenses in the second effect, generating another kilogram there, and so on. The *economy* of a multiple-effect, evaporator depends on heat-balance considerations and not on the rate of heat transfer. The *capacity,* on the other hand, is reduced by the boiling-point elevation. The capacity of a double-effect evaporator concentrating a solution with a boiling-point elevation is generally less than one-half the capacity of two single effects, each operating with the same overall temperature drop. The capacity of a triple effect is generally less than one-third that of three single effects with the same terminal temperatures.

Optimum number of effects. The cost of each effect of an evaporator per square meter or square foot of surface is a function of its total area and decreases with area, approaching an asymptote for very large installations. Thus the investment required for an N-effect evaporator is about N times that for a single-effect evaporator of the same capacity. The optimum number of effects must be found from an economic balance between the savings in steam obtained by multiple-effect operation and the added investment required.

Multiple-effect calculations

In designing a multiple-effect evaporator the results usually desired are the amount of steam consumed, the area of the heating surface required, the approximate temperatures in the various effects, and the amount of vapor leaving the last effect. As in a single-effect evaporator, these quantities are found from material balances, enthalpy balances, and the capacity equation (16.1). In a multiple-effect evaporator, however, a trial-and-error method is used in place of a direct algebraic solution.

Consider, for example, a triple-effect evaporator. There are seven equations which may be written: an enthalpy balance for each effect, a capacity equation for each effect, and the known total evaporation, or the difference between the thin- and thick-liquor rates. If the amount of heating surface in each effect is assumed to be the same, there are seven unknowns in these equations: (1) the rate of steam flow to the first effect, (2) to (4) the rate of flow from each effect, (5) the boiling temperature in the first effect, (6) the boiling temperature in the second effect, and (7) the heating surface per effect. It is possible to solve these equations for the seven unknowns, but the method is tedious and involved. Another method of calculation is as follows:

1. Assume values for the boiling temperatures in the first and second effects.
2. From enthalpy balances find the rates of steam flow and of liquor from effect to effect.
3. Calculate the heating surface needed in each effect from the capacity equations.
4. If the heating areas so found are not nearly equal, estimate new values for the boiling temperatures and repeat items 2 and 3 until the heating surfaces are equal.

In practice these calculations are done by computer. Results of a typical calculation are shown in the following example.

EXAMPLE 16.3. A triple-effect forced-circulation evaporator is to be fed with 60,000 lb/h (27,215 kg/h) of 10% caustic soda solution at a temperature 180°F (82.2°C). The concentrated liquor is to be 50% NaOH. Saturated steam at 50 lb$_f$/in.^{2} (3.43 atm) abs is to be used, and the condensing temperature of vapor from the third effect is to be 100°F (37.8°C). The feed order is II, III, I. Radiation and undercooling of condensate may be neglected. Estimated overall coefficients corrected for boiling-point elevation are given in Table 16.2. Calculate (a) the heating surface required in each effect, assuming equal surfaces in each, (b) the steam consumption, and (c) the steam economy.

Solution. The total rate of evaporation is calculated from an overall material balance, assuming that the solids go through the evaporator without loss (Table 16.3).

Repeated calculations lead to the temperatures, enthalpies, and flow rates shown in Table 16.4. Note that the steam fed to I becomes the condensate from I, the vapor from I becomes the condensate from II, and the vapor from II becomes the condensate from III. From these results the answers to the problems are found to be

    (a) Area per effect: 719 ft^2 (66.8 m^2)
    (b) Steam consumption: 19,370 lb/h (8,786 kg/h)
    (c) Economy: $48,000/19,370 = 2.48$

TABLE 16.2

Effect	Overall coefficient	
	Btu/ft$^2 \cdot$ h $\cdot$ °F	W/m$^2 \cdot$ °C
I	700	3,970
II	1,000	5,680
III	800	4,540

TABLE 16.3

Material	Flow rate, lb/h		
	Total	**Solid**	**Water**
Feed solution	60,000	6,000	54,000
Thick liquor	12,000	6,000	6,000
Water evaporated	48,000		48,000

TABLE 16.4
Temperatures, enthalpies, and flow rates for Example 16.3

Stream	Temperature °F	Saturation temperature, °F	Concentration, weight fraction	Enthalpy, Btu/lb	Flow rate, lb/h
Steam	281	281		1,174	19,370
Feed to I	113		0.228	68	26,300
Vapor from I	245	170		1,170	14,300
Condensate from I	281			249	19,370
Thick liquor from I	246		0.50	249	12,000
Raw feed to II	180		0.10	135	60,000
Vapor from II	149	142		1,126	16,340
Liquid from II	149		0.137	101	43,660
Condensate from II	170			138	14,300
Vapor from III	114	100		1,111	17,360
Condensate from III	142			110	16,340

VAPOR RECOMPRESSION

The energy in the vapor evolved from a boiling solution can be used to vaporize more water, provided there is a temperature drop for heat transfer in the desired direction. In a multiple-effect evaporator this temperature drop is created by progressively lowering the boiling point of the solution in a series of evaporators through the use of lower absolute pressures. The desired driving force can also be obtained by increasing the pressure (and, therefore, the condensing temperature) of the evolved vapor by mechanical or thermal recompression.[1] The compressed vapor is then condensed in the steam chest of the evaporator from which it came.

Mechanical recompression

The principle of mechanical vapor recompression is illustrated in Fig. 16.10. Cold feed is preheated almost to its boiling point by exchange with hot liquor and is pumped through a heater as in a conventional forced-circulation evaporator. The vapor evolved, however, is not condensed; instead it is compressed to a somewhat higher pressure by a positive-displacement or centrifugal compressor and becomes the "steam" which is fed to the heater. Since the saturation temperature of the compressed vapor is higher than the boiling point of the feed, heat flows from the vapor to the solution, generating more vapor. A small amount of makeup steam may be necessary. The optimum temperature drop for a typical system is about 5°C. The energy utilization of such a system is very good: Based on the steam equivalent of the

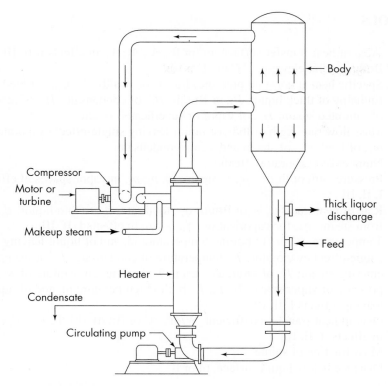

FIGURE 16.10
Mechanical recompression applied to forced-circulation evaporator.

power required to drive the compressor, the economy corresponds to that of an evaporator containing 10 to 15 effects. Important applications of mechanical compression evaporation include production of distilled water from seawater, evaporation of black liquor in the paper industry,[6] evaporation of heat-sensitive materials such as fruit juices, and crystallization of salts having inverted solubility curves[4] (see Chap. 27).

Falling-film evaporators are especially well suited for operation with vapor recompression systems.[13]

Thermal recompression

In a thermal recompression system the vapor is compressed by acting on it with high-pressure steam in a jet ejector. This results in more steam than is needed for boiling the solution, so that excess steam must be vented or condensed. The ratio of motive steam to the vapor from the solution depends on the evaporation pressure; for many low-temperature operations, with steam at 8 to 10 atm pressure, the ratio of steam required to the mass of water evaporated is about 0.5.

Since steam jets can handle large volumes of low-density vapor, thermal recompression is better suited than mechanical recompression to vacuum evaporation. Jets are cheaper and easier to maintain than blowers and compressors. The chief disadvantages of thermal recompression are the low efficiency of the jets and lack of flexibility in the system toward changed operating conditions.

SYMBOLS

A	Area of heat-transfer surface, m^2 or ft^2; A_1, A_2, A_3, in effects I, II, III
BPE	Boiling-point elevation, °C or °F
c_p	Specific heat at constant pressure, $J/g \cdot °C$ or $Btu/lb \cdot °F$; C_{pf}, of feed
H	Enthalpy of thick liquor, J/g or Btu/lb; H_c, of condensate; H_f, of feed; H_s, of saturated steam; H_v, of vapor or superheated steam
$\dot{m}$	Mass flow rate, kg/h or lb/h; of liquor leaving single-effect evaporator; $\dot{m}_f$, of feed; $\dot{m}_s$, of steam and steam condensate
N	Number of evaporator effects
p	Pressure, atm or lb_f/ft^2; p_s, of steam; p_1, p_2, p_3, in vapor spaces of effects I, II, III
q	Rate of heat transfer, W or Btu/h; q_T, total rate; q_f, to thin liquor; q_s, from steam; q_v, for vaporization; q_1, q_2, q_3, in effects I, II, III
T	Temperature, °C or °F; boiling temperature in, and of liquor leaving, single-effect evaporator; T_c, temperature of condensate; T_f, of feed; T_h, of entering steam; T_s, of saturated steam; T', boiling temperature of water at pressure of vapor space; T_1, T_2, T_3, boiling temperature in, and of liquor leaving, effects I, II, III
U	Overall heat-transfer coefficient, $W/m^2 \cdot °C$ or $Btu/ft^2 \cdot h \cdot °F$; U_1, U_2, U_3, in effects, I, II, III
x	Mass fraction of solute
Z	Distance below liquid surface, m or ft

Greek Letters

Γ	Liquid loading on tube, kg/h per meter of perimeter or lb/h per foot of perimeter
ΔT	Temperature drop, °C or °F; total overall corrected temperature drop, all effects; $\Delta T_1, \Delta T_2, \Delta T_3$, temperature drop in effects I, II, III
λ	Latent heat, J/g or Btu/lb; λ_s, latent heat of condensation of steam; λ_v, of vaporization from thick liquor
μ	Absolute viscosity, cP or $lb/ft \cdot h$

PROBLEMS

16.1. A solution of organic colloids in water is to be concentrated from 8 to 45% solids in a single-effect evaporator. Steam is available at a gauge pressure of 1.03 atm (120.5°C). A pressure of 102 mm Hg abs is to be maintained in the vapor space. The feed rate to the evaporator is 20,000 kg/h. The overall heat-transfer coefficient can be taken as 2,800 $W/m^2 \cdot °C$. The solution has a negligible elevation in boiling point and a negligible heat of dilution. Calculate the steam consumption, the economy, and the heating surface required if the temperature of the feed is (*a*) 51.7°C, (*b*) 21.1°C, (*c*) 93.3°C. The specific heat of the feed solution is 3.77 $J/g \cdot °C$, and the latent heat of vaporization of the solution may be taken equal to that of water. Radiation losses may be neglected.

16.2. A solution of organic colloids is to be concentrated from 15 to 50% solids in a vertical-tube evaporator. The solution has a negligible elevation in boiling point, and the specific heat of the feed is 0.93. Saturated steam is available at 0.8 atm abs, and the pressure in the condenser is 100 mm Hg abs. The feed enters at 15°C. The overall coefficient is 1,700 W/m²·°C. The evaporator must evaporate 25,000 kg of water per hour. How many square meters of surface are required, and what is the steam consumption in kilograms per hour?

16.3. A forced-circulation evaporator is to concentrate 60,000 kg/h of 44% NaOH to 65% using steam at 3 atm pressure. The feed temperature and the condensing temperature are both 40°C. The density of the feed solution is 1,450 kg/m³. If the overall heat-transfer coefficient is 2,000 W/m²·°C, calculate (*a*) the steam requirement, in kilograms per hour; (*b*) the heat-transfer area required.

16.4. Vertical tubes 100 mm in diameter and 6 m long are used to concentrate a dilute aqueous solution in a once-through falling-film evaporator. The flow rate per tube is 3,000 kg/h, and the viscosity of the solution at the initial boiling point is 2.5 cP. (*a*) What would be the average residence time of the liquid in the tube if no evaporation occurred? (*b*) What fraction of the water would be evaporated if the overall ΔT were 90°C and the overall heat-transfer coefficient were 3,500 W/m²·°C? (*c*) If the same fraction evaporated could be achieved in an upflow evaporator of the same size, what would be the average residence time? (Assume the fluid in the tube is two-thirds liquid and one-third vapor.)

16.5. A triple-effect evaporator of the long-tube type is to be used to concentrate 35,000 gal/h of a 17% solution of dissolved solids to 38% dissolved solids. The feed enters at 60°F and passes through three tube-and-shell heaters, *a*, *b*, and *c*, in series and then through the three effects in order II, III, I. Heater *a* is heated by vapor taken from the vapor line between the third effect and the condenser, heater *b* with vapor from the vapor line between the second and third effects, and heater *c* with vapor from the line between the first effect and the second. In each heater the warm-end temperature approach is 10°F. Other data are given below and in Table 16.5.

> Steam to I, 230°F, dry and saturated.
> Vacuum on III, 28 in., referred to a 30-in. barometer.
> Condensates leave steam chests at condensing temperatures.
> Boiling-point elevations: 1°F in II, 5°F in III, 15°F in I.
> Coefficients, in Btu/ft²·h·°F, corrected for boiling-point elevation, 450 in I, 700 in II, 500 in III.
> All effects have equal areas of heating surface.

TABLE 16.5

Concentration solids, %	Specific gravity	Specific heat, Btu/lb·°F
10	1.02	0.98
20	1.05	0.94
30	1.10	0.87
35	1.16	0.82
40	1.25	0.75

Calculate (a) the steam required in pounds per hour, (b) the heating surface per effect, (c) the economy in pounds per pound of steam, and (d) the latent heat to be removed in the condenser.

16.6. An agitated-film evaporator 120 mm in inside diameter contains a four-blade agitator turning at 400 rpm. It is vaporizing ethylene dichloride from a polymer solution with an average viscosity of 100 cP. The properties of the liquid film are: $k = 0.15$ Btu/h·ft·°F; $\rho = 60$ lb/ft^3; $c_p = 0.3$ Btu/lb·°F. (a) Estimate the heat-transfer coefficient of the agitated liquid film. (b) Compare your result with the data in Table 16.1 and comment on the discrepancy, if any.

16.7. A triple-effect evaporator is to be used to produce a 50% NaOH solution from a feed containing 25% NaOH. Steam is available at 320°F, and the vapor from the last stage is condensed at 120°F. Backward feed is used. (a) If equal amounts of water are removed in each effect, what will be the concentrations in the intermediate effects, the boiling-point elevation in each effect, and the net temperature differences available for heat transfer? (b) With the same terminal temperatures and more than three effects, what is the maximum number of effects that could be used?

16.8. A triple-effect standard vertical-tube evaporator, each effect of which has 140 m^2 of heating surface, is to be used to concentrate from 4% solids to 35% solids a solution possessing negligible boiling-point elevation. Forward feed is to be used. Steam is available at 120°C, and the vacuum in the last effect corresponds to a boiling temperature of 40°C. The overall coefficients, in W/m^2·°C, are 2,950 in I, 2,670 in II, and 1,360 in III; all specific heats may be taken as 4.2 J/g·°C; and radiation is negligible. Condensates leave at condensing temperature. The feed enters at 90°C. Calculate (a) the kilograms of 4% liquor that can be concentrated per hour and (b) the steam consumption in kilograms per hour.

16.9. A vapor recompression evaporator is to concentrate a very dilute aqueous solution. The feed rate is to be 30,000 lb/h; the evaporation rate will be 20,000 lb/h. The evaporator will operate at atmospheric pressure, with the vapor mechanically compressed as shown in Fig. 16.10 except that a natural-circulation calandria will be used. If steam costs $8 per 1,000 lb, electricity costs 3 cents per kilowatthour, and heat-transfer surface in the heater costs $70 per square foot, calculate the optimum pressure to which the vapor should be compressed. The overall compressor efficiency is 72 percent. Assume all other costs are independent of the pressure of the compressed vapor. To how many effects will this evaporator be equivalent?

REFERENCES

1. Beagle, M. J. *Chem. Eng. Prog.,* **58**(10):79 (1962).
2. Boarts, R. M., W. L. Badger, and S. J. Meisenburg. *Trans. AIChE,* **33**:363 (1937).
3. Foust, A. S., E. M. Baker, and W. L. Badger. *Trans. AIChE,* **35**:45 (1939).
4. King, R. J. *Chem. Eng. Prog.,* **80**(7):63 (1984).
5. Lindsey, E. *Chem. Eng.,* **60**(4):227 (1953).
6. Logsdon, J. D. *Chem. Eng. Prog.,* **79**(9):36 (1983).

7. McAdams, W. H. *Heat Transmission,* 3rd ed. New York: McGraw-Hill, 1954, pp. 398ff.
8. McCabe, W. L. *Trans. AIChE,* **31:**129 (1935).
9. Moore, J. G., and W. E. Hesler. *Chem. Eng. Prog.,* **59**(2):87 (1963).
10. Perry, J. H. (ed.). *Chemical Engineers' Handbook,* 6th ed. New York: McGraw-Hill, 1984, p. **10**–35.
11. Perry, R. H., and D. W. Green (eds.). *Perry's Chemical Engineers' Handbook,* 7th ed. New York: McGraw-Hill, 1997, p. **11**–109.
12. Sinek, J. R., and E. H. Young. *Chem. Eng. Prog.,* **58**(12):74 (1962).
13. Weimer, L. D., H. R. Dolf, and D. A. Austin. *Chem. Eng. Prog.,* **76**(11):70 (1980).

Mass Transfer and Its Applications

A group of operations for separating the components of mixtures is based on the transfer of material from one homogeneous phase to another. Unlike purely mechanical separations, these methods utilize differences in vapor pressure, solubility, or diffusivity, not density or particle size. The driving force for transfer is a concentration difference or a difference in activity, much as a temperature difference or a temperature gradient provides the driving force for heat transfer. These methods, covered by the term *mass-transfer operations,* include such techniques as distillation, gas absorption, dehumidification, adsorption, liquid extraction, leaching, crystallization, membrane separations, and a number of others not discussed in this book.

The function of *distillation* is to separate, by vaporization, a liquid mixture of miscible and volatile substances into individual components or, in some cases, into groups of components. The separation of a mixture of alcohol and water into its components; that of liquid air into nitrogen, oxygen, and argon; and that of crude petroleum into gasoline, kerosene, fuel oil, and lubricating stock are examples of distillation.

In *gas absorption* a solute gas is absorbed from an inert gas into a liquid in which the solute is more or less soluble. The washing of ammonia from a mixture of ammonia and air by means of liquid water is a typical example. The solute is subsequently recovered from the liquid by distillation, and the absorbing liquid can be either discarded or reused. When a solute is transferred from the solvent liquid to the gas phase, the operation is known as *desorption* or *stripping*. In *dehumidification* a pure liquid is partially removed from an inert or carrier gas by condensation. Usually the carrier gas is virtually insoluble in the liquid. Removal of water vapor from air by condensation on a cold surface and the condensation of an organic vapor such as carbon tetrachloride out of a stream of nitrogen are examples of dehumidification. In humidification operations the direction of transfer is from the liquid to the gas phase. In the *drying* of solids, a liquid, usually water, is separated by the use of hot, dry gas (usually air) and so is coupled with the humidification of the gas phase.

In *membrane separations,* including gas separations, reverse osmosis, and ultrafiltration, one component of a liquid or gaseous mixture passes through a selective membrane more readily than the other components. The fundamental driving force is the difference in thermodynamic activity, but in many cases the driving force can be expressed as a difference in concentration or partial pressure. In *adsorption* a solute is removed from either a liquid or a gas through contact with a solid adsorbent, the surface of which has a special affinity for the solute.

In *liquid extraction,* sometimes called *solvent extraction,* a mixture of two components is treated by a solvent that preferentially dissolves one or more of the components in the mixture. The mixture so treated is called the *raffinate,* and the solvent-rich phase is called the *extract.* The component transferred from raffinate to extract is the *solute,* and the component left behind in the raffinate is the *diluent.* The solvent in the extract leaving the extractor is usually recovered and reused. In extraction of solids, or *leaching,* soluble material is dissolved from its mixture with an inert solid by means of a liquid solvent. The dissolved material, or solute, can then be recovered by crystallization or evaporation. *Crystallization* is used to obtain materials in attractive and uniform crystals of good purity, separating a solute from a melt or a solution and leaving impurities behind.

The quantitative treatment of mass transfer is based on material and energy balances, equilibria, and rates of heat and mass transfer. Certain concepts applicable generally are discussed here. The individual operations are discussed in the following chapters.

TERMINOLOGY AND SYMBOLS

It is convenient to refer generally to the two streams in any one operation as the L phase and the V phase. It is also customary to choose the stream having the higher density as the L phase and the one having the lower density as the V phase. An exception may appear in liquid extraction, where the raffinate always is taken as the L phase and the extract as the V phase, even when the raffinate happens to be lighter than the extract. In drying, the L phase is the stream consisting of the solid and the liquid retained in or on the solid. Table A shows how the streams are designated in the various operations.

Note on concentrations

Strictly speaking, concentration means mass per unit volume. Mass may be in kilograms or pounds and volume in cubic meters or cubic feet. Kilogram moles or pound moles are often used as a measure of the amount of solute present. In this book the context will make clear what quantity—mole or ordinary mass—is used. It is sometimes convenient to express concentrations in mole or mass fractions. The mole fraction is the ratio of the moles of the component to the total number of moles in the mixture, with a corresponding definition for mass fraction. By definition, all mole or mass fractions in a mixture sum to unity. If there are r components, $r - 1$ of the mole fractions may be chosen independently; the mole fraction of the remaining component is thereby fixed and equals 1 less the sum of the others.

TABLE A
Terminology for streams in mass-transfer operations

Operation	V phase	L phase
Distillation	Vapor	Liquid
Gas absorption, dehumidification	Gas	Liquid
Membrane separations	Gas or liquid	Gas or liquid
Adsorption	Gas or liquid	Solid
Liquid extraction	Extract	Raffinate
Leaching	Liquid	Solid
Crystallization	Mother liquor	Crystals
Drying	Gas (usually air)	Wet solid

The relation between concentration and mole or mass fraction of a component i is

$$c_i = \rho x_i$$

where x_i = mole or mass fraction of component i
ρ = molar or mass density of mixture
c_i = corresponding concentration of component i

General symbols are needed for flow rates and concentrations. For all operations, use V and L for the flow rates of V and L phases, respectively. Use A, B, C, etc., to refer to the individual components. If only one component is transferred between phases, choose component A as that component. Use x for the concentration of a component in the L phase, and y for the concentration in the V phase. Thus, y_A is the concentration of component A in a V phase, and x_B is that of component B in an L phase. When only two components are present in a phase, the concentration of component A is x or y, and that of component B is $1 - x$ or $1 - y$, and the subscripts A and B are unnecessary.

Terminal quantities

Since in steady-flow mass-transfer operations there are two streams and each must enter and leave, there are four terminal quantities. To identify them, use subscript a to refer to that end of the process where the L phase enters and b to refer to that end where the L phase leaves. Then, for *countercurrent* flow, the terminal quantities are as shown in Table B. If there are only two components in a stream, the subscript A can be dropped from the concentration terms.

TABLE B
Terminal quantities for countercurrent flow

Stream	Flow rate	Concentration component A
L phase, entering	L_a	x_{Aa}
L phase, leaving	L_b	x_{Ab}
V phase, entering	V_b	y_{Ab}
V phase, leaving	V_a	y_{Aa}

DIFFUSIONAL PROCESSES AND EQUILIBRIUM STAGES

Mass-transfer processes can be modeled by two distinctly different methods, one based on diffusion rate processes, the other using the concept of equilibrium stages. The choice of the method depends on the kind of equipment in which the operation is carried out. Calculations involving gas absorption, liquid-liquid extraction, and other operations are modeled as diffusion processes when they are performed in packed towers and similar equipment. When distillation, leaching, and extraction are carried out in equipment such as plate towers, diffusion batteries, and mixer-settler trains, which have distinct stages, they are commonly solved by equilibrium-stage calculations. All mass-transfer calculations, however, involve a knowledge of the equilibrium relationships between phases.

Phase Equilibria

A limit to mass transfer is reached if the two phases come to equilibrium and the net transfer of material ceases. For a practical process, which must have a reasonable transfer rate, equilibrium must be avoided, as the rate of mass transfer at any point is proportional to the driving force, which is the departure from equilibrium at that point. To evaluate driving forces, a knowledge of equilibria between phases is therefore of basic importance. Several kinds of equilibria are important in mass transfer. In nearly all situations two phases are involved, and all combinations are found except two solid phases. In phases in bulk the controlling variables are the intensive properties of temperature, pressure, and concentrations. Equilibrium data can be shown in tables, equations, or graphs. For most operations considered in this text, the pertinent equilibrium relationships can be shown graphically.

Classification of equilibria

To classify equilibria and to establish the number of independent variables or degrees of freedom available in a specific situation, the phase rule is useful. It is

$$\mathscr{F} = \mathscr{C} - \mathscr{P} + 2$$

where $\mathscr{F}$ = number of degrees of freedom
$\mathscr{C}$ = number of components
$\mathscr{P}$ = number of phases

In the following paragraphs the equilibria used in mass transfer are analyzed in terms of the phase rule.

The number of degrees of freedom, or variance $\mathscr{F}$, is the number of independent intensive variables—temperature, pressure, and concentrations—that must be fixed to define the equilibrium state of the system. If fewer than $\mathscr{F}$ variables are fixed, an infinite number of states fit the assumptions; if too many are arbitrarily chosen, the system will be overspecified. When there are only two phases, as is usually the case, $\mathscr{F} = \mathscr{C}$; in systems of two components, therefore, $\mathscr{F} = 2$. If the pressure is fixed, only one variable—the liquid-phase concentration, for example—can be changed independently; the temperature and gas-phase composition (if the two phases are liquid and gas) must follow. For such systems equilibrium data are presented in temperature-composition diagrams which apply at constant pressure, or by plotting y_e, the V-phase concentration, against x_e, the L-phase composition. Such plots are called *equilibrium curves*. If there are more than two components, the equilibrium relationship cannot be represented by a single curve.

Applications of the phase rule to three typical operations follow.

Gas absorption

Assume that only one component is transferred between phases. There are three components, and $\mathscr{F} = 3$. There are four variables: pressure, temperature, and the concentrations of component A in liquid and gas. The temperature and pressure may be fixed. One concentration may be chosen as the remaining independent variable. The other concentration is then determined, and an equilibrium curve of y_e versus x_e plotted. All points on the curve pertain to the same temperature and pressure. Equilibrium data for various temperatures may also be presented in the form of solubility charts, in which the partial pressure of the solute in the gas phase is plotted as the ordinate.

Distillation

Assume that there are two components, so $\mathscr{C} = 2$, $\mathscr{P} = 2$, and $\mathscr{F} = 2$. Both components are found in both phases. There are four variables of interest: pressure, temperature, and the mole fractions of component A in the liquid and vapor phases (the mole fractions of component B are unity less those of component A). If the pressure is fixed, only one variable, for example, liquid-phase mole fraction, can be changed independently, and temperature and vapor-phase mole fraction follow.

If there are three components, $\mathscr{F} = 3$, and fixing the temperature and liquid concentrations x_A and x_B defines the system. Pressure P, liquid concentration x_C, and vapor concentrations y_A, y_B, and y_C are then all determined.

Liquid extraction

The minimum number of components is 3, so $\mathscr{F} = 3$. All three components may appear in both phases. The variables are temperature, pressure, and four concentrations. Either temperature or pressure may be taken as a constant, and two or more concentrations chosen as independent variables. The pressure usually is assumed constant, and the temperature then varies somewhat. The relations between these variables are given by various graphical methods, examples of which are shown in Figs. 23.8 to 23.11.

CHAPTER 17

Principles of Diffusion and Mass Transfer Between Phases

Diffusion is the movement, under the influence of a physical stimulus, of an individual component through a mixture. The most common cause of diffusion is a concentration gradient of the diffusing component. A concentration gradient tends to move the component in such a direction as to equalize concentrations and destroy the gradient. When the gradient is maintained by constantly supplying the diffusing component to the high-concentration end of the gradient and removing it at the low-concentration end, there is a steady-state flux of the diffusing component. This is characteristic of many mass-transfer operations. For example, when ammonia is removed from a gas by absorption in water in a packed column, at each point in the column a concentration gradient in the gas phase causes ammonia to diffuse to the gas-liquid interface, where it dissolves, and a gradient in the liquid phase causes it to diffuse into the bulk liquid. In stripping a solute from a liquid the gradients are reversed; here diffusion brings solute from the bulk liquid to the interface and from there into the gas phase. In some other mass-transfer operations such as leaching and adsorption, unsteady-state diffusion takes place, and the gradients and fluxes decrease with time as equilibrium is approached.

Although the usual cause of diffusion is a concentration gradient, diffusion can also be caused by an activity gradient, as in reverse osmosis, by a pressure gradient, by a temperature gradient, or by the application of an external force field, as in a centrifuge.[1] Molecular diffusion induced by temperature is *thermal diffusion,* and that from an external field is *forced diffusion*. Both are uncommon in chemical engineering. Only diffusion under a concentration gradient is considered in this chapter.

Diffusion is not restricted to molecular transfer through stagnant layers of solid or fluid. It also takes place when fluids of different compositions are mixed. The first step in mixing is often mass transfer caused by the eddy motion characteristic of turbulent flow. This is called *eddy diffusion*. The second step is molecular diffusion between and inside the very small eddies. Sometimes the diffusion process is accompanied by bulk flow of the mixture in a direction parallel to the direction of diffusion.

Role of diffusion in mass transfer

In all the mass-transfer operations, diffusion occurs in at least one phase and often in both phases. In distillation, the low boiler diffuses through the liquid phase to the interface and away from the interface into the vapor. The high boiler diffuses in the reverse direction and passes through the vapor into the liquid. In leaching, diffusion of solute through the solid phase is followed by diffusion into the liquid. In liquid extraction, the solute diffuses through the raffinate phase to the interface and then into the extract phase. In crystallization, solute diffuses through the mother liquor to the crystals and deposits on the solid surfaces. In humidification or dehumidification there is no diffusion through the liquid phase because the liquid phase is pure and no concentration gradient through it can exist; but the vapor diffuses to or from the liquid-gas interface into or out of the gas phase. In membrane separations diffusion occurs in all there phases: in the fluids on either side of the membrane and in the membrane itself.

THEORY OF DIFFUSION

In this section quantitative relationships for diffusion are discussed. Attention is focused on diffusion in a direction *perpendicular* to the interface between the phases and at a definite location in the equipment. Steady state is assumed, and the concentrations at any point do not change with time. This discussion is restricted to binary mixtures.

Comparison of diffusion and heat transfer

There are similarities between the conduction of heat and the transfer of mass by diffusion. Heat flux is proportional to a temperature gradient, and mass flux is proportional to a concentration gradient. The net heat flux determines the rate of temperature change of a material, and mass flux determines the rate of change in concentration. These similarities permit the solutions to the equations for heat conduction to be adapted to problems of diffusion in solids or fluids. This is especially useful for unsteady-state diffusion problems because of the many published solutions for unsteady-state heat transfer (see Chap. 10).

Differences between heat transfer and mass transfer result from the fact that heat is not a substance but energy in transit, whereas diffusion is the physical flow of material. Furthermore, all molecules of a mixture are at the same temperature at a given point in space, so heat transfer in a given direction is based on one temperature gradient and the average thermal conductivity. With mass transfer there are different concentration gradients for each component and often different diffusivities.

The material nature of diffusion and the resulting flow lead to four types of situations:

1. Only one component A of the mixture is transferred to or from the interface, and the total flow is the same as the flow of A. Absorption of a single component from a gas into a liquid is an example of this type.

2. The diffusion of component A in a mixture is balanced by an equal and opposite molar flow of component B, so that there is no net molar flow. This is generally the case in distillation, and it means there is no net volume flow in the gas phase. However, there is generally a net volume or mass flow in the liquid phase because of the difference in molar densities.

3. Diffusion of A and B takes place in opposite directions, but the molar fluxes are unequal. This situation often occurs in diffusion of chemically reacting species to and from a catalyst surface, but the equations are not covered in this text.

4. Two or more components diffuse in the same direction but at different rates, as in some membrane separations and adsorption processes.

Diffusion quantities

Five interrelated concepts are used in diffusion theory:

1. Velocity u, defined as usual by length/time.
2. Flux across a plane N, mol/area·time.
3. Flux relative to a plane of zero velocity J, mol/area·time.
4. Concentration c and molar density ρ_M, mol/volume (mole fraction may also be used).
5. Concentration gradient dc/db, where b is the length of the path perpendicular to the area across which diffusion is occurring.

Appropriate subscripts are used when needed. The equations apply equally well to SI, cgs, and fps units. In some applications, mass, rather than molal units, may be used in flow rates and concentrations.

Velocities in diffusion

Several velocities are needed to describe the movements of individual substances and of the total phase. Since absolute motion has no meaning, any velocity must be based on an arbitrary state of rest. In this discussion *velocity* without qualification refers to the velocity relative to the interface between the phases and is that apparent to an observer at rest with respect to the interface.

The individual molecules of any one component in the mixture are in random motion. If the instantaneous velocities of the components are summed, resolved in the direction perpendicular to the interface, and divided by the number of molecules of the substance, the result is the macroscopic velocity of that component. For component A, for instance, this velocity is denoted by u_A.

Molal flow rate, velocity, and flux

If the total molal flux, in moles per unit time per unit area in a direction perpendicular to a stationary plane, is denoted by N and the volumetric average velocity by u_0, then

$$N = \rho_M u_0 \tag{17.1}$$

where ρ_M is the molar density of the mixture.

For components A and B crossing a stationary plane, the molal fluxes are

$$N_A = c_A u_A \tag{17.2}$$

$$N_B = c_B u_B \tag{17.3}$$

Diffusivities are defined, not with respect to a stationary plane, but relative to a plane moving at the volume-average velocity u_0.[†] By definition there is no net volumetric flow across this reference plane, although in some cases there is a net molar flow or a net mass flow. The molar flux of component A through this reference plane is a diffusion flux designated J_A and is equal to the flux of A for a stationary plane [Eq. (17.2)] minus the flux due to the total flow at velocity u_0 and concentration c_A:

$$J_A = c_A u_A - c_A u_0 = c_A(u_A - u_0) \tag{17.4}$$

$$J_B = c_B u_B - c_B u_0 = c_B(u_B - u_0) \tag{17.5}$$

The diffusion flux J_A is assumed to be proportional to the concentration gradient dc_A/db, and the diffusivity of component A, in its mixture with component B, is denoted by D_{AB}. Thus,

$$J_A = -D_{AB} \frac{dc_A}{db} \tag{17.6}$$

A similar equation applies for component B:

$$J_B = -D_{BA} \frac{dc_B}{db} \tag{17.7}$$

Equations (17.6) and (17.7) are statements of Fick's first law of diffusion for a binary mixture. Note that this law is based on three decisions:

1. The flux is in moles per unit area per unit time.
2. The diffusion velocity is relative to the volume-average velocity.
3. The driving potential is in terms of the molar concentration (moles of component A per unit volume).

The dimensions of D_{AB} are length squared divided by time, usually given as square meters per second or square centimeters per second.

Relations between diffusivities

The relationship between D_{AB} and D_{BA} is easily determined for ideal gases, since the molar density does not depend on the composition:

$$c_A + c_B = \rho_M = \frac{P}{RT} \tag{17.8}$$

[†] Recall that linear velocity equals volumetric flow rate per unit area. Some authors define diffusivity using molar-average velocity, but the volume-average velocity is more suitable for both gases and liquids.

For diffusion of A and B in a gas at constant temperature and pressure,

$$dc_A + dc_B = d\rho_M = 0 \tag{17.9}$$

Choosing the reference plane for which there is zero volume flow, we can set the sum of the molar diffusion fluxes of A and B to zero, since the molar volumes are the same:

$$-D_{AB}\frac{dc_A}{db} - D_{BA}\frac{dc_B}{db} = 0 \tag{17.10}$$

Since $dc_A = -dc_B$, the diffusivities must be equal; that is,

$$D_{AB} = D_{BA} \tag{17.11}$$

When we are dealing with liquids, the same result is found if all mixtures of A and B have the same mass density.

$$c_A M_A + c_B M_B = \rho = \text{const} \tag{17.12}$$

$$M_A dc_A + M_B dc_B = 0 \tag{17.13}$$

For no volume flow across the reference plane, the sum of the volumetric flows due to diffusion is zero. The volumetric flow is the molar flow times the molar volume M/ρ and

$$-D_{AB}\frac{dc_A}{db}\frac{M_A}{\rho} - D_{BA}\frac{dc_B}{db}\frac{M_B}{\rho} = 0 \tag{17.14}$$

Substituting Eq. (17.13) into Eq. (17.14) gives

$$D_{AB} = D_{BA} \tag{17.15}$$

Other equations can be derived for diffusion in liquids where the density changes, but for most practical applications, equal diffusivities are assumed when we are dealing with binary mixtures. In the following equations a volumetric diffusivity D_v is used, with the subscript v used as a reminder that the driving force for diffusion is based on concentration differences in moles per volume. A common form of the diffusion equation gives the total flux relative to a fixed plane:

$$N_A = c_A u_0 - D_v \frac{dc_A}{db} \tag{17.16}$$

For gases it is often convenient to use mole fractions rather than molar concentrations, and since $c_A = \rho_M y_A$ and $u_0 = N/\rho_M$, Eq. (17.16) becomes

$$N_A = y_A N - D_v \rho_M \frac{dy_A}{db} \tag{17.17}$$

Equation (17.17) is sometimes applied to liquids, although it is only approximate if the molar density is not constant.

Interpretation of diffusion equations

Equation (17.16) is the basic equation for mass transfer in a nonturbulent fluid phase. It accounts for the amount of component A carried by the convective bulk flow of the fluid and the amount of A being transferred by molecular diffusion. The vector nature of the fluxes and concentration gradients must be understood, since these quantities are characterized by directions and magnitudes. As derived, the positive sense of the vectors is in the direction of increasing b, which may be either toward or away from the interface. As shown in Eq. (17.6), the sign of the gradient is opposite to the direction of the diffusion flux, since diffusion is in the direction of lower concentrations, or "downhill," like the flow of heat "down" a temperature gradient.

There are several types of situations covered by Eq. (17.16). The simplest case is zero convective flow and equimolal counterdiffusion of A and B, as occurs in the diffusive mixing of two gases. This is also the case for the diffusion of A and B in the vapor phase for distillations that have constant molal overflow. The second common case is the diffusion of only one component of the mixture, where the convective flow is caused by the diffusion of that component. Examples include evaporation of a liquid with diffusion of the vapor from the interface into a gas stream and condensation of a vapor in the presence of a noncondensable gas. Many examples of gas absorption also involve diffusion of only one component, which creates a convective flow toward the interface. These two types of mass transfer in gases are treated in the following sections for the simple case of steady-state mass transfer through a stagnant gas layer or film of known thickness. The effects of transient diffusion and laminar or turbulent flow are taken up later.

Equimolal diffusion

For equimolal diffusion in gases, the net volumetric and molar flows are zero, and Eq. (17.16) or Eq. (17.17) can be used with the convective term set to zero, which makes them equivalent to Eq. (17.6). Equation (17.17) is integrated over a film thickness B_T, assuming a constant flux N_A and zero total flux:

$$-D_v \rho_M \int_{y_{Ai}}^{y_A} dy_A = N_A \int_0^{B_T} db \tag{17.18}$$

where y_A = mole fraction of A at outer edge of film

$\quad\quad y_{Ai}$ = mole fraction of A at interface or inner edge of film

Integration of Eq. (17.18) and rearrangement give

$$N_A = J_A = \frac{D_v \rho_M}{B_T}(y_{Ai} - y_A) \tag{17.19}$$

or
$$N_A = J_A = \frac{D_v}{B_T}(c_{Ai} - c_A) \tag{17.20}$$

The concentration gradient for A is linear in the film, and the gradient for B has the same magnitude but the opposite sign, as shown in Fig. 17.1a. Note that for equimolal diffusion, $N_A = J_A$.

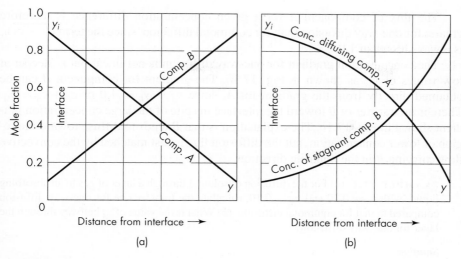

FIGURE 17.1
Concentration gradients for equimolal and unicomponent diffusion: (*a*) components *A*
and *B* diffusing at same molal rates in opposite directions; (*b*) component A diffusing,
component B stationary with respect to interface.

One-component mass transfer (one-way diffusion)

When only component A is being transferred, the total molal flux to or away
from the interface N is the same as N_A, and Eq. (17.17) becomes

$$N_A = y_A N_A - D_v \rho_M \frac{dy_A}{db} \tag{17.21}$$

Rearranging and integrating, we have

$$N_A(1 - y_A) = -D_v \rho_M \frac{dy_A}{db} \tag{17.22}$$

$$\frac{N_A B_T}{D_v \rho_M} = -\int_{y_{Ai}}^{y_A} \frac{dy_A}{1 - y_A} = \ln \frac{1 - y_A}{1 - y_{Ai}} \tag{17.23}$$

or

$$N_A = \frac{D_v \rho_M}{B_T} \ln \frac{1 - y_A}{1 - y_{Ai}} \tag{17.24}$$

Equation (17.24) can be rearranged by using the logarithmic mean of $1 - y_A$ for
easier comparison with Eq. (17.19) for equimolal diffusion. Since the driving force
$y_{Ai} - y_A$ can be written $(1 - y_A) - (1 - y_{Ai})$, the logarithmic mean becomes

$$(\overline{1 - y_A})_L = \frac{y_{Ai} - y_A}{\ln[(1 - y_A)/(1 - y_{Ai})]} \tag{17.25}$$

Combining Eqs. (17.24) and (17.25) gives

$$N_A = \frac{D_v \rho_M}{B_T} \frac{y_{Ai} - y_A}{(1 - y_A)_L} \tag{17.26}$$

The flux of component A for a given concentration difference is therefore greater for one-way diffusion than for equimolal diffusion, since the term $\overline{(1 - y_A)_L}$ is always less than 1.0.

The concentration gradient for one-way diffusion is not linear but is steeper at low values of y_A, as shown in Fig. 17.1b. The gradient for component B can be obtained directly from the gradient for A, since $y_A + y_B = 1.0$ or $c_A + c_B = \rho_M$. There is no transfer of B toward the interface in spite of the large concentration gradient shown in Fig. 17.1b. The explanation is that B tends to diffuse toward the region of lower concentration, but the diffusion flux is just matched by the convective flow carrying B in the opposite direction.

EXAMPLE 17.1. (a) For the diffusion of solute A through a layer of gas to an absorbing liquid, with $y_A = 0.20$ and $y_{Ai} = 0.10$, calculate the transfer rate for one-way diffusion compared to that for equimolal diffusion. (b) What is the value of y_A halfway through the layer for one-way diffusion?

Solution

(a) From Eq. (17.19) for equimolal diffusion,

$$N_A = J_A = \frac{D_v \rho_M}{B_T}(0.20 - 0.10)$$

From Eq. (17.24) for one-way diffusion,

$$N_A = \frac{D_v \rho_M}{B_T} \ln \frac{0.9}{0.8} = \frac{D_v \rho_M}{B_T}(0.1178)$$

(The concentration terms in both equations are reversed to make the flux toward the interface positive.) The ratio of the fluxes is $0.1178/0.10 = 1.18$. In this case the transfer rate with one-way diffusion is about 18 percent greater than that with equimolal diffusion.

(b) When $b = B_T/2$,

$$\ln \frac{1 - y_A}{0.8} = \frac{B_T}{2} \frac{N_A}{D_v \rho_M} = \frac{0.1178}{2} = 0.0589$$

$$1 - y_A = 0.8485 \qquad y_A = 0.1515$$

The concentration at the midpoint is only slightly greater than if the gradient were linear ($y_A = 0.150$).

PREDICTION OF DIFFUSIVITIES

Diffusivities are best estimated by experimental measurements, and where such information is available for the system of interest, it should be used directly. Often the desired values are not available, however, and they must be estimated from published correlations. Sometimes a value is available for one set of conditions of temperature and pressure; the correlations are then useful in predicting, from the known value, the desired values for other conditions.

Diffusion in gases

Values of D_v for some common gases diffusing in air at 0°C and 1 atm are given in App. 18. A simple theory for gases shows that D_v is proportional to the product of the average molecular velocity $\bar{u}$ and the mean free path λ.

$$D_v \cong \tfrac{1}{3}\bar{u}\lambda \tag{17.27}$$

Since the mean free path for ideal gases varies inversely with pressure, D_v also varies inversely with pressure, and the product $D_v P$ can be considered constant up to about 10 atm. The mean molecular velocity depends on $T^{0.5}$, and since the mean free path increases with $T^{1.0}$, the simple theory predicts D_v varies with $T^{1.5}$, a term that appears in some empirical equations for diffusivity. A more rigorous approach based on modern kinetic theory allows for the different sizes and velocities of the molecules and the mutual interactions as they approach one another. Using the Lennard–Jones (6-12) potential[19a] with parameters ϵ and σ leads to the following equation for binary diffusion.

$$D_{AB} = \frac{0.001858 T^{3/2}[(M_A + M_B)/M_A M_B]^{1/2}}{P \sigma_{AB}^2 \Omega_D} \tag{17.28}$$

where
D_{AB} = diffusivity, cm²/s
T = temperature, K
M_A, M_B = molecular weights of components A and B
P = pressure, atm
$\sigma_{AB} = (\sigma_A + \sigma_B)/2$ = effective collision diameter, Å
Ω_D = collision integral = $f(\mathbf{k}T/\epsilon_{AB})$
$\mathbf{k}$ = Boltzmann's constant
ϵ = Lennard–Jones force constant for common gases
$\epsilon_{AB} = \sqrt{\epsilon_A \epsilon_B}$

Tables of σ, ϵ, and Ω_D for common gases are given in App. 19. Equation (17.28) is known as the Chapman–Enskog equation.[18]

The collision integral Ω_D decreases with increasing temperature, which makes D_{AB} increase with more than the 1.5 power of the absolute temperature. The change in Ω_D with temperature is not very great, and for diffusion in air at temperatures from 300 to 1,000 K, D_v varies with about $T^{1.7-1.8}$, and $T^{1.75}$ can be used to extrapolate from room temperature data.

Diffusion in small pores. When gases diffuse in very small pores of a solid, as may occur during adsorption, drying of porous solids, or some membrane separation processes, the diffusivity is less than the normal value because of molecular collisions with the pore walls. When the pore size is much smaller than the normal mean free path, the diffusion process is called *Knudsen diffusion,* and the diffusivity for a cylindrical pore is

$$D_K = 9,700 r \sqrt{\frac{T}{M}} \tag{17.29}$$

where D_K = Knudsen diffusivity, cm²/s
   T = temperature, K
   M = molecular weight
   r = pore radius, cm

For pores of intermediate size, collisions with the wall and with other molecules are both important, and the diffusivity in the pore is predicted by combining the reciprocals of the bulk and Knudsen diffusivities.

$$\frac{1}{D_{\text{pore}}} = \frac{1}{D_{AB}} + \frac{1}{D_K} \tag{17.30}$$

EXAMPLE 17.2. Predict the volumetric diffusivity for benzene in air at 100°C and 2 atm by using the rigorous equation and by extrapolating from the published value for 0°C and 1 atm.

Solution. From App. 19, the force constants are as follows:

	ϵ/k	σ	M
Benzene	412.3	5.349	78.1
Air	78.6	3.711	29

Thus
$$\sigma_{AB} = \frac{5.349 + 3.711}{2} = 4.53$$

$$\epsilon_{AB}/K = (412.3 \times 78.6)^{0.5} = 180$$

$$\frac{kT}{\epsilon} = \frac{373}{180} = 2.072$$

From App. 19, $\Omega_D = 1.062$. From Eq. (17.28),

$$D_{AB} = \frac{0.001858 \times 373^{1.5}[(78.1 + 29)/78.1 \times 29]^{0.5}}{2 \times 4.53^2 \times 1.062} = 0.0668 \text{ cm}^2/\text{s}$$

From App. 18 at standard temperature and pressure,

$$D_{AB} = 0.299 \text{ ft}^2/\text{h} = 0.0772 \text{ cm}^2/\text{s}$$

At 373 K and 2 atm,

$$D_{AB} \cong 0.0772\left(\frac{1}{2}\right)\left(\frac{373}{273}\right)^{1.75} = 0.0666 \text{ cm}^2/\text{s}$$

Agreement with the value calculated from Eq. (17.28) is very good.

Diffusion in liquids

   The theory of diffusion in liquids is not as advanced or the experimental data as plentiful as for gas diffusion. Diffusivities in liquids are generally 4 to 5 orders of magnitude smaller than in gases at atmospheric pressure. Diffusion in liquids occurs by random motion of the molecules, but the average distance traveled between collisions is less than the molecular diameter, in contrast to gases, where the mean free path is orders of magnitude greater than the size of the molecule.

Diffusivities for dilute liquid solutions can be calculated approximately from the Wilke–Chang equation[24]

$$D_v = 7.4 \times 10^{-8} \frac{(\psi_B M_B)^{1/2} T}{\mu V_A^{0.6}} \tag{17.31}$$

where D_v = diffusivity, cm^2/s
　　　T = absolute temperature, K
　　　μ = viscosity of solution, cP
　　　V_A = molar volume of solute as liquid at its normal boiling point, cm^3/g mol
　　　ψ_B = *association parameter* for solvent
　　　M_B = molecular weight of solvent

The recommended values of ψ_B are 2.6 for water, 1.9 for methanol, 1.5 for ethanol, and 1.0 for benzene, heptane, ether, and other unassociated solvents. Equation (17.31) is valid only at low solute concentrations and does not apply when the solution has been thickened by addition of high-molecular-weight polymers. Small amounts of polymer can raise the solution viscosity more than 100-fold or even gel the solution, but the diffusivity of small solutes is only slightly reduced, because the polymer chains are too far apart to obstruct the movement of the solute molecules.[4]

For dilute aqueous solutions of nonelectrolytes a simpler equation can be used[15]

$$D_v = \frac{13.26 \times 10^{-5}}{\mu_B^{1.14} V_A^{0.589}} \tag{17.32}$$

where μ_B = viscosity of water, cP.

For values of V_A larger than 500 cm^3/mol, the Stokes–Einstein equation [Eq. (29.53) on p. 1028] should be used for the diffusivity.

The diffusion coefficient for dilute solutions of completely ionized univalent electrolytes is given by the Nernst equation

$$D_v = \frac{2RT}{(1/\lambda_+^0 + 1/\lambda_-^0) F_a^2} \tag{17.33}$$

where λ_+^0, λ_-^0 = limiting (zero-concentration) ionic conductances, $A/cm^2 \cdot (V/cm) \cdot (g$ equivalent$/cm^3)$
　　　R = gas constant, 8.314 $J/K \cdot g$ mol
　　　F_a = Faraday constant, = 96,500 Coulombs/g equivalent

Table 17.1 lists values of λ^0 at 25°C. Values for higher temperatures can be estimated from the change in T/μ.

Note that unlike the case for binary gas mixtures the diffusion coefficient for a dilute solution of A in B is not the same as for a dilute solution of B in A, since μ, M_B, and V_A will be different when the solute and solvent are exchanged. For intermediate concentrations, an approximate value of D_v is sometimes obtained by interpolation between the dilute solution values, but this method can lead to large errors for nonideal solutions.

TABLE 17.1
Limiting ionic conductances in water at 25°C[19b]

Cation	λ_+^0	Anion	λ_-^0
H^+	349.8	OH^-	197.6
Li^+	38.7	Cl^-	76.3
Na^+	50.1	Br^-	78.3
K^+	73.5	I^-	76.8
NH_4^+	73.4	NO_3^-	71.4

Schmidt number

The ratio of the kinematic viscosity to the molecular diffusivity is known as the *Schmidt number*, designated by Sc.

$$\text{Sc} = \frac{\nu}{D_v} = \frac{\mu}{\rho D_v}$$

The Schmidt number is analogous to the Prandtl number, which is the ratio of the kinematic viscosity to the thermal diffusivity (see Chap. 12).

$$\text{Pr} = \frac{\nu}{\alpha} = \frac{\mu}{\rho\alpha} = \frac{\mu}{\rho[k/(\rho c_p)]} = \frac{c_p\mu}{k}$$

Schmidt numbers for gases in air at 0°C and 1 atm are given in App. 18. Most of the values are between 0.5 and 2.0. The Schmidt number is independent of pressure when the ideal gas law applies, since the viscosity is independent of pressure, and the effects of pressure on ρ and D_v cancel. Temperature has only a slight effect on the Schmidt number because μ and ρD_v both change with about $T^{0.7-0.8}$.

The Schmidt numbers for liquids range from about 10^2 to 10^5 for typical mixtures. For small solutes in water at 20°C where $D_v \cong 10^{-5}$ cm²/s, Sc $\cong 10^3$. The Schmidt number decreases markedly with increasing temperature because of the decreasing viscosity and the increase in the diffusivity.

EXAMPLE 17.3. Estimate the diffusivity of benzene in toluene and toluene in benzene at 110°C. The physical properties are as follows:

	M	Boiling point, °C	V_A at boiling point, cm³/mol	μ at 110°C, cP
Benzene	78.11	80.1	96.5	0.24
Toluene	92.13	110.6	118.3	0.26

Solution. Equation (17.31) will be used. For benzene in toluene,

$$D_v = \frac{7.4 \times 10^{-8}(92.13)^{1/2}(383)}{0.26(96.5)^{0.6}} = 6.74 \times 10^{-5} \text{ cm}^2/\text{s}$$

For toluene in benzene,

$$D_v = \frac{7.4 \times 10^{-8}(78.11)^{1/2}(383)}{0.24(118.3)^{0.6}} = 5.95 \times 10^{-5} \text{ cm}^2/\text{s}$$

Turbulent diffusion

In a turbulent stream the moving eddies transport matter from one location to another, just as they transport momentum and heat energy. By analogy with Eqs. (3.17) and (12.46) for momentum transfer and heat transfer in turbulent streams, the equation for mass transfer is

$$J_{A,t} = -\varepsilon_N \frac{dc}{db} \tag{17.34}$$

where $J_{A,t}$ = molal flux of A, *relative to phase as a whole,* caused by turbulent action
$\quad\ \varepsilon_N$ = eddy diffusivity

The total molal flux, relative to the entire phase, becomes

$$J_A = -(D_v + \varepsilon_N) \frac{dc}{db} \tag{17.35}$$

The eddy diffusivity depends on the fluid properties but also on the velocity and position in the flowing stream. Therefore Eq. (17.35) cannot be directly integrated to determine the flux for a given concentration difference. This equation is used with theoretical or empirical relationships for ε_N in fundamental studies of mass transfer, and similar equations are used for heat or momentum transfer in developing analogies between the transfer processes. Such studies are beyond the scope of this text, but Eq. (17.35) is useful in helping to understand the form of some empirical correlations for mass transfer.

MASS-TRANSFER THEORIES

For steady-state mass transfer through a stagnant layer of fluid Eq. (17.19) or Eq. (17.24) can be used to predict the mass-transfer rate, provided B_T is known. However, this is not a common situation, because in most mass-transfer operations turbulent flow is desired to increase the rate of transfer per unit area or to help disperse one fluid in another and create more interfacial area. Furthermore, mass transfer to a fluid interface is often of the unsteady-state type, with continuously changing concentration gradients and mass-transfer rates. In spite of these differences, mass transfer in most cases is treated using the same type of equations, which feature a *mass-transfer coefficient k.* This coefficient is defined as the rate of mass transfer per unit area per unit concentration difference and is usually based on equal molal flows. The concentrations can be expressed in moles per volume or mole fractions, with subscript c indicating concentration and y or x mole fractions in the vapor or liquid phase:

$$k_c = \frac{J_A}{c_{Ai} - c_A} \tag{17.36}$$

or

$$k_y = \frac{J_A}{y_{Ai} - y_A} \tag{17.37}$$

Since k_c is a molar flux divided by a concentration difference, it has the units of velocity, such as centimeters per second or meters per second:

$$k_c = \frac{\text{mol}}{\text{s} \cdot \text{cm}^2 \cdot \text{mol/cm}^3} = \text{cm/s}$$

For k_y or k_x the units are the same as for J_A, moles per area per time, since the mole fraction driving force is dimensionless. It is apparent that k_c and k_y are related by the molar density as follows:

$$k_y = k_c \rho_M = \frac{k_c P}{RT} \tag{17.38}$$

$$k_x = k_c \rho_M = \frac{k_c \rho_x}{\bar{M}} \tag{17.39}$$

Gas-phase coefficients are sometimes based on the partial pressure driving force and are denoted by k_g.

$$k_g = \frac{J_A}{P_{Ai} - P_A} \tag{17.40}$$

Then

$$k_g = \frac{k_y}{P} = \frac{k_c}{RT} \tag{17.41}$$

The significance of k_c is brought out by combining Eq. (17.36) with Eq. (17.20) for steady-state equimolal diffusion in a stagnant film. This gives

$$k_c = \frac{J_A}{c_{Ai} - c_A} = \frac{D_v(c_{Ai} - c_A)}{B_T} \frac{1}{c_{Ai} - c_A}$$

$$= \frac{D_v}{B_T} \tag{17.42}$$

Thus the coefficient k_c is the molecular diffusivity divided by the thickness of the stagnant layer. When we are dealing with unsteady-state diffusion or diffusion in flowing streams, Eq. (17.42) can still be used to give an effective film thickness from known values of k_c and D_v.

Film Theory

The basic concept of the film theory is that the resistance to diffusion can be considered equivalent to that in a stagnant film of a certain thickness. The implication is that the coefficient k_c varies with the first power of D_v, which is rarely true, but this does not detract from the value of the theory in many applications. The film theory is often used as a basis for complex problems of multicomponent diffusion or diffusion plus chemical reaction.

For example, consider mass transfer from a turbulent gas stream to the wall of a pipe, with the concentration gradient as shown in Fig. 17.2. There is a laminar layer near the wall, where mass transfer is mainly by molecular diffusion, and the

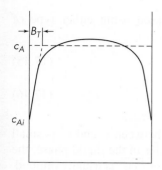

FIGURE 17.2
Concentration gradient for mass transfer in a pipe with turbulent flow of gas.

concentration gradient is almost linear. As the distance from the wall increases, turbulence becomes stronger, and the eddy diffusivity increases, which means that a lower gradient is needed for the same flux [see Eq. (17.35)]. The value of c_A is a maximum at the center of the pipe, but this value is *not* used in mass-transfer calculations. Instead the driving force is taken as $c_A - c_{Ai}$, where c_A is the concentration reached if the stream were thoroughly mixed. This is the same as a flow-weighted average concentration and is also the concentration to be used in material-balance calculations. (This is analogous to the usage in heat transfer, where the average temperature of a stream is used in defining h.)

If the gradient near the wall is linear, it can be extrapolated to c_A, and the distance from the wall at this point is the effective film thickness B_T. Generally, the resistance to mass transfer is mainly in the laminar boundary layer very close to the wall, and B_T is only slightly greater than the thickness of the laminar layer. However, as will be brought out later, the value of B_T depends on the diffusivity D_v and not just on flow parameters, such as the Reynolds number. The concept of an effective film thickness is useful, but values of B_T must not be confused with the actual thickness of the laminar layer.

Effect of one-way diffusion

As shown previously, when only component A is diffusing through a stagnant film, the rate of mass transfer for a given concentration difference is greater than if component B is diffusing in the opposite direction. From Eqs. (17.19) and (17.26), the ratio of the fluxes is

$$\frac{N_A}{J_A} = \frac{1}{(1 - y_A)_L} = \frac{1}{(y_B)_L} \tag{17.43}$$

This relationship, derived for molecular diffusion in a stagnant film, is assumed to hold reasonably well for unsteady-state diffusion or a combination of molecular and eddy diffusion. Sometimes the mass-transfer coefficient for one-way transfer is denoted by k'_c or k'_y, and the coefficients then follow the same relationship as the fluxes in Eq. (17.43):

$$\frac{k'_c}{k_c} = \frac{k'_y}{k_y} = \frac{1}{(1 - y_A)_L} \tag{17.44}$$

The rate of one-way mass transfer can be expressed using either type of coefficient:

$$N_A = k'_y(y_{Ai} - y_A) \tag{17.45}$$

$$N_A = \frac{k_y(y_{Ai} - y_A)}{(1 - y_A)_L} \tag{17.46}$$

When the value of y_A is 0.10 or less, the difference between k_y and k'_y is small and often ignored in design calculations. For mass transfer in the liquid phase, the corresponding correction term for one-way diffusion $(1 - x_A)_L$ is usually omitted, because the correction is small compared to the uncertainty in the diffusivity and the mass-transfer coefficient.

Boundary Layer Theory

Although there are few examples of diffusion through a stagnant fluid film, mass transfer often takes place in a thin boundary layer near a surface where the fluid is in laminar flow. If the velocity gradient in the boundary layer is linear and the velocity is zero at the surface, the equations for flow and diffusion can be solved to give the concentration gradient and the average mass-transfer coefficient. The coefficient depends on the two-thirds power of the diffusivity and decreases with increasing distance along the surface in the direction of flow, because an increase either in distance or in D_v makes the concentration gradient extend farther from the surface, which decreases the gradient dc_A/db at the surface.

For flow over a flat plate or around a cylinder or sphere, the velocity profile is linear near the surface, but the gradient decreases as the velocity approaches that of the main stream at the outer edge of the boundary layer. Exact calculations show that the mass-transfer coefficient still varies with $D_v^{2/3}$ if D_v is low or the Schmidt number $\mu/\rho D_v$ is 10 or larger. For Schmidt numbers of about 1, typical for gases, the predicted coefficient varies with a slightly lower power of D_v. For boundary layer flows, no matter what the shape of the velocity profile or value of the physical properties, the transfer rate cannot increase with the 1.0 power of the diffusivity, as implied by the film theory. Boundary layer theory can be used to estimate k_c for some situations, but when the boundary layer becomes turbulent or separation occurs, exact predictions of k_c cannot be made, and the theory serves mainly as a guide in developing empirical correlations. The analogy between heat and mass transfer in boundary layers permits correlations developed for heat transfer to be used for mass transfer.

Penetration Theory

The penetration theory makes use of the expression for the transient rate of diffusion into a relatively thick mass of fluid with a constant concentration at the surface. The change in concentration with distance and time is governed by the following equation:

$$\frac{\partial c_A}{\partial t} = D_v \frac{\partial^2 c_A}{\partial b^2} \tag{17.47}$$

The boundary conditions are

$$c_A = \begin{cases} c_{A0} & \text{for } t = 0 \\ c_{Ai} & \text{at } b = 0, t > 0 \end{cases}$$

The particular solution of Eq. (17.47) is the same as that for transient heat conduction to a semi-infinite solid, Eq. (10.36).

The instantaneous flux at time t is given by the analogous form of Eq. (10.39):

$$J_A = \sqrt{\frac{D_v}{\pi t}}(c_{Ai} - c_A) \tag{17.48}$$

The average flux over the time interval 0 to t_T is

$$\bar{J}_A = \frac{1}{t_T}\int_0^{t_T} J_A \, dt = \frac{c_{Ai} - c_A}{t_T}\sqrt{\frac{D_v}{\pi}}\int_0^{t_T}\frac{dt}{t^{1/2}}$$

$$= 2\sqrt{\frac{D_v}{\pi t_T}}(c_{Ai} - c_A) \tag{17.49}$$

Combining Eqs. (17.36) and (17.49) gives the average mass-transfer coefficient over time t_T:

$$\bar{k}_c = 2\sqrt{\frac{D_v}{\pi t_T}} = 1.13\sqrt{\frac{D_v}{t_T}} \tag{17.50}$$

Higbie[16] was the first to apply this equation to gas absorption in a liquid, showing that diffusing molecules will not reach the other side of a thin layer if the contact time is short. The depth of penetration, defined as the distance at which the concentration change is 1 percent of the final value, is $3.6\sqrt{D_v t_T}$. For $D_v = 10^{-5}$ cm^2/s and $t_T = 10$ s, the depth of penetration is only 0.036 cm. In gas absorption equipment, drops and bubbles often have very short lifetimes because of coalescence, and the penetration theory is likely to apply.

An alternate form of the penetration theory was developed by Danckwerts,[5] who considered the case where elements of fluid at a transfer surface are randomly replaced by fresh fluid from the bulk stream. An exponential distribution of ages or contact times results, and the average transfer coefficient is given by

$$\bar{k}_c = \sqrt{D_v s} \tag{17.51}$$

where s is the fractional rate of surface renewal, in s^{-1}.

Both Eqs. (17.50) and (17.51) predict that the coefficient varies with the one-half power of the diffusivity and give almost the same value for a given average contact time. A modified version of the penetration theory,[12] which assumes that eddies from a turbulent bulk fluid come to within random distances of the surface, gives slightly higher exponents for the diffusivity, which indicates that this theory might apply for mass transfer to pipe walls or flat surfaces such as a pool of liquid.

The various forms of the penetration theory can be classified as surface renewal models, implying either formation of new surface at frequent intervals or replacement of fluid elements at the surface with fresh fluid from the bulk. The time t_T and its reciprocal, the average rate of renewal, are functions of the fluid velocity, the fluid properties, and the geometry of the system and can be accurately predicted in only a few special cases. However, even if t_T must be determined empirically, the surface renewal models give a sound basis for correlation of mass-transfer data in many situations, particularly for transfer to drops and bubbles. The similarity between Eqs. (17.50) and (15.22) is an example of the close analogy between heat and mass transfer. It is often reasonable to assume that t_T is the same for both processes and thus to estimate rates of heat transfer from measured mass-transfer rates or vice versa.

Two-Film Theory

In many separation processes, material must diffuse from one phase into another phase, and the rates of diffusion in both phases affect the overall rate of mass transfer. In the two-film theory, proposed by Whitman[23] in 1923, equilibrium is assumed at the interface, and the resistances to mass transfer in the two phases are added to get an overall resistance, just as is done for heat transfer. The reciprocal of the overall resistance is an overall coefficient, which is easier to use for design calculations than the individual coefficients.

What makes mass transfer between phases more complex than heat transfer is the discontinuity at the interface, which occurs because the concentration or mole fraction of diffusing solute is hardly ever the same on opposite sides of the interface. For example, in distillation of a binary mixture, y_A^* is greater than x_A, and the gradients near the surface of a bubble might be as shown in Fig. 17.3a. For the absorption of a very soluble gas, the mole fraction in the liquid at the interface would be greater than that in the gas, as shown in Fig. 17.3b.

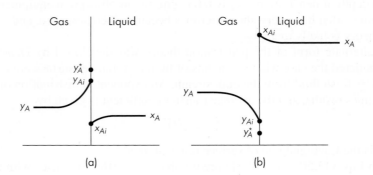

(a)   (b)

FIGURE 17.3
Concentration gradients near a gas-liquid interface: (a) distillation; (b) absorption of a very soluble gas.

In the two-film theory, the rate of transfer to the interface is set equal to the rate of the transfer from the interface:

$$r = k_x(x_A - x_{Ai}) \qquad (17.52)$$

$$r = k_y(y_{Ai} - y_A) \qquad (17.53)$$

The rate is also set equal to an overall coefficient K_y times an overall driving force $y_A^* - y_A$, where y_A^* is the composition of the vapor that would be in equilibrium with the bulk liquid of composition x_A:

$$r = K_y(y_A^* - y_A) \qquad (17.54)$$

To get K_y in terms of k_y and k_x, Eq. (17.54) is rearranged and the term $y_A^* - y_A$ replaced by $(y_A^* - y_{Ai}) + (y_{Ai} - y_A)$:

$$\frac{1}{K_y} = \frac{y_A^* - y_A}{r} = \frac{y_A^* - y_{Ai}}{r} + \frac{y_{Ai} - y_A}{r} \qquad (17.55)$$

Equations (17.52) and (17.53) are now used to replace r in the last two terms of Eq. (17.55):

$$\frac{1}{K_y} = \frac{y_A^* - y_{Ai}}{k_x(x_A - x_{Ai})} + \frac{y_{Ai} - y_A}{k_y(y_{Ai} - y_A)} \qquad (17.56)$$

Figure 17.4 shows typical values of the composition at the interface, and it is apparent that $(y_A^* - y_{Ai})/(x_A - x_{Ai})$ is the slope of the equilibrium curve. This slope is denoted by m. The equation can then be written

$$\frac{1}{K_y} = \frac{m}{k_x} + \frac{1}{k_y} \qquad (17.57)$$

The term $1/K_y$ can be considered an overall resistance to mass transfer, and the terms m/k_x and $1/k_y$ are the resistances in the liquid and gas films. These "films"

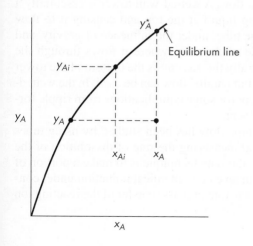

FIGURE 17.4
Bulk and interface concentrations typical of distillation.

Equilibrium line

need not be stagnant layers of a certain thickness in order for the two-film theory to apply. Mass transfer in either film may be by diffusion through a laminar boundary layer or by unsteady-state diffusion, as in the penetration theory, and the overall coefficient is still obtained from Eq. (17.57). For some problems, such as transfer through a stagnant film into a phase where the penetration theory is thought to apply, the penetration theory coefficient is slightly changed because of the varying concentration at the interface, but this effect is only of academic interest.

The essential part of the two-film theory is the method of allowing for the partition of solute between phases at the interface while combining individual coefficients to get an overall coefficient. This approach is used in the analysis of laboratory data and in equipment design for many types of mass-transfer operations, including absorption, extraction, and distillation. The same principle of adding resistances with the appropriate distribution factors also applies to membrane separations, in which there are three resistances in series.

MASS-TRANSFER COEFFICIENTS

In view of the complexity of mass transfer in actual equipment, fundamental equations for mass transfer in actual equipment are rarely available, and empirical methods, guided by dimensional analysis and by semitheoretical analogies, are relied upon to give workable equations. The approach to the problem has been made in several steps in the following manner.

Experimental Measurements

1. The coefficient k has been studied in experimental devices in which the area of contact between phases is known and where boundary layer separation does not take place. The wetted-wall tower shown in Fig. 17.5, which is sometimes used in practice, is one device of this type. It has given valuable information on mass transfer to and from fluids in turbulent flow. A wetted-wall tower is essentially a vertical tube with means for admitting liquid at the top and causing it to flow downward along the inside wall of the tube, under the influence of gravity, and means for admitting gas to the inside of the tube, where it flows through the tower in contact with the liquid. Generally the gas enters the bottom of the tower and flows countercurrent to the liquid, but parallel flow can be used. In the wetted-wall tower, the interfacial area A, except for some complications from ripple formation, is known, and form drag is absent.

2. Mass transfer to liquids in turbulent pipe flow has been studied by using tubes made from a slightly soluble solid and measuring the rate of dissolution of the solid for various liquid flow rates. An alternate technique is to make a portion of the tube wall an electrode and carry out an electrochemical reduction under conditions where the current is limited by the rate of mass transfer of the reacting ion to the wall.

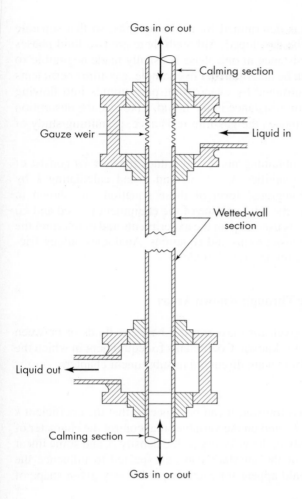

Gas in or out

Calming section

Gauze weir

Liquid in

Wetted-wall section

Liquid out

Calming section

Gas in or out

FIGURE 17.5
Wetted-wall tower.

3. External mass transfer, such as diffusion to particles or to the outside of pipes or cylinders, requires different correlations from those for internal mass transfer, because there is boundary layer flow over part of the surface, and boundary layer separation is common. The mass-transfer coefficients can be determined by studying evaporation of liquid from porous wet solids. However, it is not easy to ensure that there is no effect of internal mass-transfer resistance. Complications from diffusion in the solid are eliminated if the solid is made from a slightly soluble substance that dissolves in the liquid or sublimes into a gas. This method also permits measurement of local mass-transfer coefficients for different points on the solid particle or cylinder.

4. Finally, experiments are made with actual mass-transfer devices such as packed towers, sieve trays, and bubble columns, for which the mass-transfer area varies with operating conditions. The mass-transfer rates are converted first to a volumetric mass-transfer coefficient ka, where a is the transfer area per unit volume

of equipment. Sometimes a is determined from photographs, so that separate correlations for a and k can be developed. Although there are two fluid phases present in most cases, the resistance in one phase is usually made negligible so that ka for the other phase can be determined. For example, gas-film coefficients in a packed column are determined by evaporating pure liquids into flowing gases, and there is no diffusion resistance in the liquid. Likewise the absorption of a pure gas into water eliminates the gas-film resistance, permitting study of the liquid-film coefficient.

Experiments conducted for obtaining numerical values for k or ka consist of measuring experimentally the quantities N_A, A, y_i, and y and calculating k by Eq. (17.37) or (17.46) or an integrated form of these equations, as shown in Chap. 18. If A or a is not known, the total volume of the equipment is used and ka calculated. Dimensional analysis is used to plan the experiments and to interpret the results in the form of dimensionless groups and equations. Analogies among friction, heat transfer, and mass transfer are useful guides.

Coefficients for Mass Transfer Through Known Areas

In this section correlations are given for mass transfer between fluids or between fluids and solids where the area A is known. Coefficients for equipment in which the area between the phases is not known are discussed in subsequent chapters.

Dimensional Analysis

From the mechanism of mass transfer, it can be expected that the coefficient k would depend on the diffusivity D_v and on the variables that control the character of the fluid flow, namely, the velocity u, the viscosity μ, the density ρ, and some linear dimension D. Since the shape of the interface can be expected to influence the process, a different relation should appear for each shape. For any given shape of transfer surface

$$k = \psi(D_v, D, u, \mu, \rho)$$

Dimensional analysis gives

$$\frac{k_c D}{D_v} = \psi_1\left(\frac{DG}{\mu}, \frac{\mu}{\rho D_v}\right) \tag{17.58}$$

where $G = u\rho$.

The dimensionless group $k_c D/D_v$ is called the *Sherwood number* and is denoted by Sh. The other groups are, of course, the Reynolds number $\text{Re} = DG/\mu$ and the Schmidt number $\text{Sc} = \mu/(\rho D_v)$.

When the mass-transfer coefficient depends on the two-thirds power of the diffusivity, as is the case for some boundary layer flows, the coefficients are often presented in the form of a j_M factor analogous to the j_H factor of Eq. (12.53):

$$j_M \equiv \frac{k_c}{u}\left(\frac{\mu}{\rho D_v}\right)^{2/3} \tag{17.59}$$

Mass transfer with flow inside pipes

Correlations for mass transfer to the inside wall of a pipe are of the same form as those for heat transfer, since the basic equations for diffusion and conduction are similar. For laminar flow the Sherwood number shows the same trends as the Nusselt number, with a limiting value of 3.66 for a constant wall concentration and a one-third dependence on flow rate for short tubes. When the wall concentration is a function of axial position, as in a counterflow membrane separator, the limiting Sherwood number is somewhat higher than for a constant concentration at the wall (see Fig. 12.2), but there may be little difference with high Graetz numbers Gz. The recommended equation for moderate Graetz numbers is the theoretical equation

$$Sh = 1.62 \, Gz'^{1/3} \tag{17.60}$$

where

$$Gz' = \frac{\dot{m}}{D_v L \rho} = \frac{\pi}{4} \, Re \, Sc \, \frac{D}{L} \tag{17.61}$$

Equation (17.60) has been used to predict the internal mass-transfer resistance for separation processes using hollow-fiber membranes. The recommended equation for heat transfer, Eq. (12.25), has an empirical coefficient of 2.0, and this higher value is probably due to natural convection. There are not enough data to determine an empirical equation for mass transfer in laminar flow.

One equation for turbulent-flow mass transfer to pipe walls is a modification of Eq. (12.33) with the Nusselt and Prandtl numbers replaced by the Sherwood and Schmidt numbers.

$$Sh = 0.023 \, Re^{0.8} \, Sc^{1/3} \left(\frac{\mu}{\mu_w} \right)^{0.14} \tag{17.62}$$

This is the simplest equation that gives a fairly good fit to the published data over a wide range of Reynolds numbers and Schmidt numbers. An alternate form of the correlation is obtained by dividing Eq. (17.62) by $Re \times Sc^{1/3}$ to give the j_M factor, which was shown by Chilton and Colburn[3] to be the same as j_H and also the same as $f/2$. The term $\mu/\mu_w^{0.14}$ is usually about 1.0 for mass transfer and is omitted:

$$j_M = j_H = \tfrac{1}{2} f = 0.023 \, Re^{-0.2} \tag{17.63}$$

The analogy shown in this equation is general for heat and mass transfer in the same equipment.

Extending the analogy to include friction loss is possible for pipes only because all the loss comes from skin friction. The analogy does not apply to total friction loss when there is form drag from separation of flow, as occurs in flow around objects.

In general, j_M is a function of Re. For gas-phase mass transfer, other forms of j_M can be used:

$$j_M = \frac{k_y RT}{Pu} \, Sc^{2/3} = \frac{k_y \bar{M}}{G} \, Sc^{2/3} \tag{17.64}$$

$$j_M = \frac{k_g \bar{M} P}{G} \, Sc^{2/3} \tag{17.65}$$

Coefficient k_g is discussed in Chap. 18.

Slightly more accurate correlations for pipe flow have been presented for different ranges of the Schmidt number. Data for evaporation of several liquids in wetted-wall towers (Fig. 17.5) were correlated with slightly higher exponents for both the Reynolds and Schmidt numbers[8]

$$Sh = 0.023 \, Re^{0.81} \, Sc^{0.44} \tag{17.66}$$

The Schmidt numbers were varied from 0.60 to 2.5, and over this narrow range the difference between the exponents of 0.44 in Eq. (17.66) and 0.33 in Eq. (17.62) has only a small effect on the coefficient. The difference in exponents may have fundamental significance, since transfer to a liquid surface, which can have waves or ripples, should differ somewhat from transfer to a smooth rigid surface.

A correlation for mass transfer at high Schmidt numbers (430 to 100,000) was obtained by measuring the rate of solution of tubes of benzoic acid in water and viscous liquids[13]

$$Sh = 0.0096 \, Re^{0.913} \, Sc^{0.346} \tag{17.67}$$

The difference between the exponent on the Schmidt number and the usual value of $\frac{1}{3}$ may not be significant, but the exponent for the Reynolds number is definitely greater than 0.80. Other studies of heat transfer with large Prandtl numbers[7] have also shown an exponent of about 0.9 for the Reynolds number. Various empirical equations that cover the entire range of Sc or Pr with good accuracy are available.[21a]

EXAMPLE 17.4. (a) What is the effective thickness of the gas film for the evaporation of water into air in a 2-in.-diameter wetted-wall column at a Reynolds number of 10,000 and a temperature of 40°C? (b) Repeat the calculation for the evaporation of ethanol under the same conditions. At 1 atm the diffusivities are 0.288 cm²/s for water in air and 0.145 cm²/s for ethanol in air.

Solution. For air at 40°C,

$$\rho = \frac{29}{22,410} \times \frac{273.16}{313.16} = 1.129 \times 10^{-3} \text{ g/cm}^3$$

$$\mu = 0.0186 \text{ cP} \qquad \text{(App. 8)}$$

$$\frac{\mu}{\rho} = \frac{1.86 \times 10^{-4}}{1.129 \times 10^{-3}} = 0.165 \text{ cm}^2/\text{s}$$

(a) For the air-water system,

$$Sc = \frac{0.165}{0.288} = 0.573$$

From Eq. (17.66),

$$Sh = 0.023(10,000)^{0.81}(0.573)^{0.44} = 31.3$$

In the film theory, $k_c = D_v/B_T$, and since $Sh = k_c D/D_v$,

$$Sh = \frac{D}{B_T} \qquad \text{or} \qquad B_T = \frac{2.0}{31.3} = 0.064 \text{ in.}$$

(b) For the air-ethanol system, $Sc = 0.165/0.145 = 1.14$:

$$Sh = 0.023(10,000)^{0.81}(1.14)^{0.44} = 42.3$$

$$B_T = \frac{2.0}{42.3} = 0.047 \text{ in.}$$

Thickness B_T becomes smaller as the diffusivity decreases because k_c varies with only the 0.56 power of the diffusivity instead of the 1.0 power implied by the film theory. If Eq. (17.62) were used, the corresponding values of B_T would be 0.066 and 0.053 in. These are somewhat closer to each other because $D_v^{2/3}$ is used instead of $D_v^{0.56}$.

Flow outside tubes parallel to axis

Some membrane separators have bundles of hollow fibers in a shell-and-tube arrangement with liquid or gas flowing parallel to the tube axis on the outside of the tubes. The external flow passages are irregular in shape and not uniform, since the fibers are not held in position as are the tubes in a heat exchanger. Empirical correlations such as Eq. (17.62) for the external mass-transfer coefficient have been proposed, using an equivalent diameter to calculate the Reynolds number. For a bundle of fibers with diameter d packed in a shell with ϵ void fraction, the equivalent diameter is

$$d_e = 4 \times \frac{\text{flow area}}{\text{wetted perimeter}} = \frac{4\epsilon}{(4/d)(1-\epsilon)} = d\frac{\epsilon}{1-\epsilon} \qquad (17.68)$$

Flow normal to cylinders

A correlation of j_M versus Re for flow of air perpendicular to single cylinders is shown in Fig. 17.6.[21] The dashed line shows values of j_H calculated from Eq. (12.63), which was based on data for liquids. The data for heat transfer to air,

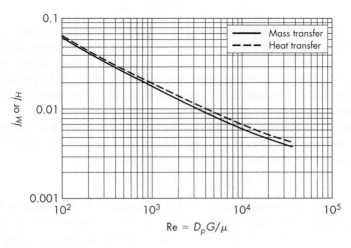

FIGURE 17.6
Heat and mass transfer, flow past single cylinders.

taken from Fig. 17.6, fall slightly below the dashed line and very close to the mass-transfer data. The good agreement shows that the analogy between mass and heat transfer holds very well for external flows as well as for flows inside pipes.

An approximate equation for mass transfer to a single cylinder can be used for Re values of 10 to 10^4:

$$Sh = 0.61 \, Re^{1/2} \, Sc^{1/3} \tag{17.69}$$

Flow normal to tube bundle

For flow normal to a bundle of tubes, the average mass-transfer coefficient is higher than for a single tube at the same superficial velocity, but the factor of increase depends on the tube spacing and the Reynolds number. Based on heat transfer to tube banks, the increase could be 20 to 40 percent, but there are few mass-transfer data for low Reynolds numbers, which is the region of interest for hollow-fiber membrane devices.

Tests of a module with 2,100 fibers and $\epsilon = 0.6$ gave $Sh = 0.24$, independent of flow rate.[25] Such a low Sherwood number indicates very uneven flow distribution, since the minimum value of Sh for ideal flow would be in the range of 2 to 4.

The following empirical equation for external mass transfer[14] was developed from data for oxygen removal from water using radial flow modules with close-packed hollow fibers.

$$Sh = 1.28 \, Re^{0.4} \, Sc^{0.33} \tag{17.70}$$

Flow past single spheres

For mass transfer to an isolated sphere, the Sherwood number approaches a lower limit of 2.0 as the Reynolds number approaches zero. A simple equation that is fairly accurate for Reynolds numbers up to 1,000 is a modification of the Frössling equation[20] [compare to Eq. (12.64) for Nu]:

$$Sh = 2.0 + 0.6 \, Re^{1/2} \, Sc^{1/3} \tag{17.71}$$

Data for high Reynolds numbers show a gradual increase of slope on a plot of Sh versus Re, as shown in Fig. 17.7. The exponent of $\frac{1}{2}$ is consistent with boundary layer theory which applies to the front portion of the sphere, where most of the transfer takes place at moderate Reynolds numbers. At high Reynolds numbers, mass transfer in the turbulent region becomes more important, and the effect of flow rate increases.

The correlation in Fig. 17.7 gives values that are too low for "creeping flow," where the Reynolds number is low and the Peclet number Pe is high ($Pe = Re \times Sc = D_p u_0 / D_v$). For this case the recommended equation is[2]

$$Sh = (4.0 + 1.21 \, Pe^{2/3})^{1/2} \tag{17.72}$$

The limiting Sherwood number of 2.0 corresponds to an effective film thickness of $D_p/2$ if the mass-transfer area is taken as the external area of the sphere. The concentration gradients actually extend out to infinity in this case, but the mass-transfer area also increases with distance from the surface, so the effective film

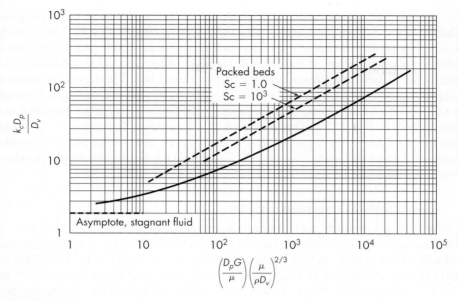

FIGURE 17.7
Heat and mass transfer, flow past single spheres (solid line) and in packed beds (dashed lines).

thickness is much less than might be estimated from the shape of the concentration profile.

Mass transfer in packed beds

There have been a great many studies of mass transfer and heat transfer from gases or liquids to particles in packed beds. The coefficients increase with about the square root of the mass velocity and the two-thirds power of the diffusivity, but the correlations presented by different workers differ appreciably, in contrast to the close agreement found in studies of single spheres. An equation that fairly well represents most of the data is[21b]

$$j_M = \frac{k_c}{u_0} \, \mathrm{Sc}^{2/3} = 1.17 \left(\frac{D_p G}{\mu} \right)^{-0.415} \tag{17.73}$$

This is equivalent to the equation

$$\mathrm{Sh} = 1.17 \, \mathrm{Re}^{0.585} \, \mathrm{Sc}^{1/3} \tag{17.74}$$

Equations (17.73) and (17.74) are recommended for spheres or roughly spherical solid particles that form a bed with about 40 to 45 percent voids. For cylindrical particles these equations can be used with the diameter of the cylinder in Re and Sh. For beds with higher void fractions or for hollow particles, such as rings, other correlations are available.[9]

To compare mass transfer in packed beds with transfer to a single particle, Sherwood numbers calculated from Eq. (17.74) are plotted in Fig. 17.7 along with the correlation for isolated spheres. The coefficients for packed beds are 2 to 3 times those for a single sphere at the same Reynolds number. Most of this difference is due to the higher actual mass velocity in the packed bed. The Reynolds number is based for convenience on the superficial velocity, but the average mass velocity is G/ϵ, and the local velocity at some points in the bed is even higher. Note that the dashed lines in Fig. 17.7 are not extended to low values of Re, since it is unlikely that the coefficients for a packed bed would ever be lower than those for single particles. Therefore Eqs. (17.73) and (17.74) should not be used for Reynolds numbers less than about 10.

Mass transfer to suspended particles

When solid particles are suspended in an agitated liquid, as in a stirred tank, a minimum estimate of the transfer coefficient can be obtained by using the terminal velocity of the particle in still liquid to calculate Re in Eq. (17.71). The effect of particle size and density difference on this minimum coefficient k_{cT} is shown in Fig. 17.8. Over a wide range of sizes, there is little change in the coefficient, because the increase in terminal velocity and Reynolds number makes the Sherwood number nearly proportional to particle diameter.

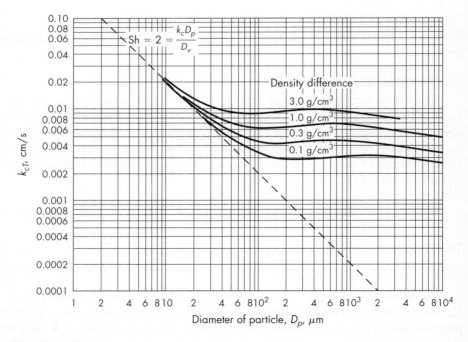

FIGURE 17.8
Mass-transfer coefficients for particles falling in water[10] (viscosity $\mu = 1$ cP, diffusivity $D_v = 10^{-5}$ cm²/s).

The actual coefficient is greater than k_{cT} because frequent acceleration and de-celeration of the particle raises the average slip velocity and because small eddies in the turbulent liquid penetrate close to the particle surface and increase the local rate of mass transfer. However, if the particles are fully suspended, the ratio k_c/k_{cT} falls within the relatively narrow range of 1.5 to 5 for a wide range of particle sizes and agitation conditions.[10] The effects of particle size, diffusivity, and viscosity follow the trends predicted for k_{cT}, but the density difference has almost no effect until it exceeds 0.3 g/cm^3. For suspended particles, k_c varies with only the 0.1 to 0.15 power of the power dissipation per unit volume, the higher exponent being for large particles. Empirical correlations for predicting k_c based on power consumption are available,[17] but at the same power, the coefficients are higher for larger ratios of ag-itator diameter to tank diameter D_a/D_t.

Power per unit volume is a satisfactory basis for scaleup when we are dealing with suspended particles, provided geometric similarity is maintained.

Mass transfer to drops and bubbles

When small drops of liquid are falling through a gas, surface tension tends to make the drops nearly spherical, and the coefficients for mass transfer to the drop surface are often quite close to those for solid spheres. The shear caused by the fluid moving past the drop surface, however, sets up toroidal circulation currents in the drop that decrease the resistance to mass transfer both inside and outside the drop. The extent of the change depends on the ratio of the viscosities of the internal and external fluids and on the presence or absence of substances such as surfactants that concentrate at the interface.[11]

For a low-viscosity drop falling through a viscous liquid with no surface-active material present, the velocity boundary layer in the external fluid almost disappears. Fluid elements are exposed to the drop for short times, and the mass transfer is governed by the penetration theory. It can be shown that the effective contact time is the time for the drop to fall a distance equal to its own diameter, and application of the penetration theory leads to the equation for the external coefficient

$$\bar{k}_c = 2\sqrt{\frac{D_v u_0}{\pi D_p}} \tag{17.75}$$

Multiplying through by D_p/D_v gives

$$\text{Sh} = \frac{2}{\sqrt{\pi}} \left(\frac{D_p u_0 \rho}{\mu} \frac{\mu}{\rho D_v} \right)^{1/2}$$

$$= 1.13 \, \text{Re}^{1/2} \, \text{Sc}^{1/2} \tag{17.76}$$

Comparing Eq. (17.76) with Eq. (17.71) for a rigid sphere shows that internal circulation can increase k_c by a factor of about 1.88 Sc$^{1/6}$, or 5.9, when Sc $= 10^3$.

Coefficients in agreement with Eq. (17.75) have been found for some drops in free fall, but in many cases the high drop viscosity or impurities in the drop reduce the circulation currents and lead to values only slightly greater than those for rigid spheres. For drops suspended in an agitated liquid, as in a stirred-tank extractor, the coefficients generally fall between those for a solid sphere and those for a completely circulating drop. The coefficients for drops increase with the 1.0 to 1.2 power of the stirrer speed, in contrast to the 0.4 to 0.5 power found with solid particles, because eddies in the suspending liquid can penetrate closer to a drop with a deformable surface than to a solid particle.[6]

It is difficult to predict k_c for a practical application, and the mass-transfer calculations are generally based on a volumetric mass-transfer coefficient $k_c a$ estimated from laboratory or pilot-plant tests.

The same uncertainties arise when we are dealing with mass transfer from bubbles of gas rising through liquid. The gas in the bubbles should circulate rapidly because of the low gas viscosity, but impurities often interfere, giving coefficients between those for rigid spheres and those for freely circulating bubbles. Bubbles 1 mm in diameter or smaller often behave as rigid spheres, and those 2 mm or larger as freely circulating bubbles. Bubbles larger than a few millimeters in diameter, however, are flattened in shape and may oscillate as they rise, making mass-transfer predictions more difficult. As with transfer to liquid drops, design correlations for bubbling systems are usually based on a volumetric coefficient.

With drops and bubbles the resistance to mass transfer in both phases may be significant. Diffusion inside a stagnant (noncirculating) drop is an unsteady-state process, and for diffusion the analogue of Eq. (10.19) for unsteady-state heat transfer is

$$\frac{\partial c}{\partial t} = D_v \frac{\partial^2 c}{\partial b^2} \qquad (17.77)$$

Equation (17.77) may be integrated in the same way as Eq. (10.19) to give an equation for a sphere analogous to Eq. (10.25); but for convenience in combining coefficients, an effective internal coefficient can be used, as was done for heat transfer in spheres [p. 305]:

$$k_{ci} = \frac{10 D_v}{D_p} \qquad (17.78)$$

where k_{ci} = effective internal mass-transfer coefficient
D_v = diffusivity inside drop
D_p = drop diameter

If the drop has a short lifetime, the internal coefficient will be greater than that given by Eq. (17.78), since the concentration gradient will not extend very far into the drop. If the lifetime is known, the penetration theory [Eq. (17.50)] can be used, but the breakup and coalescence of the drops in agitated systems make it hard to predict the drop lifetime. Measurements of the internal mass-transfer coefficient k_{ci} for drops of an organic liquid in a stirred extractor were consistent with the penetration theory and with drop lifetimes one-third to one-tenth as long as the batch time.[22]

SYMBOLS

A	Area perpendicular to direction of mass transfer, m^2 or ft^2
a	Area of interface between phases per unit volume of equipment, m^{-1} or ft^{-1}
B_T	Thickness of layer through which diffusion occurs, m or ft

b	Distance from phase boundary in direction of diffusion, m or ft
c	Concentration, kg mol/m^3 or lb mol/ft^3; c_A, of component A; c_{Ai}, of component A at interface; c_{A0}, at time zero; c_B, of component B
c_p	Specific heat at constant pressure, J/g·°C or Btu/lb·°F
D	Linear dimension or diameter, m or ft; D_a, of agitator; D_p, of bubble, drop, or particle; D_t, of tank
D_{AB}	Diffusivity of component A in component B; D_{BA}, of B in A
D_v	Volumetric diffusivity, m^2/h, cm^2/s, or ft^2/h; D_K, Knudsen diffusivity; D_{pore}, diffusivity in pore
d	Diameter of hollow fiber, m or ft; d_e, equivalent diameter
F_a	Faraday constant, 96,500 Coulombs/g equivalent
Fo	Fourier number [see Eq. (10.20)]
f	Fanning friction factor, dimensionless
G	Mass velocity, kg/m^2·s or lb/ft^2·h
Gz'	Graetz number for mass transfer, $\dot{m}/(D_v L \rho)$
h	Individual heat-transfer coefficient, W/m^2·°C or Btu/ft^2·h·°F
J	Mass flux relative to a plane of zero velocity, kg mol/m^2·s or lb mol/ft^2·h; J_A, J_B, of components A and B, respectively; $\bar{J}_A$, average value; $J_{A,t}$, of component A, caused by turbulent action
j_H	Colburn j factor for heat transfer, $(h/c_p G)(c_p \mu/k)^{2/3}$, dimensionless
j_M	Colburn j factor for mass transfer, $(k_y \bar{M}/G)(\mu/\rho D_v)^{2/3}$, dimensionless
K_y	Overall mass-transfer coefficient in gas phase, kg mol/m^2·s·unit mole fraction or lb mol/ft^2·h·unit mole fraction
k	Boltzmann constant, 1.380×10^{-23} J/K
k	Individual mass-transfer coefficient; k_c, cm/s or ft/s; k_{cT}, minimum coefficient for suspended particle (Fig. 17.8); k_{ci}, effective internal coefficient [Eq. (17.78)]; $\bar{k}_c$, average value over time t_T; k_g, in gas phase based on partial pressure driving force; k_x, k_y, in liquid phase and gas phase, respectively, based on mole-fraction differences, kg mol/m^2·s·unit mole fraction or lb mol/ft^2·h·unit mole fraction; also thermal conductivity, W/m·°C or Btu/ft·h·°F
k'	Effective mass-transfer coefficient in one-way diffusion; k_c', cm/s or ft/s; k_y', in gas phase, kg mol/m^2·s·unit mole fraction or lb mol/ft^2·h·unit mole fraction
L	Length of pipe or tube, m or ft
M	Molecular weight; M_A, M_B, of components A and B, respectively; $\bar{M}$, average value
m	Slope of equilibrium curve
$\dot{m}$	Mass flow rate, kg/s or lb/s
N	Mass-transfer flux across a plane or boundary, kg mol/m^2·s or lb mol/ft^2·h; N_A, N_B, of components A and B, respectively
Nu	Nusselt number, hD_p/k
P	Pressure, atm or lb$_f$/ft^2; P_A, partial pressure of component A; P_{Ai}, at interface
Pe	Peclet number, $D_p u_0/D_v$
Pr	Prandtl number, $c_p \mu/k$
R	Gas constant, 8,314 J/g mol·K or 1,545 ft·lb$_f$/lb mol·°R
Re	Reynolds number, DG/μ

r	Radius of pore, cm; also, rate of mass transfer, kg mol/m^2·s or lb mol/ft^2·h
Sc	Schmidt number, $\mu/(\rho D_v)$
Sh	Sherwood number, $k_c D/D_v$
s	Fractional rate of surface renewal, s^{-1}
T	Temperature, °C, K, °F, or °R
t	Time, s or h; t_T, residence time on transfer surface
u	Velocity, m/s or ft/s; u_A, u_B, of components A and B, respectively; u_0, volume-average velocity of phase; also velocity past a suspended bubble, drop, or particle; also superficial velocity in packed bed
$\bar{u}$	Average molecular velocity
V_A	Molar volume of solute as liquid at its normal boiling point, cm^3/g mol
x	Mole fraction in liquid or L phase; x_A, of component A; x_{Ai}, of component A at interface
y	Mole fraction in gas or V phase; y_A, of component A; y_{Ai}, of component A at interface; y_A^*, of vapor in equilibrium with liquid of composition x_A; y_B, of component B; y_i, at interface
z	Distance in vertical direction, m or ft

Greek Letters

α	Thermal diffusivity, $k/\rho c_p$, m^2/s or ft^2/h
ϵ	Void fraction, dimensionless; also Lennard–Jones force constant; ϵ_A, ϵ_B, for components A and B; $\epsilon_{AB} = \sqrt{\epsilon_A \epsilon_B}$
ε_N	Eddy diffusivity of mass, m^2/h, cm^2/s, or ft^2/h
λ	Mean free path of molecules
λ^0	Limiting ionic conductance, A/cm^2·(V/cm)·(g equivalent/cm^3); λ_+^0, of cation; λ_-^0, of anion
μ	Viscosity, Pa·s, cP, or lb/ft·s; μ_B, viscosity of water [Eq. (17.32)]; μ_w, value at wall
ν	Kinematic viscosity, μ/ρ, m^2/h, or ft^2/h
ρ	Density, kg/m^3 or lb/ft^3
ρ_M	Molar density, g mol/m^3 or lb mol/ft^3; ρ_{Mx}, of liquid; ρ_{My}, of gas
ψ	Function; ψ_1, in Eq. (17.58)
ψ_B	Association parameter for solvent [Eq. (17.31)]
Ω_D	Collision integral, $f(kT/\epsilon_{AB})$

PROBLEMS

17.1. Carbon dioxide is diffusing through nitrogen in one direction at atmospheric pressure and 0°C. The mole fraction of CO_2 at point A is 0.2; at point B, 3 m away, in the direction of diffusion, it is 0.02. Diffusivity D_v is 0.144 cm^2/s. The gas phase as a whole is stationary; that is, nitrogen is diffusing at the same rate as the carbon dioxide, but in the opposite direction. (*a*) What is the molal flux of CO_2, in kilogram moles per square meter per hour? (*b*) What is the *net* mass flux, in kilograms per square meter per hour? (*c*) At what speed, in meters per second, would an observer have to move from one point to the other so that the net mass flux, *relative to him or her*, would be zero? (*d*) At what speed would the observer have to move so that, relative to him or her, the *nitrogen* is stationary? (*e*) What would be the molal flux of carbon dioxide relative to the observer under condition (*d*)?

17.2. An open circular tank 8 m in diameter contains *n*-propanol at 25°C exposed to the atmosphere in such a manner that the liquid is covered with a stagnant air film estimated to be 5 mm thick. The concentration of propanol beyond the stagnant film is negligible. The vapor pressure of propanol at 25°C is 20 mm Hg. If propanol is worth $1.20 per liter, what is the value of the loss of propanol from this tank in dollars per day? The specific gravity of propanol is 0.80.

17.3. Ethanol vapor is being absorbed from a mixture of alcohol vapor and water vapor by means of a nonvolatile solvent in which alcohol is soluble but water is not. The temperature is 97°C, and the total pressure is 760 mm Hg. The alcohol vapor can be considered to be diffusing through a film of alcohol–water vapor mixture 0.1 mm thick. The mole percent of the alcohol in the vapor at the outside of the film is 80 percent, and that on the inside, next to the solvent, is 10 percent. The volumetric diffusivity of alcohol–water vapor mixtures at 25°C and 1 atm is 0.15 cm²/s. Calculate the rate of diffusion of alcohol vapor in kilograms per hour if the area of the film is 10 m².

17.4. An ethanol–water vapor mixture is being rectified by contact with an alcohol–water liquid solution. Alcohol is being transferred from gas to liquid and water from liquid to gas. The molal flow rates of alcohol and water are equal but in opposite directions. The temperature is 95°C and the pressure 1 atm. Both components are diffusing through a gas film 0.1 mm thick. The mole percentage of the alcohol at the outside of the film is 80 percent, and that on the inside is 10 percent. Calculate the rate of diffusion of alcohol and of water in pounds per hour through a film area of 10 m².

17.5. A wetted-wall column operating at a total pressure of 518 mm Hg is supplied with water and air, the latter at a rate of 120 g/min. The partial pressure of the water vapor in the airstream is 76 mm Hg, and the vapor pressure of the liquid-water film on the wall of the tower is 138 mm Hg. The observed rate of vaporization of water into the air is 13.1 g/min. The same equipment, now at a total pressure of 820 mm, is supplied with air at the same temperature as before and at a rate of 100 g/min. The liquid vaporized is *n*-butyl alcohol. The partial pressure of the alcohol is 30.5 mm, and the vapor pressure of the liquid alcohol is 54.5 mm. What rate of vaporization, in grams per minute, may be expected in the experiment with *n*-butyl alcohol?

17.6. Air at 40°C and 2.0 atm is passed through a shallow bed of naphthalene spheres 12 mm in diameter at a rate of 2 m/s, based on the empty cross section of the bed. The vapor pressure of naphthalene is 0.35 mm Hg. How many kilograms per hour of naphthalene will evaporate from 1 m³ of bed, assuming a bed porosity of 40 percent?

17.7. Diffusion coefficients for vapors in air can be determined by measuring the rate of evaporation of a liquid from a vertical glass tube. For a tube 0.2 cm in diameter filled with *n*-heptane at 21°C, calculate the expected rate of decrease of the liquid level when the meniscus is 1 cm from the top based on the published diffusivity of 0.071 cm²/s. At 21°C the vapor pressure and density of *n*-heptane are 0.050 atm and

0.66 g/cm^3, respectively. Would there be any advantage in using a larger-diameter tube?

17.8. Estimate the liquid-film mass-transfer coefficient for O$_2$ diffusing from an air bubble rising through water at 20°C. Choose a bubble size of 4.0 mm, assume a spherical shape, and assume rapid circulation of gas inside the bubble. Neglecting the change of bubble size with distance traveled, calculate the fraction of oxygen absorbed from the air in 1 m of travel if the water contains no dissolved oxygen.

17.9. Small spheres of solid benzoic acid are dissolved in water in an agitated tank. If the Sherwood number is nearly constant at a value of 4.0, show how the time for complete dissolution varies with the initial size of the particle. How much time would be required for 100-μm particles to dissolve completely in pure water at 25°C? Solubility: 0.43 g/100 g H$_2$O. $D_v = 1.21 \times 10^{-5}$ cm^2/s.

17.10. Estimate the diffusion coefficient and the Schmidt number for water vapor in air at 50°C and 20 mm Hg absolute pressure.

17.11. Predict the diffusivity of helium in natural gas at 5 atm and 20°C or 250°C. What is the exponent for this temperature range?

17.12. By what factor would the diffusion coefficient for sucrose in water increase as the temperature was raised from 0 to 100°C? Use two different equations and compare your results.

17.13. Predict the diffusion coefficients for dilute solutions of HCl and NaCl in water at 50°C.

17.14. Predict the terminal velocity for a 1-mm drop of nitrobenzene in water, and estimate the external mass-transfer coefficient if the drop is circulating or if the drop is stagnant.

17.15. Water saturated with air at 20°C is passed through hydrophobic hollow fibers at 50 cm/s. The fibers are 1 m long with an inner diameter of 500 μm, and vacuum is applied on the outside to remove oxygen as fast as it diffuses to the fiber wall. Estimate the mass-transfer coefficient for oxygen.

17.16. In the absorption of ammonia from air into water at 20°C, the slope of the equilibrium line is about 1.0. Estimate the fraction of the total resistance in the gas phase, assuming that the penetration theory applied to both phases.

17.17. If the liquid film resistance is 5 times that of the gas film for a gas absorption process, by how much would the rate of absorption change if the liquid film coefficient could be doubled without changing other parameters? What would be the effect of doubling the gas-film coefficient?

REFERENCES

1. Bird, R. B. *Advances in Chemical Engineering,* vol. 1. New York: Academic, 1956, pp. 156–239.
2. Brian, P. L. T., and H. B. Hales. *AIChE J.* **15:**419 (1969).

3. Chilton, T. H., and A. P. Colburn. *Ind. Eng. Chem.* **26:**1183 (1934).
4. Clough, S. B., H. E. Read, A. B. Metzner, and V. C. Behn. *AIChE J.* **8:**346 (1962).
5. Danckwerts, P. V. *Ind. Eng. Chem.* **43:**1460 (1951).
6. Davies, J. T. *Turbulence Phenomena.* New York: Academic, 1972, p. 240.
7. Friend, W. L., and A. B. Metzner. *AIChE J.* **4:**393 (1958).
8. Gilliland, E. R., and T. K. Sherwood. *Ind. Eng. Chem.* **26:**516 (1935).
9. Gupta, A. S., and G. Thodos. *Chem. Eng. Prog.* **58**(7):58 (1962).
10. Harriott, P. *AIChE J.* **8:**93 (1962).
11. Harriott, P. *Can. J Chem. Eng.* **40:**60 (1962).
12. Harriott, P. *Chem. Eng. Sci.* **17:**149 (1962).
13. Harriott, P., and R. M. Hamilton. *Chem. Eng. Sci.* **20:**1073 (1965).
14. Harriott, P., and S. V. Ho. *J. Membr. Sci.* **135:**55 (1997).
15. Hayduk, W., and H. Laudie. *AIChE J.* **20:**611 (1974).
16. Higbie, R. *Trans. AIChE* **31:**365 (1935).
17. Levins, D. J., and J. R. Glastonbury. *Trans. Inst. Chem. Eng. Lond.* **50:**132 (1972).
18. Perry, R. H., and D. W. Green (eds.). *Perry's Chemical Engineers' Handbook,* 7th ed. New York: McGraw-Hill, 1997, p. 5-48.
19. Reid, R. C., J. M. Prausnitz, and B. E. Poling. *The Properties of Gases and Liquids,* 4th ed. New York: McGraw-Hill, 1987, a, p. 582, b, p. 620.
20. Schlichting, H. *Boundary Layer Theory,* 7th ed. New York: McGraw-Hill, 1979, pp. 303–4.
21. Sherwood, T. K., R. L. Pigford, and C. R. Wilke. *Mass Transfer.* New York: McGraw-Hill, 1975: (*a*) p. 169, (*b*) p. 242.
22. Skelland, A. H. P., and H. Xien. *Ind. Eng. Chem. Res.* **29:**415 (1990).
23. Whitman, W. G. *Chem. Met. Eng.* **29:**146 (1923).
24. Wilke, C. R., and P. Chang. *AIChE J.* **1:**264 (1955).
25. Yang, M. C., and E. L. Cussler. *AIChE J.* **32:**1910 (1986).

CHAPTER 18

Gas Absorption

This chapter deals with the mass-transfer operations known as *gas absorption* and *stripping,* or *desorption.* In gas absorption a soluble vapor is absorbed from its mixture with an inert gas by means of a liquid in which the solute gas is more or less soluble. A major application of absorption technology is the removal of CO_2 and H_2S from natural gas or synthesis gas by absorption in solutions of amines or alkaline salts. Another example is the washing of ammonia from a mixture of ammonia and air by means of liquid water. The solute is subsequently recovered from the liquid by distillation, and the absorbing liquid can be either discarded or reused. Sometimes a solute is removed from a liquid by bringing the liquid into contact with an inert gas; such an operation, the reverse of gas absorption, is desorption or gas stripping.

PACKINGS AND PACKED TOWER DESIGN

A common apparatus used in gas absorption and certain other operations is the packed tower, an example of which is shown in Fig. 18.1. The device consists of a cylindrical column, or tower, equipped with a gas inlet and distributing space at the bottom; a liquid inlet and distributor at the top; gas and liquid outlets at the top and bottom, respectively; and a supported mass of inert solid shapes, called *tower packing.* The packing support is typically a screen, corrugated to give it strength, with a large open area so that flooding does not occur at the support. The inlet liquid, which may be pure solvent or a dilute solution of solute in the solvent and which is called the *weak liquor,* is distributed over the top of the packing by the distributor and, in ideal operation, uniformly wets the surfaces of the packing. The distributor shown in Fig. 18.1 is a set of perforated pipes. In large towers, spray nozzles or distributor plates with overflow weirs are more common. For very large towers, up to 9 m (30 ft) in diameter, Nutter Engineering advertises a plate distributor with individual drip tubes.

546

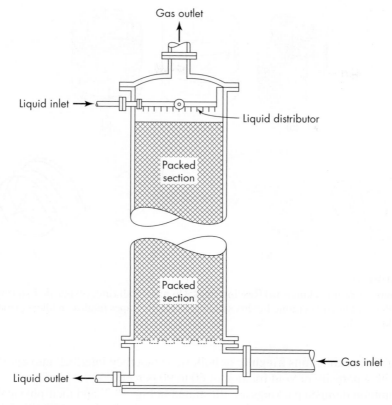

FIGURE 18.1
Packed tower.

The solute-containing gas, or rich gas, enters the distributing space below the packing and flows upward through the interstices in the packing countercurrent to the flow of the liquid. The packing provides a large area of contact between the liquid and gas and encourages intimate contact between the phases. The solute in the rich gas is absorbed by the fresh liquid entering the tower, and dilute, or lean, gas leaves the top. The liquid is enriched in solute as it flows down the tower, and concentrated liquid, called the *strong liquor,* leaves the bottom of the tower through the liquid outlet.

Tower packings are divided into three principal types: those that are dumped at random into the tower, those that must be stacked by hand, and those known as structured or ordered packings. Dumped packings consist of units 6 to 75 mm ($\frac{1}{4}$ to 3 in.) in major dimension; packings smaller than 25 mm are used mainly in laboratory or pilot-plant columns. In stacked packings the units are 50 to 200 mm (2 to 8 in.) in size. They are much less commonly used than dumped packings and are not discussed here.

Dumped tower packings are made of cheap, inert materials such as clay, porcelain, or various plastics. Thin-walled metal rings of steel or aluminum are sometimes used. High void spaces and large passages for the fluids are achieved by

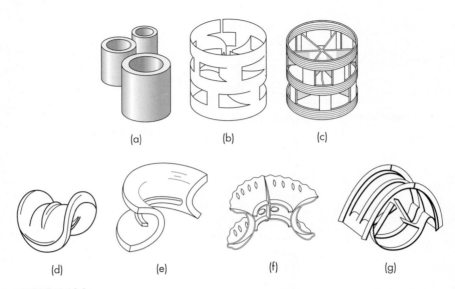

(a)   (b)   (c)

(d)   (e)   (f)   (g)

FIGURE 18.2
Common tower packings: (*a*) Raschig rings; (*b*) metal Pall ring; (*c*) plastic Pall ring;
(*d*) Berl saddle; (*e*) ceramic Intalox saddle; (*f*) plastic Super Intalox saddle; (*g*) metal
Intalox saddle.

making the packing units irregular or hollow, so that they interlock into open structures with a porosity or void fraction of 60 to 90 percent.

Common dumped packings are illustrated in Fig. 18.2, and their physical characteristics are given in Table 18.1. Ceramic Berl saddles and Raschig rings are older
types of packing that are not much used now, although they were a big improvement
over ceramic spheres or crushed stone when first introduced. Intalox saddles are
somewhat like Berl saddles, but the shape prevents pieces from nesting closely
together, and this increases the bed porosity. Super Intalox saddles are a slight variation with scalloped edges; they are available in plastic or ceramic form. Pall rings
are made from thin metal with portions of the wall bent inward or from plastic with
slots in the wall and stiffening ribs inside. Hy-pak metal packing and Flexirings
(not shown) are similar in shape and performance to metal Pall rings. Beds of Pall
rings have over 90 percent void fraction and lower pressure drop than most other
packings of the same nominal size. Norton's new IMTP (Intalox Metal Tower
Packing) has a very open structure and even lower pressure drop than Pall rings.
Additional pressure-drop packing factors for many commercial packings are given
by Robbins[13] and, in SI units, by Perry.[12]

Structured packings with ordered geometry evolved from the Stedman packing
of the late 1930s,[20] but they found few industrial uses until the Sulzer packing was
developed about 1965.[18] Early structured packings were fabricated from wire gauze;
most current ones are made of sheets of perforated corrugated metal, with adjacent
sheets arranged so that liquid spreads over their surfaces while vapor flows through
channels formed by the corrugations. The channels are set at an angle of 45° with
the horizontal; the angle alternates in direction in successive layers, as shown

TABLE 18.1
Characteristics of dumped tower packings [8, 11b, 21]

Type	Material	Nominal size, in.	Bulk density,[†] lb/ft³	Total area,[†] ft²/ft³	Porosity ε	Packing factors[‡] F_p	f_p
Raschig rings	Ceramic	$\frac{1}{2}$	55	112	0.64	580	1.52§
		1	42	58	0.74	155	1.36§
		$1\frac{1}{2}$	43	37	0.73	95	1.0
		2	41	28	0.74	65	0.92§
Pall rings	Metal	1	30	63	0.94	56	1.54
		$1\frac{1}{2}$	24	39	0.95	40	1.36
		2	22	31	0.96	27	1.09
	Plastic	1	5.5	63	0.90	55	1.36
		$1\frac{1}{2}$	4.8	39	0.91	40	1.18
Berl saddles	Ceramic	$\frac{1}{2}$	54	142	0.62	240	1.58§
		1	45	76	0.68	110	1.36§
		$1\frac{1}{2}$	40	46	0.71	65	1.07§
Intalox saddles	Ceramic	$\frac{1}{2}$	46	190	0.71	200	2.27
		1	42	78	0.73	92	1.54
		$1\frac{1}{2}$	39	59	0.76	52	1.18
		2	38	36	0.76	40	1.0
		3	36	28	0.79	22	0.64
Super Intalox saddles	Ceramic	1	—	—	—	60	1.54
		2	—	—	—	30	1.0
IMTP	Metal	1	—	—	0.97	41	1.74
		$1\frac{1}{2}$	—	—	0.98	24	1.37
		2	—	—	0.98	18	1.19
Hy-Pak	Metal	1	19	54	0.96	45	1.54
		$1\frac{1}{2}$	—	—	—	29	1.36
		2	14	29	0.97	26	1.09

[†]Bulk density and total area are given per unit volume of column.
[‡]Factor F_p is a pressure drop factor and f_p a relative mass-transfer coefficient. Factor f_p is discussed on page 584 in the paragraph "Performance of Other Packings." Its use is illustrated in Example 18.7.
§Based on NH_3–H_2O data; other factors based on CO_2–NaOH data.

schematically in Fig. 18.3. Each layer is a few inches thick. Various proprietary packings differ in the size and arrangement of the corrugations and the treatment of the packing surfaces.[3] Typically the triangular corrugations are 25 to 40 mm across the base, 17 to 25 mm on the side, and 10 to 15 mm high. The porosity ranges from 0.93 to 0.97, and the specific surface area from 60 to 76 ft²/ft³ (200 to 250 m²/m³). The Sulzer BX packing, fabricated from metal gauze, provides a specific surface area of 152 ft²/ft³ (500 m²/m³) with a porosity of 0.90.

Contact between liquid and gas

The requirement of good contact between liquid and gas is the hardest to meet, especially in large towers. Ideally the liquid, once distributed over the top of the packing, flows in thin films over all the packing surface all the way down the

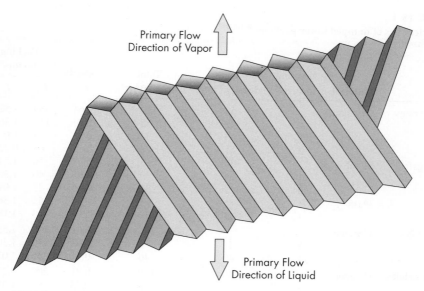

Primary Flow
Direction of Vapor

Primary Flow
Direction of Liquid

FIGURE 18.3
Schematic drawing of structured packing.

tower. Actually the films tend to grow thicker in some places and thinner in others, so that the liquid collects into small rivulets and flows along localized paths through the packing. Especially at low liquid rates much of the packing surface may be dry or, at best, covered by a stagnant film of liquid. This effect is known as *channeling;* it is the chief reason for the poor performance of large packed towers.

Channeling is severe in towers filled with stacked packing, which is the main reason they are not much used. It is less severe in dumped packings. In towers of moderate size, channeling can be minimized by having the diameter of the tower at least 8 times the packing diameter. If the ratio of tower diameter to packing diameter is less than 8 to 1, the liquid tends to flow out of the packing and down the walls of the column. Even in small towers filled with packings that meet this requirement, however, liquid distribution and channeling have a major effect on column performance.[5] In large towers the initial distribution is especially important,[10] but even with good initial distribution it is necessary to include redistributors for the liquid every 5 to 10 m in the tower, immediately above each packed section. Improved liquid distribution has made possible the effective use of packed towers as large as 9 m (30 ft) in diameter.

Pressure drop and limiting flow rates

Figure 18.4 shows typical data for the pressure drop in a packed tower. The pressure drop per unit packing depth comes from fluid friction; it is plotted on

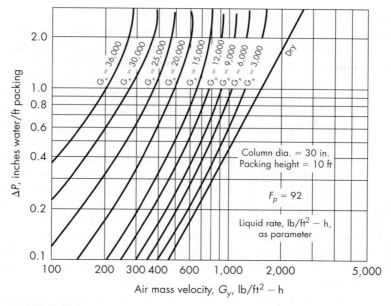

FIGURE 18.4

Pressure drop in a packed tower for air-water system with 1-in. Intalox saddles. $(1{,}000\ \text{lb/ft}^2 \cdot \text{h} = 1.356\ \text{kg/m}^2 \cdot \text{s}; 1\ \text{in.}\ H_2O/\text{ft} = 817\ \text{Pa/m})$

logarithmic coordinates versus the gas flow rate G_y, expressed in mass of gas per hour per unit of cross-sectional area, based on the empty tower. Therefore, G_y is related to the superficial gas velocity by the equation $G_y = u_0\,\rho_y$, where ρ_y is the density of the gas. When the packing is dry, the line so obtained is straight and has a slope of about 1.8. The pressure drop therefore increases with the 1.8 power of the velocity. If the packing is irrigated with a constant flow of liquid, the relationship between pressure drop and gas flow rate initially follows a line parallel to that for dry packing. The pressure drop is greater than that in dry packing, because the liquid in the tower reduces the space available for gas flow. The void fraction, however, does not change with gas flow. At moderate gas velocities the line for irrigated packing gradually becomes steeper, because the gas now impedes the downflowing liquid and the liquid holdup increases with gas rate. The point at which the liquid holdup starts to increase, as judged by a change in the slope of the pressure drop line, is called the *loading point*. However, as is evident from Fig. 18.4, it is not easy to get an accurate value for the loading point.

With still further increase in gas velocity, the pressure drop rises even more rapidly, and the lines become almost vertical when the pressure drop is about 2 to 3 in. of water per foot of packing (150 to 250 mm of water per meter). In local regions of the column, the liquid becomes the continuous phase, and the flooding point has been reached. Higher gas flows can be used temporarily, but then liquid rapidly accumulates, and the entire column may fill with liquid.

The gas velocity in an operating packed column must obviously be lower than the flooding velocity. However, as flooding is approached, most of or all the packing surface is wetted, maximizing the contact area between gas and liquid. The designer must choose a velocity far enough from the flooding velocity to ensure safe operation but not so low as to require a much larger column. Lowering the design velocity increases the tower diameter without much change in the required height, since lower gas and liquid velocities lead to a nearly proportional reduction in mass-transfer rate. Decreased pressure drop is one benefit of low gas velocity, but the cost of power consumed is usually not a major factor in optimizing the design. The gas velocity is sometimes chosen as one-half the predicted flooding velocity obtained from a generalized correlation. This might seem too conservative, but there is considerable scatter in published data for flooding velocities, and the generalized correlations are not very accurate. A closer approach to flooding may be used if detailed performance data are available for the packing selected. Packed towers may also be designed on the basis of a definite pressure drop per unit height of packing.

The flooding velocity depends strongly on the type and size of packing and the liquid mass velocity. Figure 18.5 shows data for Intalox saddles taken from Fig. 18.4 and similar curves for other sizes. Flooding was assumed to occur at a pressure drop of 2.0 in. H_2O/ft of packing, since the pressure drop curves are vertical or nearly so at this point. For low liquid rates the flooding velocity varies with about the -0.2 to -0.3 power of the liquid rate and the 0.6 to 0.7 power of the packing size. The effects of liquid rate and packing size become more pronounced at high liquid mass velocities.

Several generalized correlations have been proposed for the pressure drop and flooding velocity in packed columns. Most of these use a log-log plot with $(G_x/G_y)(\rho_y/\rho_x)^{0.5}$ on the abscissa and a function containing G_y^2 on the ordinate. Usually the flow ratio G_x/G_y is set from equilibrium and economic considerations as explained later in this chapter, and G_y can then be determined directly, whereas trial-and-error solution is needed if G_y and G_x are on separate axes, as in Fig. 18.5. The packing characteristics are accounted for by a packing factor F_p, which decreases with increasing packing size or increasing void fraction. Packing factors cannot be predicted from theory using the Ergun equation [Eq. (7.22)] because of the complex shapes, and they are determined empirically. Unfortunately, no single correlation for pressure drop gives a good fit to all packings, and values of F_p based on fitting the data for low pressure drops may differ significantly from values obtained by fitting the data for high pressure drops or by fitting the flooding data.

A widely used correlation for estimating pressure drops in dumped packings is given in Fig. 18.6 where G_x and G_y are in lb/ft$^2\cdot$s, μ_x is in cP, ρ_x and ρ_y are in lb/ft^3, and g_c is 32.174 lb $\cdot$ ft/lb$_f\cdot$s^2. Earlier versions of this correlation included a flooding line above the line for $\Delta P = 1.5$ in. H_2O/ft of packing, but recent studies show flooding at pressure drops of only 0.7 to 1.5 in. H_2O/ft of packing for 2- or 3-in. packings. An empirical equation for the limiting pressure drop is[6]

$$\Delta P_{\text{flood}} = 0.115 F_p^{0.7} \tag{18.1}$$

where ΔP_{flood} = pressure drop at flooding, in. H_2O/ft of packing

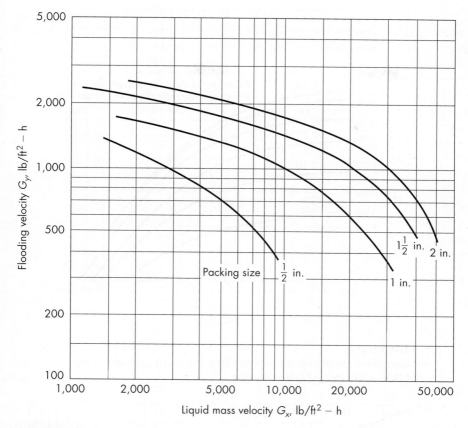

FIGURE 18.5
Flooding velocities in ceramic Intalox saddles, air-water system. ($1,000 \text{ lb/ft}^2 \cdot \text{h} = 1.356 \text{ kg/m}^2 \cdot \text{s}$).

$F_p = $ packing factor, dimensionless

Equation (18.1) can be used for packing factors from 10 to 60. For higher values of F_p, the pressure drop at flooding can be taken as 2.0 in H_2O/ft.

An alternate correlation for the pressure drop in packed columns was proposed by Strigle[21] and is shown in Fig. 18.7. The abscissa is essentially the same as for Fig. 18.6, but the ordinate includes the capacity factor $C_s = u_0\sqrt{\rho_y/(\rho_x - \rho_y)}$, where u_0 is the superficial velocity in feet per second. The kinematic viscosity of the liquid v is in centistokes. The semilog plot permits easier interpolation than the log-log plot, though both correlations are based on the same set of data.

EXAMPLE 18.1. A tower packed with 1-in. (25.4-mm) ceramic Intalox saddles is to be built to treat 25,000 ft^3 (708 m^3) of entering gas per hour. The ammonia content of the entering gas is 2% by volume. Ammonia-free water is used as absorbent. The temperature is 68°F (20°C), and the pressure is 1 atm. The ratio of gas flow to liquid flow is 1 lb of gas per pound of liquid. (*a*) If the gas velocity is to be one-half the flooding velocity,

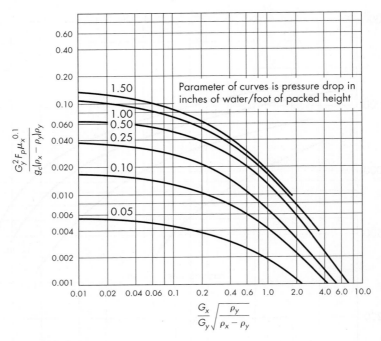

FIGURE 18.6

Generalized correlation for pressure drop in packed columns. (1 in. $H_2O/ft = 817$ Pa/m) (*After Eckert.*[2])

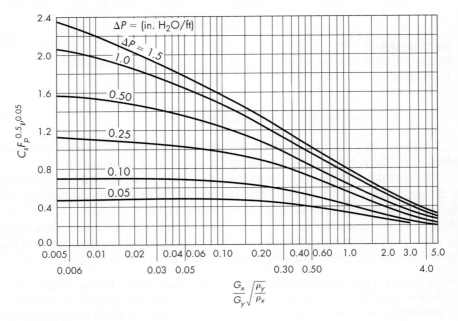

FIGURE 18.7

Alternate generalized pressure drop correlation. (1 in. $H_2O/ft = 817$ Pa/m)

what should be the diameter of the tower? (b) What is the pressure drop if the packed section is 20 ft (6.1 m) high?

Solution. The average molecular weight of the entering gas is $29 \times 0.98 + 0.02 \times 17 = 28.76$. Then

$$\rho_y = \frac{28.76 \times 492}{359(460 + 68)} = 0.07465 \text{ lb/ft}^3$$

(a) Use Fig. 18.5. When $G_y = G_x$ for 1-in. Intalox saddles, $G_y = 1{,}700$ lb/ft$^2 \cdot$ h or 0.472 lb/ft$^2 \cdot$ s. The design value is then $0.472/2 = 0.236$ lb/ft$^2 \cdot$ s. The total gas flow is $25{,}000 \times 0.07465/3{,}600 = 0.518$ lb/s. The cross-sectional area of the tower is

$$S = \frac{0.518}{0.236} = 2.19 \text{ ft}^2$$

The diameter of the tower is $\sqrt{2.19/0.7854} = 1.67$ ft (509 mm).

(b) Use Fig. 18.4. The pressure drop for $G_x = G_y = 850$ lb/ft$^2 \cdot$ h is 0.35 in. H_2O/ft. The total pressure drop is then $20 \times 0.35 = 7.0$ in. H_2O (13.1 mm Hg).

This may be checked using Fig. 18.6. The additional quantities needed are

$$\rho_x = 62.3 \text{ lb/ft}^3 \qquad\qquad \mu_x = 1 \text{ cP}$$

$$g_c = 32.174 \text{ ft} \cdot \text{lb/lb}_f \cdot \text{s}^2 \qquad \frac{G_x}{G_y} = 1$$

For 1-in. ceramic Intalox saddles, $F_p = 92$ (Table 18.1). Also

$$\frac{G_x}{G_y} \sqrt{\frac{\rho_y}{\rho_x - \rho_y}} = \sqrt{\frac{0.07465}{62.3 - 0.07}} = 0.0346$$

Note that the ordinate value in Fig. 18.6 is not dimensionless, and the stated units must be used. The ordinate value in Fig. 18.6 is

$$\frac{G_y^2 F_p \mu_x^{0.1}}{g_c(\rho_x - \rho_y)\rho_y} = \frac{0.236^2 \times 92 \times 1^{0.1}}{32.174 \times (62.3 - 0.07465) \times 0.07465}$$

$$= 0.0343$$

For these conditions the predicted pressure drop is about 0.23 in. H_2O per foot of packed height; the total pressure drop is $20 \times 0.23 = 4.6$ in. H_2O (8.6 mm Hg). The actual pressure drop would probably be close to the higher value found from Fig. 18.4, which is based on direct measurements for the specific packing used. (The generalized correlation in Fig. 18.6 is in closer agreement with Fig. 18.4 when $G_x/G_y \gg 1$.)

Structured packings

Flooding velocities in several commercial structured packings are shown in Fig. 18.8. The pressure drop in structured packings may be predicted from some rather complicated equations given by Fair and Bravo,[3] but relatively little experimental information is available. Towers containing structured packing are best designed in collaboration with the packing manufacturer. Spiegel and Meier[19] state that most structured packings reach their maximum capacity at a pressure drop of

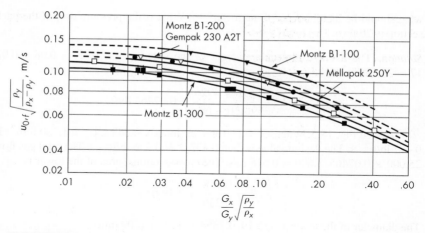

FIGURE 18.8
Flooding velocities in structured packings.[3]

about 1,000 (N/m²)/m (1.22 in. H₂O per ft), at a vapor velocity of 90 to 95 percent of the flooding velocity.

EXAMPLE 18.2. Substitute Gempak 230 A2T packing, manufactured by Glitsch, for the 1-in. Intalox saddles in the tower of Example 18.1. (*a*) What increase in capacity is expected? (*b*) Will the pressure drop at 50 percent of the flooding velocity be greater or less than in the Intalox saddles?

Solution

(*a*) Use Fig. 18.8. From Example 18.1, $G_x/G_y = 1$ and

$$\frac{G_x}{G_y}\sqrt{\frac{\rho_y}{\rho_x}} = \sqrt{\frac{0.07465}{62.3}} = 0.0346$$

From Fig. 18.8, $u_{0,f}\sqrt{(\rho_y/(\rho_x - \rho_y))} = 0.11$. The superficial vapor velocity at flooding is therefore

$$u_{0,f} = 0.11\sqrt{\frac{62.3 - 0.07}{0.07465}} = 3.175 \text{ m/s}$$

The allowable vapor velocity, at 50 percent of flooding, is

$$u_0 = \frac{3.175}{2} = 1.588 \text{ m/s or } 5.21 \text{ ft/s}$$

The corresponding mass velocity is

$$G_y = 5.21 \times 0.07465 = 0.389 \text{ lb/ft}^2 \cdot \text{s}$$

The allowable mass velocity in Example 18.1 was 0.236 lb/ft²·s. The increase by using structured packing is $0.389/0.236 - 1 = 0.65$, or 65 percent.

(b) Using Spiegel and Meier's value of 1.22 in. H_2O/ft at 90 percent of flooding, the pressure drop in 20 ft of packing at 50 percent of flooding would be no greater than

$$\Delta P = 20 \times 1.22 \times \left(\frac{0.50}{0.90}\right)^{1.8}$$

$$= 8.47 \text{ in. } H_2O$$

This is 1.2 times the pressure drop of 7 in. H_2O in the Intalox saddles. The actual pressure drop in the structured packing should be less than this, but it probably would still be somewhat greater than in the saddles.

PRINCIPLES OF ABSORPTION

As shown in the previous section, the diameter of a packed absorption tower depends on the quantities of gas and liquid handled, their properties, and the ratio of one stream to the other. The height of the tower, and hence the total volume of packing, depends on the magnitude of the desired concentration changes and on the rate of mass transfer per unit of packed volume. Calculations of the tower height, therefore, rest on material balances, enthalpy balances, and estimates of driving force and mass-transfer coefficients.

Material balances

In a differential-contact plant such as the packed absorption tower illustrated in Fig. 18.9, variations in composition are continuous from one end of the equipment to the other. Material balances for the portion of the column above an arbitrary section, as shown by the dashed line in Fig. 18.9, are as follows:

Total material: $\qquad\qquad L_a + V = L + V_a \qquad\qquad\qquad$ (18.2)

Component A: $\qquad\qquad L_a x_a + Vy = Lx + V_a y_a \qquad\qquad$ (18.3)

where V is the molal flow rate of the gas phase and L that of the liquid phase at the same point in the tower. The L-phase and V-phase concentrations x and y apply to this same location.

The overall material-balance equations, based on the terminal streams, are

Total material: $\qquad\qquad L_a + V_b = L_b + V_a \qquad\qquad\qquad$ (18.4)

Component A: $\qquad\qquad L_a x_a + V_b y_b = L_b x_b + V_a y_a \qquad\qquad$ (18.5)

The relationship between x and y at any point in the column, obtained by rearranging Eq. (18.3), is called the operating-line equation

$$y = \frac{L}{V}x + \frac{V_a y_a - L_a x_a}{V} \qquad\qquad (18.6)$$

The operating line can be plotted on an arithmetic graph along with the equilibrium curve, as shown in Fig. 18.10. The operating line must lie above the equilibrium line in order for absorption to take place, since this gives a positive driving force $y - y^*$ for absorption.

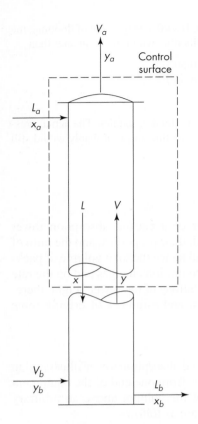

FIGURE 18.9
Material-balance diagram for packed column.

In Eq. (18.6), x and y represent the bulk compositions of the liquid and gas, respectively, in contact with each other at any given section through the column. It is assumed that the compositions at a given elevation are independent of position in the packing. The absorption of a soluble component from a gas mixture makes the total gas rate V decrease as the gas passes through the column, and the flow of liquid L increases. These changes make the operating line slightly curved, as shown in Fig. 18.10. For dilute mixtures, containing less than 10% of soluble gas, the effect of changes in total flow is usually ignored and the design is based on the average flow rates.

Limiting gas-liquid ratio

Equation (18.6) shows that the average slope of the operating line is L/V, the ratio of the molal flows of liquid and gas. Thus, for a given gas flow, a reduction in liquid flow decreases the slope of the operating line. Consider the operating line ab in Fig. 18.10. Assume that the gas rate and the terminal concentrations x_a, y_a, and y_b are held constant and the liquid flow L decreased. The upper end of the operating line then shifts in the direction of the equilibrium line, and x_b, the concentration of the strong liquor, increases. The maximum possible liquor concentration and the minimum possible liquid rate are obtained when the operating line just touches the equilibrium line, as shown by line ab' in Fig. 18.10. At this condition, an infinitely

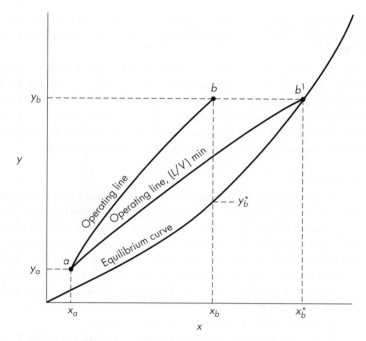

FIGURE 18.10
Limiting gas-liquid ratio.

deep packed section is necessary, as the concentration difference for mass transfer becomes zero at the bottom of the tower. In any actual tower the liquid rate must be greater than this minimum to achieve the specified change in gas composition.

The L/V ratio is important in the economics of absorption in a countercurrent column. The driving force for mass transfer is $y - y^*$, which is proportional to the vertical distance between the operating line and the equilibrium line on a diagram such as Fig. 18.10. Increasing L/V increases the driving force everywhere in the column except at the very top, and the absorption column does not need to be as tall. However, using a larger amount of liquid gives a more dilute liquid product, which makes it more difficult to recover the solute by desorption or stripping. The energy cost for stripping is often a major part of the total cost of an absorption-stripping operation. The optimum liquid rate for absorption is found by balancing the operating costs for both units against the fixed costs of the equipment. In general, the liquid rate for the absorber should be between 1.1 and 1.5 times the minimum rate, unless the liquid is to be discarded and not regenerated.

The conditions at the top of the absorber are often design variables that also have to be set, considering the balance between equipment and operating costs. For example, if tentative specifications call for 98 percent recovery of a product from a gas stream, the designer might calculate how much taller the column would have to be to get 99 percent recovery. If the value of the extra product recovered exceeds the extra costs, the optimum recovery is at least 99 percent, and the calculation should be repeated for even higher recovery. If the unremoved solute is a pollutant, its concentration in the vent gas may be set by emission standards, and the required percent recovery may exceed the optimum value based on product value and operating costs.

The diagram in Fig.18.10 shows a significant concentration of solute in the liquid fed to the column, and 99 percent removal from the gas would not be possible for this case. However, a lower value of x_a could be obtained by better stripping or more complete regeneration of the absorbing liquid. The value of x_a could be optimized, considering the extra equipment and operating costs for more complete regeneration and the savings from better operation of the absorber.

Rate of absorption

The rate of absorption can be expressed in four different ways using individual coefficients or overall coefficients based on the gas or liquid phases. Volumetric coefficients are used for most calculations, because it is more difficult to determine the coefficients per unit area and because the purpose of the design calculation is generally to determine the total absorber volume. In the following treatment the correction factors for one-way diffusion are omitted for simplicity, and the changes in gas and liquid flow rates are neglected. The equations are strictly valid only for lean gases but can be used with little error for mixtures with up to 10% solute. Absorption from rich gases is treated later as a special case.

The rate of absorption per unit volume of packed column is given by any of the following equations, where y and x refer to the mole fraction of the component being absorbed:

$$r = k_y a(y - y_i) \tag{18.7}$$

$$r = k_x a(x_i - x) \tag{18.8}$$

$$r = K_y a(y - y^*) \tag{18.9}$$

$$r = K_x a(x^* - x) \tag{18.10}$$

The individual coefficients $k_y a$ and $k_x a$ are based on a unit volume, as are the overall coefficients $K_y a$ and $K_x a$. The a in all these coefficients is the interfacial area per unit volume of the packed column or other device. It is hard to measure or to predict a, but in most cases it is not necessary to know its actual value since design calculations can be based on the volumetric coefficients.

The interface composition (y_i, x_i) can be obtained from the operating-line diagram using Eqs. (18.7) and (18.8):

$$\frac{y - y_i}{x_i - x} = \frac{k_x a}{k_y a} \tag{18.11}$$

Thus a line drawn from the operating line with a slope $-k_x a / k_y a$ will intersect the equilibrium line at (y_i, x_i), as shown in Fig. 18.11. Usually it is not necessary to know the interface compositions, but these values are used for calculations involving rich gases or when the equilibrium line is strongly curved.

The overall driving forces are easily determined as vertical or horizontal lines on the yx diagram. The overall coefficients are obtained from $k_y a$ and $k_x a$, using the

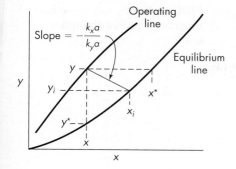

FIGURE 18.11
Location of interface compositions.

local slope of the equilibrium curve m, as was shown in Chap.17 [Eq. (17.57)]:

$$\frac{1}{K_y a} = \frac{1}{k_y a} + \frac{m}{k_x a} \tag{18.12}$$

$$\frac{1}{K_x a} = \frac{1}{k_x a} + \frac{1}{m k_y a} \tag{18.13}$$

In Eq. (18.12), the terms $1/(k_y a)$ and $m/(k_x a)$ are the resistances to mass transfer in the gas film and the liquid film, respectively. When the coefficients $k_y a$ and $k_x a$ are of the same order of magnitude,[†] and m is very much greater than 1.0, the liquid-film resistance is said to be controlling. This means that any change in $k_x a$ has a nearly proportional effect on both $K_y a$ and $K_x a$ and on the rate of absorption, whereas a change in $k_y a$ has little effect. For example, Henry's law coefficient for CO_2 in water at 20°C is 1,430 atm/mol fraction, which corresponds to $m = 1,430$ for absorption at 1 atm and $m = 143$ for absorption at 10 atm. Under these conditions the absorption of CO_2 in water is clearly liquid-film controlled. Increasing the gas velocity would increase $k_y a$ but have a negligible effect on $K_y a$. Increasing the liquid velocity would increase the interfacial area a and probably also increase k_x, both leading to an increase in $k_x a$ and $K_y a$.

When the solubility of the gas is very high, such as with HCl in water, m is very small and the gas-film resistance controls the rate of absorption. With gases of intermediate solubility both resistances are important, but the term *controlling resistance* is sometimes used for the larger resistance. The absorption of NH_3 in water is often cited as an example of gas-film control, since the gas film has about 80 to 90 percent of the total resistance.

Calculation of tower height

An absorber can be designed using any of the four basic rate equations, but the gas-film coefficients are often used, and the use of $K_y a$ will be emphasized here. Choosing a gas-film coefficient does not require any assumption about the controlling resistance. Even if the liquid film controls, a design based on $K_y a$ is as simple and accurate as one based on $K_x a$ or $k_x a$.

[†] In packed columns $k_x a$ is generally 3 to 20 times $k_y a$.

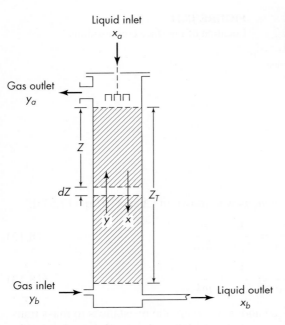

FIGURE 18.12
Diagram of packed absorption tower.

Consider the packed column shown in Fig. 18.12. The cross section is S, and the differential volume in height dZ is $S\,dZ$. If the change in molar flow rate V is neglected, the amount absorbed in section dZ is $-V\,dy$, which equals the absorption rate times the differential volume:

$$-V\,dy = K_y a(y - y^*)S\,dZ \tag{18.14}$$

This equation is rearranged for integration, grouping the constant factors V, S, and $K_y a$ with dZ and reversing the limits of integration to eliminate the minus sign:

$$\frac{K_y a S}{V} \int dZ = \frac{K_y a S Z_T}{V} = \int_a^b \frac{dy}{y - y^*} \tag{18.15}$$

The right-hand side of Eq. (18.15) can be integrated directly for certain cases, or it can be determined numerically. We will examine some of these cases.

Number of transfer units

The equation for column height can be written as follows:

$$Z_T = \frac{V/S}{K_y a} \int_a^b \frac{dy}{y - y^*} \tag{18.16}$$

The integral in Eq. (18.16) represents the change in vapor concentration divided by the average driving force and is called the *number of transfer units* (NTU) N_{Oy}. This is analogous to the number of heat-transfer units N_H, defined in Eq. (15.9). The subscripts show that N_{Oy} is based on the overall driving force for the gas phase. The other part of Eq. (18.16) has the units of length and is called the *height of a transfer unit* (HTU) H_{Oy}. Thus a simple design method is to determine

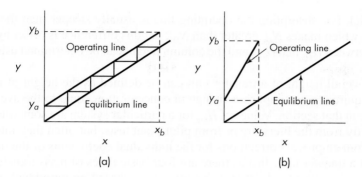

FIGURE 18.13
Relationship between number of transfer units (NTU) and number
of theoretical plates (NTP): (a) NTU = NTP; (b) NTU > NTP.

N_{Oy} from the yx diagram and multiply it by H_{Oy} obtained from the literature or calculated from mass-transfer correlations:

$$Z_T = H_{Oy} N_{Oy} \tag{18.17}$$

The number of transfer units is somewhat like the number of ideal stages, discussed in Chap. 20, but the values are equal only if the operating line and equilibrium line are straight and parallel, as in Fig. 18.13a. For this case,

$$N_{Oy} = \frac{y_b - y_a}{y - y^*} \tag{18.18}$$

In Fig. 18.13a there are about four ideal stages and four transfer units.

When the operating line is straight but steeper than the equilibrium line, as in
Fig. 18.13b, the number of transfer units is greater than the number of ideal stages.
Note that for the example shown, the driving force at the bottom is $y_b - y_a$, the
same as the change in vapor concentration across the tower, which has one ideal,
stage. However, the driving force at the top is y_a, which is several-fold smaller, so
the average driving force is much less than $y_b - y_a$. The proper average can be
shown to be the logarithmic mean of the driving forces at the two ends of the
column.

For straight operating and equilibrium lines, the number of transfer units is the
change in concentration divided by the logarithmic mean driving force:

$$N_{Oy} = \frac{y_b - y_a}{\overline{\Delta y_L}} \tag{18.19}$$

where $\overline{\Delta y_L}$ is the logarithmic mean of $y_b - y_b^*$ and $y_a - y_a^*$. Equation (18.19) is
based on the gas phase. The corresponding equation based on the liquid phase is

$$N_{Ox} = \frac{x_b - x_a}{\overline{\Delta x_L}} \tag{18.20}$$

The number of liquid-phase transfer units N_{Ox} is not the same as the number of
gas-phase transfer units N_{Oy} unless the operating and equilibrium lines are straight

and parallel. For absorption the operating line is usually steeper than the equilibrium line, which makes N_{Oy} greater than N_{Ox}, but this difference is offset by the difference between H_{Oy} and H_{Ox}, and the column height can be determined using either approach.

The overall height of a transfer unit can be defined as the height of a packed section required to accomplish a change in concentration equal to the average driving force in that section. Values of H_{Oy} for a particular system are sometimes available directly from the literature or from pilot-plant tests, but often they must be estimated from empirical correlations for the individual coefficients or the individual heights of a transfer unit. Just as there are four basic types of mass-transfer coefficients, there are four kinds of transfer units, those based on individual or overall driving forces for the gas and liquid phases. These are as follows:

Gas film: $$H_y = \frac{V/S}{k_y a} \qquad N_y = \int \frac{dy}{y - y_i} \qquad (18.21)$$

Liquid film: $$H_x = \frac{L/S}{k_x a} \qquad N_x = \int \frac{dx}{x_i - x} \qquad (18.22)$$

Overall gas: $$H_{Oy} = \frac{V/S}{K_y a} \qquad N_{Oy} = \int \frac{dy}{y - y^*} \qquad (18.23)$$

Overall liquid: $$H_{Ox} = \frac{L/S}{K_x a} \qquad N_{Ox} = \int \frac{dx}{x^* - x} \qquad (18.24)$$

Alternate forms of transfer coefficients

The gas-film coefficients reported in the literature are often based on a partial-pressure driving force instead of a mole-fraction difference and are written as $k_g a$ or $K_g a$. Their relationships to the coefficients used heretofore are simply $k_g a = k_y a/P$ and $K_g a = K_y a/P$, where P is the total pressure. The units of $k_g a$ and $K_g a$ are commonly $\text{mol/ft}^3 \cdot \text{h} \cdot \text{atm}$. Similarly liquid-film coefficients may be given as $k_L a$ or $K_L a$, where the driving force is a volumetric concentration difference; k_L is therefore the same as k_c defined by Eq. (17.36). Thus $k_L a$ and $K_L a$ are equal to $k_x a/\rho_M$ and $K_x a/\rho_M$, respectively, where ρ_M is the molar density of the liquid. The units of $k_L a$ and $K_L a$ are usually $\text{mol/ft}^3 \cdot \text{h} \cdot (\text{mol/ft}^3)$ or h^{-1}.

If G_y/M or G_M is substituted for V/S in Eqs. (18.21) and (18.23), and G_x/M for L/S in Eqs. (18.22) and (18.24), the equations for the height of a transfer unit may be written (since $M\rho_M = \rho_x$, the density of the liquid)

$$H_y = \frac{G_M}{k_g a P} \qquad \text{and} \qquad H_{Oy} = \frac{G_M}{K_g a P} \qquad (18.25)$$

$$H_x = \frac{G_x/\rho_x}{k_L a} \qquad \text{and} \qquad H_{Ox} = \frac{G_x/\rho_x}{K_L a} \qquad (18.26)$$

The terms H_G, H_L, N_G, and N_L often appear in the literature instead of H_y, H_x, N_y, and N_x, as well as the corresponding terms for overall values, but here the different subscripts do not signify any difference in either units or magnitude.

If a design is based on N_{Oy}, the value of H_{Oy} can be calculated either from $K_y a$ or from values of H_y and H_x, as shown below. Starting with the equation for overall resistance, Eq. (18.12), each term is multiplied by G_M, and the last term is multiplied by L_M/L_M, where $L_M = L/S = G_x/M$, the molar mass velocity of the liquid:

$$\frac{G_M}{K_y a} = \frac{G_M}{k_y a} + \frac{m G_M}{k_x a}\frac{L_M}{L_M} \tag{18.27}$$

From the definitions of HTU in Eqs. (18.21) to (18.23),

$$H_{Oy} = H_y + m\frac{G_M}{L_M}H_x \tag{18.28}$$

$$H_{Ox} = H_x + \frac{L_M}{m G_M}H_y \tag{18.29}$$

EXAMPLE 18.3. A gas stream containing 3.0% A is passed through a packed column to remove 99 percent of the A by absorption in water. The absorber will operate at 25°C and 1 atm, and the gas and liquid rates are to be 20 mol/h·ft^2 and 100 mol/h·ft^2, respectively. Mass-transfer coefficients and equilibrium data are given below:

$$y^* = 3.1x \qquad \text{at } 25°C$$

$$k_x a = 60 \text{ mol/h·ft}^3\text{·unit mol fraction}$$

$$k_y a = 15 \text{ mol/h·ft}^3\text{·unit mol fraction}$$

(a) Find N_{Oy}, H_{Oy}, and Z_T, assuming isothermal operation and neglecting changes in gas and liquid flow rates. What percent of the total resistance is in the gas phase?
(b) Calculate Z_T, using N_{Ox} and H_{Ox}.

Solution
   (a) Assume $x_a = 0$. Since $G_M \Delta y = L_M \Delta x$,

$$x_b = \frac{20 \times 0.03 \times 0.99}{100} = 0.00594$$

$$y_b^* = 3.1 \times 0.0054 = 0.01841$$

At the bottom of the column,

$$y_b - y_b^* = 0.03 - 0.01841 = 0.01159$$

At the top, $y_a - y_a^* = y_a = 0.0003$

Then $\overline{\Delta y_L} = \dfrac{0.01159 - 0.0003}{\ln(0.01159/0.0003)} = 0.00309$

$$N_{Oy} = \frac{\Delta y}{\Delta y_L} = \frac{0.03 \times 0.99}{0.00309} = 9.61$$

$$\frac{1}{K_y a} = \frac{1}{15} + \frac{3.1}{60} = 0.11833 \qquad K_y a = 8.45$$

$$H_{Oy} = \frac{20}{8.45} = 2.37 \text{ ft.}$$

$$Z_T = 2.37 \times 9.61 = 22.7 \text{ ft}$$

The relative gas-film resistance is $\frac{1}{15}/(1/8.45) = 0.56$, or 56 percent.
(b) At the bottom of the column,

$$x^* = \frac{0.03}{3.1} = 0.009677$$

$$\Delta x = 0.009677 - 0.00594 = 0.003737$$

At the top, $\qquad x^* = \dfrac{0.0003}{3.1} = 9.677 \times 10^{-5} \qquad x = 0$

$$\overline{\Delta x_L} = \frac{0.00374 - 0.000097}{\ln[0.00374/(9.677 \times 10^{-5})]} = 9.96 \times 10^{-4}$$

$$N_{Ox} = \frac{0.00594}{9.96 \times 10^{-4}} = 5.96$$

$$\frac{1}{K_x a} = \frac{1}{60} + \frac{1}{3.1 \times 15} = 0.03817 \qquad K_x a = 26.2$$

$$H_{Ox} = \frac{100}{26.2} = 3.817 \text{ ft}$$

$$Z_T = 5.96 \times 3.817 = 22.7 \text{ ft}$$

Effect of pressure

Absorption columns are often operated under pressure to give increased capacity and higher rates of mass transfer. The equilibrium partial pressure of the solute depends only on the liquid composition and the temperature, so the equilibrium mole fraction in the gas varies inversely with the total pressure

$$y_A = \frac{p_A}{P} \tag{18.30}$$

If the gas and liquid rates are kept constant so the operating line is not changed, going to a higher pressure increases the mole-fraction driving force, as shown in Fig. 18.14, and reduces the number of transfer units. At high pressure the minimum liquid rate is smaller, so the operating line could be changed to give a richer product, as indicated by the dashed line in Fig. 18.14, and about the same number of transfer units as before.

Temperature variations in packed towers

When rich gas is fed to an absorption tower, the temperature in the tower varies appreciably from bottom to top. The heat of absorption of the solute raises the solution temperature, but evaporation of the solvent tends to lower the temperature. Usually the overall effect is an increase in the liquid temperature, but sometimes the temperature goes through a maximum near the bottom of the column. The shape of the temperature profile depends on the rates of solute absorption, evaporation or condensation of solvent, and heat transfer between the phases. Lengthy computations are needed to get the exact temperature profiles for liquid and gas,[15,22] and in this text only simplified examples are presented. When the gas inlet temperature is close to the exit temperature of the liquid and the incoming gas is saturated, there is little effect on solvent evaporation, and the rise in liquid temperature is roughly proportional to the amount of solute absorbed. The equilibrium line is then curved

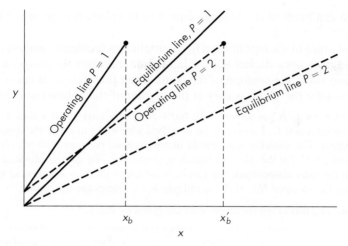

FIGURE 18.14
Effect of pressure on absorption.

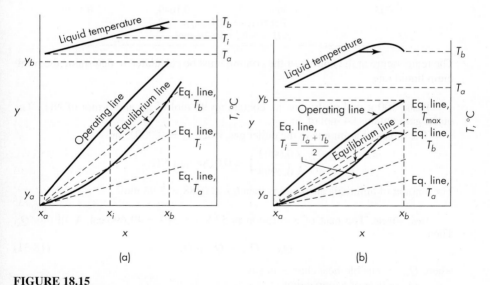

FIGURE 18.15
Temperature profiles and equilibrium lines for adiabatic absorption: (*a*) no solvent evaporation: (*b*) significant solvent evaporation or cold gas feed.

gradually upward, as shown in Fig. 18.15*a*, with increasing values of x corresponding to higher temperatures.

When the gas enters the columns 10 to 20°C below the exit liquid temperature and the solvent is volatile, evaporation will cool the liquid in the bottom part of the column, and the temperature profile may have a maximum, as shown in Fig. 18.15*b*. When the feed gas is saturated, the temperature peak is not very pronounced, and for an approximate design, either the exit temperature or the estimated maximum temperature can be used to calculate equilibrium values for the lower half of the column.

The curvature of the equilibrium line complicates the determination of the minimum liquid rate, since decreasing the liquid rate increases the temperature rise of the liquid and shifts the position of the equilibrium line. For most cases, it is satisfactory to assume the pinch occurs at the bottom of the column to calculate L_{min}.

EXAMPLE 18.4. A gas stream with 6.0% NH_3 (dry basis) and a flow rate of 4,500 SCFM (ft^3/min at 0°C, 1 atm) is to be scrubbed with water to lower the concentration to 0.02 percent. The absorber will operate at atmospheric pressure with inlet temperatures of 20 and 25°C for the gas and liquid, respectively. The gas is saturated with water vapor at the inlet temperature and can be assumed to leave as a saturated gas at 25°C. Calculate the value of N_{Oy} if the liquid rate is 1.25 times the minimum.

Solution. The following solubility data are given by Perry.[11a]

x	$y_{20°C}$	$y_{30°C}$	$y_{40°C}$
0.0308	0.0239	0.0389	0.0592
0.0406	0.0328	0.0528	0.080
0.0503	0.0417	0.0671	0.1007
0.0735	0.0658	0.1049	0.1579

For $NH_3 \rightarrow NH_3(aq)$,
$\Delta H = -8.31$ kcal/g mol

The temperature at the bottom of the column must be calculated to determine the minimum liquid rate.

Basis. 100 g mol of dry gas in, containing 94 mol of air and 6 mol of NH_3. The outlet gas contains 94 mol of air.

The moles of ammonia in the outlet gas, since $y_a = 0.0002$, are

$$94\left(\frac{0.0002}{0.9998}\right) = 0.0188 \text{ mol } NH_3$$

The amount of ammonia absorbed is then $6 - 0.0188 = 5.98$ mol.

Heat effects. The heat of absorption is $5.98 \times 8,310 = 49,690$ cal. Call this Q_a. Then

$$Q_a = Q_{sy} + Q_v + Q_{sx} \tag{18.31}$$

where Q_{sy} = sensible-heat change in gas
  Q_v = heat of vaporization
  Q_{sx} = sensible heat change in liquid

The sensible-heat changes in the gas are

$$Q_{air} = 94 \text{ mol} \times 7.0 \text{ cal/mol} \cdot °C \times 5°C = 3,290 \text{ cal}$$

$$Q_{H_2O} = 2.4 \times 8.0 \times 5 = 96 \text{ cal}$$

$$Q_{sy} = 3,290 + 96 = 3,390 \text{ cal}$$

The amount of vaporization of water from the liquid is found as follows. At 20°C, $p_{H_2O} = 17.5$ mm Hg; at 25°C, $p_{H_2O} = 23.7$ mm Hg. The amount of water in the inlet gas is

$$100 \times \frac{17.5}{742.5} = 2.36 \text{ mol}$$

In the outlet gas it is

$$94.02 \times \frac{23.7}{736.3} = 3.03 \text{ mol}$$

The amount of water vaporized is therefore $3.03 - 2.36 = 0.67$ mol. Since the heat of vaporization $\Delta H_v = 583$ cal/g,

$$Q_v = 0.67 \times 583 \times 18.02 = 7,040 \text{ cal}$$

Solving Eq. (18.31) for Q_{sx}, the sensible-heat change in the liquid, gives

$$Q_{sx} = 49,690 - 3,390 - 7,040 = 39,260 \text{ cal}$$

The outlet temperature of the liquid T_b is found by trial. Assume that for the solution $C_p = 18$ cal/g mol·°C; guess that $T_b = 40$°C and $x_{max} = 0.031$, as estimated from the equilibrium solubility lines on Fig. 18.16. Then the total moles of liquid out L_b are

$$L_b = \frac{5.98}{0.031} = 192.9 \text{ mol}$$

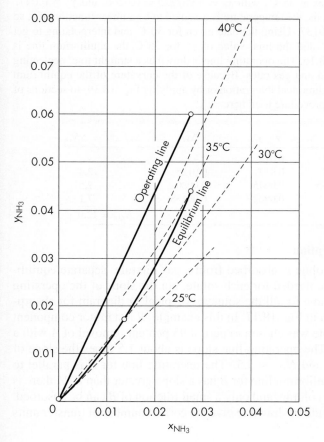

FIGURE 18.16
The yx diagram for Example 18.4.

Since $T_a = 25°C$,

$$192.9 \times 18(T_b - 25) = 39{,}260$$
$$T_b = 36.3°C$$

For a revised estimate of $T_b = 37°C$, $x_{\max} = 0.033$,

$$L_b = \frac{5.98}{0.033} = 181 \text{ mol}$$

$$T_b - 25 = \frac{39{,}260}{181 \times 18} = 12.1$$

$$T_b = 37°C$$

This procedure gives the minimum liquid rate; the minimum amount of water is

$$L_{\min} = 181 - 6 = 175 \text{ mol } H_2O$$

For a water rate 1.25 times the minimum, $L_a = 1.25 \times 175 = 219$ mol, and $L_b = 219 + 6 = 225$ mol. Then, the temperature rise of the liquid is

$$T_b - 25 = \frac{39{,}260}{225 \times 18} = 9.7°C$$

The liquid therefore leaves at 35°C, with $x_b = 5.98/225 = 0.0266$ and $y^* \approx 0.044$.

To simplify the analysis, the temperature is assumed to be a linear function of x, so that $T \approx 30°C$ at $x = 0.0137$. Using the data given for 30°C and interpolating to get the initial slope for 25°C and the final value of y^* for 35°C, the equilibrium line is drawn as shown in Fig. 18.16. The operating line is drawn as a straight line, neglecting the slight change in liquid and gas rates. Because of the curvature of the equilibrium line, N_{Oy} is evaluated by numerical integration or by applying Eq. (18.19) to sections of the column, which is the procedure used here.

y	y^*	$y - y^*$	$\overline{\Delta y_L}$	$\Delta y / \overline{\Delta y_L} = \Delta N_{Oy}$
0.06	0.048	0.012	—	—
0.03	0.017	0.013	0.0125	2.4
0.01	0.0055	0.0045	0.0080	2.5
0.0002	0	0.0002	0.00138	7.1
				$N_{Oy} = 12.0$

Multicomponent absorption

When more than one solute is absorbed from a gas mixture, separate equilibrium and operating lines are needed for each solute, but the slope of the operating line, which is L/V, is the same for all the solutes. A typical yx diagram for absorption of two solutes is shown in Fig. 18.17. In this example, B is a minor component of the gas, and the liquid rate was chosen to permit 95 percent removal of A with a reasonable packed height. The operating-line slope is about 1.5 times the slope of the equilibrium line for A, and $N_{Oy} \approx 5.5$. The operating line for B is parallel to that for A, and since the equilibrium line for B has a slope greater than L/V, there is a pinch at the bottom of the column, and only a small fraction of B can be absorbed. The operating line for B should be drawn to give the correct number of transfer units

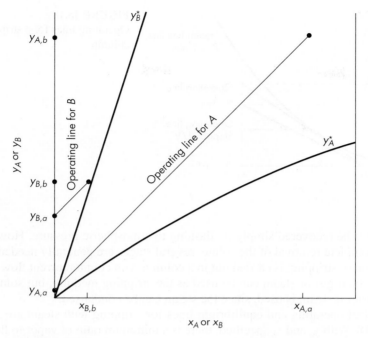

FIGURE 18.17
Equilibrium and operating lines for multicomponent absorption.

for B, which is generally about the same as N_{Oy} for A. However, in this example, $x_{B,b}$ is practically the same as x_B^*, the equilibrium value for $y_{B,b}$, and the fractional removal of B can be calculated directly from a material balance:

$$V(y_{B,b} - y_{B,a}) = L(x_{B,b}^* - x_{B,a}) \qquad (18.32)$$

If nearly complete absorption of B is required, the operating line will have to be made steeper than the equilibrium line for B. Then the operating line will be much steeper than the equilibrium line for A, and the concentration of A in the gas will be reduced to a very low value. Examples of multicomponent absorption are the recovery of light hydrocarbon gases by absorption in heavy oil, the removal of CO_2 and H_2S from natural gas or coal gasifier products by absorption in methanol or alkaline solutions, and water scrubbing to recover organic products produced by partial oxidation. For some cases, the dilute solution approach presented here may have to be corrected for the change in molar flow rates or the effect of one solute on the equilibria for other gases, as shown in the analysis of a natural gasoline absorber.[15]

Desorption or stripping

In many cases, a solute that is absorbed from a gas mixture is desorbed from the liquid to recover the solute in more concentrated form and regenerate the absorbing solution. To make conditions more favorable for desorption, the temperature may be increased or the total pressure reduced, or both these changes may be made. If the absorption is carried out under high pressure, a large fraction of the solute can

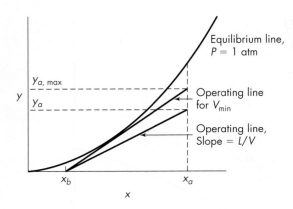

FIGURE 18.18
Operating lines for a stripping column.

sometimes be recovered simply by flashing to atmospheric pressure. However, for nearly complete removal of the solute, several stages are generally needed, and the desorption or stripping is carried out in a column with countercurrent flow of liquid and gas. Inert gas or steam can be used as the stripping medium, but solute recovery is easier if steam is used, since the steam can be condensed.

Typical operating and equilibrium lines for stripping with steam are shown in Fig. 18.18. With x_a and x_b specified, there is a minimum ratio of vapor to liquid corresponding to the operating line that just touches the equilibrium line at some point. The pinch may occur in the middle of the operating line if the equilibrium line is curved upward, as in Fig. 18.18, or it may occur at the top of the column, at (y_a, x_a). For simplicity the operating line is shown as a straight line, though it would generally be slightly curved because of the change in vapor and liquid rates.

In an overall process of absorption and stripping, the cost of steam is often a major expense, and the process is designed to use as little steam as possible. The stripping column is operated at close to the minimum vapor rate, and some solute is left in the stripped solution, rather than try for complete recovery. When the equilibrium line is curved upward, as in Fig. 18.18, the minimum steam rate becomes much higher as x_b approaches zero.

The height of a stripping column can be calculated from the number of transfer units and the height of a transfer unit, using the same equations as for absorption. Often attention is focused on the liquid-phase concentration, and N_{Ox} and H_{Ox} are used:

$$Z_T = H_{Ox}N_{Ox} = H_{Ox} \int \frac{dx}{x^* - x} \qquad (18.33)$$

where H_{Ox} is given in Eq. (18.29).

Stripping with air is used in some cases to remove small amounts of gases such as ammonia or organic solvents from water. If there is no need to recover the solute in concentrated form, the optimum amount of air used may be much greater than the minimum, since it does not cost much to provide more air, and the column height is considerably reduced. The following example shows the effect of air rate in a stripping operation.

EXAMPLE 18.5. Water containing 6 ppm trichloroethylene (TCE) is to be purified by stripping with air at 20°C. The product must contain less than 4.5 ppb TCE to meet emission standards. Calculate the minimum air rate in standard cubic meters of air per cubic meter of water and the number of transfer units if the air rate is 1.5 to 5 times the minimum value.

Solution. Henry's law coefficient for TCE in water[7] at 20°C is 0.0075 $m^3 \cdot$ atm/mol. This can be converted to the slope of the equilibrium line in mole-fraction units as follows, since $P = 1$ atm and 1 m^3 of liquid weighs 10^6g:

$$m = 0.0075 \frac{\text{atm} \cdot m^3}{\text{mol}} \times \frac{1}{1 \text{ atm}} \times \frac{10^6 \text{ mol } H_2O}{18} \frac{}{m^3} = 417$$

With this large value of m, the desorption is liquid-phase controlled. At the minimum air rate, the exit gas will be in equilibrium with the incoming solution. The molecular weight of TCE is 131.4, and

$$x_a = \frac{6 \times 10^{-6} \text{ mol TCE}}{131.4 \quad \text{g } H_2O} \times \frac{18 \text{ g}}{\text{mol } H_2O} = 8.22 \times 10^{-7}$$

$$y_a = 417(8.22 \times 10^{-7}) = 3.43 \times 10^{-4}$$

Per cubic meter of solution fed, the TCE removed is

$$V_{TCE} = \frac{10^6(6 \times 10^{-6} - 4.5 \times 10^{-9})}{131.4}$$

$$= 4.56 \times 10^{-2} \text{ mol}$$

The total amount of gas leaving is

$$V = \frac{4.56 \times 10^{-2}}{3.43 \times 10^{-4}} = 132.9 \text{ mol}$$

Since 1 g mol $= 0.0224$ std m^3 and since the change in gas flow rate is very small,

$$F_{min} = 132.9 \times 0.0224 = 2.98 \text{ std } m^3$$

The density of air at standard conditions is 1.295 kg/m^3, so the minimum rate on a mass basis is

$$\left(\frac{G_y}{G_x}\right)_{min} = \frac{2.98 \times 1.295}{1,000} = 3.86 \times 10^{-3} \text{ kg air/kg water}$$

If the air rate is 1.5 times the minimum value, then

$$y_a = \frac{3.43 \times 10^{-4}}{1.5} = 2.29 \times 10^{-4}$$

$$x_a^* = \frac{2.29 \times 10^{-4}}{417} = 5.49 \times 10^{-7}$$

$$C_a^* = 5.49 \times 10^{-7} \times \frac{131.4}{18} = 4.01 \times 10^{-6} \text{ g/g} = 4.01 \text{ ppm}$$

$$C_a - C_a^* = \Delta C_a = 6.0 - 4.01 = 1.99$$

At bottom,

$$C_b = 0.0045 \text{ ppm} \qquad C_b^* = 0 \qquad \Delta C_b = 0.0045 \text{ ppm}$$

$$\overline{(C - C^*)_L} = \frac{1.99 - 0.0045}{\ln(1.99/0.0045)} = 0.3259 \text{ ppm}$$

Using concentrations in parts per million to calculate N_{Ox},

$$N_{Ox} = \int \frac{dC}{C - C^*} = \frac{C_a - C_b}{\overline{(C - C^*)_L}}$$

$$= \frac{6 - 0.0045}{0.3259} = 18.4$$

Similar calculations for other multiples of the minimum flow rate give the following values. The packed height is based on an estimated value of $H_{Ox} = 3$ ft; this is somewhat greater than the values reported for 1-in. plastic Pall rings.

Air rate	N_{Ox}	Z, ft
$1.5V_{min}$	18.4	55.2
$2V_{min}$	13.0	39
$3V_{min}$	10.2	30.6
$5V_{min}$	8.7	26.1

Going from 1.5 to $2V_{min}$ or from 2 to $3V_{min}$ decreases the tower height considerably, and the reduction in pumping work for water is more than the additional energy needed to force air through the column. Further increase in V does not change Z very much, and the optimum air rate is probably in the range 3 to $5V_{min}$. Typical flow rates at $V = 3V_{min}$ might be $G_x = 10,000$ lb/ft²·h (49,000 kg/m²·h) and $G_y = 116$ lb/ft²·h (566 kg/m²·h).

ABSORPTION FROM RICH GASES

When the solute being absorbed is present at moderate or high concentrations in the gas, there are several additional factors to consider in design calculations. The decrease in total gas flow and the increase in liquid flow must be accounted for in the material balance, and the correction factor for one-way diffusion should be included. Also, the mass-transfer coefficients will not be constant because of the changes in flow rate, and there may be an appreciable temperature gradient in the column, which will change the equilibrium line.

The amount of solute absorbed in a differential height dZ is $d(Vy)$, since both V and y decrease as the gas passes through the tower.

$$dN_A = d(Vy) = V\, dy + y\, dV \tag{18.34}$$

If only A is being transferred, dN_A is the same as dV, so Eq. (18.34) becomes

$$dN_A = V\, dy + y\, dN_A \tag{18.35}$$

or

$$dN_A = \frac{V\, dy}{1 - y} \tag{18.36}$$

The effect of one-way diffusion in the gas film is to increase the mass-transfer rate for the gas film by the factor $1/(1 - Y)_L$, as shown by Eq. (17.46), so the effective overall coefficient $K'_y a$ is somewhat larger than the normal value of $K_y a$:

$$\frac{1}{K'_y a} = \frac{\overline{(1 - y)_L}}{k_y a} + \frac{m}{k_x a} \tag{18.37}$$

In this treatment, the effect of one-way diffusion in the liquid film is neglected. The basic mass-transfer equation is then

$$dN_A = \frac{V \, dy}{1 - y} = K'_y a S \, dZ \, (y - y^*) \tag{18.38}$$

The column height can be found by a graphical integration, allowing for changes in V, $1 - y$, $y - y^*$, and $K'_y a$:

$$Z_T = \frac{1}{S} \int_a^b \frac{V \, dy}{(1 - y)(y - y^*)(K'_y a)} \tag{18.39}$$

If the process is controlled by the rate of mass transfer through the gas film, a simplified equation can be developed. The term $\overline{(1 - y)_L}$, which strictly applies only to the gas film, as shown in Eq. (18.37), is assumed to apply to the overall coefficient, since the gas film is controlling. The coefficient $K'_y a$ in Eq. (18.39) is replaced by $K_y a/(1 - y)_L$, which leads to

$$Z_T = \frac{1}{S} \int_a^b \frac{V \overline{(1 - y)_L} \, dy}{K_y a (1 - y)(y - y^*)} \tag{18.40}$$

Since $k_y a$ varies with about $V^{0.7}$, and $K_y a$ will show almost the same variation when the gas film controls, the ratio $V/K_y a$ does not change much. This term can be taken outside the integral and evaluated at the average flow rate, or the values at the top and bottom of the tower can be averaged. The term $\overline{(1 - y)_L}$ is the logarithmic mean of $1 - y$ and $1 - y_i$, which is usually only slightly larger than $1 - y$. Therefore the terms $\overline{(1 - y)_L}$ and $1 - y$ are assumed to cancel, and Eq. (18.40) becomes

$$Z_T = \overline{\left(\frac{V/S}{K_y a}\right)} \int_b^a \frac{dy}{y - y^*} \tag{18.41a}$$

$$Z_T = H_{Oy} N_{Oy} \tag{18.41b}$$

This is the same as Eq. (18.16) for dilute gases, except that the first term, which is H_{Oy}, is an average value for the column rather than a constant. Note that $K_y a$ from Eq. (18.12) is to be used here, and not $K'_y a$, since the $\overline{(1 - y)_L}$ term was included in the derivation.

If the liquid film has the controlling resistance to mass transfer, gas-film coefficients could still be used for design calculations following Eq. (18.39). If liquid-film coefficients are used, and if the factor $\overline{(1 - x)_L}$ is introduced to allow for one-way diffusion in the liquid, an equation similar to Eq. (18.41a) can be derived:

$$Z_T = \overline{\left(\frac{L/S}{K_x a}\right)} \int_b^a \frac{dx}{x^* - x} \tag{18.42}$$

or

$$Z_T = H_{Ox} N_{Ox} \tag{18.43}$$

When the gas-film and liquid-film resistances are comparable in magnitude, there is no simple method of dealing with absorption of a rich gas. The recommended method is to base the design on the gas phase and use Eq. (18.39). Several values of y between y_a and y_b are chosen, and values of V, $K_y'a$, and $y - y^*$ are calculated. As a check, the value of Z_T from the integration should be compared with that based on the simple formula in Eq. (18.41), since the difference should not be great.

EXAMPLE 18.6. A tower packed with 1-in. (25.4-mm) rings is to be designed to absorb sulfur dioxide from air by scrubbing the gas with water. The entering gas is 20% SO_2 by volume, and the leaving gas is to contain not more than 0.5% SO_2 by volume. The entering H_2O is SO_2-free. The temperature is 80°F, and the total pressure is 2 atm. The water flow is to be twice the minimum. The airflow rate (SO_2-free basis) is to be 200 lb/ft^2·h (976 kg/m^2·h). What depth of packing is required?

The following equations are available[23] for the mass-transfer coefficients for absorption of SO_2 at 80°F in towers packed with 1-in. rings:

$$k_L a = 0.038 G_x^{0.82}$$

$$k_g a = 0.028 G_y^{0.7} G_x^{0.25}$$

where $k_L a$ is in h^{-1}, $k_g a$ is in mol/ft^3·h·atm, and G_x and G_y are in lb/ft^2·h.

The solubility of SO_2 in water is not quite proportional to the pressure, because the H_2SO_3 that forms partially dissociates to H^+ and HSO_3^-, and the fraction dissociation decreases with increasing concentration. Henry's law applies to the undissociated H_2SO_3, and the total SO_2 dissolved is the sum of the H_2SO_3 and the HSO_3^-. (The SO_3^{2-} formed in water is negligible.) A few points for the total SO_2 dissolved at 80°F are given below.

p_{SO_2}, atm	0.04	0.08	0.12	0.16	0.20
C_{SO_2}, mol/ft^3	0.0044	0.0082	0.0117	0.0152	0.0186
x_{SO_2}, mole fraction	0.00127	0.00237	0.00338	0.00439	0.00538

Solution. The coefficients are first converted to mole-fraction units. At 80°F, ρ_M for water is $62.2/18.02 = 3.45$ mol/ft^3. Therefore

$$k_x a = 0.038 G_x^{0.82} \times 3.45 = 0.131 G_x^{0.82} \text{ mol/ft}^3 \cdot \text{h}$$

Since the coefficients were measured at atmospheric pressure, and $k_y a = k_g a P$, the value of $k_g a$ is equal to $k_y a$. However, the measured coefficients were not corrected for one-way diffusion and thus represent $k_y a/(1 - y)_L$. The SO_2 concentration for the published tests ranged from 3 to 17%, and the average value of $(1 - y)_L$ was about 0.9. Therefore the published correlation is multiplied by 0.9 to give

$$k_y a = 0.025 G_y^{0.7} G_x^{0.25}$$

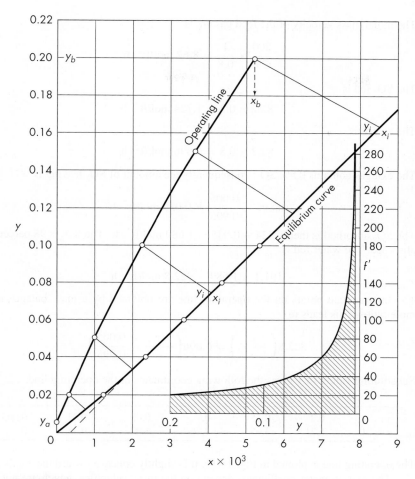

FIGURE 18.19
Diagram for Example 18.6.

The next step is to plot the equilibrium curve. Since the pressure is 2 atm, $y = p_{SO_2}/2$. The curve is shown in Fig. 18.19. The slope of the linear portion is 20.1, and the initial slope is 15.6. The lower end of the operating line is at $y = 0.005$, $x = 0$, and at the minimum liquid rate, the upper end of this line will touch the equilibrium curve at $y = 0.20$, $x = 10.36 \times 10^{-3}$. At twice the minimum liquid rate, the ratio of water to SO_2 in the liquid will be twice as great. The number of moles of water fed per mole of SO_2 is

$$2 \times \frac{0.98964}{0.01036} = 191.1$$

and

$$x_b = \frac{1}{191.1 + 1} = 0.00521$$

The molal mass velocity of the feed gas G_M is

$$\frac{200}{29} \times \frac{1}{0.8} = 8.62 \text{ mol/ft}^2 \cdot \text{h}$$

The SO_2 fed is

$$8.62 \times 0.2 = 1.724 \text{ mol/ft}^2 \cdot \text{h}$$

The air fed is

$$8.62 \times 0.8 = 6.896 \text{ mol/ft}^2 \cdot \text{h}$$

The exit gas contains 0.5% SO_2; consequently the amount of SO_2 leaving is

$$6.896 \times \frac{0.005}{0.995} = 0.035 \text{ mol/ft}^2 \cdot \text{h}$$

The SO_2 absorbed is then $1.724 - 0.035 = 1.689 \text{ mol/ft}^2 \cdot \text{h}$. The SO_2 is 98 percent absorbed. The water fed is

$$191.1 \times 1.689 = 322.8 \text{ mol/ft}^2 \cdot \text{h}$$

Intermediate points on the operating line are obtained by a mass balance using mole ratios. This leads to

$$322.8\left(\frac{x}{1-x}\right) = 6.896\left(\frac{y}{1-y} - \frac{0.005}{0.995}\right)$$

Substituting values of y gives the following coordinates of the operating line:

y	0.2	0.15	0.1	0.05	0.02	0.005
$x \times 10^3$	5.21	3.65	2.26	1.02	0.33	0

The operating line is plotted in Fig. 18.19. It is slightly concave toward the x axis.

The mass-transfer coefficients depend on the mass velocities, which are not constant. The change in liquid rate is very small, and the liquid-film coefficient is based on an average flow rate. The molecular weight of SO_2 is 64.1, and the average mass velocity is

$$\overline{G_x} = 322.8 \times 18.02 + \frac{1.689 \times 64.1}{2}$$

$$= 5,871 \text{ lb/ft}^2 \cdot \text{h}$$

The coefficient, assumed constant for the tower, is

$$k_x a = 0.131 \times 5,871^{0.82} = 161 \text{ mol/ft}^3 \cdot \text{h}$$

The gas-film coefficients are calculated for the bottom and top of the tower:

At bottom:
$$G_y = 6.9 \times 29 + 1.724 \times 64.1$$
$$= 310.6 \text{ lb/ft}^2 \cdot \text{h}$$
$$k_y a = 0.025 \times 310.6^{0.7} \times 5,871^{0.25}$$
$$= 12.15 \text{ mol/ft}^3 \cdot \text{h}$$

At top:
$$G_y = 6.9 \times 29 + 0.035 \times 64.1$$
$$= 202.3 \text{ lb/ft}^2 \cdot \text{h}$$
$$k_y a = 0.025 \times 202.3^{0.7} \times 5{,}871^{0.25}$$
$$= 9.0 \text{ mol/ft}^3 \cdot \text{h}$$

The gas-film coefficients are only 6 to 8 percent as large as the liquid-film coefficient, but the slope of the equilibrium line makes the two resistances comparable in magnitude. Values of y_i are obtained by drawing lines of slope $-k_x a \overline{(1-y)}_L / k_y a$ from the operating line to the equilibrium line. This requires a preliminary estimate of $\overline{(1-y)}_L$, but the lines are nearly parallel, and the first trial is usually accurate enough. If $\overline{(1-y)}_L$ is assumed to be about 0.82 at the bottom,

$$\frac{k_x a \overline{(1-y)}_L}{k_y a} = \frac{161 \times 0.82}{12.15} = 10.87$$

A line from (y_b, x_b) with a slope of -10.9 gives $y_i = 0.164$, and $1 - y_i = 0.836$. The logarithmic mean of 0.80 and 0.836 is 0.818, very close to the estimated value. From Eq. (18.37),

$$\frac{1}{K'_y a} = \frac{0.818}{12.15} + \frac{20.1}{161} = 0.192$$

$$K'_y a = \frac{1}{0.192} = 5.21 \text{ mol/ft}^3 \cdot \text{h}$$

The fraction of the total resistance that is in the liquid is $(20.1/161)/0.192 = 0.65$, or 65 percent.

Similar calculations are made for intermediate values of y, with the results given in Table 18.2. The values of V/S are found from $V/S = 6.896/(1-y)$, and $k_y a$ is calculated from the mass velocity. Note that $V/S = G_M$. When $y > 0.05$, slope m is taken as 20.1; when y is <0.05, $m = 15.6$.

The column height is found as follows. From Eq. (18.39),

$$dZ = \frac{(V/S)\,dy}{(1-y)(y-y^*)K'_y a} = f'\,dy$$

Values of f' are calculated for several values of y and tabulated in Table 18.2. The total height may be found by graphical integration of a plot of y versus f', as shown by the inset in Fig. 18.19, to give a total height Z_T of 8.9 ft.

TABLE 18.2
Integration for Example 18.6

y	$y - y^*$	y_i	$\overline{(1-y)}_L$	$K'_y a$	$\dfrac{V}{S}$	$f' = \dfrac{V/S}{(1-y)(y-y^*)K'_y a}$	ΔZ
0.20	0.103	0.164	0.818	5.21	8.62	20.1	
0.15	0.084	0.118	0.866	4.95	8.12	23.0	1.08
0.10	0.062	0.074	0.913	4.71	7.67	29.2	1.31
0.05	0.034	0.034	0.958	4.46	7.26	50.4	1.99
0.02	0.015	0.012	0.984	4.87	7.04	98.3	2.06
0.005	0.005	0.002	0.996	4.82	6.93	289.0	2.41
							$Z_T = 8.85$

Since the major resistance is in the liquid film, Eq. (18.43) can be used for a rough check on the column height. The equation analogous to Eq. (18.37) is

$$\frac{1}{K'_x a} = \frac{1}{k_x a} + \frac{\overline{(1-y)_L}}{mk_y a}$$

Hence

$$\frac{1}{K'_x a} = \frac{1}{161} + \frac{0.91}{20.1 \times 11} = 1.04 \times 10^{-2}$$

$$K'_x a = 96.2 \text{ mol/ft}^3 \cdot \text{h} \cdot \text{unit mole-fraction difference}$$

From Eq. (18.24),

$$H_{Ox} = \frac{322.8}{96.2} = 3.36 \text{ ft}$$

Ignoring the curvature of the equilibrium and operating lines,

$$N_{Ox} = \frac{x_b - x_a}{(x^* - x)_L}$$

$$= \frac{0.00522}{0.00174} = 3.0$$

From Eq. (18.43),

$$Z_T = 3.36 \times 3.0 = 10.1 \text{ ft}$$

MASS-TRANSFER CORRELATIONS

To predict the overall mass-transfer coefficient or the height of a transfer unit, separate correlations are required for the gas phase and the liquid phase. Such correlations are generally based on experimental data for systems in which one phase has the controlling resistance, since it is difficult to separate the two resistances accurately when they are of comparable magnitude. The liquid-phase resistance can be determined from the rate of desorption of oxygen or carbon dioxide from water. The low solubility of these gases makes the gas-film resistance negligible, and the values of H_{Ox} are essentially the same as H_x.

More accurate values of H_x are obtained from desorption measurements than from absorption tests, because the operating lines at typical gas and liquid rates have slopes much less than the slope of the equilibrium line. For oxygen in water at 20°C, the equilibrium partial pressure is 4.01×10^4 atm per mole fraction, and L/V might range from 1 to 100. For absorption of oxygen from air in pure water, a "pinch" would develop at the bottom of the packed column, as shown in Fig. 18.20. Very accurate measurements of x_b and the temperature (to determine x_b^*) would be needed to determine the driving force $x_b - x_b^*$. For desorption of oxygen from a saturated solution into nitrogen, the concentration x_b is small, but N_{Ox} can be determined with reasonable accuracy since x_b^* is zero.

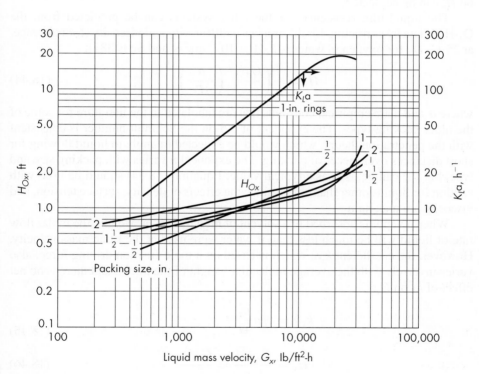

FIGURE 18.20

Typical operating lines for absorption or desorption of a slightly soluble gas.

FIGURE 18.21

Height of a transfer unit for desorption of oxygen from water at 25°C with Raschig ring packing. (*Note:* In this system $H_{Ox} \approx H_x$.)

Liquid-film coefficients

Values of H_x for the system O_2-H_2O with ceramic Raschig rings[14] are shown in Fig. 18.21. For liquid mass velocities in the intermediate range, 500 to 10,000 lb/ft²·h, H_x increases with $G_x^{0.4}$ for $\frac{1}{2}$-in. rings but with $G_x^{0.2}$ for the larger sizes. Thus for 1-, $1\frac{1}{2}$-, and 2-in. rings $k_L a$ varies with $G_x^{0.8}$. Much of the increase in $k_L a$ is due to the increasing interfacial area a, and the rest comes from an increase in k_L. At high mass velocities, the packing is nearly completely wetted, and there is only a slight increase in $k_L a$ with G_x, which makes H_x nearly proportional to G_x. Note that as far as mass transfer is concerned, the small packings are only slightly better than the large ones in the intermediate range of flows, even though the total area varies inversely with the packing size. The larger packings are generally preferred for commercial operation because of the much higher capacity (higher flooding velocity).

The data in Fig. 18.21 were taken with gas flow rates of 100 to 230 lb/ft$^2 \cdot$ h, and there was no effect of G_y in this range. For gas flow rates between loading and flooding, H_x is slightly lower because of the increased holdup of liquid. However, for a column designed to operate at one-half the flooding velocity, the effect of G_y on H_x can be neglected.

The liquid-film resistance for the other systems can be predicted from the O_2-H_2O data by correcting for differences in diffusivity and viscosity (for reference, at 25°C, D_v for oxygen in water is 2.41×10^{-5} cm^2/s and Sc is 381]:

$$H_x = \frac{1}{\alpha}\left(\frac{G_x}{\mu}\right)^n \left(\frac{\mu}{\rho D_v}\right)^{0.5} \tag{18.44}$$

where α and n are empirical constants that are tabulated in the literature for some of the older packing types. The exponent of 0.5 on the Schmidt number is consistent with the penetration theory, which would be expected to apply to liquid flowing for short distances over pieces of packing. The exponent n varies with packing size and type, but 0.3 can be used as a typical value. Equation (18.44) should be used with caution for liquids other than water, since the effects of density, surface tension, and viscosity are uncertain.

When a vapor is absorbed in a solvent of high molecular weight, the molar flow rate of liquid will be much less than if water were used at the same mass velocity. However, the coefficient $k_x a$, which is based on a mole-fraction driving force, also varies inversely with the average molecular weight of the liquid, and there is no net effect of M on H_x.

$$k_x a = k_L a \frac{\rho_x}{\overline{M}} \tag{18.45}$$

$$H_x = \frac{G_x/\overline{M}}{k_x a} = \frac{G_x/\rho_x}{k_L a} \tag{18.46}$$

The coefficient $k_L a$ depends mainly on the volumetric flow rate, diffusivity, and viscosity but not on the molecular weight, so in this respect general correlations for $k_L a$ or H_x are simpler than those for $k_x a$.

Gas-film coefficients

The absorption of ammonia in water has been used to get data on $k_g a$ or H_y, since the liquid-film resistance is only about 10 percent of the overall resistance and can be easily allowed for. Data for H_{Oy} and corrected values of H_y for $1\frac{1}{2}$-in. Raschig rings are given in Fig. 18.22. For mass velocities up to 600 lb/ft$^2 \cdot$ h, H_y varies with about the 0.3 to 0.4 power of G_y, which means $k_g a$ increases with $G_y^{0.6-0.7}$, in reasonable agreement with data for mass transfer to particles in packed

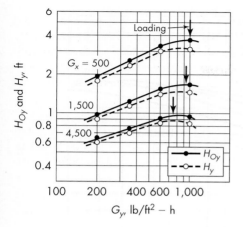

FIGURE 18.22
Height of a transfer unit for the absorption of ammonia in water with $1\frac{1}{2}$-in. ceramic Raschig rings.

beds. The slopes of the H_y plots decrease in the loading region because of the increase in interfacial area. The values of H_y vary with the -0.7 to -0.4 power of the liquid rate, reflecting the large effect of liquid rate on interfacial area.

The following equation is recommended[15] to estimate H_y for absorption of other gases in water. The Schmidt number for the NH_3-air-H_2O system is 0.66 at 25°C:

$$H_y = H_{y,NH_3}\left(\frac{Sc}{0.66}\right)^{1/2}$$ (18.47)

There are few data to support an exponent of $\frac{1}{2}$ for the diffusivity or the Schmidt number, and an exponent of $\frac{2}{3}$ has been suggested based on boundary layer theory and data for packed beds. However, the Schmidt numbers for gases do not differ widely, and the correction term is often small. There is more uncertainty about the effect of liquid properties on H_y if liquids other than water are used as solvents.

For the vaporization of pure liquid into a gas stream, there is no mass-transfer resistance in the liquid phase, and vaporization tests might seem a good method of developing a correlation for gas-film resistance. However, tests with water and other liquids give H_y values about one-half those for ammonia at the same mass velocities. The difference is attributed to pockets of nearly stagnant liquid that contribute steadily to vaporization but soon become saturated in a gas absorption test.[17] The stagnant pockets correspond to the *static holdup,* liquid that remains in the column long after the flow is shut off. The rest of the liquid constitutes the *dynamic holdup,* which increases with liquid flow rate. Correlations for the static and dynamic holdup and the corresponding interfacial areas have been developed[16] and can be used to correlate vaporization and gas absorption results.

Performance of other packings

Several packings have been developed that have high capacity and better mass-transfer characteristics than Raschig rings and Berl saddles, but comprehensive data on the gas and liquid resistances are not available. Many of these packings have been tested for the absorption of CO_2 in NaOH solution, a system where the liquid

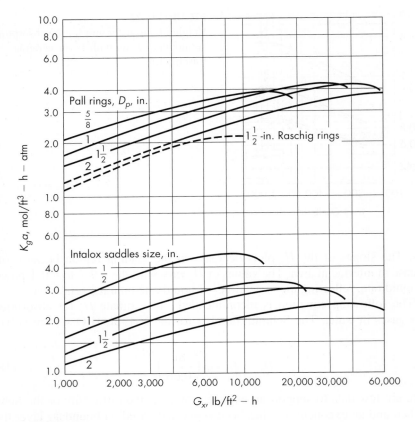

FIGURE 18.23
Mass-transfer coefficients for the absorption of CO_2 in 4% NaOH with metal Pall rings or ceramic Intalox saddles ($G_y = 500$ lb/ft$^2 \cdot$ h).

film has the controlling resistance, but the gas-film resistance is not negligible. The $K_g a$ values are 20 to 40 times the normal values for CO_2 absorption in water, because the chemical reaction between CO_2 and NaOH takes place very close to the interface, making the concentration gradient for CO_2 much steeper.

Although the $K_g a$ values for the CO_2-NaOH system cannot be used directly to predict the performance with other systems, they can be used for comparison between packings. Data for several sizes of Intalox saddles and Pall rings are shown in Fig. 18.23, along with some results for Raschig rings. The ratio of $K_g a$ for a given packing to that for $1\frac{1}{2}$-in. Raschig rings, evaluated at $G_x = 1{,}000$ lb/ft$^2 \cdot$h and $G_y = 500$ lb/ft$^2 \cdot$ h, is taken as a measure of performance f_p and listed in Table 18.1. The value of f_p is a relative measure of the total interfacial area, since the absorption of CO_2 into NaOH solution is an irreversible reaction that can take place in the static as well as the dynamic holdup. Packings that have a relatively high total interfacial area probably have a large dynamic holdup as well and a large area for normal physical absorption. For a rough estimate of the performance of the new packings for physical absorption, the value of f_p can be applied to H_{Oy} or to the overall coefficient calculated for 1.5-in. Raschig rings. The overall coefficient would be based on data for NH_3 and for O_2 and corrected for changes in diffusivity, viscosity, and flow rate.

Large columns sometimes have higher apparent values of H_{Oy} than small columns using the same packing, and various empirical correlations for the effects of column diameter and packed height have been presented.[11c] These effects probably resulted from uneven liquid distribution, which tends to make the gas flow uneven and results in local values of the operating-line slope quite far from the average. The penalty for maldistribution is greatest when the operating line is only slightly steeper than the equilibrium line and when a large number of transfer units are needed. For these cases it is especially important to provide very good liquid distribution, and for tall columns, as mentioned previously, it is advisable to pack the column in 5- to 10-m sections, with redistribution of the liquid between the sections.

EXAMPLE 18.7. Gas from a reactor has 3.0% ethylene oxide (EO) and 10% CO_2, with the rest mostly nitrogen, and 98 percent of the EO is to be recovered by scrubbing with water. The absorber will operate at 20 atm, using water with 0.04 mole percent EO at 30°C, and the gas enters at 30°C, saturated with water. How many transfer units are needed if 1.4 mol H_2O is used per mole of dry gas? Estimate the diameter of the column and the packed height if $1\frac{1}{2}$-in. Pall rings are used and the total gas feed rate is 10,000 mol/h.

Solution. Equilibrium data[4] for 30 and 40°C are shown in Fig. 18.24. By a heat balance similar to that of Example 18.4, the temperature rise of the liquid was estimated to be 12.5°C, which makes the equilibrium line for the column curve upward. The terminal points of the operating line are determined by material balance.

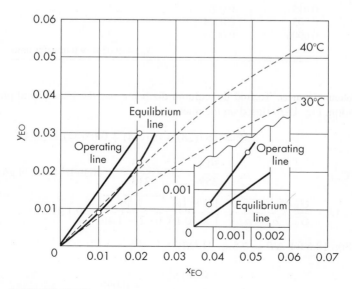

FIGURE 18.24
The yx diagram for Example 18.7.

Basis. 100 mol dry gas in; 140 mol solution in.

In	Out
87 N_2	87 N_2
10 CO_2	10 CO_2
3 EO	0.06 EO ($= 3 \times 0.02$)
100	97.06

Assume negligible CO_2 absorption and neglect the effect of H_2O on gas composition.

At top: $\qquad\qquad\qquad x = 0.0004 \qquad y = \dfrac{0.06}{97.06} = 0.00062$

Moles of EO absorbed: $\qquad\qquad 3 \times 0.98 = 2.94$

Moles of EO in water: $\qquad\qquad 140 \times 0.0004 = 0.056$

At bottom: $\qquad\qquad\qquad x = \dfrac{2.94 + 0.056}{140 + 2.94} = 0.0210$

$$y = 0.030$$

$$N_{Oy} = \int \frac{dy}{y - y^*} = \sum \frac{\Delta y}{(y - y^*)_L}$$

y	$y - y^*$	ΔN_{Oy}
0.03	0.008	—
0.015	0.006	2.14
0.005	0.0024	2.55
0.0006	0.0003	4.36
		$N_{Oy} = \overline{9.05} = 9.0$ transfer units

Column diameter. To find the column diameter, use the generalized pressure drop correlation, Fig. 18.6. Based on the inlet gas,

$$\bar{M} = 0.87(28) + 0.1(44) + 0.03(44) = 30.1$$

At 40°C, $\qquad\qquad \rho_y = \dfrac{30.1}{359} \times 20 \times \dfrac{273}{313} = 1.46 \text{ lb/ft}^3 \ (0.0234 \text{ g/cm}^3)$

$$\frac{G_x}{G_y}\sqrt{\frac{\rho_y}{\rho_x - \rho_y}} = \frac{1.4 \times 18}{1 \times 30.1}\sqrt{\frac{1.46}{62.2 - 1.46}} = 0.130$$

From Fig. 18.6 for $\Delta P = 0.5$ in. H_2O/ft,

$$\frac{G_y^2 F_p \mu_x^{0.1}}{\rho_y(\rho_x - \rho_y)g_c} = 0.045$$

From Table 18.1, $F_p = 40$. At 40°C, $\mu = 0.656$ cP. Therefore,

$$G_y^2 = \frac{0.045(1.46)(62.2 - 1.46)(32.2)}{40(0.656)^{0.1}} = 3.35$$

$$G_y = 1.83 \text{ lb/ft}^2 \cdot \text{s} = 6,590 \text{ lb/ft}^2 \cdot \text{h}$$

$$G_x = \frac{1.4 \times 18}{1 \times 30.1} \times 6,590 = 5,520 \text{ lb/ft}^2 \cdot \text{h}$$

For a feed rate of 10,000 mol/h $\times$ 30.1 = 3.01 $\times$ 10⁵ lb/h,

$$S = \frac{3.01 \times 10^5}{6,590} = 45.7 \text{ ft}^2 \qquad D = 7.6 \text{ ft}$$

Use an 8.0-ft-diameter column.

Column height. Find H_y and H_x from the data for ammonia-water and oxygen-water, with $1\frac{1}{2}$-in. Raschig rings. From Fig. 18.22 at $G_y = 500$ and $G_x = 1,500$,

$$H_{y,\text{NH}_3} = 1.4 \text{ ft}$$

The gas viscosity is assumed to be that of N_2 at 40°C and 1 atm, which is 0.0181 cP (App. 8). The diffusivity of EO in the gas is calculated from Eq. (17.28) to be, at 40°C and 20 atm,

$$D_v = 7.0 \times 10^{-3} \text{ cm}^2/\text{s}$$

$$\text{Sc} = \frac{\mu}{\rho D_v} = \frac{1.81 \times 10^{-4}}{(7.0 \times 10^{-3})(2.34 \times 10^{-2})} = 1.10$$

From Table 18.1, $f_p = 1.36$ for 1.5-in. Pall rings, so H_y is lower by this factor. The quantity H_y is assumed to vary with $G_y^{0.3}$ and $G_x^{-0.4}$. Thus

$$H_{y,\text{EO}} = 1.4\left(\frac{1.10}{0.66}\right)^{1/2} \frac{1}{1.36}\left(\frac{6,590}{500}\right)^{0.3}\left(\frac{1,500}{5,520}\right)^{0.4} = 1.71 \text{ ft}$$

From Fig. 18.21, $H_{x,\text{O}_2} = 0.9$ ft at $G_x = 1,500$. Using Eq. (17.32), $D_v = 2.15 \times 10^{-5}$ cm²/s for EO in H_2O at 40°C:

$$\text{Sc} = \frac{0.00656}{1.0 \times 2.15 \times 10^{-5}} = 305$$

Using Eq. (18.44) with the correction factor f_p and Sc = 381 for O_2 in water at 25°C,

$$H_{x,\text{EO}} = 0.9\left(\frac{5,520/0.656}{1,500/0.894}\right)^{0.3}\left(\frac{305}{381}\right)^{0.5}\frac{1}{1.36} = 0.96 \text{ ft}$$

From Fig. 18.24, the average value of m is about 1.0, and from Eq. (18.28)

$$H_{Oy} = H_y + \frac{mG_M}{L_M}H_x = 1.71 + \frac{1.0 \times 0.96}{1.4} = 2.40 \text{ ft}$$

For 9 transfer units, you need 22 ft. Use 24 ft of packing in two 12-ft sections, with redistribution of the liquid.

Absorption in Plate Columns

Gas absorption can be carried out in a column equipped with sieve trays or other types of plates normally used for distillation. A column with trays is sometimes chosen instead of a packed column to avoid the problem of liquid distribution in a large-diameter tower and to decrease the uncertainty in scaleup. Methods of determining the number of theoretical plates and estimating the average plate efficiency are discussed in Chap. 21.

ABSORPTION WITH CHEMICAL REACTION

Absorption followed by reaction in the liquid phase is often used to get more complete removal of a solute from a gas mixture. For example, a dilute acid solution can be used to scrub NH_3 from gas streams, and basic solutions are used to remove CO_2 and other acid gases. Reaction in the liquid phase reduces the equilibrium partial pressure of the solute over the solution, which greatly increases the driving force for mass transfer. If the reaction is essentially irreversible at absorption conditions, the equilibrium partial pressure is zero, and N_{Oy} can be calculated just from the change in gas composition. For $y^* = 0$,

$$N_{Oy} = \int_a^b \frac{dy}{y} = \ln \frac{y_b}{y_a} \qquad (18.48)$$

To illustrate the effect of a chemical reaction, consider the absorption of NH_3 in dilute HCl with a 300-fold reduction in gas concentration (6 to 0.02%). From Eq. (18.48), $N_{Oy} = \ln 300 = 5.7$, which can be compared with $N_{Oy} = 12$ for the same change in concentration using water at the conditions of Example 18.4.

A further advantage of absorption plus reaction is the increase in the mass-transfer coefficient. Some of this increase comes from a greater effective interfacial area, since absorption can now take place in the nearly stagnant regions (static holdup) as well as in the dynamic liquid holdup. For NH_3 absorption in H_2SO_4 solutions, $K_g a$ was 1.5 to 2 times the value for absorption in water.[17] Since the gas-film resistance is controlling, this effect must be due mainly to an increase in effective area. The values of $K_g a$ for NH_3 absorption in acid solutions were about the same as those for vaporization of water, where all the interfacial area is also expected to be effective. The factors $K_g a_{vap}/K_g a_{abs}$ and $K_g a_{react}/K_g a_{abs}$ decrease with increasing liquid rate and approach unity when the total holdup is much larger than the static holdup.

The factor $K_g a_{react}/K_g a_{abs}$ also depends on the concentration of reactant and is smaller when only a slight excess of reagent is present in the solution fed to the column. Data on liquid holdup and effective area have been published for Raschig rings and Berl saddles,[16] but similar results for newer packings are not available.

When the liquid-film resistance is dominant, as in the absorption of CO_2 or H_2S in aqueous solutions, a rapid chemical reaction in the liquid can lead to a very large increase in the mass-transfer coefficient. The coefficients shown in Fig. 18.23 for

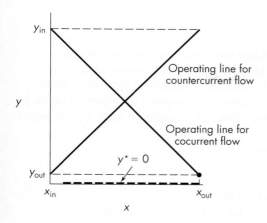

FIGURE 18.25
Operating lines for countercurrent flow and cocurrent flow for absorption plus an irreversible chemical reaction.

the CO_2-H_2O-NaOH system range from 1 to 4 mol/ft$^3 \cdot$ atm $\cdot$ h, compared to typical values for CO_2 in water of 0.05 to 0.2 mol/ft$^3 \cdot$ atm $\cdot$ h. The rapid reaction consumes much of the CO_2 very close to the gas-liquid interface, which makes the gradient for CO_2 steeper and enhances the process of mass transfer in the liquid. The ratio of the apparent value of k_L to that for physical absorption defines an enhancement factor ϕ, which ranges from 1.0 to 1,000 or more. Methods of predicting ϕ from kinetic and mass-transfer data are given in specialized texts.[1,15] When the value of ϕ is very large, the gas film may become the controlling resistance.

When absorption is accompanied by a very slow reaction, the apparent values of $K_g a$ may be lower than with absorption alone. An example is the absorption of Cl_2 in water followed by hydrolysis of the dissolved chlorine. The slow hydrolysis reaction essentially controls the overall rate of absorption.

Cocurrent flow operation

When the chemical reaction is essentially irreversible and the equilibrium partial pressure of the solute is zero, the number of transfer units for a given separation is the same for countercurrent operation or for cocurrent flow of liquid and gas. Figure 18.25 shows typical operating lines for both cases. In this diagram x is the total solute absorbed and reacted, and not the amount of solute present in the original form. For cocurrent operation with the feeds at the top, the gas leaving at the bottom is exposed to rich liquid, which has absorbed a lot of solute; but if $y^* = 0$, the driving force is just y, and N_{Oy} is calculated from Eq. (18.48), as for countercurrent flow.

The advantage of cocurrent operation is that there is no flooding limitation, and a gas flow rate much greater than normal can be used. This reduces the required column diameter, and the corresponding increase in liquid and gas mass velocity gives high mass-transfer coefficients. Liquid rates as high as 50,000 to 100,000 lb/ft$^2 \cdot$ h (70 to 140 kg/m$^2 \cdot$ s) can be used. The values of $K_g a$ or H_{Oy} can be estimated approximately by extrapolating counterflow data for the same system to the higher mass velocities.

SYMBOLS

a	Area of interface per unit packed volume, m^2/m^3 or ft^2/ft^3; a_{abs}, for absorption; a_{react}, for chemical reaction; a_{vap}, for vaporization
C	Mass concentration in liquid phase, g/g, $kg\ mol/m^3$, ppm, or $lb\ mol/ft^3$; C_a, at liquid inlet; C_b, at liquid outlet; C^*, liquid-phase concentration in equilibrium with gas of composition y; C_a^*, at liquid inlet; C_b^*, at liquid outlet
C_p	Molal heat capacity, $cal/g\ mol \cdot °C$
C_s	Capacity factor, $u_0\sqrt{\rho_y/(\rho_x - \rho_y)}$
D	Column diameter, m or ft
D_p	Diameter of packing unit, mm or in.
D_v	Diffusivity, m^2/s, cm^2/s, or ft^2/h; D_{vx}, in liquid; D_{vy}, in gas
F	Volume of gas fed to tower, m^3; F_{min}, minimum value
F_p	Packing factor for pressure drop
f_p	Relative mass-transfer coefficient (Table 18.1)
f'	Factor $(V/S)/(1 - y)(y - y^*)(K_y'a)$ in Example 18.6
G	Mass velocity based on total tower cross section, $kg/m^2 \cdot h$ or $lb/ft^2 \cdot h$; G_x, of liquid stream; $\bar{G}_x$, average value; G_y, of gas stream
G_M	Molal mass velocity, $kg\ mol/m^2 \cdot h$ or $lb\ mol/ft^2 \cdot h$
g_c	Newton's law proportionality factor, $32.174\ ft \cdot lb/lb_f \cdot s^2$
H	Height of a transfer unit, m or ft; H_G, alternate form of H_y; H_L, alternate form of H_x; H_{Ox}, overall, based on liquid phase; H_{Oy}, overall, based on gas phase; H_x, individual, based on liquid phase; H_y, individual, based on gas phase
Ka	Overall volumetric mass-transfer coefficient, $kg\ mol/m^3 \cdot h \cdot$ unit mole fraction or $lb\ mol/ft^3 \cdot h \cdot$ unit mole fraction; K_xa, based on liquid phase; K_ya, based on gas phase; $K_x'a$, $K_y'a$, including one-way diffusion factors, for liquid and gas phases, respectively
K_La	Overall volumetric mass-transfer coefficient for liquid phase, based on concentration difference, h^{-1}
K_ga	Overall volumetric mass-transfer coefficient for gas phase, based on partial-pressure driving force, $kg\ mol/m^3 \cdot h \cdot atm$ or $lb\ mol/ft^3 \cdot h \cdot atm$
k_L	Individual mass transfer coefficient for liquid phase, based on concentration difference, m/h or ft/h
ka	Individual volumetric mass-transfer coefficient, $kg\ mol/m^3 \cdot h \cdot$ unit mole fraction or $lb\ mol/ft^3 \cdot h \cdot$ unit mole fraction; k_xa, for liquid phase; k_ya, for gas phase
k_La	Individual volumetric mass-transfer coefficient for liquid phase, based on concentration difference, h^{-1}
k_ga	Individual volumetric mass-transfer coefficient for gas phase, based on partial-pressure driving force, $kg\ mol/m^3 \cdot h \cdot atm$ or $lb\ mol/ft^3 \cdot h \cdot atm$
L	Molal flow rate of liquid, mol/h; L_a, at liquid inlet; L_b, at liquid outlet; L_{min}, minimum value
L_M	Molal mass velocity of liquid, $kg\ mol/m^2 \cdot h$ or $lb\ mol/ft^2 \cdot h$
M	Molecular weight; $\bar{M}$, average value

m	Slope of equilibrium curve
N	Number of transfer units; N_G, alternate form of N_y; N_L, alternate form of N_x; N_{Ox}, overall, based on liquid phase; N_{Oy}, overall, based on gas phase; N_x, individual, liquid phase; N_y, individual, gas phase
N_A	Mass-transfer flux of component A, kg mol/m$^2 \cdot$s or lb mol/ft$^2 \cdot$h
N_H	Number of heat-transfer units
NTP	Number of theoretical plates
NTU	Number of transfer units
n	Exponent in Eq. (18.44)
P	Total pressure, atm; P_A', vapor pressure of component A
p_A	Partial pressure of component A
Q	Quantity of heat, cal; Q_a, heat of absorption; Q_{sx}, Q_{sy}, sensible-heat changes in liquid and gas, respectively; Q_v, heat of vaporization
r	Rate of absorption per unit volume, kg mol/m$^3 \cdot$h or lb mol/ft$^3 \cdot$h
S	Cross-sectional area of tower, m^2 or ft^2
Sc	Schmidt number, $\mu/\rho D_v$
T	Temperature, °C or °F; T_a, at liquid inlet; T_b, at liquid outlet; T_i, at intermediate point; T_{max}, maximum value
u_0	Superficial gas velocity, based on empty tower, m/s or ft/s; $u_{0,f}$, at flooding
V	Molal flow rate of gas, mol/h; V_a, at outlet; V_b, at inlet; V_{min}, minimum value
x	Mole fraction of solute (component A) in liquid; x_a, at liquid inlet; x_b, at liquid outlet; x_i, at gas-liquid interface; x_{max}, maximum value; x^*, equilibrium concentration corresponding to gas-phase composition y; x_b^*, in equilibrium with y_b
x_B	Mole fraction of component B in liquid; x_{Ba}, at liquid inlet; x_{Bb}, at liquid outlet
$\overline{(1-x)_L}$	One-way diffusion factor in liquid phase
y	Mole fraction of solute (component A) in gas; y_a, at gas outlet; $y_{a,max}$, maximum value; y_b, at gas inlet; y_i, at gas-liquid interface; y^*, equilibrium concentration corresponding to liquid-phase composition x; y_a^* in equilibrium with x_a; y_b^*, with x_b
y_B	Mole fraction of component B in gas; y_{Ba}, at gas outlet; y_{Bb}, at gas inlet; y_B^*, in equilibrium with x_B
$\overline{(1-y)_L}$	One-way diffusion factor in gas phase
Z	Vertical distance below top of packing, m or ft; Z_T, total height of packed section

Greek Letters

α	Constant in Eq. (18.44)
γ_A	Activity coefficient of component A
ΔC	Concentration driving force, g/g or ppm; ΔC_a, at liquid inlet; ΔC_b, at liquid outlet

ΔH	Heat of solution, kcal/g mol; ΔH_v, heat of vaporization
ΔP	Pressure drop, in. water/ft packing; ΔP_{flood}, at flooding
Δx_L	Logarithmic mean of $x_b^* - x_b$ and $x_a^* - x_a$
Δy_L	Logarithmic mean of $y_b - y_b^*$ and $y_a - y_a^*$
ε	Porosity or void fraction in packed section
μ	Viscosity, cP or lb/ft $\cdot$ h; μ_x, of liquid
ν	Kinematic viscosity, μ/ρ, m^{2}s or ft^2/s
ρ	Density, kg/m^3 or lb/ft^3; ρ_x, of liquid; ρ_y, of gas
ρ_M	Molar density, kg mol/m^3 or lb mol/ft^3; ρ_{Mx}, of liquid; ρ_{My}, of gas
ϕ	Enhancement factor in absorption with chemical reaction, dimensionless

PROBLEMS

18.1. A plant design calls for an absorber that is to recover 95 percent of the acetone in an air stream, using water as the absorbing liquid. The entering air contains 14 mole percent acetone. The absorber has cooling and operates at 80°F and 1 atm and is to produce a product containing 5.0 mole percent acetone. The water fed to the tower contains 0.02 mole percent acetone. The tower is to be designed to operate at 50 percent of the flooding velocity. (a) How many pounds per hour of water must be fed to the tower if the gas rate is 500 ft^3/min, measured at 1 atm and 32°F? (b) How many transfer units are needed, based on the overall gas-phase driving force? (c) If the tower is packed with 1-in. Raschig rings, what should be the packed height?

For equilibrium. Assume that $p_A = P_A' \gamma_A x$, where $\ln \gamma_A = 1.95(1 - x)^2$. The vapor pressure of acetone at 80°F is 0.33 atm.

18.2. An absorber is to recover 99 percent of the ammonia in the air-ammonia stream fed to it, using water as the absorbing liquid. The ammonia content of the air is 20 mole percent. Absorber temperature is to be kept at 30°C by cooling coils; the pressure is 1 atm. (a) What is the minimum water rate? (b) For a water rate 40 percent greater than the minimum, how many overall gas-phase transfer units are needed?

18.3. A soluble gas is absorbed in water using a packed tower. The equilibrium relationship may be taken as $y_e = 0.06x_e$. Terminal conditions are as follows:

	Top	Bottom
x	0	0.08
y	0.001	0.009

If $H_x = 0.24$ m and $H_y = 0.36$ m, what is the height of the packed section?

18.4. A mixture of 5% butane and 95% air is fed to a sieve-plate absorber containing eight ideal plates. The absorbing liquid is a heavy, nonvolatile oil having a molecular weight of 250 and a specific gravity of 0.90. The absorption takes place at 1 atm and 15°C. The butane is to be recovered to the extent of 95 percent. The vapor pressure of butane at 15°C is 1.92 atm, and liquid butane has a density of 580 kg/m^3 at

15°C. (*a*) Calculate the cubic meters of fresh absorbing oil per cubic meter of butane recovered. (*b*) Repeat, on the assumption that the total pressure is 3 atm and all other factors remain constant. Assume that Raoult's and Dalton's laws apply.

18.5. An absorption column is fed at the bottom with a gas containing 5% benzene and 95% air. At the top of the column a nonvolatile absorption oil is introduced, which contains 0.2% benzene by weight. Other data are as follows:

Feed, 2,000 kg of absorption oil per hour
Total pressure, 1 atm
Temperature (constant), 26°C
Molecular weight of absorption oil, 230
Viscosity of absorbing oil, 4.0 cP
Vapor pressure of benzene at 26°C, 100 mm Hg
Volume of entering gas, 0.3 m^3/s
Tower packing, Intalox saddles, 1-in. nominal size
Fraction of entering benzene absorbed, 0.90
Mass velocity of entering gas, 1.1 kg/m$^2 \cdot$s

Calculate the height and diameter of the packed section of this tower. Assume Raoult's law applies.

18.6. A vapor stream containing 3.0 mole percent benzene is scrubbed with wash oil in a packed absorber to reduce the benzene concentration in the gas to 0.02 percent. The oil has an average molecular weight of about 250 and a density of 54.6 lb/ft^3 and contains 0.015% benzene. The gas flow is 1,500 ft^3/min at 25°C and 1 atm. (*a*) If the scrubber operates isothermally at 25°C with a liquid rate of 14,000 lb/h, how many transfer units are required? (*b*) If the scrubber operates adiabatically, how many transfer units are needed? (*c*) What would be the major effect of operating with an oil of lower molecular weight, say, $M = 200$?

18.7. An aqueous waste stream containing 1.0 weight percent NH_3 is to be stripped with air in a packed column to remove 99 percent of the NH_3. What is the minimum air rate, in kilograms of air per kilogram of water, if the column operates at 20°C? How many transfer units are required at twice the minimum air rate?

18.8. An 8-ft-diameter column packed with 20 ft of 1-in. Berl saddles has air at 1.5 atm and 40°C flowing through it. The tower is apparently close to flooding, since $\Delta p = 24$ in. of water. The mass velocity of the liquid is 8.5 times that of the gas. (*a*) If the tower were repacked with $1\frac{1}{2}$-in. Intalox saddles, what would the pressure drop be? (*b*) How much higher flow rates could be used if the pressure drop were the same as it was with the Berl saddles?

18.9. An absorber is to remove 99 percent of solute A from a gas stream containing 4 mol percent A. Solutions of A in the solvent follow Henry's law, and the temperature rise of the liquid can be neglected. (*a*) Calculate N_{Oy} for operation at 1 atm using solute-free liquid at a rate of 1.5 times the minimum value. (*b*) For the same liquid rate, calculate N_{Oy} for operation at 2 atm and at 4 atm. (*c*) Would the effect of pressure on N_{Oy} be partly offset by a change in H_{Oy}?

18.10. A gas containing 2% A and 1% B is to be scrubbed with a solvent in which A is 5 times as soluble as B. Show that using two columns in series with separate regeneration of the liquid from each column would permit recovery of A and B in relatively pure form. Use a yx diagram to show the equilibrium and operating lines for the simultaneous absorption of A and B, and estimate the ratio of A and B in the liquid from the first absorber.

18.11. An absorber packed with 1-in. Intalox saddles operates at 50°C and 10 atm with a liquid mass velocity 5 times the gas mass velocity. Assuming the gas and liquid are similar to air and water, what gas mass velocity will give a pressure drop of 0.5 in. H_2O/ft packing? Use the generalized correlations to get the effect of changed physical properties, and apply a correction to the data of Fig. 18.4.

18.12. A column packed with Mellapak 250Y is operated at 1 atm with air and an aqueous solution at $G_x = 8G_y$ and 60 percent of the flooding velocity (see Fig. 18.8). (a) What are the superficial gas velocity and the mass velocity? (b) If the pressure were changed to 2 atm, what superficial velocity and mass velocity would be used?

18.13. A packed column is used to remove toluene from water by stripping with air. The water contains 60 ppm toluene (weight basis), and the concentration must be reduced to 2 ppm. The column will operate at 20°C and an average pressure of 1.1 atm. The equilibrium relationship is

$$P_{tol} = 256x$$

where P_{tol} = partial pressure of toluene, atm, in the gas phase, and x = mol fraction of toluene in the liquid. (a) Determine the minimum air rate in g mol/min for a water rate of 100 L/min. (b) Calculate N_{Ox} for an air rate twice the minimum. (c) Determine the packed height required if $H_y = 0.7$ m and $H_x = 0.6$ m.

18.14. If a column packed with 1-in. Berl saddles is repacked with 1-in. metal Pall rings, what will be the major changes in operating characteristics?

REFERENCES

1. Danckwerts, P. V. *Gas-Liquid Reactions.* New York: McGraw-Hill, 1970.
2. Eckert, J. S. *Chem. Eng. Prog.* **66**(3):39 (1970).
3. Fair, J. F., and J. L. Bravo. *Chem. Eng. Prog.* **86**(1):19 (1990).
4. Gmehling, J., U. Onken, and W. Arlt. *Vapor-Liquid Equilibria Data Collection,* vol. 1. Dechema, Frankfurt/Main, 1979.
5. Harriott, P. *Environ. Sci. Tech.* **23**:309 (1989).
6. Kister, H. Z., and D. R. Gill. *Chem. Eng. Prog.* **87**(2):32 (1991).
7. Lincoff, A. H., and J. M. Gossett. *International Symposium on Gas Transfer at Water Surfaces,* Ithaca, NY: Cornell University, June 1983.
8. Norton Chemical Process Products Division, Akron, Ohio, 1987.
9. O'Connell, H. E. *Trans. AIChE* **42**:741 (1946).
10. Perry, D., D. E. Nutter, and A. Hale. *Chem. Eng. Prog.* **86**(1):30 (1990).
11. Perry, J. H. *Chemical Engineers' Handbook,* 6th ed. New York: McGraw-Hill, 1984, (a) p. **3**–101, (b) p. **18**–23, (c) p. **18**–39.
12. Perry, R. H., and D. W. Green (eds.). *Perry's Chemical Engineers' Handbook,* 7th ed. New York: McGraw-Hill, 1997, p. **14**–43.

13. Robbins, L. A. *Chem. Eng. Prog.* **87**(5):87 (1991).
14. Sherwood, T. K., and F. A. L. Holloway. *Trans. AIChE* **36**:21, 39 (1940).
15. Sherwood, T. K., R. L. Pigford, and C. R. Wilke. *Mass Transfer,* New York: McGraw-Hill, 1975, p. 442.
16. Shulman, H. L., C. F. Ullrich, A. Z. Proulx, and J. O. Zimmerman. *AIChE J.* **1**:253 (1955).
17. Shulman, H. L., C. F. Ullrich, and N. Wells. *AIChE J.* **1**:247 (1955).
18. Sperandio, A., M. Richard, and M. Huber. *Chem. Ing. Tech.* **37**:322 (1965).
19. Spiegel, L., and W. Meier. *I. Chem. E. Symp. Ser.* **104**:A203 (1987).
20. Stedman, D. F. *Trans. AIChE* **33**:153 (1937).
21. Strigle, R. F., Jr. *Random Packings and Packed Towers.* Houston, TX: Gulf Publishing, 1987.
22. VonStockar, U., and C. R. Wilke. *Ind. Eng. Chem. Fund.* **16**:88, 94 (1977).
23. Whitney, R. P., and J. E. Vivian. *Chem. Eng. Prog.* **45**:323 (1949).

CHAPTER 19

Humidification Operations

Humidification and dehumidification involve the transfer of material between a pure liquid phase and a fixed gas that is nearly insoluble in the liquid. These operations are somewhat simpler than those for absorption and stripping, for when the liquid contains only one component, there are no concentration gradients and no resistance to mass transfer in the liquid phase. On the other hand, both heat transfer and gas-phase mass transfer are important, and they influence each other. In previous chapters they have been treated separately; here and in drying of solids (discussed in Chap. 24) they occur together, and concentration and temperature change simultaneously.

DEFINITIONS

In humidification operations, especially as applied to the system air-water, a number of rather special definitions are in common use. The usual basis for engineering calculations is a unit mass of vapor-free gas, where *vapor* means the gaseous form of the component that is also present as liquid and *gas* is the component present only in gaseous form. In this discussion a basis of a unit mass of vapor-free gas is used. In the gas phase the vapor will be referred to as component A and the fixed gas as component B. Because the properties of a gas-vapor mixture vary with total pressure, the pressure must be fixed. Unless otherwise specified, a total pressure of 1 atm is assumed. Also, it is assumed that mixtures of gas and vapor follow the ideal gas laws.

Humidity $\mathcal{H}$ is the mass of vapor carried by a unit mass of vapor-free gas. So defined, humidity depends only on the partial pressure of the vapor in the mixture when the total pressure is fixed. If the partial pressure of the vapor is p_A atm, the molal ratio of vapor to gas at 1 atm is $p_A/(P - p_A)$. The humidity is therefore

$$\mathcal{H} = \frac{M_A p_A}{M_B(P - p_A)} \tag{19.1}$$

where M_A and M_B are the molecular weights of components A and B, respectively.

The humidity is related to the mole fraction in the gas phase by the equation

$$y = \frac{\mathscr{H}/M_A}{1/M_B + \mathscr{H}/M_A} \tag{19.2}$$

Since $\mathscr{H}/M_A$ is usually small compared with $1/M_B$, often y may be considered to be directly proportional to $\mathscr{H}$.

Saturated gas is gas in which the vapor is in equilibrium with the liquid at the gas temperature. The partial pressure of vapor in saturated gas equals the vapor pressure of the liquid at the gas temperature. If $\mathscr{H}_s$ is the saturation humidity and P'_A the vapor pressure of the liquid,

$$\mathscr{H}_s = \frac{M_A P'_A}{M_B(P - P'_A)} \tag{19.3}$$

Relative humidity $\mathscr{H}_R$ is defined as the ratio of the partial pressure of the vapor to the vapor pressure of the liquid at the gas temperature. It is usually expressed on a percentage basis, so 100 percent humidity means saturated gas and 0 percent humidity means vapor-free gas. By definition

$$\mathscr{H}_R = 100\frac{p_A}{P'_A} \tag{19.4}$$

Percentage humidity $\mathscr{H}_A$ is the ratio of the actual humidity $\mathscr{H}$ to the saturation humidity $\mathscr{H}_s$ at the gas temperature, also on a percentage basis, or

$$\mathscr{H}_A = 100\,\frac{\mathscr{H}}{\mathscr{H}_s} = 100\,\frac{p_A/(P - p_A)}{P'_A/(P - P'_A)} = \mathscr{H}_R\,\frac{P - P'_A}{P - p_A} \tag{19.5}$$

At all humidities other than 0 or 100 percent, the percentage humidity is less than the relative humidity.

Humid heat c_s is the heat energy necessary to increase the temperature of 1 g or 1 lb of gas plus whatever vapor it may contain by 1°C or 1°F. Thus

$$c_s = c_{pB} + c_{pA}\mathscr{H} \tag{19.6}$$

where c_{pB} and c_{pA} are the specific heats of gas and vapor, respectively.

Humid volume v_H is the total volume of a unit mass of vapor-free gas plus whatever vapor it may contain at 1 atm and the gas temperature. From the gas laws and the values of standard molar volume (see Chap. 1, page 10), v_H in SI units is related to humidity and temperature by the equation

$$v_H = \frac{0.0224T}{273}\left(\frac{1}{M_B} + \frac{\mathscr{H}}{M_A}\right) \tag{19.7a}$$

where v_H is in cubic meters per gram and T is in kelvins. In fps units the equation is

$$v_H = \frac{359T}{492}\left(\frac{1}{M_B} + \frac{\mathscr{H}}{M_A}\right) \tag{19.7b}$$

where v_H is in cubic feet per pound and T is in degrees Rankine. For vapor-free gas $\mathscr{H} = 0$, and v_H is the specific volume of the fixed gas. For saturated gas $\mathscr{H} = \mathscr{H}_s$, and v_H becomes the *saturated volume*.

Dew point is the temperature to which a vapor-gas mixture must be cooled (at constant humidity) to become saturated. The dew point of a saturated gas phase equals the gas temperature.

Total enthalpy H_y is the enthalpy of a unit mass of gas plus whatever vapor it may contain. To calculate H_y, two reference states must be chosen, one for gas and one for vapor. Let T_0 be the datum temperature chosen for both components, and base the enthalpy of component A on liquid A at T_0. (Temperature $T_0 = 32°F$ for most air-water problems.) Let the temperature of the gas be T and the humidity $\mathcal{H}$. The total enthalpy is the sum of three items: the sensible heat of the vapor, the latent heat of the liquid at T_0, and the sensible heat of the vapor-free gas. Then

$$H_y = c_{pB}(T - T_0) + \mathcal{H}\lambda_0 + c_{pA}\mathcal{H}(T - T_0) \tag{19.8}$$

where λ_0 is the latent heat of the liquid at T_0. From Eq. (19.6) this becomes

$$H_y = c_s(T - T_0) + \mathcal{H}\lambda_0 \tag{19.9}$$

Phase equilibria

In humidification and dehumidification operations the liquid phase is a single pure component. The equilibrium partial pressure of solute in the gas phase is therefore a unique function of temperature when the total pressure on the system is held constant. Also, at moderate pressures the equilibrium partial pressure is almost independent of total pressure and is virtually equal to the vapor pressure of the liquid. By Dalton's law the equilibrium partial pressure may be converted to the equilibrium mole fraction y_e in the gas phase. Since the liquid is pure, x_e is always unity. Equilibrium data are often presented as plots of y_e versus temperature at a given total pressure, as shown for the system air-water at 1 atm in Fig. 19.1. The equilibrium mole fraction y_e is related to the saturation humidity by Eq. (19.2); thus

$$y_e = \frac{\mathcal{H}_s/M_A}{1/M_B + \mathcal{H}_s/M_A} \tag{19.10}$$

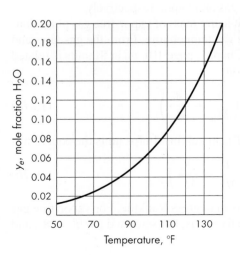

FIGURE 19.1
Equilibria for the system air-water at 1 atm.

Adiabatic saturator

Water is often sprayed into a stream of gas in a pipe or spray chamber to bring the gas to saturation. The pipe or chamber is insulated so that the process is adiabatic. The gas, with an initial humidity $\mathcal{H}$ and temperature T, is cooled and humidified. If not all the water evaporates and there is sufficient time for the gas to come to equilibrium with the water, the exit temperature of the gas is called the *adiabatic saturation temperature* T_s. The remaining liquid is also at T_s and can be recirculated to the spray nozzles. The value of T_s depends on the temperature and initial humidity of the air and to a minor extent on the initial water temperature. To simplify the analysis, the water is often assumed to enter at T_s.

An enthalpy balance can be written over this process. Pump work is neglected, and the enthalpy balance is based on temperature T_s as a datum. Then the enthalpy of the makeup liquid is zero, and the total enthalpy of the entering gas equals that of the leaving gas. Since the latter is at datum temperature, its enthalpy is simply $\mathcal{H}_s \lambda_s$, where $\mathcal{H}_s$ is the saturation humidity and λ_s is the latent heat, both at T_s. From Eq. (19.9) the total enthalpy of the entering gas is $c_s(T - T_s) + \mathcal{H}\lambda_s$, and the enthalpy balance is

$$c_s(T - T_s) + \mathcal{H}\lambda_s = \mathcal{H}_s\lambda_s$$

or

$$\frac{\mathcal{H}_s - \mathcal{H}}{T - T_s} = \frac{c_s}{\lambda_s} = \frac{c_{pB} + c_{pA}\mathcal{H}}{\lambda_s} \qquad (19.11)$$

To find the adiabatic saturation temperature for gases other than air, a heat balance similar to Eq. (19.11) is used. It may be more convenient, however, to use molar heat capacities, as in the following example.

Equation (19.11) cannot be solved directly for the adiabatic saturation temperature T_s, since $\mathcal{H}_s$, c_s, and λ_s are all functions of T_s. Thus T_s is obtained by a trial-and-error calculation, or, for the air-water system, by using humidity charts.

EXAMPLE 19.1. Flue gas at 320°F and 1 atm is to be cooled by a water spray. The gas contains 14% CO_2, 7% H_2O, 3% O_2, and 76% N_2. (*a*) Calculate the adiabatic saturation temperature if the water spray enters at 80°F. (*b*) Repeat for water entering at T_s.

Solution

(*a*) Basis: 100 mol of gas. Guess that T_s is about 120°F and evaluate the molar heat capacity C_p for each gas at $(320 + 120)/2 = 220$°F.

Gas	Number of moles n	Molar specific heat C_p	nC_p
CO_2	14	9.72	136.08
H_2O	7	8.11	56.77
O_2	3	7.14	21.42
N_2	76	6.98	530.48
	$\sum n = 100$		$\sum nC_p = 744.75$

Make a heat balance for z moles of water evaporated:

$$\sum nC_p(T - T_s) = z\lambda_s + 18z(120 - 80)$$

$$= z(\lambda_s + 720)$$

At 120°F, from App. 7,

$$\lambda_s = 1{,}025.5 \times 18 = 18{,}459 \text{ Btu/lb mol}$$

Then $744.75(320 - 120) = z(18{,}459 + 720) = 19{,}179z$

$$z = 7.77$$

Total moles of water in exit gas: $7 + 7.77 = 14.77$

Mole fraction of water in exit gas:

$$y = \frac{14.77}{107.77} = 0.137$$

From Fig. 19.1, the saturation value y_s at 120°F is 0.115. Therefore the saturation temperature must be greater than 120°F, since a higher estimate for T_s will reduce y and increase y_s.

Use Fig. 19.1 for a new estimate of T_s. For $y_e = 0.137$, $T_s = 126$°F. Then $\lambda_s = 1{,}022.1 \times 18 = 18{,}398$ Btu/lb mol. Neglecting changes in $\sum nC_p$, we have

$$744.75(320 - 126) = z[18{,}398 + 18(126 - 80)] = 19{,}226z$$

$$z = 7.51$$

Total moles of water: $7 + 7.51 = 14.51$

$$y = \frac{14.51}{107.51} = 0.135$$

This is close enough to 0.137, so $T_s \cong 126$°F.

(b) The last term of the heat balance is dropped if $T_{in} = T_s$. For $T_s = 126$°F,

$$744.75(320 - 126) = z(18{,}398)$$

$$z = 7.85$$

$$y = \frac{7.85 + 7}{107.51} = 0.138$$

The saturation temperature must be slightly higher than 126°F, but the difference is negligible. More accurate vapor pressure data would be needed to find the saturation temperature to within 0.1°F.

HUMIDITY CHART

A convenient diagram showing the properties of mixtures of a permanent gas and a condensable vapor is the humidity chart. A chart for mixtures of air and water at 1 atm is shown in Fig. 19.2. Many forms of such charts have been proposed. Figure 19.2 is based on the Grosvenor[2] chart.

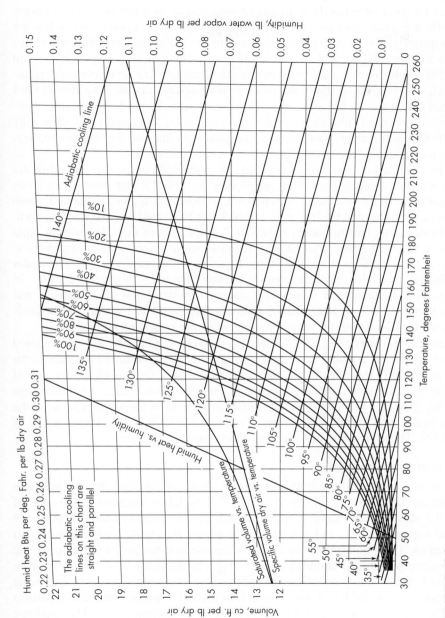

FIGURE 19.2
Humidity chart. Air-water at 1 atm.

601

On Fig. 19.2 temperatures are plotted as abscissas and humidities as ordinates. Any point on the chart represents a definite mixture of air and water. The curved line marked 100 percent gives the humidity of saturated air as a function of air temperature. By using the vapor pressure of water, the coordinates of points on this line are found from Eq. (19.3). Any point above and to the left of the saturation line represents a mixture of saturated air and liquid water. This region is important only in checking fog formation. Any point below the saturation line represents undersaturated air, and a point on the temperature axis represents dry air. The curved lines between the saturation line and the temperature axis marked in even percents represent mixtures of air and water of definite *percentage humidities*. As shown by Eq. (19.5), linear interpolation between the saturation line and the temperature axis can be used to locate the lines of constant percentage humidity.

The slanting lines running downward and to the right of the saturation line are called *adiabatic cooling lines*. They are plots of Eq. (19.11), each drawn for a given constant value of the adiabatic saturation temperature. For a given value of T_s, both H_s and λ_s are fixed, and the line of $\mathcal{H}$ versus T can be plotted by assigning values to $\mathcal{H}$ and calculating corresponding values of T. Inspection of Eq. (19.11) shows that the slope of an adiabatic cooling line, if drawn on truly rectangular coordinates, is $-c_s/\lambda_s$, and by Eq. (19.6), this slope depends on the humidity. On rectangular coordinates, then, the adiabatic cooling lines are neither straight nor parallel. In Fig. 19.2 the ordinates are sufficiently distorted to straighten the adiabatics and render them parallel, so interpolation between them is easy. The ends of the adiabatics are identified with the corresponding adiabatic saturation temperatures.

Lines are shown on Fig. 19.2 for the specific volume of dry air and the saturated volume. Both lines are plots of volume versus temperature. Volumes are read on the scale at the left. Coordinates of points on these lines are calculated by use of Eq. (19.7b). Linear interpolation between the two lines, based on percentage humidity, gives the humid volume of unsaturated air. Also, the relation between the humid heat c_s and humidity is shown as a line on Fig. 19.2. This line is a plot of Eq. (19.6). The scale for c_s is at the top of the chart.

Use of humidity chart

The usefulness of the humidity chart as a source of data on a definite air-water mixture can be shown by reference to Fig. 19.3, which is a portion of the chart of Fig. 19.2. Assume, for example, that a given stream of undersaturated air is known to have a temperature T_1 and a percentage humidity $\mathcal{H}_{A1}$. Point *a* represents this air on the chart. This point is the intersection of the constant-temperature line for T_1 and the constant-percentage-humidity line for $\mathcal{H}_{A1}$. The humidity $\mathcal{H}_1$ of the air is given by point *b*, the humidity coordinate of point *a*. The dew point is found by following the constant-humidity line through point *a* to the left to point *c* on the 100 percent line. The dew point is then read at point *d* on the temperature axis. The adiabatic saturation temperature is the temperature applying to the adiabatic cooling line through point *a*. The humidity at adiabatic saturation is found by following the adiabatic line through point *a* to its intersection *e* on the 100 percent line and reading humidity $\mathcal{H}_s$ at point *f* on the humidity scale. Interpolation between the adiabatic lines may be necessary. The adiabatic saturation temperature T_s is given by point *g*.

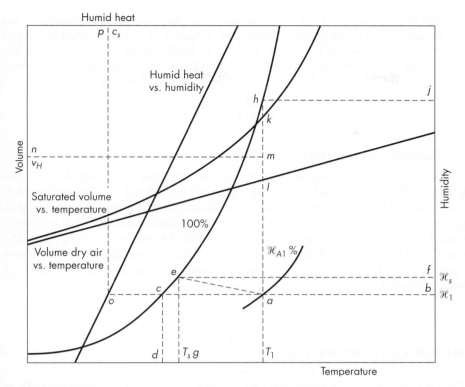

FIGURE 19.3
Use of humidity chart.

If the original air is subsequently saturated at constant temperature, the humidity after saturation is found by following the constant-temperature line through point a to point h on the 100 percent line and reading the humidity at point j.

The humid volume of the original air is found by locating points k and l on the curves for saturated and dry volumes, respectively, corresponding to temperature T_1. Point m is then found by moving along line lk a distance $(\mathcal{H}_A/100)\overline{kl}$ from point l, where $\overline{kl}$ is the line segment between points l and k. The humid volume v_H is given by point n on the volume scale. The humid heat of the air is found by locating point o, the intersection of the constant-humidity line through point a and the humid heat line, and reading the humid heat c_s at point p on the scale at the top.

EXAMPLE 19.2. The temperature and dew point of the air entering a certain dryer are 150 and 60°F (65.6 and 15.6°C), respectively. What additional data for this air can be read from the humidity chart?

Solution. The dew point is the temperature coordinate on the saturation line corresponding to the humidity of the air. The saturation humidity for a temperature of 60°F is 0.011 lb of water per pound (0.011 g/g) of dry air, and this is the humidity of the air. From the temperature and humidity of the air, the point on the chart for this air is located. At $\mathcal{H} = 0.011$ and $T = 150°F$, the percentage humidity $\mathcal{H}_A$ is found by interpolation to be 5.2 percent. The adiabatic cooling line through this point intersects the

100 percent line at 85°F (29.4°C), and this is the adiabatic saturation temperature. The humidity of saturated air at this temperature is 0.026 lb of water per pound (0.026 g/g) of dry air. The humid heat of the air is 0.245 Btu/lb dry air·°F (1.03 J/g·°C). The saturated volume at 150°F is 20.7 ft³/lb (1.29 m³/kg) of dry air, and the specific volume of dry air at 150°F is 15.35 ft³/lb (0.958 m³/kg). The humid volume is, then,

$$v_H = 15.35 + \frac{0.011 \times 359}{18} \left(\frac{610}{492} \right) = 15.62 \text{ ft}^3/\text{lb dry air } (0.978 \text{ m}^3/\text{kg})$$

Humidity charts for systems other than air-water

A humidity chart may be constructed for any system at any desired total pressure. The data required are the vapor pressure and latent heat of vaporization of the condensable component as a function of temperature, the specific heats of pure gas and vapor, and the molecular weights of both components. If a chart on a mole basis is desired, all equations can easily be modified to the use of molal units. If a chart at a pressure other than 1 atm is wanted, obvious modifications in the above equations may be made. Charts for several common systems besides air-water have been published.[5]

WET-BULB TEMPERATURE

The properties discussed above and those shown on the humidity charts are static or equilibrium quantities. Equally important are the rates at which mass and heat are transferred between the gas and liquid phases that are not in equilibrium. The driving forces for mass and heat transfer are concentration and temperature differences, which can be predicted by using a quantity called the wet-bulb temperature.

The wet-bulb temperature is the steady-state, nonequilibrium temperature reached by a small mass of liquid exposed under adiabatic conditions to a continuous stream of gas. Because the gas flow is continuous, the properties of the gas are constant and are usually evaluated at inlet conditions. If the gas is not saturated, some liquid evaporates, cooling the remaining liquid until the rate of heat transfer to the liquid just balances the heat needed for evaporation. The liquid temperature when steady state is reached is the wet-bulb temperature.

The method of measuring the wet-bulb temperature is shown in Fig. 19.4a. A thermometer or other temperature-measuring device, such as a thermocouple, is covered by a wick, which is saturated with pure liquid and immersed in a stream of gas having a definite temperature T and humidity $\mathcal{H}$. Assume that initially the temperature of the liquid is about that of the gas. Since the gas is not saturated, liquid evaporates; and because the process is adiabatic, the latent heat is supplied at first by cooling the liquid. As the temperature of the liquid decreases below that of the gas, sensible heat is transferred to the liquid. Ultimately a steady state is reached at such a liquid temperature that the heat needed to evaporate the liquid and heat the vapor to gas temperature is exactly balanced by the sensible heat flowing from the gas to the liquid. It is this steady-state temperature, denoted by T_w, that is

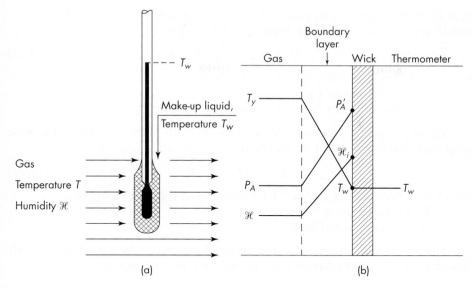

FIGURE 19.4
(*a*) Wet-bulb thermometer. (*b*) Gradients in the gas boundary layer.

called the *wet-bulb temperature*. It is a function of both T and $\mathscr{H}$. The temperature and concentration gradients at steady state are shown in Fig. 19.4*b*.

To measure the wet-bulb temperature with precision, three precautions are necessary: (1) The wick must be completely wet, so no dry areas of the wick are in contact with the gas; (2) the velocity of the gas should be large enough (at least 5 m/s) to ensure that the rate of heat flow by radiation from warmer surroundings to the bulb is negligible in comparison with the rate of sensible heat flow by conduction and convection from the gas to the bulb; (3) if makeup liquid is supplied to the bulb, it should be at the wet-bulb temperature. When these precautions are taken, the wet-bulb temperature is independent of gas velocity over a wide range of flow rates.

The wet-bulb temperature superficially resembles the adiabatic saturation temperature T_s. Indeed, for air-water mixtures the two temperatures are nearly equal. This is fortuitous, however, and is not true of mixtures other than air and water. The wet-bulb temperature differs fundamentally from the adiabatic saturation temperature. The temperature and humidity of the gas vary during adiabatic saturation, and the endpoint is a true equilibrium rather than a dynamic steady state.

Commonly, an uncovered thermometer is used along with the wet bulb to measure T, the actual gas temperature, and the gas temperature is usually called the *dry-bulb temperature*.

Theory of wet-bulb temperature

At the wet-bulb temperature the rate of heat transfer from the gas to the liquid may be equated to the product of the rate of vaporization and the sum of the latent heat of evaporation at temperature T_w and the sensible heat of the vapor. Since radiation may be neglected, this balance may be written

$$q = M_A N_A [\lambda_w + c_{pA}(T - T_w)] \tag{19.12}$$

where q = rate of sensible heat transfer to liquid
  N_A = molal rate of vaporization
  λ_w = latent heat of liquid at wet-bulb temperature T_w

The rate of heat transfer may be expressed in terms of the area, the temperature drop, and an effective heat-transfer coefficient in the usual way, or

$$q = h_y(T - T_i)A \tag{19.13}$$

where h_y = heat-transfer coefficient between gas and surface of liquid
  T_i = temperature at interface
  A = surface area of liquid

The rate of mass transfer may be expressed in terms of the mass-transfer coefficient, the area, and the driving force in mole fraction of vapor, or

$$N_A = \frac{k_y}{(1 - y)_L}(y_i - y)A \tag{19.14}$$

where N_A = molal rate of transfer of vapor
  y_i = mole fraction of vapor at interface
  y = mole fraction of vapor in airstream
  k_y = mass-transfer coefficient, mole per unit area per unit mole fraction
  $\overline{(1 - y)_L}$ = one-way diffusion factor

If the wick is completely wet and no dry spots show, the entire area of the wick is available for both heat and mass transfer and the areas in Eqs. (19.13) and (19.14) are equal. Since the temperature of the liquid is constant, no temperature gradients are necessary in the liquid to act as driving forces for heat transfer within the liquid, the surface of the liquid is at the same temperature as the interior, and the surface temperature of the liquid T_i equals T_w. Since the liquid is pure, no concentration gradients exist, and granting interfacial equilibrium, y_i is the mole fraction of vapor in saturated gas at temperature T_w. It is convenient to replace the mole fraction terms in Eq. (19.14) by humidities through the use of Eq. (19.2), noting that y_i corresponds to $\mathcal{H}_w$, the saturation humidity at the wet-bulb temperature. [See Eq. (19.10).] Following this by substituting q from Eq. (19.13) and N_A from Eq. (19.14) into Eq. (19.12) gives

$$h_y(T - T_w) = \frac{k_y}{(1 - y)_L}\left(\frac{\mathcal{H}_w}{1/M_B + \mathcal{H}_w/M_A} - \frac{\mathcal{H}}{1/M_B + \mathcal{H}/M_A}\right)$$
$$\times [\lambda_w + c_{pA}(T - T_w)] \tag{19.15}$$

Equation (19.15) may be simplified without serious error in the usual range of temperatures and humidities as follows: (1) the factor $(1 - y)_L$ is nearly unity and can be omitted; (2) the sensible-heat item $c_{pA}(T - T_w)$ is small in comparison with λ_w

and can be neglected; (3) the terms $\mathcal{H}_w/M_A$ and $\mathcal{H}/M_A$ are small in comparison with $1/M_B$ and may be dropped from the denominators of the humidity terms. With these simplifications Eq. (19.15) becomes

$$h_y(T - T_w) = M_B k_y \lambda_w (\mathcal{H}_w - \mathcal{H})$$

or

$$\frac{\mathcal{H} - \mathcal{H}_w}{T - T_w} = -\frac{h_y}{M_B k_y \lambda_w} \tag{19.16}$$

For a given wet-bulb temperature, both λ_w and $\mathcal{H}_w$ are fixed. The relation between $\mathcal{H}$ and T then depends on the ratio h_y/k_y. The close analogy between mass transfer and heat transfer provides considerable information on the magnitude of this ratio and the factors that affect it.

It has been shown in Chap. 12 that heat transfer by conduction and convection between a stream of fluid and a solid or liquid boundary depends on the Reynolds number DG/μ and the Prandtl number $c_p\mu/k$. Also, as shown in Chap. 17, the mass-transfer coefficient depends on the Reynolds number and the Schmidt number $\mu/\rho D_v$. As discussed in Chap. 17, the rates of heat and mass transfer, when these processes are under the control of the same boundary layer, are given by equations that are identical in form. For turbulent flow of the gas stream these equations are

$$\frac{h_y}{c_p G} = b \, \text{Re}^n \, \text{Pr}^m \tag{19.17}$$

and

$$\frac{\bar{M}k_y}{G} = b \, \text{Re}^n \, \text{Sc}^m \tag{19.18}$$

where b, n, m = constants
$\bar{M}$ = average molecular weight of gas stream

Substitution of h_y from Eq. (19.17) and k_y from Eq. (19.18) in Eq. (19.16), assuming $\bar{M} = M_B$, gives

$$\frac{\mathcal{H} - \mathcal{H}_w}{T - T_w} = -\frac{h_y}{M_B k_y \lambda_w} = -\frac{c_p}{\lambda_w}\left(\frac{\text{Sc}}{\text{Pr}}\right)^m \tag{19.19}$$

and

$$\frac{h_y}{M_B k_y} = c_p\left(\frac{\text{Sc}}{\text{Pr}}\right)^m \tag{19.20}$$

If m is taken as $\frac{2}{3}$, the predicted value of $h_y/M_B k_y$ for air in water is $0.24(0.62/0.71)^{2/3}$, or 0.22 Btu/lb·°F (0.92 J/g·°C). The experimental value[6] is 0.26 Btu/lb·°F (1.09 J/g·°C), somewhat larger than predicted, because of heat transfer by radiation. For organic liquids in air it is larger, in the range 0.4 to 0.5 Btu/lb·°F (1.6 to 2.0 J/g·°C). The difference, as shown by Eq. (19.20), is the result of the differing ratios of Prandtl and Schmidt numbers for water and for organic vapors.

Psychrometric line and Lewis relation

For a given wet-bulb temperature, Eq. (19.19) can be plotted on the humidity chart as a straight line having a slope of $-h_y/M_B k_y \lambda_w$ and intersecting the 100 percent line at T_w. This line is called the *psychrometric line*. When both a psychrometric line, from Eq. (19.19), and an adiabatic saturation line, from Eq. (19.11), are plotted for the same point on the 100 percent curve, the relation between the lines depends on the relative magnitudes of c_s and $h_y/M_B k_y$.

For the system air-water at ordinary conditions the humid heat c_s is almost equal to the specific heat c_p, and the following equation is nearly correct:

$$\frac{h_y}{M_B k_y} \cong c_s \tag{19.21}$$

Equation (19.21) is known as the *Lewis relation*.[4] When this relation holds, the psychrometric line and the adiabatic saturation line become essentially the same. In Fig. 19.2 for air-water, therefore, the same line may be used for both. For other systems separate lines must be used for psychrometric lines. With nearly all mixtures of air and organic vapors, the psychrometric lines are steeper than the adiabatic saturation lines, and the wet-bulb temperature of any mixture other than a saturated one is higher than the adiabatic saturation temperature.

Measurement of humidity

The humidity of a stream or mass of gas may be found by measuring either the dew point or the wet-bulb temperature or by direct absorption methods.

Dew-point methods. If a cooled, polished disk is inserted into gas of unknown humidity and the temperature of the disk gradually lowered, the disk reaches a temperature at which mist condenses on the polished surface. The temperature at which this mist just forms is the temperature of equilibrium between the vapor in the gas and the liquid phase. It is therefore the dew point. A check on the reading is obtained by slowly increasing the disk temperature and noting the temperature at which the mist just disappears. From the average of the temperatures of mist formation and disappearance, the humidity can be read from a humidity chart.

Psychrometric methods. A very common method of measuring the humidity is to determine simultaneously the wet-bulb and dry-bulb temperatures. From these readings the humidity is found by locating the psychrometric line intersecting the saturation line at the observed wet-bulb temperature and following the psychrometric line to its intersection with the ordinate of the observed dry-bulb temperature.

Direct methods. The vapor content of a gas can be determined by direct analysis, in which a known volume of gas is drawn through an appropriate analytical device.

COOLING TOWERS

When warm liquid is brought into contact with unsaturated gas, part of the liquid evaporates and the liquid temperature drops. The most important application of this principle is in the use of cooling towers to lower the temperature of recirculated water used for condensers and heat exchangers in chemical plants, power plants, and air conditioning units. Cooling towers are large-diameter columns with special types of packing designed to give good gas-liquid contact with low pressure drop. Warm water is distributed over the packing by spray nozzles or a grid of notched troughs or pipes. Air is passed through the packing by forced-draft or induced-draft fans, or in some designs it is drawn through by natural convection.

Two of the major types of cooling towers are shown in Fig. 19.5. In the cross-flow tower, which is rectangular in cross section, air passes horizontally through the inclined beds of packing or fill, while the water flows downward. Slanted louvers prevent the escape of water drops to the outside, and angled baffles called drift eliminators capture most of the droplets that are entrained in the exit air. Air is drawn through the tower by a propeller-type fan with several blades. In large units, the pitch of the blades can be adjusted to vary the flow of air. The fan is often located at the throat of a venturi-shaped cylinder that promotes smooth flow of air to the fan and gives some pressure recovery in the expanding section. The cylinder also discharges the moist air farther above ground level, which decreases the chance of recirculation to the air inlet.[3] This type of pressure recovery cylinder is used mainly with large cooling towers.

In the counterflow tower, air enters below the layer of fill and passes upward countercurrent to the flow of descending water. This is a more efficient arrangement for heat transfer and permits a closer temperature approach, as was shown in comparing counterflow and crossflow heat exchangers in Chap. 15. The counterflow tower can be used with a forced-draft fan at the base of the tower, but a large space below the fill is needed for good air distribution.

In older towers, the packing consisted of horizontal redwood or cypress slats spaced so that water falling on the slats would splash and the droplets be intercepted by the next layer of packing. Splash-type fill is still used in some crossflow towers, but V-shaped bars of polyvinyl chloride are used instead of wooden slats.[8] The pressure drop is low because airflow is parallel to the bars or slats, and the open structure makes inspection and cleaning relatively easy. Splash-type packing is not recommended for counterflow towers.

The most common type of packing for new installations is cellular fill or film-type packing, which consists of corrugated sheets of plastic similar to those used in plate-type heat exchangers. Water flows over the surface of the packing, giving more transfer area per unit volume than splash-type packings. The plastic sheets are spaced 3/4 to 1.0 in. (18 to 25 mm) apart to permit high flow rates of air and water with only moderate pressure drop. The depth of fill may be only a few feet, a small fraction of the total height of the unit. With cellular fill, it is especially important to get good distribution of the water at the top, since redistribution does not occur naturally, as it does with random dumped packings.

The reduction in water temperature in the cooling tower comes mainly from evaporation, although when the air temperature is low, there is also some sensible-heat transfer to the air. However, even when the air is warmer than the water, water can be cooled by evaporation if the wet-bulb temperature is below that of the water. In practice, the discharge temperature of the water is 5 to 15°F (3 to 8°C) above the wet-bulb temperature, and this difference is known as the *approach*. The change in

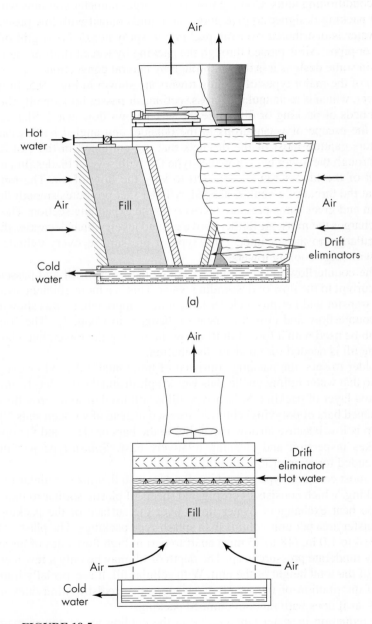

FIGURE 19.5
Typical cooling towers: (*a*) crossflow tower; (*b*) counterflow
tower.

water temperature from inlet to exit is known as the *range,* and the range is generally 10 to 30°F (6 to 17°C).[7]

The loss of water by evaporation during cooling is small. Since about 1,000 Btu is required to vaporize 1 lb of water, and 50 lb must be cooled 20°F to provide 1,000 Btu, a range of 20°F means an evaporation loss of 2 percent. In addition there are losses of spray droplets, referred to as drift or windage, but in a well-designed tower these amount to only about 0.2 percent. The total makeup water supplied must equal the evaporation and drift losses plus the amount of purge or blowdown needed to limit the buildup of dissolved salts.

Cooling towers are generally selected after consultation with equipment suppliers and consideration of factors such as the average and maximum heat duty, the required temperature range, the availability and quality of makeup water, and the local weather conditions. The tower is often sized to satisfy all but the most extreme conditions, such as when the wet-bulb temperature exceeds a certain limit for a few days per year. Detailed weather data for design purposes for all regions of the United States are available from the Marley Cooling Tower Company[3] or from government agencies.

Theory of Counterflow Cooling Towers

When the wet-bulb temperature is measured, heat transfer and mass transfer take place at steady state with gradients as shown in Fig. 19.4*b.* The flow of heat to the interface just matches that needed for evaporation of the water that diffuses as vapor into the bulk gas. There is no significant gradient in the liquid, which remains at a constant temperature. By contrast in a cooling tower, the water temperature changes as droplets pass through the tower, and it is necessary to consider heat flow in the liquid phase as well as heat and mass transfer in the gas.

Typical gradients at the bottom and top of a cooling tower are sketched in Fig. 19.6. At the bottom, the air temperature can be greater than the water temperature (Fig. 19.6*a*), but the water is being cooled because the interface temperature T_i

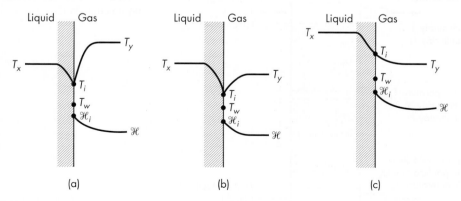

(a)   (b)   (c)

FIGURE 19.6
Conditions in cooling tower: (*a*), (*b*) at bottom of tower, (*c*) at top of tower.

is lower than the bulk water temperature T_x. The humidity at the interface is greater than that in the bulk gas, which provides a driving force for mass transfer of water vapor. If the inlet air temperature is less than the exit water temperature, as in Fig. 19.6b, the gradients are similar in shape, but there is less sensible-heat transfer through the gas film. In all cases, the interface temperature must be above the wet-bulb temperature, since if $T_x = T_w$, all the heat for vaporization would come from the gas, and there would be no temperature gradient in the water and no cooling of the water.

As the air passes up through the tower, the air temperature might decrease for a short distance, but it will eventually increase as the air contacts warmer and warmer water. At the top, the gradients might be as shown in Fig. 19.6c. Heat transferred from the water to the interface is used to warm the air as well as to provide heat of vaporization, although the water cooling due to evaporation is much greater than that from sensible-heat transfer to the air. The exit gas temperature is usually within a few degrees Fahrenheit of the inlet water temperature.

Equations for cooling-tower analysis

Consider the counterflow cooling tower shown in Fig. 19.7. Air at humidity $\mathcal{H}_b$ and temperature T_{yb} enters the bottom of the tower and leaves at the top with humidity $\mathcal{H}_a$ and temperature T_{ya}. Water enters the top at temperature T_{xa} and leaves at the bottom at temperature T_{xb}. The mass velocity of the air is G_y', the mass of vapor-free air per hour per unit cross section of tower. The mass velocities of water at the inlet and outlet are, respectively, G_{xa} and G_{xb}. At a distance Z from the bottom of the contact zone, the air and water temperatures are T_y and T_x, and the humidity is $\mathcal{H}$. At the gas-liquid interface, the temperature is T_i and the humidity $\mathcal{H}_i$. Assume for convenience that the interface temperature is greater than the gas temperature, as in Fig. 19.6c. (The following derivation is still valid if $T_i < T_y$.)

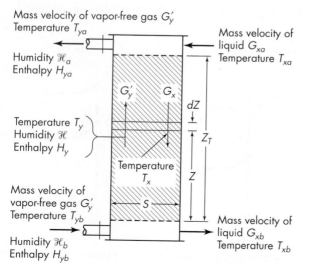

Mass velocity of vapor-free gas G_y'
Temperature T_{ya}

Humidity $\mathcal{H}_a$
Enthalpy H_{ya}

Mass velocity of liquid G_{xa}
Temperature T_{xa}

G_y' G_x

dZ

Z_T

Temperature T_y
Humidity $\mathcal{H}$
Enthalpy H_y

Temperature T_x

Z

Mass velocity of vapor-free gas G_y'
Temperature T_{yb}

S

Humidity $\mathcal{H}_b$
Enthalpy H_{yb}

Mass velocity of liquid G_{xb}
Temperature T_{xb}

FIGURE 19.7
Flow diagram of countercurrent gas-liquid contactor.

An enthalpy balance for a short section of tower dZ is

$$G_y' \, dH_y = d(G_x H_x) \tag{19.22}$$

Since the change in liquid rate in the tower is only 1 to 2 percent, G_x is assumed to be constant.

$$G_y' \, dH_y = G_x c_L \, dT_x \tag{19.23}$$

The change in gas enthalpy is the change in sensible heat plus the change in humidity times the heat of vaporization.

$$dH_y = c_s \, dT_y + \lambda_0 \, d\mathcal{H} \tag{19.24}$$

where λ_0 = heat of vaporization at 32°F.

The enthalpy of saturated air is

$$H_{y,\text{sat}} = c_s (T_y - 32) + \lambda_0 \mathcal{H}_s \tag{19.25}$$

The overall energy balance for the tower is

$$G_y'(H_a - H_b) = G_x c_L (T_{xa} - T_{xb}) \tag{19.26}$$

At an intermediate point in the tower, the enthalpy balance is

$$G_y'(H_a - H_y) = G_x c_L (T_{xa} - T_x) \tag{19.27}$$

Equation (19.27) is the operating line for the tower, and it is shown as a straight line of slope $G_x c_L / G_y'$ on a plot of air enthalpy versus water temperature in Fig. 19.8. The equilibrium line gives the enthalpy of air saturated with water vapor [Eq. (19.25)] as a function of temperature. The enthalpy-temperature diagram for the cooling tower is similar to that for a stripping column, but energy rather than a solute is being transferred from the water to the air. For given water temperatures and inlet air conditions, there is a minimum air rate corresponding to the operating line that just touches the equilibrium line, as shown in Fig. 19.8. Because of the curvature of the equilibrium line, the minimum air rate is sometimes determined by a line tangent to the curve. The air rate is generally chosen to be 1.2 to 2.0 times the minimum value.

The required packed height for a cooling tower can be determined by using the operating line–equilibrium line diagram and an overall coefficient based on an enthalpy driving force. To show why this is true, the rate equations for the air-water system are examined.

The rate of sensible-heat transfer from the water to the interface is

$$G_x c_L \, dT_x = h_x a(T_x - T_i) \, dZ \tag{19.28}$$

where $h_x a$ = volumetric heat-transfer coefficient for liquid. The rate of heat transfer from the interface to the gas is

$$G_y' c_s \, dT_y = h_y a(T_i - T_y) \, dZ \tag{19.29}$$

where $h_y a$ = volumetric heat-transfer coefficient for gas.

The rate of mass transfer of water vapor through the gas film is written first in the normal way, using a molar flow rate and a mole fraction driving force. A dilute gas is assumed so that $(1 - y)_L \cong 1.0$.

$$G_M \, dy = k_y a(y_i - y) \, dZ \tag{19.30}$$

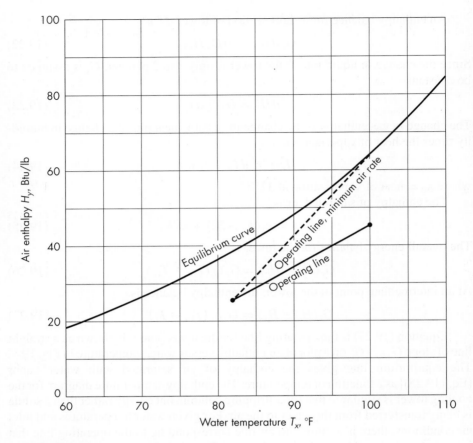

FIGURE 19.8
Operating diagram for cooling tower; plot of enthalpy of the air versus water temperature.

Since for low-humidity air, $G_M \cong G_y'/M_B$, where M_B is the molecular weight of the inert gas (air), y is approximately proportional to $\mathscr{H}$ and $\mathscr{H}/M_A \ll 1/M_B$, Eq. (19.30) can be changed to

$$G_y' \, d\mathscr{H} = k_y a M_B (\mathscr{H}_i - \mathscr{H}) \, dZ \qquad (19.31)$$

The a terms in $h_x a$, $h_y a$, and $k_y a$ are assumed to be identical.

Equation (19.31) is converted to an energy basis by multiplying by λ_0.

$$G_y' \lambda_0 \, d\mathscr{H} = k_y a M_B \lambda_0 (\mathscr{H}_i - \mathscr{H}) \, dZ \qquad (19.32)$$

Combining Eq. (19.32) and Eq. (19.29) gives

$$G_y' (\lambda_0 \, d\mathscr{H} + c_s \, dT_y) = [k_y a M_B \lambda_0 (\mathscr{H}_i - \mathscr{H}) + h_y a (T_i - T_y)] \, dZ \quad (19.33)$$

The Lewis relationship $h_y = c_s M_B k_y$ is used to replace $h_y a$ in the bracketed term of Eq. (19.33).

$$G_y' (\lambda_0 \, d\mathscr{H} + c_s \, dT_y) = k_y a M_B [\lambda_0 (\mathscr{H}_i - \mathscr{H}) + c_s (T_i - T_y)] \, dZ \quad (19.34)$$

Since the term in parentheses on the left-hand side is the differential change in enthalpy and the bracketed term is an enthalpy difference, Eq. (19.34) becomes

$$G_y' \, dH_y = k_y a M_B (H_i - H_y) \, dZ \qquad (19.35)$$

Thus the rate of change of gas enthalpy is proportional to the difference between the enthalpy at the interface and that in the bulk gas, and the transfer coefficient is the normal gas-film transfer coefficient multiplied by M_B, since G_y' and H are based on mass, not moles.

To determine conditions at the interface, the heat-transfer rate in the liquid film is equated to the change of gas enthalpy.

$$h_x a (T_x - T_i) \, dZ = k_y a M_B (H_i - H_y) \, dZ \qquad (19.36)$$

or

$$\frac{H_i - H_y}{T_i - T_x} = -\frac{h_x a}{k_y a M_B} \qquad (19.37)$$

Thus, a tie line from point H_i, T_i on the equilibrium curve to point H_y, T_x on the operating line has a slope $-h_x a / k_y a \, M_B$. By constructing tie lines of this slope for different values of H_i, Eq. (19.35) can be integrated to get the overall height.

$$\int \frac{dH_y}{H_i - H_y} = \frac{k_y a M_B Z_T}{G_y'} \qquad (19.38)$$

However, for most packings, there are no published correlations for $h_x a$ and $k_y a$, and a simpler approach based on an overall coefficient and an overall enthalpy driving force can be used.

$$G_y' \, dH_y = K_y a (H_y^* - H_y) \, dZ \qquad (19.39)$$

where $\dfrac{1}{K_y a} = \dfrac{1}{k_y a M_B} + \dfrac{m}{h_x a}$

$\qquad m = \dfrac{dH^*}{dT} = $ slope of equilibrium line

$\qquad H_y^* = $ enthalpy of gas in equilibrium with liquid at temperature T_x

The number of transfer units and the height of a transfer unit can be defined in the same way as for gas absorption.

$$\int \frac{dH_y}{H_y^* - H_y} = N_{Oy} = \frac{Z_T}{H_{Oy}} \qquad (19.40)$$

where $H_{Oy} = G_y' / (K_y a)$.

The use of an overall gas-phase coefficient may introduce some error in the design or analysis of cooling tower performance, because the slope of the equilibrium line changes with temperature. As shown in Fig. 19.9, the slope increases by 30 to 40 percent for a 10°F change in temperature. However, the gas film has the major resistance, so the change in $K_y a$ with temperature may be relatively small, With cellular fill at the normal flow rates, the value of H_{Oy} is typically 2 to 3 ft (0.6 to 1 m). For older towers with wood slats H_{Oy} may be 10 to 20 ft (3 to 6 m).[1]

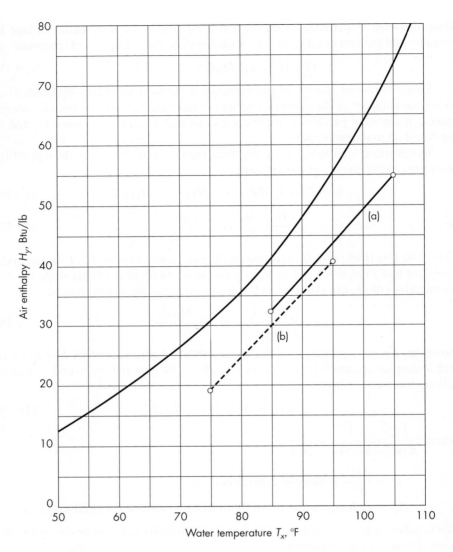

FIGURE 19.9
Operating diagram for Example 19.3.

Although the detailed design of cooling towers is usually left to specialists, the effects of changes in weather conditions on the performance of an existing tower are readily predicted. From data for normal conditions, an energy balance is made as a check, and the number of overall transfer units is calculated using an enthalpy diagram. A new operating line is then located by trial to give the same number of transfer units. This is illustrated in Example 19.3.

EXAMPLE 19.3. A counterflow induced-draft cooling tower operates with inlet and exit water temperatures of 105 and 85°F when the air has dry-bulb and wet-bulb temperatures, respectively, of 90 and 76°F. The tower has 4 ft of stacked plastic fill, and the flow rates are $G_y = 2,000$ lb/h·ft^2 and $G_x = 2,200$ lb/h·ft^2. (a) Determine the number of transfer units, the height of a transfer unit based on the overall gas-phase driving force, and the temperature approach. (b) If the cooling load remains the same but the air temperature drops to 70°F with a wet-bulb temperature of 60°F, predict the water temperature and the temperature approach.

Solution

(a) From the psychrometric chart (Fig. 19.2) the inlet humidity and humid heat are determined.

$$\mathcal{H}_b = 0.017 \text{ lb water/lb air} \qquad \mathcal{H}_R = \frac{0.017}{0.031} \times 100 = 55\%$$

$$c_s = 0.248 \text{ Btu/lb}\cdot°\text{F}$$

$$H_b = 0.248(90 - 32) + 1,075(.017) = 32.7 \text{ Btu/lb}$$

$$2,200(1.0)(105 - 85) = 2,000(H_a - 32.7)$$

$$H_a = 54.7 \text{ Btu/lb}$$

The points $T_{xa} = 105$, $H_a = 54.7$ and $T_{xb} = 85$, $H_b = 32.7$ are plotted in Fig. 19.9 as the endpoints of the operating line. The number of transfer units is obtained by determining the driving force at the middle of the column and using the log mean ΔH to get the number of transfer units for each section.

T_x	H^*	H	$H^* - H$	$(H^* - H)_L$	ΔN
85	41.5	32.7	8.8		
95	55.5	43.7	11.8	10.2	1.08
105	73	54.7	18.3	14.8	0.74
					$N_{Oy} = 1.82$

$$H_{Oy} = \frac{4}{1.82} = 2.2 \text{ ft}$$

The temperature approach is $85 - 76 = 9°F$.

(b) For $T_y = 70°F$ and $T_w = 60°F$,

$$\mathcal{H}_b = 0.009 \qquad c_s = 0.244$$

$$H_b = 0.244(70 - 32) + 1,075(0.009) = 18.9 \text{ Btu/lb}$$

For the same cooling load and constant water and airflows

$$2,200(1.0)(20) = 2,000(H_a - 18.9)$$

$$H_a = 18.9 + 22 = 40.9$$

The operating line is located by trial to give the same number of transfer units. For $T_{xa} = 95°F$, $T_{xb} = 75°F$, $N_{Oy} = 1.78$, which is close enough to 1.82. The operating line is shown as a dashed line in Fig. 19.9. The approach to the wet-bulb temperature is

$$T_{xb} - T_w = 75 - 60 = 15°F$$

The position of the operating line depends very much on the wet-bulb temperature of the entering air, since this fixes the air enthalpy. Changes in dry-bulb temperature have almost no effect. If the air is 100 percent saturated, cooling will still take place as long as the water temperature is greater than the wet-bulb temperature. In practice, towers are not designed to operate with an approach of less than 5°F (2.8°C), and approaches of 10 to 15°F (5.6 to 8.3°C) are more typical. As shown in Example 19.3, decreasing the wet-bulb temperature lowers the outlet water temperature, but the approach is increased because of the curvature of the equilibrium line.

If the gas rate or liquid rate changes, some change in the number of transfer units is expected. The gas-film coefficient increases with about the 0.8 power of the gas rate, so if the gas film controls, H_{Oy} goes up with the 0.2 power of G_y'. Decreasing the liquid rate may lead to an increase in H_{Oy} because of incomplete wetting of the packing.

SYMBOLS

A	Surface area of liquid, m^2 or ft^2
a	Transfer area, m^2/m^3 or ft^2/ft^3
b	Constant in Eqs. (19.17) and (19.18)
C_p	Molar specific heat, $J/g \, mol \cdot °C$ or $Btu/lb \, mol \cdot °F$
c_L	Specific heat of liquid, $J/g \cdot °C$ or $Btu/lb \cdot °F$
c_p	Specific heat, $J/g \cdot °C$ or $Btu/lb \cdot °F$; c_{pA}, c_{pB}, of components A and B, respectively
c_s	Humid heat, $J/g \cdot °C$ or $Btu/lb \cdot °F$
D	Diameter, m or ft
D_v	Diffusivity, m^2/h, cm^2/s, or ft^2/h
G	Mass velocity, $kg/m^2 \cdot h$ or $lb/ft^2 \cdot h$; G_x, of liquid at any point; G_{xa}, of liquid at entrance; G_{xb}, of liquid at exit; G_y', of gas, mass of vapor-free gas per unit area of tower cross section per hour
G_M	Molal mass velocity, $kg \, mol/m^2 \cdot h$ or $lb \, mol/ft^2 \cdot h$
H	Enthalpy, J/g or Btu/lb; H_x, of liquid; H_y, of gas; H_{ya}, H_{yb}, of gas at entrance and exit, respectively H^*, equilibrium value; H_y^*, of gas in equilibrium with liquid
H_{Oy}	Height of a transfer unit, m or ft, overall, based on gas phase
$\mathcal{H}$	Humidity, mass of vapor per unit mass of vapor-free gas; $\mathcal{H}_a$, at top of contactor; $\mathcal{H}_b$, at bottom of contactor; $\mathcal{H}_i$, at gas-liquid interface; $\mathcal{H}_s$, saturation humidity; $\mathcal{H}_w$, saturation humidity at wet-bulb temperature
$\mathcal{H}_A$	Percentage humidity, $100\mathcal{H}/\mathcal{H}_s$
$\mathcal{H}_R$	Relative humidity, $100 p_A/P_A'$
h	Heat-transfer coefficient, $W/m^2 \cdot °C$ or $Btu/ft^2 \cdot h \cdot °F$; h_x, liquid side; h_y, gas side
$K_y a$	Overall volumetric mass-transfer coefficient based on gas phase
k	Thermal conductivity, $W/m \cdot °C$ or $Btu/ft \cdot h \cdot °F$

k_y	Mass-transfer coefficient, g mol/m$^2 \cdot$ h $\cdot$ unit mole fraction or lb mol/ft$^2 \cdot$ h $\cdot$ unit mole fraction
M	Molecular weight; M_A, M_B, of components A and B, respectively; $\bar{M}$, average molecular weight of gas stream
m	Exponent in Eqs. (19.17) and (19.18); also, slope of equilibrium line
N	Number of transfer units; N_{Oy}, overall, based on gas phase
N_A	Rate of transfer or vaporization of liquid, mol/h
n	Exponent in Eqs. (19.17) and (19.18); also number of moles (Example 19.1)
P	Pressure, atm; P_A', vapor pressure of liquid
Pr	Prandtl number, $c_p \mu / k$
p_A	Partial pressure of vapor, atm
q	Rate of sensible-heat transfer to liquid, W or Btu/h
Re	Reynolds number, DG/μ
Sc	Schmidt number, $\mu/(\rho D_v)$
T	Temperature, K, °C, °R or °F; T_i, at gas-liquid interface; T_s, adiabatic saturation temperature; T_w, wet-bulb temperature; T_x, of bulk of liquid; T_{xa}, of liquid at top of contactor; T_{xb}, of liquid at bottom of contactor; T_y, of bulk of gas; T_{ya}, of gas at top of contactor; T_{yb}, of gas at bottom of contactor; T_0, datum for computing enthalpy
v_H	Humid volume, m^3/kg or ft^3/lb
x	Mole fraction of gas component in liquid stream; x_e, equilibrium value
y	Mole fraction of liquid component in gas stream; y_e, equilibrium value; y_i, at gas-liquid interface; y_s, saturation value
$\overline{(1-y)_L}$	One-way diffusion factor
Z	Distance from bottom of contact zone, m or ft; Z_T, total height of contact section
z	Number of moles of water evaporated (Example 19.1)

Greek Letters

λ	Latent heat of vaporization, J/g or Btu/lb; λ_s, at T_s; λ_w; at T_w; λ_0, at T_0
μ	Viscosity, cP or lb/ft $\cdot$ h
ρ	Density of gas, kg/m^3 or lb/ft^3

PROBLEMS

19.1. One method of removing acetone from cellulose acetate is to blow an airstream over the cellulose acetate fibers. To know the properties of the air-acetone mixtures, the process control department requires a humidity chart for air-acetone. After investigation, it was found that an absolute humidity range of 0 to 6.0 and a temperature range of 5 to 55°C would be satisfactory. Construct the following portions of such a humidity chart for air-acetone at a total pressure of 760 mm Hg: (*a*) percentage humidity

TABLE 19.1
Properties of acetone

Temperature, °C	Vapor pressure, mm Hg	Latent heat, J/g	Temperature, °C	Vapor pressure, mm Hg	Latent heat, J/g
0		564	50	620.9	
10	115.6		56.1	760.0	521
20	179.6	552	60	860.5	517
30	281.0		70	1,189.4	
40	420.1	536	80	1,611.0	495

lines for 50 and 100 percent, (*b*) saturated volume versus temperature, (*c*) latent heat of acetone versus temperature, (*d*) humid heat versus humidity, (*e*) adiabatic cooling lines for adiabatic saturation temperatures of 20 and 40°C, and (*f*) wet-bulb temperature (psychrometric) lines for wet-bulb temperatures of 20 and 40°C. The necessary data are given in Table 19.1. For acetone vapor, $c_p = 1.47$ J/g·°C and $h/(M_B k_y) = 1.7$ J/g·°C.

19.2. A mixture of air and benzene vapor is to be cooled from 70 to 15°C in a tubular cooler condenser. The humidity at the inlet is 0.7 kg benzene vapor per kilogram of air. Calculate (*a*) the wet-bulb temperature of the entering gas, (*b*) the humidity at the outlet, and (*c*) the total amount of heat to be transferred per kilogram of air.

19.3. A counterflow cooling tower with 30 in. of cellular fill was designed for an approach of 10°F and a cooling range of 17°F when the wet-bulb temperature is 75°F. The tower cross section is 36 × 36 ft, and the airflow and water flow rates are 523,000 cfm (at 90°F) and 6,000 gpm. (*a*) What are the inlet air enthalpy and the change in enthalpy across the tower? (*b*) Calculate N_{Oy} and H_{Oy} for these conditions. (*c*) How much more fill would be needed if the wet-bulb temperature were 78°F and the desired approach were 7°F?

19.4. For the cooling tower described in Example 19.3, predict the effect of increasing the air rate by 20 percent with the same liquid rate and a wet-bulb temperature of 75°F.

19.5. (*a*) Show that for small drops of water evaporating in warm air, the evaporation time is proportional to the square of the droplet size. (*b*) Calculate the evaporation time for 50-μm drops in air at 140°F. (*c*) Evaluate the volumetric heat-transfer coefficient for a spray of 50-μm drops that contains 1 percent of drops by volume.

19.6. Air at 27°C and 60 percent relative humidity is circulated past 1.5-cm-OD tubes through which water is flowing at 60 cm/s and 15°C. The air velocity approaching the tubes is 1.5 m/s. (*a*) Will water condense on the tubes? (*b*) What are the wall temperature and the interface temperature if condensation occurs?

19.7. Use the penetration theory, with the same contact time for gas and liquid, to find approximate values of $k_y a$ and $h_x a$. What fraction of the total resistance is in the gas phase?

REFERENCES

1. Burger, R. *Hydrocarbon Proc.,* **70**(3):59 (1991).
2. Grosvenor, W. M. *Trans. AIChE* **1:**184 (1908).
3. Hensley, J. C. (ed.). *Cooling Tower Fundamentals,* 2nd ed. Overland Park, KS: Marley Cooling Tower Co., 1998.
4. Lewis, W. K. *Trans. AIME* **44:**325 (1922).
5. Perry, R. H., and D. W. Green (eds.). *Perry's Chemical Engineers' Handbook,* 7th ed. New York: McGraw-Hill, 1997, pp. **12**–29, **12**–30.
6. Sherwood, T. K., and R. L. Pigford. *Absorption and Extraction,* 2nd ed. New York: McGraw-Hill, 1952, pp. 97–101.
7. Strigle, R. F., Jr. *Random Packing and Packed Towers.* Houston, TX: Gulf Publication Co., 1987.
8. Willa, J. L. *Chem. Eng.* **104**(11):92 (1997).

CHAPTER 20

Equilibrium-Stage Operations

One class of mass-transfer devices consists of assemblies of individual units, or stages, interconnected so that the materials being processed pass through each stage in turn. The two streams move countercurrently through the assembly; in each stage they are brought into contact, mixed, and then separated. Such multistage systems are called *cascades*. For mass transfer to take place, the streams entering each stage must not be in equilibrium with one another, for it is the departure from equilibrium conditions that provides the driving force for transfer. The leaving streams are usually not in equilibrium either but are much closer to being so than the entering streams are. The closeness of the approach to equilibrium depends on the effectiveness of mixing and mass transfer between the phases. To simplify the design of a cascade, the streams leaving each stage are often assumed to be in equilibrium, which, by definition, makes each stage *ideal*. A correction factor or efficiency is applied later to account for any actual departures from equilibrium.

To illustrate the principle of an equilibrium-stage cascade, two typical countercurrent multistage devices are described here, one for distillation or gas absorption, where the stages are arranged one above the other in a vertical column, and one for solid-liquid contacting as in leaching, where the stages are a series of stirred tanks on the same level. Other types of mass-transfer equipment are discussed in later chapters.

EQUIPMENT FOR STAGE CONTACTS

Typical distillation equipment

Equipment for continuous distillation is shown in Fig. 20.1. Column C is fed continuously with the liquid mixture to be distilled, and the liquid in reboiler A is partially converted to vapor by heat transferred from the heating element B. The

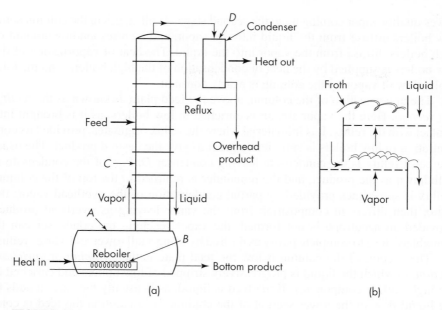

FIGURE 20.1
(*a*) Reboiler with fractionating column: *A*, reboiler; *B*, heating element; *C*, column; *D*, condenser. (*b*) Detail of sieve plate.

vapor stream from the still is brought into intimate countercurrent contact with a descending stream of boiling liquid in the column, or tower, *C*. This liquid must be rich enough in the low boiler that there is mass transfer of the low boiler from the liquid to the vapor at each stage of the column. Such a liquid can be obtained simply by condensing the overhead vapors and returning some of the liquid to the top of the column. This return liquid is called *reflux*. The use of reflux increases the purity of the overhead product, but not without some cost, since the vapor generated in the reboiler must provide both reflux and overhead product, and this energy cost is a large part of the total cost of separation by distillation.

The reflux entering the top of the column is often at the boiling point; but if it is cold, it is almost immediately heated to its boiling point by the vapor. Throughout the rest of the column, the vapor at any stage is at the same temperature as the liquid, which is at its boiling point. The temperature increases on going down the column because of the increase in pressure and the increasing concentration of high boiling components.

The vapor is enriched at each stage because the vapor coming to a stage contains less low boilers than the vapor that would be in equilibrium with the liquid fed to that stage. If, as is usual, the overhead vapor is totally condensed, it has the same composition as the product and the reflux. The reflux, however, has an equilibrium vapor composition which is richer than the vapor coming up to the top stage. This vapor is therefore enriched in low boilers at the expense of the reflux liquid. This partially depletes the reflux of low boilers, but if the flow rates have been set correctly, the liquid passing down to the second stage is still able to enrich the lower-quality vapor coming up to the second stage. At all stages in the column some low boilers diffuse from the liquid into the vapor, and a corresponding amount of high boilers diffuse from the vapor into the liquid. The heat of vaporization of the low boilers is supplied by the heat of condensation of the high boilers, and the total molal flow of vapor up the column is nearly constant.

The upper section of the column, above the feed plate, is known as the *rectifying section.* Here the vapor stream is enriched in low boilers, as it is brought into contact with the reflux. It is immaterial where the reflux originates, provided its concentration in low boilers is sufficiently great to give the desired product. The usual source of reflux is the condensate leaving condenser *D.* Part of the condensate is withdrawn as the product, and the remainder is returned to the top of the column. Reflux is sometimes provided by partial condensation of the overhead vapor; the reflux then differs in composition from the vapor leaving as overhead product. Provided an azeotrope is not formed, the vapor reaching the condenser can be brought as close to complete purity as desired by using a tall tower and a large reflux.

The section of the column below the feed plate is the *stripping* or *enriching section,* in which the liquid is progressively stripped of low boilers and enriched in the high boiling components. If the feed is liquid, as is usually the case, it adds to the liquid flow in the lower section of the column. If in addition the feed is cold, additional vapor must be provided by the reboiler to raise the temperature of the feed to the boiling point. To accomplish this, the additional vapor is condensed when it comes in contact with the feed, adding still more to the liquid flowing down through the stripping section. Details of column performance and the effects of feed condition are discussed in Chap. 21.

From the reboiler, liquid is withdrawn which contains most of the high boiling components and usually only a little of the low boilers. This liquid is called the *bottom product* or *bottoms.*

The column shown in Fig. 20.1*a* contains a number of plates, or trays, stacked one above the other. Often these plates are perforated and are known as sieve plates, details of which are shown in Fig. 20.1*b.* They consist of horizontal trays carrying a number of holes and a vertical plate which acts as a downcomer and a segmental weir. Sometimes, as discussed in Chap. 21, the holes contain valves or plugs which are lifted as vapor passes through them. The downcomer from a given plate reaches nearly to the tray below. Liquid flows over the weirs from plate to plate down the column, passing across the plates where the rising vapor causes it to froth. The vapor space above the froth contains a mist of fine droplets formed by the collapsing bubbles. Most of the drops fall back into the liquid, but some are entrained by the vapor and are carried to the plate above. See Chap. 21 for a discussion of the effect of this entrainment on column performance.

Typical leaching equipment

In leaching, soluble material is dissolved from its mixture with an inert solid by means of a liquid solvent. A diagrammatic flow sheet of a typical countercurrent leaching plant is shown in Fig. 20.2. It consists of a series of units, in each of which the solid from the previous unit is mixed with the liquid from the succeeding unit, and the mixture is allowed to settle. The solid is then transferred to the

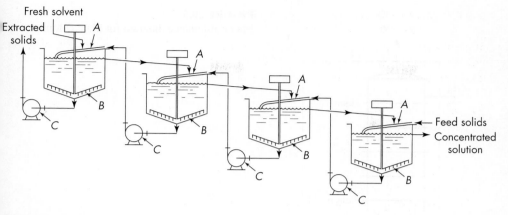

FIGURE 20.2
Countercurrent leaching plant: A, launder; B, rake; C, slurry pump.

next succeeding unit, and the liquid to the previous unit. As the liquid flows from unit to unit, it becomes enriched in solute, and as the solid flows from unit to unit in the reverse direction, it becomes impoverished in solute. The solid discharged from one end of the system is well extracted, and the solution leaving at the other end is strong in solute. The thoroughness of the extraction depends on the amount of solvent and the number of units. In principle, the unextracted solute can be reduced to any desired amount if enough solvent and a sufficient number of units are used.

Any suitable mixer and settler can be chosen for the individual units in a countercurrent leaching system. In those shown in Fig. 20.2 mixing occurs in the feed troughs (known as "launders") leading to each tank and also in the upper parts of the tanks themselves. Rakes B move settled solids to the discharge, and pumps C move the slurry from tank to tank.

PRINCIPLES OF STAGE PROCESSES

In the sieve-plate tower and the countercurrent leaching plant shown in Figs. 20.1 and 20.2, the cascade consists of a series of interconnected units, or stages. Study of the assembly as a whole is best made by focusing attention on the streams passing between the individual stages. An individual unit in a cascade receives two streams, one a V phase and one an L phase, from the two units adjacent to it, brings them into close contact, and delivers L and V phases, respectively, to the same adjacent units. The fact that the contact units may be arranged either one above the other, as in the sieve-plate column, or side by side, as in a stage leaching plant, is important mechanically and may affect some of the details of operation of individual stages. The same material-balance equations, however, can be used for either arrangement.

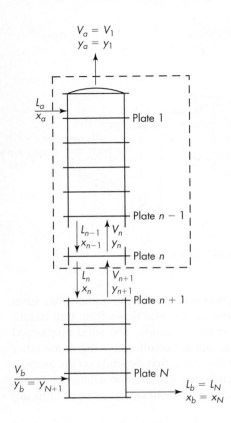

$V_a = V_1$
$y_a = y_1$

FIGURE 20.3
Material-balance diagram for plate column.

Terminology for stage-contact plants

The individual contact units in a cascade are numbered serially, starting from one end. In this book, the stages are numbered in the direction of flow of the L phase, and the last stage is that discharging the L phase. A general stage in the system is the nth stage, which is number n counting from the entrance of the L phase. The stage immediately ahead of stage n in the sequence is stage $n - 1$, and that immediately following it is stage $n + 1$. Using a plate column as an example, Fig. 20.3 shows how the units in a cascade are numbered. The total number of stages is N, and the last stage in the plant is therefore the Nth stage.

To designate the streams and concentrations pertaining to any one stage, all streams originating in that stage carry the number of the unit as a subscript. Thus, for a two-component system, y_{n+1} is the mole fraction of component A in the V phase leaving stage $n + 1$, and L_n is the molal flow rate of the L phase leaving the nth stage. The streams entering and leaving the cascade and those entering and leaving stage n in a plate tower are shown in Fig. 20.3. Quantities V_a, L_b, y_a, and x_b in Table B, page 507, are equal to V_1, L_N, y_1, and x_N, respectively. This can be seen by reference to Fig. 20.3.

Material balances

Consider the portion of the cascade that includes stages 1 through n, as shown by the section enclosed by the dashed line in Fig. 20.3. The total input of material to this section is $L_a + V_{n+1}$ mol/h, and the total output is $L_n + V_a$ mol/h. Since, under steady flow, there is neither accumulation nor depletion, the input and the output are equal and

$$L_a + V_{n+1} = L_n + V_a \tag{20.1}$$

Equation (20.1) is a total material balance. Another balance can be written by equating input to output for component A. Since the number of moles of this component in a stream is the product of the flow rate and the mole fraction of A in the stream, the input of component A to the section under study, for a two-component system, is $L_a x_a + V_{n+1} y_{n+1}$ mol/h, the output is $L_n x_n + V_a y_a$ mol/h, and

$$L_a x_a + V_{n+1} y_{n+1} = L_n x_n + V_a y_a \tag{20.2}$$

A material balance can also be written for component B, but such an equation is not independent of Eqs. (20.1) and (20.2), since if Eq. (20.2) is subtracted from Eq. (20.1), the result is the material-balance equation for component B. Equations (20.1) and (20.2) yield all the information that can be obtained from material balances alone written over the chosen section.

Overall balances covering the entire cascade are found in the same manner:

Total material balance: $$L_a + V_b = L_b + V_a \tag{20.3}$$

Component A balance: $$L_a x_a + V_b y_b = L_b x_b + V_a y_a \tag{20.4}$$

Enthalpy balances

In many equilibrium-stage processes the general energy balance can be simplified by neglecting mechanical potential energy and kinetic energy. If, in addition, the process is workless and adiabatic, a simple enthalpy balance applies. Then, for a two-component system, for the first n stages,

$$L_a H_{L,a} + V_{n+1} H_{V,n+1} = L_n H_{L,n} + V_a H_{V,a} \tag{20.5}$$

where H_L and H_V are the enthalpies per mole of the L phase and V phase, respectively. For the entire cascade,

$$L_a H_{L,a} + V_b H_{V,b} = L_b H_{L,b} + V_a H_{V,a} \tag{20.6}$$

Graphical methods for two-component systems

For systems containing only two components it is possible to solve many mass-transfer problems graphically. The methods are based on material balances and equilibrium relationships; some more complex methods require enthalpy balances as well. These more complex methods will be discussed in Chap. 21. The principles underlying the simple graphical methods are discussed in the following paragraphs. Their detailed applications to specific operations are covered in later chapters.

Operating-line diagram

For a binary system, the compositions of the two phases in a cascade can be shown on an arithmetic graph where x is the abscissa and y the ordinate. As shown by Eq. (20.2), the material balance at an intermediate point in the column involves x_n, the concentration of the L phase *leaving* stage n, and y_{n+1}, the concentration of the V phase *entering* that stage. Equation (20.2) can be written to show the relationship more clearly:

$$y_{n+1} = \frac{L_n}{V_{n+1}}x_n + \frac{V_a y_a - L_a x_a}{V_{n+1}} \qquad (20.7)$$

Equation (20.7) is the operating-line equation for the column; if the points x_n and y_{n+1} for all the stages are plotted, the line through these points is called the *operating line*. Note that if L_n and V_{n+1} are constant throughout the column, the equation is that of a straight line with slope L/V and intercept $y_a - (L/V)x_a$, and the line is easily located. For this case the operating line can also be drawn as a straight line connecting the terminal compositions (x_a, y_a) and (x_b, y_b). To understand why this is true, extend the dashed rectangle in Fig. 20.3 to include plate N, and consider that the stream V_b coming to the bottom stage is equivalent to a stream from a hypothetical stage $N + 1$, so that y_b corresponds to y_{N+1} and x_b to x_N. Similarly, the stream L_a at the top of the column can be considered to come from a hypothetical stage numbered 0 so that the point (x_0, y_1) or (x_a, y_a) locates the true upper end of the operating line.

In a packed column or other nonstagewise contacting device, such as the one described in Chap. 18 for gas absorption, x and y are continuous functions of height Z, whereas in staged towers x and y have only discrete values. The equilibrium line is of course continuous, but the operating line is drawn to connect a series of compositions, relating y_{n+1} to x_n. Usually, however, we are not sure of the values of y_n and x_n, so we draw the operating line as a solid line.

When the flow rates are not constant in the column, the operating line on a simple arithmetic plot is not straight. The terminal compositions may still be used to locate the ends of the line, and material-balance calculations over sections of the column are made to establish a few intermediate points. Often only one or two other points are needed because usually the operating line is only slightly curved.

The position of the operating line relative to the equilibrium line determines the direction of mass transfer and how many stages are required for a given separation. The equilibrium data are found by experiment, by thermodynamic calculations, or from published sources,[2] and the equilibrium line is just a plot of the equilibrium values of x_e and y_e. For rectification in a distillation column, the operating line must lie below the equilibrium line, as shown in Fig. 20.4a. Then the vapor coming to any plate contains less of the low boiler than the vapor in equilibrium with the liquid leaving the plate, so that vapor passing through the liquid will be enriched in the low-boiling component. The relative slopes of the lines are not important as long as the lines do not touch; the operating line could be less steep than the equilibrium line, and progressive enrichment of the vapor would still take place. The driving force for mass transfer is the difference $y_e - y_{n+1}$, as shown in Fig. 20.4a.

When one component is to be transferred from the V phase to the L phase, as in the absorption of soluble material from an inert gas, the operating line must lie above the equilibrium line, as in Fig. 20.4b. The driving force for mass transfer is now $y_{n+1} - y_e$, or the difference between the actual vapor composition and the vapor composition in equilibrium with the liquid for that position in the column. In

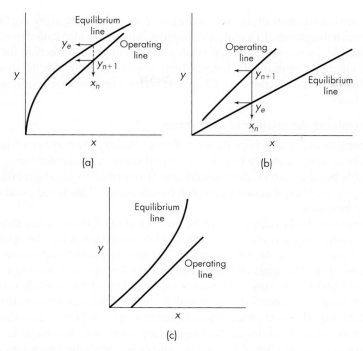

FIGURE 20.4
Operating and equilibrium lines: (*a*) for rectification, (*b*) for gas absorption, (*c*) for desorption.

the design of gas absorbers the liquid rate is usually chosen to make the operating line somewhat steeper than the equilibrium line, which gives a moderately large driving force in the bottom part of the column and permits the desired separation to be made with relatively few stages.

In absorbing one component of the gas into a nonvolatile solvent, the total gas rate decreases and the total liquid rate increases as the two phases pass through the column. Therefore the operating line is usually curved, though the percentage change in slope, or L/V, is not as great as the change in either L or V, since both L and V are largest at the bottom of the column and smallest at the top. A method of calculating intermediate points on the operating line is shown later in Example 20.1.

The reverse of gas absorption is called desorption or stripping, an operation carried out to recover valuable solute from the absorbing solution and regenerate the solution. The operating line must then lie below the equilibrium line, as in Fig. 20.4c. Usually the temperature or pressure is changed to make the equilibrium curve much steeper than for the absorption process.

Ideal contact stages

The ideal stage is a standard to which an actual stage may be compared. In an ideal stage, the V phase leaving the stage is in equilibrium with the L phase leaving the same stage. For example, if plate n in Fig. 20.3 is an ideal stage, concentrations

x_n and y_n are coordinates of a point on the curve of x_e versus y_e showing the equilibrium between the phases. In a plate column ideal stages are also called *perfect plates*.

To use ideal stages in design, it is necessary to apply a correction factor, called the *stage efficiency* or *plate efficiency,* which relates the ideal stage to an actual one. Plate efficiencies are discussed in Chap. 21, and the present discussion is restricted to ideal stages.

Determining the number of ideal stages

A problem of general importance is that of finding the number of ideal stages required in an actual cascade to cover a desired range of concentration x_a to x_b or y_a to y_b. If this number can be determined, and if information on stage efficiencies is available, the number of actual stages can be calculated. This is the usual method of designing cascades.

A simple method of determining the number of ideal stages when there are only two components in each phase is a graphical construction using the operating-line diagram. Figure 20.5 shows the operating line and the equilibrium curve for a typical gas absorber. The ends of the operating line are point a, having coordinates (x_a, y_a), and point b, having coordinates (x_b, y_b). The problem of determining the number of ideal stages needed to accomplish the gas-phase concentration change y_b to y_a and the liquid-phase concentration change x_a to x_b is solved as follows.

The concentration of the gas leaving the top stage, which is stage 1, is y_a, or y_1. If the stage is ideal, the liquid leaving is equilibrium with the vapor leaving, so the point (x_1, y_1) must lie on the equilibrium curve. This fact fixes point m, found by moving horizontally from point a to the equilibrium curve. The abscissa of point m is x_1. The operating line is now used. It passes through all points having coordinates of the type (x_n, y_{n+1}), and since x_1 is known, y_2 is found by moving vertically from point m to the operating line at point n, the coordinates of which are (x_1, y_2). The step, or triangle, defined by points a, m, and n represents one ideal stage, the first one in this column. The second stage is located graphically on the diagram by repeating the same construction, passing horizontally to the equilibrium curve at

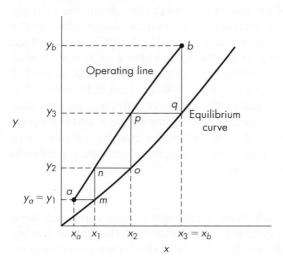

FIGURE 20.5
Operating-line diagram for gas absorber.

point o, having coordinates (x_2, y_2), and vertically to the operating line again at point p, having coordinates (x_2, y_3). The third stage is found by again repeating the construction, giving triangle pqb. For the situation shown in Fig. 20.5, the third stage is the last, as the concentration of the gas leaving that stage is y_b, and the liquid leaving it is x_b, which are the desired terminal concentrations. Three ideal stages are required for this separation.

The same construction can be used for determining the number of ideal stages needed in any cascade, whether it is used for gas absorption, distillation, leaching, or liquid extraction. The graphical step-by-step construction utilizing alternately the operating and equilibrium lines to find the number of ideal stages was first applied to the design of distillation columns, and is known as the *McCabe–Thiele method.*[4] The construction can be started at either end of the column, and in general the last step will not exactly meet the terminal concentrations, as was the case in Fig. 20.5. A fractional step may be assigned, or the number of ideal stages may be rounded up to the nearest whole number.

EXAMPLE 20.1. By means of a plate column, acetone is absorbed from its mixture with air in a nonvolatile absorption oil. The entering gas contains 30 mole percent acetone, and the entering oil is acetone-free. Of the acetone in the air 97 percent is to be absorbed, and the concentrated liquor at the bottom of the tower is to contain 10 mole percent acetone. The equilibrium relationship is $y_e = 1.9x_e$. Plot the operating line and determine the number of ideal stages.

Solution. Choose 100 mol of entering gas as a basis, and set this equal to V_b. The acetone entering is then $0.3 \times 100 = 30$ mol; the air entering is $100 - 30 = 70$ mol. With 97 percent absorbed, the acetone leaving is $0.03 \times 30 = 0.9$ mol and $y_a = 0.9/70.9 = 0.0127$; the acetone absorbed is $30 - 0.9 = 29.1$ mol. With 10% acetone in the leaving solution and no acetone in the entering oil, $0.1L_b = 29.1$, and $L_b = 291$ mol. Then $L_a = 291 - 29.1 = 261.9$ mol.

To find an intermediate point on the operating line, make an acetone balance around the top part of the tower, assuming a particular value of yV, the moles of acetone left in the gas. For 10 mol left in the gas,

$$y = \frac{10}{10 + 70} = 0.125$$

The moles of acetone lost by the gas in this section, $10 - 0.9$, or 9.1, must equal the moles gained by the liquid. Hence where $y = 0.125$,

$$x = \frac{9.1}{261.9 + 9.1} = 0.0336$$

Similar calculations for $yV = 20$ give $y = 20/90 = 0.222$ and $x = 19.1/(261.9 + 19.1) = 0.068$.

The operating line is plotted in Fig. 20.6. Note that it is only slightly curved, even though the gas flow rate changes almost 30 percent.

The number of ideal stages is 4 and a fraction. Based on the required change in x relative to the change that would be made in a full step, the fraction is l_1/l_2, or 0.27. A similar construction based on changes in y gives the fraction 0.33; the values differ because the operating and equilibrium lines are not parallel. The answer would be given as 4.3 stages.

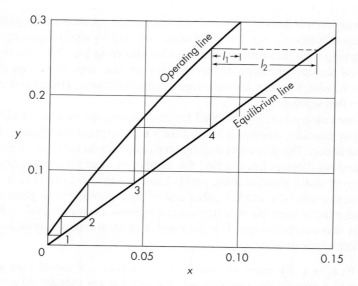

FIGURE 20.6
Diagram for Example 20.1.

Absorption factor method for calculating the number of ideal stages

When the operating and equilibrium lines are both straight over a given concentration range x_a to x_b, the number of ideal stages can be calculated directly, and graphical construction is unnecessary. Formulas for this purpose are derived as follows.

Let the equation of the equilibrium line be

$$y_e = mx_e + B \tag{20.8}$$

where, by definition, m and B are constant. If stage n is ideal,

$$y_n = mx_n + B \tag{20.9}$$

Substitution for x_n into Eq. (20.7) gives, for ideal stages and constant L/V,

$$y_{n+1} = \frac{L(y_n - B)}{mV} + y_a - \frac{Lx_a}{V} \tag{20.10}$$

It is convenient to define an absorption factor A by the equation

$$A \equiv \frac{L}{mV} \tag{20.11}$$

The absorption factor is the ratio of the slope of the operating line L/V to that of the equilibrium line m. It is a constant when both of these lines are straight. Equation (20.10) can be written

$$y_{n+1} = A(y_n - B) + y_a - Amx_a$$
$$= Ay_n - A(mx_a + B) + y_a \tag{20.12}$$

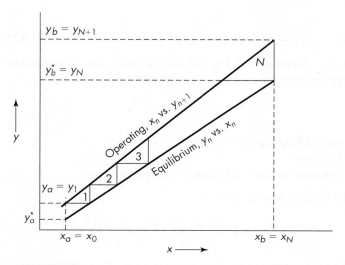

FIGURE 20.7
Derivation of absorption factor equation.

Normally A is made greater than 1.0 to permit nearly complete removal of solute from the V phase. The quantity $mx_a + B$ is, by Eq. (20.8), the concentration of the vapor that is in equilibrium with the inlet L phase, the concentration of which is x_a. This can be seen from Fig. 20.7. The symbol y^* is used to indicate the concentration of a V phase in equilibrium with a specified L phase. Then

$$y_a^* = mx_a + B \tag{20.13}$$

and Eq. (20.12) becomes

$$y_{n+1} = Ay_n - Ay_a^* + y_a \tag{20.14}$$

Equation (20.14) can be used to calculate, step by step, the value of y_{n+1} for each stage, starting with stage 1. The method may be followed with the aid of Fig. 20.7.

For stage 1, using $n = 1$ in Eq. (20.14) and noting that $y_1 = y_a$ gives

$$y_2 = Ay_a - Ay_a^* + y_a = y_a(1 + A) - Ay_a^*$$

For stage 2, using $n = 2$ in Eq. (20.14) and eliminating y_2 give

$$y_3 = Ay_2 - Ay_a^* + y_a = A[y_a(1 + A) - Ay_a^*] - Ay_a^* + y_a$$
$$= y_a(1 + A + A^2) - y_a^*(A + A^2)$$

These equations can be generalized for the nth stage, giving

$$y_{n+1} = y_a(1 + A + A^2 + \cdots + A^n) - y_a^*(A + A^2 + \cdots + A^n) \tag{20.15}$$

For the entire cascade, $n = N$, the total number of stages, and

$$y_{n+1} = y_{N+1} = y_b$$

Then

$$y_b = y_a(1 + A + A^2 + \cdots + A^N) - y_a^*(A + A^2 + \cdots + A^N) \quad (20.16)$$

The sums in the parentheses of Eq. (20.16) are both sums of geometric series. The sum of such a series is

$$s_n = \frac{a_1(1 - r^n)}{1 - r}$$

where s_n = sum of first n terms of series
$\quad a_1$ = first term
$\quad r$ = constant ratio of each term to preceding term

Equation (20.16) can then be written

$$y_b = y_a \frac{1 - A^{N+1}}{1 - A} - y_a^* A \frac{1 - A^N}{1 - A} \quad (20.17)$$

Equation (20.17) is a form of the Kremser equation.[3] It can be used as such or in the form of a chart relating N, A, and the terminal concentrations.[1,5] It can also be put into a simpler form by the following method.

Equation (20.14) is, for stage N,

$$y_b = Ay_N - Ay_a^* + y_a \quad (20.18)$$

Figure 20.7 shows that $y_N = y_b^*$, and Eq. (20.18) can be written

$$y_a = y_b - A(y_b^* - y_a^*) \quad (20.19)$$

Collecting terms in Eq. (20.17) containing A^{N+1} gives

$$A^{N+1}(y_a - y_a^*) = A(y_b - y_a^*) + y_a - y_b \quad (20.20)$$

Substituting $y_a - y_b$ from Eq. (20.19) into Eq. (20.20) gives

$$A^N(y_a - y_a^*) = y_b - y_a^* - y_b^* + y_a^* = y_b - y_b^* \quad (20.21)$$

Taking logarithms of Eq. (20.21) and solving for N give

$$N = \frac{\ln\left[(y_b - y_b^*)/(y_a - y_a^*)\right]}{\ln A} \quad (20.22)$$

and from Eq. (20.19)

$$\frac{y_b - y_a}{y_b^* - y_a^*} = A \quad (20.23)$$

Equation (20.22) can be written

$$N = \frac{\ln\left[(y_b - y_b^*)/(y_a - y_a^*)\right]}{\ln\left[(y_b - y_a)/(y_b^* - y_a^*)\right]} \quad (20.24)$$

The various concentration differences in Eq. (20.24) are shown in Fig. 20.8.

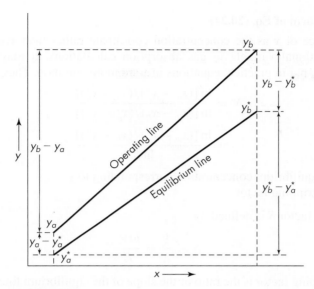

FIGURE 20.8
Concentration differences in Eq. (20.24).

When the operating line and the equilibrium line are parallel, A is unity and Eqs. (20.22) and (20.24) are indeterminate. In this case the number of steps is just the overall change in concentration divided by the driving force, which is constant. Thus,

$$N = \frac{y_b - y_a}{y_a - y_a^*} = \frac{y_b - y_a}{y_b - y_b^*} \tag{20.25}$$

If the operating line has a lower slope than the equilibrium line, A is less than 1.0, but Eqs. (20.22) and (20.24) can still be used by inverting both terms to give

$$N = \frac{\ln\left[(y_a - y_a^*)/(y_b - y_b^*)\right]}{\ln(1/A)} \tag{20.26}$$

or
$$N = \frac{\ln\left[(y_a - y_a^*)/(y_b - y_b^*)\right]}{\ln\left[(y_b^* - y_a^*)/(y_b - y_a)\right]} \tag{20.27}$$

In the design of an absorber, as stated earlier, the liquid rate is usually chosen to make the operating line steeper than the equilibrium line or to make A greater than unity. Values of A less than 1.0 can arise when we are dealing with two or more absorbable components. If the value of A is slightly greater than 1.0 for the major solute, a second component with a much lower solubility (higher value of m) will have a value of A appreciably less than 1.0. If the gas stream and the solution are dilute, the preceding equations can be applied to each component independently.

L-phase form of Eq. (20.24)

The choice of y as the concentration coordinate rather than x is arbitrary. It is the conventional variable in gas absorption calculations. It may be used for stripping also, but in practice, equations in x are more common. They are

$$N = \frac{\ln\left[(x_a - x_a^*)/(x_b - x_b^*)\right]}{\ln\left[(x_a - x_b)/(x_a^* - x_b^*)\right]}$$

$$= \frac{\ln\left[(x_a - x_a^*)/(x_b - x_b^*)\right]}{\ln S} \qquad (20.28)$$

where $x^* =$ equilibrium concentration corresponding to y
  $S =$ stripping factor

The stripping factor S is defined by

$$S \equiv \frac{1}{A} = \frac{mV}{L} \qquad (20.29)$$

The stripping factor is the ratio of the slope of the equilibrium line to that of the operating line, and the conditions are usually chosen to make S greater than unity. The concentration differences in Eq. (20.28) are shown in Fig. 20.9.

As shown in the derivations, it is not assumed that the linear extension of the equilibrium line passes through the origin. It is only necessary that the line be linear in the range where the steps representing the stages touch the line, as shown by line

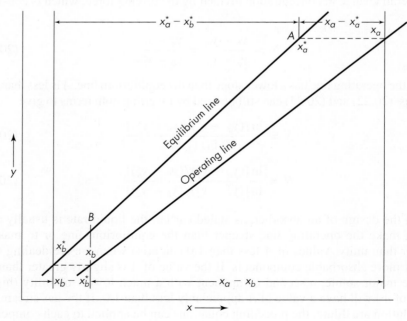

FIGURE 20.9
Concentration differences in Eq. (20.28).

AB in Fig. 20.9. Thus an equilibrium line that is almost linear near the origin but curves at higher concentrations is sometimes fitted by a straight line over part of its range to permit use of the Kremser equation.

The various forms of the Kremser equation were derived using concentrations in mole fractions, which is the usual choice for distillation or absorption. For some operations, including extraction and leaching, the concentrations may be expressed using mole ratios or mass ratios, defined as the amount of diffusing component divided by the amount of inert nondiffusing components. If this choice of units gives straight equilibrium and operating lines, the same Kremser equations can be used to find the number of ideal stages.

In the design of a plant, N is calculated from the proposed terminal concentrations and a selected value of A or S. Equation (20.22) or (20.24) is used for absorption and Eq. (20.28) for stripping. In estimating the effect of a change in operating conditions of an existing plant, Eq. (20.21) is used for absorption or its analog, Eq. (20.30), for stripping.

$$S^N = \frac{x_a - x_a^*}{x_b - x_b^*} \tag{20.30}$$

EXAMPLE 20.2. Ammonia is stripped from a dilute aqueous solution by countercurrent contact with air in a column containing seven sieve trays. The equilibrium relationship is $y_e = 0.8x_e$, and when the molar flow of air is 1.5 times that of the solution, 90 percent of the ammonia is removed. (a) How many ideal stages does the column have, and what is the stage efficiency? (b) What percentage removal would be obtained if the air rate were increased to 2.0 times the solution rate?

Solution

(a) The stripping factor is

$$S = \frac{mV}{L} = 0.8 \times 1.5 = 1.2$$

All concentrations can be expressed in terms of x_a, the mole fraction of NH_3 in the entering solution:

$$x_b = 0.1x_a \qquad x_b^* = 0 \qquad \text{since } y_b = 0$$

From an ammonia balance, $V \, \Delta y = V y_a = L \, \Delta x = L(0.9x_a)$. Hence

$$y_a = \frac{L}{V}(0.9x_a) = \frac{0.9}{1.5}x_a = 0.6x_a$$

Also,

$$x_a^* = \frac{y_a}{m} = \frac{0.6x_a}{0.8} = 0.75x_a$$

From Eq. (20.28),

$$N = \frac{\ln\left[(x_a - 0.75x_a)/(0.1x_a - 0)\right]}{\ln S}$$

$$= \frac{\ln\left(0.25x_a/0.1x_a\right)}{\ln 1.2} = 5.02$$

The separation corresponds to 5.02 ideal stages, so the stage efficiency is $5.02/7 = 72$ percent.

(b) If V/L is increased to 2.0 and the number of ideal stages N does not change (same stage efficiency), $S = 0.8 \times 2.0 = 1.6$. Then from Eq. (20.30)

$$\ln \frac{x_a - x_a^*}{x_b} = 5.02 \ln 1.6 = 2.36$$

$$\frac{x_a - x_a^*}{x_b} = 10.59$$

Let f be the fraction of NH_3 removed. Then $x_b = (1 - f)x_a$. By a material balance,

$$y_a = \frac{L}{V}(x_a - x_b)$$

$$= \tfrac{1}{2}[x_a - (1 - f)x_a] = \tfrac{1}{2}f x_a$$

$$x_a^* = \frac{y_a}{m} = \frac{0.5 f x_a}{0.8} = 0.625 f x_a$$

Thus
$$x_a - x_a^* = (1 - 0.625 f)x_a$$

Also,
$$x_a - x_a^* = 10.59 x_b = 10.59(1 - f)x_a$$

From these, $f = 0.962$, or 96.2 percent is removed.

The conditions for the original case and the new case are sketched in Fig. 20.10.

Equilibrium-Stage Calculations for Multicomponent Systems

For systems containing more than two or three components, graphical procedures are ordinarily of little value, and the number of ideal stages required in a given problem must be found by algebraic calculations. These involve knowledge and application of the equilibrium relationships; material balances [Eqs. (20.1) and (20.2)]; and (sometimes) enthalpy balances [Eqs. (20.5) and (20.6)]. The calculations are begun at some point in the cascade where conditions are known. On the basis of certain assumptions the conditions on succeeding stages which satisfy the equilibrium requirements and the material and energy balances are found mathematically, usually by iteration. The calculations are continued, stage by stage, until the desired terminal conditions are reached or, as often happens, it becomes evident that they cannot be reached. If this occurs, the underlying assumptions are modified and the entire calculation is repeated until the problem is solved. For preliminary calculations on

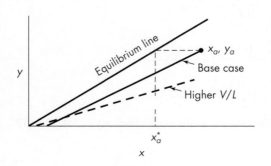

FIGURE 20.10
Diagram for Example 20.2.

multicomponent systems, certain approximate methods are available which greatly reduce the labor involved, but the rigorous computation of multicomponent cascades is now done by computer. (See Chap. 22.)

SYMBOLS

A	Absorption factor, L/mV, dimensionless
a_1	First term of geometric series
B	Constant in Eq. (20.8)
f	Fraction of ammonia absorbed (Example 20.2)
H	Specific enthalpy, J/g or Btu/lb; H_L, of L phase; $H_{L,a}$, at entrance; $H_{L,b}$, at exit; $H_{L,n}$, of L phase leaving stage n; H_V, of V phase; $H_{V,a}$, at exit; $H_{V,b}$, at entrance; $H_{V,n+1}$, of V phase leaving stage $n + 1$
L	Flow rate of L phase, kg mol/h or lb mol/h; L_N, from final stage of cascade; L_a, at entrance; L_b, at exit; L_n, from stage n
l_1, l_2	Lengths of line segments in Fig. 20.6
m	Slope of equilibrium curve, dy_e/dx_e
N	Total number of ideal stages
n	Serial number of ideal stage, counting from inlet of L phase
r	Ratio of succeeding terms of geometric series
S	Stripping factor, mV/L, dimensionless
s_n	Sum of first n terms of geometric series
V	Flow rate of V phase, kg mol/h or lb mol/h; V_a, at exit; V_b, at entrance; V_{n+1}, from stage $n + 1$; V_1, leaving first stage of cascade
x	Mole fraction in L phase; used for component A when only two components are present; x_N, in L phase from final stage of cascade; x_a, at entrance; x_b, at exit; x_e, at equilibrium; x_n, mole fraction in L phase from stage n; x^*, mole fraction in L phase in equilibrium with specified stream of V phase; x_a^*, in equilibrium with y_a; x_b^*, in equilibrium with y_b; x_0, in L phase entering first stage of cascade; x_1, x_2, in L phase leaving first and second stages, respectively
y	Mole fraction in V phase; used for component A when only two components are present; y_{N+1}, in V phase entering stage N of cascade; y_a, at exit; y_b, at entrance; y_e, at equilibrium; y_n, in V phase from stage n; y^*, mole fraction in V phase in equilibrium with specified stream of L phase; y_a^*, in equilibrium with x_a; y_b^*, in equilibrium with x_b; y_1, y_2, in V phase leaving first and second stages, respectively

PROBLEMS

20.1. Calculate the number of ideal stages for the system described in Example 20.1 if the conditions are changed to the following:

Acetone in entering gas, 25 mole percent
Acetone in entering oil, 1.5 mole percent

Acetone in bottoms liquor, 8 mole percent
Acetone absorbed, 90 percent

20.2. What are the effects on the concentrations of the exit gas and liquid streams of the following changes in the operating conditions of the column of Example 20.2? (*a*) A drop in the operating temperature that changes the equilibrium relationship to $y_e = 0.6x_e$. Unchanged from the original design: N, L/V, y_b, and x_a. (*b*) A reduction in the L/V ratio from 1.5 to 1.25. Unchanged from original design: temperature, N, y_b, and x_a. (*c*) An increase in the number of ideal stages from 5.02 to 8. Unchanged from original design: temperature, L/V, y_b, and x_a.

20.3. A column with eight plates and an estimated plate efficiency of 75 percent is used to remove component A from a dilute gas by absorption in water. By what factor must the L/V ratio exceed the slope of the equilibrium line to achieve 95 percent removal of component A?

20.4. If an ammonia absorber has 2% ammonia in the inlet gas and no ammonia in the incoming water, what fraction of the ammonia could be absorbed with an absorption factor of 0.9 and $N = 5$ or $N = 10$?

20.5. A toxic hydrocarbon is stripped from water with air in a column with eight ideal stages. (*a*) What stripping factor is needed for 98 percent removal? (*b*) What percentage removal could be achieved with a stripping factor of 2.0?

20.6. For the conditions of Example 20.1, how many ideal stages would be needed for 97 percent adsorption of the acetone of the incoming oil contained 0.005 mol fraction acetone?

REFERENCES

1. Brown, G. G., M. Souders, Jr., and H. V. Nyland. *Int. Eng. Chem.* **24**:522 (1932).
2. Gmehling, J., U. Onken, et al. *Vapor Liquid Equilibria Data Collection.* Frankfurt: DECHEMA, 1977.
3. Kremser, A. *Natl. Petr. News* **22**(21):42 (May 21, 1930).
4. McCabe, W. L., and E. W. Thiele. *Ind. Eng. Chem.* **17**:605 (1925).
5. Perry, R. H., and D. W. Green (eds.). *Perry's Chemical Engineers' Handbook,* 7th ed. New York: McGraw-Hill, 1997, p. **13**–38.

CHAPTER 21

Distillation

In practice, distillation may be carried out by either of two principal methods. The first method is based on the production of a vapor by boiling the liquid mixture to be separated and condensing the vapors without allowing any liquid to return to the still. There is then no reflux. The second method is based on the return of part of the condensate to the still under such conditions that this returning liquid is brought into intimate contact with the vapors on their way to the condenser. Either of these methods may be conducted as a continuous process or as a batch process. The first sections of this chapter deal with continuous steady-state distillation processes, including single-stage partial vaporization without reflux (flash distillation) and continuous distillation with reflux (rectification) for systems containing only two components. (Multicomponent distillation is discussed in Chap. 22.) Later sections are concerned with the design and performance of distillation equipment such as sieve tray columns and with batch distillation.

FLASH DISTILLATION

Flash distillation consists of vaporizing a definite fraction of the liquid in such a way that the evolved vapor is in equilibrium with the residual liquid, separating the vapor from the liquid, and condensing the vapor. Figure 21.1 shows the elements of a flash distillation plant. Feed is pumped by pump a through heater b, and the pressure is reduced through valve c. An intimate mixture of vapor and liquid enters the vapor separator d, in which sufficient time is allowed for the vapor and liquid portions to separate. Because of the intimacy of contact of liquid and vapor before separation, the separated streams are in equilibrium. Vapor leaves through line e and liquid through line g.

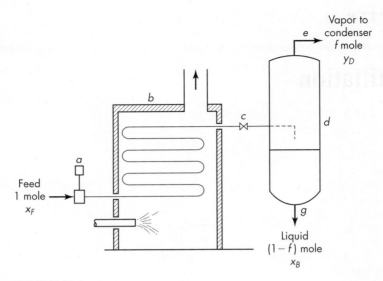

FIGURE 21.1
Plant for flash distillation.

Flash distillation of binary mixtures

Flash distillation is used extensively in petroleum refining, in which petroleum fractions are heated in pipe stills and the heated fluid is flashed into vapor and residual liquid streams, each containing many components. Liquid from an absorber is often flashed to recover some of the solute; liquid from a high-pressure reactor may be flashed to a lower pressure, causing some vapor to be evolved.

Consider 1 mol of a two-component mixture fed to the equipment shown in Fig. 21.1. Let the concentration of the feed be x_F, in mole fraction of the more volatile component. Let f be the molal fraction of the feed that is vaporized and withdrawn continuously as vapor. Then $1 - f$ is the molal fraction of the feed that leaves continuously as liquid. Let y_D and x_B be the concentrations of the vapor and liquid, respectively. By a material balance for the more volatile component, based on 1 mol of feed, all that component in the feed must leave in the two exit streams, or

$$x_F = f y_D + (1 - f)x_B \qquad (21.1)$$

There are two unknowns in Eq. (21.1): x_B and y_D. To use the equation, a second relationship between the unknowns must be available. Such a relationship is provided by the equilibrium curve, as y_D and x_B are coordinates of a point on this curve. If x_B and y_D are replaced by x and y, respectively, Eq. (21.1) can be written

$$y = -\frac{1-f}{f}x + \frac{x_F}{f} \qquad (21.2)$$

The fraction f is not fixed directly but depends on the enthalpy of the hot incoming liquid and the enthalpies of the vapor and liquid leaving the flash chamber.

For a given feed condition, the fraction f can be increased by flashing to a lower pressure.

Equation (21.2) is the equation of a straight line with a slope of $-(1 - f)/f$, and it can be plotted on the equilibrium diagram. The coordinates of the intersection of the line and the equilibrium curve are $x = x_B$ and $y = y_D$. The intersection of this material-balance line and the diagonal $x = y$ can be used conveniently as a point on the line. Letting $x = x_F$ in Eq. (21.2) gives

$$y = -\frac{1 - f}{f}x_F + \frac{x_F}{f}$$

from which $y = x_F = x$. The material-balance line crosses the diagonal at $x = x_F$ for all values of f.

EXAMPLE 21.1. A mixture of 50 mole percent benzene and 50 mole percent toluene is subjected to flash distillation at a separator pressure of 1 atm. The vapor-liquid equilibrium curve and boiling-point diagram are shown in Figs. 21.2 and 21.3. Plot the following quantities, all as functions of f, the fractional vaporization: the temperature in the separator, the composition of the liquid leaving the separator, and the composition of the vapor leaving the separator.

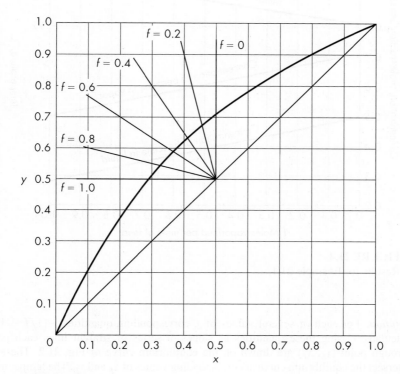

FIGURE 21.2
Equilibrium curve, system benzene-toluene. Graphical construction for Example 21.1.

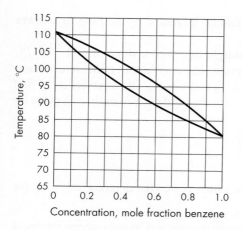

FIGURE 21.3
Boiling-point diagram (system benzene-toluene at 1 atm).

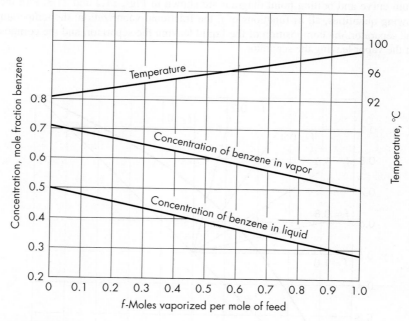

FIGURE 21.4
Results for Example 21.1.

Solution. For each of several values of f corresponding quantities $-(1/f - 1)$ are calculated. Using these quantities as slopes, a series of straight lines each passing through point (x_F, x_F) are drawn on the equilibrium curve of Fig. 21.2. These lines intersect the equilibrium curve at corresponding values of x_B and y_D. The temperature of each vaporization is then found from Fig. 21.3. The results are shown in Table 21.1 and plotted in Fig. 21.4. The limits for 0 and 100 percent vaporization are the *bubble* and *dew points*, respectively.

TABLE 21.1
Data for Example 21.1

Fraction vaporized f	Slope $-\dfrac{1-f}{f}$	Concentration mol fraction C_6H_6		Temperature, °C
		Liquid x_B	Vapor y_D	
0	∞	0.50	0.71	92.2
0.2	-4	0.455	0.67	93.7
0.4	-1.5	0.41	0.63	95.0
0.6	-0.67	0.365	0.585	96.5
0.8	-0.25	0.325	0.54	97.7
1.0	0	0.29	0.50	99.0

CONTINUOUS DISTILLATION WITH REFLUX

Flash distillation is used most for separating components that boil at widely different temperatures. It is not effective in separating components of comparable volatility, which requires the use of distillation with reflux. For large-scale production, continuous distillation, as described in this section, is far more common than batch distillation which is discussed later in this chapter.

Action on an ideal plate

On an ideal plate, by definition, the liquid and vapor leaving the plate are brought into equilibrium. Consider a single plate in an ideal cascade, such as plate n in Fig. 21.5. Assume that the plates are numbered serially from the top down and that

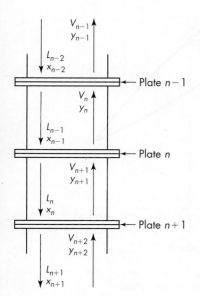

FIGURE 21.5
Material-balance diagram for plate n.

the plate under consideration is the nth plate from the top. Then the plate immediately above plate n is plate $n - 1$, and that immediately below it is plate $n + 1$. Subscripts are used on all quantities to show the point of origin of the quantity.

Two fluid streams enter plate n, and two leave it. A stream of liquid L_{n-1} mol/h from plate $n - 1$ and a stream of vapor V_{n+1} mol/h from plate $n + 1$ are brought into intimate contact. A stream of vapor V_n mol/h rises to plate $n - 1$, and a stream of liquid L_n mol/h descends to plate $n + 1$. Since the vapor streams are the V phase, their concentrations are denoted by y. The liquid streams are the L phase, and their concentrations are denoted by x. Then the concentrations of the streams entering and leaving the nth plate are as follows:

Vapor leaving plate y_n
Liquid leaving plate x_n
Vapor entering plate y_{n+1}
Liquid entering plate x_{n-1}

Figure 21.6 shows the boiling-point diagram for the mixture being treated. The four concentrations given above are shown in this figure. By definition of an ideal plate, the vapor and liquid leaving plate n are in equilibrium, so x_n and y_n represent equilibrium concentrations. This is shown in Fig. 21.6. The vapor is enriched in more volatile component A as it travels up the column, and the liquid is depleted of

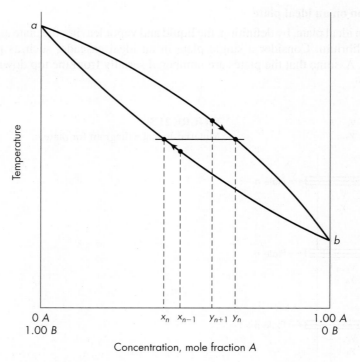

FIGURE 21.6
Boiling-point diagram showing rectification on ideal plate.

A as it flows downward. Thus the concentration of A in both phases increases with the height of the column; x_{n-1} is greater than x_n, and y_n is greater than y_{n+1}. Although the streams leaving the plate are in equilibrium, those entering it are not. This can be seen from Fig. 21.6. When the vapor from plate $n + 1$ and the liquid from plate $n - 1$ are brought into intimate contact, their concentrations tend to move toward an equilibrium state, as shown by the arrows in Fig. 21.6. Some of the more volatile component A is vaporized from the liquid, decreasing the liquid concentration from x_{n-1} to x_n; and some of the less volatile component B is condensed from the vapor, increasing the vapor concentration from y_{n+1} to y_n. Since the liquid streams are at their bubble points and the vapor streams at their dew points, the heat necessary to vaporize component A must be supplied by the heat released in the condensation of component B. Each plate in the cascade acts as an interchange apparatus in which component A is transferred to the vapor stream and component B to the liquid stream. Also, since the concentration of A in both liquid and vapor increases with column height, the temperature decreases, and the temperature of plate n is greater than that of plate $n - 1$ and less than that of plate $n + 1$.

Combination rectification and stripping

To produce nearly pure products at both the top and bottom of the distillation column, the feed is admitted to a plate in the central portion of the column. If the feed is liquid, it flows down the column to the reboiler and is stripped of component A by the vapor rising from the reboiler. By this means a bottom product can be produced which is nearly pure B.

A typical continuous fractionating column equipped with the necessary auxiliaries and containing rectifying and stripping sections is shown in Fig. 21.7. Column A is fed near its center with a steady flow of feed of definite concentration. Assume that the feed is a liquid at its boiling point. The action in the column is not dependent on this assumption, and other conditions of the feed will be discussed later. The plate on which the feed enters is called the *feed plate*. All plates above the feed plate constitute the rectifying section, and all plates below the feed, *including the feed plate itself,* constitute the stripping section. The feed flows down the stripping section to the bottom of the column, in which a definite level of liquid is maintained. Liquid flows by gravity to reboiler B. This is a steam-heated vaporizer that generates vapor and returns it to the bottom of the column. The vapor passes up the entire column. At one end of the reboiler is a weir. The bottom product is withdrawn from the pool of liquid on the downstream side of the weir and flows through the cooler G. This cooler also preheats the feed by heat exchange with the hot bottoms.

The vapors rising through the rectifying section are completely condensed in condenser C, and the condensate is collected in accumulator D, in which a definite liquid level is maintained. Reflux pump F takes liquid from the accumulator and delivers it to the top plate of the tower. This liquid stream is called *reflux*. It provides the downflowing liquid in the rectifying section that is needed to act on the upflowing vapor. Without the reflux, no rectification would occur in the rectifying section, and the concentration of the overhead product would be no greater than that of the vapor rising from the feed plate. Condensate not picked up by the reflux pump is

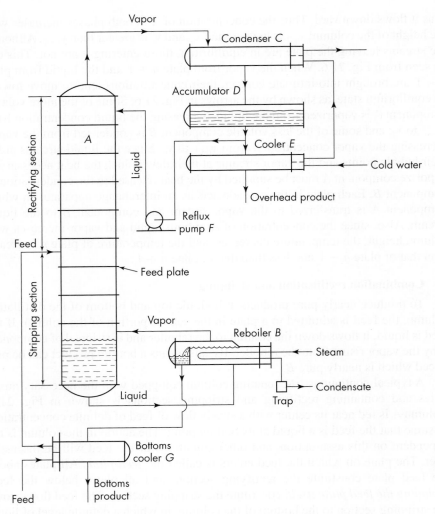

FIGURE 21.7
Continuous fractionating column with rectifying and stripping sections.

cooled in heat exchanger *E*, called the *product cooler*, and withdrawn as the overhead product. If no azeotropes are encountered, both overhead and bottom products may be obtained in any desired purity if enough plates and adequate reflux are provided.

The plant shown in Fig. 21.7 is often simplified for small installations. In place of the reboiler, a heating coil may be placed in the bottom of the column to generate vapor from the pool of liquid there. The condenser is sometimes placed above the top of the column, and the reflux pump and accumulator are omitted. Reflux then returns to the top plate by gravity. A special valve, called a *reflux splitter*, may be used to control the rate of reflux return. The remainder of the condensate forms the overhead product.

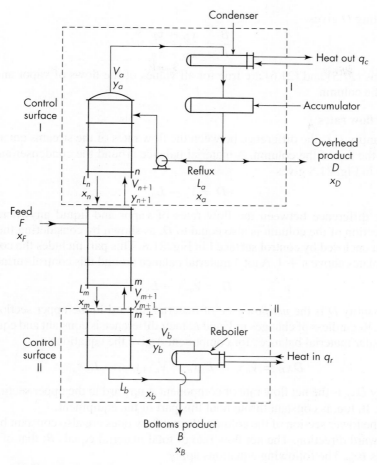

FIGURE 21.8
Material-balance diagram for continuous fractionating column.

Material Balances in Plate Columns

Overall material balances for two-component systems

Figure 21.8 is a material-balance diagram for a typical continuous distillation plant. The column is fed with F mol/h of concentration x_F and delivers D mol/h of overhead product of concentration x_D and B mol/h of bottom product of concentration x_B. Two independent overall material balances can be written.

Total-material balance $$F = D + B \tag{21.3}$$

Component A balance $$Fx_F = Dx_D + Bx_B \tag{21.4}$$

Eliminating B from these equations gives

$$\frac{D}{F} = \frac{x_F - x_B}{x_D - x_B} \tag{21.5}$$

Eliminating D gives

$$\frac{B}{F} = \frac{x_D - x_F}{x_D - x_B} \tag{21.6}$$

Equations (21.5) and (21.6) are true for all values of the flows of vapor and liquid within the column.

Net flow rates

Quantity D is the difference between the flow rates of the streams entering and leaving the top of the column. A material balance around the condenser and accumulator in Fig. 21.8 gives

$$D = V_a - L_a \tag{21.7}$$

The difference between the flow rates of vapor and liquid anywhere in the upper section of the column is also equal to D, as shown by considering the part of the plant enclosed by control surface I in Fig. 21.8. This part includes the condenser and all plates above $n + 1$. A total material balance around this control surface gives

$$D = V_{n+1} - L_n \tag{21.8}$$

Thus quantity D is the *net flow rate* of material upward in the upper section of the column. Regardless of changes in V and L, their difference is constant and equal to D.

Similar material balances for component A give the equations

$$D x_D = V_a y_a - L_a x_a = V_{n+1} y_{n+1} - L_n x_n \tag{21.9}$$

Quantity $D x_D$ is the net flow rate of component A upward in the upper section of the column. It, too, is constant throughout this part of the equipment.

In the lower section of the column, the net flow rates are also constant but are in a downward direction. The net flow rate of total material equals B; that of component A is $B x_B$. The following equations apply:

$$B = L_b - V_b = L_m - V_{m+1} \tag{21.10}$$

$$B x_B = L_b x_b - V_b y_b = L_m x_m - V_{m+1} y_{m+1} \tag{21.11}$$

Subscript m is used in place of n to designate a general plate in the stripping section.

Operating lines

Because there are two sections in the column, there are also two operating lines, one for the rectifying section and the other for the stripping section. Consider first the rectifying section. As shown in Chap. 20 [Eq. (20.7)], the operating line for this section is

$$y_{n+1} = \frac{L_n}{V_{n+1}} x_n + \frac{V_a y_a - L_a x_a}{V_{n+1}} \tag{21.12}$$

Substitution for $V_a y_a - L_a x_a$ from Eq. (21.9) gives

$$y_{n+1} = \frac{L_n}{V_{n+1}} x_n + \frac{D x_D}{V_{n+1}} \tag{21.13}$$

The slope of the line defined by Eq. (21.13) is, as usual, the ratio of the flow of the liquid stream to that of the vapor stream. For further analysis it is convenient to eliminate V_{n+1} from Eq. (21.13) by Eq. (21.8), giving

$$y_{n+1} = \frac{L_n}{L_n + D} x_n + \frac{D x_D}{L_n + D} \tag{21.14}$$

For the section of the column below the feed plate, a material balance over control surface II in Fig. 21.8 gives

$$V_{m+1} y_{m+1} = L_m x_m - B x_B \tag{21.15}$$

In a different form, this becomes

$$y_{m+1} = \frac{L_m}{V_{m+1}} x_m - \frac{B x_B}{V_{m+1}} \tag{21.16}$$

This is the equation for the operating line in the stripping section. Again the slope is the ratio of the liquid flow to the vapor flow. Eliminating V_{m+1} from Eq. (21.16) by Eq. (21.10) gives

$$y_{m+1} = \frac{L_m}{L_m - B} x_m - \frac{B x_B}{L_m - B} \tag{21.17}$$

Equation (21.14) shows that the slope of the operating line in the rectifying section is always less than 1.0; in the stripping section, as shown by Eq. (21.17), the slope is always greater than 1.0.

Number of Ideal Plates; McCabe–Thiele Method

When the operating lines represented by Eqs. (21.14) and (21.17) are plotted with the equilibrium curve on the xy diagram, the McCabe–Thiele step-by-step construction can be used to compute the number of *ideal* plates needed to accomplish a definite concentration difference in either the rectifying or the stripping section.[10] Equations (21.14) and (21.17), however, show that unless L_n and L_m are constant, the operating lines are curved and can be plotted only if the change in these internal streams with concentration is known. Enthalpy balances are required in the general case to determine the position of a curved operating line, and a method of doing this is described later in this chapter.

Constant molal overflow

For most distillations, the molar flow rates of vapor and liquid are nearly constant in each section of the column, and the operating lines are almost straight. This results from nearly equal molar heats of vaporization, so that each mole of high boiler that condenses as the vapor moves up the column provides energy to vaporize about 1 mol of low boiler. For example, the molar heats of vaporization of toluene and benzene are 7,960 and 7,360 cal/mol, respectively, so that 0.92 mol of toluene corresponds to 1.0 mol of benzene.[†] The changes in enthalpy of the liquid and vapor streams and heat losses from the column often require slightly more vapor to be formed at the bottom, so the molar ratio of vapor flow at the bottom of a column section to that at the top is even closer to 1.0. In designing columns, therefore, the concept of *constant molal overflow* is generally used, which means simply that in Eqs. (21.8) to (21.17), subscripts n, $n + 1$, m, and $m + 1$ on L and V may be dropped, and L and V now refer to flows in the upper part of the column, and $\bar{L}$ and $\bar{V}$ denote flows in the lower section. In this simplified model the material-balance equations are linear and the operating lines straight. An operating line can be plotted if the coordinates of two points on it are known. Then the McCabe–Thiele method is used without requiring enthalpy balances.

Reflux ratio

The analysis of fractionating columns is facilitated by the use of a quantity called the *reflux ratio*. Two such quantities are used. One is the ratio of the reflux to the overhead product, and the other is the ratio of the reflux to the vapor. Both ratios refer to quantities in the rectifying section. The equations for these ratios are

$$R_D = \frac{L}{D} = \frac{V - D}{D} \quad \text{and} \quad R_V = \frac{L}{V} = \frac{L}{L + D} \tag{21.18}$$

In this text only R_D will be used.

If both numerator and denominator of the terms on the right-hand side of Eq. (21.14) are divided by D, the result is, for constant molal overflow,

$$y_{n+1} = \frac{R_D}{R_D + 1} x_n + \frac{x_D}{R_D + 1} \tag{21.19}$$

Equation (21.19) is an equation for the operating line of the rectifying section. Its slope is $R_D/(R_D + 1)$; by substitution of $L = V - D$ from Eq. (21.18), it can be shown to be equal to L/V. The y intercept of this line is $x_D/(R_D + 1)$. The value of x_D is set by conditions of the design, and R_D, the reflux ratio, is an operating variable that can be controlled at will by adjusting the split between reflux and overhead product or by changing the amount of vapor formed in the reboiler for a given flow rate of the overhead product. A point at the upper end of the operating line can be obtained by setting x_n equal to x_D in Eq. (21.19):

$$y_{n+1} = \frac{R_D}{R_D + 1} x_D + \frac{x_D}{R_D + 1} = \frac{x_D(R_D + 1)}{R_D + 1} = x_D \tag{21.20}$$

The operating line for the rectifying section then intersects the diagonal at point (x_D, x_D). This is true for either a partial or a total condenser. (Partial condensers are discussed in the next section.)

[†]For benzene-toluene and for many other pairs of similar hydrocarbons, the heat of vaporization per unit *mass* is higher for the low boiler, but the ratio is still close to 1.0, and operating lines based on mass fraction would be almost straight. For systems such as ethanol-water, however, the heats of vaporization are about the same per mol but quite different per unit mass, so the use of molar quantities is advantageous for distillation calculations.

Condenser and top plate

The McCabe–Thiele construction for the top plate does depend on the action of the condenser. Figure 21.9 shows material-balance diagrams for the top plate and the condenser. The concentration of the vapor from the top plate is y_1, and that for the reflux to the top plate is x_c. In accordance with the general properties of operating lines, the upper terminus of the line is at the point (x_c, y_1).

The simplest arrangement for obtaining reflux and liquid product, and one that is frequently used, is the single total condenser shown in Fig. 21.9b, which condenses all the vapor from the column and supplies both reflux and product. When such a single total condenser is used, the concentrations of the vapor from the top plate, of the reflux to the top plate, and of the overhead product are equal and can all be denoted by x_D. The operating terminus of the operating line becomes point (x_D, x_D), which is the intersection of the operating line with the diagonal. Triangle abc in Fig. 21.10a then represents the top plate in the column.

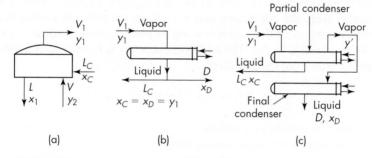

FIGURE 21.9
Material-balance diagrams for top plate and condenser: (a) top plate; (b) total condenser; (c) partial and final condensers.

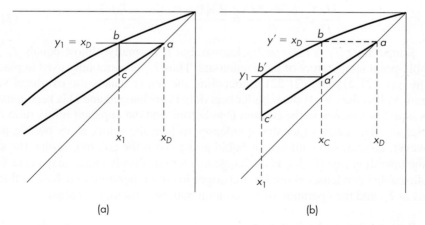

FIGURE 21.10
Graphical construction for top plate: (a) using total condenser; (b) using partial and final condensers.

When a partial condenser is used, the liquid reflux does not have the same composition as the overhead product; that is, $x_c \neq x_D$. Sometimes two condensers are used in series, first a partial condenser to provide reflux, then a final condenser to provide liquid product. Such an arrangement is shown in Fig. 21.9c. Vapor leaving the partial condenser has a composition y', which is the same as x_D. Under these conditions the diagram in Fig. 21.10b applies. The operating line passes through the point (x_D, x_D) on the diagonal, but as far as the column is concerned, the operating line ends at point a', which of course has the coordinates (x_c, y_1). Triangle $a'b'c'$ in Fig. 21.10b represents the top plate in the column. Since the vapor leaving a partial condenser is normally in equilibrium with the liquid condensate, the vapor composition y' is the ordinate value of the equilibrium curve where the abscissa is x_c, as shown in Fig. 21.10b. The partial condenser, represented by the dotted triangle aba' in Fig. 21.10b, is therefore equivalent to an additional theoretical stage in the distillation apparatus.

In the preceding treatment it is assumed that the condenser removes latent heat only and that the condensate is liquid at its bubble point. Then the reflux L is equal to L_c, the reflux from the condenser, and $V = V_1$. If the reflux is cooled below the bubble point, a portion of the vapor coming to plate 1 must condense to heat the reflux; so $V_1 < V$ and $L > L_c$. The additional amount ΔL that is condensed inside the column is found from the equation

$$\Delta L = \frac{L_c c_{pc}(T_1 - T_c)}{\lambda_c} \tag{21.21}$$

where c_{pc} = specific heat of condensate
T_1 = temperature of liquid on top plate
T_c = temperature of returned condensate
λ_c = heat of vaporization of condensate

The actual reflux ratio in the column is then

$$\frac{L}{D} = \frac{L_c + \Delta L}{D} = \frac{L_c[1 + c_{pc}(T_1 - T_c)/\lambda_c]}{D} \tag{21.22}$$

Temperature T_1 is not usually known, but it normally almost equals T_{bc}, the bubble-point temperature of the condensate. Thus T_{bc} is commonly used in place of T_1 in Eqs. (21.21) and (21.22). Subcooling the reflux makes the overhead vapor stream V_1 less than V_n. If the reboiler heat duty is unchanged and D is kept constant, less liquid is returned to the column than before, and the apparent reflux ratio R_D is decreased. The extra vapor that condenses to heat the reflux to its bubble point, however, increases the amount of liquid going down the column, so that the slope of the operating line $(L/V)_n$ is unchanged. A serious disadvantage may occur when an air-cooled condenser is used, for changes in air temperature may lead to fluctuations in T_c, and the operation of the column may be difficult to control.

Bottom plate and reboiler

The action at the bottom of the column is analogous to that at the top. Thus, Eq. (21.17), written for constant molal overflow, becomes, with $\bar{L}$ and later $\bar{V}$ used

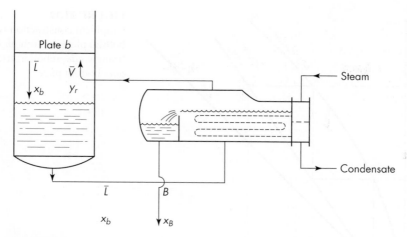

FIGURE 21.11
Material-balance diagram for bottom plate and reboiler.

to denote flow rates in this section,

$$y_{m+1} = \frac{\bar{L}}{\bar{L} - B}x_m - \frac{Bx_B}{\bar{L} - B} \qquad (21.23)$$

If x_m is set equal to x_B in Eq. (21.23), y_{m+1} is also equal to x_B, so the operating line for the stripping section crosses the diagonal at point (x_B, x_B). This is true no matter what type of reboiler is used, as long as there is only one bottom product. The lower operating line could then be constructed using the point (x_B, x_B) and the slope $\bar{L}/(\bar{L} - B)$, which equals $\bar{L}/\bar{V}$, but a more convenient method is described in the discussion on feed plates in the next section.

The material-balance diagram for the bottom plate and the reboiler is shown in Fig. 21.11. The lowest point on the operating line for the column itself is the point for the bottom plate (x_b, y_r), where x_b and y_r are the concentrations in the liquid leaving the bottom plate and the vapor coming from the reboiler. However, as shown earlier, the operating line can be extended to cross the diagonal at point (x_B, x_B).

In the common type of reboiler shown in Figs. 21.7 and 21.11, the vapor leaving the reboiler is in equilibrium with the liquid leaving as bottom product. Then x_B and y_r are coordinates of a point on the equilibrium curve, and the reboiler acts as an ideal plate. Figure 21.12 shows the graphical construction for the reboiler (triangle cde) and the bottom plate (triangle abc). Such a reboiler is called a *partial reboiler*. Its construction is shown in detail in Fig. 13.8.

Feed plate

At the plate where the feed is admitted, the liquid rate or the vapor rate or both may change, depending on the thermal condition of the feed. Figure 21.13 shows diagrammatically the liquid and vapor streams into and out of the feed plate for various feed conditions. In Fig. 21.13a, cold feed is assumed, and the entire feed

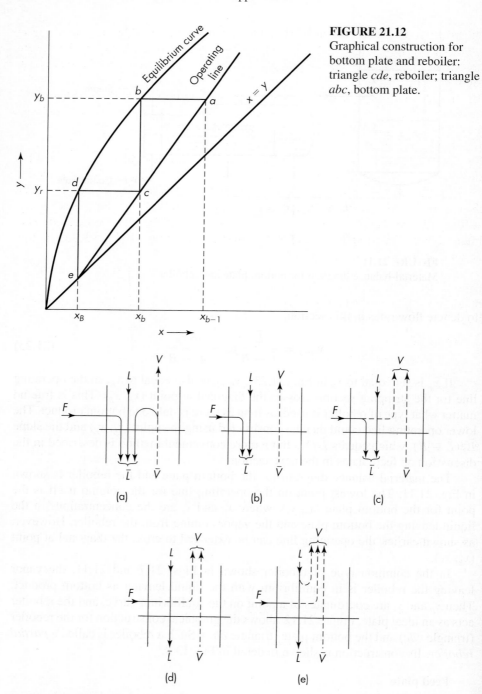

FIGURE 21.12
Graphical construction for bottom plate and reboiler: triangle *cde*, reboiler; triangle *abc*, bottom plate.

FIGURE 21.13
Flow through feed plate for various feed conditions: (*a*) feed cold liquid; (*b*) feed saturated liquid; (*c*) feed partially vaporized; (*d*) feed saturated vapor; (*e*) feed superheated vapor.

stream adds to the liquid flowing down the column. In addition, some vapor condenses to heat the feed to the bubble point; this makes the liquid flow even greater in the stripping section and decreases the flow of vapor to the rectifying section.

In Fig. 21.13b the feed is assumed to be at its bubble point. No condensation is required to heat the feed, so $V = \bar{V}$ and $\bar{L} = F + L$. If the feed is partly vapor, as shown in Fig. 21.13c, the liquid portion of the feed becomes part of $\bar{L}$ and the vapor portion becomes part of V. If the feed is saturated vapor, as shown in Fig. 21.13d, the entire feed becomes part of V, so $L = \bar{L}$ and $V = F + \bar{V}$. Finally, if the feed is superheated vapor, as shown in Fig. 21.13e, part of the liquid from the rectifying column is vaporized to cool the feed to a state of saturated vapor. Then the vapor in the rectifying section consists of (1) the vapor from the stripping section, (2) the feed, and (3) the extra moles vaporized in cooling the feed. The liquid flow to the stripping section is less than that in the rectifying section by the amount of additional vapor formed.

All five of the feed types can be characterized by the use of a single factor, denoted by q and defined as the moles of liquid flow in the stripping section that result from the introduction of each mole of feed. Then q has the following numerical limits for the various conditions:

Cold feed, $q > 1$
Feed at bubble point (saturated liquid), $q = 1$
Feed partially vapor, $0 < q < 1$
Feed at dew point (saturated vapor), $q = 0$
Feed superheated vapor $q < 0$

If the feed is a mixture of liquid and vapor, q is the fraction that is liquid. Such a feed may be produced by an equilibrium flash operation, so $q = 1 - f$, where f is the fraction of the original stream vaporized in the flash.

The value of q for cold-liquid feed is found from the equation

$$q = 1 + \frac{c_{pL}(T_b - T_F)}{\lambda} \tag{21.24}$$

For superheated vapor the equation is

$$q = -\frac{c_{pV}(T_F - T_d)}{\lambda} \tag{21.25}$$

where c_{pL}, c_{pV} = specific heats of liquid and vapor, respectively
T_F = temperature of feed
T_b, T_d = bubble point and dew point of feed, respectively
λ = heat of vaporization

Feed line

The value of q obtained from Eq. (21.24) or (21.25) can be used with the material balances to find the locus for all points of intersection of the operating lines. The equation for this line of intersections can be found as follows.

The contribution of the feed stream to the internal flow of liquid is qF, so the total flow rate of reflux in the stripping section is

$$\bar{L} = L + qF \qquad \text{and} \qquad \bar{L} - L = qF \tag{21.26}$$

Likewise, the contribution of the feed stream to the internal flow of vapor is $F(1 - q)$, and so the total flow rate of vapor in the rectifying section is

$$V = \bar{V} + (1 - q)F \qquad \text{and} \qquad V - \bar{V} = (1 - q)F \tag{21.27}$$

For constant molal overflow, the material-balance equations for two sections are

$$V y_n = L x_{n+1} + D x_D \tag{21.28}$$

$$\bar{V} y_m = \bar{L} x_{m+1} - B x_B \tag{21.29}$$

To locate the point where the operating lines intersect, let $y_n = y_m$ and $x_{n+1} = x_{m+1}$ and subtract Eq. (21.29) from Eq. (21.28):

$$y(V - \bar{V}) = (L - \bar{L})x + D x_D + B x_B \tag{21.30}$$

From Eq. (21.4), the last two terms in Eq. (21.30) can be replaced by $F x_F$. Also, substituting for $L - \bar{L}$ from Eq. (21.26) and for $V - \bar{V}$ from Eq. (21.27) and simplifying lead to the result

$$y = -\frac{q}{1 - q} x + \frac{x_F}{1 - q} \tag{21.31}$$

Equation (21.31) represents a straight line, called the *feed line*, on which all intersections of the operating lines must fall. The position of the line depends only on x_F and q. The slope of the feed line is $-q/(1 - q)$, and as can be demonstrated by substituting x for y in Eq. (21.31) and simplifying, the line crosses the diagonal at $x = x_F$.

Construction of operating lines

The simplest method of plotting the operating lines is to (1) locate the feed line; (2) calculate the y-axis intercept $x_D/(R_D + 1)$ of the rectifying line and plot that line through the intercept and the point (x_D, x_D); (3) draw the stripping line through point (x_B, x_B) and the intersection of the rectifying line with the feed line. The operating lines in Fig. 21.14 show the result of this procedure.

In Fig. 21.14 are plotted operating lines for various types of feed, on the assumption that x_F, x_B, x_D, L, and D are all constant. The corresponding feed lines are shown. If the feed is a cold liquid, the feed line slopes upward and to the right; if the feed is a saturated liquid, the line is vertical; if the feed is a mixture of liquid and vapor, the line slopes upward and to the left, and the slope is the negative of the ratio of the liquid to the vapor; if the feed is a saturated vapor, the line is horizontal; and finally, if the feed is superheated vapor, the line slopes downward and to the left.

Feed plate location

After the operating lines have been plotted, the number of ideal plates is found by the usual step-by-step construction, as shown in Fig. 21.15. The construction can

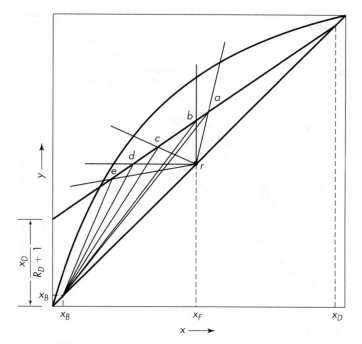

FIGURE 21.14
Effect of feed condition on feed line: *ra*, feed cold liquid; *rb*,
feed saturated liquid; *rc*, feed partially vaporized; *rd*, feed
saturated vapor; *re*, feed superheated vapor.

start either at the bottom of the stripping line or at the top of the rectifying line. In the following it is assumed that the construction begins at the top and that a total condenser is used. As the intersection of the operating lines is approached, it must be decided when the steps should transfer from the rectifying line to the stripping line. The change should be made in such a manner that the maximum enrichment per plate is obtained, so that the number of plates is as small as possible. Figure 21.15 shows that this criterion is met if the transfer is made immediately after a value of x is reached that is less than the x coordinate of the intersection of the two operating lines. The feed plate is always represented by the triangle that has one corner on the rectifying line and one corner on the stripping line. At the optimum position, the triangle representing the feed plate straddles the intersection of the operating lines.

The transfer from one operating line to the other, and hence the feed plate location, can be made at any location between points a and b in Fig. 21.15; but if the feed plate is placed anywhere but at the optimum point, an unnecessarily large number of plates are called for. For example, if the feed plate in Fig. 21.15 is number 7, the smaller steps shown by the dashed lines make the number of ideal plates needed about 8 plus a reboiler, instead of 7 plus a reboiler when the feed is on plate number 5. Note that the liquid on the feed plate does not have the same

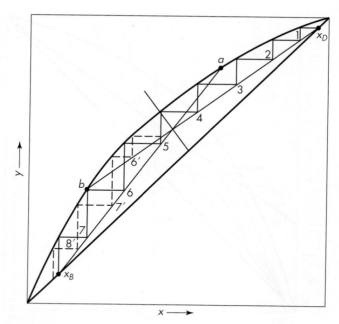

FIGURE 21.15
Optimum feed plate location: ———, with feed on plate 5
(optimum location); ------, with feed on plate 7.

composition as the feed except by coincidence, even when the feed plate location is optimum.

When we are analyzing the performance of a real column, the switch from one operating line to another must be made at a real feed plate. Because of changes in feed composition and uncertainties in plate efficiency, large columns are often operated with the feed entering a few plates above or below the optimum location. If large changes in feed composition are anticipated, alternate feed locations can be provided.

In an actual column with a fixed number of plates, feeding on the wrong plate may seriously affect the column performance. Feeding on too low a plate, for example, close to point *b* in Fig. 21.15, increases the number of plates in the rectifying section; but many of them are now operating in a pinched region where the driving force is small. These plates do very little separation. The diagram, therefore, must change, lowering the quality of both the top and bottom products to reflect the poorer performance of the plates. Feeding too high in the column leads to similar consequences.

Heating and cooling requirements

Heat loss from a large insulated column is relatively small, and the column itself is essentially adiabatic. The heat effects of the entire unit are confined to the condenser and the reboiler. If the average molal latent heat is λ and the total sensible heat change in the liquid streams is small, the heat added in the reboiler q_r is $\bar{V}\lambda$, either in watts or Btu per hour. When the feed is liquid at the bubble point ($q = 1$), the heat supplied in the reboiler is approximately equal to that removed in the condenser; but for other values of q this is not true. (See enthalpy balances on page 674.)

If saturated steam is used as the heating medium, the steam required at the reboiler is

$$\dot{m}_s = \frac{\bar{V}\lambda}{\lambda_s} \qquad (21.32)$$

where $\dot{m}_s$ = steam consumption
$\bar{V}$ = vapor rate from reboiler
λ_s = latent heat of steam
λ = molal latent heat of mixture

If water is used as the cooling medium in the condenser and the condensate is not subcooled, the cooling water requirement is

$$\dot{m}_w = \frac{V\lambda}{T_2 - T_1} \qquad (21.33)$$

where $\dot{m}_w$ = water consumption
$T_2 - T_1$ = temperature rise of cooling water

EXAMPLE 21.2. A continuous fractionating column is to be designed to separate 30,000 kg/h of a mixture of 40% benzene and 60% toluene into an overhead product containing 97% benzene and a bottom product containing 98% toluene. These percentages are by weight. A reflux ratio of 3.5 mol to 1 mol of product is to be used. The molal latent heats of benzene and toluene are 7,360 and 7,960 cal/g mol, respectively. Benzene and toluene form a nearly ideal system with a relative volatility of about 2.5; the equilibrium curve is shown in Fig. 21.16. The feed has a boiling point of 95°C at a pressure of 1 atm. (*a*) Calculate the moles of overhead product and bottom product per hour. (*b*) Determine the number of ideal plates and the position of the feed plate (*i*) if the feed is liquid and at its boiling point; (*ii*) if the feed is liquid and at 20°C (specific heat 0.44 cal/g · °C); (*iii*) if the feed is a mixture of two-thirds vapor and one-third liquid. (*c*) If steam at 20 lb$_f$/in.^{2} (1.36 atm) gauge is used for heating, how much steam is required per hour for each of the above three cases, neglecting heat losses and assuming the reflux is a saturated liquid? (*d*) If cooling water enters the condenser at 25°C and leaves at 40°C, how much cooling water is required, in cubic meters per hour?

Solution

(*a*) The molecular weight of benzene is 78, and that of toluene is 92. The concentrations of feed, overhead, and bottoms in mole fraction of benzene are

$$x_F = \frac{\frac{40}{78}}{\frac{40}{78} + \frac{60}{92}} = 0.440 \qquad x_D = \frac{\frac{97}{78}}{\frac{97}{78} + \frac{3}{92}} = 0.974$$

$$x_B = \frac{\frac{2}{78}}{\frac{2}{78} + \frac{98}{92}} = 0.0235$$

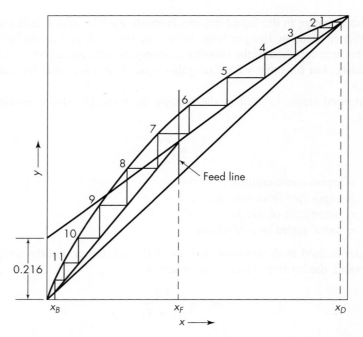

FIGURE 21.16
Example 21.2, part (*b*)(*i*).

The average molecular weight of the feed is

$$\frac{100}{\frac{40}{78} + \frac{60}{92}} = 85.8$$

The average heat of vaporization of the feed is

$$\lambda = 0.44(7{,}360) + 0.56(7{,}960) = 7{,}696 \text{ cal/g mol}$$

The feed rate F is $30{,}000/85.8 = 350$ kg mol/h. By an overall benzene balance, using Eq. (21.5),

$$D = 350\left(\frac{0.440 - 0.0235}{0.974 - 0.0235}\right) = 153.4 \text{ kg mol/h}$$

$$B = 350 - 153.4 = 196.6 \text{ kg mol/h}$$

(*b*) Next we determine the number of ideal plates and position of the feed plate.

(*i*) The first step is to plot the equilibrium diagram and on it erect verticals at x_D, x_F, and x_B. These should be extended to the diagonal of the diagram. Refer to Fig. 21.16. The second step is to draw the feed line. Here $f = 0$, and the feed line is vertical and is a continuation of line $x = x_F$. The third step is to plot the operating lines. The intercept of the rectifying line on the y axis is, from Eq. (21.19), $0.974/(3.5 + 1) = 0.216$. This point is connected with the point x_D on the yx reference line. From the intersection of the rectifying operating line and the feed line, the stripping line is drawn.

The fourth step is to draw the rectangular steps between the two operating lines and the equilibrium curve. In drawing the steps, the transfer from the rectifying line to the

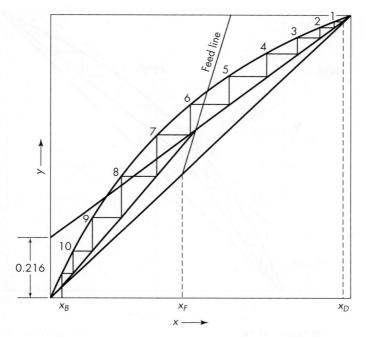

FIGURE 21.17
Example 21.2, part (*b*)(*ii*).

stripping line is at the seventh step. By counting steps it is found that, besides the reboiler, 11 ideal plates are needed and feed should be introduced on the seventh plate from the top.[†]

(*ii*) The latent heat of vaporization of the feed λ is $7{,}696/85.8 = 89.7$ cal/g. Substitution in Eq. (21.24) gives

$$q = 1 + \frac{0.44(95 - 20)}{89.7} = 1.37$$

From Eq. (21.31) the slope of the feed line is $-1.37/(1 - 1.37) = 3.70$. When steps are drawn for this case, as shown in Fig. 21.17, it is found that a reboiler and 10 ideal plates are needed and that the feed should be introduced on the sixth plate.

(*iii*) From the definition of q it follows that for this case $q = \frac{1}{3}$ and the slope of the feed line is -0.5 The solution is shown in Fig. 21.18. It calls for a reboiler and 12 plates, with the feed entering on the seventh plate.

More plates are needed when the feed is largely vapor, partly because the feed line slants to the left and a few more plates are required in the rectifying section. The major

---

[†]To fulfill the conditions of the problem literally, the last step, which represents the reboiler, should reach the concentration x_B exactly. This is nearly true in Fig. 21.16. Usually, x_B does not correspond to an integral number of steps. An arbitrary choice of the four quantities x_D, x_F, x_B, and R_D is not necessarily consistent with an integral number of steps. An integral number can be obtained by a slight adjustment of one of the four quantities, but in view of the fact that a plate efficiency must be applied before the actual number of plates is established, there is little reason for making this adjustment.

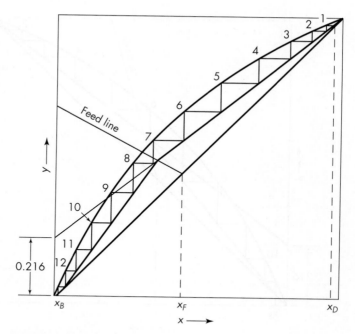

FIGURE 21.18
Example 21.2, part (b)(iii).

reason, however, is that partially vaporized feed contributes less liquid to the stripping section than does a totally liquid feed, and the reflux ratio in the stripping section is reduced.

(c) The vapor flow V in the rectifying section, which must be condensed in the condenser, is 4.5 mol per mole of overhead product, or $4.5 \times 153.4 = 690$ kg mol/h. From Eq. (21.27)

$$\bar{V} = 690 - 350(1 - q)$$

Using the heat of vaporization of toluene rather than that of benzene to be slightly conservative in design, $\lambda = 7,960$ cal/g mol. The heat from 1 lb of steam at 20 $lb_f/in.^2$ gauge, from App. 7, is 939 Btu/lb; hence $\lambda_S = 939/1.8 = 522$ cal/g. The steam required, from Eq. (21.32), is

$$\dot{m}_s \frac{7,960}{522} \bar{V} = 15.25[690 - 350(1 - q)] \qquad \text{kg/h}$$

The results are given in Table 21.2.

Note: Using the equation $\alpha = 2.34 + 0.27x$ for the relative volatility of benzene-toluene mixtures, a computer program indicated 10.59 plates required for part (*i*), 10.17 plates for part (*ii*), and 11.85 for part (*iii*).

(d) The cooling water needed, which is the same in all cases, is, from Eq. (21.33),

$$\dot{m}_w = \frac{7,960 \times 690}{40 - 25} = 366,160 \text{ kg/h}$$

The density of water at 25°C (77°F), from App. 6, is 62.24 lb/ft³, or 996.3 kg/m³. The water requirement is $366,160/996.3 = 367.5$ m³/h.

TABLE 21.2
Solution to Example 21.2, part (c)

Case	q	Reboiler steam $\dot{m}_s$, kg/h	Number of ideal plates
(i)	1.0	10,520	11
(ii)	1.37	12,500	10
(iii)	0.333	6,960	12

The use of cold feed, case (ii), requires the smallest number of plates but the greatest amount of reboiler steam. The total energy requirement for the reboiler and the preheater is about the same for all three cases. The reasons for preheating the feed, in most cases, are to keep the vapor flow rate about the same in both sections of the column and to make use of the energy in a hot-liquid stream such as the bottom product.

Minimum number of plates

Since the slope of the rectifying line is $R_D/(R_D + 1)$, the slope increases as the reflux ratio increases until, when R_D is infinite, $V = L$ and the slope is 1. The operating lines then both coincide with the diagonal. This condition is called *total reflux*. At total reflux the number of plates is a minimum, but the rates of feed and of both the overhead and bottom products are zero. Total reflux represents one limiting case in the operation of fractionating columns. The minimum number of plates required for a given separation may be found by constructing steps on an xy diagram between compositions x_D and x_B, using the 45° line as the operating line for both sections of the column. Since there is no feed in a column operating under total reflux, there is no discontinuity between the upper and lower sections.

For the special case of ideal mixtures, a simple method is available for calculating the value of $N_{\min}$ from the terminal concentrations x_B and x_D. This is based on the relative volatility of the two components α_{AB}, which is defined in terms of the equilibrium concentrations

$$\alpha_{AB} = \frac{y_{Ae}/x_{Ae}}{y_{Be}/x_{Be}} \qquad (21.34)$$

A useful form of Eq. (21.34) for finding y_e from x_e is

$$y_e = \frac{\alpha_{AB} x_e}{1 + (\alpha_{AB} - 1)x_e} \qquad (21.35)$$

An ideal mixture follows Raoult's law, and the relative volatility is the ratio of the vapor pressures. Thus

$$p_A = P_A' x_A \qquad y_A = \frac{p_A}{P}$$

$$p_B = P_B' x_B \qquad y_B = \frac{p_B}{P}$$

$$\alpha_{AB} = \frac{y_A/x_A}{y_B/x_B} = \frac{P_A'/P}{P_B'/P} = \frac{P_A'}{P_B'} \qquad (21.36)$$

The ratio P_A'/P_B' does not change much over the range of temperatures encountered in a typical column, so the relative volatility is taken as constant in the following derivation.

For a binary system y_A/y_B and x_A/x_B may be replaced by $y_A/(1-y_A)$ and $x_A/(1-x_A)$, so Eq. (21.34) can be written for plate $n+1$ as

$$\frac{y_{n+1}}{1-y_{n+1}} = \alpha_{AB}\frac{x_{n+1}}{1-x_{n+1}} \tag{21.37}$$

Since at total reflux $D=0$ and $L/V=1$, $y_{n+1}=x_n$. See Eq. (21.13), and note that the operating line is the 45° line; this leads to

$$\frac{x_n}{1-x_n} = \alpha_{AB}\frac{x_{n+1}}{1-x_{n+1}} \tag{21.38}$$

At the top of the column, if a total condenser is used, $y_1 = x_D$, so Eq. (21.37) becomes

$$\frac{x_D}{1-x_D} = \alpha_{AB}\frac{x_1}{1-x_1} \tag{21.39}$$

Writing Eq. (21.38) for a succession of n plates gives

$$\frac{x_1}{1-x_1} = \alpha_{AB}\frac{x_2}{1-x_2}$$

$$\cdots\cdots\cdots\cdots\cdots\cdots \tag{21.40}$$

$$\frac{x_{n-1}}{1-x_{n-1}} = \alpha_{AB}\frac{x_n}{1-x_n}$$

If Eq. (21.39) and all the equations in the set of Eqs. (21.40) are multiplied together and all the intermediate terms canceled, then

$$\frac{x_D}{1-x_D} = (\alpha_{AB})^n\frac{x_n}{1-x_n} \tag{21.41}$$

To reach the bottom discharge from the column, N_{min} plates and a reboiler are needed, and Eq. (21.41) gives

$$\frac{x_D}{1-x_D} = (\alpha_{AB})^{N_{min}+1}\frac{x_B}{1-x_B}$$

Solving the equation for N_{min} by logarithms gives

$$N_{min} = \frac{\ln\left[x_D(1-x_B)/x_B(1-x_D)\right]}{\ln \alpha_{AB}} - 1 \tag{21.42}$$

Equation (21.42) is the *Fenske equation,* which applies when α_{AB} is constant. If the change in the value of α_{AB} from the bottom of the column to the top is moderate, a geometric mean of the extreme values is recommended for α_{AB}.

Minimum reflux

At any reflux less than total, the number of plates needed for a given separation is larger than at total reflux and increases continuously as the reflux ratio is decreased. As the ratio becomes smaller, the number of plates becomes very large, and at a definite minimum, called the *minimum reflux ratio,* the number of plates becomes infinite. All actual columns producing a finite amount of desired top and bottom products must operate at a reflux ratio between the minimum, at which the number of plates is infinity, and infinity, at which the number of plates is a minimum. If L_a/D is the operating reflux ratio and $(L_a/D)_{min}$ is the minimum reflux ratio, then

$$\left(\frac{L_a}{D}\right)_{min} < \frac{L_a}{D} < \infty \qquad (21.43)$$

The minimum reflux ratio can be found by following the movement of the operating lines as the reflux is reduced. In Fig. 21.19 both operating lines coincide with the diagonal *afb* at total reflux. For an actual operation lines *ae* and *eb* are typical operating lines. As the reflux is further reduced, the intersection of the operating lines moves along the feed line toward the equilibrium curve, the area on the diagram available for steps shrinks, and the number of steps increases. When either one or both of the operating lines touch the equilibrium curve, the number of steps

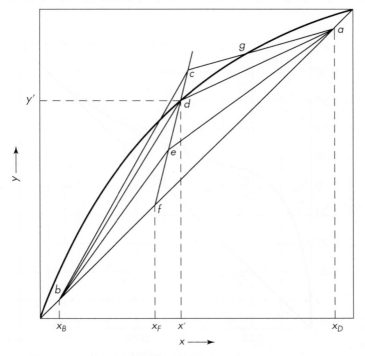

FIGURE 21.19
Minimum reflux ratio.

necessary to cross the point of contact becomes infinite. The reflux ratio corresponding to this situation is, by definition, the minimum reflux ratio.

For the normal type of equilibrium curve, which is concave downward throughout its length, the point of contact, at minimum reflux, of the operating and equilibrium lines is at the intersection of the feed line with the equilibrium curve, as shown by lines *ad* and *db* in Fig. 21.19. A further decrease in reflux brings the intersection of the operating lines outside of the equilibrium curve, as shown by lines *agc* and *cb*. Then even an infinite number of plates cannot pass point *g*, and the reflux ratio for this condition is less than the minimum.

The slope of operating line *ad* in Fig. 21.19 is such that the line passes through the points (x', y') and (x_D, x_D), where x' and y' are the coordinates of the intersection of the feed line and the equilibrium curve. Let the minimum reflux ratio be R_{Dm}.

Then
$$\frac{R_{Dm}}{R_{Dm} + 1} = \frac{x_D - y'}{x_D - x'}$$

or
$$R_{Dm} = \frac{x_D - y'}{y' - x'} \tag{21.44}$$

Equation (21.44) cannot be applied to all systems. Thus, if the equilibrium curve has a concavity upward, for example, the curve for ethanol and water shown in Fig. 21.20, it is clear that the rectifying line first touches the equilibrium curve

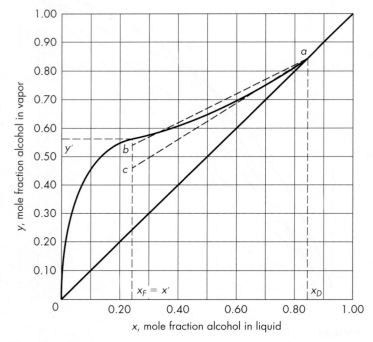

FIGURE 21.20
Equilibrium diagram (system ethanol-water).

between abscissas x_F and x_D and line ac corresponds to minimum reflux. Operating line ab is drawn for a reflux less than the minimum, even though it does intersect the feed line below point (x', y'). In such a situation the minimum reflux ratio must be computed from the slope of the operating line ac that is tangent to the equilibrium curve.

Invariant zone

At minimum reflux ratio an acute angle is formed at the intersection of an operating line and the equilibrium curve, as shown at point d in Fig. 21.19 or at the point of tangency in Fig. 21.20. In each angle an infinite number of steps are called for, representing an infinite number of ideal plates, in all of which there is no change in either liquid or vapor concentrations from plate to plate, so $x_{n-1} = x_n$ and $y_{n+1} = y_n$. The term *invariant zone* is used to describe these infinite sets of plates. The more descriptive term *pinch point* also is used.

With a normal equilibrium curve it is seen from Fig. 21.19 that at minimum reflux ratio the intersection of the q line and the equilibrium curve gives the concentrations of liquid and vapor at the feed plate (and at an infinite number of plates on either side of that plate). So an invariant zone forms at the bottom of the rectifying section and a second one at the top of the stripping section. The two zones differ only in that the liquid-vapor ratio is L/V in one and $\bar{L}/\bar{V}$ in the other.

Optimum reflux ratio

As the reflux ratio is increased from the minimum, the number of plates decreases, rapidly at first and then more and more slowly until, at total reflux, the number of plates is a minimum. It will be shown later that the cross-sectional area of a column usually is approximately proportional to the flow rate of vapor. As the reflux ratio increases, both V and L increase for a given production, and a point is reached where the increase in column diameter is more rapid than the decrease in the number of plates. The cost of the unit is roughly proportional to the total plate area, that is, the number of plates times the cross-sectional area of the column; so the fixed charges for the column first decrease and then increase with reflux ratio. Fixed charges on the heat-exchange equipment—the reboiler and condenser—increase steadily with the reflux ratio. Curve 2 in Fig. 21.21 shows the total fixed charges, which drop sharply at first, then pass through a very shallow minimum.

Also important are the costs of heating and cooling, shown by curve 1, which rises linearly with the reflux ratio. At the optimum reflux ratio the steam cost is often about two-thirds of the total annual cost, that is, the sum of the fixed charges and the costs of heating and cooling.[14] The total cost is shown by the topmost curve. It is a minimum at a definite reflux ratio not much greater than the minimum reflux. This is the point of most economical operation, and this ratio is called the *optimum reflux ratio*. Figure 21.21 is based on a study of benzene-toluene distillation where the product compositions were 92% benzene and 95% toluene, and the optimum reflux ratio was 1.1 times R_{Dm}.[14] In a similar optimization study where the products were very pure—99.97% benzene and 99.83% toluene—the optimum reflux ratio was $1.25R_{Dm}$.[4] The optimum reflux ratio depends on the cost of energy; it will be closer to R_{Dm} when energy costs are relatively high, and farther from R_{Dm} when distillation

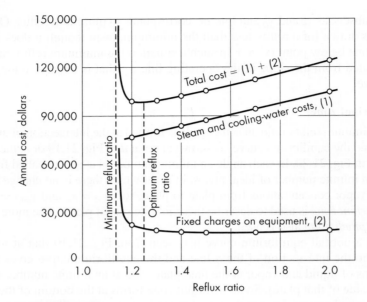

FIGURE 21.21
Optimum reflux ratio. (*By permission from M. S. Peters and K. D. Timmerhaus, Plant Design and Economics for Chemical Engineers, 3rd ed., 1980, McGraw-Hill.*)

equipment is made of expensive alloys. Actually, most plants are operated at reflux ratios somewhat above the optimum, because the total cost is not very sensitive to reflux ratio in this range, and better operating flexibility is obtained if a reflux ratio greater than the optimum is used.[†]

EXAMPLE 21.3. What are (*a*) the minimum reflux ratio and (*b*) the minimum number of plates for cases (*b*)(*i*), (*b*)(*ii*), and (*b*)(*iii*) of Example 21.2?

Solution

(*a*) For minimum reflux ratio use Eq. (21.44). Here $x_D = 0.974$. The results are given in Table 21.3.

(*b*) For minimum number of plates, the reflux ratio is infinite, the operating lines coincide with the diagonal, and there are no differences between the three cases. The plot is given in Fig. 21.22. A reboiler and eight ideal plates are needed.

Nearly pure products

When either the bottom or overhead product is nearly pure, a single diagram covering the entire range of concentrations is impractical as the steps near $x = 0$ and $x = 1$ become small. Auxiliary diagrams for the ends of the construction range may be prepared, on a large scale, so that the individual steps are large enough to be drawn. In practice, however, the calculations are usually done by computer, and the

---

[†]Because of deviations in the column from constant molal overflow, the true minimum reflux ratio may be larger than that predicted from the McCabe–Thiele diagram. (See p. 678.)

TABLE 21.3

Case	x'	y'	R_{Dm}
(b)(i)	0.440	0.658	1.45
(b)(ii)	0.521	0.730	1.17
(b)(iii)	0.300	0.513	2.16

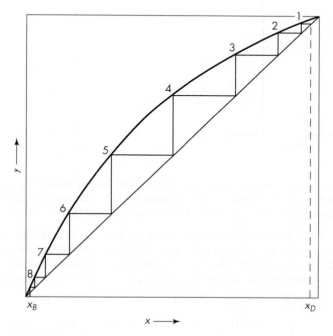

FIGURE 21.22
Example 21.3, part (b).

scale may easily be expanded to cover the desired range. (With computer calculations, of course, it is not necessary to display the McCabe–Thiele diagram to find
the number of plates needed, but it is often helpful to do so in order to visualize the
solution to the problem.)

Another method of treating nearly pure products is based on the principle that
Raoult's law applies to the major component and Henry's law to the minor component at each end of the equilibrium curve. In these regions, therefore, both the equilibrium and the operating lines are straight, so Eq. (20.27) can be used, and no graphical construction is required. The same equation may be used anywhere in the
concentration range where both the operating and equilibrium lines are straight or
nearly so.

EXAMPLE 21.4. A mixture of 2 mol percent ethanol and 98 mol percent water is to be
stripped in a plate column to a bottom product containing not more than 0.01 mol
percent ethanol. Steam, admitted through an open coil in the liquid on the bottom plate,

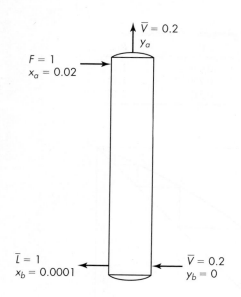

FIGURE 21.23
Material-balance diagram for Example 21.4.

is to be used as a source of vapor. The feed is at its boiling point. The steam flow is to be 0.2 mol per mole of feed. For dilute ethanol-water solutions, the equilibrium line is straight and is given by $y_e = 9.0x_e$. How many ideal plates are needed?

Solution. Since both equilibrium and operating lines are straight, Eq. (20.27) rather than a graphical construction may be used. The material-balance diagram is shown in Fig. 21.23. No reboiler is needed, as the steam enters as a vapor. Also, the liquid flow in the tower equals the feed entering the column. By conditions of the problem

$$F = \bar{L} = 1 \qquad \bar{V} = 0.2 \qquad y_b = 0 \qquad x_a = 0.02$$

$$x_b = 0.0001 \qquad m = 9.0 \qquad y_a^* = 9.0 \times 0.02 = 0.18$$

$$y_b^* = 9.0 \times 0.0001 = 0.0009$$

To use Eq. (20.27), y_a, the concentration of the vapor leaving the column, is needed. This is found by an overall ethanol balance

$$\bar{V}(y_a - y_b) = \bar{L}(x_a - x_b) \qquad 0.2(y_a - 0) = 1(0.02 - 0.0001)$$

from which $y_a = 0.0995$. Substituting into Eq. (20.27) gives

$$N = \frac{\ln\left[(0.0995 - 0.18)/(0 - 0.0009)\right]}{\ln\left[(0.0009 - 0.18)/(0 - 0.0995)\right]}$$

$$= \frac{\ln 89.4}{\ln 1.8} = 7.6 \text{ ideal plates}$$

ENTHALPY BALANCES

The actual variations in the V and L streams in a distillation column depend on the enthalpies of the vapor and liquid mixtures. The limitations imposed by assuming constant molal overflow can be removed by enthalpy balances used in conjunction

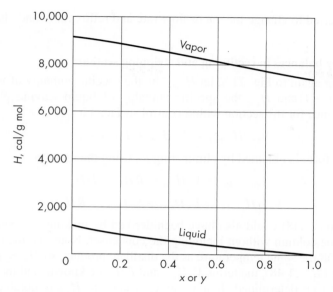

FIGURE 21.24
Enthalpy–concentration diagram for benzene-toluene at 1 atm.

with material balances and phase equilibria. The enthalpy data may be available from an enthalpy–concentration diagram, such as the one in Fig. 21.24. Since benzene-toluene solutions are ideal, this diagram was constructed using molar average heat capacities and heats of vaporization. Some sample calculations are given later in Example 21.5. The reference temperature was taken as 80°C, the boiling point of benzene, to simplify further calculations.

The enthalpy values in Fig. 21.24 are for liquid mixtures at the bubble point and for vapor mixtures at the dew point, both at 1.0 atm. Thus each point on the line is for a different temperature between 110.6 and 80°C, but the temperature for $x = 0.5$ is not the same as for $y = 0.5$, as shown by the difference between bubble and dew points in Fig. 21.3. The slight curvature in the enthalpy–concentration graphs is due to the nonlinear change in bubble point and dew point with mole fraction benzene.

Consider an overall enthalpy balance for the system shown in Fig. 21.8. In addition to the quantities shown in the figure, let H_F, H_D, and H_B represent the specific enthalpies of the feed, overhead product, and bottom product, respectively, all in energy per mole. The enthalpy balance for the entire system is

$$F H_F + q_r = D H_D + B H_B + q_c \qquad (21.45)$$

When the feed is liquid at the boiling point, H_F is between H_D and H_B, and the terms $F H_F$ and $D H_D + B H_B$ nearly cancel, making the heat supplied at the reboiler q_r about equal to that removed in the condenser q_c.

For given feed and product streams, only one of heat effects, q_r or q_c, is independent and subject to choice by the designer or operator. In designing a column, q_c is usually chosen to correspond to the desired reflux ratio and moles of overhead vapor. Then q_r can be calculated from Eq. (21.45). However, in operating a column,

q_r is often varied to change the vapor flow rate and reflux ratio, and changes in q_c then follow.

Enthalpy balances in rectifying and stripping sections

Referring again to Fig. 21.8, let $H_{y,n+1}$ be the specific enthalpy of vapor rising from plate $n + 1$ and $H_{x,n}$ the specific enthalpy of liquid leaving plate n. The enthalpy balance for the section within control surface I is

$$V_{n+1} H_{y,n+1} = L_n H_{x,n} + DH_D + q_c \tag{21.46}$$

An alternate form is obtained by eliminating q_c, using the relation

$$q_c = V_a H_{y,a} - RH_D - DH_D \tag{21.47}$$

$$V_{n+1} H_{y,n+1} = L_n H_{x,n} + V_a H_{y,a} - RH_D \tag{21.48}$$

Equation (21.48) could also have been derived by making a balance over the top part of the column without including the condenser. Note that the reflux is assumed to be at the same temperature as the distillate product, so $H_R = H_D$.

In using Eq. (21.48), the terms $V_a H_{y,a}$ and RH_D are known, and the terms Y_{n+1} and L_n are to be determined. If a value x_n is chosen, $H_{x,n}$ is obtained from an enthalpy–concentration diagram, or it is calculated from the average specific heat and the bubble point. The value of $H_{y,n+1}$ depends on y_{n+1}, which is not known until the operating line has been drawn on the McCabe–Thiele diagram or values of Y_{n+1} and L_n specified for the operating-line equation.

An exact value of V_{n+1} calls for a trial-and-error solution using Eq. (21.48), the enthalpy–concentration diagram, and the following equations for the individual and overall material balances:

$$y_{n+1} = \frac{L_n x_n}{V_{n+1}} + \frac{D x_D}{V_{n+1}} \tag{21.49}$$

$$V_{n+1} = L_n + D \tag{21.50}$$

However, a satisfactory value of V_{n+1} can usually be obtained on the first trial by using the flows at the top of the column L_a and V_a to calculate y_{n+1} from x_n in Eq. (21.49) (this corresponds to using a straight operating line based on constant molal overflow). Then $H_{y,n+1}$ is evaluated from y_{n+1}, and Eq. (21.48) is solved for V_{n+1} after L_n is replaced by $V_{n+1} - D$.

Only a few values of V_{n+1} and L_n are needed to establish a slightly curved operating line. For a plate-by-plate calculation, the values of V_{n+1} and L_n would be obtained by using the corresponding values for the previous plate V_n and L_{n-1} in the calculation of y_{n+1} from Eq. (21.49).

In the stripping section of the column, the flow rates at an intermediate plate m are calculated using an enthalpy balance for control surface II in Fig. 21.8:

$$V_{m+1} H_{y,m+1} = L_m H_{x,m} + q_r - BH_B \tag{21.51}$$

$$y_{m+1} = \frac{L_m}{V_{m+1}} x_m - \frac{B x_B}{V_{m+1}} \tag{21.52}$$

$$L_m = V_{m+1} + B \tag{21.53}$$

Following the same approach as before, a value of x_m is chosen and y_{m+1} is calculated from Eq. (21.52) using L_b and V_b to approximate L_m and V_{m+1}. Then V_{m+1} is calculated from Eq. (21.51), using the specific enthalpies $H_{x,m}$ and $H_{y,m+1}$ and substituting $V_{m+1} + B$ for L_m.

EXAMPLE 21.5. A mixture of 50 mol percent benzene and toluene is to be separated by distillation at atmospheric pressure into products of 98 percent purity using a reflux ratio 1.2 times the minimum value. The feed is liquid at the boiling point. Use enthalpy balances (Table 21.4) to calculate the flows of liquid and vapor at the top, middle, and bottom of the column, and compare these values with those based on constant molal overflow. Estimate the difference in the number of theoretical plates for the methods.

Solution

$$x_F = 0.50 \qquad x_D = 0.98 \qquad x_B = 0.02$$

From Eq. (21.5)

$$\frac{D}{F} = \frac{x_F - x_B}{x_D - x_B} = \frac{0.5 - 0.02}{0.98 - 0.02} = 0.50$$

Basis: $\qquad F = 100 \text{ mol} \qquad D = 50 \text{ mol} \qquad B = 50 \text{ mol}$

From Eq. (21.3)

$$R_{Dm} = \frac{x_D - y'}{y' - x'} \qquad \text{based on } \frac{L}{V} = \text{const}$$

For this feed, $q = 1.0$ and $x' = x_F = 0.50$. From the equilibrium curve, $y' = 0.72$, and

$$R_{Dm} = \frac{0.98 - 0.72}{0.72 - 0.50} = 1.18$$

$$R_D = 1.2(1.18) = 1.42$$

$$R = 1.42(50) = 71 \text{ mol}$$

At top of column,

$$V_1 = R + D = 71 + 50 = 121 \text{ mol}$$

Enthalpy balance. Choose 80°C as a reference temperature so that the reflux and distillate product at 80°C have zero enthalpy. For benzene vapor, the enthalpy is the heat

TABLE 21.4
Data for Example 21.5

Component	Enthalpy of vaporization, cal/g mol	Specific heat at constant pressure, cal/g mol · °C		Boiling point, °C
		Liquid	Vapor	
Benzene	7,360	33	23	80.1
Toluene	7,960	40	33	110.6

of vaporization at 80°C plus the sensible heat of the vapor:

$$H_y = 7{,}360 + 23(T - 80) \text{ cal/mol}$$

For toluene vapor, the enthalpy of vaporization at $80°C$ is calculated from the value at the boiling point:

$$\Delta H_v = \Delta H_{v,b} + (C_{p,l} - C_{p,v})(T_b - T)$$

For toluene at $T = 80°C$, $\Delta H_v = 7{,}960 + (40 - 33)(30.6) = 8{,}174$ cal/mol. For toluene at $T°C$, $H_y = 8{,}174 + 33(T - 80)$.

From Eq. (21.48) with $H_D = 0$ and $V_a = V_1$

$$V_{n+1} H_{y,n+1} = L_n H_{x,n} + V_1 H_{y,1}$$

Assume the top plate temperature is about 80°C to evaluate $H_{y,1}$. Since $y_1 = x_D = 0.98$,

$$H_y = 0.98(7{,}360) + 0.02(8{,}174) = 7{,}376 \text{ cal/mol}$$

Pick $x_n = 0.5$, and get $T_b = 92°C$ from Fig. 21.3. Then

$$H_{x,n} = (0.5 \times 33 + 0.5 \times 40)(92 - 80)$$
$$= 438 \text{ cal/mol}$$

To estimate y_{n+1}, use the operating line for constant molal overflow (dashed line in Fig. 21.25):

$$y_{n+1} \approx 0.70 \qquad T_d \approx 93°C \qquad \text{from Fig. 21.3}$$

$$H_{y,n+1} = 0.7(7{,}360 + 23 \times 13) + 0.3(8{,}174 + 33 \times 13)$$
$$= 7{,}942 \text{ cal/mol}$$

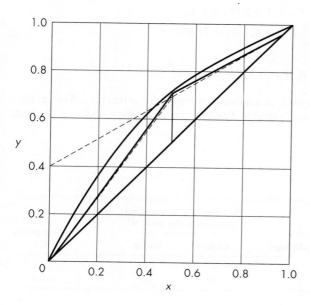

FIGURE 21.25
McCabe–Thiele diagram for Example 21.5 benzene-toluene distillation: ---------, based on constant molal overflow; ————, based on enthalpy balance.

Since $L_n = V_{n+1} - D$, from Eq. (21.48)

$$V_{n+1}(7,942) = (V_{n+1} - 50)(438) + 121(7,376)$$

$$V_{n+1} = \frac{870,596}{7,504} = 116.0 \text{ mol} \qquad L_n = 66.0 \text{ mol}$$

From Eq. (21.49),

$$y_{n+1} = \frac{66}{116}(0.50) + \frac{50(0.98)}{116} = 0.707$$

This is close enough to 0.70.

A similar calculation for $x_n = 0.7$ gives

$$V_{n+1} = 118 \qquad L_n = 68 \qquad y_{n+1} = 0.818$$

The operating line, shown in Fig. 21.25 as a solid line, is almost straight but lies above the operating line based on constant molal overflow (dashed line).

To get the vapor rate at the reboiler, an overall balance is made to get q_r:

$$F H_F + q_r = D H_D + B H_B + q_c$$

For feed at 92°C,

$$H_F = (0.5 \times 33 + 0.5 \times 40)(92 - 80)$$
$$= 438 \text{ cal/mol}$$

For bottoms at 111°C,

$$H_B = (0.02 \times 33 + 0.98 \times 40)(111 - 80)$$
$$= 1,236 \text{ cal/mol}$$

$$q_c = 121 \times 7,376 = 892,496 \text{ cal}$$

$$q_r = 50 \times 0 + 50 \times 1,236 + 892,496 - 100 \times 438$$
$$= 910,496 \text{ cal}$$

An enthalpy balance around the reboiler is then made:

$$q_r + L_b H_{x,b} = V_b H_{y,b} + B H_b$$

The vapor from the reboiler is about 5% benzene at 111°C, and

$$H_{y,b} = 0.05(7,360 + 23 \times 31) + 0.95(8,174 + 33 \times 31)$$
$$= 9,141 \text{ cal/mol}$$

The liquid to the reboiler is about 4% benzene at 110°C, and

$$H_{x,b} = 0.04(33 \times 30) + 0.96(40 \times 30) = 1,192 \text{ cal/mol}$$

Since $L_b = V_b + 50$, and $H_{y,b} - H_{x,b} = 9,141 - 1,192 = 7,949 \text{ cal/mol}$,

$$V_b = \frac{910,496 + 50(1,192) - 50(1,236)}{7,949} = 114.3$$

$$L_b = 114.3 + 50 = 164.3$$

Approximately the same value for V_b could have been obtained from q_r and the heat of vaporization of toluene:

$$V_b \approx \frac{q_r}{\Delta H_v} = \frac{910,496}{7,960} = 114.4$$

Use Eq. (21.51) to get an intermediate value of V_{m+1}:

$$V_{m+1} H_{y,m+1} = L_m H_{x,m} + q_r - B H_B$$

For $x_m = 0.4$, $Y_{m+1} = 0.55$ (from the operating line in Fig. 21.25); also

$$T_m = 95°C \qquad T_{m+1} = 97°C$$

$$H_{y,m+1} = 0.55[7,360 + 23(97 - 80)] + 0.45[8,174 + 33(17)] = 8,194 \text{ cal/mol}$$

$$H_{x,m} = (0.4 \times 33 + 0.6 \times 40)(95 - 80)$$

$$= 558 \text{ cal/mol}$$

$$L_m = V_{m+1} + 50$$

From Eq. (21.51),

$$8,194 V_{m+1} = 558(V_{m+1} + 50) + 910,496 - 1,236 \times 50$$

$$V_{m+1} = 114.8 \text{ mol}$$

$$L_m = 164.8 \text{ mol}$$

Note that in this case there is almost no change in L and V in the stripping section, in contrast to the 7 percent decrease in L in the rectifying section. The lower operating line can be drawn as a straight line to the intersection of the upper operating line and the q line.

Counting steps, about 27 ideal stages are required for this separation, compared to 21 based on the assumption of constant molal overflow. The difference would be smaller if a higher reflux ratio were used. The calculations were based on 1.2 times the nominal value of R_{min}, but this really corresponds to about 1.1 times the true minimum reflux, as can be seen from Fig. 21.25.

In Example 21.5, the molar liquid rate decreased about 7 percent in going from the top plate to the feed plate, mainly because of the higher molar heat of vaporization of toluene. The terms for the change in sensible heat of the liquid and vapor streams nearly cancel, since the liquid has a higher heat capacity but a lower flow rate than the vapor. In the stripping section of the column, there was almost no change in liquid rate, although the vapor composition changed even more than in the rectifying section. The liquid flow rate in the stripping section is always greater than the vapor rate; and in Example 21.5, the product of flow rate and heat capacity for the liquid was 1.73 times that for the vapor. Only part of the energy needed to heat the liquid from the feed plate temperature to the reboiler temperature could be supplied by cooling the vapor, and the rest came from condensation of vapor in the column. The difference between the heat of condensation of toluene and the heat of vaporization of benzene was nearly used up in providing the extra energy needed to heat the liquid, so that there was only a slight increase in vapor flow from the reboiler to the feed plate.

Similar changes in L and V are likely to be observed for other ideal mixtures.[†] The more volatile component has a lower molal heat of vaporization, since the heat of vaporization is roughly proportional to the normal boiling point (Trouton's rule). The change in V will be greatest in the upper section of the column, where L is less than V, and may be almost zero in the lower section, where $L/V > 1.0$. The percentage change in L/V will be smaller than the change in L or V, but the slight upward shift of the operating lines may be important when operating close to the minimum reflux ratio, as was the case in Example 21.5. For operation at twice the minimum reflux ratio or greater, the effect of operating-line curvature would be very small.

The slope of a straight operating line is L/V for that section of the column, but the local slope of a curved operating line is not equal to the local value of L/V. Starting with the equation for the rectifying section, Eq. (21.49), the following equations can be obtained by replacing first L_n and then V_{n+1} with $V_{n+1} - D$ and $L_n + D$:

or
$$V_{n+1}y_{n+1} = (V_{n+1} - D)x_n + Dx_D$$
$$V_{n+1}(y_{n+1} - x_n) = D(x_D - x_n) \tag{21.54}$$

or
$$(L_n + D)y_{n+1} = L_nx_n + Dx_D$$
$$L_n(y_{n+1} - x_n) = D(x_D - y_{n+1}) \tag{21.55}$$

Dividing Eq. (21.55) by Eq. (21.54), we have

$$\frac{L_n}{V_{n+1}} = \frac{x_D - y_{n+1}}{x_D - x_n} \tag{21.56}$$

Thus L_n/V_{n+1} is the slope of the chord connecting points (x_D, x_D) and (y_{n+1}, x_n). A similar derivation for the stripping section shows that

$$\frac{L_m}{V_{m+1}} = \frac{y_{m+1} - x_B}{x_m - x_B} \tag{21.57}$$

Multiple feeds and side-stream drawoffs

Sometimes a column is fed with two (or even more) streams, admitted to the column on different plates. The yx diagram for a column with two liquid feeds F_1 and F_2 is illustrated in Fig. 21.26a. The upper operating line, above where F_1 enters, is not affected; the middle line has a larger slope than the upper line, because of the liquid added from F_1; and the line for the lowest section, where the amount of liquid is greatest, has the largest slope. The diagram is constructed (or the computer program prepared) from the relative amounts of liquid and vapor in each section of the column. Plates are stepped off in the usual way.

Similar procedures are used when a side stream is drawn off below the top of the column, with a composition x_D'. The drawoff reduces the amount of liquid passing to the section below, so the slope of the operating line in this region is less than that

---

[†]For aqueous systems, the molar heats of vaporization may be quite different, leading to significant changes in L and V.

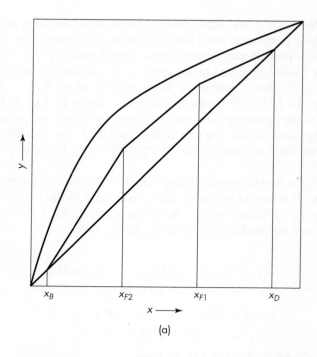

(a)

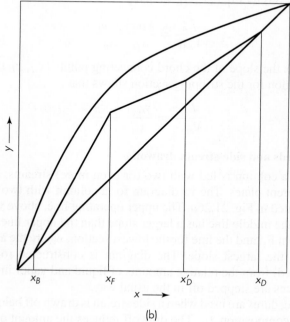

(b)

FIGURE 21.26
Complex distillation columns: (a) with two liquid feeds;
(b) with side-stream drawoff.

in the topmost section. This is illustrated in Fig. 21.26b. As before, the slope of the operating line equals the ratio of liquid to vapor L/V everywhere in the column.

DESIGN OF SIEVE-PLATE COLUMNS

To translate ideal plates to actual plates, a correction for the efficiency of the plates must be applied. There are other important decisions, some at least as important as fixing the number of plates, that must be made before a design is complete. A mistake in these decisions results in poor fractionation, lower-than-desired capacity, poor operating flexibility, and with extreme errors, an inoperative column. Correcting such errors after a plant has been built can be costly. Since many variables that influence plate efficiency depend on the design of the individual plates, the fundamentals of plate design are discussed first.

The extent and variety of rectifying columns and their applications are enormous. The largest units are usually in the petroleum industry, but large and very complicated distillation plants are encountered in fractionating solvents, in treating liquefied air, and in general chemical processing. Tower diameters may range from 300 mm (1 ft) to more than 9 m (30 ft) and the number of plates from a few to about a hundred. Plate spacings may vary from 150 mm or less to 1 or 2 m. Formerly bubble-cap plates were most common; today most columns contain sieve trays or lift-valve plates. Many types of liquid distribution are specified. Columns may operate at high pressures or low, from temperatures of liquid gases up to 900°C reached in the recification of sodium and potassium vapors. The materials distilled can vary greatly in viscosity, diffusivity, corrosive nature, tendency to foam, and complexity of composition. Plate towers are as useful in absorption as in rectification, and the fundamentals of plate design apply to both operations.

Designing fractionating columns, especially large units and those for unusual applications, is best done by experts. Although the number of ideal plates and the heat requirements can be computed quite accurately without much previous experience, other design factors are not precisely calculable, and a number of equally sound designs can be found for the same problem. In common with most engineering activities, sound design of fractionating columns relies on a few principles, on a number of empirical correlations (which are in a constant state of revision), and on much experience and judgment.

The following discussion is limited to the usual type of column, equipped with sieve plates, operating at pressures not far from atmospheric, and treating mixtures having ordinary properties.

Normal operation of sieve plate

A sieve plate is designed to bring a rising stream of vapor into intimate contact with a descending stream of liquid. The liquid flows across the plate and passes over a weir to a downcomer leading to the plate below. The flow pattern on each plate is therefore crossflow rather than countercurrent flow, but the column as a whole is still considered to have countercurrent flow of liquid and vapor. The fact that there

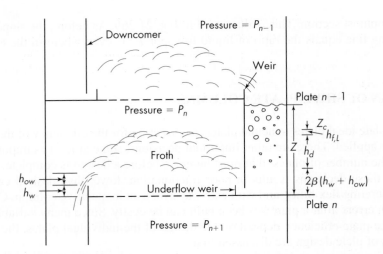

FIGURE 21.27
Normal operation of sieve plate.

is crossflow of liquid on the plate is important in analyzing the hydraulic behavior of the column and in predicting the plate efficiency.

Figure 21.27 shows a plate in a sieve-tray column in normal operation. The downcomers are the segment-shaped regions between the curved wall of the column and the straight chord weir. Each downcomer usually occupies 10 to 15 percent of the column cross section, leaving 70 to 80 percent of the column area for bubbling or contacting. In small columns the downcomer may be a pipe welded to the plate and projecting up above the plate to form a circular weir. For very large columns, additional downcomers may be provided at the middle of the plate to decrease the length of the liquid flow path. In some cases an underflow weir or tray inlet weir is installed as shown in Fig. 21.27 to improve the liquid distribution and to prevent vapor bubbles from entering the downcomer.

The vapor passes through the perforated region of the plate, which occupies most of the space between downcomers. The holes are usually 5 to 12 mm ($\frac{3}{16}$ to $\frac{1}{2}$ in.) in size and arranged in a triangular pattern. One or two rows of holes may be omitted near the overflow weir to permit some degassing of the liquid before it passes over the weir. Some holes may also be omitted near the liquid inlet to keep vapor bubbles out of the downcomer. Under normal conditions, the vapor velocity is high enough to create a frothy mixture of liquid and vapor that has a large surface area for mass transfer. The average density of the froth may be as low as 0.2 of the liquid density, and the froth height is then several times the value corresponding to the amount of liquid actually on the plate.

Vapor pressure drop

The flow of vapor through the holes and the liquid on the plate requires a difference in pressure. The pressure drop across a single plate is usually 50 to 70 mm H_2O, and the pressure drop over a 40-plate column is then about 2 to 3 m H_2O.

The pressure required is automatically developed by the reboiler, which generates vapor at a pressure sufficient to overcome the pressure drop in the column and condenser. The overall pressure drop is calculated to determine the pressure and temperature in the reboiler, and the pressure drop per plate must be checked to make sure the plate will operate properly, without weeping or flooding.

The pressure drop across the plate can be divided into two parts, the friction loss in the holes and the pressure drop due to the holdup of liquid on the plate. The pressure drop is usually given as an equivalent head in millimeters or inches of liquid:

$$h_t = h_d + h_l \tag{21.58}$$

where h_t = total pressure drop per plate, mm of liquid
h_d = friction loss for dry plate, mm of liquid
h_l = equivalent head of liquid on plate, mm of liquid

The pressure drop through the holes can be predicted from a modification of Eq. (8.39) for flow through an orifice:

$$h_d = \left(\frac{u_0^2}{C_0^2}\right)\left(\frac{\rho_V}{2g\rho_L}\right) = 51.0\left(\frac{u_0^2}{C_0^2}\right)\left(\frac{\rho_V}{\rho_L}\right) \tag{21.59}$$

where u_0 = vapor velocity through holes, m/s
ρ_V = vapor density
ρ_L = liquid density
C_0 = orifice coefficient

Equation (21.59) gives h_d in millimeters of liquid. The coefficient comes from

$$\frac{1,000 \text{ mm/m}}{2 \times 9.8 \text{ m/s}^2} = 51.0$$

If u_0 is expressed in feet per second and h_d in inches, the coefficient becomes

$$\frac{12}{2 \times 32.2} = 0.186$$

Orifice coefficient C_0 depends on the fraction open area (the ratio of the total cross-sectional area of the holes to the column cross section) and on the ratio of tray thickness to hole diameter, as shown in Fig. 21.28. The increase in C_0 with open area is similar to the change in C_0 for single orifices as the ratio of orifice diameter to pipe diameter increases. The coefficients vary with plate thickness, but for most sieve plates the thickness is only 0.1 to 0.3 times the hole size. For these thicknesses and the typical fraction open area of 0.08 to 0.10, the value of C_0 is 0.66 to 0.72.

The amount of liquid on the plate increases with the weir height and with the flow rate of liquid; but it decreases slightly with increasing vapor flow rate, because this decreases the density of the froth. The liquid holdup also depends on the physical properties of liquid and vapor, and only approximate methods of predicting the holdup are available. A simple method of estimating h_l uses the weir height h_w, the calculated height of clear liquid over the weir h_{ow}, and an empirical correlation

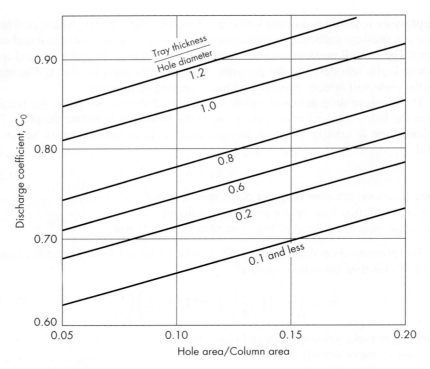

FIGURE 21.28
Discharge coefficients for vapor flow, sieve trays. [*I. Liebson, R. E. Kelley, and L. A. Bullington, Petrol. Refin.,* **36**(2):127, 1957; **36**(3):288, 1957.]

factor β:

$$h_l = \beta(h_w + h_{ow}) \tag{21.60}$$

The height over the weir is calculated from a form of the Francis equation, which for a straight segmental weir is

$$h_{ow} = 43.4\left(\frac{q_L}{L_w}\right)^{2/3} \tag{21.61}$$

where h_{ow} = height, mm
q_L = flow rate of clear liquid, m³/min
L_w = length of weir, m

If q_L/L_w is in gallons per minute per inch, a coefficient of 0.48 in Eq. (21.61) gives h_{ow} in inches.

The actual height of froth over the weir is greater than h_{ow}, since the vapor has only partially separated from the liquid, making the volumetric flow rate at the weir greater than that of the liquid alone. However, the actual height over the weir is not needed in estimating h_l, since the effect of froth density is included in the correlation factor β. For typical weir heights of 25 to 50 mm (1 to 2 in.) and the normal range of vapor velocities, values of β are 0.4 to 0.7. The change in β with flow rates of vapor and liquid is complex, and there is no generally accepted correlation. For design purposes, a value of $\beta = 0.6$ can be used in Eq. (21.60), and some error can be tolerated since most of the pressure drop at high vapor flows is due to the holes.

When h_{ow} is small relative to h_w, Eq. (21.60) shows that h_l may be less than h_w, which means less liquid on the tray than corresponds to the weir height. This is a fairly common situation.

Downcomer level

The level of liquid in the downcomer must be considerably greater than that on the plate because of the pressure drop across the plate. Referring to Fig. 21.27, note that the top of the downcomer for plate n is at the same pressure as plate $n - 1$. Therefore, the equivalent level in the downcomer must exceed that on the plate by an amount h_t plus any friction losses in the liquid $h_{f,L}$. The total height of clear liquid Z_c is

$$Z_c = \beta(h_w + h_{ow}) + h_t + h_{f,L} \tag{21.62}$$

Using Eqs. (21.58) and (21.60) for h_t gives

$$Z_c = 2\beta(h_w + h_{ow}) + h_d + h_{f,L} \tag{21.63}$$

The contributions to Z_c are shown in Fig. 21.27. Note that an increase in h_w or h_{ow} comes in twice, since it increases the level of liquid on the plate and increases the pressure drop for vapor flow. The term for $h_{f,L}$ is usually small, corresponding to one to two velocity heads based on the liquid velocity under the bottom of the downcomer.

The actual level of aerated liquid Z in the downcomer is greater than Z_c because of the entrained bubbles. If the average volume fraction liquid is ϕ_d, the level is

$$Z = \frac{Z_c}{\phi_d} \tag{21.64}$$

When the height of aerated liquid becomes as great as or greater than the plate spacing, the flow over the weir on the next plate is hindered, and the column becomes flooded. For conservative design, a value of $\phi_d = 0.5$ is assumed, and the plate spacing and operating conditions are chosen so that Z is less than the plate spacing.

Operating limits for sieve trays

At low vapor velocities, the pressure drop is not great enough to prevent liquid from flowing down through some of the holes. This condition is called *weeping* and is more likely to occur if there is a slight gradient in liquid head across the plate. With such a gradient, vapor will tend to flow through the region where there is less liquid and therefore less resistance to flow, and liquid will flow through the section where the depth is greatest. Weeping decreases the plate efficiency, since some liquid passes to the next plate without contacting the vapor. The lower limit of operation could be extended by using smaller holes or a lower fraction open area, but these changes would increase the pressure drop and reduce the maximum flow rate. A sieve tray can usually be operated over a three- to fourfold range of flow rates between the weeping and flooding points.

If a greater range is desired, other types of plates such as valve trays can be used. (See page 689.)

The upper limit of the velocity in a sieve-tray column is determined by the flooding point or by the velocity at which entrainment becomes excessive. Flooding occurs when the liquid in the downcomer backs up to the next plate, and this is determined mainly by the pressure drop across the plate and the plate spacing. Near the flooding point, most of the pressure drop comes from the term h_d in Eq. (21.58), so the total pressure drop varies approximately with the square of the velocity and the ratio of vapor and liquid densities. A long-established empirical correlation consistent with the above reasoning states that the maximum permissible vapor velocity is proportional to $\sqrt{(\rho_L - \rho_V)/\rho_V}$. Thus

$$u_c = K_v' \sqrt{\frac{\rho_L - \rho_V}{\rho_V}} \tag{21.65}$$

where u_c = maximum permissible velocity based on bubbling or active area
$\quad\quad\;\; K_v'$ = empirical coefficient

Coefficient K_v' has been evaluated from plant data and correlated with operating variables such as plate spacing and flow rates. A correlation[3] that includes the effect of surface tension is given in [Eq. (21.66)]; values of K_v for use with this equation are shown in Fig. 21.29. A value of $\sigma = 20$ dyn/cm is typical of organic liquids, and the correlation shows that the flooding velocity with such liquids is about 20 percent

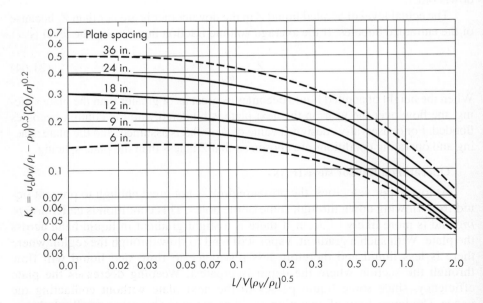

FIGURE 21.29
Values of K_v at flooding conditions for sieve plates; L/V = ratio of mass flow rate of liquid to vapor, u is in feet per second, and σ is in dynes per centimeter. [*J. R. Fair, Petrol. Chem. Eng.,* **33**(10):45, 1961. *Courtesy Petroleum Engineer.*]

lower than that for water, for which σ is about 72 dyn/cm. The correlation is not recommended for liquids of very low surface tension or for systems that foam easily:

$$u_c = K_v \sqrt{\frac{\rho_L - \rho_V}{\rho_V}} \left(\frac{\sigma}{20}\right)^{0.2} \tag{21.66}$$

The term $\rho_L - \rho_V$ appears in Eqs. (21.65) and (21.66) because it was once thought that entrainment of liquid drops determined the limiting vapor velocity, and the setting velocity of drops depends on this density difference. The term $\rho_L - \rho_v$ is practically the same as ρ_L in most cases; the effect of vapor density is reflected primarily in the terms involving $\sqrt{\rho_V}$, which show that the flooding velocity is appreciably lower in columns operated at high pressure than those at atmospheric pressure or below.

EXAMPLE 21.6. A sieve-plate column operating at atmospheric pressure is to produce nearly pure methanol from an aqueous feed containing 40 mol percent methanol. The distillate product rate is 5,800 kg/h. (a) For a reflux ratio of 3.5 and a plate spacing of 18 in., calculate the allowable vapor velocity and the column diameter. (b) Calculate the pressure drop per plate if each sieve tray is $\frac{1}{8}$ in. thick with $\frac{1}{4}$-in. holes on a $\frac{3}{4}$-in. triangular spacing and a weir height of 2 in. (c) What is the froth height in the downcomer?

Solution

Physical properties of methanol. Molecular weight is 32, normal boiling point is 65°C, and the density of vapor is

$$\rho_V = \frac{32 \times 273}{22.4 \times 338} = 1.15 \text{ kg/m}^3$$

From Perry, *Chemical Engineers' Handbook,* 6th ed., page **3**–188, the density of liquid methanol is 810 kg/m³ at 0°C and 792 kg/m³ at 20°C. At 65°C, the estimated density ρ_L is 750 kg/m³. Lange's *Handbook of Chemistry,* 9th ed., 1956, page 1650, gives the surface tension of methanol at 20 and 100°C. By interpolation, at 65°C, $\sigma = 19$ dyn/cm.

(a) Vapor velocity and column diameter. In Fig. 21.29 the abscissa is

$$\frac{L}{V}\left(\frac{\rho_V}{\rho_L}\right)^{1/2} = \frac{3.5}{4.5}\left(\frac{1.15}{750}\right)^{1/2} = 3.04 \times 10^{-2}$$

For 18-in. plate spacing,

$$K_v = 0.29 = u_c \left(\frac{\rho_V}{\rho_V - \rho_L}\right)^{1/2} \left(\frac{20}{\sigma}\right)^{0.2}$$

Allowable vapor velocity:

$$u_c = 0.29 \left(\frac{750 - 1.15}{1.15}\right)^{1/2} \left(\frac{19}{20}\right)^{0.2}$$

$$= 7.32 \text{ ft/s or } 2.23 \text{ m/s}$$

Vapor flow rate:

$$V = D(R + 1) = 4.5D$$

$$= \frac{5,800 \times 4.5}{3,600 \times 1.15} = 6.30 \text{ m}^3/\text{s}$$

Cross-sectional area of column:

$$\text{Bubbling area} = \frac{6.30}{2.23} = 2.83 \text{ m}^2$$

If the bubbling area is 0.7 of the total column area,

$$\text{Column area} = \frac{2.83}{0.7} = 4.04 \text{ m}^2$$

Column diameter:

$$D_c = \left(\frac{4 \times 4.04}{\pi}\right)^{1/2} = 2.27 \text{ m}$$

(b) Pressure drop. The plate area of one unit of three holes on a triangular $\frac{3}{4}$-in. pitch is $\frac{1}{2} \times \frac{3}{4}(\frac{3}{4} \times \sqrt{3}/2) = 9\sqrt{3}/64 \text{ in.}^2$. The hole area in this section (half a hole) is $\frac{1}{2} \times \pi/4 \times (\frac{1}{4})^2 = \pi/128 \text{ in.}^2$. Thus the hole area is $\pi/128 \times 64/9\sqrt{3} = 0.1008$, or 10.08 percent of the bubbling area.

Vapor velocity through holes:

$$u_0 = \frac{2.23}{0.1008} = 22.1 \text{ m/s}$$

Use Eq. (21.59) for the pressure drop through the holes. From Fig. 21.28, $C_0 = 0.73$. Hence

$$h_d = \frac{51.0 \times 22.1^2 \times 1.15}{0.73^2 \times 750} = 71.7 \text{ mm methanol}$$

Head of liquid on plate:

Weir height: $h_w = 2 \times 25.4 = 50.8$ mm

Height of liquid above weir: Assume the downcomer area is 15 percent of the column area on each side of the column. From Perry, 6th ed., page 1–26, the chord length for such a segmental downcomer is 1.62 times the radius of the column, so

$$L_w = 1.62 \times 2.27/2 = 1.84 \text{ m}$$

Liquid flow rate:

$$q_L = \frac{5,800 \times 3.5}{750 \times 60} = 0.45 \text{ m}^3/\text{min}$$

From Eq. (21.61),

$$h_{ow} = 43.4 \left(\frac{0.45}{1.84}\right)^{2/3} = 17.0 \text{ mm}$$

From Eq. (21.60), with $\beta = 0.6$,

$$h_l = 0.6(50.8 + 17.0) = 40.7 \text{ mm}$$

Total head of liquid [from Eq. (21.58)]:

$$h_t = 71.7 + 40.7 = 112.4 \text{ mm}$$

(c) Froth height in downcomer. Use Eq. (21.63). Estimate $h_{f,L} = 10$ mm methanol. Then

$$Z_c = 2 \times 40.7 + 71.7 + 10 = 163.1 \text{ mm}$$

From Eq. (21.64),

$$Z = \frac{163.1}{0.5} = 326 \text{ mm } (12.8 \text{ in.})$$

Valve-tray columns

In a valve-tray column the openings in the plate are quite large, typically 38 mm ($1\frac{1}{2}$ in.) in diameter. The openings are covered with lids or "valves" which rise and fall as the vapor rate varies, providing a variable area for vapor flow. Downcomers and crossflow of the liquid are used as with ordinary sieve trays. Valve trays are more expensive than conventional trays but have the advantage of a large *turndown ratio* (the ratio of the maximum allowable vapor velocity to the lowest velocity at which the column will operate satisfactorily), up to 10 or more, so the operating range of the column is large. Recent developments in valve trays have increased the available bubbling area and improved the distribution of vapor flowing through the plates. Information on the design and limitations of valve-tray columns is available from the Norton Chemical Process Products Corporation.[13]

PLATE EFFICIENCIES

To translate ideal plates to actual plates, the plate efficiency must be known. The following discussion applies to columns for gas absorption as well as to those for distillation.

Types of plate efficiency

Three kinds of plate efficiency are used: (1) overall efficiency, which concerns the entire column; (2) Murphree efficiency, which has to do with a single plate; and (3) local efficiency, which pertains to a specific location on a single plate.

The *overall efficiency* η_o is simple to use but is the least fundamental. It is defined as the ratio of the number of ideal plates needed in an entire column to the number of actual plates.[8] For example, if six ideal plates are called for and the overall efficiency is 60 percent, the number of actual plates is $6/0.60 = 10$.

The *Murphree efficiency*[12] η_M is defined by

$$\eta_M = \frac{y_n - y_{n+1}}{y_n^* - y_{n+1}} \tag{21.67}$$

where y_n = actual concentration of vapor leaving plate n
y_{n+1} = actual concentration vapor entering plate n
y_n^* = concentration of vapor in equilibrium with liquid leaving downpipe from plate n

The Murphree efficiency is therefore the change in vapor composition from one plate to the next divided by the change that would have occurred if the vapor leaving were in equilibrium with the *liquid leaving*. The liquid leaving is generally not the same as the average liquid on the plate, and this distinction is important in comparing local and Murphree efficiencies.

The Murphree efficiency is defined using vapor concentrations as a matter of custom, but the measured efficiencies are rarely based on analysis of the vapor phase because of the difficulty in getting reliable samples. Instead, samples are taken of the liquid on the plates, and the vapor compositions are determined from a McCabe–Thiele diagram. A plate efficiency can be defined using liquid concentrations, but this is used only occasionally for desorption or stripping calculations.

Columns operated at high velocity will have significant entrainment, and this reduces the plate efficiency, because the drops of entrained liquid are less rich in the more volatile component than is the vapor. Although methods of allowing for entrainment have been published,[1] most empirical correlations for the Murphree efficiency are based on liquid samples from the plates, and this includes the effect of entrainment.

The *local efficiency* η' is defined by

$$\eta' = \frac{y_n' - y_{n+1}'}{y_{en}' - y_{n+1}'} \tag{21.68}$$

where y_n' = concentration of vapor leaving specific location on plate n
  y_{n+1}' = concentration of vapor entering plate n at same location
  y_{en}' = concentration of vapor in equilibrium with liquid at same location

Since y_n' cannot be greater than y_{en}', a local efficiency cannot be greater than 1.00, or 100 percent.

Relation between Murphree and local efficiencies. In small columns, the liquid on a plate is sufficiently agitated by vapor flow through the perforations for there to be no measurable concentration gradients in the liquid as it flows across the plate. The concentration of the liquid in the downpipe x_n is that of the liquid on the entire plate. The change from concentration x_n to x_{n+1} occurs right at the exit from the downcomer, as liquid leaving the downcomer is vigorously mixed with the liquid on plate $n + 1$. Since the concentration of the liquid on the plate is everywhere the same, so is that of the vapor from the plate and no gradients exist in the vapor streams. A comparison of the quantities in Eqs. (21.67) and (21.68) shows that $y_n = y_n'$, $y_{n+1} = y_{n+1}'$, and $y_n^* = y_{en}'$. Then $\eta_M = \eta'$, and the local and Murphree efficiencies are equal.

In larger columns, liquid mixing in the direction of flow is not complete, and a concentration gradient does exist in the liquid on the plate. The maximum possible variation is from a concentration of x_{n-1} at the liquid inlet to a concentration of x_n at the liquid outlet. To show the effect of such a concentration gradient, consider a portion of the McCabe–Thiele diagram, as shown in Fig. 21.30. This diagram corresponds to a Murphree efficiency of about 0.9, with y_n almost equal to y_n^*. However, if there is no horizontal mixing of the liquid, the vapor near the liquid inlet

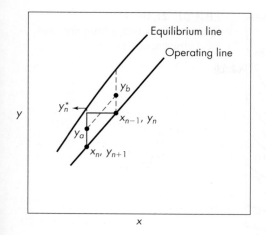

FIGURE 21.30
Local and average vapor compositions for an unmixed plate.

would contact liquid of composition x_{n-1} and be considerably richer than vapor contacting liquid of composition x_n near the exit. To be consistent with an average vapor composition y_n, the local vapor composition must range from y_a near the liquid exit to y_b near the liquid inlet. The local efficiency is therefore considerably lower than the Murphree efficiency, and η' would be about 0.6 for this example.

When the local efficiency is high, say, 0.8 or 0.9, the presence of concentration gradients in the liquid sometimes gives average vapor concentration greater than y_n^*, and the Murphree efficiency is then greater than 100 percent. An example of this is shown later in Fig. 21.35.

The relation between η_M and η' depends on the degree of liquid mixing and whether or not the vapor is mixed before going to the next plate. Calculations have shown[9] only a small difference in efficiency for completely mixed vapor or unmixed vapor, but the effect of no liquid mixing can be quite large. Most studies have assumed complete vapor mixing in order to simplify the calculations for various degrees of liquid mixing. A correlation based on plug flow of liquid across the plate with eddy diffusion in the liquid phase was developed by workers at the University of Delaware[1] and is given in Fig. 21.31. The abscissa is $(mV/L)\eta'$, and the parameter on the graphs is a Peclet number for axial dispersion:

$$\text{Pe} = \frac{Z_l^2}{D_E t_L} \tag{21.69}$$

where Z_l = length of liquid flow path, m
       D_E = eddy diffusivity, m²/s
       t_L = residence time of liquid on plate, s

For distillation at atmospheric pressure in a column 0.3 m (1 ft) in diameter the Peclet number is about 10, based on empirical correlations for dispersion on bubble-cap and sieve trays.[1,5] This is in the range where significant enhancement of the efficiency should result because of gradients on the plate. For a column 1 m or larger in diameter, the Peclet number would be expected to be greater than 20, and the efficiency should be almost as high as it would be with no mixing in the

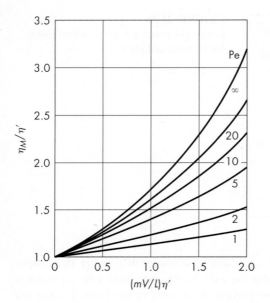

FIGURE 21.31
Relationship between Murphree and local efficiencies.

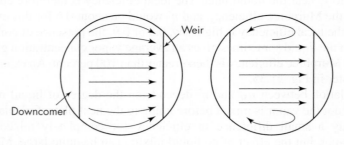

FIGURE 21.32
Possible liquid flow patterns in a large column.

direction of flow. Tests on very large columns, however, sometimes show lower plate efficiencies than for medium-size columns, probably because of departure from plug flow. With a large column and segmental downcomers the liquid flowing around the edge of the bubbling area, as shown in Fig. 21.32, has an appreciably longer flow path than liquid crossing the middle; and a wide distribution of residence times or even some backflow of liquid may result. These effects can be minimized by special plate design.

Use of Murphree efficiency

When the Murphree efficiency is known, it can readily be used in the McCabe–Thiele diagram. The diagram for an actual plate as compared with that for an ideal plate is shown in Fig. 21.33. Triangle acd represents the ideal plate and triangle abe the actual plate. The actual plate, instead of enriching the vapor from y_{n+1} to y_n^*, shown by line segment ac, accomplished a lesser enrichment $y_n - y_{n+1}$,

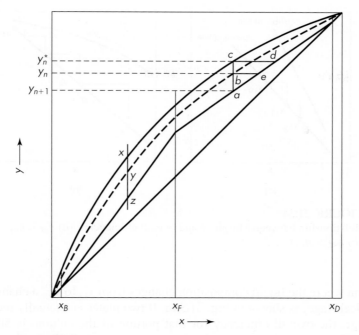

FIGURE 21.33
Use of Murphree efficiency on xy diagram. Dashed line is effective equilibrium curve, y'_e, versus x_e for $\eta_M = 0.60$; $ba/ca = yz/xz = 0.60$.

shown by line segment ab. By definition of η_M, the Murphree efficiency is given by the ratio ab/ac. To apply a known Murphree efficiency to an entire column, it is necessary only to replace the true equilibrium curve y_e versus x_e by an effective equilibrium curve y'_e versus x_e, whose ordinates are calculated from the equation

$$y'_e = y + \eta_M(y_e - y) \tag{21.70}$$

In Fig. 21.33 an effective equilibrium curve for $\eta_M = 0.60$ is shown. Note that the position of the y'_e versus x_e curve depends on both the operating line and the true equilibrium curve. Once the effective equilibrium curve has been plotted, the usual step-by-step construction is made and the number of actual plates determined. The reboiler is not subject to a discount for plate efficiency, and the true equilibrium curve is used for the last step in the stripping section.

Relation between Murphree and overall efficiencies. The overall efficiency of a column is not the same as the average Murphree efficiency of the individual plates. The relation between these efficiencies depends on the relative slopes of the equilibrium line and the operating line. When the equilibrium line is steeper than the operating line, which is typical for stripping columns, the overall efficiency is greater than the Murphree efficiency if η_M is less than 1.0. Consider a portion of

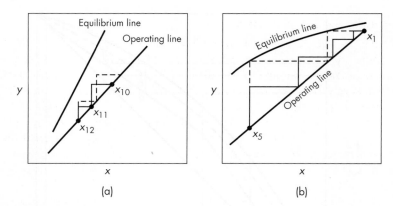

FIGURE 21.34
Relationship between Murphree and overall efficiencies: (a) $\eta_M < \eta_o$; (b) $\eta_M > \eta_o$.

the column where the liquid composition changes from x_{12} to x_{10}, a change requiring 1.0 ideal stage, as shown in Fig. 21.34a. If two plates are actually required for this change, the overall efficiency for that portion of the column is 50 percent. However, if two partial steps are drawn assuming $\eta_M = 0.50$, as shown by the dashed line, then the predicted value of x_{10} is too high, since the first step would go halfway from x_{12} to x_{10} and the second step would be a much larger change. The correct value of η_M is about 0.40, as shown in the steps drawn with a solid line.

When the equilibrium line is less steep than the operating line, as usually occurs near the top of the rectifying section, the overall efficiency is less than the Murphree efficiency, as illustrated in Fig. 21.34b. For this case, two ideal plates are needed to go from x_1 to x_5, and if four actual plates are required, the overall efficiency is 0.50. By trial and error, a Murphree efficiency of 0.6 is found to give the correct value of x_5 after four partial steps.

For columns with both a stripping and a rectifying section, the overall value of η_o may be fairly close to the average value of η_M, since the higher value of η_o in the stripping section, where $mV/L > 1$, tends to offset the lower value of η_o in the rectifying section, where $mV/L < 1$. For this reason, the difference between η_o and η_M is sometimes ignored in designing a column. However, in analyzing the performance of a real column or a section of the column, where the composition change over several plates is measured, the correct value of η_M should be determined by trial rather than just determining η_o and assuming $\eta_o = \eta_M$.

For the special case where the equilibrium and operating lines are straight, the following equation can be applied:

$$\eta_o = \frac{\ln\left[1 + \eta_M(mV/L - 1)\right]}{\ln(mV/L)} \tag{21.71}$$

where m is the slope of the equilibrium line. Note that when $mV/L = 1.0$ or when $\eta_M \approx 1.0$, $\eta_M = \eta_o$.

Factors influencing plate efficiency

Although thorough studies of plate efficiency have been made,[7,11,16] the estimation of efficiency is largely empirical. Sufficient data are at hand, however, to show the major factors involved and to provide a basis for estimating the efficiencies for conventional types of columns operating on mixtures of common substances.

The most important requirement for obtaining satisfactory efficiencies is that the plates operate properly. Adequate and intimate contact between vapor and liquid is essential. Any misoperation of the column, such as excessive foaming or entrainment, poor vapor distribution, or short-circuiting, weeping, or dumping of liquid, lowers the plate efficiency.

Plate efficiency is a function of the rate of mass transfer between liquid and vapor. The prediction of mass-transfer coefficients in sieve trays and their relationship to plate efficiency are discussed later. Some published values of the plate efficiency of a 1.2-m column are shown in Fig. 21.35. This column had sieve trays with 12.7-mm holes and 8.32 percent open area, a 51-mm weir height, and 0.61-m tray spacing. The data are plotted against a flow parameter F, which tends to cover about the same range for different total pressures, since the flooding velocity varies inversely with $\sqrt{\rho_V}$, as shown by Eq. (21.66). Parameter F, generally known as the *F factor*, is defined as follows:

$$F \equiv u\sqrt{\rho_V} \tag{21.72}$$

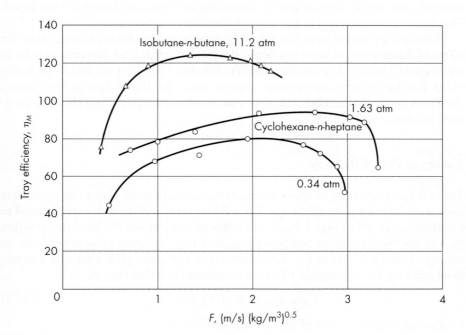

FIGURE 21.35
Efficiency of sieve trays in a 1.2-m column. (*From M. Sakata and T. Yanagi, 3rd Int. Symp. Dist.,* p. 3.2/21, *ICE, 1979.*)

The normal range of F for sieve trays is 1 to 3 (m/s) $(kg/m^3)^{0.5}$, or 0.82 to 2.46 (ft/s) $(lb/ft^3)^{0.5}$.

EXAMPLE 21.7. What is the F factor at the top of the column in Example 21.6 if the vapor velocity is the maximum allowable?

Solution. From Eq. (21.72) the F factor is

$$F = u_c\sqrt{\rho_V}$$

$$= 2.23\sqrt{1.15} = 2.39 \ \frac{m}{s} \left(\frac{kg}{m^3}\right)^{0.5}$$

As shown by Fig. 21.35, this is a reasonable value for a sieve-plate column at 1 atm pressure.

The efficiency does not change much with vapor rate in the range between the weeping point and the flooding point. The increase in vapor flow increases the froth height, creating more mass-transfer area, so that the total mass transferred goes up about as fast as the vapor rate. The data for Fig. 21.35 were obtained at total reflux, so the increase in liquid rate also contributed to the increase in interfacial area. The sharp decrease in efficiency near the flooding point is due to entrainment.

The data for cyclohexane-n-heptane show a lower efficiency for operation at a lower pressure, which has been confirmed by tests on other systems. Lowering the pressure decreases the concentration driving force in the vapor phase but increases the vapor diffusivity. It also lowers the temperature, which increases the liquid viscosity and surface tension and decreases the diffusion coefficients in the liquid. The decrease in efficiency is due to a combination of these effects.

Efficiencies greater than 100 percent for the isobutane-butane system show the effect of liquid concentration gradients; the local efficiency for this case may be in the range 0.7 to 0.9. A somewhat smaller difference between η_M and η' probably exists for the cyclohexane-heptane system, since the liquid flow is less, which decreases the Peclet number and makes the conditions on the plate closer to complete mixing.

Theory of Plate Efficiency

The two-film theory can be applied to mass transfer on a sieve plate to help correlate and extend data for plate efficiency. The bubbles formed at the holes are assumed to rise through a pool of liquid that is vertically mixed and has the local composition x_A. The bubbles change in composition as they rise, and there is assumed to be no mixing of the gas phase in the vertical direction. For a unit plate area with a super-ficial velocity $\bar{V}_s$, the number of moles transferred in a thin slice dz is

$$\bar{V}_s\rho_M \, dy_A = K_y a(y_A^* - y_A) \, dz \tag{21.73}$$

Integrating over the height of the aerated liquid Z gives

$$\int_{y_{A1}}^{y_{A2}} \frac{dy_A}{y_A^* - y_A} = \ln \frac{y_A^* - y_{A1}}{y_A^* - y_{A2}} = \frac{K_y a Z}{\bar{V}_s \rho_M} \tag{21.74}$$

or
$$\frac{y_A^* - y_{A2}}{y_A^* - y_{A1}} = \exp\left(-\frac{K_y a Z}{\bar{V}_s \rho_M}\right) \tag{21.75}$$

The local efficiency η' is given by

$$\eta' = \frac{y_{A2} - y_{A1}}{y_A^* - y_{A1}} \tag{21.76}$$

and
$$1 - \eta' = \frac{y_A^* - y_{A1} - y_{A2} + y_{A1}}{y_A^* - y_{A1}} \tag{21.77}$$

From Eq. (21.75),

$$1 - \eta' = \exp\left(-\frac{K_y a Z}{\bar{V}_s \rho_M}\right) = e^{-N_{Oy}} \tag{21.78}$$

where N_{Oy} is the number of overall gas-phase transfer units. For distillation of low-viscosity liquids such as water, alcohol, or benzene at about 100°C, the value of N_{Oy} is about 1.5 to 2, nearly independent of the gas velocity over the normal operating range of the column. This gives a local efficiency of 78 to 86 percent, and the Murphree plate efficiency will be slightly higher or lower, depending on the degree of lateral mixing on the plate and on the amount of entrainment.

The relative importance of the gas and liquid resistances can be estimated by assuming that the penetration theory applies to both phases and with the same contact time. Since the penetration theory [Eq. (17.50)] gives k_c, and k_y and k_x equal $k_c \rho_{My}$ and $k_c \rho_{Mx}$, respectively,

$$\frac{k_y}{k_x} = \left(\frac{D_{vy}}{D_{vx}}\right)^{1/2} \frac{\rho_{My}}{\rho_{Mx}} \tag{21.79}$$

EXAMPLE 21.8. (*a*) Use the penetration theory to estimate the fraction of the total resistance that is in the gas film in the distillation of a benzene-toluene mixture at 110°C and 1 atm pressure. The liquid viscosity μ is 0.26 cP. The diffusivities and densities are, for liquid,

$$D_{vx} = 6.74 \times 10^{-5} \text{ cm}^2/\text{s} \qquad \rho_{Mx} = 8.47 \text{ mol/L}$$

and, for vapor,

$$D_{vy} = 0.0494 \text{ cm}^2/\text{s} \qquad \rho_{My} = 0.0318 \text{ mol/L}$$

(*b*) How would a fourfold reduction in total pressure change the local efficiency and the relative importance of the gas-film and liquid-film resistances?

Solution

(*a*) Substitution into Eq. (21.79) gives

$$\frac{k_y}{k_x} = \left(\frac{0.0494}{6.74 \times 10^{-5}}\right)^{1/2} \frac{0.0318}{8.47} = 0.102$$

Thus the gas-film coefficient is predicted to be only 10 percent of the liquid-film coefficient, and if $m = 1$, about 90 percent of the overall resistance to mass transfer would be in the gas film.

(b) Assume that the column is operated at the same F factor and that this gives the same interfacial area a and froth height Z. The boiling temperature of toluene at 0.25 atm is 68°C, or 341 K, compared to 383 K at 1 atm.

Gas film. Since $D_{vy} \propto T^{1.81}/P$, the new value of D_{vy} is

$$D'_{vy} = \left(\frac{341}{383}\right)^{1.81} \frac{D_{vy}}{0.25} = 3.24 \text{ times old value}$$

Assuming that the penetration theory holds with the same t_T, k_c increases by $\sqrt{3.24}$, or 1.8, but at 0.25 atm and 68°C, ρ_{My} is 0.00894 mol/L, so k_y changes by $1.8 \times 0.00894/0.0318 = 0.506$.

Liquid film. Here $D_{vx} \propto T/\mu$, and since $\mu = 0.35$ cP at 68°C, the new value of D_{vx} is

$$D'_{vx} = \frac{341}{383}\left(\frac{0.26 D_{vx}}{0.35}\right) = 0.66 \text{ times old value}$$

Thus k_c decreases by $\sqrt{0.66} = 0.81$, and considering the small change in molar density to 8.92 mol/L, k_x changes by $0.81 \times 8.92/8.47 = 0.86$.

If the local efficiency at 1 atm pressure is 0.86, corresponding to two transfer units, and if the relative values of k_x and k_y are estimated as in part (a), the new value of K'_y is obtained as follows:

$$k'_y = 0.506 k_y \qquad k'_x = 0.86 k_x$$

At 1 atm, $k_y = 0.102 k_x$ and $K_y = 0.907 k_y$. Thus

$$k'_x = \frac{0.86}{0.102} k_y = 8.43 k_y$$

For $m = 1$,

$$\frac{1}{K'_y} = \frac{1}{k'_y} + \frac{1}{k'_x} = \frac{1}{0.506 k_y} + \frac{1}{8.43 k_y} = \frac{2.10}{k_y}$$

$$K'_y = 0.476 k_y$$

The ratio of the number of transfer units is the ratio of the overall coefficients divided by the molar flow rate. If the column is operated at the same F factor, $\sqrt{\rho_y}$ changes by $[(383 \times 0.25)/341]^{0.5} = 0.53$, and $\bar{V}_s$ changes by $1/0.53$. If a, the area per unit volume, is the same, the new value of N'_{Oy} is

$$N'_{Oy} = 2 \times \frac{0.476}{0.53} = 1.80 \qquad \eta' = 1 - e^{-1.80} = 0.83$$

Thus the local efficiency is predicted to drop from 86 to 83 percent, with 94 percent of the total resistance in the gas phase. Close agreement with the actual values of efficiency is not expected because of the assumptions made to simplify the analysis, but the trend is correct, as shown in Fig. 21.35, and it is clear that the gas-film resistance is increasingly important at low pressures. For distillation at high pressures, k_y and k_x are more nearly equal.

DISTILLATION IN PACKED COLUMNS

Packed columns are often used for distillation when the separation is relatively easy and the required column diameter is not very large. They are generally less expensive than plate columns and have a lower pressure drop. The main disadvantage is the difficulty in getting good liquid distribution, particularly for large-diameter columns or very tall columns. Even if liquid is spread evenly over the packing at the top of the column, liquid tends to move toward the wall and to flow through the packing in preferred channels. Regions of high liquid flow tend to have low vapor flow, and the local variations in L/V decrease the separation that can be achieved.[6] To minimize this effect, tall columns are often split into sections, with redistributors every 3 to 4 m.

The column height is usually based on the number of theoretical plates and the height equivalent to a theoretical plate (HETP). In most cases the upper operating line is slightly steeper than the equilibrium line, and the reverse is true for the lower operating line, which makes the average HETP about the same as H_{Oy}. The $1\frac{1}{2}$- or 2-in. sizes of the common packings have about the same capacity (allowable vapor velocity) as a sieve tray, and the packed height equivalent to a theoretical plate is generally in the range of 1 to 2 ft (0.3 to 0.6 m). The smaller packings have lower values of HETP, sometimes less than 1.0 ft, but they also have lower capacity and are not likely to be used in a large column. The pressure drop per equivalent theoretical plate is generally less than that for a sieve or bubble-cap tray, which is an important advantage for vacuum operation.

Packed distillation towers can often be operated over a moderate range of flow rates at nearly constant separation efficiency. Data for isooctane-toluene separation at total reflux are shown in Fig. 21.36. The three Intalox metal tower (IMTP) packings numbered 25, 40, and 50 correspond to nominal sizes of 1, 1.5, and 2 in., respectively. As the capacity factor increases, both the liquid rate and the vapor rate increase, which explains why the HETP is nearly constant. The gas film has the controlling resistance to mass transfer, and H_{Oy} increases with the 0.3 to 0.4 power of G_y, but this is offset by the decrease in H_{Oy} with increasing G_x, as shown in Fig. 18.22. The net effect is a nearly constant HTU or HETP over a 2- to 2.5-fold range of flows. The HETP decreases in the loading region because of the more rapid increase in liquid holdup and wetted area, but HETP then increases sharply as flooding is approached. With these high-capacity packings, the rise in HETP is believed due to entrainment of the liquid. The recommended design velocity for these packings is 20 percent less than the velocity at which HETP starts to rise rapidly.

Structured sheet-metal packings give slightly better separation than dumped packings, and the HETP values are nearly constant over a three- to fourfold range of flow rates. Even more efficient are the gauze-type packings, with HETP values that increase from about 3 to 6 in. (0.1 to 0.2 m) over a fourfold range of velocities. The gauze is completely wetted even at low flow rates, and the HETP can be predicted from fundamental equations for mass transfer in wetted channels.[2] There is no corresponding theory for other structured packings or for dumped packings,

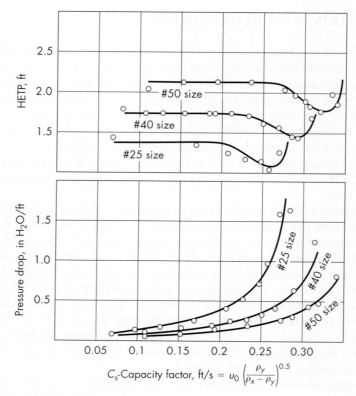

FIGURE 21.36
HETP and pressure drop in the distillation of isooctane and
toluene in Intalox metal tower packings.[15]

because it is difficult to predict the wetted area. Low surface tension tends to increase the wetted area, but too low values may lead to foaming, which increases
HETP.

BATCH DISTILLATION

In some small plants, volatile products are recovered from liquid solution by batch
distillation. The mixture is charged to a still or reboiler, and heat is supplied through
a coil or through the wall of the vessel to bring the liquid to the boiling point and
then vaporize part of the batch. In the simplest method of operation, the vapors are
taken directly from the still to a condenser, as shown in Fig. 21.37. The vapor
leaving the still at any time is in equilibrium with the liquid in the still, but since the
vapor is richer in the more volatile component, the compositions of liquid and vapor
are not constant.

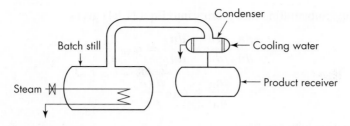

FIGURE 21.37
Simple distillation in a batch still.

To show how the compositions change with time, consider what happens if n_0 moles are charged to a batch still. Let n be the moles of liquid left in the still at a given time and y and x be the vapor and liquid compositions. The total moles of component A left in the still n_A will be

$$n_A = xn \tag{21.80}$$

If a small amount of liquid dn is vaporized, the change in the moles of component A is $y\, dn$, or dn_A. Differentiating Eq. (21.80) gives

$$dn_A = d(xn) = n\, dx + x\, dn \tag{21.81}$$

Hence

$$n\, dx + x\, dn = y\, dn$$

By rearrangement,

$$\frac{dn}{n} = \frac{dx}{y - x} \tag{21.82}$$

Equation (21.82) is integrated between the limits of x_0 and x_1, the initial and final concentrations,

$$\int_{n_0}^{n_1} \frac{dn}{n} = \int_{x_0}^{x_1} \frac{dx}{y - x} = \ln \frac{n_1}{n_0} \tag{21.83}$$

Equation (21.83) is known as the Rayleigh equation. The function $dx/(y - x)$ can be integrated graphically or numerically using tabulated equilibrium data or an equilibrium curve.

A simple alternative to the Rayleigh equation can be derived for an ideal mixture based on the relative volatility. Although the temperature in the still increases during a batch distillation, the relative volatility, which is the ratio of vapor pressures, does not change much, and an average value can be used. From Eq. (21.34)

$$\frac{y_A}{y_B} = \alpha_{AB} \frac{x_A}{x_B} \tag{21.84}$$

If the mixture has n_A mol of A and n_B mol of B, the ratio n_A/n_B is equal to x_A/x_B; when dn mol is vaporized, the change in A is $y_A\, dn$ or dn_A, and the change in B is

$y_B \, dn$ or dn_B. Substituting these terms into Eq. (21.84) gives

$$\frac{dn_A/dn}{dn_B/dn} = \frac{dn_A}{dn_B} = \alpha_{AB}\frac{n_A}{n_B}$$

or

$$\frac{dn_A}{n_A} = \alpha_{AB}\frac{dn_B}{n_B} \qquad (21.85)$$

After integration between limits

$$\ln\frac{n_A}{n_{0A}} = \alpha_{AB}\ln\frac{n_B}{n_{0B}} \qquad (21.86)$$

or

$$\frac{n_B}{n_{0B}} = \left(\frac{n_A}{n_{0A}}\right)^{1/\alpha_{AB}} \qquad (21.87)$$

Equation (21.87) can be plotted as a straight line on logarithmic coordinates to help follow the course of a batch distillation, or it can be used directly if the recovery of one of the components is specified.

EXAMPLE 21.9. A batch of crude pentane contains 15 mole percent n-butane and 85 percent n-pentane. If a simple batch distillation at atmospheric pressure is used to remove 90 percent of the butane, how much pentane will be removed? What will be the composition of the remaining liquid?

Solution. The final liquid is nearly pure pentane, and its boiling point is 36°C. The vapor pressure of butane at this temperature is 3.4 atm, giving a relative volatility of 3.4. For the initial conditions, the boiling point is about 27°C, and the relative volatility is 3.6. Therefore, an average value of 3.5 is used for α_{AB}.

   Basis: 1 mol feed

$$n_{0A} = 0.15 \text{ (butane)} \qquad n_A = 0.015 \qquad n_{0B} = 0.85 \text{ (pentane)}$$

From Eq. (21.87)

$$\frac{n_B}{0.85} = 0.1^{1/3.5} = 0.518 \qquad n_B = 0.518(0.85) = 0.440$$

$$n = 0.44 + 0.015 = 0.455 \text{ mol} \qquad x_A = \frac{0.015}{0.455} = 0.033$$

Batch distillation with reflux

Batch distillation with only a simple still does not give a good separation unless the relative volatility is very high. In many cases, a rectifying column with reflux is used to improve the performance of the batch still. If the column is not too large, it may be mounted on top of the reboiler, as shown in Fig. 20.1, or it may be supported independently, with connecting pipes for the vapor and liquid streams.

   The operation of a batch still and column can be analyzed using a McCabe–Thiele diagram, with the same operating-line equation that was used for the rectifying section of a continuous distillation [Eq. (21.19)]:

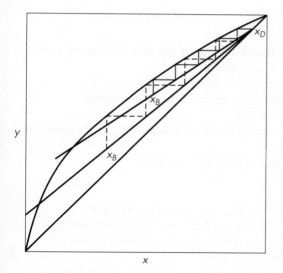

FIGURE 21.38
McCabe–Thiele diagrams for a batch distillation. Upper operating line and solid lines: initial conditions; lower operating line and dashed lines: after one-third of the charge has been removed.

$$y_{n+1} = \frac{R_D}{R_D + 1} x_n + \frac{x_D}{R_D + 1}$$

The system may be operated to keep the top composition constant by increasing the reflux ratio as the composition of the liquid in the reboiler changes. The McCabe–Thiele diagram for this case would have operating lines of different slope positioned such that the same number of ideal stages was used to go from x_D to x_B at any time. A typical diagram is shown in Fig. 21.38 for a still with five ideal stages including the reboiler. The upper operating line is for the initial conditions, when the concentration of low boiler in the still is about the same as the charge composition. (The concentration x_B is slightly lower than x_F because of the holdup of liquid on the plates.) The lower operating line and the dashed line steps show conditions when about one-third of the charge has been removed as overhead product.

To determine the reflux ratio needed for a constant x_D and given x_B requires a trial-and-error calculation, since the last step on the assumed operating line must end exactly at x_B. However, once the initial reflux ratio is chosen by this method, the value of x_B for a later stage in the distillation can be obtained by assuming a value of R_D, constructing the operating line, and making the correct number of steps ending at x_B. By a material balance, Eqs. (21.5) and (21.6), the amount of product and remaining charge, can be calculated.

An alternative method of running a batch distillation is to fix the reflux ratio and let the overhead product purity vary with time, stopping the distillation when the amount of product or the average concentration in the total product reaches a certain value. To calculate the performance of the still, operating lines of constant slope are drawn starting at different values of x_D, and the actual number of stages is stepped off to determine x_B. The total number of moles left in the still is then calculated by integration of Eq. (21.83), where x_D is equal to y and x_I is equal to x_B.

SYMBOLS

a	Area of interface between liquid and vapor per unit volume of packing, m^{-1} or ft^{-1}
B	Flow rate of bottoms product, mol/h, kg/h, or lb/h
C_0	Discharge coefficient, flow through perforations of sieve plate
C_s	Capacity factor, $u_o\sqrt{\rho_y/(\rho_x - \rho_y)}$
c_p	Specific heat at constant pressure, J/g·°C or Btu/lb·°F; c_{pc}, of condensate; c_{pL}, of liquid; c_{pV}, of vapor
D	Flow rate of overhead product, mol/h, kg/h, or lb/h
D_c	Diameter of column, m or ft
D_E	Eddy diffusivity, m^2/s
D_v	Diffusivity, m^2/s, cm^2/s, or ft^2/h; D_{vx}, in liquid; D_{vy}, in vapor
F	Feed rate, mol/h, kg/h, or lb/h; F_1, F_2, feeds to column with multiple feeds; also factor for estimating column capacity, defined by Eq. (21.72)
f	Fraction of feed that is vaporized
G	Mass velocity based on total tower cross section; kg/m^2·h or lb/ft^2·h; G_x, of liquid stream; G_y, of vapor
g	Acceleration of gravity, m/s^2 or ft/s^2
H	Enthalpy, energy per mole or per unit mass; H_B, of bottom product; H_D, of overhead product; H_F, of feed; H_R, of reflux; H_x, of saturated liquid; H_{xm}, of liquid from plate m of stripping column; H_{xn}, of liquid from plate n of rectifying column; H_y, of saturated vapor; $H_{y,a}$, of vapor entering column; $H_{y,m+1}$, of vapor from plate $m + 1$ of stripping column; $H_{y,n+1}$, of vapor from plate $n + 1$ of rectifying column; H_{y1}, of vapor from top plate
H_{Oy}	Overall height of a transfer unit, based on vapor phase
HETP	Height equivalent to a theoretical plate
HTU	Height of a transfer unit
h	Pressure drop or head, mm of liquid; h_d, for dry plate; $h_{f,L}$, friction loss in liquid; h_l, equivalent head of liquid on plate; h_{ow}, height of clear liquid over weir; h_t, total drop per plate; h_w, height of weir
K_v	Coefficient in Eq. (21.66); K_v', in Eq. (21.65)
K_y	Overall mass-transfer coefficient, based on vapor phase; K_y', new value in Example 21.8
k	Individual mass-transfer coefficient; k_c, based on concentration difference, m/h or ft/h; k_x, k_y, in liquid and vapor phases, respectively, based on mole-fraction differences, kg mol/m^2·s·unit mole fraction or lb mol/ft^2·h·unit mole fraction; k_x', k_y', new values in Example 21.8
L	Flow rate of liquid in general or in rectifying column, mol/h, kg/h, or lb/h; L_a, entering top of column; L_b, leaving bottom of column; L_c, of reflux from condenser; L_m, from plate m of stripping column; L_n, from plate n of rectifying column; $\bar{L}$, in stripping column
L_w	Length of weir, m
m	Serial number of plate in stripping column, counting from feed plate; also, slope of equilibrium curve, dy_e/dx_e

$\dot{m}$	Mass flow rate, kg/h or lb/h; $\dot{m}_s$, of steam to reboiler; $\dot{m}_w$, of cooling water to condenser
N	Number of ideal plates; $N_{\min}$, minimum number of ideal plates
N_{Oy}	Overall number of transfer units based on vapor phase; N'_{Oy}, new value in Example 21.8
n	Serial number of plate in rectifying column, counting from top; also, number of moles in still or mixture; n_A, n_B, of components A and B, respectively; n_0, moles charged to still; n_{0A}, n_{0B}, of components A and B, respectively; n_1, final value; also exponent in Eq. (21.41)
P	Pressure, N/m^2 or lb$_f$/ft^2; P_{n-1}, P_n, P_{n+1}, in vapor space above plates $n-1$, n, and $n+1$, respectively; P', vapor pressure; P'_A, P'_B, of components A and B, respectively
Pe	Peclet number for axial dispersion, $Z_l^2/D_E t_L$
p_A, p_B	Partial pressure of components A and B, respectively, N/m^2 or lb$_f$/ft^2
q	Rate of heat flow, W or Btu/h; q_c, heat removed in condenser; q_r, heat added at reboiler; also, moles of liquid to stripping section of column per mole of feed
q_L	Volumetric flow rate of liquid in downpipe, m^3/s or ft^3/s
R	Reflux ratio; $R_D = L/D$; $R_V = L/V$; R_{Dm}, minimum reflux ratio
T	Temperature, °C or °F; T_F, of feed; T_b, bubble point; T_{bc}, bubble point of condensate; T_c, of condensate; T_d, dew point; T_1, of liquid on top plate; also, entering temperature of cooling water; $T_2 - T_1$, temperature rise of cooling water
t	Time, s or h; t_L, residence time of liquid on plate; t_T, residence time on transfer surface in penetration theory
u	Linear velocity, m/s or ft/s; u_c, maximum permissible vapor velocity, based on area of bubbling section; u_0, vapor velocity through perforations in sieve plate or superficial velocity in packed column
V	Flow rate of vapor, in general or in rectifying column, mol/h, kg/h or lb/h; V_a, from top of column; V_b, entering bottom of column; V_{m+1}, from plate $m+1$ in stripping column; V_n, V_{n+1}, from plates n and $n+1$, respectively, in rectifying column; V_1, from top plate to condenser; $\bar{V}$, in stripping column
$\bar{V}_s$	Superficial velocity, m/s or ft/s
x	Mole fraction or mass fraction in liquid; x_A, of component A; x_{Ae}, in equilibrium with vapor of concentration y_{Ae}; x_B, in bottoms product, also, of component B in liquid; x_{Be}, in equilibrium with vapor of concentration y_{Be}; x_D, in overhead product; x'_D, in side-stream drawoff; x_F, in feed; x_a, in liquid entering single-section column; x_b, in liquid leaving single-section column; x_c, in reflux from condenser; x_e, in equilibrium with vapor of composition y_e; x_m, in liquid from plate m of stripping column; x_{n-1}, x_n, in liquid from plates $n-1$ and n, respectively, of rectifying column; x', at intersection of feed line and equilibrium curve; x_0, x_1, initial and final values in batch distillation
y	Mole fraction or mass fraction in vapor; y_A, y_B, of components A and B, respectively; y_{Ae}, y_{Be}, in equilibrium with liquids of concentration x_{Ae}, x_{Be},

respectively; y_D, in vapor overhead product; y_a, in vapor leaving single-section column; y_b, in vapor entering single-section column; y_e, in equilibrium with liquid of concentration x_e; y_{m+1}, of vapor from plate $m + 1$ in stripping column; y_n, y_{n+1}, in vapor from plates n and $n + 1$, respectively, in rectifying column; y_r, from reboiler; y^*, in vapor in equilibrium with specific stream of liquid; y_a^*, in equilibrium with x_a; y_b^*, in equilibrium with x_b; y_n^*, in equilibrium with x_n; y_1, in vapor from top plate; y', at intersection of feed line and equilibrium curve; also, of vapor leaving partial condenser; y_e', pseudoequilibrium value [Eq. (21.70)]; y_{en}' in equilibrium with liquid at a specific location on plate n; y_n', in vapor leaving a specific location on plate n; y_{n+1}', entering plate n at same location as for y_n',

Z $\qquad$ Height of liquid in downcomer, m or ft; actual height of aerated liquid; Z_c, height of clear liquid

Z_l $\qquad$ Length of liquid flow path, m

Greek Letters

α_{AB} $\qquad$ Relative volatility, component A relative to component B

β $\qquad$ Correction factor in Eq. (21.60)

ΔH_v $\qquad$ Enthalpy of vaporization, cal/mol; $\Delta H_{v,b}$, at boiling point

ΔL $\qquad$ Additional liquid condensed in column from cold condensate [Eq. (21.21)]

η $\qquad$ Efficiency; η_M, Murphree plate efficiency; η_o, overall plate efficiency; η', local plate efficiency

λ $\qquad$ Latent heat of vaporization, energy per unit mass; λ_c, of condensate; λ_s, of steam

μ $\qquad$ Viscosity, Pa · s, cP, or lb/ft·s

ρ $\qquad$ Density, kg/m^3 or lb/ft^3; ρ_L, of liquid; ρ_V, of vapor

ρ_M $\qquad$ Molar density, kg mol/m^3 or lb mol/ft^3; ρ_{Mx}, of liquid; ρ_{My}, of vapor

σ $\qquad$ Surface tension, dyn/cm

ϕ_d $\qquad$ Volume fraction liquid in aerated mixture

PROBLEMS

21.1. A liquid containing 25 mole percent toluene, 40 mole percent ethylbenzene, and 35 mole percent water is subjected to a continuous flash distillation at a total pressure of 0.5 atm. Vapor pressure data for these substances are given in Table 21.5. Assuming that mixtures of ethylbenzene and toluene obey Raoult's law and that the hydrocarbons are completely immiscible in water, calculate the temperature and compositions of liquid and vapor phases (*a*) at the bubble point, (*b*) at the dew point, (*c*) at the 50 percent point (one-half of the feed leaves as vapor and the other half as liquid).

TABLE 21.5

Vapor pressures of ethylbenzene, toluene, and water

Temperature, °C	Vapor pressure, mm Hg		
	Ethylbenzene	Toluene	Water
50	35.2		92.5
60	55.5	139.5	149.4
70	84.8	202.4	233.7
80	125.8	289.4	355.1
90	181.9	404.6	525.8
100	257.0	557.2	760.0
110	353.3		
110.6		760.0	
120	481.8		

TABLE 21.6

Equilibrium data for methanol-water

x	0.1	0.2	0.3	0.4	0.5	0.6	0.7	0.8	0.9	1.0
y	0.417	0.579	0.669	0.729	0.780	0.825	0.871	0.915	0.959	1.0

TABLE 21.7

System acetone-methanol

Temperature, °C	Mole fraction acetone		Temperature, °C	Mole fraction acetone	
	Liquid	Vapor		Liquid	Vapor
64.5	0.00	0.000	56.7	0.50	0.586
63.6	0.05	0.102	56.0	0.60	0.656
62.5	0.10	0.186	55.3	0.70	0.725
60.2	0.20	0.322	55.05[†]	0.80	0.80
58.65	0.30	0.428	56.1	1.00	1.00
57.55	0.40	0.513			

[†]Azeotrope.

21.2. A plant must distill a mixture containing 75 mole percent methanol and 25 percent water. The overhead product is to contain 99.99 mole percent methanol and the bottom product 0.002 mole percent. The feed is cold, and for each mole of feed 0.15 mol of vapor is condensed at the feed plate. The reflux ratio at the top of the column is 1.4, and the reflux is at its bubble point. Calculate (*a*) the minimum number of plates; (*b*) the minimum reflux ratio; (*c*) the number of plates using a total condenser and a reboiler, assuming an average Murphree plate efficiency of 72 percent; (*d*) the number of plates using a reboiler and a partial condenser operating with the reflux in equilibrium with the vapor going to a final condenser. Equilibrium data are given in Table 21.6.

21.3. The boiling point–equilibrium data for the system acetone-methanol at 760 mm Hg are given in Table 21.7. A column is to be designed to separate a feed analyzing

25 mole percent acetone and 75 mole percent methanol into an overhead product containing 78 mole percent acetone and a bottom product containing 1.0 mole percent acetone. The feed enters as an equilibrium mixture of 30 percent liquid and 70 percent vapor. A reflux ratio equal to twice the minimum is to be used. An external reboiler is to be used. Bottom product is removed from the reboiler. The condensate (reflux and overhead product) leaves the condenser at 25°C, and the reflux enters the column at this temperature. The molal latent heats of both components are 7,700 g cal/g mol. The Murphree plate efficiency is 70 percent. Calculate (a) the number of plates required above and below the feed; (b) the heat required at the reboiler, in Btu per pound mole of overhead product; (c) the heat removed in the condenser, in Btu per pound mole of overhead product.

21.4. An equimolal mixture of benzene and toluene is to be separated in a bubble-plate tower at the rate of 100 kg mol/h at 1 atm pressure. The overhead product must contain at least 98 mole percent benzene. The feed is saturated liquid. A tower is available containing 24 plates. Feed may be introduced either on the 11th or the 17th plate from the top. The maximum vaporization capacity of the reboiler is 120 kg mol/h. The plates are about 50 percent efficient. How many moles per hour of overhead product can be obtained from this tower?

21.5. An aqueous solution of a volatile component A containing 7.94 mole percent A preheated to its boiling point is to be fed to the top of a continuous stripping column operated at atmospheric pressure. Vapor from the top of the column is to contain 11.25 mole percent A. No reflux is to be returned. Two methods are under consideration, both calling for the same expenditure of heat, namely, a vaporization of 0.562 mol per mole feed in each case. Method 1 is to use a still at the bottom of a plate column, generating vapor by use of steam condensing inside a closed coil in the still. In method 2 the still and heating coil are omitted, and live steam is injected directly below the bottom plate. Equilibrium data are given in Table 21.8. The usual simplifying assumptions may be made. What are the advantages of each method?

21.6. A tower containing six ideal plates, a reboiler, and a total condenser is used to separate, partially, oxygen from air at 65 $lb_f/in.^2$ gauge. It is desired to operate at reflux ratio (reflux to product) of 2.6 and to produce a bottom product containing 51 weight percent oxygen. The air is fed to the column at 65 $lb_f/in.^2$ gauge and 30 percent vapor by mass. The enthalpy of oxygen-nitrogen mixtures at this pressure is given in Table 21.9. Compute the composition of the overhead if the vapors are just condensed but not cooled.

21.7. A rectifying column containing the equivalent of three ideal plates is to be supplied continuously with a feed consisting of 0.4 mole percent ammonia and 99.6 mole

TABLE 21.8
Equilibrium data in mole fraction A

x	0.0035	0.0077	0.0125	0.0177	0.0292	0.0429	0.0590	0.0784
y	0.0100	0.0200	0.0300	0.0400	0.0600	0.0800	0.1000	0.1200

TABLE 21.9
Enthalpy of oxygen-nitrogen at 65 $lb_f/in.^2$ gauge

Temperature, °C	Liquid		Vapor	
	N_2, wt%	H_x, cal/g mol	N_2, wt%	H_y, cal/g mol
−163	0.0	420	0.0	1,840
−165	7.5	418	19.3	1,755
−167	17.0	415	35.9	1,685
−169	27.5	410	50.0	1,625
−171	39.0	398	63.0	1,570
−173	52.5	378	75.0	1,515
−175	68.5	349	86.0	1,465
−177	88.0	300	95.5	1,425
−178	100.0	263	100.0	1,405

percent water. Before entering the column, the feed is converted wholly to saturated vapor, and it enters between the second and third plates from the top of the column. The vapors from the top plate are totally condensed but not cooled. Per mole of feed, 1.35 mol of condensate is returned to the top plate as reflux, and the remainder of the distillate is removed as overhead product. The liquid from the bottom plate overflows to a reboiler, which is heated by closed steam coils. The vapor generated in the reboiler enters the column below the bottom plate, and bottom product is continuously removed from the reboiler. The vaporization in the reboiler is 0.7 mol per mole of feed. Over the concentration range involved in this problem, the equilibrium relation is given by the equation

$$y = 12.6x$$

Calculate the mole fraction of ammonia in (*a*) the bottom product from the reboiler, (*b*) the overhead product, (*c*) the liquid reflux leaving the feed plate.

21.8. It is desired to produce an overhead product containing 80 mole percent benzene from a feed mixture of 68 mole percent benzene and 32 percent toluene. The following methods are considered for this operation. All are to be conducted at atmospheric pressure. For each method calculate the moles of product per 100 mol of feed and the number of moles vaporized per 100 mol feed. (*a*) Continuous equilibrium distillation. (*b*) Continuous distillation in a still fitted with a partial condenser, in which 55 mole percent of the entering vapors is condensed and returned to the still. The partial condenser is so constructed that vapor and liquid leaving it are in equilibrium and holdup in it is negligible.

21.9. The operation of a fractionating column is circumscribed by two limiting reflux ratios: one corresponding to the use of an infinite number of plates and the other a total reflux, or infinite reflux, ratio. Consider a rectifying column fed at the bottom with a constant flow of a binary vapor having a constant composition, and assume also that the column has an infinite number of plates. (*a*) What happens in such a column operating at total reflux? (*b*) Assume that a product is withdrawn at a constant rate from the top of this column. What happens as more and more product is withdrawn in successive steps if each step achieves steady state between changes?

21.10. A laboratory still is charged with 10 L of a methanol-water mixture containing 0.70 mole fraction methanol. This is to be distilled batchwise without reflux at 1 atm pressure until 5 L of liquid remains in the still, that is, 5 L has been boiled off. The rate of heat input is constant at 4 kW. The partial molar volumes are 40.5 cm^3/g mol for methanol and 18 cm^3/g mol for water. Neglecting any volume changes on mixing and using an average heat of vaporization of 40 kJ/g mol, calculate (*a*) the time t_T required to boil off 5 L; (*b*) the mole fraction of methanol left in the still at times $t_T/2$, $3t_T/4$, and t_T; (*c*) the average composition of the total distillate at time t_T. Equilibrium data for the system methanol-water are given in Table 21.6.

21.11. An equimolal mixture of A and B with a relative volatility of 2.3 is to be separated into a distillate product with 98.5 percent A, a bottoms product with 2 percent A, and an intermediate liquid product that is 80 percent A and has 40 percent of the A fed. (*a*) Derive the equation for the operating line in the middle section of the column, and sketch the three operating lines on a McCabe–Thiele diagram. (*b*) Calculate the amounts of each product per 100 mol of feed, and determine the minimum reflux rate if the feed is liquid at the boiling point. (*c*) How much greater is the minimum reflux rate because of the withdrawal of the side-stream product?

21.12. Distillation is used to prepare 99 percent pure products from a mixture of *n*-butane and *n*-pentane. Vapor pressures are given below.

	Temperature, °C	
P, atm	$n\text{-}C_4H_{10}$	$n\text{-}C_5H_{12}$
0.526	−16.3	18.5
1	−0.5	36.1
2	18.8	58.0
5	50.0	92.4
10	79.5	124.7
20	116.0	164.3

(*a*) Plot the vapor pressures in a form that permits accurate interpolation, and determine the average relative volatility for columns operating at 1, 2, and 8 atm. (*b*) Determine the minimum number of ideal plates for the separation at these three pressures. What is the main advantage of carrying out the separation at above atmospheric pressure?

21.13. Ethyl benzene (boiling point 136.2°C) and styrene (boiling point 145.2°C) are separated by continuous distillation in a column operated under a vacuum to keep the temperature under 110°C and to avoid styrene polymerization. The feed is 30,000 kg/h with 54% ethyl benzene and 46% styrene (weight percent), and the products have 97% and 0.2% ethyl benzene. The relative volatility is 1.37, and with a reflux ratio of 6.15, about 70 plates are needed. The top of the column operates at 50 mm Hg and 58°C, and the average pressure drop per tray is 2.5 mm Hg. (*a*) If the column is designed to have an F factor of 2.8 (m/s)(kg/m^3)$^{0.5}$ at the top, what diameter column is needed? (*b*) For a uniform diameter column, what would be the F factor at the bottom of the column? (*c*) If the column was built in two sections, with a smaller diameter for the bottom section, what diameter should be used so that F is never greater than 2.8? (See C. J. King, *Separation Processes,* McGraw-Hill, New York, 1971, p. 608, for more about this system.)

21.14. A plant has two streams containing benzene and toluene, one with 37% benzene and one with 68% benzene. About equal amounts of the two streams are available, and a distillation tower with two feed points is proposed to produce 98% benzene and 99% toluene in the most efficient manner. However, combining the two streams and feeding at one point would be a simpler operation. For the same reflux rate, calculate the number of ideal stages required for the two cases.

21.15. Toluene saturated with water at 30°C has 680 ppm H_2O and is to be dried to 0.3 ppm H_2O by fractional distillation. The feed is introduced to the top plate of the column, and the overhead vapor is condensed, cooled to 30°C, and separated into two layers. The water layer is removed, and the toluene layer, saturated with water, is recycled. The average relative volatility of water to toluene is 120. How many theoretical stages are required if 0.25 mol of vapor is used per mole of liquid feed? (Neglect the change in L/V in the column.)

21.16. A sieve-tray column with 15 plates is used to prepare 99% methanol from a feed containing 40% methanol and 60% water (mole percent). The plates have 8 percent open area, $\frac{1}{4}$-in. holes, and 2-in. weirs with segmental downcomers. (*a*) If the column is operated at atmospheric pressure, estimate the flooding limit based on conditions at the top of the column. What are the F factor and the pressure drop per plate at this limit? (*b*) For the flow rate calculated in part (*a*), determine the F factor and the pressure drop per plate near the bottom of the column. Which section of the column will flood first as the vapor rate is increased?

21.17. A distillation column is separating methanol and water with the reflux returned to the column at its bubble point. A larger condenser is now installed, significantly cooling the reflux stream. Control devices keep the apparent (or external) reflux ratio constant. If the same amount of product is to be withdrawn as before, what will happen to (*a*) the slope of the operating line, (*b*) the reboiler heat duty, and (*c*) the purity of the overhead product?

REFERENCES

1. *Bubble Tray Design Manual,* New York: American Institute of Chemical Engineers, 1958.
2. Bravo, J. L., J. A. Rocha, and J. R. Fair. *Hydrocarbon Proc.* **64**(1):91 (1985).
3. Fair, J. R. *Petrol. Chem. Eng.* **33**(10):45 (1961).
4. Fisher, W. R., M. F. Doherty, and J. M. Douglas. *Ind. Eng. Chem. Proc. Des. Dev.* **24**:955 (1985).
5. Gerster, J. A. *Ind. Eng. Chem.* **52**:645 (1960).
6. Harriott, P. *Environ. Sci. Technol.* **23**:309 (1988).
7. Jones, J. B., and C. Pyle. *Chem. Eng. Prog.* **51**:424 (1955).
8. Lewis, W. K. *Ind. Eng. Chem.* **14**:492 (1922).

9. Lewis, W. K., Jr. *Ind. Eng. Chem.* **28**:399 (1936).
10. McCabe, W. L., and E. W. Thiele. *Ind. Eng. Chem.* **17**:605 (1925).
11. McFarland, S. A., P. M. Sigmund, and M. Van Winkle. *Hydro. Proc.* **51**(7):111 (1972).
12. Murphree, E. V. *Ind. Eng. Chem.* **17**:747 (1925).
13. *Valve Tray Design Manual,* Akron, OH: Norton Chemical Process Products Corp., 1996.
14. Peters, M. S., and K. D. Timmerhaus. *Plant Design and Economics for Chemical Engineers,* 3rd ed. New York: McGraw-Hill, 1980, p. 387.
15. Strigle, R. F., Jr., and F. Rukovena, Jr. *Chem. Eng. Prog.* **75**(3):87 (1979).
16. Vital, T. J., S. S. Grossel, and P. I. Olsen. *Hydro. Proc.* **63**(11):147 (1984).

CHAPTER 22

Introduction to Multicomponent Distillation

In multicomponent distillation, as in the distillation of binary mixtures, the calculation of equilibrium stages uses mass and enthalpy balances and vapor-liquid equilibria. A mass balance can be written for each component for the column as a whole or for a single stage, but there is only one enthalpy balance for the column or for each stage. The phase equilibria are much more complex than for binary systems, because of the several components and because the equilibria depend on temperature, which changes from stage to stage. In binary systems the temperature and equilibria also change from stage to stage, but except with azeotropes, the more volatile component is more volatile than the other component throughout the column. In multicomponent mixtures one component may be more volatile than the average in one part of the column and less volatile than the average in another part, which leads to complex concentration profiles.

In practice, the field is dominated by the use of digital computers because of the mass of numbers needed to quantify the operating and engineering variables and the many iterations required to obtain convergence of the solutions to the equations. This chapter does not include computer programs, but all such programs are based squarely on the principles to which this text is devoted.

PHASE EQUILIBRIA IN MULTICOMPONENT DISTILLATION

The vapor-liquid equilibria for a mixture are described by distribution coefficients or K factors, where K for each component is the ratio of mole fractions in the vapor and liquid phases at equilibrium:

$$K_i \equiv \frac{y_{ie}}{x_{ie}} \tag{22.1}$$

If Raoult's law and Dalton's law hold, values of K_i can be calculated from the vapor pressure and the total pressure of the system:

$$p_i = x_i P_i' \tag{22.2}$$

$$y_i = \frac{p_i}{P} \tag{22.3}$$

$$K_i = \frac{x_i P_i'}{P x_i} = \frac{P_i'}{P} \tag{22.4}$$

Raoult's law is a good approximation for mixtures of similar compounds, such as the paraffins found in the low-boiling fractions of petroleum or the aromatics recovered from coke production. However, at high pressures, K factors do not vary exactly inversely with total pressure because of compressibility effects.

The K factors are strongly temperature-dependent because of the change in vapor pressure, but the relative values of K for two components change only moderately with temperature. The ratio of K factors is the same as the relative volatility of the components:

$$\alpha_{ij} = \frac{y_i/x_i}{y_j/x_j} = \frac{K_i}{K_j} \tag{22.5}$$

When Raoult's law applies,

$$\alpha_{ij} = \frac{P_i'}{P_j'} \tag{22.6}$$

As will be shown later, the average relative volatility of a key component in the overhead or distillate product to that of a key component in the bottoms product can be used to estimate the minimum number of stages for a multicomponent distillation.

Bubble-point and dew-point calculation

Determination of the bubble point (initial boiling point of a liquid mixture) or the dew point (initial condensation temperature) is required for a flash distillation calculation and for each stage of a multicomponent distillation. The basic equations are, for the bubble point,

$$\sum_{i=1}^{N_c} y_i = \sum_{i=1}^{N_c} K_i x_i = 1.0 \tag{22.7}$$

and, for the dew point,

$$\sum_{i=1}^{N_c} x_i = \sum_{i=1}^{N_c} \frac{y_i}{K_i} = 1.0 \tag{22.8}$$

where N_c is the number of components.

To use Eq. (22.7), a temperature is assumed, and values of K_i are obtained from published tables or from vapor pressure data and the known total pressure. If the summation of $K_i x_i$ exceeds 1.0, a lower temperature is chosen and the calculation repeated until Eq. (22.7) is satisfied. If the bubble-point temperature is determined exactly ($\sum K_i x_i = 1.00$), the composition of the vapor in equilibrium with this liquid is given directly by the terms $K_i x_i$. However, when the summation is close to 1.0, the vapor composition can be determined with little error from the relative contribution of each term to the summation:

$$y_i = \frac{K_i x_i}{\sum\limits_{i=1}^{N_c} K_i x_i} \tag{22.9}$$

A similar procedure is used to determine the dew point of a vapor mixture and the composition of the liquid in equilibrium with this mixture.

EXAMPLE 22.1. Find the bubble-point and the dew-point temperatures and the corresponding vapor and liquid compositions for a mixture of 33 mole percent *n*-hexane, 37 mole percent *n*-heptane, and 30 mole percent *n*-octane at 1.2 atm total pressure.

Solution. Plot the vapor pressures of the three components as a semilogarithmic plot of log P versus T (Fig. 22.1) or log P versus $1/T_{abs}$, where T_{abs} is the absolute temperature in kelvins.

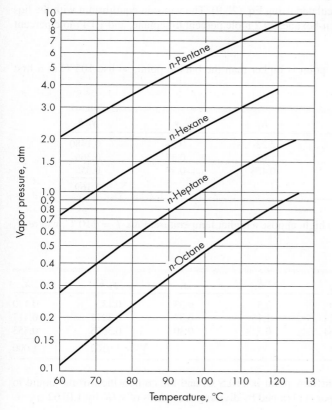

FIGURE 22.1
Diagram for Example 22.1.

Bubble point. Choose $T = 105°C$, where the vapor pressure of heptane, the middle component, is 1.2 atm.

Component	P'_i 105°C, atm	$K_i = P'_i/1.2$	x_i	$y_i = K_i x_i$
1. Hexane	2.68	2.23	0.33	0.7359
2. Heptane	1.21	1.01	0.37	0.3737
3. Octane	0.554	0.462	0.30	0.1386
				$\sum = 1.248$

Since $\sum y_i$ is too large, try a lower temperature. Since the major contribution comes from the first term, pick a temperature where K_i is lower by a factor $1/1.24$. Choose $T = 96°C$, where $P'_i = 2.16$ atm.

Component	P'_i at 96°C	K_i	x_i	$K_i x_i$	y_i
1	2.16	1.8	0.33	0.5940	0.604
2	0.93	0.775	0.37	0.2868	0.292
3	0.41	0.342	0.30	0.1025	0.104
				$\sum = 0.9833$	1.000

Since $\sum K_i x_i = 0.9833$, $y_i = K_i x_i/0.9833$.

By interpolation, the bubble point is 97°C, close enough to 96°C so that the vapor compositions can be calculated using Eq. (22.9). The vapor in equilibrium with the liquid is 60.4 mole percent n-hexane, 29.2 mole percent n-heptane, and 10.4 mole percent n-octane.

Dew point. The dew point is higher than the bubble point, so use 105°C as a first guess.

Component	K_i	y_i	y_i/K_i
1	2.23	0.33	0.1480
2	1.01	0.37	0.366
3	0.458	0.30	0.655
			$\sum = 1.169$

Since the sum is too high, choose a higher temperature. Pick $T = 110°C$, where K_3 is 17 percent higher.

Component	P'_i	K_i	y_i	y_i/K_i	x_i
1	3.0	2.5	0.33	0.132	0.130
2	1.38	1.15	0.37	0.3217	0.317
3	0.64	0.533	0.30	0.5625	0.553
				$\sum = 1.0162$	1.000

By extrapolation, the dew point is 110.5°C, and the composition of the liquid in equilibrium with the vapor is obtained by dividing the values of y_i/K_i by 1.0162.

FLASH DISTILLATION OF MULTICOMPONENT MIXTURES

Equation (21.2) can be written for each component in a flash distillation in the form

$$y_{Di} = \frac{x_{Fi}}{f} - \frac{1-f}{f} x_{Bi} \tag{22.10}$$

Since the distillate and bottom streams are in equilibrium, this equation may be changed to

$$\frac{y_{Di}}{x_{Bi}} = K_i = \frac{1}{f}\left(\frac{x_{Fi}}{x_{Bi}} + f - 1\right) \tag{22.11}$$

Solving Eq. (22.11) for x_{Bi} and summing over N_c components give

$$\sum_{i=1}^{N_c} x_{Bi} = 1 = \sum_{i=1}^{N_c} \frac{x_{Fi}}{f(K_i - 1) + 1} \tag{22.12}$$

This equation is solved by iteration in the same manner as the dew-point calculation using Eq. (22.8), and the final values of T and K_i are used to calculate the compositions of the product streams.

> **EXAMPLE 22.2.** The mixture of Example 22.1 is subjected to a flash distillation at 1.2 atm pressure, and 60 percent of the feed is vaporized. (a) Find the temperature of the flash and the composition of the liquid and vapor products. (b) To what temperature must the feed liquid be heated for 60 percent vaporization on flashing?
>
> **Solution**
> (a) The flash temperature must lie between the bubble point (97°C) and the dew point (110.5°C). Assume $T = 105$°C, which is $97 + 0.6(110.5 - 97)$. From Fig. 22.1, $K_1 = 2.68/1.2 = 2.23$, $K_2 = 1.21/1.2 = 1.01$, and $K_3 = 0.554/1.2 = 0.462$. The value of f is 0.6. The right-hand side of Eq. (22.12) becomes
>
> $$\frac{0.33}{0.6(2.23 - 1) + 1} + \frac{0.37}{0.6(1.01 - 1) + 1} + \frac{0.30}{0.6(0.462 - 1) + 1}$$
> $$= 0.190 + 0.368 + 0.443 = 1.001$$
>
> The flash temperature is 105°C. The composition of the liquid product is n-hexane, 19.0 mole percent; n-heptane, 36.8 mole percent; and n-octane, 44.2 mole percent.
>
> The composition of the vapor product is computed from the values of K and x:
>
> | | |
> |---|---|
> | n-hexane, $y = 0.190(2.23)$ | $= 0.424$ |
> | n-heptane, $y = 0.368(1.01)$ | $= 0.372$ |
> | n-octane, $y = 0.442(0.462)$ | $= 0.204$ |
> | | $\overline{1.000}$ |
>
> (b) To determine the temperature of the feed before flashing, an enthalpy balance is made using 105°C as the reference temperature. The heats of vaporization at 105°C and the average heat capacities of the liquid from 105 to 200°C are obtained from the literature.

	C_p, cal/mol · °C	ΔH_v, cal/mol
n-hexane	62	6,370
n-heptane	70	7,510
n-octane	78	8,560

Based on liquid at 105°C, the enthalpies of the product are

$$H_{vapor} = 0.6(0.424 \times 6,370 + 0.372 \times 7,510 + 0.204 \times 8,560)$$

$$H_{vapor} = 4,345 \text{ cal} \qquad H_{liquid} = 0$$

For the feed,

$$\bar{C}_p = 0.33 \times 62 + 0.37 \times 70 + 0.30 \times 78$$

$$= 69.8 \text{ cal/mol} \cdot °C$$

$$69.8(T_0 - 105) = 4,345$$

$$T_0 = 167°C = \text{preheat temperature}$$

For a more accurate answer, the liquid heat capacities could be reevaluated for the range 105 to 170°C.

FRACTIONATION OF MULTICOMPONENT MIXTURES

As in the fractionation of binary mixtures, ideal plates are assumed in the design of cascades, and the number of stages is subsequently corrected for plate efficiencies. The two limiting conditions of total reflux and minimum reflux are also encountered.

Calculations for distillation plants are made by either of two methods. In the first, a desired separation of components is assumed, and the numbers of plates above and below the feed are calculated from a chosen reflux ratio. In the second, the number of plates above and below the feed is assumed, and the separation of components is calculated using assumed flows of reflux from the condenser and vapor from the reboiler. In binary distillation the first method is the more common; in multicomponent cases the second approach is preferred, especially in computer calculations.

In final computer calculations, neither constant molal overflow nor temperature-independent K factors are assumed, and plate efficiencies are also introduced, but in preliminary estimates the simplifying assumptions are common. When the activity coefficients are assumed to be temperature-independent, group methods are used in which the number of ideal stages in a cascade is the dependent variable. The calculation gives this number without solving for either the plate temperature or the compositions of the interstreams between plates. If α values are temperature-dependent, these simple methods are not used and plate-by-plate calculations are necessary. The temperature and liquid composition for plate $n + 1$ are calculated by trial from those already known for plate n, and the calculation proceeds from plate to plate up or down the column.

In the rest of this chapter these methods are sampled in the following way: The estimates of the minimum number of plates at infinite reflux and of minimum reflux with an infinite number of plates are made by group methods on the assumption of constant relative volatilities and are based on a design point of view. An empirical relation for the number of plates at an operating reflux is described.

Key components

The objective of distillation is the separation of the feed into streams of nearly pure products. In binary distillation, the purity is usually defined by specifying x_D and x_B, the mole fraction of light component in the distillate and bottoms products. As shown by Eq. (21.5), fixing these concentrations fixes the amounts of both products per unit of feed. The reflux ratio is then chosen, and the number of theoretical stages is calculated.

In multicomponent distillation, there are three or more components in the products, and specifying the concentrations of one component in each does not fully characterize these products. However, if the concentrations of two out of three or three out of four components are specified for the distillate and bottoms products, it is generally impossible to meet these specifications exactly. An increase in reflux ratio or number of plates would increase the sharpness of the separation, and the desired concentration of one component in each product could be achieved, but it would be a coincidence if the other concentrations exactly matched those specified beforehand. The designer generally chooses two components whose concentrations or fractional recoveries in the distillate and bottoms products are a good index of the separation achieved. After these components are identified, they are called key components. Since the keys must differ in volatility, the more volatile, identified by subscript L, is called the *light key,* and the less volatile, identified by subscript H, is called the *heavy key.*

Having chosen the keys, the designer arbitrarily assigns small numbers to x_H in the distillate (x_{DH}) and to x_L in the bottoms (x_{BL}), just as small numbers are assigned to x_{DB} and x_{BA} in binary distillation. Choosing small values for x_{BL} and x_{DH} means that most of the light key ends up in the distillate and most of the heavy key in the bottoms. The distillate may be nearly pure light key if the keys are the two most volatile components, since components heavier than the heavy key will tend to concentrate in the liquid phase and not be carried much above the feed plate. Often there are components lighter than the light key, and these are nearly completely recovered in the distillate. Any components heavier than the heavy key are usually completely recovered in the bottoms. The exceptions to these generalizations are encountered in distillation of very close boiling materials, such as mixtures of isomers.

Unlike the binary case, the choice of two keys does not give determinate mass balances, because not all other mole fractions are calculable by mass balances alone and equilibrium calculations are required to calculate the concentrations of the dew-point vapor from the top plate and the bubble-point liquid leaving the reboiler.

Although any two components can be nominated as keys, usually they are adjacent in the rank order of volatility. Such a choice is called a *sharp separation.* In sharp separations the keys are the only components that appear in both products in appreciable concentrations.

Minimum number of plates

The Fenske equation (21.42) applies to any two components i and j in a conventional plant at infinite reflux ratio. In this case, the equation has the form

$$N_{\min} = \frac{\ln\left[(x_{Di}/x_{Bi})/(x_{Dj}/x_{Bj})\right]}{\ln\bar{\alpha}_{ij}} - 1 \qquad (22.13)$$

$$\bar{\alpha}_{ij} = \sqrt[3]{\alpha_{Dij}\alpha_{Fij}\alpha_{Bij}} \qquad (22.14)$$

The subscripts D, F, and B in Eq. (22.14) refer to the temperatures of the distillate, feed plate, and bottoms in the column.

EXAMPLE 22.3. A mixture with 33% n-hexane, 37% n-heptane, and 30% n-octane is to be distilled to give a distillate product with 0.01 mole fraction n-heptane and a bottoms product with 0.01 mole fraction n-hexane. The column will operate at 1.2 atm with 60 percent vaporized feed. Calculate the complete product compositions and the minimum number of ideal plates at infinite reflux.

Solution. The n-hexane is the light key (LK), the n-heptane is the heavy key (HK), and the n-octane is a heavy nonkey (HNK), which goes almost entirely to the bottoms. The product compositions are found by mass balance assuming no n-octane and 0.99 mole fraction n-hexane in the distillate. Basing the calculations on a feed rate of 100 mol/h,

$$F = D + B = 100$$

For hexane,

$$Fx_F = Dx_D + Bx_B$$

$$100 \times 0.33 = 0.99D + (100 - D)(0.01)$$

$$D = \frac{32}{0.98} = 32.65 \text{ mol/h}$$

$$B = 100 - D = 67.35 \text{ mol/h}$$

The amount of hexane in the overhead is

$$Dx_D = 32.65 \times 0.99 = 32.32 \text{ mol/h}$$

The composition of the bottoms product can be calculated directly since this stream contains all the octane, all but $0.01D$ of the heptane, or $37 - 0.01(32.65) = 36.67$ mol/h, and 0.68 mol/h hexane. Table 22.1 gives the compositions.

TABLE 22.1

Component	Feed, mol	Distillate		Bottoms		K at 105 °C, 1.2 atm
		Mol	x	Mol	x	
LK n-hexane	33	32.32	0.99	0.68	0.010	2.23
HK n-heptane	37	0.33	0.01	36.67	0.544	1.01
HNK n-octane	30	0	0	30	0.446	0.462
	100	32.65		67.35		

The minimum number of plates is obtained from the Fenske equation [Eq. (22.13)] using the relative volatility of the light key to the heavy key, which is the ratio of the K factors. The K values at the flash temperature were taken from Example 22.2 and are given in Table 22.1:

$$\alpha_{LK, HK} = \frac{2.23}{1.01} = 2.21$$

$$N_{min} = \frac{\ln\left[(0.99/0.01)/(0.01/0.544)\right]}{\ln 2.21} - 1 = 10.8 - 1 = 9.8$$

The minimum number of ideal stages is 9.8 plus a reboiler.

A more accurate estimate of N_{min} can be obtained using a mean relative volatility based on values at the top, middle, and bottom of the column. The top temperature is about 75°C, the boiling point of n-hexane at 1.2 atm, and the relative volatility is 2.53 from the vapor pressures in Fig. 22.1. The bottom temperature is about 115°C, by a bubble-point calculation for the bottoms product, giving a relative volatility of 2.15. From Eq. (22.14)

$$\bar{\alpha}_{LK, HK} = \sqrt[3]{2.53 \times 2.21 \times 2.15} = 2.29$$

Using ln 2.29 in the denominator of Eq. (22.13) gives $N_{min} = 9.4$.

To check the assumption of no octane in the distillate, Eq. (22.13) can be applied to heptane and octane using $\alpha = K_2/K_3 = 1.01/0.462 = 2.19$:

$$N_{min} + 1 = 10.4 = \frac{\ln[(0.01/0.544)/(x_{D3}/0.446)]}{\ln 2.19}$$

from which $x_{D3} = 2.4 \times 10^{-6}$, which is negligible.

Minimum reflux ratio

The minimum reflux ratio for a multicomponent distillation has the same significance as for binary distillation; at this reflux ratio, the desired separation is just barely possible, but an infinite number of plates are required. The minimum reflux ratio is a guide in choosing a reasonable reflux ratio for an operating column and in estimating the number of plates needed for a given separation at certain values of the reflux ratio.

For a multicomponent system, the desired separation usually refers to the amount of light key recovered in the distillate and the amount of heavy key recovered in the bottoms. For example, the specifications might call for 98 percent recovery of the light key in the distillate and 99 percent recovery of the heavy key in the bottoms. The actual mole fractions of the key components in the products are not usually specified, since they depend on the amounts of nonkey components in the feed. Small changes in these nonkey components in the feed would change the product compositions without significantly affecting the basic separation of the light and heavy keys.

Although the separation achieved in a column depends to some extent on all components in the feed, an approximate value of the minimum reflux ratio can be obtained by treating the mixture as a pseudobinary. Taking only the moles of light key and heavy key to make a new pseudofeed, product compositions could be calculated along with a vapor-liquid equilibrium curve based on $\alpha_{LK\text{-}HK}$. Then R_{Dm}

could be obtained by using Eq. (21.44) as illustrated in Fig. 21.19. An alternate equation for a *saturated liquid feed*[2] gives the minimum ratio of liquid rate to feed rate for a binary mixture of A and B:

$$\frac{L_{min}}{F} = \frac{(Dx_{DA}/Fx_{FA}) - \alpha_{AB}(Dx_{DB}/Fx_{FB})}{\alpha_{AB} - 1} \tag{22.15}$$

The terms in parentheses in Eq. (22.15) are the fractional recovery of A and B in the distillate product. For a multicomponent mixture, these terms would be the specified recovery of light key in the distillate and the fraction of heavy key in the feed that is allowed in the distillate. Note that the minimum value of L depends mainly on the relative volatility. Changing the recovery of light key from 0.95 to 0.99 or even 0.999 changes L_{min}/F only about 4 to 5 percent, since the term $\alpha_{AB}(Dx_{DB}/Fx_{FB})$ is usually quite small. The feed composition has little effect in Eq. (22.15); but when x_{FA} is low, D will be small and the reflux ratio L/D will be greater than that for a richer feed.

Equation (22.15) gives a good approximation for multicomponent mixtures if the key components make up 90 percent or more of the feed. It generally overestimates the value of L needed for these cases, since components more volatile than the light key or heavier than the heavy key are more easily separated than the keys themselves. For other mixtures, the distribution of nonkey components in the products must be estimated as a first step in a more rigorous calculation of the minimum reflux ratio. The complete composition of the products cannot be specified beforehand, and the amounts of nonkey components in the products change with the reflux ratio, even when the number of plates is adjusted to maintain the desired separation of the key components. To help estimate the product compositions at minimum reflux, the concepts of *distributed* and *undistributed* components are introduced.

Distributed and undistributed components

A distributed component is found in both the distillate and bottoms products, whereas an undistributed component is found in only one product. The light key and heavy key are always distributed, as are any components having volatilities between those two keys. Components more volatile than the light key are almost completely recovered in the distillate, and those less volatile than the heavy key are found almost completely in the bottoms. Whether such components are called distributed or undistributed depends on the interpretation of the definition. For a real column with a finite number of plates, all components are theoretically present in both products, though perhaps some are at concentrations below the detectable limit. If the mole fraction of a heavy nonkey component in the distillate is 10^{-6} or less, the component may be considered undistributed from a practical standpoint. However, to start a plate-by-plate calculation to get the number of plates for the column, this small but finite value needs to be estimated.

For the case of minimum reflux, the distinction between distributed and undistributed components is clearer, since heavy nonkey components are generally absent from the distillate, and light nonkey components are not present in the bottoms. The concentrations of these species can go to zero because of an infinite number of plates in the column and conditions that lead to a progressive reduction in concentration for each plate beyond the feed plate.

Consider what is required for a heavy component to be completely absent from the distillate. If x_D is zero and constant molal overflow is assumed, the material-balance equation for the upper part of the column [Eq. (22.14)] becomes

$$y_{n+1} = \left(\frac{L}{V}\right)_n x_n \tag{22.16}$$

For an ideal stage, $y_n = Kx_n$, and the ratio of vapor concentrations for successive plates is

$$\frac{y_n}{y_{n+1}} = \frac{KV}{L} \tag{22.17}$$

If K for the component being considered is less than L/V, y_n will be smaller than y_{n+1}; and if this is true for all plates above the feed, an infinite number of plates will make y go to zero. Of course, K is a function of temperature, but if K is less than L/V at the feed-plate temperature, it will be even smaller for plates above the feed, where the temperature is lower, and the decrease in y from plate to plate will be more rapid. Heavy components will generally be undistributed if K is more than 10 percent below the K for the heavy key or if the relative volatility based on the heavy key is less than 0.9.

Light components are undistributed at minimum reflux if the value of K is high enough for plates below the feed plate. If x_B is assumed zero, Eq. (21.16) leads to

$$y_{m+1} = \frac{\bar{L}}{\bar{V}} x_m \tag{22.18}$$

Thus,

$$\frac{y_{m+1}}{y_m} = \frac{\bar{L}}{K\bar{V}} \tag{22.19}$$

If K is always greater than $\bar{L}/\bar{V}$ in the bottom section of the column, y will become zero, justifying the assumption that $x_B = 0$. In practice, a K value or relative volatility 10 percent greater than that of the light key nearly always means an undistributed component. For a feed with components only slightly more volatile than the light key or slightly less volatile than the heavy key, there are techniques for calculating whether such components distribute and for estimating their concentrations in the products.[4]

Calculation of minimum reflux ratio

At the minimum reflux ratio, there are invariant zones above and below the feed plate where the compositions of the liquid and vapor do not change from plate to plate. These zones are similar to the "pinch" regions shown in Fig. 21.19, but they do not necessarily occur at the feed plate, as they do in binary distillation. If there are undistributed components in the feed, their concentrations change from plate to plate near the feed, and the concentrations have been reduced to zero when the invariant zone is reached. Thus for a feed with both light and heavy undistributed components, the liquid in the upper invariant zone will have all components except the heavy undistributed ones.

At the lower invariant zone, all components except the light undistributed ones will be present. The two invariant zones will be at different temperatures and have different liquid and vapor compositions because of the undistributed components. If the undistributed components are a small fraction of the feed, the temperatures of the two invariant zones are nearly the same, and the calculation of the minimum reflux ratio is relatively easy. When these temperatures differ considerably, exact calculation of the minimum reflux is difficult because the relative volatilities in the two zones are different. The following analysis is intended to give the concepts underlying the determination of the minimum reflux ratio and a convenient approximate equation. The complete derivation of the equation for R_{Dm} is beyond the scope of this text.

The material-balance equation for each component in the upper section of the column [Eq. (21.14)] can be written with y_n/K in place of x_n, which assumes perfect equilibrium between vapor and liquid:

$$V_{n+1, i} y_{n+1, i} = \frac{L_n y_{ni}}{K_i} + D x_{Di} \qquad (22.20)$$

In the invariant zone, there is no change in composition from plate to plate, so $y_{n+1, i} = y_{ni}$ and is designated $y_{\infty i}$. The subscript ∞ denotes an infinite number of plates. Equation (22.20) then becomes

$$V_\infty y_{\infty i} = \frac{L_\infty y_{\infty i}}{K_{\infty i}} + D x_{Di} \qquad (22.21)$$

Rearrangement of this equation leads to

$$y_{\infty i} = \frac{D x_{Di}}{V_\infty - L_\infty / K_{\infty i}} \qquad (22.22)$$

or

$$y_{\infty i} = \frac{D}{V_\infty} \left(\frac{x_{Di}}{1 - L_\infty / V_\infty K_{\infty i}} \right) \qquad (22.23)$$

Equation (22.23) is summed for all components appearing in the distillate, and the sum must equal 1.0:

$$\sum y_{\infty i} = 1.0 = \frac{D}{V_\infty} \sum \frac{x_{Di}}{1 - L_\infty / V_\infty K_{\infty i}} \qquad (22.24)$$

A similar treatment for the lower section of the column leads to

$$\bar{V}_\infty y_{\infty i} = \frac{\bar{L}_\infty y_{\infty i}}{\bar{K}_{\infty i}} - B x_{Bi} \qquad (22.25)$$

$$y_{\infty i} = -\frac{B x_{Bi}}{\bar{V}_\infty - \bar{L}_\infty / \bar{K}_{\infty i}} \qquad (22.26)$$

$$y_{\infty i} = -\frac{B}{\bar{V}_\infty} \left(\frac{x_{Bi}}{1 - \bar{L} / \bar{V}_\infty \bar{K}_{\infty i}} \right) \qquad (22.27)$$

Equation (22.27) is summed over all the components appearing in the bottoms product, and the signs are changed to make the denominator positive:

$$\sum y_{\infty i} = 1.0 = \frac{B}{V_\infty} \sum \frac{x_{Bi}}{\bar{L}_\infty / \bar{V}_\infty \bar{K}_{\infty i} - 1} \tag{22.28}$$

To determine the minimum reflux ratio using Eq. (22.24), a value of R_D is assumed, which gives L/V and D/V. The temperature at which Eq. (22.24) is satisfied with all terms positive is determined by trial. Other sets of K values will give a sum equal to 1.0 but with some negative terms, which have no physical significance.

The flow rates in the lower section of the column are then calculated from the feed condition: $\bar{L} = L + qF$, $\bar{V} = V - (1 - q)F$, where q is the number of moles of liquid entering the stripping section per mole of feed [see Eqs. (21.24) and (21.25)]. The temperature that satisfies Eq. (22.28) with all terms positive is found. The temperature in the lower invariant zone should be higher than that found for the upper invariant zone if there are some undistributed components. If the calculated temperatures are the same or are in the wrong order, the assumed R_D is incorrect, and the calculations are repeated for a lower value of R_D. Figure 22.2 shows how this procedure would apply to a binary system. For any selected value of R_D or L/V, a temperature would be found that corresponds to a pinch, where the operating line touches the equilibrium line. For a higher R_D, the upper pinch occurs at a lower value of x or a higher temperature, and the lower pinch occurs at a higher x and lower temperature. For a binary, the two pinch points coincide at the true minimum reflux, but for a multicomponent feed, the invariant zones differ in temperature and composition. Unfortunately, there is no simple method of determining the temperature separation, and plate-to-plate calculations are needed in the region between these zones to get the exact value of R_{Dm}.

An approximate but fairly accurate method of determining R_{Dm} was developed by Underwood.[5] The relative volatility for each component is taken to be the same

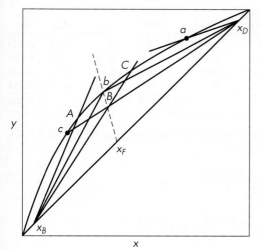

FIGURE 22.2
Invariant zones in a binary system.

in the upper and lower invariant zones, and constant molal overflow is assumed. The equations for the invariant zones are written in terms of the relative volatility α_i, where $\alpha_i = K_i/K_{ref}$, with the heavy key generally taken as the reference component. The two equations are combined with an overall material balance and the feed quality equation to give an equation that must be solved by trial. The correct root ϕ of this equation lies between the values of α for the keys. Other values of ϕ satisfy the equation but have no physical significance. The equation is

$$1 - q = \sum \frac{\alpha_i x_{Fi}}{\alpha_i - \phi} = \sum f_i \tag{22.29}$$

The value of ϕ is then used to get V_{min}/D:

$$\frac{V_{min}}{D} = R_{Dm} + 1 = \sum \frac{\alpha_i x_{Di}}{\alpha_i - \phi} \tag{22.30}$$

Note that all components of the feed are included in the summation of Eq. (22.29), but only those found in the distillate are included in Eq. (22.30). If there are one or more compounds in the feed between the light and heavy keys, there are two or more values of ϕ between the α values of the keys that will satisfy Eq. (22.29). The correct value of ϕ must then be found by solving Eqs. (22.29) and (22.30) simultaneously.

EXAMPLE 22.4. A mixture with 4% n-pentane, 40% n-hexane, 50% n-heptane, and 6% n-octane is to be distilled at 1 atm with 98% of the hexane and 1% of the heptane recovered in the distillate. (a) What is the minimum reflux ratio for a liquid feed at the boiling point? (b) What are the temperatures and compositions in the upper and lower invariant zones?

Solution. The keys are n-hexane and n-heptane, and the other components are sufficiently different in volatility to be undistributed. Below are given the moles in the products per 100 mol of feed along with K values at 80°C.

		x_F	Fx_F	Moles in D	x_D	Moles in B	x_B	$K_{80°}$	Kx_F
	$n\text{-}C_5$	0.04	4	4	0.092	0	0	3.62	0.145
LK	$n\text{-}C_6$	0.40	40	39.2	0.897	0.8	0.014	1.39	0.556
HK	$n\text{-}C_7$	0.50	50	0.5	0.011	49.5	0.879	0.56	0.280
	$n\text{-}C_8$	0.06	6	0	0	6	0.107	0.23	0.014
				$D = 43.7$		$B = 56.3$			0.995

(a) The bubble point is 80°C, and at this temperature $\alpha_{LK\text{-}HK}$ is $1.39/0.56 = 2.48$. For an approximate solution, use Eq. (22.15):

$$\frac{L_{min}}{F} = \frac{0.98 - 2.48(0.01)}{2.48 - 1} = 0.645$$

$$\frac{L_{min}}{D} = \frac{L_{min}}{F}\frac{F}{D} = 0.645\left(\frac{1}{0.437}\right) = 1.48$$

To use the Underwood method, the K values at 80°C are converted to relative volatilities, and the root of Eq. (22.29) between 1 and 2.48 is found by trial. Since $q = 1.0$, the terms must sum to zero.

	α_i	x_{Fi}	$f_i, \phi = 1.5$	$f_i, \phi = 1.48$
$n\text{-}C_5$	6.46	0.04	0.052	0.052
$n\text{-}C_6$	2.48	0.40	1.012	0.992
$n\text{-}C_7$	1.0	0.50	-1.00	-1.042
$n\text{-}C_8$	0.41	0.06	-0.023	-0.023
			0.041	-0.021

By further trials or interpolation, $\phi = 1.487$. From Eq. (22.30),

$$R_{Dm} + 1 = \sum \frac{\alpha_i x_{Di}}{\alpha_i - 1.487}$$

$$= \frac{6.46(0.092)}{6.64 - 1.487} + \frac{2.48(0.897)}{2.48 - 1.487} + \frac{1(0.011)}{1 - 1.487}$$

$$= 0.120 + 2.24 - 0.023 = 2.337$$

$$R_{Dm} = 1.34$$

Note this is 10 percent less than the approximate value obtained using Eq. (22.15).

(b) To get the conditions in the upper invariant zone, use Eq. (22.24) with the following flow ratios:

$$\frac{V}{D} = R_D + 1 = 2.34 \qquad \frac{D}{V} = 0.427$$

$$\frac{V}{F} = \frac{V}{D}\frac{D}{F} = 2.34 \times 0.437 = 1.02 \qquad \frac{L}{V} = \frac{R_D}{R_D + 1} = \frac{1.34}{2.34} = 0.573$$

$$y_i = \frac{D}{V}\left(\frac{x_{Di}}{1 - L/VK_i}\right)$$

		x_{Di}	$K_{80°}$	y_i	$K_{81°}$	y_i	$K_{81.2°}$	y_i	y_i at 81.1°C
	$n\text{-}C_5$	0.092	3.62	0.047	3.72	0.046	3.74	0.046	0.046
LK	$n\text{-}C_6$	0.897	1.39	0.652	1.43	0.639	1.44	0.636	0.637
HK	$n\text{-}C_7$	0.011	0.56	-0.202	0.58	0.389	0.584	0.249	0.317
						1.074		0.931	1.00

For an assumed $T = 80°C$, the calculated y for heptane is negative, so the temperature must be slightly higher (so that $K_i > L/V$). The term for heptane is very sensitive to the assumed temperature, and the K values would have to be given to four significant figures to make the summation 1.00. From the above values

$$T \text{ upper zone} \approx 81.1°C$$

The vapor compositions in this zone (y_i in the final column) are corrected to the correct sum by making most of the adjustment to the value for heptane.

The vapor composition and temperature in the lower invariant zone are obtained using Eq. (22.28) with the following flow ratios. For $q = 1.0$,

$$V = \bar{V} \qquad \bar{L} = L + F$$

$$\frac{B}{\bar{V}} = \frac{B}{F}\frac{F}{\bar{V}} = \frac{0.563}{1.02} = 0.552 \qquad \frac{\bar{L}}{\bar{V}} = \frac{L}{V} + \frac{F}{V} = 0.573 + \frac{1}{1.02} = 1.55$$

$$y_i = \frac{B}{\bar{V}}\left(\frac{x_{Bi}}{\bar{L}/\bar{V}K_i - 1}\right)$$

		x_{Bi}	$K_{83°}$	y_i	$K_{83.2°}$	y_i	y_i at 83.3°C
LK	n-C$_6$	0.014	1.52	0.392	1.53	0.591	0.662
HK	n-C$_7$	0.879	0.618	0.322	0.622	0.325	0.326
	n-C$_8$	0.107	0.258	0.012	0.26	0.012	0.012
				0.726		0.928	1.000

Here the term for hexane changes most rapidly with temperature, and the final values of y_i are adjusted accordingly:

$$T \text{ lower zone} \approx 83.3°C$$

The liquid compositions in the invariant zones are calculated from $x_i = y_i/K_i$.

		Lower zone	Upper zone
T, °C		83.3	81.1
LK	x	0.433	0.442
	y	0.662	0.637
HK	x	0.524	0.543
	y	0.326	0.317
	$\alpha_{\text{LK-HK}}$	2.46	2.47
	$y_{\text{LK}}/y_{\text{HK}}$	2.03	2.01

Between the lower and upper invariant zones, the mole fraction of both keys in the vapor phase decreases, and the ratio of light key to heavy key decreases. This region of the column serves to remove the light nonkey components from the liquid flowing down and the heavy nonkey component from the material that will flow up and form the distillate. The small amount of *reverse fractionation* shown for the key components is an interesting phenomenon that is often found in real columns operating at close to the minimum reflux ratio.

Calculation of required reflux ratio and concentration profiles

The number of plates needed for a specified separation at a selected reflux ratio can be determined by a plate-by-plate calculation called the Lewis–Matheson method.[3] The amount of all components in the products must be specified to start the calculation. From the composition of the distillate (which is the same as the vapor from the top if a total condenser is used), the temperature and liquid composition on the top plate can be determined by a dew-point calculation from

Eq. (22.8):

$$\sum x_i = 1.0 = \sum \frac{y_i}{K_i}$$

The K factors are stored as a table of values or calculated from empirical equations for a given temperature and pressure. If the mixtures are nonideal, equations for the activity coefficients are also required.

From the liquid composition on the top plate and the distillate composition, material-balance equations are used to get the composition of the vapor from plate 2:

$$y_{2i} V_2 = L_1 x_{11} + D x_{Di} \tag{22.31}$$

Equal molal overflow could be assumed, but if the calculations are done by computer, an enthalpy balance would probably be made and the change in pressure from stage to stage would also be allowed for. The calculations are continued in this fashion, alternating the use of equilibrium and material-balance relationships, until the composition is close to that of the feed. Similar calculations are carried out for the lower section of the column, starting with an estimated reboiler or bottoms composition. The next step is to match the compositions at the feed stage for the two sets of calculations. Based on the differences for individual components, the product compositions are adjusted and the calculations repeated until all errors fall below a specified value. In some procedures, the number of plates and the feed plate are fixed beforehand, and the calculations are repeated for different reflux ratios until the desired match is obtained at the designated feed plate.

Convergence to the specified conditions at the feed plate is easy when the non-keys are all heavy or all light.[7a] In other situations it may be very difficult, even when constant molal overflow or constant relative volatilities may be assumed. For the general case it is necessary to use rather elaborate matrix methods that form the basis of commercially available computer software. These methods are discussed in Ref. 7b.

Concentration profiles calculated[6] for a depropanizer operating at 300 lb/in.^{2} absolute are shown in Fig. 22.3. There are 40 stages counting the reboiler and condenser, with feed entering as liquid on stage 20. The reflux ratio is 2.62, which is $1.25 R_{Dm}$. The concentration profiles are characteristic of systems with components both lighter and heavier than the keys. The maxima shown for the light key and the heavy key and the shape of the other profiles can be better understood by examining the operating lines and the equilibrium relationships for individual components on a yx diagram.

In the upper section of the column, L/V is nearly constant, and the operating line for ethane is

$$y_{n+1} = 0.724 x_n + 0.061$$

The equilibrium relationship $y = Kx$ is shown in Fig. 22.4a as a family of straight lines, with the slope increasing as n increases. Each of the straight lines is used only once for the appropriate plate number. Starting from $x_D = 0.222$, only a few plates are needed to reduce x to about 0.05, which results in a "pinch." The pinch shifts to lower values of x as the temperature increases and K increases, but the change from plate to plate is very small, as shown in Fig. 22.3. At the feed plate, the calculation

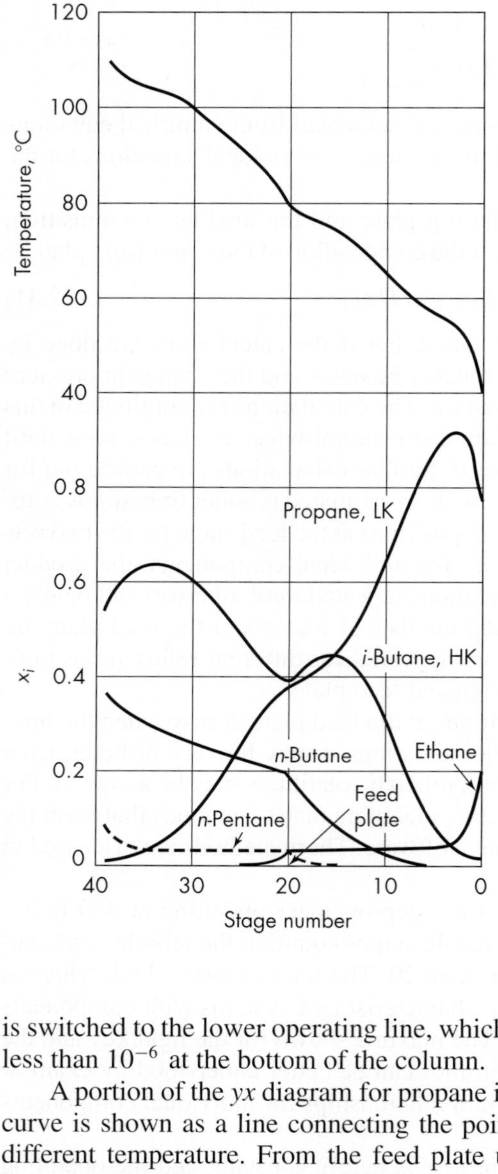

FIGURE 22.3
Temperature and concentration profiles for a depropanizer.

is switched to the lower operating line, which gives a rapid decrease in x and values less than 10^{-6} at the bottom of the column.

A portion of the yx diagram for propane is shown in Fig. 22.4b. The equilibrium curve is shown as a line connecting the points for the individual plates, each at a different temperature. From the feed plate up to plate 6, the temperature is high enough so that K exceeds 1.0, and the vapor is richer in propane than the liquid. The enrichment from plate to plate is nearly constant in this region. For plate 5 and above, K is less than 1.0, and the increase in x and y per stage becomes smaller. At plate 3, the equilibrium curve intersects the operating line, which in a binary mixture would mean no further change in concentration. However, in this multicomponent system, the further decrease in temperature puts the equilibrium curve below the operating line, and more steps can be made between the two lines, which brings x back to lower values and gives a product with 77 percent propane. The top plates of the column thus serve to enrich the distillate in ethane, mainly by reducing the amount of propane. It would not be possible to get 90 percent propane just by eliminating these few plates, since

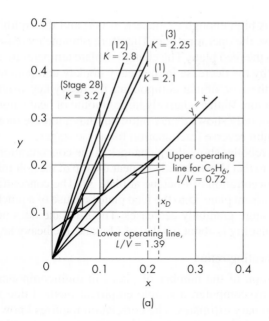

(a)

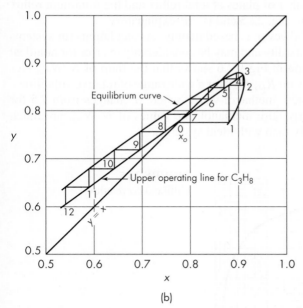

(b)

FIGURE 22.4
Operating-line diagrams for
individual components in a
depropanizer; (*a*) for C_2H_6;
(*b*) for C_3H_8.

all the ethane ends up in the distillate product, and the peak propane concentration would just shift a few plates closer to the feed. It would be possible to take a side-stream product a few plates from the top that would be richer in propane, but for pure products, it would be better to send the crude propane to a deethanizer column.

The concentration profile for the heavy key *i*-butane seems normal from the top of the column to plate 15. In this region K is less than 0.7, and the operating line lies above the equilibrium curve, so x decreases going up the column. From the feed plate to plate 15 the temperature is high enough to make $K > 0.7$, and the equilibrium line shifts to give values above the operating line. Therefore i-butane increases in concentration on going up from the feed plate. The change in plate temperature in this region is strongly influenced by the decrease in the heavy nonkey components n-butane and n-pentane. Without these or similar components the heavy key would not show any maximum concentration. With a relatively large amount of heavy nonkey components, the heavy key not only shows a maximum but may increase more rapidly than the light key and exhibit reverse fractionation for a few stages.

The first few plates above the reboiler show sharp changes in the concentrations of n-pentane and n-butane, similar to that shown by ethane near the top of the column, and this results in a maximum concentration of i-butane at plate 34. The concentration of n-pentane is nearly constant from plate 35 to the feed plate because of a pinch, and changes slowly as the temperature gradually decreases. The n-butane does not show such a plateau because its volatility is about 0.8 that of i-butane, the heavy key.

Number of ideal plates at operating reflux

Although the precise calculation of the number of plates in multicomponent distillation is best accomplished by computer, a simple empirical method due to Gilliland[1] is much used for preliminary estimates. The correlation requires knowledge only of the minimum number of plates at total reflux and the minimum reflux ratio. The correlation is given in Fig. 22.5 and is self-explanatory.

The Gilliland correlation, however, is based mainly on calculations for systems with nearly constant relative volatility and may be considerably in error for nonideal systems. An alternate correlation in Fig. 22.6 shows that the ratio of N/N_{min} (for binary systems) depends mainly on R_D/R_{Dm} for a wide range of relative volatilities. On the other hand, for the system methanol-water, where α changes from 7.5 for dilute solutions to 2.7 for nearly pure methanol, the values of N/N_{min} are much greater and change more rapidly than with ideal systems.

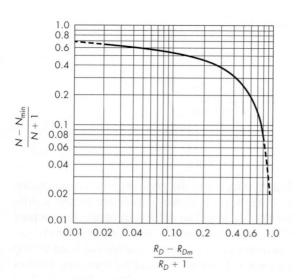

FIGURE 22.5
Gilliland correlation.

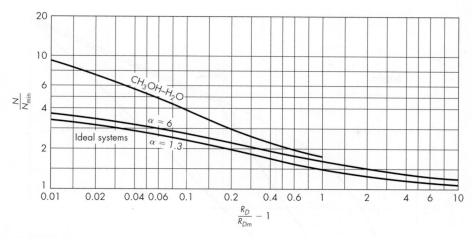

FIGURE 22.6
Alternate correlation for $N/N_{\min}$.

The McCabe–Thiele diagram in Fig. 22.7 shows why this is so. The upper part of the equilibrium line is virtually linear and parallel to the operating line, and many plates are needed in the rectifying section of the column. Similar situations can arise in multicomponent distillation, where changes in activity coefficients with concentration may result in a small driving force in a considerable portion of the column.

EXAMPLE 22.5. Estimate the number of ideal plates required for the separation specified in Example 22.3 if the reflux ratio is $1.5R_{Dm}$.

Solution. From Example 22.3, the minimum number of ideal stages is 9.4 plus a reboiler, or 10.4. The value of R_{Dm} is obtained by the Underwood method.

		x_F	x_D	K	α
LK	n-Hexane	0.33	0.99	2.23	2.21
HK	n-Heptane	0.37	0.01	1.01	1.0
	n-Octane	0.30	0	0.462	0.457

For a liquid feed, $q = 1$,

$$\sum \frac{\alpha_i x_{Fi}}{\alpha_i - \phi} = 0$$

By trial, $\phi = 1.45$:

$$R_{Dm} + 1 = \sum \frac{\alpha_i x_{Di}}{\alpha_i - 1.45} = \frac{2.21(0.99)}{2.21 - 1.45} + \frac{1.0(0.01)}{1 - 1.45} = 2.86$$

$$R_{Dm} = 1.86$$

$$R_D = 1.5 \times 1.86 = 2.79$$

$$\frac{R_D - R_{Dm}}{R_D + 1} = \frac{2.79 - 1.86}{3.79} = 0.245$$

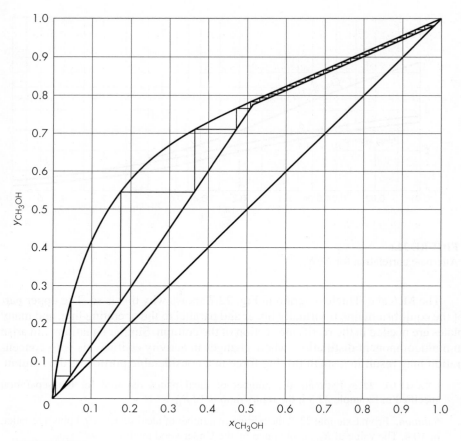

FIGURE 22.7
McCabe–Thiele diagram for methanol-water at $R_D = 1.1 R_{dm}$.

From Fig. 22.5,

$$\frac{N - N_{min}}{N + 1} = 0.41$$

$$N - 10.4 = 0.41N + 0.41$$

$$N = \frac{10.81}{0.59} = 18.3 \text{ stages}$$

From Fig. 22.6, at $R_D/R_{Dm} - 1 = 0.5$, $N/N_{min} = 1.8$

$$N = 10.4(1.8) = 18.7 \text{ stages}$$

AZEOTROPIC AND EXTRACTIVE DISTILLATION

The separation of components that have nearly the same boiling points is difficult by simple distillation even if the mixtures are ideal, and complete separation may be impossible because of azeotrope formation. For such systems the separation can often be improved by adding a third component to alter the relative volatility of the original components. The added component may be a higher-boiling liquid or "solvent" that is miscible with both of the key components but is chemically more similar to one of them. The key component that is more like the solvent will have a lower activity coefficient in the solution than the other component, so the separation is enhanced. This process is called *extractive distillation* and is like liquid-liquid extraction with an added vapor phase.

One example of extractive distillation is the use of furfural to permit the separation of butadiene from a mixture containing butane and butenes. Furfural, which is a highly polar solvent, lowers the activity of butadiene more than it does for butenes or butane, and the butadiene is concentrated in the furfural-rich stream from the bottom of the column. Butadiene is distilled from the furfural, which is returned to the top of the extractive distillation column. This column would operate with a reflux containing butane and butenes, but the total liquid rate in the top section of the column would be the reflux rate plus the flow rate of furfural.

Separation of the original mixture may also be enhanced by adding a solvent that forms an azeotrope with one of the key components. This process is called *azeotropic distillation*. The azeotrope forms the distillate or bottoms product from the column and is later separated into solvent and key component. Usually the material added forms a low-boiling azeotrope and is taken overhead, and such materials are called *entrainers*. The azeotrope will of course contain some of all components in the feed, but it will have a much different ratio of the keys than the feed.

An example of azeotropic distillation is the use of benzene, heptane, or cyclohexane to permit the separation of ethanol and water, which form a minimum-boiling azeotrope with 95.6 weight percent alcohol. The alcohol-water mixture with about 95 percent alcohol is fed near the middle of the dehydrating column, and nearly pure alcohol is removed as the bottoms product. The overhead vapor is a ternary azeotrope, which is condensed and separated into two phases. The upper organic layer is returned to the top of the dehydrating column as reflux, and the water layer is sent to a stripping column, where alcohol and the hydrocarbon entrainer are taken overhead and returned to the first column. The stripper bottom is an aqueous stream that is discharged as wastewater or sent to a third column to recover some of the alcohol.

SYMBOLS

B	Flow rate of heavy or bottom product, mol/h
C_p	Molar heat capacity, cal/g mol·°C; $\bar{C}_p$, average value
D	Flow rate of light or overhead product, mol/h
F	Feed rate, mol/h
f	Fraction of feed that is vaporized; f_i, of component i
H	Enthalpy of stream, cal/g mol or Btu/lb mol
K	Equilibrium ratio, y_e/x_e; K_i, K_j, of components i and j; K_{ref}, of reference component; K_∞, for an infinite number of plates; $\bar{K}$, stripping section

L Flow rate of liquid, in general or in rectifying section, mol/h; L_{min}, minimum flow rate [Eq. (22.15)]; L_n, on plate n L_∞, for an infinite number of plates; $\bar{L}$, in stripping section

m Plate number in stripping section

N Number of ideal plates; N_{min}, minimum number at total reflux

N_c Number of components

n Plate number in rectifying section

P Total pressure, atm or lb_f/ft^2; P', vapor pressure; P'_i, P'_j, of components i and j

p Partial pressure, atm or lb_f/ft^2; p_i, of component i

q Moles of liquid to stripping section per mole of feed

R_D Reflux ratio, L/D; R_{Dm}, minimum value

T Temperature, °C or °F; T_{abs}, absolute temperature, K or °R; T_0, preheat temperature (Example 22.2b)

V Flow rate of vapor, in general or in rectifying section, mol/h; V_{min}, at minimum reflux ratio [Eq. (22.30)]; V_{n+1}, from plate $n+1$; V_∞, for an infinite number of plates; $\bar{V}$, in stripping section

x Mole fraction of component in liquid phase; x_B, in bottoms; x_D, in overhead; x_{DA}, x_{DB}, of components A and B in overhead; x_F, in feed; x_H, of heavy key; x_L, of light key; x_e, of liquid in equilibrium with vapor of composition y_e; x_i, of component i; x_{ie}, equilibrium value for component i; x_j, of component j; x_m, x_n, on plates m and n

y Mole fraction of component in vapor phase; y_D, in overhead; y_e, of vapor in equilibrium with liquid of composition x_e; y_i, y_j, of components i and j; y_{ie}, equilibrium value for component i; y_m, y_n, from plates m and n; y_{ni}, of component i from plate n; $y_{\infty i}$, for an infinite number of plates

Greek Letters

α Relative volatility, dimensionless; α_{AB}, of component A relative to component B in binary system; α_B, α_D, α_F, in bottoms, overhead, and feed, respectively; $\alpha_{LK,HK}$, of light key relative to heavy key; α_i, of component i, defined as K_i/K_{ref}; α_{ij}, of component i relative to component j; $\bar{\alpha}_{ij}$, average value, defined by Eq. (22.14)

ΔH_v Heat of vaporization, cal/mol

ϕ Root of Eq. (22.29)

PROBLEMS

22.1. The feed to a conventional distillation column and the relative volatilities are shown in Table 22.2. The recovery of component 2 in the distillate is 99 percent, and 98 percent of component 3 is to leave in the bottoms. Calculate the minimum number of plates.

TABLE 22.2

Component	x_{Fi}	α_i
1	0.05	2.1
2	0.42	1.7
3	0.46	1.0
4	0.07	0.65

TABLE 22.3

	Component	x_{Fi}	α_i
	1	0.06	2.6
LK	2	0.40	1.9
	3	0.05	1.5
HK	4	0.42	1.0
	5	0.07	0.6

TABLE 22.4

	Component	Boiling point, °C	x_{Fi}	α_i	% recovery in D
	Ethyl benzene	136.2	0.054	1.23	
	p-Xylene	138.5	0.221	1.15	
LK	m-Xylene	139.1	0.488	1.13	99.0
HK	o-Xylene	144.4	0.212	1.0	3.8
	n-Propyl benzene	159.3	0.025	0.70	

22.2. For the conditions of Prob. 22.1, estimate the minimum reflux ratio if the feed is liquid at the bubble point. About how many plates would be required at a reflux ratio 1.3 times the minimum?

22.3. The feed to a distillation column operating at 270 lb$_f$/in.^{2} absolute contains 10% ethane, 45% propane, 30% i-butane, and 15% n-butane. Calculate the bubble point of the feed and the fraction vaporized when the liquid enters the column if the feed is heated as a liquid 20°C above the bubble point.

22.4. A five-component mixture is to be distilled with 99 percent recovery of the light and heavy keys in the distillate and bottoms (Table 22.3). Calculate the product compositions for the case of infinite reflux. Explain how these concentrations would shift as the reflux ratio was decreased, using the compositions at minimum reflux as a guide.

22.5. A mixture of xylenes plus other aromatics is separated in a large fractionating column operating at atmospheric pressure. Calculate the minimum number of plates and the minimum reflux ratio for the conditions in Table 22.4. Use the Gilliland correlation to estimate the reflux ratio that will permit the separation to occur in 100 ideal stages. The relative volatilities are calculated for 18 lb$_f$/in.^{2} absolute and 150°C, the estimated conditions near the feed tray.

22.6. A mixture of 30% benzene, 25% toluene, and 45% ethyl benzene is to be separated by distillation at atmospheric pressure, with 98 percent of the benzene and only 1 percent of the toluene in the distillate. (*a*) Calculate the minimum number of ideal plates and the approximate compositions of the products. (*b*) Estimate the concentration of ethyl benzene in the distillate. (*c*) If the distillation were carried out at 0.2 atm, would there be any change in N_{min} or in the heat duty?

REFERENCES

1. Gilliland, E. R. *Ind. Eng. Chem.* **32:**110 (1940).
2. King, C. J. *Separation Processes,* 2nd ed. New York: McGraw-Hill, 1980, p. 416.
3. Lewis, W. L., and G. L. Matheson. *Ind. Eng. Chem.* **24:**494 (1932).
4. Shiras, R. N., D. N. Hanson, and C. H. Gibson. *Ind. Eng. Chem.* **42:**871 (1950).
5. Underwood, A. J. V. *Chem. Eng. Prog.* **44:**603 (1948).
6. Vorhis, F. H. Chevron Research Co., private communication, 1983.
7. Wankat, P. C. *Equilibrium Staged Separations.* New York: Elsevier, 1988; (*a*) p. 243, (*b*) pp. 251–63.

CHAPTER 23

Leaching and Extraction

This chapter discusses the methods of removing one constituent from a solid or liquid by means of a liquid solvent. These techniques fall into two categories. The first, called *leaching* or *solid extraction,* is used to dissolve soluble matter from its mixture with an insoluble solid. The second, called *liquid extraction,* is used to separate two miscible liquids by the use of a solvent that preferentially dissolves one of them. Although the two processes have certain common fundamentals, the differences in equipment and, to some extent, in theory are sufficient to justify separate treatment.

LEACHING

Leaching differs very little from the washing of filtered solids, as discussed in Chap. 29, and leaching equipment strongly resembles the washing section of various filters. In leaching, the amount of soluble material removed is often greater than in ordinary filtration washing, and the properties of the solids may change considerably during the leaching operation. Coarse, hard, or granular feed solids may disintegrate into pulp or mush when their content of soluble material is removed.

Leaching Equipment

When the solids form an open, permeable mass throughout the leaching operation, solvent may be percolated through an unagitated bed of solids. With impermeable solids or materials that disintegrate during leaching, the solids are dispersed into the solvent and are later separated from it. Both methods may be either batch or continuous.

Leaching by percolation through stationary solid beds

Stationary solid-bed leaching is done in a tank with a perforated false bottom to support the solids and permit drainage of the solvent. Solids are loaded into the tank, sprayed with solvent until their solute content is reduced to the economical minimum, and excavated. In some cases the rate of solution is so rapid that one passage of solvent through the material is sufficient, but countercurrent flow of solvent through a battery of tanks is more common. In this method, fresh solvent is fed to the tank containing the solid that is most nearly extracted; it flows through the several tanks in series and is finally withdrawn from the tank that has been freshly charged. Such a series of tanks is called an *extraction battery*. The solid in any one tank is stationary until it is completely extracted. The piping is arranged so that fresh solvent can be introduced to any tank and strong solution withdrawn from any tank, making it possible to charge and discharge one tank at a time. The other tanks in the battery are kept in countercurrent operation by advancing the inlet and drawoff tanks one at a time as the material is charged and removed. Such a process is sometimes called a *Shanks process*.

In some solid-bed leaching the solvent is volatile, necessitating the use of closed vessels operated under pressure. Pressure is also needed to force solvent through beds of some less permeable solids. A series of such pressure tanks operated with countercurrent solvent flow is known as a *diffusion battery*.

Moving-bed leaching[5]

In the machines illustrated in Fig. 23.1 the solids are moved through the solvent with little or no agitation. The Bollman extractor (Fig. 23.1*a*) contains a bucket elevator in a closed casing. There are perforations in the bottom of each bucket. At the top right-hand corner of the machine, as shown in the drawing, the buckets are loaded with flaky solids such as soybeans and are sprayed with appropriate amounts of *half miscella* as they travel downward. Half miscella is the intermediate solvent containing some extracted oil and some small solid particles. As solids and solvent flow concurrently down the right-hand side of the machine, the solvent extracts more oil from the beans. Simultaneously the fine solids are filtered out of the solvent, so that clean *full miscella* can be pumped from the right-hand sump at the bottom of the casing. As the partially extracted beans rise through the left side of the machine, a stream of pure solvent percolates countercurrently through them. It collects in the left-hand sump and is pumped to the half-miscella storage tank. Fully extracted beans are dumped from the buckets at the top of the elevator into a hopper from which they are removed by paddle conveyors. The capacity of typical units is 50 to 500 tons of beans per 24-h day.

In the Rotocel extractor, illustrated in Fig. 23.1*b*, a horizontal basket is divided into walled compartments with a floor that is permeable to the liquid. The basket rotates slowly about a vertical axis. Solids are admitted to each compartment at the feed point; the compartments then successively pass a number of solvent sprays, a drainage section, and a discharge point at which the floor of the compartment opens to discharge the extracted solids. The empty compartment moves to the feed point to receive its next load of solids. To give countercurrent extraction, fresh solvent is fed only to the last compartment before the discharge point, and

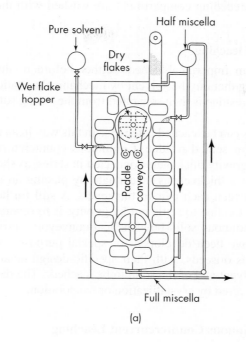

Pure solvent

Half miscella

Dry flakes

Wet flake hopper

Paddle conveyor

Full miscella

(a)

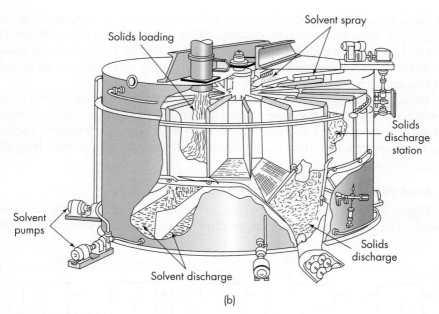

Solids loading

Solvent spray

Solids discharge station

Solids discharge

Solvent pumps

Solvent discharge

(b)

FIGURE 23.1
Moving-bed leaching equipment: (*a*) Bollman extractor; (*b*) Rotocel extractor[11b] (*by permission of McGraw-Hill, Inc.*).

the solids in each preceding compartment are washed with the effluent from the succeeding one.

Dispersed-solid leaching

Solids that form impermeable beds, either before or during leaching, are treated by dispersing them in the solvent by mechanical agitation in a tank or flow mixer. The leached residue is then separated from the strong solution by settling or filtration.

Small quantities can be leached batchwise in this way in an agitated vessel with a bottom drawoff for settled residue. Continuous countercurrent leaching is obtained with several gravity thickeners connected in series, as shown in Fig. 23.2, or when the contact in a thickener is inadequate by placing an agitator tank in the equipment train between each pair of thickeners. A still further refinement, used when the solids are too fine to settle out by gravity, is to separate the residue from the miscella in continuous solid-bowl helical-conveyor centrifuges. Many other leaching devices have been developed for special purposes, such as the solvent extraction of various oilseeds, with their specific design details governed by the properties of the solvent and of the solid to be leached.[5] The dissolved material, or solute, is often recovered by crystallization or evaporation.

Principles of Continuous Countercurrent Leaching

The most important method of leaching is the continuous countercurrent method using stages. Even in an extraction battery, where the solid is not moved physically from stage to stage, the charge in any one cell is treated by a succession of liquids of constantly decreasing concentration as if it were being moved from stage to stage in a countercurrent system.

Because of its importance, only the continuous countercurrent method is discussed here. Also, since the stage method is normally used, the differential-contact method is not considered. In common with other stage cascade operations, leaching may be considered, first, from the standpoint of ideal stages and, second, from that of stage efficiencies.

Ideal stages in countercurrent leaching

Figure 23.2 shows a material-balance diagram for a continuous countercurrent cascade. The stages are numbered in the direction of flow of the solid. The V phase

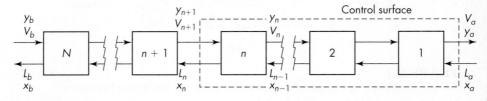

FIGURE 23.2
Countercurrent leaching cascade.

is the liquid that overflows from stage to stage in a direction counter to that of the flow of the solid, dissolving solute as it moves from stage N to stage 1. The L phase is the solid flowing from stage 1 to stage N. Exhausted solids leave stage N, and concentrated solution overflows from stage 1.

It is assumed that the solute-free solid is insoluble in the solvent and that the flow rate of this solid is constant throughout the cascade. The solid is porous and inert (no adsorption) and carries with it an amount of solution that may or may not be constant. Let L refer to the flow of this retained liquid and V to the flow of the overflow solution. The flows V and L may be expressed in mass per unit time or may be based on a definite flow of dry solute-free solid. Also, in accordance with standard nomenclature, the terminal concentrations are as follows:

Solution retained by entering solid x_a
Solution retained by leaving solid x_b
Fresh solvent entering the system y_b
Concentrated solution leaving the system y_a

As in absorption and distillation, the quantitative performance of a countercurrent system can be analyzed by utilizing an equilibrium line and an operating line, and as before, the method to be used depends on whether these lines are straight or curved.

Equilibrium

In leaching, provided sufficient solvent is present to dissolve all the solute in the entering solid and there is no adsorption of solute by the solid, equilibrium is attained when the solute is completely dissolved and the concentration of the solution so formed is uniform. Such a condition may be obtained simply or with difficulty, depending on the structure of the solid. These factors are considered when stage efficiency is discussed. At present, it is assumed that the requirements for equilibrium are met. Then the concentration of the liquid retained by the solid leaving any stage is the same as that of the liquid overflow from the same stage. The equilibrium relationship is simply $x_e = y$.

Operating line

The equation for the operating line is obtained by writing material balances for that portion of the cascade consisting of the first n units, as shown by the control surface indicated by the dashed lines in Fig. 23.2. These balances are

Total solution:
$$V_{n+1} + L_a = V_a + L_n \tag{23.1}$$

Solute:
$$V_{n+1}y_{n+1} + L_a x_a = L_n x_n + V_a y_a \tag{23.2}$$

Solving for y_{n+1} gives the operating-line equation, which is the same as that derived earlier for the general case of an equilibrium-stage cascade [Eq. (20.7)]:

$$y_{n+1} = \left(\frac{L_n}{V_{n+1}}\right) x_n + \frac{V_a y_a - L_a x_a}{V_{n+1}} \tag{23.3}$$

As usual, the operating line passes through the points (x_a, y_a) and (x_b, y_b), and if the flow rates are constant, the slope is L/V.

Constant and variable underflow

Two cases are to be considered. If the density and viscosity of the solution change considerably with solute concentration, the solids from the lower-numbered stages may retain more liquid than those from the higher-numbered stages. Then, as shown by Eq. (23.3), the slope of the operating line varies from unit to unit. If, however, the mass of solution retained by the solid is independent of concentration, then L_n is constant and the operating line is straight. This condition is called *constant solution underflow*. If the underflow is constant, so is the overflow. Constant underflow and variable underflow are given separate consideration.

Number of ideal stages for constant underflow

When the operating line is straight, a McCabe–Thiele construction can be used to determine the number of ideal stages; but since in leaching the equilibrium line is always straight, Eq. (20.24) can be used directly for constant underflow. The use of this equation is especially simple here because $y_a^* = x_a$ and $y_b^* = x_b$.

Equation (20.24) cannot be used for the entire cascade if L_a, the solution entering with the unextracted solids, differs from L, the underflows within the system. Equations have been derived for this situation,[1,7] but it is easy to calculate, by material balances, the performance of the first stage separately and then to apply Eq. (20.24) to the remaining stages.

Number of ideal stages for variable underflow

When the underflow and overflow vary from stage to stage, a modification of the McCabe–Thiele graphical method may be used for calculations. The terminal points on the operating line are determined using material balances. Assuming the amount of underflow L is known as a function of underflow composition, an intermediate value of x_n is chosen to fix L_n, and V_{n+1} is calculated from Eq. (23.1). The composition of the overflow y_{n+1} is then calculated from Eq. (23.2), and the point (x_n, y_{n+1}) is plotted along with the terminal compositions to give the curved operating line. Unless there is a large change in L and V or the operating line is very close to the equilibrium line, only one intermediate point need be calculated.

EXAMPLE 23.1. Oil is to be extracted from meal by means of benzene using a continuous countercurrent extractor. The unit is to treat 1,000 kg of meal (based on completely exhausted solid) per hour. The untreated meal contains 400 kg of oil and is contaminated with 25 kg of benzene. The fresh solvent mixture contains 10 kg of oil and 655 kg of benzene. The exhausted solids are to contain 60 kg of unextracted oil. Experiments carried out under conditions identical with those of the projected battery show that the solution retained depends on the concentration of the solution, as shown in Table 23.1. Find (*a*) the concentration of the strong solution, or extract; (*b*) the concentration of the solution adhering to the extracted solids; (*c*) the mass of solution leaving with the extracted meal; (*d*) the mass of extract; (*e*) the number of stages required. All quantities are given on an hourly basis.

TABLE 23.1
Data for Example 23.1

Concentration, kg oil/kg solution	Solution retained, kg/kg solid	Concentration, kg oil/kg solution	Solution retained, kg/kg solid
0.0	0.500	0.4	0.550
0.1	0.505	0.5	0.571
0.2	0.515	0.6	0.595
0.3	0.530	0.7	0.620

Solution. Let x and y be the mass fractions of oil in the underflow and overflow solutions. *At the solvent inlet,*

$$V_b = 10 + 655 = 665 \text{ kg solution/h}$$

$$y_b = \frac{10}{665} = 0.015$$

Determine the amount and composition of the solution in the spent solids by trial. If $x_b = 0.1$, the solution retained, from Table 23.1, is 0.505 kg/kg. Then

$$L_b = 0.505(1{,}000) = 505 \text{ kg/h}$$

$$x_b = \frac{60}{505} = 0.119$$

From Table 23.1, the solution retained is 0.507 kg/kg:

$$L_b = 0.507(1{,}000) = 507$$

$$x_b = \frac{60}{507} = 0.118 \text{ (close enough)}$$

Benzene in the underflow at L_b is $507 - 60 = 447$ kg/h. *At the solid inlet,*

$$L_a = 400 + 25 = 425 \text{ kg solution/h}$$

$$x_a = \frac{400}{425} = 0.941$$

Oil in extract = oil in $-60 = 10 + 400 - 60 = 350$ kg/h. Benzene in extract $= 655 + 25 - 447 = 233$ kg/h.

$$V_a = 350 + 233 = 583 \text{ kg/h}$$

$$y_a = \frac{350}{583} = 0.600$$

The answers to parts (a) to (d) are

    (a) $y_a = 0.60$
    (b) $x_b = 0.118$
    (c) $L_b = 507$ kg/h
    (d) $V_a = 583$ kg/h

(e) Determine the inlet and exit concentrations for the first stage and locate the operating line for the remaining stages. Since $x_1 = y_a = 0.60$, solution retained is 0.595 kg/kg solid.

$$L_1 = 0.505(1,000) = 595$$

Overall material balance:

$$V_2 = L_1 + V_a - V_2 = 595 + 583 - 425 = 753 \text{ kg/h}$$

Oil balance:

$$L_a x_a + V_2 y_2 = L_1 x_1 + V_a y_a$$

$$V_2 y_2 = 595(0.60) + 583(0.60) - 425(0.941) = 307$$

$$y_2 = \frac{307}{753} = 0.408$$

The point $x_1 = 0.60$, $y_2 = 0.408$ is at one end of the operating line for the remaining stages. To determine an intermediate point on the operating line, choose $x_n = 0.30$.

$$L_n = \text{solution retained} = 0.53(1,000) = 530 \text{ kg/h}$$

By an overall balance,

$$V_{n+1} = 530 + 583 - 425 = 688 \text{ kg/h}$$

An oil balance gives

$$V_{n+1} y_{n+1} = L_n x_n + V_a y_a - L_a x_a$$

$$= 530(0.30) + 583(0.60) - 400 = 108.8$$

$$y_{n+1} = \frac{108.8}{688} = 0.158$$

The points x_n, y_{n+1}, x_a, y_a, and x_b, y_b define a slightly curved operating line, as shown in Fig. 23.3. Four ideal stages are required.

Saturated concentrated solution

A special case of leaching is encountered when the solute is of limited solubility and the concentrated solution reaches saturation. This situation can be treated by the above methods.[6] The solvent input to stage N should be the maximum that is consistent with a saturated overflow from stage 1, and all liquids except that adhering to the underflow from stage 1 should be unsaturated. If too little solvent is used and saturation is attained in stages other than the first, all but one of the "saturated" stages are ineffective, and the solute concentration in the underflow from stage N is higher than it would be if more fresh solvent were used.

Stage efficiencies

In most leaching operations the solute is distributed through a more or less permeable solid. The rate of leaching is largely governed by the rate of diffusion through the solid, as discussed in Chap. 17, and the actual number of stages required

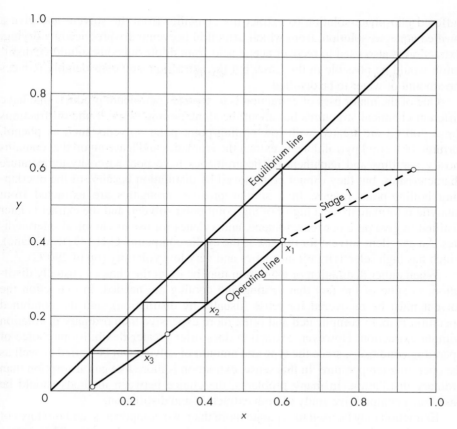

FIGURE 23.3
McCabe–Thiele diagram for leaching (Example 23.1).

may be considerably greater than the number of ideal stages. In washing impermeable solids, where the solute is confined to a film of strong solution on the solid
surfaces, the approach to equilibrium is rapid, and the stage efficiency may be taken
as unity.

LIQUID EXTRACTION

When separation by distillation is ineffective or very difficult, liquid extraction is
one of the main alternatives to consider. Close-boiling mixtures or substances that
cannot withstand the temperature of distillation, even under a vacuum, may often be
separated from impurities by extraction, which utilizes chemical differences instead
of vapor pressure differences. For example, penicillin is recovered from the fermentation broth by extraction with a solvent such as butyl acetate, after lowering
the pH to get a favorable partition coefficient. The solvent is then treated with a

buffered phosphate solution to extract the penicillin from the solvent and give a purified aqueous solution, from which penicillin is eventually produced by drying. Extraction is also used to recover acetic acid from dilute aqueous solutions; distillation would be possible in this case, but the extraction step considerably reduces the amount of water to be distilled.

One of the major uses of extraction is to separate petroleum products that have different chemical structures but about the same boiling range. Lube oil fractions (bp > 300°C) are treated with low-boiling-point polar solvents such as phenol, furfural, or methyl pyrrolidone to extract the aromatics and leave an oil that contains mostly paraffins and naphthenes. The aromatics have poor viscosity-temperature characteristics, but they cannot be removed by distillation because of the overlapping boiling-point ranges. In a similar process, aromatics are extracted from catalytic reformate using a high-boiling-point polar solvent, and the extract is later distilled to give pure benzene, toluene, and xylenes for use as chemical intermediates. An excellent solvent for this use is the cyclic compound $C_4H_8SO_2$ (Sulfolane), which has high selectivity for aromatics and very low volatility (bp of 290°C).

When either distillation or extraction may be used, the choice is usually distillation, in spite of the fact that heating and cooling are needed. In extraction the solvent must be recovered for reuse (usually by distillation), and the combined operation is more complicated and often more expensive than ordinary distillation without extraction. However, extraction does offer greater flexibility in choice of operating conditions, since the type and amount of solvent can be varied as well as the operating temperature. In this sense, extraction is more like gas absorption than ordinary distillation. In many problems, the choice between methods should be based on a comparative study of both extraction and distillation.

Extraction may be used to separate more than two components; and mixtures of solvents, instead of a single solvent, are needed in some applications. These more complicated methods are not treated in this text.

Extraction Equipment[11a]

In liquid-liquid extraction, as in gas absorption and distillation, two phases must be brought into good contact to permit transfer of material and then separated. In absorption and distillation, the mixing and separation are easy and rapid. In extraction, however, the two phases have comparable densities, so that the energy available for mixing and separation—if gravity flow is used—is small, much smaller than when one phase is a liquid and the other is a gas. The two phases are often hard to mix and harder to separate. The viscosities of both phases also are relatively high, and linear velocities through most extraction equipment are low. In some types of extractors, therefore, energy for mixing and separation is supplied mechanically.

Extraction equipment may be operated batchwise or continuously. A quantity of feed liquid may be mixed with a quantity of solvent in an agitated vessel, after which the layers are settled and separated. The extract is the layer of solvent plus extracted solute, and the raffinate is the layer from which solute has been removed.

The extract may be lighter or heavier than the raffinate, and so the extract may be shown coming from the top of the equipment in some cases and from the bottom in others. The operation may, of course, be repeated if more than one contact is required; but when the quantities involved are large and several contacts are needed, continuous flow becomes economical. Most extraction equipment is continuous with either successive stage contacts or differential contacts. Representative types are mixer-settlers, vertical towers of various kinds that operate by gravity flow, agitated tower extractors, and centrifugal extractors. The characteristics of various types of extraction equipment are listed in Table 23.2. Liquid-liquid extraction can also be carried out using porous membranes as described in Chap. 26. This method has promise for difficult separations.

Mixer-settlers

For batchwise extraction the mixer and settler may be the same unit. A tank containing a turbine or propeller agitator is most common. At the end of the mixing cycle the agitator is shut off, the layers are allowed to separate by gravity, and extract and raffinate are drawn off to separate receivers through a bottom drain line carrying a sight glass. The mixing and settling times required for a given extraction can be determined only by experiment; 5 min for mixing and 10 min for settling are typical, but both shorter and much longer times are common.

For continuous flow the mixer and settler are usually separate pieces of equipment. The mixer may be a small agitated tank provided with inlets and a drawoff line and baffles to prevent short-circuiting, or it may be a motionless mixer or other flow mixer. The settler is often a simple continuous gravity decanter. With liquids that emulsify easily and have nearly the same density it may be necessary to pass

TABLE 23.2
Performance of commercial extraction equipment

Type	Liquid capacity of combined streams, ft³/ft²·h[†]	HTU, ft	Plate or stage efficiency, %	Spacing between plates or stages, in.	Typical applications
Mixer-settler			75–100		Duo-Sol lube-oil process
Spray column	50–250	10–20			Ammonia extraction of salt from caustic soda
Packed column	20–150	5–20			Phenol recovery
Perforated-plate column	10–200	1–20	6–24	30–70	Furfural lube-oil process
Baffle column	60–105	4–6	5–10	4–6	Acetic acid recovery
Agitated tower	50–100	1–2	80–100	12–24	Pharmaceuticals and organic chemicals

[†] ft² is the total cross-sectional area.

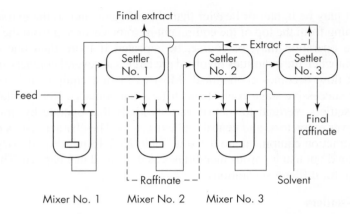

FIGURE 23.4
Mixer-settler extraction system.

the mixer discharge through a screen or pad of glass fiber to coalesce the droplets of the dispersed phase before gravity settling is feasible. For even more difficult separations, tubular or disk-type centrifuges are employed.

If, as is usual, several contact stages are required, a train of mixer-settlers is operated with countercurrent flow, as shown in Fig. 23.4. The raffinate from each settler becomes the feed to the next mixer, where it meets intermediate extract or fresh solvent. The principle is identical with that of the continuous countercurrent stage leaching system shown in Fig. 20.2.

Spray and packed extraction towers

These tower extractors give differential contacts, not stage contacts, and mixing and settling proceed simultaneously and continuously. In the spray tower shown in Fig. 23.5, the lighter liquid is introduced at the bottom and distributed as small drops by the nozzles A. The drops of light liquid rise through the mass of heavier liquid, which flows downward as a continuous stream. The drops are collected at the top and form the stream of light liquid leaving the top of the tower. The heavy liquid leaves the bottom of the tower. In Fig 23.5, light phase is dispersed and heavy phase is continuous. This may be reversed, and the heavy stream sprayed into the light phase at the top of the column, to fall as dispersed phase through a continuous stream of light liquid. The choice of dispersed phase depends on the flow rates, viscosities, and wetting characteristics of both phases and is usually based on experience. The phase with the higher flow rate may be dispersed to give a greater interfacial area, but if there is a significant difference in viscosities, the more viscous phase may be dispersed to give a higher settling rate. Some say that in packed towers the continuous phase should wet the packing, but this need not be true for good performance. Whichever phase is dispersed, the movement of drops through the column constantly brings the liquid in the dispersed phase into fresh contact with the other phase to give the equivalent of a series of mixer-settlers.

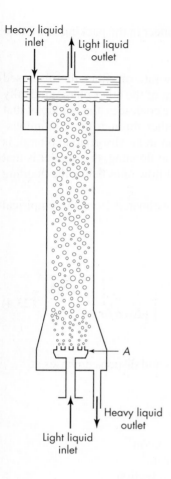

Heavy liquid inlet

Light liquid outlet

A

Heavy liquid outlet

Light liquid inlet

FIGURE 23.5
Spray tower; *A*, nozzle to distribute light liquid.

There is continuous transfer of material between phases, and the composition of each phase changes as it flows through the tower. At any given level, of course, equilibrium is not reached; indeed, it is the departure from equilibrium that provides the driving force for material transfer. The rate of mass transfer is relatively low compared to distillation or absorption, and a tall column may be equivalent to only a few perfect stages.

In actual spray towers, contact between the drops and the continuous phase often appears to be most effective in the region where the drops are formed. This could be due to a higher rate of mass transfer in the newly formed drops or to back-mixing of the continuous phase. In any case, adding height does not give a proportional increase in the number of stages; it is much more effective to redisperse the drops at frequent intervals throughout the tower. This can be done by filling the tower with packing, such as rings or saddles. The packing causes the drops to coalesce and reform and, as shown in Table 23.2, may increase the number of stages in a given height of column. Packed towers approach spray towers in simplicity and can be made to handle almost any problem of corrosion or pressure at a reasonable cost. Their chief disadvantage is that solids tend to collect in the packing and cause channeling.

Flooding velocities in packed towers. If the flow rate of either the dispersed phase or the continuous phase is held constant and that of the other phase gradually increased, a point is reached where the dispersed phase coalesces, the holdup of that phase increases, and finally both phases leave together through the continuous-phase outlet. The effect, like the corresponding action in an absorption column, is called flooding. The larger the flow rate of one phase at flooding, the smaller is that of the other. A column obviously should be operated at flow rates below the flooding point.

Flooding velocities in packed columns can be estimated from the empirical equation[9]

$$\bar{V}_{s,c}(1 + R^{0.5})^2 \left(\frac{a_v}{g}\right)^{0.5}$$

$$= \alpha C_1 \varepsilon^{1.54} \left(\frac{\Delta\rho}{\rho_d}\right)^{0.41} \left[\frac{1}{a_v}\left(\frac{\Delta\rho^2 g}{\mu_c^2}\right)^{1/3}\right]^{0.3} \left[\frac{\mu_c}{(\Delta\rho\,\sigma/a_v)^{0.5}}\right]^{0.15} \quad (23.4)$$

where
$$R = \bar{V}_{s,d}/\bar{V}_{s,c}$$
$\bar{V}_{s,c},\ \bar{V}_{s,d} =$ superficial velocities of continuous and dispersed phases at flooding, respectively, m/s
$\mu_c =$ viscosity of continuous phase, Pa · s
$\sigma =$ interfacial tension between phases, N/m
$\rho_d =$ density of dispersed phase, kg/m³
$\Delta\rho =$ density difference between phases, kg/m³
$a_v =$ specific surface area of packing, m²/m³
$\varepsilon =$ fraction voids or porosity of packed section
$g =$ acceleration of gravity, m/s²
$\alpha =$ 1.0 for continuous-phase wetting, 1.2 for dispersed-phase wetting
$C_1 =$ function of packing type

The function C_1 is 0.28 for Raschig rings or Intalox saddles. Values for other packings range[9] from 0.204 to 0.42. For a given system the term $\bar{V}_{s,c}(1 + R^{0.5})^2$ is constant.

Perforated-plate towers

Redispersion of liquid drops is also done by transverse perforated plates like those in the sieve-plate distillation tower described in Chap. 20. The perforations in an extraction tower are $1\frac{1}{2}$ to $4\frac{1}{2}$ mm in diameter. Plate spacings are 150 to 600 mm (6 to 24 in.). Usually the light liquid is the dispersed phase, and downcomers carry the heavy continuous phase from one plate to the next. As shown in Fig. 23.6a, light liquid collects in a thin layer beneath each plate and jets into the thick layer of heavy liquid above. A modified design is shown in Fig. 23.6b, in which the perforations

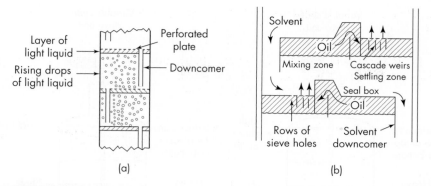

FIGURE 23.6
Perforated-plate extraction towers: (*a*) perforations in horizontal plates;
(*b*) cascade weir tray with mixing and settling zones. (*After Bushell* and
Fiocco.[4])

are on one side of the plate only, alternating from left to right from one plate to the
next. Nearly all the extraction takes place in the mixing zone above the perforations,
with the light liquid (oil) rising and collecting in a space below the next-higher
plate, then flowing transversely over a weir to the next set of perforations. The
continuous-phase heavy liquid (solvent) passes horizontally from the mixing zone
to a settling zone in which any tiny drops of light liquid have a chance to separate
and rise to the plate above. This design often greatly reduces the quantity of oil carried downward by the solvent and increases the effectiveness of the extractor.

Baffle towers

These extraction towers contain sets of horizontal baffle plates. Heavy liquid
flows over the top of each baffle and cascades to the one beneath; light liquid flows
under each baffle and sprays upward from the edge through the heavy phase.
The most common arrangements are disk-and-doughnut baffles and segmental, or
side-to-side, baffles. In both types the spacing between baffles is 100 to 150 mm
(4 to 6 in.).

Baffle towers contain no small holes to clog or be enlarged by corrosion. They
can handle dirty solutions containing suspended solids; one modification of the
disk-and-doughnut towers even contains scrapers to remove deposited solids from
the baffles. Because the flow of liquid is smooth and even, with no sharp changes in
velocity or direction, baffle towers are valuable for liquids that emulsify easily. For
the same reason, however, they are not effective mixers, and each baffle is equivalent to only a 0.05 to 0.1 ideal stage.[15]

Agitated tower extractors

Mixer-settlers supply mechanical energy for mixing the two liquid phases, but
the tower extractors so far described do not. They depend on gravity flow both for
mixing and for separation. In some tower extractors, however, mechanical energy is
provided by internal turbines or other agitators, mounted on a central rotating shaft.

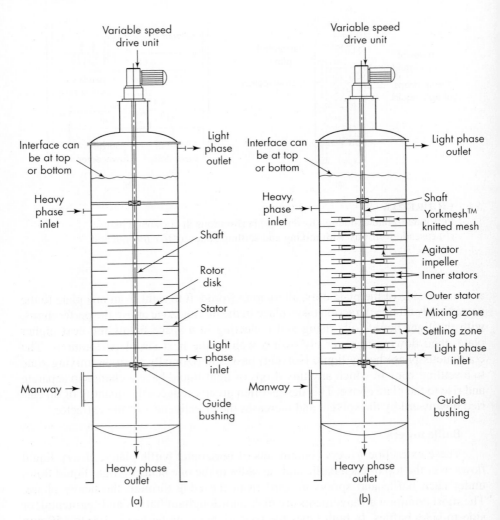

FIGURE 23.7
Agitated extraction towers: (*a*) rotating-disk unit; (*b*) York-Scheibel extractor.

In the rotating-disk contactor shown in Fig. 23.7*a*, flat disks disperse the liquids and impel them outward toward the tower wall, where stator rings create quiet zones in which the two phases can separate. In other designs, sets of impellers are separated by calming sections to give, in effect, a stack of mixer-settlers one above the other. In the York-Scheibel extractor illustrated in Fig. 23.7*b*, the regions surrounding the agitators are packed with wire mesh to encourage coalescence and separation of the phases. Most of the extraction takes place in the mixing sections, but some also occurs in the calming sections, so that the efficiency of each mixer-settler unit is sometimes greater than 100 percent. Typically each mixer-settler is 300 to 600 mm (1 to 2 ft) high, which means that several theoretical contacts can be provided in a reasonably short column. The problem of maintaining the internal moving parts, however, particularly where the liquids are corrosive, may be a serious disadvantage.

Pulse columns

Agitation may also be provided by external means, as in a pulse column. A reciprocating pump "pulses" the entire contents of the column at frequent intervals, so that a rapid reciprocating motion of relatively small amplitude is superimposed on the usual flow of the liquid phases. The tower may contain ordinary packing or special sieve plates. In a packed tower the pulsation disperses the liquids and eliminates channeling, and the contact between the phases is greatly improved. In sieve-plate pulse towers the holes are smaller than in nonpulsing towers, ranging from 1.5 to 3 mm in diameter, with a total open area in each plate of 6 to 23 percent of the cross-sectional area of the tower. Such towers are used almost entirely for processing highly corrosive radioactive liquids. No downcomers are used. Ideally the pulsation causes light liquid to be dispersed into the heavy phase on the upward stroke and the heavy phase to jet into the light phase on the downward stroke. Under these conditions the stage efficiency may reach 70 percent. This is possible, however, only when the volumes of the two phases are nearly the same and when there is almost no volume change during extraction. In the more usual case, the successive dispersions are less effective, and there is backmixing of one phase in one direction. The plate efficiency then drops to about 30 percent. Nevertheless, in both packed and sieve-plate pulse columns, the height required for a given number of theoretical contacts is often less than one-third that required in an unpulsed column.[13]

Centrifugal extractors

The dispersion and separation of the phases may be greatly accelerated by centrifugal force, and several commercial extractors make use of this. In the Podbielniak extractor a perforated spiral ribbon inside a heavy metal casing is wound about a hollow horizontal shaft through which the liquids enter and leave. Light liquid is pumped to the outside of the spiral at a pressure between 3 and 12 atm to overcome the centrifugal force; heavy liquid is fed to the center. The liquids flow countercurrently through the passage formed by the ribbons and the casing walls. Heavy liquid moves outward along the outer face of the spiral; light liquid is forced by displacement to flow inward along the inner face. The high shear at the liquid-liquid interface results in rapid mass transfer. In addition, some liquid sprays through the perforations in the ribbon and increases the turbulence. Up to 20 theoretical contacts may be obtained in a single machine, although 3 to 10 contacts are more common. Centrifugal extractors are expensive and find relatively limited use. They have the advantages of providing many theoretical contacts in a small space and of very short holdup times—about 4 s. Thus they are valuable in the extraction of sensitive products such as vitamins and antibiotics.

Auxiliary equipment

The dispersed phase in an extraction tower is allowed to coalesce at some point into a continuous layer from which one product stream is withdrawn. The interface between this layer and the predominant continuous phase is set in an open section at the top or bottom of a packed tower; in a sieve-plate tower it is set in an open section near the top of the tower when the light phase is dispersed. If the heavy phase is dispersed, the interface is kept near the bottom of the tower. The interface level may be automatically controlled by a vented overflow leg for the heavy phase, as in a continuous gravity decanter. In large columns the interface is often held at the desired point by a level controller actuating a valve in the heavy-liquid discharge line.

In liquid-liquid extraction the solvent must nearly always be removed from the extract or raffinate or both. Thus auxiliary stills, evaporators, heaters, and condensers form an essential part of most extraction systems and often cost much more than the extraction device itself. As mentioned at the beginning of this section, if a given separation can be done by either extraction or distillation, economic considerations usually favor distillation. Extraction provides a solution to problems that cannot be solved by distillation alone but does not usually eliminate the need for distillation or evaporation in some part of the separation system.

Principles of Extraction

Since most continuous extraction methods use countercurrent contacts between two phases, one a light liquid and the other a heavier one, many of the fundamentals of countercurrent gas absorption and of distillation carry over into the study of liquid extraction. Thus questions about ideal stages, stage efficiency, minimum ratio between the two streams, and size of equipment have the same importance in extraction as in distillation.

Extraction of dilute solutions

For batch or multistage extraction of dilute solutions, where changes in flow rate can be neglected and the distribution coefficient[†] K_D is constant, it is convenient to use an extraction factor E which is equivalent to the stripping factor S defined by Eq. (20.29)

$$E \equiv \frac{K_D V}{L} \tag{23.5}$$

For a single-stage extraction with pure solvent the fraction of solute remaining is $1/(1 + E)$ and the fraction recovered is $E/(1 + E)$. The various forms of the Kremser equation [Eqs. (20.24), (20.25), (20.26), and (20.28)] can be used for a countercurrent extraction cascade.

EXAMPLE 23.2. Penicillin F is recovered from a dilute aqueous fermentation broth by extraction with amyl acetate, using 6 volumes of solvent per 100 volumes of the aqueous phase. At pH = 3.2 the distribution coefficient K_D is 80. (a) What fraction of the penicillin would be recovered in a single ideal stage? (b) What would be the recovery with two-stage extraction using fresh solvent in both stages? (c) How many ideal stages

---

[†]The distribution coefficient is the ratio of the equilibrium concentration of the solute in the extract to that in the raffinate.

would be needed to give the same recovery as in part (b) if a counterflow cascade were used with $V/L = 0.06$?

Solution

(a) By a material balance, since $y_0 = 0$,

$$L(x_0 - x_1) = Vy_1$$

$$y_1 = K_D x_1$$

$$x_1 \left(\frac{VK_D}{L} + 1 \right) = x_0$$

The extraction factor is

$$E = \frac{VK_D}{L} = \frac{6 \times 80}{100} = 4.8$$

Thus

$$\frac{x_1}{x_0} = \frac{1}{1 + E} = \frac{1}{5.8} = 0.172$$

Recovery is $1 - 0.172 = 0.828$, or 82.8 percent.

(b) With the same value of E,

$$\frac{x_2}{x_1} = \frac{1}{1 + E}$$

$$\frac{x_2}{x_0} = \frac{1}{(1 + E)^2} = 0.0297$$

Recovery is $1 - 0.0297 = 0.9703$, or 97.0 percent.

(c) With K_D and V/L constant, the number of ideal stages can be calculated from the stripping form of the Kremser equation [Eq. (20.28)], using E in place of its equivalent, the stripping factor S.

$$N = \frac{\ln[(x_a - x_a^*)/(x_b - x_b^*)]}{\ln E}$$

Let $x_a = x_0 = 100$ Then $x_b = 3.0$, and $y_a = 97(100)/6 = 1{,}617$.

$$x_a^* = \frac{y_a}{K_D} = \frac{1{,}617}{80} = 20.2$$

$$x_b^* = 0$$

$$N = \frac{\ln[(100 - 20.2)/3]}{\ln 4.8} = 2.09$$

Using a counterflow process requires only a slightly larger number of ideal stages than in part (b), but uses one-half as much solvent and increases the concentration of the extract.

Extraction of concentrated solutions; phase equilibria

In the extraction of concentrated solutions, the equilibrium relationships are more complicated than in other kinds of separation, because there are three or more components present and some of each component is present in each phase. The

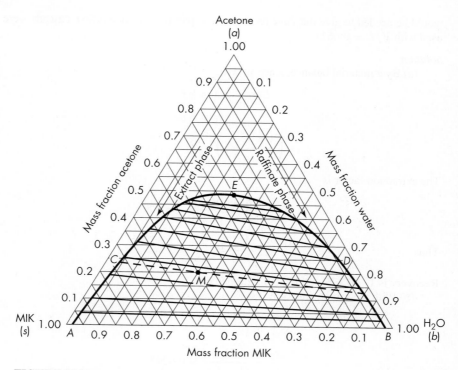

FIGURE 23.8

System acetone-MIK-water at 25°C. (*After Othmer, White, and Trueger.*[10])

equilibrium data are often presented on a triangular diagram, such as those shown in Figs. 23.8 and 23.9. The system acetone–water–methyl isobutyl ketone (MIK), Fig. 23.8, is an example of a type I system, which shows partial miscibility of the solvent (MIK) and the diluent (water) but complete miscibility of the solvent and the component to be extracted (acetone). Aniline–*n*-heptane–methylcyclohexane (MCH) forms a type II system (Fig. 23.9), where the solvent (aniline) is only partially miscible with both the other components.

Some of the features of an extraction process can be illustrated by using Fig. 23.8. When solvent is added to a mixture of acetone and water, the composition of the resulting mixture lies on a straight line between the point for pure solvent and the point for the original binary mixture. When enough solvent is added so that the overall composition falls under the dome-shaped curve, the mixture separates into two phases. The points representing the phase compositions can be joined by a straight tie line, which passes through the overall mixture composition. For clarity, only a few such tie lines are shown, and others can be obtained by interpolation. The line *ACE* shows compositions of the MIK layer (extract), and line *BDE* shows compositions of the water layer (raffinate). As the overall acetone content of the mixture increases, the compositions of the two phases approach each other, and they become equal at point *E*, the *plait point*.

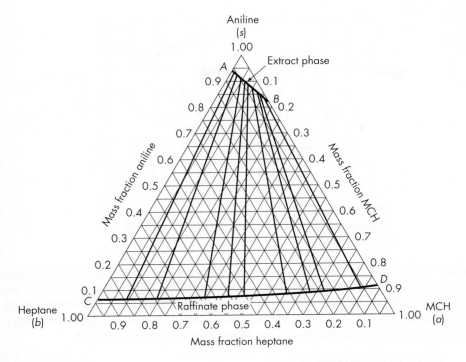

FIGURE 23.9
System aniline–*n*-heptane–MCH at 25°C: *a*, solute, MCH: *b*, diluent, *n*-heptane;
s, solvent, aniline. (*After Varteressian and Fenske.*[16])

The tie lines in Fig. 23.8 slope up to the left, and the extract phase is richer in acetone than the raffinate phase. This suggests that most of the acetone could be extracted from the water phase using only a moderate amount of solvent. If the tie lines were horizontal or sloped up to the right, extraction would still be possible, but more solvent would have to be used, since the final extract would not be as rich in acetone.

The ratio of desired product (acetone) to diluent (water) should be high for a practical extraction process. The solubility of water in MIK solvent is only 2 percent, but as the acetone concentration increases, the water content of the extract phase also increases. The data from Fig. 23.8 are replotted in Fig. 23.10 to show the gradual increase in water content y_{H_2O} with acetone content y_A. The ratio y_A/y_{H_2O} goes through a maximum at about 27 weight percent acetone in the extract phase. A higher concentration of acetone could be obtained, but the greater amount of water in the extract product would probably make operation at these conditions undesirable.

The phase compositions resulting from a single-stage extraction are easily obtained using the triangular diagram. For example, if a mixture with 40% acetone and 60% water is contacted with an equal mass of MIK solvent, the overall mixture is represented by point *M* in Fig. 23.8. A new tie line is drawn to show that the extract

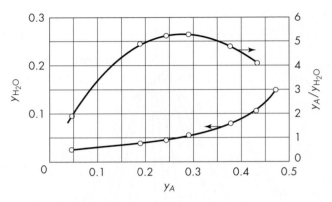

FIGURE 23.10
Composition of extract phase for MIK-acetone-H$_2$O.

phase would be 0.232 acetone, 0.043 water, and 0.725 MIK. The raffinate phase would be 0.132 acetone, 0.845 water, and 0.023 MIK. Repeated contacting of the raffinate phase with fresh solvent, a process called crosscurrent extraction, would permit recovery of most of the acetone; but this would be less efficient than using a countercurrent cascade because of the large volume of solvent needed.

Use of McCabe–Thiele method

The separation achieved for a given number of ideal stages in a counterflow cascade can be determined by using a triangular diagram and special graphical techniques, but a modified McCabe–Thiele method, which is the approach used here, is simple to use and has satisfactory accuracy for most cases. The method focuses on the concentration of solute in the extract and raffinate phases, and the diagram does not show the concentration of the diluent in the extract or the concentration of solvent in the raffinate. However, these minor components of both phases are accounted for in determining the total flow of extract and raffinate, which affects the position of the operating line.

To apply the McCabe–Thiele method to extraction, the equilibrium data are shown on a rectangular graph, where the mass fraction of solute in the extract or V phase is plotted as the ordinate and the mass fraction of solute in the raffinate phase as the abscissa. For a type I system, the equilibrium line ends with equal compositions at the plait point. The use of only one concentration to characterize a ternary mixture may seem strange, but if the phases leaving a given stage are in equilibrium, only one concentration is needed to fix the compositions of both phases.

The operating line for the extraction diagram is based on Eq. (23.3), which gives the relationship between the solute concentration leaving stage n in the L phase and that coming from stage $n + 1$ in the V phase. The terminal points on the operating line (x_a, y_a) and (x_b, y_b) are usually determined by an overall material balance, taking into account the ternary equilibrium data. Because of the decrease in the raffinate phase (L) and the increase in the extract phase (V) as they pass through the column, the operating line is curved. A material balance over a portion of the cascade is made to establish one or more intermediate points on the operating line. The number of ideal stages is then determined by drawing steps in the normal manner.

If the number of ideal stages is specified, the fraction of solute extracted and the final compositions are determined by trial and error. The fraction extracted or the final extract composition is assumed, and the curved operating line is constructed. If too many stages are required, a smaller fraction extracted is assumed and the calculations are repeated. Such calculations are generally done by computer.

EXAMPLE 23.3. A countercurrent extraction plant is used to extract acetone (A) from its mixture with water by means of methyl isobutyl ketone (MIK) at a temperature of 25°C. The feed consists of 40% acetone and 60% water. Pure solvent equal in mass to the feed is used as the extracting liquid. How many ideal stages are required to extract 99 percent of the acetone fed? What is the extract composition after removal of the solvent?

Solution. Use the data in Fig. 23.8 to prepare a plot of the equilibrium relationship y_A versus x_A, which is the upper curve in Fig. 23.11. The terminal points for the operating line are determined by material balances with allowance for the amounts of water in the extract phase and MIK in the raffinate phase. Basis: $F = 100$ mass units per hour.

Let n = mass flow rate of H_2O in extract

m = mass flow rate of MIK in raffinate

For 99 percent recovery of A, the extract has $0.99 \times 40 = 39.6A$, and the raffinate has $0.4A$. The total flows are

At the top, $L_a = F = 100 = 40A + 60\ H_2O$

$V_a = 39.6A + n\ H_2O + (100 - m)MIK = 139.6 + n - m$

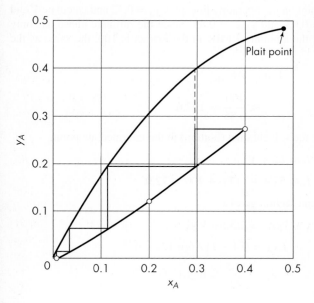

FIGURE 23.11
McCabe–Thiele diagram for extraction (Example 23.3).

At the bottom, $V_b = 100$ MIK

$$L_b = 0.4A + (60 - n)H_2O + mMIK = 60.4 + m - n$$

Since n and m are small and tend to cancel in the summations for V_a and L_a, the total extract flow V_a is about 140, which would make $y_{A,a} \approx 39.6/140 = 0.283$. The value of $X_{A,b}$ is about $0.4/60 = 0.0067$. These estimates are adjusted after calculating values of n and m.

From Fig. 23.8 for $y_A = 0.283$, $y_{H_2O} = 0.049$,

$$n = \frac{0.049}{1 - 0.049}(39.6 + 100 - m)$$

If m is very small, $n \approx (0.049/0.951)(139.6) = 7.2$.
From Fig. 23.8 for $x_A = 0.007$, $x_{MIK} = 0.02$,

$$m = \frac{0.02}{1 - 0.02}(0.4 + 60 - n)$$

$$\approx \frac{0.02}{0.98}(0.4 + 52.8) = 1.1$$

Revised $n = (0.049/0.951)(139.6 - 1.1) = 7.1$:

$$V_a = 139.6 + 7.1 - 1.1 = 145.6$$

$$y_{A,a} = \frac{39.6}{145.6} = 0.272$$

$$L_b = 60.4 + 1.1 - 7.1 = 54.4$$

$$x_{A,b} = \frac{0.4}{54.4} = 0.0074$$

Plot points (0.0074, 0) and (0.40, 0.272) to establish the ends of the operating line.

For an intermediate point on the operating line, pick $y_A = 0.12$ and calculate V and L. From Fig. 23.8, $y_{H_2O} = 0.03$, and $y_{MIK} = 0.85$. Since the raffinate phase has only 2 to 3% MIK, assume that the amount of MIK in the extract is 100, the same as the solvent fed:

$$100 \approx V y_{MIK}$$

$$V \approx \frac{100}{0.85} = 117.6$$

By an overall balance from the solvent inlet (bottom) to the intermediate point,

$$V_b + L = L_b + V$$

$$L \approx 54.4 + 117.6 - 100 = 72.0$$

A balance on A over the same section gives x_A:

$$Lx_A + V_b y_b = L_b x_b + V y_A$$

$$Lx_A \approx 0.4 + 117.6(0.12) - 0$$

$$x_A \approx \frac{14.5}{72} = 0.201$$

This value is probably accurate enough, but corrected values of V, L, and x_A can be determined. For $x_A = 0.201$, $x_{MIK} \approx 0.03$ (Fig. 23.8). A balance on MIK from the solvent inlet to the intermediate point gives

$$V_b + Lx_{MIK} = L_b x_{MIK,b} + V y_{MIK}$$

$$V y_{MIK} = 100 + 72(0.03) - 1.1$$

$$\text{Revised } V = \frac{101.1}{0.85} = 118.9$$

$$\text{Revised } L = 54.4 + 118.9 - 100 = 73.3$$

$$\text{Revised } x_A = \frac{0.4 + 118.9(0.12)}{73.3} = 0.200$$

Plot $x_A = 0.20$, $y_A = 0.12$, which gives a slightly curved operating line. From Fig. 23.11, $N = 3.4$ stages.

Countercurrent extraction of type II systems using reflux

Just as in distillation, reflux can be used in countercurrent extraction to improve the separation of the components in the feed. This method is especially effective in treating type II systems, because with a center-feed cascade and the use of reflux, the two feed components can be separated into nearly pure products.

A flow diagram for countercurrent extraction with reflux is shown in Fig. 23.12. To emphasize the analogy between this method and fractionation, it is assumed that the cascade is a plate column. Any other kind of cascade, however, may be used. The method requires that sufficient solvent be removed from the extract leaving the cascade to form a raffinate, part of which is returned to the cascade as reflux, the remainder being withdrawn from the plant as a product. Raffinate is withdrawn from the cascade as bottoms product, and fresh solvent is admitted directly to the bottom of the cascade. None of the bottom raffinate needs to be returned as reflux, for the number of stages required is the same whether or not any of the raffinate is recycled to the bottom of the cascade.[14] The situation is not the same as in continuous distillation, in which part of the bottoms must be vaporized to supply heat to the column.

The solvent separator, which is ordinarily a still, is shown in Fig. 23.12. As also shown in Fig. 23.12, solvent may be removed from both products by stripping, or in some cases by water washing, to give solvent-free products.

The close analogy between distillation and extraction, both using reflux, is shown in Table 23.3. Note that the solvent plays the same part in extraction that heat does in distillation.

Limiting reflux ratios

Just as in distillation, two limiting cases exist in operating a countercurrent extractor with reflux. As the reflux ratio R_D becomes very great, the number of stages approaches a minimum, and as R_D is reduced, a minimum value of the reflux ratio is reached where the number of stages becomes infinite. The minimum number of

TABLE 23.3
Comparison of extraction with distillation, both using reflux

Distillation	Extraction
Vapor flow in cascade V	Extract flow in cascade V
Liquid flow in cascade L	Raffinate flow in cascade L
Overhead product D	Extract product D
Bottom product B	Raffinate product B
Condenser	Solvent separator
Bottom-product cooler	Raffinate solvent stripper
Overhead-product cooler	Extract solvent stripper
Heat to reboiler q_r	Solvent to cascade s_B
Heat removal in condenser q_c	Solvent removal in separator s_D
Reflux ratio $R_D = L_a/D$	Reflux ratio $R_D = L_a/D$
Rectifying section	Extract-enriching section
Stripping section	Raffinate-stripping section

FIGURE 23.12
Countercurrent extraction with reflux.

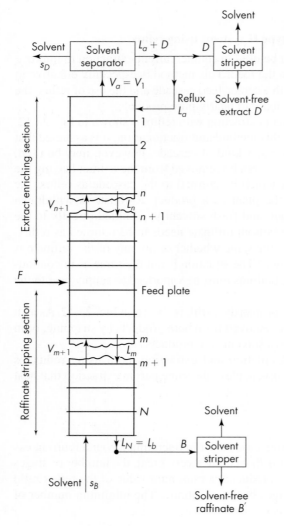

stages and the minimum reflux ratio are found by exactly the same methods used to determine the same quantities in distillation.

Practical examples of extraction with reflux

There are few, if any, practical examples of reflux in the simple manner shown in Fig. 23.12. For systems such as aniline-heptane-methylcyclohexane (Fig. 23.9), the ratio of MCH to heptane in the extract is only modestly greater than in the raffinate, so a great many stages would be needed for high-purity products. Furthermore, the low solubility of both solutes in aniline would mean a very large flow of solvent to be handled. However, a modification of the reflux concept has been applied in several industrial processes for extractive separation. Enrichment of the extract is accomplished by countercurrent washing with another liquid, chosen so that the small amounts of this liquid that dissolve in the extract can be easily removed. The Sulfolane process for extraction of aromatics is an example of this type.

Sulfolane process. A flow sheet for the Sulfolane process[3] is shown in Fig. 23.13. The hydrocarbon feed is introduced near the middle of the extractor, and the heavy solvent is fed at the top. In the top section, nearly all the aromatics are extracted from the raffinate, but the solvent at this point also contains a few percent paraffins and naphthenes. The boiling point ranges overlap, so preparation of pure

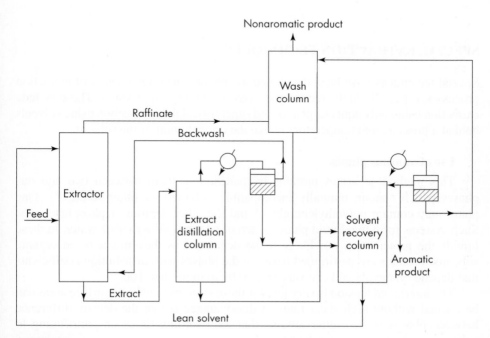

FIGURE 23.13
Sulfolane extraction process. (*After Broughton and Asselin.*[3])

aromatics by distillation of this material is not practical. Also, Sulfolane and the hydrocarbons form a type I system with a plait point, so paraffin-free aromatics cannot be obtained by refluxing some of the solvent-free extract product.

In the lower section of the extractor, medium- and high-boiling paraffins are displaced from the extract phase by contact with a low-boiling hydrocarbon fraction prepared by distillation of the extract. Water is present in this system and forms a low-boiling azeotrope with the lighter hydrocarbons, so the distillation is actually an azeotropic distillation (see Chap. 22). Vapors from the column are condensed and separated into two phases, and the hydrocarbon phase is returned to the extractor as backwash. In the lower section of the extractor, enough stages are provided for nearly complete transfer of the medium and heavy paraffins to the raffinate phase. The extract leaves saturated with light hydrocarbons, but these are removed in the azeotropic distillation column.

Solvent is recovered in the second distillation column, which is a vacuum steam distillation column, with reflux of the organic phase to get high-purity aromatics. The final column is a multistage extraction column, where water is used to wash the solvent from the raffinate. There is a closed cycle for water to minimize solvent loss, and a small amount of water can be tolerated in the extraction solvent.

The backwash stream is sometimes called countersolvent or reflux, but the term *reflux* should be reserved for streams having the same composition as the product. The backwash need not be a low-molecular-weight material; in one version of the Sulfolane process, the backwash is a heavy paraffinic fraction, which is easily removed from both extract and raffinate.

SPECIAL EXTRACTION TECHNIQUES

Special techniques have been developed to enhance the effectiveness of extraction processes, especially in the recovery of sensitive biological products. These include extraction using only aqueous phases and supercritical fluid extraction using solvents held at a pressure and temperature above the critical point of the solvent.

Use of aqueous phases

This method separates proteins by partitioning them between two aqueous phases which contain mutually incompatible polymers or other solutes.[8,17] One phase may contain polyethylene glycol and the other dextran or phosphate salts. Such systems form two liquid phases, each of which is 80% to 90% water. In these liquids the proteins are not likely to be denatured as they might be in organic solvents. Proteins are partitioned between the phases with distribution coefficients that depend on the pH and can vary from 0.01 to more than 100.

The interfacial tension is very low for these systems, and large surface areas can be created without high shear rates. A disadvantage is that the density difference between phases is also quite low, so that long settling times or centrifuging to enhance the settling rate may be needed.

Supercritical Fluid Extraction

Certain compounds may be separated from solids or liquids by extraction with a supercritical solvent. In addition to its selective dissolving power, a supercritical fluid has several advantages over conventional liquid solvents. Its density and viscosity are low, and the diffusivities of solutes in it are high—about 100 times greater than those in ordinary liquids. Consequently, the supercritical fluid easily penetrates porous or fibrous solids. Solutes can be recovered from a supercritical fluid merely by changing the temperature or the pressure, although other ways are also used. The chief disadvantage of supercritical fluid extraction is the high pressure required.

Phase equilibria

A useful solvent for supercritical extraction, especially in food processing, is carbon dioxide, which has a critical point of 31.06°C and 73.8 bars (1,070 lb$_f$/in.^{2}). The phase diagram for pure CO_2 (Fig. 23.14) shows the equilibrium regions of solid, liquid, and gas and the conditions under which a supercritical fluid exists. In the supercritical region there is no distinction between liquid and gas and no phase transition from one to the other; the supercritical fluid acts as a very dense gas or a light, mobile liquid.

Solubility and selectivity in a supercritical fluid are strong functions of temperature and pressure. For nearly total extraction of solutes by supercritical CO_2, as indicated in Fig. 23.14, the highest pressure is used, since here the solubilities are

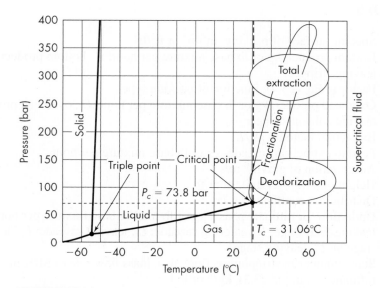

FIGURE 23.14
Phase diagram for carbon dioxide, illustrating regions useful for food processing applications.[12]

highest. The loaded solvent may then be passed through a series of separation vessels in which the temperature or pressure is changed stepwise, to give what is called multistage fractionation. For selective removal of odor-producing volatile components, conditions close to the critical point are advantageous. Here the solubilities are smaller, but the selectivity for the most volatile compounds is much higher.

Equilibrium phase diagrams for one or more solutes in supercritical CO_2 and other solvents can be very complicated.[2] Relatively little equilibrium information has been published; this lack coupled with the cost of high-pressure equipment and the difficulties of scaleup have to date limited the commercial applications of supercritical fluid extraction.

Commercial process[12]

A practical example of a supercritical fluid extraction process is the decaffeination of coffee. Coffee beans are first soaked in water to make the extraction more selective and then are loaded into an extraction vessel through which supercritical CO_2 is circulated to dissolve the caffeine. In a separate scrubbing vessel the caffeine is transferred from the CO_2 to water, also at high pressure. Extraction is continued until the caffeine content of the beans, originally 0.7 to 3 percent, is reduced to less than 0.02 percent. At the end of the batch cycle the water is depressurized and distilled away from the caffeine. In an alternative process the caffeine is separated from the CO_2 by adsorption on activated carbon. In either case the supercritical fluid extraction is said not to affect the characteristic flavor and aroma of the coffee.

SYMBOLS

a_v	Specific surface of packing, m^2/m^3 or ft^2/ft^3
B	Bottom product, kg or lb base mixture per hour; B', bottom product leaving stripper
C_1	Factor in Eq. (23.4), function of packing type
D	Overhead product, kg or lb base mixture per hour; D', overhead product leaving solvent stripper
E	Extraction factor, $K_D V/L$
F	Feed to extraction cascade, kg or lb base mixture per hour
g	Acceleration of gravity, m/s^2
HTU	Height of a transfer unit, m or ft
K_D	Distribution coefficient, y_e/x_e
L	Underflow, or raffinate phase, kg or lb total or base mixture per hour; L_N, leaving final stage; L_a, entering cascade; L_b, leaving cascade; L_m, leaving stage m; L_n, leaving stage n
m	Stage number in stripping section; also, mass flow rate of MIK in raffinate (Example 23.3), kg/h or lb/h
N	Number of ideal stages
n	Stage number in rectifying section; also mass flow rate of water in extract (Example 23.3), kg/h or lb/h

P_c	Critical pressure, bars
q	Heat added, J/g or Btu/lb; $-q_c$, to condenser; q_r, to reboiler
R	Ratio of superficial velocities, $\bar{V}_{s,d}/\bar{V}_{s,c}$
R_D	Reflux ratio
S	Stripping factor, defined by Eq. (20.29)
s	Rate of solvent flow, kg/h or lb/h; s_B, added to bottom product; s_D, from overhead product
T_c	Critical temperature, °C
V	Overflow, or extract phase, mass or moles base mixture per hour; V_a, leaving cascade; V_b, entering cascade; V_{m+1}, leaving stage $m+1$; V_{n+1}, leaving stage $n+1$
$\bar{V}_s$	Superficial velocity, based on cross section of column, m/h or ft/h; $\bar{V}_{s,c}$ of continuous phase; $\bar{V}_{s,d}$ of dispersed phase
x	Mass fraction of solute in underflow or L phase; x_A, x_{MIK}, mass fraction of acetone, MIK, respectively, based on entire L phase; x_a, at entrance; x_b, at exit; x_e, equilibrium value; x_n, leaving stage n
x^*	Concentration of underflow solution in equilibrium with a specific overflow solution; x_a^*, in equilibrium with y_a; x_b^*, in equilibrium with y_b
y	Mass fraction of solute in overflow or V phase; y_A, y_{H_2O}, y_{MIK}, mass fraction of acetone, water, MIK, respectively, based on entire V phase; y_a, at exit; y_b, at entrance; y_e, equilibrium value; y_{n+1}, leaving stage $n+1$
y^*	Concentration of overflow solution in equilibrium with specific underflow solution; y_a^*, in equilibrium with x_a; y_b^*, in equilibrium with x_b

Greek Letters

α	Factor in Eq. (23.4)
$\Delta\rho$	Density difference between phases, kg/m^3 or lb/ft^3
ε	Fraction voids in packed section
μ_c	Viscosity of continuous phase, kg/m·s or lb/ft·h
ρ_d	Density of dispersed phase, kg/m^3 or lb/ft^3
σ	Interfacial tension, dyn/cm or lb$_f$/ft

PROBLEMS

23.1. Roasted copper ore containing the copper as $CuSO_4$ is to be extracted in a counter-current stage extractor. Each hour a charge consisting of 10 tons of inert solids, 1.2 tons of copper sulfate, and 0.5 ton of water is to be treated. The strong solution produced is to consist of 90% H_2O and 10% $CuSO_4$ by weight. The recovery of $CuSO_4$ is to be 98 percent of that in the ore. Pure water is to be used as the fresh solvent. After each stage, 1 ton of inert solids retains 2 tons of water plus the copper sulfate dissolved in that water. Equilibrium is attained in each stage. How many stages are required?

23.2. A five-stage countercurrent extraction battery is used to extract the sludge from the reaction

$$Na_2CO_3 + CaO + H_2O \rightarrow CaCO_3 + 2NaOH$$

The $CaCO_3$ carries with it 1.5 times its weight of solution in flowing from one unit to another. It is desired to recover 99 percent of the NaOH. The products from the reaction enter the first unit with no excess reactants but with 0.6 kg of H_2O per kilogram of $CaCO_3$. (*a*) How much wash water must be used per kilogram of calcium carbonate? (*b*) What is the concentration of the solution leaving each unit, assuming that $CaCO_3$ is completely insoluble?

23.3. In Prob. 23.2 it is found that the sludge retains solution varying with the concentration as shown in Table 23.4. If a 12% solution of the NaOH is to be produced, how many stages must be used to recover 97 percent of the NaOH?

TABLE 23.4

NaOH, wt %	0	5.	10	15	20
Kg solution/kg $CaCO_3$	1.50	1.75	2.20	2.70	3.60

23.4. Oil is to be extracted from halibut livers by means of ether in a countercurrent extraction battery. The entrainment of solution by the granulated liver mass was found by experiment to be as shown in Table 23.5. In the extraction battery, the charge per cell is to be 100 lb, based on completely exhausted livers. The unextracted livers contain 0.043 gal of oil per pound of exhausted material. A 95 percent recovery of oil is desired. The final extract is to contain 0.65 gal of oil per gallon of extract. The ether fed to the system is oil-free. (*a*) How many gallons of ether are needed per charge of livers? (*b*) How many extractors are needed?

TABLE 23.5

Solution retained by 1 lb exhausted livers, gal	Solution concentration, gal oil/gal solution	Solution retained by 1 lb exhausted livers, gal	Solution concentration, gal oil/gal solution
0.035	0	0.068	0.4
0.042	0.1	0.081	0.5
0.050	0.2	0.099	0.6
0.058	0.3	0.120	0.68

23.5. In a continuous countercurrent train of mixer-settlers, 100 kg/h of a 40:60 acetone-water solution is to be reduced to 10% acetone by extraction with pure 1,1,2-trichloroethane at 25°C. (*a*) Find the minimum solvent rate. (*b*) At 1.8 times the minimum (solvent rate)/(feed rate), find the number of stages required. (*c*) For conditions of part (*b*) find the mass flow rates of all streams. Data are given in Table 23.6.

TABLE 23.6
Equilibrium data

	Limiting solubility curve	
$C_2H_3Cl_3$, wt %	Water, wt %	Acetone, wt %
94.73	0.26	5.01
79.58	0.76	19.66
67.52	1.44	31.04
54.88	2.98	42.14
38.31	6.84	54.85
24.04	15.37	60.59
15.39	26.28	58.33
6.77	41.35	51.88
1.72	61.11	37.17
0.92	74.54	24.54
0.65	87.63	11.72
0.44	99.56	0.00

	Tie lines				
	Weight % in water layer			Weight % in trichloroethane layer	
$C_2H_3Cl_3$	Water	Acetone	$C_2H_3Cl_3$	Water	Acetone
0.52	93.52	5.96	90.93	0.32	8.75
0.73	82.23	17.04	73.76	1.10	25.14
1.02	72.06	26.92	59.21	2.27	38.52
1.17	67.95	30.88	53.92	3.11	42.97
1.60	62.67	35.73	47.53	4.26	48.21
2.10	57.00	40.90	40.00	6.05	53.95
3.75	50.20	46.05	33.70	8.90	57.40
6.52	41.70	51.78	26.26	13.40	60.34

23.6. A spray-tower extractor operates at 30°C with hydrocarbon drops dispersed in water. The density of the hydrocarbon phase is 53 lb/ft^3. If the average drop size is 2.0 mm, calculate the terminal velocity of the drops. If the slip velocity is assumed to be independent of the holdup and if the dispersed-phase volumetric flow rate is taken to be twice that of the continuous phase, at what flow rate of the dispersed phase would the fraction holdup be 0.30 (close to flooding)? For different ratios of flow rates, is either the sum of the flow rates or the sum of their square roots nearly constant at a given holdup? [See Eq. (23.4)]

23.7. Estimate the flooding velocity for an extraction column packed with 15-mm spheres and operating with water and toluene at 30°C. Toluene is dispersed and has a flow rate twice that of the water phase.

23.8. A mixture containing 40 weight percent acetone and 60 weight percent water is contacted with an equal amount of MIK. (a) What fraction of the acetone can be extracted in a single-stage process? (b) What fraction of the acetone could be extracted if the fresh solvent were divided into two parts and two successive extractions used?

23.9. An antibiotic that has been extracted from a fermentation broth using amyl acetate at low pH is to be extracted back into clean water at pH = 6, where $K_D = 0.15$. If the water flow rate is set at 0.45 times the solvent rate, how many ideal stages would be needed for 98 percent recovery of the antibiotic in a countercurrent cascade?

REFERENCES

1. Baker, E. M. *Trans. AIChE* **32:**62 (1936).
2. Brennecke, J. F., and C. A. Eckert. *AIChE J.* **35:**1409 (1989).
3. Broughton, D. B., and G. F. Asselin. *Proc. Seventh World Petroleum Congress,* vol. 4, New York: Elsevier, 1967, p. 65.
4. Bushnell, J. D., and R. J. Fiocco. *Hydrocarbon Proc.* **59**(5):119 (1980).
5. Cofield, E. P., Jr. *Chem. Eng.* **58**(1):127 (1951).
6. Elgin, J. C. *Trans. AIChE* **32:**451 (1936).
7. Grosberg, J. A. *Ind. Eng. Chem.* **42:**154 (1950).
8. King, R. S., H. W. Blanch, and J. M. Prausnitz. *AIChE J.* **34:**1585 (1988).
9. Kumar, A., and S. Hartland. *Trans. Inst. Chem. Engrs.* **72A:**89 (1994).
10. Othmer, D. F., R. E. White, and E. Trueger. *Ind. Eng. Chem.* **33:**1240 (1941).
11. Perry, R. H., and D. W. Green (eds.). *Perry's Chemical Engineers' Handbook,* 7th ed. New York: McGraw-Hill, 1997; (*a*) pp. **15**-22 to **15**-47; (*b*) p. **18**-56.
12. Rivzi, S. S., A. L. Benado, J. A. Zollweg, and J. A. Daniels. *Food Tech.* **40**(6):55 (1986).
13. Sage, G., and F. W. Woodfield. *Chem. Eng. Prog.* **50:**396 (1954).
14. Skelland, A. H. P. *Ind. Eng. Chem.* **53:**799 (1961).
15. Treybal, R. E. *Liquid Extraction,* 2nd ed. New York: McGraw-Hill, 1963.
16. Varteressian, K. A., and M. R. Fenske. *Ind. Eng. Chem.* **29:**270 (1937).
17. Walter, H., D. E. Brooks, and D. Fisher. *Partitioning in Aqueous Two-Phase Systems.* New York: Academic, 1985.

CHAPTER 24

Drying of Solids

In general, drying a solid means the removal of relatively small amounts of water or other liquid from the solid material to reduce the content of residual liquid to an acceptably low value. Drying is usually the final step in a series of operations, and the product from a dryer is often ready for final packaging.

Water or other liquids may be removed from solids mechanically by presses or centrifuges or thermally by vaporization. This chapter is restricted to drying by thermal vaporization. It is generally cheaper to remove liquid mechanically than thermally, and thus it is advisable to reduce the liquid content as much as practicable before feeding the material to a heated dryer.

The liquid content of a dried substance varies from product to product. Occasionally the product contains no liquid and is called *bore-dry*. More commonly, the product does contain some liquid. Dried table salt, for example, contains about 0.5% water, dried coal about 4%, and dried casein about 8%. Drying is a relative term and means merely that there is a reduction in liquid content from an initial value to some acceptable final value.

The solids to be dried may be in many different forms—flakes, granules, crystals, powders, slabs, or continuous sheets—and may have widely differing properties. The liquid to be vaporized may be on the surface of the solid, as in drying salt crystals; it may be entirely inside the solid, as in solvent removal from a sheet of polymer; or it may be partly outside and partly inside. The feed to some dryers is a liquid in which the solid is suspended as particles or is in solution. The dried product may be able to stand rough handling and a very hot environment, or it may require gentle treatment at low or moderate temperatures. Consequently a multitude of types of dryers are on the market for commercial drying. They differ chiefly in the way the solids are moved through the drying zone and in the way heat is transferred.

Classification of dryers

There is no simple way of classifying drying equipment. Some dryers are continuous, and others operate batchwise; some agitate the solids, and others are essentially unagitated. Operation under vacuum may be used to reduce the drying temperature. Some dryers can handle almost any kind of material, while others are severely limited in the type of feed they can accept.

Major divisions may be made among (1) dryers in which the solid is directly exposed to a hot gas (usually air), (2) dryers in which heat is transferred to the solid from an external medium such as condensing steam, usually through a metal surface with which the solid is in contact,[5] and (3) dryers that are heated by dielectric, radiant, or microwave energy. Dryers that expose the solids to a hot gas are called *adiabatic* or *direct dryers;* those in which heat is transferred from an external medium are known as *nonadiabatic* or *indirect dryers.* Some units combine adiabatic and nonadiabatic drying; they are known as *direct-indirect* dryers.

Solids handling in dryers

Most industrial dryers handle particulate solids during part of or all the drying cycle, although some, of course, dry large individual pieces such as ceramic ware or sheets of polymer. The properties of particulate solids are discussed in Chap. 28. Here it is important only to describe the different patterns of motion of solid particles through dryers as a basis for understanding the principles of drying discussed in the next section.

In adiabatic dryers the solids are exposed to the gas in the following ways:

1. Gas is blown across the surface of a bed or slab of solids or across one or both faces of a continuous sheet or film. This process is called *cross-circulation drying* (Fig. 24.1a).
2. Gas is blown through a bed of coarse granular solids that are supported on a screen. This is known as *through-circulation drying.* As in cross-circulation drying, the gas velocity is kept low to avoid any entrainment of solid particles (Fig. 24.1b).
3. Solids are showered downward through a slowly moving gas stream, often with some undesired entrainment of fine particles in the gas (Fig. 24.1c).
4 Gas passes through the solids at a velocity sufficient to fluidize the bed, as discussed in Chap. 7. Inevitably there is some entrainment of finer particles (Fig. 24.1d).
5. The solids are all entrained in a high-velocity gas stream and are pneumatically conveyed from a mixing device to a mechanical separator (Fig. 24.1e).

In nonadiabatic dryers the only gas to be removed is the vaporized water or solvent, although sometimes a small amount of "sweep gas" (often air or nitrogen) is passed through the unit. Nonadiabatic dryers differ chiefly in the ways in which the solids are exposed to the hot surface or other source of heat.

1. Solids are spread over a stationary or slowly moving horizontal surface and "cooked" until dry. The surface may be heated electrically or by a heat-transfer

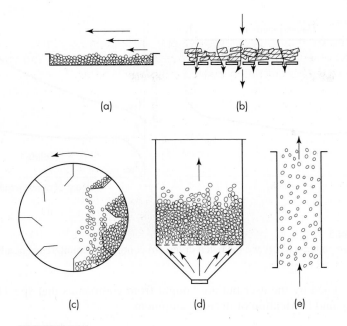

FIGURE 24.1
Patterns of gas-solid interaction in dryers: (*a*) gas flow across
a static bed of solids; (*b*) gas passing through a bed of
preformed solids; (*c*) showering action in a rotary dryer;
(*d*) fluidized solids bed; (*e*) cocurrent gas-solid flow in a
pneumatic-conveyor flash dryer.

fluid such as steam or hot water. Alternatively, heat may be supplied by a radiant
heater above the solid.

2. Solids are moved over a heated surface, usually cylindrical, by an agitator or a
   screw or paddle conveyor.
3. Solids slide by gravity over an inclined heated surface or are carried upward with
   the surface for a time and then slide to a new location. (See "Rotary Dryers,"
   p. 797.)

PRINCIPLES OF DRYING

Because of the wide variety of materials that are dried in commercial equipment and
the many types of equipment used, there is no single theory of drying that covers all
materials and dryer types. Variations in shape and size of stock, in moisture equilibria, in the mechanism of flow of moisture through the solid, and in the method of
providing the heat required for the vaporization—all prevent a unified treatment.
General principles used in a semiquantitative way are relied upon. Dryers are

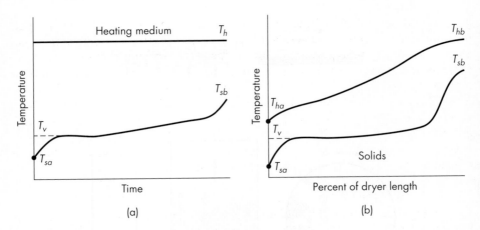

FIGURE 24.2
Temperature patterns in dryers: (*a*) batch dryer; (*b*) continuous countercurrent adiabatic dryer.

seldom designed by the user but are bought from companies that specialize in the engineering and fabrication of drying equipment.

Temperature patterns in dryers

The way in which temperatures vary in a dryer depends on the nature and liquid content of the feedstock, the temperature of the heating medium, the drying time, and the allowable final temperature of the dry solids. The pattern of variation, however, is similar from one dryer to another. Typical patterns are shown in Fig. 24.2.

In a batch dryer with a heating medium at constant temperature (Fig. 24.2*a*), the temperature of the wet solids rises rather quickly from its initial value T_{sa} to the initial vaporization temperature T_v. In a nonadiabatic dryer with no sweep gas, T_v is essentially the boiling point of the liquid at the pressure prevailing in the dryer. If a sweep gas is used or if the dryer is adiabatic, T_v is at or near the wet-bulb temperature of the gas (which equals the adiabatic saturation temperature if the gas is air and water is the liquid being evaporated). Drying may occur at T_v for a considerable time, but often after a short time the temperature of the wet solids gradually rises as a zone of dry solids forms near the surface. The vaporization temperature then depends on the heat- and mass-transfer resistances in the dry zone, as well as on those in the external boundary layer. In the final stages of drying, the solids temperature rises rapidly to some higher value T_{sb}.

The drying time indicated in Fig. 24.2*a* may range from a few seconds to many hours. The solids may be at T_v for most of the drying cycle or for only a small fraction of it. The temperature of the heating medium is often constant, as shown, or it may be programmed to change as drying proceeds.

In an ideal continuous dryer, each particle or element of the solid passes through a cycle similar to that shown in Fig. 24.2*a* on its way from the inlet to the outlet of the dryer. In steady-state operation the temperature at any given point in a continuous dryer is constant, but it varies along the length of the dryer. Figure 24.2*b* shows a temperature pattern for an adiabatic countercurrent dryer. The solids inlet and gas outlet are on the left; the gas inlet and solids outlet are on the right. Again the solids are quickly heated from T_{sa} to T_v. The vaporization temperature may change as drying proceeds, even though the wet-bulb temperature remains the same. Near the gas inlet the solids may be heated to a temperature well above T_v in a relatively short length of the dryer, since the energy needed to heat the dry solids is small compared to that needed for vaporization. For heat-sensitive materials the dryer would be designed to keep T_{sb} close to T_v. Hot gas enters the dryer at T_{hb}, usually with low humidity. The temperature profile of the gas may have a complex shape because of the variation of the temperature driving force and the change in overall heat-transfer coefficient during the drying process.

Heat transfer in dryers

Drying of wet solids is by definition a thermal process. While it is often complicated by diffusion in the solid or through a gas, it is possible to dry many materials merely by heating them above the boiling point of the liquid—perhaps well above, to free the last traces of adsorbed material. Wet solids, for example, can be dried by exposure to highly superheated steam. Here there are no diffusion limitations; the problem is solely one of heat transfer. In most adiabatic drying, of course, diffusion may occur in the solid or gas phase, but often drying rates are more dependent on heat-transfer coefficients than on mass-transfer coefficients, and the principles given in Chaps. 10 and 14 can be used in dryer calculations. Many dryers are designed on the basis of heat-transfer considerations alone.

Calculation of heat duty. Heat must be applied to a dryer to accomplish the following:

1. Heat the feed (solids and liquid) to the vaporization temperature.
2. Vaporize the liquid.
3. Heat the solids to their final temperature.
4. Heat the vapor to its final temperature.

Items 1, 3, and 4 are often negligible compared with item 2. In the general case the total rate of heat transfer may be calculated as follows. If $\dot{m}_s$ is the mass of bone-dry solids to be dried per unit time and X_a and X_b are the initial and final liquid contents in mass of liquid per unit mass of bone-dry solid, then the quantity of heat transferred per unit mass of solid $q_T/\dot{m}_s$ is

$$\frac{q_T}{\dot{m}_s} = c_{ps}(T_{sb} - T_{sa}) + X_a c_{pL}(T_v - T_{sa}) + (X_a - X_b)\lambda$$

$$+ X_b c_{pL}(T_{sb} - T_v) + (X_a - X_b)c_{pv}(T_{va} - T_v) \qquad (24.1)$$

where
T_{sa} = feed temperature
T_v = vaporization temperature
T_{sb} = final solids temperature
T_{va} = final vapor temperature
λ = heat of vaporization
c_{ps}, c_{pL}, c_{pv} = specific heats of solid, liquid, and vapor, respectively

Equation (24.1) is based on average specific heats for the temperature range from inlet to outlet and on the heat of vaporization at T_v. However, if vaporization occurs over a range of temperatures, Eq. (24.1) still applies, because the total enthalpy change is independent of the path followed from the initial to the final state.

In an adiabatic dryer the heat transferred to the solids, liquid, and vapor, as found from Eq. (24.1), comes from the cooling of the gas; for a continuous adiabatic dryer the heat balance gives

$$q_T = \dot{m}_g c_{sb}(T_{hb} - T_{ha}) \tag{24.2}$$

where $\dot{m}_g$ = mass rate of dry gas
c_{sb} = humid heat of gas at inlet humidity

Heat-transfer coefficients. In dryer calculations the basic heat-transfer equation, a form of Eq. (11.14), applies:

$$q_T = UA \, \overline{\Delta T} \tag{24.3}$$

where U = overall coefficient
A = heat-transfer area
$\overline{\Delta T}$ = average temperature difference (not necessarily logarithmic mean)

Sometimes A and $\overline{\Delta T}$ are known and the capacity of the dryer can be estimated from a calculated or measured value of U. For tray dryers and moving-belt dryers, A is the area of the horizontal surface carrying the wet solids. For drum dryers A is the active surface area of the drum, and for through-circulation dryers it may be taken as the total surface area of the particles. In some dryers such as screw-conveyor dryers or rotary dryers, the effective area for heat and mass transfer is hard to determine. Such dryers are designed on the basis of a *volumetric* heat-transfer coefficient Ua, where a is the (unknown) heat-transfer area per unit dryer volume. The governing equation is

$$q_T = Ua \, V\overline{\Delta T} \tag{24.4}$$

where Ua = volumetric heat-transfer coefficient, Btu/ft$^3 \cdot$ h $\cdot ^\circ$F or W/m$^3 \cdot ^\circ$C
V = dryer volume, ft^3 or m^3

For most dryers no general heat-transfer correlations are available, and coefficients must be found experimentally. Empirical coefficients are often based on more or less arbitrary definitions of heat transfer and average temperature difference. Examples of empirical correlations are given later in this chapter under the discussion of the particular kind of dryer to which they apply. When such correlations are used, the definitions and units of U and A must be carefully checked.

Heat-transfer units

Some adiabatic dryers, especially rotary dryers, are conveniently rated in terms of the number of heat-transfer units they contain. Heat-transfer units are discussed in Chap. 15, p. 442. For a two-fluid exchanger, heat-transfer units are generally based on the stream with the lower capacity [see Eq. (15.9)], but for dryers they are always based on the gas. The number of transfer units in a dryer is given by

$$N_t = \int_{T_{ha}}^{T_{hb}} \frac{dT_h}{T_h - T_s}$$ (24.5)

or

$$N_t = \frac{T_{hb} - T_{ha}}{\overline{\Delta T}}$$ (24.6)

When the initial liquid content of the solids is high and most of the heat transferred is for vaporization, $\overline{\Delta T}$ may be taken as the logarithmic mean difference between the dry-bulb and wet-bulb temperatures. Then

$$\overline{\Delta T} = \overline{\Delta T}_L = \frac{T_{hb} - T_{wb} - (T_{ha} - T_{wa})}{\ln[(T_{hb} - T_{wb})/(T_{ha} - T_{wa})]}$$ (24.7)

For the system water-air $T_{wb} = T_{wa}$ and Eq. (24.6) becomes

$$N_t = \ln \frac{T_{hb} - T_{wb}}{T_{ha} - T_{wb}}$$ (24.8)

In Eq. (24.8) it is assumed that $T_v = T_{wb}$, which may or may not be true. For purposes of calculation, however, this assumption is often made, since T_v is usually not known.

The length of a transfer unit and the number of transfer units appropriate for good design are discussed later under "Drying Equipment."

Phase Equilibria

Equilibrium data for moist solids are commonly given as relationships between the relative humidity of the gas and the liquid content of the solid, in mass of liquid per unit mass of bone-dry solid.[†] Examples of equilibrium relationships are shown in Fig. 24.3. Curves of this type are nearly independent of temperature. The abscissas of such curves are readily converted to absolute humidities, in mass of vapor per unit mass of dry gas.

The remainder of the discussion in this section is based on the air-water system, but it should be remembered that the underlying principles apply equally well to other gases and liquids.

When a wet solid is brought into contact with air of lower humidity than that corresponding to the moisture content of the solid, as shown by the humidity-equilibrium curve, the solid tends to lose moisture and dry to equilibrium with the air. When the air is more humid than the solid in equilibrium with it, the solid absorbs moisture from the air until equilibrium is attained.

[†] A liquid content expressed in this way is said to be on a *dry basis;* it may, and often does, exceed 100 percent.

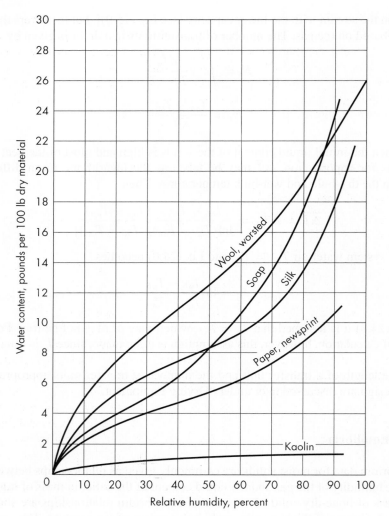

FIGURE 24.3
Equilibrium-moisture curves at 25°C.

In fluid phases diffusion is governed by concentration differences expressed in mole fractions. In a wet solid, however, the term *mole fraction* may have little meaning, and for ease in drying calculations the moisture content is nearly always expressed in mass of water per unit mass of bone-dry solid. This practice is followed throughout this chapter.

Equilibrium moisture and free moisture

The air entering a dryer is seldom completely dry but contains some moisture and has a definite relative humidity. For air of definite humidity, the moisture content of the solid leaving the dryer cannot be less than the equilibrium moisture content corresponding to the humidity of the entering air. That portion of the water in the wet solid that cannot be removed by the inlet air, because of the humidity of the latter, is called the *equilibrium moisture.*

The free water is the difference between the total water content of the solid and the equilibrium water content. Thus, if X_T is the total moisture content and if X^* is the equilibrium moisture content, the free moisture X is

$$X = X_T - X^*$$

It is X, rather than X_T, that is of interest in drying calculations.

Bound and unbound water

If an equilibrium curve like those in Fig. 24.3 is continued to its intersection with the axis for 100 percent humidity, the moisture content so defined is the minimum moisture this material can carry and still exert a vapor pressure as great as that exerted by liquid water at the same temperature. If such a material contains more water than that indicated by this intersection, it can still exert only the vapor pressure of water at the solids temperature. This makes possible a distinction between two types of water held by a given material. The water corresponding to concentrations lower than that indicated by the intersection of the curves in Fig. 24.3 with the line for 100 percent humidity is called *bound water,* because it exerts a vapor pressure less than that of liquid water at the same temperature. Water corresponding to concentrations greater than that indicated by the intersections is called *unbound water.* Substances containing bound water are often called *hygroscopic* substances.

Bound water may exist in several conditions. Liquid water in fine capillaries exerts an abnormally low vapor pressure because of the highly concave curvature of the surface; moisture in cell or fiber walls may suffer a vapor pressure lowering because of solids dissolved in it; water in natural organic substances is in physical and chemical combination, the nature and strength of which vary with the nature and moisture content of the solid. Unbound water, on the other hand, exerts its full vapor pressure and is largely held in the voids of the solid. Large nonporous particles, such as coarse sand, contain only unbound water.

The terms employed in this discussion may be clarified by reference to Fig. 24.3. Consider, for instance, curve 2 for worsted yarns. This intersects the curve for 100 percent humidity at 26 percent moisture; consequently, any sample of wool that contains less than 26 percent water contains only bound water. Any moisture that a sample may contain above 26 percent is unbound water. If the sample contains 30 percent water, for example, 4 percent of this water is unbound and 26 percent is bound. Assume, now, that this sample is to be dried with air of 30 percent relative humidity. Curve 2 shows that the lowest moisture content that can be reached under these conditions is 9 percent. This, then, is the equilibrium moisture content for this particular set of conditions. If a sample containing 30 percent total moisture is to be dried with air at 30 percent relative humidity, it contains 21 percent free water and 9 percent equilibrium moisture. Any amount up to a concentration of 26 percent is still bound water, but most of this can be evaporated into the air and hence is free water. Thus water can be both bound and free at the same time—partially bound to the solid but free to be evaporated. The distinction between bound and unbound water depends on the material itself, while the distinction between free and equilibrium moisture depends on the drying conditions.

CROSS-CIRCULATION DRYING

When both heat and mass transfer are involved, the mechanism of drying depends on the nature of the solids and on the method of contacting the solids and gas. Solids are of three kinds: crystalline, porous, and nonporous. Crystalline particles contain no interior liquid, and drying occurs only at the surface of the solid. A bed of such particles, of course, can be considered a highly porous solid. Truly porous solids, such as catalyst pellets, contain liquid in interior channels. Nonporous solids include colloidal gels such as soap, glue, and plastic clay; dense cellular solids such as wood and leather; and many polymeric materials.

Mass transfer between the solid surface and the gas is covered by the relations discussed in Chap. 17. The drying rate of solids containing internal liquid, however, depends on the way the liquid moves and on the distance it must travel to reach the surface. This is especially important in cross-circulation drying of slabs or beds of solids. Drying by this method is slow, is usually done batchwise, and has been displaced by other faster methods in most large-scale drying operations; it remains important, however, in the production of pharmaceuticals and fine chemicals, especially when drying conditions must be carefully controlled.

Constant drying conditions.   Consider a bed of wet solids, perhaps 50 to 75 mm (2 to 3 in.) deep, over which air is circulated. Assume that the temperature, humidity, and velocity and direction of flow of the air across the drying surface are constant. This is called drying under *constant drying conditions*. Note that only the conditions in the airstream are constant, as the moisture content and other factors in the solid are changing.

Rates of drying

As time passes, the moisture content X_T typically falls as shown by graph A in Fig. 24.4. After a short period during which the feed material is heated to the vaporization temperature, the graph becomes nearly linear, then curves toward the horizontal, and finally levels off. The drying rate, which is the derivative of curve A, is shown by graph B; the rate is constant or decreasing slightly for a considerable period. This time is often referred to as the *constant-rate period,* even though the drying rate may decrease somewhat. Next comes the *falling-rate period,* in which the drying rate may decrease linearly with time or may give plots that are concave upward or concave downward, depending on the nature of the solid and the mechanism of the internal moisture flow. Sometimes, as in Fig. 24.6 for drying a porous ceramic plate, there are two falling-rate periods.

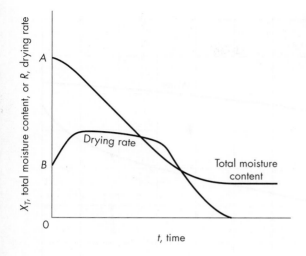

FIGURE 24.4
Typical plots of total moisture content and drying rate versus drying time.

A true constant-rate period is expected if the solid is so wet that a continuous film of liquid exists over the entire external surface. The rate of evaporation is then the same as that from a pool of liquid, and the gas-liquid interface and the solid surface are at the wet-bulb temperature. As drying proceeds, a truly constant-rate period can be maintained only if some mechanism brings water from the interior fast enough to keep the entire surface wet. Some solids such as soap are swollen when very wet and shrink during drying. The shrinkage helps keep the surface wet and prolongs the constant-rate period. However, rapid drying may also make the surface of the shrunken material hard and perhaps impervious, enclosing the bulk of the solid so the interior moisture cannot be easily removed. This effect is called *case hardening.*

When drying granular solids that are spread in a horizontal layer, as in a tray dryer, water lost by evaporation is replaced by air (or vapor), and the air-water interface recedes to below the solid surface. If the solid has a range of pore sizes, capillary action may keep some small pores full of liquid while larger pores become dry. The drying rate will then depend on the fraction of the surface that is wet, and will be at least somewhat less than the initial drying rate. The effect will depend on the relative heat- and mass-transfer resistances in the boundary layer above the solid. If the mass-transfer resistance in the boundary layer is relatively large, the drying rate may be reduced by only a small amount even when one-half of the surface is dry. This is because water vapor can spread by diffusion from the wet areas to the region above the larger pores, when the width or diameter of the dry areas is small compared to the thickness of the gas boundary layer.

In a true constant-rate period the surface temperature is the same as the wet-bulb temperature, provided there is no heat transferred by radiation or by conduction through the solid. In practice, however, material in a tray dryer is likely to receive significant radiation from the tray above and conduction from the tray below, making the initial vaporization temperature higher than T_{wb} and increasing

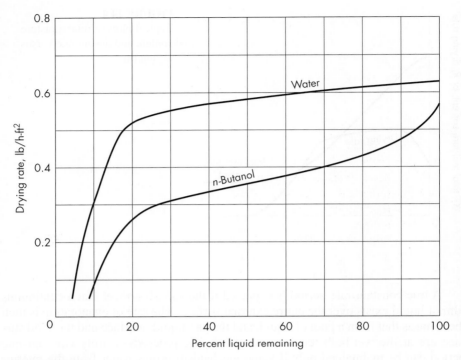

FIGURE 24.5
Drying rate in beds of glass beads, 88 to 105 μm in diameter, for water and n-butanol. (*After Morgan and Yerazunis.*[8])

the drying rate by increasing the driving force for vapor diffusion. However, since it is difficult to determine T_v, heat-transfer coefficients for dryers are nearly always calculated using $T_h - T_{wb}$ as the driving force.

Figure 24.5 shows experimental data for drying beds of glass beads in a small tunnel dryer, with water or n-butanol as the liquid being vaporized.[8] The drying rate is plotted against the water content, defined as the percentage of void volume in the solid bed that is filled with liquid. The curve for water shows a slowly decreasing drying rate as the moisture content falls to about 25 percent, then a rapidly falling rate thereafter. For n-butanol an initial steep drop in the drying rate is followed by more gradual but still significant decline, with a change to a sharply falling rate at a butanol content of about 25 percent. In neither case is there a period in which the drying rate is truly constant. Much of the difference between the two curves results from the differences in interfacial tension or the rates of surface diffusion.

During a true constant-rate period, where the interface temperature T_i can be considered equal to the wet-bulb temperature T_{wb}, the drying rate per unit area R_c can be estimated with fair precision from correlations developed for evaporation from a free liquid surface. The calculations may be based on mass transfer

[Eq. (24.9)] or on heat transfer [Eq. (24.10)] as follows:

$$\dot{m}_v = \frac{M_v k_y (y_i - y) A}{(1 - y)_L} \tag{24.9}$$

or
$$\dot{m}_v = \frac{h_y (T - T_i) A}{\lambda_i} \tag{24.10}$$

where $\dot{m}_v$ = rate of evaporation
  A = drying area
  h_y = heat-transfer coefficient
  k_y = mass-transfer coefficient
  M_v = molecular weight of vapor
  T = temperature of gas
  T_i = temperature at interface
  y = mole fraction of vapor in gas
  y_i = mole fraction of vapor at interface
  λ_i = latent heat at temperature T_i

To estimate the coefficient for air flowing parallel with the surface of a solid in the absence of experimental information, the following *dimensional equation* is recommended[10]

$$h_y = \frac{8.8 G^{0.8}}{D_e^{0.2}} \tag{24.11}$$

where h_y = heat-transfer coefficient, W/m^2·°C
  G = mass velocity, kg/s·m^2
  D_e = equivalent diameter of the airflow channel, m

Equation (24.11) is based on the properties of air at 95°C; it applies for Reynolds numbers between 2,600 and 22,000.

When the flow is perpendicular to the surface, at air velocities between 0.9 and 4.5 m/s, the equation is[16]

$$h_y = 24.2 G^{0.37} \tag{24.12}$$

In fps units, with h in Btu/ft^2·h·°F, G in lb/ft^2·h, and D_e in ft, the coefficient in Eq. (24.11) is 0.01 and that in Eq. (24.12) is 0.37.

The constant drying rate R_c is simply

$$R_c = \frac{\dot{m}_v}{A} = \frac{h_y (T - T_i)}{\lambda_i} \tag{24.13}$$

Critical moisture content

The point at which the constant-rate period ends (whether or not the drying rate is truly constant) is called the *critical moisture content*. Sometimes it is clearly identifiable, as shown by point B in Fig. 24.6; more often it is approximate. In Fig. 24.5 it is approximately 25 percent (of the void volume in the solids) for both water and *n*-butanol. It represents the moisture content below which insufficient liquid can be

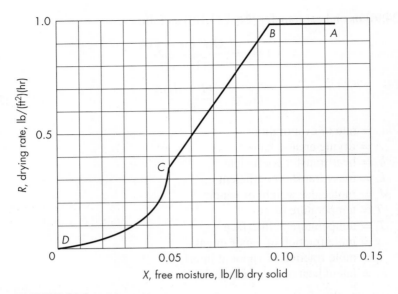

FIGURE 24.6
Drying rate curve for porous ceramic plate. (*After Sherwood and Comings.*[15])

transferred from the interior of the solid to maintain a continuous or nearly continuous liquid film on the surface.

If the initial moisture content of the solid is below the critical value, there will be no constant-rate period.

The critical moisture content is not only a property of the material being dried. It also varies with the thickness of the material, the rate of drying, and the resistances to heat and mass transfer within the solid. Decreasing the thickness of the material gives a lower critical moisture content, because the internal resistances become relatively small and the external resistances control the drying rate for a longer period.

EXAMPLE 24.1. A filter cake 24 in. (610 mm) square and 2 in. (51 mm) thick, supported on a screen, is dried from both sides with air at a wet-bulb temperature of 80°F (26.7°C) and a dry-bulb temperature of 160°F (71.1°C). The air flows parallel with the faces of the cake at a velocity of 8 ft/s (2.44 m/s). The dry density of the cake is 120 lb/ft³ (1,922 kg/m³). The equilibrium moisture content is negligible. Under the conditions of drying the critical moisture is 9 percent, dry basis. (*a*) What is the drying rate during the constant-rate period? (*b*) How long would it take to dry this material from an initial moisture content of 20 percent (dry basis) to a final moisture content of 10 percent? Equivalent diameter D_e is equal to 6 in. (153 mm). Assume that heat transfer by radiation or by conduction is negligible.

Solution

(*a*) The interface temperature T_i is the wet-bulb temperature of the air, 80°F. Then λ_i is, by App. 7, 1,049 Btu/lb. The mass velocity of the air is

$$G = \frac{8 \times 29 \times 492 \times 3,600}{359(460 + 120)} = 1,972 \text{ lb/ft}^2 \cdot \text{h}$$

The coefficient h_y, by Eq. (24.11), in fps units, is

$$h = 0.01 \times \frac{1{,}972^{0.8}}{(0.5)^{0.2}} = 4.97 \text{ Btu/ft}^2 \cdot \text{h} \cdot {}^\circ\text{F}$$

Substituting in Eq. (24.13) gives

$$R_c = \frac{4.97(160 - 80)}{1{,}049} = 0.379 \text{ lb/ft}^2 \cdot \text{h}$$

(b) Since drying is from both faces, area A is $2 \times (24/12)^2 = 8$ ft². The rate of drying $\dot{m}_v$ is therefore

$$\dot{m}_v = 0.379 \times 8 = 3.03 \text{ lb/h}$$

The volume of the cake is $(24/12)^2 \times \frac{2}{12} = 0.667$ ft³, and the mass of bone-dry solids is $120 \times 0.667 = 80$ lb. The quantity of moisture to be vaporized is $80(0.20 - 0.10) = 8$ lb. Drying time t_T is therefore $8/3.03 = 2.64$ h.

Drying in the falling-rate period

Methods of estimating drying rates in the falling-rate period depend on whether the solid is porous or nonporous. In a nonporous material, once there is no more superficial moisture, further drying can occur only at a rate governed by diffusion of internal moisture to the surface. In a porous material other mechanisms appear, and vaporization may even take place inside the solid instead of at the surface.

Nonporous solids. The moisture distribution in a typical nonporous solid is qualitatively consistent with that called for by assuming that the moisture flows by diffusion through the solids, in accordance with Eq. (17.46). It differs somewhat from the theoretical distribution, mainly because the diffusivity varies considerably with moisture content and is especially sensitive to shrinkage.

Equation (17.46) has long been used in estimating the rate of drying of nonporous solids.[9,14] Materials drying in this way are said to be drying by diffusion, although the actual mechanism is probably considerably more complicated than simple diffusion.

Diffusion is characteristic of slow-drying materials. The resistance to mass transfer of water vapor from the solid surface to the air is often negligible, and diffusion in the solid controls the overall drying rate. The moisture content at the surface, therefore, is at or very near the equilibrium value. The velocity of the air has little or no effect, and the humidity of the air influences the process primarily through its effect on the equilibrium moisture content. Since diffusivity increases with temperature, the rate of drying increases with the temperature of the solid.

Porous solids. The moisture distribution in a porous solid during drying differs greatly from that indicated by diffusion in a nonporous material. Moisture flows through porous solids by capillarity[1,2,4] and to some extent by surface diffusion (see Chap. 25, p. 828). A porous material contains a complicated network of interconnecting pores and channels, the cross sections of which vary greatly. At the surface are the mouths of pores of various sizes. As water is removed by vaporization, a meniscus across each pore is formed, which sets up capillary forces by the interfacial tension between the water and the solid. The capillary forces possess components in the direction perpendicular to the surface of the solid. It is these forces that provide the driving force for the movement of water through the pores toward the surface.

The strength of capillary forces at a given point in a pore depends on the curvature of the meniscus, which is a function of the pore cross section. Small pores develop greater capillary forces than large ones, and small pores, therefore, can pull water out of the large pores. As the water at the surface is depleted, the large pores tend to empty first. Air must displace the water so removed. This air enters either through the mouths of the larger pores at the drying surface or from the sides and back of the material if drying is from one side only.

The rate-of-drying curve for a typical porous solid having small pores is shown in Fig. 24.6. As long as the delivery of water from the interior to the surface is sufficient to keep the surface completely wet, the drying rate is constant. The pores are progressively depleted of water, and at the critical point the surface layer of water begins to recede into the solid. This starts with the larger pores. The high points on the surface of the solid begin to emerge from the liquid, and the area available for mass transfer from the solid into the air decreases. Then, although the rate of evaporation per unit *wetted* area remains unchanged, the rate based on the *total* surface, including both wet and dry areas, is less than that in the constant-rate period. The rate continues to decrease as the fraction of dry surface increases.

The first section of the falling-rate period is shown by line *BC* in Fig. 24.6. The rate of drying during this period depends on the same factors that are active during the constant-rate period, since the mechanism of evaporation is unchanged and the vaporization zone is at or near the surface. The water in the pores is the continuous phase, and the air is the dispersed phase. In the first falling-rate period, the rate-of-drying curve is usually linear.

As the water is progressively removed from the solid, the fraction of the pore volume that is occupied by air increases. When the fraction reaches a certain limit, there is insufficient water left to maintain continuous films across the pores and the pores fill with air, which now becomes the continuous phase. The remaining water is relegated to small isolated pools in the corners and interstices of the pores. When this state appears, the rate of drying again suddenly decreases, as shown by line *CD* in Fig. 24.6. The moisture content at which this break appears, shown by point *C* in Fig. 24.6, is called the second critical point, and the period that it initiates is called the second falling-rate period.

In this final drying period, the vaporization rate is practically independent of the velocity of the air. The water vapor must diffuse through the solid, and the heat of vaporization must be transmitted to the vaporization zones by conduction through the solid. Temperature gradients are set up in the solid, and the temperature of the solid surface approaches the dry-bulb temperature of the air. For fine pores, the rate-of-drying curve during the second falling-rate period conforms to the diffusion model, and the drying-rate curve is concave upward.

Calculation of drying time under constant drying conditions

For drying under constant drying conditions, the time of drying can be determined from the rate-of-drying curve if it can be constructed. Often the only source of this curve is an experiment on the material to be dried, and this gives the drying time directly. Drying rate curves for one set of conditions often may be modified to other conditions, and then working back from the drying rate curve to drying time is useful. By definition,

$$R = -\frac{dm_v}{A\,dt} = -\frac{m_s}{A}\frac{dX}{dt} \tag{24.14}$$

Integrating Eq. (24.14) between X_1 and X_2, the initial and final free-moisture contents, respectively, gives

$$t_T = \frac{m_s}{A}\int_{X_2}^{X_1}\frac{dX}{R} \tag{24.15}$$

where t_T is the total drying time. Equation (24.15) may be integrated numerically from the rate-of-drying curve or analytically if equations are available giving R as a function of X.

In the constant-rate period $R = R_c$ and the drying time is simply

$$t_c = \frac{m_s(X_1 - X_2)}{A R_c} \tag{24.16}$$

If R is linear in X, as with many porous solids, during the falling-rate period,

$$R = aX + b \tag{24.17}$$

where a and b are constants, and $dR = a\,dX$. Substitution for dX in Eq. (24.15) gives, for the time required in the falling-rate period,

$$t_f = \frac{m_s}{aA}\int_{R_2}^{R_1}\frac{dR}{R} = \frac{m_s}{aA}\ln\frac{R_1}{R_2} \tag{24.18}$$

where R_1 and R_2 are the drying rates corresponding to the initial and final moisture contents. The constant a is the slope of the drying rate curve. If there are two falling-rate periods, as in Fig. 24.6, a may be written as

$$a = \frac{R_c - R'}{X_c - X'} \tag{24.19}$$

where R_c = rate at first critical point
R' = rate at second critical point
X_c = free-moisture content at first critical point
X' = free-moisture content at second critical point

Substitution of a from Eq. (24.19) into Eq. (24.18) gives

$$t_f = \frac{m_s(X_c - X')}{A(R_c - R')}\ln\frac{R_1}{R_2} \tag{24.20}$$

When the drying process covers both a constant-rate period and a falling-rate period, X_2 of Eq. (24.16) equals X_c, and R_1 of Eq. (24.20) equals R_c. The total time of drying t_T is then

$$t_T = t_c + t_f = \frac{m_s}{A}\left(\frac{X_1 - X_c}{R_c} + \frac{1}{a}\ln\frac{R_c}{R_2}\right) \tag{24.21}$$

Here X_1 is the moisture content at the start of the entire process, and R_2 is the drying rate at the end of the process.

In some situations, a single straight line passing through the origin adequately represents the entire falling-rate period. The point (X_c, R_c) lies on this line. When this approximation may be made, Eq. (24.21) may be simplified by noting that $a = R_c/X_c$ and that $R_c/R_2 = X_c/X_2$. Equation (24.21) then becomes

$$t_T = \frac{m_s}{A R_c} \left(X_1 - X_c + X_c \ln \frac{X_c}{X_2} \right)$$

(24.22)

Here X_2 is the moisture content at the end of the entire process.

THROUGH-CIRCULATION DRYING

If the particles of wet solid are large enough, gas may be passed through the bed instead of across it, usually with a significant increase in drying rate. Even if the individual particles are too small to permit this, the material in many cases may be "preformed" into a condition suitable for through-circulation drying. Filter cake, for example, may be granulated or extruded into "biscuits" or spaghettilike cylinders, perhaps 6 mm in diameter and several centimeters long. Preforms usually retain their shape during drying and form a permeable bed of fairly high porosity.

The rates of heat and mass transfer to particle surfaces in a bed of solids can be calculated from Eq. (17.74), as illustrated in Example 24.2 below. The results of such calculations, however, are best used for preliminary estimates only, for it is always advisable, and usually necessary, to make experimental tests with the actual material to be dried.

EXAMPLE 24.2. The filter cake of Example 24.1 is extruded onto a screen in the form of cylinders $\frac{1}{4}$ in. in diameter and 3 in. long. The solids loading is 8 lb of dry solids per square foot of screen surface. The bed porosity is 45 percent. Air at 160°F (dry-bulb) and with a wet-bulb temperature of 80°F is passed through the bed at a superficial velocity of 4 ft/s. (The air velocity is lower than that in Example 24.1 to minimize the pressure drop and avoid dusting of the dry material.) The critical moisture content of the solids, as before, is 9 percent. How long will it take to dry the solids from 20 to 10 percent moisture?

Solution. Since the critical moisture content is less than 10 percent, all drying takes place in the constant-rate period, and the vaporization temperature, as before, is 80°F and λ is 1,049 Btu/lb. Per square foot of screen surface, therefore, the mass of water to be evaporated is $8(0.20 - 0.10) = 0.8$ lb, and the quantity of heat to be transferred Q_T is $0.8 \times 1,049 = 839$ Btu.

The mass of dry solids in one cylinder is

$$m_p = \frac{\pi \times \left(\frac{1}{4}\right)^2}{4 \times 144} \times \frac{3}{12} \times 120 = 0.0102 \text{ lb}$$

Neglecting the ends of the pieces, the surface area of one cylinder is

$$A_p = \frac{\pi \times \frac{1}{4}}{12} \times \frac{3}{12} = 0.0164 \text{ ft}^2$$

The total area exposed by 8 lb of solids is

$$A = \frac{8}{0.0102} \times 0.0164 = 12.9 \text{ ft}^2$$

The heat-transfer coefficient is found from the equivalent form of Eq. (17.74):

$$\frac{hD}{k} = 1.17 \, \text{Re}^{0.585} \, \text{Pr}^{1/3}$$

For air at 1 atm and 160°F, the properties are

$$\rho = \frac{29}{359} \times \frac{492}{620} = 0.0641 \text{ lb/ft}^3 \qquad \mu = 0.020 \text{ cP} \qquad \text{(App. 8)}$$

$$k = 0.0171 \text{ Btu/ft} \cdot \text{h} \cdot {}^\circ\text{F} \qquad \text{(App. 12)} \qquad c_p = 0.25 \text{ Btu/lb} \cdot {}^\circ\text{F} \qquad \text{(App. 14)}$$

The Reynolds number, based on the diameter of the particles, is

$$\text{Re} = \frac{D_p G}{\mu} = \frac{D_p \bar{V} \rho}{\mu}$$

$$= \frac{\frac{1}{48} \times 4 \times 0.0641}{0.020 \times 6.72 \times 10^{-4}} = 397$$

The Prandtl number is

$$\text{Pr} = \frac{0.25 \times 0.020 \times 2.42}{0.0171} = 0.71$$

From Eq. (17.74),

$$h = \frac{0.0171 \times 1.17 \times 397^{0.585} \times 0.71^{1/3}}{\frac{1}{48}}$$

$$= 28.4 \text{ Btu/ft}^2 \cdot \text{h} \cdot {}^\circ\text{F}$$

The heat transferred from the gas to a thin section of bed is

$$\dot{m}_g c_s \, dT_h = h \, d A (T_h - T_w)$$

which integrates to

$$\ln \frac{T_{hb} - T_w}{T_{ha} - T_w} = \frac{h A}{\dot{m}_g c_s}$$

Per square foot of screen surface, the mass flow rate of the air is

$$\dot{m}_g = 4 \times 3{,}600 \times 0.0641 = 923 \text{ lb/h}$$

From Fig. 19.2, $c_s = 0.245$. Then

$$\ln \frac{T_{hb} - T_w}{T_{ha} - T_w} = \frac{28.4 \times 12.9}{923 \times 0.245} = 1.62$$

Since $T_{hb} - T_w = 160 - 80 = 80°F$,

$$T_{ha} - T_w = 15.83°F$$

$$\overline{\Delta T_L} = \frac{80 - 15.83}{\ln(80/15.83)} = 39.6°F$$

If q_T is the rate of heat transfer and t_T is the drying time,

$$q_T = \frac{Q_T}{t_T} = hA \,\overline{\Delta T_L}$$

$$= 28.4 \times 12.9 \times 39.6$$

$$= 14{,}500 \text{ Btu/h}$$

From this, since $Q_T = 839$ Btu,

$$t_T = \frac{839}{14{,}500} = 0.058 \text{ h, or 3.47 min}$$

Note that the final moisture content of 10 percent in Example 24.2 is the average value for all the solids. The top layer, exposed to the hottest air, would be dried to 10 percent moisture in about 1.7 min, so that at the end of the 3.47-min drying time the top layer would be drier than the average and the bottom layer would be somewhat wetter. Note also that the drying time in Example 24.2 is less than 2 percent of the time required with cross-circulation drying (Example 24.1). Although the solids loading in Example 24.2 is only 40 percent of that in Example 24.1, the amount of solids dried per unit time with through-circulation drying is more than 20 times as much as with cross-circulation drying. It is true that this is for constant-rate drying, but drying rates in the falling-rate period are also enhanced, primarily because of the reduction in diffusion distance.

Drying of Suspended Particles

The rate of heat transfer from a gas to a single particle may be estimated from Eq. (12.64), provided the velocity difference between the particle or gas is known. Rates of mass transfer from the surface of a solid particle or drop may be found from Eq. (17.71) or from Fig. 17.7. Often, as in a tower dryer or rotary dryer, only part of the drying is done while the particles are being showered through the gas, and such dryers are therefore designed using empirical equations (see "Rotary Dryers," page 797).

Transfer to suspended particles is estimated from an equation like Eq. (17.71); it is often appropriate simply to assume that the Nusselt or Sherwood number is 2.0, the limit for a relative velocity of zero. As discussed later under fluid-bed dryers, however, it is not always possible to calculate an average temperature difference or mass-transfer driving force between the gas and solid, and drying rates must be found by experiment.

Internal diffusion in spherical particles may be estimated from an equation analogous to Eq. (10.32).

The time required to dry individual particles is usually very short, so short, in fact, that the terms *constant rate* and *falling rate* have no significance. In flash dryers and some kinds of spray dryers, the drying is all finished in $\frac{1}{2}$ to 5 s.

FREEZE-DRYING

Lyophilization, or freeze drying, is a process for drying foods, vitamins, and other heat-sensitive products at temperatures below 0°C.[7,13] The material to be dried is quickly frozen and placed in thin layers on trays in special vacuum chambers, or it can be frozen in place by refrigerant circulating inside the hollow trays. Full vacuum is applied and fluid is passed through the trays to supply the heat of sublimation. As drying proceeds, the ice front retreats from the surface, leaving behind a zone of nearly dry porous solid. For most of the drying period, the temperature at the ice front is nearly constant at a value well below 0°C. This temperature is determined by a balance between the rate of heat transfer through the dry zone and the external gas film and the rate of mass transfer of water vapor through the same regions. This is similar to the balance reached in measuring the wet-bulb temperature of a gas, but in freeze-drying, the resistance to heat transfer and mass transfer increase with time.

Temperature and partial pressure profiles are sketched in Fig. 24.7 for a material being dried from both sides. Screens or ribs between the material and the heating surface permit escape of water vapor, but they add some resistance to heat transfer.

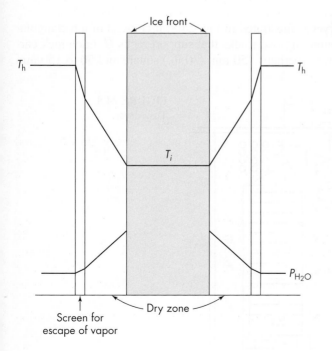

FIGURE 24.7
Gradients in freeze-drying.

The major resistances to heat and mass transfer are in the dry zone of the material, and the drying time is nearly proportional to the square of the sample thickness. Drying times are generally several hours long, because the heat required for sublimation is much greater than the change in enthalpy of the dry solid, and the temperature differences are smaller than in other types of drying.

DRYING EQUIPMENT

Of the many types of commercial dryers available,[11a,17] only a small number of important types are considered here. The first and larger group comprises dryers for rigid or granular solids and semisolid pastes; the second group consists of dryers that can accept slurry or liquid feeds.

Dryers for Solids and Pastes

Typical dryers for solids and pastes include tray and screen-conveyor dryers for materials that cannot be agitated and tower, rotary, screw-conveyor, fluid-bed, and flash dryers where agitation is permissible. In the following treatment these types are ordered, as far as possible, according to the degree of agitation and the method of exposing the solid to the gas or contacting it with a hot surface, as discussed at the beginning of this chapter. The ordering is complicated, however, by the fact that some types of dryers may be either adiabatic or nonadiabatic or a combination of both.

Tray dryers

A typical batch tray dryer is illustrated in Fig. 24.8. It consists of a rectangular chamber of sheet metal containing two trucks that support racks H. Each rack carries a number of shallow trays, perhaps 750 mm (30 in.) square and 50 to 150 mm

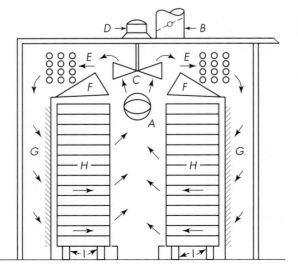

FIGURE 24.8
Tray dryer.

(2 to 6 in.) deep, that are loaded with the material to be dried. Heated air is circulated at 2 to 5 m/s (7 to 15 ft/s) between the trays by fan C and motor D and passes over heaters E. Baffles G distribute the air uniformly over the stack of trays. Some moist air is continuously vented through exhaust duct B; makeup fresh air enters through inlet A. The racks are mounted on truck wheels I, so that at the end of the drying cycle the trucks can be pulled out of the chamber and taken to a tray-dumping station.

Tray dryers are useful when the production rate is small. They can dry almost anything, but because of the labor required for loading and unloading, they are expensive to operate. They find most frequent application for valuable products like dyes and pharmaceuticals. Drying by circulation of air across stationary layers of solid is slow, and drying cycles are long: 3 to 48 h per batch. Occasionally through-circulation drying is used, but this is usually neither economical nor necessary in batch dryers because shortening the drying cycle does not reduce the labor required for each batch. Energy savings may be significant, however.

Tray dryers may be operated under vacuum, often with indirect heating. The trays may rest on hollow metal plates supplied with steam or hot water or may themselves contain spaces for a heating fluid. Vapor from the solid is removed by an ejector or vacuum pump.

Screen-conveyor dryers

A typical through-circulation screen-conveyor dryer is shown in Fig. 24.9. A layer 25 to 150 mm (1 to 6 in.) thick of material to be dried is slowly carried on a traveling metal screen through a long drying chamber or tunnel. The chamber consists of a series of separate sections, each with its own fan and air heater. At the inlet end of the dryer, the air usually passes upward through the screen and the solids; near the discharge end, where the material is dry and may be dusty, air is passed downward through the screen. The air temperature and humidity may differ in the various sections, to give optimum conditions for drying at each point.

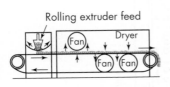

(a) Path of travel of permeable bed through a 3-unit through-circulation dryer

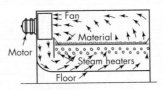

(b) Air flow in wet end

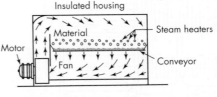

(c) Air flow in dry end

FIGURE 24.9
Through-circulation screen-conveyor dryer.

Screen-conveyor dryers are typically 2 m (6 ft) wide and 4 to 50 m (12 to 150 ft) long, giving drying times of 5 to 120 min. The minimum screen size is about 30-mesh. Coarse granular, flaky, or fibrous materials can be dried by through circulation without any pretreatment and without loss of material through the screen. Pastes and filter cakes of fine particles, however, must be preformed before they can be handled on a screen-conveyor dryer. The aggregates usually retain their shape while being dried and do not dust through the screen except in small amounts. Provision is sometimes made for recovering any fines that do sift through the screen.

Screen-conveyor dryers handle a variety of solids continuously and with a very gentle action; their cost is reasonable, and their steam consumption is low, typically 2 kg of steam per kilogram of water evaporated. Air may be recirculated through, and vented from, each section separately or passed from one section to another countercurrently to the solid. These dryers are particularly applicable when the drying conditions must be appreciably changed as the moisture content of the solid is reduced. They are designed by methods similar to that illustrated in Example 24.2.

Tower dryers

A tower dryer contains a series of circular trays mounted one above the other on a central rotating shaft. Solid feed dropped on the topmost tray is exposed to a stream of hot air or gas that passes across the tray. The solid is then scraped off and dropped to the tray below. It travels in this way through the dryer, discharging as dry product from the bottom of the tower. The flow of solids and gas may be either parallel or countercurrent.

The *turbodryer* illustrated in Fig. 24.10 is a tower dryer with internal recirculation of the heating gas. Turbine fans circulate the air or gas outward between some

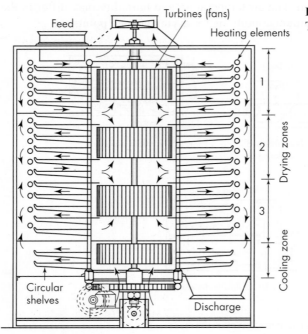

FIGURE 24.10
Turbodryer.

of the trays, over heating elements, and inward between other trays. Gas velocities are commonly 0.6 to 2.4 m/s (2 to 8 ft/s). The bottom two trays of the dryer shown in Fig. 24.10 constitute a cooling section for dry solids. Preheated air is usually drawn in the bottom of the tower and discharged from the top, giving countercurrent flow. A turbodryer functions partly by cross-circulation drying, as in a tray dryer, and partly by showering the particles through the hot gas as they tumble from one tray to another.

Rotary dryers

A rotary dryer consists of a revolving cylindrical shell, horizontal or slightly inclined toward the outlet. Wet feed enters one end of the cylinder; dry material discharges from the other. As the shell rotates, internal flights lift the solids and shower them down through the interior of the shell. Rotary dryers are heated by direct contact of gas with the solids, by hot gas passing through an external jacket, or by steam condensing in a set of longitudinal tubes mounted on the inner surface of the shell. The last of these types is called a steam-tube rotary dryer. In a direct-indirect rotary dryer, hot gas first passes through the jacket and then through the shell, where it comes into contact with the solids.

A typical adiabatic countercurrent air-heated rotary dryer is shown in Fig. 24.11. A rotating shell A made of sheet steel is supported on two sets of rollers B and driven by a gear and pinion C. At the upper end is a hood D, which connects through fan E to a stack and a spout F, which brings in wet material from the feed hopper. Flights G, which lift the material being dried and shower it down through the current of hot air, are welded inside the shell. At the lower end the dried product discharges into a screw conveyor H. Just beyond the screw conveyor is a set of steam-heated extended-surface pipes that preheat the air. The air is moved through the dryer by a fan, which may, if desired, discharge into the air heater so that the whole system is under a positive pressure. Alternatively, the fan may be placed in the stack as shown, so that it draws air through the dryer and keeps the system under a slight

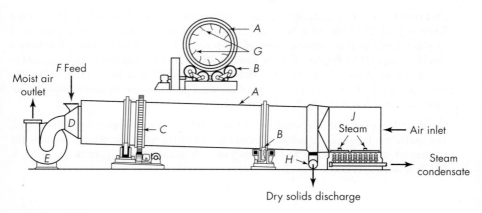

FIGURE 24.11
Countercurrent air-heated rotary dryer: A, dryer shell; B, shell-supporting rolls; C, drive gear; D, air discharge hood; E, discharge fan; F, feed chute; G, lifting flights; H, product discharge; J, air heater.

vacuum. This is desirable when the material tends to dust. Rotary dryers of this kind are widely used for salt, sugar, and all kinds of granular and crystalline materials that must be kept clean and may not be directly exposed to very hot flue gases.

The allowable mass velocity of the gas in a direct-contact rotary dryer depends on the dusting characteristics of the solid being dried and ranges from 2,000 to 25,000 kg/m²·h (400 to 5,000 lb/ft²·h) for coarse particles. Inlet gas temperatures are typically 120 to 175°C (250 to 350°F) for steam-heated air and 550 to 800°C (1,000 to 1,500°F) for flue gas from a furnace. Dryer diameters range from 1 to 3 m (3 to 10 ft); the peripheral speed of the shell is commonly 20 to 25 m/min (60 to 75 ft/min).

Direct-contact rotary dryers are designed on the basis of heat transfer. An empirical *dimensional* equation for the rate of heat transfer q_T in Btu/h is[11b]

$$q_T = \frac{0.5G^{0.67}}{D} V \, \overline{\Delta T}$$

$$= 0.125\pi \, DLG^{0.67} \, \overline{\Delta T} \tag{24.23}$$

where V = dryer volume, ft³
L = dryer length, ft
$\overline{\Delta T}$ = average temperature difference, taken as logarithmic mean of wet-bulb depressions at inlet and outlet of dryer
G = mass velocity, lb/ft²·h
D = dryer diameter, ft

The volumetric heat-transfer coefficient Ua in Btu/ft³·h·°F, from Eq. (24.23), is

$$Ua = \frac{0.5G^{0.67}}{D} \tag{24.24}$$

The proper outlet gas temperature is a matter of economics; it may be estimated from Eq. (24.6) or (24.8) since it has been found empirically[11b] that rotary dryers are operated most economically when N_t is between 1.5 and 2.5.

EXAMPLE 24.3. Calculate the diameter and length of an adiabatic rotary dryer to dry 2,800 lb/h (1,270 kg/h) of a heat-sensitive solid from an initial moisture content of 15 percent to a final moisture content of 0.5 percent, both dry basis. The solids have a specific heat of 0.52 Btu/lb·°F; they enter at 80°F (26.7°C) and must not be heated to a temperature above 125°F (51.7°C). Heating air is available at 260°F (126.7°C) and a humidity of 0.01 lb of water per pound of dry air. The maximum allowable mass velocity of the air is 700 lb/ft²·h (3,420 kg/m²·h).

Solution. In view of the heat sensitivity of the solids, concurrent operation will be used. The outlet gas temperature is found from Eq. (24.8) for adiabatic drying. Assume the number of transfer units is 1.5. The inlet wet-bulb temperature T_{wb}, from Fig. 19.2, is 102°F. Since T_{hb} is 260°F, Eq. (24.8) gives

$$N_t = 1.5 = \ln \frac{260 - 102}{T_{ha} - 102}$$

From this, $T_{ha} = 137°F$ and T_{sb} may reasonably be set at the maximum allowable value, 125°F.

Other quantities needed are

$$\lambda \text{ at } 102°F = 1,036 \text{ Btu/lb} \qquad \text{(App. 7)}$$

Specific heats, in Btu/lb · °F, are

$$c_{ps} = 0.52 \qquad c_{pv} = 0.45 \qquad \text{(App. 14)} \qquad c_{pL} = 1.0$$

Also

$$X_a = 0.15 \qquad X_b = 0.005 \qquad \dot{m}_s = 2,800 \text{ lb/h}$$

The rate of mass transfer,

$$\dot{m}_v = \dot{m}_s(X_a - X_b) = 2,800(0.15 - 0.005) = 406 \text{ lb/h}$$

The heat duty is found from substitution in Eq. (24.1):

$$\frac{q_T}{\dot{m}_s} = 0.52(125 - 80) + 0.15 \times 1.0(102 - 80)$$

$$+ (0.15 - 0.005)(1,036) + 0.005 \times 1.0(125 - 102)$$
$$+ 0.145 \times 0.45(137 - 102)$$
$$= 23.4 + 3.3 + 150.2 + 0.1 + 2.3 = 179.3 \text{ Btu/lb}$$

Only the first and third terms are significant. From this, $q_T = 179.3 \times 2,800 = 502,040$ Btu/h.

The flow rate of entering air is found from a heat balance and the humid heat c_{sb}. From Fig. 19.2, $c_{sb} = 0.245$ Btu/lb · °F. Thus

$$\dot{m}_g(1 + \mathcal{H}_b) = \frac{q_T}{c_{sb}(T_{hb} - T_{ha})}$$

$$= \frac{502,040}{0.245(260 - 137)} = 16,660 \text{ lb/h}$$

Since $\mathcal{H}_b = 0.01$, $\dot{m}_g = 16,660/1.01 = 16,495$ lb/h of dry air.
The outlet humidity

$$\mathcal{H}_a = \mathcal{H}_b + \dot{m}_v/\dot{m}_g = 0.01 + \frac{406}{16,495} = 0.0346 \text{ lb/lb}$$

At a dry-bulb temperature T_{ha} of 137°F, the wet-bulb temperature T_{wa} for $\mathcal{H}_a = 0.0346$ is 102°F, the same as T_{wb} (as it should be for adiabatic drying).

The dryer diameter is found from the allowable mass velocity and the flow rate of the entering air. For $G = 700$ lb/ft² · h, the cross-sectional area of the dryer must be $16,660/700 = 23.8$ ft², and the dryer diameter is

$$D = \left(\frac{4 \times 23.8}{\pi}\right)^{0.5} = 5.50 \text{ ft (1.68 m)}$$

The dryer length is given by Eq. (24.23):

$$L = \frac{q_T}{0.125\pi \, DG^{0.67} \, \Delta T}$$

The logarithmic mean temperature difference is

$$\overline{\Delta T} = \frac{260 - 102 - (137 - 102)}{\ln[(260 - 102)/(137 - 102)]} = 81.6°F$$

Therefore,

$$L = \frac{502,040}{0.125\pi \times 5.5 \times 700^{0.67} \times 81.6}$$
$$= 35.4, \text{ say, } 36 \text{ ft } (11.0 \text{ m})$$

This gives a ratio L/D of $36/5.5 = 6.54$, a reasonable value for rotary dryers.

Screw-conveyor dryers

A screw conveyor is a continuous indirect-heat dryer, consisting essentially of a horizontal screw conveyor (or paddle conveyor) enclosed in a cylindrical jacketed shell. Solid fed in one end is conveyed slowly through the heated zone and discharges from the other end. The vapor evolved is withdrawn through pipes set in the roof of the shell. The shell is 75 to 600 mm (3 to 24 in.) in diameter and up to 6 m (20 ft) long; when greater length is required, several conveyors are set one above another in a bank. Often the bottom unit in such a bank is a cooler in which water or another coolant in the jacket lowers the temperature of the dried solids before they are discharged.

The rate of rotation of the conveyor is slow, from 2 to 30 rpm. Heat-transfer coefficients are based on the entire inner surface of the shell, even though the shell runs only 10 to 60 percent full. The coefficient depends on the loading in the shell and on the conveyor speed. It ranges, for many solids, between 15 and 60 $W/m^2 \cdot °C$ (3 and 10 $Btu/ft^2 \cdot h \cdot °F$).

Screw-conveyor dryers handle solids that are too fine and too sticky for rotary dryers. They are completely enclosed and permit recovery of solvent vapors with little or no dilution by air. When provided with appropriate feeders, they can be operated under moderate vacuum. Thus they are adaptable to the continuous removal and recovery of volatile solvents from solvent-wet solids, such as spent meal from leaching operations. For this reason they are sometimes known as *desolventizers*.

A related type of equipment is described later under "Thin-Film Dryers."

Fluid-bed dryers

Dryers in which the solids are fluidized by the drying gas find applications in a variety of drying problems.[3] The particles are fluidized by air or gas in a boiling-bed unit, as shown in Fig. 24.12. Mixing and heat transfer are very rapid. Wet feed is admitted to the top of the bed; dry product is taken out from the side, near the bottom. In the dryer shown in Fig. 24.12 there is a random distribution of residence times; the average time a particle stays in the dryer is typically 30 to 120 s when only surface liquid is vaporized and up to 15 to 30 min if there is also internal diffusion. Small particles are heated essentially to the exit dry-bulb temperature of the fluidizing gas; consequently, thermally sensitive materials must be dried in a relatively cool suspending medium. Even so, the inlet gas may be hot, for it mixes so rapidly that the temperature is virtually uniform, at the exit gas temperature, throughout the

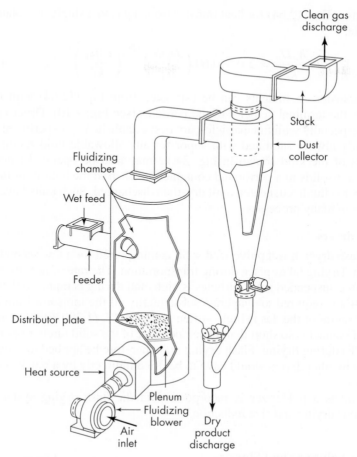

FIGURE 24.12
Continuous fluid-bed dryer.

bed. If fine particles are present, either from the feed or from particle breakage in the fluidized bed, there may be considerable solids carryover with the exit gas, and cyclones and bag filters are needed for fines recovery.

Some rectangular fluid-bed dryers have separately fluidized compartments through which the solids move in sequence from inlet to outlet. These are known as *plug flow dryers;* in them the residence time is almost the same for all particles. Drying conditions can be changed from one compartment to another, and often the last compartment is fluidized with cold gas to cool the solids before discharge.

Because of the rather complex temperature patterns, the true average temperature difference for the dryer as a whole is not easy to define. Usually, in fact, the outlet temperatures of solids and gas are so nearly the same that the difference between them cannot be measured. Heat-transfer coefficients are therefore hard to estimate and may be of limited utility. One general equation that is useful in drying calculations is Eq. (12.64) for heat transfer from a gas to a single or isolated spherical particle:

$$\frac{h_o D_p}{k_f} = 2.0 + 0.60 \left(\frac{D_p G}{\mu_f}\right)^{0.50} \left(\frac{c_p \mu_f}{k_f}\right)^{1/3} \tag{12.64}$$

Heat-transfer coefficients can be estimated from Eq. (12.64) with $G = \rho u_t$, where u_t is the terminal velocity of the particles (see Fig. 7.10). Dryer capacities, however, especially with fine particles, are best established by experiment. The exit gas is nearly always saturated with vapor for any allowable fluidization velocity.

Dryers of the kind shown in Fig. 24.12 may also be operated batchwise. A charge of wet solids in a perforated container attached to the bottom of the fluidizing chamber is fluidized, heated until dry, then discharged. Such units have replaced tray dryers in many processes.

Flash dryers

In a flash dryer, a wet pulverized solid is transported for a few seconds in a hot gas stream. Drying takes place during transportation. The rate of heat transfer from the gas to the suspended solid particles is high, and drying is rapid, so that no more than 3 or 4 s is required to evaporate substantially all the moisture from the solid. The temperature of the gas is high—often about 650°C (1,200°F) at the inlet—but the time of contact is so short that the temperature of the solid rarely rises more than 50°C (90°F) during drying. Flash drying may therefore be applied to sensitive materials that in other dryers would have to be dried indirectly by a much cooler heating medium.

Sometimes a pulverizer is incorporated in the flash drying system to give simultaneous drying and size reduction.

Dryers for Solutions and Slurries

A few types of dryers evaporate solutions and slurries entirely to dryness by thermal means. Typical examples are spray dryers, thin-film dryers, and drum dryers.

Spray dryers

In a spray dryer, a slurry or liquid solution is dispersed into a stream of hot gas in the form of a mist of fine droplets. Moisture is rapidly vaporized from the droplets, leaving residual particles of dry solid, which are then separated from the gas stream. The flow of liquid and gas may be cocurrent, countercurrent, or a combination of both in the same unit.

Droplets are formed inside a cylindrical drying chamber by pressure nozzles, two-fluid nozzles, or, in large dryers, high-speed spray disks. In all cases it is essential to prevent the droplets or wet particles of solid from striking solid surfaces before drying has taken place, so that drying chambers are necessarily large. Diameters of 2.5 to 9 m (8 to 30 ft) are common.

In the typical spray dryer shown in Fig. 24.13, the chamber is a cylinder with a short conical bottom. Liquid feed is pumped into a spray-disk atomizer set in the

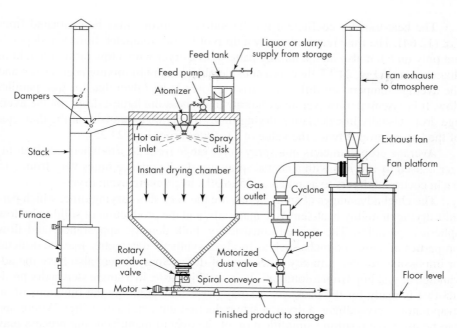

FIGURE 24.13
Spray dryer with parallel flow.

roof of the chamber. In this dryer the spray disk is about 300 mm (12 in.) in diameter and rotates at 5,000 to 10,000 rpm. It atomizes the liquid into tiny drops, which are thrown radially into a stream of hot gas entering near the top of the chamber. Cooled gas is drawn by an exhaust fan through a horizontal discharge line set in the side of the chamber at the bottom of the cylindrical section. The gas passes through a cyclone separator where any entrained particles of solid are removed. Much of the dry solid settles out of the gas into the bottom of the drying chamber, from which it is removed by a rotary valve and screw conveyor and combined with any solid collected in the cyclone.

An equation for the volume-surface mean diameter $\bar{D}_s$ of the drops from a disk atomizer is[6]

$$\frac{\bar{D}_s}{r} = 0.4 \left(\frac{\Gamma}{\rho_L n r^2} \right)^{0.6} \left(\frac{\mu}{\Gamma} \right)^{0.2} \left(\frac{\sigma \rho_L L_p}{\Gamma^2} \right)^{0.1} \tag{24.25}$$

where $\bar{D}_s$ = average drop diameter, m or ft
r = disk radius, m or ft
Γ = spray mass rate per unit length of disk periphery, kg/s·m or lb/s·ft
σ = surface tension of liquid, N/m or lb$_f$/ft
ρ_L = density of liquid, kg/m^3 or lb/ft^3
n = disk speed, r/s
μ = viscosity of liquid, Pa·s or lb/ft·s
L_p = disk periphery, $2\pi r$, m or ft

The heat-transfer coefficient for the individual drops may be estimated from Eq. (12.64). The time required to dry a drop of known diameter, however, depends not only on h but also on the mixing pattern. In a dryer with a low ratio of height to diameter, as in Fig. 24.13, there is a considerable amount of mixing near the top and the average temperature-difference driving force is less than that for true parallel flow. It is essential, however, to evaporate to dryness the largest drop, not the average drop. Usually in calculating drying times it may be assumed that the diameter of the largest drop is twice the value of $\bar{D}_s$ found from Eq. (24.25).

Average drop diameters in a spray dryer range from 20 μm when a disk atomizer is used to 180 μm with a coarse spray nozzle. Residence times vary from 3 to 6 s in cocurrent dryers to as much as 25 to 30 s in countercurrent dryers.

The chief advantages of spray dryers are the very short drying time, which permits drying of highly heat-sensitive materials, and the production of solid or hollow spherical particles. The desired consistency, bulk density, appearance, and flow properties of some products, such as foods or synthetic detergents, may be difficult or impossible to obtain in any other type of dryer. Spray dryers also have the advantage of yielding from a solution, slurry, or thin paste, in a single step, a dry product that is ready for the package. A spray dryer may combine the functions of an evaporator, a crystallizer, a dryer, a size reduction unit, and a classifier. Where one can be used, the resulting simplification of the overall manufacturing process may be considerable.

Considered as dryers alone, spray dryers are not highly efficient. Much heat is ordinarily lost in the discharged gases. They are bulky and very large, often 25 m (80 ft) or more high, and are not always easy to operate. The bulk density of the dry solid—a property of especial importance in packaged products—is often difficult to keep constant, for it may be highly sensitive to changes in the solids content of the feed, to the inlet gas temperature, and to other variables.[11c]

In spray-drying solutions, the evaporation from the surface of the drops leads to initial deposition of solute at the surface before the interior of the drop reaches saturation. The rate of diffusion of solute back into the drop is slower than the flow of water from the interior to the surface, and the entire solute content accumulates at the surface. The final dry particles are often hollow, and the product from a spray dryer is quite porous.

Thin-film dryers

Competitive with spray dryers in some situations are thin-film dryers which can accept a liquid or a slurry feed and produce a dry, free-flowing solid product. They normally are built in two sections, the first of which is a vertical agitated evaporator-dryer similar to the device illustrated in Fig. 16.2. Here most of the liquid is removed from the feed, and partially wet solid is discharged to the second section, as illustrated in Fig. 24.14, in which the residual liquid content of the material from the first section is reduced to the desired value.

The thermal efficiency of thin-film dryers is high, and there is little loss of solids, since little or no gas needs to be drawn through the unit. They are useful in removing and recovering solvents from solid products. They are relatively expensive and are somewhat limited in heat-transfer area. With both aqueous and

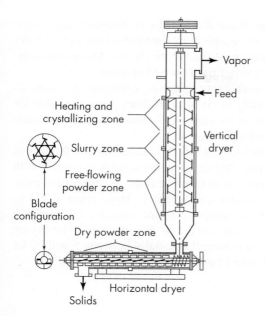

FIGURE 24.14
"Combi" thin-film dryer. (*LCI Corp.*)

solvent feeds the acceptable feed rate is usually between 100 and 200 kg/m²·h (20 and 40 lb/ft²·h).

Drum dryers

A drum dryer consists of one or more heated metal rolls on the outside of which a thin layer of liquid is evaporated to dryness. Dried solid is scraped off the rolls as they slowly revolve.

A typical drum dryer—a double-drum unit with center feed—is shown in Fig. 24.15. Liquid is fed from a trough or perforated pipe into a pool in the space above and between the two rolls. The pool is confined there by stationary end plates. Heat is transferred by conduction to the liquid, which is partly concentrated in the space between the rolls. Concentrated liquid issues from the bottom of the pond as

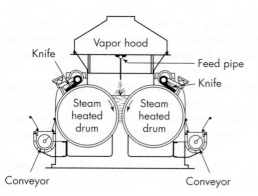

FIGURE 24.15
Double-drum dryer with center feed.

a viscous layer covering the remainder of the drum surfaces. Substantially all the liquid is vaporized from the solid as the drums turn, leaving a thin layer of dry material that is scraped off by doctor blades into conveyors below. Vaporized moisture is collected and removed through a vapor head above the drums.

Double-drum dryers are effective with dilute solutions, concentrated solutions of highly soluble materials, and moderately heavy slurries. They are not suitable for solutions of salts with limited solubility or for slurries of abrasive solids that settle out and create excessive pressure between the drums.

The rolls of a drum dryer are 0.6 to 3 m (2 to 10 ft) in diameter and 0.6 to 4 m (2 to 14 ft) long, revolving at 1 to 10 rpm. The time that the solid is in contact with hot metal is 6 to 15 s, which is short enough to result in little decomposition even of heat-sensitive products. The heat-transfer coefficient is high, from 1,200 to 2,000 W/m$^2 \cdot$ °C (220 to 360 Btu/ft$^2 \cdot$ h $\cdot$ °F) under optimum conditions, although it may be only one-tenth of these values when conditions are adverse.[12] The drying capacity is proportional to the active drum area; it is usually between 5 and 50 kg per square meter of drying surface per hour (1 and 10 lb/ft$^2 \cdot$ h).

Selection of Drying Equipment

The first consideration in selecting a dryer is its operability; above all else, the equipment must produce the desired product in the desired form at the desired rate. Despite the variety of commercial dryers on the market, the various types are largely complementary, not competitive, and the nature of the drying problem dictates the type of dryer that must be used, or at least limits the choice to perhaps two or three possibilities. The final choice is then made on the basis of capital and operating costs. Attention must be paid, however, to the costs of the entire isolation system, not just the drying unit alone.

General considerations

There are some general guidelines for selecting a dryer, but it should be recognized that the rules are far from rigid and exceptions not uncommon. Batch dryers, for example, are most often used when the production rate of dried solid is less than 150 to 200 kg/h (300 to 400 lb/h); continuous dryers are nearly always chosen for production rates greater than 1 or 2 tons/h. At intermediate production rates other factors must be considered. Thermally sensitive materials must be dried at low temperature under vacuum, with a low-temperature heating medium, or very rapidly as in a flash or spray dryer. Fragile crystals must be handled gently as in a tray dryer, a screen-conveyor dryer, or tower dryer.

The dryer must also operate reliably, safely, and economically. Operating and maintenance costs must not be excessive; pollution must be controlled; energy consumption must be minimized. As with other equipment, these considerations may conflict with one another, and a compromise must be reached in finding the optimum dryer for a given service.

As far as the drying operation itself is concerned, adiabatic dryers are generally less expensive than nonadiabatic dryers, in spite of the lower thermal efficiency of adiabatic units. Unfortunately there is usually a lot of dust carryover from adiabatic dryers, and these entrained particles must be removed almost quantitatively from the drying gas. Elaborate particle removal equipment may be needed, equipment that may cost as much as the dryer itself. This often makes adiabatic dryers less economical than a "buttoned-up" nonadiabatic system in which little or no gas is used. Rotary dryers are an example; they were once the most common type of continuous dryer, but because of the inevitable entrainment, other types of dryers which avoid the problem of dust carryover would now, if possible, be selected in their place. Nonadiabatic dryers are always chosen for very fine particles or for solids that are too chemically reactive to be exposed to a stream of gas. They are also widely used for solvent removal and recovery.

Complete isolation train; evaporator-dryers

Many industrial processes involve the isolation of solids from solution in water or other solvents to yield a dry, purified granular product suitable for packaging and sale. The process steps needed to accomplish this are typically evaporation, crystallization, filtration or centrifuging, drying, size reduction, and classification or screening. Five or six pieces of equipment are needed. Sometimes an evaporator-dryer can eliminate the need for many of these. Spray dryers, drum dryers, and thin-film dryers can accept liquid feed and convert it directly into dry product ready for packaging. The added cost of removing moisture thermally instead of mechanically is more than made up by the economies of installing and operating one piece of equipment instead of many. It may, of course, be necessary to treat the liquid feed prior to drying to remove impurities that cannot be permitted to appear in the dried product. It is essential, therefore, that the engineer pay attention to the whole isolation process and not focus narrowly on the drying step alone.

SYMBOLS

A Area for drying, m^2 or ft^2; A_p, surface area of a single particle

a Surface area per unit volume, m^2/m^3 or ft^2/ft^3; also slope of drying rate curve [Eq. (24.17)]

b Constant in Eq. (24.17)

c_p Specific heat at constant pressure, $J/g \cdot °C$ or $Btu/lb \cdot °F$; c_{pL}, of liquid; c_{ps}, of solids; c_{pv}, of vapor

c_s Humid heat, $J/g \cdot °C$ or $Btu/lb \cdot °F$; c_{sb}, at gas inlet

D Diameter, m or ft; D_e, equivalent diameter of channel, m or ft; D_p, of particle; $\bar{D}_s$, volume-surface mean diameter

D_v Volumetric diffusivity of liquid through solid, m^2/h, cm^2/s, or ft^2/h

G Mass velocity of gas, $kg/m^2 \cdot h$ or $lb/ft^2 \cdot h$

$\mathcal{H}$ Humidity of gas, mass of vapor per unit mass of dry gas; $\mathcal{H}_a$, at outlet; $\mathcal{H}_b$, at inlet

h Heat-transfer coefficient, $W/m^2 \cdot °C$ or $Btu/ft^2 \cdot h \cdot °F$; h_o, between gas and particle; h_y, between gas and surface of slab

k Thermal conductivity, W/m·°C or Btu/ft·h·°F; k_f, at mean film temperature

k_y Mass-transfer coefficient, kg mol/m²·h·unit mole-fraction difference or lb mol/ft²·h·unit mole-fraction difference

L Length of dryer, m or ft; L_p, perimeter of atomizing disk

M_v Molecular weight of vapor

m Mass, kg or lb; m_p, of single particle; m_s, of solids; m_v, of vapor

$\dot{m}$ Mass rate of flow, kg/h or lb/h; $\dot{m}_g$, of gas; $\dot{m}_s$, of bone-dry solids; $\dot{m}_v$, rate of vaporization

N_t Number of transfer units

n Rotational speed of disk, r/s

Pr   Prandtl number, $c_p \mu / k$

Q_T Quantity of heat transferred, J or Btu

q_T Rate of heat transfer, W or Btu/h

R Rate of drying, kg/m²·h or lb/ft²·h; R_c, in constant-rate period; R_1, at start of drying; R_2, at end of drying; R', at second critical moisture content

Re   Reynolds number, DG/μ

r Radius of atomizing disk, m or ft

T Temperature, °C or °F; T_h, of heating medium; T_{ha}, at outlet; T_{hb}, at inlet; T_i, at interface; T_s, of solids; T_{sa}, at inlet; T_{sb}, at outlet; T_v, vaporization temperature; T_{va}, vapor temperature at outlet; T_w, wet-bulb temperature; T_{wa}, at outlet; T_{wb}, at inlet

t Drying time, h or s; t_T, total drying time; t_c, in constant-rate period; t_f, in falling-rate period

U Overall heat-transfer coefficient, W/m²·°C or Btu/ft²·h·°F

Ua Volumetric heat-transfer coefficient, W/m³·°C or Btu/ft³·h·°F

u_t Terminal velocity of particle, m/s or ft/s

V Volume of dryer, m³ or ft³

$\bar{V}$ Superficial velocity of gas, m/s or ft/s

X Free-moisture content, mass of water (or other liquid) per unit mass of dry solid; X_T, total moisture content; X_{T1}, X_a, initial total moisture content; X_b, final total moisture content; X_c, free-moisture content at first critical point; X_1, initial value; X_2, final value; X^*, equilibrium value; X', at second critical point

y Mole fraction of vapor in gas phase; y_i, at interface; $\overline{(1 - y)_L}$, log mean value of $1 - y$ and $1 - y_i$

Greek Letters

Γ Liquid rate per foot of disk periphery, kg/s·m or lb/s·ft

$\overline{\Delta T}$ Average temperature difference; $\overline{\Delta T_L}$, log mean value

λ Latent heat of vaporization, J/g or Btu/lb; λ_i, at interface temperature T_i

μ Viscosity, Pa·s, lb/ft·s or lb/ft·h; μ_f, at mean film temperature

ρ Density, kg/m³ or lb/ft³; ρ_L, of liquid

σ Surface tension of liquid, N/m or lb_f/ft

PROBLEMS

24.1. Fluorspar (CaF_2) is to be dried from 6 to 0.4 percent moisture (dry basis) in a countercurrent adiabatic rotary dryer at a rate of 18,000 lb/h of bone-dry solids. The heating air enters at 1000°F with a humidity of 0.03 and a wet-bulb temperature of 150°F. The solids have a specific heat of 0.48 Btu/lb·°F; they enter the dryer at 70°F and leave at 200°F. The maximum allowable mass velocity of the air is 2,000 lb/ft²·h. (a) Assuming Eq. (24.8) applies, what would be the diameter and length of the dryer if $N_t = 2.2$? Is this a reasonable design? (b) Repeat part (a) with $N_t = 1.8$.

24.2. A porous solid is dried in a batch dryer under constant drying conditions. Seven hours are required to reduce the moisture content from 35 to 10 percent. The critical moisture content was found to be 20 percent and the equilibrium moisture 4 percent. All moisture contents are on the dry basis. Assuming that the rate of drying during the falling-rate period is proportional to the free-moisture content, how long should it take to dry a sample of the same solid from 35 to 5 percent under the same drying conditions?

24.3. A slab with a wet weight of 5 kg originally contains 50 percent moisture (wet basis). The slab is 600 by 900 by 75 mm thick. The equilibrium moisture content is 5 percent of the total weight when in contact with air of 20°C and 20 percent humidity. The drying rate is given in Table 24.1 for contact with air of the above quality at a definite velocity. Drying is from one face. How long will it take to dry the slab to 15 percent moisture content (wet basis)?

24.4. A continuous countercurrent dryer is to be designed to dry 800 lb of wet porous solid per hour from 140 percent moisture to 20 percent, both on the dry basis. Air at 120°F dry-bulb and 70°F wet-bulb temperature is to be used. The exit humidity is to be 0.012. The average equilibrium moisture content is 5 percent of the dry weight. The total moisture content (dry basis) at the critical point is 40 percent. The stock may be assumed to remain at a temperature 3°F above that of the wet-bulb temperature of the air throughout the dryer. The heat-transfer coefficient is 12 Btu/ft²·h·°F. The area exposed to the air is 1.1 ft² per pound of dry solids. How long must the solids remain in the dryer?

24.5. Show that the dimensional equation for h_y, Eq. (24.11), is consistent with the dimensionless equation for turbulent flow, Eq. (12.32), for a certain range of dryer conditions.

24.6. (a) Show that for a small drop of dilute slurry evaporating in a spray dryer, the drying time is proportional to the square of the initial size of the droplet. (b) What is the approximate drying time for a 50-μm droplet when the average temperature driving force is 50°C?

TABLE 24.1
Data for Prob. 24.3

Wet-slab weight, kg	9.1	7.2	5.3	4.2	3.3	2.9	2.7
Drying rate, kg/m²·h	4.9	4.9	4.4	3.9	3.4	2.0	1.0

24.7. A filter cake 2 in. thick in a tray dryer is exposed to a stream of air at 160°F. Estimate the effective coefficient for radiant heat transfer from the tray above, and compare this with the convective coefficient in Example 24.1, where the average air velocity was 8 ft/s. By what factor would the radiant heat transfer increase the initial drying rate? Would the total drying time be reduced by the same factor?

24.8. Spherical alumina catalyst particles with 30 percent moisture and an average size of 300 μm are dried in a batch fluidized bed with air entering at 250°F, a superficial velocity of 1.2 ft/s, and a humidity of 0.016 lb H_2O/lb dry air. (*a*) If the settled bed depth is 4 ft and the initial solids temperature is 80°F, estimate the exit temperature of the air. (*b*) What is the temperature of the solids during the constant-rate period? (*c*) What is the maximum drying rate, in pounds of water removed per hour per pound of catalyst?

24.9. (*a*) Calculate the mean droplet diameter for a spray dryer with a 6-in. rotary atomizer turning at 10,000 rpm. The feed rate is 30 lb/min at 120°F. Assume the slurry density is 70 lb/ft^3 and the surface tension is that of water. (*b*) In scaling up to a larger dryer with a 12-in. atomizer, what rotation rate should be used to obtain the same droplet size?

24.10. A food product containing 65 percent moisture is to be freeze-dried in a vacuum dryer operating at 0.5 mm Hg and a condensing temperature of −25°C. The material is 1.0 cm thick, and drying is to take place from both sides, with a 1-mm gap between the sample and the heating surfaces, which are at −5°C. The thermal conductivities of the dry solid and the vapor film are about 1.2×10^{-4} and 4.8×10^{-5} cal/s·cm·K, respectively. (*a*) Allowing for radiation, what is the coefficient for heat transfer from the hot surface to the exterior of the solid? (*b*) When one-half the moisture has been removed, what is the overall heat-transfer coefficient? (*c*) Using this value as an average coefficient, estimate the required time for drying. (*d*) If the material was in layers 0.5 cm thick, by what factor would the drying time be reduced?

REFERENCES

1. Ceaglske, N. H., and O. A. Hougen. *Trans. AIChE* **33**:283 (1937).
2. Ceaglske, N. H., and F. C. Kiesling. *Trans. AIChE* **36**:211 (1940).
3. Clark, W. E. *Chem. Eng.* **74**(6):177 (1967).
4. Comings, E. W., and T. K. Sherwood. *Ind. Eng. Chem.* **26**:1096 (1934).
5. Dittman, F. W. *Chem. Eng.* **84**(2):106 (1977).
6. Friedman, S. J., F. A. Gluckert, and W. R. Marshall, Jr. *Chem. Eng. Progr.* **48**:181 (1952).
7. King, C. J. "Freeze Drying." In *Unit Operations Handbook*, ed. J. J. McKetta. New York: Marcel Dekker, 1993.
8. Morgan, R. P., and S. Yerazunis. *Chem. Eng. Prog. Symp. Ser.* **67**(79):1 (1967).
9. Newman, A. B. *Trans. AIChE* **27**:203, 310 (1931).
10. Perry, R. H., and D. W. Green (eds.). *Perry's Chemical Engineers' Handbook*, 6th ed. New York: McGraw-Hill, 1984, p. **20**-20.

11. Perry, R. H., and D. W. Green (eds.). *Perry's Chemical Engineers' Handbook,* 7th ed. New York: McGraw-Hill, 1997; (*a*) pp. **12**-36 to **12**-90; (*b*) p. **12**-54; (*c*) p. **12**-85.
12. Riegel, E. R. *Chemical Process Machinery,* 2nd ed. New York: Reinhold, 1953, chap. 17.
13. Sandall, O. C., C. J. King, and C. R. Wilke. *AIChEJ.* **13**:428 (1967).
14. Sherwood, T. K. *Ind. Eng. Chem.* **21**:12 (1929).
15. Sherwood, T. K., and E. W. Comings. *Trans. AIChE* **28**:118 (1932).
16. Treybal, R. E. *Mass Transfer Operations,* 3rd ed. New York: McGraw-Hill, 1980, p. 675.
17. Wallace, S. M. *Chemical Process Equipment.* Stoneham, MA: Butterworths, 1988, pp. 237–277.

CHAPTER 25

Fixed-Bed Separations

In several processes for separating gaseous or liquid mixtures, the fluid is contacted with small particles of a porous solid which selectively adsorbs or complexes with certain components of the feed. In adsorption, the solid is usually held in a fixed bed, and fluid is passed continuously through the bed until the solid is nearly saturated. The flow is then switched to a second bed, and the saturated bed is replaced or regenerated. Ion exchange is another process that is carried out in this semibatch fashion in a fixed bed. Water to be softened or deionized is passed over beds of ion-exchange resin until the resin becomes nearly saturated. For both adsorption and ion exchange, continuous countercurrent operation could be achieved by moving the solids through the bed and continuously replacing spent particles with fresh ones. However, this method of operation is not often used, because it is difficult to get uniform flow of solids.

Chromatography is a process similar to adsorption in that gas or liquid mixtures are passed through a bed of porous particles, but the feed is introduced in small pulses, rather than continuously. The individual components move through the bed at different rates and are collected at the exit. The bed is continuously regenerated by passage of the carrier gas or liquid and can be operated for long periods, but only small amounts of the feed mixture are separated at a time.

For all these processes, the performance depends on fluid-solid equilibria and on mass-transfer rates, which are discussed in the following sections.

ADSORPTION

Adsorbents and adsorption processes

Most adsorbents are highly porous materials, and adsorption takes place primarily on the walls of the pores or at specific sites inside the particle. Because the pores are generally very small, the internal surface area is orders of magnitude greater than the external area and is often 500 to 1,000 m^2/g. Separation occurs because differences in molecular weight, shape, or polarity cause some molecules to be held more strongly on the surface than others or because the pores are too small to admit the larger molecules. In many cases, the adsorbing component (or adsorbate) is held strongly enough to permit complete removal of that component from the fluid with very little adsorption of other components. Regeneration of the adsorbent can then be carried out to obtain the adsorbate in concentrated or nearly pure form.

Applications of vapor-phase adsorption include the recovery of organic solvents used in paints, printing inks, and solutions for film casting or fabric coating. The solvent-laden air may first be sent to a water-cooled or refrigerated condenser to collect some of the solvent, but it is generally impractical to cool the gas far below ambient temperature in an attempt to eliminate solvent losses. The air with a small amount of solvent is passed through a bed of carbon adsorbent particles, which can reduce the solvent concentration to less than 1 ppm. The concentration may be set by government emission standards rather than by the economics of solvent recovery. Adsorption on carbon is also used to remove pollutants such as H_2S, CS_2, and other odorous compounds from air circulating in ventilation systems, and canisters of carbon are placed in most new automobiles to prevent gasoline vapors from being vented to the air.

Drying of gases is often carried out by adsorbing the water on silica gel, alumina, or other inorganic porous solids. The zeolites, or molecular sieves, which are natural or synthetic aluminosilicates with a very regular, fine pore structure, are especially effective in preparing gases with low dew points ($-75°C$). Adsorption on molecular sieves can also be used to separate oxygen and nitrogen, to prepare pure hydrogen from synthesis gas, and to separate normal paraffins from branched paraffins and aromatics.

Adsorption from the liquid phase is used to remove organic components from drinking water or aqueous wastes, colored impurities from sugar solutions and vegetable oils, and water from organic liquids. Adsorption can also be used to recover reaction products that are not easily separated by distillation or crystallization. Some of the same types of solids are used for both vapor-phase and liquid-phase adsorption, though often adsorbents with larger pores are preferred for use with liquids.

ADSORPTION EQUIPMENT

Fixed-bed adsorbers

A typical system used for adsorption of solvent vapors is shown in Fig. 25.1. The adsorbent particles are placed in a bed 0.3 to 1.2 m (1 to 4 ft) deep supported on a screen or perforated plate. The feed gas passes down through one of the beds while the other is being regenerated. Downflow is preferred because upflow at high rates might fluidize the particles, causing attrition and loss of fines. When the concentration of solute in the exit gas reaches a certain value or at a scheduled time,

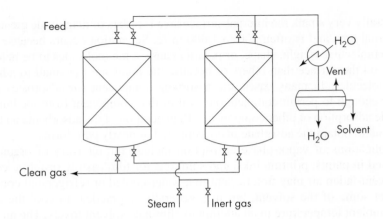

FIGURE 25.1
Vapor-phase adsorption system.

valves are automatically switched to direct the feed to the other bed and initiate the regeneration sequence.

Regeneration can be carried out with hot inert gas, but steam is usually preferred if the solvent is not miscible with water. Steam condenses in the bed, raising the temperature of the solid and providing the energy for desorption. The solvent is condensed, separated from the water, and perhaps dried before reuse. The bed may then be cooled and dried with inert gas, but it is not necessary to lower the entire bed to ambient temperature. If some water vapor can be tolerated in the clean gas, evaporation of water during the adsorption cycle will help cool the bed and partially offset the heat of adsorption.

The size of the adsorbent bed is determined by the gas flow rate and the desired cycle time. Usually the cross-sectional area is calculated to give a superficial velocity of 0.15 to 0.45 m/s (0.5 to 1.5 ft/s), which results in a pressure drop of a few inches of water per foot when using typical adsorbents (4 by 10 mesh or 6 by 16 mesh). For very large flow rates, a rectangular bed may be installed in the middle of a horizontal cylinder rather than use a vertical tank with a diameter much greater than the bed depth.

The bed depth and flow rate are generally chosen to provide an adsorption cycle of 2 to 24 h. By using a longer bed, the adsorption cycle could be extended to several days, but the higher pressure drop and the greater capital cost of the adsorber would probably make this uneconomical. Using a bed only 0.3 m (1 ft) or less in depth is sometimes recommended to decrease the pressure drop and the size of the adsorber, but shallow beds do not give as complete a separation and require more energy for regeneration.

Gas-drying equipment. The equipment for drying gases is similar to that shown in Fig. 25.1, but hot gas is used for regeneration. The moist gas from the bed being regenerated may be vented, or much of the water may be removed in a condenser and the gas recirculated through a heater to the bed. For small dryers, electric heaters are sometimes installed inside the bed to provide the energy for regeneration.

When regeneration is carried out at a much lower pressure than adsorption, it may not be necessary to supply heat, since low pressure favors desorption. If a gas dryer operates at several atmospheres pressure during the adsorption cycle, nearly complete regeneration can be accomplished by passing part of the dry gas through the bed at atmospheric pressure without preheating. Some of the heat of adsorption, which was stored in the bed as sensible heat of the solid, becomes available for desorption, and the bed cools during regeneration. The amount of gas needed for regeneration is only a fraction of the gas fed in the adsorption cycle, since the gas leaving at 1 atm will have a much higher mole fraction of water than the feed gas. Applying the same principle, vacuum regeneration offers an alternative to steam or hot-gas regeneration when adsorption is carried out at atmospheric pressure.

Pressure-swing adsorption. Although adsorption is most often used as a purification process to remove small amounts of material, a number of applications involve separations of gas mixtures with moderate to high concentration of adsorbates. These are called bulk separations, and they often use different operating procedures than for gas purification. Pressure-swing adsorption (PSA) is a bulk separation process that is used for small-scale air separation plants and for concentration of hydrogen in process streams.

A simple PSA scheme for air separation uses two beds of molecular sieves, with one adsorbing at several atmospheres pressure and the other being regenerated at 1 atm. For the same concentration, nitrogen is adsorbed 3 to 4 times as strongly as oxygen, so nearly pure oxygen can be produced as the product. However, the adsorption time is quite short (less than a minute) because of the high concentration of N_2 in air and the low capacity of the sorbent. Therefore the holdup of gas in the bed is significant relative to the amount adsorbed, and this gas volume must be considered in designing the adsorption cycle. After adsorption, the bed is depressurized, which removes most of the gas holdup and some of the adsorbed gas. The bed is purged at 1 atm with part of the product gas from the other adsorber to finish the desorption. Then the bed is pressurized with product gas before being switched to air feed. More complex schemes use three or four beds with only one bed adsorbing and the others being depressurized, purged, or repressurized, all under control of a sequence timer.

Although PSA is usually based on differences in the equilibrium amount adsorbed, some separations utilize differences in the adsorption kinetics. Nearly pure nitrogen can be prepared from air by using carbon molecular sieves,[17c] materials that have very small pore mouths which admit oxygen molecules much more readily than the slightly larger nitrogen molecules. The equilibrium amounts adsorbed are about the same, but the effective diffusivity of O_2 is 100 times greater than that of N_2, and very little N_2 is adsorbed during the short adsorption period.[17c]

Adsorption from liquids. An important example of adsorption from the liquid phase is the use of activated carbon to remove pollutants from aqueous wastes. Carbon adsorbents are also used to remove trace organics from municipal water supplies, which improves the taste and reduces the chance of forming toxic compounds in the chlorination step. For these uses the carbon beds are many feet in diameter and up to 10 m (30 ft) tall, and there may be several beds operating in parallel. Tall beds are needed to ensure adequate treatment, because the rate of adsorption from liquids is much slower than from gases. Also the spent carbon is usually removed from the bed for regeneration, and so relatively long periods between regeneration are desirable.

An alternative method of treating wastewaters is to add powdered carbon to a tank of solution using mechanical stirrers or air spargers to keep the particles suspended. With the fine particles, the adsorption is much more rapid than with granular carbon, but large equipment is needed to remove the spent carbon by sedimentation or filtration. The treatment with powdered carbon can be done batchwise, or it can be carried out continuously, with metered addition of carbon to the waste stream and continuous removal of the spent carbon.

Equilibria; Adsorption Isotherms

The adsorption isotherm is the equilibrium relationship between the concentration in the fluid phase and the concentration in the adsorbent particles at a given temperature. For gases, the concentration is usually given in mole percent or as a partial pressure. For liquids, the concentration is often expressed in mass units, such as mg/L (ppm) or μg/L (ppb). The concentration of adsorbate on the solid is given as mass adsorbed per unit mass of original adsorbent.

Types of isotherms

Some typical isotherm shapes are shown as arithmetic graphs in Fig. 25.2. The linear isotherm goes through the origin, and the amount adsorbed is proportional to the concentration in the fluid. Isotherms that are convex upward are called favorable, because a relatively high solid loading can be obtained at low concentration in the fluid. The Langmuir isotherm is given by the relation $W = W_{max}[Kc/(1 + Kc)]$, where W is the adsorbate loading, c is the concentration in the fluid, and K is the adsorption constant. This isotherm is of the favorable type. When K is large and

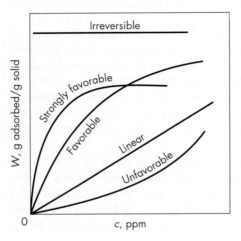

FIGURE 25.2
Adsorption isotherms.

$Kc \gg 1$, the isotherm is strongly favorable; and when $Kc < 1$, the isotherm is nearly linear. The Langmuir isotherm is derived assuming a uniform surface—not a valid assumption—but the relation works fairly well for gases that are weakly adsorbed. For strongly favorable isotherms the empirical Freundlich equation $W = bc^{m}$, where b and m are constants and $m < 1$, is often a better fit, particularly for adsorption from liquids.

The limiting case of a very favorable isotherm is irreversible adsorption, where the amount adsorbed is independent of concentration down to very low values. All systems show a decrease in the amount adsorbed with an increase in temperature, and of course adsorbate can be removed by raising the temperature even for the cases labeled "irreversible." However, desorption requires a much higher temperature when the adsorption is strongly favorable or irreversible than when the isotherms are linear.

An isotherm that is concave upward is called unfavorable because relatively low solid loadings are obtained and because it leads to quite long mass-transfer zones in the bed. Isotherms of this shape are rare, but they are worth studying to help understand the regeneration process. If the adsorption isotherm is favorable, mass transfer from the solid back to the fluid phase has characteristics similar to those for adsorption with an unfavorable isotherm.

To show the variety of isotherm shapes for a single adsorbate, data for water adsorbed from air on three dessicants are given in Fig. 25.3. Silica gel has a nearly linear isotherm up to 50 percent relative humidity, and the ultimate capacity is about twice that for the other solids. At high humidity, the small pores become filled with liquid by capillary condensation, and the total amount adsorbed depends on the volume of the small pores and not just the surface area. Water is held most strongly

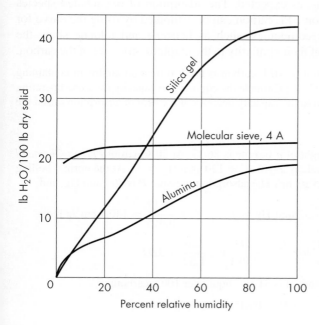

FIGURE 25.3
Adsorption isotherms for water in air at 20 to 50°C.

by molecular sieves, and the adsorption is almost irreversible, but the pore volume is not as great as for silica gel. The curves in Fig. 25.3 are based on relative humidity, which makes the isotherms for a range of temperatures fall on a single curve. Note that except for the molecular sieves, the amount adsorbed at a given partial pressure decreases strongly with increased temperature. For air with 1 percent H_2O at 20°C, $\mathscr{H}_R = 7.6$ mm Hg/$17.52 \times 100 = 43.4$ percent, and the amount adsorbed on silica gel is $W = 0.26$ lb/lb. For the same concentration at 40°C, $\mathscr{H}_R = 7.6/55.28 \times 100 = 13.7$ percent, and $W = 0.082$ lb/lb.

The adsorption data for hydrocarbon vapors on activated carbon are sometimes fitted to Freundlich isotherms, but data for a wide range of pressures show the isotherm slopes gradually decrease as the pressure is increased. The amount adsorbed depends primarily on the ratio of the partial pressure of the adsorbate in the gas to the vapor pressure of the liquid at the same conditions and on the surface area of the carbon. Generalized correlations have been developed based on the concept of the adsorption potential,[10,14] and some results for a typical coal-based carbon are shown in Fig. 25.4. For a given class of materials, the amount adsorbed depends on $(T/V) \log (f_s/f)$, where T is the adsorption temperature, in kelvins, V is the molar volume of the liquid at the boiling point, f_s is the fugacity of the saturated liquid at adsorption temperature, and f is the fugacity of the vapor. For adsorption at atmospheric pressure, the partial pressure and vapor pressure can be used for the fugacities. The volume adsorbed is converted to mass by assuming the adsorbed liquid has the same density as liquid at the boiling point.

The curves for n-paraffins and sulfur compounds were adapted from similar plots by Grant and Manes,[10,11] who used a different definition of the molar volume. Limited data show that highly chlorinated hydrocarbons are more strongly adsorbed than the sulfur compounds, as shown in Fig. 25.4, but data for vinyl chloride fall closer to the line for paraffins, as expected. The adsorption of oxygenated species such as ketones and alcohols on BPL carbon can be estimated by using the curve for sulfur compounds. Aromatic compounds such as benzene and toluene show the strongest adsorption because of their similarity to the graphitic structure of the carbon.

EXAMPLE 25.1. Adsorption on BPL carbon is used to treat an airstream containing 0.2 percent n-hexane at 20°C. (*a*) Estimate the equilibrium capacity for a bed operating at 20°C. (*b*) How much would the capacity decrease if the heat of adsorption raised the bed temperature to 40°C?

Solution

(*a*) The molecular weight of n-hexane is 86.17. At 20°C (from *Perry's Chemical Engineers' Handbook*, 7th ed., p. 2-70) $P' = 120$ mm Hg $\approx f_s$. At the normal boiling point (68.7°C), $\rho_L = 0.615$ g/cm³. The adsorption pressure P is 760 mm Hg, and

$$p = 0.002 \times 760 = 1.52 \text{ mm Hg} \approx f \qquad V = \frac{86.17}{0.615} = 140.1 \text{ cm}^3/\text{g mol}$$

$$\frac{T}{V} \log \frac{f_s}{f} = \frac{293}{140.1} \log \frac{120}{1.52} = 3.97$$

From Fig. 25.4, volume adsorbed is 31 cm³ liquid per 100 g carbon:

$$W = 0.31 \times 0.615 = 0.19 \text{ g/g carbon}$$

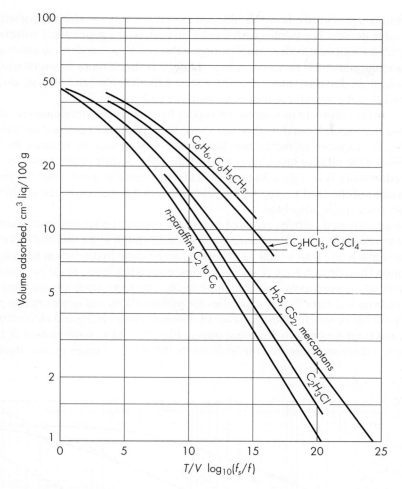

FIGURE 25.4
Generalized adsorption correlation[10,11] for Calgon BPL carbon
(1,040 m²/g).

(*b*) At 40°C, $P' = 276$ mm Hg:

$$\frac{T}{V} \log \frac{f_s}{f} = \frac{313}{140.1} \log \frac{276}{1.52} = 5.05$$

Volume adsorbed, from Fig. 25.4, is 27 cm³ per 100 g carbon:

$$W = 0.27 \times 0.615 = 0.17 \text{ g/g carbon}$$

Adsorption isotherms are generally presented for a single component, but many applications involve multicomponent mixtures. The Langmuir isotherm is easily modified for multiple adsorbates by adding terms to the denominator:

$$W_1 = W_{\max} \left(\frac{K_1 c_1}{1 + K_1 c_1 + K_2 c_2 + \cdots} \right)$$

As mentioned previously, however, this equation is not very satisfactory for strongly adsorbed materials. For solutes with similar properties, the generalized correlations of Fig. 25.4 could be used by using average values of V, ρ_L, and f_s to estimate the total amount adsorbed. However, when solute A is much more strongly adsorbed than solute B, separate mass-transfer zones will develop, and A will displace adsorbed B as the zone for A moves through the bed.

A common problem in removal of vapors from air is the simultaneous adsorption of water vapor. Even though water vapor is only weakly adsorbed on activated carbons, the capacity of the carbon for organic vapors may be reduced by 50 to 70 percent if the relative humidity is 80 percent. There is little effect, however, if the relative humidity is less than 30 percent. This may make it worthwhile to raise the temperature of the feed by 10 to 20°C, even though this decreases the equilibrium capacity for a single adsorbate.

When activated carbons are used to remove organic pollutants from water, the hydrocarbons have to compete with water for adsorption sites, and the capacity is generally much less than would be expected for the same hydrocarbon fugacity and adsorption from dry air. New types of carbonaceous adsorbents, made by controlled pyrolysis of macroreticular ion-exchange resins, are more hydrophobic than activated carbons and have higher capacities for hydrocarbons and other organic pollutants dissolved in water.[16] Isotherms for trichloroethylene (TCE) and chloroform on Rohm and Haas Ambersorb 563 are shown in Fig. 25.5. At a concentration of 1 ppm or lower, the amount adsorbed on Ambersorb 563 is 6 to 8 times greater than on a

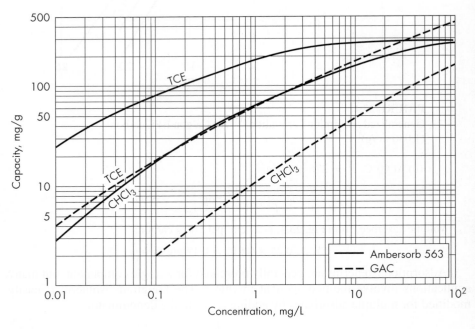

FIGURE 25.5
Adsorption isotherms for trichloroethylene (TCE) and chloroform ($CHCl_3$) in water solution, on Ambersorb 563 and granular activated carbon (GAC).

typical granular activated carbon. At high concentrations the trend is reversed, because the granular carbons have greater micropore volume. A typical application for Ambersorb adsorbents is to treat groundwater containing about 1 ppm TCE and reduce the concentration to the water quality standard of 0.005 ppm or 5 ppb.

For both adsorbents shown in Fig. 25.5 the amount of TCE adsorbed is several times that of chloroform, although the difference in vapor-phase adsorption is less than twofold at the same concentration. The higher solubility of chloroform in water (some 8 times higher than that of TCE) makes the activity at a given concentration in the water much less for chloroform than for TCE.

Principles of Adsorption

Concentration patterns in fixed beds

In fixed-bed adsorption, the concentrations in the fluid phase and the solid phase change with time as well as with position in the bed. At first, most of the mass transfer takes place near the inlet of the bed, where the fluid first contacts the adsorbent. If the solid contains no adsorbate at the start, the concentration in the fluid drops exponentially with distance essentially to zero before the end of the bed is reached. This concentration profile is shown by curve t_1 in Fig. 25.6a, where c/c_0 is the concentration in the fluid relative to that in the feed. After a few minutes, the solid near the inlet is nearly saturated, and most of the mass transfer takes place

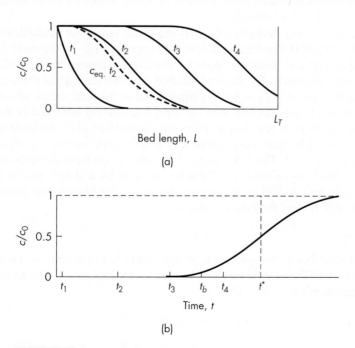

FIGURE 25.6
(a) Concentration profiles and (b) breakthrough curve for adsorption in a fixed bed.

farther from the inlet. The concentration gradient becomes S-shaped, as shown by curve t_2. The region where most of the change in concentration occurs is called the mass-transfer zone, and the limits are often taken as c/c_0 values of 0.95 to 0.05.

With time, the mass-transfer zone moves down the bed, as shown by profiles t_3 and t_4. Similar profiles could be drawn for the average concentration of adsorbate on the solid, showing nearly saturated solid at the inlet, a large change in the region of the mass-transfer zone, and zero concentration at the end of the bed. Instead of plotting the actual concentration on the solid, the concentration in the fluid phase in equilibrium with the solid is shown as a dashed line for time t_2. This concentration must always be less than the actual fluid concentration, and the difference in concentrations, or driving force, is large where the concentration profile is steep and mass transfer is rapid.

Breakthrough curves

Few fixed beds have internal probes that would permit measurement of profiles such as those in Fig. 25.6a. However, these profiles can be predicted and used in calculating the curve of concentration versus time for fluid leaving the bed. The curve shown in Fig. 25.6b is called a breakthrough curve. At times t_1 and t_2, the exit concentration is practically zero, as shown also in Fig. 25.6a. When the concentration reaches some limiting permissible value, or break point, the flow is stopped or diverted to a fresh adsorbent bed. The break point is often taken as a relative concentration of 0.05 or 0.10, and since only the last portion of fluid processed has this high a concentration, the average fraction of solute removed from the start to the break point is often 0.99 or higher.

If adsorption were continued beyond the break point, the concentration would rise rapidly to about 0.5 and then more slowly approach 1.0, as shown in Fig. 25.6b. This S-shaped curve is similar to those for the internal concentration profiles, and it is often nearly symmetric. By material balance, it can be shown that the area between the curve and a line at $c/c_0 = 1.0$ is proportional to the total solute adsorbed if the entire bed comes to equilibrium with the feed. The amount adsorbed is also proportional to the rectangular area to the left of the dashed line at t^*, the ideal adsorption time for a vertical breakthrough curve. For a symmetric curve, t^* is also the time when c/c_0 reaches 0.5. The movement of the adsorption front through the bed and the effect of process variables on t^* can be obtained by a simple material balance.

For a unit area of bed cross section, the solute feed rate is the product of the superficial velocity and the concentration:

$$F_A = u_0 c_0 \tag{25.1}$$

For an ideal breakthrough curve, all the solute fed in time t^* is adsorbed, and the concentration on the solid has increased from the initial value W_0 to the equilibrium or saturation value W_{sat}. Thus

$$u_0 c_0 t^* = L\rho_b (W_{sat} - W_0) \tag{25.2}$$

or
$$t^* = \frac{L\rho_b (W_{sat} - W_0)}{u_0 c_0} \tag{25.3}$$

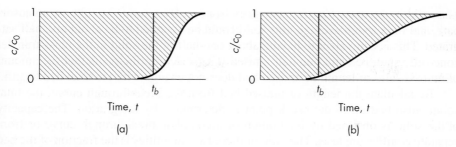

FIGURE 25.7
Breakthrough curves for (*a*) a narrow and (*b*) a wide mass-transfer zone.

where L and ρ_b are the length and bulk density of the bed, respectively. For fresh carbon or completely regenerated carbon, $W_0 = 0$; but complete regeneration is often too costly.

The break-point time t_b is always less than t^*, and the actual amount of solute adsorbed at the break point can be determined by integrating the breakthrough curve up to time t_b, as shown in Fig. 25.7. If the mass-transfer zone is narrow relative to the bed length, the breakthrough curve will be rather steep, as in Fig. 25.7*a*, and most of the capacity of the solid will be utilized at the break point. When the mass-transfer zone is almost as long as the bed, the breakthrough curve is greatly extended, as in Fig. 25.7*b*, and less than one-half of the bed capacity is utilized. A narrow mass-transfer zone is desirable to make efficient use of the absorbent and to reduce the energy costs in regeneration. In the ideal case of no mass-transfer resistance and no axial dispersion, the mass-transfer zone would be of infinitesimal width, and the breakthrough curve would be a vertical line from 0 to 1.0 when all the solid was saturated.

Scaleup

The width of the mass-transfer zone depends on the mass-transfer rate, the flow rate, and the shape of the equilibrium curve. Methods of predicting the concentration profiles and zone width have been published, but lengthy computations are often required, and the results may be inaccurate because of uncertainties in the mass-transfer correlations. Usually adsorbers are scaled up from laboratory tests in a small-diameter bed, and the large unit is designed for the same particle size and superficial velocity. The bed length need not be the same, as shown in the next section.

Length of unused bed. For systems with a favorable isotherm, the concentration profile in the mass-transfer zone soon acquires a characteristic shape and width that do not change as the zone moves down the bed. Thus, tests with different bed lengths give breakthrough curves of the same shape; but with longer beds, the mass-transfer zone is a smaller fraction of the bed length, and a greater fraction of the bed is utilized. At the break point, the solid between the inlet of the bed and the start of the mass-transfer zone is completely saturated (at equilibrium with the feed).

The solid in the mass-transfer zone goes from nearly saturated to almost no adsorbate, and for a rough average, this solid could be assumed to be about one-half saturated. This is equivalent to having about one-half of the solid in the mass-transfer zone fully saturated and one-half unused. The scaleup principle is that the amount of unused solid or length of unused bed does not change with the total bed length.[5]

To calculate the length of unused bed from the breakthrough curve, the total solute adsorbed up to the break point is determined by integration. The capacity of the solid is obtained by integration of a complete breakthrough curve or from separate equilibrium tests. The ratio of these two quantities is the fraction of the bed capacity utilized at the break point, and 1.0 minus this ratio is the unused fraction. The unused fraction is converted to an equivalent length of unused bed (LUB), which is assumed to be constant. The break-point time is calculated from the ideal time and the fraction of bed utilized:

$$t_b = t^* \left(1 - \frac{\text{LUB}}{L} \right) \tag{25.4}$$

As an example, if a 20-cm-deep bed gives 60 percent utilization at the break point, the length of the unused bed is 8 cm. Increasing the bed length to 40 cm makes the unused portion $\frac{8}{40}$, or 20 percent. Therefore the break-point time is increased by a factor $40/20 \times 0.8/0.6 = 2.67$, because of the longer bed and the greater fraction used.

EXAMPLE 25.2. The adsorption of n-butanol from air was studied[9] in a small fixed bed (10.16-cm diameter) with 300 and 600 g carbon, corresponding to bed lengths of 8 and 16 cm. (a) From the following data for effluent concentration, estimate the saturation capacity of the carbon and the fraction of the bed used at $c/c_0 = 0.05$. (b) Predict the break-point time for a bed length of 32 cm.

Data for n-butanol on Columbia JXC 4/6 carbon are as follows:

$$u_0 = 58 \text{ cm/s} \qquad D_p = 0.37 \text{ cm}$$
$$c_0 = 365 \text{ ppm} \qquad S = 1{,}194 \text{ m}^2/\text{g}$$
$$T = 25°C \qquad \rho_b = 0.461 \text{ g/cm}^3$$
$$P = 737 \text{ mm Hg} \qquad \varepsilon = 0.457$$

300 g		600 g	
t, h	c/c₀	t, h	c/c₀
1	0.005	5	0.0019
1.5	0.01	5.5	0.003
2	0.027	6	0.0079
2.4	0.050	6.5	0.018
2.8	0.10	7	0.039
3.3	0.20	7.5	0.077
4	0.29	8	0.15
5	0.56	8.5	0.24

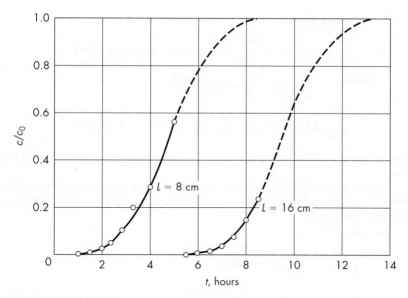

FIGURE 25.8
Breakthrough curves for Example 25.2.

Solution

(a) The concentration profiles are plotted in Fig. 25.8 and extended to $c/c_0 = 1.0$, assuming the curves are symmetric about $c/c_0 = 0.5$.

Per square centimeter of bed cross section, the solute feed rate is

$$F_A = u_0 c_0 M$$

$$= 58 \ \frac{\text{cm}}{\text{s}} \left(\frac{365 \times 10^{-6}}{22{,}400} \times \frac{273}{298} \times \frac{737}{760} \right) \frac{\text{mol}}{\text{cm}^3} \times 74.12 \ \text{g/mol}$$

$$= 6.22 \times 10^{-5} \ \text{g/cm}^2 \cdot \text{s} \quad \text{or} \quad 0.224 \ \text{g/cm}^2 \cdot \text{h}$$

The total solute adsorbed is the area above the graph multiplied by F_A. For the 8-cm bed, the area is

$$\int_0^{8.5} \left(1 - \frac{c}{c_0} \right) dt = 4.79 \ \text{h}$$

This area corresponds to the ideal time that would be required to adsorb the same amount if the breakthrough curve were a vertical line. The mass of carbon per unit cross-sectional area of bed is $8 \times 0.461 = 3.69 \ \text{g/cm}^2$. Thus

$$W_{\text{sat}} = \frac{0.224 \times 4.79}{3.69} = 0.291 \ \text{g solute/g carbon}$$

At the break point, where $c/c_0 = 0.05$, and $t = 2.4$ h

$$\int_0^{2.4} \left(1 - \frac{c}{c_0} \right) dt = 2.37 \ \text{h}$$

The amount adsorbed up to the break point is then

$$W_b = \frac{0.224 \times 2.37}{3.69} = 0.144 \text{ g solute/g carbon}$$

$$\frac{W_b}{W_{\text{sat}}} = \frac{0.144}{0.291} = 0.495$$

$$\text{LUB} = L\left(1 - \frac{W_b}{W_{\text{sat}}}\right) = 8(0.505) = 4.04 \text{ cm}$$

Thus 50 percent of the bed capacity is unused, which can be represented by a length of 4 cm.

For the 16-cm bed the breakthrough curve has the same initial slope as the curve for the 8-cm bed, and although data were not taken beyond $c/c_0 = 0.25$, the curves are assumed to be parallel.

For the entire bed,

$$\int_0^{13}\left(1 - \frac{c}{c_0}\right)dt = 9.59 \text{ h}$$

$$W_{\text{sat}} = \frac{0.224 \times 9.59}{16 \times 0.461} = 0.291 \text{ g solute/g carbon}$$

At $c/c_0 = 0.05$, $t = 7.1$ h, and

$$\int_0^{7.1}\left(1 - \frac{c}{c_0}\right)dt = 7.07 \text{ h}$$

$$W_b = \frac{0.224 \times 7.07}{16 \times 0.461} = 0.215 \text{ g solute/g carbon}$$

$$\frac{W_b}{W_{\text{sat}}} = \frac{0.215}{0.291} = 0.739$$

At the break point, 74 percent of the bed capacity is used, which corresponds to an unused section of length $0.26 \times 16 = 4.2$ cm. Within experimental error, the lengths of unused bed agree, and 4.1 cm is the expected value for a still longer bed.

(b) For $L = 32$ cm, the expected length of the fully used bed is $32 - 4.1 = 27.9$ cm. The fraction of the bed used is

$$\frac{W_b}{W_{\text{sat}}} = \frac{27.9}{32} = 0.872$$

The break-point time is

$$t_b = \frac{L(W_b/W_{\text{sat}})\rho_b W_{\text{sat}}}{F_A} = \frac{27.9 \times 0.461 \times 0.291}{0.224} = 16.7 \text{ h}$$

Summary

L, cm	8	16	32
t_b, h	2.4	7.1	16.7
W_b/W_{sat}	0.50	0.74	0.87

Effect of feed concentration. The effect of moderate changes in feed concentration on the breakthrough curve can be predicted, since the width of the mass-transfer zone does not change. The equilibrium capacity is determined from the adsorption isotherm, and the break-point time is proportional to the capacity of the solid and to the reciprocal of the feed concentration [Eqs. (25.3) and (25.4)]. Laboratory tests using higher than expected concentrations of a pollutant may be made to shorten the time for a breakthrough test. Very large differences in concentration may lead to errors in scaleup because of a change in the mass-transfer coefficient or because of temperature effects.

Adsorption is an exothermic process, and a bed temperature rise of 10 to 50°C may result when treating vapors with only 1 percent adsorbable component. In small-diameter beds, heat loss will limit the temperature rise, but a large unit will operate almost adiabatically, and significant differences in performance could result. In such cases, a large-diameter pilot column should be used, or detailed calculations made to account for heat release and heat transfer in the bed.

Basic Equations for Adsorption

Although adsorbers are generally designed from laboratory data, the approximate performance can sometimes be predicted from equilibrium data and mass-transfer calculations. In this section the basic equations are presented for isothermal adsorption in fixed beds, and solutions are given for some limiting cases. This material can give insight into factors affecting the width of the mass-transfer zone in an adsorber.

Rate of mass transfer

Equations for mass transfer in fixed-bed adsorption are obtained by making a solute material balance for a section dL of the bed, as shown in Fig. 25.9. The rate of accumulation in the fluid and in the solid is the difference between input and output flows. The change in superficial velocity is neglected:

$$\varepsilon \, dL \, \frac{\partial c}{\partial t} + (1 - \varepsilon) \, dL \, \rho_p \, \frac{\partial W}{\partial t} = u_0 c - u_0(c + dc) \tag{25.5}$$

or
$$\varepsilon \, \frac{\partial c}{\partial t} + (1 - \varepsilon)\rho_p \, \frac{\partial W}{\partial t} = -u_0 \, \frac{\partial c}{\partial L} \tag{25.6}$$

The term ε is the external void fraction of the bed, and solute dissolved in the pore fluid is included with the particle fraction $1 - \varepsilon$. For adsorption from a gas or a

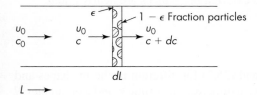

FIGURE 25.9
Mass balance for a section of a fixed bed.

dilute solution, the first term in Eq. (25.6), the accumulation in the fluid, is usually negligible compared to the accumulation on the solid.

The mechanism of transfer to the solid includes diffusion through the fluid film around the particle and diffusion through the pores to internal adsorption sites. The actual process of physical adsorption is practically instantaneous, and equilibrium is assumed to exist between the surface and the fluid at each point inside the particle. The transfer process is approximated using an overall volumetric coefficient and an overall driving force:

$$\rho_p(1-\varepsilon)\frac{\partial W}{\partial t} = K_c a(c - c^*) \tag{25.7}$$

The mass-transfer area a is taken as the external surface of the particles, which is $6(1-\varepsilon)/D_p$ for spheres. The concentration c^* is the value that would be in equilibrium with the average concentration W in the solid.

Internal and external mass-transfer coefficients

The overall coefficient K_c depends on the external coefficient $k_{c,\text{ext}}$ and on an effective internal coefficient $k_{c,\text{int}}$. Diffusion within the particle is actually an unsteady-state process, and the value of $k_{c,\text{int}}$ decreases with time, as solute molecules must penetrate farther and farther into the particle to reach adsorption sites. Using Eq. (17.77), an average effective coefficient can be used to give an approximate fit to uptake data for spheres:

$$k_{c,\text{int}} \approx \frac{10 D_e}{D_p}$$

This leads to

$$\frac{1}{K_c} \approx \frac{1}{k_{c,\text{ext}}} + \frac{D_p}{10 D_e} \tag{25.8}$$

The effective diffusion coefficient D_e depends on the particle porosity, the pore diameter, the tortuosity, and the nature of the diffusing species. For gas-filled pores, the above factors can be allowed for to make a reasonable estimate of the effective diffusivity in the gas phase. However, diffusion of adsorbed molecules along the pore walls, called surface diffusion, often contributes much more to the total flux than diffusion in the gas phase. This is particularly evident in the adsorption of water vapor on silica gel and the adsorption of hydrocarbon vapors on carbon, where the measured values of K_c correspond to internal and external coefficients of comparable magnitude or even to external film control. In the adsorption of solutes from aqueous solutions, surface diffusion also occurs, but its effect is hard to predict. In some cases surface diffusion is slow, and the internal diffusion resistance dominates, but in others the internal and external resistances are nearly equal.

Solutions to Mass-Transfer Equations

There are many solutions to Eq. (25.6) and (25.7) for different isotherm shapes and controlling steps, and all solutions involve a dimensionless time τ and a parameter N

representing the overall number of transfer units:

$$\tau \equiv \frac{u_0 c_0 (t - L\varepsilon/u_0)}{\rho_p (1 - \varepsilon) L (W_{sat} - W_0)} \tag{25.9}$$

$$N \equiv \frac{K_c a L}{u_0} \tag{25.10}$$

The term $L\varepsilon/u_0$ in Eq. (25.9) is the time to displace fluid from external voids in the bed, which is normally negligible, and $\rho_p(1 - \varepsilon)$ is the bed density ρ_b. Then τ is the ratio of the time to the ideal time t^* from Eq. (25.3):

$$\tau = \frac{t}{t^*} \tag{25.11}$$

If there were no mass-transfer resistance, the adsorber could be operated with complete removal of solute up to $\tau = 1.0$, and then the concentration would jump from 0 to $c/c_0 = 1.0$. With a finite rate of mass transfer, breakthrough occurs at $\tau < 1.0$, and the steepness of the breakthrough curve depends on the parameter N and on the shape of the equilibrium curve.

Irreversible adsorption

Irreversible adsorption with a constant mass-transfer coefficient is the simplest case to consider, since the rate of mass transfer is then just proportional to the fluid concentration. A truly constant coefficient is obtained only when all resistance is in the external film, but a moderate internal resistance does not change the breakthrough curve very much. Strongly favorable adsorption gives almost the same results as irreversible adsorption, because the equilibrium concentration in the fluid is practically zero until the solid concentration is over one-half the saturation value. If the accumulation term for the fluid is neglected, Eqs. (25.6) and (25.7) are combined to give

$$-u_0 \frac{\partial c}{\partial L} = K_c a c \tag{25.12}$$

The initial shape of the concentration profile is obtained by integration of Eq. (25.12), which gives

$$\ln \frac{c}{c_0} = -\frac{K_c a L}{u_0} \tag{25.13}$$

Since the term $K_c a L/u_0$ is defined as N in Eq. (25.10), the concentration at the end of the bed is given by

$$c = c_0 e^{-N} \tag{25.14}$$

If the bed contained only three transfer units, the exit concentration would be $0.05c_0$ right at the start of the test; but usually N will be 10 or more, and c at the exit will be a very small fraction of c_0, small enough to be considered zero.

The rate of mass transfer to the first layer of particles is assumed to be constant until the particles reach equilibrium with the fluid, and until this happens, the

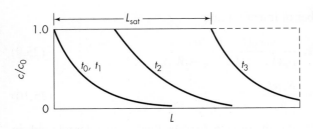

FIGURE 25.10
Concentration profiles for irreversible adsorption with a constant coefficient.

concentration profile in the bed remains constant. The time to saturate the first portion of the bed t_1 is the equilibrium capacity divided by the initial transfer rate ($W_0 = 0$ to simplify the analysis):

$$t_1 = \frac{W_{sat}\rho_p(1-\varepsilon)}{K_c a c_0} \qquad (25.15)$$

After this time, the concentration profile moves steadily down the bed, keeping the same shape. The transfer zone moves at a velocity v_z, which is equal to the amount of solute removed per unit time divided by the amount retained on the solid per unit length of bed:

$$v_z = \frac{u_0 c_0}{\rho_p(1-\varepsilon)W_{sat}} \qquad (25.16)$$

The concentration is constant at c_0 for the saturated portion of the bed and then falls exponentially in the mass-transfer zone, as shown in Fig. 25.10.

$$\ln\frac{c}{c_0} = \frac{-K_c a(L - L_{sat})}{u_0} \qquad (25.17)$$

To predict the break point, Eq. (25.17) is applied for a bed of length L with c/c_0 set at 0.05 or another selected value. The length of the saturated bed is the product of transfer zone velocity and the time since the zone started to move:

$$L_{sat} = v_z(t - t_1) \qquad (25.18)$$

$$L_{sat} = \frac{u_0 c_0}{\rho_p(1-\varepsilon)W_{sat}}\left[t - \frac{W_{sat}\rho_p(1-\varepsilon)}{K_c a c_0}\right] \qquad (25.19)$$

Substituting the equation for L_{sat} into Eq. (25.17) and using the dimensionless terms τ and N [Eqs. (25.9) and (25.10)] give

$$\ln\frac{c}{c_0} = -N + N\tau - 1 \qquad (25.20)$$

or

$$\ln\frac{c}{c_0} = N(\tau - 1) - 1 \qquad (25.21)$$

The predicted breakthrough curve is shown as a solid line in Fig. 25.11. The slope increases with time, and c/c_0 becomes 1.0 at $N(\tau - 1) = 1.0$. In practice, the

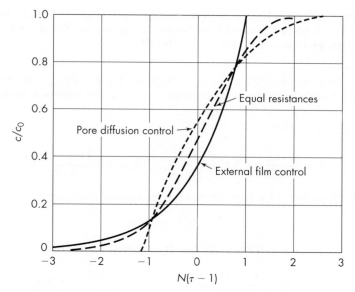

FIGURE 25.11
Breakthrough curves for irreversible adsorption.

breakthrough curves are usually S-shaped, because the internal diffusion resistance is not negligible, and it increases somewhat when the solid becomes nearly saturated.

If pore diffusion controls the rate of adsorption, the breakthrough curve has the opposite shape from that for external-film control. The corresponding line in Fig. 25.11 was taken from the work of Hall et al.,[12] who presented breakthrough curves for several cases of irreversible adsorption. For pore diffusion control the initial slope of the curve is high, because the solid near the front of the mass-transfer zone has almost no adsorbate, and the average diffusion distance is a very small fraction of the particle radius. The curve has a long tail because the final molecules adsorbed have to diffuse almost to the center of the particle.

When both internal and external resistances are significant, the breakthrough curve is S-shaped, as shown by the dashed line in Fig. 25.11. For this plot, the value of N is based on the overall mass-transfer coefficient given by Eq. (25.8), or it can be expressed in Hall's terminology as

$$\frac{1}{N} = \frac{1}{N_f} + \frac{1}{N_p} \tag{25.22}$$

where

$$N_f = \frac{k_{c,\text{ext}} a L}{u_0}$$

$$N_p = \frac{10 D_e a L}{D_p u_0}$$

EXAMPLE 25.3. (*a*) Use the breakthrough data in Example 25.2 to determine N and $K_c a$ for the 8-cm bed, assuming irreversible adsorption. (*b*) Compare $K_c a$ with the predicted $k_c a$ for the external film.

Solution

(*a*) From Example 25.2, at $c/c_0 = 0.05$, $W/W_{sat} = 0.495$, $\tau = 0.495$, $\tau - 1 = -0.505$. Assume equal internal and external resistances to determine N from Fig. 25.11:

$$N(\tau - 1) = -1.6 \qquad \text{at} \quad \frac{c}{c_0} = 0.05$$

$$N = \frac{-1.6}{-0.505} = 3.17 = \frac{K_c a L}{u_0} \qquad K_c a = \frac{3.17 \times 58 \text{ cm/s}}{8 \text{ cm}} = 23.0 \text{ s}^{-1}$$

(*b*) Prediction of $k_c a$ from Re, Sc (k_c is the external coefficient):

$$D_p = 0.37 \text{ cm}$$

At 25°C, 1 atm, $\mu/\rho = 0.152$ cm²/s and $D_v = 0.0861$ cm²/s. Then

$$\text{Re} = \frac{0.37(58)}{0.152} = 141 \qquad \text{Sc} = \frac{0.152}{0.0861} = 1.765$$

From Eq. (17.74)

$$\text{Sh} = 1.17(141)^{0.585}(1.765)^{1/3} = 25.6$$

$$k_c = \frac{25.6(0.0861)}{0.37} = 5.96 \text{ cm/s}$$

$$a = \frac{6(1 - \varepsilon)}{D_p} = \frac{6(1 - 0.457)}{0.37} = 8.81 \text{ cm}^2/\text{cm}^3$$

$$k_c a = 5.96 \times 8.81 = 52.5 \text{ s}^{-1}$$

Since $K_c a$ is slightly less than one-half the predicted value of $k_c a$, the external resistance is close to one-half the total resistance, and the calculated value of N need not be revised. The internal coefficient can be obtained from

$$\frac{1}{k_{c,\text{int}}} = \frac{1}{K_c} - \frac{1}{k_{c,\text{ext}}}$$

$$K_c = \frac{23.0}{8.81} = 2.61 \text{ cm/s}$$

$$k_{c,\text{int}} = \frac{1}{1/2.61 - 1/5.96} = 4.64 \text{ cm/s}$$

If diffusion into the particle occurred only in the gas phase, the maximum possible value of D_e would be about $D_v/4$, which leads to

$$k_{c,\text{int}} = \frac{10D_e}{D_p} = \frac{10 \times 0.0861}{4 \times 0.37} = 0.58 \text{ cm/s}$$

Since the measured value of $k_{c,\text{int}}$ is an order of magnitude greater than this value, surface diffusion must be the dominant transfer mechanism.

Linear isotherm

Adsorption with a linear isotherm is another limiting case for which solutions for Eqs. (25.6) and (25.7) are readily available. With a linear isotherm, the equations are of the same form as those for passage of a temperature wave through a fixed bed:

$$\varepsilon c_p \frac{\partial T_g}{\partial t} + (1 - \varepsilon)\rho_p c_s \frac{\partial T_s}{\partial t} = -u_0 \rho c_p \frac{\partial T_g}{\partial L} \tag{25.23}$$

$$(1 - \varepsilon)\rho_p c_s \frac{\partial T_s}{\partial t} = Ua(T_g - T_s) \tag{25.24}$$

Solutions for the heat-transfer problem, which arises in the use of packed beds as direct-contact recuperative heat exchangers, were presented by Furnas[8] in 1930. The parameter N_H is the number of heat-transfer units. For heat transfer, the dimensionless time τ is the heat capacity of the gas times the amount of gas that has passed through the bed divided by the total bed capacity. For $N_H = \infty$, the breakthrough curve of T_g/T_0 versus τ would be a vertical line at $\tau = 1.0$, just as for mass transfer. The defining equations are

$$\tau = \frac{u_0 \rho c_p t}{\rho_s c_s (1 - \varepsilon)L} \tag{25.25}$$

$$N_H = \frac{UaL}{\rho c_p u_0} \tag{25.26}$$

The breakthrough curves are complex expressions that are best presented graphically, as in Fig. 25.12. These curves become steeper relative to bed length for

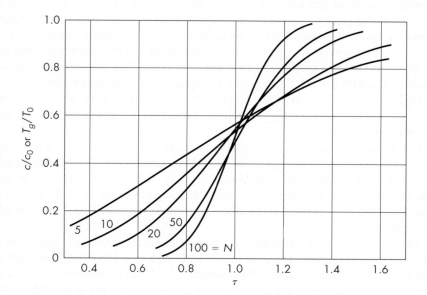

FIGURE 25.12
Breakthrough curves for adsorption with a linear isotherm or for heat transfer, for which $N = N_H$.

high values of N, but the absolute width of the transfer zone actually increases with $L^{1/2}$. For adsorption with a linear isotherm, a very long bed is required to make the transfer zone a small fraction of the bed width, in contrast to adsorption with a favorable isotherm. For example, if $N = 10$, Fig. 25.12 shows the break point would be reached at $\tau = 0.35$, and only about 35 percent of the bed capacity would be utilized. Doubling the bed length to make $N = 20$ raises τ at the break point to only 0.50, and N would have to be more than 100 to utilize 80 percent of the bed capacity. For irreversible adsorption with equal internal and external resistances, $N = 10$ gives $N(\tau - 1) = -1.6$ or $\tau = 0.84$ at the break point.

Favorable adsorption

For favorable adsorption, the break point occurs between the values predicted for linear adsorption and irreversible adsorption. Solutions are available for certain isotherm shapes and different values of internal and external resistances.[12,22] These solutions have found use for the design of ion exchangers, where the solid-fluid equilibria and the internal diffusivities are more readily characterized than for adsorption.

Adsorber Design

The design of an adsorber for gas or liquid purification involves choosing the adsorbent and the particle size, selecting an appropriate velocity to get the bed area, and either determining the bed length for a given cycle time or calculating the breakthrough time for a chosen length. Using a shorter bed length means a smaller inventory of sorbent and lower pressure drop in the bed. However, the shorter bed means more frequent regeneration and higher regeneration costs, since a smaller fraction of the bed is saturated at breakthrough. For gas purification, 4×6- or 4×10-mesh carbon is often chosen, but smaller sizes can be used when better mass transfer is needed and pressure drop is not a problem. The gas velocity is usually between 15 and 60 cm/s (0.5 and 2 ft/s). Because the external area varies with $1/D_p$ and both $k_{c,\text{ext}}$ and $k_{c,\text{int}}$ increase as D_p decreases, $k_c a$ is expected to vary with the -1.5 to -2.0 power of D_p. Therefore a reduction in particle size gives a much steeper breakthrough curve.

For adsorption from liquids, smaller particle sizes are chosen, and the fluid velocity is much lower than with gases. Typical conditions are 20×50-mesh carbon ($D_p = 0.3$ to 0.8 mm) and a superficial velocity of 0.3 cm/s (0.01 ft/s or about 4 gal/min · ft²). Even with these conditions $K_c a / u_0$ is smaller than for typical gas adsorption, and LUB may be 10 to 20 cm or even as much as 1 m if internal diffusion controls.

There have been few published studies of regeneration kinetics, since it is difficult to match conditions for a large adiabatic bed in a laboratory adsorber and modeling is complicated because the desorption isotherm is nonlinear and unfavorable. For steam regeneration of carbon, reverse flow is generally used with steam at 130 to 150°C. Regeneration is stopped soon after the temperature front reaches the top of the bed, and the typical steam consumption[19] is 0.2 to 0.4 lb steam/lb carbon.

Although a significant amount of absorbate remains, regeneration is most complete near the bottom of the bed, so that a high percentage removal can still be achieved in the next adsorption period. If extremely high solute removal is needed, the bed can be almost completely regenerated by prolonging the steaming period or using higher-temperature steam.

For some processes where adsorption is used to recover a valuable product, regeneration by thermal-swing desorption is not feasible, because the high temperature needed for desorption would degrade the product. An alternative is to displace the adsorbate by a gas or liquid which is moderately strongly adsorbed and can be easily separated from the product. A number of materials may be suitable for displacement regeneration, since an adsorbate can be nearly completely displaced by another species even if the displacing substance is less strongly adsorbed.

The Exxon process for separation of medium-molecular-weight linear paraffins is an example of an adsorption cycle with displacement regeneration.[1] The linear molecules are removed from the liquid feed by adsorption at 300°C in the cavities of 5Å molecular sieves, while the branched and cyclic isomers are excluded. The linear paraffins are desorbed by gaseous ammonia, which is moderately strongly adsorbed in spite of its low molecular weight. The hydrocarbons are easily removed by condensation, and the ammonia is reused. The complete cycle time is 12 to 30 min depending on the concentration of linear paraffins in the feed.[17b]

EXAMPLE 25.4. Adsorption on activated carbon is being considered to treat a process airstream that has 0.12 volume percent methyl ethyl ketone (MEK). The gas is at 25°C and 1 atm, and the flow is 16,000 ft³/min. The pressure drop across the bed should not exceed 12 in. H_2O. (*a*) If BPL 4 × 10-mesh carbon is used, predict the saturation capacity and the working capacity if the average bed temperature is 35°C and the regeneration is stopped when $W = \frac{1}{3} W_{sat}$. (*b*) What gas velocity and bed size could be used to give a reasonable cycle time if the length of unused bed is 0.5 ft? How much carbon is needed?

Solution

(*a*) From the handbooks, $P' = f_s = 151$ mm Hg at 35°C and $\rho_L = 0.805$ g/cm³ at 20°C. The normal boiling point is 79.6°C, and the estimated density at this temperature is $\rho_L = 0.75$ g/cm³. The molecular weight is 72.1.

$$V = \frac{72.1}{0.75} = 96.1 \qquad p = 0.0012 \times 760 = 0.912 \text{ mm Hg} = f$$

At 35°C,

$$\frac{T}{V} \log \frac{f_s}{f} = \frac{308}{96.1} \log \frac{151}{0.912} = 7.11$$

From Fig. 25.4, the volume adsorbed is 24 cm³ per 100 g carbon:

$$W_{sat} = 24(0.75) = 18 \text{ g}/100 \text{ g carbon}$$

$$W_0 = \frac{1}{3} W_{sat} = 6 \text{ g}/100 \text{ g carbon}$$

Working capacity $= W_{sat} - W_0 = 12$ g/100 g carbon $= 0.12$ lb/lb carbon.

(b) Try $u_0 = 1$ ft/s:

$$A = \frac{16,000 \text{ ft}^3}{60} \times \frac{1}{\text{s}} \times \frac{1}{1 \text{ ft/s}} = 267 \text{ ft}^2$$

For a circular cross section, $D = 18.4$ ft. A rectangular bed 10 ft $\times$ 27 ft might be more suitable if the bed depth is only 3 to 4 ft. Try $L = 4$ ft. From Eq. (25.3)

$$t^* = \frac{L\rho_b(W_{sat} - W_0)}{u_0 c_0}$$

At 25°C,

$$c_0 = \frac{0.0012}{359} \times \frac{273}{298} \times 72.1 = 2.21 \times 10^{-4} \text{ lb/ft}^3$$

$$\rho_p(1 - \varepsilon) = \rho_b \cong 30 \text{ lb/ft}^3 \qquad t^* = \frac{30 \times 4 \times 0.12}{1 \times 2.21 \times 10^{-4}} = 6.52 \times 10^4 \text{ s} = 18.1 \text{ h}$$

If the length of unused bed is 0.5 ft, 3.5 ft is used, and

$$t_b = \frac{3.5}{4.0} \,(18.1) = 15.8 \text{ h}$$

If the bed length is 3 ft with 2.5 ft used,

$$t_b = \frac{2.5}{4} \,(18.1) = 11.3 \text{ h}$$

Allowing for uncertainties in the calculations, a bed length of 3 ft would be satisfactory with regeneration once per 8-h shift.

Check ΔP *using Eq. (7.22).* For granular carbon, assume $\Phi_s = 0.7$ (see Table 7.1). Assume external void fraction $\varepsilon = 0.35$ (see Table 7.2). From handbooks, the properties of air at 25°C are

$$\mu = 0.018 \text{ cP} = 1.21 \times 10^{-5} \text{ lb/ft} \cdot \text{s} \qquad \rho = 0.074 \text{ lb/ft}^3$$

From Perry, 7th ed., p. **19**-20, for 4×10-mesh carbon,

$$D_p = \frac{4.76 + 2.0}{2} = 3.38 \text{ mm} = 1.108 \times 10^{-2} \text{ ft}$$

$$\frac{\Delta P}{L} = \frac{150 \times 1.0 \times 1.21 \times 10^{-5}}{32.2 \times 0.7^2 \times (1.108 \times 10^{-2})^2} \frac{0.65^2}{0.35^3} + \frac{1.75 \times 0.074 \times 1.0^2}{32.2 \times 0.7 \times 0.01108} \frac{0.65}{0.35^3}$$

$$= 9.23 + 7.86 = 17.09 \text{ lb}_f/\text{ft}^2 \cdot \text{ft}$$

$$= \frac{17.09 \times 12}{62.4} = 3.29 \text{ in. H}_2\text{O/ft}$$

For $L = 3$ ft, $\Delta P = 9.9$ in. H_2O, which is satisfactory.

A velocity of 1.5 ft/s would give $\Delta P/L = 6.06$ in. H_2O/ft and require $L \leq 2$ ft to keep $\Delta P < 12$ in. H_2O. However, the breakthrough time would be reduced to $11.3/1.5 \times (1.5/2.5) = 4.5$ h, and the bed would have to be regenerated twice each shift. This design might be satisfactory but does not give as great a margin for error.

The recommended design is for two beds $10 \times 27 \times 3$ ft placed in horizontal cylinders. The total inventory of carbon is

$$m_C = 2(270 \times 3) \text{ ft}^3 \times 30 \text{ lb/ft}^3 = 48{,}600 \text{ lb}$$

EXAMPLE 25.5. Water contaminated with 1.2 ppm TCE is to be purified in a fixed bed of 20×50-mesh Ambersorb 563. (a) For a bed length of 2 ft and a flow rate of 4.5 gal/min · ft², estimate the breakthrough time if the length of the unused bed is 0.6 ft. (b) What is the effective capacity in volume treated per unit bed volume? The adsorbent will be regenerated by steam to remove 85 percent of the TCE. The bulk density of the adsorbent is 0.53 g/cm³.

Solution

(a) From Fig. 25.5,

$$W_{\text{sat}} = 200 \text{ mg/g or } 0.2 \text{ lb/lb}$$

$$W_0 = (1 - 0.85)W_{\text{sat}} = 0.03$$

$$u_0 = \frac{4.5}{7.48} \times 60 = 36.1 \text{ ft/h (0.01 ft/s)}$$

From Eq. (25.3),

$$t^* = \frac{2(0.53 \times 62.4)(0.2 - 0.03)}{36.1(1.2 \times 10^{-6} \times 62.4)} = 4.16 \times 10^3 \text{ h}$$

$$\frac{t_b}{t^*} = 1 - \frac{\text{LUB}}{L} = 1 - \frac{0.6}{2} = 0.7$$

Breakthrough time is

$$t_b = 0.7(4.16 \times 10^3) = 2{,}910 \text{ h}$$

(b) Bed volumes treated,

$$\frac{u_0 t_b}{L} = \frac{36.1 \times 2{,}910}{2} = 5.25 \times 10^4$$

Continuous Operation

When a fixed-bed adsorber is operated in a cyclic fashion, the mass-transfer zone is often a small fraction of the bed length, and the rest of the bed is not being used for mass transfer. More efficient utilization of the adsorbent is possible with continuous countercurrent flow of adsorbent and fluids. Two beds could be arranged vertically as in Fig. 25.13, and solids would move by gravity through the desorber and the adsorber and then be carried by pneumatic conveyor or bucket elevator to the top column. The flow rates of feed, solid, and desorbent fluid would be set to provide a reasonable average driving force in each bed, and the number of ideal stages or the number of transfer units would be determined from a McCabe–Thiele diagram. For the example in Fig. 25.13, the temperature and pressure are assumed to be the same in both units, and a common equilibrium line is shown. If conditions were different, a separate equilibrium line would be drawn for the desorber.

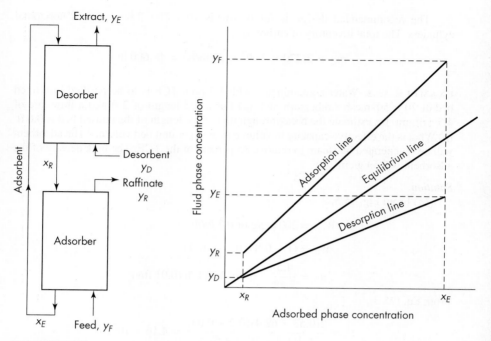

FIGURE 25.13
McCabe–Thiele diagram for a countercurrent adsorber.

Large-diameter moving beds of adsorbent carbon are used in some water treatment plants. Spent carbon is removed at intervals from the bottom of the bed, and fresh carbon is added at the top. The solid phase does not move in ideal plug flow, but this is not critical, since the mass-transfer zone is several feet long. Spent carbon is regenerated at high temperature in a rotary kiln or is sent back to the supplier for credit. A more complex moving-bed process called hypersorption was developed by Union Oil Co. for recovery of ethylene from refinery gas streams. Activated carbon particles moved slowly downward through a tall column with sections for cooling, adsorption, rectifying, and stripping.[2] The process was a technical success, but it was more costly than cryogenic separation methods and is no longer used.

A novel approach to continuous counterflow operation is the Sorbex process developed by UOP, which features a simulated moving bed. A tall column of adsorbent is divided into 12 segments, each with a flow distributor for addition or removal of liquid. A schematic diagram of the column is given in Fig. 25.14. The inlet and exit points for the four streams are moved to the adjacent segments of the bed with each new position of the rotary valve, and the effect is nearly the same as for the countercurrent flow of solids and fluid. Sorbex units are widely used for separation of xylene isomers, mixtures of olefins and paraffins, and branched and linear paraffins. For optimum performance in separating a binary or pseudobinary mixture, the desorbent fluid is chosen to have greater affinity for the solid than the weakly bound species of the feed but less affinity than the more strongly bound species.[3,4]

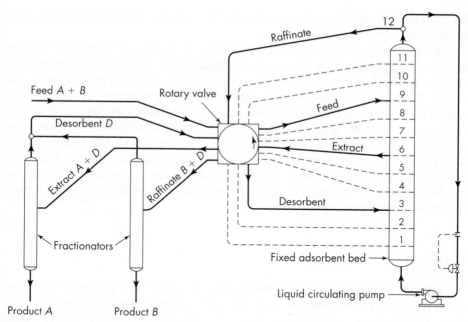

FIGURE 25.14
Schematic diagram of Sorbex process with simulated moving bed.[3]

Ion Exchange

Ion exchange is a process in which solid particles containing exchangeable cations or anions are contacted with an electrolyte solution to change the composition of the solution. Major applications are seen in softening water by exchanging calcium ions for sodium ions and demineralization of water by removing both cations and anions. Other applications include recovery of metals from dilute solutions and separation of products from bioreactors.

Ion-exchange capability exists in some natural materials such as clays and zeolites, but most processes use synthetic ion-exchange resins. These are prepared from organic polymers such as cross-linked polystyrene to which ionizable groups have been added. Cation exchangers include strong-acid resins with sulfonic acid groups ($-SO_3^-$), weak-acid resins with carboxylic acid groups ($-COO^-$), and other types with intermediate acid strength. Anion exchangers can have strong-base quaternary ammonium groups [$-N^+(CH_3)_3$] or weak-base amine groups ($-N^+H_3$).

In both cation and anion exchangers, the acid or base groups are chemically bonded to the resin matrix, and the resins have a high concentration of fixed negative or positive charges. These are balanced by mobile counterions such as H^+, Na^+, or Ca^{2+} for cation resins and Cl^-, OH^-, or NO_3^- for anion resins, so that electrical neutrality is always maintained in the resin particles. Ion exchange takes place when the activity of ions in the external solution differs from that of the mobile ions in the resin phase. For example, exposing a hydrogen-form resin HR to a solution with

Na$^+$ and H$^+$ will result in diffusion of some Na$^+$ ions into the resin and diffusion of some H$^+$ ions into the solution.

Ion-exchange resins are insoluble in water, but they swell in aqueous solution to an extent determined by the degree of cross-linking, the concentration of fixed charges, and the concentration of electrolytes in the solution. For a sulfonic acid resin with a moderate degree of cross-linking (8% divinylbenzene), the swollen volume of the Na-form resin in dilute solution is about 1.8 times the volume of the dry resin, and the porosity is 45 percent. This resin swells an additional 8 percent when converted to HR in dilute acid solutions. Some swelling is desirable to increase diffusion rates inside the particles, but swelling also decreases the capacity of the resin per unit volume of particle or per unit volume of bed. Resins are available as spherical beads with sizes between 0.3 and 1.2 mm and are generally used in fixed beds similar to those for adsorption from liquids.

Equilibria

The capacity of an ion exchanger is the number of exchangeable groups per unit mass of dry resin. For cation resins, the capacity is often expressed as milliequivalents per gram of dry hydrogen-form resin, or meq/g H$^+$. For anion resins, the basis is a gram of dry chlorine-form resin. For practical purposes, the capacity may be given in equivalents per liter of bed. The theoretical capacity for a styrene-divinylbenzene copolymer is about 5.4 meq/g H$^+$ when each of the aromatic rings has one sulfonic acid group. The corresponding bed capacity is about 2.5 equivalents per liter.

Ion exchange is a reversible reaction in which the counterion in the resin is replaced by a different ion from the external solution. For cation exchange of a sodium-form resin with a monovalent ion A$^+$, the reaction is

$$A^+ + NaR \rightleftharpoons AR + Na^+ \tag{25.27}$$

The equilibrium constant for the reaction is expressed in terms of activities or concentrations times activity coefficients

$$K_{eq} = \frac{c_{Na^+} c_{AR}}{c_{A^+} c_{NaR}} \times \frac{\gamma_{Na^+} \gamma_{AR}}{\gamma_{A^+} \gamma_{NaR}} \tag{25.28}$$

For dilute solutions, the activity coefficients do not change much with concentration, and a simple concentration-based equilibrium constant can be used.

$$K' = \frac{c_{Na^+} c_{AR}}{c_{A^+} c_{NaR}} \tag{25.29}$$

When a monovalent ion is replaced by a divalent ion, each new counterion balances two charged sites in the resin.

$$B^{2+} + 2NaR \rightleftharpoons BR_2 + 2Na^+ \tag{25.30}$$

$$K' = \frac{(c_{Na^+})^2 c_{BR_2}}{c_{B^{2+}} (c_{NaR})^2} \tag{25.31}$$

TABLE 25.1

Equilibrium constants for polystyrene cation and anion exchangers, 8% DVB cross-linking[†]

Strong-acid sulfonated resin		Strong-base trimethyl amine resin	
Counterion	K'	Counterion	K'
Li^+	1.00	Iodide	8.7
H^+	1.27	Nitrate	3.8
Na^+	1.98	Bromide	2.8
NH_4^+	2.55	Cyanide	1.6
K^+	2.90	Chloride	1.0
Mg^{2+}	3.29	Bicarbonate	0.3
Cu^{2+}	3.47	Acetate	0.2
Ni^{2+}	3.93	Sulfate	0.15
Ca^{2+}	5.16	Fluoride	0.09
Ba^{2+}	11.5	Hydroxide	0.05–0.07

[†]By permission, from R. H. Perry, and D. W. Green (eds.): *Perry's Chemical Engineers' Handbook*, 7th ed., McGraw-Hill, New York, 1997, p. **16**-14.

Values of K' for typical strong-acid and strong-base resins are given in Table 25.1. For cation exchangers, Li^+ is the reference ion, and K' for any reaction can be calculated from the ratio of K' values. For example, for the reaction of Na^+ with HR, $K' = 1.98/1.27 = 1.56$. The K' values for monovalent cations increase from Li^+ to K^+ because of the decreasing size of the hydrated ions. The average hydration number for Li^+ is about 3.3 and for K^+ is about 0.6.[13a] For more strongly cross-linked resins, which do not swell as much, the effect of ion size on K' values is somewhat greater than on the values in Table 25.1.

Divalent ions tend to have somewhat higher K' values than monovalent ions, but in most applications, the preference of the exchanger for divalent ions is greater than suggested by the equilibrium constants. For very low solution concentrations, an exchange reaction of the type shown in Eq. (25.30) can go almost completely to the right because the concentration of the monovalent ion is squared in Eq. (25.31).

EXAMPLE 25.6. For the Cu^{2+}/Na^+ exchange with a strong-acid resin, show how the fraction CuR_2 in the resin varies with the fraction Cu^{2+} in solution for total solution normalities of 0.5, 0.1, and 0.01. Assume a resin normality of 2 meq/mL.

Solution

$$Cu^{2+} + 2NaR \rightleftharpoons CuR_2 + 2Na^+$$

$$K' = K'_{Cu^{2+}}/K'_{Na^+} = 3.47/1.98 = 1.75$$

For $Cu^{2+} = 0.02\ M$, $Na^+ = 0.06\ M$

$$c_{total} = 2Cu^{2+} + Na^+ = 0.04 + 0.06 = 0.10\ N$$

Equivalent fraction Cu^{2+} in solution: $0.04/0.10 = 0.40$. From Eq. (25.31)

$$\frac{c_{CuR_2}}{(c_{NaR})^2} = \frac{1.75 \times 0.02}{0.06^2} = 9.72$$

Resin normality = 2.0, so

$$2.0 = c_{NaR} + 2c_{CuR} = c_{NaR} + 2(9.72)(c_{NaR})^2$$

$$19.44(c_{NaR})^2 + c_{NaR} - 2 = 0$$

Using the quadratic formula,

$$c_{NaR} = \frac{-1 + [1 + 8(19.44)]^{0.5}}{2 \times 19.94} = 0.296$$

$$c_{CuR_2} = \frac{2 - 0.296}{2} = 0.852$$

On an equivalent basis, fraction Cu^{2+} in resin $= 2 \times 0.852/2 = 0.852$.

Similar computations give the results plotted in Fig. 25.15. The curves are similar in shape to adsorption isotherms of the favorable type, and the exchange reaction is very strongly favored at 0.01 N total concentration. The predicted results are quite close to experimental values for a similar cation exchanger.

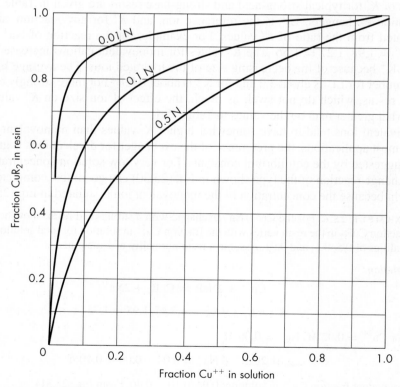

FIGURE 25.15
Predicted selectivity curves for Cu^{2+}/Na^+ exchange.

Mass-Transfer Rates

The rates of ion-exchange reactions are limited by the rates of diffusion of the counterions in the external solution and in the pores of the resin particles. Equilibrium can be assumed to exist at any point inside each particle, and predicting the concentration changes with time would require solution of the unsteady-state diffusion equations. For engineering applications, the mass-transfer rate can be approximated using an overall coefficient $K_c a$ and a driving force based on the concentration in the solution and the concentration c^* that would be in equilibrium with the average particle concentration. The rate for a packed bed is proportional to this driving force.

$$r = K_c a(c - c^*) \tag{25.32}$$

where $\dfrac{1}{K_c} = \dfrac{1}{k_{c,\text{ext}}} + \dfrac{1}{m k_{c,\text{int}}}$

a = external area per unit bed volume
m = ratio of internal to external concentration

The external mass-transfer coefficient can be predicted from the correlation for packed beds, given in Chap. 17,

$$\text{Sh} = 1.17\,\text{Re}^{0.585}\,\text{Sc}^{1/3} \tag{17.74}$$

The area per unit volume of packing for a bed of spheres is $6(1 - \varepsilon)/D_p$, where ε is the external void fraction. The internal coefficient is based on the effective diffusivity and a diffusion distance equal to one-tenth of the particle diameter [see Eqs. (10.30) and (17.77)].

$$k_{c,\text{int}} = \frac{10 D_e}{D_p} \tag{25.33}$$

Diffusivities for ions in the resin are lower than in the external solution because of the limited internal void fraction, the tortuosity, and to some extent the hindered diffusion in small pores. For monovalent ions, the effective diffusivity is about one order of magnitude lower than normal, but for divalent ions, it may be lower by two orders of magnitude.[13b] Fortunately, the internal resistance to diffusion is often no larger than the external resistance and may even be negligible in spite of the low effective diffusivity, because a low concentration in the external solution makes m very large. For a solution that is 10^{-3} M, m would be about 2,000, and the external resistance would be controlling even with a 10- or 100-fold lower internal diffusivity.

With favorable equilibrium curves of the type in Fig. 25.15, the mass-transfer zone soon acquires a constant length, and the shape of the breakthrough curve does not change with bed length. The steepness of this curve depends on the number of mass-transfer units and the nature of the controlling resistance, as is true for adsorption (see Fig. 25.11). The exact shape of the breakthrough curve and the value of t_b are hard to predict, and the design of large units is usually based on experience or tests with a small column. If the particle size and liquid velocity are kept constant, the length of unused bed (LUB) should not change in going to a larger column. With a favorable equilibrium curve, the LUB may be only a few inches, a small fraction of the normal bed length.

Regeneration of an exhausted bed is carried out by passing a concentrated solution of acid, base, or salt through the bed. By using a concentrated solution, the minimum regeneration time is decreased, and the strong preference of the resin for multivalent ions is greatly reduced, as illustrated in Fig. 25.15. When the equilibrium curve is almost linear, the width of the mass-transfer zone increases with bed length (see Fig. 25.12). The time for complete (or nearly complete) regeneration may be much greater than the minimum regeneration time, but it is still considerably shorter than the breakthrough time for the solute removal part of the cycle.

EXAMPLE 25.7. Laboratory tests for the removal of metal ions from a 0.008 M solution using a short bed of ion-exchange beads indicated an effective bed capacity of 0.75 mmol/mL at breakthrough ($c = 0.02c_0$) compared to an equilibrium capacity of 1.15 mmol/mL. The bed was 2 cm in diameter and 30 cm long; the superficial velocity was 0.40 cm/s (5.9 gal/min · ft^2). (a) A large ion-exchange column will be operated at 3 gal/min · ft^2 with a bed 0.6 m in diameter and 1.5 m long. Assuming the same LUB as in the laboratory tests, predict t_b and the effective bed capacity at breakthrough. (b) How might the change in velocity affect the LUB and the effective capacity? (c) If regeneration is carried out with a 2 M NaCl solution at 3 gal/min · ft^2, what is the time for regeneration if twice the minimum amount of solution is used?

Solution

(a) Length of unused bed in laboratory column,

$$\frac{t_b}{t^*} = \frac{0.75}{1.15} = 0.652 = 1 - \frac{LUB}{L}$$

$$LUB = (1 - 0.652)(30) = 10.4 \text{ cm}$$

For the same LUB in the large bed,

$$\frac{t_b}{t^*} = 1 - \frac{10.4}{150} = 0.931$$

The effective capacity is $0.931(1.15) = 1.07$ mmol/mL (or mol/L). From Eq. (25.2)

$$t^* = \frac{\rho_b L W_{sat}}{u_0 c_0}$$

$$\rho_b W_{sat} = 1.15 \text{ mol/L}$$

$$u_0 = \frac{3 \text{ gal}}{\text{min} \cdot \text{ft}^2} \times \frac{3{,}785 \text{ cm}^3/\text{gal}}{60 \text{ s/min}} \times \frac{1}{930 \text{ cm}^2/\text{ft}^2} = 0.204 \text{ cm/s}$$

$$t^* = \frac{1.15 \text{ mol}}{L} \times \frac{150 \text{ cm}}{0.204 \text{ cm/s} \times 0.008 \text{ mol/L}} \times \frac{1}{3{,}600} = 29.4 \text{ h}$$

$$t_b = 0.931(29.4) = 27.4 \text{ h}$$

(b) For very favorable equilibrium, the breakthrough curves in Fig. 25.11 show that $N(\tau - 1) = -2.9$ when the external film controls. Since $N = k_c a L/u_0$, and $k_c a$

increases with about $u_0^{1/2}$, going to about one-half the laboratory velocity decreases $k_c a$ but increases N by $\sqrt{2}$, making LUB $\cong 7$ cm and $t_b \cong 0.95 t^*$. If both internal and external resistances are important, $k_c a$ will not decrease as much, and LUB will be even shorter.

(c)

$$t^* = \frac{1.15 \text{ mol}}{\text{L}} \times \frac{150 \text{ cm}}{0.204 \text{ cm/s} \times 2.0 \text{ mol/L}} \times \frac{1}{60} = 7.05 \text{ min}$$

$$t = 2t^* = 15.1 \text{ min for regeneration}$$

Operation of Ion Exchangers

An ion-exchange system for semicontinuous operation requires at least two columns, so that one can always be online. When the time needed for regeneration is quite short, it may save money to use two beds in series for ion removal. The first bed is switched to regeneration when breakthrough occurs at the second bed, and the previously regenerated bed is made the second bed in the series. The inventory of resin is reduced, and only fully saturated beds are regenerated, which is more efficient.

The feed solution is normally passed downward through the bed at a flow rate of 1 to 5 gal/min·ft^2 (1 gpm/ft$^2 = 0.67$ ft/min). After breakthrough, the bed is taken off-line and backwashed with upflow of water to remove small particles or sediment trapped in the bed. During backwash, the bed fluidizes and expands 50 to 100 percent, so sufficient headroom must be provided above the resin bed. Regenerating solution is passed either upflow or downflow through the bed at 2 to 5 gal/min·ft^2. Using counterflow of regeneration solution is more effective in removing strongly held counterions, since these tend to concentrate near the feed end of the bed. However, a cocurrent arrangement with downflow of feed and regenerant is commonly used.

For preparation of very pure water, both cation- and anion-exchange resins are used. A strong cation resin replaces the cations with H$^+$ ions, and in the next column, an anion-exchange resin removes anions and produces OH$^-$ ions, which react to form H$_2$O. Sometimes three or four beds are used in series with a combination of strong-acid, weak-acid, strong-base, and weak-base resins arranged in various ways to facilitate regeneration. Deionization can also be achieved with a mixed bed of cation and anion resins. The beads are separated before regeneration by gentle fluidization, since the anion resins are lighter and more easily suspended.[7]

CHROMATOGRAPHY

Chromatography is the name given to a class of processes for the separation of multicomponent mixtures of gases or liquids. It uses a bed of solids or immobilized liquid as a stationary phase, and intermittent feed of the material to be separated. Components of the mixture are moved through the bed or eluted by continuous flow of a carrier gas or liquid, which becomes the mobile phase. Feed components partition between the mobile and stationary phases and move at different rates through the bed because of different distribution coefficients. If the bed or column is long enough, all the components emerge sequentially as separate pulses, and an analyzer at the exit shows the concentration of each component in the mobile phase.

The term *chromatography* arose from the bands of color seen when a glass column was used to separate liquid mixtures of plant cell pigments. It is now applied to other similar separations. The column, analyzer, and associated equipment for controlling the flow rate and temperature are called a *chromatograph*. The plot of the analyzer signal showing the peaks for each component is a *chromatogram*. A typical chromatogram is shown in Fig. 25.16.

Chromatographs are classified by the nature of the mobile and stationary phases. Gas chromatography (GC) includes GSC, where the stationary phase is a solid adsorbent, and GLC, where a liquid of high molecular weight, held in a porous solid, is the stationary phase. GLC is more common the GSC, since components of the feed usually partition between the phases in constant proportions, which leads to symmetric patterns in the chromatogram. In GSC, if one component is strongly adsorbed with a nonlinear isotherm, the resulting peak is asymmetric with a long tail, which makes the separation from other peaks difficult. In GLC the feed can be a gas, but it is often a liquid injected as a small sample into a vaporizer and swept into the column by the carrier gas. Helium or hydrogen is commonly used as the carrier gas because very low concentrations of other components can be determined by measuring the thermal conductivity or density, or by using flame ionization detectors.

The separation of liquid mixtures by chromatography (LC), can be carried out in a column containing a solid stationary phase, LSC, or with an immiscible liquid as the stationary phase, LLC. Diffusion resistances are important in liquid chromatography, and performance is improved by using very small particles in the stationary phase, even though this may lead to column pressure drops of several atmospheres. The term *high-performance liquid chromatography* (HPLC) is applied to separations carried out at high pressure with very fine particles and high flow rates.

A different separation principle is used in *gel permeation chromatography* (GPC). A mixture of high-molecular-weight polymers or biomaterials dissolved in a liquid is fractionated because the largest molecules are excluded from some of the pores in the stationary phase. The molecules are separated by size, with the largest ones coming out first and the smallest eluted last. This is the opposite of the order of elution in other chromatographic processes.

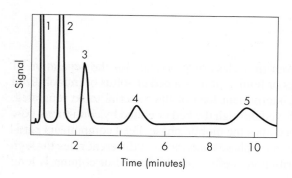

FIGURE 25.16
Chromatogram for C_1 to C_5 *n*-alkyl bromides. (*By permission from S. Dal Nogare and R. S. Juvet, Jr.: Gas-Liquid Chromatography, Theory and Practice, Interscience, New York, 1962.*)

Chromatography is used mainly as an analytical tool, because the intermittent feed and small pulses of feed material make the production rate from the column quite low, and the separation cost per unit of product is high. There are also problems in scaling up large columns, as discussed later. Large-scale or preparative chromatography is justified for some high-value products or where separation by other means is very difficult.

Gas-liquid chromatography

In gas-liquid chromatography, each component of the feed sample moves through the column at a rate proportional to the velocity of the carrier gas and inversely proportional to the capacity of the stationary phase plus the mobile phase. The equilibrium relationship for component A, expressed in mole fractions, is

$$y_A = \frac{p_A}{P} = \frac{\gamma_A x_A P'_A}{P} \tag{25.34}$$

Converting to mass concentrations in the gas and liquid phases gives

$$c_{A,l} = x_A \left(\frac{\rho_l}{M_l} \right) M_A = x_A \rho_{M,l} M_A \tag{25.35}$$

$$c_{A,g} = y_A \rho_{M,g} M_A \tag{25.36}$$

Combining Eqs. (25.34) to (25.36), we get

$$c_{A,l} = \frac{P}{\gamma_A P'_A} \left(\frac{\rho_{M,l}}{\rho_{M,g}} \right) c_{A,g} \tag{25.37}$$

The mass of solvent phase per unit volume w equals the bed density times the fraction solvent loading s.

$$w = \rho_b s \tag{25.38}$$

The equilibrium mass of component A in a unit volume of bed is the amount in the pore volume plus the amount dissolved in the stationary phase.

$$m_A = \varepsilon c_{A,g} + \left(\rho_b s \frac{P}{\gamma_A P'_A} \frac{\rho_{M,l}}{\rho_{M,g}} \right) c_{A,g} \tag{25.39}$$

To get good separation of the peaks, the feed pulse is quite small. This makes x_A small and γ_A nearly constant. For isothermal operation, then, the term in parentheses is constant and becomes a partition coefficient K_A. Pressure changes in the column do not change K_A, since $\rho_{M,g}$ is proportional to pressure.

$$m_A = c_{A,g}(\varepsilon + K_A) \tag{25.40}$$

When a carrier gas is passed through the column at a superficial velocity u_0, the actual average velocity is u_0/ε, and a component that is not absorbed by the solvent leaves the column at time t.

$$t = \frac{L\varepsilon}{u_0} \tag{25.41}$$

A component that partitions between the mobile and stationary phases moves at a lower velocity, and the midpoint of the pulse leaves the column at the *retention time t^**

$$t_A^* = \frac{L(\varepsilon + K_A)}{u_0} \tag{25.42}$$

Equation (25.42) is similar to Eq. (25.3) for the breakthrough curve in adsorption.

For most GLC analyses, $K_A \gg \varepsilon$, and the retention time is essentially proportional to K_A. For the separation of a binary mixture the ratio of retention times must be large enough that the peaks on the chromatogram do not overlap. When ε can be neglected, this ratio is

$$\frac{t_A^*}{t_B^*} = \frac{K_A}{K_B} = \frac{\gamma_B P_B'}{\gamma_A P_A'} \tag{25.43}$$

When the liquid mixture is ideal, the ratio of retention times depends only on the ratio of the vapor pressures, and the lower-boiling component or components exit first. However, when the components are quite different in polarity, the proper choice of solvent can make the activity coefficients different enough to permit a high boiler to come out before a low boiler.

Peak broadening

As it passes through the column, each individual pulse becomes broader because of mass-transfer resistances, axial diffusion, and departure from plug flow. If the injected sample is quite small, and if the partition coefficients are constant, the pulse becomes Gaussian in shape, and the resulting peaks on the chromatogram can be characterized by the mean retention time, the peak height, and the standard deviation. For a standard distribution, tangents to the points of inflection intersect the baseline at $t^* \pm 2\sigma$, so the width Δt in Fig. 25.17 corresponds to 4 standard

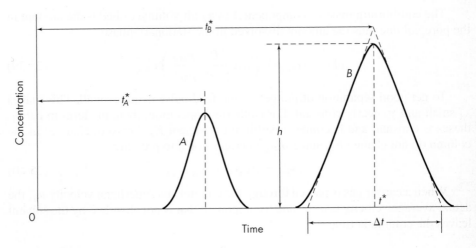

FIGURE 25.17
Typical chromatogram for a binary mixture.

deviations. The area under the curve is a measure of the amount of that component; the exact value is determined using a calibration factor.

In the early studies of chromatography the column was considered equivalent to an array of equilibrium stages, and the number of ideal stages or plates, NTP or N, was found from the relative width of a peak on the chromatogram.[6]

$$N = 16\left(\frac{t^*}{\Delta t}\right)^2 \tag{25.44}$$

Other measures of peak spreading can be used to determine N. The volume of carrier gas used up to the midpoint of a peak is called the *retention volume* V_R, and N can be calculated from the relation[14]

$$N = \left(\frac{V_R}{\sigma}\right)^2 \tag{25.45}$$

For peak B in Fig. 25.17, $t^*/\Delta t = 2.65$, and from Eq. (25.44), $N = 112$. About the same value of N is obtained using peak A. The number of ideal plates, however, is not exactly the same for different components, since differences in diffusivity affect the peak broadening. The change in N is more apparent when the components of the feed have a wide range of molecular weight and retention times.

Since for this example $t_B^* \cong 2t_A^*$, the separation is easy and a relatively short column can be used. When the partition coefficients differ by only 10 to 20 percent, a much larger value of N is needed. For good separation of two components, the difference in retention times should be equal to or greater than one-half the sum of the peak widths.

$$t_B^* - t_A^* \geq \frac{\Delta t_B}{2} + \frac{\Delta t_A}{2} \tag{25.46}$$

The ratio of retention times is called α, which is approximately the ratio of the partition coefficients [Eq. (25.43)]. If N is the same for both components, Eqs. (25.43), (25.44), and (25.46) are combined to give

$$t_A^* + \frac{2t_A^*}{\sqrt{N}} = t_B^* - \frac{2t_B^*}{\sqrt{N}} = \alpha t_A^*\left(1 - \frac{2}{\sqrt{N}}\right) \tag{25.47}$$

Rearrangement of Eq. (25.47) leads to an equation for the minimum number of plates

$$N_{\min} = 4\left(\frac{1+\alpha}{\alpha - 1}\right)^2 \tag{25.48}$$

A low value of α requires a large number of plates and a relatively long column. The retention time is proportional to the column length, but the peak width increases with only the square root of N or L, so separation improves with column length. Fortunately a GLC column only 1 or 2 m long may contain more than 1,000 theoretical plates, so separation is possible even with quite low values of α. The height of a theoretical plate, HETP or H, is often 0.1 cm or less when small particles are used. Some data[15] showing the effect of particle size and gas velocity on H are

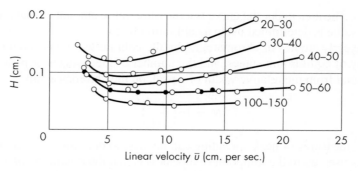

FIGURE 25.18
Experimental data for the dependence of H upon $\bar{u}$ for packed
columns of different d_p. B.S.S. mesh range quoted. (Closed circles
are duplicate column data.) (*by permission from H. Purnell, Gas
Chromatography, Wiley, 1962*).

shown in Fig. 25.18. The lowest H is obtained with the finest particles and a gas
velocity of 5 to 10 cm/s. The general shape of the curves agrees with the theory of
van Deemter et al.[21]

$$H = A + \frac{B}{u} + Cu \qquad (25.49)$$

The A term gives the peak broadening caused by the irregular nature of the flow
channels in the packed bed; it is usually a small multiple of D_p. The term B/u gives
the effect of axial molecular diffusion. This term increases with D_v. At low gas
velocities the use of He or H_2 as a carrier gas gives higher values of H than N_2. The
third term shows the spreading due to mass transfer between the mobile phase and
stationary phase; it includes the resistance to diffusion in the external gas film and
in the open pores of the solid. The C term also includes the effect of diffusion into
the liquid phase. The relative importance of diffusion in the gas and liquid phases is
hard to predict, since it depends on the size and distribution of the elements of liquid
in the pores. If the liquid loading is only 2 to 5 percent of the solids, the thickness
of the liquid elements and the resistance to diffusion are small. If the pores are
nearly full of liquid, the average diffusion distance in the liquid and that in the gas
are comparable, and the liquid resistance dominates because of the much lower
diffusivity.

The small value of HETP for GLC columns may seem surprising, since the
HTU values for gas absorption or gas adsorption are much larger. A typical ammonia absorber with $D_p \cong 4$ cm and $u_0 \cong 100$ cm/s has $H_{Og} \cong 30$ to 60 cm (see
Fig. 18.22). An adsorber has smaller particles, which increases the mass-transfer
area, and a lower gas velocity which also contributes to a lower HTU. Referring to
Examples 25.2 and 25.3, a butanol adsorber with $D_p = 0.37$ cm and $u_0 = 58$ cm/s
had $H_{Og} = L/N = 8.0/3.17 = 2.5$ cm, more than 10 times smaller than the
absorber in spite of a significant internal mass-transfer resistance. For GLC with
still smaller particles (40- to 50-mesh, $D_p = 0.04$ cm), and $u_0 \cong 5$ cm/s, H_{Og}

would be at least 100 times smaller than for adsorption, or about 0.02 cm based on predicted mass-transfer rates. The reported HETP for GLC at these conditions is about 0.08 cm, as shown in Fig. 25.18. The values of HETP and H_{Og} would be about the same if the peak broadening were due only to mass transfer between the mobile and stationary phases. The difference here reflects the importance of the axial dispersion terms A and B/u in the van Deemter equation.

Preparative chromatography

When a chromatographic process is scaled up for commercial production, increasing the column diameter usually increases the peak broadening caused by uneven flow distribution and wall effects. The HETP may increase severalfold unless great care is taken in packing the column and getting uniform distribution of the feed. Using a narrow range of particle sizes and vibrating the column while it is being packed help keep the HETP close to the laboratory values. Baffles may be introduced at regular intervals to redistribute the flow, but even with such baffles in place the HETP still increases with column diameter.[17a]

Large-scale chromatography is mainly used for the separation of liquids. One example is the Asahi process for recovering p-xylene from mixed xylenes using a zeolite adsorbent.[18] The feed is introduced intermittently for short periods, and the isomers are moved through the main column by the continuous flow of a desorbent liquid. The peaks are not Gaussian in shape because of the size of the feed pulse and the interaction between components. Four fractions are collected as shown in Fig. 25.19. The first fraction with weakly adsorbed m-xylene and o-xylene is returned to the isomerization unit to produce more p-xylene. The next two fractions are sent to separate auxiliary columns to produce some ethyl benzene, p-xylene, and mixed xylenes. The final fraction is distilled to recover the solvent and produce nearly pure p-xylene. Complete separation of all the components could probably be achieved in one long column with small feed pulses, but the hourly capacity would be reduced and the production cost increased.

It is interesting to compare the number of theoretical plates needed for chromatographic separation with the number required to produce 99 percent pure

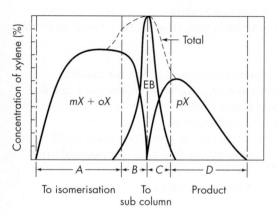

FIGURE 25.19
Large-scale chromatography of mixed xylenes.[18]

TABLE 25.2
Theoretical plates required for binary mixture separation

α	$N_{min, dist}$	$N_{min} \times 2$	$N_{chromatograph}$
4	6.63	13.3	11
2	13.26	26.5	36
1.5	22.67	45.3	100
1.2	50.4	101	484
1.1	96.4	193	1,764
1.05	188	377	6,724

products by fractional distillation. The number of plates for distillation is taken as twice the minimum number calculated from the Fenske equation [Eq. (21.42)]. As shown in Table 25.2, there is not much difference when α is large, but as α approaches 1.0, N for the chromatograph rises much more rapidly than does N_{min} for distillation.

SYMBOLS

A	Cross-sectional area of bed, m^2 or ft^2; also, constant in Eq. (25.49)
a	External surface area per unit volume of bed, m^{-1} or ft^{-1}
B	Constant in Eq. (25.49)
C	Constant in Eq. (25.49)
c	Concentration, g/cm^3 or ppm (mg/L); c_A, of component A; $c_{A,g}$, in gas; $c_{A,l}$, in liquid; c_0, in feed; c_1, c_2, of components 1 and 2 of multicomponent mixture; c^*, in equilibrium with concentration in solid
c_p	Heat capacity of fluid, $J/g \cdot °C$ or $Btu/lb \cdot °F$
c_s	Heat capacity of solid, $J/g \cdot °C$ or $Btu/lb \cdot °F$
D	Diameter, m or ft; D_p, of particle
D_e	Effective diffusivity, cm^2/s or ft^2/h
D_v	Volumetric diffusivity, cm^2/s or ft^2/h
F_A	Feed rate of adsorbate per unit cross-sectional area of bed, $g/cm^2 \cdot s$
f	Fugacity, atm or mm Hg; f_s, of saturated liquid
H	Height of a theoretical plate or transfer unit, m or ft; H_{Og}, based on overall coefficient in gas phase
$\mathscr{H}_R$	Relative humidity
K	Constant in equation for adsorption isotherm; K_1, K_2, for multiple adsorbates; also, partition coefficient; K_A, K_B, for components A and B
K_c	Overall mass-transfer coefficient, m/s or ft/s
$K_c a$	Volumetric overall mass-transfer coefficient, s^{-1}
K_{eq}	Equilibrium constant in chemical reaction; K', based on concentrations
k_c	Individual mass-transfer coefficient, m/s or ft/s; $k_{c,ext}$, for external film; $k_{c,int}$, for internal diffusion
$k_c a$	Volumetric individual mass-transfer coefficient, s^{-1}

L Distance through bed, m or ft; L_T, total bed length; L_{sat}, length of saturated portion of bed

M Molecular weight; M_A, of component A; M_l, of liquid

m Exponent in Freundlich equation

m_A Equilibrium mass of component A per unit bed volume, kg/m^3 or lb/ft^3

m_C Total mass of carbon in bed, kg or lb

N Number of mass-transfer units or theoretical plates; N_f, based on external film; N_p, based on pore diffusion; N_{min}, minimum number of transfer units or theoretical plates

N_H Number of heat-transfer units

P Total pressure, atm or mm Hg; P', vapor pressure; P'_A, P'_B, of components A and B

p Partial pressure, atm or mm Hg; p_A, of component A

Re   Reynolds number, $D_p u_0 \rho / \mu$

S Internal surface area of solid, m^2/g or ft^2/lb

Sc   Schmidt number, $\mu / \rho D_v$

Sh   Sherwood number, $k_c D_p / D_v$

s Fractional solvent loading

T Temperature, °C, K, °F or °R; T_g, of gas; T_s, of solid; T_0, initial value

t Time, s or h; t_b, at break point; t_0, at start of adsorption; t_1, time to saturate first portion of bed; t^*, ideal adsorption time for vertical breakthrough curve; t_A^*, t_B^*, of components A and B

U Overall heat-transfer coefficient, W/m$^2 \cdot$°C or Btu/ft$^2 \cdot$h$\cdot$°F

Ua Volumetric overall heat-transfer coefficient, W/m$^3 \cdot$°C or Btu/ft$^3 \cdot$h$\cdot$°F

u_0 Superficial velocity of fluid, cm/s, m/s, or ft/s

V Molar volume at normal boiling point, cm^3/g mol

v_z Velocity of transfer zone, cm/s or ft/s

W Adsorbate loading, g/g solid; W_b, at break point; W_{max}, maximum value as $c \to \infty$; W_{sat}, at equilibrium with the fluid; W_0, initial value; W_1, for component 1 of multiple adsorbates

w Mass of solvent per unit bed volume, kg/m^3 or lb/ft^3

x Mole fraction in liquid; x_A, of component A; x_E, in extract; x_R, in raffinate

y Mole fraction in gas; y_A, of component A; y_E, in extract; y_F, in feed; y_R, in raffinate

Abbreviations

GC   Gas chromatography; GLC, gas-liquid chromatography; GSC, gas-solid chromatography

GPC   Gel permeation chromatography

HETP   Height equivalent to a theoretical plate

HTU   Height of a transfer unit

LUB   Length of unused bed

PSA   Pressure-swing adsorption

Greek Letters

α	Ratio of retention times, t_B^* / t_A^*
γ	Activity coefficient; γ_A, γ_B, of components A and B
ΔP	Pressure drop in adsorption column, lb_f/ft^2 or in. H_2O
Δt	Width of peak in chromatogram; Δt_A, Δt_B, of components A and B
ε	External void fraction of bed
μ	Absolute viscosity, Pa·s, cP, or lb/ft·h
ρ	Density, kg/m^3 or lb/ft^3; ρ_L, of liquid; ρ_b, bulk density of bed; ρ_p, density of particle
ρ_M	Molar density, mol/m^3 or mol/ft^3; $\rho_{M,g}$, of gas; $\rho_{M,l}$, of liquid
σ	Standard deviation, units of time or volume
τ	Throughput parameter defined by Eq. (25.9), dimensionless
Φ_s	Sphericity, dimensionless

PROBLEMS

25.1. Adsorption on 6×10-mesh activated carbon is being considered to recover methyl ethyl ketone (MEK) from an airstream at 25°C and 1 atm. The airflow is 12,000 std ft^3/min, and the air has 0.40 lb MEK/1,000 std ft^3. If the superficial velocity is 0.5 ft/s, and an adsorption cycle of at least 8 h is desired, about what bed dimensions should be used? Assume the bulk density of the carbon is 30 lb/ft^3.

25.2. Granular carbon is used to remove phenol from an aqueous waste. If 10×20-mesh carbon is used with a superficial velocity of 0.03 m/s, estimate the number of transfer units in a bed 4 m deep. The effective diffusivity in the particles can be taken as 0.2 times the bulk diffusivity.

25.3. Air with relative humidity of 50 percent at 20°C is compressed to 8 atm, cooled to 30°C to condense some of the water, and dried in a silica-gel dryer. What fraction of the water in the air is removed in the condenser? Calculate the equilibrium capacity of the silica gel if the adsorption is carried out at 30°C and if the average bed temperature rises to 50°C.

25.4. The preliminary design for a vapor-phase adsorber specified a bed of 6×16-mesh carbon 6 ft in diameter and 4 ft deep, with a superficial velocity of 60 ft/min. However, the estimated pressure drop of 16 in. H_2O is more than twice the desired value. If the same amount of carbon is used in a wider and shallower bed, how will that affect the pressure drop and the steepness of the breakthrough curve? Suggest an appropriate design for the bed.

25.5. Data for drying nitrogen with molecular sieve type 4A are given by Collins.[5] Calculate the saturation capacity from the breakthrough curve, and determine the length of unused bed based on a break-point concentration c/c_0 of 0.05:

$T = 79°F$	$L_T = 1.44$ ft
$P = 86$ psia	$\rho_b = 44.5$ lb/ft^3
N_2 feed $= 29.2$ mol/h·ft^2	$c_0 = 1,490$ ppm

t, h	0	10	15	15.4	15.6	15.8	16	16.2
c, ppm	<1	<1	<1	5	26	74	145	260
t, h	16.4	16.6	16.8	17	17.2	17.6	18	18.5
c, ppm	430	610	798	978	1,125	1,355	1,465	1,490

25.6. (a) Use the correlation in Fig. 25.4 to predict the amount of n-butanol adsorbed on carbon for the conditions of Example 25.2, and compare with the experimental data. (b) Estimate the saturation capacity of the carbon for a butanol concentration of 20 ppm in air at 30°C and 1 atm.

25.7. A molecular sieve adsorbent used for air separation has a maximum capacity for N_2 of 0.046 g/g at 30°C, and the data can be fitted to a Langmuir isotherm with $K = 0.50$ atm^{-1}. (a) What is the saturation capacity when the adsorbent is exposed to air at 10 atm and 30°C? (b) Calculate the ideal absorption time for a bed 4 ft deep operating with a superficial velocity of 20 ft/min. The pellet density is 1.3 g/cm^3.

25.8. Granular activated carbon impregnated with sulfur is used to remove mercury vapor from natural gas and from air. Because chemical reaction converts the mercury to mercuric sulfide, the adsorption is irreversible, and the capacity is as high as 20 weight percent. (a) Assuming that the adsorption rate for removal of mercury from air is controlled by external mass transfer, calculate the overall coefficient for a bed of 4×6-mesh carbon at 20°C and a superficial velocity of 75 cm/s. (b) If the inlet concentration is $10\ \mu g/Nm^3$ (micrograms per normal cubic meter, with the gas volume evaluated at standard conditions) and the treated gas must contain less than $10^{-3}\ \mu g/Nm^3$, what is the minimum bed length? (c) For a bed 50 cm deep, what is the predicted time to breakthrough?

25.9. An adsorber for removing vinyl chloride monomer (VCM) from air has two beds of BPL carbon that are 4 ft deep, and at breakthrough, the beds have 0.09 lb VCM/lb carbon. The average bed temperature is 80°F. The beds are regenerated with air at 200°F and 120 ft/min, and the exhaust gas is incinerated. Regeneration at higher temperature is being considered. (a) How fast does the temperature front move through the bed if the bed operates adiabatically and the heat of desorption is 1.5 times the heat of vaporization? (b) If the inlet temperature is increased to 300°F with the same mass velocity, what is the rate of movement of the temperature front? (c) If the bed temperature could be increased very rapidly to 200°F by using electric heaters in the bed, what would be the equilibrium concentration of VCM in the vapor phase? If airflow were then started at 120 ft/min and 200°F how rapidly would the desorption front advance through the bed? (d) What is the main advantage of increasing the regeneration temperature?

25.10. A water softener has a resin bed 1 ft in diameter and 1.5 ft long. If the water contains 20 ppm Ca^{2+}, 10 ppm Mg^{2+}, and 300 ppm Na^+, how many gallons can be treated before the resin needs regeneration?

REFERENCES

1. Asher, W. J., M. L. Campbell, W. R. Epperly, and J. L. Robertson. *Hydrocarbon Proc.* **481**:134 (1969).

2. Berg, C. *Trans. AIChE.* **42:**665 (1946).
3. Broughton, D. B. *Chem. Eng. Prog.* **64**(8):60 (1968).
4. Broughton, D. B., R. W. Neuzil, J. M. Pharis, and C. S. Brearly. *Chem. Eng. Prog.* **66**(9):70 (1970).
5. Collins, J. J. *AIChE Symp. Ser.* **63**(74):31 (1967).
6. Dal Nogare, S., and R. S. Juvet, Jr. *Gas-Liquid Chromatography, Theory and Practice.* New York: Interscience, 1962.
7. Dickert, C., in R. E. Kirk, D. F. Othmer, J. I. Kroschwitz, and M. Howe-Grant (eds.), *Encyclopedia of Chemical Technology,* Vol. 14, 4th ed. New York: Wiley, 1992, p. 749.
8. Furnas, C. C. *Trans. AIChE* **24:**142 (1930).
9. Golovy, A., and J. Braslaw. *Environ. Progr.* **1:**89 (1982).
10. Grant, R. J., M. Manes, and S. B. Smith. *AIChE J.* **8:**403 (1962).
11. Grant, R. J., and M. Manes. *Ind. Eng. Chem. Fund.* **3:**221 (1964).
12. Hall, K. R., L. C. Eagleton, A. Acrivos, and T. Vermeulen. *Ind. Eng. Chem. Fund.* **5:**212 (1966).
13. Helfferich, F. *Ion Exchange.* New York: Dover, 1995; (*a*) p. 106, (*b*) p. 306.
14. Lewis, W. K., E. R. Gilliland, B. Chertow, and W. P. Cadogan. *Ind. Eng. Chem.* **42:**1326 (1950).
15. Purnell, H. *Gas Chromatography.* New York: Wiley, 1962.
16. Rohm and Haas Corp. *Technical Notes, Ambersorb Carbonaceous Adsorbents,* Philadelphia, 1992.
17. Ruthven, D. M. *Principles of Adsorption and Adsorption Processes.* New York: Wiley, 1984; (*a*) p. 328, (*b*) p. 375, (*c*) p. 372.
18. Seko, M., T. Miyake, and K. Inada. *Ind. Eng. Chem. Prod. Res. Dev.* **18:**263 (1979).
19. Shuliger, W. G. Calgon Corp., personal communication.
20. Subba Rao, H. C., and M. M. David. *AIChEJ.* **3:**187 (1957).
21. van Deemter, J. J., F. J. Zuiderweg, and A. Klinkenberg. *Chem. Eng. Sci.* **5:**271 (1956).
22. Vermeulen, T., G. Klein, and N. K. Hiester, in J. H. Perry (ed.), *Chemical Engineers' Handbook,* 5th ed., sec. 16. New York: McGraw-Hill, 1973.

CHAPTER 26

Membrane Separation Processes

Many processes for separation of gaseous or liquid mixtures use semipermeable membranes that allow one or more constituents of the mixture to pass through more readily than the others. The membranes may be thin layers of a rigid material such as porous glass or sintered metal, but more often they are flexible films of synthetic polymers prepared to have a high permeability for certain types of molecules. This chapter deals with processes in which separation is due to differences in solubilities and diffusion rates through the membrane. Separation of gas mixtures with porous or nonporous membranes is treated first. Then several processes for liquid mixtures are discussed, including pervaporation, liquid-liquid extraction, reverse osmosis, and dialysis. The process of ultrafiltration, where separation is accomplished mainly by excluding large molecules or colloidal particles from the pores of the membrane, is treated as a special case of filtration in Chap. 29.

SEPARATION OF GASES

Porous membranes

When a gas mixture is allowed to diffuse through a porous membrane to a region of lower pressure, the gas permeating the membrane is enriched in the lower-molecular-weight components, since they diffuse more rapidly. When the pores are much smaller than the mean free path in the gas phase (about 1,000 Å at standard conditions), the gases diffuse independently by Knudsen diffusion, and the diffusivity in the pore is proportional to the pore size and the average molecular velocity, which varies inversely with the square root of the molecular weight M. For Knudsen diffusion of gas A in cylindrical pores [see Eq. (17.29)]

$$D_A = 9,700r \left(\frac{T}{M_A} \right)^{0.5} \tag{26.1}$$

In Eq. (26.1), r is the mean pore radius in centimeters, T is the absolute temperature in kelvins, and D_A is in square centimeters per second.

The flux per unit membrane area depends on an effective diffusivity D_e that is lower than the pore diffusivity by the factor ε/τ, where ε is the porosity and τ the tortuosity. For membranes with about 50 percent porosity, this factor is generally 0.2 to 0.3:

$$D_{eA} = \frac{D_A \varepsilon}{\tau} \cong \frac{1}{4} D_A \qquad (26.2)$$

The flux of each gas is proportional to the concentration gradient, which is linear if the membrane structure is uniform and the gases do not interact. Usually the gradient is expressed as a partial pressure gradient, and ideal gases are assumed:

$$J_A = D_{eA}\left(\frac{\Delta c_A}{\Delta z}\right) = D_{eA}\left(\frac{\Delta p_A/RT}{\Delta z}\right) \qquad (26.3)$$

The composition of the permeate depends on the fluxes of all species. For a binary system, the mole fraction of A in the permeate is

$$y_A = \frac{J_A}{J_A + J_B} \qquad (26.4)$$

Typical pressure gradients for a membrane exposed to an equimolar binary mixture are shown in Fig. 26.1. For this case, gas A is assumed to have a diffusivity 2 times that of gas B (helium and methane, e.g.), and the upstream and downstream pressures are 2.4 and 1.0 atm, respectively. The permeate has 60 percent A, which is only a slight enrichment over the 50 percent A on the feed side. The enrichment makes the gradient for A less than that for B ($\Delta p_A = 1.2 - 0.6 = 0.6$; $\Delta p_B = 1.2 - 0.4 = 0.8$) so the flux of A is only $2 \times (0.6/0.8) = 1.5$ times that for B, giving a permeate with 60 percent A. The permeate would be slightly richer in A if a higher feed pressure were used or if the permeate side of the membrane were at less than atmospheric pressure. A smaller amount of purer product could be obtained by compressing the permeate and sending it to another membrane unit. A cascade of

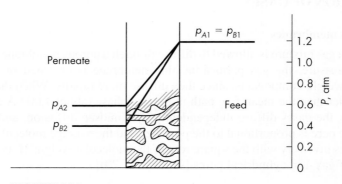

FIGURE 26.1
Typical pressure gradients in a porous membrane.

countercurrent stages with recycle could be designed to produce nearly pure products at high recovery, but the membrane cost and the cost of compression at each stage generally make such processes too expensive.

A well-known example of gas separation by porous membranes, and perhaps the only large-scale application, is the separation of uranium isotopes using the hexafluorides $^{235}UF_6$ and $^{238}UF_6$. Since natural uranium has only 0.7 percent ^{235}U and the diffusivities of these hexafluorides differ by only 0.4 percent, more than a thousand stages are needed to get a product with 4 percent ^{235}U and a residue with 0.25 percent ^{235}U.[13]

Polymer membranes

The transport of gases through dense (nonporous) polymer membranes occurs by a solution diffusion mechanism. The gas dissolves in the polymer at the high-pressure side of the membranes, diffuses through the polymer phase, and desorbs or evaporates at the low-pressure side. The rate of mass transfer depends on the concentration gradient in the membrane, which is proportional to the partial pressure gradient across the membrane if the solubility is proportional to the pressure. Typical gradients for a binary mixture are shown in Fig. 26.2. Henry's law is assumed to apply for each gas, and equilibrium is assumed at the interface. The gas-film resistances are neglected for this case, so the partial pressures at the gas-polymer interface are the same as those in the bulk. The flux for gas A is

$$J_A = -D_A\left(\frac{dc_A}{dz}\right) = D_A\left(\frac{c_{A1} - c_{A2}}{z}\right) \tag{26.5}$$

The concentrations are related to the partial pressures by a solubility coefficient S, which has units such as mol/cm$^3 \cdot$ atm (S is the reciprocal of the Henry law coefficient):

$$c_A = p_A S_A \qquad c_B = p_B S_B \tag{26.6}$$

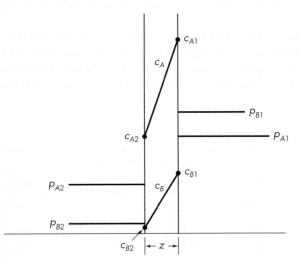

FIGURE 26.2
Gradients in a dense polymer membrane.

Using Eq. (26.6) to replace the concentration gradient with a pressure gradient gives

$$J_A = \frac{D_A S_A (p_{A1} - p_{A2})}{z} \tag{26.7}$$

The product $D_A S_A$ is the flux per unit pressure gradient, which is called the permeability coefficient q_A and is often expressed in Barrers, where 1 Barrer = 10^{-10} cm^3 (STP)·cm/cm^2·s·cm Hg. Since the actual membrane thickness is not always known or specified for commercial membranes, it is customary to use the flux per unit pressure difference, which will be called the permeability Q_A:

$$J_A = \frac{q_A (p_{A1} - p_{A2})}{z} = Q_A (p_{A1} - p_{A2}) \tag{26.8}$$

Convenient units of Q_A might be std ft^3/ft^2·h·atm or L(STP)/m^2·h·atm. Units must be checked carefully in using published values of permeability since different definitions are in use.

The ratio of permeabilities for a binary mixture is the membrane selectivity α (also called the ideal separation factor):

$$\alpha = \frac{Q_A}{Q_B} = \frac{D_A S_A}{D_B S_B} \tag{26.9}$$

A high selectivity can be obtained from either a favorable diffusivity ratio or a large difference in solubilities. The diffusivities in the membrane depend more strongly on the size and shape of the molecules than do gas-phase diffusivities, and large differences may exist for molecules of almost the same size. For example, the ratio D_{O_2}/D_{N_2} is between 1.5 and 2.5 for several polymers,[6] though the O_2 molecule is only 10 percent smaller than the N_2 molecule. Values of the diffusivity vary widely with the type of polymer, with lowest values for glassy or crystalline polymers and high values for polymers that are above their glass transition temperature. The diffusivity may vary with the concentration of solute in the polymer, particularly when there is appreciable swelling of the polymer. A few values of diffusivity are given in Table 26.1.

The gas solubility also varies widely with the gas and the type of polymer. The solubility is low for gases that have a low boiling point or critical temperature, but the similarity of the gas and the polymer is also important. Polar gases tend to be more soluble in polymers with a high concentration of polar groups, and the

TABLE 26.1
Diffusion coefficients in selected polymers[6]

Polymer	$D \times 10^9$ at 25°C, cm^2/s			
	O_2	N_2	CO_2	CH_4
Polyethylene terephthalate	3.6	1.4	0.54	0.17
Polyethylene ($\rho = 0.964$ g/cm^3)	170	93	124	57
Polyethylene ($\rho = 0.914$ g/cm^3)	460	320	372	193
Natural rubber	1,580	1,110	1,110	890

solubility of water vapor is high in materials that can form hydrogen bonds with water molecules. With a wide range of diffusivities and solubilities, it is not surprising that some membranes have quite high selectivities for certain gas mixtures. For silicone rubber, the selectivity is 4.9 for CO_2/H_2 and 5.4 for CO_2/O_2. For Kapton, an aromatic polyether diimide and a glassy polymer, the permeabilities are two to four orders of magnitude lower than for silicone rubber, and the order of permeabilities is altered.[14] The selectivities for Kapton are 0.18 for CO_2/H_2 and 1.8 for CO_2/O_2. A selectivity of 4 or greater is generally needed for a good separation, as shown in a later section.

For most gases, the permeability increases with temperature, because the increase in diffusivity more than offsets any decrease in solubility. The change in permeability is often correlated with the exponential equation $Q = a \exp(-E/RT)$, with activation energies E that range from 1 to 5 kcal/mol. However, an increase in temperature usually decreases the membrane selectivity, so the operating temperature is determined by balancing the needs for high flux and high selectivity.

Membrane structure

The flux through a dense polymer film is inversely proportional to the thickness [Eq. (26.7)], so there is a strong incentive to make the membrane as thin as possible without having holes or weak spots in it. Gas separation processes operate with pressure differences of 1 to 20 atm, so the thin membrane must be supported by a porous structure capable of withstanding such pressures but offering little resistance to the flow of gas. The support is made from a porous ceramic, metal, or polymer, and it should have a porosity of about 50 percent. The pore size should be comparable to the thickness of the thin selective film that covers the support. However, handling a thin layer and bonding it to the support without tearing are difficult, and most gas separation membranes are prepared with the support as an integral part of the membrane. Special methods of casting are used to prepare asymmetric membranes, which have a thin, dense layer or skin on one side and a highly porous substructure over the rest of the membrane. A picture of such a membrane is shown in Fig. 26.3.

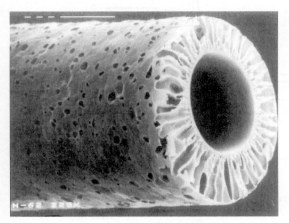

FIGURE 26.3
Capillary ultrafiltration membrane. Electron micrograph ($150\times$) of a DIAFLO™ hollow fiber. (*Courtesy of Millipore Corporation*)

   Typical asymmetric membranes are 50 to 200 μm thick with a 0.1- to 1-μm skin.[22] New techniques may permit production of commercial membranes with a skin thinner than 0.1 μm. Membranes with a very thin skin are more likely to have pinholes, and since flow through such flaws is very rapid compared to diffusion through dense polymer, only a few pinholes per unit area can lower the selectivity appreciably. One solution to this problem is to coat the membrane with a very permeable but nonselective polymer, which fills the pinholes and does not greatly reduce the permeability of the rest of the membrane.[9] The asymmetric membranes can be prepared in the form of flat sheets, tubes, or hollow fibers as small as 40 μm in diameter. The small, hollow fibers are strong enough to withstand high pressures without any additional support, but the flat sheets need additional supports and spacers.

   The concentration gradients in an asymmetric membrane are complex, because transport through the skin layer is by molecular diffusion in the dense polymer, but transport through the porous support is by diffusion plus laminar flow of gas in the tortuous pores. For high-flux membranes there may also be significant mass-transfer resistances in the boundary layers on both sides, where transport is also by diffusion plus flow.

   Figure 26.4 shows pressure and concentration gradients for an asymmetric membrane. For this example, the permeability of A is much greater than that of B, and the flux of A is several times that of B. The sketch shows a slight pressure gradient for A in the feed boundary layer, but the large drop in c_A shows that the

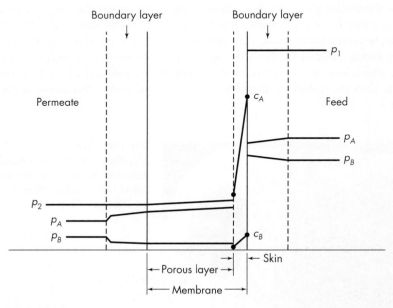

FIGURE 26.4
Pressure and concentration gradients for an asymmetric membrane with boundary layer resistance.

skin has most of the resistance to mass transfer. Note that the gradient for B in the boundary layer is negative, and B is carried against its concentration gradient by the total flow, which is mostly A.

The gases are assumed to be in equilibrium with the polymer phase on both sides of the skin layer. The composition of the gas in the pores next to the skin is generally not the same as the bulk composition of the permeate at that point. The bulk composition depends on the flow arrangement of the separator, and the bulk gas could have either more A or less A than the gas in the porous layer. The diagram in Fig. 26.4 shows a case in which the bulk permeate is about 70 percent A and the gas leaving the skin layer is about 90 percent A.

Flow patterns in membrane separators

There are several ways of arranging the surface area in a gas separator, and some of these are illustrated in Fig. 26.5 for hollow-fiber membranes with an external skin. Only a few fibers are shown, and their size is greatly exaggerated for clarity. A commercial separator has up to a million fibers in a shell several inches in diameter. The fibers are sealed into a tube sheet with an epoxy potting compound at one or both ends of the unit, to keep the feed and permeate separated.

Figure 26.5a shows a separator arranged for countercurrent flow with the feed gas on the shell side. The fibers are closed at one end, so the permeate flow increases from zero at the closed end to the final value at the discharge end. The feed gas has to flow across some of the fibers near the inlet and exit, so the flow is not always parallel to the axis, as it would be in ideal countercurrent flow. Good distribution of flow on the shell side is a design problem for large-diameter units. The feed is sometimes introduced to the lumen side or inside of the fibers to improve the flow distribution.

In some separators, both ends of the fibers are open, as in Fig. 26.5b, and permeate flows from the center toward each end. This makes the flow countercurrent in half of the separator and parallel in the other half. This arrangement decreases the pressure drop for permeate flow inside the fibers or permits longer units to be made with the same pressure drop. Sometimes there is little difference in the permeate composition for parallel or counterflow operation, since the fluxes depend mainly on the feed-side partial pressures when the permeate pressure is much less than the feed pressure. Counterflow operation is necessary, however, to make a high-purity residue stream, as in the production of nearly pure N_2 from air.

The problem of getting good flow distribution on the shell side is alleviated by using a crossflow arrangement, as in Fig. 26.5c. The fibers are bundled around a perforated discharge pipe, and feed gas flows radially from the outside of the shell to the central pipe. With flow radially inward and the flow decreasing as gas permeates the fibers, there is not much change in velocity past the fibers. Some commercial separators are arranged with feed at the center and radial flow outward, even though this makes the change in velocity greater from inlet to exit. The fibers can be sealed in tube sheets at one or both ends of the unit. Crossflow also exists in spiral-wound separators made with flat-sheet membranes, as shown in Fig. 26.19. With crossflow, the predicted separation is not quite as good as with counterflow, but it is better than with parallel flow.[33]

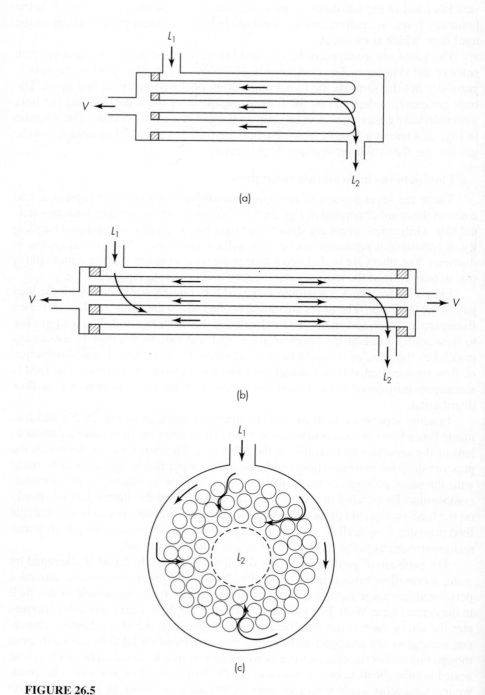

FIGURE 26.5
Flow arrangements for hollow-fiber membranes: (*a*) countercurrent flow; (*b*) parallel
and countercurrent flow; (*c*) radial crossflow.

Product purity and yield

The compositions of the permeate and the residue depend on many variables, including the pressure difference across the membrane, the composition and flow rate of the feed, the permeability of the various species, the total membrane area, and the flow arrangement. An important dependent variable is the fraction of feed that is recovered as permeate. This is called the *stage cut*. Data for permeate and residue compositions are often shown as functions of the stage cut, which can be varied over a wide range by changing the pressure or the feed rate.

The equations derived here for the separation of a binary mixture with an asymmetric membrane are based on the assumption of negligible resistances in the porous sublayer and the boundary layers. The frictional pressure drops on the feed side and the permeate side are also assumed to be negligible. The terminology is similar to that for distillation, with x and L representing composition and flow rate on the feed side, and y and V the composition and flow rate of the permeate. Here, x and y refer to mole fractions of the more permeable species, which is A in an A-B binary mixture. Unlike their behavior in distillation, L and V change appreciably as the gases pass through the separator, and by material balance the decrease in L is equal to the increase in V. The feed can be inside or outside the hollow fibers and is assumed to be in plug flow, so x gradually changes with length along the fiber axis, as shown in Fig. 26.6.

For the permeate, also assumed to be in plug flow, it is necessary to distinguish between y, the average composition of the gas at some axial position, and y', the local composition of the gas leaving the membrane surface at the same axial position. The

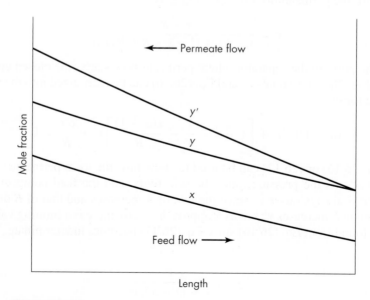

FIGURE 26.6
Local and average permeate compositions and residue composition for a counterflow separator.

values of y and y' are identical at the end of the separator where the permeate flow starts, but they change at different rates as permeate accumulates. As shown in Fig. 26.6, y' increases as the permeate flows toward the discharge end because x is increasing. The concentration y does not rise as rapidly because y is an average of the y' values for the permeate formed up to the point.

The relationship between x and y' is not an equilibrium one but depends on the relative permeabilities Q_A and Q_B and on differences in partial pressure. The fluxes J_A and J_B are given by the following equations, where P_1 is the feed pressure and P_2 the permeate pressure:

$$J_A = Q_A(P_1 x - P_2 y) \tag{26.10}$$

$$J_B = Q_B[P_1(1 - x) - P_2(1 - y)] \tag{26.11}$$

The ratio of absolute pressures R is introduced into the flux equations to eliminate P_2:

$$R \equiv \frac{P_2}{P_1} \tag{26.12}$$

$$J_A = Q_A P_1(x - Ry) \tag{26.13}$$

$$J_B = Q_B P_1[1 - x - R(1 - y)] \tag{26.14}$$

The local permeate composition y' depends on the flux ratio at that point:

$$y' = \frac{J_A}{J_A + J_B} = \frac{Q_A P_1(x - Ry)}{Q_A P_1(x - Ry) + Q_B P_1[1 - x - R(1 - y)]} \tag{26.15}$$

Using α for the permeability ratio Q_A/Q_B gives

$$y' = \frac{x - Ry}{x - Ry + (1 - x - R + Ry)/\alpha} \tag{26.16}$$

At the point in the separator where permeate flow starts (the closed end of the fibers in Fig. 26.5a), y equals y', and Eq. (26.16) can be rearranged into the standard quadratic form:

$$(\alpha - 1)(y')^2 + \left[1 - \alpha - \frac{1}{R} - \frac{x(\alpha - 1)}{R}\right]y' + \frac{\alpha x}{R} = 0 \tag{26.17}$$

Equation (26.16) or (26.17) can be used to show how the local permeate composition depends on the pressure ratio, the selectivity, and the feed composition. An increase in x always raises y', since the flux of A increases and that of B decreases. A decrease in R increases y'; but as R approaches zero, there is a limiting value of y', which is found from Eq. (26.16) since Eq. (26.17) becomes indeterminate. Then for $R = 0$,

$$y' = \frac{x}{x + (1 - x)/\alpha} \tag{26.18}$$

or

$$y' = \frac{\alpha x}{1 + (\alpha - 1)x}$$

When $R = 1.0$, no separation occurs in a binary system, since there is no driving force for diffusion. If a third component is added to the permeate side as a sweep gas, the partial pressures of A and B are lowered, and A and B can be separated even when $R = 1.0$. Sometimes a sweep gas is added to the permeate side to improve the separation.

At a fixed pressure the separation improves with increasing selectivity, but there is sometimes an upper limit to y'. The partial pressure of A in the permeate can never exceed that in the feed, and the maximum value of y' is found by equating the partial pressures.

$$P_1 x \geq P_2 y' = P_1 R y'$$

$$y'_{max} = \frac{x}{R} \tag{26.19}$$

For example, for a feed with 40 percent A and a pressure ratio $R = 0.5$, the highest permeate concentration is 0.80 even for a very selective membrane. However, if the pressure ratio is decreased to 0.20, nearly pure A can be obtained if the membrane has a very high selectivity. The variation of y' with α is shown in Fig. 26.7 for two values of R and $x = 0.4$ or 0.2.

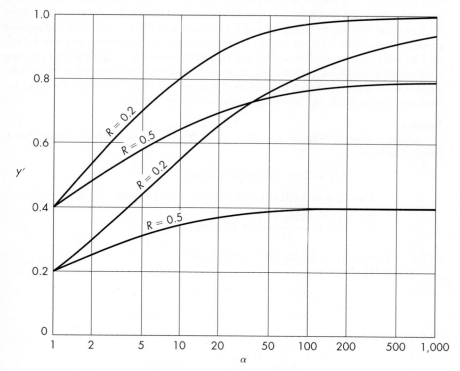

FIGURE 26.7
Effects of selectivity and pressure ratio on local permeate composition.

The performance of a separator can be predicted by combining the flux equations with material balances and integrating numerically from one end of the separator to the other. For an approximate prediction or an analysis of data from an operating unit, it may be satisfactory to use average values of the permeate composition or the driving force.

The overall and component material balances for the separator are

$$L_{in} = L_{out} + V_{out} \tag{26.20}$$

$$L_{in}x_{in} = L_{out}x_{out} + V_{out}y_{out} \tag{26.21}$$

If the change in x is not very large, the permeate composition is estimated by averaging the values of the local permeate composition at the ends of the separator.

$$\bar{y} \cong \frac{y'_{inlet} + y'_{outlet}}{2} \tag{26.22}$$

The membrane area required is obtained from the flux equation for the more permeable gas.

$$A \cong \frac{V_{out}y_{out}}{Q_A(P_1x - P_2y)_{ave}} \tag{26.23}$$

EXAMPLE 26.1. A hollow-fiber separator with polysulfone membranes was arranged for counterflow and tested for air separation. At a feed rate of 20 L/min (STP), it gave 3.1 L/min of 40% O_2 as permeate and a residue with 17% O_2. The feed and residue gauge pressures were 40.0 and 39.5 $lb_f/in.^2$, respectively; the membrane area was 5.2 m^2. (a) Calculate the permeabilities of O_2 and N_2 and the selectivity. (b) If the feed pressure is increased to 80 $lb_f/in.^2$, what oxygen concentration could be obtained in the permeate? (c) What is the maximum oxygen concentration that could be obtained with a very large feed rate at 80 $lb_f/in.^2$ or with a vacuum on the permeate side?

Solution

(a) By a material balance, $L = F - V = 20 - 3.1 = 16.9$ L/min. Check the O_2 balance. In feed,

$$Fx_F = 20(0.209) = 4.18 \text{ L/min}$$

$$Vy + Lx = 3.1(0.40) + 16.9(0.17) = 4.11 \text{ L/min} \qquad \text{OK}$$

$$P_1 = \frac{40 + 14.7}{14.7} = 3.72 \text{ atm}$$

At the feed end, $p_{1,O_2} = 3.72(0.209) = 0.777$ atm.
   Assume $P_2 = 1.0$ atm

$$p_{2,O_2} = P_2y = 0.40 \text{ atm}$$

$$\Delta p_{O_2} = 0.777 - 0.40 = 0.377 \text{ atm}$$

At the residue end,

$$P_1 = \frac{39.5 + 14.7}{14.7} = 3.69 \text{ atm}$$

$$p_{1,O_2} = 3.69(0.17) = 0.627 \text{ atm}$$

Neglecting the frictional pressure drop, $P_2 \cong 1.0$ atm, but y is not yet known.

As shown by Fig. 26.6, the initial value of y is less than the final value, but not much less in this case since x decreases only from 0.209 to 0.17. Guess that $y \cong 0.9(0.40) = 0.36$.

$$p_{2,O_2} \cong 0.36 \text{ atm}$$

$$\Delta p_{O_2} = 0.627 - 0.36 = 0.267 \text{ atm}$$

Average difference in O_2 partial pressure:

$$\overline{\Delta p_{O_2}} = \frac{0.377 + 0.267}{2} = 0.322 \text{ atm}$$

From Eq. (26.23),

Permeability of O_2:
$$Q_{O_2} = \frac{3.1 \times 0.40}{5.2 \times 0.322} = 0.741 \text{ L/min} \cdot \text{m}^2 \cdot \text{atm}$$

For N_2 at the feed end,
$$p_{1,N_2} = 0.791 \times 3.72 = 2.94 \text{ atm}$$

$$p_{2,N_2} = 1 - 0.40 = 0.60 \text{ atm}$$

$$\Delta p_{N_2} = 2.94 - 0.60 = 2.34 \text{ atm}$$

At the residue end
$$p_{1,N_2} = 3.69(1 - 0.17) = 3.06 \text{ atm}$$

$$p_{2,N_2} = 1 - y = 1 - 0.36 = 0.64 \text{ atm}$$

$$\Delta p_{N_2} = 3.06 - 0.64 = 2.42 \text{ atm}$$

Average difference
$$\overline{\Delta p_{N_2}} = \frac{2.34 + 2.42}{2} = 2.38 \text{ atm}$$

Permeability of N_2:
$$Q_{N_2} = \frac{3.1 \times 0.60}{5.2 \times 2.38} = 0.150 \text{ L/min} \cdot \text{m}^2 \cdot \text{atm}$$

Selectivity:
$$\alpha = \frac{0.741}{0.150} = 4.94$$

If $\alpha = 4.94$, the initial value of y can be checked from Eq. (26.17), using $R = 1/3.72 = 0.269$.

$$3.94(y')^2 + [1 - 4.94 - 3.72 - (0.17 \times 3.94 \times 3.72)]y' + 4.94 \times 0.17 \times 3.72 = 0$$

$$3.94(y')^2 - 10.15y' + 3.12 = 0$$

$$y' = 0.357, \text{ close to } 0.36 \text{ assumed}$$

(b) Feed pressure:
$$P_1 = \frac{80 + 14.7}{14.7} = 6.44 \text{ atm}$$

$$P_2 = 1.0 \text{ atm} \qquad R = \frac{1}{6.44} = 0.1553$$

If the flow is adjusted to keep $x = 0.17$ in the residue, y' at the residue end can be obtained from Eq. (26.17).

$$3.94(y')^2 + [1 - 4.94 - 6.44 - (0.17 \times 3.94 \times 6.44)]y' + 4.94 \times 0.17 \times 6.44 = 0$$

$$3.94(y')^2 - 14.69y' + 5.408 = 0$$

$$\text{Initial } y' = 0.414$$

At the feed end, the value of y' is obtained from Eq. (26.16), which includes y, the final permeate concentration. Assume

$$y = \frac{0.414 + y'}{2} = 0.207 + 0.5y'$$

$$y' = \frac{0.209 - 0.1553(0.207 + 0.5y')}{0.209 - 0.1553(0.207 + 0.5y') + [1 - 0.209 - 0.1553 + 0.1553(0.207 + 0.5y')]/4.94}$$

$$= \frac{0.17685 - 0.07765y'}{0.31204 - 0.06193y'}$$

$$0.06193(y')^2 - 0.38969y' - 0.17685 = 0$$

$$y' = 0.492$$

$$y = \frac{0.414 + 0.492}{2} = 0.453 \quad \text{or} \quad 45\% \ O_2$$

(c) For a very large feed rate, the stage cut approaches zero and the residue composition approaches that of air (20.9% O_2). From Eq. (26.17) with $R = 1/6.44$,

$$3.94(y')^2 + [1 - 4.94 - 6.44 - (0.209 \times 3.94 \times 6.44)]y' + 4.94 \times 0.209 \times 6.44 = 0$$

$$3.94(y')^2 - 15.683y' + 6.649 = 0$$

$$y' = 0.482 \quad \text{or} \quad 48\% \ O_2$$

If a vacuum is applied, R approaches zero, and from Eq. (26.18) with $x = 0.209$,

$$y' = \frac{4.94 \times 0.209}{1 + 3.94 \times 0.209} = 0.566 \quad \text{or} \quad 57\% \ O_2$$

As Example 26.1 shows, the concentration of oxygen in the permeate is increased by raising the upstream pressure, even though this increases the flux of both oxygen and nitrogen. The oxygen flux increases more because the partial pressure of oxygen in the permeate has a large effect on the oxygen driving force. With $P = 40 \ lb_f/in.^2$ or 3.72 atm, the driving force for oxygen at the inlet is $3.72(0.209) - 0.40 = 0.377$ atm, and at $80 \ lb_f/in.^2$ it is $6.44(0.209) - 0.45 = 0.896$ atm, a 2.4-fold increase. The change in the driving force for N_2 is only 1.94-fold on going from 40 to $80 \ lb_f/in.^2$. Because the flux of each component depends on a partial pressure difference, the total flux is not proportional to the total pressure difference except in the limit as R approaches zero.

When one is using a stepwise solution to determine the performance of a counterflow separator, it is best to start at the residue end with a specified value of x_{out} and an arbitrary value of L_{out}, say, 100 L/min. The initial value of y' is found from Eq. (26.17), and the local fluxes of A and B are calculated using Eqs. (26.10) and (26.11). Then ΔV and ΔL are obtained for a small increment of surface area.

$$\Delta V = \Delta L = (J_A + J_B) \Delta A \tag{26.24}$$

New values of x and y are calculated from the material balance and flux equations, and the computations are continued until x reaches the feed composition. The

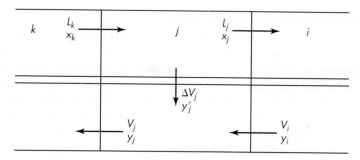

FIGURE 26.8
Flows in a counterflow separator.

stage cut is then V/L. A differential length of the separator is shown in Fig. 26.8 with the inlet and outlet flows. The other equations used are

$$L_k = L_j + \Delta V_j \tag{26.25}$$

$$V_j = V_i + \Delta V_j \tag{26.26}$$

$$L_k x_k = L_j x_j + \Delta V_j y_j' \tag{26.27}$$

Replacing L_j with $L_k - \Delta V_j$ in Eq. (26.27) leads to an equation for the incremental change in x.

$$L_k(x_k - x_j) = \Delta V_j (y_j' - x_j) \tag{26.28}$$

The new value of y comes from

$$V_j y_j = V_i y_i + \Delta V_j y_j' \tag{26.29}$$

where

$$y_j' = \frac{J_A}{J_A + J_B} \tag{26.30}$$

For greater accuracy the value of y_j' is taken as the average of the values at the beginning and end of each increment.

Multicomponent mixtures. To predict the performance of a membrane separator for a multicomponent mixture, the flux equations for each component are written in the same form as Eq. (26.10), and a trial-and-error procedure is used to get the amount of permeate and the local permeate compositions for each increment of area. To start the calculation, an approximate residue composition can be selected, and computations carried out going toward the feed end; or alternatively an approximate permeate composition can be used to permit calculations starting at the feed end. In either case, the procedure is repeated until the terminal compositions are correct. As with other multicomponent separations, the complete composition of the product streams cannot be specified, and the design goal is either a certain concentration of a key component in the permeate or residue, a given percent recovery of one component, or a specified stage cut, which is the molar ratio of permeate to feed.

Pressure drop and mass-transfer effects

A rigorous analysis of separator performance must include the friction pressure drops and diffusion resistances on both sides of the membrane. For permeate flow inside hollow fibers, the internal pressure gradient is zero at the closed end of the fibers, and increases gradually to a maximum value at the tube sheet. If the pressure at the closed end is as much as 20 to 30 percent above the discharge pressure, the effect on the fluxes is usually significant. The pressure drop on the shell side is generally much smaller and is often neglected, since the flow area is greater and the volumetric flow rate smaller because of the high pressure.

The frictional pressure drop inside the fibers affects the choice of fiber size, since large diameters are needed when the permeability is high. Very long fibers are not practical, since the pressure drop changes with about the square of the fiber length because of the increase in permeate velocity.

In separating low-molecular-weight gases with polysulfone or other glassy polymers, the mass-transfer resistances in the support layer and the boundary layers are usually negligible. These resistances are important, however, with new ultrathin membranes or membranes made of more permeable polymers. The local composition on the downstream side of the selective layer can then differ significantly from the bulk composition, and the flux equations have to be modified to account for flow and diffusion in the support layer and the boundary layer. Mass-transfer effects have been observed in studies of gas separation with high-flux silicone membranes.[18]

Applications

The major applications of gas separation membranes are to make products that are enriched in one or more components but are not of very high purity. Products of equal or greater purity can usually be obtained by liquefaction and distillation at low temperature, but the membrane processes have the advantage of operation at or near room temperature.

Several companies supply permeators for air separation using membranes with characteristics similar to those in Example 26.1. One of the main applications is to provide nitrogen at purities of 95 to 99 percent, which is sufficient for many inert-gas requirements. The air is usually fed to the lumen side to get a more uniform residence time distribution and to lower the friction pressure drop on the permeate side. Figure 26.9 shows the residual oxygen concentration as a function of L_{out}/L_{in} for an ideal counterflow separator. A low oxygen concentration requires a small value of R and a large stage cut. Commercial units with capacities of 3 to 1,000 m^3/h are available, and for moderate nitrogen purities their performance is close to that predicted for $\alpha = 5$ and $R = 0.1$. For 99 percent purity or higher, the fraction recovered is somewhat less than predicted.

If the goal is to make enriched oxygen for medical uses or to improve combustion, only a small stage cut is used, and a permeate with 40% to 50% O_2 is obtained. For enriched oxygen production, the feed can be at atmospheric pressure, and a vacuum pump is used to get a low value of R. Then the only work needed is to compress the permeate to atmospheric pressure instead of compressing a much larger flow of air to several atmospheres. Although purer oxygen can be made by using

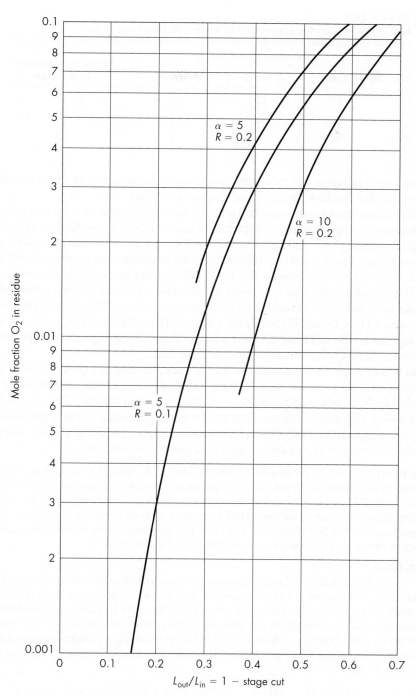

FIGURE 26.9
Nitrogen production from air in a counterflow separator.

TABLE 26.2
Gas compositions for separation of ammonia plant purge gas[23]

Gas	Gas composition, mol %			Percentage recovery in permeate
	Feed	Residue	Permeate	
H_2	59	12	86	93
N_2	21	43	8	24
Ar	6	11	3	32
CH_4	14	34	3	14

several stages in series, the recompression costs make this uneconomical compared to oxygen from liquid-air plants or adsorption processes.

A combination of membrane and other technologies might be attractive for some applications. Using membranes to prepare a gas with 50% O_2 as feed to a pressure-swing adsorption plant would greatly increase the capacity of the adsorber. Also, a gas with only a few ppm O_2 can be produced by using membranes to get down to 0.5% O_2 and then removing the oxygen with added hydrogen in a catalytic combustor.[4]

There are many installations using membranes to recover hydrogen from purge streams in ammonia, methanol, and hydrogenation plants. The selectivities for H_2 relative to CH_4, CO, and N_2 are in the range of 10 to 100, and permeate considerably enriched in H_2 can be obtained in a single stage. The compositions reported for a petrochemical application are shown in Table 26.2.

In this example the feed pressure was 72 bars, the residue pressure was 71 bars, and the permeate pressure was 31 bars. Most of the hydrogen is recovered and recycled to the synthesis plant, and the residue is burned for fuel. There is no need to get a very high concentration of hydrogen in the permeate. In other applications, hydrogen concentrations of 95% to 98% can be obtained by taking a lower stage cut (smaller recovery of H_2) or operating with a lower pressure ratio.

EXAMPLE 26.2. A gas containing 70% H_2, 24% CH_4, and 6% C_2H_6 is to be separated into a nearly pure H_2 stream and a fuel gas using a hollow-fiber permeator with a selectivity of 100 for H_2/CH_4. (*a*) If the upstream and downstream pressures are 600 and 300 lb$_f$/in.^{2} absolute, what fraction of H_2 in the feed is recovered in the permeate if the permeate is 96% H_2? What is the stage cut? (*b*) How much more H_2 could be recovered by lowering the permeate pressure to 180 lb$_f$/in.^{2} absolute?

Solution

(*a*) The mixture is treated as a pseudobinary, since the permeability of C_2H_6 is only slightly less than that of CH_4. A stepwise solution is used because the high value of the downstream pressure ($R = 0.5$) makes the driving force for H_2 change several-fold from inlet to outlet. To simplify the calculations, the local permeate compositions are obtained from Eq. (26.17) for $\alpha = 100$ and $R = 0.5$. This method slightly underestimates the values of y' for counterflow operation but is nearly correct for crossflow. At the feed inlet, $x = 0.70$, and

$$99y_i^2 + [1 - 100 - 2 - 2(99)(0.7)]y_i + 100(0.7)(2) = 0$$
$$99y_i^2 - 239.6y_i + 140 = 0$$
$$y_i = 0.9860$$

The calculation is repeated for other values of x using increments of 0.05. For $x = 0.65$

$$99y_j^2 - 229.7y_j + 130 = 0$$

$$y_j = 0.9793$$

For the first increment, $\bar{y} = (0.9860 + 0.9793)/2 = 0.9826$. From Eq. (26.28) with $L_{in} = 100$

$$\Delta V = \frac{L_i(x_i - x_j)}{\bar{y} - x_j} = \frac{100(0.70 - 0.65)}{0.9826 - 0.65} = 15.03$$

$$H_2 \text{ in permeate} = \Delta V\,\bar{y} = 15.03(0.9826) = 14.77$$

$$H_2 \text{ recovery} = 14.77/70 = 0.211 = 21.1\%$$

Calculations are continued until the total H_2 recovery is over 95 percent. The results are plotted in Fig. 26.10.

For $R = 0.5$, the H_2 purity is 96 percent at a recovery of 62.5 percent.
The H_2 recovered per 100 mol total feed is $0.625(70) = 43.75$.
The permeate flow $V = 43.75/0.96 = 45.6$.
The stage cut is 45.6 percent, and the residue contains 48.3% H_2. The local value of y' goes from 0.986 at the feed end to 0.877 at the residue end, and a simple average of these values does not give the correct permeate composition.

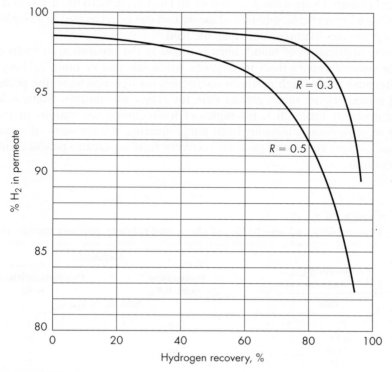

FIGURE 26.10
Permeate purity for H_2/CH_4 separation with $\alpha = 100$ and $x_0 = 0.70$.

(b) $R = 180/600 = 0.3$. At $x = 0.70$

$$99y_i^2 + \left[1 - 100 - \frac{1}{0.3} - \frac{99}{0.3}(0.7)\right]y_i + \frac{100(0.7)}{0.3} = 0$$

$$y_i = 0.9926$$

Continuing the calculations as in part (a) gives $y = 0.96$ at 88 percent H_2 recovery. The stage cut is $0.88(70)/0.96 = 64.2$ percent.

Another industrial application of gas separation membranes is the removal of carbon dioxide from natural gas. The CO_2/CH_4 selectivity is about 20 to 30 for polycarbonate, polysulfone, and cellulose acetate membranes at 35°C and 40 atm. A selectivity of over 60 can be obtained with Kapton, but this polymer is much less permeable than the others. Increasing the temperature raises the permeability of most polymers but generally causes a slight decrease in selectivity. The operating temperature is chosen to be somewhat above the dew point of the residue gas. There is considerable CO_2 absorbed in the membranes at high CO_2 partial pressures, and the plasticization effect of CO_2 increases the effective diffusion coefficients for all gases and makes the selectivity less than that based on pure-gas data.[11] Methods of allowing for such nonlinear effects have been presented.[15]

The separation of helium from natural gas is a potential application of membrane technology. Using a membrane with an He/CH_4 selectivity of 190 and a feed with only 0.82 percent He, one-half of the helium can be recovered in a single stage, giving a permeate 30 times richer than the feed.[24]

A new application of membranes is found in the separation of hydrocarbon gases or solvent vapors from inert gases, using silicone or other rubbery polymers. Because of the high partition coefficient for hydrocarbons, the permeability and selectivity are much greater than for glassy membranes. Some selectivities are given in Table 26.3. Examples of hydrocarbon separations include the control of refrigerant emissions from purge streams, the recovery of monomers from reactor vent gases, and gasoline recovery from airstreams produced in loading operations.[3]

TABLE 26.3
Permeability and selectivities of glassy and rubbery polymer membranes

Permeability Q_{N_2}, L (STP)/min·m²·atm	Membrane	
	Polysulfone 0.05–0.5	Polydimethylsiloxane 5–50
Gas	Selectivity, α_{gas/N_2}	
O_2	5–6	3
CO_2	17	6
H_2O	280	—
CH_4	20	3
C_3H_6	—	10
C_4H_{10}	—	30
Toluene	—	80

Separator arrangement

Most applications of membranes for gas or liquid separation require multiple units, since the largest units are only about 1 ft (0.3 m) in diameter and 10 to 15 ft (3 to 5 m) long. A hollow-fiber module of this size might have many thousand square feet of membrane area and be able to process several hundred cubic feet of gas per minute. For treating the much larger flows in a refinery or chemical plant, several units could be arranged in parallel, as shown in Fig. 26.11a. Care must be

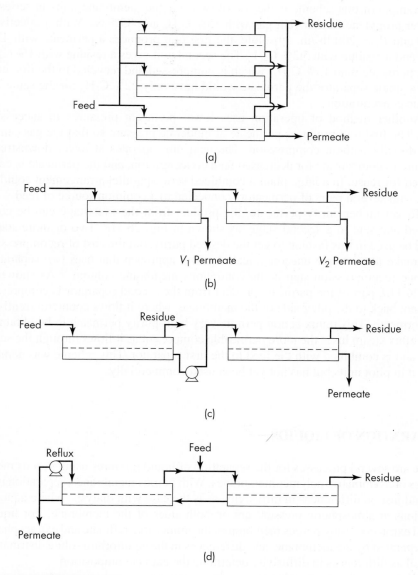

FIGURE 26.11
Separator arrangements: (a) parallel flow; (b) series flow; (c) two-stage flow; (d) continuous membrane column.

taken in designing the feed distribution system to ensure the same flow to all units. When operating at low capacity, some of the units can be shut off to keep the flow per module about the same. If all units were kept in service, the greater permeate recovery at low flow rate could lead to condensation of liquid on the feed side.[21]

Separators are sometimes arranged in series, as shown in Fig. 26.11b. The friction pressure drop on the feed side is usually small (<1 atm), so two or three units can be put in series without having to recompress the feed. The permeate streams differ in purity and may be used for different purposes or they may all be combined. An example of this scheme is the use of two silicone membrane units in series to recover propylene from a vent gas with 15% C_3H_6 and 85% N_2. With a selectivity of 10 and $P_1 = 200$ lb$_f$/in.^{2} absolute, the first unit produces a permeate with 43% C_3H_6 and a residue with 5.3% C_3H_6. The second unit gives a residue with 1% C_3H_6 and a permeate with 15% C_3H_6, which is compressed and recycled to the first unit. With a single separator the permeate would have only 22% C_3H_6 for the same 1% residue concentration.[2]

Another method of operation uses lower permeate pressures in successive units. The first unit produces permeate at moderate pressure so that the gas can be used directly without compression. The next unit operates at lower downstream pressure to compensate for decreased feed concentration, and the permeate is compressed for reuse. In a large plant a combined series-parallel arrangement could be used, with several pairs of permeators connected to a common source of feed.

To get higher-purity permeate, the product from the first stage can be compressed and sent to a second stage, as shown in Fig. 26.11c. Two or more stages could be used in this fashion to get the desired purity, but the cost of recompression may make this scheme uneconomical. A novel approach that uses two separators and one recompression step is the continuous membrane column.[12] As shown in Fig. 26.11d, part of the permeate product from the second separator is compressed and sent back to the other side of the membrane, where it flows countercurrently to the permeate. This reflux action permits very high-purity permeate to be obtained. The reflux steam loses the more permeable component as it flows through the separator and is combined with the feed to the first separator. This scheme was demonstrated in pilot units but has not yet been used commercially.

SEPARATION OF LIQUIDS

There are several processes for the separation of liquid mixtures using porous membranes or asymmetric polymer membranes. With porous membranes, separation may depend just on differences in diffusivity, as is the case with dialysis, where aqueous solutions at atmospheric pressure are on both sides of the membrane. For liquid-liquid extraction using porous membranes, the immiscible raffinate and extract phases are separated by the membrane, and differences in the equilibrium solute distribution as well as differences in diffusivity determine the extract composition.

With asymmetric membranes or dense polymer membranes, permeation of liquids occurs by a solution diffusion mechanism. The selectivity depends on the solubility ratio as well as the diffusivity ratio, and these ratios are very dependent on the chemical structure of the polymer and the liquids. The driving force for transport is the activity gradient in the membrane, but in contrast to gas separation, the driving force cannot be changed over a wide range by increasing the upstream pressure, since pressure has little effect on activity in the liquid phase. In pervaporation, one side of the membrane is exposed to the feed liquid at atmospheric pressure, and vacuum or sweep gas is used to form a vapor phase on the permeate side. This lowers the partial pressure of the permeating species and provides an activity driving force for permeation. In reverse osmosis, the permeate is nearly pure water at about 1 atm, and very high pressure is applied to the feed solution to make the activity of the water slightly greater than that in the permeate. This provides an activity gradient across the membrane even though the concentration of water in the product is higher than that in the feed.

Dialysis

Thin porous membranes are used in dialysis, a process for selectively removing low-molecular-weight solutes from a solution by allowing them to diffuse into a region of lower concentration. There is little or no pressure difference across the membrane, and the flux of each solute is proportional to the concentration difference. Solutes of high molecular weight are mostly retained in the feed solution, because their diffusivity is low and because diffusion in small pores is greatly hindered when the molecules are almost as large as the pores.

Concentration gradients for a typical dialysis process are shown in Fig. 26.12. The feed is assumed to contain a low-molecular-weight solute A, a solute of intermediate size B, and a colloid C. There are concentration boundary layers on both

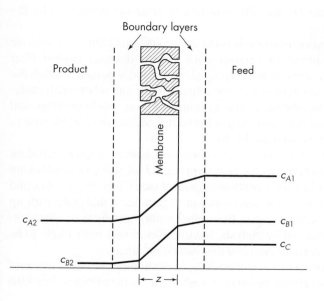

FIGURE 26.12
Concentration gradients in dialysis.

sides of the membrane, and these may contribute significantly to the overall resistance if the membrane is thinner than the boundary layers. The gradient for A or B in the membrane is steeper than in the boundary layers, because the effective diffusivity is less than the bulk value, and at steady state, the flux through the membrane equals that through the boundary layers. The values of c_A and c_B in the membrane are the concentrations in the pore fluid and not the concentrations based on total membrane volume. In the pore fluid and in the product, $c_C = 0$ because the colloidal particles are larger than the pore size.

The general equations for solute flux allow for three resistances in series:

$$J_A = K_A(c_{A1} - c_{A2}) \tag{26.31}$$

$$\frac{1}{K_A} = \frac{1}{k_{1A}} + \frac{1}{k_{mA}} + \frac{1}{k_{2A}} \tag{26.32}$$

The coefficients k_1 and k_2 for the feed and product depend on the flow rates, physical properties, and membrane geometry, and they can be predicted using the correlations in Chap. 17. The membrane coefficient depends on the effective diffusivity D_e and the membrane thickness z:

$$k_m = \frac{D_e}{z} \tag{26.33}$$

A theoretical equation for D_e is based on λ, the ratio of molecular size to pore size[30]

$$D_e = \frac{D_v \varepsilon}{\tau}(1 - \lambda)^2(1 - 2.104\lambda + 2.09\lambda^2 - 0.95\lambda^3) \tag{26.34}$$

The term $(1 - \lambda)^2$ is the volume fraction available for a spherical molecule in a cylindrical pore, and the last term in Eq. (26.34) is a hindered diffusion term. For $\varepsilon = 0.5$, $\tau = 2$, and $\lambda = 0.1$, $D_e = 0.164 D_v$, and for $\lambda = 0.5$, $D_e = 0.022 D_v$. Since D_e is much lower than D_v, the diffusion flux is generally controlled by the membrane resistance.

The best-known application of dialysis is the use of artificial kidneys to remove waste products from the blood of persons with kidney disease. Hollow-fiber cellulosic or polysulfone membranes are employed, and blood is passed through the fibers while saline solution is circulated on the outside. Urea and other small molecules diffuse through the membrane to the external solution, while proteins and cells are retained in the blood. The dialyzing solution has added salts and glucose to prevent loss of these materials from the blood.

An industrial application of dialysis is the recovery of caustic from hemicellulose solutions produced in making rayon by the viscose process. Flat-sheet membranes are placed parallel to each other in a filter press arrangement (see Chap. 29, p. 994), and water is passed countercurrent to the feed solution to produce a dialyzate with up to 6% NaOH. Recovery of salts or sugars from other natural products or other colloidal solutions could be achieved by dialysis, but ultrafiltration is more likely to be used because of the higher permeation rates that can be obtained.

Many large-scale applications of electrodialysis use ion-selective membranes and a potential gradient to speed migration of ions through the membranes. Brackish

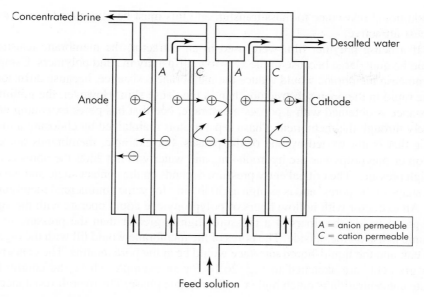

FIGURE 26.13
Schematic diagram of an electrodialysis cell.

water can be made potable by passage through an array with alternate cation- and anion-permeable membranes, as shown in Fig. 26.13. In one-half of the spaces, cations move out to one side and anions move out to the other side, leaving purer water. Solution in the alternate spaces becomes more concentrated and is eventually discarded. Similar units are used to concentrate salt solutions in a variety of processes. One example is the use of electrodialysis to treat the reject salt solution from a reverse-osmosis system.[28] The salt concentration is increased eightfold, which reduces the disposal cost, and the reclaimed water is recycled to the reverse-osmosis plant. In this application, the electrode polarity is reversed at regular intervals to minimize scale problems at the high salt concentration.

Membranes for Liquid-Liquid Extraction

Extraction of a solute from water to an organic liquid, or vice versa, can be carried out using membranes to separate the phases and provide a high surface area for mass transfer. Hollow-fiber or flat-sheet membranes can be used, and the mass-transfer area is then fixed by the design and does not depend on variables such as flow rate, viscosity, and surface tension, which affect the area of liquid-liquid dispersions. The membrane extractor can be arranged to have counterflow of the two phases with no flooding limitations, unlike the situation in packed or spray columns. A further advantage is that there is no need for a settling tank or deemulsifier, because the phases are kept separate by the membrane. However, the membrane does introduce an additional resistance to mass transfer, and this must be minimized to make the process attractive.

If a dense polymer film were used in an extractor, the membrane resistance would be quite large because of the very low diffusivity in solid polymers. Using an asymmetric membrane would reduce the membrane resistance, because diffusion is more rapid in the open substructure than in the dense skin. However, the minimum resistance is obtained with a porous membrane, one that has pores extending completely through the membrane. Phase separation is maintained by choosing a membrane that is not wetted by one of the phases. For example, membranes made of Teflon or polypropylene are hydrophobic, and water will not enter the pores except at high pressure. The critical entry pressure depends on the contact angle and the size and shape of the pores[7] and is as high as 50 $lb_f/in.^2$ for some commercial membranes.

An extractor with hollow fibers of polypropylene could operate with the aqueous phase inside the fibers at a pressure slightly greater than the pressure of the organic phase on the outside. The pores of the membrane would fill with the organic solvent, and the liquid-liquid interface would be at the pore mouths. The concentration gradients are sketched in Fig. 26.14 for an example where the equilibrium solute concentration is much higher in the organic phase. The overall resistance for this case is

$$\frac{1}{K_w} = \frac{1}{k_w} + \frac{1}{m}\left(\frac{1}{k_o} + \frac{z}{D_{e,0}}\right) \tag{26.35}$$

The coefficients for the water phase, k_w, the organic phase, k_o, and the membrane, $D_{e,0}/z$, are generally about the same magnitude; but if the distribution coefficient m is large, most of the resistance is in the water phase. Here m is the ratio of the solute concentration in the organic phase to that in the water phase.

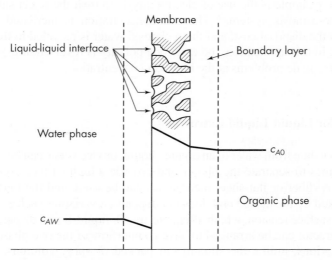

FIGURE 26.14
Liquid-liquid extraction with a porous hydrophobic membrane.

If a hydrophilic membrane is used, the pores fill with the water phase, and the organic phase must be kept at a high pressure to prevent water from passing through the pores and forming drops in the organic phase. For the system shown in Fig. 26.14, use of a hydrophilic membrane would mean two water-phase resistances and a lower overall coefficient, as shown by the equation

$$\frac{1}{K_w} = \frac{1}{k_w} + \frac{z}{D_{e,w}} + \frac{1}{mk_o} \tag{26.36}$$

If the distribution coefficient for the solute strongly favors the water phase ($m \ll 1$), the organic phase has the controlling resistance, and a hydrophilic membrane might be selected to make the membrane resistance smaller.

Selective extraction of a solute from an aqueous phase into another aqueous phase can be accomplished with a hydrophobic membrane, the pores of which are filled with a polymeric liquid that has a high partition coefficient for the solute.[10]

The overall mass-transfer coefficient K_w for such a process is found from the equation

$$\frac{1}{K_w} = \frac{1}{k_{w1}} + \frac{1}{k_m K_P} + \frac{1}{k_{w2}} \tag{26.37}$$

where k_m is the mass-transfer coefficient for the membrane and K_P is the partition coefficient for the solute distributed between water and the polymeric liquid. The recovery and concentration of phenol from a dilute salt solution using this process have been demonstrated using commercial radial-flow modules.[8]

Other hollow-fiber extractors have been tested in the laboratory,[26,36] and the mass-transfer rates are generally consistent with theory, although satisfactory correlations for the external coefficient may not be available (see Chap. 17). These devices should find practical application for systems where it is difficult to get a good dispersion or where emulsification makes the final phase separation difficult.

Pervaporation

Pervaporation is a separation process in which one or more components of a liquid mixture diffuse through a selective membrane, evaporate under low pressure on the downstream side, and are removed by a vacuum pump or a chilled condenser. Composite membranes are used with the dense layer in contact with the liquid and the porous supporting layer exposed to the vapor. The phase change occurs in the membrane, and the heat of vaporization is supplied by the sensible heat of the liquid conducted through the thin dense layer. The decrease in temperature of the liquid as it passes through the separator lowers the rate of permeation, and this usually limits the application of pervaporation to removal of small amounts of feed, typically 2 to 5 percent for one-stage separation. If a greater removal is needed, several stages are used in series with intermediate heaters. Commercial units generally use flat-sheet membranes stacked in a filter press arrangement, with spacers acting as product channels, although spiral-wound membranes could also be used. Hollow-fiber membranes are not as suitable because of the pressure drop from the permeate flow through the small-bore fibers.

The flux of each component is proportional to the concentration gradient and the diffusivity in the dense layer. However, the concentration gradient is often nonlinear because the membrane swells appreciably as it absorbs liquid, and the diffusion coefficient in the fully swollen polymer may be 10 to 100 times the value in the dense unswollen polymer. Furthermore, when the polymer is swollen mainly by absorption of one component, the diffusivity of other components is increased also. This interaction makes it difficult to develop correlations for membrane permeability and selectivity.

For permeation of a single species, the diffusivity can be expressed as an exponential function of the concentration:

$$D = D_0 e^{\beta c} \tag{26.38}$$

where β is a constant and D_0 is the diffusivity at infinite dilution. At steady state, the flux across a film of thickness z is

$$J = \frac{D_0}{\beta z}(e^{\beta c_1} - e^{\beta c_2}) \tag{26.39}$$

If βc_1 is larger than 1.0 and c_2 is much smaller than c_1, there is a several-fold decrease in diffusivity across the membrane, and the flux is not proportional to the concentration difference. For example, if $\beta c_1 = 2.0$ and $c_2 = 0.5c_1$, the flux is $(e^2 - e)D_0/\beta z = 4.67D_0/\beta z$; but lowering c_2 to zero, which doubles the concentration difference, only increases the flux to $6.39D_0/\beta z$, a 37 percent increase. However, increasing c_1 by 50 percent to double the driving force raises the flux to $17.4D_0/\beta z$, a 3.7-fold increase.

In an early study of pervaporation[5] using n-heptane at 1 atm and 99°C, the flux was inversely proportional to the thickness of the dense polymer film, as expected, but the flux increased only slightly as the downstream pressure was decreased from 500 to 50 mm Hg. This is consistent with Eq. (26.39) and a value of 5 or more for βc_1. Other studies with pure feed liquids have given similar results, and direct measurements have shown very nonlinear concentration profiles in the membrane.[1] However, in commercial applications of pervaporation, the liquid feed usually has a low concentration of the more permeable species, so the swelling of the membrane and the resulting nonlinear effects are not as pronounced as when testing pure liquids or solutions of high concentration.

The gradients at steady state for a typical system are shown in Fig. 26.15. The feed is rich in B, but A is assumed to be much more soluble in the polymer, and there is a high concentration of A in the membrane at the upstream side. The gradient of A is small near this boundary because of the high diffusivity, but D_A decreases as c_A decreases. The gradient of B has a similar shape, reflecting the change in D_B through the membrane. The concentrations c_{A2} and c_{B2} are likely to be proportional to the downstream partial pressure p_{A2} and p_{B2}, but Henry's law may not hold at the upstream side, where the membrane is highly swollen by solvent.

The concentration of a solvent in the membrane at the upstream side depends on its solubility in the polymer at the operating temperature and the activity of the solvent in the liquid. The feed is usually heated, because the membrane permeability

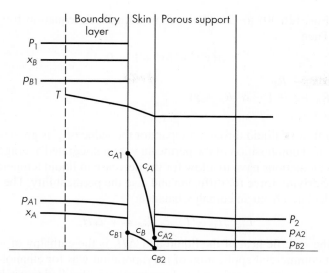

FIGURE 26.15
Gradients in a pervaporation membrane.

increases 20 to 40 percent for a 10°C increase in temperature without much loss in selectivity. The upstream pressure is 1 atm or slightly higher to prevent partial vaporization of the feed. It is not worthwhile to use very high pressure on the upstream side, since the activity in the liquid phase increases only a small amount for a large increase in pressure. The downstream pressure is kept as low as possible to provide a large driving force for diffusion across the membrane. The driving force can be expressed as a partial-pressure difference $(\gamma_A x_A P'_A - y_A P_2)$, where γ is the activity coefficient, P'_A the vapor pressure of the pure component, and x and y the mole fractions in the liquid and vapor. If the nonlinear effects are neglected and average values of permeability are used, the equations for a binary mixture become

$$J_A = Q_A(\gamma_A x_A P'_A - y_A P_2) \tag{26.40}$$

$$J_B = Q_B[\gamma_B(1 - x_A)P'_B - (1 - y_A)P_2] \tag{26.41}$$

$$y_A = \frac{J_A}{J_A + J_B} \tag{26.42}$$

These equations are similar to those for gas separation, but the pressure ratio R is replaced by modified ratios R_A and R_B, which include the activity coefficients and vapor pressures:

$$R_A = \frac{P_2}{\gamma_A P'_A} \qquad R_B = \frac{P_2}{\gamma_B P'_B} \tag{26.43}$$

The activity coefficients and vapor pressures are also included in a modified selectivity,

$$\alpha' = \frac{Q_A \gamma_A P'_A}{Q_B \gamma_B P'_B} \tag{26.44}$$

Combining Eqs. (26.40) to (26.44) leads to a quadratic equation for y', the local value of y_A. Then

$$a(y')^2 + by' + c = 0 \qquad (26.45)$$

where $a = \alpha' R_A - R_B$
$b = R_B + x - 1 - \alpha'(R_A + x)$
$c = \alpha' x$

For plug flow of liquid through a separator the value of y' is greatest at the feed end, and the final composition of the permeate y is an integrated average of the local values. The calculations have to allow for the decrease in liquid temperature, which decreases the driving force for diffusion and also the permeability. The temperature change is calculated from an enthalpy balance:

$$L_{in}c_p T_{in} - L_{out}c_p T_{out} = V\,\Delta H_v \qquad (26.46)$$

where c_p is the specific heat of the liquid and ΔH_v is the enthalpy of vaporization.

The first commercial application of pervaporation was for ethanol-water separation. The dilute solution produced by fermentation is distilled to produce an overhead product with 90% to 95% alcohol (close to the azeotrope), and this solution is fed to the membrane unit to give nearly pure ethanol (99.9 percent). The permeate stream with about 20% to 40% alcohol is recyled to the distillation column. The vapor-liquid equilibrium curve and the vapor and liquid compositions for a poly(vinyl alcohol) membrane are shown in Fig. 26.16.[34] The membrane gives a permeate that is always richer in water than the liquid, in contrast to distillation, where alcohol is the more volatile component over most of the range. The shape of the curve for permeate composition indicates strongly nonlinear behavior. Between 40% and 80% alcohol, the alcohol content of the vapor decreases as the concentration in the liquid increases, but not all membranes show this type of behavior. The membrane is most selective between 80% and 85% alcohol, and the permeate is only about 5% alcohol in this range. The curve for permeate composition would be displaced upward at higher permeate pressures because the driving force for water permeation is more sensitive to changes in P_2 than is the driving force for alcohol.

EXAMPLE 26.3. Laboratory tests of a pervaporation membrane exposed to liquid with 90 weight percent ethanol and 10% water at 60°C showed a flux of 0.20 kg/m²·h and a permeate composition of 7.1% ethanol when the downstream pressure was 15 mm Hg. (a) Calculate the permeability of the membrane to ethanol and to water at the test conditions and the selectivity for water. (b) Predict the local permeate composition for 90% ethanol and 60°C if the downstream pressure is kept at 30 mm Hg by a water-cooled condenser. What is the condensing temperature? (c) Calculate the local permeate composition for 95%, 99%, and 99.9% ethanol at 60°C and 30 mm Hg, assuming the permeabilities are the same as for part (a).

Solution
(a) The flux of each component is calculated from the total flux and the permeate composition with $A = H_2O$ and $B = C_2H_5OH$:

$$J_A = 0.20(0.929) = 0.1858 \text{ kg/m}^2 \cdot \text{h}$$

$$J_B = 0.20(0.071) = 0.0142 \text{ kg/m}^2 \cdot \text{h}$$

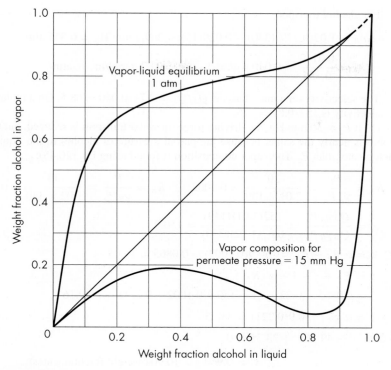

FIGURE 26.16
Permeate compositions for pervaporation with a PVA membrane and an ethanol-water feed at 60°C.

The vapor pressures at 60°C are $P'_A = 149$ mm Hg and $P'_B = 340$ mm Hg. For 10 weight percent H_2O, $x = (10/18)/(10/18 + 90/46) = 0.221$ mole fraction H_2O. The activity coefficients can be estimated from the Margules equations[25] using activity coefficients applicable at the boiling point and neglecting the change in γ with temperature.

For H_2O, $\ln \gamma_A = (0.7947 + 1.615 x_A)x_B^2$:

$$\text{For } x_A = 0.221, x_B = 0.779, \gamma_A = 2.01$$

For alcohol, $\ln \gamma_B = (1.6022 - 1.615 \, x_B)x_A^2$

$$\gamma_B = 1.02$$

The permeate is $100 - 7.1$, or 92.9 weight percent H_2O:

$$y = \frac{92.9/18}{92.9/18 + 7.1/46} = 0.971 \text{ mole fraction } H_2O$$

The driving force for water transport is

$$\Delta p_A = 2.01(0.221)(149) - 0.971(15) = 51.6 \text{ mm Hg} = 0.0679 \text{ atm}$$

$$Q_A = \frac{0.1858}{0.0679} = 2.74 \text{ kg/m}^2 \cdot \text{h} \cdot \text{atm} = 152 \text{ g mol/m}^2 \cdot \text{h} \cdot \text{atm}$$

For alcohol transport

$$\Delta p_B = 1.02(0.779)(340) - 0.029(15) = 269.7 \text{ mm Hg} = 0.355 \text{ atm}$$

$$Q_B = \frac{0.0142}{0.355} = 0.040 \text{ kg/m}^2 \cdot \text{h} \cdot \text{atm} = 0.87 \text{ g mol/m}^2 \cdot \text{h} \cdot \text{atm}$$

The selectivity on a mass basis is $Q_A/Q_B = 2.74/0.040 = 68.5$. On a molar basis, the selectivity is $152/0.87 = 175$.

(b) If $P_2 = 30$ mm Hg, the driving force for water transport is reduced, but that for alcohol is nearly the same, because the partial pressure of alcohol in the liquid is so much greater than P_2. The vapor composition is found using Eq. (26.45):

$$R_A = \frac{P_2}{\gamma_A P_A'} = \frac{30}{2.01 \times 149} = 0.1002 \qquad R_B = \frac{P_2}{\gamma_B P_B'} = \frac{30}{1.02 \times 340} = 0.0865$$

$$\alpha' = \frac{Q_A \gamma_A P_A'}{Q_B \gamma_B P_B'} = \frac{152(2.01)(149)}{0.87(1.02)(340)} = 150.9$$

$$a = \alpha' R_A - R_B = 150.9(0.1002) - 0.0865 = 15.03$$

$$b = R_B + x - 1 - \alpha'(R_A + x)$$

$$= 0.0865 + 0.221 - 1 - 150.9(0.1002 + 0.221) = -49.16$$

$$c = \alpha' x = 150.9(0.221) = 33.35$$

$$15.03(y')^2 - 49.16y' + 33.35 = 0$$

$$y' = 0.960 \qquad (0.904 \text{ weight fraction water})$$

This small increase in alcohol content of the permeate (from 7.1 to 9.6 percent) might seem tolerable, but the change becomes much greater as the feed solution becomes richer in alcohol. In practice, pressures even lower than 15 mm Hg are used to minimize the alcohol content of the permeate.

Since the vapor is mostly water, the condensing temperature at 30 mm Hg is estimated from the vapor pressure of water to be 29°C.

(c) For 95%, 99%, and 99.9% alcohol, the concentrations are converted to mole fractions, and new values of the activity coefficients and other parameters are calculated. The results are given in Table 26.4.

Reducing the water content of the ethanol from 1% to 0.1% is difficult if $P_2 = 30$ mm Hg. At 0.1% H_2O the partial pressure of water over the liquid is only 0.84 mm Hg,

TABLE 26.4

Permeate compositions for pervaporation of ethanol-water mixtures, Example 26.3

Wt % alcohol	x_{H_2O}	γ_A	γ_B	R_A	R_B	α'	y'	wt % H_2O
95	0.1186	2.15	1.0	0.0936	0.0882	164.6	0.915	80.8
99	0.0252	2.21	1.0	0.0911	0.0882	169.2	0.256	11.9
99.9	0.00255	2.21	1.0	0.0911	0.0882	169.2	0.026	1.0
99.9[†]	0.00255	2.21	1.0	0.00911	0.00865	169.2	0.159	6.8

[†]For $P_2 = 3$ mm Hg

and the local permeate contains only 1 weight percent H_2O. If P_2 could be reduced to 3 mm Hg, the increased driving force would raise y' to 0.159 and the local permeate would contain 6.8 weight percent H_2O. Therefore much less alcohol would be removed in going to 99.9% alcohol.

Although membranes could be used for the entire alcohol purification process, a great many stages would be needed, and it is cheaper to use distillation for most of the separation. If a membrane selectively permeable to alcohol could be developed, it might be feasible to separate alcohol from dilute solutions entirely by membranes.

The membranes used for ethanol purification are also suitable for dehydration of many other organic solvents, including methanol, isopropanol, butanol, methyl ethyl ketone, acetone, and chlorinated solvents. Commercial units use up to 12 stages with reheating between stages, and product water contents lower than 100 ppm can be obtained.

Another application of pervaporation is the removal of volatile organic contaminants from water using silicone rubber or other organophilic polymers for the membrane. Substances such as chlorinated solvents or gasoline components that are only slightly soluble in water have very high activity coefficients in aqueous solution. Therefore the equilibrium concentration in the membrane may be moderately large even with only a few ppm in the solution, and nearly complete removal of the organic compounds can be achieved at a low stage cut (not much removal of water). Purification of water containing traces of trichloroethylene (TCE) was demonstrated using silicone hollow fibers at 20°C, and up to 90 percent TCE removal was obtained in a single stage.[27]

At low concentrations of nearly insoluble organic materials, the mass-transfer resistance in the water phase may control the rate of permeation. For water flow inside hollow fibers at low Reynolds numbers, the limiting Sherwood number is about 4 [see Fig. 12.2], corresponding to a "film thickness" of $\frac{1}{4}d_i$, where d_i is the inside diameter of the tube. The membrane thickness ranges from $\frac{1}{10}d_i$ to $\frac{1}{2}d_i$, about the same magnitude as the inside film thickness. The solute diffusivity in the membrane is lower than in water, but this is more than offset by the much higher concentration of solute in the membrane. As a result, the mass-transfer resistance for solute is often greater in the water phase than in the membrane. By contrast, the transfer of water through the membrane is completely controlled by the membrane resistance, since the water phase contains more than 99% H_2O. This leads to an interesting optimization problem, since doubling the membrane thickness halves the water flux but only slightly decreases the flux of organic solute. The lower water flux means lower costs for vapor pumping and condensation and less water in the permeate, but the membrane area for a given solute removal is slightly increased. In a study by Lipski and Coté,[16] the optimum conditions for removal of volatile organics were calculated for hollow fibers with inside flow or transverse flow of the feed. The optimum membrane thickness was 75 μm for flow inside 500-μm fibers but only 30 μm for flow transverse to the fibers, which gave a higher mass-transfer coefficient. Boundary layer effects are also very important when spiral-wound membranes are used for solvent recovery by pervaporation, and the selectivity can be increased by selecting turbulence-promoting spacer materials.[35]

Reverse Osmosis

When miscible solutions of different concentration are separated by a membrane that is permeable to the solvent but nearly impermeable to the solute, diffusion of solvent occurs from the less concentrated to the more concentrated solution, where the solvent activity is lower. The diffusion of solvent is called *osmosis,* and osmotic transfer of water occurs in many plant and animal cells. The transfer of solvent can be stopped by increasing the pressure of the concentrated solution until the activity of the solvent is the same on both sides of the membrane. If pure solvent is on one side of the membrane, the pressure required to equalize the solvent activities is the osmotic pressure of the solution π. If pressure higher than the osmotic pressure is applied, solvent will diffuse from the concentrated solution through the membrane into the dilute solution. This phenomenon is called *reverse osmosis,* because the solvent flow is opposite to the normal osmotic flow.

Reverse osmosis is used primarily to prepare pure water from dilute aqueous solutions, although it could be used for purifying organic solvents. The main advantages of the process are that separation can take place at room temperature and there is no phase change, which would require supplying and removing large amounts of energy. The energy for the separation of saltwater into pure water and a concentrated brine comes from the work of pressurizing the feed, and some of this energy can be recovered with a turbine, so the thermodynamic efficiency of the process is relatively high.

Several polymers have a high permeability for water and low permeabilities for dissolved salts. Cellulose acetate is outstanding in these respects and relatively inexpensive. In early work by Reid and Breton,[29] dense membranes made from cellulose acetate with 40% acetyl content gave salt rejections of 95 to 98 percent at pressures of 50 to 90 atm, but the water fluxes were impractically low. Membranes with higher acetyl content gave higher salt rejection but even lower fluxes. Reverse osmosis became a practical process with the discovery of asymmetric cellulose-acetate membranes by Loeb and Sourirajan.[17] With a skin or dense layer thickness less than 1 μm and a porous substructure, fluxes of 10 to 20 gal/ft$^2 \cdot$ day are possible with high salt rejection. Asymmetric cellulose acetate membranes can now be prepared by different casting techniques as flat sheets, hollow fibers, or a coating on a porous pipe to make a tubular membrane. Hollow-fiber membranes made by du Pont from an aromatic polyamide are also used for water purification,[20] and the fluxes are 1 to 3 gal/ft$^2 \cdot$ day (0.04 to 0.12 m^3/m$^2 \cdot$ day).

The mechanism of water and salt transport in reverse osmosis is not completely understood. One theory is that water and solutes diffuse separately through the polymer by a solution diffusion mechanism. The concentration of water in the dense polymer is assumed to be proportional to the activity of water in the solution. On the low-pressure side of the dense layer, the activity is essentially unity if nearly

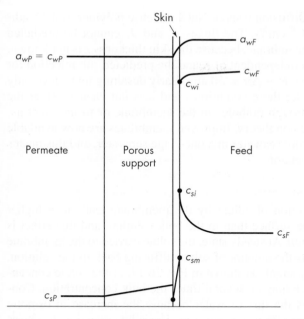

FIGURE 26.17
Concentration gradients for a reverse-osmosis membrane.

pure water is produced at 1 atm. On the high-pressure side, the activity would be slightly less than 1.0 at atmospheric pressure (0.97 for a 5% NaCl solution), 1.0 at the osmotic pressure, and slightly greater than 1.0 at higher pressures. The upstream pressure is generally set at 20 to 50 atm above the osmotic pressure of the feed solution. At these pressures, the activity of the water a_w is only a few percent greater than for pure water at 1 atm, and the change in activity and concentration across the membrane are small, as shown in Fig. 26.17. The driving force for water transport is the difference in activity, which is proportional to the pressure difference ΔP minus the difference in osmotic pressures of the feed and product $\Delta \pi$. The equation for water flux is[19]

$$J_w = \frac{c_w D_w v_w}{RT}\left(\frac{\Delta P - \Delta \pi}{z}\right) \tag{26.47}$$

In Eq. (26.47) D_w is the diffusivity in the membrane, c_w is the average water concentration in g/cm^3, and v_w is the partial molar volume of water.

The flux of solute is assumed proportional to the difference in solution concentrations, the diffusivity, and a solubility or distribution coefficient:

$$J_s = D_s S_s\left(\frac{\Delta c_s}{z}\right) \tag{26.48}$$

Equation (26.47) shows that the water flux increases strongly with the pressure difference ΔP, and the selectivity increases also, since the salt flux does not depend on ΔP. Experiments confirm these trends, but the salt rejection with cellulose acetate is not as high as predicted. The water content c_w is about 0.2 g/cm^3, and tracer tests show $D_w \cong 10^{-6}\,\text{cm}^2/\text{s}$. Diffusion tests of NaCl in dense polymer films[19] indicate $S_s = 0.035$ and $D_s = 10^{-9}$ cm²/s. The fluxes J_w and J_s cannot be predicted accurately for an asymmetric membrane, because the skin thickness z is not known. However, the ratio of fluxes is independent of z, and the predicted salt rejection for seawater is 99.6 percent when $\Delta P - \Delta \pi = 50$ atm. Early desalting units gave only 97 to 98 percent rejection under these conditions, and it is not clear whether the difference was due to flow through pinholes in the membrane or to incorrect assumptions in the solution diffusion theory. Improved membranes are now available for producing potable water from seawater in a one-stage separator, and this means greater than 99 percent salt rejection.

Concentration polarization

The nearly complete rejection of solute by the membrane leads to a higher concentration at the membrane surface than in the bulk solution, and this effect is called *concentration polarization*. At steady state, the solute carried to the membrane by the water flux almost equals the amount of solute diffusing back to the solution. The gradient may be relatively small, as shown in Fig. 26.17, or the solute concentration at the membrane surface may be several times the bulk concentration. Concentration polarization reduces the flux of water because the increase in osmotic pressure reduces the driving force for water transport. The solute rejection decreases both because of the lower water flux and because the greater salt concentration at the surface increases the flux of solute.

Equations for concentration polarization have been derived for simple cases such as laminar flow of feed solution between parallel plates or inside hollow fibers.[32] Numerical solutions were required because of the developing concentration boundary layer and the gradual decrease in solution flow rate as permeation occurs. Exact solutions are not available for the more important cases of flow outside hollow fibers or in the channels of a spiral-wound module, but an approximate analysis may still be helpful.

Consider a membrane with a water flux of J_w cm³/s·cm² when the bulk solute concentration is c_s g/cm³ and f is the fraction of solute rejected. Diffusion of solute away from the membrane surface is characterized by a mass-transfer coefficient k_c and driving force $c_{si} - c_s$, where c_{si} is the solute concentration at the surface. At steady state, the diffusion flux equals the amount of solute rejected per unit area:

$$J_w c_s f = k_c(c_{si} - c_s) \tag{26.49}$$

The polarization factor Γ is defined as the relative concentration difference

$$\Gamma \equiv \frac{c_{si} - c_s}{c_s} = \frac{J_w f}{k_c} \tag{26.50}$$

The mass-transfer coefficient can be predicted from correlations such as those in Chap. 17. If the polarization factor Γ is less than 0.1, its effect can be neglected. If Γ is large, the change in rejection and water flux can be estimated using Eqs. (26.47), (26.48), and (26.50) or the more exact expression for the concentration profile, Eq. (29.51), can be used. However, a large value of Γ is an opportunity to improve the performance by changing the dimensions or velocities in the separator to give better mass transfer.

EXAMPLE 26.4. A hollow-fiber permeator with $d_o = 300\ \mu m$ and $d_i = 200\ \mu m$ gives a water flux of 10 gal/day·ft^2 with 0.1 M NaCl solution at 20°C, and the salt rejection is 97 percent. Feed solution flows normal to the fibers at an average superficial velocity of 0.5 cm/s. Is concentration polarization significant?

Solution. For 10 gal/day·ft^2

$$J_w = 10 \times \frac{231 \times 16.3871}{24 \times 3,600 \times 929} = 4.72 \times 10^{-4}\ \text{cm/s}$$

$$\text{Re} = \frac{3 \times 10^{-2}\ \text{cm} \times 0.5\ \text{cm/s} \times 1\ \text{g/cm}^3}{0.01\ \text{g/cm} \cdot \text{s}} = 1.5$$

$$D_s = 1.6 \times 10^{-5}\ \text{cm}^2/\text{s}$$

$$\text{Sc} = \frac{0.01}{1 \times 1.6 \times 10^{-5}} = 625$$

For flow normal to fibers in a bundle, use Eq. (17.70):

$$\text{Sh} = 1.28\,\text{Re}^{0.4}\,\text{Sc}^{0.33}$$

$$= 1.28(1.5)^{0.4}(625)^{0.33} = 12.6$$

$$k_c = \frac{12.6(1.6 \times 10^{-5})}{0.03} = 6.72 \times 10^{-3}\ \text{cm/s}$$

From Eq. (26.50),

$$\Gamma = \frac{4.72 \times 10^{-4}(0.97)}{6.72 \times 10^{-3}} = 0.068$$

A concentration difference of 6.8 percent between the surface and the bulk solution would not have much effect on the salt rejection or the water flux. However, if good flow distribution is not maintained, sections receiving little flow might have significant polarization.

Friction pressure drop

Hollow-fiber membranes are generally made with the skin on the outside, and a bundle with thousands of closely packed fibers is sealed in a metal cylinder. Feed solution passes radially across the fibers or flows parallel to the fibers on the shell side, and product water is collected from the fiber lumens at one or both ends of the bundle. The diameter and length of the fibers are chosen so that the pressure drop for product flow inside the fibers is not large relative to the driving force for water permeation. For some units with low water flux, fibers with d_o and d_i as small as 50 and 25 μm are used, but larger diameters are employed with more permeable membranes. For high production rates, feed water is passed in parallel through a large number of permeators, and the residue streams may be combined and passed through another set of permeators, as shown in Fig. 26.18. With this arrangement, the velocity on the shell side is kept high to get good flow distribution and to mini-mize concentration polarization.

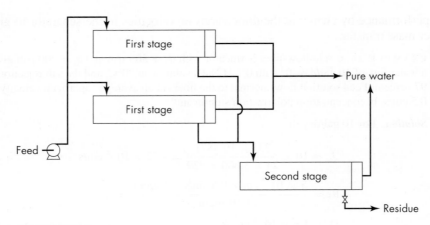

FIGURE 26.18
Two-stage reverse-osmosis system.

The flow inside the fibers is laminar, and the pressure gradient caused by skin friction dp_s/dL is given by the derivative form of the Hagen-Poiseuille equation, Eq. (5.20):

$$\frac{dp_s}{dL} = \frac{32\bar{V}\mu}{D^2} \tag{26.51}$$

where $\bar{V}$ is the average velocity, μ is the viscosity, and D is the tube diameter. The velocity increases with distance from the closed end of the fiber, and the incremental change in flow rate is the flux per unit wall area times the incremental area:

$$\frac{\pi D^2}{4}\,d\bar{V} = J_w\pi D\,dL \tag{26.52}$$

$$\frac{d\bar{V}}{dL} = \frac{4J_w}{D} \tag{26.53}$$

The water flux J_w changes along the length of the separator, since increasing the concentration of salt increases $\Delta\pi$, and pressure buildup inside the fibers decreases ΔP. For an approximate solution, J_w is assumed constant, and Eq. (26.53) is integrated directly:

$$\bar{V} = \frac{4J_wL}{D} \tag{26.54}$$

Substituting Eq. (26.54) into Eq. (26.51) and integrating give

$$\frac{dp_s}{dL} = \frac{128J_w\mu L}{D^3} \tag{26.55}$$

$$\Delta p_s = \frac{128J_w\mu}{D^3}\frac{L^2}{2} \tag{26.56}$$

Note that the pressure drop is just one-half the value that would be calculated if the water velocity $\bar{V}$ were constant at the exit value $4J_wL/D$. This is shown by rearranging Eq. (26.56) to give, for comparison with Eq. (26.51),

$$\Delta p_s = \frac{1}{2} \frac{32\mu L}{D^2} \frac{4J_w L}{D} \tag{26.57}$$

EXAMPLE 26.5. (a) For the permeator of Example 26.4 estimate the exit velocity and the pressure drop within the fibers if the fiber length is 3 m and the average water flux is 10 gal/day · ft² based on the external area. (b) What is the pressure drop if the fibers are open at both ends?

Solution.

(a) Convert the flux to J_w based on inside area. Using $d_i = 200\ \mu$m and $d_o = 300\ \mu$m, and the conversion factor from Example 26.4,

$$J_w = (4.72 \times 10^{-4})\frac{300}{200} = 7.08 \times 10^{-4}\ \text{cm/s} = 7.08 \times 10^{-6}\ \text{m/s}$$

Assume $\mu = 1\ \text{cP} = 10^{-3}\ \text{Pa} \cdot \text{s}$.

$$D = d_i = 200 \times 10^{-6}\ \text{m}$$

From Eq. (26.54),

$$\bar{V} = \frac{4(7.08 \times 10^{-6})(3)}{200 \times 10^{-6}} = 0.425\ \text{m/s}$$

From Eq. (26.57),

$$\Delta p_s = \frac{0.425(32)10^{-3}(3)}{(2 \times 10^{-4})^2}\frac{1}{2} = 5.1 \times 10^5\ \text{Pa} = 5.03\ \text{atm}$$

This is a significant pressure drop, but if the feed is at 50 atm and $\Delta\pi$ goes from 5 atm in the feed to 10 atm at the discharge, the driving force for water transport $\Delta P - \Delta\pi$ has a maximum value of $50 - 5 = 45$ and a minimum value of $45 - 10 = 35$, so the assumption of constant flux is not greatly in error.

(b) If the fibers are open at both ends, the effective length is 1.5 m and the exit velocity is one-half as great. The pressure drop is one-fourth as large as it was:

$$\Delta P = \frac{5.03}{4} = 1.26\ \text{atm}$$

Flat-sheet membranes for reverse osmosis are usually used in spiral-wound modules. The membrane is folded over a porous spacer sheet, through which product drains, and the edges are sealed. A plastic screen is placed on top to serve as a feed distributor, and the sandwich is rolled in a spiral around a small perforated drainpipe. The module is inserted in a small pressure vessel, and many units are installed in parallel. The mass-transfer area is a few hundred ft²/ft³, an order of magnitude lower than for typical hollow-fiber separators,[31] and the pressure drop due to permeate flow is generally negligible. Diagrams of spiral-wound and hollow-fiber units are shown in Fig. 26.19.

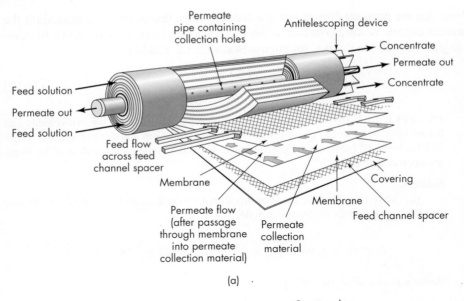

(a)

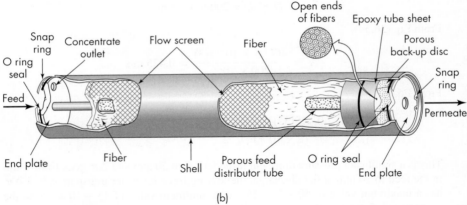

(b)

FIGURE 26.19
Cutaway views of permeators: (*a*) spiral-wound separator [*by permission, from W. Eykamp and J. Steen, in Handbook of Separation Process Technology, R. W. Rousseau (ed.), Wiley, 1987, p. 838*]; (*b*) PERMASEP hollow-fiber separator (*by permission, from Du Pont brochure, PERMASEP Permeators, 1990*).

SYMBOLS

A	Area, m^2 or ft^2
a	Constant
a_w	Activity of water; a_{wF}, in feed; a_{wP}, in permeate
b	Constant in Eq. (26.45)
c	Concentration, g mol/cm^3, kg mol/m^3, or lb mol/ft^3; c_A, of component A; c_B, of component B; c_C, of colloid; c_s, of solute; c_{sF}, in feed; c_{sP}, in

permeate; c_{si}, at membrane surface; c_{sm}, in membrane; c_w, of water; c_{wF}, in feed; c_{wP}, in permeate; c_{wi}, at membrane surface; c_1, c_2, average concentrations in feed and permeate, respectively; also, constant in Eq. (26.45)

c_p — Specific heat of liquid, J/g·°C or Btu/lb·°F

D — Volumetric diffusivity, cm²/s, m²/h, or ft²/h; D_A, D_B, of components A and B; D_e, effective diffusivity; D_{eA}, of component A; $D_{e,o}$, in organic phase; $D_{e,w}$, in water phase; D_s, of salt; D_v, bulk value; D_w, of water in membrane; D_0, at infinite dilution; also, diameter, m or ft

d — Tube or fiber diameter, μm; d_i, inside diameter; d_o, outside diameter

E — Activation energy, kcal/mol

e — Base of natural logarithms, 2.71828 $\cdots$

F — Feed rate, mol/h, kg/h, or lb/h

f — Fraction of solute rejected

J — Molar flux, mass flux, or volume flux, mol/m²·h, kg/m²·h, or m³/m²·h; J_A, J_B, of components A and B; J_s, of salt; J_w, of water

K — Overall mass-transfer coefficient, kg mol/m²·s·(kg mol/m³) or m/s; K_A, of component A; K_w, based on water phase

K_P — Partition coefficient for solute distributed between water and polymeric liquid

k — Individual mass-transfer coefficient; k_c, based on concentration [Eq. (26.49)]; k_m, for membrane; k_o, for organic phase; k_w, for water phase; k_1, k_2, for phases 1 and 2

L — Flow rate of feed or residue, mol/h, kg/h, lb/h, std ft³/h, or L/h; L_i, at station i; L_j, at station j; L_k, at station k; L_1, at entrance; L_2, at discharge; also, length, m or ft

M — Molecular weight; M_A, of component A

m — Distribution coefficient for solute (ratio of concentration in organic phase to that in water phase)

P — Total pressure, atm or lb$_f$/ft²; P_1, upstream or of feed; P_2, downstream or of permeate

P' — Vapor pressure, atm, lb$_f$/ft² or mm Hg; P'_A, of component A; P'_B, of component B

p — Partial pressure; p_A, of component A; p_B, of component B; dp_s/dL, pressure gradient due to skin friction

Q — Permeability [flux per unit pressure difference, Eq. (26.8)], L/m²·h·atm or ft³/ft²·h·(lb$_f$/ft²); Q_A, Q_B, of components A and B

q — Permeability coefficient (flux per unit pressure gradient; $q = DS$, where S is the solubility coefficient, or Qz), cm³/cm²·s·(cm Hg/cm), L/m²·h·(atm/m), or ft³/ft³·h·[(lb$_f$/ft²)/ft]; q_A, of component A

R — Gas constant, 8.314 J/g mol·K or 1,545 ft·lb$_f$/lb mol·°R; also, pressure ratio P_2/P_1; R_A, R_B, modified pressure ratios [Eq. (26.43)]

Re — Reynolds number, $d\bar{V}\rho/\mu$

r — Pore radius, μm or cm

S — Solubility coefficient, mol/cm³·atm [Eq. (26.6)]; S_A, S_B, of components A and B; also distribution coefficient [Eq. (26.48)]; S_s, of solute

Sc — Schmidt number, $\mu/\rho D$

Sh	Sherwood number, $k_c d / D$
T	Absolute temperature, K
V	Flow rate of permeate, mol/h, kg/h, lb/h, std ft^3/h or L/h; V_j, at station j; V_k, at station k
$\bar{V}$	Average fluid velocity, m/s or ft/s
v_w	Partial molar volume of water, cm^3/g
x	Mole fraction in feed or residue; x_A, of component A; x_B, of component B; x_i, at station i; x_j, at station j; x_0, at feed inlet
y	Mole fraction of more permeable species in permeate; y_A, of component A; y_i, at station i; y_j, at station j; y_k, at station k; y', local value; $\bar{y}$, average value over incremental length or in separator [Eq. (26.22)]
z	Membrane thickness, cm or μm; also, distance normal to surface

Greek Letters

α	Membrane selectivity for gas separation, dimensionless [Eq. (26.9)]; α', modified selectivity for pervaporation [Eq. (26.44)]
β	Exponent in Eq. (26.38)
Γ	Polarization factor, dimensionless [Eq. (26.49)]
γ	Activity coefficient; γ_A, γ_B, of components A and B
ΔA	Increment of area, m^2 or ft^2
Δc	Concentration difference; Δc_A, of component A; Δc_B, of component B
ΔH_v	Enthalpy of vaporization, cal/mol, J/g, or Btu/lb
ΔL	Incremental change in flow rate of residue
ΔP	Pressure difference, atm or lb$_f$/in.^{2}
Δp	Partial-pressure difference; Δp_A, of component A; Δp_B, of component B; $\overline{\Delta p}$, average value
Δp_s	Pressure drop due to skin friction, atm or lb$_f$/ft^2
ΔV	Incremental change in flow rate of permeate; ΔV_j, at station j
Δz	Increment of distance, m or ft
$\Delta \pi$	Difference in osmotic pressures, atm or lb$_f$/ft^2
ε	Porosity or void fraction, dimensionless
λ	Ratio of molecular size to pore size
μ	Viscosity, cP, Pa·s, or lb/ft·s
π	Osmotic pressure, atm or lb$_f$/ft^2
ρ	Density, g/cm^3, kg/m^3, or lb/ft^3
τ	Tortuosity, dimensionless

PROBLEMS

26.1. (a) If a membrane has a selectivity of 8 for air separation, what is the maximum oxygen concentration that can be obtained for a single-stage device? (b) What is the approximate permeate composition if 60 percent of the oxygen in the feed is recovered in the permeate?

26.2. Laboratory tests of a membrane for H_2/CH_4 separation gave a permeate composition of 80 percent H_2 and a residue of 42 percent H_2 when the feed had 50 percent H_2 and the feed and permeate absolute pressures were 100 and 15 $lb_f/in.^2$. The permeate flow was 20 percent of the feed flow. (*a*) What is the membrane selectivity? (*b*) About what permeate composition would have been obtained if a vacuum had been used on the downstream side?

26.3. Dialysis is being considered to recover a product A with molecular weight 150 from a dilute aqueous stream. The principal contaminant is a polymer B with molecular weight 15,000. Predict the initial fluxes of A and B if the membrane has a porosity of 45 percent, a mean pore size of 0.05 μm, and a thickness of 30 μm, and the feed solution contains 1% A and 1% B. Neglect boundary layer resistances, and assume pure water on the product side.

26.4. A hollow-fiber separator used for reverse osmosis is suspected of having flaws in the 0.1-μm dense layer, since the salt rejection is only 97 percent when tested with seawater at 1,000 $lb_f/in.^2$ abs, and the predicted rejection is 99.5 percent. The measured product flux is 6.5 gal/day·ft^2. (*a*) If the flaws are 0.01-μm pinholes in the dense layer, how many holes per square centimeter would be needed to account for the lower salt rejection? (*b*) What is the corresponding fraction hole area in the dense layer? (*c*) What would be the product flux if the pinholes were sealed without increasing the thickness of the membrane?

26.5. Show how the water flux and salt rejection would change with upstream pressure for a perfect cellulose acetate asymmetric membrane and a feed with an osmotic pressure of 20 atm. Use the diffusivity and solubility values for NaCl.

26.6. (*a*) Calculate the internal pressure drop for an oxygen-nitrogen hollow-fiber separator that has 600-μm outside-diameter and 400-μm inside-diameter fibers 1.0 m long. When $P_1 = 75$ $lb_f/in.^2$ abs and $P_2 \cong 15$ $lb_f/in.^2$ abs, the permeate flux is 2.0 L/min·m^2. (*b*) What would be the internal pressure drop for $L = 5$ m?

26.7. Pervaporation is used to produce nearly pure ethanol from a feed with 90% ethanol and 10% water. The feed enters at 80°C, and the design calls for reheating the liquid after the temperature has dropped to 70°C. About how many stages and how many heaters are needed?

26.8. The permeation rates of CH_4 and CO_2 in a GASEP membrane at 80°F were given as 0.00205 and 0.0413 ft^3 (std conditions)/ft^2·h·($lb_f/in.^2$), respectively. At 100°F the corresponding values were 0.00290 and 0.0425. (*a*) Calculate the apparent activation energies for permeation of CH_4 and CO_2. (*b*) Predict the permeabilities and the CO_2/CH_4 selectivity at 130°F.

26.9. A three-component mixture at 600 $lb_f/in.^2$ abs is fed to a membrane separator with a permeate pressure of 120 $lb_f/in.^2$ abs. The mixture has 50 percent A, 30 percent B, and 20 percent C, and the permeabilities are $Q_A = 0.4$, $Q_B = 0.1$, and $Q_C = 0.15$, all in ft^3 (std)/ft^2·h·atm. Estimate the local composition of the permeate at the inlet end of the separator.

26.10. (*a*) Use the data in Table 26.2 to estimate the membrane selectivity for N_2/CH_4 based on the average driving force. (*b*) Explain why the same method cannot be used to estimate the H_2/CH_4 selectivity. (*c*) How could the H_2/CH_4 selectivity be determined from these data?

26.11. The hollow-fiber separator of Example 26.1 has 300×600 μm fibers 1 m long and an external area of 5.2 m². (*a*) With a permeate flow of 3.1 L (STP)/min, what are the exit velocity and the pressure drop in the fiber lumen? (*b*) If a separator were made with the same membrane in the form of 150×300 μm fibers, what would be the pressure drop?

26.12. A membrane system treating air with 1% CFC-11 produces a residue with 0.1% CFC-11 and a permeate of 4.9% CFC-11 with a stage cut of 19 percent. $R = 0.05$. (*a*) What is the selectivity? (*b*) If the permeate is sent to a smaller second unit operating at the same value of R, what permeate composition could be reached if the residue from the second unit contains 1% CFC-11?

26.13. A brochure from the Permea Corporation claims that increasing the operating pressure from 90 to 180 $lb_f/in.^2$ raises the nitrogen productivity of their air separation membrane units by a factor of 2.3. The nitrogen purity is kept the same at 97 percent. Why isn't the permeate flow rate proportional to the pressure difference across the membrane?

REFERENCES

1. Aptel, P., and J. Ne'el: in P. M. Bungay, H. K. Lonsdale, and M. N. dePinho (eds.). *Synthetic Membranes: Science, Engineering, and Applications.* Boston: Dordrecht, 1986, p. 403.
2. Baker, R. W., J. G. Wijmans, and J. H. Kaschemekat. *J. Membrane Sci.* **151:**55 (1998).
3. Baker, R. W., and J. G. Wijmans: in D. R. Paul and Y. P. Yampol'skii (eds.). *Polymeric Gas Separation Membranes.* Boca Raton, FL: CRC Press, 1994.
4. Beaver, E. R., and P. V. Bhat. *AIChE Symp. Ser.* **84**(261):113 (1988).
5. Binning, R. C., R. J. Lee, J. F. Jennings, and E. C. Martin. *Ind. Eng. Chem.* **53:**45 (1961).
6. Crank, J., and G. S. Park (eds.). *Diffusion in Polymers.* New York: Academic, 1968.
7. Harriott, P., and B. Kim. *J. Colloid Interface Sci.* **115:**1 (1987).
8. Harriott, P., and S. V. Ho. *J. Membrane Sci.* **135:**55 (1997).
9. Henis, J. M. S., and M. K. Tripodi. *Separation Sci. Technol.* **15:**1059 (1980).
10. Ho, S. V., P. W. Sheridan, and E. Krupetsky. *J. Membrane Sci.* **112:**13 (1996).
11. Hogsett, J. E., and W. H. Mazur. *Hydrocarbon Proc.* **62**(8):52 (1983).
12. Hwang, S. T., and J. M. Thorman: *AIChE J.* **26:**558 (1980).
13. Kirk, R. E., and D. F. Othmer (eds.). *Encyclopedia of Chemical Technology,* 3rd ed., vol. 7. New York: Wiley, 1979, p. 639.
14. Koros, W. J., and R. T. Chern: in R. W. Rousseau (ed.). *Handbook of Separation Process Technology.* New York: Wiley, 1987, p. 862.
15. Lee, S. Y., and B. S. Minhas. *AIChE Symp. Ser.* **84**(261):93 (1988).
16. Lipski, C., and P. Coté. *Environ. Prog.* **9:**254 (1990).
17. Loeb, S., and S. Sourirajan. *Adv. Chem. Ser.* **38:**117 (1962).

18. Lokhandwala, K. A., S. Segelke, P. Nguyen, R. W. Baker, T. T. Su, and L. Pinnau. *Ind. Eng. Chem. Res.* **38:**3606 (1999).

19. Lonsdale, H. K.: in U. Merten (ed.). *Desalination by Reverse Osmosis.* Cambridge, MA: MIT Press, 1966, p. 93.

20. Lonsdale, H. K.: in P. M. Bungay, H. K. Lonsdale, and M. N. dePinho (eds.). *Synthetic Membranes: Science, Engineering, and Applications.* Boston: Dordrecht, 1986, p. 307.

21. MacLean, D. L., D. J. Stookey, and T. R. Metzger. *Hydrocarbon Proc.* **62**(8):47 (1983).

22. Matson, S. L., J. Lopez, and J. A. Quinn. *Chem. Eng. Sci.* **38:**503 (1983).

23. Medal Membrane Separation System brochure, Du Pont-Air Liquide, 1989.

24. Pan, C. Y. *AIChE J.* **29:**545 (1983).

25. Perry, R. H., and D. W. Green (eds.). *Perry's Chemical Engineers' Handbook,* 7th ed. New York: McGraw-Hill, 1997, p. **4**-22.

26. Prasad, R., and K. K. Sirkar. *AIChE J.* **33:**1057 (1987).

27. Psaume, R., P. Aptel, Y. Aurelle, J. C. Mora, and J. L. Bersillon. *J. Membrane Sci.* **36:**373 (1988).

28. Reahl, E. R. *Desalination,* **78:**77 (1990).

29. Reid, C. E., and E. J. Breton. *J. Appl. Polym. Sci.* **1:**133 (1959).

30. Renkin, E. M. *J. Gen. Physiol.* **38:**225 (1954).

31. Schell, W. J., and C. D. Houston: in T. E. Whyte, Jr. C. M. Yon, and E. H. Wagener (eds.). *Industrial Gas Separations, Am. Chem. Soc. Symp. Ser.* **223:**125 (1983).

32. Sherwood, T. K., P. L. T. Brian, R. E. Fisher, and L. Dresser. *Ind. Eng. Chem. Fund.* **4:**113 (1965).

33. Walawender, W., and S. A. Stern. *Separation Sci.* **7:**553 (1972).

34. Wesslein, M., A. Heintz, and R. N. Lichtenthaler: *J. Membrane Sci.,* **51:**169 (1990).

35. Wijmans, J. G., A. L. Athayde, R. Daniels, J. H. Ly, H. D. Kumaruddin, and I. Pinnau. *J. Membrane Sci.* **109:**135 (1996).

36. Yang, M. C., and E. L. Cussler. *AIChE J.* **32:**1910 (1986).

CHAPTER 27

Crystallization

Crystallization is the formation of solid particles within a homogeneous phase. It may occur as the formation of solid particles in a vapor, as in snow; as solidification from a liquid melt, as in the manufacture of large single crystals; or as crystallization from liquid solution. This chapter deals mainly with the last situation. The concepts and principles described here equally apply to the crystallization of a dissolved solute from a saturated solution and to the crystallization of part of the solvent itself, as in freezing ice crystals from seawater or other dilute salt solutions.

Crystallization from solution is important industrially because of the variety of materials that are marketed in the crystalline form. Its wide use has a twofold basis: A crystal formed from an impure solution is itself pure (unless mixed crystals occur), and crystallization affords a practical method of obtaining pure chemical substances in a satisfactory condition for packaging and storing.

Magma. In industrial crystallization from solution, the two-phase mixture of mother liquor and crystals of all sizes, which occupies the crystallizer and is withdrawn as product, is called a *magma*.

Purity of product

A sound, well-formed crystal itself is nearly pure, but it retains mother liquor when removed from the final magma, and if the crop contains crystalline aggregates, considerable amounts of mother liquor may be occluded within the solid mass. When retained mother liquor of low purity is dried on the product, contamination results, the extent of which depends on the amount and degree of impurity of the mother liquor retained by the crystals.

In practice, much of the retained mother liquor is separated from the crystals by filtration or centrifuging, and the balance is removed by washing with fresh solvent. The effectiveness of these purification steps depends on the size and uniformity of the crystals.

Importance of crystal size

Clearly, good yield and high purity are important objectives in crystallization, but the appearance and size range of a crystalline product also are significant. If the crystals are to be further processed, reasonable size and size uniformity are desirable for filtering, washing, reacting with other chemicals, transporting, and storing the crystals. If the crystals are to be marketed as a final product, customer acceptance requires individual crystals to be strong, nonaggregated, uniform in size, and noncaking in the package. For these reasons, *crystal size distribution (CSD)* must be under control; it is a prime objective in the design and operation of crystallizers.

CRYSTAL GEOMETRY

A crystal is the most highly organized type of nonliving matter. It is characterized by the fact that its constituent particles, which may be atoms, molecules, or ions, are arranged in orderly three-dimensional arrays called space lattices. As a result of this arrangement of the particles, when crystals are allowed to form without hindrance from other crystals or outside bodies, they appear as polyhedrons having sharp corners and flat sides, or faces. Although the relative sizes of the faces and edges of various crystals of the same material may be widely different, the angles made by corresponding faces of all crystals of the same material are equal and are characteristic of that material.

Crystallographic systems

Since all crystals of a definite substance have the same interfacial angles in spite of wide differences in the extent of development of individual faces, crystal forms are classified on the basis of these angles. The seven classes are cubic, hexagonal, trigonal, tetragonal, orthorhombic, monoclinic, and triclinic. A given material may crystallize in two or more different classes depending on the conditions of crystallization. Calcium carbonate, for example, occurs most commonly in nature in the hexagonal form (as calcite) but also occurs in the orthorhombic form (aragonite).

Invariant crystals

Under ideal conditions, a growing crystal maintains geometric similarity during growth. Such a crystal is called *invariant*. Figure 27.1 shows cross sections of an invariant crystal during growth. Each of the polygons in the figure represents the outline of the crystal at a different time. Since the crystal is invariant, these

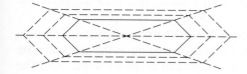

FIGURE 27.1
Growth of an invariant crystal.

polygons are geometrically similar, and the dotted lines connecting the corners of the polygons with the center of the crystal are straight. The center point may be thought of as the location of the original nucleus from which the crystal grew. The rate of growth of any face is measured by the velocity of translation of the face away from the center of the crystal in a direction perpendicular to the face. Unless the crystal is a regular polyhedron, the rates of growth of the various faces of an invariant crystal are not equal.

The size of a crystal may be specified by its characteristic length L, defined as $\Phi_s D_p$. Hence, from a rearrangement of Eq. (7.10),

$$L = \frac{6v_p}{s_p} \tag{27.1}$$

where v_p and s_p are the volume and the total surface area of the crystal, respectively, and Φ_s is the sphericity. This is satisfactory with regular solids for which Φ_s is close to 1.0, but not for disks or needles with very small values of Φ_s. In practice, L is usually taken as equal to the size determined by screening.

The concept of invariant growth is useful in analyzing the crystallization process, even though in most crystallizers the conditions are far from ideal and growth is often far from invariant. In extreme cases one face may grow much more rapidly than any of the others, giving rise to long needlelike crystals. Slow growth of one face may give rise to thin-plate or disk-shaped crystals, typical of ice formed from aqueous solution.

Even with invariant crystals the different faces of a growing crystal normally have different translational velocities. This can greatly alter the shape and appearance of the crystal. By what is called the *overlapping principle,* faces with low translational velocities can dominate the growth process to the extent that faces with high velocities diminish and ultimately disappear. Only faces with the lowest velocities survive. Conversely, when a crystal is being dissolved, faces with low translational velocities disappear, and only those with the highest velocities survive.[12c]

EQUILIBRIA AND YIELDS

Equilibrium in crystallization processes is reached when the solution is saturated, and the equilibrium relationship for bulk crystals is the solubility curve. (As shown later, the solubility of extremely small crystals is greater than that of crystals of ordinary size.) Solubility data are given in standard tables.[12a,21] Curves showing solubility as a function of temperature are given in Fig. 27.2. Most materials follow curves similar to curve 1 for KNO_3; that is, their solubility increases more or less rapidly with temperature. A few substances follow curves like curve 2 for NaCl, with little change in solubility with temperature; others have what is called an *inverted solubility curve* (curve 3 for $MnSO_4 \cdot H_2O$), which means that their solubility decreases as the temperature is raised.

Many important inorganic substances crystallize with water of crystallization. In some systems, several different hydrates are formed, depending on the concentration and temperature, and phase equilibria in such systems can be quite complicated. The

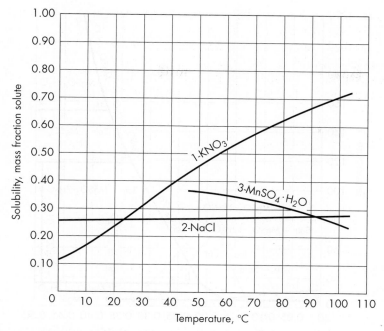

FIGURE 27.2
Solubility curves for (1) KNO_3, (2) NaCl, and (3) $MnSO_4 \cdot H_2O$ in aqueous solution.

phase diagram for the system magnesium sulfate–water is shown in Fig. 27.3. The equilibrium temperature in degrees Fahrenheit is plotted against the concentration in mass fraction of anhydrous magnesium sulfate. The entire area above and to the left of the broken solid line represents undersaturated solutions of magnesium sulfate in water. The broken line *eagfhij* represents complete solidification of the liquid solution to form various solid phases. The area *pae* represents mixtures of ice and saturated solution. Any solution containing less than 16.5% $MgSO_4$ precipitates ice when the temperature reaches line *pa*. Broken line *abcdq* is the solubility curve. Any solution more concentrated than 16.5% precipitates, on cooling, a solid when the temperature reaches this line. The solid formed at point *a* is called a *eutectic*. It consists of an intimate mechanical mixture of ice and $MgSO_4 \cdot 12H_2O$. Between points *a* and *b* the crystals are $MgSO_4 \cdot 12H_2O$; between *b* and *c* the solid phase is $MgSO_4 \cdot 7H_2O$ (epsom salt); between *c* and *d* the crystals are $MgSO_4 \cdot 6H_2O$; and above point *d* they are $MgSO_4 \cdot H_2O$. In the area *cihb*, the system at equilibrium consists of mixtures of saturated solution and crystalline $MgSO_4 \cdot 7H_2O$. In area *dkjc*, the mixture consists of saturated solution and crystals of $MgSO_4 \cdot 6H_2O$. In area *qdk*, the mixture is saturated solution and $MgSO_4 \cdot H_2O$.

Yields

In many industrial crystallization processes, the crystals and mother liquor are in contact long enough to reach equilibrium, and the mother liquor is saturated at the

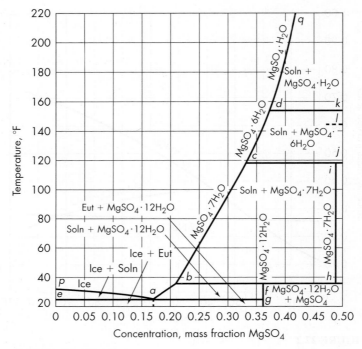

FIGURE 27.3
Phase diagram, system $MgSO_4 \cdot H_2O$. [*By permission, from J. H. Perry (ed.), Chemical Engineers' Handbook, 4th ed. Copyright, 1963, McGraw-Hill Book Company.*]

final temperature of the process. The yield of the process can then be calculated from the concentration of the original solution and the solubility at the final temperature. If appreciable evaporation occurs during the process, this must be known or estimated.

When the rate of crystal growth is slow, considerable time is required to reach equilibrium. This is especially true when the solution is viscous or where the crystals collect in the bottom of the crystallizer so there is little crystal surface exposed to the supersaturated solution. In such situations, the final mother liquor may retain appreciable supersaturation, and the actual yield will be less than that calculated from the solubility curve.

If the crystals are anhydrous, calculation of the yield is simple, as the solid phase contains no solvent. When the crop contains water of crystallization, account must be taken of the water accompanying the crystals, since this water is not available for retaining solute in solution. Solubility data are usually given either in parts by mass of anhydrous material per 100 parts by mass of total solvent or in mass fraction anhydrous solute. These data ignore water of crystallization. The key to calculations of yields of hydrated solutes is to express all masses and concentrations in terms of hydrated salt and free water. Since it is this latter quantity that remains in the liquid phase during crystallization, concentrations or amounts based on free water can be subtracted to give a correct result.

EXAMPLE 27.1. A solution consisting of 30% $MgSO_4$ and 70% H_2O is cooled to 60°F. During cooling, 5 percent of the total water in the system evaporates. How many kilograms of crystals are obtained per 1,000 kg of original mixture?

Solution. From Fig. 27.3 note that the crystals are $MgSO_4 \cdot 7H_2O$ and that the concentration of the mother liquor is 24.5% anhydrous $MgSO_4$ and 75.5% H_2O. Per 1,000 kg of original solution, the total water is $0.70 \times 1,000 = 700$ kg. The evaporation is $0.05 \times 700 = 35$ kg. The molecular weights of $MgSO_4$ and $MgSO_4 \cdot 7H_2O$ are 120.4 and 246.5, respectively; so the total $MgSO_4 \cdot 7H_2O$ in the batch is $1,000 \times 0.30(246.5/120.4) = 614$ kg, and the free water is $1,000 - 35 - 614 = 351$ kg. In 100 kg of mother liquor, there is $24.5(246.5/120.4) = 50.16$ kg of $MgSO_4 \cdot 7H_2O$ and $100 - 50.16 = 49.84$ kg of free water. The $MgSO_4 \cdot 7H_2O$ in the mother liquor, then, is $(50.16/49.84)351 = 353$ kg. The final crop is $614 - 353 = 261$ kg.

Enthalpy balances

In heat-balance calculations for crystallizers, the heat of crystallization is important. This is the latent heat evolved when solid forms from a solution. Ordinarily, crystallization is exothermic, and the heat of crystallization varies with both temperature and concentration. The heat of crystallization is equal to the heat absorbed by crystals dissolving in a saturated solution, which may be found from the heat of solution in a very large amount of solvent and the heat of dilution of the solution from saturation to high dilution. Data on heats of solution and of dilution are available,[2] and these, together with data on the specific heats of the solutions and of the crystals, can be used to construct enthalpy-concentration charts like those of Fig. 16.6 but extended to include solid phases. The diagram is especially useful in calculating enthalpy balances for crystallization processes. An H-x diagram, showing enthalpies of solid phases, for the system $MgSO_4$ and H_2O is given in Fig. 27.4. This diagram is consistent with the phase diagram of Fig. 27.3. As before, enthalpies are given in Btu per pound. They refer to 1 lb of total mixture regardless of the number of phases in the mixture. The area above line *pabcdq* represents enthalpies of undersaturated solutions of $MgSO_4$ in H_2O, and the isotherms in this area have the same significance as those in Fig. 16.6. The area *eap* in Fig. 27.4 represents all equilibrium mixtures of ice and freezing $MgSO_4$ solutions. Point *n* represents ice at 32°F. The isothermal (25°F) triangle *age* gives the enthalpies of all combinations of ice with partly solidified eutectic or of partly solidified eutectic with $MgSO_4 \cdot 12H_2O$. Area *abfg* gives the enthalpy-concentration points for all magmas consisting of $MgSO_4 \cdot 12H_2O$ crystals and mother liquor. The isothermal (35.7°F) triangle *bhf* shows the transformation of $MgSO_4 \cdot 7H_2O$ to $MgSO_4 \cdot 12H_2O$, and this area represents mixtures consisting of a saturated solution containing 21% $MgSO_4$, solid $MgSO_4 \cdot 7H_2O$, and solid $MgSO_4 \cdot 12H_2O$. The area *cihb* represents all magmas of $MgSO_4 \cdot 7H_2O$ and mother liquor. Isothermal (118.8°F) triangle *cji* represents mixtures consisting of a saturated solution containing 33% $MgSO_4$, solid $MgSO_4 \cdot 6H_2O$, and solid $MgSO_4 \cdot 7H_2O$. Area *dljc* gives enthalpies of $MgSO_4 \cdot 6H_2O$ and mother liquor. The isothermal (154.4°F) triangle *dkl* represents mixtures of a saturated solution containing 37% $MgSO_4$, solid $MgSO_4 \cdot H_2O$, and solid $MgSO_4 \cdot 6H_2O$. Area *qrkd* is part of the field representing saturated solutions in equilibrium with $MgSO_4 \cdot H_2O$.

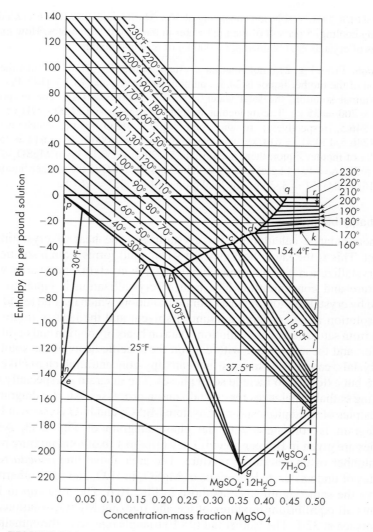

FIGURE 27.4
Enthalpy-concentration diagram, system $MgSO_4 \cdot H_2O$. Datum is
liquid water at 32°F (0°C). [*By permission, from J. H. Perry (ed.),
Chemical Engineers' Handbook, 4th ed. Copyright 1963,
McGraw-Hill Book Company.*]

EXAMPLE 27.2. A 32.5% solution of $MgSO_4$ at 120°F (48.9°C) is cooled, without
appreciable evaporation, to 70°F (21.1°C) in a batch water-cooled crystallizer. How
much heat must be removed from the solution per ton of crystals?

Solution. The initial solution is represented by the point on Fig. 27.4 at a concentration
of 0.325 in the undersaturated solution field on a 120°F isotherm. The enthalpy coordinate of this point is −33.0 Btu/lb. The point for the final magma lies on the 70°F

isotherm in area *cihb* at concentration 0.325. The enthalpy coordinate of this point is -78.4. Per 100 lb of original solution the change in enthalpy of the solution is

$$100(33.0 - 78.4) = -4,540 \text{ Btu}$$

This is a heat evolution of 4,540 Btu/100 lb (1.06×10^5 J/kg).

The split of the final slurry between crystals and mother liquor can be found from conventional mass balances, or by what is called the "center-of-gravity principle," which says that the masses of the two phases, in a two-phase mixture, are inversely proportional to the differences between their concentrations and that of the overall mixture. This principle is applied to the 70°F isotherm in either Fig. 27.3 or 27.4. The concentration of the mother liquor is 0.259, and that of the crystals is 0.488. Then the crystals are

$$100\left(\frac{0.325 - 0.259}{0.488 - 0.259}\right) = 28.8 \text{ lb/100 lb slurry}$$

The heat evolved per ton of crystals is $(4,540/28.8)(2,000) = 315,000$ Btu/ton (3.66×10^5 J/kg).

Supersaturation

Mass and enthalpy balances shed no light on the CSD of the product from a crystallizer. Conservation laws are unchanged if the product is one huge crystal or a mush of small ones.

In the formation of a crystal two steps are required: (1) the birth of a new particle and (2) its growth to macroscopic size. The first step is called *nucleation*. In a crystallizer, the CSD is determined by the interaction of the rates of nucleation and growth, and the overall process is complicated kinetically. The driving potential for both rates is supersaturation, and neither crystal growth nor formation of nuclei from the solution can occur in a saturated or unsaturated solution. Of course, very small crystals can be formed by attrition in a saturated solution, and these may act just as new nuclei as sites for further growth if the solution later becomes supersaturated.

In the theories of nucleation and growth, mole units are used in place of mass units.

Supersaturation may be generated by one or more of three methods. If the solubility of the solute increases strongly with increase in temperature, as is the case with many common inorganic salts and organic substances, a saturated solution becomes supersaturated by simple cooling and temperature reduction. If the solubility is relatively independent of temperature, as is the case with common salt, supersaturation may be generated by evaporating a portion of the solvent. If neither cooling nor evaporation is desirable, as when the solubility is very high, supersaturation may be generated by adding a third component. The third component may act physically by forming, with the original solvent, a mixed solvent in which the solubility of the solvent is sharply reduced. This process is called salting. Or, if a nearly complete precipitation is required, a new solute may be created chemically by adding a third component that will react with the original solute and form an insoluble substance. This process is called precipitation. By the addition of a third component, the rapid creation of very large supersaturations is possible.

Units for supersaturation. Supersaturation is the concentration difference between that of the supersaturated solution in which the crystal is growing and that of a solution in equilibrium with the crystal. The two phases are very nearly at the same temperature. Concentrations may be defined either in mole fraction of the solute, denoted by y, or in moles of solute in unit volume of solution, denoted by c. Since only one component is transferred across phase boundaries, component subscripts are omitted. The two supersaturations are defined by the equations

$$\Delta y \equiv y - y_s \tag{27.2}$$

$$\Delta c \equiv c - c_s \tag{27.3}$$

where Δy = supersaturation, mole fraction of solute
$\quad y$ = mole fraction of solute in solution
$\quad y_s$ = mole fraction of solute in saturated solution
$\quad \Delta c$ = molar supersaturation, moles per unit volume
$\quad c$ = molar concentration of solute in solution
$\quad c_s$ = molar concentration of solute in saturated solution

The supersaturations defined by Eqs. (27.2) and (27.3) are related by the equation

$$\Delta c = \rho_M y - \rho_s y_s \tag{27.4}$$

where ρ_M and ρ_s are the molar densities of the solution and the saturated solution, respectively. In general, since supersaturations in crystallizers are small, the densities ρ_M and ρ_s may be considered equal and ρ_M used to designate both quantities. Then

$$\Delta c = \rho_M \, \Delta y \tag{27.5}$$

The concentration ratio α and the fractional supersaturation s are defined by

$$\alpha \equiv \frac{c}{c_s} = 1 + \frac{\Delta c}{c_s} = \frac{y}{y_s} = 1 + \frac{\Delta y}{y_s} \equiv 1 + s \tag{27.6}$$

The quantity $100s$ is the percentage of supersaturation. In practice, it is usually less than about 2 percent.

Temperature difference as a potential. When the solubility increases appreciably with temperature, the supersaturation can be expressed as an equivalent temperature difference instead of a concentration difference. The relation between these driving potentials is shown in Fig. 27.5, which contains a small section of the solubility curve in molar concentrations. The field above the line represents unsaturated solutions and that below the line, supersaturated solutions. Point A refers to a saturated solution at temperature T_c, which is the temperature of the growing crystal, and point D to the supersaturated solution at temperature T. Since heat is evolved by the crystal as it grows, T_c is slightly larger than T, providing the driving force ΔT_h for heat transfer from the crystal to the liquid. This temperature difference is typically of the order of 0.01 to 0.02°C. The supersaturation α is normally based on the bulk temperature and, as shown by points E and D, is slightly greater than the actual supersaturation.

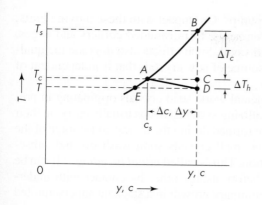

FIGURE 27.5
Supersaturation and temperature
potentials.

Point B refers to a saturated solution of the same composition as the supersaturated solution in which the crystal is growing. It would be at a temperature T_s, where $T_s > T$. Point C refers to temperature T_c and a concentration equal to that of the supersaturated solution.

From Eqs. (27.2) and (27.3) the supersaturation potential is represented by line segment $\overline{AC}$. The equivalent temperature driving potential is shown by line segment $\overline{BC}$. Segment $\overline{AB}$ of the solubility curve can be considered linear over the small concentration spanned by line $\overline{AC}$ and the temperature potential defined by

$$\Delta T_c \equiv T_s - T_c = \kappa(y - y_s) = \kappa \, \Delta y = \frac{\kappa}{\rho_M} \Delta c \qquad (27.7)$$

where κ = slope of T-versus-y line
ρ_M = molar density

In crystallization from aqueous solutions, the temperature potential is only a little less than that calculated from the actual temperature T of the solution and its saturation temperature T_s. Since ΔT_h is so small, the difference between $T_s - T$ and $T_s - T_c$ is usually insignificant.

NUCLEATION

The rate of nucleation is the number of new particles formed per unit time per unit volume of magma or solids-free mother liquor. This quantity is the first kinetic parameter controlling the CSD.

Origins of crystals in crystallizers

If all sources of particles are subsumed under the term *nucleation,* a number of kinds of nucleation may occur. Many of these are important only as methods to be avoided. They may be classified into three groups: spurious nucleation, primary nucleation, and secondary nucleation.

One origin of crystals is macroscopic attrition, which is more akin to comminution than to real nucleation. Circulating magma crystallizers have internal propeller agitators or external rotary circulating pumps. On impact with these moving parts, soft or weak crystals can break into fragments, form rounded corners and edges, and so give new crystals, both large and small. Such effects also degrade the quality of the product. Attrition is the only source of new crystals that is independent of supersaturation.

Occasionally, especially in experimental work, seed crystals originating in previous crystallizations are added to crystallizing systems. Seeds usually carry on their surfaces many small crystals which were formed during the drying and storage of the seeds after their own manufacture. The small crystals soon wash off and subsequently grow in the supersaturated solution. This is called *initial breeding*.[24] It can be prevented by curing the seed crystals before using, either by contact with undersaturated solution or solvent or by a preliminary growth in a stagnant supersaturated solution.

Growth-related spurious nucleation occurs at large supersaturations or accompanies poor magma circulation. It is characterized by abnormal needlelike and whiskerlike growths from the ends of the crystals, which, under these conditions, may grow much faster than the sides. The spikes are imperfect crystals, which are bound to the parent crystal by weak forces and which break off to give crystals of poor quality. This is called *needle breeding*.[24]

Another growth-related imperfection, unrelated to nucleation, is called *veiled growth* and occurs at moderate supersaturations. It is the result of the occlusion of mother liquor into the crystal face, giving a milky surface and an impure product. The cause of veiled growth is too rapid crystal growth, which traps mother liquor into the crystal faces.

Figures 27.6 and 27.7, which apply to $MgSO_4 \cdot 7H_2O$, show the appearance of good and inferior crystals and the variation in quality of product at various supersaturations.

All forms of spurious nucleation can be avoided by growing crystals at low supersaturations and by using only well-designed and operated pumps and agitators.

Primary nucleation

In scientific usage, nucleation refers to the birth of very small bodies of a new phase within a supersaturated homogeneous existing phase. Basically, the phenomenon of nucleation is the same for crystallization from solution, crystallization from a melt, condensation of fog drops in a supercooled vapor, and generation of bubbles in a superheated liquid. In all instances, nucleation is a consequence of rapid local fluctuations on a molecular scale in a homogeneous phase that is in a state of metastable equilibrium.[7] The basic phenomenon is called *homogeneous nucleation,* which is further restricted to the formation of new particles within a phase uninfluenced in any way by solids of any sort, including the walls of the container or even the most minute particles of foreign substances.

A variant of homogeneous nucleation occurs when solid particles of foreign substances do influence the nucleation process by catalyzing an increase of nucleation rate at a given supersaturation or giving a finite rate at a supersaturation where homogeneous nucleation would occur only after a vast time. This is called *heterogeneous nucleation.*

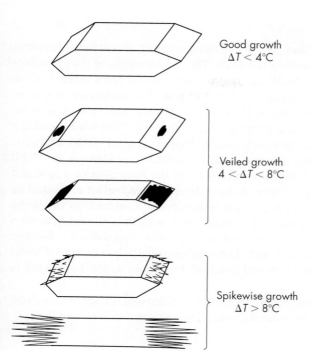

FIGURE 27.6
Illustration of growth regimes.
(*After Clontz and McCabe.*[4])

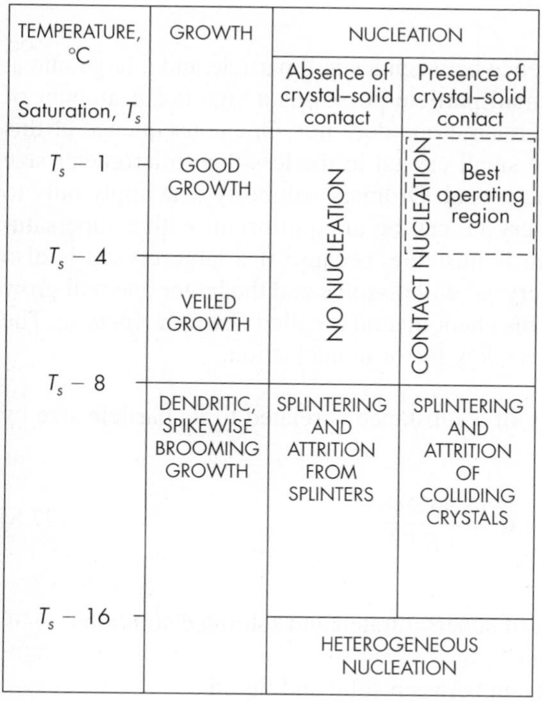

Good growth
$\Delta T < 4°C$

Veiled growth
$4 < \Delta T < 8°C$

Spikewise growth
$\Delta T > 8°C$

FIGURE 27.7
Effect of supersaturation on crystal growth quality and type of nucleation for $MgSO_4 \cdot 7H_2O$. (*Adapted from Ref. 5.*)

TEMPERATURE, °C Saturation, T_s	GROWTH	NUCLEATION	
		Absence of crystal–solid contact	Presence of crystal–solid contact
$T_s - 1$	GOOD GROWTH	NO NUCLEATION	CONTACT NUCLEATION — Best operating region
$T_s - 4$			
	VEILED GROWTH		
$T_s - 8$			
	DENDRITIC, SPIKEWISE BROOMING GROWTH	SPLINTERING AND ATTRITION FROM SPLINTERS	SPLINTERING AND ATTRITION OF COLLIDING CRYSTALS
$T_s - 16$		HETEROGENEOUS NUCLEATION	

Homogeneous nucleation

In crystallization from solution, homogeneous nucleation almost never happens, except perhaps in some precipitation reactions. The fundamentals of the phenomenon, however, are important in understanding the more useful types of nucleation.

Crystal nuclei may form from various kinds of particles: molecules, atoms, or ions. In aqueous solutions these may be hydrated. Because of their random motion, in any small volume several of these particles may associate to form what is called a *cluster*—a rather loose aggregation which usually disappears quickly. Occasionally, however, enough particles associate into what is known as an *embryo,* in which there are the beginnings of a lattice arrangement and the formation of a new and separate phase. For the most part, embryos have short lives and revert to clusters or individual particles, but if the supersaturation is large enough, an embryo may grow to such a size that it is in thermodynamic equilibrium with the solution. It is then called a *nucleus,* which is the smallest assemblage of particles that will not redissolve and can therefore grow to form a crystal. The number of particles needed for a stable nucleus ranges from a few to several hundred. For liquid water the number is about 80.

Nuclei are in a state of unstable equilibrium: If a nucleus loses units, it dissolves; if it gains units, it grows and becomes a crystal. The sequence of stages in the evolution of a crystal is, then,

$$Cluster \rightarrow embryo \rightarrow nucleus \rightarrow crystal$$

Equilibrium

Thermodynamically, the difference between a small particle and a large one at the same temperature is that the small particle possesses a significant amount of surface energy per unit mass and the large one does not. One consequence of this difference is that the solubility of a small crystal in the less-than-micrometer size range is larger than that of the large crystal. Ordinary solubility data apply only to moderately large crystals. A small crystal can be in equilibrium with a supersaturated solution. Such an equilibrium is unstable, because if a large crystal is also present in the solution, the smaller crystal will dissolve and the larger one will grow until the small crystal disappears. This phenomenon is called *Ostwald ripening.* The effect of particle size on solubility is a key factor in nucleation.

Kelvin equation. The solubility of a substance is related to its particle size by the Kelvin equation

$$\ln \alpha = \frac{4V_M \sigma}{\nu RTL} \tag{27.8}$$

where L = crystal size
  α = ratio of concentrations of supersaturated and saturated solutions
  V_M = molar volume of crystal
  σ = average interfacial tension between solid and liquid
  ν = number of ions per molecule of solute (for molecular crystals $\nu = 1$)

Since $\alpha = 1 + s$, Eq. (27.8) shows that a very small crystal of size L can exist in equilibrium with a solution having a supersaturation of s, relative to a saturated solution in equilibrium with large crystals.

Rate of nucleation

The rate of nucleation, from the theory of chemical kinetics, is given by the equation

$$B° = C \exp\left[-\frac{16\pi\sigma^3 V_M^2 N_a}{3v^2 (RT)^3 (\ln\alpha)^2} \right]$$
(27.9)

where $B°$ = nucleation rate, number/cm$^3 \cdot$ s
N_a = Avogadro constant, 6.0222×10^{23} molecules/g mol
R = gas constant, 8.3143×10^7 ergs/g mol $\cdot$ K
C = frequency factor

The factor C is a statistical measure of the rate of formation of embryos that reach the critical size. It is proportional to the concentration of the individual particles and to the rate of collision of these particles with an embryo of the critical size required to form a stable nucleus. Its value for nucleation from solutions is not known. From analogy with nucleation of water drops from supersaturated water vapor,[6] it is of the order of 10^{25} nuclei/cm$^3 \cdot$ s. Its accurate value is not important, because the kinetics of nucleation is dominated by the ln α term in the exponent.

Numerical values for σ also are uncertain. The experimental determination of solid-liquid interfacial tensions is difficult, and few values are available. They may be estimated from solid-state theory, using lattice energies. For ordinary salts, σ is on the order[31] of 80 to 100 ergs/cm^2. When the above values of C and σ are used in Eq. (27.9), a value of s can be calculated that will correspond to a nucleation rate of one nucleus per second per cubic centimeter, or $B° = 1$. The calculation gives a very large value of s and one that is impossible for materials of usual solubility. This is one reason for the conclusion that homogeneous nucleation in ordinary crystallization from solution never occurs and all actual nucleations in these situations are heterogeneous. Equation (27.9), then, does not give actual rates of nucleation.

In precipitation reactions, where y_s is very small and where large supersaturation ratios can be generated rapidly, homogeneous nucleation probably occurs.[31]

Heterogeneous nucleation

The catalytic effect of solid particles on nucleation rate is the reduction of the energy required for nucleation. One theory of this effect holds that if the nucleus "wets" the surface of the catalyst, the work of nucleus formation is reduced by a factor that is a function of the angle of wetting between the nucleus and the catalyst. Experimental data on the heterogeneous nucleation of potassium chloride solutions[14] show that the nucleation of this substance is consistent with an apparent value of the interfacial tension in the range of 2 to 3 ergs/cm^2 for both catalyzed nucleation and nucleation without an added catalyst. If the latter situation was actually a secondary nucleation self-catalyzed by microscopic seeds, the value of σ for seeded solutions of KCl would be 2.8 ergs/cm^2 at a temperature of 300 K. If σ_a is used to denote the apparent interfacial tension, if C is taken as 10^{25}, and if the mathematical approximation $\ln \alpha = \alpha - 1 = s$ is accepted for small values of $\alpha - 1$, Eq. (27.9) may be written

$$B^\circ = 10^{25} \exp\left[-\frac{16\pi V_M^2 N_a \sigma_a^3}{3(RT)^3 v^2 s^2}\right] \tag{27.10}$$

This equation, although it rests on incomplete data, gives results of the correct order and does reflect the very strong effect of supersaturation on nucleation.

EXAMPLE 27.3. Assuming that the rate of heterogeneous nucleation of potassium chloride is consistent with an apparent interfacial tension of 2.5 ergs/cm², determine the nucleation rate as a function of s at a temperature of 80°F (300 K).

Solution. Use Eq. (27.10). The molecular weight of KCl is 74.56. The density of the crystal is 1.988 g/cm³. Since KCl dissociates into two ions K$^+$ and Cl$^-$, $v = 2$. Then

$$V_M = \frac{74.56}{1.988} = 37.51 \text{ cm}^3/\text{g mol} \qquad \sigma_a = 2.5 \text{ ergs/cm}^2$$

The exponent in Eq. (27.10) is

$$-\frac{16\pi (37.51)^2 \times 6.0222 \times 10^{23} \times 2.5^3}{3(300 \times 8.3134 \times 10^7)^3 (2^2 s^2)} = -\frac{0.03575}{s^2}$$

For $B^\circ = 1$, the value of s is given by

$$1 = 10^{25} e^{-0.03575/s^2} = e^{57.565} e^{-0.03575/s^2}$$

$$\frac{0.03575}{s^2} = 57.565 \qquad s = \sqrt{\frac{0.03575}{57.565}} = 0.02492$$

From the equation

$$B^\circ = e^{57.565} e^{-0.03575/s^2}$$

the value of B° can be calculated for magnitudes of s around 0.025. The results are shown in Table 27.1. The explosive increase in B° as s increases is apparent.

EXAMPLE 27.4. What would be the size of a nucleus in equilibrium with a supersaturation of 0.029, under the conditions of Example 27.3?

Solution. This is solved from the Kelvin equation. Here $\alpha = 1 + 0.029 = 1.029$. Substitution of the values for V_M, σ_a, v, R, and T from Example 27.3 into Eq. (27.8) gives

$$\ln 1.029 = \frac{4 \times 37.51 \times 2.5}{2 \times 8.3143 \times 10^7 \times 300L}$$

TABLE 27.1

s	B°	s	B°
0.023	4.47×10^{-5}	0.0255	13.3
0.024	1.11×10^{-2}	0.027	5.04×10^3
0.02492	1	0.029	3.46×10^6

From this,

$$L = 2.63 \times 10^{-7} \text{ cm, or } 2.63 \text{ nm}$$

Note that this is consistent with the size of a nucleus containing a few hundred particles about 0.3 nm in diameter. If σ were 80 ergs/cm^2, α would be 2.5 for a 2.63-nm particle, and the supersaturation required would be 150 percent—clearly an impossible value for a soluble salt like KCl.

Secondary nucleation

The formation of nuclei attributable to the influence of the existing macroscopic crystals in the magma is called *secondary nucleation*.[9,28] Two kinds are known, one attributable to fluid shear and the other to collisions between existing crystals with one another or with the walls of the crystallizer and rotary impellers or agitator blades.

Fluid-shear nucleation. This type is known to take place under certain conditions and is suspected in others. When supersatured solution moves past the surface of a growing crystal at a substantial velocity, the shear stresses in the boundary layer may sweep away embryos or nuclei that would otherwise be incorporated into the growing crystal and so appear as new crystals. This has been reported in work on sucrose crystallization.[13] It also has been demonstrated[25] in the nucleation of $MgSO_4 \cdot 7H_2O$, if the solution is subjected to shear at the crystal face at one supersaturation and then quickly cooled to a higher supersaturation and allowed to stand while nuclei grow to macroscopic size.

Contact nucleation

It has been known for a long time that secondary nucleation is influenced by the intensity of agitation, but it was not until the 1970s that the phenomenon of contact nucleation was isolated and studied experimentally. It is the most common type of nucleation in industrial crystallizers, for it occurs at low supersaturations where the growth rate of the crystals is at an optimum for good quality. It is proportional to the first power of the supersaturation, instead of the 20th or greater power for primary nucleation, so that control is comparatively easy without unstable operation.

The energy at which a crystal must be struck is amazingly low, on the order of a few hundred ergs, and no visible effect is observable on the crystal surface. The position of contact nucleation in the fields of crystal growth and nucleation is shown by the dashed area (the "best operating region") in Fig. 27.7.

In experimental studies of this phenomenon[5,6,23] small rods are made to strike, with known amounts of impact energy, selected faces of individual crystals positioned in a flowing supersaturated solution, and the number of resulting nuclei is measured. This number has been shown to depend only on the supersaturation and the energy of impact. For inorganic crystals the number of nuclei per contact N is proportional to the supersaturation s; for some organic crystals, ln N is proportional to s. For organic and hydrated inorganic crystals, the number N is directly proportional to the contact energy E over a range that is significant in practice; but for anhydrous inorganic crystals the energy required is larger than for other crystals, and a threshold energy is required before nucleation will take place at all. For higher contact energies N is proportional[27] to exp E.

An appreciable healing time, usually several seconds long, is needed between contacts if successive contacts are to give the same nucleation.[6]

Contact nucleation is probably a combination of breakage of microscopic dendritic growths on the surface of the growing crystal and an interference by the contacting object with clusters of solute particles moving to become organized into the crystal. It has been hypothesized[24] that the action of the contacting object deflects or dislodges particles ranging in size from embryos to small crystals much larger than L, that is, the maximum size of a crystal that can exist in equilibrium with the supersaturated solution, as called for by the Kelvin equation (27.8). Particles at least as large as L survive and grow as new nuclei, while smaller ones dissolve.

In commercial crystallization equipment, nucleation results from both homogeneous and contact nucleation. The overall rate $B°$ can be written[12d]

$$B° = B_{ss} + B_e + B_c \qquad (27.11)$$

where B_{ss} = rate of homogeneous nucleation due to supersaturation driving force
B_e = nucleation rate from contact of crystals with impeller
B_c = nucleation rate from crystal-crystal contact

With high supersaturations and no agitation, homogeneous nucleation may be significant. In commercial equipment, with low supersaturations and agitation to keep the crystals in suspension, contact nucleation is the predominant mechanism.

CRYSTAL GROWTH

Crystal growth is a diffusional process, modified by the effect of the solid surfaces on which the growth occurs. Solute molecules or ions reach the growing faces of a crystal by diffusion through the liquid phase. The usual mass-transfer coefficient k_y applies to this step. On reaching the surface, the molecules or ions must be accepted by the crystal and organized into the space lattice. The reaction occurs at the surface at a finite rate, and the overall process consists of two steps in series. Neither the diffusional nor the interfacial step will proceed unless the solution is supersaturated.

Individual and overall growth coefficients

In mass-transfer operations it is generally assumed that equilibrium exists at the interface between phases. If this were true in crystallization, the concentration of the solution at the face of the crystal would be the saturation value y_s, and the total driving force for mass transfer would be $y - y_s$, where y is the concentration at a distance from the crystal face. Because of the surface reaction, however, a driving force is needed for the interfacial step, and the concentration at the interface is therefore y', where $y_s < y' < y$. Only $y - y'$ remains as the driving force for mass transfer. This is illustrated in Fig. 27.8.

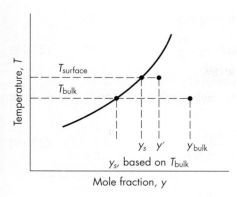

FIGURE 27.8
Temperatures and concentrations in crystallization.

The coefficients for both mass transfer and surface reaction differ from one crystal face to another, but it is nearly always adequate to consider only average values for the entire crystal. The equation for mass transfer can then be written

$$N_A = \frac{\dot{m}}{s_p} = k_y(y - y')$$ (27.12)

where N_A = molar flux, moles per unit time per unit area
$\dot{m}$ = rate of mass transfer, mol/h
s_p = surface area of crystal
k_y = mass-transfer coefficient defined by Eq. (17.37)

The equation for the surface reaction using the coefficient k_s is

$$\frac{\dot{m}}{s_p} = k_s(y' - y_s)$$ (27.13)

The resistances for the two steps may be added to give an overall coefficient K defined by

$$K \equiv \frac{\dot{m}}{s_p(y - y_s)}$$ 27.14)

Elimination of y' from Eqs. (27.12) and (27.13), followed by substitution in Eq. (27.14), gives

$$K = \frac{1}{1/k_y + 1/k_s}$$ (27.15)

Growth rate. For an invariant crystal the volume of the crystal v_p is proportional to the cube of its characteristic length L; that is,

$$v_p = aL^3$$ (27.16)

where a is a constant. If ρ_M is the molar density, the mass of the crystal m is then

$$m = v_p\rho_M = aL^3\rho_M$$ (27.17)

Differentiating Eq. (27.17) with respect to time gives

$$\dot{m} = \frac{dm}{dt} = 3aL^2 \rho_M \frac{dL}{dt} \tag{27.18}$$

The growth rate dL/dt is denoted by the symbol G.

From Eq. (27.1), $s_p = 6v_p/L = 6aL^2$. Substituting this for s_p and $\dot{m}$ from Eq. (27.18) into Eq. (27.14) gives

$$K = \frac{3aL^2 \rho_M G}{6aL^2(y - y_s)} \tag{27.19}$$

From this

$$G = \frac{2K(y - y_s)}{\rho_M} \tag{27.20}$$

Mass-transfer coefficients. For spheres or crystals with a shape factor close to 1.0, Eq. (17.71) may be used to predict the mass-transfer coefficient k_y. As discussed on page 539, for suspended particles in an agitated system the coefficient will be 1.5 to 5 times that calculated from the terminal setting velocity of the crystals.

Surface-growth coefficients. Much research on the growth of crystals by the interfacial reaction has been done and reported in standard monographs on crystallization.[3,22,29] Although a coherent theory of crystal growth has evolved, numerical data on k_s of a kind that can be used in design are scarce.

One theory is based on the concept that growth occurs layer by layer on the crystal face, and that each new layer begins as a two-dimensional nucleus attached to the face. This theory predicts that growth does not start until an appreciable threshold supersaturation is reached and that the rate of growth then increases rapidly until, at some fairly high value of supersaturation, it becomes linear with supersaturation. Actually, the growth rate of most crystals is linear with supersaturation at all supersaturations, even at very low values. There seems to be no threshold value required.

The difference between the observed and theoretical growth rates has been reconciled by the Frank screw dislocation theory. Actual space lattices of real crystals are far from perfect, and crystals have imperfections called *dislocations*. Planes of particles on the surfaces and within the crystals are displaced, and several kinds of dislocations are known. One common dislocation is a screw dislocation (Fig. 27.9), where the individual particles are shown as cubical building blocks. The dislocation is in a shear plane perpendicular to the surface of the crystal, and the slipping of the crystal creates a ramp. The edge of the ramp acts as a portion of a two-dimensional nucleus and provides a kink into which particles can easily fit. A complete face never can form, and no nucleation is necessary. As growth continues, the edge of the ramp becomes a spiral, and the continued deposit of particles along the edge of the spiral constitutes the mechanism of crystal growth.

Since this mechanism was first suggested, many examples of screw dislocations have actually been observed by electron microscopy and other methods of very high magnification.

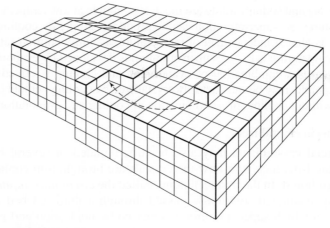

FIGURE 27.9
Screw dislocation in crystal surface and movement of particle into kink.

The ΔL law of crystal growth[8]

If all crystals in magma grow in a uniform supersaturation field and at the same temperature and if all crystals grow from birth at a rate governed by the supersaturation, then all crystals are not only invariant but also have the same growth rate that is independent of size. This generalization is called the ΔL *law*. When applicable, $G \neq f(L)$, the total growth of each crystal in the magma during the same time interval Δt is the same, and

$$\Delta L = G \Delta t \tag{27.21}$$

The model underlying Eq. (27.21) is highly idealized and is certainly not realistic in all situations. Figure 17.8 shows that the mass-transfer coefficient is approximately constant only over a limited range of particle diameters; when the mass-transfer resistance is significant, therefore, a constant growth rate would be expected only for crystals between about 50 and 500 μm. When Eq. (27.21) is not applicable, $G = f(L)$ and the growth is called *size-dependent*. Perhaps because of cancellation of errors, the ΔL law is sufficiently precise in many situations to be usable, and so it greatly simplifies the calculation of industrial processes.

CRYSTALLIZATION EQUIPMENT[12f]

Commercial crystallizers may operate either continuously or batchwise. Except for special applications, continuous operation is preferred. The first requirement of any crystallizer is to create a supersaturated solution, because crystallization cannot occur without supersaturation. Three methods are used to produce supersaturation, depending primarily on the nature of the solubility curve of the solute. (1) Solutes like potassium nitrate and sodium sulfite are much less soluble at low temperatures than at high temperatures, so supersaturation can be produced simply by cooling. (2) When the solubility is almost independent of temperature, as with common salt, or diminishes as the temperature is raised, supersaturation is developed by evaporation. (3) In intermediate cases a combination of evaporation and cooling is effective. Sodium nitrate, for example, may be satisfactorily crystallized by cooling without evaporation, evaporation without cooling, or a combination of cooling and evaporation.

Variations in crystallizers

Commercial crystallizers may also be differentiated in several other ways. One important difference is in how the crystals are brought into contact with the supersaturated liquid. In the first technique, called the *circulating-liquid method,* a stream of supersaturated solution is passed through a fluidized bed of growing crystals, within which supersaturation is released by nucleation and growth. The saturated liquid then is pumped through a cooling or evaporating zone, in which supersaturation is generated, and finally the supersaturated solution is recycled through the crystallizing zone. In the second technique, called the *circulating-magma method,* the entire magma is circulated through both crystallization and supersaturation steps without separating the liquid from the solid. Supersaturation as well as crystallization occurs in the presence of crystals. In both methods feed solution is added to the circulating stream between the crystallizing and supersaturating zones.

One type of crystallizer uses size classification devices designed to retain small crystals in the growth zone for further growth and to allow only crystals of a specified minimum size to leave the unit as product. Ideally, such a crystallizer would produce a classified product of a single uniform size. Other crystallizers are designed to maintain a thoroughly mixed suspension in the crystallizing zone, in which crystals of all sizes from nuclei to large crystals are uniformly distributed throughout the magma. Ideally, the size distribution in the product from a mixed suspension unit is identical to that in the crystallizing magma itself.

To make the average crystal size larger than that for a mixed-suspension unit, some crystallizers are equipped with devices that segregate and remove most of the fine crystals from the crystallizing zone. These small crystals are redissolved and returned to the crystallizer. Other crystallizers have two takeoff lines, one for large crystals and another for small ones. The two product streams, which may differ considerably in volume, are combined and sent to a filter or other separation unit. These techniques of enhancing crystal size are discussed later.

Most crystallizers utilize some form of agitation to improve growth rate, to prevent segregation of supersaturated solution that causes excessive nucleation, and to keep crystals in suspension throughout the crystallizing zone. Internal propeller agitators may be used, often equipped with draft tubes and baffles, and external pumps also are common for circulating liquid or magma through the supersaturating or crystallizing zones. The latter method is called *forced circulation.* One advantage of forced-circulation units with external heaters is that several identical units can be connected in multiple effect by using the vapor from one unit to heat the next in line. Systems of this kind are *evaporator-crystallizers.*

Vacuum crystallizers

Most modern crystallizers fall in the category of vacuum units in which adiabatic evaporative cooling is used to create supersaturation. In its original and simplest form, such a crystallizer is a closed vessel in which a vacuum is maintained by a condenser, usually with the help of a steam-jet vacuum pump, or booster, placed between the crystallizer and the condenser. A warm saturated solution at a temperature well above the boiling point at the pressure in the crystallizer is fed to the vessel. A magma volume is maintained by controlling the level of the liquid and crystallizing solid in the vessel, and the space above the magma is used for release of vapor and elimination of entrainment. The feed solution cools spontaneously to the equilibrium temperature; since both the enthalpy of cooling and the enthalpy of crystallization appear as enthalpy of vaporization, a portion of the solvent evaporates. The supersaturation generated by both cooling and evaporation causes nucleation and growth. Product magma is drawn from the bottom of the crystallizer. The theoretical yield of crystals is proportional to the difference between the concentration of the feed and the solubility of the solute at equilibrium temperature.

Figure 27.10 shows a continuous vacuum crystallizer with the conventional auxiliary units for feeding the unit and processing the product magma. The essential

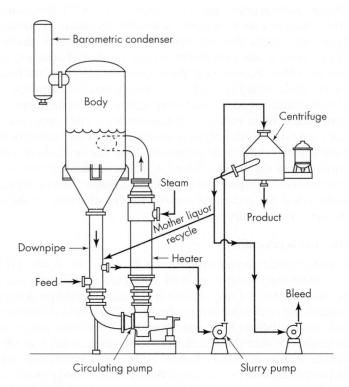

FIGURE 27.10
Continuous crystallizer. (*By permission, from R. C. Bennett and M. Van Buren, Chem. Eng. Prog. Symp. Ser. 95,* **65:**38, 1969.)

action of a single body is much like that of a single-effect evaporator, and in fact these units can be operated in multiple effect. The magma circulates from the cone bottom of the crystallizer body through a downpipe to a low-speed low-head circulating pump, passes upward through a vertical tubular heater with condensing steam in the shell, and thence into the body. The heated stream enters through a tangential inlet just below the level of the magma surface. This imparts a swirling motion to the magma, which facilitates flash evaporation and equilibrates the magma with the vapor through the action of an adiabatic flash. The supersaturation thus generated provides the driving potential for nucleation and growth. The volume of the magma divided by the volumetric flow rate of magma through the slurry pump gives the average residence time.

Feed solution enters the downpipe before the suction of the circulating pump. Mother liquor and crystals are drawn off through a discharge pipe upstream from the feed inlet in the downpipe. Mother liquor is separated from the crystals in a continuous centrifuge; the crystals are taken off as a product or for further processing, and the mother liquor is recycled to the downpipe. Some of the mother liquor is bled from the system by a pump to prevent accumulation of impurities.

The simple form of vacuum crystallizer has serious limitations from the standpoint of crystallization. Under the low pressure existing in the unit, the effect of static head on the boiling point is important; for example, water at 7°C has a vapor pressure of 7.6 mm Hg, which is a pressure easily obtainable by steam-jet boosters. A static head of 300 mm increases the absolute pressure to 30 mm Hg, where the boiling point of water is 29°C. Feed at this temperature would not flash if admitted at any level more than 300 mm below the surface of the magma. Admission of the feed at a point where it does not flash, as in Fig. 27.10, is advantageous in controlling nucleation.

Because of the effect of static head, evaporation and cooling occur only in the liquid layer near the magma surface, and concentration and temperature gradients near the surface are formed. Also crystals tend to settle to the bottom of the crystallizer, where there may be little or no supersaturation. The crystallizer will not operate satisfactorily unless the magma is well agitated, to equalize concentration and temperature gradients and suspend the crystals. The simple vacuum crystallizer provides no good method for nucleation control, for classification, or for removal of excess nuclei and very small crystals.

Draft tube–baffle crystallizer

A more versatile and effective equipment is the draft tube–baffle (DTB) crystallizer. The crystallizer body is equipped with a draft tube, which also acts as a baffle to control the circulation of the magma, and an upward-directed propeller agitator to provide a controllable circulation within the crystallizer. An additional circulation system, outside the crystallizer body and driven by a circulating pump, contains the heater and feed inlet. Product slurry is removed through an outlet near the bottom of the conical lower section of the crystallizer body. For a given feed rate, both the internal and external circulations are independently variable and provide controllable variables for obtaining the required CSD.

Draft tube–baffle crystallizers can be equipped with an elutriation leg below the body to classify the crystals by size and may also be equipped with a baffled settling

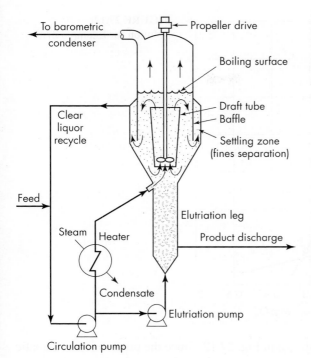

FIGURE 27.11
Draft tube–baffle crystallizer with internal system for fines separation and removal. (*By permission, from A. D. Randolph, Chem. Eng., May 1970, p. 86.*)

zone for fines removal. An example of such a unit is shown in Fig. 27.11. Part of the circulating liquor is pumped to the bottom of the leg and used as a hydraulic sorting fluid to carry small crystals back into the crystallizing zone for further growth. The action here is that of hindered-settling classification described in Chap. 29. The discharge slurry is withdrawn from the lower part of the elutriation leg and sent to a filter or centrifuge, and the mother liquor is returned to the crystallizer.

Unwanted nuclei are removed by providing an annular space, or jacket, by enlarging the cone bottom and using the lower wall of the crystallizer body as a baffle. The annular space provides a settling zone, in which hydraulic classification separates fine crystals from larger ones by floating them in an upward-flowing stream of mother liquor, which is withdrawn from the top of the settling zone. The fine crystals so withdrawn are 60-mesh in size or smaller, and although their number is huge, their mass is small, so that the stream from the jacket is nearly solids free. When this stream, called the *clear liquor recycle,* is mixed with the fresh feed and pumped through a steam heater, the solution becomes unsaturated and most of the tiny crystals dissolve. The liquor, now essentially clear, is rapidly mixed with the slurry circulating in the main body of the crystallizer.

By removing a large fraction of the mother liquor from the jacket in this fashion, the magma density is sharply increased. Magma densities of 30 to 50 percent, based on the ratio of the volume of settled crystals to that of the total magma, are achieved.

Yield of vacuum crystallizer. The yield from a vacuum crystallizer can be calculated by enthalpy and material balances. A graphical calculation based on the

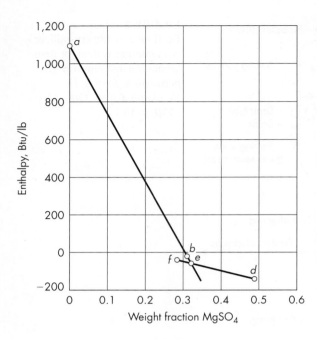

FIGURE 27.12
Solution for Example 27.5.

enthalpy chart of Fig. 27.4 is shown in Fig. 27.12. Since the process is an adiabatic split of the feed into product magma and vapor, points b for the feed, a for the vapor, and e for the magma lie on a straight line. The isotherm connecting point d for the crystals with point f for the mother liquor also passes through point e, and this point is located by the intersection of lines ab and df. From the line segments and the center-of-gravity principle, the ratios of the various streams are calculated.

> **EXAMPLE 27.5.** A continuous vacuum crystallizer is fed with a 31% $MgSO_4$ solution. The equilibrium temperature of the magma in the crystallizer is 86°F (30°C), and the boiling-point elevation of the solution is 2°F (1.11°C). A product magma containing 5 tons (4,536 kg) of $MgSO_4 \cdot 7H_2O$ per hour is obtained. The volume ratio of solid to magma is 0.15; the densities of the crystals and mother liquor are 105 and 82.5 lb/ft³, respectively. What are the temperature of the feed, the feed rate, and the rate of evaporation?

Solution. Figure 27.12 shows the graphical solution of the problem. The vapor leaves the crystallizer at the pressure corresponding to 84°F and carries 2°F superheat, which may be neglected. From steam tables, the enthalpy of the vapor is that of saturated steam at 0.5771 $lb_f/in.^2$, and the coordinates of point a are $H = 1,098$ Btu/lb and $c = 0$. The enthalpy and average concentration of the product magma are calculated from data given by Fig. 27.4. The straight line fd is the 86°F isotherm in the area $bcih$ of Fig. 27.4. The coordinates for its terminals are, for point f, $H = -43$ Btu/lb and $c = 0.285$, and for point d, $H = -149$ Btu/lb, $c = 0.488$. The mass ratio of crystals to mother liquor is

$$\frac{0.15 \times 105}{0.85 \times 82.5} = 0.224$$

The rate of production of mother liquor is $10,000/0.224 = 44,520$ lb/h, and the total magma produced is $10,000 + 44,520 = 54,520$ lb/h. The average concentration of

$MgSO_4$ in the magma is

$$\frac{0.224 \times 0.488 + 0.285}{1.224} = 0.322$$

The enthalpy of the magma is

$$\frac{0.224(-149) + (-43)}{1.224} = -62.4 \text{ Btu/lb}$$

These are the coordinates of point e. The point for the feed must lie on the straight line ae. Since the feed concentration is 0.31, the enthalpy of the feed is the ordinate of point b, or -21 Btu/lb. Point b is on the 130°F (94.4°C) isotherm, so this temperature is that of the feed. By the center-of-gravity principle, the evaporation rate is

$$54,520 \frac{-21 - (-62.4)}{1,098 - (-21)} = 2,017 \text{ lb/h} \ (915 \text{ kg/h})$$

The total feed rate is $54,520 + 2,017 = 56,537$ lb/h (25,645 kg/h).

CRYSTALLIZER DESIGN: CRYSTAL SIZE DISTRIBUTION

Once the theoretical yield from a crystallizer has been calculated from mass and energy balances, there remains the problem of estimating the CSD of the product from the kinetics of nucleation and growth. An idealized crystallizer model, called the *mixed suspension–mixed product removal (MSMPR) model,* has served well as a basis for identifying the kinetic parameters and showing how knowledge of them can be applied to calculate the performance of such a crystallizer.[17,18,20]

MSMPR Crystallizer

Consider a continuous crystallizer that operates in conformity with the following stringent requirements.[15]

1. The operation is steady state.
2. At all times the crystallizer contains a mixed-suspension magma, with no product classification.
3. At all times uniform supersaturation exists throughout the magma.
4. The ΔL law of crystal growth applies.
5. No size-classified withdrawal system is used.
6. There are no crystals in the feed.
7. The product magma leaves the crystallizer in equilibrium, so the mother liquor in the product magma is saturated.
8. No crystal breakage into finite particle size occurs.

The process is called *mixed suspension–mixed product removal* crystallization. Because of the above restraints, the nucleation rate, in number of nuclei generated in unit time and unit volume of mother liquor, is constant at all points in the magma; the

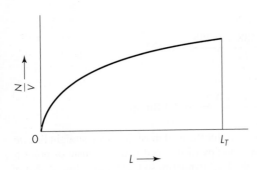

FIGURE 27.13
Cumulative number density versus length.

rate of growth, in length per unit time, is constant and independent of crystal size and location; all volume elements of mother liquor contain a mixture of particles ranging in size from nuclei to large crystals; and the particle size distribution is independent of location in the crystallizer and is identical to the size distribution in the product.

By the use of a generalized population balance the MSMPR model[†] is extended to account for unsteady-state operation, classified product removal, crystals in the feed, crystal fracture, variation in magma volume, and time-dependent growth rate.[18] These variations are not included in the following derivations.

Population density function

The basic quantity in the theory of the CSD is the population density. To understand the meaning of this variable, assume that a distribution function of the cumulative number of crystals in the magma, in number per unit volume of mother liquor, is known as a function of L, the crystal size. Such a function is plotted in Fig. 27.13. The abscissa is L, and the ordinate is N/V, where V is the volume of mother liquor in the magma and N is the number of crystals of size L and smaller in the magma. At $L = 0$, $N = 0$; the total number of crystals is N_T, corresponding to the length L_T of the largest crystal in the magma.

The population density n is defined as the slope of the cumulative distribution curve at size L, or

$$n \equiv \frac{d(N/V)}{dL} = \frac{1}{V}\frac{dN}{dL} \tag{27.22}$$

By assumption 1, V is a constant. The function $n = f(L)$ has a maximum value $n°$ where $L = 0$ and is zero where $L = L_T$. In the MSMPR model, the functions of N/V and n with L are invariant in both time and location in the magma. The dimensions of n are number/volume · length.

A growing crystal moves with time along the size axis in the direction of increasing L. A dissolving crystal moves in the other direction. For computations in the MSMPR model, a relation between population density n and size L is needed.

Consider the $n\,dL$ crystals between sizes L and $L + dL$ per unit volume of magma in the crystallizer. In the MSMPR model, each crystal of length L has the same age, and if t_m is the age of a crystal,

$$L = Gt_m \tag{27.23}$$

[†]Walas[30a] has proposed the term *continuous-stirred-tank crystallizer* (CSTC) in place of MSMPR.

Assume, now, that of the $n\,dL$ crystals per unit volume of liquid, $\Delta n\,dL$ are withdrawn as product during time increment Δt. Since the operation is in steady state, withdrawal of product does not affect the size distribution in either magma or product, and since in the MSMPR model the discharge is accurately representative of the magma, it follows that the fraction of particles withdrawn is identical to the ratio of the volume of product liquid taken out in time Δt to the total volume of liquid in the crystallizer. Then if Q is the volumetric flow rate of liquid in the product and V_c is the total volume of liquid in the crystallizer,

$$-\frac{\Delta n\,dL}{n\,dL} = -\frac{\Delta n}{n} = \frac{Q\,\Delta t}{V_c} \tag{27.24}$$

The time interval Δt also is a period in the life of each crystal, and during this time, by Eq. (27.23), the growth of each crystal is

$$\Delta L = G\,\Delta t \tag{27.25}$$

Eliminating Δt from Eqs. (27.24) and (27.25) gives

$$-\frac{\Delta n}{\Delta L} = \frac{Qn}{GV_c}$$

Letting $\Delta L \to 0$, so that

$$\lim_{\Delta L \to 0} \frac{\Delta n}{\Delta L} = \frac{dn}{dL}$$

leads to

$$-\frac{dn}{dL} = \frac{Qn}{GV_c}$$

The retention time τ of the magma in the crystallizer is defined by

$$\tau \equiv \frac{V_c}{Q}$$

and

$$-\frac{dn}{n} = \frac{1}{G\tau}\,dL \tag{27.26}$$

Integration of Eq. (27.26) gives the function of cumulative population versus length, where $n°$ is the population density at $L = 0$ and is assumed to represent the nuclei. Thus

$$\int_{n°}^{n} \frac{dn}{n} = -\frac{1}{G\tau}\int_{0}^{L} dL$$

$$\ln \frac{n°}{n} = \frac{L}{G\tau} \tag{27.27}$$

The quantity $L/G\tau$ is dimensionless. It may be replaced by z, called the *dimensionless length,* which is defined by

$$z \equiv \frac{L}{G\tau} \tag{27.28}$$

Equation (27.27) may be written

$$n = n^\circ e^{-z} \tag{27.29}$$

When plotted on semilogarithmic coordinates, the linear graph gives n° when $z = 0$ and has a slope of $-1/G\tau$. Such a graph is shown in Fig. 27.15.

Moment equations

Equation (27.29) is the fundamental relation of the MSMPR crystallizer. From it differential and cumulative equations can be derived for crystal population, crystal length, crystal area, and crystal mass. Also, the kinetic coefficients G and B° are embedded in these equations.

These calculations use moments of the n-versus-z relation of Eq. (27.29). The normalized jth moment is defined by

$$\mu_j \equiv \frac{\int_0^z n z^j \, dz}{\int_0^\infty n z^j \, dz} \tag{27.30}$$

Using Eq. (27.29) gives

$$\mu_j = \frac{\int_0^z z^j e^{-z} \, dz}{\int_0^\infty z^j e^{-z} \, dz} \tag{27.31}$$

Integrating this equation for values of $j = 0$ through $j = 3$ gives

$$\mu_0 = 1 - e^{-z} \tag{27.32}$$

$$\mu_1 = 1 - (1 - z)e^{-z} \tag{27.33}$$

$$\mu_2 = 1 - \left(1 + z + \tfrac{1}{2}z^2\right)e^{-z} \tag{27.34}$$

$$\mu_3 = 1 - \left(1 + z + \tfrac{1}{2}z^2 + \tfrac{1}{6}z^3\right)e^{-z} \tag{27.35}$$

The differential distributions are

$$\frac{d\mu_0}{dz} = e^{-z} \tag{27.36}$$

$$\frac{d\mu_1}{dz} = z e^{-z} \tag{27.37}$$

$$\frac{d\mu_2}{dz} = \frac{z^2 e^{-z}}{2} \tag{27.38}$$

$$\frac{d\mu_3}{dz} = \frac{z^3 e^{-z}}{6} \tag{27.39}$$

Equations (27.32) to (27.35) give directly the distribution of the crystals from the idealized MSMPR plant. Thus, μ_0 is the number distribution x_n, μ_1 the size distribution x_L, μ_2 the area distribution x_a, and μ_3 the mass distribution x_m.

Likewise, the differential distributions for number, size, area, and mass are given in Eqs. (27.36) to (27.39). Plots of x_n, x_m, dx_n/dz, and dx_m/dz are given in Fig. 27.14.

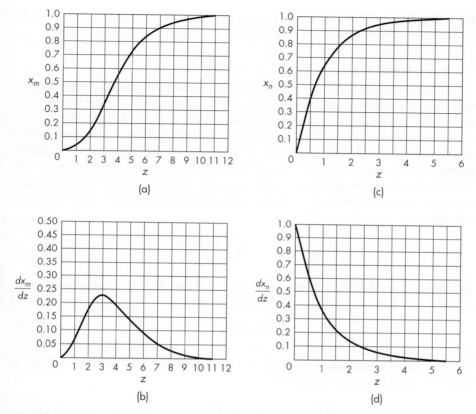

FIGURE 27.14

Size distribution relations, mixed suspension: (*a*) cumulative mass distribution; (*b*) differential mass distribution; (*c*) cumulative population distribution; (*d*) differential population distribution.

Figure 27.14*a* relates the cumulative mass fraction x_m to the dimensionless length z. For given values of G and τ, z may be converted by Eq. (27.28) to the crystal size L, and Fig. 27.14*a* then becomes a plot of the cumulative screen analysis. This procedure is illustrated in Example 27.6.

Since both z and μ are dimensionless, Eqs. (27.32) to (27.39) are universally applicable to MSMPR crystallizers.

Predominant crystal size

Figure 27.14*b*, the differential mass distribution, shows a node where the value of dx_m/dz is a maximum. Differentiating the right-hand side of Eq. (27.39) with respect to z and setting the result equal to zero show that the node is at abscissa $z_{pr} = 3$. The predominant crystal size in the product occurs where $L_{pr}G\tau = 3$, in the sense that more of the mass is in the size interval dz at $z = 3$ than in any other size interval of width dz. The most populous size in terms of number, as shown by Fig. 27.14*c* and *d*, is the extremely small size where z approaches zero. Differentiating

Eqs. (27.37) and (27.38) shows that nodes appear in the differential curves for the size and area distributions (not illustrated) at $z = 1$ and $z = 2$, respectively.

Kinetic coefficients

Clearly, to use Eq. (27.27) and its consequences in calculations, three parameters must be evaluated: the growth rate G, the nucleus density $n°$, and the retention time τ. The last parameter is under the control of the designer, but the other two are dependent variables generated by the crystallizer and its contents and are functions of the engineering details of the plant, the action of the impeller or circulating pump, and the nucleating and growth characteristics of the crystallizing system. There is an important equation, readily derivable from the MSMPR model, which connects the nucleation rate $B°$ with the zero-size particle density $n°$. Thus, by formal calculus,

$$\lim_{L\to 0} \frac{dN}{dt} = \lim_{L\to 0}\left(\frac{dL}{dt}\frac{dN}{dL}\right)$$

By definition of the terms in this equation, the limits at $L = 0$ are $dN/dt \to B°$, $dL/dt \to G$, and $dN/dL \to n°$. Therefore,

$$B° = Gn° \tag{27.40}$$

Number of crystals per unit mass. The number of crystals n_c in a unit volume of liquid in either magma or product is

$$n_c = \int_0^\infty n\,dL = n°\tau G \int_0^\infty e^{-z}\,dz = n°\tau G \tag{27.41}$$

The total mass of product crystals in a unit volume of liquid is

$$m_c = \int_0^\infty mn\,dL = a\rho_c(G\tau)^3 n° \int_0^\infty z^3 e^{-z}\,dz$$
$$= 6a\rho_c n°(G\tau)^4$$

where ρ_c is the density of the crystals. The number of crystals per unit mass is

$$\frac{n_c}{m_c} = \frac{n°G\tau}{6a\rho_c n°(G\tau)^4} = \frac{1}{6a\rho_c(G\tau)^3} \tag{27.42}$$

If the predominant size is used as a design parameter, then from the equation $L_{pr} = 3G\tau$ and Eq. (27.42),

$$\frac{n_c}{m_c} = \frac{9}{2a\rho_c L_{pr}^3} \tag{27.43}$$

The nucleation rate must be just sufficient to generate one nucleus for each crystal in the product. If C is the mass production rate of crystals, the required nucleation, in number per unit time and volume of mother liquor, is, from Eqs. (27.42) and (27.43),

$$B° = \frac{Cn_c}{m_c V_c} = \frac{C}{6a\rho_c(G\tau)^3 V_c} = \frac{9C}{2a\rho_c V_c L_{pr}^3} \tag{27.44}$$

EXAMPLE 27.6. In the crystallizer of Example 27.5, a growth rate G of 0.0018 ft/h (0.00055 m/h) is anticipated, and a predominant crystal size of 20-mesh is desired. How large must the magma volume in the crystallizer be; what nucleation rate $B°$ is necessary; and what is the screen analysis of the product, assuming that the operation conforms to all the requirements of the mixed suspension–mixed product crystallization?

Solution. The screen opening of a 20-mesh standard screen is, from App. 5, 0.0328 in., or 0.00273 ft. This dimension can be used for L, and the factor a [Eq. (27.16)] is assumed to be unity. From Example 27.5, the volume flow rate of mother liquor in the product magma is

$$Q = \frac{44,520}{82.5} = 540 \text{ ft}^3/\text{h}$$

Since, when $z = 3$, $L_{pr} = 0.00273$, Eq. (27.28) gives for the drawdown time and the volume of liquid in the crystallizer

$$\tau = \frac{L_{pr}}{3G} = \frac{0.00273}{3 \times 0.0018} = 0.506 \text{ h} \qquad V_c = 0.506 \times 540 = 273 \text{ ft}^3$$

The total magma volume is $273/0.85 = 321$ ft^3, or 2,400 gal.
    The nucleation rate is, from Eq. (27.44), since $C = 10,000$ lb/h,

$$B° = \frac{9 \times 10,000}{2 \times 105 \times 273 \times 0.00273^3}$$

$$= 7.72 \times 10^7 \text{ nuclei/ft}^3 \cdot \text{h} \ (2.74 \times 10^9 \text{ nuclei/m}^3 \cdot \text{h})$$

By Eq. (27.40), the zero-size particle density is $n° = 7.72 \times 10^7/0.0018 = 4.289 \times 10^{10}$ nuclei/ft^4. The value of $L/G\tau$ is $L/(0.0018)(0.506) = 1.1 \times 10^3 L$. The equation for the number-density distribution is, from Eq. (27.27),

$$\log n = \log n° - \frac{1.1 \times 10^3 L}{2.3026} = 10.632 - 4.777 \times 10^2 L$$

This equation is plotted in Fig. 27.15.
    The screen analysis is found by reading ordinates from Fig. 27.14c for values of z corresponding to mesh openings. For example, for the 20-mesh point, where $z = 3$, x_m is 0.35. In general,

$$z = \frac{L}{\tau G} = \frac{L}{0.506 \times 0.0018} = 1,098L$$

Details are given in Table 27.2.

TABLE 27.2

Mesh	Size				Screen analysis, %	
	ft	mm	z		Cumulative	Differential
8	0.0078	2.37	8.5		97	3
9	0.0065	1.98	7.1		93	4
10	0.0054	1.65	5.9		84	9
12	0.0046	1.40	5.0		74	10
14	0.0038	1.16	4.2		61	13
16	0.0033	1.01	3.6		48	13
20	0.0027	0.82	3.0		35	13
24	0.0023	0.70	2.5		25	10
28	0.0019	0.58	2.1		17	8
32	0.0016	0.49	1.8		11	6
35	0.0014	0.43	1.5		6	5
42	0.0011	0.34	1.2		4	2

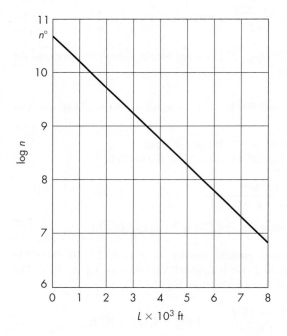

FIGURE 27.15
Population density versus length (Example 27.6).

Increasing the crystal size

Two techniques are used to make crystals larger than those from an MSMPR crystallizer. The first, known as fines destruction with solute recycle, has been described in the discussion of DTB crystallizers. The objective is to reduce sharply the number of nuclei present so that the remaining ones can grow. The presence of too many nuclei leads to a large surface area and makes the supersaturation small, so the growth rate of individual crystals is low. Fines destruction is most useful in systems involving highly soluble materials, where the tiny crystals can easily be redissolved by heating or by adding a small amount of solvent.

With the second technique, known as *double drawoff* (DDO), a large overflow stream is removed through a classifier so that crystals larger than a certain cut size L_F are left behind. A smaller underflow stream is taken from the well-mixed region of the crystallizer, and the large crystals that leave with this stream have a longer average residence time than they would have in a mixed crystallizer with a single drawoff. The overflow stream is typically 8 to 12 times as large as the underflow stream, and the residence time for large crystals is thus increased by about 10 times.

The classification can be accomplished by removing the overflow through a vertical pipe or baffled segment of the tank, with the area chosen to give an upward velocity slightly greater than the terminal velocity of crystals of size L_F. Alternatively, the overflow can be drawn directly from the tank and passed through a hydroclone (see Chap. 29), with oversize particles being returned to the tank. The overflow and underflow streams are combined outside the crystallizer and sent to a filter or other separator.

The population distribution in the combined product is strongly bimodal, with one peak at about the same size as with MSMPR operation. The second peak size may be several times as large as the first one, and the average size 2 to 3 times that obtainable from an MSMPR crystallizer. This may dramatically change the filtration characteristics of the product. In the disposal of waste gypsum, for example, ordinary crystallization produced an unfilterable slime, but the product from a DDO crystallizer, despite its high content of small crystals, filtered "like sand."[4, 16]

Calculation of the optimum fines cut size in a DDO crystallizer is discussed by White and Randolph.[32] The optimum cut size for fines destruction with solute recycle, with design charts for DDO and fines-destruction crystallizers, is given by Sutradhar and Randolph.[26]

Contact nucleation in crystallizers

The contact nucleation discussed on previous pages is based on the single-particle experiments of Clontz and McCabe,[5] which determined the number of crystals generated by a single contact at known supersaturation, energy, and area of contact. The results have been applied to the construction of nucleation models for the design of magma crystallizers.[1, 11] A correlation developed by Bennett et al.[1] is based on the following ideas:

1. The total generation of nuclei is proportional to the sum of the nucleation from all crystals of all sizes each time they pass through the impeller.
2. The driving potential is the supersaturation, which is proportional to the growth rate G.
3. The energy imparted to a crystal of size L and mass $\rho_c L^3$ is that necessary to accelerate the particle from the speed of the flowing magma to the speed of the tip of the impeller.
4. The area of contact is proportional to L^2.

These assumptions lead to the following equation:

$$B° = K_N n° G \frac{u_T^2}{t_{To}} \rho_c (G\tau)^5 \qquad (27.45)$$

where $B°$ = nucleation rate per unit volume of liquid
$\quad\quad K_N$ = dimensional constant
$\quad\quad u_T$ = tip speed of impeller
$\quad\quad t_{To}$ = turnover time

Use of Eq. (27.45) in practice requires empirical data from a pilot plant or actual operation of crystallizers of the same design. In the crystallization of KCl, for example, Randolph et al.[19] found that $B° \propto G^{2.77} m_c^{0.91}$, where m_c is the solids concentration in the crystallizer, in mass of crystals per unit volume.

Crystallization of organic chemicals

All the examples given in this chapter are for inorganic salts. The principles of crystallization from solution, however, apply equally to organic compounds. Crystallization data for a number of organic compounds are reported in the literature.[12e] Organic compounds are also often purified by crystallization from the melt, as discussed in the following section.

CRYSTALLIZATION FROM MELTS

The final step in making some organic compounds is to solidify the product from a purified molten stream. This is commonly done by freezing on a drum flaker or a cooled vibrating conveyor or in a jacketed agitated vessel.[12b] Sometimes the liquid can be subcooled 20 or even 50°C without freezing, and mechanical agitation or seeding of the liquid may be needed to induce crystallization.

Binary and ternary organic mixtures, especially of aromatic isomers, may be purified by simple cooling. Unless solid solutions are formed, virtually pure crystals of one isomer can often be produced. Mixtures of xylene isomers, for example, are processed on a commercial scale at temperatures approaching −60°C in a two-stage crystallization system. Crystals of p-xylene from each stage are separated in a filtering centrifuge (see Chap. 29); the first crop, which is not sufficiently pure, is remelted and recrystallized, and the second crop is washed with toluene or other solvent to remove adhering mother liquor.

In other processing systems the pure product is melted, and part of it is refluxed to wash and purify the crystals. A typical unit of this type is the *Brodie purifier*, illustrated in Fig. 27.16. This consists of several scraped-surface coolers set in series and followed by a purification column. Molten feed enters the middle of the last cooler. Liquid flows toward the residue end of the system where the temperature is lowest. Crystals are moved by the conveyors toward the product end where they discharge into a vertical purifier column. At the bottom of the column they are melted; part of the resulting liquid is withdrawn as product, and part flows upward through the column as reflux. In the purifier the reflux washes the crystal surfaces to remove impurities; in addition, the crystals are held at their melting point for a time to permit occluded impurities to migrate into the reflux. By this means p-dichlorobenzene with a purity of greater than 99.9 percent can be produced from a mixture of chlorobenzene

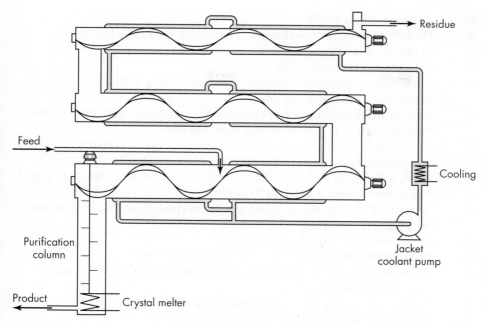

FIGURE 27.16
The Brodie purifier countercurrent cooling crystallizer. (*C. W. Nofsinger Company.*)

isomers. The Brodie purifier is also used commercially to reduce the thionaphthenate content of naphthalene from 2 to 0.2 to 0.4 percent, even though these compounds form solid solutions instead of a eutectic mixture.[10]

Other countercurrent devices and melt-refining techniques are described in the literature.[12g,30b]

SYMBOLS

a	Shape factor, defined by Eq. (27.16), dimensionless
$B°$	Nucleation rate, number/cm$^3 \cdot$ s or number/ft$^3 \cdot$ h; B_c, rate from crystal-crystal contact; B_e, rate from contact of crystals with impeller; B_{ss}, rate of homogeneous nucleation
C	Frequency factor in nucleation, number/cm$^3 \cdot$ s; also, mass production rate of crystals, kg/h or lb/h
c	Concentration of solution, mol/unit volume or g mol/m^3; c_s, in saturated solution
D_p	Diameter of particle, m or ft
E	Energy of contact in contact nucleation, ergs
f	Function of
G	Growth rate of crystal, m/h or ft/h

H	Enthalpy, J/g or Btu/lb
K	Overall mass-transfer coefficient, g mol/m$^2 \cdot$h$\cdot$unit mole fraction or lb mol/ft$^2 \cdot$h$\cdot$unit mole fraction
K_N	Dimensional constant in Eq. (27.45)
k_s	Coefficient of interfacial reaction, g mol/m$^2 \cdot$h$\cdot$unit mole fraction or lb mol/ft$^2 \cdot$h$\cdot$unit mole fraction
k_y	Mass-transfer coefficient from solution to crystal face, g mol/m$^2 \cdot$h$\cdot$unit mole fraction or lb mol/ft$^2 \cdot$h$\cdot$unit mole fraction
L	Linear dimension or size of crystal, m or ft; L_F, cut size in DDO crystallizer; L_T, maximum size; L_{pr}, predominant size
m	Mass, g mol or lb mol; m_c, total mass of crystals per unit volume of liquid
$\dot{m}$	Molal growth rate, g mol/h or lb mol/h
N	Number of crystals of size L and smaller in crystallizer; N_T, total number of crystals in crystallizer
N_A	Molar flux, g mol/m$^2 \cdot$h or lb mol/ft$^2 \cdot$h
$\mathbf{N}_a$	Avogadro's constant, 6.0222×10^{23} molecules/g mol
n	Population density defined by Eq. (27.22) number/m^4 or number/ft^4; $n°$, maximum value, for nuclei
n_c	Number of crystals per unit volume of crystals
Q	Volumetric flow of liquid in product, m^3/h or ft^3/h
R	Gas constant, 8.3143×10^7 ergs/g mol$\cdot$K
s	Fractional supersaturation, defined by Eq. (27.7); $100s$, percent supersaturation
s_p	Surface area of crystal, m^2 or ft^2
T	Temperature, °C or °F; T_c, of growing crystal; T_s, of saturated solution
t	Time; t_{T_0}, turnover time between passages of a given crystal between contacts with rotating impeller
t_m	Age of crystal
u_T	Tip speed of impeller, m/s or ft/s
V	Volume, m^3 or ft^3; volume of liquid in magma; V_c, total volume of liquid in crystallizer
V_M	Molar volume, $1/\rho_M$, cm^3/g mol or ft^3/lb mol
v_p	Volume of crystal, m^3 or ft^3
x	Mass fraction; also, distribution relation; x_L, size distribution; x_a, area distribution; x_m, mass distribution; x_n, number distribution
y	Mole fraction of solute in solution, at a distance from crystal face; y_s, in saturated solution; y', at interface between crystal and liquid
z	Dimensionless length, $L/G\tau$; z_{pr}, predominant value

Abbreviations

CSD	Crystal size distribution
DDO	Double drawoff
DTB	Draft tube–baffle, type of crystallizer
MSMPR	Mixed suspension–mixed product removal

Greek Letters

α Concentration ratio, defined by Eq. (27.7)

Δc Supersaturation, g mol/m^3 or lb mol/ft^3

ΔL Increase in crystal size in time increment Δt, m or ft

Δn Increment of population density

ΔT Temperature driving potential, °C or °F; ΔT_c, for crystallization; ΔT_h, for heat transfer from crystal to liquid

Δt Time increment, h or s

Δy Supersaturation, mole fraction of solute

κ Slope of temperature-concentration line

μ Distribution of crystals from idealized MSMPR plant; μ_0, number distribution; μ_1, size distribution; μ_2, area distribution; μ_3, mass distribution

μ_j Normalized jth moment of crystal distribution, defined by Eq. (27.30)

ν Number of ions per molecule of solute

ρ_M Molar density of solution, g mol/cm^3 or lb mol/ft^3; ρ_c, of crystal; ρ_s, of saturated solution

σ Interfacial energy, ergs/cm^2

σ_a Apparent interfacial tension between nucleus and catalyst, ergs/cm^2

τ Retention time of magma in crystallizer, h

Φ_s Sphericity

PROBLEMS

27.1. $CuSO_4 \cdot H_2O$ containing 3.5 percent of a soluble impurity is dissolved continuously in sufficient water and recycled mother liquor to make a saturated solution at 80°C. The solution is then cooled to 25°C, and crystals of $CuSO_4 \cdot 5H_2O$ are thereby obtained. These crystals carry 10 percent of their dry weight as adhering mother liquor. The crystals are then dried to zero free water ($CuSO_4 \cdot 5H_2O$). The allowable impurity in the product is 0.6 percent. Calculate (a) the weight of water and of recycled mother liquor required per 100 kg of impure copper sulfate; (b) the percentage recovery of copper sulfate, assuming that the mother liquor not recycled is discarded. The solubility of $CuSO_4 \cdot 5H_2O$ at 80°C is 120 g per 100 g of free H_2O and at 25°C is 40 g per 100 g of free H_2O.

27.2. A solution of $MgSO_4$ containing 43 g of solid per 100 g of water is fed to a vacuum crystallizer at 220°F. The vacuum in the crystallizer corresponds to an H_2O boiling temperature of 43°F, and a saturated solution of $MgSO_4$ has a boiling-point elevation of 2°F. How much solution must be fed to the crystallizer to produce 900 kg of epsom salt ($MgSO_4 \cdot 7H_2O$) per hour?

27.3. An ideal product classification in a continuous vacuum crystallizer would achieve the retention of all crystals within the crystallizer until they attained a desired size and then discharge them from the crystallizer.[20] The size distribution of the product would be uniform, and all crystals would have the same value of D_p. Such a process conforms to the other constraints for the mixed suspension–mixed product crystallizer except that the magma in the unit is classified by size and each crystal has the same retention or growth time. For such a process show that

$$m_c = \frac{a\rho_c B^\circ L^4}{4G} \qquad \text{and} \qquad \tau = \frac{L_{pr}}{4G}$$

27.4. Assume that $CuSO_4 \cdot 5H_2O$ is to be crystallized in an ideal product-classifying crystallizer. A 1.4-mm product is desired. The growth rate is estimated to be 0.2 μm/s. The geometric constant a is 0.20, and the density of the crystal is 2,300 kg/m^3. A magma consistency of 0.35 m^3 of crystals per cubic meter of mother liquor is to be used. What is the production rate, in kilograms of crystals per hour per cubic meter of mother liquor? What rate of nucleation, in number per hour per cubic meter of mother liquor, is needed?

27.5. An MSMPR crystallizer produces 1 ton of product per hour having a predominant size of 35-mesh. The volume of crystals per unit volume of magma is 0.15. The temperature in the crystallizer is 120°F, and the retention time is 2.0 h. The densities of crystals and mother liquor are 105 and 82.5 lb/ft^3, respectively. (a) Plot the cumulative screen analysis of the theoretical product. (b) Determine the required growth rate G and the necessary nucleation rate B°.

27.6. Results have been reported on the performance of a crystallizer operating on sodium chloride.[1] Results from one experiment are

$$\text{Tip speed } u_T = 1{,}350 \text{ ft/min} \qquad \text{Retention time} = 1.80 \text{ h}$$

$$\text{Time between passages, or turnover time} = 35 \text{ s}$$

Size distribution parameters are

$$n^\circ = 1.46 \times 10^6 \text{ number/cm}^4 \qquad B^\circ = 1.84 \text{ number/cm}^3 \cdot \text{s}$$

$$\text{Crystal density} = 2.163 \text{ g/cm}^3$$

Calculate the dimensional constant K_N, using the units specified above, and the growth rate in millimeters per hour.

27.7. A potassium chloride crystallizer that operates at 100°F has a fines dissolution loop where the slurry temperature is 130°F.[19] The solubility of KCl is 39.3 g/100 g H_2O at 100°F and 44 g/100 g H_2O at 130°F. (a) If the solution is initially supersaturated by 1.0 g/100 g and the temperature is suddenly increased to 130°F, how long will it take to dissolve 10-μm crystals or 50-μm crystals if the change in solution concentration is neglected? (b) If the average crystal size in the fines loop is 200 μm and the slurry has 30 g solid/L, how close to saturation will the solution be after 10 s at 130°F?

27.8. Crystallization of gypsum takes place in a 200,000-gal stirred tank that is part of a process for flue gas desulfurization. With a single outlet to a settling tank and filter, the slurry has 15 weight percent $CaSO_4 \cdot 2H_2O$ and a mean particle size of about 30 μm. Pilot-plant tests under similar conditions[4] but with double drawoff and an overflow-underflow ratio of 3.5 showed a bimodal size distribution with peaks at 20 and 50 μm. The classification was made with hydroclones with a cut size of 30 μm. (a) What is the average residence time for large crystals with DDO operation relative to the average residence time for MSMPR operation? (b) What evidence is there that the supersaturation is different in the two modes of operation? (*c*) Estimate the concentrations in the overflow and underflow streams and the concentration of the combined streams.

27.9. Use the moment equations for constant growth rate in an MSMPR crystallizer to calculate (*a*) the surface-volume mean size and (*b*) the mass average size. (*c*) Compare these values with the sizes where a maximum occurs in the corresponding distribution curves.

REFERENCES

1. Bennett, R. C., H. Fiedelman, and A. D. Randolph. *Chem. Eng. Prog.* **69**(7):86 (1973).
2. Bichowsky, F. R., and F. D. Rossini. *Thermochemistry of Chemical Substances.* New York: Reinhold, 1936.
3. Buckley, H. E. *Crystal Growth.* New York: Wiley, 1951.
4. Chang, J. C. S., and T. G. Brna. *Chem. Eng. Prog.* **82**(11):51 (1986).
5. Clontz, N. A., and W. L. McCabe. *AIChE Symp. Ser., No.* 110, **67**:6 (1971).
6. Johnson, R. T., R. W. Rousseau, and W. L. McCabe. *AIChE Symp. Ser., No.* 121, **68**:31 (1972).
7. La Mer, V. K. *Ind. Eng. Chem.* **44**:1270 (1952).
8. McCabe, W. L. *Ind. Eng. Chem.* **21**:30, 121 (1929).
9. McCabe, W. L.: in J. C. Perry (ed.). *Chemical Engineers' Handbook,* 3rd ed. New York: McGraw-Hill, 1950, p. 1056.
10. Meyer, D. W. *Chem. Proc.* **53**(1):50 (1990).
11. Ottens, E. P. K. *Nucleation in Continuous Agitated Crystallizers.* Delft, Netherlands: Technological University, 1972.
12. Perry, R. H., and D. W. Green (eds.). *Perry's Chemical Engineers' Handbook,* 7th ed. New York: McGraw-Hill, 1997; (*a*) pp. **2**-120 to **2**-124, (*b*) pp. **11**-58 to **11**-67, (*c*) p. **18**-38, (*d*) p. **18**-40, (*e*) p. **18**-44, (*f*) pp. **18**-44 to **18**-54, (*g*) pp. **22**-3 to **22**-13.
13. Powers, H. E. C. *Ind. Chem.* **39**:351 (1963).
14. Preckshot, G. W., and G. G. Brown. *Ind. Eng. Chem.* **44**:1314 (1952).
15. Randolph, A. D. *AIChE J.* **11**:424 (1965).
16. Randolph, A. D. Private communication, 1991.
17. Randolph, A. D., and M. A. Larson. *AIChE J.* **8**:639 (1962).
18. Randolph, A. D., and M. A. Larson. *Theory of Particulate Processes.* New York: Academic, 1971.
19. Randolph, A. D., E. T. White, and C. -C. D. Low. *Ind. Eng. Chem. Proc. Des. Dev.* **20**:496 (1981).
20. Saeman, W. C. *AIChE J.* **2**:107 (1956).
21. Seidell, A. *Solubilities,* 3rd ed. Princeton, NJ: Van Nostrand, 1940 (supplement, 1950).
22. Strickland-Constable, R. F. *Kinetics and Mechanism of Crystallization.* New York: Academic, 1968.
23. Strickland-Constable, R. F., and R. E. A. Mason. *Nature,* **197**:4870 (1963).
24. Strickland-Constable, R. F. *AIChE Symp. Ser., No.* 121, **68**:1 (1972).
25. Sung, C. Y., J. Estrin, and G. R. Youngquist. *AIChE J.* **19**:957 (1973).
26. Sutradhar, B. C., and A. D. Randolph. Unpublished manuscript. Chemical Engineering Department, University of Arizona, Tucson, 1991.

27. Tai, C. Y., W. L. McCabe, and R. W. Rousseau. *AIChE J.* **21**:351 (1975).
28. Ting, H. H., and W. L. McCabe. *Ind. Eng. Chem.* **26**:1201 (1934).
29. VanHook, A. *Crystallization, Theory and Practice,* New York: Wiley, 1951.
30. Walas, S. M. *Chemical Process Equipment.* Stoneham, MA: Butterworths, 1988; (*a*) p. 533; (*b*) pp. 543–8.
31. Walton, A. G. *Science* **148**:601 (1965).
32. White, E. T., and A. D. Randolph. *Ind. Eng. Chem. Res.* **28**(3):276 (1989).

Properties and Handling of Particulate Solids

Of all the shapes and sizes that may be found in solids, the most important from a chemical engineering standpoint is the small particle. An understanding of the characteristics of masses of particulate solids is necessary in designing processes and equipment for dealing with streams containing such solids.

CHARACTERIZATION OF SOLID PARTICLES

Individual solid particles are characterized by their size, shape, and density. Particles of homogeneous solids have the same density as the bulk material. Particles obtained by breaking up a composite solid, such as a metal-bearing ore, have various densities, usually different from the density of the bulk material. Size and shape are easily specified for regular particles, such as spheres and cubes, but for irregular particles (such as sand grains or mica flakes) the terms *size* and *shape* are not so clear and must be arbitrarily defined.

Particle shape

As discussed in Chap. 7, the shape of an individual particle is conveniently expressed in terms of the sphericity Φ_s, which is independent of particle size. For a spherical particle of diameter D_p, $\Phi_s = 1$; for a nonspherical particle, the sphericity is defined by the relation

$$\Phi_s \equiv \frac{6v_p}{D_p s_p} \tag{28.1}$$

where D_p = equivalent diameter or nominal diameter of particle
s_p = surface area of one particle
v_p = volume of one particle

The equivalent diameter is sometimes defined as the diameter of a sphere of equal volume. For fine granular materials, however, it is difficult to determine the exact volume and surface area of a particle, and D_p is usually taken to be the nominal size based on screen analyses or microscopic examination. The surface area may be found from adsorption measurements or from the pressure drop in a bed of particles, and Eq. (28.1) used to calculate Φ_s. For many crushed materials Φ_s is between 0.6 and 0.8, as shown in Table 7.1, but for particles rounded by abrasion Φ_s may be as high as 0.95.

Particle size

In general, "diameters" may be specified for any equidimensional particle. Particles that are not equidimensional, that is, that are longer in one direction than in others, are sometimes characterized by the *second*-longest major dimension. By convention, particle sizes are expressed in different units depending on the size range involved. Coarse particles are measured in inches or millimeters; fine particles in terms of screen size; very fine particles in micrometers or nanometers. Ultrafine particles are sometimes described in terms of their surface area per unit mass, usually in square meters per gram.

Mixed particle sizes and size analysis

In a sample of uniform particles of diameter D_p the total volume of the particles is m/ρ_p, where m and ρ_p are the total mass of the sample and the density of the particles, respectively. Since the volume of one particle is v_p, the number of particles in sample N is

$$\Phi_s = \frac{6\,v_p}{D_p\,s_p}$$

$$N = \frac{m}{\rho_p v_p} \tag{28.2}$$

The total surface area of the particles is, from Eqs. (28.1) and (28.2),

$$A = N s_p = \frac{6m}{\Phi_s \rho_p D_p} \tag{28.3}$$

To apply Eqs. (28.2) and (28.3) to mixtures of particles having various sizes and densities, the mixture is sorted into fractions, each of constant density and approximately constant size. Each fraction can then be weighed, or the individual particles in it can be counted or measured by any of a number of methods. Equations (28.2) and (28.3) can then be applied to each fraction and the results added.

Information from such a particle size analysis is tabulated to show the mass or number fraction in each size increment as a function of the average particle size (or size range) in the increment. An analysis tabulated in this way is called a *differential analysis*. The results are often presented as a histogram, as shown in Fig. 28.1a, with a continuous curve like the dashed line used to approximate the distribution. A second way to present the information is through a *cumulative analysis* obtained by adding, consecutively, the individual increments, starting with that containing the smallest particles, and tabulating or plotting the cumulative sums against the maximum particle diameter in the increment. Figure 28.1b is a cumulative analysis plot

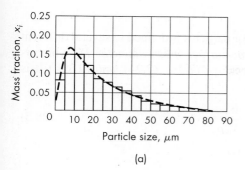
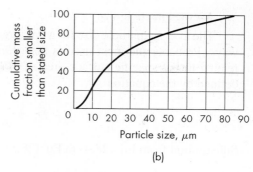

FIGURE 28.1
Particle size distribution for powder: (*a*) differential analysis; (*b*) cumulative analysis.

of the distribution shown in Fig. 28.1*a*. In a cumulative analysis the data may appropriately be represented by a continuous curve.

Cumulative plots may also be made on semilogarithmic paper or, more commonly, on logarithmic probability paper on which the abscissa scale is divided in accordance with a Gaussian probability distribution. Size analyses from a crusher or grinder often give linear plots on such paper, at least over much of the particle size range.

Calculations of average particle size, specific surface area, or particle population of a mixture may be based on either a differential or a cumulative analysis. In principle, methods based on the cumulative analysis are more precise than those based on the differential analysis, since when the cumulative analysis is used, the assumption that all particles in a single fraction are equal in size is not needed. The accuracy of particle size measurements, however, is rarely great enough to warrant the use of the cumulative analysis, and calculations are nearly always based on the differential analysis.

Specific surface of mixture

- If the particle density ρ_p and sphericity Φ_s are known, the surface area of the particles in each fraction may be calculated from Eq. (28.3) and the results for all fractions added to give A_w, the *specific surface* (the total surface area of a unit mass of particles). If ρ_p and Φ_s are constant, A_w is given by

$$A_w = \frac{6x_1}{\Phi_s \rho_p \bar{D}_{p1}} + \frac{6x_2}{\Phi_s \rho_p \bar{D}_{p2}} + \cdots + \frac{6x_n}{\Phi_s \rho_p \bar{D}_{pn}}$$

$$= \frac{6}{\Phi_s \rho_p} \sum_{i=1}^{n} \frac{x_i}{\bar{D}_{pi}} \tag{28.4}$$

where subscripts = individual increments
$\quad\quad\quad\quad\; x_i$ = mass fraction in a given increment
$\quad\quad\quad\quad\; n$ = number of increments
$\quad\quad\quad\quad\; \bar{D}_{pi}$ = average particle diameter, taken as arithmetic average of smallest and largest particle diameters in increment

Average particle size

The average particle size for a mixture of particles is defined in several different ways. Probably the most used is the *volume-surface mean diameter* $\bar{D}_s$, which is related to the specific surface area A_w. [See Eqs. (7.23), (7.24), and (9.39).] It is defined by the equation

$$\bar{D}_s \equiv \frac{6}{\Phi_s A_w \rho_p} \tag{28.5}$$

Substitution from Eq. (28.4) in Eq. (28.5) gives

$$\bar{D}_s = \frac{1}{\sum\limits_{i=1}^{n} (x_i / \bar{D}_{pi})} \tag{28.6}$$

This is the same as Eq. (7.24).

If the number of particles in each fraction N_i is known instead of the mass fraction, $\bar{D}_s$ is given by Eq. (7.23).

Other averages are sometimes useful. The *arithmetic mean diameter* $\bar{D}_N$ is

$$\bar{D}_N = \frac{\sum\limits_{i=1}^{n} (N_i \bar{D}_{pi})}{\sum\limits_{i=1}^{n} N_i} = \frac{\sum\limits_{i=1}^{n} (N_i \bar{D}_{pi})}{N_T} \tag{28.7}$$

where N_T is the number of particles in the entire sample.

The *mass mean diameter* $\bar{D}_w$ is found from the equation

$$\bar{D}_w = \sum_{i=1}^{n} x_i \bar{D}_{pi} \tag{28.8}$$

Dividing the total volume of the sample by the number of particles in the mixture (see below) gives the average volume of a particle. The diameter of such a particle is the *volume mean diameter* $\bar{D}_V$, which is found from the relation

$$\bar{D}_V = \left[\frac{1}{\sum\limits_{i=1}^{n} (x_i / \bar{D}_{pi}^3)} \right]^{1/3} \tag{28.9}$$

For samples consisting of uniform particles these average diameters are, of course, all the same. For mixtures containing particles of various sizes, however, the several average diameters may differ widely from one another.

Number of particles in mixture

To calculate, from the differential analysis, the number of particles in a mixture, Eq. (28.2) is used to compute the number of particles in each fraction, and N_w, the total population in one mass unit of sample, is obtained by summation over all the fractions. For a given particle shape, the volume of any particle is proportional to its "diameter" cubed, or

$$v_p = a D_p^3 \tag{28.10}$$

where a is the *volume shape factor*. From Eq. (28.2), then, assuming that a is independent of size,

$$N_w = \frac{1}{a\rho_p} \sum_{i=1}^{n} \frac{x_i}{\bar{D}_{pi}^3} = \frac{1}{a\rho_p \bar{D}_V^3} \tag{28.11}$$

The specific surface area, the various average diameters, and the number of particles are readily calculated from the particle size analyses through the use of simple computer programs. Many measuring instruments for very fine particles are programmed to report these quantities directly.

Screen analysis; standard screen series

Standard screens are used to measure the size (and size distribution) of particles in the size range between about 3 and 0.0015 in. (76 mm and 38 μm). Testing sieves are made of woven wire screens, the mesh and dimensions of which are carefully standardized. The openings are square. Each screen is identified in meshes per inch. The actual openings are smaller than those corresponding to the mesh numbers, however, because of the thickness of the wires. The characteristics of one common series, the Tyler standard screen series, are given in App. 5. This set of screens is based on the opening of the 200-mesh screen, which is established at 0.074 mm. The area of the openings in any one screen in the series is exactly twice that of the openings in the next-smaller screen. The ratio of the actual mesh dimension of any screen to that of the next-smaller screen is, then, $\sqrt{2} = 1.41$. For closer sizing, intermediate screens are available, each of which has a mesh dimension $\sqrt[4]{2}$, or 1.189, times that of the next-smaller standard screen. Ordinarily these intermediate screens are not used.

In making an analysis, a set of standard screens is arranged serially in a stack, with the smallest mesh at the bottom and the largest at the top. The sample is placed on the top screen and the stack shaken mechanically for a definite time, perhaps 20 min. The particles retained on each screen are removed and weighed, and the masses of the individual screen increments are converted to mass fractions or mass percentages of the total sample. Any particles that pass the finest screen are caught in a pan at the bottom of the stack.

The results of a screen analysis are tabulated to show the mass fraction of each screen increment as a function of the mesh size range of the increment. Since the particles on any one screen are passed by the screen immediately ahead of it, two numbers are needed to specify the size range of an increment, one for the screen through which the fraction passes and the other on which it is retained. Thus, the notation 14/20 means "through 14 mesh and on 20 mesh."

A typical screen analysis is shown in Table 28.1. The first two columns give the mesh size and width of opening of the screens; the third column is the mass fraction of the total sample that is retained on the designated screen. This is x_i, where i is the

TABLE 28.1
Screen analysis

Mesh	Screen opening D_{pi}, mm	Mass fraction retained, x_i	Average particle diameter in increment, $\bar{D}_{pi}$, mm	Cumulative fraction smaller than $\bar{D}_{pi}$
4	4.699	0.0000	—	1.0000
6	3.327	0.0251	4.013	0.9749
8	2.362	0.1250	2.845	0.8499
10	1.651	0.3207	2.007	0.5292
14	1.168	0.2570	1.409	0.2722
20	0.833	0.1590	1.001	0.1132
28	0.589	0.0538	0.711	0.0594
35	0.417	0.0210	0.503	0.0384
48	0.295	0.0102	0.356	0.0282
65	0.208	0.0077	0.252	0.0205
100	0.147	0.0058	0.178	0.0147
150	0.104	0.0041	0.126	0.0106
200	0.074	0.0031	0.089	0.0075
Pan	—	0.0075	0.037	0.0000

number of the screen starting at the bottom of the stack; thus $i = 1$ for the pan, and screen $i + 1$ is the screen immediately above screen i. The symbol D_{pi} means the particle diameter equal to the mesh opening of screen i.

The last two columns in Table 28.1 show the average particle diameter $\bar{D}_{pi}$ in each increment and the cumulative fraction smaller than each value of D_{pi}. In screen analyses cumulative fractions are sometimes written starting at the top of the stack and are expressed as the fraction *larger than* a given size.

Size measurements with fine particles

Dry screening is useful for sizing particles with diameters greater than about 44 μm (325-mesh); wet-screen analysis can be used for diameters down to 10 μm. Particles finer than this can be measured in a variety of ways. Optical microscopy and gravity sedimentation are used with particles 1 to 100 μm in diameter, as is the Coulter counter, a device that measures changes in the resistivity of an electrolyte as it carries the particles one by one through a small orifice. Light-scattering techniques, sedimentation in centrifuges or ultracentrifuges, photon correlation spectroscopy, and electron microscopy are useful with even finer particles.[12a]

EXAMPLE 28.1. The screen analysis shown in Table 28.1 applies to a sample of crushed quartz. The density of the particles is 2,650 kg/m³ (0.00265 g/mm³), and the shape factors are $a = 2$ and $\Phi_s = 0.571$. For the material between 4-mesh and 200-mesh in particle size, calculate (a) A_w in square millimeters per gram and N_w in particles per gram, (b) $\bar{D}_V$, (c) $\bar{D}_s$, (d) $\bar{D}_w$, and (e) N_i for the 150/200-mesh increment. (f) What fraction of the total number of particles is in the 150/200-mesh increment?

Solution. To find A_w and N_w, Eq. (28.4) can be written

$$A_w = \frac{6}{0.571 \times 0.00265} \sum \frac{x_i}{\bar{D}_{pi}} = 3,965 \sum \frac{x_i}{\bar{D}_{pi}}$$

and Eq. (28.11) as

$$N_w = \frac{1}{2 \times 0.00265} \sum \frac{x_i}{\bar{D}_{pi}^3} = 188.7 \sum \frac{x_i}{\bar{D}_{pi}^3}$$

(a) For the 4/6-mesh increment $\bar{D}_{pi}$ is the arithmetic mean of the mesh openings of the defining screens; or, from Table 28.1, $(4.699 + 3.327)/2 = 4.013$ mm. For this increment $x_i = 0.0251$; hence $x_i/\bar{D}_{pi} = 0.0251/4.013 = 0.0063$ and $x_i \bar{D}_{pi}^3 = 0.0004$. Corresponding quantities are calculated for the other 11 increments and summed to give $\sum x_i/\bar{D}_{pi} = 0.8284$ and $\sum x_i/\bar{D}_{pi}^3 = 8.8296$. Since the pan fraction is excluded, the specific surface and number of particles per unit mass of particles 200-mesh or larger are found by dividing the results from Eqs. (28.4) and (28.11) by $1 - x_1$ (since $i = 1$ for the pan), or $1 - 0.0075 = 0.9925$. Then

$$A_w = \frac{3,965 \times 0.8284}{0.9925} = 3,309 \text{ mm}^2/\text{g}$$

$$N_w = \frac{188.7 \times 8.8296}{0.9925} = 1,679 \text{ particles/g}$$

(b) From Eq. (28.9),

$$\bar{D}_V = \frac{1}{8.8296^{1/3}} = 0.4238 \text{ mm}$$

(c) The volume-surface mean diameter is found from Eq. (28.6):

$$\bar{D}_s = \frac{1}{0.8284} = 1.207 \text{ mm}$$

(d) Mass mean diameter $\bar{D}_w$ is obtained from Eq. (28.8). For this, from the data in Table 28.1,

$$\sum x_i \bar{D}_{pi} = \bar{D}_w = 1.677 \text{ mm}$$

(e) The number of particles in the 150/200-mesh increment is found from Eq. (28.11):

$$N_2 = \frac{x_2}{a\rho_p \bar{D}_{p2}^3} = \frac{0.0031}{2 \times 0.00265 \times 0.089^3}$$

$$= 836 \text{ particles/g}$$

This is $836/1,679 = 0.498$, or 49.8 percent of the particles in the top 12 increments. For the material in the pan fraction, the number of particles and specific surface area are enormously greater than for the coarser material, but they cannot be accurately estimated from the data in Table 28.1.

PROPERTIES OF MASSES OF PARTICLES

Masses of solid particles, especially when the particles are dry and not sticky, have many of the properties of a fluid. They exert pressure on the sides and walls of a container; they flow through openings or down a chute. They differ from liquids and gases in several ways, however, because the particles interlock under pressure and cannot slide over one another until the applied force reaches an appreciable magnitude. Unlike most fluids, granular solids and solid masses permanently resist distortion when subjected to a moderate distorting force. When the force is large enough, failure occurs and one layer of particles slides over another; but between the layers on each side of the failure there is appreciable friction.

Solid masses have the following distinctive properties:

1. The pressure is not the same in all directions. In general, a pressure applied in one direction creates some pressure in other directions, but it is always smaller than the applied pressure. It is a minimum in the direction at right angles to the applied pressure. In a homogeneous mass the ratio of the normal pressure to the applied pressure is a constant K', which is characteristic of the material. It depends on the shape and interlocking tendencies of the particles, on the stickiness of the grain surfaces, and on how tightly the material is packed. It is nearly independent of particle size until the grains become very small and the material is no longer free-flowing.
2. A shear stress applied at the surface of a mass is transmitted throughout a static mass of particles unless failure occurs.
3. The density of the mass may vary, depending on the degree of packing of the grains. The density of a fluid is a unique function of temperature and pressure, as is that of each individual solid particle; but the bulk density of the mass is not. The bulk density is a minimum when the mass is "loose." It rises to a maximum when the mass is packed by vibrating or tamping.
4. Before a mass of tightly packed particles can flow, it must increase in volume to permit interlocking grains to move past one another. Without such dilation flow is not possible.
5. When granular solids are piled up on a flat surface, the sides of the pile are at a definite reproducible angle with the horizontal. This angle, α_r, is called the *angle of repose* of the material. For free-flowing granular solids α_r is often between 15° and 30°.

Depending on their flow properties, particulate solids are divided into two classes, *cohesive* and *noncohesive*. Noncohesive materials such as grain, dry sand, and plastic chips flow freely out of a storage bin or silo. For these solids K' is often between 0.35 and 0.6. Cohesive solids, such as wet clay, are characterized by their reluctance to flow through openings. For them, the value of K' approaches zero.

Storage and Conveying of Solids

Bulk storage

Coarse solids such as gravel and coal are stored outside in large piles, unprotected from the weather. When hundreds or thousands of tons of material are involved, this is the most economical method. The solids are removed from the pile by dragline or tractor shovel and delivered to a conveyor or to the process. Outdoor

storage can lead to environmental problems such as dusting or leaching of soluble material from the pile. Dusting may necessitate a protective cover of some kind for the stored solid; leaching can be controlled by covering the pile or by locating it in a shallow basin with an impervious floor from which the runoff may be safely withdrawn.

Bin storage

Solids that are too valuable or too soluble to expose in outdoor piles are stored in bins, hoppers, or silos. These are cylindrical or rectangular vessels of concrete or metal. A silo is tall and relatively small in diameter; a bin is not so tall and usually fairly wide. A hopper is a small vessel with a sloping bottom, for temporary storage before feeding solids to a process. All these containers are loaded from the top by some kind of elevator; discharging is ordinarily done from the bottom. As discussed later, a major problem in bin design is to provide satisfactory discharge.

Pressures in bins and silos. When granular solids are placed in a bin or silo, the lateral pressure exerted on the walls at any point is less than predicted from the head of material above that point. Furthermore there usually is friction between the wall and the solid grains, and because of the interlocking of the particles, the effect of this friction is felt throughout the mass. The friction force at the wall tends to offset the weight of the solid and reduces the pressure exerted by the mass on the floor of the container.

The vertical pressure on the vessel floor or the packing support is much smaller than that exerted by a column of liquid of the same density and height. The actual pressure from the solids depends on the value of K' for the solids, the coefficient of friction between the solids and the vessel wall, and the way the solids are placed in the vessel. In general, when the height of the solids column is greater than about 3 times the diameter of the container, additional solids have no effect on the pressure at the base. The total mass, of course, increases if more solids are added, but the additional mass is carried by the walls and foundation, not by the floor of the vessel.

In granular solids a high pressure does not always increase the tendency of the material to flow, as it does in a liquid; instead, increased pressure packs the grains more tightly together and makes flow more difficult. In extreme cases the combination of gravitational and friction forces at some point in the container causes the solids to arch or bridge, so that they do not fall even when the material below them is removed. Nearly all large bins contain an *archbreakr*, an upward-pointing shallow metal cone set near the bottom to keep the solids at the discharge opening from becoming tightly packed. Granular solids, especially with angular particles, must be loose in order to flow.

Flow out of bins

Solids tend to flow out of any opening near the bottom of a bin but are best discharged through an opening in the floor. Flow through a side opening tends to be uncertain and increases the lateral pressure on the other side of the bin while the solids are flowing. A bottom outlet is less likely to clog and does not induce abnormally high pressures on the wall at any point.

Factors influencing the flow of solids out of bins have been studied by Jenike et al.[5] When the outlet at the bottom of a bin containing free-flowing solids is opened, the material immediately above the opening begins to flow. One of two flow patterns will develop, depending on the steepness of the walls in the bottom section of the bin and on the coefficient of friction between the solids and the bin walls.[9b] *Mass flow* occurs in cone-bottomed bins with a tall, steep cone; all the material moves downward uniformly from the top of the bin. *Tunnel flow* develops in bins with a shallow cone angle or with vertical walls and a central opening in the floor. Here a vertical column of solids above the opening moves downward without disturbing the material at the sides. Eventually lateral flow begins, first from the topmost layer of solids. A conical depression is formed in the surface of the mass. The solids at the bin floor, at or near the walls, are the last to leave. The material slides laterally into the central column at an angle approximating the angle of internal friction of the solids. If additional material is added at the top of the bin at the same rate as material is flowing out the bottom, the solids near the bin walls remain stagnant and do not discharge, no matter how long flow persists.

The rate of flow of granular solids by gravity through a circular opening in the bottom of a bin depends on the diameter of the opening and on the properties of the solid. Within wide limits it does not depend on the height of the bed of solids. With free-flowing particles the rate of solids flow $\dot{m}$ varies approximately with D_o^3, where D_o is the diameter of the discharge opening.[9,15]

With cohesive solids it is often hard to start flow. Once flow does start, however, it again begins in the material directly above the discharge opening. Frequently the column of solids above the outlet moves out as a plug, leaving a "rathole" with nearly vertical sides. Sticky solids and even some dry powders adhere strongly to vertical surfaces and have enough shear strength to support a plug of considerable diameter above an open discharge. Thus to get flow started and to keep the material moving, vibrators on the bin walls, internal plows near the bin floor, or jets of air in the discharge opening are often needed.

The discharge opening should be small enough to be readily closed when solids are flowing, yet not so small that it will clog. It is best to make the opening large enough to pass the full desired flow when half open. It can then be opened further to clear a partial choke. If the opening is too large, however, the shutoff valve may be hard to close and control of the flow rate will be poor.

Conveyors

Pneumatic conveyors for solids are described in Chap. 7. Other common devices for transportation include belt conveyors and bucket elevators, closed-belt conveyors with zipperlike fasteners, and various kinds of drag and flight conveyors. These all include a return leg that carries the empty belt or chain back from the discharge to the loading point. Vibrating conveyors and screw conveyors have no return leg but only operate over relatively short distances. Pneumatic conveyors also have no return leg and are not so limited as to distance of travel. Conveyors for solids are discussed in Ref. 12*i*.

MIXING OF SOLIDS

The mixing of solids, whether free-flowing or cohesive, resembles to some extent the mixing of low-viscosity liquids. Both processes intermingle two or more separate components to form a more or less uniform product. Some of the equipment normally used for blending liquids may, on occasion, be used to mix solids.

Yet there are significant differences between the two processes. Liquid blending depends on the creation of flow currents, which transport unmixed material to the mixing zone adjacent to the impeller. In heavy pastes or masses of particulate solids no such currents are possible, and mixing is accomplished by other means. In consequence, much more power is normally required in mixing pastes and dry solids than in blending liquids.

Another difference is that in blending liquids a "well-mixed" product usually means a truly homogeneous liquid phase, from which random samples, even of very small size, all have the same composition. In mixing pastes and powders the product often consists of two or more easily identifiable phases, each of which may contain individual particles of considerable size. From a "well-mixed" product of this kind small random samples will differ markedly in composition; in fact, samples from any given such mixture must be larger than a certain critical size (several times the size of the largest individual particle in the mix) if the results are to be significant.

Measures of mixer performance

Mixing is harder to define with solids and pastes than it is with liquids. Quantitative measures of mixing, based on statistical procedures, are sometimes used to evaluate mixer performance. These procedures are based on analyses of spot samples taken from the mix at various times. A mixture in which one component is randomly distributed through another is said to be completely mixed.

For granular noncohesive solids, multiple small samples containing about the same number of particles are used. Consider a mixture of components A and B from which N spot samples, each containing n particles, are taken and analyzed. The standard deviations s is estimated from the analytical results by the equation

$$s = \sqrt{\frac{\sum\limits_{i=1}^{N}(x_i - \bar{x})^2}{N - 1}} = \sqrt{\frac{\sum\limits_{i=1}^{N}x_i^2 - \bar{x}\sum\limits_{i=1}^{N}x_i}{N - 1}} \tag{28.12}$$

where x_i = number fraction of A in each sample
$\bar{x}$ = average value of measured number fractions

Even if the mixture is completely mixed, the value of x_i in the various spot samples will not be the same; there is always some chance that a sample drawn from a random mixture will contain a larger (or smaller) proportion of one kind of particle than the population from which it was taken. The theoretical standard deviation σ_e for a completely random mixture is given by

$$\sigma_e = \sqrt{\frac{\mu_p(1 - \mu_p)}{n}} \tag{28.13}$$

For cohesive solids, mass fractions instead of number fractions are used. The standard deviation of the spot analyses is estimated as before from Eq. (28.12). The standard deviation for the mixture σ_0, before mixing begins, is

$$\sigma_0 = \sqrt{\mu(1 - \mu)} \qquad (28.14)$$

where μ is the overall mass fraction of component A in the mix.

Sometimes s is used directly as a measure of the degree of mixing, but more often the standard deviation (or its square, the variance) is compared with the theoretical standard deviations or variances found from Eqs. (28.13) and (28.14).[6,10,14]

In actual practice, the proof of a mixer is in the properties of the mixed material it produces. A well-mixed product is one that does what is required and has the necessary property—visual uniformity, high strength, uniform burning rate, or other desired characteristic. A good mixer is one that produces this well-mixed product at the lowest overall cost.

Mixing heavy pastes, plastic solids, and rubber is more of an art than a science. The properties of the materials to be mixed vary enormously from one process to another. Even in a single material they may be widely different at various times during the mixing operation. A batch may start as a dry, free-flowing powder, become pasty on the addition of liquid, stiff and gummy as a reaction proceeds, and then perhaps dry, granular, and free-flowing once more. Indeterminate properties of the material such as stiffness, tackiness, and wettability are as significant in these mixing problems as viscosity and density. Mixers for pastes and plastic masses must, above all, be versatile. In a given problem the mixer chosen must handle the material when in its worst condition and may not be so effective as other designs during other parts of the mixing cycle. As with other equipment, the choice of a mixer for heavy materials is often a compromise.

Mixers for Noncohesive Solids

Mixers for dry powders include some machines that are also used for heavy pastes and some machines that are restricted to free-flowing powders. Mixing is by slow-speed agitation of the mass with an impeller, by tumbling, or by centrifugal smearing and impact. These mixers are of fairly light construction, and their power consumption per unit mass of material mixed is moderate.

Ribbon blenders

A ribbon blender consists of a horizontal trough containing a central shaft and a helical ribbon agitator. A typical mixer is shown in Fig. 28.2. Two counteracting ribbons are mounted on the same shaft, one moving the solid slowly in one direction, the other moving it quickly in the other. The ribbons may be continuous or interrupted. Mixing results from the "turbulence" induced by the counteracting agitators, not from mere motion of the solids through the trough. Some ribbon blenders operate batchwise, with the solids charged and mixed until satisfactory; others mix continuously,

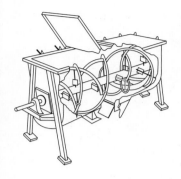

FIGURE 28.2
Ribbon mixer.

with solids fed in one end of the trough and discharged from the other. The trough is open or lightly covered for light duty and closed and heavy-walled for operation under pressure or vacuum. Ribbon blenders are effective mixers for thin pastes and for powders that do not flow readily. Some batch units are very large, holding up to 34 m³ (9,000 gal) of material. The power they require is moderate.

Internal screw mixers

Free-flowing grains and other light solids are often mixed in a vertical tank containing a helical conveyor that elevates and circulates the material. Many different designs are commercially available. In one common type the double-motion helix orbits about the central axis of a conical vessel, visiting all parts of the mix. Mixing is generally slower than in ribbon blenders, but the power required is somewhat less.

Tumbling mixers

Many materials are mixed by tumbling them in a partly filled container rotating about a horizontal axis. The ball mills described later in this chapter are often used as mixers. Most tumbling mixers, however, do not contain grinding elements. Tumbling barrels, for example, resemble ball mills without the balls; they effectively mix suspensions of dense solids in liquids and heavy dry powders. Other tumbling blenders, such as those illustrated in Fig. 28.3, handle lighter dry solids only. The double-cone mixer shown at (*a*) is a popular mixer for free-flowing dry powders. A batch is charged into the body of the machine from above until it is 50 to 60 percent full. The ends of the container are closed and the solids tumbled for 5 to 20 min. The machine is stopped; mixed material is dropped out the bottom of the container into a conveyor or bin. The twin-shell blender shown at (*b*) is made from two cylinders joined to form a V and rotated about a horizontal axis. Like a double-cone blender, it may contain internal sprays for introducing small amounts of liquid into the mix or mechanically driven devices for breaking up agglomerates of solids. Twin-shell blenders are more effective in some blending operations than double-cone blenders. Tumbling mixers are made in a wide range of sizes and materials of construction. They draw a little less power, ordinarily, than ribbon blenders.

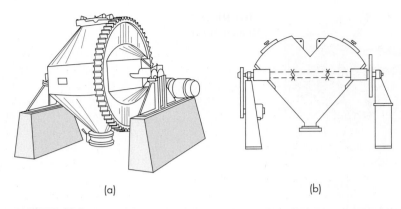

(a)   (b)

FIGURE 28.3
Tumbler mixers: (*a*) double-cone mixer; (*b*) twin-shell blender.

One method of scaling up such mixers is based upon keeping the Froude number $n^2 L/g$ constant, where L is a characteristic length of the equipment, n is the rate of rotation, and g is the gravitational acceleration.[4] Scaleup procedures for tumbling mixers are discussed in detail by Wang and Fan.[17]

Mixing in tumbling blenders is initially rapid, but never complete. In this type of mixer the components are never blended in a completely random manner. After a time the quality of the mix levels off, fluctuates, and may even decrease. Unblending forces, usually electrostatic, are always at work in a dry solids blender, and their effects are especially noticeable here. These forces often prevent the mix from becoming completely blended; when the mixing time is long, they may lead to a considerable degree of unmixing and segregation.[18]

Impact wheels

Fine, light powders such as insecticides may be blended continuously by spreading them out in a thin layer under centrifugal action. A premix of the several dry ingredients is fed continuously near the center of a high-speed spinning disk 250 to 700 mm (10 to 27 in.) in diameter, which throws it outward into a stationary casing. The intense shearing forces acting on the powders during their travel over the disk surface thoroughly blend the various materials. The disk in some machines is vertical; in others it is horizontal. The attrition mill shown in Fig. 28.12 is an effective mixer of this type. In some devices, designed for mixing and not size reduction, the premix is dropped onto a horizontal double rotor carrying short vertical pins near its periphery to increase the mixing effectiveness. A 350-mm (14-in.) disk turns at 1,750 rpm for easy problems and 3,500 rpm for materials that are hard to mix. Sometimes several passes through the same machine or through machines in series are necessary. For good results the premix fed to an impact wheel must be fairly uniform, for there is almost no holdup of material in the mixer and no chance for recombining material that has passed through with that which is entering. Impact wheels blend 1 to 25 tons/h of light free-flowing powders.

Mixers for Cohesive Solids

Some of the most difficult of all mixing problems involve cohesive solids such as pastes, plastic materials, and rubber. In some ways these substances resemble liquids, but their enormously high viscosity means that the mixing equipment must be different from and much more powerful than the mixers described in Chap. 9. With cohesive solids the mixing elements cannot generate flow currents; instead they shear, fold, stretch, and compress the material to be mixed. The mechanical energy is applied by moving parts directly to the mass of material. In the closed type, such as Banbury mixers, the inner wall of the casing acts as part of the mixing means, and all mixing action occurs close to the moving parts. Clearances between mixing arms, rotors, and wall of casing are small. The forces generated in these mixers are large, the machinery must be ruggedly built, and the power consumption is high. The heat evolved per unit mass of material may be sufficient to require cooling to prevent the temperature from reaching a level dangerous to the equipment or the material.

Change-can mixers

These devices blend viscous liquids or light pastes, as in food processing or paint manufacture. A small removable can of 5 to 100 gal in size holds the material to be mixed. In the *pony mixer* shown in Fig. 28.4a the agitator consists of several vertical blades or fingers held on a rotating head and positioned near the wall of the can. The blades are slightly twisted. The agitator is mounted eccentrically with respect to the axis of the can. The can rests on a turntable driven in a direction opposite to that of the agitator, so that during operation all the liquid or paste in the can is brought to the blades to be mixed. When the mixing is complete, the agitator head is raised, lifting the blades out of the can; the blades are wiped clean; and the can is replaced with another containing a new batch.

In the beater mixer in Fig. 28.4b the can or vessel is stationary. The agitator has a planetary motion; as it rotates, it precesses, so that it repeatedly visits all parts of the vessel. Beaters are shaped to pass with close clearance over the side and bottom of the mixing steel

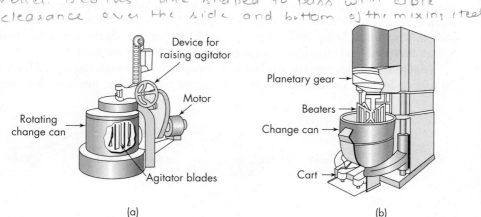

Device for raising agitator

Motor

Planetary gear →

Rotating change can →

Beaters →

Change can →

Agitator blades

Cart →

(a)

(b)

FIGURE 28.4
Double-motion paste mixers: (*a*) pony mixer; (*b*) beater mixer.

the vessel. Beaters are shaped to pass with close clearance over the side and bottom of the mixing vessel.

Kneaders, dispersers, and masticators

Kneading is a method of mixing used with deformable or plastic solids. It involves squashing the mass flat, folding it over on itself, and squashing it once more. Most kneading machines also tear the mass apart and shear it between a moving blade and a stationary surface. Considerable energy is required even with fairly thin materials, and as the mass becomes stiff and rubbery, the power requirements become very large.

A *two-arm kneader* handles suspensions, pastes, and light plastic masses. Typical applications are in the compounding of lacquer bases from pigments and carriers and in shredding cotton linters into acetic acid and acetic anhydride to form cellulose acetate. A *disperser* is heavier in construction and draws more power than a kneader; it works additives and coloring agents into stiff materials. A *masticator* is still heavier and draws even more power. It can disintegrate scrap rubber and compound the toughest plastic masses that can be worked at all. Masticators are often called *intensive mixers*.

In all these machines the mixing is done by two heavy blades on parallel horizontal shafts turning in a short trough with a saddle-shaped bottom. The blades turn toward each other at the top, drawing the mass downward over the point of the saddle, then shearing it between the blades and the wall of the trough. The circles of rotation of the blades are usually tangential, so that the blades may turn at different speeds in any desired ratio. The optimum ratio is about $1\frac{1}{2}:1$. In some machines the blades overlap and turn at the same speed or with a speed ratio of 2:1.

Designs of mixing blades for various purposes are shown in Fig. 28.5. The common sigma blade shown at the left is used for general-purpose kneading. Its edges may be serrated to give a shredding action. The double-naben, or fishtail, blade in the center is particularly effective with heavy plastic materials. The dispersion blade at the right develops the high shear forces needed to disperse powders or liquids into plastic or rubbery masses. Masticator blades are even heavier than those shown, sometimes being little larger in diameter than the shafts that drive them. Spiral, flattened, and elliptical designs of masticator blades are used.

Material to be kneaded or worked is dropped into the trough and mixed for 5 to 20 min or longer. Sometimes the mass is heated while in the machine, but more

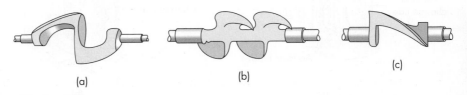

FIGURE 28.5
Kneader and disperser blades: (*a*) sigma blade; (*b*) double-naben blade; (*c*) disperser blade.

commonly it must be cooled to remove the heat generated by the mixing action. The trough is often unloaded by tilting it so that its contents spill out. In kneaders and some dispersers only one agitator blade is directly driven; the other is turned by timing gears. In masticators both shafts are independently driven, sometimes from both ends, so that the trough cannot be tilted and must be unloaded through an opening in the floor.

In many kneading machines the trough is open, but in some designs, known as *internal mixers,* the mixing chamber is closed during the operating cycle with a cover, the underside of which conforms to the volume swept out by the blades. Such mixers do not tilt. They are used for dissolving rubber and for making dispersions of rubber in liquids. The most common internal mixer is the *Banbury mixer,* shown in Fig. 28.6. This is a heavy-duty two-arm mixer in which the agitators are in the form of interrupted spirals. The shafts turn at 30 to 40 rpm. Solids are charged in from above and held in the trough during mixing by an air-operated piston under a pressure of 1 to 10 atm. Mixed material is discharged through a sliding door in the bottom of the trough. Banbury mixers compound rubber and plastic solids, masticate crude rubber, devulcanize rubber scrap, and make water dispersions and rubber solutions. They also accomplish the same tasks as kneaders but in a shorter time and with smaller batches. The heat generated in the material is removed by cooling water sprayed on the walls of the mixing chamber and circulated through the hollow agitator shafts.

Continuous kneaders

The machines just described operate batchwise on relatively small amounts of material. The more difficult the material is to mix, the smaller the batch size must be. Many industrial processes are continuous, with steady uniform flow into and out of units of equipment; into such processes batch equipment is not readily incorporated. Continuous kneading machines have been developed that can handle light to fairly heavy materials. In a typical design a single horizontal shaft slowly turning in a mixing chamber carries rows of teeth arranged in a spiral pattern to move the material through the chamber. The teeth on the rotor pass with close clearance between stationary teeth set in the wall of the casing. The shaft turns and also reciprocates in the axial direction. Material between the meshing teeth is therefore smeared in an axial or longitudinal direction as well as being subjected to radial shear. Solids enter the machine near the driven end of the rotor and discharge through an opening surrounding the shaft bearing in the opposite end of the mixing chamber. The chamber is an open trough with light solids, a closed cylinder with plastic masses. These machines can mix several tons per hour of heavy, stiff, or gummy materials.

Mixer-extruders

If the discharge opening of a continuous kneader is restricted by covering it with an extrusion die, the pitched blades of the rotor build up considerable pressure in the material. The mix is cut and folded while in the mixing chamber and subjected to additional shear as it flows through the die. Other mixer-extruders function

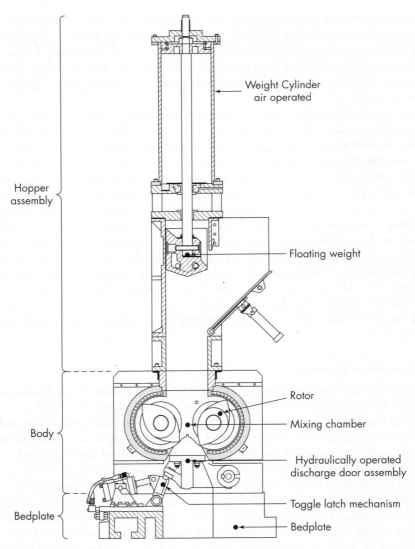

FIGURE 28.6
Banbury internal mixer (*Farrel Corp., Ansonia, CT.*)

in the same way. They contain one or two horizontal shafts, rotating but not reciprocating, carrying a helix or blades set in a helical pattern. Pressure is built up by reducing the pitch of the helix near the discharge, reducing the diameter of the mixing chamber, or reducing both. Mixer-extruders continuously mix, compound, and work thermoplastics, doughs, clays, and other hard-to-mix materials. Some also carry a heating jacket and vapor discharge connections to permit removal of water or solvent from the material as it is being processed.

Mixing rolls

Another way of subjecting pastes and deformable solids to intense shear is to pass them between smooth metal rolls turning at different speeds. By repeated passes between such mixing rolls, solid additives can be thoroughly dispersed into pasty or stiff plastic materials. Continuous mills for mixing pastes contain three to five horizontal rolls set one above the other in a vertical stack; the paste passes from the slower rolls to successively faster ones. Rubber products and pastes can be compounded on batch roll mills with two rolls set in the same horizontal plane. Solids are picked up on the faster roll, cut at an angle by the operator, and folded back into the "bite" between the rolls. Additives are sprinkled on the material as it is being worked. Batch roll mills require long mixing times and careful attention by the operator and have largely been displaced by internal mixers and continuous kneaders.

Muller mixers

A muller gives a distinctly different mixing action from that of other machines. Mulling is a smearing or rubbing action similar to that in a mortar and pestle. In large-scale processing this action is given by the wide, heavy wheels of the mixer shown in Fig. 28.7. In this particular design of muller the pan is stationary, and the central vertical shaft is driven, causing the muller wheels to roll in a circular path over a layer of solids on the pan floor. The rubbing action results from the slip of the wheels on the solids. Plows guide the solids under the muller wheels or to an opening in the pan floor at the end of the cycle when the mixer is being discharged. In another design the axis of the wheels is held stationary, and the pan is rotated; in still another the wheels are not centered in the pan but are offset, and both the pan and the wheels are driven. Mixing plows may be substituted for the muller wheels to give what is called a *pan mixer*. Mullers are good mixers for batches of heavy solids and pastes; they are especially effective in uniformly coating the particles of granular solid with a small amount of liquid. Continuous muller mixers with two mixing pans connected in series are also available.

Pugmills

In a pugmill the mixing is done by blades or knives set in a helical pattern on a horizontal shaft turning in an open trough or closed cylinder. Solids continuously enter one end of the mixing chamber and discharge from the other. While in the

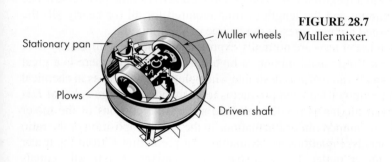

Stationary pan

Plows

Muller wheels

Driven shaft

FIGURE 28.7
Muller mixer.

chamber, they are cut, mixed, and moved forward to be acted upon by each succeeding blade. Single-shaft mills utilize an enclosed mixing chamber; open-trough
double-shaft mills are used where more rapid or more thorough mixing is required.
The chamber of most enclosed mills is cylindrical, but in some it is polygonal in
cross section to prevent sticky solids from being carried around with the shaft. Pug-
mills blend and homogenize clays, break up agglomerates in plastic solids, and mix
liquids with solids to form thick, heavy slurries. Sometimes they operate under vacuum to remove air from clay or other materials. They are built with jackets for heating or cooling.

Mixing effectiveness

The performance of a mixer for cohesive solids is judged by the time required,
the power load, and the properties of the product. These criteria vary widely from
one problem to another: Sometimes a very high degree of uniformity is required, at
other times rapid mixing, and at still other times a minimum amount of power.

As discussed earlier, the degree of mixing may be measured by analyzing spot
samples and comparing the estimated standard deviation s with the standard deviation at zero mixing σ_0. Studies have shown[10] that the degree of mixing in two-arm
kneaders and muller mixers rises rapidly at first, then levels off at a value characteristic of the material, with sandy, granular solids mixed best and heavy sticky
pastes much more poorly. In contrast, some continuous kneaders mix plastic materials more effectively than they do granular free-flowing solids.

Axial mixing

In the helical element mixer described in Chap. 9, page 264, two fluids are
being mixed radially at any given cross section, but there is little mixing in an axial
or longitudinal direction. The fluid behavior approximates that in plug flow, in
which there is no axial mixing whatever. In some continuous paste mixers there is
also little axial mixing, a desirable characteristic in certain mixing operations or
chemical reactions; in others the axial mixing may be significant.

In paste mixers the degree of axial mixing is measured by the injection of a
tracer, over a very short time, into the feed, followed by monitoring the concentration of tracer in the outlet stream. Typically the tracer appears at the outlet a little
earlier than expected from the mean residence time of the mixer contents. Its outlet
concentration rises to a maximum, then decays toward zero as time progresses. The
height of the maximum and the length of time required for all (or nearly all) the
tracer to be discharged are measures of the degree of axial mixing.

Results of such tracer tests are normally expressed in terms of a diffusivity E. A
low diffusivity means little axial mixing; a high diffusivity means there is a great
deal. Obviously a small value of E is desirable when plug flow is best, as in chemical
reactors in which mixing of feed and product is to be avoided. A large value of E is
desirable when axial mixing is needed to blend successive portions of the mixer
feed, for example, to dampen minor fluctuations in the feed composition or the ratio
of the feed components. Equations are available[16] for predicting E from the tracer
time data at the mixer outlet. For two-shaft paddle mixers E typically equals

0.02UL to 0.2UL, where U is the longitudinal velocity of the material in the mixer and L is the mixer length.

The ratio UL/E is known as the *Peclet number* Pe. Thus for paddle mixers Pe ranges from 5 to 50. With some agitator designs Pe is large and falls as the rotor speed increases; with other designs it is small and virtually independent of rotor speed.[16]

SIZE REDUCTION

The term *size reduction* is applied to all the ways in which particles of solids are cut or broken into smaller pieces. Throughout the process industries, solids are reduced by different methods for different purposes. Chunks of crude ore are crushed to workable size; synthetic chemicals are ground into powder; sheets of plastic are cut into tiny cubes or diamonds. Commercial products must often meet stringent specifications regarding the size and sometimes the shape of the particles they contain. Reducing the particle size also increases the reactivity of solids; it permits separation of unwanted ingredients by mechanical methods; it reduces the bulk of fibrous materials for easier handling and for waste disposal.

Solids may be broken in many different ways, but only four are commonly used in size-reduction machines: (1) compression, (2) impact, (3) attrition, or rubbing, and (4) cutting. A nutcracker, a hammer, a file, and a pair of shears exemplify these four types of action. Sometimes size reduction results from the attrition of a particle by one or more other particles or from intense shear in the supporting fluid. In general, compression is used for coarse reduction of hard solids, to give relatively few fines; impact gives coarse, medium, or fine products; attrition yields very fine products from soft, nonabrasive materials. Cutting gives a definite particle size and sometimes a definite shape, with few or no fines.

Criteria for comminution

Comminution is a generic term for size reduction; crushers and grinders are types of comminuting equipment. An ideal crusher or grinder would (1) have a large capacity, (2) require a small power input per unit of product, and (3) yield a product of the single size or the size distribution desired. The usual method of studying the performance of process equipment is to set up an ideal operation as a standard, compare the characteristics of the actual equipment with those of the ideal unit, and account for the difference between the two. When this method is applied to crushing and grinding equipment, the differences between the ideal and the actual are very great, and despite extensive study the gaps have not been completely accounted for. On the other hand, useful empirical equations for predicting equipment performance have been developed from the incomplete theory now at hand.

Characteristics of comminuted products

The objective of crushing and grinding is to produce small particles from larger ones. Smaller particles are desired either because of their large surface or because of their shape, size, and number. One measure of the efficiency of the operation is based on the energy required to create new surface, since the surface area of a unit mass of particles increases greatly as the particle size is reduced.

Unlike an ideal crusher or grinder, an actual unit does not yield a uniform product, whether the feed is uniformly sized or not. The product always consists of a mixture of particles, ranging from a definite maximum size to very small particles. Some machines, especially in the grinder class, are designed to control the magnitude of the largest particles in their products, but the fine sizes are not under control. In some types of grinders, fines are minimized, but they are not eliminated. If the feed is homogeneous, both in the shapes of the particles and in chemical and physical structure, then the shapes of the individual units in the product may be quite uniform; otherwise, the grains in the various sizes of a single product may differ considerably in shape.

The ratio of the diameters of the largest and smallest particles in a comminuted product is on the order of 10^4. Because of this extreme variation in the sizes of the individual particles, relationships adequate for uniform sizes must be modified when applied to such mixtures. The term *average size,* for example, is meaningless until the method of averaging is defined, and as discussed earlier in this chapter, several different average sizes can be calculated.

Unless they are smoothed by abrasion after crushing, comminuted particles resemble polyhedrons with nearly plane faces and sharp edges and corners. The particles may be compact, with length, breadth, and thickness nearly equal; or they may be platelike or needlelike.

Energy and power requirements in comminution[5]

The cost of power is a major expense in crushing and grinding, so the factors that control this cost are important. During size reduction, the particles of feed material are first distorted and strained. The work necessary to strain them is stored temporarily in the solid as mechanical energy of stress, just as mechanical energy can be stored in a coiled spring. As additional force is applied to the stressed particles, they are distorted beyond their ultimate strength and suddenly rupture into fragments. New surface is generated. Since a unit area of solid has a definite amount of surface energy, the creation of new surface requires work, which is supplied by the release of energy of stress when the particle breaks. By conservation of energy, all energy of stress in excess of the new surface energy created must appear as heat.

Efficiency. Size reduction is one of the least energy-efficient of all the unit operations. Laboratory studies of crushing have shown that less than 1 percent of the energy delivered to the solids is used to create new surface; the rest is dissipated as heat. In operating machines energy must also be supplied to overcome friction in the bearings and other moving parts. The mechanical efficiency, the ratio of the energy delivered to the solids to the total energy input to the machine, ranges from 25 to 60 percent.[12b]

Crushing laws and work index

Crushing laws proposed many years ago by Rittinger and Kick have been shown to apply only over a very limited range of conditions. A more realistic way of estimating the power required for crushing and grinding was proposed by Bond.[3] Bond postulated that the work required to form particles of size D_p from very large feed is proportional to the square root of the surface-to-volume ratio of the product s_p/v_p. By Eq. (28.1), $s_p/v_p = 6/\Phi_s D_p$, from which it follows that

$$\frac{P}{\dot{m}} = \frac{K_b}{\sqrt{D_p}} \tag{28.15}$$

where K_b is a constant that depends on the type of machine and on the material being crushed. To use Eq. (28.15), a work index W_i is defined as the gross energy requirement in kilowatthours per ton (2,000 lb) of feed needed to reduce a very large feed to such a size that 80 percent of the product passes a 100-μm screen. This definition leads to a relation between K_b and W_i. If D_p is in millimeters, P in kilowatts, and $\dot{m}$ in tons per hour,

$$K_b = \sqrt{100 \times 10^{-3}} \ W_i = 0.3162 W_i \tag{28.16}$$

If 80 percent of the feed passes a mesh size of D_{pa} mm and 80 percent of the product a mesh of D_{pb} mm, it follows from Eqs. (28.15) and (28.16) that

$$\frac{P}{\dot{m}} = 0.3162 W_i \left(\frac{1}{\sqrt{D_{pb}}} - \frac{1}{\sqrt{D_{pa}}} \right) \tag{28.17}$$

The work index includes the friction in the crusher, and the power given by Eq. (28.17) is gross power.

Table 28.2 gives typical work indices for some common minerals. These data do not vary greatly among different machines of the same general type and apply to

TABLE 28.2
Work indices for dry crushing[†] or wet grinding[‡]

Material	Specific gravity	Work index W_i
Bauxite	2.20	8.78
Cement clinker	3.15	13.45
Cement raw material	2.67	10.51
Clay	2.51	6.30
Coal	1.4	13.00
Coke	1.31	15.13
Granite	2.66	15.13
Gravel	2.66	16.06
Gypsum rock	2.69	6.73
Iron ore (hematite)	3.53	12.84
Limestone	2.66	12.74
Phosphate rock	2.74	9.92
Quartz	2.65	13.57
Shale	2.63	15.87
Slate	2.57	14.30
Trap rock	2.87	19.32

[†]For dry grinding, multiply by $\frac{4}{3}$.
[‡]From Allis-Chalmers. Solids Processing Equipment Div., Appleton, Wisconsin, by permission.

dry crushing or to wet grinding. For dry grinding, the power calculated from Eq. (28.17) is multiplied by $\frac{4}{3}$.

EXAMPLE 28.2. What is the power required to crush 100 ton/h of limestone if 80 percent of the feed passes a 2-in. screen and 80 percent of the product a $\frac{1}{8}$-in. screen?

Solution. From Table 28.2, the work index for limestone is 12.74. Other quantities for substitution into Eq. (28.17) are

$$\dot{m} = 100 \text{ ton/h}$$

$$D_{pa} = 2 \times 25.4 = 50.8 \text{ mm} \qquad D_{pb} = 0.125 \times 25.4 = 3.175 \text{ mm}$$

The power required is

$$P = 100 \times 0.3162 \times 12.74 \left(\frac{1}{\sqrt{3.175}} - \frac{1}{\sqrt{50.8}} \right)$$

$$= 169.6 \text{ kW (227 hp)}$$

Computer Simulation of Milling Operations

The size distribution of products from various types of size reduction equipment can be predicted by a computer simulation of the comminution process.[12c,13] This makes use of two basic concepts, that of a *grinding rate function* S_u and a *breakage function* $\Delta B_{n,u}$. The material in a mill or crusher at any time is made up of particles of many different sizes, and they all interact with one another during the size reduction process; but for purposes of computer simulation the material is imagined to be divided into a number of discrete fractions (such as the ones retained on the various standard screens), and particle breakage occurs in each fraction more or less independently of the other fractions.

Consider a stack of n_T standard screens, and let n be the number of a particular screen in the stack. Here it is convenient to number the screens from the top down, beginning with the coarsest screen. (In discussing Table 28.1 the numbering began at the bottom of the stack.) For any given value of n, let the upper screens, coarser than screen n, be designated by the subscript u. (Note that $u < n$.) The grinding rate function S_u is the fraction of the material of a given size, coarser than that on screen n, which is broken in a given time. If x_u is the mass fraction retained on one of the upper screens, its rate of change by breakage to smaller size is

$$\frac{dx_u}{dt} = -S_u x_u \tag{28.18}$$

Suppose, for example, that the coarsest material in the charge to a grinding mill is 4/6-mesh, that the mass fraction of this material x_1 is 0.05, and that one-hundredth of this material is broken every second. Then S_u will be 0.01 s^{-1}, and x_1 will diminish at the rate of $0.01 \times 0.05 = 0.0005$ s^{-1}.

The breakage function $\Delta B_{n,u}$ gives the size distribution resulting from the breakage of the upper material. Some of the 4/6-mesh material, after breaking, would be fairly coarse, some very small, and some in between. Probably very little would be as large as 6/8-mesh, and only a small amount as small as 200-mesh. One would expect sizes in the intermediate range to be favored. Consequently $\Delta B_{n,u}$ varies with both n and u. Furthermore it varies with the composition of the material in the mill, since coarse particles may break differently in the presence of large amounts of fines than they do in the absence of fines. In a batch mill, therefore, $\Delta B_{n,u}$ (and S_u also) would be expected to vary with time as well as with all the other milling variables.

If $\Delta B_{n,u}$ and S_u are known or can be assumed, the rate of change of any given fraction can be found as follows. For any fraction except the coarsest, the initial amount is diminished by breakage to smaller sizes and simultaneously augmented by the creation of new particles from breakage of all coarser fractions. If the input and output from a given screen are at equal rates, the fraction retained on that screen remains constant. Usually, however, this is not the case, and the mass fraction retained on screen n changes according to the equation

$$\frac{dx_n}{dt} = -S_n x_n + \sum_{u=1}^{n-1} x_u S_u \, \Delta B_{n,u} \tag{28.19}$$

Equation (28.19) can be simplified if it is assumed that S_u and $\Delta B_{n,u}$ are constant, and analytical and matrix solutions are available for this case,[12c] but these assumptions are highly unrealistic. In crushing coal, for particles larger than about 28-mesh, S_u has been found to vary with the cube of the particle size[1] and the breakage function to depend on the reduction ratio $\bar{D}_n/\bar{D}_u$ according to the equation

$$B_{n,u} = \left(\frac{\bar{D}_n}{\bar{D}_u}\right)^{\beta} \tag{28.20}$$

where the exponent, β may be constant or may vary with the value of B.

In Eq. (28.20), $B_{n,u}$ is the *total* mass fraction smaller than size $\bar{D}_n$. It is a cumulative mass fraction, in contrast with $\Delta B_{n,u}$, which is the fraction of size $\bar{D}_n$ (retained between screens n and $n+1$) resulting from breakage of particles of size $\bar{D}_u$. If β in Eq. (28.20) is constant, this equation says that the particle size distribution of the crushed material is the same for all sizes of the initial material. The value of $\Delta B_{n,u}$ in crushing 4/6-mesh material to 8/10-mesh will be the same as in crushing 6/8-mesh particles to 10/14-mesh, since the size reduction ratio is the same.

Usually Eq. (28.19) is solved by the Euler method of numerical approximation, in which the changes in all fractions during successively short time intervals Δt (say, 30 s) are calculated by the approximation $dx_n/dt = \Delta x_n/\Delta t$. Changes in S_u and $\Delta B_{n,u}$ with screen size and (if known) with time can be incorporated.

Equipment for Size Reduction

Size reduction equipment is divided into crushers, grinders, ultrafine grinders, and cutting machines. *Crushers* do the heavy work of breaking large pieces of solid material into small lumps. A primary crusher operates on run-of-mine material, accepting anything that comes from the mine face and breaking it into 150- to 250-mm (6- to 10-in.) lumps. A secondary crusher reduces these lumps to particles perhaps

6 mm ($\frac{1}{4}$ in.) in size. *Grinders* reduce crushed feed to powder. The product from an intermediate grinder might pass a 40-mesh screen; most of the product from a fine grinder would pass a 200-mesh screen with a 74-μm opening. An *ultrafine grinder* accepts feed particles no larger than 6 mm ($\frac{1}{4}$ in.); the product size is typically 1 to 50 μm. *Cutters* give particles of definite size and shape, 2 to 10 mm in length.

These machines do their work in distinctly different ways. Compression is the characteristic action of crushers. Grinders employ impact and attrition, sometimes combined with compression; ultrafine grinders operate principally by attrition. A cutting action is of course characteristic of cutters, dicers, and slitters.

Crushers

Crushers are slow-speed machines for coarse reduction of large quantities of solids. The main types are jaw crushers, gyratory crushers, smooth-roll crushers, and toothed-roll crushers. The first three operate by compression and can break large lumps of very hard materials, as in the primary and secondary reduction of rocks and ores. Toothed-roll crushers tear the feed apart as well as crushing it; they handle softer feeds such as coal, bone, and soft shale.

Jaw crushers. In a jaw crusher feed is admitted between two jaws, set to form a V open at the top. One jaw, the fixed, or anvil, jaw, is nearly vertical and does not move; the other, the swinging jaw, reciprocates in a horizontal plane. It makes an angle of 20° to 30° with the anvil jaw. It is driven by an eccentric so that it applies great compressive force to lumps caught between the jaws. The jaw faces are flat or slightly bulged; they may carry shallow horizontal grooves. Large lumps caught between the upper parts of the jaws are broken, drop into the narrower space below, and are recrushed the next time the jaws close. After sufficient reduction they drop out the bottom of the machine. The jaws open and close 250 to 400 times per minute. Some machines with a 1.8- by 2.4-m (72- by 96-in.) feed opening can accept rocks 1.8 m (6 ft) in diameter and crush 1,200 tons/h to a maximum product size of 250 mm (10 in.). Smaller secondary crushers reduce the particle size of precrushed feed to 6 to 50 mm ($\frac{1}{4}$ to 2 in.) at much lower rates of throughput.

Gyratory crushers. A gyratory crusher may be looked upon as a jaw crusher with circular jaws, between which material is being crushed at some point at all times. A conical crushing head gyrates inside a funnel-shaped casing, open at the top. As shown in Fig. 28.8, the crushing head is carried on a heavy shaft pivoted at the top of the machine. An eccentric drives the bottom end of the shaft. At any point on the periphery of the casing, therefore, the bottom of the crushing head moves toward, and then away from, the stationary wall. Solids caught in the V-shaped space between the head and the casing are broken and rebroken until they pass out the bottom. The crushing head is free to rotate on the shaft and turns slowly because of friction with the material being crushed.

The speed of the crushing head is typically 125 to 425 gyrations per minute. Because some part of the crushing head is working at all times, the discharge from a gyratory crusher is continuous instead of intermittent as in a jaw crusher. The load on the motor is nearly uniform; less maintenance is required than with a jaw crusher;

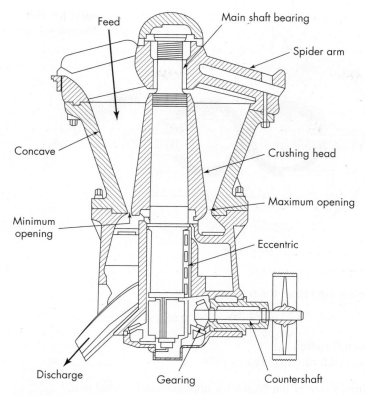

FIGURE 28.8
Gyratory crusher.

and the power requirement per ton of material crushed is smaller. Big gyratories handle up to 4,500 tons/h. The capacity of a gyratory crusher varies with the jaw setting, the impact strength of the feed, and the speed of gyration of the machine. The capacity is almost independent of the compressive strength of the material being crushed.

Roll crushers. Heavy, smooth-faced metal rolls turning on parallel horizontal axes are the working elements of the smooth-roll crusher illustrated in Fig. 28.9. Particles of feed caught between the rolls are broken in compression and drop out below. The rolls turn toward each other at the same speed. They have relatively narrow faces and are large in diameter so that they can "nip" moderately large lumps. Typical rolls are 600 mm (24 in.) in diameter with a 300-mm (12-in.) face to 2,000 mm (78 in.) in diameter with a 914-mm (36-in.) face. Roll speeds range from 50 to 300 rpm. Smooth-roll crushers are secondary crushers, with feeds 12 to 75 mm ($\frac{1}{2}$ to 3 in.) in size and products 12 mm ($\frac{1}{2}$ in.) to about 1 mm.

The limiting size $D_{p,\text{max}}$ of particles that can be nipped by the rolls depends on the coefficient of friction between the particle and the roll surface, but in most cases

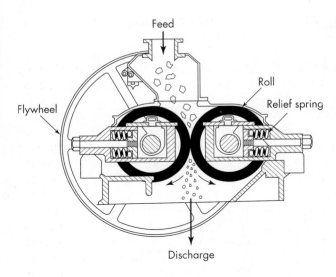

Feed

Roll

Flywheel

Relief spring

Discharge

FIGURE 28.9
Smooth-roll crusher.

it can be estimated from the relation

$$D_{p,\max} = 0.08R + d \tag{28.21}$$

where R = roll radius
d = width of gap between rolls

The maximum size of the product is approximately equal to d.

The particle size of the product depends on the spacing between the rolls, as does the capacity of a given machine. Smooth-roll crushers operate most effectively when set to give a reduction ratio of 3 or 4 to 1; that is, the maximum particle diameter of the product is one-third or one-fourth that of the feed.

In many roll crushers the roll faces carry corrugations, breaker bars, or teeth. Such crushers may contain two rolls, or only one roll working against a stationary curved breaker plate. Machines known as *disintegrators* contain two corrugated rolls turning at different speeds, which tear the feed apart, or a small high-speed roll with transverse breaker bars on its face turning toward a large slow-speed smooth roll. Some crushing rolls for coarse feeds carry heavy pyramidal teeth. Other designs utilize a large number of thin-toothed disks that saw through slabs or sheets of material. Toothed-roll crushers are much more versatile than smooth-roll crushers, within the limitation that they cannot handle very hard solids. They operate by compression, impact, and shear, not by compression alone, as do smooth-roll machines. They are not limited by the problem of nip inherent with smooth rolls and can therefore reduce much larger particles. The particle size of the feed to these machines may be as great as 500 mm (20 in.); their capacity ranges up to 500 tons/h.

Grinders

The term *grinder* refers to a variety of size reduction machines for intermediate duty. Product from a crusher is often fed to a grinder for further reduction.

Commercial grinders described here are hammer mills and impactors, rolling-compression machines, attrition mills, and tumbling mills.

Hammer mills and impactors. These mills all contain a high-speed rotor turning inside a cylindrical casing. Usually the shaft is horizontal. Feed dropped into the top of the casing is broken and falls out through a bottom opening. In a hammer mill, the particles are broken by sets of swing hammers pinned to a rotor disk. A particle of feed entering the grinding zone cannot escape being struck by the hammers. It shatters into pieces, which fly against a stationary anvil plate inside the casing and break into still smaller fragments. These in turn are rubbed into powder by the hammers and pushed through a grate or screen that covers the discharge opening.

Several rotor disks, 150 to 450 mm (6 to 18 in.) in diameter and each carrying four to eight swing hammers, are often mounted on the same shaft. The hammers may be straight bars of metal with plain or enlarged ends or with ends sharpened to a cutting edge. Intermediate hammer mills yield a product 25 mm (1 in.) to 20-mesh in particle size. In hammer mills for fine reduction, the peripheral speed of the hammer tips may reach 110 m/s (360 ft/s); they reduce 0.1 to 15 tons/h to sizes finer than 200-mesh. Hammer mills grind almost anything—tough fibrous solids like bark or leather, steel turnings, soft wet pastes, sticky clay, hard rock. For fine reduction they are limited to the softer materials.

The capacity and power requirement of a hammer mill vary greatly with the nature of the feed and cannot be estimated with confidence from theoretical considerations. They may be found from published information[12d] or better from small-scale or full-scale tests of the mill with a sample of the actual material to be ground. Commercial mills typically reduce 60 to 240 kg of solid per kilowatthour (100 to 400 lb/hp·h) of energy consumed.

An *impactor,* illustrated in Fig. 28.10, resembles a heavy-duty hammer mill except that it contains no grate or screen. Particles are broken by impact alone, without the rubbing action characteristic of a hammer mill. Impactors are often primary-reduction machines for rock and ore, processing up to 600 tons/h. The rotor in an impactor, as in many hammer mills, may be run in either direction to prolong the life of the hammers.

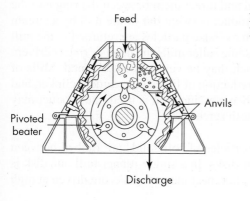

Feed

Pivoted
beater

Anvils

Discharge

FIGURE 28.10
Impactor.

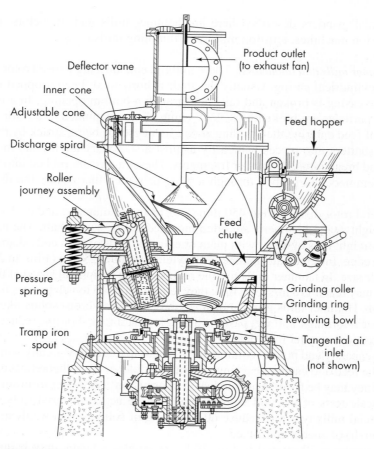

FIGURE 28.11
Roller mill.

Rolling-compression machines. In the roller mill illustrated in Fig. 28.11 solids are caught and crushed between vertical cylindrical rollers and a stationary anvil ring or bull ring. The rollers are driven at moderate speeds in a circular path. Plows lift the solid lumps from the floor of the mill and direct them between the ring and the rolls, where the reduction takes place. Product is swept out of the mill by a stream of air to a classifier separator, from which oversize particles are returned to the mill for further reduction. In a bowl mill and some roller mills the bowl or ring is driven; the rollers rotate on stationary axes, which may be vertical or horizontal. Mills of this kind find greatest application in the reduction of limestone, cement clinker, and coal. They pulverize up to 50 tons/h. When classification is used, the product may be as fine as 99 percent through a 200-mesh screen.

Attrition mills. In an attrition mill particles of soft solids are rubbed between the grooved flat faces of rotating circular disks. In a single-runner mill one disk is stationary and one rotates; in a double-runner machine both disks are driven at high

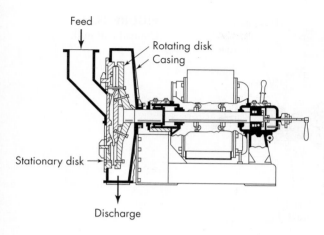

FIGURE 28.12
Attrition mill.

speed in opposite directions. Feed enters through an opening in the hub of one of the disks; it passes outward through the narrow gap between the disks and discharges from the periphery into a stationary casing. The width of the gap, within limits, is adjustable. At least one grinding plate is spring-mounted so that the disks can separate if unbreakable material gets into the mill. Mills with different patterns of grooves, corrugations, or teeth on the disks perform a variety of operations, including grinding, cracking, granulating, shredding, and sometimes blending.

A single-runner attrition mill is shown in Fig. 28.12. Single-runner mills contain disks of buhrstone or rock emery for reducing solids such as clay and talc, or metal disks for solids such as wood, starch, insecticide powders, and carnauba wax. Metal disks are usually of white iron, although for corrosive materials disks of stainless steel are sometimes necessary. Double-runner mills, in general, grind to finer products than single-runner mills but process softer feeds. Air is often drawn through the mill to remove the product and prevent choking. The disks may be cooled with water or refrigerated brine.

The disks of a single-runner mill are 250 to 1,400 mm (10 to 54 in.) in diameter, turning at 350 to 700 rpm. Disks in double-runner mills turn faster, at 1,200 to 7,000 rpm. The feed is precrushed to a maximum particle size of about 12 mm ($\frac{1}{2}$ in.) and must enter at a uniform controlled rate. Attrition mills grind from $\frac{1}{2}$ to 8 tons/h to products that will pass a 200-mesh screen. The energy required depends strongly on the nature of the feed and the degree of reduction accomplished and is much higher than in the mills and crushers described so far. Typical values are between 8 and 80 kWh (10 and 100 hp·h) per ton of product.

Tumbling mills. A typical tumbling mill is shown in Fig. 28.13. A cylindrical shell slowly turning about a horizontal axis and filled to about one-half its volume with a solid grinding medium forms a tumbling mill. The shell is usually steel, lined with high-carbon steel plate, porcelain, silica rock, or rubber. The grinding medium is metal rods in a rod mill, lengths of chain or balls of metal, rubber, or wood in a ball mill, flint pebbles or porcelain or zircon spheres in a pebble mill. For intermediate and fine reduction of abrasive materials, tumbling mills are unequaled.

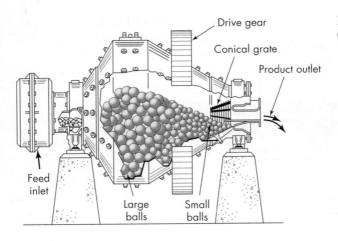

FIGURE 28.13
Conical ball mill.

Unlike the mills previously discussed, all of which require continuous feed, tumbling mills may be continuous or batch. In a batch machine a measured quantity of the solid to be ground is loaded into the mill through an opening in the shell. The opening is then closed and the mill turned on for several hours; it is then stopped, and the product is discharged. In a continuous mill the solid flows steadily through the revolving shell.

In all tumbling mills, the grinding elements are carried up the side of the shell nearly to the top, from whence they fall on the particles underneath. The energy expended in lifting the grinding units is utilized in reducing the size of the particles. In some tumbling mills, as in a *rod mill,* much of the reduction is done by rolling compression and by attrition as the rods slide downward and roll over one another. The grinding rods are usually steel, 25 to 125 mm (1 to 5 in.) in diameter, with several sizes present at all times in any given mill. Rod mills are intermediate grinders, reducing a 20-mm ($\frac{3}{4}$-in.) feed to perhaps 10-mesh, often preparing the product from a crusher for final reduction in a ball mill. They yield a product with little oversize and a minimum of fines.

In a *ball mill* or *pebble mill,* most of the reduction is done by impact as the balls or pebbles drop from near the top of the shell. In a large ball mill the shell might be 3 m (10 ft) in diameter and 4.25 m (14 ft) long. The balls are 25 to 125 mm (1 to 5 in.) in diameter; the pebbles in a pebble mill are 50 to 175 mm (2 to 7 in.). A *tube mill* is a continuous mill with a long cylindrical shell, in which material is ground for 2 to 5 times as long as in the shorter ball mill. Tube mills are excellent for grinding to very fine powders in a single pass where the amount of energy consumed is not of primary importance. Putting slotted transverse partitions in a tube mill converts it into a *compartment mill.* One compartment may contain large balls, another small balls, and a third pebbles. This segregation of the grinding media into elements of different size and weight aids considerably in avoiding wasted work, for the large, heavy balls break only the large particles, without interference by the fines.

Segregation of the grinding units in a single chamber is a characteristic of the *conical ball mill* illustrated in Fig. 28.13. Feed enters from the left through a 60° cone into the primary grinding zone, where the diameter of the shell is a maximum. Product leaves through the 30° cone to the right. A mill of this kind contains balls of different sizes, all of which wear and become smaller as the mill is operated. New large balls are added periodically. As the shell of such a mill rotates, the large balls move toward the point of maximum diameter, and the small balls migrate toward the discharge. The initial breaking of the feed particles, therefore, is done by the largest balls dropping the greatest distance; small particles are ground by small balls dropping a much smaller distance. The amount of energy expended is suited to the difficulty of the breaking operation, increasing the efficiency of the mill.

The load of balls in a ball or tube mill is normally such that when the mill is stopped, the balls occupy about one-half the volume of the mill. The void fraction in the mass of balls, when at rest, is typically 0.40. The grinding may be done with dry solids, but more commonly the feed is a suspension of the particles in water, increasing both the capacity and the efficiency of the mill.

When the mill is rotated, the balls are picked up by the mill wall and carried nearly to the top, where they break contact with the wall and fall to the bottom to be picked up again. Centrifugal force keeps the balls in contact with the wall and with one another during the upward movement. While in contact with the wall, the balls do some grinding by slipping and rolling over one another, but most of the grinding occurs at the zone of impact, where the free-falling balls strike the bottom of the mill.

The faster the mill is rotated, the farther the balls are carried up inside the mill and the greater the power consumption and the capacity of the mill. If the speed is too high, however, the balls are carried over and the mill is said to be centrifuging. The speed at which centrifuging occurs is called the *critical speed*. From a balance between the gravitational and centrifugal forces, the critical speed n_c may be found from the equation

$$n_c = \frac{1}{2\pi}\sqrt{\frac{g}{R - r}} \qquad (28.22)$$

where g is the acceleration of gravity, R is the radius of the mill, and r is the radius of the grinding elements.

The operating speed n must be less than n_c. Tumbling mills run at 65 to 80 percent of the critical speed, with the lower values for wet grinding in viscous suspensions.[12e]

ULTRAFINE GRINDERS

Many commercial powders must contain particles averaging 1 to 20 μm in size, with substantially all particles passing a standard 325-mesh screen that has openings 44 μm wide. Mills that reduce solids to such fine particles are called *ultrafine grinders*. Ultrafine grinding of dry powder is done by grinders, such as high-speed hammer mills, provided with internal or external classification, and by fluid-energy or jet mills. Ultrafine wet grinding is done in agitated mills.

Classifying hammer mills. In a hammer mill with internal classification a set of swing hammers is held between two rotor disks as in a conventional machine, but in addition to the hammers the rotor shaft carries two fans, which draw air through the mill inward toward the drive shaft and then discharge into ducts leading to collectors for the product. On the rotor disks are short radial vanes for separating oversize particles from those of acceptable size. Acceptably fine particles are carried past the radial vanes; particles that are too large are thrown back for further reduction in the grinding chamber. The maximum particle size of the product is varied by changing the rotor speed or the size and number of the separator vanes. Mills of this kind reduce 1 or 2 tons/h to an average particle size of 1 to 20 μm, with an energy requirement of about 40 kWh/t (50 hp · h/ton).

Fluid energy mills

In these mills the particles are suspended in a high-velocity gas stream. In some designs the gas flows in a circular or elliptical path; in others there are jets that oppose one another or vigorously agitate a fluidized bed. Some reduction occurs when the particles strike or rub against the walls of the confining chamber, but most of the reduction is believed to be caused by interparticle attrition. Internal classification keeps the larger particles in the mill until they are reduced to the desired size.

The suspending gas is usually compressed air or superheated steam, admitted at a pressure of 7 atm (100 lb$_f$/in.^{2}) through energizing nozzles. In the mill shown in Fig. 28.14 the grinding chamber is an oval loop of pipe 25 to 200 mm (1 to 8 in.) in

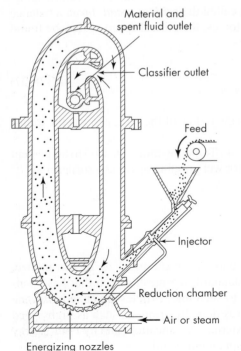

Material and
spent fluid outlet

Classifier outlet

Feed

Injector

Reduction chamber

Air or steam

Energizing nozzles

FIGURE 28.14
Fluid-energy mill. (*By permission, Fluid Energy Processing and Equipment Co.*)

diameter and 1.2 to 2.4 m (4 to 8 ft) high. Feed enters near the bottom of the loop through a venturi injector. Classification of the ground particles takes place at the upper bend of the loop. As the gas stream flows around this bend at high speed, the coarser particles are thrown outward against the outer wall while the fines congregate at the inner wall. A discharge opening in the inner wall at this point leads to a cyclone separator and a bag collector for the product. The classification is aided by the complex pattern of swirl generated in the gas stream at the bend in the loop of pipe.[2] Fluid-energy mills can accept feed particles as large as 12 mm ($\frac{1}{2}$ in.) but are more effective when the feed particles are no larger than 100-mesh. They reduce up to 1 ton/h of nonsticky solid to particles averaging $\frac{1}{2}$ to 10 μm in diameter, using 1 to 4 kg of steam or 6 to 9 kg of air per kilogram of product. Loop mills can process up to 6,000 kg/h.

Agitated mills. For some ultrafine grinding operations, small batch nonrotary mills containing a solid grinding medium are available. The medium consists of hard solid elements such as balls, pellets, or sand grains. These mills are vertical vessels 4 to 1,200 L (1 to 300 gal) in capacity, filled with liquid in which the grinding medium is suspended. In some designs the charge is agitated with a multiarmed impeller; in others, used especially for grinding hard materials (such as silica or titanium dioxide), a reciprocating central column "vibrates" the vessel contents at about 20 Hz. A concentrated feed slurry is admitted at the top, and product (with some liquid) is withdrawn through a screen at the bottom. Agitated mills are especially useful in producing particles 1 μm in size or finer.[12f]

Colloid mills.[12g] In a colloid mill, intense fluid shear in a high-velocity stream is used to disperse particles or liquid droplets to form a stable suspension or emulsion. The final size of the particles or droplets is usually less than 5 μm. Often there is little actual size reduction in the mill; the principal action is the disruption of lightly bonded clusters or agglomerates. Syrups, milk, purees, ointments, paints, and greases are typical products processed in this way. Chemical additives are often useful for stablizing the dispersion.

In most colloid mills the feed liquid is pumped between closely spaced surfaces one of which is moving relative to the other at speeds of 50 m/s or more. In one design the liquid passes through the narrow spaces between a disk-shaped rotor and its casing. The clearances are adjustable down to 25 μm. Often cooling is required to remove the heat generated. The capacities of colloid mills are relatively low, ranging from 2 or 3 L/min (30 to 50 gal/h) for small mills up to 440 L/min (7,000 gal/h) for the largest units.

Cutting machines

In some size reduction problems the feed stocks are too tenacious or too resilient to be broken by compression, impact, or attrition. In other problems the feed must be reduced to particles of fixed dimensions. These requirements are met by machines known as *granulators,* which yield more or less irregular pieces, and *cutters,* which produce cubes, thin squares, or diamonds. These devices find application in many manufacturing processes, but are especially well adapted to size reduction problems in making rubber and plastics. They find important applications in recycling paper and plastic materials.[8]

Typical rotary knife cutters contain a horizontal rotor turning at 200 to 900 rpm in a cylindrical chamber. On the rotor are 2 to 12 flying knives with edges of tempered steel or stellite, passing with close clearance over 1 to 7 stationary bed knives. Feed particles entering from above may be cut several times before they are small enough to pass through a bottom screen with 5- to 8-mm openings. Other rotary cutters and granulators are similar in design.

Equipment operation

In selecting and operating size reduction machinery, attention must be given to many details of procedure and of auxiliary equipment. A crusher, grinder, or cutter cannot be expected to perform satisfactorily unless (1) the feed is of suitable size and enters at a uniform rate, (2) the product is removed as soon as possible after the particles are of the desired size, (3) unbreakable material is kept out of the machine, and (4) in the reduction of low-melting or heat-sensitive products, the heat generated in the mill is removed. Heaters and coolers, metal separators, pumps and blowers, and constant-rate feeders are therefore important adjuncts to the size reduction unit. Factors to consider in specifying a size reduction system, including energy efficiency and environmental issues, are discussed by Kukla[8] and Hixon.[7]

Open-circuit and closed-circuit operation.   In many mills the feed is broken into particles of satisfactory size by passing it once through the mill. When no attempt is made to return oversize particles to the machine for further reduction, the mill is said to be operating in *open circuit*. This may require excessive amounts of power, for much energy is wasted in regrinding particles that are already fine enough. Thus it is often economical to remove partially ground material from the mill and pass it through a size separation device. The undersize becomes the product, and the oversize is returned to be reground. The separation device is sometimes inside the mill, as in ultrafine grinders; more commonly it is outside the mill. *Closed-circuit operation* is the term applied to the action of a mill and separator connected so that the oversize particles are returned to the mill. Energy must be supplied to drive the conveyors and separators in a closed-circuit system, but despite this, the reduction in total energy requirement over open-circuit grinding often reaches 25 percent.

Energy consumption.   Enormous amounts of energy are consumed in size-reduction operations, especially in manufacturing cement; crushing coal, rock, and shale; and preparing ores for making steel and copper.[7] Size reduction is probably the most inefficient of all unit operations: Over 99 percent of the energy goes to operating the equipment, producing undesirable heat and noise, leaving less than 1 percent for creating new surface. As processes have been developed that require finer and finer particles as feed to a kiln or reactor, the total energy requirement has increased, for reduction to very fine sizes is much more costly in energy than simple crushing to relatively coarse products. This is illustrated in Fig. 28.15, which also shows typical amounts of energy consumed per unit mass of product by the various kinds of size reduction equipment.

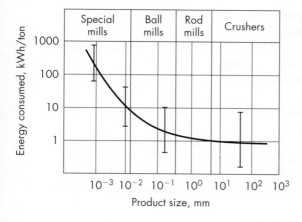

FIGURE 28.15
Energy consumption versus product size in size reduction equipment. (*By permission, from Comminution and Energy Consumption, NMAB-264, National Academy Press, 1981.*)

Heat removal. Since only a very small fraction of the energy supplied to the solid is used in creating new surface, the bulk of the energy is converted to heat, which may raise the temperature of the solid by many degrees. The solid may melt, decompose, or explode unless this heat is removed. For this reason cooling water or refrigerated brine is often circulated through coils or jackets in the mill. Sometimes the air blown through the mill is refrigerated, or solid carbon dioxide (dry ice) is admitted with the feed. Still more drastic temperature reduction is achieved with liquid nitrogen, to give grinding temperatures below $-75°C$. The purpose of such low temperatures is to alter the breaking characteristics of the solid, usually by making it more friable. In this way substances such as lard and beeswax become hard enough to shatter in a hammer mill; tough plastics, which stall a mill at ordinary temperatures, become brittle enough to be ground without difficulty.

Size enlargement

To improve handling properties, reduce the pressure drop in packed beds, increase rates of settling or drying, or prevent dusting, small particles are often bonded together to form lumps of the desired size. This may be accomplished by forming small pellets or larger briquets by compaction; by extrusion through a die under high pressure; by solidification of liquid drops in a spray dryer or prilling tower; by flocculation and coagulation of particles in a dilute slurry; or by heat bonding in nodulization and sintering processes. It is difficult to predict the behavior of a given material without actual tests. A binder is often added when making pellets, briquets, or extrudates so that the particles will stick together. Equipment for these size enlargement operations is described by Perry.[12h]

SYMBOLS

A	Area, m^2 or ft^2; total surface area of particles
A_w	Specific surface area of particles, m^2/g or ft^2/lb
a	Volume shape factor [Eq. (28.10)]

$B_{n,u}$ Total mass fraction of particles smaller than size $\bar{D}_{pn}$ resulting from breakage of particles of size $\bar{D}_{pu}$

D_p Particle size, mm or ft; D_{pa}, of feed; D_{pb}, of product; $D_{p,\max}$, maximum particle size nipped by rolls

D_{pn} Mesh opening in screen n, mm or ft; $D_{p(n+1)}$, in screen $n+1$; D_{pu}, in screen u

D Diameter, ft or mm; D_o, diameter of bin opening; D_p, particle size D_{pi}, mesh opening in screen i

$\bar{D}$ Average particle size, mm, μm, or ft; $\bar{D}_N$, arithmetic mean diameter [Eq. (28.7)]; $\bar{D}_V$, volume mean diameter [Eq. (28.9)]; $\bar{D}_n$, arithmetic average of D_{pn} and $D_{p(n+1)}$; $\bar{D}_{pi}$, arithmetic mean of D_{pi} and D_{pi+1}; $\bar{D}_s$, mean volume-surface diameter [Eq. (28.6)]; $\bar{D}_w$, mass mean diameter [Eq. (28.8)]

d Width of gap between rolls, m or ft

E Diffusivity in axial mixing, m^2/s or ft^2/s

g Gravitational acceleration, m/s^2 or ft/s^2

i Number of fraction or increment; also, screen number, counting from smallest size

K' Ratio of pressures, p_V/p_L

L Length, ft or m

m Mass of sample, g or lb

$\dot{m}$ Mass flow rate, lb/min

N Number of particles; N_T, total number; N_i, number in fraction i; also, number of spot samples

n Number of increments or screens; number of particles in spot sample; also speed of ball mill, r/s; n_c, critical speed

P Power, kW or hp

Pe   Peclet number, UL/E

R Radius of crushing rolls or ball mill, m or ft

r Radius of balls in ball mill, m or ft

S Grinding rate function, s^{-1}; S_n, for screen n; S_u, for screen u

u Number of screen coarser than screen n

v_p Volume of particle, m^3 or ft^3

W_i Bond work index, kWh/ton

x Mass fraction; also measured fraction of component A in spot sample; x_i, in increment i; x_n, on screen n; x_u, on screen u; x_1, on coarsest screen; $\bar{x}$, average measured fraction of component A

Greek Letters

α_r Angle of repose

β Exponent in Eq. (28.20)

$\Delta B_{n,u}$ Breakage function, fraction of particles of size $\bar{D}_u$ that are broken to size $\bar{D}_n$

Δt	Time increment, s
Δx_n	Change in x_n in time Δt
ρ_p	Density of particle, kg/m^3 or lb/ft^3
Φ_s	Sphericity [Eq. (28.1)]

PROBLEMS

28.1. Calculate the arithmetic mean diameter $\bar{D}_N$ for the −4- to +200-mesh fractions of the material analyzed in Table 28.1. How does $\bar{D}_N$ differ qualitatively from the volume mean diameter $\bar{D}_V$?

28.2. Plot the cumulative distribution given in Table 28.1 on logarithmic-probability paper. Is the plot linear over any range of particle sizes? How does the amount of fine material (smaller than 20-mesh) differ from what would be predicted from the size distribution of the coarser material?

28.3. A large Banbury mixer masticates 1,800 lb of scrap rubber with a density of 70 lb/ft^3. The power load is 6,000 hp per 1,000 gal of rubber. How much cooling water, in gallons per minute, is needed to remove the heat generated in the mixer if the temperature of the water is not to rise more than 15°F?

28.4. Data on the rate of mixing of sand and salt particles in an air-fluidized bed are given in Table 28.3. The number of particles in each spot sample was about 100. (*a*) For each run, calculate the estimated standard deviation s and the theoretical standard deviation for complete mixing σ_e. (*b*) How close was the mixture to being completely mixed after 87 s?

TABLE 28.3
Data on mixing of 35/48-mesh salt in sand in a 2-in. air-fluidized mixer[11]

Run no.	Mixing time, s	Number fraction of sand in spot samples									
1	45	0.64	0.68	0.74	0.63	0.73	0.81	0.59	0.65	0.62	0.70
		0.66	0.64	0.77	0.70	0.67	0.58	0.60	0.65	0.87	0.60
		0.49	0.52	0.49	0.54	0.64	0.38	0.32	0.34	0.49	0.52
		0.25	0.32	0.33	0.35	0.48	0.23	0.16	0.32	0.44	0.39
		0.26	0.26	0.21	0.32	0.38	0.22	0.24	0.22	0.15	0.36
2	87	0.53	0.54	0.60	0.60	0.60	0.55	0.56	0.60	0.69	0.63
		0.48	0.67	0.65	0.63	0.62	0.46	0.63	0.58	0.48	0.59
		0.49	0.53	0.46	0.49	0.58	0.34	0.52	0.45	0.50	0.47
		0.42	0.35	0.43	0.49	0.59	0.38	0.39	0.45	0.52	0.39
		0.35	0.36	0.37	0.49	0.48	0.37	0.49	0.32	0.32	0.36

28.5. Trap rock is crushed in a gyratory crusher. The feed is nearly uniform 2-in. spheres. The differential screen analysis of the product is given in column (1) of Table 28.4. The power required to crush this material is 400 kW. Of this 10 kW is needed to

TABLE 28.4
Data for Prob. 28.5

Mesh	Product	
	First grind (1)	Second grind (2)
4/6	3.1	
6/8	10.3	3.3
8/10	20.0	8.2
10/14	18.6	11.2
14/20	15.2	12.3
20/28	12.0	13.0
28/35	9.5	19.5
35/48	6.5	13.5
48/65	4.3	8.5
−65	0.5	
65/100		6.2
100/150		4.0
−150		0.3

operate the empty mill. By reducing the clearance between the crushing head and the cone, the differential screen analysis of the product becomes that given in column (2) in Table 28.4. The feed rate is 110 tons/h. Using the Bond method, estimate the power necessary per ton of rock in both the first and second grinds.

28.6. What rotational speed, in revolutions per minute, would you recommend for a ball mill 1,200 mm in diameter charged with 75-mm balls?

REFERENCES

1. Arbiter, N., and C. C. Harris. *Br. Chem. Eng.* **10**:240 (1965).
2. Berry, C. E. *Ind. Eng. Chem.* **38**:672 (1946).
3. Bond, F. C. *Trans. AIME*, TP-3308B, and *Mining Eng.* May 1952.
4. Clump, C. W.: in V. W. Uhl and J. B. Gray (eds.). *Mixing, Theory and Practice,* vol. 2. New York: Academic, 1967, p. 284.
5. Galanty, H. E. *Ind. Eng. Chem.* **55**(1):46 (1963).
6. Harnby, N., M. F. Edwards, and A. W. Nienow. *Mixing in the Process Industries.* London: Butterworths, 1985, pp. 24–38, 91–3.
7. Hixon, L. M. *Chem. Eng. Prog.* **87**(5):36 (1991).
8. Kukla, R. J. *Chem. Eng. Prog.* **87**(5):23 (1991).
9. Laforge, R. M., and B. K. Boruff. *Ind. Eng. Chem.* **56**(2):42 (1964).
10. Michaels, A. S., and V. Puzinauskis. *Chem. Eng. Prog.* **50**:604 (1954).
11. Nicholson, W. J. "The Blending of Dissimilar Particles in a Gas-Fluidized Bed," Ph.D. thesis. Ithaca, NY: Cornell University, 1965.
12. Perry, R. H., and D. W. Green (eds.). *Perry's Chemical Engineers' Handbook,* 7th ed. New York: McGraw-Hill, 1997; (*a*) pp. **20**-7 to **20**-10, (*b*) p. **20**-14, (*c*) pp. **20**-18 to **20**-22,

(*d*) pp. **20**-24 to **20**-48, (*e*) p. **20**-32, (*f*) p. **20**-38, (*g*) p. **20**-45, (*h*) pp. **20**-56 to **20**-89, (*i*) pp. **21**-4 to **21**-27.

13. Reid, K. J. *Chem. Eng. Sci.* **20**(11):953 (1965).
14. Smith, J. C. *Ind. Eng. Chem.* **47**:2240 (1955).
15. Smith, J. C., and U. S. Hattiangadi. *Chem. Eng. Commun.* **6:**105 (1980).
16. Todd, D. B., and H. F. Irving. *Chem. Eng. Prog.* **56**(9):84 (1969).
17. Wang, R. H., and L. T. Fan. *Chem. Eng.* **81**(11):88 (1974).
18. Weidenbaum, S. S., and C. F. Bonilla. *Chem. Eng. Prog.* **51**:27-J (1955).

CHAPTER 29

Mechanical Separations

Separations are extremely important in chemical manufacture—in fact, much processing equipment is devoted to separating one phase or one material from another. Separations are divided into two classes. One class, known as diffusional operations, involves the transfer of material between phases, as discussed in Chaps. 17 to 27. Another class, known as mechanical separations, is the subject of this chapter.

Mechanical separations are applicable to heterogeneous mixtures, not to homogeneous solutions. The techniques are based on physical differences between the particles such as size, shape, or density. They are applicable to separating solids from gases, liquid drops from gases, solids from solids, and solids from liquids. Two general methods are the use of a sieve, septum, or membrane, such as a screen or a filter, which retains one component and allows the other to pass; and the utilization of differences in the rate of sedimentation of particles or drops as they move through a liquid or gas. For special problems other methods, not discussed here, are used. These special methods exploit differences in the wettability or the electrical or magnetic properties of the substances.

SCREENING

Screening is a method of separating particles according to size alone. In industrial screening the solids are dropped on, or thrown against, a screening surface. The undersize particles, or *fines*, pass through the screen openings; oversize particles, or *tails*, do not. A single screen can make but a single separation into two fractions. These are called unsized fractions, because although either the upper or lower limit of the particle sizes they contain is known, the other limit is unknown. Material passed through a series of screens of different sizes is separated into sized fractions,

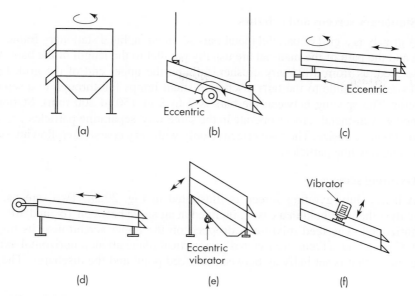

FIGURE 29.1
Motions of screens. (*a*) gyrations in horizontal plane; (*b*) gyrations in vertical plane; (*c*) gyrations at one end, shaking at other; (*d*) shaking; (*e*) mechanically vibrated; (*f*) electrically vibrated.

that is, fractions in which both the maximum and minimum particle sizes are known. Screening is done wet occasionally but much more commonly dry.

Industrial screens are made from woven wire, silk or plastic cloth, metal bars, perforated or slotted metal plates, or wires that are wedge-shaped in cross section. Various metals are used, with steel and stainless steel the most common. Standard screens range in mesh size from 4 in. to 400-mesh, and woven metal screens with openings as small as 1 μm are commercially available.[†] Screens finer than about 150-mesh are not commonly used, however, because with very fine particles other methods of separation are usually more economical.

Screening Equipment

Many varieties of screens are available for different purposes, and only a few representative types are discussed here. In most screens the particles drop through the openings by gravity; in a few designs they are pushed through the screen by a brush or centrifugal force. Coarse particles drop easily through large openings in a stationary surface, but with fine particles the screen surface must be agitated in some way, such as by shaking, gyrating, or vibrating it mechanically or electrically. Typical screen motions are illustrated in Fig. 29.1.

[†]Standard screens are discussed in Chap. 28; screen sizes are tabulated in App. 5.

Stationary screens and grizzlies

A grizzly is a grid of parallel metal bars set in an inclined stationary frame. The slope and the path of the material are usually parallel to the length of the bars. Very coarse feed, as from a primary crusher, falls on the upper end of the grid. Large chunks roll and slide to the tails discharge; small lumps fall through to a separate collector. The spacing between the bars is 2 to 8 in. (50 to 200 mm). Stationary inclined woven-metal screens operate in the same way, separating particles $\frac{1}{2}$ to 4 in. (12 to 100 mm) in size. They are effective only with very coarse, free-flowing solids containing few fine particles.

Gyrating screens

A heavy-duty gyrating screen is illustrated in Fig. 29.2a. Two screens, one above the other, are held in a casing inclined at an angle between 16° and 30° with the horizontal. The feed mixture is dropped on the upper screen near its highest point. Casing and screens are gyrated in a vertical plane about a horizontal axis by an eccentric that is set halfway between the feed point and the discharge. The rate

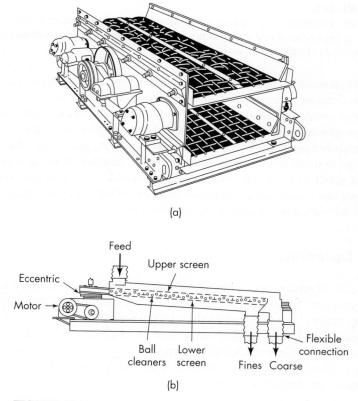

(a)

(b)

FIGURE 29.2

(a) Heavy-duty vertically gyrated screen; (b) Horizontally gyrated screen.

of gyration is between 600 and 1,800 rpm. The screens are rectangular and fairly long, typically $1\frac{1}{2}$ by 4 ft (0.5 to 1.2 m) to 5 by 14 ft (1.5 to 4.3 m). Oversize particles fall from the lower ends of the screens into collecting ducts; fines pass through the bottom screen into a discharge chute.

Finer screens are usually gyrated at the feed end in a horizontal plane. The discharge end reciprocates but does not gyrate. This combination of motions stratifies the feed, so that fine particles travel downward to the screen surface, where they are pushed through by the larger particles on top. Often the screening surface is double; as shown in Fig. 29.2b, between the two screens are rubber balls held in separate compartments. As the screen operates, the balls strike the screen surface and free the openings of any material that tends to plug them. Dry, hard, rounded or cubical grains ordinarily pass without trouble through screens, even fine screens; but elongated, sticky, flaky, or soft particles do not. Under the screening action such particles may become wedged into the openings and prevent other particles from passing through. A screen plugged with solid particles is said to be *blinded.*

Vibrating screens

Screens that are rapidly vibrated with small amplitude are less likely to blind than are gyrating screens. The vibrations may be generated mechanically or electrically. Mechanical vibrations are usually transmitted from high-speed eccentrics to the casing of the unit and from there to steeply inclined screens. Electrical vibrations from heavy-duty solenoids are transmitted to the casing or directly to the screens. Ordinarily no more than three decks are used in vibrating screens. Between 1,800 and 3,600 vibrations per minute are usual. A 48- by 120-in. (1.2- to 3-m) screen draws about 4 hp (3 kW).

Comparison of ideal and actual screens

The objective of a screen is to accept a feed containing a mixture of particles of various sizes and separate it into two fractions, an underflow that is passed through the screen and an overflow that is rejected by the screen. Either one, or both, of these streams may be a product.

An ideal screen would sharply separate the feed mixture in such a way that the smallest particle in the overflow would be just larger than the largest particle in the underflow. Such an ideal separation defines a cut diameter D_{pc} that marks the point of separation between the fractions. Usually D_{pc} is chosen to be equal to the mesh opening of the screen. Actual screens do not give a perfect separation about the cut diameter. The closest separations are obtained with spherical particles on standard testing screens, but even here there is an overlap between the smallest particles in the overflow and the largest ones in the underflow. The overlap is especially pronounced when the particles are needlelike or fibrous or where the particles tend to aggregate into clusters that act as large particles. Commercial screens usually give poorer separations than testing screens of the same mesh opening operating on the same mixture.

Material balances over screen

Simple material balances can be written over a screen which are useful in calculating the ratios of feed, oversize, and underflow from the screen analyses of the three streams and knowledge of the desired cut diameter. Let F, D, and B be the mass flow rates of the feed, overflow, and underflow, respectively, and x_F, x_D, and x_B be the mass fraction of oversize material in these three streams. The mass fractions of undersize material in the feed, overflow, and underflow are $1 - x_F$, $1 - x_D$, and $1 - x_B$.

Since the total material fed to the screen must leave it either as underflow or as overflow,

$$F = D + B \tag{29.1}$$

The material A in the feed must also leave in these two streams and

$$Fx_F = Dx_D + Bx_B \tag{29.2}$$

Elimination of B from Eqs. (29.1) and (29.2) gives

$$\frac{D}{F} = \frac{x_F - x_B}{x_D - x_B} \tag{29.3}$$

Elimination of D gives[†]

$$\frac{B}{F} = \frac{x_D - x_F}{x_D - x_B} \tag{29.4}$$

Screen efficiency

The efficiency of a screen is a measure of the success of a screen in closely separating materials. If the screen functioned perfectly, all the oversize material would be in the overflow and all the undersize material would be in the underflow. A common measure of screen efficiency is the ratio of oversize material that is actually in the overflow to the amount entering with the feed. These quantities are Dx_F and Fx_F, respectively. Thus

$$E_A = \frac{Dx_D}{Fx_F} \tag{29.5}$$

where E_A is the screen efficiency based on the oversize. Similarly, an efficiency E_B based on the undersize materials is given by

$$E_B = \frac{B(1 - x_B)}{F(1 - x_F)} \tag{29.6}$$

Although various formulas have been proposed, a uniform method of figuring overall efficiency has never been established.[30g]

Screen capacity

The capacity of a screen is measured by the mass of material that can be fed per unit time to a unit area of the screen. Capacity and efficiency are opposing factors.

[†]Note the identity of Eqs. (29.3) and (29.4) with Eqs. (21.5) and (21.6) for distillation. Although they are not alike physically, both operations are separation operations, and the same overall material-balance equations apply to them.

To obtain maximum efficiency, the capacity must be small, and large capacity is obtainable only at the expense of a reduction in efficiency. In practice, a reasonable balance between capacity and efficiency is desired.

The capacity of a screen is controlled simply by varying the rate of feed to the unit. The efficiency obtained for a given capacity depends on the nature of the screening operation. The overall chance of passage of a given undersize particle is a function of the number of times the particle strikes the screen surface and the probability of passage during a single contact. If the screen is overloaded, the number of contacts is small and the chance of passage on contact is reduced by the interference of the other particles. The improvement of efficiency attained at the expense of reduced capacity is a result of more contacts per particle and better chances for passage on each contact.

The probability of passage of a particle through a screen depends on the fraction of the total surface represented by openings, on the ratio of the diameter of the particle to the width of an opening in the screen, and on the number of contacts between the particle and the screen surface. The maximum capacity of a given screen is roughly proportional to the screen opening D_{pc}. Curves for estimating the screen capacity as a function of screen opening are given by Perry.[30h] For coarse screens, with openings $\frac{1}{4}$ to 4 in. (6 to 100 mm) in size, the capacity ranges from 1 to 8 tons/h·ft² (2.7 to 22 kg/s·m²), depending on the density of the material being screened; for screens with 0.05- to 0.25-in. (1- to 6-mm) openings it is 0.1 to 1.0 ton/h·ft² (0.27 to 2.7 kg/s·m²).

As the particle size is reduced, screening becomes progressively more difficult, and both the capacity and efficiency are, in general, low for particle sizes smaller than 150-mesh (0.1 mm).

FILTRATION; GENERAL CONSIDERATIONS

Filtration is the removal of solid particles from a fluid by passing the fluid through a filtering medium, or *septum,* on which the solids are deposited. Industrial filtrations range from simple straining to highly complex separations. The fluid may be a liquid or a gas; the valuable stream from the filter may be the fluid, or the solids, or both. Sometimes it is neither, as when waste solids must be separated from waste liquid prior to disposal. In industrial filtration the solids content of the feed ranges from a trace to a very high percentage. Often the feed is modified in some way by pretreatment to increase the filtration rate, as by heating, recrystallizing, or adding a "filter aid" such as cellulose or diatomaceous earth. Because of the enormous variety of materials to be filtered and the widely differing process conditions, many types of filters have been developed,[30d, 37a] a few of which are described in Fig. 29.3.

Fluid flows through a filter medium by virtue of a pressure differential across the medium. Filters are also classified, therefore, into those that operate with a pressure above atmospheric on the upstream side of the filter medium and those that operate with atmospheric pressure on the upstream side and a vacuum on the downstream side. Pressures above atmospheric may be developed by the force of gravity acting on a column of liquid, by a pump or blower, or by centrifugal force. Centrifugal filters are discussed in a later section of this chapter. In a gravity filter the filter medium can be no finer than a coarse screen or a bed of coarse particles like sand. Gravity filters are therefore restricted in their industrial applications to the draining of liquor from very coarse crystals, the clarification of potable water, and the treatment of wastewater.

Most industrial filters are pressure filters, vacuum filters, or centrifugal separators. They are also either continuous or discontinuous, depending on whether the discharge of filtered solids is steady or intermittent. During much of the operating cycle of a discontinuous filter, the flow of fluid through the device is continuous, but it must be interrupted periodically to permit discharging the accumulated solids. In a continuous filter the discharge of both solids and fluid is uninterrupted as long as the equipment is in operation.

Filters are divided into three main groups: cake filters, clarifying filters, and crossflow filters. Cake filters separate relatively large amounts of solids as a cake of crystals or sludge, as illustrated in Fig. 29.3a. Often they include provisions for washing the cake and for removing some of the liquid from the solids before discharge. Clarifying filters remove small amounts of solids to produce a clean gas or sparkling clear liquids such as beverages. The solid particles are trapped inside the filter medium, as shown in Fig. 29.3b, or on its external surfaces. Clarifying filters differ from screens in that the pores of the filter medium are much larger in diameter than the particles to be removed. In a crossflow filter the feed suspension flows

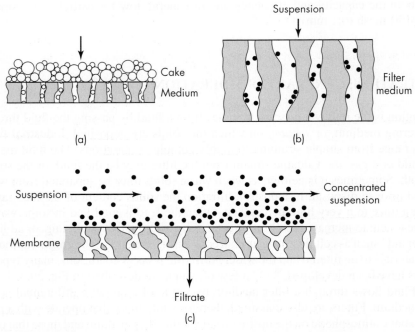

FIGURE 29.3
Mechanisms of filtration: (a) cake filter; (b) clarifying filter; (c) crossflow filter.

under pressure at a fairly high velocity across the filter medium (Fig. 29.3c). A thin layer of solids may form on the surface of the medium, but the high liquid velocity keeps the layer from building up. The filter medium is a ceramic, metal, or polymer membrane with pores small enough to exclude most of the suspended particles. Some of the liquid passes through the medium as clear filtrate, leaving a more concentrated suspension behind. As discussed later, an ultrafilter is a crossflow unit containing a membrane with extremely small openings, used for the separation and concentration of colloidal particles and large molecules.

CAKE FILTERS

At the start of filtration in a cake filter, some solid particles enter the pores of the medium and are immobilized, but soon others begin to collect on the septum surface. After this brief initial period the cake of solids does the filtration, not the septum; a visible cake of appreciable thickness builds up on the surface and must be periodically removed. Except as noted under bag filters for gas cleaning, cake filters are used almost entirely for liquid-solid separations. As with other filters they may operate with above-atmospheric pressure upstream from the filter medium or with vacuum applied downstream. Either type can be continuous or discontinuous, but because of the difficulty of discharging the solids against a positive pressure, most pressure filters are discontinuous.

Discontinuous pressure filters

Pressure filters can apply a large pressure differential across the septum to give economically rapid filtration with viscous liquids or fine solids. The most common types of pressure filters are filter presses and shell-and-leaf filters.

Filter press. A filter press contains a set of plates designed to provide a series of chambers or compartments in which solids may collect. The plates are covered with a filter medium such as canvas. Slurry is admitted to each compartment under pressure; liquor passes through the canvas and out a discharge pipe, leaving a wet cake of solids behind.

The plates of a filter press may be square or circular, vertical or horizontal. Most commonly the compartments for solids are formed by recesses in the faces of molded polypropylene plates. In other designs, they are formed as in the *plate-and-frame press* shown in Fig. 29.4, in which square plates 6 to 78 in. (150 mm to 2 m) on a side alternate with open frames. The plates are $\frac{1}{4}$ to 2 in. (6 to 50 mm) thick, the frames $\frac{1}{4}$ to 8 in. (6 to 200 mm) thick. Plates and frames sit vertically in a metal rack, with cloth covering the face of each plate, and are squeezed tightly together by a screw or a hydraulic ram. Slurry enters at one end of the assembly of plates and frames. It passes through a channel running lengthwise through one corner of the assembly. Auxiliary channels carry slurry from the main inlet channel into each frame. Here the solids are deposited on the cloth-covered faces of the plates. Liquor passes through the cloth, down grooves or corrugations in the plate faces, and out of the press.

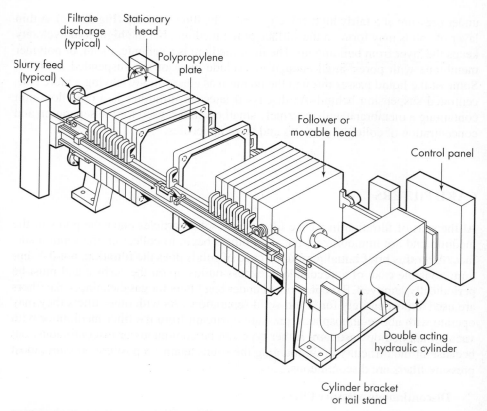

FIGURE 29.4
Filter press equipped for automatic operation. (*Shriver Filters, Eimco Process Equipment Co.*)

After assembly of the press, slurry is admitted from a pump or pressurized tank usually at a pressure of 3 to 10 atm. Filtration is continued until liquor no longer flows out the discharge or the filtration pressure suddenly rises. These occur when the frames are full of solid and no more slurry can enter. The press is then said to be *jammed*. Wash liquid may then be admitted to remove soluble impurities from the solids, after which the cake may be blown with steam or air to displace as much residual liquid as possible. The press is then opened, and the cake of solids is removed from the filter medium and dropped to a conveyor or storage bin. In many filter presses these operations are carried out automatically, as in the press illustrated in Fig. 29.4.

Thorough washing in a filter press may take several hours, for the wash liquid tends to follow the easiest paths and to bypass tightly packed parts of the cake. If the cake is less dense in some parts than in others, as is usually the case, much of the wash liquid will be ineffective. If washing must be exceedingly good, it may be best to reslurry a partly washed cake with a large volume of wash liquid and refilter it or to use a shell-and-leaf filter, which permits more effective washing than a plate-and-frame press.

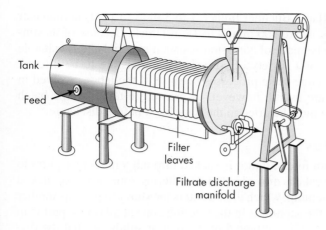

FIGURE 29.5
Horizontal-tank pressure leaf filter.

Shell-and-leaf filters. For filtering under higher pressures than are possible in a plate-and-frame press, to economize on labor, or where better washing of the cake is needed, a shell-and-leaf filter may be used. In the horizontal-tank design shown in Fig. 29.5, a set of vertical leaves is held on a retractable rack. The unit shown in the figure is open for discharge; during operation the leaves are inside the closed tank. Feed enters through the side of the tank; filtrate passes through the leaves into a discharge manifold. The design shown in Fig. 29.5 is widely used for filtrations involving filter aids, as discussed later in this chapter.

Automatic belt filter

The Larox belt filter is a discontinuous pressure filter that separates, compresses, washes, and automatically discharges the cake. Filtration takes place in from 2 to 20 horizontal chambers, set one above the other. A belt of filter cloth passes through the filter chambers in turn. With the belt held stationary, each chamber is filled with solids during the filtration cycle. High-pressure water is then pumped behind a flexible diaphragm in the chamber ceiling, squeezing the cake and mechanically expressing some of the liquid. With the diaphragm released, wash water may be passed through the cake and the cake recompressed by the diaphragm if desired. Finally, air is blown through the cake to remove additional liquid.

The chambers are then opened hydraulically so that the belt may be moved a distance somewhat greater than the length of one chamber. This action discharges cake from both sides of the filter. At the same time, part of the belt passes between spray nozzles for washing. After all the cake has been discharged, the belt is stopped, the chambers are closed, and the filtration cycle is repeated. All the steps are actuated automatically by impulses from a control panel. Filter sizes range from 0.8 m^2 (8.6 ft^2) to 31.5 m^2 (339 ft^2). The overall cycle is relatively short, typically 10 to 30 min, so that these filters can be used in continuous processes.

Discontinuous vacuum filters

Pressure filters are usually discontinuous; vacuum filters are usually continuous. A discontinuous vacuum filter, however, is sometimes a useful tool. A vacuum *nutsche* is little more than a large Büchner funnel, 1 to 3 m (3 to 10 ft) in diameter and forming a layer of solids 100 to 300 mm (4 to 12 in.) thick. Because of its simplicity, a nutsche can readily be made of corrosion-resistant material and is valuable where experimental batches of a variety of corrosive materials are to be filtered. Nutsches are uncommon in large-scale processes because of the labor involved in digging out the cake; they are useful, however, as pressure filters in some batch operations in which the cake must be dried in the filter itself before discharge.[29]

Continuous vacuum filters

In all continuous vacuum filters, liquor is sucked through a moving septum to deposit a cake of solids. The cake is moved out of the filtering zone, washed, sucked dry, and dislodged from the septum, which then reenters the slurry to pick up another load of solids. Some part of the septum is in the filtering zone at all times, part is in the washing zone, and part is being relieved of its load of solids, so that the discharge of both solids and liquids from the filter is uninterrupted. The pressure differential across the septum in a continuous vacuum filter is not high, ordinarily between 250 and 500 mm Hg. Various designs of filter differ in the method of admitting slurry, the shape of the filter surface, and the way in which the solids are discharged. Most all, however, apply vacuum from a stationary source to the moving parts of the unit through a rotary valve.

Rotary-drum filter. The most common type of continuous vacuum filter is the rotary-drum filter illustrated in Fig. 29.6. A horizontal drum with a slotted face turns at 0.1 to 2 rpm in an agitated slurry trough. A filter medium, such as canvas, covers the face of the drum, which is partly submerged in the liquid. Under the slotted cylindrical face of the main drum is a second, smaller drum with a solid surface. Between the two drums are radial partitions dividing the annular space into separate compartments, each connected by an internal pipe to one hole in the rotating plate of the rotary valve. Vacuum and air are alternately applied to each compartment as the drum rotates. A strip of filter cloth covers the exposed face of each compartment to form a succession of panels.

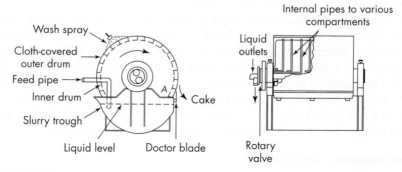

FIGURE 29.6
Continuous rotary vacuum filter.

Consider now the panel shown at *A* in Fig. 29.6. It is just about to enter the slurry in the trough. As it dips under the surface of the liquid, vacuum is applied through the rotary valve. A layer of solids builds up on the face of the panel as liquid is drawn through the cloth into the compartment, through the internal pipe, through the valve, and into a collecting tank. As the panel leaves the slurry and enters the washing and drying zone, vacuum is applied to the panel from a separate system, sucking wash liquid and air through the cake of solids. As shown in the flow sheet of Fig. 29.7, wash liquid is drawn through the filter into a separate collecting tank. After the cake of solids on the face of the panel has been sucked as dry as possible, the panel leaves the drying zone, vacuum is cut off, and the cake is removed by scraping it off with a horizontal knife known as a *doctor blade*. A little air is blown in under the cake to belly out the cloth. This cracks the cake away from the cloth and makes it unnecessary for the knife to scrape the drum face itself. Once the cake is dislodged, the panel reenters the slurry and the cycle is repeated. The operation of any given panel, therefore, is cyclic; but since some panels are in each part of the cycle at all times, the operation of the filter as a whole is continuous.

Many variations of the rotary-drum filter are commercially available. In some designs there are no compartments in the drum; vacuum is applied to the entire inner surface of the filter medium. Filtrate and wash liquid are removed together through a dip pipe; solids are discharged by airflow through the cloth from a stationary shoe

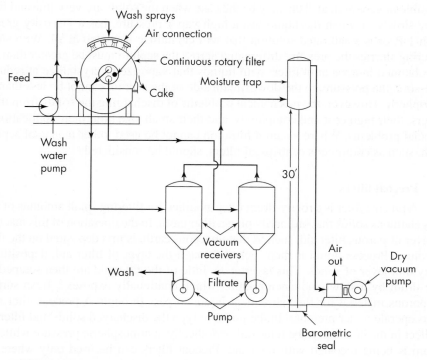

FIGURE 29.7
Flow sheet for continuous vacuum filtration.

inside the drum, bellying out the filter cloth and cracking off the cake. In other models the cake is lifted from the filter surface by a set of closely spaced parallel strings or by separating the filter cloth from the drum surface and passing it around a small-diameter roller. The sharp change in direction at this roller dislodges the solids. The cloth may be washed as it returns from the roller to the underside of the drum. Wash liquid may be sprayed directly on the cake surface; or, with cakes that crack when air is drawn through them, it may be sprayed on a cloth blanket that travels with the cake through the washing zone and is tightly pressed against its outer surface.

The amount of submergence of the drum is also variable. Most bottom-feed filters operate with about 30 percent of their filter area submerged in the slurry. When high filtering capacity and no washing are desired, a high-submergence filter, with 60 to 70 percent of its filter area submerged, may be used. The capacity of any rotary filter depends strongly on the characteristics of the feed slurry and particularly on the thickness of the cake that may be deposited in practical operation. The cakes formed on industrial rotary vacuum filters are 3 to about 40 mm ($\frac{1}{8}$ to $1\frac{1}{2}$ in.) thick. Standard drum sizes range from 0.3 m (1 ft) in diameter with a 0.3-m (1-ft) face to 3 m (10 ft) in diameter with a 4.3-m (14-ft) face.

Rotary-drum pressure filters

Continuous rotary vacuum filters are sometimes adapted to operate under positive pressures up to about 15 atm for situations in which vacuum filtration is not feasible or economical. This may be the case when the solids are very fine and filter very slowly or when the liquid has a high vapor pressure, has a viscosity greater than 1 P, or is a saturated solution that will crystallize if cooled at all. With slow-filtering slurries the pressure differential across the septum must be greater than can be obtained in a vacuum filter; with liquids that vaporize or crystallize at reduced pressure, the pressure on the downstream side of the septum cannot be less than atmospheric. However, the mechanical problems of discharging the solids from these filters, their high cost and complexity, and their small size limit their application to special problems. Where vacuum filtration cannot be used, other means of separation, such as continuous centrifugal filters, should be considered.

Precoat filters

A precoat filter is a rotary-drum filter modified for filtering small amounts of fine or gelatinous solids that ordinarily plug a filter cloth. In the operation of this machine a layer of porous filter aid, such as diatomaceous earth, is first deposited on the filter medium. Process liquid is then sucked through the layer of filter aid, depositing a very thin layer of solids. This layer and a little of the filter aid are then scraped off the drum by a slowly advancing knife, which continually exposes a fresh surface of porous material for the subsequent liquor to pass through. A precoat filter may also operate under pressure. In the pressure type the discharged solids and filter aid collect in the housing, to be removed periodically at atmospheric pressure while the drum is being recoated with filter aid. Precoat filters can be used only where the solids are to be discarded or where their admixture with large amounts of filter aid

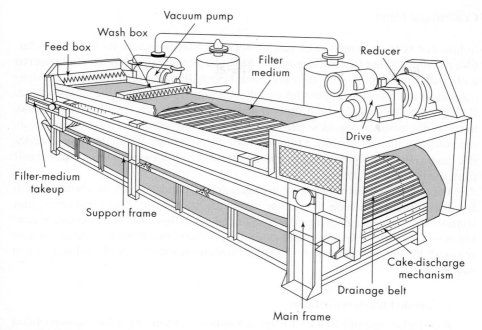

FIGURE 29.8
Horizontal belt filter.

introduces no serious problem. The usual submergence of a precoat-filter drum is 50 percent.

Horizontal belt filter. When the feed contains coarse fast-settling particles of solid, a rotary-drum filter works poorly or not at all. The coarse particles cannot be suspended well in the slurry trough, and the cake that forms often will not adhere to the surface of the filter drum. In this situation a top-fed horizontal filter may be used. The moving belt filter shown in Fig. 29.8 is one of several types of horizontal filter; it resembles a belt conveyor, with a transversely ridged support or drainage belt carrying the filter cloth, which is also in the form of an endless belt. Central openings in the drainage belt slide over a longitudinal vacuum box, into which the filtrate is drawn. Feed slurry flows onto the belt from a distributor at one end of the unit; filtered and washed cake is discharged from the other.

Belt filters are especially useful in waste treatment, since the waste often contains a very wide range of particle sizes.[37a] They are available in sizes from 0.6 to 5.5 m (2 to 18 ft) wide and 4.9 to 33.5 m (16 to 110 ft) long, with filtration areas up to 110 m² (1,200 ft²). Some models are "indexing" belt filters, similar in action to the Larox pressure filter described earlier; in these the vacuum is intermittently cut off and reapplied. The belt is moved forward one-half a meter or so when the vacuum is off and is held stationary while vacuum is applied. This avoids the difficulty of maintaining a good vacuum seal between the vacuum box and a moving belt.

Centrifugal Filters

Solids that form a porous cake can be separated from liquids in a filtering centrifuge. Slurry is fed to a rotating basket having a slotted or perforated wall covered with a filter medium such as canvas or metal cloth. Pressure resulting from the centrifugal action forces the liquor through the filter medium, leaving the solids behind. If the feed to the basket is then shut off and the cake of solids spun for a short time, much of the residual liquid in the cake drains off the particles, leaving the solids much "drier" than those from a filter press or vacuum filter. When the filtered material must subsequently be dried by thermal means, considerable savings may result from the use of a centrifuge.

The main types of filtering centrifuges are suspended batch machines, which are discontinuous in their operation; automatic short-cycle batch machines; and continuous conveyor centrifuges. In suspended centrifuges the filter media are canvas or other fabric or woven metal cloth. In automatic machines fine metal screens are used; in conveyor centrifuges the filter medium is usually the slotted wall of the basket itself.

Suspended batch centrifuges

A common type of batch centrifuge in industrial processing is the top-suspended centrifuge shown in Fig. 29.9. The perforated baskets range from 750 to 1,200 mm (30 to 48 in.) in diameter and from 450 to 750 mm (18 to 30 in.) deep and turn at speeds between 600 and 1,800 rpm. The basket is held at the lower end of a free-swinging vertical shaft driven from above. A filter medium lines the perforated wall of the basket. Feed slurry enters the rotating basket through an inlet pipe or chute. Liquor drains through the filter medium into the casing and out a discharge pipe; the solids form a cake 50 to 150 mm (2 to 6 in.) thick inside the basket. Wash liquid may be sprayed through the solids to remove soluble material. The cake is then spun as dry as possible, sometimes at a higher speed than during the loading and washing steps. The motor is shut off, and the basket nearly stopped by means of a brake. With the basket slowly turning, at perhaps 30 to 50 rpm, the solids are discharged by cutting them out with an unloader knife, which peels the cake off the filter medium and drops it through an opening in the basket floor. The filter medium is rinsed clean, the motor turned on, and the cycle repeated.

Top-suspended centrifuges are used extensively in sugar refining, where they operate on short cycles of 2 to 3 min per load and produce up to 5 tons/h of crystals per machine. Automatic controls are often provided for some of or all the steps in the cycle. In most processes where large tonnages of crystals are separated, however, other automatic or continuous conveyor centrifuges are used.

Another type of batch centrifuge is driven from the bottom, with the drive motor, basket, and casing all suspended from vertical legs mounted on a base plate. Solids are unloaded by hand through the top of the casing or plowed out through openings in the floor of the basket as in top-suspended machines. Except in sugar refining, suspended centrifuges usually operate on cycles of 10 to 30 min per load, discharging solids at a rate of 300 to 1,800 kg/h (700 to 4,000 lb/h).

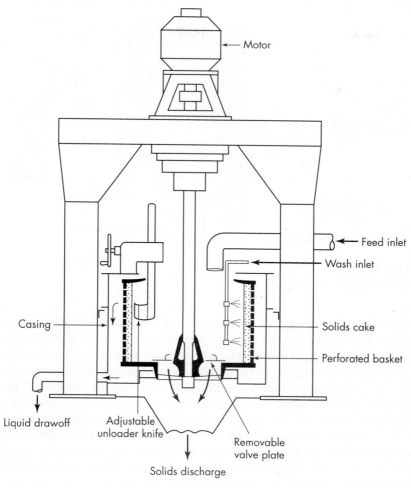

FIGURE 29.9
Top-suspended basket centrifuge.

Automatic batch centrifuges

A short-cycle automatic batch centrifuge is illustrated in Fig. 29.10. In this machine the basket rotates at constant speed about a horizontal axis. Feed slurry, wash liquid, and screen rinse are successively sprayed into the basket at appropriate intervals for controlled lengths of time. The basket is unloaded while turning at full speed by a heavy knife that rises periodically and cuts the solids out with considerable force through a discharge chute. Cycle timers and solenoid-operated valves control the various parts of the operation: feeding, washing, spinning, rinsing, and unloading. Any part of the cycle may be lengthened or shortened as desired.

The basket in these machines is between 500 and 1,100 mm (20 and 42 in.) in diameter. Automatic centrifuges have high productive capacity with free-draining

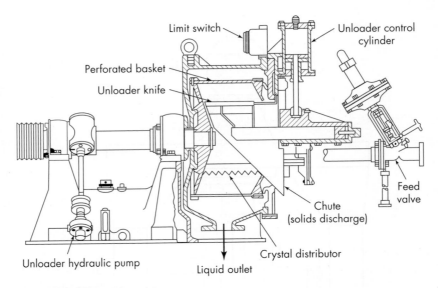

FIGURE 29.10
Automatic batch centrifuge.

crystals. Usually they are not used when the feed contains many particles finer than 150-mesh. With coarse crystals the total operating cycle ranges from 35 to 90 s, so that the hourly throughput is large. Because of the short cycle and the small amount of holdup required for feed slurry, filtrate, and discharged solids, automatic centrifuges are easily incorporated into continuous manufacturing processes. The small batches of solid can be effectively washed with small amounts of wash liquid, and— as in any batch machine—the amount of washing can be temporarily increased to clean up off-quality material should it become necessary. Automatic centrifuges cannot handle slow-draining solids, which would give uneconomically long cycles, or solids that do not discharge cleanly through the chute. There is also considerable breakage or degradation of the crystals by the unloader knife.

Continuous filtering centrifuges

A continuous centrifugal separator for coarse crystals is the reciprocating-conveyor centrifuge shown in Fig. 29.11. A rotating basket with a slotted wall is fed through a revolving feed funnel. The purpose of the funnel is to accelerate the feed slurry gently and smoothly. Feed enters the small end of the funnel from a stationary pipe at the axis of rotation of the basket. It travels toward the large end of the funnel, gaining speed as it goes, and when it spills off the funnel onto the wall of the basket, it is moving in the same direction as the wall and at very nearly the same speed. Liquor flows through the basket wall, which may be covered with a woven metal cloth. A layer of crystals 25 to 75 mm (1 to 3 in.) thick is formed. This layer is moved over the filtering surface by a reciprocating pusher. Each stroke of the pusher moves the crystals a few inches toward the lip of the basket; on the return stroke a space is opened on the filtering surface in which more cake can be deposited. When the

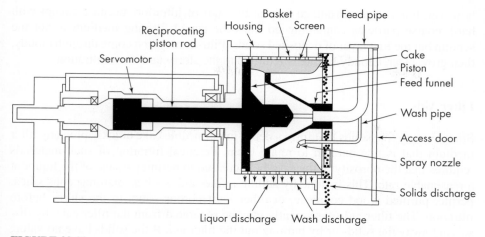

FIGURE 29.11
Reciprocating-conveyor continuous centrifuge.

crystals reach the lip of the basket, they fly outward into a large casing and drop into a collector chute. Filtrate and any wash liquid that is sprayed on the crystals during their travel leave the casing through separate outlets. The gentle acceleration of the feed slurry and deceleration of the discharged solids minimize breakage of the crystals. Multistage units that minimize the distance of travel of the crystals in each stage are used with solid cakes that do not "convey" properly in a single-stage machine. Reciprocating centrifuges are made with baskets ranging in diameter from 300 to 1,200 mm (12 to 48 in.). They dewater and wash 0.3 to 25 tons/h of solids containing no more than about 10 percent by weight of material finer than 100-mesh.

Filter Media

The septum in any filter must meet the following requirements:

1. It must retain the solids to be filtered, giving a reasonably clear filtrate.
2. It must not plug or blind.
3. It must be resistant chemically and strong enough physically to withstand the process conditions.
4. It must permit the cake formed to discharge cleanly and completely.
5. It must not be prohibitively expensive.

In industrial filtration a common filter medium is canvas cloth, either duck or twill weave. Many different weights and patterns of weave are available for different services. Corrosive liquids require the use of other filter media, such as woolen cloth, metal cloth of monel or stainless steel, glass cloth, or paper. Synthetic fabrics such as nylon, polypropylene, and various polyesters are also highly resistant chemically.[7,19]

In a cloth of a given mesh size, smooth synthetic or metal fibers are less effective than the more ragged natural fibers in removing very fine particles. Ordinarily, however, this is a disadvantage only at the start of filtration, because except with hard, coarse particles containing no fines the actual filtering medium is not the septum but the first layer of deposited solids. Filtrate may first come through cloudy, then grow clear. Cloudy filtrate is returned to the slurry tank for refiltration.

Filter Aids

Slimy or very fine solids that form a dense impermeable cake quickly plug any filter medium that is fine enough to retain them. Practical filtration of such materials requires that the porosity of the cake be increased to permit passage of the liquor at a reasonable rate. This is done by adding a filter aid, such as diatomaceous silica, perlite, purified wood cellulose, or other inert porous solid, to the slurry before filtration. The filter aid subsequently may be separated from the filter cake by dissolving away the solids or by burning out the filter aid. If the solids have no value, they and the filter aid are discarded together.

Another way of using a filter aid is by precoating, that is, depositing a layer of it on the filter medium before filtration. In batch filters the precoat layer is usually thin; in a continuous precoat filter, as described previously, the layer of precoat is thick, and the top of the layer is continually scraped off by an advancing knife to expose a fresh filtering surface. Precoats prevent gelatinous solids from plugging the filter medium and give a clearer filtrate. The precoat is really a part of the filter medium rather than of the cake.

Principles of Cake Filtration

Filtration is a special example of flow through porous media, which was discussed in Chap. 7 for cases in which the resistances to flow are constant. In filtration the flow resistances increase with time as the filter medium becomes clogged or a filter cake builds up, and the equations given in Chap. 7 must be modified to allow for this. The chief quantities of interest are the flow rate through the filter and the pressure drop across the unit. As time passes during filtration, either the flow rate diminishes or the pressure drop rises. In what is called *constant-pressure filtration* the pressure drop is held constant and the flow rate allowed to fall with time; less commonly, the pressure drop is progressively increased to give what is called *constant-rate filtration*.

In cake filtration the liquid passes through two resistances in series: that of the cake and that of the filter medium. The filter medium resistance, which is the only resistance in clarifying filters, is normally important only during the early stages of cake filtration. The cake resistance is zero at the start and increases with time as filtration proceeds. If the cake is washed after it is filtered, both resistances are constant during the washing period and that of the filter medium is usually negligible.

The overall pressure drop at any time is the sum of the pressure drops over medium and cake. If p_a is the inlet pressure, p_b the outlet pressure, and p' the pressure at the boundary between cake and medium, then

$$\Delta p = p_a - p_b = (p_a - p') + (p' - p_b) = \Delta p_c + \Delta p_m \qquad (29.7)$$

where Δp = overall pressure drop
Δp_c = pressure drop over cake
Δp_m = pressure drop over medium

Pressure drop through filter cake

Figure 29.12 shows diagrammatically a section through a filter cake and filter medium at a definite time t from the start of the flow of filtrate. At this time the thickness of the cake, measured from the filter medium, is L_c. The filter area, measured perpendicularly to the direction of flow, is A. Consider the thin layer of cake of thickness dL lying in the cake at a distance L from the medium. Let the pressure at this point be p. This layer consists of a thin bed of solid particles through which the filtrate is flowing. In a filter bed the velocity is sufficiently low to ensure laminar flow. Accordingly, as a starting point for treating the pressure drop through the cake, an equation of the form of Eq. (7.17) can be used, noting that $\Delta p/L = dp/dL$. If the superficial velocity of the filtrate is designated as u, Eq. (7.17) becomes

$$\frac{dp}{dL} = \frac{150\mu u(1-\varepsilon)^2}{(\Phi_s D_p)^2 \varepsilon^3} \tag{29.8}$$

Often the pressure drop is expressed as a function of the surface-volume ratio instead of the particle size. Substitution of $6(v_p/s_p)$ for $\Phi_s D_p$ [Eq. (7.10)] gives

$$\frac{dp}{dL} = \frac{4.17\mu u(1-\varepsilon)^2(s_p/v_p)^2}{\varepsilon^3} \tag{29.9}$$

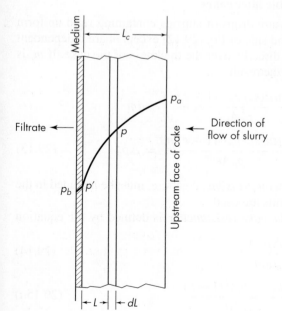

FIGURE 29.12
Section through filter medium and cake, showing pressure gradients; p, fluid pressure; L, distance from filter medium.

where $\frac{dp}{dL}$ = pressure gradient at thickness L
  μ = viscosity of filtrate
  u = linear velocity of filtrate, based on filter area
  s_p = surface of single particle
  v_p = volume of single particle
  ε = porosity of cake

The linear velocity u is given by the equation

$$u = \frac{dV/dt}{A} \tag{29.10}$$

where V is the volume of filtrate collected from the start of the filtration to time t. Since the filtrate must pass through the entire cake, V/A is the same for all layers and u is independent of L.

The volume of solids in the layer is $A(1 - \varepsilon)\,dL$, and if ρ_p is the density of the particles, the mass dm of solids in the layer is

$$dm = \rho_p(1 - \varepsilon)A\,dL \tag{29.11}$$

Elimination of dL from Eqs (29.9) and (29.11) gives

$$dp = \frac{k_1 \mu u (s_p/v_p)^2(1 - \varepsilon)}{\rho_p A \varepsilon^3}\,dm \tag{29.12}$$

where k_1 is used in place of the coefficient 4.17 in Eq. (29.9). If fps units are used, the Newton law proportionality factor g_c must be included in the denominators of Eqs. (29.9) and (29.12).

Compressible and incompressible filter cakes

In the filtration under low-pressure drops of slurries containing rigid uniform particles, all factors on the right-hand side of Eq. (29.12) except m are independent of L, and the equation is integrable directly, over the thickness of the cake. If m_c is the total mass of solids in the cake, the result is

$$\int_{p'}^{p_a} dp = \frac{k_1 \mu u (s_p/v_p)^2(1 - \varepsilon)}{\rho_p A \varepsilon^3} \int_0^{m_c} dm$$

$$p_a - p' = \frac{k_1 \mu u (s_p/v_p)^2(1 - \varepsilon)m_c}{\rho_p A \varepsilon^3} = \Delta p_c \tag{29.13}$$

Filter cakes of this type are called *incompressible*. Again, g_c must be included in the denominator of Eq. (29.13) if fps units are used.

For use in Eq. (29.13) a *specific cake resistance* α is defined by the equation

$$\alpha \equiv \frac{\Delta p_c\, A}{\mu u m_c} \tag{29.14}$$

where

$$\alpha = \frac{k_1 (s_p/v_p)^2(1 - \varepsilon)}{\varepsilon^3 \rho_p} \tag{29.15a}$$

The cake resistance α may also be expressed in terms of the particle size D_p, with a new coefficient k_2:

$$\alpha = \frac{k_2(1 - \varepsilon)}{(\Phi_s D_p)^2 \varepsilon^3 \rho_p} \qquad (29.15b)$$

For incompressible cakes α is independent of the pressure drop and of position in the cake. The dimensions of α are $\bar{L}M^{-1}$.

From Eq. (29.14), α is the resistance of a cake that gives a unit pressure drop when μ, u, and m_c/A all equal 1.0. Equation (29.15b) shows that α is influenced solely by the physical properties of the cake, especially the particle size D_p and the porosity ε.

Most cakes encountered industrially are not made up of individual rigid particles. The usual slurry is a mixture of agglomerates, or flocs, consisting of loose assemblies of very small particles, and the resistance of the cake depends on the properties of the flocs rather than on the geometry of the individual particles.[18] The flocs are deposited from the slurry on the upstream face of the cake and form a complicated network of channels to which Eq. (29.13) does not precisely apply. The resistance of such a cake is sensitive to the method used in preparing the slurry and to the age and temperature of the material. Also, the flocs are distorted and broken down by the forces existing in the cake, and the factors ε, k_2, and s_p/v_p vary from layer to layer.

Such a filter cake is called compressible. In a compressible cake, α varies with distance from the septum, since the cake nearest the septum is subject to the greatest compressive force and has the lowest void fraction. This makes the pressure gradient nonlinear, as shown in Fig. 29.12. The local value of α may also vary with time. In consequence, Eq. (29.13) does not strictly apply. In practice, however, the variations in α with time and location are ignored. An average value is obtained experimentally for the material to be filtered, using Eq. (29.14) for the calculation.

Filter medium resistance

A filter medium resistance R_m can be defined by analogy with the cake resistance $\alpha m_c/A$. The equation is

$$R_m \equiv \frac{p' - p_b}{\mu u} = \frac{\Delta p_m}{\mu u} \qquad (29.16)$$

The dimension of R_m is $\bar{L}^{-1}$.

The filter medium resistance R_m may vary with the pressure drop, since the higher liquid velocity caused by a large pressure drop may force additional particles of solid into the filter medium. Resistance R_m also varies with the age and cleanliness of the filter medium; but since it is important only during the early stages of filtration, it is nearly always satisfactory to assume that it is constant during any given filtration and to determine its magnitude from experimental data. When R_m is treated as an empirical constant, it also includes any resistance to flow that may exist in the pipes leading to and from the filter.

From Eqs. (29.14) and (29.16),

$$\Delta p = \Delta p_c + \Delta p_m = \mu u \left(\frac{m_c \alpha}{A} + R_m \right) \tag{29.17}$$

Strictly, the cake resistance α is a function of Δp_c rather than of Δp. During the important stage of the filtration, when the cake is of appreciable thickness, Δp_m is small in comparison with Δp_c, and the effect on the magnitude of α of carrying the integration of Eq. (29.13) over a range Δp instead of Δp_c can be safely ignored. In Eq. (29.17), then α is taken as a function of Δp.

In using Eq. (29.17) it is convenient to replace u, the linear velocity of the filtrate, and m_c, the total mass of solid in the cake, by functions of V, the total volume of filtrate collected to time t. Equation (29.10) relates u and V, and a material balance relates m_c and V. If c is the mass of the particles deposited in the filter per unit volume of filtrate,[†] the mass of solids in the filter at time t is Vc and

$$m_c = Vc \tag{29.19}$$

Substituting u from Eq. (29.10) and m_c from Eq. (29.19) in Eq. (29.17) gives

$$\frac{dt}{dV} = \frac{\mu}{A \, \Delta p} \left(\frac{\alpha c V}{A} + R_m \right) \tag{29.20}$$

Constant-pressure filtration

When Δp is constant, the only variables in Eq. (29.20) are V and t. When $t = 0$, $V = 0$ and $\Delta p = \Delta p_m$; hence

$$\frac{\mu R_m}{A \, \Delta p} = \left(\frac{dt}{dV} \right)_0 = \frac{1}{q_0} \tag{29.21}$$

Equation (29.20) may therefore be written

$$\frac{dt}{dV} = \frac{1}{q} = K_c V + \frac{1}{q_0} \tag{29.22}$$

where

$$K_c = \frac{\mu c \alpha}{A^2 \, \Delta p} \tag{29.23}$$

[†]The concentration of solid in the slurry fed to the filter is slightly less than c, since the wet cake includes sufficient liquid to fill its pores, and V, the actual volume of filtrate, is slightly less than the total liquid in the original slurry. Correction for this retention of liquid in the cake can be made by material balances if desired. Thus, let m_F be the mass of the wet cake, including the filtrate retained in its voids, and m_c be the mass of the dry cake obtained by washing the cake free of soluble material and drying. Also, let ρ be the density of the filtrate. Then if c_F is the concentration of solids in the slurry in kilograms per cubic meter of liquid fed to the filter, material balances give

$$c = \frac{c_F}{1 - (m_F/m_c - 1)c_s/\rho} \tag{29.18}$$

Integration of Eq. (29.22) between the limits $(0, 0)$ and (t, V) gives

$$\frac{t}{V} = \left(\frac{K_c}{2}\right)V + \frac{1}{q_0} \tag{29.24}$$

Thus a plot of t/V versus V will be linear, with a slope equal to $K_c/2$ and an intercept of $1/q_0$. From such a plot and Eqs. (29.21) and (29.23), the values of α and R_m may be calculated as shown in Example 29.1.

Empirical equations for cake resistance

By conducting constant-pressure experiments at various pressure drops, the variation of α with Δp may be found. If α is independent of Δp, the cake is incompressible. Ordinarily α increases with Δp, as most cakes are at least to some extent compressible. For highly compressible cakes, α increases rapidly with Δp.

Empirical equations may be fitted to observed data for Δp versus α, the most common of which is

$$\alpha = \alpha_0(\Delta p)^s \tag{29.25}$$

where α_0 and s are empirical constants. Constant s is the *compressibility coefficient* of the cake. It is zero for incompressible cakes and positive for compressible ones. It usually falls between 0.2 and 0.8. Equation (29.25) should not be used in a range of pressure drops much different from that used in the experiments conducted to evaluate α_0 and s.

EXAMPLE 29.1. Laboratory filtrations conducted at constant pressure drop on a slurry of $CaCO_3$ in H_2O gave the data shown in Table 29.1. The filter area was 440 cm², the mass of solid per unit volume of filtrate was 23.5 g/L, and the temperature was 25°C. Evaluate the quantities α and R_m as a function of pressure drop, and fit an empirical equation to the results for α.

TABLE 29.1
Volume-time data[35] for Example 29.1[†]

Filtrate volume V, L	Test I t, s	t/V	Test II t, s	t/V	Test III t, s	t/V	Test IV t, s	t/V	Test V t, s	t/V
0.5	17.3	34.6	6.8	13.6	6.3	12.6	5.0	10.0	4.4	8.8
1.0	41.3	41.3	19.0	19.0	14.0	14.0	11.5	11.5	9.5	9.5
1.5	72.0	48.0	34.6	23.1	24.2	16.13	19.8	13.2	16.3	10.87
2.0	108.3	54.15	53.4	26.7	37.0	18.5	30.1	15.05	24.6	12.3
2.5	152.1	60.84	76.0	30.4	51.7	20.68	42.5	17.0	34.7	13.88
3.0	201.7	67.23	102.0	34.0	69.0	23.0	56.8	18.7	46.1	15.0
3.5			131.2	34.49	88.8	25.37	73.0	20.87	59.0	16.86
4.0			163.0	40.75	110.0	27.5	91.2	22.8	73.6	18.4
4.5					134.0	29.78	111.0	24.67	89.4	19.87
5.0					160.0	32.0	133.0	26.6	107.3	21.46
5.5							156.8	28.51		
6.0							182.5	30.42		

[†]Δp, in $lb_f/in.^2$: I, 6.7; II, 16.2; III, 28.2; IV, 36.3; V, 49.1.

Solution. The first step is to prepare plots, for each of the five constant-pressure experiments, of t/V versus V. The data are given in Table 29.1, and the plots are shown in Fig. 29.13. The slope of each line is $K_c/2$, in seconds per liter per liter. To convert to seconds per cubic foot per cubic foot, the conversion factor is $28.31^2 = 801$. The intercept of each line on the axis of ordinates is $1/q_0$, in seconds per liter. The factor to convert this to seconds per cubic foot is 28.31. The slopes and intercepts, in the observed and converted units, are given in Table 29.2.

The viscosity of water is, from App. 6, 0.886 cP, or $0.886 \times 6.72 \times 10^{-4} = 5.95 \times 10^{-4}$ lb/ft·s. The filter area is $440/30.48^2 = 0.474$ ft². The concentration c is $(23.5 \times 28.31)/454 = 1.47$ lb/ft³.

From the values of $K_c/2$ and $1/q_0$ in Table 29.2, corresponding values of R_m and α are found from Eqs. (29.21) and (29.23). Thus, with g_c included in the denominator

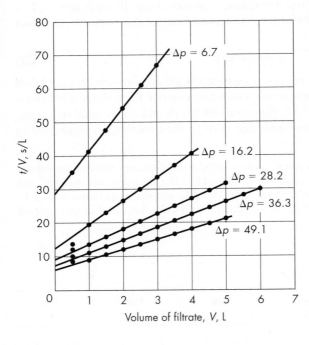

FIGURE 29.13
Plot of t/V versus V for Example 29.1.

TABLE 29.2
Values for K_c, $1/q_0$, R_m, and α for Example 29.1

Test	Pressure drop Δp		Slope $K_c/2$		Intercept $1/q_0$		R_m,	α,
	lb$_f$/in.²	lb$_f$/ft²	s/L²	s/ft⁶	s/L	s/ft³	ft⁻¹ × 10⁻¹⁰	ft/lb × 10⁻¹¹
I	6.7	965	13.02	10,440	28.21	800	1.98	1.66
II	16.2	2,330	7.24	5,800	12.11	343	2.05	2.23
III	28.2	4,060	4.51	3,620	9.43	267	2.78	2.43
IV	36.3	5,230	3.82	3,060	7.49	212	2.84	2.64
V	49.1	7,070	3.00	2,400	6.35	180	3.26	2.80

of Eq. (29.21),

$$R_m = \frac{A \, \Delta p \, g_c (1/q_0)}{\mu} = \frac{0.474 \times 32.17 \, \Delta p (1/q_0)}{5.95 \times 10^{-4}}$$

$$= 2.56 \times 10^4 \, \Delta p \, \frac{1}{q_0}$$

$$\alpha = \frac{A^2 \, \Delta p \, g_c K_c}{c\mu} = \frac{0.474^2 \times 32.17 \, \Delta p K_c}{5.95 \times 10^{-4} \times 1.47}$$

$$= 8.26 \times 10^3 \, \Delta p \, K_c$$

Table 29.2 shows the values of $K_c/2$ and $1/q_0$ for each test, calculated by the method of least squares. In all but test I the first point, which does not fall on the linear graph, was omitted. Also in Table 29.2 are the values of α and R_m. Figure 29.14 is a plot of R_m versus Δp. The filter medium resistance increases linearly with pressure drop, but rises by only 50 percent for a 6-fold increase in Δp.

Figure 29.15 is a logarithmic plot of α versus Δp. The points closely define a straight line, so Eq. (29.25) is suitable as an equation for α as a function of Δp. The slope of the line, which is the value of s for this cake, is 0.26. The cake is only slightly compressible.

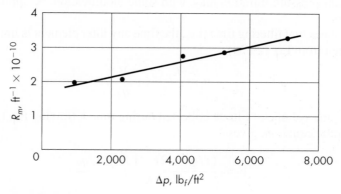

FIGURE 29.14
Plot of R_m versus Δp for Example 29.1.

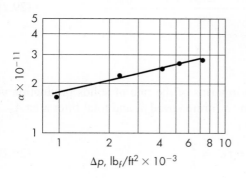

FIGURE 29.15
Log-log plot of α versus Δp for Example 29.1.

Constant α_0 can be calculated by reading the coordinates of any convenient point on the line of Fig. 29.15 and calculating α_0 by Eq. (29.25). For example, when $\Delta p = 1,000$, $\alpha = 1.75 \times 10^{11}$, and

$$\alpha_0 = \frac{1.75 \times 10^{11}}{1,000^{0.26}} = 2.90 \times 10^{10} \text{ ft/lb } (1.95 \times 10^{10} \text{ m/kg})$$

Equation (29.25) becomes for this cake

$$\alpha = 2.90 \times 10^{10} \, \Delta p^{0.26}$$

Continuous filtration

In a continuous filter, say, of the rotary-drum type, the feed, filtrate, and cake move at steady constant rates. For any particular element of the filter surface, however, conditions are not steady but transient. Follow, for example, an element of the filter cloth from the moment it enters the pond of slurry until it is scraped clean once more. It is evident that the process consists of several steps in series—cake formation, washing, drying, and discharging—and that each step involves progressive and continual change in conditions. The pressure drop across the filter during cake formation is, however, held constant. Thus the foregoing equations for discontinuous constant-pressure filtration may, with some modification, be applied to continuous filters.

If t is the actual filtering time (i.e., the time any filter element is immersed in the slurry), then from Eq. (29.24)

$$t = \frac{K_c V^2}{2} + \frac{V}{q_0} \tag{29.26}$$

where V is the volume of filtrate collected during time t. Solving Eq. (29.26) for V, as a quadratic equation, gives

$$V = \frac{\left(1/q_0^2 + 2K_c t\right)^{1/2} - 1/q_0}{K_c} \tag{29.27}$$

Substitution for $1/q_0$ and K_c from Eqs. (29.21) and (29.23), followed by division by tA, leads to the equation

$$\frac{V}{tA} = \frac{[2 \, \Delta p \, c\alpha / \mu t + (R_m/t)^2]^{1/2} - R_m/t}{c\alpha} \tag{29.28}$$

where $\dfrac{V}{t}$ = rate of filtrate collection

A = submerged area of filter

Equation (29.28) may be written in terms of the rate of solids production $\dot{m}_c$ and the filter characteristics: cycle time t_c, drum speed n, and total filter area A_T. If the fraction of the drum submerged is f,

$$t = f t_c = \frac{f}{n} \tag{29.29}$$

The rate of solids production, from Eq. (29.19), is

$$\dot{m}_c = c\frac{V}{t} \tag{29.30}$$

Since $A/A_T = f$, the rate of cake production, divided by the total area of the filter, is

$$\frac{\dot{m}_c}{A_T} = \frac{[2c\alpha\,\Delta p\,fn/\mu + (nR_m)^2]^{1/2} - nR_m}{\alpha} \tag{29.31}$$

The filter medium resistance R_m includes that of any cake not removed by the discharge mechanism and carried through the next cycle. When the filter medium is washed after the cake is discharged, R_m is usually negligible and Eq. (29.31) becomes

$$\frac{\dot{m}_c}{A_T} = \left(\frac{2c\,\Delta p\,fn}{\alpha\mu}\right)^{1/2} \tag{29.32}$$

If the specific cake resistance varies with pressure drop according to Eq. (29.25), then Eq. (29.32) may be modified to

$$\frac{\dot{m}_c}{A_T} = \left(\frac{2c\,\Delta p^{1-s}\,fn}{\alpha_0\mu}\right)^{1/2} \tag{29.33}$$

Equations (29.31) and (29.32) apply both to continuous vacuum filters and to continuous pressure filters. When R_m is negligible, Eq. (29.32) predicts that the filtrate flow rate varies inversely with the square root of the viscosity and of the cycle time. This has been observed experimentally with thick cakes and long cycle times.[28] With short cycle times, however, this is not true, and the more complicated relationship shown in Eq. (29.31) must be used.[34] In general, the filtration rate increases as the drum speed increases and the cycle time t_c diminishes, because the cake formed on the drum face is thinner than at low drum speeds. At speeds above a certain critical value, however, the filtration rate no longer increases with speed but remains constant, and the cake tends to become wet and difficult to discharge.

The filter area required for a given filtration rate is calculated as shown in Example 29.2.

EXAMPLE 29.2. A rotary drum filter with 30 percent submergence is to be used to filter a concentrated aqueous slurry of $CaCO_3$ containing 14.7 lb of solids per cubic foot of water (236 kg/m³). The pressure drop is to be 20 in. Hg. If the filter cake contains 50 percent moisture (wet basis), calculate the filter area required to filter 10 gal/min of slurry when the filter cycle time is 5 min. Assume that the specific cake resistance is the same as that in Example 29.1 and that the filter medium resistance R_m is negligible. The temperature is 20°C.

Solution. Equation (29.33) will be used. The quantities needed for substitution are

$$\Delta p = 20\left(\frac{14.69}{29.92}\right) \times 144 = 1{,}414\ \text{lb}_f/\text{ft}^2$$

$$f = 0.30 \qquad t_c = 5 \times 60 = 300\ \text{s} \qquad n = \tfrac{1}{300}\,\text{s}^{-1}$$

From Example 29.1

$$\alpha_0 = 2.90 \times 10^{10} \text{ ft/lb} \qquad s = 0.26$$

Also $\qquad \mu = 1 \text{ cP} = 6.72 \times 10^{-4} \text{ lb/ft} \cdot \text{s} \qquad \rho = 62.3 \text{ lb/ft}^3$

Quantity c is found from Eq. (29.18). The slurry concentration c_F is 14.7 lb/ft³. Since the cake contains 50 percent moisture, $m_F/m_c = 2$. Substitution of these quantities in Eq. (29.18) gives

$$c = \frac{14.7}{1 - (2 - 1)(14.7/62.3)} = 19.24 \text{ lb/ft}^3$$

Solving Eq. (29.33) for A_T including g_c as needed gives

$$A_T = \dot{m}_c \left(\frac{\alpha_0 \mu}{2c \, \Delta p^{1-s} \, g_c f n} \right)^{1/2} \qquad (29.34)$$

The solids production rate $\dot{m}_c$ equals the slurry flow rate times its concentration c_F. Thus since the density of $CaCO_3$ is 168.8 lb/ft³,

$$\dot{m}_c = \frac{10}{60} \frac{1}{7.48} \left(\frac{1}{14.7/168.8 + 1} \right) 14.7 = 0.302 \text{ lb/s}$$

Substitution in Eq. (29.34) gives

$$A_T = 0.302 \left(\frac{2.90 \times 10^{10} \times 6.72 \times 10^{-4}}{2 \times 19.24 \times 1{,}414^{0.74} \times 32.17 \times 0.30 \times \frac{1}{300}} \right)^{1/2}$$

$$= 81.7 \text{ ft}^2 \ (7.59 \text{ m}^2)$$

Constant-rate filtration

If filtrate flows at a constant rate, the linear velocity u is constant and

$$u = \frac{dV/dt}{A} = \frac{V}{At} \qquad (29.35)$$

Equation (29.14) can be written, after substituting m_c from Eq. (29.19) and u from Eq. (29.35), as

$$\frac{\Delta p_c}{\alpha} = \frac{\mu c}{t} \left(\frac{V}{A} \right)^2 \qquad (29.36)$$

The specific cake resistance α is retained on the left-hand side of Eq. (29.36) because it is a function of Δp for compressible sludges.[†]

If α is known as a function of Δp_c and if Δp_m, the pressure drop through the filter medium, can be estimated, then Eq. (29.36) can be used directly to relate the overall pressure drop to time when the rate of flow of filtrate is constant. A more

---

[†]The concentration c may also vary somewhat with pressure drop. In operation, c_s rather than c is constant and, by Eq. (29.18) since m_F/m_c changes with pressure, c also changes when $(m_F/m_c - 1)(c_s/\rho)$ is appreciable in comparison with unity. Any such variation in c with pressure drop can be ignored, in view of the other approximations made in the general theory of filtration.

direct use of this equation can be made, however, if Eq. (29.25) is accepted to relate α and Δp_c.[23] If α from Eq. (29.25) is substituted in Eq. (29.36), and if $\Delta p - \Delta p_m$ is substituted for Δp_c, the result is

$$\Delta p_c^{1-s} = \alpha_0 \mu c t \left(\frac{V}{At}\right)^2 = (\Delta p - \Delta p_m)^{1-s} \tag{29.37}$$

Again, the simplest method of correcting the overall pressure drop for the pressure drop through the filter medium is to assume the filter medium resistance is constant during a given constant-rate filtration. Then, by Eq. (29.16), Δp_m is also constant in Eq. (29.37). Since the only variables in Eq. (29.37) are Δp and t, the equation can be written

$$(\Delta p - \Delta p_m)^{1-s} = K_r t \tag{29.38}$$

where K_r is defined by

$$K_r = \mu u^2 c \alpha_0 \tag{29.39}$$

Principles of centrifugal filtration

The basic theory of constant-pressure filtration can be modified to apply to filtration in a centrifuge. The treatment applies after the cake has been deposited and during flow of clear filtrate or fresh water through the cake. Figure 29.16 shows such a cake. In this figure,

$$r_1 = \text{radius of inner surface of liquid}$$
$$r_i = \text{radius of inner face of cake}$$
$$r_2 = \text{inside radius of basket}$$

The following simplifying assumptions are made: The effects of gravity and of changes in kinetic energy of the liquid are neglected, and the pressure drop from centrifugal action equals the drag of the liquid flowing through the cake; the cake is completely filled with liquid; the flow of the liquid is laminar; the resistance of the

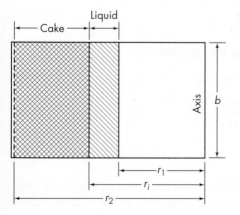

Liquid
Cake

Axis b

r_1
r_i
r_2

FIGURE 29.16
Centrifugal filter.

filter medium is constant; and the cake is nearly incompressible, so an average specific resistance can be used as a constant.

In light of these assumptions, the flow rate of liquid through the cake is predicted as follows. Assume first that area A for flow does not change with radius, as would nearly be true with a thin cake in a large-diameter centrifuge. The linear velocity of the liquid is then given by

$$u = \frac{dV/dt}{A} = \frac{q}{A} \tag{29.40}$$

where q is the volumetric flow rate of liquid. Substitution from Eq. (29.40) into Eq. (29.17) gives

$$\Delta p = q\mu \left(\frac{m_c \alpha}{A^2} + \frac{R_m}{A} \right) \tag{29.41}$$

The pressure drop from centrifugal action, from Eq. (2.8), is

$$\Delta p = \frac{\rho \omega^2 \left(r_2^2 - r_1^2 \right)}{2} \tag{29.42}$$

where ω = angular velocity, rad/s
ρ = density of liquid

Combining Eqs. (29.41) and (29.42) and solving for q give

$$q = \frac{\rho \omega^2 \left(r_2^2 - r_1^2 \right)}{2\mu (\alpha m_c / A^2 + R_m / A)} \tag{29.43}$$

When the change in A with radius is too large to be neglected, it can be shown that Eq. (29.43) should be written[18]

$$q = \frac{\rho \omega^2 \left(r_2^2 - r_1^2 \right)}{2\mu (\alpha m_c / \bar{A}_L \bar{A}_a + R_m / A_2)} \tag{29.44}$$

where A_2 = area of filter medium (inside area of centrifuge basket)
$\bar{A}_a$ = arithmetic mean of cake area
$\bar{A}_L$ = logarithmic mean of cake area

The average areas $\bar{A}_a$ and $\bar{A}_L$ are defined by the equations

$$\bar{A}_a \equiv (r_i + r_2)\pi b \tag{29.45}$$

$$\bar{A}_L \equiv \frac{2\pi b(r_2 - r_i)}{\ln (r_2/r_i)} \tag{29.46}$$

where b = height of basket
r_i = inner radius of cake

Note that Eq. (29.44) applies to a cake of definite mass and is *not* an integrated equation over an entire filtration starting with an empty centrifugal. The cake resistance

α in Eqs. (29.43) and (29.44) is generally somewhat greater than that found in a pressure or vacuum filter under comparable conditions. Especially with compressible cakes, α increases with the applied centrifugal force.

Washing filter cakes

To wash soluble material that may be retained by the filter cake after a filtration, a solvent miscible with the filtrate may be used as a wash. Water is the most common wash liquid. The rate of flow of the wash liquid and the volume of liquid needed to reduce the solute content of the cake to a desired degree are important in the design and operation of a filter. Although the following general principles apply to the problem, these questions cannot be completely answered without experiment.[10]

The volume of wash liquid required is related to the concentration-time history of the wash liquid leaving the filter. During the first part of the wash period, the concentration is nearly constant. The effluent consists essentially of the filtrate that was left on the filter, which is swept out by the first wash liquid without appreciable dilution. This stage of washing, called *displacement wash,* is the ideal method of washing a cake. Under favorable conditions where the particle size of the cake is small, as much as 90 percent of the solute in the cake can be recovered during this stage. The volume of wash liquid needed for a displacement wash is equal to the volume of filtrate left in the cake, or $\bar{\varepsilon}AL$, where L is the cake thickness and $\bar{\varepsilon}$ is the average porosity of the cake. The second stage of washing is characterized by a rapid drop in concentration of the effluent. The volume of wash liquid used in this stage is also on the order of magnitude of that used in the first stage. In the third stage the concentration of solute in the effluent is low, and the remaining solute is slowly leached from the cake. If sufficient wash liquid is used, the residual solute in the cake can be reduced to any desired point; but once it is acceptably low, any further washing should be stopped when the value of the unrecovered solute is less than the cost of recovering it.

In most filters the wash liquid follows the same path as the filtrate.[†] The rate of flow of the wash liquid is, in principle, equal to that of the last of the filtrate, provided the pressure drop remains unchanged in passing from filtration to washing. If the viscosities of filtrate and wash liquid differ, correction for this difference can be made. This rule is only approximate, however, as during washing the liquid does not actually follow exactly the path of the filtrate because of channeling, formation of cracks in the cake, and short-circuiting.

When a cake of porous particles is washed, solute in the pores must first diffuse to the surface of the particles. This is a slow process compared to displacement from the external channels in the cake. The curves of concentration versus time are similar to the breakthrough curves for adsorption or other fixed-bed processes, and the equations in Chap. 25 can be applied to predict the effects of particle size, fluid velocity, and other variables.

[†]In a filter press the wash passes through the entire thickness of the cake. The last filtrate may pass through only one-half the final cake.

CLARIFYING FILTERS

Clarifying filters remove small amounts of solids or liquid droplets from either liquids or gases. The particles are trapped inside the filter medium or on its surfaces. Clarification differs from screening in that pores in the filter medium are larger—sometimes much larger—than the particles to be removed. The particles are caught by surface forces and immobilized on the surfaces or within the flow channels, where they reduce the effective diameter of the channels but usually do not block them completely.

Liquid Clarification

Clarifying filters for liquids are especially important in "polishing" materials such as beverages, pharmaceutical products, fuel oil, lubricants, and electroplating solutions and are essential in cleaning the feed in processes of fiber spinning and film extrusion.[30e] They include the gravity-bed filters for water treatment mentioned earlier and various disk and plate presses and cartridge clarifiers. The feed to such units generally contains no more than 0.10 percent solids. Some cake filters, especially tank filters and continuous precoat filters, are used extensively for clarification, as are many of the crossflow filters described in the next section.

In a batch unit, the filtration rate and solids removal efficiency are typically almost constant for a considerable period of operation, but eventually the solids contents of the filtrate rises to an unacceptable "breakthrough" value, and backwashing of the filter element becomes necessary.

A pack of disks for a clarifying *disk filter* is illustrated in Fig. 29.17. The disks are made of asbestos and cellulose fibers. In operation the assembly shown in the figure is sealed into a pressure case, usually operating at a gauge pressure less than

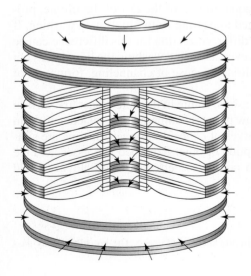

FIGURE 29.17
Pack of disks for clarifying filter.

345 kPa (50 $lb_f/in.^2$). Liquid flows inward through the disks and into a central manifold with a central or peripheral discharge. Individual units deliver as much as 378 L/min (6,000 gal/h) of low-viscosity liquid.[30e]

Gas Cleaning

Filters for gas cleaning include pad filters for atmospheric dust and granular beds and bag filters for process dusts. Air is cleaned by passing it through pads of cellulose pulp, cotton, felt, glass fiber, or metal screening; the pad material may be dry or coated with a viscous oil to act as a dust holder. For light duty the pads are disposable, but in large-scale gas cleaning they are frequently rinsed and recoated with oil.

Granular bed filters contain stationary or moving beds of granules ranging from 30-mesh to 8-mesh in size in some designs, to 12 to 40 mm ($\frac{1}{2}$ to $1\frac{1}{2}$ in.) in others. A bag filter contains one or more large bags of felt or thin woven fabric, mounted inside a metal housing. Dust-laden gas usually enters the bag at the bottom and passes outward, leaving the dust behind, although sometimes the flow is inward. Efficiencies are typically 99 percent even with extremely fine particles—far finer than the openings in the bag material. Periodically the flow is automatically cut off, and clean gas is blown back or the bag mechanically shaken to dislodge the dust for recovery or disposal. In most cases bag filters act as clarifiers, with particles trapped within the fabric of the bag, but with heavy dust loadings a thin but definite cake of dust is allowed to build up before it is discharged.

In all these filters most of the separation is by impingement, as discussed below.

Principles of Clarification

If the solid particles being removed completely plug the pores of the filter medium and the rate of plugging is constant with time, the mechanism is known as *direct sieving*. Direct sieving is rarely encountered. Much more commonly the particles partially block the pores, giving a gradual reduction in pore size; this is called *standard blocking*. Grace[19] gives equations for flow rate through a clarifying filter as a function of time.

Most often, especially in cleaning gases, the separation is by impingement of the particles against a solid surface placed in the flowing stream. The particles, because of their inertia, are expected to cross the streamlines of the fluid and strike and adhere to the solid, from which they can subsequently be removed. The principle of impingement separation is shown in Fig. 29.18. The solid lines are the streamlines passing around a sphere, and the dotted lines show the paths followed by the particles. Particles initially moving along the streamlines between A and B strike the solid and can be removed if they adhere to the wall and are not reentrained. Particles initially following streamlines outside lines A and B do not strike the solid and cannot be removed from the gas stream by impingement. The *target efficiency* η_t is defined as the fraction of the particles in the gas stream directly approaching the separator element that strike the solid. For particles that would settle through

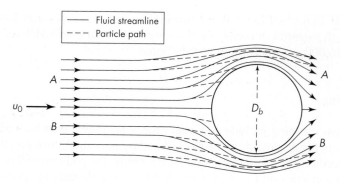

FIGURE 29.18
Principle of impingement. [*By permission, from J. H. Perry (ed.), Chemical Engineers' Handbook, 6th ed., p. 20–81. Copyright 1984, McGraw-Hill Book Company, New York.*]

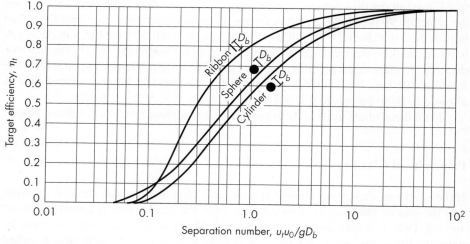

FIGURE 29.19
Target efficiency of spheres, cylinders, and ribbons. [*By permission, from J. H. Perry (ed.), Chemical Engineers' Handbook, 6th ed., p. 20–83. Copyright 1984, McGraw-Hill Book Company, New York.*]

still fluid in the Stokes law range, the target efficiencies for ribbons, spheres, and cylinders are shown in Fig. 29.19. The abscissa is the dimensionless group N_s, called the *separation number* $u_t u_0 / g D_b$, where u_t is the terminal velocity of the particle in still fluid, u_0 is the velocity of the fluid approaching the solid, g is the acceleration of gravity, and D_b is the width of the ribbon or the diameter of the sphere or cylinder.

In settling in the Stokes law range, the terminal velocity u_t is proportional to D_p^2. Thus the smaller the particle, the lower the target efficiency. The target efficiency, however, can be increased by decreasing the target size D_b, and for the collection of very small particles filter pads are made using extremely fine fibers of glass, metal, or polymers. The overall collection efficiency depends on the target efficiency, the solid fraction of the filter, and the depth of the filter. Even when the target efficiency is not very high, nearly complete removal of the particles can be accomplished by using a deep bed of fibers to give repeated chances for particle interception.

For very small drops 0.1 μm or so in diameter (as in a mist of H_2SO_4), the target efficiency rises as the particle diameter decreases, contrary to what is predicted by impingement theory. The tiny droplets move across the streamlines by diffusion and are immobilized on the solid surfaces. Diffusion is more and more effective as the particles become smaller, and the target efficiency is correspondingly increased.

CROSSFLOW FILTRATION; MEMBRANE FILTERS

The principle of crossflow filtration can be applied to concentrate suspensions of fine particles or collodial material or to fractionate solutions of macromolecules. The term *microfiltration* is generally used for particles in the size range of 0.5 to 5 μm. Cake filtration could be used for some of these materials, but a cake of 1 μm particles would have a high resistance to flow, and the filtration rate would be very low. *Ultrafiltration* (UF) covers a wider size range, from 0.5-μm particles down to molecules about 10^{-3} μm in size ($M \cong 300$). The term *hyperfiltration* is sometimes used for separation of small molecules or ions, but reverse osmosis is a more descriptive term, because the osmotic pressure has a major effect on the flux. Furthermore, the separation in *reverse osmosis* occurs by a solution diffusion mechanism in the dense polymer rather than by a screening action at the membrane surface (see Chap. 26).

Types of Membranes

The ideal membrane for crossflow filtration would have a high porosity and a narrow pore size distribution, with the largest pores slightly smaller than the particles or molecules to be retained. An asymmetric membrane is preferred, with a thin selective skin supported on a thicker layer with large pores to decrease the hydraulic resistance. Some commercial membranes have a continuous gradation in pore size, as shown in Fig. 29.20. Several kinds of polymers are used for ultrafiltration membranes, including cellulose acetate, polyacrylonitrile, polysulfone, polyamide, and polyimide. The skin thickness and average pore size can be varied by changing the casting conditions or postcasting treatment.

Ultrafiltration membranes have a range of pore sizes in the selective layer, and they are often characterized by a molecular weight cutoff based on measurements of fraction rejected versus molecular weight.[15] Molecules larger than the cutoff size are almost completely rejected, but there is a wide range of sizes for which partial rejection occurs. Rejection curves for some membranes are shown in Fig. 29.21, but such data should be used with caution, since the fraction rejected for a given

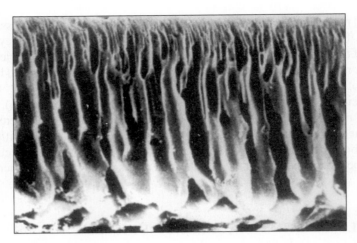

FIGURE 29.20
Cross section of an anisotropic ultrafiltration membrane. (Courtesy of Millipore Corporation.)

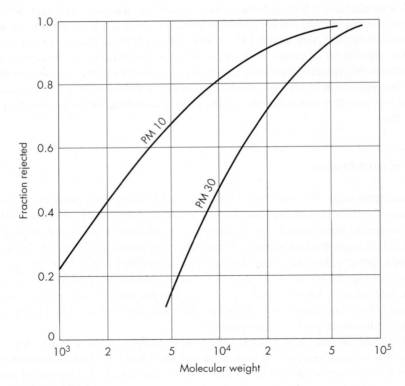

FIGURE 29.21
Solute rejection curves for Amicon UF membranes. (*After Porter.*[32])

molecular weight varies with the molecular shape, the solvent permeation rate, the shear rate near the surface, and the extent of membrane fouling.

Ultrafiltration and microfiltration membranes are also made from sintered stainless steel or other metals and from porous carbon or alumina. The usual methods of preparation give membranes with fairly large pores (1 to 100 μm), but partially filling the surface layer with zirconium oxide or other inorganic material produces asymmetric membranes with controlled pore size. The advantages of inorganic membranes over polymer membranes are that they can operate at higher temperatures and can better stand chemical cleaning or sterilization treatment.

Several types of equipment are available for laboratory use and for commercial UF installations.[5] Small samples may be processed with flat circular membranes supported on a porous disk in a cylindrical cell. A magnet-driven stirrer above the membrane provides crossflow of the solution. Rectangular cells can also be used with solution recirculated through the small gap between the membrane and the top of the cell. These devices are useful for preliminary tests, but performance data are not directly applicable to large units.

For industrial applications, large areas are obtained using modules with many tubular or hollow-fiber membranes or by using large sheets in a filter press arrangement or a spiral-wound unit. Tubular membranes are 5 to 25 mm in inside diameter and up to 3 m long, and velocities of 1 to 5 m/s are used to give turbulent flow and good mass transfer. The polymer membranes are supported by porous metal or ceramic tubes, and several tubes are clustered in a cylindrical shell. Alumina membranes are available as single tubes or as monolithic multichannel elements of the type shown in Fig. 29.22. Separation occurs in the thin layer of oxide particles bound to the alumina, and the pores in the support are large enough to have little resistance to permeate flow. Tubular membranes have the advantages of easy membrane replacement and the feasibility of mechanical cleaning, but the ratio of surface area to volume is relatively low.

Hollow-fiber UF membranes have diameters of 0.2 to 2.0 mm, and hundreds or thousands of fibers are sealed in each cylindrical module. Flow through the fibers is usually laminar, but the velocity is kept high to give high shear at the wall and

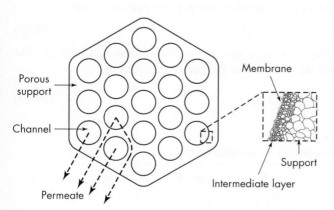

FIGURE 29.22
Cross-sectional schematic of a monolithic multichannel element. (*After Hsieh.*[22])

Porous support

Channel

Permeate

Membrane

Support

Intermediate layer

improve the flux. The area per unit volume is greatest for the smallest fibers, but these are most susceptible to plugging by suspended matter in the feed, and the fibers used are generally larger than those in reverse-osmosis units.

Spiral-wound modules of the type used for reverse osmosis (see Fig. 26.19) are widely used for UF. They are not as prone to plugging as hollow-fiber units, since the entrance is a narrow slit about 1 mm wide, but prefiltration of the feed solution is recommended. The velocity in the feed channels corresponds to laminar flow, but the flow disturbances caused by the spacers make the pressure drop and the mass transfer greater than for true laminar flow.

Permeate Flux for Ultrafiltration

The performance of a UF membrane can be characterized by the permeate flux, the percent rejection, and the concentration of solute in the retentate stream. The permeate flux often decreases with time because of membrane fouling, but fouling may increase the rejection. We deal first with the performance of a clean membrane. The water is assumed to pass by laminar flow through the small pores of the selective layer, and the driving force is the pressure difference Δp minus the difference in osmotic pressures across the membrane $\Delta \pi$. The flux is proportional to the void fraction ε and the square of the average pore size, D. The pores are at random angles to the surface, and the nominal thickness of the active layer L is multiplied by a tortuosity factor τ. A modified Hagen–Poiseuille equation gives the volume flux v, which is the superficial permeate velocity normal to the surface:

$$v = \frac{(\Delta p - \Delta \pi)D^2 \varepsilon}{32 L \tau \mu} \tag{29.47}$$

where ε is the void fraction of the membrane. In SI units, the flux has units of m^3/s, m^2, or m/s, but more common units are L/m$^2 \cdot$h or gal/ft$^2 \cdot$day. Conversion factors are shown in Table 29.3.

It is difficult to get independent measurements of ε, D, τ, and L for use in Eq. (29.47), but these characteristics are incorporated in the membrane permeability Q_m, the flux of pure water at room temperature per unit pressure drop:

$$v = Q_m \Delta p \tag{29.48}$$

For UF use, the permeate flux for a clean membrane can be predicted from Q_m, the driving force $\Delta p - \Delta \pi$, and the viscosity of the permeate solution:

$$v = Q_m(\Delta p - \Delta \pi)\frac{\mu_{H_2O}}{\mu} \tag{29.49}$$

TABLE 29.3
Conversion factors for permeate flux

m/s	m/h	L/m$^2 \cdot$h	gal/ft$^2 \cdot$day
2.78×10^{-7}	10^{-3}	1	0.589
4.72×10^{-7}	1.698×10^{-3}	1.698	1

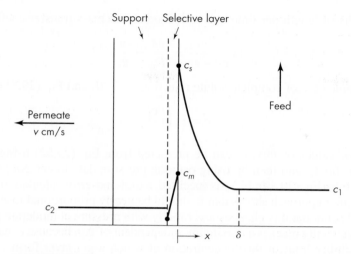

FIGURE 29.23
Concentration gradients in a UF membrane.

Concentration Polarization

The osmotic pressure difference $\Delta\pi$ depends on the solute concentration at the membrane surface, which is often much greater than the bulk concentration, particularly if the permeate flux is high and the solute diffusivity is low. Figure 29.23 shows concentration gradients for a UF system with partial rejection of the solute. The concentration just inside the pores c_m is lower than the concentration at the surface c_s by the factor K, the equilibrium partition coefficient. There is a similar discontinuity in solute concentration where the selective layer joins the large-pore support. For cylindrical pores and spherical molecules, $K = (1 - \lambda)^2$, where λ is the ratio of molecular size to pore size. A low value of K contributes to a high rejection, and there is a further effect of wall friction on the solute diffusivity, which may make the rejection quite high even when the molecules are less than one-half the size of the pores [see Eq. (26.34)].

In Chap. 26, concentration polarization in reverse osmosis was treated using a simple mass-transfer equation, Eq. (26.49), which is satisfactory where the surface concentration is only moderately higher than the bulk concentration. For UF, the large change in concentration near the surface requires integration to get the concentration profile. The basic equation states that the flux of solute due to convection plus diffusion is constant in the boundary layer and equal to the flux of solute in the permeate[4]

$$vc + D_v \frac{dc}{dx} = vc_2 \tag{29.50}$$

Equation (29.50) is integrated with the boundary conditions $c = c_s$ at $x = 0$ and $c = c_1$ at $x = \delta$, the thickness of the concentration boundary layer. The term D_v/δ

has the units of length per time and is defined as the mass-transfer coefficient k_c:

$$\ln \frac{c_s - c_2}{c_1 - c_2} = \frac{v\delta}{D_v} = \frac{v}{k_c} \qquad (29.51)$$

For the simple case of complete solute rejection, $c_2 = 0$, and Eq. (29.51) becomes

$$v = k_c \ln \frac{c_s}{c_1} \qquad (29.52)$$

For low values of flux, c_s can be predicted from Eq. (29.52) using standard correlations for k_c, and then by using osmotic pressure data to get $\Delta \pi$, Δp can be calculated from Eq. (29.49). If Δp is specified, a trial-and-error calculation is needed to get v. Either approach shows that v should be nearly proportional to Δp for low values of Δp, but the flux changes less rapidly with pressure at moderate pressures. The flux eventually reaches a maximum, independent of Δp, because c_s has reached c_g, the solubility limit or the concentration at which a gel layer forms. A further sudden increase in Δp gives a temporary increase in permeate flow, but this decreases to the maximum steady-state flux as the gel layer becomes thicker and increases the hydraulic resistance. Figure 29.24 shows some performance curves for ultrafiltration of skim milk, a process used for concentrating the protein fraction. The limiting flux increases with the feed velocity, and the limit is reached at pressure differences of 10 to 20 $lb_f/in.^2$ (69 to 138 kPa).

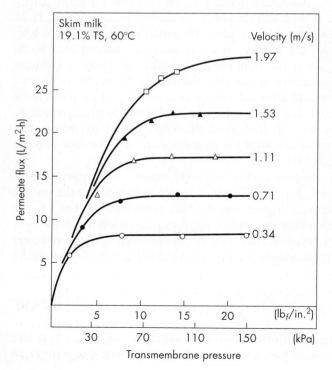

FIGURE 29.24
Ultrafiltration of skim milk in a hollow-fiber UF module (Romicon HF 15–43–PM50). (*After Cheryan and Chiang.*[6])

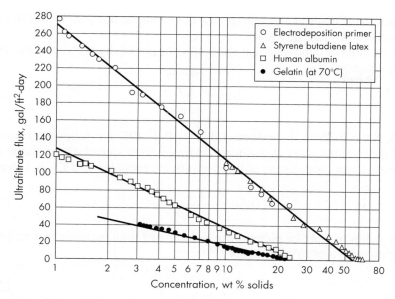

FIGURE 29.25
Effect of solute concentration of flux. (*After Porter.*[32])

Ultrafiltration systems generally operate close to the limiting flux, which occurs at pressure differences of only 1 to 5 atm. The flux increases with velocity past the membrane surface because of the increase in k_c, and the flux is lower for high bulk concentration, as indicated by Eq. (29.52). The change in flux with concentration is illustrated in Fig. 29.25, and semilog plots of this type can be extrapolated to zero flux to determine c_g. The gel concentration ranges from 10 to 50 weight percent for various proteins and is about 70 weight percent for latex suspensions.

For solutions of polymers and proteins up to about 300,000 in molecular weight, some tests with tubular membranes and thin-channel cells gave limiting fluxes in reasonable agreement (± 30 percent) with those predicted using standard correlations for mass transfer,[24,31] such as Eq. (17.60) for laminar flow and Eq. (17.67) for turbulent flow. Others have found fluxes considerably lower than expected using the same correlations and published diffusivities.[17,39] Some of the differences may be due to the higher viscosity near the wall or to the variation in diffusivity with concentration, factors not allowed for in the simple theory.

A more dramatic departure from theory occurs in the ultrafiltration of colloidal suspensions. The limiting flux is generally larger than for protein solutions, even though the diffusivity of the colloidal particles is an order of magnitude lower! Figure 29.25 shows that the flux for a latex suspension, which has particles about 0.2 μm in diameter, is 2 to 3 times greater than for albumin ($M = 170,000$, $D_p = 7.2 \times 10^{-3}$ μm) and 5 times greater than for gelatin ($M = 350,000$). Tests with whole blood, milk, and other suspensions give similar results. This phenomenon has been called the *flux paradox*. Something other than normal diffusion must be responsible for the transfer of particles from the surface of the membrane or the gel layer back to the mainstream. Current theories about the flux paradox are discussed in the section on microfiltration.

For particles or large spherical molecules, the diffusivity is given by the Stokes-Einstein equation

$$D_v = \frac{kT}{6\pi r_0 \mu} \tag{29.53}$$

where k is the Boltzmann constant, 1.380×10^{-23} J/K. The following convenient form of the equation gives D_v in square centimeters per second when r_0 is the radius of the particles in centimeters, μ' the viscosity in centipoise, and T is in kelvins:

$$D_v = \frac{7.32 \times 10^{-16} T}{r_0 \mu'} \tag{29.54}$$

The radius of globular proteins can be estimated from the molecular weight and density, using Avogadro's number $\mathbf{N}_a$.

$$r_0 = \left(\frac{3M}{4\pi \mathbf{N}_a \rho_p} \right)^{1/3} \tag{29.55}$$

For proteins with a density of 1.4 g/cm³,

$$r_0 = 6.57 \times 10^{-9} \, M^{1/3} \tag{29.56}$$

For random-coil polymer molecules, r_0 should be taken as the radius of gyration, which is generally 2 to 3 times the radius calculated from the molar volume.

EXAMPLE 29.3. A tubular membrane with a diameter of 2 cm and a water permeability of 250 L/m²·h·atm is being used for UF of cheese whey. The whey proteins have an average diffusivity of 4×10^{-7} cm²/s, and the osmotic pressure in atmospheres is given by Jonsson's equation[24]

$$\pi = 4.4 \times 10^{-3} c - 1.7 \times 10^{-6} c^2 + 7.9 \times 10^{-8} c^3$$

where c is the protein concentration in grams per liter. (a) Calculate the effect of Δp on the flux for a clean membrane if the solution velocity is 1.5 m/s and the protein concentration is 10, 20, or 40 g/L. Assume the gel concentration is 400 g/L and the rejection is 100 percent. (b) If the membrane permeability is reduced 5-fold by plugging, what is the effect on the permeate flux?

Solution

(a) Assume the bulk solutions have the same density and viscosity as water:

$$D = 2 \text{ cm} \qquad \bar{V} = 150 \text{ cm/s} \qquad \rho = 1 \text{ g/cm}^3 \qquad \mu = 0.01 \text{ g/cm} \cdot \text{s}$$

$$\text{Re} = 2 \times 150 \times 1/0.01 = 30,000 \qquad \text{Sc} = \frac{0.01}{1 \times 4 \times 10^{-7}} = 25,000$$

From Eq. (17.67),

$$\text{Sh} = 0.0096 \times 30,000^{0.913} \times 25,000^{0.346} = 3.9 \times 10^3$$

$$k_c = \frac{3.9 \times 10^3 (4 \times 10^{-7})}{2} = 7.8 \times 10^{-4} \text{ cm/s}$$

For $c_1 = 10$ g/L, pick $v = 10^{-3}$ cm/s or 36 L/m$^2 \cdot$ h. From Eq. (29.52)

$$\ln \frac{c_s}{c_1} = \frac{v}{k_c} = \frac{10^{-3}}{7.8 \times 10^{-4}} = 1.282$$

$$c_s = 3.60c_1 = 36 \text{ g/L}$$

At c_s, $\pi = 4.4 \times 10^{-3} \times 36 - 1.7 \times 10^{-6}(36)^2 + 7.9 \times 10^{-8}(36)^3 = 0.16$ atm. For complete rejection of protein $\Delta\pi = \pi = 0.16$:

$$Q_m = \frac{250 \text{ L}}{\text{m}^2 \cdot \text{h} \cdot \text{atm}} \times \frac{1}{36{,}000} = 6.94 \times 10^{-3} \text{ cm/s} \cdot \text{atm}$$

From Eq. (29.49)

$$\Delta p - \Delta\pi = \frac{10^{-3}}{6.94 \times 10^{-3}} = 0.144 \text{ atm}$$

$$\Delta p = 0.144 + 0.16 = 0.304 \text{ atm}$$

Note that over one-half of the driving force is needed to overcome the osmotic pressure difference caused by concentration polarization.

The maximum flux is obtained from Eq. (29.52) with $c_s = 400$:

$$v_{\max} = 7.8 \times 10^{-4} \ln \frac{400}{10} = 2.88 \times 10^{-3} \text{ cm/s} = 104 \text{ L/m}^2 \cdot \text{h}$$

At this point,

$$\Delta p - \Delta\pi = \frac{2.88 \times 10^{-3}}{6.94 \times 10^{-3}} = 0.41 \text{ atm}$$

For $c = 400$, $\pi = 6.54$ atm,

$$\Delta p = 6.54 + 0.41 = 6.95 \text{ atm}$$

Similar calculations are made for other values of v from 10^{-4} to $v_{\max}$, and the flux is plotted against the pressure drop in Fig. 29.26. The predicted flux is constant for $\Delta p > 6.95$, and a gel layer of increasing thickness is presumed to form as the pressure increases; but in practice, the flux might decrease slightly because of compression of the gel layer.

The curves for the three concentrations are similar in shape, but the flux is zero until the pressure difference exceeds the osmotic pressure of the solution, and the intercept is more noticeable for the higher concentrations.

(b) If $Q_m = 250/5 = 50$ L/m$^2 \cdot$ h$\cdot$atm, $v_{\max}$ is not changed, but a greater Δp is needed for any value of v. For example, at $v = 10^{-3}$ cm/s and $c = 40$ g/L,

$$Q_m = 50 \times \frac{1}{36{,}000} = 1.39 \times 10^{-3} \text{ cm/s} \cdot \text{atm}$$

$$\Delta p - \Delta\pi = \frac{10^{-3}}{1.39 \times 10^{-3}} = 0.719 \text{ atm}$$

$$\frac{c_s}{c_1} = 3.6 \qquad \text{as in part } (a)$$

$$c_s = 3.6 \times 40 = 144$$

$$\pi = 0.834 = \Delta\pi$$

$$\Delta p = 0.719 + 0.834 = 1.55 \text{ atm}$$

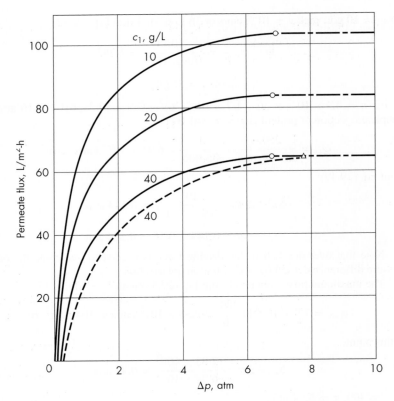

FIGURE 29.26
Effects of pressure drop and concentration on flux for Example 29.3:
(——) $Q_m = 250$ L/m$^2 \cdot$h$\cdot$atm [part (a)]; (-------) $Q_m = 50$ L/m$^2 \cdot$h$\cdot$atm
[part (b)]; (-- --) extrapolation of top three curves.

The dashed line in Fig. 29.26 shows that the largest effect of the lower membrane permeability is a 30 percent reduction in flux at low pressure drops.

Partial rejection of solutes. In many applications of UF, the membrane selected has some pores that are larger than the solute molecules, and the solute is only partially rejected. The fraction rejected R is sometimes defined using the feed and permeate concentrations:

$$R_F \equiv 1 - \frac{c_P}{c_F} \tag{29.57}$$

Since the concentrations change along the length of a separator, a more fundamental definition of R is based on the local retentate and permeate compositions c_1 and c_2 (see Fig. 29.23):

$$R \equiv 1 - \frac{c_2}{c_1} \tag{29.58}$$

The rejection R depends mainly on the ratio of solute size to pore size λ, which determines the partition coefficient K, and on the ratio v/k_c, which determines the concentration polarization effect. From Eqs. (29.51) and (29.58)

$$c_s = c_1\left(1 - R + R\exp\frac{v}{k_c}\right) \tag{29.59}$$

If diffusion in the membrane is negligible and solute is carried through the pores by the permeate flow, the permeate concentration is the same as that in equilibrium with c_s:

$$c_2 = Kc_s \tag{29.60}$$

Combining Eqs. (29.58), (29.59), and (29.60) leads to

$$\frac{1 - R}{R} = \frac{K}{1 - K}\exp\frac{v}{k_c} \tag{29.61}$$

When Eq. (29.61) applies, the rejection approaches $1 - K$ as v/k_c approaches zero, and the rejection decreases with increasing flux because of concentration polarization. For UF of a very dilute dextran solution with a tubular membrane,[17] the rejection varied from 77 to 93 percent with decreasing flux, in good agreement with Eq. (29.61) and $K = 0.044$.

When filtering solutions of moderate or high concentration, the maximum flux is low, and molecular diffusion in the pores may become important. The basic equation for solute flux in the selective layer is similar to Eq. (29.50) for mass transfer in the boundary layer, but the diffusion term adds to the convective term:

$$vc_2 = vc - D_e\frac{dc}{dy} \tag{29.62}$$

where

$$D_e = \frac{D_{\text{pore}}\varepsilon}{\tau} \tag{29.63}$$

y = distance from membrane surface

Integrating Eq. (29.62) and assuming the same partition factor at both boundaries of the selective layer leads to

$$\frac{c_2}{c_s} = \frac{K\exp(vL/D_e)}{K - 1 + \exp(vL/D_e)} \tag{29.64}$$

where L is the thickness of the selective layer.

Where the term vL/D_e is less than 2.0, solute diffusion has a significant effect on the rejection. Because diffusion lowers the rejection at low permeate flux and concentration polarization is important at high flux, the fraction rejected is predicted to go through a maximum with permeate flux.[20] The diffusion effect should be quite pronounced if the active layer is very thin, say, 0.1 to 0.2 μm, but there are not enough data to confirm this.

When one is dealing with a mixture of solutes or colloidal particles, a gel layer formed from one material may increase the rejection for a solute of lower molecular weight. Also, after a gel layer has formed, operation at high pressure may compress this layer and make it less permeable to smaller solutes as well as to solvent.

Membrane fouling. A common problem in the operation of ultrafiltration units is the gradual decrease in the permeate flux caused by membrane fouling. On start-up, there is usually a high initial flux corresponding to the pure water permeability and then a rapid decline to a much lower value, limited by concentration polarization. This is often followed by a gradual decline in the rate of permeate flow over a period of several hours, resulting from some type of membrane fouling. Molecules or particles slightly smaller than the pore size may enter the membrane but become blocked at constrictions in the pores. Some molecules in the feed may be strongly adsorbed on the pore walls, reducing the diameter of the channels and increasing the membrane resistance. Deposits can also form at the surface of the membrane by precipitation or by adsorption.

In the flux decline is caused by surface fouling or by compaction of the gel layer, the membrane can usually be cleaned by reversing the flow, or, for tubular membranes, by mechanical cleaning. For internal fouling, cleaning with chemical solutions may be required. If the mechanism of fouling can be identified, changes in the solution chemistry, pretreatment of the membrane surface, or careful choice of operating conditions may greatly prolong the period of satisfactory operation. For example, a change in pH may decrease the amount of adsorption, or operating with a relatively low pressure drop may give a smaller initial flux but a greater average flux because of less plugging of the pores.

EXAMPLE 29.4. Ultrafiltration tests with a 1.5-cm tubular membrane at Re = 25,000 gave a permeate flux of 40 L/m²·h and 75 percent rejection for a 5% polymer solution. The polymer has an average molecular weight of 30,000, and the estimated diffusivity is 5×10^{-7} cm/s. (*a*) Neglecting the effect of molecular diffusion in the pores, predict the fraction rejected for a flux of 20 L/m²·h, and predict the maximum rejection. (*b*) Estimate the fraction rejected for the low-molecular-weight fraction of the polymer with $M \cong 10,000$. (*c*) If the selective layer thickness is 0.2 μm, does molecular diffusion have a significant effect on the rejection for case (*a*)?

Solution
(*a*) Base case:

$$v = 40 \times 2.78 \times 10^{-5} = 1.112 \times 10^{-3} \text{ cm/s}$$

$$Sc = \frac{0.01}{5 \times 10^{-7}} = 20,000$$

From Eq. (17.67)

$$Sh = 0.0096(25,000)^{0.913}(20,000)^{0.346} = 3,060$$

$$k_c = \frac{3,060(5 \times 10^{-7})}{1.5} = 1.02 \times 10^{-3} \text{ cm/s}$$

$$\frac{1-R}{R} = \frac{0.25}{0.75} = \frac{K}{1-K} \exp \frac{1.112 \times 10^{-3}}{1.02 \times 10^{-3}}$$

$$\frac{K}{1-K} = 0.112$$

$$K = \frac{0.112}{1.112} = 0.101$$

If the flux is reduced to $20 \text{ L/m}^2 \cdot \text{h}$ or 0.556×10^{-3} cm/s,

$$\frac{1-R}{R} = \frac{0.101}{0.899} \exp \frac{0.556}{1.02} = 0.194$$

$$R = \frac{1}{1.194} = 0.84$$

As the flux approaches zero, R approaches $1 - K$:

$$R_{\max} = 1 - 0.101 = 0.90$$

(b) Use Fig. 29.21 for a rough estimate. Locate the point $R_1 = 0.75$ and $M_1 = 30,000$ on the graph, and draw a line similar to that for PM 30. At $M_2 = 10,000$, $R_2 \cong 0.35$. For an independent calculation, predict K and k_c. If $K_1 = 0.101 = (1 - \lambda_1)^2$,

$$\lambda_1 = 0.682 = \frac{D_1}{D_{\text{pore}}}$$

$$D_2 \cong D_1 \left(\frac{10,000}{30,000}\right)^{1/3} = 0.694 D_1$$

$$\lambda_2 = 0.682(0.694) = 0.473$$

$$K_2 = (1 - 0.473)^2 = 0.278$$

The diffusivity of large molecules varies with $-\frac{1}{3}$ power of the size, and k_c varies with D_v^{0-65} or $M^{-0.22}$:

$$k_{c_2} = k_{c_1} \times 3^{0.22} = 1.02 \times 10^{-3} \times 1.27 = 1.29 \times 10^{-3} \text{ cm/s}$$

At $v = 1.112 \times 10^{-3}$ cm/s,

$$\frac{1 - R_2}{R_2} = \frac{0.278}{1 - 0.278} \exp \frac{1.112}{1.29} = 0.912$$

$$R_2 = \frac{1}{1.912} = 0.52$$

This is appreciably higher than the estimate of 0.35, but in any case, a sharp separation is not possible for molecules differing only threefold in molecular weight.

(c) For $M = 30,000$ and $D_v = 5 \times 10^{-7}$ cm^2/s, estimate

$$D_{\text{pore}} = 1 \times 10^{-7} \text{ cm}^2/\text{s} \qquad \varepsilon = 0.5 \qquad \tau = 2$$

$$D_e = 2.5 \times 10^{-8} \text{ cm}^2/\text{s}$$

$$L = 0.2 \ \mu\text{m} = 2 \times 10^{-5} \text{ cm}$$

$$\frac{vL}{D_e} = \frac{(5.56 \times 10^{-4})(2 \times 10^{-5})}{2.5 \times 10^{-8}} = 0.445$$

From Eq. (29.64) with $K = 0.101$,

$$\frac{c_2}{c_s} = \frac{0.101 \exp 0.445}{0.101 - 1 + \exp 0.445} = 0.24$$

Diffusion in the membrane makes the permeate concentration about twice as high as it would be if $c_2 = Kc_s = 0.101c_s$. This indicates that the partition coefficient is lower than that estimated in part (a).

Microfiltration

There is not a sharp dividing line between microfiltration, which treats suspensions of small particles, and ultrafiltration, which usually deals with solutions of large molecules. For very small particles, such as 0.1-μm spheres of polymer in a latex paint, either term could be applied. However, for particles of this size or larger, the osmotic pressure can be neglected, and the molecular diffusivity is too low to account for significant mass transfer away from the membrane surface. The solvent flux in microfiltration is generally much less than the flux with pure water, indicating that a layer of solids has formed at the surface. The limiting flux depends on the type and concentration of the particles, membrane characteristics, and flow rate of the suspension. Theories for suspensions of uniform spherical particles have been developed to help guide the selection of operating conditions.

Fundamental studies of particles in motion have been made for laminar flow of dilute suspensions in thin channels or tubes. Neutrally buoyant particles in a nonporous tube tend to move away from the wall because of an inertial lifting force, giving a particle-free zone near the wall. This is called the tubular-pinch effect. Where the wall is porous, as in microfiltration, flow toward the wall exerts a drag on the particles opposing the lifting force. For particles 10 μm or greater in size, the inertial lifting force is often large enough to keep the particle from reaching the wall.[1] Measured trajectories for isolated particles agree with predictions, but the theory has not yet been extended to concentrated suspensions or turbulent flow.[3]

For 1-μm particles, the inertial force is quite small relative to the drag force, and rapid formation of a particle or gel layer is predicted to occur a short distance from the channel entrance.[33] The layer reaches steady-state thickness when the number of particles brought to the gel layer by permeate flow is balanced by the outward motion of particles from the top of the layer. The high shear rate near the surface causes a tumbling motion of the particles, which expands the layer and leads to particle migration away from the wall.[11] This *shear-induced dispersion* and the particle movement toward regions of lower concentration can be modeled with a particle diffusivity D_s that is proportional to the shear rate and the square of the particle size. An empirical equation for particle dispersion in laminar flow was obtained by tracking radioactive particles in a Couette flow device[14]

$$D_s = 0.03\gamma r^2 \tag{29.65}$$

where D_s = shear-induced diffusivity, cm^2/s
  γ = shear rate, s^{-1}
  r = particle radius, cm

This equation, when combined with Eq. (29.52) and the equation for laminar flow mass transfer [Eq. (17.60)], gave a reasonable fit to limiting flux data in several studies of microfiltration.[41] More work is needed to understand the effects of particle shape and size distribution, and to extend the correlation to turbulent flow.

Although the permeate flux increases with crossflow velocity, velocities greater than a few meters per second are usually impractical because of the large pressure drop. Another way of providing high shear at the membrane surface is to use a rotating filter plate or a rotating disk very close to the filter plate. Several devices of this type have been tested and shown[27] to give fluxes of 100 to 300 L/m$^2 \cdot$h.

GRAVITY SEDIMENTATION PROCESSES

Many mechanical separations are based on the sedimentation of solid particles or liquid drops through a fluid, impelled by the force of gravity or by centrifugal force. This section deals with gravity settling and the next with centrifugal sedimentation. The fluid may be a gas or liquid; it may be flowing or at rest. In some situations the objective of the process is to remove particles from a stream in order to eliminate contaminants from the fluid or to recover the particles, as in the elimination of dust and fumes from air or flue gas or the removal of solids from liquid wastes. In other problems, particles are deliberately suspended in fluids to obtain separations of the particles into fractions differing in size or density. The fluid is then recovered, sometimes for reuse, from the fractionated particles.

The principles of particle mechanics that underlie the separations described here are discussed in Chap. 7. If a particle starts at rest with respect to the fluid in which it is immersed and is then moved through the fluid by an external force, its motion can be divided into two stages. The first stage is a short period of acceleration, during which the velocity increases from zero to the terminal velocity. The second stage is the period during which the particle is at its terminal velocity.

Since the period of initial acceleration is short, usually on the order of tenths of a second or less, initial-acceleration effects are short-range. Terminal velocities, on the other hand, can be maintained as long as the particle is under treatment in the equipment. Equations such as (7.30) and (7.32) apply during the acceleration period, and equations such as (7.40) and (7.43) during the terminal-velocity period. Some separation methods, such as jigging and tabling, depend on differences in particle behavior during the acceleration period. Most common methods, however, including all those described here, make use of the terminal-velocity period only.

Particles heavier than the suspending fluid may be removed from a gas or liquid in a large settling box or settling tank, in which the fluid velocity is low and the particles have ample time to settle out. Simple devices of this kind, however, have limited usefulness because of the incompleteness of the separation and the labor required to remove the settled solids from the floor of the vessel.

Industrial separators nearly all provide for the continuous removal of settled solids. The separation may be partial or very nearly complete. A settler that removes virtually all the particles from a liquid is known as a *clarifier*, whereas a device that separates the solids into two fractions is called a *classifier*. The same principles of sedimentation apply to both kinds of equipment.

Gravity classifiers

Most classifiers in chemical processing separate particles on the basis of size, in situations in which the density of the fine particles is the same as that of the larger ones. The elutriation leg of the crystallizer shown in Fig. 27.11 is an example. By adjusting the upward velocity of the liquid so that it is smaller than the terminal settling velocity of acceptably large crystals, this device carries unwanted fine crystals back to the crystallizing zone for further growth.

Mechanical classifiers are used in closed-circuit grinding, especially in metallurgical operations. Here the relatively coarse particles are called *sands,* and the slurry of fine particles is called *slimes.* Sufficient time is provided to allow the sands to settle to the bottom of the device; the slimes leave in the effluent liquid.

In a typical mechanical classifier the settling vessel is a semicylindrical trough set at an angle of about 12° with the horizontal, with a liquid overflow at the lower end. Slurry is fed continuously to the middle of the trough. The flow rate is adjusted so that fines do not have time to settle but are carried out with the overflow liquid. Larger particles sink to the floor of the trough. A rotating helical conveyor moves the settled solids upward along the floor of the trough, out of the pool of liquid, and up to the sands discharge chute. Such a classifier works well with coarse particles where exact splits are not required. Typical applications are in connection with ball or rod mills for reduction to particle sizes between 8- and 20-mesh.

Sorting classifiers

Devices that separate particles of differing densities are known as *sorting classifiers.* They use one or the other of two principal separation methods: sink-and-float and differential settling.

Sink-and-float methods. A sink-and-float method uses a liquid sorting medium, the density of which is intermediate between that of the light material and that of the heavy. Then the heavy particles settle through the medium, and the lighter ones float, and a separation is thus obtained. This method has the advantage that, in principle, the separation depends only on the difference in the densities of the two substances and is independent of the particle size. This method is also called *heavy-fluid separation.*

Heavy-fluid processes are used to treat relatively coarse particles, usually greater than 10-mesh. The first problem is to choose a liquid medium of the proper density to allow the light material to float and the heavy to sink. True liquids can be used, but since the specific gravity of the medium must be in the range of 1.3 to 3.5 or greater, there are only a few liquids that are sufficiently heavy, cheap, nontoxic, and noncorrosive to be practicable. Halogenated hydrocarbons and $CaCl_2$ solutions have been used for this purpose. A more common choice of medium is a pseudoliquid consisting of a suspension in water of fine particles of a heavy mineral. Magnetite (specific gravity 5.17), ferrosilicon (specific gravity 6.3 to 7.0), and galena (specific gravity 7.5) are used. The ratio of mineral to water can be varied to give a wide range of medium densities. Provision must be made for feeding the mixture to be separated, for removing overflow and underflow, and for recovering the separating fluid, which may be expensive relative to the value of the materials being treated.

Hindered settling is used. Cleaning coal, and concentrating ores are common applications of sink and float. Under proper conditions, clean separations between materials differing in specific gravity by only 0.1 have been claimed.[38]

Differential settling methods. Differential settling methods utilize the difference in terminal velocities that can exist between substances of different density. The density of the medium is less than that of either substance. The disadvantage of the method is that since the mixture of materials to be separated covers a range of particle sizes, the larger, light particles settle at the same rate as the smaller, heavy ones and a mixed fraction is obtained.

In differential settling, both light and heavy materials settle through the same medium. This method brings in the concept of equal-settling particles. Consider particles of two materials A and B settling through a medium of density ρ. Let material A be the heavier; for example, component A might be galena (specific gravity 7.5) and component B quartz (specific gravity 2.65). The terminal velocity of a particle of size D_p and of density ρ_p settling under gravity through a medium of density ρ is given by Eq. (7.40) for settling in the Stokes' law regime. This equation can be written, for a galena particle of density ρ_{pA} and diameter D_{pA}, as

$$u_{tA} = \frac{g D_{pA}^2 (\rho_{pA} - \rho)}{18\mu} \tag{29.66}$$

For a quartz particle of density ρ_{pB} and diameter D_{pB},

$$u_{tB} = \frac{g D_{pB}^2 (\rho_{pB} - \rho)}{18\mu} \tag{29.67}$$

For equal-settling particles $u_{tA} = u_{tB}$, therefore

$$\frac{D_{pA}}{D_{pB}} = \sqrt{\frac{\rho_{pB} - \rho}{\rho_{pA} - \rho}} \tag{29.68}$$

For settling in the Newton's law range, the diameters of equal-settling particles, from Eq. (7.43), are related by the equation

$$\frac{D_{pA}}{D_{pB}} = \frac{D_{pB} - \rho}{\rho_{pA} - \rho} \tag{29.69}$$

The significance in a separation process of the equal-settling ratio of diameters is shown by Fig. 29.27, in which curves of u_t versus D_p are plotted for components A and B, for settling in the intermediate regime between Stokes'-law and Newton's-law settling. Assume that the diameter range for both substances lies between points D_{p1} and D_{p4} on the size axis. Then all particles of the light component B having diameters between D_{p1} and D_{p2} will settle more slowly than any particle of the heavy substance A and in theory can be obtained as a pure fraction. Likewise, any particles of substance A having diameters between D_{p3} and D_{p4} settle faster than any particle of substance B and can also be obtained as a pure fraction. But any light

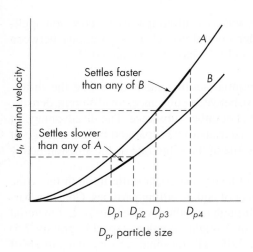

FIGURE 29.27
Equal-settling particles.

particle having a diameter between D_{p2} and D_{p4} settles at the same speed as a particle of substance A in the size range between D_{p1} and D_{p3}, and all particles in these size ranges form a mixed fraction.

Equations (29.68) and (29.69) show that the sharpness of separation is improved if the density of the medium is increased. It is also clear from Fig. 29.27 that the mixed fraction can be reduced or eliminated by closer sizing of the feed. For example, if the size range of the feed is from D_{p3} to D_{p4} in Fig. 29.27, complete separation is possible.

Clarifiers and thickeners

Gravity separation under hindered settling conditions is often used to convert a dilute slurry of fine particles into a clarified liquid and a concentrated suspension. This process is carried out in large open tanks called thickeners or clarifiers. The concentrated suspension, or sludge, may have to be filtered to produce a drier product, but the cost of the filtration step is much lower than if the original slurry had been filtered directly. The clarified liquid is free or nearly free of suspended particles, and it may be reused as process water or discharged as waste.

Flocculation. If the solids in a suspension are mainly individual particles only a few micrometers in diameter, the gravity settling rate will be very low and perhaps too low for practical operation. Fortunately, in many fine suspensions, the particles form agglomerates or clusters of particles that settle at reasonable rates. Agglomeration is sometimes promoted by adding flocculating agents including strong electrolytes, which reduce the repulsive forces between the charged particles, or polymeric flocculants that may be cationic, anionic, or nonionic in character. Flocculation is also carried out by adding inexpensive materials such as lime, alumina, or sodium silicate, which form loose agglomerates that carry fines down with them.

Flocculated particles have different settling characteristics from suspensions of dispersed dense solids. The aggregates have a high porosity and retain a considerable

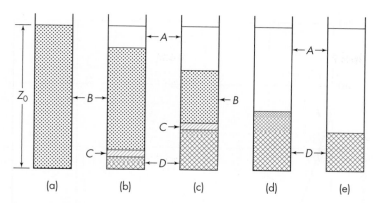

FIGURE 29.28
Batch sedimentation.

amount of water that accompanies the flocs when they settle. The aggregates are also loosely bonded, and the sludge at the bottom of the settler compresses under the weight of additional solids. Because the size, shape, and effective density of the flocs are not readily definable, it is not possible to predict the settling rate or the sludge density from theories or general correlations. The thickener design is generally based on measurements of the settling rates obtained from batch tests in the laboratory.

Batch sedimentation. These are several stages in the settling of a flocculated suspension, and different zones are formed as sedimentation proceeds. Usually, the concentration of solids is high enough that sedimentation of individual particles or flocs is hindered by other solids to such an extent that all solids at a given level settle at a common velocity. At first, the solid is uniformly distributed in the liquid, as shown in Fig. 29.28a. The total depth of the suspension is Z_0. After a short time, the solids have settled to give a zone of clear liquid, zone A in Fig. 29.28b, and a zone D of settled solids. Above zone D is a transition layer, zone C, in which the solids content varies from that in the original pulp to that in zone D. In zone B, the concentration is uniform and equal to the original concentration, since the settling rate is the same throughout this zone. The boundaries between zones D and C and between C and B may not be distinct, but the boundary between zones A and B is usually sharp.

As settling continues, the depths of zones D and A increase. The depth of zone C remains nearly constant, and that of zone B decreases. This is shown in Fig. 29.28c. Eventually zone B disappears, and all the solids are in zones C and D (see Fig. 29.28d). Meanwhile, the gradual accumulation of solid puts stress on the material at the bottom, which compresses solids in layer D. Compression breaks down the structure of the flocs or aggregates, and liquid is expelled into the upper zones. Sometimes liquid in the flocs spurts out of zone D as small geysers as layer D compresses. Finally, when the weight of the solid is balanced by the compressive strength of the flocs, the settling process stops, as shown in Fig. 29.28e. The entire process shown in Fig. 29.28 is called sedimentation.

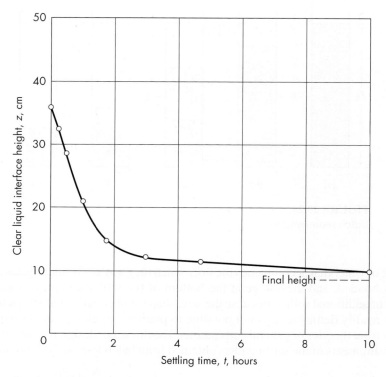

FIGURE 29.29
Batch settling of a limestone slurry, $c_0 = 236$ g/L. (*From Foust et al.,*
"Principles of Unit Operations," 2nd ed., Wiley, New York, 1980.
Copyright © 1980 by John Wiley & Sons, Inc. Reprinted by permission
of John Wiley & Sons, Inc.)

Rate of sedimentation. A typical plot of interface height (the boundary between
zones A and B) versus time is shown in Fig. 29.29. During the early stages of settling
the velocity is constant, as shown by the first portion of the curve. When zone B
disappears, the rate of settling starts to decrease and steadily drops until the ultimate
height is reached. For the example shown, the interface height was still decreasing at
20 h, and the ultimate height was only estimated.

Slurries vary greatly in their settling rates and in the relative heights of various
zones during settling. The initial rate is a function of the feed concentration, but in
the late stages, the settling rate also depends on the initial height Z_0, since compression effects are more important with the thicker sludge layers. Experimental studies
of the settling rate for different initial heights and concentrations are needed for the
design of a thickener.

Equipment for sedimentation; thickeners

Industrially, the above process is conducted on a large scale in equipment called
thickeners. For relatively fast-settling particles, a batch settling tank or continuous

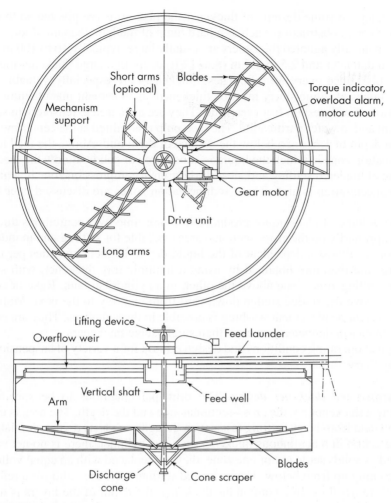

FIGURE 29.30
Gravity thickener. (*Eimco Corp.*)

settling cone may be adequate. For many duties, however, a mechanically agitated thickener like that shown in Fig. 29.30 must be employed. This is a large, fairly shallow tank with slow-moving radial rakes driven from a central shaft. Its bottom may be flat or a shallow cone. Dilute feed slurry flows from an inclined trough or launder into the center of the thickener. The feed slurry, being more dense than water, tends to flow downward until it reaches a zone of equal density. Then it moves radially outward at a constantly decreasing velocity, and the flow gradually divides between the downward-moving suspension and the upward-moving flow that is nearly free of solids. Liquor moves radially at a constantly decreasing velocity, allowing the solids to settle to the bottom of the tank. Clear liquor spills over the edge of the tank into a launder. The rake arms gently agitate the sludge and move it to the center of the tank, where it flows through a large opening to the inlet of a sludge pump. In some designs of thickener the rake arms are pivoted so that they can ride over an obstruction, such as a hard lump of mud, on the tank floor.

Mechanically agitated thickeners are usually large, typically 10 to 100 m (30 to 300 ft) in diameter and 2.5 to 3.5 m (8 to 12 ft) deep. In a large thickener the rakes may revolve once every 30 min. These thickeners are especially valuable when large volumes of dilute slurry must be thickened, as in cement manufacture or the production of magnesium from seawater. They are also used extensively in sewage treatment and in water purification. The feed pulp is admitted at the centerline of the unit at a depth of 1 m or so below the surface of the liquid. Above the feed level is a clarification zone in which the liquid is almost free of solids. Below the feed level is a zone of hindered settling and, near the bottom, a compression zone in which the solids concentration is high. These sedimentation zones are discussed later in this section.

The volume of clear liquor produced in a unit time by a continuous thickener depends primarily on the cross-sectional area available for settling and in industrial separators is almost independent of the liquid depth. Higher capacities per unit of floor area are therefore obtained by using a multiple-tray thickener, with several shallow settling zones, one above the other, in a cylindrical tank. Rake or scraper agitators move the settled sludge downward from one tray to the next. Multistage countercurrent displacement washing is possible in these devices. They are considerably smaller in diameter, however, than single-stage thickeners.

Typical operating conditions and design criteria for a variety of suspensions are given by Perry.[30c]

Clarifier and thickener design. The principal quantities to be specified in designing a thickener are the cross-sectional area and the depth. The area is usually based on data from batch settling tests, even though such tests do not simulate very well the action in a continuous thickener. In a test cylinder, there is no net volume flow, and as solids settle out of one zone, they are replaced with an equal volume of liquid coming up from below. The measured settling rates are valid for a reference frame of no net flow. In a continuous thickener, the depth of the layers is usually constant, at least for a short time, but some liquid is flowing downward with the solid, and the rest is flowing upward in the clarification zone. The design is based on a one-dimensional analysis assuming upward flow in the clarifier and downward flow in the settling zone.

In a continuous thickener, the total downward solids flux is made up of two parts: the flux of solids carried by the downflowing liquid and the additional flux resulting from the settling of the solids through the liquid. The first is called the transport flux G_t and is the product of the solids concentration c and the downward velocity u. The second is the settling flux G_s, which is the product of the solids concentration and the settling rate dZ/dt, as measured in batch tests,

$$G = G_t + G_s = uc + \frac{dZ}{dt} c \qquad (29.70)$$

The settling flux goes through a maximum with increasing concentration, since the settling rate is nearly constant at very low concentrations but decreases rapidly at

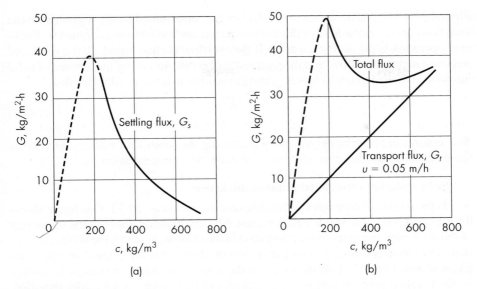

FIGURE 29.31
Fluxes for continuous thickening of a limestone slurry: (*a*) settling flux; (*b*) transport flux and total flux for $u = 0.05$ m/h.

high concentrations. Figure 29.31*a* shows data for limestone slurries derived from the settling curve in Fig. 29.29. These data cover only the declining flux portion of the curve, and the dashed line shows the approximate shape of the curve for low concentrations. Usually a thickener operates at a high underflow concentration, and the design is based on the declining flux part of the curve.

Different methods have been proposed for establishing the settling flux curve, and they often give different results. In the method of Coe and Clevenger,[8] the initial settling velocity is measured for several suspensions with concentrations between the feed and desired underflow concentrations, and the settling velocity is multiplied by the initial concentration to get the settling flux. This assumes that the settling rate depends only on concentration, but in practice, the settling rate may also depend on the concentration-time history of the suspension. Only one batch settling curve is needed for the Kynch method,[25] which was used to prepare Fig. 29.31. For each of several times, a tangent is drawn to the settling curve to get the settling rate, and the corresponding solids concentration is estimated by the following procedure. The tangent is extended to the ordinate axis to give an intercept Z_i, and the initial concentration is multiplied by Z_0/Z_i to get the concentration at the top of the settling zone. While only one laboratory test is needed, the measured settling rates may depend on the initial concentration and height of the suspension because of compression effects. Other methods of analyzing batch tests are described in the literature.[9,13,15,30b]

The transport flux increases linearly with concentration and is much lower than the settling flux at low concentrations, but it becomes the major factor at high concentrations. Because of this trend and the maximum in G_s, the total flux goes through a maximum and then a minimum with increasing concentration, as shown in

Fig. 29.31b. The required area of the thickener is determined by the minimum in the total flux curve, since a zone of this concentration tends to form in a continuous thickener operating at the design capacity. If the overflow is clear liquid, all the solids are removed in the underflow, and the required area is determined by the amount of solid introduced in the feed, which is Fc_0, and the minimum downward solids flux:

$$A = \frac{Fc_0}{(G_t + G_s)_{\min}} \tag{29.71}$$

The required area can be decreased by increasing the downward velocity, increasing the solids flux, but this also decreases the underflow concentration c_u.

Sedimentation zones in continuous thickeners

Typical vertical concentration profiles are shown in Fig. 29.32. This figure shows the region of very low solids concentration (the clarification zone) in the upper part of the thickener; the hindered settling zone in the middle, in which the solids concentration is virtually constant; and the region of high solids concentration (the compression zone) near the bottom, in which the solids concentration rises as the sludge is slowly raked to the discharge. The thickness of these zones is not easily specified, since there are radial as well as axial gradients in concentration, and changes in the feed rate or solids concentration lead to gradual changes in average zone thickness. The hindered settling zone may be from 0.3 to as much as 2 or 3 m thick depending on the operating history of the unit. At steady state, the performance does not depend on the thickness of this layer, as long as it does not extend above the feed level. If the thickener is operated for a short time at feed rates above the design value, the settling zone gradually increases in thickness because the solids feed rate exceeds the limiting solids flux. If the normal concentration profile were similar to curve a in Fig. 29.32, satisfactory operation might continue for several hours as the settling zone slowly increased in height. If the original concentration profile were similar to curve b, the accumulation of solids would lead to poorer clarification and loss of solids in the overflow.

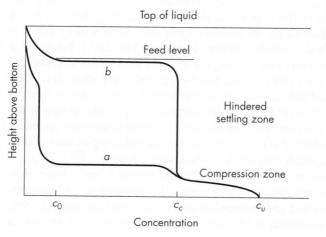

FIGURE 29.32
Concentration profiles in a continuous thickener.

CENTRIFUGAL SEDIMENTATION PROCESSES

A given particle in a given fluid settles under gravitational force at a fixed maximum rate. To increase the settling rate, the force of gravity acting on the particle may be replaced by a much stronger centrifugal force. Centrifugal separators have, to a considerable extent, replaced gravity separators in production operations because of their greater effectiveness with fine drops and particles and their much smaller size for a given capacity.

Separation of solids from gases; cyclones

Most centrifugal separators for removing particles from gas streams contain no moving parts. They are typified by the cyclone separator shown in Fig. 29.33. It consists of a vertical cylinder with a conical bottom, a tangential inlet near the top, and an outlet for dust at the bottom of the cone. The inlet is usually rectangular. The outlet pipe is extended into the cylinder to prevent short-circuiting of air from inlet to outlet.

The incoming dust-laden air travels in a spiral path around and down the cylindrical body of the cyclone. The centrifugal force developed in the vortex tends to move the particles radially toward the wall, and the particles that reach the wall slide down into the cone and are collected. The cyclone is basically a settling device in which a strong centrifugal force, acting radially, is used in place of a relatively weak gravitational force acting vertically.

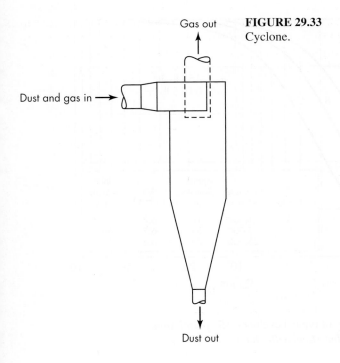

Gas out

Dust and gas in →

Dust out

FIGURE 29.33
Cyclone.

The centrifugal force F_c at radius r is equal to mu_{tan}^2/r_c, where m is the mass of the particle and u_{tan} is its tangential velocity. The ratio of the centrifugal force to the force of gravity is then

$$\frac{F_c}{F_g} = \frac{mu_{\text{tan}}^2/r}{mg} = \frac{u_{\text{tan}}^2}{rg} \tag{29.72}$$

For a cyclone 1 ft (0.3 m) in diameter with a tangential velocity of 50 ft/s (15 m/s) near the wall, the ratio F_c/F_g, called the *separation factor*, is $2{,}500/(0.5 \times 32.2) = 155$. A large-diameter cyclone has a much lower separation factor at the same velocity, and velocities above 50 to 70 ft/s (15 to 20 m/s) are usually impractical because of the high pressure drop and increased abrasive wear. Small-diameter cyclones may have separation factors as high as 2,500.[30a] To handle large gas flows, a number of small-diameter cyclones may be grouped in a single enclosure with common headers for the feed and product gases and a single dust hopper.

The dust particles entering a cyclone are accelerated radially, but the force on a particle is not constant because of the change in r and because the tangential velocity in the vortex varies with r and with distance below the inlet. Calculation of particle trajectories is difficult, and the efficiency of a cyclone is generally predicted from empirical correlations. Typical data for commercial cyclones are given in Fig. 29.34, which shows the strong effects of particle size and cyclone diameter on collection efficiency.

These three cyclones are of similar proportions with diameters of about 14, 32, and 72 in. (0.36, 0.81, and 1.83 m), and the lower efficiency of the larger cyclones

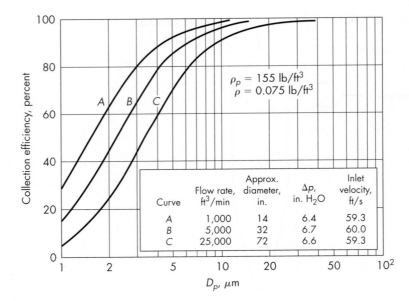

FIGURE 29.34
Collection efficiency of typical cyclones. (*By permission, Fisher-Klosterman Inc., Louisville, KY.*)

is mainly a result of the decrease in centrifugal force. For a given airflow rate and inlet velocity, however, moderate increases in cyclone diameter and length improve the collection efficiency, because the increase in surface area offsets the decreased centrifugal force. The results in Fig. 29.34 are for intermediate-size cyclones, and higher or lower efficiencies would be expected with larger or smaller units at the same flow rate and inlet velocity.

The decrease in efficiency with decreasing particle size is actually more gradual than predicted by simple theories. For small particles, the radial velocity and the collection efficiency should be a function of D_p^2, but agglomeration of the fines may occur to raise the efficiency for these particles. Because of the particle size effect, the uncollected dust leaving with the gas has a much smaller average size than the entering dust, which may be important in setting emission limits. Also, the overall efficiency is a function of the particle size distribution of the feed and cannot be predicted just from the average size.

The collection efficiency of a cyclone increases with the particle density and decreases as the gas temperature is increased because of the increase in gas viscosity. The efficiency is quite dependent on flow rate because of the u_{tan}^2 term in Eq. (29.72). The cyclone is one of the few separation devices that work better at full load than at partial load. Sometimes two identical cyclones are used in series to get more complete solids removal, but the efficiency of the second unit is less than that of the first, because the feed to the second unit has a much lower average particle size.

The pressure drop in a cyclone is proportional to the gas density and the square of the inlet velocity. It does not depend on the density of the solid particles. Surprisingly, the pressure drop actually decreases somewhat with increasing particle concentration.[21]

Liquid-solid separations; hydroclones[30h, 37b]

Cyclones are also used for separating solids from liquids, sometimes as thickeners but much more commonly as classifiers. In these services they are called hydrocyclones or hydroclones. The action in a hydroclone is shown in Fig. 29.35a. Feed enters tangentially at high velocity near the top. The liquid follows a spiral path near the vessel wall, forming a strong downward vortex. Large or heavy solid particles separate to the wall and are pushed downward and out of the cyclone as a slurry or paste. A variable-discharge orifice controls the consistency of the underflow. Most of the liquid goes back upward in an inner vortex and leaves through the central discharge pipe, which is known as a *vortex finder*.

In a hydroclone it is not possible to have both good solids removal and a high underflow concentration. In thickening operations, with nearly all the solids removed from the overflow, the underflow concentration must be less than about 12 percent by volume. When hydroclones are used for classifying, the underflow can be more concentrated, up to a maximum of about 50 percent by volume for slurries of limestone or coal. Figure 29.35b shows how the shape of a hydroclone is modified depending on the function of the unit.

The pressure drop Δp in a hydroclone varies with the feed rate raised to a power between 2.0 and 3.3. For dilute feeds the cut diameter varies with the 1.5 power of

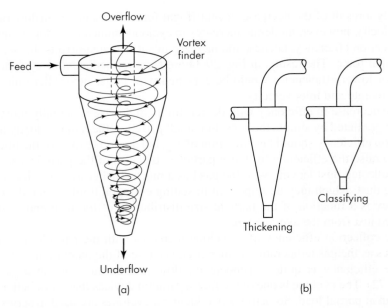

FIGURE 29.35
Hydroclones: (*a*) pattern of flow; (*b*) shape adapted to the kind of service.
(*From Walas, Chemical Process Equipment: Selection and Design,*
p. 329. Butterworths, Stoneham, MA, 1988.)

the cyclone diameter, so for a given pressure drop a small diameter gives better separation than a large one. Hydroclones are therefore small: They range in diameter from 10 mm (0.4 in.) to about 1.2 m (48 in.). To handle large flows, many small hydroclones are connected in parallel, with as many as 480 ten-mm units manifolded in a single assembly.

The cut size is a weak function of the pressure drop; for dilute feeds it varies with $\Delta p^{-0.25}$. Hence large pressure drops are not economical. Large hydroclones operate with a Δp of about 1 atm; in small ones Δp is 4 to 5 atm.

Hydroclones find applications in degritting operations in alumina production, removing carbon in upgrading gypsum for phosphoric acid manufacture, classifying pigments and crystal magmas, and similar process steps. They have largely replaced mechanical classifiers in closed-circuit grinding.

Centrifugal decanters

Immiscible liquids are separated industrially in centrifugal decanters as described in Chap. 2. The separating force is much larger than that of gravity, and it acts in the direction away from the axis of rotation instead of downward toward the earth's surface. The main types of centrifugal decanters are tubular centrifuges and disk centrifuges.

Tubular centrifuge. A tubular liquid-liquid centrifuge is shown in Fig. 29.36. The bowl is tall and narrow, 100 to 150 mm (4 to 6 in.) in diameter, and turns in a

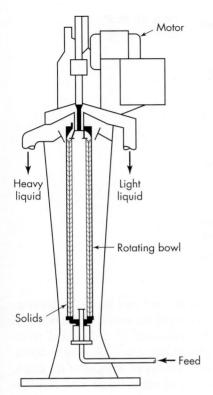

FIGURE 29.36
Tubular centrifuge.

stationary casing at about 15,000 rpm. Feed enters from a stationary nozzle inserted through an opening in the bottom of the bowl. It separates into two concentric layers of liquid inside the bowl. The inner, or lighter, layer spills over a weir at the top of the bowl; it is thrown outward into a stationary discharge cover and from there to a spout. Heavy liquid flows over another weir into a separate cover and discharge spout. The weir over which the heavy liquid flows is removable and may be replaced with another having an opening of a different size. The position of the liquid-liquid interface (the neutral zone) is maintained by a hydraulic balance as shown in Fig. 2.6 and Eq. (2.17). In some designs the liquids discharge under pressure, and the interface position is set by adjusting external valves in the discharge lines.

Disk centrifuge. For some liquid-liquid separations the disk-type centrifuge illustrated in Fig. 29.37 is highly effective. A short, wide bowl 200 to 500 mm (8 to 20 in.) in diameter turns on a vertical axis. The bowl has a flat bottom and a conical top. Feed enters from above through a stationary pipe set into the neck of the bowl. Two liquid layers are formed as in a tubular centrifuge; they flow over adjustable dams into separate discharge spouts. Inside the bowl and rotating with it are closely spaced "disks," which are actually cones of sheet metal set one above the other. Matching holes in the disks about halfway between the axis and the wall of the bowl form channels through which the liquids pass. In operation, feed liquid enters the

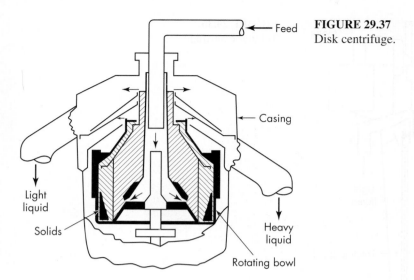

FIGURE 29.37
Disk centrifuge.

bowl at the bottom and flows into the channels and upward past the disks. Heavier liquid is thrown outward, displacing lighter liquid toward the center of the bowl. In its travel the heavy liquid very soon strikes the underside of a disk and flows beneath it to the periphery of the bowl without encountering any more light liquid. Light liquid similarly flows inward and upward over the upper surfaces of the disks. Since the disks are closely spaced, the distance a drop of either liquid must travel to escape from the other phase is short, much shorter than in the comparatively thick liquid layers in a tubular centrifuge. In addition, in a disk machine there is considerable shearing at the liquid-liquid interface, as one phase flows in one direction and the other phase in the opposite direction. This shearing helps break certain types of emulsions. Disk centrifuges are particularly valuable where the purpose of the centrifuging is not complete separation but the concentration of one fluid phase, as in the separation of cream from milk and the concentration of rubber latex.

If the liquid fed to a disk or tubular centrifuge contains dirt or other heavy solid particles, the solids accumulate inside the bowl and periodically must be discharged. This is accomplished by stopping the machine, removing and opening the bowl, and scraping out its load of solids. This becomes uneconomical if the solids are more than a few percent of the feed.

Tubular and disk centrifuges are used to advantage for removing traces of solids from lubricating oil, process liquids, ink, and beverages that must be perfectly clean. They can take out gelatinous or slimy solids that would quickly plug a filter. Usually they clarify a single liquid in a bowl provided with but a single liquid overflow; however, they also may throw down solids while simultaneously separating two liquid phases.

Nozzle-discharge centrifuge

When the feed liquid contains more than a few percent of solids, means must be provided for discharging the solids automatically. One way of doing this is shown in

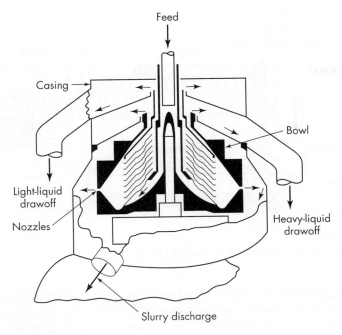

FIGURE 29.38
Nozzle-discharge centrifuge.

Fig. 29.38. This separator is a modified disk-type centrifuge with a double conical bowl. In the periphery of the bowl at its maximum diameter is a set of small holes, or nozzles, perhaps 3 mm in diameter. The central part of the bowl operates in the same way as the usual disk centrifuge, overflowing either one or two streams of clarified liquid. Solids are thrown to the periphery of the bowl and escape continuously through the nozzles, together with considerable liquid. In some designs part of the slurry discharge from the nozzles is recycled through the bowl to increase its concentration of solids; wash liquids may also be introduced into the bowl for displacement washing. In still other designs the nozzles are closed most of the time by plugs or valves that open periodically to discharge a moderately concentrated slurry.

Sludge separators

In a nozzle-discharge centrifuge the solids leave the bowl from below the liquid surface and therefore carry with them considerable quantities of liquid. For separating a feed slurry into a clear liquid fraction and a heavy "dry" sludge, the settled solids must be moved mechanically from the liquid and given a chance to drain while still under centrifugal force. This is done in continuous sludge separators, a typical example of which is illustrated in Fig. 29.39. In this helical-conveyor centrifuge a cylindrical bowl with a conical end section rotates about a horizontal axis. Feed enters through a stationary axial pipe, spraying outward into a "pond" or annular layer of liquid inside the cylindrical bowl. Clarified liquid flows through overflow ports in the

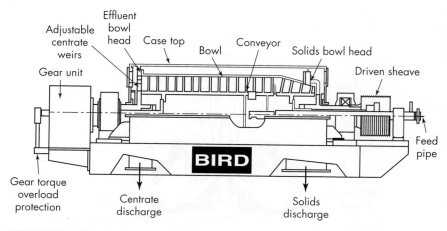

FIGURE 29.39
Helical-conveyor centrifuge (*Bird Machine Co.*).

plate covering the nonconical end of the bowl. The radial position of these ports fixes the thickness of the annular layer of liquid in the bowl. Solids settle through the liquid to the inner surface of the bowl; a helical conveyor turning slightly slower than the bowl moves the solids out of the pond and up the "beach" to the discharge openings in the small end of the cone. Wash liquid may be sprayed on the solids as they travel up the beach, to remove soluble impurities. The wash flows into the pond and discharges with the liquor. Drained sludge and clarified liquor are thrown out from the bowl into different parts of the casing, from which they leave through suitable openings.

Helical-conveyor centrifuges are made with maximum bowl diameters from 100 to 1,400 mm (4 to 54 in.). They separate large amounts of material. A 450-mm machine, for example, might handle 1 to 2 tons of solids per hour; a 1,400-mm machine, 50 tons/h. With thick feed slurries the capacity of a given machine is limited by the allowable torque on the conveyor. With dilute slurries the liquid-handling capacity of the bowl and overflow ports limits the throughput.

Practical operation of a sludge separator, of course, requires that the solids be heavier than the liquid and not be resuspended by the action of the conveyor. A modification known as an axial flow conveyor centrifuge (described later) is more effective in separating fine light solids than the full-scroll centrifuge shown in Fig. 29.39. Even so, the liquid effluent from these machines is usually not completely free from solids and may require subsequent clarification. Within these restrictions sludge separators solve a wide variety of problems. They separate fine particles from liquids, dewater and wash free-draining crystals, and are often used as classifiers.

Principles of centrifugal sedimentation

In a sedimenting centrifuge a particle of given size is removed from the liquid if sufficient time is available for the particle to reach the wall of the separator bowl. If it is assumed that the particle is at all times moving radially at its terminal

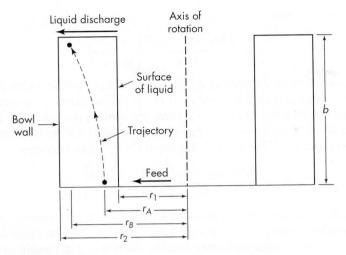

FIGURE 29.40

Particle trajectory in sedimenting centrifuge.

velocity, the diameter of the smallest particle that should just be removed can be calculated.

Consider the volume of liquid in a centrifuge bowl shown in Fig. 29.40. The feed point is at the bottom, and the liquid discharge is at the top. Assume that all the liquid moves upward through the bowl at a constant velocity, carrying solid particles with it. A given particle, as shown in the figure, begins to settle at the bottom of the bowl at some position in the liquid, say, at a distance r_A from the axis of rotation. Its settling time is limited by the residence time of the liquid in the bowl; at the end of this time let the particle be at a distance r_B from the axis of rotation. If $r_B < r_2$, the particle leaves the bowl with the liquid; if $r_B = r_2$, it is deposited on the bowl wall and removed from the liquid. If the particle settles in the Stokes'-law range, the terminal velocity at radius r is, by Eq. (7.40),

$$u_t = \frac{\omega^2 r(\rho_p - \rho)D_p^2}{18\mu}$$

Since $u_t = dr/dt$,

$$dt = \frac{18\mu}{\omega^2(\rho_p - \rho)D_p^2}\frac{dr}{r} \qquad (29.73)$$

Integrating Eq. (29.73) between the limits $r = r_A$ at $t = 0$ and $r = r_B$ at $t = t_T$ gives

$$t_T = \frac{18\mu}{\omega^2(\rho_p - \rho)D_p^2}\ln\frac{r_B}{r_A} \qquad (29.74)$$

Residence time t_T is equal to the volume of liquid in the bowl V divided by the volumetric flow rate q. Volume V equals $\pi b(r_2^2 - r_1^2)$. Substitution into Eq. (29.74) and rearrangement give

$$q = \frac{\pi b \omega^2 (\rho_p - \rho) D_p^2}{18\mu} \frac{r_2^2 - r_1^2}{\ln(r_B/r_A)} \tag{29.75}$$

A *cut point* can be defined[2] as the diameter of that particle which just reaches one-half the distance between r_1 and r_2. If D_{pc} is the cut diameter, a particle of this size moves a distance $y = (r_2 - r_1)/2$ during the settling time allowed. If a particle of diameter D_{pc} is to be removed, it must reach the bowl wall in the available time. Thus $r_B = r_2$ and $r_A = (r_1 + r_2)/2$. Equation (29.75) then becomes

$$q_c = \frac{\pi b \omega^2 (\rho_p - \rho) D_{pc}^2}{18\mu} \frac{r_2^2 - r_1^2}{\ln[2r_2/(r_1 + r_2)]} \tag{29.76}$$

where q_c is the volumetric flow rate corresponding to the cut diameter. At this flow rate most particles whose diameters are larger than D_{pc} will be eliminated by the centrifuge, and most particles having smaller diameters will remain in the liquid.

If the thickness of the liquid layer is small compared to the radius of the bowl, $r_1 \approx r_2$, and Eq. (29.76) becomes indeterminate. Under these conditions, however, the settling velocity may be considered constant and given by the equation

$$u_t = \frac{D_p^2 (\rho_p - \rho) \omega^2 r_2}{18\mu} \tag{29.77}$$

Let the thickness of the liquid layer be s and the settling distance for particles of cut diameter D_{pc} be $s/2$. Then

$$u_t = \frac{s}{2t_T} \tag{29.78}$$

where t_T is the residence time, given by

$$t_T = \frac{V}{q_c} \tag{29.79}$$

Combining Eqs. (29.75) to (29.77) and solving for q_c give

$$q_c = \frac{2Vu_t}{s} = \frac{2V D_p^2 (\rho_p - \rho) \omega^2 r_2}{18\mu s} \tag{29.80}$$

Sigma value; scaleup. For application to industrial centrifuges Eq. (29.80) is modified as follows. Radius r_2 and thickness s are replaced by r_e and s_e, respectively, which are appropriate average values of r and s for the type of centrifuge under consideration. The right-hand side of Eq. (29.80) is multiplied and divided by g, the gravitational acceleration, and all factors relating to the centrifuge are collected in one group and those relating to the solids and liquid in another. This gives

$$q_c = \frac{2V\omega^2 r_e}{gs} \frac{D_p^2 (\rho_p - \rho) g}{18\mu}$$

$$= 2\Sigma u_g \tag{29.81}$$

TABLE 29.4
Characteristics of sedimenting centrifuges[30f,40]

Type	Bowl diameter, in.	Speed, rpm	Σ value, ft$^2 \times 10^{-4}$
Tubular	4.125	15,000	2.7
Disk	9.5	6,500	21.5
	13.7	4,650	39.3
	19.5	4,240	105
Helical conveyor	14	4,000	1.34
	25	3,000	6.1
Axial flow conveyor			
No vanes	29	2,600	4.05
96 vanes	29	2,600	12.7

where Σ, the *sigma value,* is a characteristic of the centrifuge and u_g is the terminal settling velocity of the particles under gravity settling conditions. Physically, Σ is the cross-sectional area of a gravity settling tank of the same separation capacity as the centrifuge. Typical values are given in Table 29.4. A 19.5-in (0.5-m) disk centrifuge, for example, is equivalent to a gravity sedimenter with an area of over 10^6 ft^2 (10^5 m^2). In practice, the actual capacity of a centrifuge may be somewhat less than that indicated by the Σ value, because of the complicated flow patterns in a revolving centrifuge bowl and, in some designs, the resuspension of particles by an internal conveyor.

Axial flow conveyor centrifuges

In 1970 Schnittger[36] showed that the actual residence time in a conveyor centrifuge is much less than that given by Eq. (29.79). He proposed a model in which liquid flows in a thin "boundary layer" over an essentially stagnant liquid pond. The critical separation takes place in this layer, for once a particle moves from the boundary layer into the pond, it is effectively removed unless later resuspended mechanically. A thin boundary layer gives better separation than a thick one. The total pond volume is not important; the pond needs to be no deeper than necessary to ensure that the compacted solids are well below the surface.

In an axial flow conveyor centrifuge, the scroll is cut away in the feed and sedimentation sections to form a ribbon conveyor operating near the bowl wall, leaving the liquid surface undisturbed. The boundary layer can then flow in an axial direction from the feed point to the liquid discharge. Turbulence is minimized in the feed zone and at the liquid exit by the design of the flow channels. Longitudinal vanes in the sedimentation section make sure the liquid in the boundary layer rotates at full bowl speed; these vanes, set at an angle with the radial direction, also function as the disks of a disk centrifuge, reducing the distance that a particle must settle before it strikes a solid surface. Once this occurs, the particle is unlikely to be resuspended.

In a full-scroll centrifuge (Fig. 29.39) the liquid flows in a spiral path against the motion of the conveyor. Under these conditions the boundary layer is as much as 10 times as thick as it is when the layer flows only in an axial direction. Furthermore, spiral flow induces eddies and turbulence in the boundary layer, impeding sedimentation; with axial flow the boundary layer is in laminar flow, and sedimentation is unimpeded. Willis and Shapiro[40] give Σ values for axial flow centrifuges. The addition of vanes can increase Σ by 3 to 4 times.

Centrifugal classifiers

During the passage of the liquid through a centrifuge bowl, the heavier, larger solid particles are thrown out of the liquid. Finer, lighter particles may not settle in the time available and may be carried out with the liquid effluent. As in a gravity hydraulic classifier, solid particles can be sorted according to size, shape, or specific gravity. In a centrifugal classifier the separating force is on the order of 600 times the force of gravity, permitting sharp separations of particles 1 μm or less in diameter. Much coarser particles than this, however, are also classified in centrifugal machines.

The high settling force in a centrifuge means that practical settling rates can be obtained with much smaller particles than in a gravity classifier. While the higher force does not change the relative settling velocities of small particles, it does overcome the small but disturbing effects of free convection currents and Brownian motion in a gravity classifier, and it permits separation in some cases where none is possible in a gravity unit. With coarse particles the settling regime may be changed, so that a particle that settles by gravity according to Stokes's law may settle in the intermediate or Newton law regime in a centrifuge. Thus mixtures of equal-settling particles from a gravity unit may sometimes be partially separated in a centrifuge. On the other hand, loose flocs or weak agglomerates which settle rapidly in a gravity thickener are often broken up in a centrifugal classifier and settle slowly or not at all despite the increased force for sedimentation.

SYMBOLS

A	Area, m^2 or ft^2; A_T, total area of continuous filter; A_1, area of inner surface of material in centrifuge; A_2, area of outer surface of material in centrifuge; $\bar{A}_L$, logarithmic mean of A_1 and A_2; $\bar{A}_a$, arithmetic mean of A_1 and A_2
B	Underflow from screen, kg/h or lb/h
b	Width of centrifuge basket, m or ft
c	Mass of solid deposited in filter per unit volume of filtrate, kg/m^3 or lb/ft^3; also, concentration of solids in suspension, kg/m^3, g/L, or lb/ft^3; c_F, in feed; c_P, in permeate; c_c, critical concentration in thickener; c_g, concentration at which a gel layer forms in ultrafiltration; c_m, in pores of medium; c_s, in feed slurry, also, at surface in ultrafiltration; c_u, in thickener underflow; c_0, in feed to sedimenter; c_1, c_2, local retentate and permeate concentrations
D	Overflow from screen, kg/h or m/h; also, diameter or pore size, m, μm, or ft
D_b	Width or diameter of impingement target, m or ft
D_p	Particle size, m or ft; D_{pA}, of heavy particle; D_{pB}, of light particle; D_{pc}, cut diameter

D_s Shear-induced diffusivity, cm^2/s, [see Eq. (29.65)]

D_v Volumetric diffusivity, m^2/h, cm^2/s, or ft^2/h; D_e, effective diffusivity defined by Eq. (29.63); D_{pore}, diffusivity in pores

E Screen efficiency, dimensionless; E_A, based on oversize; E_B, based on undersize

F Feed rate, kg/h or lb/h; also force, N or lb$_f$; F_c, centrifugal force; F_g, force of gravity

f Fraction of filter cycle available for cake formation

G Mass flux in sedimenter, kg/m$^2 \cdot$h or lb/ft$^2 \cdot$h; G_s, settling flux; G_t, transport flux

g Acceleration of gravity, m/s^2 or ft/s^2

g_c Newton's-law proportionality factor, 32.174 ft$\cdot$lb/lb$_f \cdot$s^2

K Equilibrium partition coefficient in ultrafiltration

K_c Constant in equation for constant-pressure cake filtration, defined by Eq. (29.23)

K_r Constant in equation for constant-rate filtration, Eq. (29.38)

k   Boltzmann constant, 1.380×10^{-23} J/K

k_c Mass-transfer coefficient based on concentration, cm/s [Eq. (29.51)]

k_1, k_2 Constants in Eqs. (29.12) and (29.15b), respectively

L Distance in cake measured from filter medium, m or ft; also, thickness of selective layer in ultrafiltration; L_c, filter cake thickness

M Molecular weight

m Mass, kg or lb; m_F, mass of wet filter cake; m_c, mass of solids in filter cake

$\dot{m}$ Mass flow rate, kg/h or lb/h; $\dot{m}_c$, of solids from continuous filter

N_a Avogadro's number, 6.022×10^{23} molecules/g mol

N_s Separation number, $u_t u_0 / g D_b$

n Drum speed of continuous filter, r/s

p Pressure, atm or lb$_f$/ft^2; pressure in cake at distance L from filter medium; p_a, at inlet to filter; p_b, at discharge from filter; p', at boundary between cake and medium in filter

Q_m Membrane permeability, $v/\Delta p$ [Eq. (29.48)]

q Volumetric flow rate, m^3/s or ft^3/s; q_c, corresponding to removal of particles of cut diameter; q_0, at start of filtration

R Fraction of solute rejected in ultrafiltration, defined by Eq. (29.58); R_F, based on feed and permeate [Eq. (29.57)]

R_m Filter medium resistance, m^{-1} or ft^{-1}

Re   Reynolds number, $D u \rho / \mu$

r Radius, m or ft; r_A, initial position of particle in sedimenting centrifuge; r_B, final position; r_e, effective average value; r_i, of interface between cake and liquid layer in centrifuge; r_o, of particle; r_1, inner radius of material in centrifuge; r_2, outer radius

Sc   Schmidt number, $\mu / \rho D_v$

Sh   Sherwood number, $k_c D / D_v$

s Thickness of liquid layer in centrifuge, m or ft; s_e, effective average value; also, compressibility coefficient [Eq. (29.25)]

s_p Surface area of single particle, m^2 or ft^2

T Absolute temperature, K

t Time, h or s; t_T, residence time in centrifuge; t_c, cycle time in continuous filter

u Linear velocity, m/s or ft/s; u_g, settling velocity in gravity field; u_t, terminal setting velocity; u_{tA}, of heavy particle; u_{tB}, of light particle; u_{tan}, tangential velocity of gas in cyclone; u_o, velocity of undisturbed fluid approaching solid

V Volume, m^3, L, or ft^3; also, volume of filtrate collected to time t

$\bar{V}$ Average solution velocity, m/s or ft/s

v Volume flux (superficial permeate velocity) in ultrafiltration, m/s or ft/s; v_{max}, maximum value

v_p Volume of single particle, m^3 or ft^3

x Mass fraction of cut in mixture of particles; x_B, in underflow from screen; x_D, in overflow from screen; x_F, in feed to screen; also, distance, m or ft

y Distance from membrane surface, m, μm, or ft

Z Height of liquid-solid interface in sedimentation test, m or ft; Z_i, intercept in Kynch method for sedimenter design; Z_0, initial height

Greek Letters

α Specific cake resistance, m/kg or ft/lb; α_0, constant in Eq. (29.25)

Δp Overall pressure drop through filter, atm or lb_f/ft^2, $p_a - p_b$; Δp_c, pressure drop through cake, $p_a - p'$; Δp_m, pressure drop through filter medium, $p' - p_b$

$\Delta \pi$ Difference in osmotic pressures, atm or lb_f/ft^2

δ Thickness of concentration boundary layer, m, μm, or ft

ε Porosity or volume fraction voids in bed of solids, dimensionless; $\bar{\varepsilon}$, average porosity of filter cake

η_t Target efficiency, impingement separator

λ Ratio of molecular size to pore size

μ Viscosity, cP or lb/ft·s; μ', cP in dimensional equation [Eq. (29.54)]

π Osmotic pressure, atm or lb_f/ft^2

ρ Density, kg/m^3 or lb/ft^3; of fluid or filtrate; ρ_p, of particle; ρ_{pA}, of heavy particle; ρ_{pB}, of light particle

Σ Sigma value for scaleup of centrifuges [Eq. (29.81)]

τ Tortuosity factor, dimensionless

Φ_s Shape factor or sphericity, defined by Eq. (28.1)

ω Angular velocity, rad/s

PROBLEMS

29.1. It is desired to separate a mixture of crystals into three fractions, a coarse fraction retained on an 8-mesh screen, a middle fraction passing an 8-mesh but retained on a 14-mesh screen, and a fine fraction passing a 14-mesh. Two screens in series are used, an 8-mesh and a 14-mesh, conforming to the Tyler standard. Screen analyses of feed, coarse, medium, and fine fractions are given in Table 29.5. Assuming the analyses are accurate, what do they show about the ratio by weight of each of the three fractions actually obtained? What are the efficiencies E_A and E_B of each screen?

TABLE 29.5
Screen analyses for Prob. 29.1

Screen	Feed	Coarse fraction	Middle fraction	Fine fraction
3/4	3.5	14.0		
4/6	15.0	50.0	4.2	
6/8	27.5	24.0	35.8	
8/10	23.5	8.0	30.8	20.0
10/14	16.0	4.0	18.3	26.7
14/20	9.1		10.2	20.2
20/28	3.4		0.7	19.6
28/35	1.3			8.9
35/48	0.7			4.6
Total	100.0	100.0	100.0	100.0

29.2. The screens used in Prob. 29.1 are shaking screens with a capacity of 4 metric tons/m$^2 \cdot$h$\cdot$mm mesh size. How many square meters of screen are needed for each of the screens in Prob. 29.1 if the feed to the first screen is 100 tons/h?

29.3. The data in Table 29.6 were taken in a constant-pressure filtration of a slurry of CaCO$_3$ in H$_2$O. The filter was a 6-in. filter press with an area of 1.0 ft^2. The mass fraction of solids in the feed to the press was 0.139. Calculate the values of α, R_m, and cake thickness for each of the experiments. The temperature is 70°F.

TABLE 29.6
Data from constant-pressure filtration[†]

5-lb$_f$/in.^{2} pressure drop (1)		15-lb$_f$/in.^{2} pressure drop (2)		30-lb$_f$/in.^{2} pressure drop (3)		50-lb$_f$/in.^{2} pressure drop (4)	
Filtrate, lb	Time, s	Filtrate, lb	Time, s	Filtrate, lb	Time, s	Filtrate, lb	Time, s
0	0	0	0	0	0	0	0
2	24	5	50	5	26	5	19
4	71	10	181	10	98	10	68
6	146	15	385	15	211	15	142
8	244	20	660	20	361	20	241
10	372	25	1,009	25	555	25	368
12	524	30	1,443	30	788	30	524
14	690	35	2,117	35	1,083	35	702
16	888						
18	1,188						

[†]Mass ratio of wet cake to dry cake: (1) 1.59, (2–4), 1.47. Dry cake density: (1) 63.5, (2, 3) 73.0, (4) 73.5 lb/ft^3. From E. L. McMillen and H. A. Webber, *Trans, AIChE,* **34**:213 (1938).

29.4. The slurry of Prob. 29.3 is to be filtered in a press having a total area of 8 m² and operated at a constant pressure drop of 2 atm. The frames are 36 mm thick. Assume that the filter medium resistance in the large press is the same as that in the laboratory filter. Calculate the filtration time required and the volume of filtrate obtained in one cycle.

29.5. Assuming that the actual rate of washing is 85 percent of the theoretical rate, how long will it take to wash the cake in the press of Prob. 29.4 with a volume of wash water equal to that of the filtrate?

29.6. A continuous rotary vacuum filter operating with a pressure drop of 0.7 atm is to handle the feed slurry of Prob. 29.3. The drum submergence is to be 25 percent. What total filter area must be provided to match the overall productive capacity of the filter press described in Prob. 29.4? Drum speed is 2 rpm.

29.7. The following relation between α and Δp for superlight $CaCO_3$ has been determined[18] to be

$$\alpha = 8.8 \times 10^{10}[1 + 3.36 \times 10^{-4}(\Delta p)^{0.86}]$$

where Δp is in pounds force per square foot. This relation is followed over a pressure range from 0 to 1,000 lb_f/in.². A slurry of this material giving 3.0 lb of cake solid per cubic foot of filtrate is to be filtered at a constant pressure drop of 70 lb_f/in.² and a temperature of 70°F. Experiments on this sludge and the filter cloth to be used gave a value of $R_m = 1.2 \times 10^{10}$ ft⁻¹. A pressure filter of the tank type is to be used. How many square feet of filter surface are needed to give 1,400 gal of filtrate in a 1-h filtration? The viscosity is that of water at 70°F.

29.8. The filter of Prob. 29.7 is washed at 70°F and 70 lb_f/in.² with a volume of wash water equal to one-third that of the filtrate. The washing rate is 85 percent of the theoretical value. How long should it take to wash the cake?

29.9. The filter of Prob. 29.7 is operated at a constant rate of 0.6 gal/ft²·min from the start of the run until the pressure drop reaches 70 lb_f/in.² and then at a constant pressure drop of 70 lb_f/in.² until a total of 1,400 gal of filtrate is obtained. The operating temperature is 70°F. What is the total filtration time required?

29.10. A continuous pressure filter is to yield 1,400 gal/h of filtrate from the slurry described in Prob. 29.7. The pressure drop is limited to a maximum of 50 lb_f/in.² How much filter area must be provided if the cycle time is 3 min and the drum submergence is 50 percent?

29.11. Air carrying particles of density 1,800 kg/m³ and an average diameter of 20 μm enters a cyclone at a linear velocity of 18 m/s. The diameter of the cyclone is 600 mm. (a) What is the approximate separation factor for this cyclone? (b) What fraction of the particles will be removed from the gas stream?

29.12. The dust-laden air of Prob. 29.11 is passed through an impingement separator at a linear velocity of 8 m/s. The separator consists essentially of ribbons 25 mm wide. What is the maximum fraction of the particles that can be removed by the first row of ribbons, which covers 50 percent of the cross-sectional area of the duct?

29.13. What is the capacity in cubic meters per hour of a clarifying centrifuge operating under the following conditions?

Diameter of bowl, 600 mm	Specific gravity of liquid 1.2
Thickness of liquid layer, 75 mm	Specific gravity of solid, 1.6
Depth of bowl, 400 mm	Viscosity of liquid 2 cP
Speed, 1,200 rpm	Cut size of particles, 30 μm

29.14. A batch centrifugal filter having a bowl diameter of 750 mm and a bowl height of 450 mm is used to filter a suspension having the following properties:

Liquid, water	Final thickness of cake, 150 mm
Temperature, 25°C	Speed of centrifuge, 2,000 rpm
Concentration of solid in feed, 60 g/L	Specific cake resistance, 9.5×10^{10} ft/lb
Porosity of cake, 0.435	Filter medium resistance, 2.6×10^{10} ft^{-1}
Density of dry solid in cake, 2,000 kg/m^3	

The final cake is washed with water under such conditions that the radius of the inner surface of the liquid is 200 mm. Assuming that the rate of flow of wash water equals the final rate of flow of filtrate, what is the rate of washing in cubic meters per hour?

29.15. (a) Calculate the apparent mass-transfer coefficient for latex from the ultrafiltration results shown in Fig. 29.25. (b) Using the corresponding data for albumin as a reference, estimate the shear-induced diffusivity D_s for the latex spheres, and compare this with the normal diffusivity D_v. (c) What shear rate would be needed to explain the value of D_s?

29.16. (a) Predict the effect of pressure drop on the permeate flux for the ultrafiltration of cheese whey in a 2-cm-diameter tubular membrane at a solution velocity of 3 m/s. Assume the membrane characteristics are the same as those in Example 29.3. (b) Calculate the frictional pressure drop if the tube length is 1.5 m.

29.17. A microfiltration unit used to concentrate a suspension of 0.8-μm spherical particles has a permeate flux of 150 L/m$^2 \cdot$h at 30°C and a pressure drop of 2.1 atm. Under the same conditions, the pure water flux was 280 L/m$^2 \cdot$h. How thick a layer of particles would be needed to account for the gel-layer resistance if $\varepsilon = 0.40$?

REFERENCES

1. Altena, F. W., and G. Belfort. *Chem. Eng. Sci.* **39**:343 (1984).
2. Ambler, C. M. *Chem. Eng. Prog.* **48**:150 (1952).
3. Belfort, G. *J. Membrane Sci.* **40**:123 (1989).
4. Blatt, W. F., A. David, A. S. Michaels, and L. Nelsen: in J. E. Flinn (ed.). *Membrane Science and Technology.* New York: Plenum Press, 1970.
5. Cheryan, M. *Ultrafiltration Handbook.* Lancaster, PA: Technomic Publishing Co., 1986, p. 127.
6. Cheryan, M., and B. H. Chiang: in B. M. McKenna (ed.). *Engineering and Food.* London: Applied Science Publ., 1984.

7. Clark, J. G. *Chem. Eng. Prog.* **86**(11):45 (1990).

8. Coe, F. S., and G. H. Clevenger. *Trans. AIME* **55**:356 (1916).

9. Concha, F. A. *AIChE J.* **37**:1425 (1991).

10. Crozier, H. E., and L. E. Brownell. *Ind. Eng. Chem.* **44**:631 (1952).

11. Davis, R. S., and D. T. Leighton. *Chem. Eng. Sci.* **42**:279 (1987).

12. Dick, R. I. *Fluid/Particle Separation J.* **2**(2):77 (1989).

13. Dixon, D. C. *AIChE J.* **37**:1431 (1991).

14. Eckstein, E. C., D. G. Bailey, and A. H. Shapiro. *J. Fluid Mech.* **79**:191 (1974).

15. Fane, A. G., C. J. D. Fell, and A. G. Waters. *J. Membrane Sci.* **9**:245 (1981).

16. Fitch, B. *AIChE J.* **36**:1545 (1990).

17. Goldsmith, R. L. *Ind. Eng. Chem. Fund.* **10**:113 (1971).

18. Grace, H. P. *Chem. Eng. Prog.* **49**:303, 367, 427 (1953).

19. Grace, H. P. *AIChE J.* **2**:307, 316 (1956).

20. Harriott, P. *Separation Sci.* **8**(3):291 (1973).

21. Hoffman, A. C., A. van Santen, R. W. K. Allen, and R. Clift. *Powder Technology* **70**:83 (1992).

22. Hsieh, H. P. *Chem. Eng. Prog. Symp. Ser.* **84**(261):1 (1988).

23. Hughes, O. D., R. W. Ver Hoeve, and C. D. Luke. Paper given at meeting of AIChE, Columbus, OH, December 1950.

24. Jonsson, G. *Desalination* **51**:61 (1984).

25. Kynch, G. J. *Trans. Faraday Soc.* **48**:161 (1952).

26. Mulder, M. *Basic Principles of Membrane Technology.* Dordrecht, Netherlands: Kluwer Academic Publishers, 1991, p. 54.

27. Murkes, J., and C. -G. Carlsson. *Crossflow Filtration Theory and Practice.* New York: Wiley, 1988.

28. Nickolaus, N., and D. A. Dahlstrom. *Chem. Eng. Prog.* **52**(3):87M (1956).

29. Perlmutter, B. A. *Chem. Eng. Prog.* **87**(7):29 (1991).

30. Perry, R. H., and D. W. Green (eds.). *Perry's Chemical Engineers' Handbook,* 7th ed. New York: McGraw-Hill, 1997; (*a*) p. **17**-27, (*b*) p. **18**-61, (*c*) p. **18**-72, (*d*) pp. **18**-90 to **18**-103, (*e*) p. **18**-100, (*f*) p. **18**-116, (*g*) p. **19**-23, (*h*) p. **19**-24.

31. Porter, M. C. *Ind. Eng. Chem. Prod. Res. Devel.* **11**:234 (1972).

32. Porter, M. C.: in P. A. Schweitzer (ed.). *Handbook of Separation Techniques for Chemical Engineers.* New York: McGraw-Hill, 1979.

33. Romero, C. A., and R. H. Davis. *J. Membrane Sci.* **62**:249 (1991).

34. Rushton, A., and M. S. Hameed. *Filtr. Separation* **7**:25 (1970).

35. Ruth, B. F. Personal communication.

36. Schnittger, J. R. *Ind. Eng. Chem. Proc. Des. Dev.* **9**(3):407 (1970).

37. Svarovsky, L. *Chem. Eng.* vol. 86, 1979; (*a*) no. 14, p. 62; (*b*) no. 15, p. 101.

38. Taggart, A. F. *Handbook of Mineral Dressing: Ores and Industrial Minerals.* New York: Wiley, 1945, p. **11**-123.

39. Van den Berg, G. B., I. G. Racz, and C. A. Smolders. *J. Membrane Sci.* **47**:25 (1989).

40. Willis, F. F., and L. Shapiro. Technical Report No. 936. Warminster, PA: Alfa-Laval Separation, Inc., 1991.

41. Zydney, A. L., and C. K. Colton. *Chem. Eng. Commun.* **47**:1 (1987).

APPENDIX 1

Conversion Factors and Constants of Nature

To convert from	To	Multiply by[†]
acre	ft^2	43,560*
	m^2	4,046.85
atm	N/m^2	1.01325* $\times$ 10^5
	lb$_f$/in.^{2}	14.696
Avogadro's number	particles/g mol	6.022169 $\times$ 10^{23}
bbl (petroleum)	ft^3	5.6146
	gal (U.S.)	42*
	m^3	0.15899
bar	N/m^2	1* $\times$ 10^5
	lb$_f$/in.^{2}	14.504
Boltzmann's constant	J/K	1.380622 $\times$ 10^{-23}
Btu	cal$_{IT}$	251.996
	ft$\cdot$lb$_f$	778.17
	J	1,055.06
	kWh	2.9307 $\times$ 10^{-4}
Btu/lb	cal$_{IT}$/g	0.55556
Btu/lb$\cdot$°F	cal$_{IT}$/g$\cdot$°C	1*
Btu/ft$^2\cdot$h	W/m^2	3.1546
Btu/ft$^2\cdot$h$\cdot$°F	W/m$^2\cdot$°C	5.6783
	kcal/m$^2\cdot$h$\cdot$K	4.882
Btu$\cdot$ft/ft$^2\cdot$h$\cdot$°F	W$\cdot$m/m$^2\cdot$°C	1.73073
	kcal/m$\cdot$h$\cdot$K	1.488
cal$_{IT}$	Btu	3.9683 $\times$ 10^{-3}
	ft$\cdot$lb$_f$	3.0873
	J	4.1868*
cal	J	4.184*
cm	in.	0.39370
	ft	0.0328084
cm^3	ft^3	3.531467 $\times$ 10^{-5}
	gal (U.S.)	2.64172 $\times$ 10^{-4}

(Continued)

To convert from	To	Multiply by[†]
cP (centipoise)	kg/m·s	$1^* \times 10^{-3}$
	lb/ft·h	2.4191
	lb/ft·s	6.7197×10^{-4}
cSt (centistoke)	m^2/s	$1^* \times 10^{-6}$
faraday	C/g mol	9.648670×10^4
ft	m	0.3048*
ft·lb_f	Btu	1.2851×10^{-3}
	cal_{IT}	0.32383
	J	1.35582
ft·lb_f/s	Btu/h	4.6262
	hp	1.81818×10^{-3}
ft^2/h	m^2/s	2.581×10^{-5}
	cm^2/s	0.2581
ft^3	m^3	0.0283168
	gal (U.S.)	7.48052
	L	28.31684
ft^3·atm	Btu	2.71948
	cal_{IT}	685.29
	J	2.8692×10^3
ft^3/s	gal (U.S.)/min	448.83
gal (U.S.)	ft^3	0.13368
	$in.^3$	231*
gas law constant, R, see Table 1.2, p. 11		
gravitational constant	$N·m^2/kg^2$	6.673×10^{-11}
gravity acceleration, standard	m/s^2	9.80665*
h	min	60*
	s	3,600*
hp	Btu/h	2,544.43
	kW	0.74624
hp/1,000 gal	kW/m^3	0.197
in.	cm	2.54*
$in.^3$	cm^3	16.3871
J	erg	$1^* \times 10^7$
	ft·lb_f	0.73756
kg	lb	2.20462
kWh	Btu	3,412.1
L	m^3	$1^* \times 10^{-3}$
lb	kg	0.45359237*
lb/ft^3	kg/m^3	16.018
	g/cm^3	0.016018
$lb_f/in.^2$	N/m^2	6.89473×10^3
lb mol/ft^2·h	kg mol/m^2·s	1.3562×10^{-3}
	g mol/cm^2·s	1.3562×10^{-4}
light, speed of	m/s	2.997925×10^8
m	ft	3.280840
	in.	39.3701
m^3	ft^3	35.3147
	gal (U.S.)	264.17
N	dyn	$1^* \times 10^5$
	lb_f	0.22481
N/m^2	$lb_f/in.^2$	1.4503×10^{-4}

(*Continued*)

To convert from	To	Multiply by[†]
Planck's constant	J·s	6.626196×10^{-34}
proof (U.S.)	percent alcohol by volume	0.5
ton (long)	kg	1,016
	lb	2,240*
ton (short)	lb	2,000*
t (metric)	kg	1,000*
	lb	2,204.6
yd	ft	3*
	m	0.9144*

[†]Values that end in an asterisk are exact, by definition.

APPENDIX 2

Dimensionless Groups

Symbol	Name	Definition
Bi	Biot number	$\dfrac{hs}{k}$ for slab
		$\dfrac{hr_m}{k}$ for cylinder or sphere
C_D	Drag coefficient	$\dfrac{2F_{Dc}}{\rho u_0^2 A_p}$
Fo	Fourier number	$\dfrac{\alpha t}{r^2}$
Fr	Froude number	$\dfrac{u^2}{gL}$
f	Fanning friction factor	$\dfrac{\Delta p_{sc} D}{2L\rho \bar{V}^2}$
Gr	Grashof number	$\dfrac{L^3 \rho^2 \beta g \Delta T}{\mu^2}$
Gz	Graetz number	$\dfrac{\dot{m} c_p}{kL}$
Gz'	Graetz number for mass transfer	$\dfrac{\dot{m}}{\rho D_v L}$
j_H	Heat-transfer factor	$\dfrac{h}{c_p G}\left(\dfrac{c_p \mu}{k}\right)^{2/3}\left(\dfrac{\mu_w}{\mu}\right)^{0.14}$
j_M	Mass-transfer factor	$\dfrac{k\bar{M}}{G}\left(\dfrac{\mu}{D_v \rho}\right)^{2/3}$

(Continued)

Symbol	Name	Definition
Ma	Mach number	$\dfrac{u}{a}$
N_{Ae}	Aeration number	$\dfrac{q_g}{nD_a^3}$
N_P	Power number	$\dfrac{P_c}{\rho n^3 D^5}$
N_Q	Flow number	$\dfrac{q}{nD_a^3}$
Nu	Nusselt number	$\dfrac{hD}{k}$
Pe	Peclet number	$\dfrac{D\bar{V}}{\alpha}$ or $\dfrac{Du_o}{D_v}$
Pr	Prandtl number	$\dfrac{c_p\mu}{k}$
Re	Reynolds number	$\dfrac{DG}{\mu}$
N_s	Separation number	$\dfrac{u_t u_0}{gD_p}$
Sc	Schmidt number	$\dfrac{\mu}{D_v\rho}$
Sh	Sherwood number	$\dfrac{k_c D}{D_v}$
We	Weber number	$\dfrac{D\rho\bar{V}^2}{\sigma}$

APPENDIX 3

Dimensions, Capacities, and Weights of Standard Steel Pipe[†]

Nominal pipe size, in.	Outside diameter, in.	Schedule no.	Wall thickness, in.	Inside diameter, in.	Cross-sectional area of metal, in.^{2}	Inside sectional area, ft^2	Circumference, ft or surface, ft^2/ft of length		Capacity at 1 ft/s velocity		Pipe weight, lb/ft
							Outside	Inside	U.S. gal/min	Water, lb/h	
$\frac{1}{8}$	0.405	40	0.068	0.269	0.072	0.00040	0.106	0.0705	0.179	89.5	0.24
		80	0.095	0.215	0.093	0.00025	0.106	0.0563	0.113	56.5	0.31
$\frac{1}{4}$	0.540	40	0.088	0.364	0.125	0.00072	0.141	0.095	0.323	161.5	0.42
		80	0.119	0.302	0.157	0.00050	0.141	0.079	0.224	112.0	0.54
$\frac{3}{8}$	0.675	40	0.091	0.493	0.167	0.00133	0.177	0.129	0.596	298.0	0.57
		80	0.126	0.423	0.217	0.00098	0.177	0.111	0.440	220.0	0.74
$\frac{1}{2}$	0.840	40	0.109	0.622	0.250	0.00211	0.220	0.163	0.945	472.0	0.85
		80	0.147	0.546	0.320	0.00163	0.220	0.143	0.730	365.0	1.09
$\frac{3}{4}$	1.050	40	0.113	0.824	0.333	0.00371	0.275	0.216	1.665	832.5	1.13
		80	0.154	0.742	0.433	0.00300	0.275	0.194	1.345	672.5	1.47
1	1.315	40	0.133	1.049	0.494	0.00600	0.344	0.275	2.690	1,345	1.68
		80	0.179	0.957	0.639	0.00499	0.344	0.250	2.240	1,120	2.17
$1\frac{1}{4}$	1.660	40	0.140	1.380	0.668	0.01040	0.435	0.361	4.57	2,285	2.27
		80	0.191	1.278	0.881	0.00891	0.435	0.335	3.99	1,995	3.00
$1\frac{1}{2}$	1.900	40	0.145	1.610	0.800	0.01414	0.497	0.421	6.34	3,170	2.72
		80	0.200	1.500	1.069	0.01225	0.497	0.393	5.49	2,745	3.63
2	2.375	40	0.154	2.067	1.075	0.02330	0.622	0.541	10.45	5,225	3.65
		80	0.218	1.939	1.477	0.02050	0.622	0.508	9.20	4,600	5.02
$2\frac{1}{2}$	2.875	40	0.203	2.469	1.704	0.03322	0.753	0.647	14.92	7,460	5.79
		80	0.276	2.323	2.254	0.02942	0.753	0.608	13.20	6,600	7.66
3	3.500	40	0.216	3.068	2.228	0.05130	0.916	0.803	23.00	11,500	7.58
		80	0.300	2.900	3.016	0.04587	0.916	0.759	20.55	10,275	10.25
$3\frac{1}{2}$	4.000	40	0.226	3.548	2.680	0.06870	1.047	0.929	30.80	15,400	9.11
		80	0.318	3.364	3.678	0.06170	1.047	0.881	27.70	13,850	12.51
4	4.500	40	0.237	4.026	3.17	0.08840	1.178	1.054	39.6	19,800	10.79
		80	0.337	3.826	4.41	0.07986	1.178	1.002	35.8	17,900	14.98
5	5.563	40	0.258	5.047	4.30	0.1390	1.456	1.321	62.3	31,150	14.62
		80	0.375	4.813	6.11	0.1263	1.456	1.260	57.7	28,850	20.78
6	6.625	40	0.280	6.065	5.58	0.2006	1.734	1.588	90.0	45,000	18.97
		80	0.432	5.761	8.40	0.1810	1.734	1.508	81.1	40,550	28.57
8	8.625	40	0.322	7.981	8.396	0.3474	2.258	2.089	155.7	77,850	28.55
		80	0.500	7.625	12.76	0.3171	2.258	1.996	142.3	71,150	43.39
10	10.75	40	0.365	10.020	11.91	0.5475	2.814	2.620	246.0	123,000	40.48
		80	0.594	9.562	18.95	0.4987	2.814	2.503	223.4	111,700	64.40
12	12.75	40	0.406	11.938	15.74	0.7773	3.338	3.13	349.0	174,500	53.56
		80	0.688	11.374	26.07	0.7056	3.338	2.98	316.7	158,350	88.57

[†]Based on ANSI B36.10-1959 by permission of ASME.

APPENDIX 4

Condenser and Heat-Exchanger Tube Data[†]

Outside diameter, in.	Wall thickness BWG no.	Wall thickness in.	Inside diameter, in.	Cross-sectional area metal, in.²	Inside sectional area, ft²	Circumference, ft or surface, ft²/ft of length Outside	Circumference, ft or surface, ft²/ft of length Inside	Velocity, ft/s, for 1 U.S. gal/min	Capacity at 1 ft/s velocity U.S. gal/min	Capacity at 1 ft/s velocity Water, lb/h	Weight, lb/ft[‡]
$\frac{5}{8}$	12	0.109	0.407	0.177	0.000903	0.1636	0.1066	2.468	0.4053	202.7	0.602
	14	0.083	0.459	0.141	0.00115	0.1636	0.1202	1.938	0.5161	258.1	0.479
	16	0.065	0.495	0.114	0.00134	0.1636	0.1296	1.663	0.6014	300.7	0.388
	18	0.049	0.527	0.089	0.00151	0.1636	0.1380	1.476	0.6777	338.9	0.303
$\frac{3}{4}$	12	0.109	0.532	0.220	0.00154	0.1963	0.1393	1.447	0.6912	345.6	0.748
	14	0.083	0.584	0.174	0.00186	0.1963	0.1529	1.198	0.8348	417.4	0.592
	16	0.065	0.620	0.140	0.00210	0.1963	0.1623	1.061	0.9425	471.3	0.476
	18	0.049	0.652	0.108	0.00232	0.1963	0.1707	0.962	1.041	520.5	0.367
$\frac{7}{8}$	12	0.109	0.657	0.262	0.00235	0.2291	0.1720	0.948	1.055	527.5	0.891
	14	0.083	0.709	0.207	0.00274	0.2291	0.1856	0.813	1.230	615.0	0.704
	16	0.065	0.745	0.165	0.00303	0.2291	0.1950	0.735	1.350	680.0	0.561
	18	0.049	0.777	0.127	0.00329	0.2291	0.2034	0.678	1.477	738.5	0.432
1	10	0.134	0.732	0.364	0.00292	0.2618	0.1916	0.763	1.310	655.0	1.237
	12	0.109	0.782	0.305	0.00334	0.2618	0.2047	0.667	1.499	750.0	1.037
	14	0.083	0.834	0.239	0.00379	0.2618	0.2183	0.588	1.701	850.5	0.813
	16	0.065	0.870	0.191	0.00413	0.2618	0.2278	0.538	1.854	927.0	0.649
$1\frac{1}{4}$	10	0.134	0.982	0.470	0.00526	0.3272	0.2571	0.424	2.361	1,181	1.598
	12	0.109	1.032	0.391	0.00581	0.3272	0.2702	0.384	2.608	1,304	1.329
	14	0.083	1.084	0.304	0.00641	0.3272	0.2838	0.348	2.877	1,439	1.033
	16	0.065	1.120	0.242	0.00684	0.3272	0.2932	0.326	3.070	1,535	0.823
$1\frac{1}{2}$	10	0.134	1.232	0.575	0.00828	0.3927	0.3225	0.269	3.716	1,858	1.955
	12	0.109	1.282	0.476	0.00896	0.3927	0.3356	0.249	4.021	2,011	1.618
	14	0.083	1.334	0.370	0.00971	0.3927	0.3492	0.229	4.358	2,176	1.258
2	10	0.134	1.732	0.7855	0.0164	0.5236	0.4534	0.136	7.360	3,680	2.68
	12	0.109	1.782	0.6475	0.0173	0.5236	0.4665	0.129	7.764	3,882	2.22

[†]Condensed, by permission, from J. H. Perry (ed.), *Chemical Engineers' Handbook*, 5th ed., p. **11**-12. Copyright © 1973, McGraw-Hill Book Company, New York.

[‡]For steel; for copper, multiply by 1.14; for brass, multiply by 1.06.

APPENDIX 5

Tyler Standard Screen Scale

This screen scale has as its base an opening of 0.0029 in., which is the opening in 200-mesh 0.0021-in. wire, the standard sieve, as adopted by the National Bureau of Standards.

Mesh	Clear opening, in.	Clear opening, mm	Approximate opening, in.	Wire diameter, in.
	1.050	26.67	1	0.148
†	0.883	22.43	$\frac{7}{8}$	0.135
	0.742	18.85	$\frac{3}{4}$	0.135
†	0.624	15.85	$\frac{5}{8}$	0.120
	0.525	13.33	$\frac{1}{2}$	0.105
†	0.441	11.20	$\frac{7}{16}$	0.105
	0.371	9.423	$\frac{3}{8}$	0.092
$2\frac{1}{2}$†	0.312	7.925	$\frac{5}{16}$	0.088
3	0.263	6.680	$\frac{1}{4}$	0.070
$3\frac{1}{2}$†	0.221	5.613	$\frac{7}{32}$	0.065
4	0.185	4.699	$\frac{3}{16}$	0.065
5†	0.156	3.962	$\frac{5}{32}$	0.044
6	0.131	3.327	$\frac{1}{8}$	0.036
7†	0.110	2.794	$\frac{7}{64}$	0.0328
8	0.093	2.362	$\frac{3}{32}$	0.032
9†	0.078	1.981	$\frac{5}{64}$	0.033
10	0.065	1.651	$\frac{1}{16}$	0.035
12†	0.055	1.397		0.028
14	0.046	1.168	$\frac{3}{64}$	0.025
16†	0.0390	0.991		0.0235
20	0.0328	0.833	$\frac{1}{32}$	0.0172
24†	0.0276	0.701		0.0141
28	0.0232	0.589		0.0125
32†	0.0195	0.495		0.0118
35	0.0164	0.417	$\frac{1}{64}$ <None>	0.0122
42†	0.0138	0.351		0.0100
48	0.0116	0.295		0.0092
60†	0.0097	0.246		0.0070
65	0.0082	0.208		0.0072
80†	0.0069	0.175		0.0056
100	0.0058	0.147		0.0042
115†	0.0049	0.124		0.0038
150	0.0041	0.104		0.0026
170†	0.0035	0.088		0.0024
200	0.0029	0.074		0.0021
270	0.0021	0.053		
325	0.0017	0.044		

†These screens, for closer sizing, are inserted between the sizes usually considered as the standard series. With the inclusion of these screens the ratio of diameters of openings in two successive screens is as $1 : \sqrt[4]{2}$ instead of $1 : \sqrt{2}$.

APPENDIX 6

Properties of Liquid Water

Temperature T, °F	Viscosity[†] μ, cP	Thermal conductivity[‡] k, Btu/ft·h·°F	Density[§] ρ, lb/ft^3	$\psi_f = \left(\dfrac{k^3 \rho^2 g}{\mu^2}\right)^{1/3}$
32	1.794	0.320	62.42	1,410
40	1.546	0.326	62.43	1,590
50	1.310	0.333	62.42	1,810
60	1.129	0.340	62.37	2,050
70	0.982	0.346	62.30	2,290
80	0.862	0.352	62.22	2,530
90	0.764	0.358	62.11	2,780
100	0.682	0.362	62.00	3,020
120	0.559	0.371	61.71	3,530
140	0.470	0.378	61.38	4,030
160	0.401	0.384	61.00	4,530
180	0.347	0.388	60.58	5,020
200	0.305	0.392	60.13	5,500
220	0.270	0.394	59.63	5,960
240	0.242	0.396	59.10	6,420
260	0.218	0.396	58.53	6,830
280	0.199	0.396	57.94	7,210
300	0.185	0.396	57.31	7,510

[†]From *International Critical Tables,* vol. 5, McGraw-Hill Book Company, New York, 1929, p. 10.

[‡]From E. Schmidt and W. Sellschopp, *Forsch. Geb. Ingenieurw.,* **3:**277 (1932).

[§]Calculated from J. H. Keenan and F. G. Keyes, *Thermodynamic Properties of Steam,* John Wiley & Sons., Inc., New York, 1937.

APPENDIX 7

Properties of Saturated Steam and Water[†]

Temperature T, °F	Vapor pressure p_A, $lb_f/in.^2$	Specific volume, ft³/lb		Enthalpy, Btu/lb		
		Liquid v_x	Saturated vapor v_y	Liquid H_x	Vaporization λ	Saturated vapor H_y
32	0.08859	0.016022	3,305	0	1,075.4	1,075.4
35	0.09992	0.016021	2,948	3.00	1,073.7	1,076.7
40	0.12166	0.016020	2,445	8.02	1,070.9	1,078.9
45	0.14748	0.016021	2,037	13.04	1,068.1	1,081.1
50	0.17803	0.016024	1,704.2	18.06	1,065.2	1,083.3
55	0.2140	0.016029	1,431.4	23.07	1,062.4	1,085.5
60	0.2563	0.016035	1,206.9	28.08	1,059.6	1,087.7
65	0.3057	0.016042	1,021.5	33.09	1,056.8	1,089.9
70	0.3632	0.016051	867.7	38.09	1,054.0	1,092.0
75	0.4300	0.016061	739.7	43.09	1,051.1	1,094.2
80	0.5073	0.016073	632.8	48.09	1,048.3	1,096.4
85	0.5964	0.016085	543.1	53.08	1,045.5	1,098.6
90	0.6988	0.016099	467.7	58.07	1,042.7	1,100.7
95	0.8162	0.016114	404.0	63.06	1,039.8	1,102.9
100	0.9503	0.016130	350.0	68.05	1,037.0	1,105.0
110	1.2763	0.016166	265.1	78.02	1,031.4	1,109.3
120	1.6945	0.016205	203.0	88.00	1,025.5	1,113.5
130	2.225	0.016247	157.17	97.98	1,019.8	1,117.8
140	2.892	0.016293	122.88	107.96	1,014.0	1,121.9
150	3.722	0.016343	96.99	117.96	1,008.1	1,126.1
160	4.745	0.016395	77.23	127.96	1,002.2	1,130.1
170	5.996	0.016450	62.02	137.97	996.2	1,134.2
180	7.515	0.016509	50.20	147.99	990.2	1,138.2
190	9.343	0.016570	40.95	158.03	984.1	1,142.1
200	11.529	0.016634	33.63	168.07	977.9	1,145.9
210	14.125	0.016702	27.82	178.14	971.6	1,149.7
212	14.698	0.016716	26.80	180.16	970.3	1,150.5

(Continued)

Temperature T, °F	Vapor pressure p_A, $lb_f/in.^2$	Specific volume, ft³/lb		Enthalpy, Btu/lb		
		Liquid v_x	Saturated vapor v_y	Liquid H_x	Vaporization λ	Saturated vapor H_y
220	17.188	0.016772	23.15	188.22	965.3	1,153.5
230	20.78	0.016845	19.386	198.32	958.8	1,157.1
240	24.97	0.016922	16.327	208.44	952.3	1,160.7
250	29.82	0.017001	13.826	218.59	945.6	1,164.2
260	35.42	0.017084	11.768	228.76	938.8	1,167.6
270	41.85	0.017170	10.066	238.95	932.0	1,170.9
280	49.18	0.017259	8.650	249.18	924.9	1,174.1
290	57.53	0.017352	7.467	259.44	917.8	1,177.2
300	66.98	0.017448	6.472	269.73	910.4	1,180.2
310	77.64	0.017548	5.632	280.06	903.0	1,183.0
320	89.60	0.017652	4.919	290.43	895.3	1,185.8
340	117.93	0.017872	3.792	311.30	879.5	1,190.8
350	134.53	0.017988	3.346	321.80	871.3	1,193.1
360	152.92	0.018108	2.961	332.35	862.9	1,195.2
370	173.23	0.018233	2.628	342.96	854.2	1,197.2
380	195.60	0.018363	2.339	353.62	845.4	1,199.0
390	220.2	0.018498	2.087	364.34	836.2	1,200.6
400	247.1	0.018638	1.8661	375.12	826.8	1,202.0
410	276.5	0.018784	1.6726	385.97	817.2	1,203.1
420	308.5	0.018936	1.5024	396.89	807.2	1,204.1
430	343.3	0.019094	1.3521	407.89	796.9	1,204.8
440	381.2	0.019260	1.2192	418.98	786.3	1,205.3
450	422.1	0.019433	1.1011	430.2	775.4	1,205.6

†Abstracted from *Steam Tables*, by Joseph H. Keenan, Frederick G. Keyes, Philip G. Hill, and Joan G. Moore, John Wiley & Sons, New York, 1969, with the permission of the publisher.

APPENDIX 8

Viscosities of Gases[†]

No.	Gas	X	Y	No.	Gas	X	Y
1	Acetic acid	7.7	14.3	29	Freon-113	11.3	14.0
2	Acetone	8.9	13.0	30	Helium	10.9	20.5
3	Acetylene	9.8	14.9	31	Hexane	8.6	11.8
4	Air	11.0	20.0	32	Hydrogen	11.2	12.4
5	Ammonia	8.4	16.0	33	$3H_2 + N_2$	11.2	17.2
6	Argon	10.5	22.4	34	Hydrogen bromide	8.8	20.9
7	Benzene	8.5	13.2	35	Hydrogen chloride	8.8	18.7
8	Bromine	8.9	19.2	36	Hydrogen cyanide	9.8	14.9
9	Butene	9.2	13.7	37	Hydrogen iodide	9.0	21.3
10	Butylene	8.9	13.0	38	Hydrogen sulfide	8.6	18.0
11	Carbon dioxide	9.5	18.7	39	Iodine	9.0	18.4
12	Carbon disulfide	8.0	16.0	40	Mercury	5.3	22.9
13	Carbon monoxide	11.0	20.0	41	Methane	9.9	15.5
14	Chlorine	9.0	18.4	42	Methyl alcohol	8.5	15.6
15	Chloroform	8.9	15.7	43	Nitric oxide	10.9	20.5
16	Cyanogen	9.2	15.2	44	Nitrogen	10.6	20.0
17	Cyclohexane	9.2	12.0	45	Nitrosyl chloride	8.0	17.6
18	Ethane	9.1	14.5	46	Nitrous oxide	8.8	19.0
19	Ethyl acetate	8.5	13.2	47	Oxygen	11.0	21.3
20	Ethyl alcohol	9.2	14.2	48	Pentane	7.0	12.8
21	Ethyl chloride	8.5	15.6	49	Propane	9.7	12.9
22	Ethyl ether	8.9	13.0	50	Propyl alcohol	8.4	13.4
23	Ethylene	9.5	15.1	51	Propylene	9.0	13.8
24	Fluorine	7.3	23.8	52	Sulfur dioxide	9.6	17.0
25	Freon-11	10.6	15.1	53	Toluene	8.6	12.4
26	Freon-12	11.1	16.0	54	2,3,3-Trimethylbutane	9.5	10.5
27	Freon-21	10.8	15.3	55	Water	8.0	16.0
28	Freon-22	10.1	17.0	56	Xenon	9.3	23.0

Coordinates for use with figure on next page.

[†]By permission, from J. H. Perry (ed.), *Chemical Engineers' Handbook,* 5th ed., pp.**3**-210 and **3**-211. Copyright © 1973, McGraw-Hill Book Company, New York.

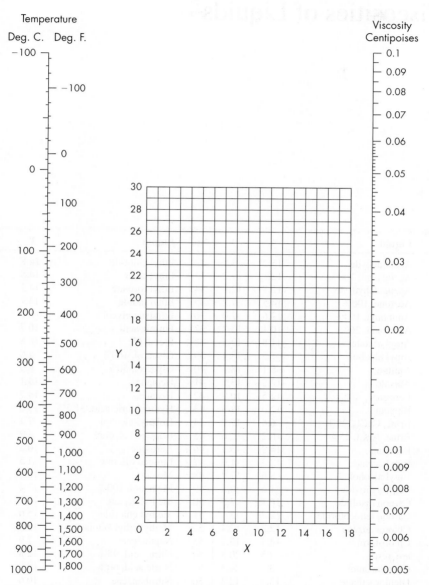

Viscosities of gases and vapors at 1 atm; for coordinates, see table on previous page.

APPENDIX 9

Viscosities of Liquids[†]

No.	Liquid	X	Y	No.	Liquid	X	Y
1	Acetaldehyde	15.2	4.8	32	Ethyl chloride	14.8	6.0
2	Acetic acid, 100%	12.1	14.2	33	Ethyl ether	14.5	5.3
3	Acetic anhydride	12.7	12.8	34	Ethyl formate	14.2	8.4
4	Acetone, 100%	14.5	7.2	35	Ethyl iodide	14.7	10.3
5	Ammonia, 100%	12.6	2.0	36	Ethylene glycol	6.0	23.6
6	Ammonia, 26%	10.1	13.9	37	Formic acid	10.7	15.8
7	Amyl acetate	11.8	12.5	38	Freon-12	16.8	5.6
8	Amyl alcohol	7.5	18.4	39	Glycerol, 100%	2.0	30.0
9	Aniline	8.1	18.7	40	Glycerol, 50%	6.9	19.6
10	Anisole	12.3	13.5	41	Heptane	14.1	8.4
11	Benzene	12.5	10.9	42	Hexane	14.7	7.0
12	Biphenyl	12.0	18.3	43	Hydrochloric acid, 31.5%	13.0	16.6
13	Brine, $CaCl_2$, 25%	6.6	15.9	44	Isobutyl alcohol	7.1	18.0
14	Brine, NaCl, 25%	10.2	16.6	45	Isopropyl alcohol	8.2	16.0
15	Bromine	14.2	13.2	46	Kerosene	10.2	16.9
16	Butyl acetate	12.3	11.0	47	Linseed oil, raw	7.5	27.2
17	Butyl alcohol	8.6	17.2	48	Mercury	18.4	16.4
18	Carbon dioxide	11.6	0.3	49	Methanol, 100%	12.4	10.5
19	Carbon disulfide	16.1	7.5	50	Methyl acetate	14.2	8.2
20	Carbon tetrachloride	12.7	13.1	51	Methyl chloride	15.0	3.8
21	Chlorobenzene	12.3	12.4	52	Methyl ethyl ketone	13.9	8.6
22	Chloroform	14.4	10.2	53	Napthalene	7.9	18.1
23	m-Cresol	2.5	20.8	54	Nitric acid, 95%	12.8	13.8
24	Cyclohexanol	2.9	24.3	55	Nitric acid, 60%	10.8	17.0
25	Dichloroethane	13.2	12.2	56	Nitrobenzene	10.6	16.2
26	Dichloromethane	14.6	8.9	57	Nitrotoluene	11.0	17.0
27	Ethyl acetate	13.7	9.1	58	Octane	13.7	10.0
28	Ethyl alcohol,100%	10.5	13.8	59	Octyl alcohol	6.6	21.1
29	Ethyl alcohol, 95%	9.8	14.3	60	Pentane	14.9	5.2
30	Ethyl alcohol, 40%	6.5	16.6	61	Phenol	6.9	20.8
31	Ethyl benzene	13.2	11.5	62	Sodium	16.4	13.9

(*Continued*)

No.	Liquid	X	Y	No.	Liquid	X	Y
63	Sodium hydroxide, 50%	3.2	25.8	70	Toluene	13.7	10.4
64	Sulfur dioxide	15.2	7.1	71	Trichloroethylene	14.8	10.5
65	Sulfuric acid, 98%	7.0	24.8	72	Vinyl acetate	14.0	8.8
66	Sulfuric acid, 60%	10.2	21.3	73	Water	10.2	13.0
67	Tetrachloroethane	11.9	15.7	74	o-Xylene	13.5	12.1
68	Tetrachloroethylene	14.2	12.7	75	m-Xylene	13.9	10.6
69	Titanium tetrachloride	14.4	12.3	76	p-Xylene	13.9	10.9

Coordinates for use with figure on next page.

[†]By permission, from J. H. Perry (ed.), *Chemical Engineers' Handbook,* 5th ed., pp. **3**-212 and **3**-213. Copyright © 1973, McGraw-Hill Book Company, New York.

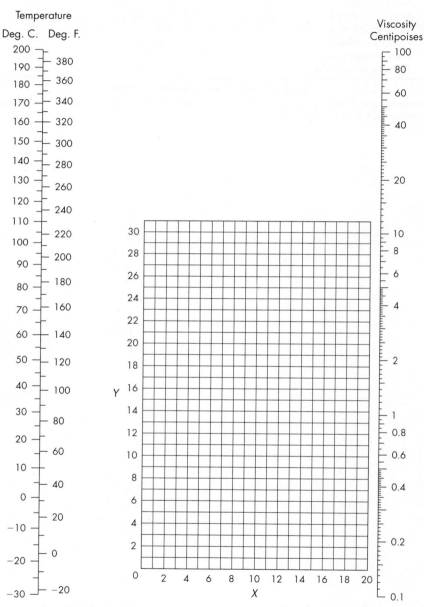

Viscosities of liquids at 1 atm. For coordinates, see table on previous page.

APPENDIX 10

Thermal Conductivities of Metals[†]

Metal	Thermal conductivity k[‡]		
	32°F	64°F	212°F
Aluminum	117		119
Antimony	10.6		9.7
Brass (70 copper, 30 zinc)	56		60
Cadmium		53.7	52.2
Copper (pure)	224		218
Gold		169.0	170.0
Iron (cast)	32		30
Iron (wrought)		34.9	34.6
Lead	20		19
Magnesium	92	92	92
Mercury (liquid)	4.8		
Nickel	36		34
Platinum		40.2	41.9
Silver	242		238
Sodium (liquid)			49
Steel (mild)			26
Steel (1% carbon)		26.2	25.9
Steel (stainless, type 304)			9.4
Steel (stainless, type 316)			9.4
Steel (stainless, type 347)			9.3
Tantalum		32	
Tin	36		34
Zinc	65		64

[†]Based on W. H. McAdams, *Heat Transmission,* 3rd ed., McGraw-Hill Book Company, New York, 1954, pp. 445–447.

[‡]k = Btu/ft·h·°F. To convert to W/m·°C, multiply by 1.73073.

APPENDIX 11

Thermal Conductivities of Various Solids and Insulating Materials[†]

Material	Apparent density ρ, lb/ft³	Temperature T, °C	Thermal conductivity k, Btu/h · ft² · (°F/ft)
Asbestos	29	−200	0.043
	36	0	0.087
	36	400	0.129
Bricks			
Alumina	—	1,315	2.7
Building brickwork	—	20	0.4
Carbon	96.7	—	3.0
Fire clay (Missouri)	—	200	0.58
	—	1,000	0.95
	—	1,400	1.02
Kaolin insulating firebrick	19	200	0.050
	19	760	0.113
Silicon carbide, recrystallized	129	600	10.7
	129	1,000	8.0
	129	1,400	6.3
Cardboard, corrugated	—	—	0.37
Concrete			
Clinker	—	—	0.20
Stone	—	—	0.54
1:4 dry	—	—	0.44
Cork, ground	9.4	30	0.025
Glass			
Borosilicate	139	30–75	0.63
Window	—	—	0.3–0.61

(*Continued*)

Material	Apparent density ρ, lb/ft^3	Temperature T, °C	Thermal conductivity k, Btu/h·ft^2·(°F/ft)
Granite	—	—	1.0–2.3
Ice	57.5	0	1.3
Insulating materials			
Fiberglass batts[‡]	6	20	0.019
	6	150	0.027
	6	200	0.035
	9	20	0.018
	9	150	0.023
Kapok	0.88	20	0.020
Polystyrene foam[§]	1	20	0.023
	2–5	20	0.020
Polyurethane foam[§] (made with fluorocarbon gas)	1.3–3.0	—	0.014
	4–8	—	0.018
Polyurethane foam[§] (made with CO_2)	1.3–3.0	—	0.018
Wall board	14.8	21	0.028
Magnesia, powdered	49.7	47	0.35
Paper	—	—	0.75
Porcelain	—	200	0.88
Rubber, soft	—	21	0.075–0.092
Snow	34.7	0	0.27
Wood (across grain)			
Oak	51.5	15	0.12
Maple	44.7	50	0.11
Pine, white	34.0	15	0.087
Wood (parallel to grain)			
Pine	34.4	21	0.20

[†]From J. H. Perry (ed.), *Chemical Engineers' Handbook,* 6th ed., McGraw-Hill, New York, p. **3**-260, except as noted.

[‡]From *Heat Transfer and Fluid Data Book,* vol. 1, Genium Publishing Corp., Schenectady, NY, 1984, sect. 515.24, p. 1.

[§]From *Modern Plastics Encyclopedia,* vol. 65, no. 11, McGraw-Hill Book Co., New York, 1988, p. 657.

APPENDIX 12

Thermal Conductivities of Gases and Vapors[†]

Substance	Thermal conductivity k[‡]	
	32°F	212°F
Acetone	0.0057	0.0099
Acetylene	0.0108	0.0172
Air	0.0140	0.0184
Ammonia	0.0126	0.0192
Benzene		0.0103
Carbon dioxide	0.0084	0.0128
Carbon monoxide	0.0134	0.0176
Carbon tetrachloride		0.0052
Chlorine	0.0043	
Ethane	0.0106	0.0175
Ethyl alcohol		0.0124
Ethyl ether	0.0077	0.0131
Ethylene	0.0101	0.0161
Helium	0.0818	0.0988
Hydrogen	0.0966	0.1240
Methane	0.0176	0.0255
Methyl alcohol	0.0083	0.0128
Nitrogen	0.0139	0.0181
Nitrous oxide	0.0088	0.0138
Oxygen	0.0142	0.0188
Propane	0.0087	0.0151
Sulfur dioxide	0.0050	0.0069
Water vapor (at 1 atm abs pressure)		0.0136

[†]Based on W. H. McAdams, *Heat Transmission,* 3rd ed., McGraw-Hill Book Company, New York, 1954, pp. 457–458.

[‡]$k = $ Btu/ft $\cdot$ h $\cdot$ °F. To convert to W/m $\cdot$ °C, multiply by 1.73073.

APPENDIX 13

Thermal Conductivities of Liquids Other Than Water[†]

Liquid	Temperature, °F	k[‡]
Acetic acid	68	0.099
Acetone	86	0.102
Ammonia (anhydrous)	5–86	0.29
Aniline	32–68	0.100
Benzene	86	0.092
n-Butyl alcohol	86	0.097
Carbon bisulfide	86	0.093
Carbon tetrachloride	32	0.107
Chlorobenzene	50	0.083
Ethyl acetate	68	0.101
Ethyl alcohol (absolute)	68	0.105
Ethyl ether	86	0.080
Ethylene glycol	32	0.153
Gasoline	86	0.078
Glycerine	68	0.164
n-Heptane	86	0.081
Kerosene	68	0.086
Methyl alcohol	68	0.124
Nitrobenzene	86	0.095
n-Octane	86	0.083
Sulfur dioxide	5	0.128
Sulfuric acid (90%)	86	0.21
Toluene	86	0.086
Trichloroethylene	122	0.080
o-Xylene	68	0.090

[†]Based on W. H. McAdams, *Heat Transmission*, 3rd ed., McGraw-Hill Book Company, New York, 1954, pp. 455–456.

[‡]$k = $ Btu/ft·h·°F. To convert to W/m·°C, multiply by 1.73073.

APPENDIX 14

Specific Heats of Gases[†]

$$c_p = \text{Specific heat} = \text{Btu/lb-°F} = \text{cal/g-°C}$$

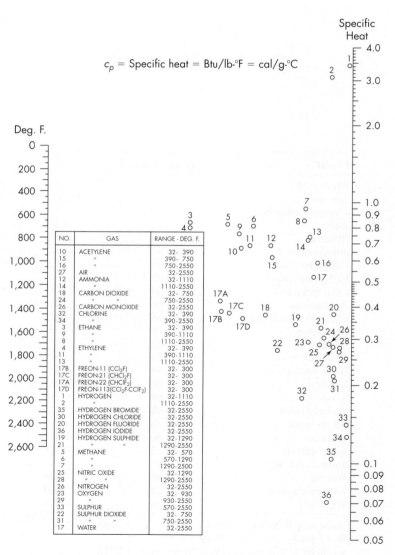

True specific heats c_p of gases and vapors at 1 atm pressure.

[†]*Courtesy of T. H. Chilton.*

Specific Heats of Liquids†

Specific heat = Btu/lb-°F = cal/g-°C

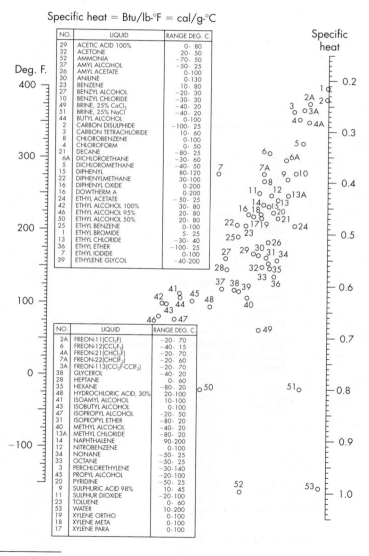

NO.	LIQUID	RANGE DEG. C.
29	ACETIC ACID 100%	0- 80
32	ACETONE	20- 50
52	AMMONIA	−70- 50
37	AMYL ALCOHOL	−50- 25
26	AMYL ACETATE	0-100
30	ANILINE	0-130
23	BENZENE	10- 80
27	BENZYL ALCOHOL	−20- 30
10	BENZYL CHLORIDE	−30- 30
49	BRINE, 25% CaCl₂	−40- 20
51	BRINE, 25% NaCl	−40- 20
44	BUTYL ALCOHOL	0-100
2	CARBON DISULPHIDE	−100- 25
3	CARBON TETRACHLORIDE	10- 60
8	CHLOROBENZENE	0-100
4	CHLOROFORM	0- 50
21	DECANE	−80- 25
6A	DICHLOROETHANE	−30- 60
5	DICHLOROMETHANE	−40- 50
15	DIPHENYL	80-120
22	DIPHENYLMETHANE	30-100
16	DIPHENYL OXIDE	0-200
16	DOWTHERM A	0-200
24	ETHYL ACETATE	− 50- 25
42	ETHYL ALCOHOL 100%	30- 80
46	ETHYL ALCOHOL 95%	20- 80
50	ETHYL ALCOHOL 50%	20- 80
25	ETHYL BENZENE	0-100
1	ETHYL BROMIDE	5- 25
13	ETHYL CHLORIDE	−30- 40
36	ETHYL ETHER	−100- 25
7	ETHYL IODIDE	0-100
39	ETHYLENE GLYCOL	−40-200

NO.	LIQUID	RANGE DEG. C.
2A	FREON-11(CCl₃F)	−20- 70
6	FREON-12(CCl₂F₂)	−40- 15
4A	FREON-21(CHCl₂F)	−20- 70
7A	FREON-22(CHClF₂)	−20- 60
3A	FREON-113(CCl₂F-CClF₂)	−20- 70
38	GLYCEROL	−40- 20
28	HEPTANE	0- 60
35	HEXANE	−80- 20
48	HYDROCHLORIC ACID, 30%	20-100
41	ISOAMYL ALCOHOL	10-100
43	ISOBUTYL ALCOHOL	0-100
47	ISOPROPYL ALCOHOL	−20- 50
31	ISOPROPYL ETHER	−80- 20
40	METHYL ALCOHOL	−40- 20
13A	METHYL CHLORIDE	−80- 20
14	NAPHTHALENE	90-200
12	NITROBENZENE	0-100
34	NONANE	−50- 25
33	OCTANE	−50- 25
3	PERCHLORETHYLENE	−30-140
45	PROPYL ALCOHOL	−20-100
20	PYRIDINE	−50- 25
9	SULPHURIC ACID 98%	10- 45
11	SULPHUR DIOXIDE	−20-100
23	TOLUENE	0- 60
53	WATER	10-200
19	XYLENE ORTHO	0-100
18	XYLENE META	0-100
17	XYLENE PARA	0-100

†*Courtesy of T. H. Chilton.*

APPENDIX 16

Prandtl Numbers for Gases at 1 atm and 100°C[†]

Gas	$Pr = \dfrac{c_p \mu}{k}$
Air	0.69
Ammonia	0.86
Argon	0.66
Carbon dioxide	0.75
Carbon monoxide	0.72
Helium	0.71
Hydrogen	0.69
Methane	0.75
Nitric oxide, nitrous oxide	0.72
Nitrogen	0.70
Oxygen	0.70
Water vapor	1.06

[†]Based on W. H. McAdams, *Heat Transmission,* 3rd ed., McGraw-Hill Book Company, New York, 1954, p. 471.

Prandtl Numbers for Liquids[†]

	$Pr = \dfrac{c_p \mu}{k}$	
Liquid	**61°F**	**212°F**
Acetic acid	14.5	10.5
Acetone	4.5	2.4
Aniline	69	9.3
Benzene	7.3	3.8
n-Butyl alcohol	43	11.5
Carbon tetrachloride	7.5	4.2
Chlorobenzene	9.3	7.0
Ethyl acetate	6.8	5.6
Ethyl alcohol	15.5	10.1
Ethyl ether	4.0	2.3
Ethylene glycol	350	125
n-Heptane	6.0	4.2
Methyl alcohol	7.2	3.4
Nitrobenzene	19.5	6.5
n-Octane	5.0	3.6
Sulfuric acid (98%)	149	15.0
Toluene	6.5	3.8
Water	7.7	1.5

[†]Based on W. H. McAdams, *Heat Transmission,* 3rd ed.,
McGraw-Hill Book Company, New York, 1954, p. 470.

APPENDIX 18

Diffusivities and Schmidt Numbers for Gases in Air at 0°C and 1 atm[†]

Gas	Volumetric diffusivity D_v, ft²/h[¶]	$Sc = \dfrac{\mu}{\rho D_v}$[‡]
Acetic acid	0.413	1.24
Acetone	0.32[§]	1.60
Ammonia	0.836	0.61
Benzene	0.299	1.71
n-Butyl alcohol	0.273	1.88
Carbon dioxide	0.535	0.96
Carbon tetrachloride	0.26[§]	1.97
Chlorine	0.43[§]	1.19
Chlorobenzene	0.24[§]	2.13
Ethane	0.49[§]	1.04
Ethyl acetate	0.278	1.84
Ethyl alcohol	0.396	1.30
Ethyl ether	0.302	1.70
Hydrogen	2.37	0.22
Methane	0.74[§]	0.69
Methyl alcohol	0.515	1.00
Naphthalene	0.199	2.57
Nitrogen	0.70[§]	0.73
n-Octane	0.196	2.62
Oxygen	0.690	0.74
Phosgene	0.31[§]	1.65
Propane	0.36[§]	1.42
Sulfur dioxide	0.44[§]	1.16
Toluene	0.275	1.86
Water vapor	0.853	0.60

[†]By permission, from T. K. Sherwood and R. L. Pigford, *Absorption and Extraction,* 2nd ed., p. 20. Copyright 1952, McGraw-Hill Book Company, New York.

[‡]The value of μ/ρ is that for pure air, 0.512 ft²/h.

[§]Calculated by Eq. (17.28).

[¶]To convert ft²/h to cm²/s, multiply by 0.2581.

APPENDIX 19

Collision Integral and Lennard–Jones Force Constants[†]

Collision integral Ω_D

$\dfrac{kT}{\varepsilon_{12}}$	Ω_D	$\dfrac{kT}{\varepsilon_{12}}$	Ω_D	$\dfrac{kT}{\varepsilon_{12}}$	Ω_D
0.30	2.662	1.65	1.153	4.0	0.8836
0.35	2.476	1.70	1.140	4.1	0.8788
0.40	2.318	1.75	1.128	4.2	0.8740
0.45	2.184	1.80	1.116	4.3	0.8694
0.50	2.066	1.85	1.105	4.4	0.8652
0.55	1.966	1.90	1.094	4.5	0.8610
0.60	1.877	1.95	1.084	4.6	0.8568
0.65	1.798	2.00	1.075	4.7	0.8530
0.70	1.729	2.1	1.057	4.8	0.8492
0.75	1.667	2.2	1.041	4.9	0.8456
0.80	1.612	2.3	1.026	5.0	0.8422
0.85	1.562	2.4	1.012	6	0.8124
0.90	1.517	2.5	0.9996	7	0.7896
0.95	1.476	2.6	0.9878	8	0.7712
1.00	1.439	2.7	0.9770	9	0.7556
1.05	1.406	2.8	0.9672	10	0.7424
1.10	1.375	2.9	0.9576	20	0.6640
1.15	1.346	3.0	0.9490	30	0.6232
1.20	1.320	3.1	0.9406	40	0.5960
1.25	1.296	3.2	0.9328	50	0.5756
1.30	1.273	3.3	0.9256	60	0.5596
1.35	1.253	3.4	0.9186	70	0.5464
1.40	1.233	3.5	0.9120	80	0.5352
1.45	1.215	3.6	0.9058	90	0.5256
1.50	1.198	3.7	0.8998	100	0.5130
1.55	1.182	3.8	0.8942	200	0.4644
1.60	1.167	3.9	0.8888	400	0.4170

Lennard–Jones force constants

Compound	ε/k (K)	σ (Å)
Acetone	560.2	4.600
Acetylene	231.8	4.033
Air	78.6	3.711
Ammonia	558.3	2.900
Argon	93.3	3.542
Benzene	412.3	5.349
Bromine	507.9	4.296
n-butane	310	5.339
i-butane	313	5.341
Carbon dioxide	195.2	3.941
Carbon disulfide	467	4.483
Carbon monoxide	91.7	3.690
Carbon tetrachloride	322.7	5.947
Carbonyl sulfide	336	4.130
Chlorine	316	4.217
Chloroform	340.2	5.389
Cyanogen	348.6	4.361
Cyclohexane	297.1	6.182
Cyclopropane	248.9	4.807
Ethane	215.7	4.443
Ethanol	362.6	4.530
Ethylene	224.7	4.163
Fluorine	112.6	3.357
Helium	10.22	2.551
n-Hexane	339.3	5.949
Hydrogen	59.7	2.827
Hydrogen cyanide	569.1	3.630
Hydrogen chloride	344.7	3.339
Hydrogen iodide	288.7	4.211
Hydrogen sulfide	301.1	3.623
Iodine	474.2	5.160
Krypton	178.9	3.655
Methane	148.6	3.758
Methanol	481.8	3.626
Methylene chloride	356.3	4.898
Methyl chloride	350	4.182
Mercury	750	2.969
Neon	32.8	2.820
Nitric oxide	116.7	3.492
Nitrogen	71.4	3.798
Nitrous oxide	232.4	3.828
Oxygen	106.7	3.467
n-Pentane	341.1	5.784
Propane	237.1	5.118
n-Propyl alcohol	576.7	4.549
Propylene	298.9	4.678
Sulfur dioxide	335.4	4.112
Water	809.1	2.641

[†]From J. O. Hirschfelder, C. F. Curtiss, and R. B. Bird, *Molecular Theory of Gases and Liquids*, New York: Wiley, 1954.

INDEX

A

Absolute velocity of fluid, 203
Absorbing gases, radiation to, 425
Absorption:
  with chemical reaction, 588–589
  multicomponent, 570–571
  in plate columns, 588
  of radiation, by opaque solids, 410–413
  rate of, 560–561
  from rich gases, 574–580
Absorption coefficient, 424
Absorption factor, 632
Absorption factor method for calculating
    number of ideal stages, 632–635
Absorption length, 423
Absorptivity, 406
  of opaque solids, 410–411
Acoustic velocity, 130
  of ideal gas, 131–132
Activity coefficients, 885
Adiabatic compression, 214–215
Adiabatic cooling lines, 602
Adiabatic dryers, 774
Adiabatic friction flow, 140–145
Adiabatic saturation temperature, 599
Adiabatic saturator, 599–600
Adiabatic wall temperature, 353
Adsorbents, 812–813
Adsorber design, 834–837
Adsorbers, continuous operation, 837–838
Adsorption, 506, 812–813
  basic equations for, 827–828
  favorable, 834
  irreversible, 829–832
  with linear isotherms, 833–834
  from liquids, 815–816
  pressure-swing, 815
  principles of, 821–827
  of water vapor, 820
Adsorption equipment, 813–816
Adsorption isotherms, 816–821
Adsorption process, 812–813
Affinity laws for pumps, 207–208
Aggregate fluidization, 175

Agitated-film evaporators, 478–479
Agitated mills, 979
Agitated tower extractors, 753–755
Agitated vessels, 239–240
  heat-transfer coefficients in, 460–462
  heat transfer in, 459–463
  transient heating or cooling in, 462
Agitation:
  of liquids, 238–285
  mixing versus, 238
  purposes of, 238–239
Agitation nozzles, 461
Agitator selection and scaleup, 278–281
Air:
  natural convection to, from vertical shapes
    and horizontal planes, 368–369
  nitrogen production from, 873
Air-cooled heat exchangers, 454–455
Air pump, 209
Air rate effect in stripping, 572–574
Amicon UF membranes, 1022
Analogy equations, 355
Angle of repose, 952
Angle of vision, 415
Angular momentum, 81
Angular-momentum equation, 81–82
Anion exchangers, 839
Annulus, laminar flow in, 102–104
Approaches, 316
Aqueous phases, use of, 766
Archbreaker, 953
Area meters, 226–228
Arithmetic mean diameter, 948
Asahi process, 851
Association parameter, 521
Asterisk condition, 132
Asymmetric membranes, concentration
    gradients for, 862
Attenuation, 423–424
Attrition mills, 974–975
Automatic batch centrifuges, 1001–1002
Automatic belt filters, 995
Average coefficient, 349
Average stream temperatures, 321
Average velocity, relations between maximum
    velocity and, 109

D

M

N

O

P

T

U

Y

Z